Topics in Geobiology

Volume 43

Topics in Geobiology series treats geobiology – the broad discipline that covers the history of life on Earth. The series aims for high quality, scholarly volumes of original research as well as broad reviews. Recent volumes have showcased a variety of organisms including cephalopods, corals, and rodents. They discuss the biology of these organisms-their ecology, phylogeny, and mode of life – and in addition, their fossil record – their distribution in time and space.

Other volumes are more theme based such as predator-prey relationships, skeletal mineralization, paleobiogeography, and approaches to high resolution stratigraphy, that cover a broad range of organisms. One theme that is at the heart of the series is the interplay between the history of life and the changing environment. This is treated in skeletal mineralization and how such skeletons record environmental signals and animal-sediment relationships in the marine environment. The series editors also welcome any comments or suggestions for future volumes.

For further volumes:
http://www.springer.com/series/6623

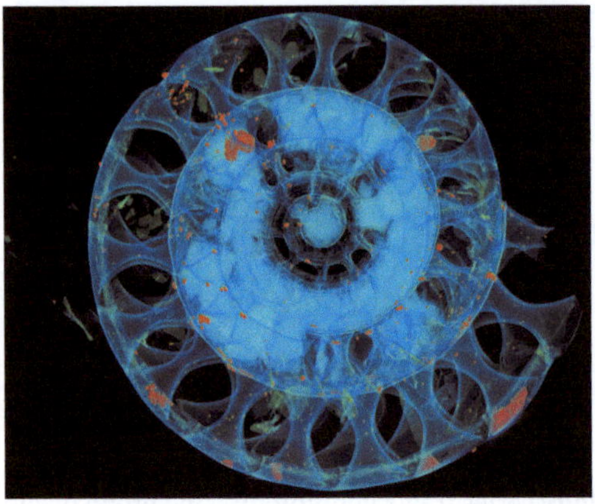

CT-scan reconstructed by René Hoffmann (Bochum) of the Carboniferous ammonoid *Arnsbergites* (Harz Mountains, Germany). This example stands for the increase in use of non-invasive imaging techniques and 3D-visualization in ammonoid paleobiology: Specimen discovered by Hartmut Knappe (Wernigerode). Image by courtesy of René Hoffmann (Bochum).

Christian Klug • Dieter Korn • Kenneth De Baets
Isabelle Kruta • Royal H. Mapes
Editors

Ammonoid Paleobiology: From Anatomy to Ecology

Springer

Editors
Christian Klug
Paläontologisches Institut und Museum
University of Zurich
Zürich
Switzerland

Dieter Korn
Leibniz-Institut für Evolutions und Biodiversitätsforschung
Museum für Naturkunde
Berlin
Germany

Kenneth De Baets
GeoZentrum Nordbayern
Friedrich-Alexander-Universität Erlangen
Erlangen
Germany

Isabelle Kruta
Division of Paleontology
American Museum of Natural History
New York
New York
USA

Royal H. Mapes
North Carolina Museum of Natural Sciences
Raleigh
USA

ISSN 0275-0120
Topics in Geobiology
ISBN 978-94-017-9629-3 ISBN 978-94-017-9630-9 (eBook)
DOI 10.1007/978-94-017-9630-9
Springer Dordrecht Heidelberg New York London

Library of Congress Control Number: 2015941022

Printed on acid-free paper

Springer is part of Springer Science+Business Media (www.springer.com)

Image courtesy W. Gerber (Tübingen)

To the memory of Adolf ("Dolf") Seilacher (24. February 1925 to 26. April 2014)
There are paleontologists, who have published hundreds of articles, but there is hardly anybody who stimulated ammonoid paleobiological research as much with insightful and innovative articles as Dolf Seilacher. Also, he published the possibly most widely cited phrase on ammonoids:
"Ammonites are for paleontologists what Drosophila is in genetics. The structural complexity of their shells, the complete ontogenetic protocol and a long and rather perfect fossil record make them the most suitable invertebrate group for macroevolutionary studies."
Seilacher (1989: p. 67)

Dolf Seilacher died peacefully on April 26th 2014 at the age of 89, short before the completion of this book. In order to acknowledge his scientific input and stimulation of research on ammonoids, we dedicate this book to his memory.
His main contributions to the field were probably the balloon-model for simple septa, the tie-point-model for complex septa, the Cartesian diver model, the use of epizoa to constrain ammonoid ecology and his works on ammonoid taphonomy and paleobiology. Below, we provide a list of his papers on ammonoids, which included ammonoid data.

Christian Klug, Dieter Korn, Kenneth De Baets, Isabelle Kruta, Royal H. Mapes

List of Contributions to Ammonoid Research by Seilacher

Seilacher A (1960) Epizoans as a key to ammonoid ecology. J. Paleontol. 34:189–193

Seilacher A (1963) Umlagerung und Rolltransport von Cephalopoden-Gehäusen. N Jahrb Geol Paläontol Mh 11:593–615

Seilacher A (1966) Lobenlibellen und Füllstrukturen bei Ceratiten. N Jahrb Geol Paläontol Abh 125:480–488

Seilacher A (1968) Sedimentationsprozesse in Ammonitengehäusen. Akad Wiss Lit Mainz, Mat-Nat Kl, Abh 1967(9):189–203

Seilacher A (1971) Preservational history of ceratite shells. Palaeontology 14:16–21.

Seilacher A, Westphal F (1971) Fossil-Lagerstätten. In: Müller G (ed) Sedimentology of parts of Central Europe. Kramer, Frankfurt

Seilacher A (1973) Fabricational noise in adaptive morphology. Syst Zoo 22:451–465

Seilacher A (1975) Mechanische Simulation und funktionelle Evolution des Ammoniten-Septums. Paläontol Z 49:268–286

Seilacher A, Andalib F, Dietl G, Gocht H (1976) Preservational history of compressed Jurassic ammonites from southern Germany. N Jahrb Geol Paläontol Abh 152:307–356

Seilacher A (1976) Ammoniten-Erhaltung. Zentralbl Geol Paläontol II 1976:355–362

Brenner K, Seilacher A (1979) New aspects about the origin of the Toarcian Posidonia Shales. N Jahrb Geol Paläontol, Abh 11–18

Seilacher A (1982) Ammonite shells as habitats in the Posidonia Shales of Holzmaden—floats or benthic islands? N Jahrb Geol Paläontol Mh 2:98–114

Seilacher A (1982) Posidonia shales (Toarcian, S. Germany)—stagnant basin model revalidated. In: Motanaro Gallitelli E (ed) Palaeontology, essential of historical geology: proceedings of the first international meeting on palaeontology, essential of historical geology. Modena, Italy: S.T.E.M. Mucchi

Seilacher A (1982) Erfolgte die Ammoniten-Besiedlung im Holzmadener Posidonienschiefer vor oder nach dem Absinken? N Jahrb Geol Paläont Abh 164:31

Seilacher A (1982) Ammonite shells as habitats–floats or benthic islands? In: Einsele G, Seilacher A (eds) Cyclic and event stratification. Springer, Heidelberg

Seilacher A (1983) Papers on cephalopod paleobiology and phylogeny. N Jahrb Geol Paläont Abh 165:327–329

Seilacher A (1984) Sedimentary structures tentatively attributed to seismic events. Mar Geol 55:1–12

Seilacher A, Reif W-E, Westphal F (1985) Sedimentological, ecological and temporal patterns of fossil Lagerstätten. Philos Trans Roy Soc London B 311:5–23

Seilacher A (1988) Why are nautiloid and ammonite sutures so different? N Jahrb. Geol Paläont Abh 177:41–69

Seilacher A, Chinzei K (1988) Selbstbildungs-Mechanismen als Prozessoren organismischer Entwicklung und Evolution. Leichtbau in Architektur und Natur. Nat Konstr 3: 55–63

Seilacher A (1990) Die Holzmadener Posidonienschiefer: Entstehung der Fossillagerstätte und eines Erdölmuttergesteins. In: Weidert WK (ed) Klassische Fundstellen der Paläontologie. Goldschneck, Korb

Seilacher A (1991) Self-organization: morphogenetic mechanisms as processors of evolution. Rev Esp Paleontol Extraordin 5:5–11

Seilacher A (1993) Ammonite aptychi: how to transform a jaw into an operculum. In: Dodson P, Gingerich P (eds) Functional morphology and evolution. New Haven, Kline Geol. Lab., Yale. Am J Sci spec vol 293A:20–32

Seilacher A (1993) Fossil-Lagerstätten im Muschelkalk. In: Hagdorn H, Seilacher A (eds) Muschelkalk, Schöntaler Symposium 1991. Sonderbände Ges Nat Württ 2:213–222

Seilacher A, Klug C (1993) Selbst-Organisation bei Kerzenmuscheln. Naturwiss Rundsch 46:132–134

Seilacher A, Chinzei K (eds, 1993) Progress in constructional morphology. Neues Jahrb Geol Paläont Abh 190:165–167

Seilacher A, Chinzei K (1993) Remote biomineralization II: fill skeletons controlling buoyancy in shelled cephalopods. N Jahrb Geol Paläont Abh 190:199–208

Seilacher A, Gunji YP (1993) Morphogenetic countdown: another view on heteromorphy shells in gastropods and ammonites. N Jahrb Geol Paläontol Abh 190:73–101

Seilacher A, LaBarbera M (1995) Ammonites as Cartesian divers. Palaios 10:493–506

Maeda H, Seilacher A (1996) Ammonoid taphonomy. In: Landman NH, Tanabe K, Davis RA (eds) Ammonoid paleobiology. Plenum, New York, Topics in Geobiol 13:543–578

Kase T, Johnston PA, Seilacher A, Boyce GB (1998) Alleged mosasaur marks on Late Cretaceous ammonites are limpet (patellogastropod) home-scars. Geology 26:947–950

Seilacher A (1999) *Oecoptychius* Rätsel. Fossilien 3:131

Keupp H, Röper M, Seilacher A (1999) Paläobiologische Aspekte von syn vivobesiedehen Ammonoideen im Plattenkalk des Ober-Kimmeridgiums von Brunn in Ostbayern. Berliner Geowiss Abh E 30:121–145

Seilacher A (2000) Ammoniten als Kartesische Taucher? Fossilien 4:231–237

Seilacher A, Keupp H (2000) Wie sind Ammoniten geschwommen? Fossilien 5:310–313

Seilacher A (2004) Trittbrettfahrer im Muschelkalkmeer. Fossilien 3:157–160
Seilacher A (2012) Cyclism revisited: extinction and 'Achilles' Heels' keep diversification in check on macroevolutionary time scales. Hist Biol 25:239–250
Seilacher A (2013) Patterns of macroevolution through the Phanerozoic. Palaeontology 56:1273–1283
Seilacher A, Gishlick AD (2014) Morphodynamics. CRC Press, London

Foreword to the New Edition

This two-volume work is a testament to the abiding interest and human fascination with ammonites. As Niles Eldredge wrote in the forward to our 1996 book "Ammonoid Paleobiology" (fondly referred to as the Red Book), ammonites are "the quintessential fossils." They have contributed to ideas about biostratigraphy, paleoecology, paleobiology, paleoenvironment, paleobiogeography, paleogeography, paleoceanography, evolution, phylogeny, and ontogeny. All of these themes are treated in the present book. The past two decades have witnessed an explosion of new information about ammonites: early life history, evolution of the buccal mass, feeding habits, soft-tissue preservation, radiation-and extinction-patterns, shell microstructure, sutures and pseudosutures, cameral membranes, mode of life, phylogeny, and habitats. Many of these discoveries have benefitted from the application of new technologies such as isotopic analysis, organic geochemistry, geographic information systems, geometric morphometrics, computerized tomography, and synchrotron imaging. They have also relied on more traditional techniques such as scanning electron microscopy and electron dispersive analysis, which continue to furnish an abundance of data. Fortunately, too, our field is constantly being re-energized by the discovery of new fossil finds that shed light on old questions and raise new ones. Given all these advances in our knowledge, this book is a comprehensive and timely "state of the art" compilation. Moreover, it also points the way for future studies to further enhance our understanding of this endlessly fascinating group of organisms.

Neil H. Landman, Kazushige Tanabe, and Richard Arnold Davis, Editors of the 1996 book "Ammonoid Paleobiology" (the original three musketeers)

Foreword to the First Edition: Ammonoids Do It All

Ammonoids are *the* quintessential fossils, seemingly covering all the major themes of paleontology. Method and theory of stratigraphic correlation using fossils? Albert Oppel, whose concepts of zonation were explicated and applied by W. J. Arkell exhaustively in his monumental works on the Jurassic System, immediately spring to mind-works based virtually exclusively on the stratigraphic distributions of ammonoid species. Evolution? W. Waagen leaps to mind, applying the term "mutation" to his ammonoid lineages, and thus introducing the word to the scientific literature well before geneticists co-opted "mutation" for their own, starkly different, use.

Extinction? Cretaceous heteromorphs were type examples of "racial senescence"-if now wholly discredited, nonetheless an important part of earlier discourse on what is one of the most compelling issues that paleobiology brings to general biological theory. I was myself stunned, when compiling data on the end-Cretaceous mass extinction in the late 1960s for a seminar conducted by Norman D. Newell, to find that the scaphitids-far from dwindling to a precious few as Cretaceous time was running out-were actually in the midst of an evolutionary radiation, an expansion of diversity cut abruptly short by whatever it was that disrupted things so badly 65 million years ago.

Indeed, though of course much remains to be learned about ammonoid phylogeny, every chart that I have seen published in the last 30 years showing the basic outlines of ammonoid evolution against the backdrop of SilurianCretaceous geologic time constitutes a stark object lesson on the resonance between evolution and extinction. The theme of early "experimentation" shows up amidst Devonian ammonoid diversity: the clymeniids constitute an arch example, with their siphuncle on the opposite side of the body from what proved to be the "normal" ammonoid condition-an experiment that failed to survive the late Devonian biotic crisis. thus forever depleting ammonoid morphological diversity. And are the goniatites, ceratites, and ammonites mere grades, as nearly everyone suspected back in the parallel-evolution-mad 1960s? Or are they, as now seems evident, genealogically coherent, monophyletic clades that represent radiations consequent to major biotic crises of the Permo-Triassic and Triassic-Jurassic boundaries? That grade-like patterns can come from evolutionary radiations following severe extinction bottlenecks is an aspect of evolutionary theory yet to be fully expounded. And it is the ammonoids that show such patterns best.

Biostratigraphy, evolution, extinction-not to mention biogeography, paleoecology, and functional morphology: of all major taxa in the fossil record, the ammonoids arguably do it best. But there is something more to them, a certain allure that makes them deserved rivals of trilobites as the most ardently desired and sought-after relics of the deep past. Ammonoids are at once exotic yet familiarly organic. Though nearly always simply the empty shells of long-dead animals, they nonetheless seem complete. They are almost always beautiful-and sometimes even colorful. It's probably the (nearly always planispiral) logarithmic spiral that, in spite of its mathematical precision, nonetheless casts an aura of intrigue and mystery to what is otherwise just another fossil. A few years back I published a lavishly illustrated book on fossils, using photographs of many of the finest specimens of all taxa from the rich paleontological collections of the American Museum of Natural History. And though I had skulls of a male and female Tertiary artiodactyl on the front cover, it is the photo on the back-of a pretty little pyritized specimen of the Jurassic ammonoid *Hecticoceras-that* attracted the most attention, and that has been subsequently reproduced over and over again.

I can only conclude that, over and above the prodigious intellectual contributions that continue to come from contemplation of these marvelous animals (as this present volume amply demonstrates), ammonoids also have that certain *je ne sais quoi* that will always keep them at the forefront of the paleontological realm. Ammonoids really do seem to have it all.

The American Museum of Natural History Niles Eldredge
New York, New York

Preface

Imagine you belong to any religion and your chief deity asks you: "Could you imagine editing the new sacred book?" This is the feeling you have as an ammonoid worker, when you are offered to take care of the new edition of 'Ammonoid Paleobiology'. Not only for us, who had the honor and burden of this gigantic task, 'Ammonoid Paleobiology' represented a comparably important book since we consulted it so often in order to better understand these organisms, which went extinct 65 million years ago.

Although many of the early ammonoid researchers of the nineteenth century have spent thoughts on the ammonoid organism and its mode of life, most of the major contributions to modern ammonoid paleobiology appeared roughly in the past half century. Looking at the scientific output of these decades, it appears like the first edition of "Ammonoid paleobiology" was a product of something like a golden age of ammonoid research. The two decades preceding its publication saw the first five international symposiums "Cephalopods—Present and Past" and many important articles by colleagues such as John Callomon, Antonio Checa, John A. Chamberlain, Larissa Doguzhaeva, Jean-Louis Dommergues, Jean Guex, Roger H. Hewitt, Michael House, David K Jacobs, Jim Kennedy, Cyprian Kulicki, Neil Landman, Ulrich Lehmann, Harry Mutvei, Takashi Okamoto, Bruce Saunders, Yasunari Shigeta, Kazushige Tanabe, Henri Tintant, Jost Wiedmann, Peter D. Ward. Gerd Westermann, Yuri Zakharov (incomplete list!) contributed essential data and interpretations, but they also stimulated further research in this field. Unfortunately, many important cephalopod workers and good colleagues have passed way in the last two decades. In 2014 alone, for example, Fabrizio Cecca, Adolf Seilacher, Helmut Hölder, **Hiromichi** Hirano, and Gerd E. G. Westermann passed away.

Due to fundamental changes in the structure of scientific communities including the dubious judgment of the value of scientific work by impact factors and citation rates, cephalopod research has changed as well. Additionally, the community of ammonoid researchers appears to have started shrinking. Nevertheless, the past decades still saw thousands of interesting contributions on representatives of this fantastic clade. And still, we have a lot of work ahead of us prior to becoming able to respond to all questions regarding ammonoid paleobiology.

So what is new? In terms of content, we have restructured the former into a two-volume work with the main parts shell, ontogeny, anatomy, habit and habitats, macroevolution, paleobiogeography, ammonoids through time, fluctuations in ammonoid diversity, and taphonomy. Most of these parts are subdivided into chapters. The great amount of 41 chapters reflects the panel of ammonoid workers present nowadays in academia, junior and senior scientists from many countries and a higher percentage of female authors compared to the previous edition. We aimed at being as up-to-date as possible, which had the consequence that some chapters also present unpublished specimens, data and results. We also included two chapters on the geochemistry of ammonoid shells, a field that still offers vast possibilities for new research. This is also reflected in the slightly different views presented therein.

Furthermore, we added an introductory for the definition of terms and with a recommendation for the description of new ammonoid taxa. We emphasized the next challenges in ammonoid research such as reconstructing ammonoid phylogeny, understanding their intraspecific variability or reconstructing the soft parts. Studying intraspecific variability has been widely neglected, but it offers a wealth of possible implications for life histories, ontogeny, reproduction and, most importantly, for evolution. In this context, another challenge is establishing a phylogeny for ammonoids, and thus, one part comprising five chapters is dedicated to ammonoid macroevolution. In our eyes, paleontological data yield the essential information for research on evolution. As pointed out already by Seilacher and Eldredge, ammonoids are of particular interest due to their accretionary shell, which has a good fossilization potential and hold a record of their life history, their high evolutionary rates, their wide geographic distribution, high taxonomic diversity and morphological disparity as well as their well-constrained stratigraphic (i.e., temporal) framework. In the case of ammonoids, however, countless homoplasies occurred throughout their evolution, thus hampering attempts to reconstruct ammonoid phylogeny. Nevertheless, a sound phylogenetic model for the ammonoid clade should be one of the central tasks in ammonoid research because the knowledge of ammonoid phylogeny is still patchy. Furthermore, although some quantitative approaches have been pioneered with ammonoids (e.g., Raup's morphospace, Okamoto's growing tube model), such methods are still too little used in many studies on ammonoid paleobiology and evolution; many studies restrict themselves to narrative discussions or qualitative assessments. For this reason, the application of several quantitative and statistical methods to study many aspects of ammonoid like biostratigraphy, biogeography, intraspecific variability, evolutionary trends, etc. are explained and demonstrated in several of the chapters of these two volumes, in the hope these methods will be used more widely in the ammonoid community.

Finally, we added new information obtained from tomographic data obtained both from computer tomography and grinding tomography. The field of virtual paleontology has just started to deliver ammonoid data, which are of special interest in the studies of shell morphology, ontogeny, buoyancy, mode of life, and ultimately evolution.

These two volumes would have been impossible without our wonderful authors, and especially the help of Neil Landman as well as Kazushige Tanabe. Additionally, we greatly appreciate the support of the army of reviewers, who are listed and thanked in the corresponding chapters. Naturally, our partners and families have been affected more or less from the additional time consumed by the preparation of the volumes, we apologize for that and thank them for all their patience, inspiration, and support.

Christian Klug
Dieter Korn
Kenneth De Baets
Isabelle Kruta
Royal H. Mapes

Contents

Contributors

Kenneth De Baets GeoZentrum Nordbayern, Fachgruppe PaläoUmwelt, Universität Erlangen, Erlangen, Germany

Didier Bert UMR-CNRS 6118 Géosciences, Université de Rennes 1, campus Beaulieu, Rennes cedex, France

Laboratoire du Groupe de recherche en paléobiologie et biostratigraphie des ammonites (GPA), La Mure-Argens, France

John Chamberlain Department of Earth and Environmental Sciences, Brooklyn College of CUNY, Brooklyn, NY, USA

Doctoral Programs in Biology and Earth and Environmental Sciences, CUNY Graduate Center, New York, NY, USA

Larisa A. Doguzhaeva Department of Palaeobiology, Swedish Museum of Natural History, Stockholm, Sweden

René Hoffmann Department of Earth Sciences, Institute of Geology, Mineralogy, and Geophysics, Ruhr-Universität Bochum, Bochum, Germany

Bernhard Hostettler Naturhistorisches Museum, Bern, Switzerland

Andrzej Kaim Institute of Paleobiology, Polish Academy of Sciences, Warszawa, Poland

Helmut Keupp Institute of Geological Sciences, Branch Palaeontology, Freie Universität Berlin, Berlin, Germany

Christian Klug Paläontologisches Institut und Museum, University of Zurich, Zurich, Switzerland

Dieter Korn Museum für Naturkunde, Leibniz-Institut für Evolutions- und Biodiversitätsforschung, Berlin, Germany

Isabelle Kruta Division of Paleontology (Invertebrates), American Museum of Natural History, New York, NY, USA

Sorbonne Universités, UPMC Université Paris 06, CR2P-UMR 7207 CNRS, MNHN, Paris, France

Cyprian Kulicki Institute of Paleobiology, Polish Academy of Sciences, Warszawa, Poland

Neil H. Landman Division of Paleontology (Invertebrates), American Museum of Natural History, New York, NY, USA

Neal L. Larson Larson Paleontology Unlimited, Keystone, SD, USA

Jens Lehmann Fachbereich Geowissenschaften, University of Bremen, Bremen, Germany

Robert Lemanis Department of Earth Sciences, Institute of Geology, Mineralogy, and Geophysics, Ruhr-Universität Bochum, Bochum, Germany

Alexander Lukeneder Department of Geology and Paleontology, Natural History Museum, Vienna, Austria

Royal H. Mapes North Carolina Museum of Natural Sciences, Raleigh, NC, USA

Claude Monnet UMR CNRS 8198 Evo-Eco-Paleo, Université de Lille, UFR Sciences de la Terre (SN5), Villeneuve d'Ascq, France

Kazuyoshi Moriya Faculty of Natural System, Institute of Science and Engineering, Kanazawa University, Kakuma-machi, Kanazawa, Japan

Harry Mutvei Department of Palaeobiology, Swedish Museum of Natural History, Stockholm, Sweden

Carole Naglik Paläontologisches Institut und Museum, University of Zurich, Zurich, Switzerland

Horacio Parent Laboratorio de Paleontología, IFG–FCEIA, Universidad Nacional de Rosario, Rosario, Argentina

Kristin Polizzotto Department of Biological Sciences, Kingsborough Community College, Brooklyn, NY, USA

Takenori Sasaki Department of Historical Geology and Paleontology, The University Museum, The University of Tokyo, Tokyo, Japan

Amane Tajika Paläontologisches Institut und Museum, University of Zurich, Zurich, Switzerland

Kazushige Tanabe Department of Historical Geology and Paleontology, The University Museum, The University of Tokyo, Tokyo, Japan

Séverine Urdy Department of Anatomy, Department of Bioengineering and Therapeutic Sciences, University of California San Francisco, San Francisco, CA, USA

Physics Institute, Disordered and Biological Soft Matter, University of Zurich, Zurich, Switzerland

Margaret M. Yacobucci Department of Geology, Bowling Green State University, Bowling Green, Ohio, USA

Michał Zatoń Faculty of Earth Sciences, University of Silesia, Sosnowiec, Poland

Part I
Conch

Chapter 1
Describing Ammonoid Conchs

Christian Klug, Dieter Korn, Neil H. Landman, Kazushige Tanabe, Kenneth De Baets and Carole Naglik

1.1 Introduction

Because ammonoid jaws are rare (Tanabe et al. 2015) and preserved soft parts as well as radulae (Klug and Lehmann 2015; Kruta et al. 2015) are even rarer, most paleontologists are limited in the available morphological information to the conch when describing ammonoids. Taking the great diversity and disparity as well as the over 300 Ma of the clade's existence into account, it becomes obvious that the different ammonoid clades have divergent sets of characters requiring descriptive procedures adapted to the requirements. For example, in the earliest ammonoids, details

C. Klug (✉) · C. Naglik
Paläontologisches Institut und Museum, University of Zurich, Karl Schmid-Strasse 6, 8006 Zurich, Switzerland
e-mail: chklug@pim.uzh.ch

C. Naglik
e-mail: carole.naglik@pim.uzh.ch

D. Korn
Museum für Naturkunde, Leibniz-Institut für Evolutions- und Biodiversitätsforschung, Invalidenstraße 43, 10115 Berlin, Germany
e-mail: dieter.korn@mfn-berlin.de

N. H. Landman
Division of Paleontology (Invertebrates), American Museum of Natural History, Central Park West at 79th St., New York 10024-5192, NY, USA
e-mail: landman@amnh.org

K. Tanabe
Department of Historical Geology and Paleontology, The University Museum, The University of Tokyo, Hongo 7-3-1, Tokyo 113-0033, Japan
e-mail: tanabe@um.u-tokyo.ac.jp

K. De Baets
GeoZentrum Nordbayern, Fachgruppe PaläoUmwelt, Universität Erlangen, Loewenichstr. 28, 91054 Erlangen, Germany
e-mail: kenneth.debaets@fau.de

C. Klug et al. (eds.), *Ammonoid Paleobiology: From Anatomy to Ecology*, Topics in Geobiology 43, DOI 10.1007/978-94-017-9630-9_1

of the suture line and the ornamentation are often less important while conch geometry yields important information. By contrast, ornamentation and sutures can be essential for the systematics of Late Paleozoic and Mesozoic ammonoid groups, while conch shape might play a lesser role. Additionally, intraspecific variability differs strongly between ammonoid clades and thus, small differences between some forms might justify the introduction of a new species whereas in other clades, such a small difference could fall within the broad range of intraspecific variability (De Baets et al. 2015).

Nevertheless, we will try to give a guideline on the optimal features that systematic descriptions of ammonoid species should take into consideration offering some suggestions which certainly go beyond the normal framework of descriptions, but which would give them a special above average quality. At the same time we are well aware that some of our suggestions would lead to some kind of 'de luxe' description, presuming all our suggestions are fully implemented.

Naturally, this is not the first attempt to produce a guideline for a more uniform and intelligible mode of ammonoid description. Many pioneers, however, did not explicitly state their strategies in describing ammonoids in their monographs, although these authors commonly followed certain rules.

Miller et al. (1957) and Arkell (1957) summarized the available morphological terms in the Treatise for Invertebrate Paleontology for Paleozoic and Mesozoic ammonoids, respectively. As far as Paleozoic ammonoids are concerned, it was Ruzhencev (1960) who set the standards for the description of Paleozoic ammonoids. His descriptions are not only well-structured but also provide the same set of information in a uniform order, accompanied by photographs of lateral and ventral views as well as suture lines and often cross sectional drawings. His introduction to conch shape and terminology in the Osnovy Paleontologii (Ruzhencev 1962, 1974) belongs to the best that have been printed.

Branco (1879–1880) described general characteristics of the early internal conch features of some ammonoids. Subsequent works with SEM (Tanabe et al. 1979; Drushchits and Doguzhaeva 1982) have demonstrated that the study of ontogenetic development of internal structures is as important as that of suture, shape and sculpture of conchs to construct an adequate scheme of major taxonomy and systematics of Ammonoidea (Kulicki et al. 2015).

In his famous books, Lehmann (1976, 1981, 1990) presented important descriptive terms with simple line drawings. However, his main focus was on paleobiological aspects of ammonoids.

Landman et al. (1996) and Westermann (1996) also defined morphological terms in a qualitative way. They distinguished various types of conch shapes for 'normal', planispirally coiled, ammonoids (with touching or overlapping whorls): cadiconic, discoconic, elliptospheroconic, planorbiconic, platyconic, serpenticonic, spheroconic. They also use specific terms to refer to "heteromorph" ammonoids, which are not planispirally coiled and/or have successive whorls in contact with one another: ancyloconic, breviconic, gyroconic, hamitoconic, orthoconic, scaphitoconic, torticonic and vermiconic. For relative terms, they use 'evolute' for more loosely coiled conchs, 'involute' was used to refer to tightly coiled conchs with a large whorl overlap and 'advolute' was used to refer to whorls, which are touching but not overlapping.

Landman et al. (1996) and Westermann (1996) also used the terms 'brevidomic', 'mesodomic' and 'longidomic' to describe body chamber lengths of approximately one-half whorl, three-fourth whorl, and a whorl or more in length, respectively. Body chamber length is usually expressed as the Body chamber angle (BCA) or the

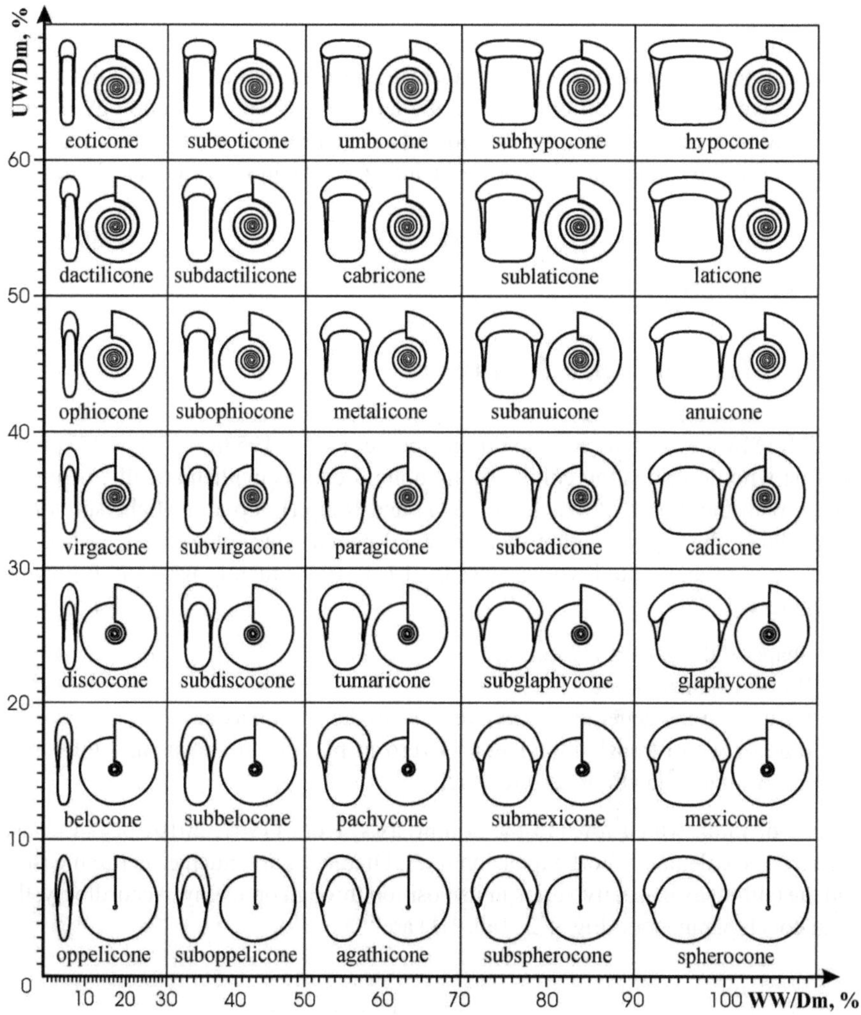

Fig. 1.1 Description of conch shapes as suggested by Kutygin (with permission, from Kutygin 1998) for Permian ammonoids. In the terminology of Arkell (1957), the following terms would be synonymous: oxycone—oppelicone; serpenticone—eoticone/dactilicone; platycone—suboppelicone/subbelocone; sphaerocone—subcadicone via mexicone and agathicone to spherocone

angular length measured from the septal neck (medial saddle of the external lobe) of the ultimate septum to the peristome (apertural edge), excluding lappets or rostra.

In several of his articles and monographs, Korn (e.g., Korn 1997; Korn and Klug 2002, 2003, 2007) quantified terms which he commonly uses to describe morphological aspects of ammonoid conchs. In order to make this more broadly known, he published "*A key for the description of Palaeozoic ammonoids*" (Korn 2010), where he listed terms, how to calculate certain ratios, and how to illustrate them properly.

Kutygin (1998) subdivided conch shapes of normally coiled ammonoids according to their umbilical width/conch diameter ratio versus whorl width/conch diameter ratio (Fig. 1.1). He outlined a theoretical morphospace of ammonoid conch-shapes, which he used to illustrate morphological change through ontogeny (Kutygin 2006).

Other examples for comprehensive definitions of terms are the monographs of Schlegelmilch (1976, 1985, 1994). He produced drawings of ribbing types, whorl cross sections, conch shapes, keels, shapes of apertures, and other conch parts.

Here, we provide an introduction to the terminology and methodology needed and/or recommended to describe ammonoids in general. There is such a wealth of terms, definitions and methods that we include only the most widely used ones.

1.2 Geometry

1.2.1 Classical Conch Parameters

Possibly, Moseley (1838) and Guido Sandberger (1851, 1953a, 1953b, 1857) were the first who described the coiling of ammonoid conchs mathematically. More recently, with the works of Trueman (1941) and Raup (Raup and Michelson 1965; Raup 1967), the quantification of ammonoid conch morphology has reached the 'high table' of ammonoid workers. Raup (1961, 1966) mainly used the following parameters:

S Shape of the generating curve;
W Whorl expansion rate;
D Position of the generating curve relative to the coiling axis;
T Rate of whorl translation. T equals zero in planispiral conchs and thus is of lesser interest in ammonoid research.

Instead of radii, which refer to the coiling axis, Korn (1997, 2010) began to use diameters to calculate whorl expansion rates. Diameters are much easier to measure and the coiling axis usually varies in its position through ontogeny. Accordingly, the main conch parameters (Fig. 1.2; Tab. 1.1) are:

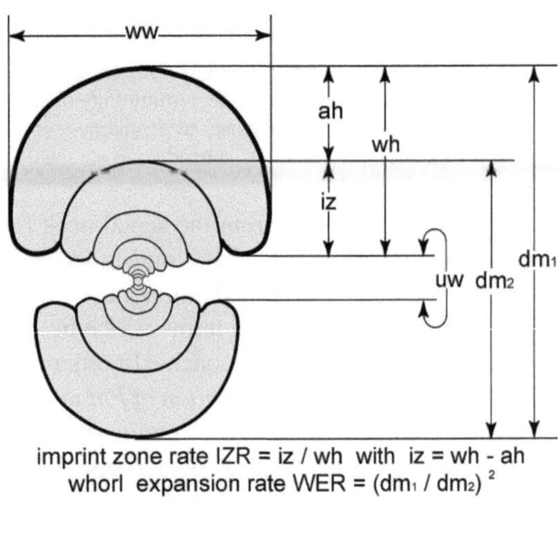

Fig. 1.2 Overview over the main conch parameters and ratios, exemplified on a cross section of the Middle Devonian ammonoid *Subanarcestes*

Table 1.1 Classification of the maximum conch diameters of individual specimens, the conch shape (ww/dm), the whorl width index (ww/wh), the umbilical width index (uw/dm), the whorl expansion rate (WER), and the whorl overlap or imprint zone rate (IZR). All values taken from Korn (2010)

Parameter	Descriptive term	Equation	Value
Max. conch diameter	Very small	dm	<25 mm
Max. conch diameter	Small	dm	25–50 mm
Max. conch diameter	Moderate	dm	50–100 mm
Max. conch diameter	Large	dm	100–200 mm
Max. conch diameter	Very large	dm	>200 mm
Conch shape	Extremely discoidal	ww/dm	<0.35
Conch shape	Discoidal	ww/dm	0.35–0.60
Conch shape	Pachyconic	ww/dm	0.60–0.85
Conch shape	Globular	ww/dm	0.85–1.10
Conch shape	Spindle-shaped	ww/dm	>1.10
Whorl width index WWI	Strongly compressed	ww/wh	<0.50
Whorl width index WWI	Weakly compressed	ww/wh	0.50–1.00
Whorl width index WWI	Weakly depressed	ww/wh	1.00–1.50
Whorl width index WWI	Moderately depressed	ww/wh	1.50–2.00
Whorl width index WWI	Strongly depressed	ww/wh	2.00–2.50
Whorl width index WWI	Very strongly depressed	ww/wh	2.50–3.00
Whorl width index WWI	Extremely depressed	ww/wh	>3.00
Umbilical width index UWI	Very narrow (involute)	uw/wh	<0.15
Umbilical width index UWI	Narrow (subinvolute)	uw/wh	0.15–0.30
Umbilical width index UWI	Moderate (subevolute)	uw/wh	0.30–0.45
Umbilical width index UWI	Wide (evolute)	uw/wh	0.45–0.60
Umbilical width index UWI	Very wide (very evolute)	uw/wh	>0.60
Whorl expansion rate WER	Very low	$[dm/(dm-ah)]^2$	<1.50
Whorl expansion rate WER	Low	$[dm/(dm-ah)]^2$	1.50–1.75
Whorl expansion rate WER	Moderate	$[dm/(dm-ah)]^2$	1.75–2.00
Whorl expansion rate WER	High	$[dm/(dm-ah)]^2$	2.00–2.25
Whorl expansion rate WER	Very high	$[dm/(dm-ah)]^2$	2.25–2.50
Whorl expansion rate WER	Extremely high	$[dm/(dm-ah)]^2$	>2.50
Imprint zone rate IZR	Weakly embracing	(wh − ah)/wh	<0.15
Imprint zone rate IZR	Moderately embracing	(wh − ah)/wh	0.15–0.30
Imprint zone rate IZR	Strongly embracing	(wh − ah)/wh	0.30–0.45
Imprint zone rate IZR	Very strongly embracing	(wh − ah)/wh	>0.45

- conch diameter: The maximum diameter is abbreviated as dm (or dm_1). In order to determine the whorl expansion rate, a second diameter value is needed, namely the diameter measured half a whorl earlier (180° behind the aperture or dm_1, respectively; dm_2). The conch diameter has often been used as proxy for size (and relative age). However, other properties like body chamber volume might be more suitable as a proxy for size because it better reflects the volume of the soft body than the conch diameter, especially when comparing forms with very different conch geometries (e.g., Bucher et al. 1996; Dommergues et al. 2002; De Baets et al. 2012, 2013a, 2015). In extant coleoids (Nixon and Young 2003; Boyle and Rodhouse 2005), mostly the (dorsal) mantle length (which would correspond with the body chamber length in ammonoids) is used as a measure of size. Other measures are also used such as weight (which would correspond to the weight of the soft tissue with or without the conch) or the total length (with arms as they can form a major part of the coleoid). Nevertheless, the diameter will always be an important parameter in ammonoids as it is easy to obtain and has been widely used and available in the literature (Bucher et al. 1996).

- whorl width: It is measured perpendicular to the plane of symmetry and abbreviated as ww. In ornamented forms, this parameter is commonly measured between the ornament, so it represents a kind of minimal value. If this measurement is taken from older ontogenetic stages (e.g., from cross sections) in half a whorl distance (each 180 degrees), the values are labeled accordingly ww_1, ww_2, ww_3. This can also be done with the following parameters.
- whorl height: This parameter, abbreviated as wh, is measured parallel to the plane of symmetry from the umbilical seam or umbilical wall to the middle of the venter.
- umbilical width: Being a secondary parameter, it can be measured from umbilical wall to umbilical wall or it can be calculated as follows:
  $uw = dm_1 - wh_1 - wh_2$
- aperture height: This value is measured from the dorsum of the preceding whorl to the dorsum of the whorl under consideration. It can also be calculated:
  $ah = dm_1 - dm_2$
- imprint zone width: This parameter describes the degree of whorl overlap and is measured from the umbilical seam of the whorl under consideration to the dorsum of the preceding whorl. It may be calculated using the following equation:
  $iz = wh_1 - ah = wh_1 - (dm_1 - dm_2)$

1.2.2 Cross Section and Ratios

An easy way to assemble a lot of morphometric data from ammonoids is to produce cross sections perpendicular to the plane of symmetry and through the initial chamber. This allows quantification of ontogenetic change in the parameters listed above and also makes changes in shell thickness and in whorl cross section visible. A peculiar aspect of conch shape, made visible by cross sections, is the umbilical lid (a continuation of the lateral conch wall partially covering the umbilicus) of the Early Devonian auguritids and the Middle Devonian pinacitids (Klug and Korn 2002; Monnet et al. 2011) as well as in Middle Devonian pharciceratids (Bockwinkel et al. 2009). In the Auguritidae and Pinacitidae, the lateral wall begins to extend over the umbilicus starting in the juvenile whorls. Although this is just an example, such cross sections can also reveal shell thickenings at the umbilicus, keels and other morphological details (e.g., Tozer 1972).

The greatest advantage of cross sections is the access to comprehensive morphometric data throughout ontogeny. In order to assure accuracy of the cross sections, the section should optimally run through the maximum diameter of the initial chamber (protoconch) and should be perpendicular to the plane of symmetry (Fig. 1.3). The values measured on complete specimens or sections can then be used to calculate the following simple ratios (Korn 2010):

- conch width index: CWI = ww/dm
- whorl width index: WWI = ww/wh
- umbilical width index: UWI = uw/dm = $(dm_1 - wh_1 - wh_2)/dm_1$

Based on the conch width index and the umbilical width index, the conch shapes and cross sections can be classified as (Fig. 1.4):

- discoidal (CWI < 0.60)
- pachyconic (0.60 ≤ CWI < 0.85)
- globular (0.85 ≤ CWI < 1.10)
- spindle-shaped (CWI ≥ 1.10)

According to the umbilical width, ammonoid conchs can be termed as

- involute (UWI < 0.15)
- subinvolute (0.15 ≤ UWI < 0.30)
- subevolute (0.30 ≤ UWI < 0.45)
- evolute (0.45 ≤ UWI < 0.60)
- very evolute to advolute (UWI ≥ 0.60)
- advolute: whorls touch but do not overlap
- heteromorphic/criocone: whorls do not touch

Cross sections also better reveal details of the conch morphology such as the vaulting of lateral or ventral walls. They help to describe the whorl cross section more correctly.

1.2.3 Expansion Rates

Due to their nearly logarithmic conch growth, most conch parameters also increase at differing rates. Caused by allometric growth, the change in certain parameters through ontogeny is not necessarily perfectly linear in a loglog-space (Kant 1973; Kant and Kullmann 1980; Klug 2001; Korn and Klug 2003; Urdy et al. 2010a, 2010b; Korn 2012; Urdy 2015). In order to quantify these changes, parameters taken from transverse cross sections or values measured in the plane of symmetry can be used.

Longitudinal (median) sections should optimally be in the plane of symmetry, i.e. the siphuncle or the siphuncular perforations should be visible completely. These sections offer the opportunity to measure parameters such as apertural height

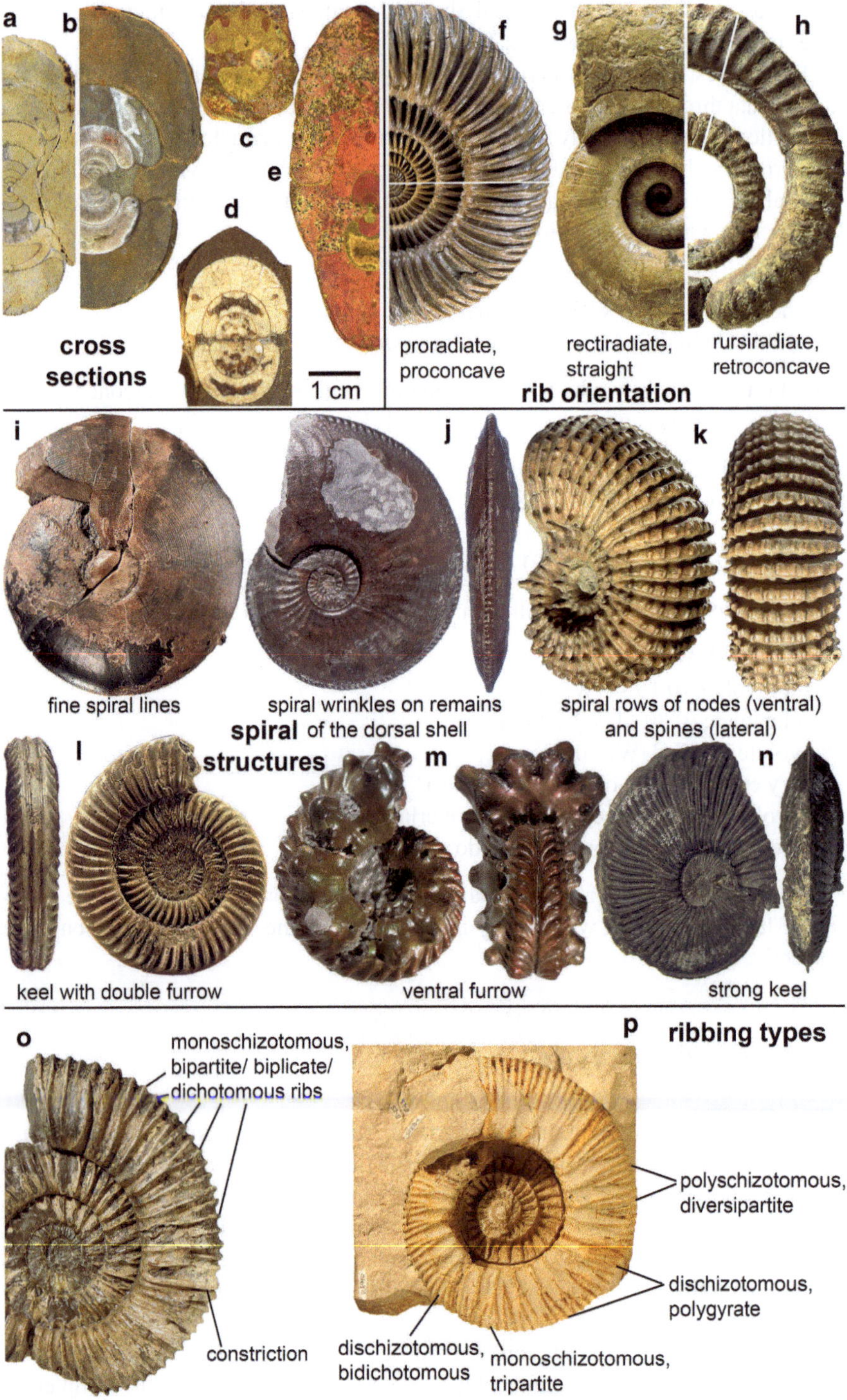

Fig. 1.3 Example of **cross sections** of various ammonoids: **a** *Sellanarcestes* cf. *tenuior*, late Emsian, Devonian, Filon 12, Tafilalt, Morocco; protoconch is visible, almost perpendicular to the

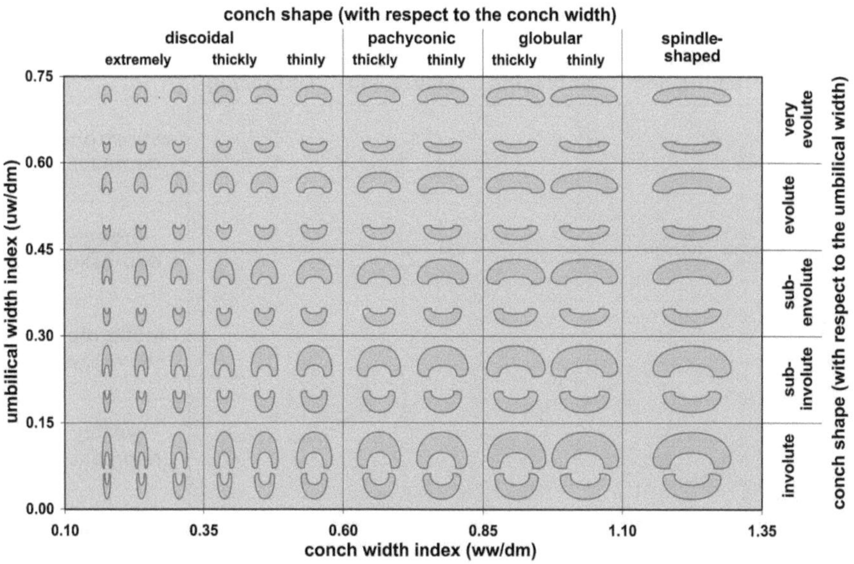

Fig. 1.4 Bivariate plot illustrating the terminology of the conch width index (ww/dm) on the x-axis and umbilical width index (uw/dm) on the y-axis (from Korn 2010)

and diameter in small increments, while whorl width, umbilical width, imprint zone width, and whorl height cannot be measured. Additionally, the angle between septa becomes measurable (Bucher et al. 1996; Kraft et al. 2008).

---

plane of symmetry (note the septa). **b** *Subanarcestes macrocephalus*, middle Eifelian, Filon 12, Tafilalt, Morocco; note the siphuncle. **c** *Goniatites multiliratus*, Visean, Elm Creek, Oklahoma, USA; note the symmetry in the septa, indicating a plane perpendicular to the plane of symmetry. **d, e** *Macrocephalites* sp., PIMUZ 19078, Callovian, Jurassic, Anwil, Switzerland; note the approximately symmetrically cut septa. **Orientation of ornament:** *F, Parkinsonia parkinsoni,* Bajocian, Port-en-Bessin, France, dm 124 mm, Staatliches Museum für Naturkunde Stuttgart. G, *Lytoceras fimbriatum,* Pliensbachian, Jurassic, Fresney-le-Puceux near Caen, France, dm 209 mm, Staatliches Museum für Naturkunde Stuttgart. H. *Erbenoceras advolvens,* GPIT 1849-2002, early Emsian, Devonian, northern Tafilalt, Morocco, dm 156 mm. (all images: W. Gerber, Tübingen; A, B reproduced from Ernst and Klug 2010). **Spiral ornamentation.** I to K: Lateral structures. I, *Maxigoniatites saourensis,* Visean, Carboniferous, near Merzouga, Tafilalt, Morocco, dm 72 mm. J. *Amaltheus margaritatus,* PIMUZ 13468, Pliensbachian, Reichenbach near Aalen, Germany. K, *Douvilleiceras mammillatum,* Albian, Cretaceous, Courcelles near Troyes, France, dm 10.8 cm, image: A. E. Richter, Augsburg. L to N: ventral structures. L, *Arietites* sp., Sinemurian, Jurassic, Mögglingen, Germany, dm 70 mm, Staatliches Museum für Naturkunde Stuttgart. (image: W. Gerber, Tübingen). M, *Euhoplites proboscideus,* PIMUZ 23108, Lower Gault, Albian, Cretaceous, Folkestone, Kent, UK, dm 40 mm. N, *Venezoliceras karsteni,* J 17830, Albian, NNE of Barbacoa, Venezuela, dm 110 mm, Naturhistorisches Museum Basel. **Ribbing patterns.** O, *Virgatisphinctes* sp., PIMUZ 16975, Unterhausen near Neuburg/Donau, Germany, dm 110 mm. P, *Pavlovia pallasioides,* Kimmeridgian, Jurassic, Kimmeridge Bay, Dorset, UK, dm 140 mm.

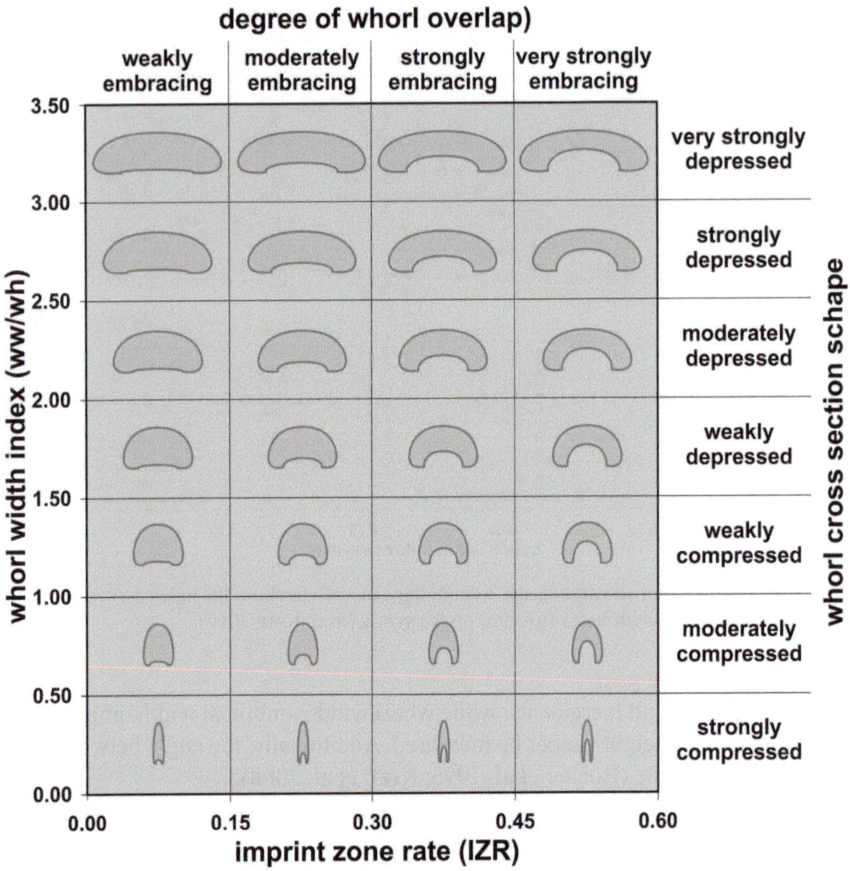

Fig. 1.5 Bivariate plot illustrating the terminology of the imprint zone rate (IZR) on the x-axis and whorl width index (ww/wh) on the y-axis (from Korn 2010)

Parameters measured through ontogeny on either kind of cross sections can be utilized to calculate the following expansion rates:

- whorl expansion rate:
  $WER_1 = (dm_1/dm_2)^2 = [dm_1/(dm_1 - ah_1)]^2$
- imprint zone rate:
  $IZR_1 = wh_1 - ah1/wh_1 = -[wh_1 \cdot (dm - dm_2)]/wh_1$

See Fig. 1.5 and Tab. 1.1 for subdivisions of whorl expansion rates and imprint zone rates.

Korn and Klug (2002) introduced a slightly different formula for the Whorl Expansion Rate (WER) than the one used by Raup (1967), which better reflects the growth of the soft-body during ontogeny (in loosely coiled forms) and which is easier and more precisely applicable than the classical equation proposed by Raup and Michelson (1965).

Parent et al. (2010, 2011) also independently arrived at a similar formula for planispirally coiled Mesozoic forms. They slightly modified the Raup-model to also include planispirally coiled forms with non-touching whorls.

1.3 Ornamentation

1.3.1 Radial Elements

All ammonoids bear fine or coarse radial elements on the conch. The finest structures are commonly the growth lines (Bucher et al. 1996), which are formed during conch growth. They form when shell is secreted discontinuously at the aperture and may be spaced at distances of around 0.1 mm (Vermeij 1993; Bucher et al. 1996). Characteristically, they are interrupted and cannot be traced around the entire whorl. To examine them, well preserved original or replacement shell is needed.

Lirae are usually much stronger; they are also formed more or less regularly with distances sometimes exceeding 1 mm. Normally, lirae can be traced around the ammonoid's circumference, but the limits between growth lines and lirae are not clearly defined. Both are simply fine and coarse traces of former apertures, recording their shape through growth.

Ribs represent even larger undulations in the conch wall and are not present in all ammonoid taxa. Their shape, arrangement, strength, etc. varies broadly and significant changes during ontogeny may be observed. Rather often, ribs continue into nodes or spines. They still carry valuable taxonomic information, although the strength of the ribs often covaries with the whorl cross section (Checa et al. 1996). Their strength, spacing and orientation may be quantified for taxonomic purposes (e.g., De Baets et al. 2013a, 2013b).

Constrictions are less frequent than the previously mentioned radial elements; they usually occur in a lower number than ribs (often between one and five per whorl) and commonly are produced at growth halts (megastriae; see Bucher et al. 1996; Urdy 2015). At least on the internal mould (steinkern), constrictions are visible as furrows. Often, constrictions are internal shell thickenings which may have made interim apertures more resistant against mechanical damage by any cause during growth halts. In some cases, the shell thickening equalized the inward bent shell surface in such a way, that it is barely visible from the outside. Since they represented growth halts, the orientation of younger radial elements tend to display an orientation differing from that of the preceding ones. In some cases, these interim apertures carried collars, spines or nodes, which can be diagnostic for certain taxa.

The orientation of the radial elements (Fig. 1.3, 1.6) can be described as rectiradiate (radial orientation), proradiate or prorsiradiate (turning toward the aperture in the ventral direction) and rursiradiate (turning away from the aperture in the ventral direction). Depending on their curvature, the ribs can be concave (vaulted away from the aperture) or convex (vaulted towards the aperture); these two terms can

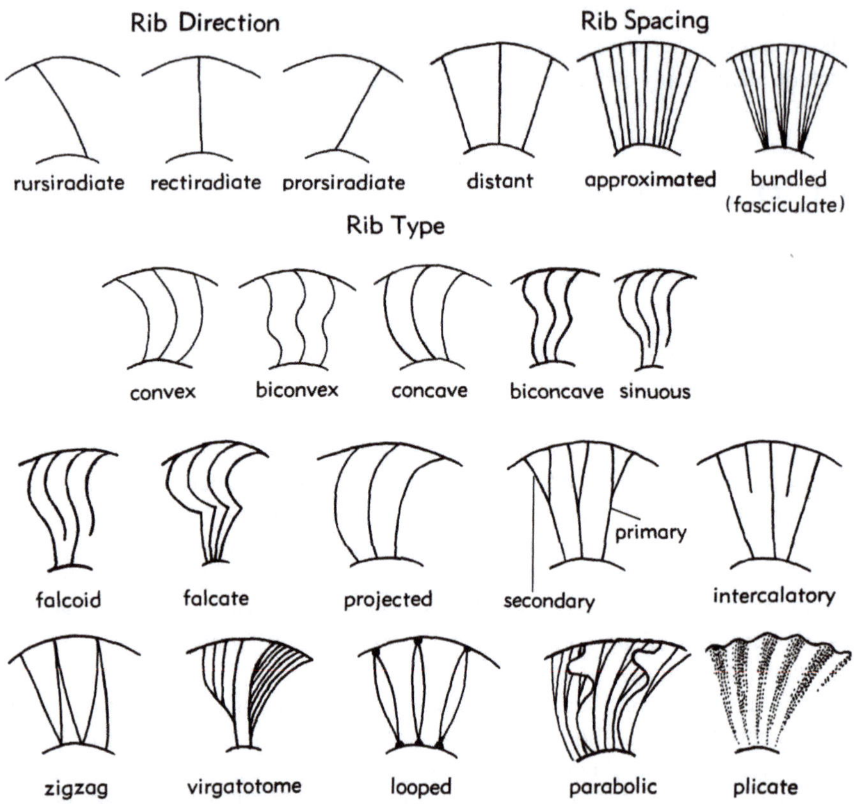

Fig. 1.6 Rib shape, spacing and course (modified from Arkell 1957)

be combined with the prefix pro- when they are inclined anteriorly (ventrad) and retro- when they are inclined posteriorly (ventrad). If the rib is partially concave and partially convex, it is called sinusoidal (sigmoid, sinusoid) and if the dorsal part of the rib is straight it is termed falcate. Ribs can split in various ways (Fig. 1.3, 1.6):

- simple (not branching)
- monoschizotomous (branching once): primary splits into two (bipartite, biplicate, dichotomous), three (tripartite) or four (quadripartite) secondary ribs
- dischizotomous (branching twice): primary splits into three (polygyrate) or four (bidichotomous) branches
- polyschizotomous (branching more than twice): branching only on one side of the primary rib (virgatipartite, virgatotomous) or branching on both sides of the primary (diversipartite)
- fibulate: ribs split and fuse again, forming a narrow ellipsis

In several Devonian and Carboniferous ammonoid species, subadult to adult specimens display the wrinkle layer (Korn et al. 2013). In all cases, these wrinkles are irregular and form a kind of fingerprint pattern on the dorsal shell. The elevation of the wrinkles varies between a fraction of a millimetre and a few millimetres. In contrast to the wrinkle layer, Ritzstreifen extend over the entire conch and occur only in the Devonian (Sandberger and Sandberger 1850; Korn et al. 2013; compare Kulicki et al. 2015).

The spacing of radial elements is usually measured per half-whorl or demi-whorl (e.g., RDW or ribs per demi-whorl = the amount of ribs counted on a half-whorl). Other parameters such a rib-indexes have also been used to quantify rib spacing more locally (or on fragments: compare Yacobucci 2004; De Baets et al. 2013a).

1.3.2 Spiral Elements

Spiral ornament can be subdivided according to position, i.e. ventral, lateral, or dorsal. Many ammonoids display spiral ornament such as spiral lines, which are usually rather weak compared to many radial structures (Fig. 1.3). Spiral lines are particularly common in Paleozoic ammonoids, where they sometimes form reticulate patterns when they occur in combination with radial lirae or ribs. Another common phenomenon is spiral rows of spines or tubercles (Fig. 1.3), which may occur laterally and ventrally.

In the Early Jurassic Amaltheidae, the dorsal conch commonly displays spiral wrinkles comparable to the radial wrinkles of the wrinkle layer known from Paleozoic ammonoids (Fig. 1.3). As far as ventral structures are concerned, keels have to be mentioned. These may be sharp or rounded, they can be connected to the flanks with a smooth transition or they can be clearly set off, they may be accompanied by a pair of furrows or a set of several parallel keels can occur (e.g. in Frasnian Beloceratidae). Families such as the Parkinsoniidae or the Hoplitidae have a midventral furrow.

1.3.3 Spines, Nodes, Tubercles

There are several kinds of ornamentation, which are neither truly radial nor spiral in orientation. Spines are pointed and elongate, while tubercles and nodes are knob-shaped. The term node is sometimes used for bigger structures, although the use is not uniform and some might consider nodes and tubercles synonymous terms. All these structures can be arranged radially and/or spirally, for example in *Douvilleiceras* (Fig. 1.3).

Some Paleozoic forms have developed deep ventral sinuses in their aperture (ventral band). At the edge of this sinus, collar-like projections developed in some genera, which formed long ventral ‘median spines’ in genera such as *Armatites* or *Kosmoclymenia* (e.g., Korn 1979, 2014).

1.4 Septa

1.4.1 Suture Line

The suture line is the line, where the septal mantle first attached the organic septal membrane and later the septum is formed by mineralization of the membrane. Its importance in systematics and taxonomy varies, depending on the researcher and also on the taxon under consideration. In Early and some Middle Devonian forms, differences in the suture line are sometimes so subtle that other conch characters are of greater use (e.g., Chlupáč and Turek 1983).

Nevertheless, the suture line yields valuable information on systematics and ultimately also phylogeny. In order to produce good drawings of suture lines, growth lines or constrictions, various techniques can be used.

1. A very simple procedure that can be applied to sutures, which lack microscopic detail, is the following: A sharp pencil is used to trace the suture line directly on the specimen. Afterwards, a strip of thin transparent duct tape is used to cover the entire suture line under consideration. After rubbing the surface of the tape, where the suture was colored before, the tape can be removed and attached to a sheet of paper. Next steps are scanning and tracing the suture line formerly copied on tape with any vector graphic software.
2. The classical method is to mount the specimen with modeling clay under a binocular microscope and then use a drawing mirror (camera lucida) or a grid within an ocular in order to transfer the suture on paper. In order to depict the entire suture, the specimen needs to be turned and mounted again in a new position on the modeling clay. The raw drawing can then also be scanned and traced with vector graphic software.

The convention is that the saddles (Klug and Hoffmann 2015) point with their convex sides towards the top (aperture). The plane of symmetry (the center of the E-lobe) is marked by an arrow, the umbilical shoulder (if present) can be indicated by a dotted line, the umbilical seam by a curve segment and the dorsal intersection with the plane of symmetry is indicated by either two straight dashes or two straight lines. If possible, mostly the right side of the suture is depicted, at least until the umbilicus and, if visible, the internal suture is also added.

In the case of suture lines, it is also very helpful, when more than one ontogenetic stage is depicted, because the change in complexity through ontogeny can be extreme, especially in Mesozoic species. It is also important to illustrate an adult suture, because usually, the adult sutures display the peak complexity.

1.4.2 The Septum in Space

In many publications, the third dimension of the septum is neglected. This is somewhat justified because normally, the septum displays the strongest folding at the suture line. The way in which the septum is folded, however, might yield additional information for the discrimination of taxa or for the reconstruction of phylogenetic relationships. Accordingly, it can be rewarding to pay special attention to the morphology of the entire septum, also because it might display soft-tissue imprints (Klug et al. 2008).

To some extent, septum shape depends on the whorl cross section. For example, in strongly compressed as well as in extremely depressed forms, there are often high numbers of sutural elements (Ruzhencev 1949). Corresponding pairs of sutural elements are often linked by bulges in the septum, namely in the case of compressed forms symmetrically arranged in lateral direction, in the case of depressed forms arranged in approximately dorsoventral direction. Depending on the orientation of this bulging (see Klug and Hoffmann 2015 for illustrations), the terms central fluting (bulges are radially arranged relative to the initial chamber), lateral fluting (bulges are perpendicular to the plane of symmetry) or radial fluting (bulges are arranged radially around the center of the septum) were introduced. Additionally, ammonoid septa may be synclastically (concave toward the aperture) or anticlastically folded (partially concave and partially convex toward the aperture).

1.5 Discriminating New Species

Naturally, the requirements for the introduction of a new species are not uniform across all taxonomic and stratigraphic boundaries. Nevertheless, some common rules apply to most groups of ammonoids. In the following, we will highlight some important aspects that can be taken into account, when new species are described. We are well aware that not all material yields all the information to perform all the studies listed below.

1.5.1 Ontogeny

A common problem with many taxa that have been introduced in the nineteenth century is that hardly anything is known about ontogenetic changes in these taxa. However, some parts of ammonoid conchs grow allometrically (Klug 2001; Korn 2012) and in the course of their growth, variability was not uniform (Ropolo 1995; De Baets et al. 2013a). This has been more generally shown for mollusk conchs by Urdy et al. (2010a, 2010b). Variability is often the lowest in the early and the latest whorls, i.e. these are the most characteristic, but still variable (De Baets et al. 2015).

If the material permits, as many of the major ontogenetic stages (embryonic conch, neanoconch, juvenile conch, preadult conch, adult conch; Westermann 1996; Klug 2001) as possible, especially of the last three stages, should be displayed and described in order to avoid that future researchers ascribe different ontogenetic stages of the same species to a different taxon.

1.5.2 Intraspecific Variability and the Quality of Characters

Discrete characters such as the absence or presence of certain structures or significant differences in numbers of lobes can be very helpful to identify species and also to justify the introduction of new species. In the case of characters, where transitions in character states between the supposedly new species and closely related species are known, intraspecific variability can be examined based on some tens of specimens of one size class in order to use such a character to explain the separation of a new species (De Baets et al. 2013a, 2015). The obvious disadvantage of the quantitative evaluation of intraspecific variability and its description as well as illustration is that they cost a lot of time and that they require a lot of printed space. Therefore, compromises are usually unavoidable and intraspecific variability cannot be examined for every single species. Nevertheless, it is a good idea to attempt to understand the intraspecific variability of the group one has to deal with, because then, the meaning of differences in any character between specimens can be more confidently interpreted with respect to its meaning, be it variation within a species or a difference in taxon.

When differences between supposedly new species in ratios such as UWI, CWI or WWI and in expansion rates such as WER and IZR (see Chap. 1.2) are evaluated, awareness of the respective intraspecific variability of the character under consideration can be of great help to both justify species separation and to avoid mistakes (e.g. by overestimating the character's meaning); however, intraspecific variability is roughly known only for a few species and genera, hampering such studies. This implies that, if time, the material, and thus the morphometric data permit, tests could be carried out to understand how the various character states are distributed through ontogeny. Are they normally distributed within a size class? Are one or several maxima present? Optimally, there should be two or more clearly separate peaks in the curve in order to make a quantitative character useful for species separation. An example, how such data can be represented, is given in Fig. 1.7. In any case, intraspecific variability of ammonoids is so poorly studied that it yields a wide array of possibilities for future studies (De Baets et al. 2015).

As stated above, intraspecific variability changes through ontogeny; it is usually the highest in middle whorls. This is partially reflected in some studies on covariation, Buckman's laws as well as in some other articles on variability (Hohenegger and Tatzreiter 1992; Dagys and Weitschat 1993; Checa et al. 1996; Korn and Klug 2007; De Baets et al. 2013a). If possible, we recommend basing descriptions on several specimens displaying several ontogenetic stages only. Furthermore, it is very helpful to include information on adult specimens, because adult conch modifications can show important diagnostic characters.

Sexual dimorphism (Klug et al. 2015) also contributes to intraspecific variability, especially in Mesozoic forms (Makowski 1962; Callomon 1963; Westermann 1964). Since dimorphism mostly applies to the last part of ontogeny, aspects of variability linked to dimorphism can be discriminated from variability within one sex.

Another poorly studied topic is differences in variability between regions; one of the inherent problems is the difficulty to reconstruct whether regional morpho-

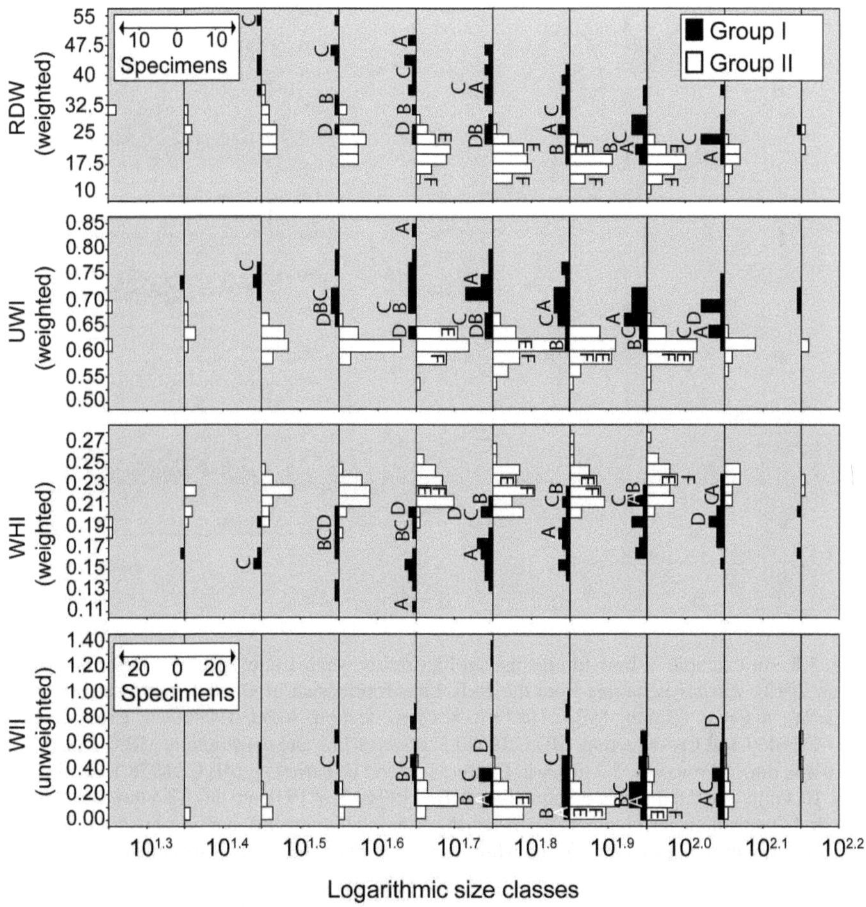

Fig. 1.7 Diagrams from De Baets et al. (2013a). Back-to-back histograms of Group I (*black*) and Group II (*white*) for ribs per half-whorl, RDW (weighted); umbilical width index, UWI (weighted); whorl height index, WHI (weighted); and whorl interspace index, WII (nonweighted)

logical differences originate in the facts that they are different species or whether these differences were caused by phenotypic plasticity or variation (Jacobs et al. 1994; Wilmsen and Mosavinia 2011). Body size may have varied geographically; intraspecific variability has certainly differed between regions, too (De Baets et al. 2015).

Although the study of intraspecific variability might appear as a nuisance, partially because it is time-consuming and partially because it is difficult to understand and describe in detail, it is actually an interesting topic for research since variation is essential for evolution, particularly heritable phenotypic variation (Hunt 2004, 2007). Furthermore, research on links between ecology and variability can also be rewarding (Jacobs et al. 1994; De Baets et al. 2015).

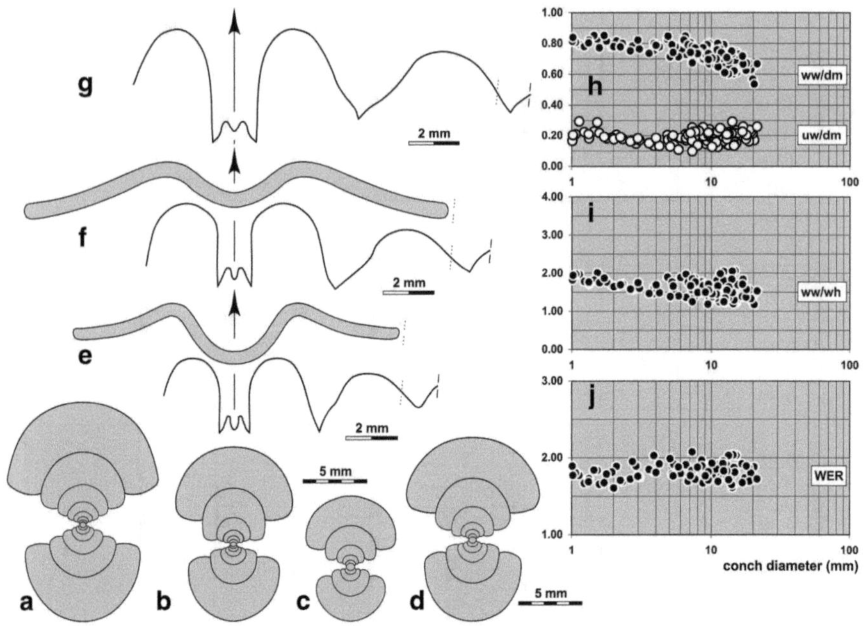

Fig. 1.8 An example of how to organise an illustration when a new species is described (Korn et al. 2010): *Eurites permutus* from the early Late Tournaisian of Oued Temertasset (Mouydir, Algeria). **a** Cross section, MB.C.18849.1. **b** Cross section, MB.C.19040.3. **c** Cross section, MB.C.19040.4. **d** Cross section, MB.C.19040.5. **e** Suture line and constriction, MB.C.18835.1, at 11.0 mm dm, 7.9 mm ww, 5.3 mm wh. **f** Suture line and constriction, MB.C.18978.1, at 14.0 mm dm, 10.4 mm ww, 7.0 mm wh. **g** Suture line, MB.C.19040.1, at 19.0 mm dm, 12.5 mm ww, 9.0 mm wh. **h–j** Ontogenetic development of the conch width index (ww/dm), umbilical width index (uw/dm), whorl width index (ww/wh), and whorl expansion rate (WER) of all specimens

1.5.3 Number of Specimens and Figures

As ammonoids may show great intraspecific variability and since many taxa display allometric growth in their conch, it is advantageous when more than one specimen is available when a new species is described. If possible, these specimens should show the major growth stages, especially the juvenile, the preadult, and the adult stage. Similarly, it is helpful when the main allometric changes, as well as a large part of the intraspecific variability of one growth stage can be illustrated.

As far as the number of figures is concerned, we recommend producing some graphs illustrating the ontogenetic change of morphometric aspects of the ornamentation (rib spacing, ornament strength, ornament orientation etc.), UWI, WWI and WER through ontogeny of several specimens. This yields an idea of intraspecific variability and allometry. Optimally, one specimen of each growth phase should be illustrated, with the adult growth phase being possibly the most important. If dimorphism is strongly expressed, a dimorphic pair can be illustrated. Furthermore, a cross section (photograph or drawing after a photograph) may yield valuable ontogenetic information and drawings of the suture as well as of growth lines, the aperture, constrictions, etc. Some of these illustrations may be meaningfully combined (see Fig. 1.8).

1.6 Organizing A Species Description

Classically, a diagnosis is given first, usually combined with information on synonymies, origin of the name, type material, its provenance, other materials used and the repository. Although slightly dated, Matthews (1973) and Bengtson (1988) are still some of the best references for how best to use synonymy lists and open nomenclature, respectively. The diagnosis should be concise and contain the major characters of the taxon. It should list the main aspects of conch morphology and ontogeny, ornament, and suture line (Korn 2010).

In the detailed description, the same topics should be addressed in the same order, but in greater detail. The task and strength of such a descriptive text is to highlight important parameters and character states. Comparisons can be listed in the "comparisons" or "remarks" paragraph.

As most ammonoids underwent more or less profound changes in morphology throughout their ontogeny, it is advantageous to provide a reasonable amount of information on these ontogenetic trajectories, at least as far as such data can be obtained. Some examples for representations of ontogenetic changes can be found in the following articles: Korn (1997); Klug (2001); Monnet et al. (2012).

Although Korn et al. (2010) replaced the descriptive paragraphs in the systematic section by tables with the main morphological information, many might want to list the characters and their states in descriptions. One can order the descriptions according to

- specimen
- ontogeny (initial chamber, ammonitella, juvenile/neanic, preadult, adult/mature/gerontic)
- character (conch shape, ornamentation, suture line)

It is the easiest for the reader, when one of these orders is chosen and adhered to throughout the entire manuscript. Naturally, several aspects will vary according to the individual style.

Acknowledgments Some of the insights grew in the course of the research projects with the numbers 200021-113956/1, 200020-25029, and 200020-132870 funded by the Swiss National Science Foundation SNF (CK, CN, KDB). We greatly appreciate the constructive reviews of Ottilia Szives (Budapest) and Christian Meister (Geneva).

References

Arkell WJ (1957) Introduction to Mesozoic Ammonoidea. In: Moore RC (ed) Treatise on Invertebrate Paleontology. Part L, Mollusca 4, Cephalopoda-Ammonoidea. GSA and University of Kansas Press, L80–L100

Bengtson P (1988) Open nomenclature. Palaeontology 31:223–227

Bockwinkel J, Becker RT, Ebbighausen V (2009) Upper Givetian ammonoids from Dar Kaoua (Tafilalt, SE Anti-Atlas, Morocco). Berl Paläobiol Abh 10:61–128

Boyle P, Rodhouse P (2005) Cephalopods: ecology and fisheries. Blackwell Publishing, Singapore

Branco W (1879–1880) Beiträge zur Entwicklungsgeschichte der fossilen Cephalopoden. Palaeontogr 26(1879):15–50 (27(1880):17–81)

Bucher H, Landman NH, Klofak, SM, Guex J (1996) Mode and rate of shell growth. In: Landman NH, Tanabe K, Davis RA (eds) Ammonoid paleobiology. Plenum, New York

Callomon JH (1963) Sexual dimorphism in Jurassic ammonites. Trans Leic Lit Philos Soc 57:21–56

Checa A, Company M, Sandoval J, Weitschat W (1996) Covariation of morphological characters in the Triassic ammonoid *Czekanowskites rieberi*. Lethaia 29:225–235

Chlupáč I, Turek V (1983) Devonian goniatites from the Barrandian area. Rozpr Ustred Ust Geol 46:1–159

Dagys AS, Weitschat W (1993) Extensive intraspecific variation in a Triassic ammonoid from Siberia. Lethaia 26:113–121

De Baets K Klug C Korn D Landman NH (2012) Early evolutionary trends in ammonoid embryonic development. Evolution 66:1788–1806

De Baets K Klug C Monnet C (2013a) Intraspecific variability through ontogeny in early ammonoids. Paleobiology 39(1):75–94

De Baets K Klug C Korn D Bartels C Poschmann M (2013b) Emsian Ammonoidea and the age of the Hunsrück Slate (Rhenish Mountains, Western Germany). Palaeontogr A 299(1–6):1–113

De Baets K, Bert D, Hofmann R, Monnet C, Yacobucci MM, Klug C (2015) Ammonoid intraspecific variability (this volume)

Dommergues J-L, Montuire S, Neige P (2002) Size patterns through time: the case of the Early Jurassic ammonite radiation. Paleobiology 28:423–434

Drushchits VV, Doguzhaeva LA (1982) Ammonites under the electron microscope. Moscow University Press, Moscow (in Russian)

Ernst HU, Klug C (2011) Perlboote und Ammonshörner weltweit. Nautilids and Ammonites worldwide. Pfeil Verlag, München

Hohenegger J, Tatzreiter F (1992) Morphometric methods in determination of ammonite species, exemplified through *Balatonites* shells (Middle Triassic). J Paleont 66:801–816.

Hunt G (2004) Phenotypic variation in fossil samples: modeling the consequences of time-averaging. Paleobiology 30:426–443

Hunt G (2007) Variation and early evolution. Science 317:459–460

Jacobs DK, Landman NH, Chamberlain JA (1994) Ammonite shell shape covaries with facies and hydrodynamics: iterative evolution as a response to changes in basinal environment. Geology 22:905–908

Kant R (1973) Allometrisches Wachstum paläozoischer Ammonoideen: Variabilität und Korrelation einiger Merkmale. Neues Jahrb Geol Paläontol Abh 143(2):153–192

Kant R, Kullmann J (1980) Umstellungen im Gehäusebau jungpaläozoischer Ammonoideen. Neues Jahrb Geol Paläontol Mh 1980(11):673–685

Klug C (2001) Life-cycles of Emsian and Eifelian ammonoids (Devonian). Lethaia 34:215–233

Klug C, Hoffmann R (2015) Ammonoid septa and sutures. (this volume)

Klug C, Korn D (2002) Occluded umbilicus in the Pinacitinae (Devonian) and its palaeoecological implications. Palaeontology 45:917–931

Klug C, Lehmann J (2015) Soft-part anatomy of ammonoids: reconstructing the animal based on exceptionally preserved specimens and actualistic comparisons. (this volume)

Klug C, Meyer E, Richter U, Korn D (2008) Soft-tissue imprints in fossil and recent cephalopod septa and septum formation. Lethaia 41:477–492

Klug C, Zatoń M, Parent H, Hostettler B, Tajika A (2015) Mature modifications and sexual dimorphism. (this volume)

Korn D (1979) Mediandornen bei *Kosmoclymenia* Schindewolf (Ammonoidea, Cephalopoda). N Jahrb Geol Paläont Mh 7:399–405

Korn D (1997) The Palaeozoic ammonoids of the South Portuguese Zone. Memórias do Instituto Geológico e Mineiro 33:1–131

Korn D (2010) A key for the description of Palaeozoic ammonoids. Foss Rec 13:5–12

Korn D (2012) Quantification of ontogenetic allometry in ammonoids. Evol Dev 14(6):501–514

Korn D (2014) *Armatites kaufmanni* n. sp., the first Late Devonian goniatite with ventral spines. Neues Jahrb Geol Paläontol Abh 271:349–352

Korn D, Klug C (2002) Ammoneae Devonicae. In: Riegraf W (ed) Fossilium catalogus. Backhuys, Leiden
Korn D, Klug C (2003) Morphological pathways in the evolution of Early and Middle Devonian ammonoids. Paleobiology 29:329–348
Korn D, Klug C (2007) Conch form analysis, variability, morphological disparity, and mode of life of the Frasnian (Late Devonian) Ammonoid *Manticoceras* from Coumiac (Montagne Noire, France). In: Landman NH, Davis RA, Mapes RH (eds) Cephalopods present and past: New insights and fresh perspectives. Springer, Dordrecht
Korn D, Klug C (2012) Palaeozoic ammonoids—diversity and development of conch morphology. In: Talent J (ed) Extinction intervals and biogeographic perturbations through time: earth and Life (International year of planet earth). Springer, Netherlands
Korn D, Ebbighausen V, Bockwinkel J, Klug C (2003) On the A-mode sutural ontogeny in prolecanitid ammonoids. Palaeontology 46:1123–1132
Korn D, Bockwinkel J, Ebbighausen V (2010) The ammonoids from the Argiles de Teguentour of Oued Temertasset (early Late Tournaisian; Mouydir, Algeria). Foss Rec 13:35–152
Korn D, Mapes RH, Klug C (2013) The coarse wrinkle layer of Palaeozoic ammonoids: new evidence from the Early Carboniferous of Morocco. Palaeontology 57:771–781. doi:10.1111/pala.12087
Kraft S, Korn D, Klug C (2008) Ontogenetic patterns of septal spacing in Carboniferous ammonoids. Neues Jahrb Geol Paläontol Abh 250:31–44
Kruta I, Landman NH, Tanabe K (2015) Ammonoid radulae. (this volume)
Kulicki C, Tanabe K, Landman NH, Kaim A (2015) Ammonoid shell microstructure. (this volume)
Kutygin RV (1998) Shell shapes of Permian ammonoids from northeastern Russia. Paleont Zh 1998 (1):20–31
Kutygin RV (2006) Methods for studying ammonoid shell shape (example of Permian Goniatitida from northeastern Asia). Russiskaja Akademija Nauk Paleontologiczeskij Institut, pp 96–98 [in Russian]
Landman NH, Tanabe K, Davis RA (eds) (1996) Ammonoid paleobiology. Plenum, New York
Lehmann U (1976) Ammoniten. Ihr Leben und ihre Umwelt. Enke, Stuttgart, p 171
Lehmann U (1981) The ammonites: their life and their world. Cambridge University Press, New York
Lehmann U (1990) Ammonoideen. Enke, Stuttgart
Makowski H (1962) Problem of sexual dimorphism in ammonites. Palaeontol Pol 12:1–92
Matthews SC (1973) Notes on open nomenclature and synonymy lists. Palaeontology 16:713–719
Miller AK, Furnish WM, Schindewolf OH (1957) Paleozoic Ammonoidea. In: Moore RC (ed) Treatise on Invertebrate Paleontology, Part L, Mollusca 4, Cephalopoda-Ammonoidea. GSA and University of Kansas Press, L11–L20
Monnet C, Klug C, De Baets K (2011) Parallel evolution controlled by adaptation and covariation in ammonoid cephalopods. BMC Evol Biol 11(115):1–21
Monnet C, Bucher H, Guex J, Wasmer M (2012) Large-scale evolutionary trends of Acrochordiceratidae Arthaber, 1911 (Ammonoidea, Middle Triassic) and Cope's rule. Palaeontology 55:87–107
Moseley H (1838) On the geometrical forms of turbinated and discoid shells. R Soc Lond Phil Trans 138:351–370
Nixon M, Young JZ (2003) The brains and lives of cephalopods. Oxford University Press, Oxford
Parent H, Greco AF, Bejas M (2010) Size-Shape relationships in the Mesozoic Planispiral Ammonites. Acta Palaeont Pol 55, 85–98
Parent H, Bejas M, Greco A, Hammer O (2011) Relationships between dimensionless models of ammonoid shell morphology. Acta Palaeont Pol 57:445–447
Raup DM (1961) The geometry of coiling in gastropods. Proc Natl Acad. Sci U S A 47:602–609
Raup DM (1966) Geometric analysis of shell coiling: general problems. J Paleontol 40(5):1178–1190
Raup DM (1967) Geometric analysis of shell coiling: coiling in ammonoids. J Paleontol 41(1):43–65
Raup DM, Michelson A (1965) Theoretical morphology of the coiled shell. Science 147:1294–1295

Ropolo P (1995) Implications of variation in coiling in some Hauterivian (Lower Cretaceous) heteromorph ammonites from the Vocontian basin, France. Mem Descr della Carta Geol Ital 51:137–165.

Ruzhencev VE (1949) Biostratigrafiya verkhnego karbona (Upper Carboniferous biostratigraphy). Dokl Akad Nauk SSSR 67(3):529–532

Ruzhencev VE (1960) Printsipy sistematiki, sistema i filogeniya paleozoyskikh ammonoidey (Principles of systematics, the system and phylogeny of Paleozoic ammonoids). Trudy Paleontol Inst Akad Nauk SSSR 133:1–331 [in Russian]

Ruzhencev VE (1962) Nadotryad Ammonoidea. Ammonoidei. Obshchaya chast' (Superorder Ammonoidea. Ammonoidei. General section). In: Orlov YA, Ruzhencev VE (eds) Osnovy Paleontologii, 5, Mollyuski: Golovonogie 1. Akademiya Nauk SSSR, Moskva

Ruzhencev VE (1974) Superorder Ammonoidea. General section. In: Orlov YA, Ruzhencev VE (eds) Fundamentals of paleontology. V. Mollusca: Cephalopoda I, Jerusalem.

Sandberger G (1851) Beobachtungen über mehrere schwierige Puncte der Organisation der Goniatiten. Jahrb Ver Nat Herzogthum Nassau 7:292–304

Sandberger G (1853a) Einige Beobachtungen über Clymenien; mit besonderer Rücksicht auf die westphälischen Arten. Verh Naturhist Ver Preuss Rheinl Westph 10:171–216

Sandberger G (1853b) Über Clymenien. Neues Jahrb Miner Geogn Geol Petrefakten-K 1853:513–523

Sandberger G (1857) Paläontologische Kleinigkeiten aus den Rheinlanden. Verh Naturhist Ver Preuss Rheinl Westph 14:140–142

Sandberger G, Sandberger F (1850–1856) Die Versteinerungen des rheinischen Schichtensystems in Nassau. Mit einer kurzgefassten Geognosie dieses Gebietes und mit steter Berücksichtigung analoger Schichten anderer Länder I-XIV, vol 1850, pp 1–72

Schlegelmilch R (1976) Die Ammoniten des süddeutschen Lias. Fischer, Stuttgart

Schlegelmilch R (1985) Die Ammoniten des süddeutschen Doggers. Fischer, Stuttgart

Schlegelmilch R (1994) Die Ammoniten des süddeutschen Malms. Fischer, Stuttgart.

Tanabe K, Kruta I, Landman NH (2015) Ammonoid buccal mass and jaw apparatus. (this volume)

Tanabe K, Obata I, Fukuda Y, Futakami M (1979) Early shell growth in some Upper Cretaceous ammonites and its implications to major taxonomy. Bull Nat Sci Mus (Tokyo) C 5:155–176

Tozer ET (1972) Observations on the shell structure of Triassic ammonoids. Palaeontology 15:637–654

Trueman AE (1941) The ammonite body chamber, with special reference to the buoyancy and mode of life of the living ammonite. Quart J Geol Soc Lond 96:339–383

Urdy S (2015) Theoretical modelling of the molluscan shell: what has been learnt from the comparison among molluscan taxa? (this volume)

Urdy S, Goudemand N, Bucher H, Chirat R (2010a) Allometries and the morphogenesis of the molluscan shell: a quantitative and theoretical model. J Exp Zool B 314:280–302

Urdy S, Goudemand N, Bucher H, Chirat R (2010b) Growth dependent phenotypic variation of molluscan shell shape: implications for allometric data interpretation. J Exp Zool B 314:303–326

Vermeij GJ (1993) A natural history of shells. Princeton University Press, Princeton

Westermann GEG (1964) Sexual-Dimorphismus bei Ammonoideen und seine Bedeutung für Taxonomie der Otoitidae (einschliesslich Sphaeroceratinae; Ammonitina, M. Jura). Palaeontogr A 124:1–3, 33–73

Westermann GEG (1996) Ammonoid life and habit. In: Landman NH, Tanabe K, Davis RA (eds) Ammonoid paleobiology. Plenum, New York

Wilmsen M, Mosavinia A (2011) Phenotypic plasticity and taxonomy of *Schloenbachia* varians (J. Sowerby, 1817) (Cretaceous Ammonoidea). Paläontol Z 85:169–184

Yacobucci MM (2004) Buckman's Paradox: variability and constraints on ammonoid ornament and shell shape. Lethaia 37:57–69

Chapter 2
Ammonoid Color Patterns

Royal H. Mapes and Neal L. Larson

2.1 Introduction and Background

Mapes and Davis (1996) provided an extensive overview of preserved ammonoid color patterns. Their report covered the general history of reported occurrences and listed three genera from the Triassic, six genera from the Jurassic, and six genera from the Cretaceous (Tab. 2.1). They also discussed the possible kinds of color patterns (transparency, achromatism, monochromatism, irregular and regular spot patterns, transverse zigzag, chevron and radial stripes and longitudinal stripes or bands), geographic occurrences, and ways to recognize biological patterns vs. false patterns produced by shell thickening and diagenetic processes. Included was a comparison of the color pattern in *Nautilus* (and *Allonautilus*), the only known externally shelled cephalopods living today. In addition to the above, they speculated on the functions of patterns and on the possible different mechanisms of taphonomic destruction of the ammonoid color patterns to explain why such patterns are so rarely preserved.

Mapes and Davis (1996) were puzzled that orthoconic and coiled nautiloids and bactritoids with color patterns had been recovered from the Late Paleozoic (Carboniferous), but none of the ammonoids co-occurring with those cephalopods retained any trace of a color pattern even though in some cases many thousands of ammonoid specimens had been recovered from the same locations and strata. This problem remains unresolved. Since 1996, some new information has come to light and is reported herein. This includes the recognition of iridescent color patterns, some new false color patterns, and some limited information on the development of specific color patterns associated with habitat and mode of life.

R. H. Mapes (✉)
North Carolina Museum of Natural Sciences, West Jones St., Raleigh, NC, USA
e-mail: mapes@ohio.edu

N. L. Larson
Larson Paleontology Unlimited, 12799 Wolframite Rd., Keystone, SD 57751, USA
e-mail: ammoniteguy@gmail.com

C. Klug et al. (eds.), *Ammonoid Paleobiology: From Anatomy to Ecology*,
Topics in Geobiology 43, DOI 10.1007/978-94-017-9630-9_2

Table 2.1 Occurrences of ammonoids with color patterns, modified after Mapes and Davis (1996)

Taxon	Author	Pattern	Quality	Country
Triassic				
Arctoceras sp.	This report	M	M	USA
Dieneroceras knechti	This report	L	D on L	USA
Dieneroceras spathi	Mapes and Sneck 1987	L	D on L	USA
Dieneroceras subquadratum	Mapes and Davis 1996	L	D on L	USA
Flemingites russelli	This report	M	M	USA
Juvenites septentrionalis	This report	R	D on L	USA
Juvenites thermarum	This report	R	D on L	USA
Kashmirites sp.	This report	L	D on L	USA
Owenites koeneni	Tozer 1972; Mapes and Sneck 1987	R	D on L	USA
Owenites sp. *cf. 0. koeneni*	Mapes and Sneck 1987	R	D on L	USA
Owenites sp.	Mapes and Sneck 1987	C	D on L	USA
Paranannites aspensis	This report	R	D on L	USA
Paranannites mulleri	This report			
Prosphingites slossi	Keupp 2000	R	D on L	USA
Preflorianites toulai	Keupp 2000	R	D on L	USA
Xenoceltites sp.	This report	L	D on L	USA
Jurassic				
Amaltheus stokesi	Spath 1935	L pathologic	D on L	England
Amaltheus subnodosus	Pinna 1972	L	D on L	Germany
Amaltheus gibbosus	Pinna 1972	L	D on L	Germany
Amaltheus margaritatus	Mapes and Davis 1996	L	D on L	England
Androgynoceras lataecosta	Spath 1935; Arkell 1957	L	D on L	England
Asteroceras stellare	Arkell 1957	L	L on D	France
Asteroceras stellare	Manley 1977	S	D on L	England
Cadoceras sp.	This report	L	I on N	Russia
Eboriceras sp.	This report	L	I on N	Russia
Kosmoceras jason	This report	L	I on N	Russia
Leioceras sp.	Greppin 1898; Arkell 1957	L	L on D	Switzerland
Pleuroceras spinatum	Schindewolf 1928, 1931; Arkell 1957	R	D on L	Germany
Pleuroceras salebrosum	Pinna 1972; Lehmann 1990	S	D on L	Germany
Pleuroceras transiens	Pinna 1972	S	D on L	Germany
Pleuroceras sp. aff. *solare*	Heller 1977	L	D on L	Germany
Proriceras sp.	This report	L	I on N	Russia
Quenstedtoceras (Lamberticeras) lamberti	This report	L	I on N	Russia

Table 2.1 (continued)

Taxon	Author	Pattern	Quality	Country
Quenstedtoceras henrici	This report	L	I on N	Poland
Tragophylloceras loscombi	Pinna 1972	L	D on L	England
Xipheroceras sp.	This report	R	D on L	Germany
Cretaceous				
Beudanticeras ambanjabese	This report	L	I on N	Madagascar
Beudanticeras beudanti	This report	L	I on N	Madagascar
Beudanticeras caseyi	This report	L	I on N	Madagascar
Calliphylloceras sp. aff. *C. demedoffi*	Bardhan et al. 1993	R	D on L	India
Cleoniceras sp.	This report	L	I on N	Madagascar
Desmoceras media	This report	L	I on N	Madagascar
Desmoceras inflatum	This report	L	I on N	Madagascar
Hoploscaphities nicolletii	This report	L	I on N	USA
Hoploscaphities (Jeletz-kytes) reesidei	This report	L	I on N	USA
Libycoceras afikpoense	Reyment 1957	s	D on L	Nigeria
Placenticeras meeki	This report			USA
Paratexanites (Parabe-vahites) serratomarginatus	Matsumoto and Hirano 1976	L	D on L	Japan
Protexanites (Protexanites) botanti shimizui	Matsumoto and Hirano 1976	L	D on L	Japan
Protexanites (Anatexanites) fukazawai	Matsumoto and Hirano 1976	L	D on L	Japan
Pseudouhligella sp.	This report	L	I on N	Madagascar
Puzosia malandiandrensis	This report	L	I on N	Madagascar
Hoploscaphites nicolletii	This report	L	I on N	USA
Submortoniceras woodsi	Kennedy et al. 1981	L	D on L	South Africa
Tetragonites glabrus	Tanabe and Kanie 1978	L	D on L	Japan
Tetragonites sp.	Tanabe and Kanie 1978	R	D on L	Japan

S spots, *L* longitudinal, *R* radial/transverse, *C* combined, *M* monochrome, *D* on *L* dark on light, *L* on *D* light on dark, *I* on *N* iridescent on nacre color

2.2 Additional Reports of Ammonoid Color Patterns

Mapes and Davis provided a relatively complete documentation of the known color patterns that had been described up to 1996. However, because there is no comprehensive listing of this biological condition for fossil cephalopods, inevitably some reports and specimens in museum collections were missed. One missed report is by Klinger and Kennedy (1981) where a specimen of *Submortoniceras woodsi* (Spath 1921) from the Cretaceous of South Africa was described and illustrated with traces of longitudinal banding. According to these authors, similar banding was described by Matsumotoi and Hirano (1976) on a stratigraphically older specimen of *Protexanites (P.) bontanti shimizui* Matsumoto 1970 from Hokkaido, Japan.

We know of only three new reports since Mapes and Davis (1996) summarized the known occurrences of ammonoids with color patterns. Probably the most significant new information is the report by Ebbighausen et al. (2007) of a Devonian goniatite with a color pattern from Morocco. The single specimen has dark transverse bands that follow the sinuous growth lines on a specimen of *Tornoceras* sp. This color-banded taxon is limited to a single specimen at this time. We accept this report with some reservation because of the limited number of specimens reported. However, presuming that this report is valid, this specimen would represent the oldest color pattern known in the ammonoid lineage, the only specimen of a goniatite with a color pattern, and it is the only ammonoid specimen with a color pattern known from the Paleozoic.

A Jurassic specimen of *Xipheroceras* sp. from Germany has been brought to our attention as having multiple longitudinal color bands (Klug, personal communication 2012). The specimen is approximately 185 mm in diameter and the relatively thin bands are dark on a light background on the outer shell. The bands are present on both the lateral, ventrolateral, and venter of the conch (Fig. 2.1). These color bands are similar to those described and illustrated by Lehman (1990, Fig. 2.1) on *Pleuroceras* sp. from the Jurassic of Germany.

A possible color pattern on *Placenticeras meeki* from an unknown locality in the Cretaceous of North America has also been discovered (Fig. 2.2). There are two types of dark bands on the specimen. One is expressed as dark bands that are widely spaced and that follow the growth lines of the specimens. These widely spaced bands are interpreted as a false color pattern and probably result from shell thickening like that seen in other false color patterns (see below). The other set of transverse bands are closely spaced, slightly darker bands than that of the shell material and are only observed on the right side of the conch on an internal whorl of

Fig. 2.1 Longitudinal color bands (see *arrows*) on a specimen of *Xipheroceras* sp. *SNMS* (= Staatliches Museum fuer Naturkunde, Stuttgart, Germany) 70085 from the Arietenkalk Formation, basal late Sinemurian, Obtusum-Zone, at Schwäbisch-Gmünd-Unterbettringen, Germany. Scale Bar = 3.0 cm

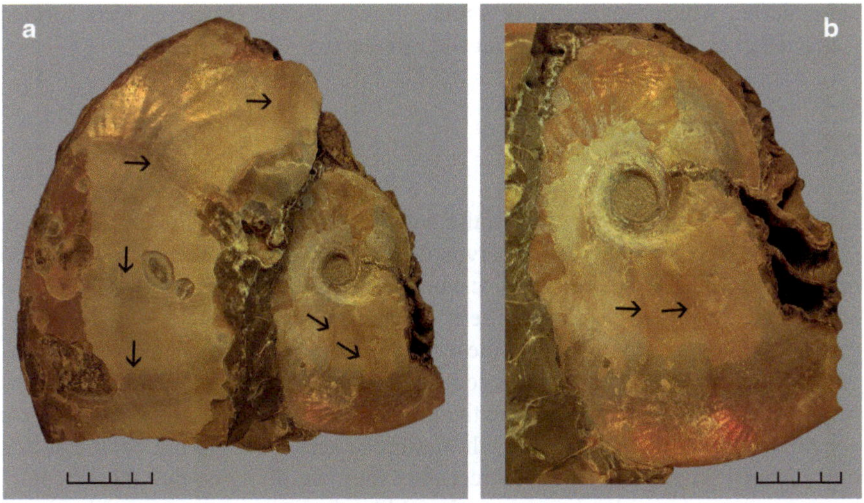

Fig. 2.2 *Placenticeras meeki,* late Campanian, Late Cretaceous, Pierre Shale, most likely South Dakota with a possible pigment emplaced color pattern. **a**. Overall view of the specimen showing several dark transverse bands that follow the growth lines on the outer whorl (at *arrows*) that are interpreted as false color patterns. Scale bar=4.0 cm. **b**. Enlarged view of the exposed inner whorl showing approximately seven closely spaced transverse dark bands on the exposed shell near the venter that are interpreted as possible color bands. Also, on this part of the specimen there are several (see *two arrows*) of the more widely spaced dark transverse false color bands similar to those seen at the arrows on the outer whorl as shown in **a**

the phragmocone. These dark bands are on a lighter surface of the external shell and appear to be confined to the outer shell layer. That this case is a legitimate biologically emplaced color pattern is uncertain because bilateral symmetry of the bands cannot be observed and the pattern is not expressed any other places on the testiferous conch. Thus, we believe that additional specimens are necessary to confirm that these are biologically emplaced color bands.

2.3 False Color Patterns

2.3.1 Background

According to Mapes and Davis (1996), false color patterns can have several different appearances ranging from random blotches to longitudinal and transverse bands on the shell. These different patterns are caused by a variety of different mechanisms. The false color pattern that Mapes and Davis (1996) indicated as most common is where thickening of the shell is present. Such places are at constrictions, pseudoconstrictions, and varices. Additional places not noted by Mapes and Davis (1996) include ribs, falcations, nodes and other places where shell thickening can occur. In any case, it appears that these false color patterns are associated with growth halts, during which megastriae or similar growth-related structures were formed (see Klug et al. 2007 for a discussion and references).

2.3.2 Reports

Since 1996, several new reports of false color patterns have been made. Klug et al. (2007) indicated that they had observed dark areas where the shell was thickened on several Carboniferous and Triassic ammonoid genera. These transversely placed bands were at regular intervals where an interim aperture position was located (megastriae). They concluded that these dark bands were the remnants of black organic material like that seen on the dorsal shell and at some injury sites on modern *Nautilus*, and that these bands were the product of some form of stress or adverse conditions during the animals growth in a broad sense. We are not confident that this explanation is adequate or correct; however, we do agree that these "color" bands are not an organic pigment emplaced like the brown transverse stripes seen on modern *Nautilus*. Thus, until additional study is done on these phenomena, we will consider them to be, with reservation, in the general category of "false color bands".

In another report Klug et al. (2012) described a situation where 17 specimens of the ammonite *Baculites* sp. (Late Cretaceous) from Germany were recovered with exceptional soft tissue preservation. One of the specimens is reported to have 11–12 false color bands on the shell. Klug et al. (2012) correctly determined that the bands were a false color pattern and indicated that these dark bands were places where growth of the animal had slowed or temporarily halted and faint ribs were deposited.

A report on a false color pattern not mentioned by Mapes and Davis (1996) is by Branson (1964). He described a specimen of *Goniatites choctawensis* (now considered to be *G. multiliratus*) from the Delaware Creek Member of the Caney Formation in south central Oklahoma as an internal mold with dark and light bands in the spaces between the septa. He considered the infilling material to be calcite with "internal mineral color". He also noted that similar specimens from the Early Carboniferous of Derbyshire, England had similar brown patterns. We agree that this coloration is a false color pattern although it could even be post mortem (i.e. diagenetic); however, little study has been done on why the calcite in the cameral chambers is colored. Under normal circumstances, one would assume that the coloration is a result of trace mineral elements in the calcite. However, sliced specimens of various Carboniferous goniatites (including ones from the Delaware Creek Member of the Caney Formation in Oklahoma), Permian prolecanitids, and Cretaceous ammonites that had colored calcite in the phragmocone chambers received some limited study (RHM, new observations). These specimens were very slowly (weeks) dissolved in very dilute acetic acid, and it was discovered that when the calcite was removed from the cameral chambers, the colored cameral filling was a fibrous-like gel that dried into a fibrous mat similar to the pelical lining seen in the chambers of modern *Nautilus*. Significantly, the goniatite specimens with clear calcite in the same acid containers with specimens that had the colored calcite did not have this fibrous gel material in the cameral spaces. The significance of this fibrous gel remains unstudied, but it is here speculated that this fibrous gel material is a remnant of the original cameral fluid and that the original cameral fluid in ammonoids was more of a gel, rather than a liquid as is seen in modern *Nautilus*. Additional study of this phenomenon is recommended.

Another kind of false color pattern is known from two cases, one is in the Devonian of Germany (Richter 2002; Richter and Fischer 2002) (Fig. 2.3), and the other is from the Early Jurassic of England (Paul 2011). In all cases, the specimens are extremely well preserved pyritized internal molds. On the surface of these molds, there are regular patterns of dark areas on the brilliant shiny surface of the internal mold of the specimen. These patterns can form transverse bands that follow the septa in phragmocones or as bilaterally symmetrical dark patches (Fig. 2.3). Richter (2002) and Richter and Fischer (2002) provided a convincing discussion as to the origin of these patterns. They observed that the darkening of these areas on the internal mold of the shell are places where the shell was rough-textured with micropyrite crystals. These rough textured areas are without the polish seen on the smooth surfaces of the internal mold of the rest of the specimens. They concluded that these dark patches are attachment scars and interpreted them as soft tissue attachment structures (Richter 2002; Richter and Fischer 2002). Paul (2011), in his analysis of the English specimens, concluded that the transverse bands he observed represented pauses in the growth of the animals he analyzed. In all of these cases, the dark bands are on the internal mold, and therefore, they are not biologically emplaced pigments in the shell of these animals. Also, these authors were convinced that the repetition of the bands is directly related to the episodic growth of the animal.

2.4 Iridescent Color Patterns

2.4.1 Background

In the last decade, hundreds of ammonite specimens with undescribed iridescent color patterns have been discovered. Keupp (2005) was the first to describe and illustrate this occurrence, although Reyment (1957) indicated he had seen such a color pattern but did not adequately document this phenomenon. It must be noted that photography of this phenomenon is difficult because this feature is usually more evident as the specimen is moved in the light, and it may for this reason that Reyment could not adequately document it in his 1956 report.

These newly recognized color patterns are preserved as iridescent longitudinal bands parallel to the venter on the external surface (presumably the outer prismatic layer) of the shell. These bands are probably the result of the prismatic shell ultrastructure selectively breaking down light into different wavelengths (i.e., different colors), but this phenomenon needs additional study to confirm this assumption and to test in how far the appearance of the color patterns was altered by diagenesis.

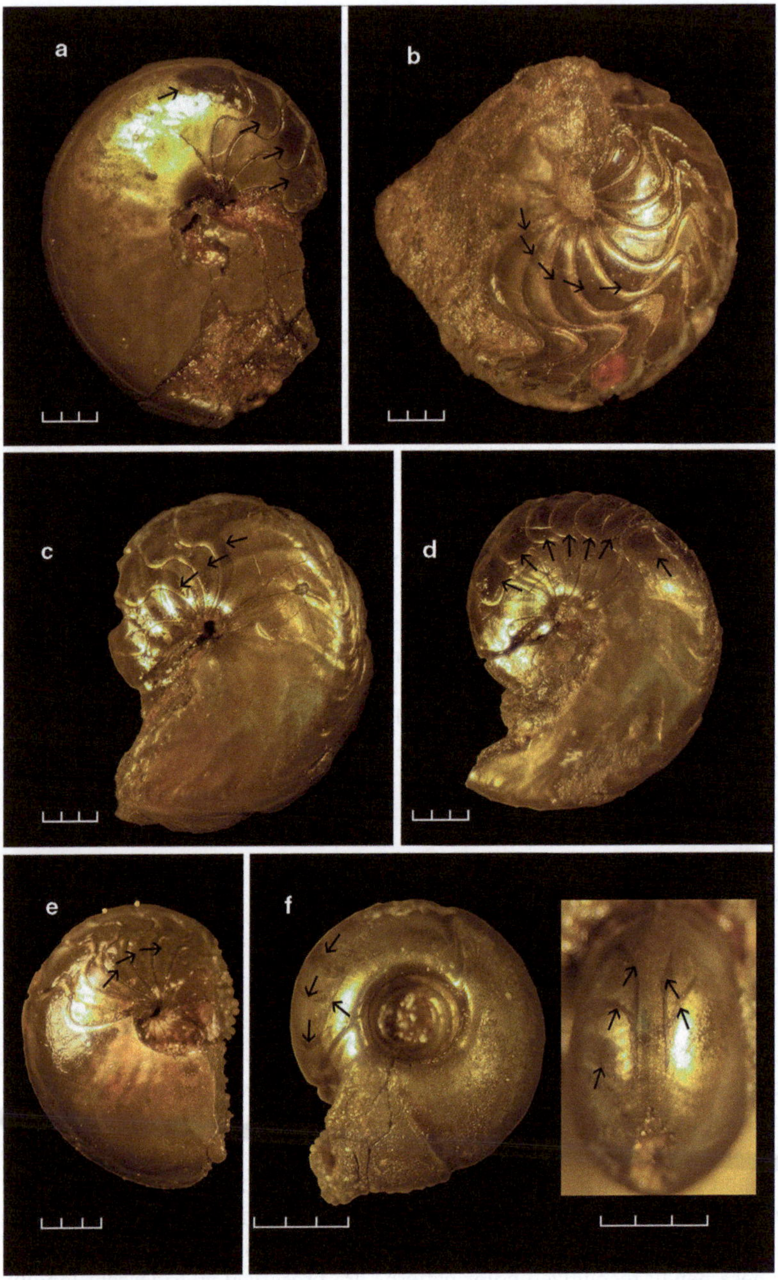

Fig. 2.3 **a–e** Pyritized internal molds of six Devonian ammonoid genera from Germany with false color patterns. The false color patterns are bilateral *black* areas on the golden pyrite of the phragmocones and sometimes the apical parts of body chambers (see *arrows*). All specimens are reposited at the Institut für Geologie und Paläeontologie der Universtät Münster (MB.C). **a** Left lateral view of *Armatites* aff. *planidorsatus* specimen MB.C 2910(272,E54), **b** Right lateral view of *Linguatornoceras haugi* specimen MB.C 3087 (265,E38), **c** Left lateral view of ? *Truyolsoceras* sp. specimen MB.C 3071 (2614,E410), **d** Left lateral view of *Cheiloceras (Puncticeras) longilobum*

Fig. 2.4 World map showing the locations around the world where ammonoids with iridescent color patterns have been recovered. In North America: *1* north-central South Dakota and *2* Coon Creek, Tennessee; Europe: *3* Normandy, France, *4* Łuków, Poland, and *5* Saratov, Russia, and Africa: *6* Mahajanga province, Madagascar

2.4.2 Locations, Description, and Ages of Specimens

Ammonoid sites that preserve an iridescent color pattern have been collected from around the world (Fig. 2.4) in the Jurassic (Fig. 2.5), the Early Cretaceous (Fig. 2.6), and the Late Cretaceous (Fig. 2.7). The locations and ages are as follows: Late Jurassic, Callovian Stage of Saratov, Russia and Łuków, Poland (Fig. 2.5); the Early Cretaceous, Albian Stage of Mahajanga Province, Madagascar and Normandy, France (Fig. 2.6); the Late Cretaceous, late Campanian Stage of south-central Tennessee, USA and the Late Cretaceous, late Maastrichtian Stage of north-central South Dakota, USA (Fig. 2.7).

2.4.2.1 Russian Occurrence – Late Jurassic

At the Saratov site (Russia, Late Jurassic) ammonoids were collected from the Dubki Quarry, a site that quarries a hard-clay used in the production of bricks. The Dubki Quarry lays in the *Quenstedtoceras* (*L.*) *lamberti* Zone – latest late Callovian of the Late Jurassic. This zone contains clays and marls deposited under oxygen

specimen MB.C 2908 (15,E66), **e** *Right* lateral view of *Falcitornoceras korni* (MB.C 3073 (2606, E402), **f–g** *Right* lateral and ventral views, respectively, of *Paratorleyoceras globosum* (MB.C 3086 (661,E33). Scale bars = 3.0 mm

Fig. 2.5 Five ammonoid genera from the Late Jurassic with iridescent longitudinal color bands. Four of the ammonoid genera are from the late Callovian, Late Jurassic, Dubki Quarry near Saratov, Russia (**a–o**) and one genus is from the late Callovian, Late Jurassic in Łuków, Poland (**p–q**).

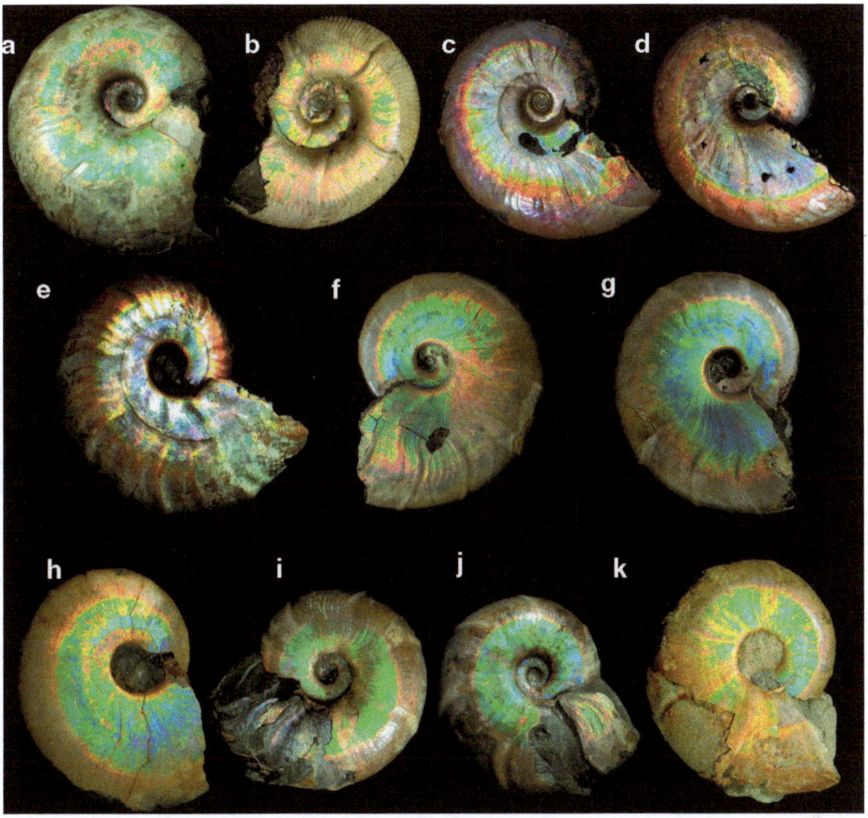

Fig. 2.6 Ammonoid specimens from the Albian stage, Early Cretaceous, Ambarimaninga, Mahajunga Province, Madagascar and from the Albian stage, Early Cretaceous, Bully, France. **a** Right side of *Pseudouhligella* sp., specimen OUZC 6192, D=55.0 mm, **b** Left side of *Puzosia malandiandrensis*, specimen OUZC 6193, D=43.5 mm, **c d** *Beudanticeras beudanti*, specimens OUZC 6194 and 6195, respectively, D=25.0 mm and 38.0 mm, respectively, **e** *Cleoniceras* sp., specimen OUZC 6196, D=28 mm, **f–g** *Beaudanticeras ambanjabese*, specimen OUZC 6197, D=72 mm, **h** *B. caseyi,* right side of specimen OUZC 6198, D=68 mm, **i–j** *Desmoceras media* specimen OUZC 6199, D=65 mm, **k** *D. inflatum*, right side of specimen OUZC 6200, D=84 mm

depleted marine conditions in the central portion of the Russian Platform (Meledina 1988). The ammonite fauna in the *Q.* (*L.*) *lamberti* Zone is also quite similar to that of the stratotype described from Europe (England and Poland). In addition to *Quenstedtoceras*, other ammonite genera include *Eboriceras*, *Grossouvria*, *Hecti-*

---

a–h, *Quenstedtoceras* (*Lamberticeras*) *lamberti* **a–b** Specimen OUZC 6180, D=70.0 mm, **c** Specimen OUZC 6181, D=69.0 mm, **d** Specimen OUZC 6282, D=50.0 mm, **e** Specimen OUZC 6183, D=51.0 mm, **f** Specimen OUZC 6183, D=51.0 mm, **g** Specimen OUZC 6185, D=105.3 mm, **h** Specimen OUZC 6186, D=103.0 mm. Note that all the *Q.* (*L.*) *lamberti* specimens have a greenish yellow or pink longitudinal band located above the umbilical shoulder, and that the remainder of the conch above and below the band is dominated by darker red. **i–l** *Proriceras* sp. **i–j** Specimen OUZC 6187, D=41.0 mm, **k–l** Specimen OUZC 6188, D=41.5 mm; *Cadoceras* sp. **m** *Left* side of specimen OUZC 6289, D=80 mm; *Eboriceras* sp. **n–o** Specimen OUZC 6190, D=64 mm, **p–q** *Quenstedtoceras henrici*, specimen OUZC 6191, D=8.8 mm

Fig. 2.7 Ammonoid specimens from the Fox Hills Formation, Late Maastrichtian Stage, Late Cretaceous (see **a**, **b**, **g**, **h**), Corson and Ziebach counties (respectively), South Dakota; from the Coon Formation, Late Campanian Stage, Late Cretaceous, McNairy County, Tennessee (see **c–d**), and from the Oxfordian Stage, Late Cretaceous, Ryazan, Russia (see **e–f**). **a–b** *Hoploscaphities nicolletii* microconch, specimen OUZC 6176, D=52.5 mm. **c–d** *Hoploscaphities (Jeletzkytes) reesidei*, specimen OUZC 6177, D=48 mm. **e–f** *Kosmoceras jason*, specimen OUZC 6178, D=5.2 mm. **g–h** *Left* side of *Hoploscaphities nicolletii* macroconch (OUZC 6179). Lateral view of the three-dimensional specimen (**g**) with an iridescent inner shell but retains no trace of a color pattern. Part of the carbonate concretion, which yielded the specimen (**h**) shows the extremely thin external shell with the iridescent color pattern attached to the matrix

coceras, *Kosmoceras*, *Peltoceras*, *Proriceras*, *Cadoceras*, and *Ruriceras* (Larson 2007). Further information on the geology of the area can also be found in Aleksejev and Repin (1986).

Ammonites are preserved in the clay and separated during the quarrying process. While their aragonite shell is well preserved, these ammonites are preserved with pyrite of both crystallized and botryoidal forms coating the interior septa or completely filling the phragmocone. Preservation is generally exceptional, with the outer prismatic layer of shell commonly preserved.

2.4.2.2. Polish Occurrence – Late Jurassic

The Łuków, Poland material (Late Jurassic) was collected by Harry Mutvei from a clay quarry that is considered to be a glacial erratic. The clay was used for bricks and the fossils were undesirable inclusions (Mutvei 1998, personal communication). The fauna comes from the *Quenstedtoceras henrici* Zone, late Callovian, Late Jurassic. Makowski (1952) described the locality and the systematics of the ammonoid fauna.

The fossils are preserved in carbonate concretions where abundant ammonoids are often present along with other mollusks. As from the Dubki Quarry in Russia, some ammonites are also preserved outside of these concretions in hard clay. Often, the ammonoids have hollow phragmocones with their septa sometimes coated with a thin layer of pyrite crystals. Ammonites that are taken out of their host rock often leave the outer layers of their shell stuck to the rock, whereas those found outside of the carbonate concretions often are able to be cleaned leaving intact all layers of their shell.

2.4.2.3. Madagascar Occurrence – Early Cretaceous

Fauna from the early Albian Stage, Early Cretaceous of Mahajanga Province, northwest Madagascar comprise unique benthic forms (generally molluscan) that are remarkably well preserved. The fauna used in this paper comes from two different zones: the *Cleoniceras besairiei* and *Douvilleiceras inaequinodom* Zones of the Early Cretaceous (Collignon 1963). These zones were assigned to the *Douvilleiceras mammillatum* Superzone of the Early Albian (Owen, 1988b). Barrabé (1929) and Collignon (1949) briefly described the geology of the region.

The ammonite material comes from a calcareous, glauconitic, sandy matrix in an extremely fossiliferous zone generally about 15 m thick (Collignon 1949). These fossils are quarried about 50 km south of the city of Mitsinjo, between the towns of Ambarimaninga and Ambinda. Infilling of the body chambers and often the entire phragmocone are common. Phragmocones tend to be filled with calcite while the body chambers are only filled with the host rock. Most ammonites are well preserved with preservation so fine many specimens can be easily removed from their matrix without the layers of shell splitting or sticking to the host rock. All of this material is collected by local people and sold to commercial dealers worldwide.

2.4.2.4. French occurrence – Early Cretaceous

In Normandy, France (Early Cretaceous), ammonoids are recovered from the late early Albian (*Douvilleiceras mammillatum* Zone) Gault Clay of Bully, Pays de Bray, Seine Maritim, France (Amédro 1992). Specimens of *Beudanticeras* are abundant from the clays at Bully. Both phragmocones and the body chambers are commonly filled with the host rock.

2.4.2.5. Tennessee Occurrence – Late Cretaceous

At the type locality of the Coon Creek Formation, Coon Creek, Tennessee (Late Cretaceous), ammonoids were recovered from diverse near-shore marine deposits of sand, glauconitic clay and isolated calcareous concretions. Mollusks dominate the fauna but well-preserved ammonoids are relatively rare. Wade (1917) made the first record of cephalopods from Coon Creek and later (1926) published a monograph on the geology and the entire fauna from this locality. Sohl (1960, 1964) described the gastropods and Cobban and Kennedy (1994) described the ammonites. Brewster and Young (2007) did an extensive report on the site regarding the history, collecting, geology, paleontology and the ongoing science investigations and research of the site. Larson (2012) did a revision of the cephalopod fauna from the type locality.

Ammonite shell preservation within the concretions is generally poor. The shell tends to adhere to the concretion and when broken open the ammonites are generally incomplete and crushed. However, nearly all of the invertebrates collected from the sands of Coon Creek Formation at its type locality are well preserved and when carefully prepared, the matrix can be removed leaving all layers of their shell. So good is the preservation that many bivalves and gastropods still retain their color pattern (Wade, 1926; Sohl, 1960, 1964). When ammonites are found in the sand, infilling of the body chambers and the phragmocones with the sand matrix are common.

2.4.2.6. South Dakota Occurrence – Late Cretaceous

Late Maastrichtian, Late Cretaceous ammonoid horizons of the Fox Hills Formation in north-central South Dakota, USA, are well known (Larson et al. 1997; Waage 1968; Landman and Waage 1993). The ammonites and associated fauna are collected from the *Hoploscaphites nicolletii* and *Jeletzkytes nebrascensis* Zones as well as several subzones (Waage 1968; Landman and Waage 1993). The Fox Hills Formation is a shallow-water, near shore deposit consisting of sands, sandstones, and shale beds with and without carbonate concretions. It was deposited near the end of the Cretaceous as the North American Western Interior Seaway was retreating.

Shell preservation is extremely good to excellent from most sites. The best-preserved ammonoids are found in carbonate concretions and occasionally outside of concretions. Infilling of the body chambers and often the entire phragmocone are common. Phragmocones tend to be filled with calcite or nothing, while the body chambers are filled with hard carbonate rock. In nearly every case, when the concretions are broken open, the outer layers of the shell adhere to the rock and the shell splits in layers. Thus, except in some rare cases, the iridescent color pattern separates from the inner shell of the ammonoid and the external color pattern is lost. An exceptional case is illustrated in Fig. 2.7g and h.

2.5 Habitat and Life Mode

In 1987, Mapes and Sneck described the oldest and most extensive color patterns on several ammonoid genera and species from the famous Early Triassic locality at Crittenden Springs, Nevada (USA). A few years later, Mapes and Davis (1996) summarized what was then known about ammonoid color patterns in the Mesozoic. In 1999, Gardner, and later, Gardner and Mapes (2001) described the color patterns on a total of 15 ammonoid genera and species discovered within an extensive collection (over 10,000 specimens composed of 29 different genera and species) from the Crittenden Springs locality. In this more extensive study of the Crittenden Springs ammonoid fauna, the systematic identifications of Kummel and Steele (1962) and Tozer (1981, 1994) were used. Within that collection, there were three

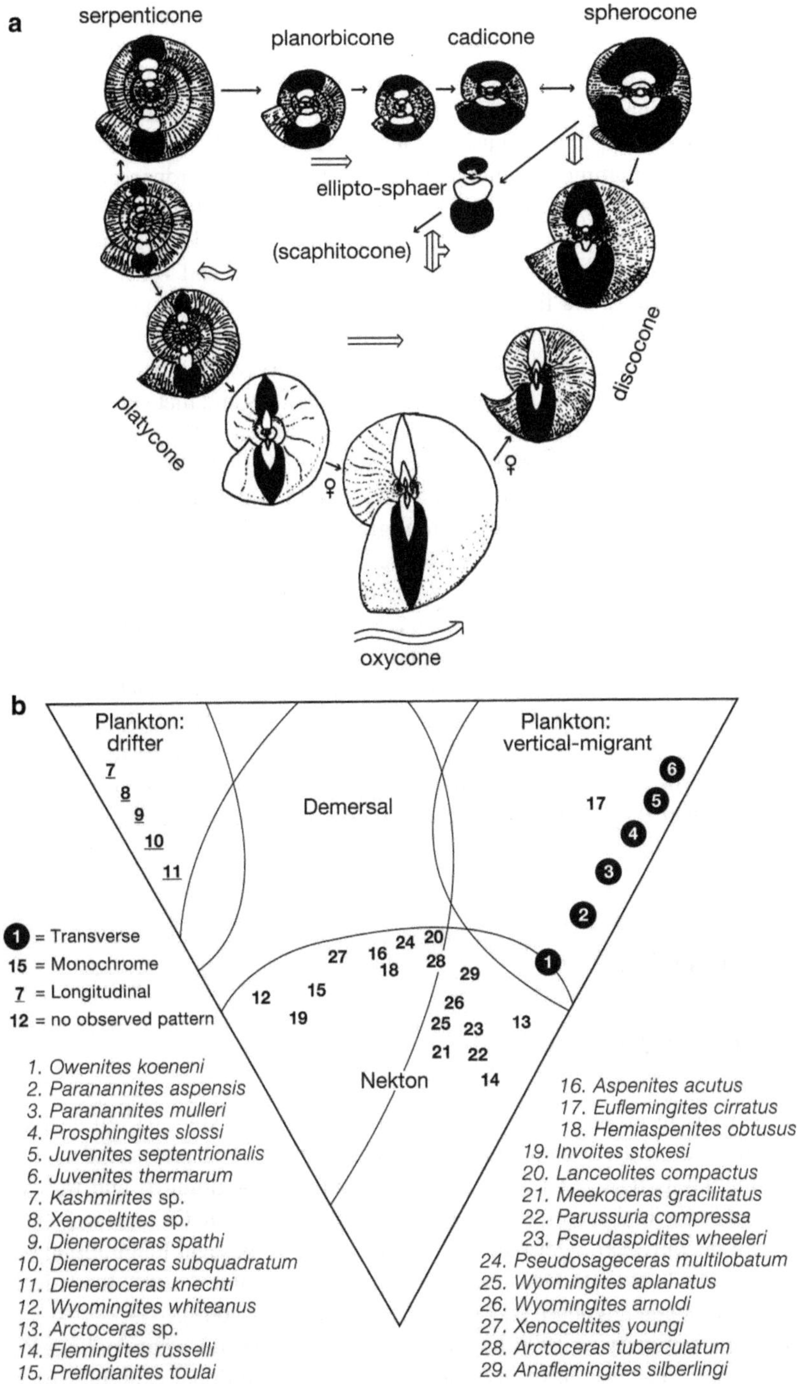

Fig. 2.8 Habitat by shell form of ammonoids as proposed by Westermann (1996). Based on shell form, the Crittenden Springs ammonoids in the Ohio University collection (now at the American Museum of Natural History in New York City, NY) with transverse color patterns are placed in the planktonic-vertical migratory habitat, those with longitudinal bands are placed in the planktonic drifter habitat. The monochromatic taxa are in the nektonic and demersal-nektonic habitats

genera that exhibited monochromatic coloration, which had not been previously confirmed as a possible color pattern by Mapes and Davis (1996) in their summary. Additionally, six genera and species had transverse bands, and six genera and species had longitudinal bands (see Tab. 1 and Fig. 2.8 for a list of the taxa).

Westermann (1996, Fig. 2.1) used conch shapes (and other features) to develop an overall model of ammonoid habitat (Fig. 2.8a). In 2001, Gardner and Mapes applied the conch morphology of the 28 ammonoid genera and species with (n=15) and without (n=14) preserved color patterns from Crittenden Springs and their identifications, as provided by Kummel and Steele (1962) and Tozer (1981, 1994) (see Tab. 2.1 and Fig. 2.8 for a list of taxa), to the Westermann model. Based on that integration of the Westermann model with the 15 ammonoid genera and species with color patterns and the 14 taxa, which had no observed color patterns, it appears that certain color patterns may represent distinct habitats and that certain groups of ammonoids in the Crittenden fauna had similar life patterns (Fig. 2.8b). All of the ammonoids with transverse bands (5 genera and species) fall in the plankton: drifter life mode, while all the longitudinal banded taxa (6 genera and species with one exception) fall in the plankton: vertical migrant life mode. The single exception is *Eufliningites cirratus*, which has no observed color pattern. The four taxa that were determined to be monochromatic fall within the nektonic life mode together with 13 genera and species that appeared to not have a color pattern. This latter distribution of the 3 monochromatic and 13 genera and species without observed color patterns suggests that the taxa without observed color patterns may have lacked coloration during life. Some modern mollusks are known to have shells that lack color patterns and in some cases the shells can be transparent. It still needs to be tested whether the color patterns are simply linked with shell shape and thus represent fabricational noise or whether they really are adaptations to certain ecological requirements of their habitat and mode of life.

2.6 Conclusions

2.6.1 General Summary

Color patterns in mollusks are usually exhibited on the exterior shell surface as shades of black, gray, brown, and red. A wide variety of shapes and designs are known including zigzag, chevron and wavy patterns, longitudinal and transverse bands, and spots. In the Ammonoidea (Early Devonian to Late Cretaceous), the oldest known unequivocal color patterns are transverse stripes and longitudinal bands from the Early Triassic in Nevada, USA. Unequivocal color patterns for Paleozoic ammonoids remain unknown (but see Ebbighausen et al. 2007).

Since the summary in 1996 by Mapes and Davis, several additional Mesozoic taxa with color patterns have become known. Also, additional examples of false color patterns have been detected.

2.6.2 *Iridescent Color Patterns*

In the last decade, numerous ammonoid specimens with an iridescent color pattern have been discovered. These new patterns are iridescent longitudinal bands on the external surface (the outer prismatic layer?) of the shell. Ammonoids from sites that preserve these iridescent color patterns (it cannot be ruled out entirely that the iridescent appearance might be secondary) have been collected from the Late Jurassic, Russia and Łuków, Poland; the Early Cretaceous, Madagascar and Normandy, France; the Late Cretaceous from Coon Creek, Tennessee and South Dakota, USA.

We are aware that the modern nautilids normally have transverse brown, red-brown and purplish-brown bands in the outer shell layer (see Mapes and Landman 2012 for an exception). And yet, those animals might have lived below the euphotic zone where visibility is extremely limited or totally lacking and detection by other visual animals may be extremely limited. Only at night when natural light is limited do these animals move into shallower water. Thus, modern nautilids have color patterns that are not used in any way that is readily apparent.

We are aware that the same could be true of the ammonoids with iridescent color patterns. If these iridescent patterns were utilized as a visual condition, these ammonoids must have lived in the euphotic zone (shallow water) as part of the nektonic food web. If the bright iridescent colors displayed on these shells were used for sexual display or as a danger warning to predators (i.e., a signal that indicates "*I am poisonous*" or "*I taste bad*"), then to be most effective, the display must be easily seen. If the marine environment was turbid with sediment and other micro particles, which reduced visibility, the ammonoid visibility was probably enhanced by the brilliant iridescent coloration of the shell, presuming they were already iridescent during their life (the iridescence might also be a diagenetic effect). Alternatively, an iridescent, color-banded shell may have also been to break up the light waves in shallow, light-filled environments to help camouflage the ammonites.

To our knowledge, iridescent color patterns are unknown in the modern taxa assigned to the phylum Mollusca (which might also point at a taphonomic problem). Also, other than the ammonoids described herein, we are not aware of any fossil mollusks that had this kind of color pattern. Given that this condition is only known for some Mesozoic ammonoids, it appears that possibly, this unique condition became extinct with the demise of the Ammonoidea at the end of the Cretaceous.

2.6.3 *Habitat and Function of Pigment Emplaced Color Patterns*

New insights into pigmented color patterns were developed by Gardner and Mapes (2001) where they integrated the ammonoid shell shape and life style model developed by Westermann (1996) with the different ammonoid color patterns (monochromatic, transverse, longitudinal and no detectable color pattern) in a collection of 28 genera and species of Early Triassic ammonoids from Nevada. The results were striking in that certain color patterns and taxa with no detectable color pattern on ammonoids were found to be specific to a preferred life style, although it could also be a link with the morphogenesis of certain shell forms.

Acknowledgements We would like to thank R. Thomas Becker (Münster, Germany) for bringing to our attention the Devonian pyritized material with the false color patterns during a visit to Berlin and to Dieter Korn (Berlin, Germany) for loaning the specimens to us for examination. He also brought to our attention the important report by Ebbighausen et al. (2007). We wish to thank Christian Klug (Zürich, Switzerland) for providing timely updates on literature sources and some photographs for this report.

References

Aleksejev V, Repin A (1986) New data on the Callovian deposits of Malinov ravine, Saratov area of the Volga region. In: Mesezhnikov MS (ed) Jurassic deposits of the Russian platform; a collection of scientific works. Geologorazved, Lenningrad

Amédro F (1992) The Albian in the Anglo-Paris basin: ammonites, phyletic zonation, séquences. Bull Cent Rech Expor-Prod Elf Aquitaine 16:187–233

Arkell WJ (1957) Introduction to Mesozoic Ammonoidea. In: Moore RC (ed) Treatise on invertebrate paleontology, part l, mollusca 4. GSA and University of Kansas Press, Lawrence

Bardhan S, Jana SK, Datta K (1993) Preserved color pattern of a phylloceratid ammonoid from the Jurassic Chari formation, Kutch, India, and its functional significance. J Paleontol 67:140–143

Barrabé L (1929) Contribution á l'étude stratigraphique et petrographique de la partie médiane du Pays sakalave (Madagascar). Mém de la Soc Géol de France, nouvelle série, V. fas:3–4

Branson C (1964) False color pattern on an Oklahoma goniatite. Okla Geol Notes 24:160

Brewster R, Young R (2007) The fossils of Coon Creek: an Upper Cretaceous Mississippi embayment marine site in McNairy County, Tennessee. Coon Creek Sci Cen, Memphis Pink Palace Family of Muse, Memphis, Tennessee, p. 66

Cobban WA, Kennedy WJ (1994) Upper Cretaceous ammonites from the Coon Creek Tongue of the Ripley Formation at its type locality in McNairy County, Tennessee. USGS Bull Short Contrib Paleontol Strat 2073B:1–12

Collignon M (1949) Recherches sur les faunes Albiennes de Madagascar. I. L'albien D'Ambarimaninga (Madagascar). Annales Geologiques du Service des Mines, Madagascar, fascicule 16:1–128

Collignon M (1963) Atlas des fossiles caractéristiques de Madagascar (Ammonites). Fascicule X (Albien). Serv Géol 184

Ebbighausen V, Becker RT, Bockwinkel J, Aboussalam ZS (2007) Givetian (Middle Devonian) brachiopod goniatite correlation in the dra valley (Anti-Atlas, Morocco) and Bergisch Gladbach Paffrath Syncline (Rhenish Massif, Germany). Geol Soc Lond Sp Pub 278:157–172

Gardner G (1999) Color patterns, habitat and repaired shell damage for lower Triassic ammonoids from Crittenden springs, Ohio University Master's thesis, Elko County

Gardner G, Mapes RH (2001) The relationships of color patterns and habitat for Lower Triassic ammonoids from Crittenden springs, Elko County. Revue de Paléobiol Genève 8:109–122.

Greppin E (1898) Description des fossiles du Bajocien supérieur des environs de Bâle. Mem Soc Paleontol Suisse 25:1–52

Heller F (1977) Ein *Pleuroceras* aff. *solare* (PHILL.) mit gut erhaltener Farbzeichnung aus den Amaltheentonen Frankens. Geol Bl Nordost-Bayern Angrenzende Geb 27(3–4):161–168

Kennedy WJ, Klinger HC, Summesberger H (1981) Cretaceous faunas from Zululand and Natal, South Africa. Additional observations on the ammonite subfamily Texanitinae Collignon, 1948. Ann S Afr Mus 86(4):115–155

Keupp H (2000) Ammoniten. Paläobiologische Erfolgsspiralen. Thorbecke, Stuttgart

Keupp H (2005) Das Geheimnis der Spiralbänder: Farbmuster auf Ammonitengehäusen. Fossilien 6:369–374

Klinger HC, Kennedy WJ (1981) Colour-banding in micromorphs of *Submortoniceras woodsi* (Spath, 1921) and their homoeomorphy with *Protexanites (P.) bontanti shimizui* Matsumoto,

1970. In: Kennedy WJ, Klinger HC, Summesberger H (eds) Cretaceous faunas from Zululand and Natal, South Africa additional observations on the ammonite subfamily Texanitidae Collignon, 1948. Ann S African Mus 86(4):115–155

Klug C, Brühwiler T, Korn D, Schweigert G, Brayard A, and Tilsley J (2007) Ammonoid shell structures of primary organic composition. Palaeontology 50:1463–1468

Klug C, Riegraf W, Lehman, J (2012) Soft-part preservation in heteromorph ammonites from the Cenomanian-Turonian boundary event (OAE 2) in north-west Germany. Palaeontology 55:1307–13331.

Kummel B, Steele D (1962) Ammonites from the *Meekoceras Gracilitatus* zone at Crittenden Spring, Elko County, Nevada. J Paleont 36:638–703

Landman NH, Waage KM (1993) Scaphitid ammonites of the Upper Cretaceous (Maastrichtian) Fox Hills Formation in South Dakota and Wyoming. Bull Am Mus Nat Hist 215:1–257

Larson NL (2007) Deformities in the late Callovian (Late, Middle Jurassic) ammonite fauna from Saratov, Russia. In: Landman NH, Davis, RA, Mapes RH (eds) Cephalopods present and past: new insights and fresh perspectives. Springer, Dordrecht

Larson NL (2012) The Late Campanian (Upper Cretaceous) cephalopod fauna of the Coon Creek Formation at its type locality. J Paleontol Sci (online), p.68

Larson NL, Jorgensen SD, Farrar RA, Larson PL (1997) Ammonites and the other cephalopods of the Pierre seaway. Geoscience Press, Tucson, p.148

Lehmann U (1990) Ammonoideen-Leben zwischen Skylla und Charybdis. Enke, Stuttgart

Makowski H (1952) La faune callovienne de Łuków en Poland. Palaeontologica Polanica 4:1–64

Manley EC (1977) Unusual pattern preservation in a Liassic ammonite from Dorset. Palaeontology 20:913–916

Mapes RH, Davis RA (1996) Color patterns in ammonoids. In: Landman NH, Davis RA, Tanabe K (eds) Ammonoid paleobiology, topics in geobiology. Plenum, NY

Mapes RH, Sneck D (1987) The oldest ammonoid "colour" patterns: description, comparison with *Nautilus*, and implications. Palaeont 30:299–309

Matsumoto T (1970) A monograph of the Collignoniceratidae from Hokkaido, Part 1. Mem Fac Sci Kyushu Univ 20:225–304

Matsumoto T, Hirano H (1976) Colour patterns in some Cretaceous ammonites from Hokkaido. Trans Proc Palaeontol Soc Japan (N. S.) 102:334–342

Meledina SV (1988) Callovian. In: Krymholts GY, Mesezhnikov MS, Westermann GEG (eds) The Jurassic ammonite zones of the Soviet Union. Geol Soc Am Spec Pap 223:33–39

Owen HG (1988) The ammonite zonal sequence and ammonite taxonomy in the *Douvilleiceras mammillatum* superzone (Lower Albian) in Europe. Bull Brit Mus Nat Hist Geol. 44:177–231

Paul CRC (2011) Dark bands on pyritic internal moulds of the Early Jurassic ammonites *Oxynoticeras* and *Cheltonia* from Glouchestershire, England: intrepretation and significance to ammonite growth analysis. Palaeontology 54:1213–1221

Pinna G (1972) Presenza di tracce di colore sul guscio di alcune ammoniti della famiglia Amaltheidae Hyatt, 1867. Soc Ital Sci Nat Mus Civ Star Nat Milano Atti 113(2):193–200

Reyment RA (1957) Über Farbspuren bei einigen Ammoniten. N Jb Geol Paläont Mh 7–8:343–351

Richter U (2002) Gewebeansatz-Structuren auf Steinkernen von Ammonoideen. Geol Beiträg Hannover 4:1–113

Richter U, Fischer R (2002) Soft tissue attachment structures on pyritized internal molds of ammonoids. Abh der Geol BA Wien 57:139–149

Schindewolf OH (1928) Über Farbstreifen bei *Amaltheus* (*Paltopleuroceras*) *spinatum* (Brug.). Paläontol Z 10:136–143

Schindewolf OH (1931) Nochmals über Farbstreifen bei *Amaltheus* (*Paltopleuroceras*) *spinatus* (Brug.). Paläontol Z 13(4):284–287

Sohl NF (1960) Archeogastopoda, Mesogastropoda and Stratigraphy of the Ripley, Owl Creek, and Prairie Bluff Formations. USGS Prof Pap 331-A:1–152

Sohl NF (1964) Neogastropoda, Opisthobranchia and Basommatophora from the Ripley, Owl Creek, and Prairie Bluff Formations. USGS Prof Pap 331-B:153–334

Spath LF (1921) On Cretaceous Cephalopoda from Zululand. Ann S African Mus 12:217–321

Spath LF (1935) On colour-markings in ammonites. Ann Mag Nat Hist Ser 10 15(87):395–398
Tanabe K, Kanie Y (1978) Colour markings in two species of tetragonitid ammonites from the upper Cretaceous of Hokkaido, Japan. Sci Rep Yokosuka City Mus 25:1–6
Tozer ET (1961) Triassic stratigraphy and faunas, Queen Elizabeth Islands, Arctic Archipelago. Geol Surv Canada Mem 316:1–116
Tozer ET (1972) Observations on the shell structure of Triassic ammonoids. Palaeontology 15:637–654
Tozer ET (1981) Triassic Ammonoidea: classification, evolution and relationship with Permian and Jurassic forms. In: House MR, Senior JR (eds) The ammonoidea. Syst Assoc Spec. 18:66–100
Tozer ET (1994) Canadian Triassic ammonoid faunas. Geol Surv Canada Bull 467:1–661
Waage KM (1968) The type Fox Hills Formation, Cretaceous (Maestrichtian), South Dakota, pt. 1, Stratigraphy and paleoenvironments. Peabody Mus Nat Hist Bull 27:1–175
Wade B (1917) A remarkable Upper Cretaceous fauna from Tennessee. John Hopkins Univ Circ 3 [new series] (271–299):73–101
Wade B (1926) The fauna of the Ripley Formation on Coon Creek, Tennessee. USGS Prof Pap 137:1–272
Westermann GEG (1996) Ammonoid life and habitat. In: Landman NH, Davis RA, Tanabe K (eds) Ammonoid paleobiology. Topics in Geobiology, vol. 13. Plenum, NY, pp. 607–707

Chapter 3
Ammonoid Septa and Sutures

Christian Klug and René Hoffmann

3.1 Introduction

Comparable to the conch of extant *Nautilus*, the ectocochleate ammonite conch is divided into two main parts: the most apertural (anterior) undivided body chamber, which contains the majority of the soft body of the living animal, and the apical (posterior) chambered phragmocone. Both are separated from each other by the latest septum. From the last septum to its earliest (embryonic) part (the protoconch), the phragmocone was internally divided in chambers by aragonitic septa. All chambers are connected by the siphuncle, which extends from the protoconch (initial chamber).

Iterative evolution from simple to more complexly folded (frilled/fluted/corrugated/vaulted) septa is well documented throughout the evolution of ammonoids (c. 335 million years) from the Early Devonian (Emsian) to the Cretaceous-Palaeogene boundary (Fig. 3.1, 3.2, 3.3 and 3.4; Seilacher 1988; Garcia-Ruiz et al. 1990; Guex 1992; Saunders 1995; Saunders and Work 1996; Checa and Garcia-Ruiz 1996; Saunders et al. 1999; Hassan et al. 2002; Allen 2007; Yacobucci and Manship 2011). Increasing complexity is also recorded during ontogeny starting with juvenile septa with a simple morphology (Schindewolf 1929, 1951, 1961–1968; Erben 1960, 1962, 1964; Ruzhencev 1962, 1974; Bayer 1977a; Wiedmann and Kullmann 1981; Hewitt 1985; Korn et al. 2003). From the center of each septum to the connection to the inner surface of the conch wall (suture), its complexity increases as well (Fig. 3.1, 3.2). Accordingly, the center of the septum is only slightly undulated or flat (Seilacher 1975; Korn 1997).

C. Klug (✉)
Paläontologisches Institut und Museum, University of Zurich, Karl Schmid-Strasse 6, 8006 Zurich, Switzerland
e-mail: chklug@pim.uzh.ch

R. Hoffmann
Department of Earth Sciences, Institute of Geology, Mineralogy, and Geophysics, Ruhr-Universität Bochum, Bochum, Germany
e-mail: rene.hoffmann@rub.de

C. Klug et al. (eds.), *Ammonoid Paleobiology: From Anatomy to Ecology*,
Topics in Geobiology 43, DOI 10.1007/978-94-017-9630-9_3

Fig. 3.1 Sutures and septa of some ammonoids. **a** *Agoniatites vanuxemi* (Hall 1879), latest Eifelian, Jebel Amessoui, Tafilalt, Morocco; Institut für Geowissenschaften (Tübingen, Germany); dm = 15 cm. **b** *Goniatites multiliratus* Gordon, 1862, middle Mississippian, Jackfork Creek, S

Major folds bulging adorally are called saddles with a convex curvature to the aperture; like saddles for horseback riding, they are commonly rather rounded. Adapically directed folds with a concave curvature to the aperture are dubbed lobes, which are usually more pointed than saddles. Secondary elements of the saddles are folioles and secondary lobes are lobules, respectively (Fig. 3.1). As a rule, the lobes have a greater amplitude than the saddles, i.e. longer distances from the center of the septum and are attached at a lower angle to the conch wall compared to saddles. Therefore, the overall orientation of the septum appears to be convex to the aperture. Maximum thickness is located at the center of the septum and its thickness continuously decreases to the septal margin (Hewitt and Westermann 1986, 1987, 1997).

According to the position of the lobes along the conch wall, Wedekind (1916) introduced the generally accepted suture line terminology used today: E for the external, A for the adventive, L for the lateral, U for the umbilical, and I for the internal lobe. Later, Kullmann and Wiedmann (1970) added I_s for the septal lobe generated by the internal lobe (see below). Accordingly, Hoffmann (2010) introduced the terminology E_s, L_s, U_{1s}, and U_{2s} for all lobes, which are attached to the preceding septum. All these lobes are regularly attached to the preceding septum and have nothing to do with the phenomenon described as septal crowding indicative for the adult stage by Hölder (1952).

Having evolved from cephalopods with orthoconic shells and bactritoids with unfolded septa, the first fold appears when the siphuncle moved to its ventral position in bactritoids (Kröger and Mapes 2007; Kröger et al. 2011). Its attachment was supported by the first lobe in an external (= midventral) position (Kröger and Mapes 2007; Kröger et al. 2011; Klug et al. 2015). Simultaneous with these phylogenetic changes in sutural complexity, folding of the septa occurred (Fig. 3.2). From orthocerids via bactritoids to the earliest ammonoids, the septa stayed simply dome-shaped with the convex side pointing apically (Korn 1997; Korn and Klug 2002). As early as in the Emsian (Early Devonian), some ammonoids evolved septa with synclastically corrugated septa, where part of the septum deviates from the simple dome-shape (e.g. *Latanarcestes*, *Sellanarcestes*). In these septa, some parts are actually vaulted in the opposite direction, i.e. adaperturally. In the course of the Devonian, the vaulting in the apertural direction increased, leading to anticlastically corrugated septa in, e.g., the Gephuroceratidae. With this innovation, the majority of the surface of the septum became vaulted adaperturally, i.e. in the opposite direction as before. This might indicate a change in the balance of hydrostatic pressure between the ambient pressure and the pressure within the newly forming chamber. In all orthocerids, all bactritoids, and most Early and Middle Devonian ammonoids, the pressure in the new chamber was speculatively lower than ambient pressure (as reflected in septum vaulting); it might have changed to nearly equal or slightly high-

of Ada, Oklahoma (USA); PIMUZ 31257 (Paläontologisches Institut und Museum, Universität Zürich); dm =43 mm. **c** lateral and ventral views of *Amphipopanoceras* cf. *medium* (McLearn 1948), SGPIMH no. 3181 (Universität Hamburg), Triassic, Spitsbergen; dm =25 mm; acid-prepared specimen with phosphatized septa and siphuncle; from Weitschat (1986). **e** *Lytoceras* sp., Aalenian (Jurassic), Heiningen near Göppingen, whorl height 84 mm, Staatliches Museum für Naturkunde Stuttgart, SMNS 23156 (after Ernst and Klug 2011). **f**, **g** entire phragmocone and detail showing the complex suture of *Pinacoceras metternichi*, MB.C 2933 (Museum für Naturkunde, Berlin), Norian, Triassic, Hallstatt (Feuerkogel?), Austria, dm ~50 cm

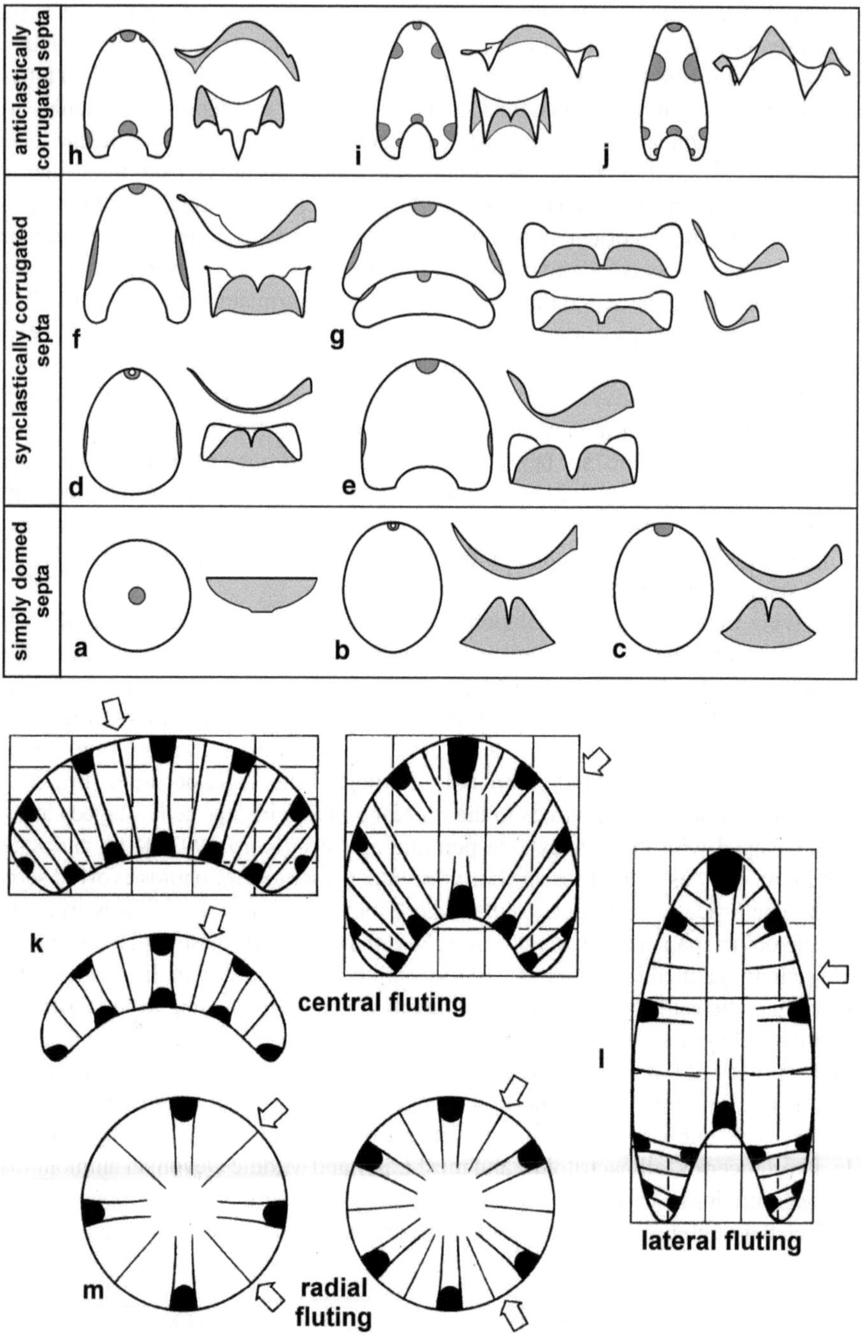

Fig. 3.2 Main shapes of septa: **a–j** Septal projections of various cephalopod septa (from Korn and Klug 2002). Note the change in vaulting from the synclastically to the anticlastically corrugated septa! In **a–j** septal view is on the right (ventral on the *top*), *bottom right* is ventral view (aperture upward), *top right* is lateral view (ventral on the *right*, aperture upward). **a** Orthocerida.

er chamber-pressures compared to ambient hydrostatic pressure in more derived ammonoids. This change occurred during phylogeny but is not visible in ontogeny.

In more complex septa, a system of regularly arranged septal folds ("*pillars*") evolved (Fig. 3.2; Westermann 1956; Schindewolf 1972; Korn and Klug 2002). Depending on the whorl cross section, these pillars can be grouped into three types: central, lateral, and radial fluting. Central fluting occurs in ammonoids with depressed shells (low aperture) and means that the folds are arranged with their axis more or less through the coiling axis. Lateral fluting is commonly found in compressed forms (narrow aperture), where the folds are arranged subparallel to the coiling axis. Radial fluting is less common and has been described from forms without imprint zone (no whorl overlap) such as the Cretaceous baculitids. In this case, the folds cross each other in the center of the septum.

Importance of the suture line for phylogenetic reconstructions of higher taxonomic units (usually at or above the generic level) was demonstrated by the outstanding monographs of, e.g., Ruzhencev (1960, 1962, 1974) and Schindewolf (1961–1968). Ammonoid evolution started with a trilobate primary suture line (ELI), which is based on the appearance of the lobes during ontogeny (Fig. 3.3, 3.4). Lobes formed between E and L in ontogenetic sequence are called adventive lobes (A), while lobes formed between L and I are called umbilical lobes (U). The idea that major Palaeozoic ammonite groups are characterized either by the A-mode (insertion of new lobes between E and L; e.g., Goniatitida) or the U-mode (insertion of new lobes between L and I; e.g., Gephuroceratina and Prolecanitida) was developed by Schindewolf (1954); this hypothesis was accepted until the sixties (Miller et al. 1957; Ruzhencev 1960). All Mesozoic ammonoids were regarded as possessing a U-mode sutural ontogeny due to their assumed derivation from the prolecanitids. Recently, this hypothesis was refuted by Korn et al. (2003), who showed that prolecanitids and their derivatives possess an A-mode sutural ontogeny, while the L-lobe was reduced during ontogeny (Fig. 3.4).

Due to the absence of complexly fluted structures, which grew in the same way as ammonoid septa, in phragmocone bearing extant cephalopods (*Sepia*, *Spirula*, *Nautilus*) and elsewhere in the animal kingdom, the ammonoid septa still remain enigmatic and fascinating to paleontologists. A similar degree in complexity of folding is reached in the walls perpendicular to the septa of the phragmocone of Sepiida, but there, these pillars are extremely low, wide and incomplete, i.e. they do not close a void completely. Thus, the differences in vaulting between the center and the margin of the septum are lower than in derived ammonoids.

Explanations for the morphogenesis of the ammonoid septa focusing on their complexly folded margin are twofold: (1) The adaptive aspect refers to the suitability of a structure to its function (Buckland 1836; Pfaff 1911; Schmidt 1925; Val-

b *Lobobactrites*. **c** *Gyroceratites*. **d** *Mimosphinctes*. **e** *Latanarcestes*. **f** *Achguigites*. **g** *Sellanarcestes*. **h** *Pseudoprobeloceras*. **i** *Manticoceras*. **j** *Probeloceras*. **j–l** the three main types of septal fluting in ammonoids; white arrows mark the transmission direction of hydrostatic pressure. (From Westermann 1956; Schindewolf 1972; Korn and Klug 2002)

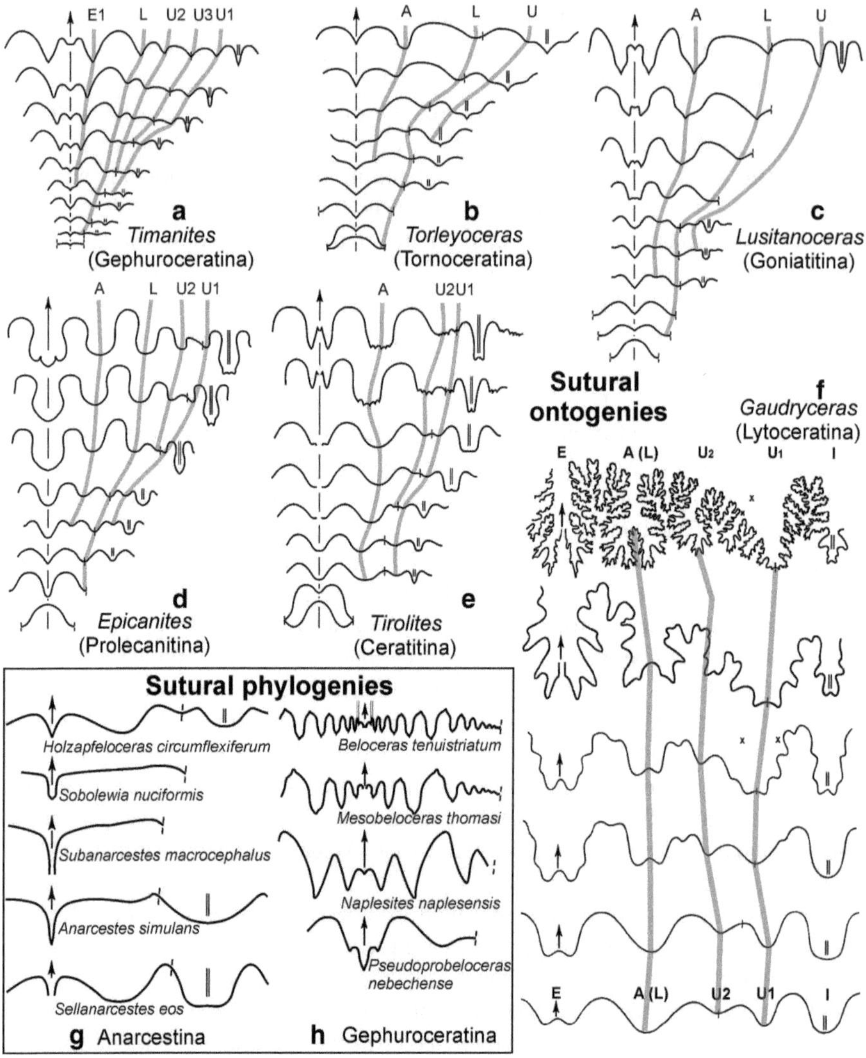

Fig. 3.3 Sutural ontogenies and phylogenies. **a–e** from Korn et al. (2003). **f** from Schindewolf (1960). **g** and **h** from Korn and Klug (2002). ***E*** external lobe, ***A*** adventive lobe, ***L*** lateral lobe, ***U*** umbilical lobe, ***I*** internal lobe

entini and Finks 1974). (2) The fabricational aspect deals with biophysical forces or processes leading to pattern formation (Seilacher 1973, 1975; Reif et al. 1985; Checa and Garcia-Ruiz 1996). The latter authors provide a review of fabricational hypotheses available for cephalopod septa such as, e.g., Owen's (1832) preseptal gas hypothesis for *Nautilus*, the gas-pressure theory presented by Tate and Blake

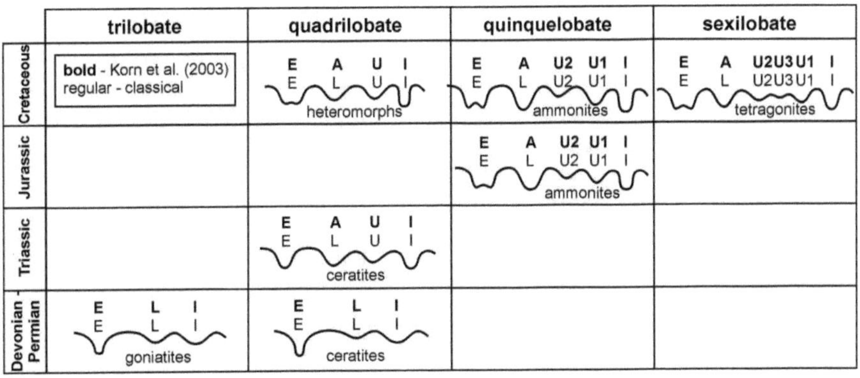

Fig. 3.4 Primary sutures of ammonoids with their traditional and reinterpreted sutural formulae; from Korn et al. (2003). In bold face: New interpretation of Korn et al. (2003). (See Fig. 3.3 for the explanation of E, A, L, U, and I)

(1876), Solger (1901), Swinnerton and Trueman (1918), Spath (1919), Schmidt (1925), Seilacher (1975), and Buckman (1909–1930), accepting that the lobes and lobules represent muscular attachment sites of an organic membrane (Klug et al. 2008). The gas-pressure theory was retained by Arkell (1949, 1957).

Much of our understanding today of the processes related to septal formation in ammonoids was improved by the findings of Denton and Gilpin-Brown (1966, 1973) and Denton (1974), who showed that the last built chamber of *Nautilus* was filled with liquid and emptied via the siphuncular osmotic pump and by the fabricational models introduced by Seilacher (1973, 1975). Accordingly, it was assumed that the ammonoid septum was an elastic membrane subject to radial tension with a series of tie points and that the mantle initially had planar adhesion to the shell wall. The presence of such radial tension on an organic membrane becomes visible by weak parallel furrows along the basis of a septum of a hollow preserved *Gaudryceras* (Fig. 3.5). Concerning the septal mantle (the part of the soft body that precipitated the septa), Westermann (1975) postulated that its adapertural part had a stiff (flexible) margin (aponeurosis) maintaining its shape during translocation and thereby caused exact duplication of the septal shape, which was also accepted by subsequent workers (Zaborski 1986; Hewitt et al. 1991). Bayer (1977a, b, 1978a, b) suggested a weakly elastic septal mantle surface that became more complicated during ontogeny, more or less representing the shape of the resulting septa. According to his articles, the mantle was attached not only at tie points but along its full length. Changing from gas to hydrostatic pressure, it was widely accepted that the pressure was subequal to or higher than that in the rear of the soft body (Checa and Garcia-Ruiz 1996). Several models for the ammonoid septal formation and their possible function(s) are available and will be described below.

Fig. 3.5 *Gaudryceras sp.*, (Coll. Y. Shigeta, Tokyo), Coniacian, Kamtchatka. **a** overview of the specimen preserved in hollow condition; scale = 0.5 cm. **b** close up of two septa to demonstrate that all lobes of the last preserved septum are in contact with the saddles of the preceding septum. **c** SEM image of lobe with weak parallel furrows along the basis of the septum documenting radial tension on an organic membrane

3.2 Ammonoid Septal Formation Models

3.2.1 Viscous Fingering Model

The morphogenetic Viscous Fingering Model was introduced by García-Ruiz et al. (1990). First, García-Ruiz and Checa (1993) used it as an alternative explanation to the Tie-Point Model of Seilacher (1973, 1975). Checa and Garcia-Ruiz (1996)

explain the fabricational process for the septal formation to some extent and compare them with fractals. In this case, septal complexity is largely independent of genetic constraints or the shape of the apical mantle (Checa 2003). The Viscous Fingering Model was applied to the septal formation because it represents a growth mechanism fulfilling the Laplace equation and deals with a fluid mechanics problem (Stanley 1987). Septal sutures were previously compared with adhesion figures by Thompson (1942) and Damiani (1986, 1990), but only for morphological reasons and less as a morphogenetic model.

Unlike mathematical fractals, natural fractals are recognized as objects not being strictly self-similar in a mathematical sense, but rather scale-invariant structures over a range of magnification. With the beginning of the seventies, a number of authors realized that fractal geometry is a good tool to quantify sutural complexity (Vicencio 1973; Guex 1981; Bayer 1985; Damiani 1986, 1990; Seilacher 1988). The fractal dimension accounts for the extent to which the septal suture fills a two-dimensional space and therefore varies between one and two (Fig. 3.6). This method was applied by Boyajian and Lutz (1992) to the sutures of more than 600 ammonoid genera. A protocol for morphogenetic studies was developed by García-Ruiz (1992). Checa and García-Ruiz (1996) published a step-by-step demonstration and the improvements done for the Viscous Fingering Model. The Viscous Fingering Model serves well as an explanation of the frilled septal margin, the distribution and/or spacing of saddles and lobes and for the attachment to the conch wall as well (Fig. 3.7). They also demonstrate that opposite elements are linked across minimal distances as was pointed out by Westermann (1956, 1958) and Seilacher (1975, 1988). Furthermore, it allows a comparison with phylogenetic or ontogenetic suture line development showing that lobes at first became more complex while saddles remained smooth, a trend recognized in both phylogeny and ontogeny of ammonoids (Seilacher 1973, 1975). An external lobe consisting of two lobules that grow symmetrically on both sides of an implemented rod was produced in all of their experiments caused by a screening effect.

By contrast, the ammonite septa contained within the ammonitella are usually not frilled along their margin. First corrugation occurs when the cross section of the conch tube increases, which often happened after the formation of the nepionic constriction (Kulicki 1974; Hewitt 1985; Checa 1991). Nevertheless, the lobes occurring during ontogeny are recognized as major elements. All subsequently occurring lobules or folioles are of lower rank. In this case, the Viscous Fingering Model failed to explain the (rare) disappearances of major lobes as demonstrated for the lateral lobe in early ceratitids by Korn et al. (2003) or the sixth lobe for *Anagaudryceras* by Krivoshapkina (1978).

3.2.2 Tie-Point Model

The model introduced by Seilacher (1973, 1975, 1988; see also Westermann 1965, 1971, 1975), which stresses fabricational processes and constraints by hydraulic forces, argues that the organic septal membrane attaches at distinct (genetically controlled) points on the inner surface of the conch wall—these are Seilacher′s tie

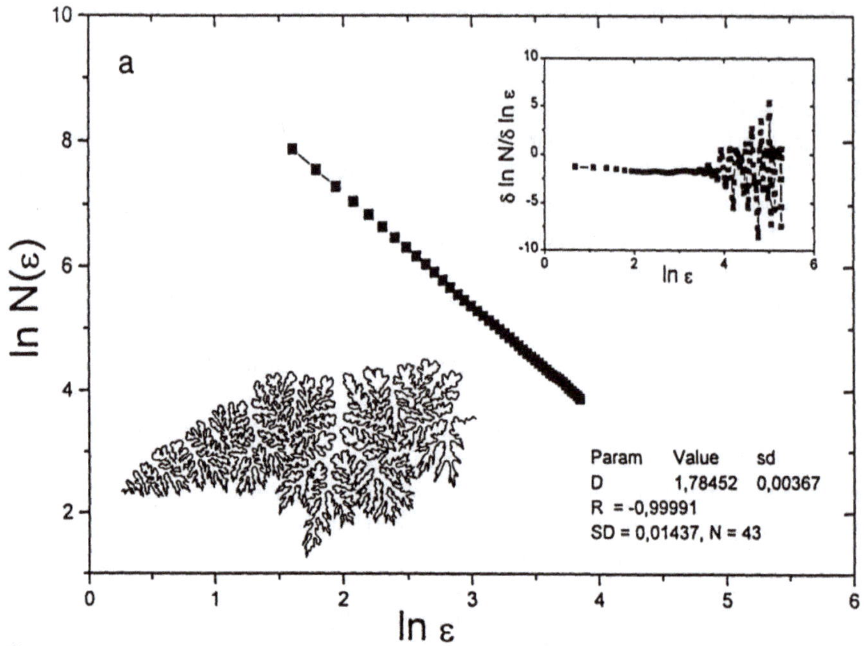

a

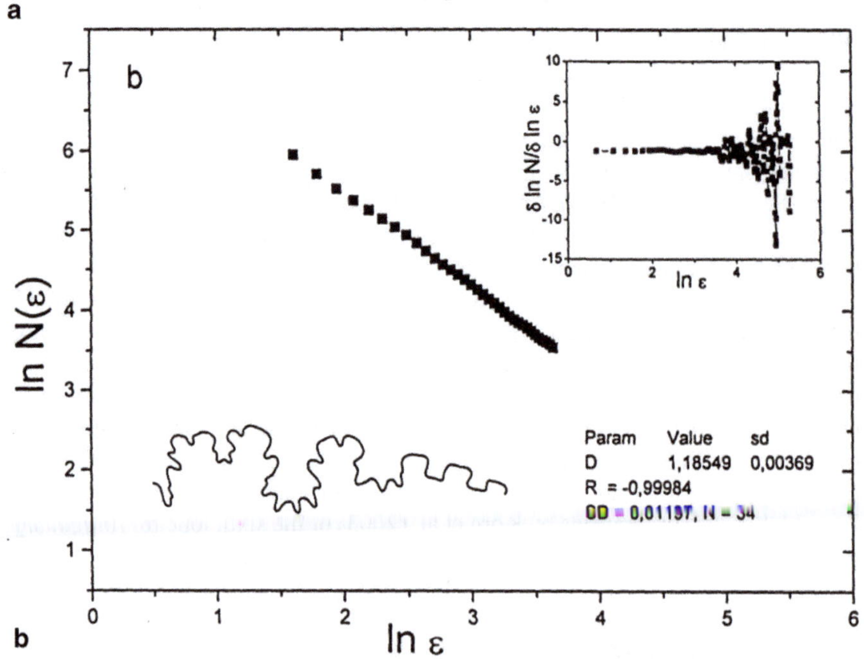

b

Fig. 3.6 Fractal dimensions of sutures. Lol-lo plots for the determination of the fractal dimension (D) of the septal sutures of two ammonoid genera. Top: *Puzosia*. Bottom: *Cottreauites* (both sutures after Arkell et al. 1957). *ε*, size of box edge (in pixels); *N(ε)*, number of boxes needed to cover the object; *sd*, standard deviation of the *D* value (fractal dimension); *SD* standard deviation of the R value, *N* number of data points. All figures reproduced from Checa and Garcia-Ruiz (1996)

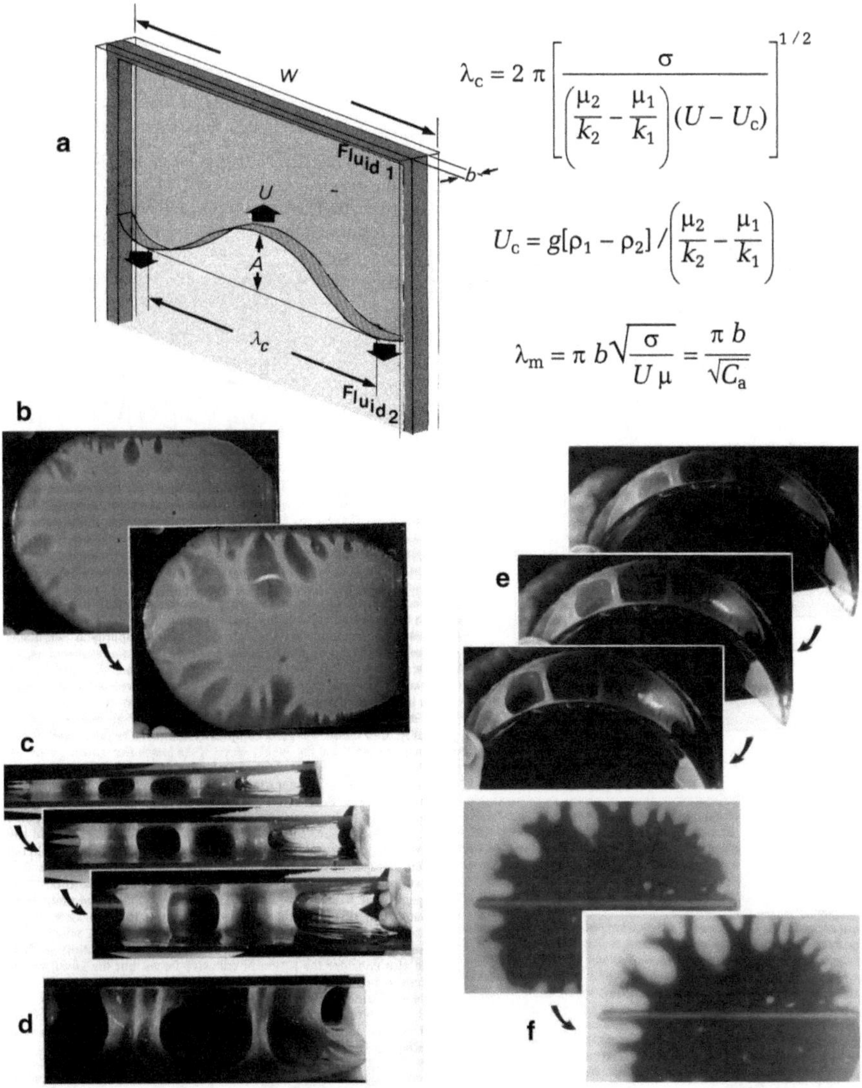

Fig. 3.7 Viscous fingering and the Hele-Shaw cell. **a** Schematic drawing of the Hele-Shaw cell. Fluids 1 and 2 (different viscosities) are injected in the cell. Without movement, the interface is horizontal. When the cell is turned upside down, fluid 1 tries to move down and deforms the interface. Then, the interface becomes unstable and forms undulations (viscous fingers) with a predictable wavelength λ_c. *A* wave amplitude, *b* cell thickness, *k* permeability of the cell, ρ, ρ_1, ρ_2 densities of the fluids, *W* cell width, *U* velocity of displacement of the interface, U_c critical velocity. b to f various views of Hele-Shaw cells illustrating the similarity of shapes formed therein to ammonite sutures. All figures reproduced from Checa and Garcia-Ruiz (1996)

points (Fig. 3.8). The tie points are first all located at the tips of what now is recognized as lobes or lobules (depending on suture complexity). After the membrane was attached, the preseptal membrane (and/or the septal mantle) became blown out

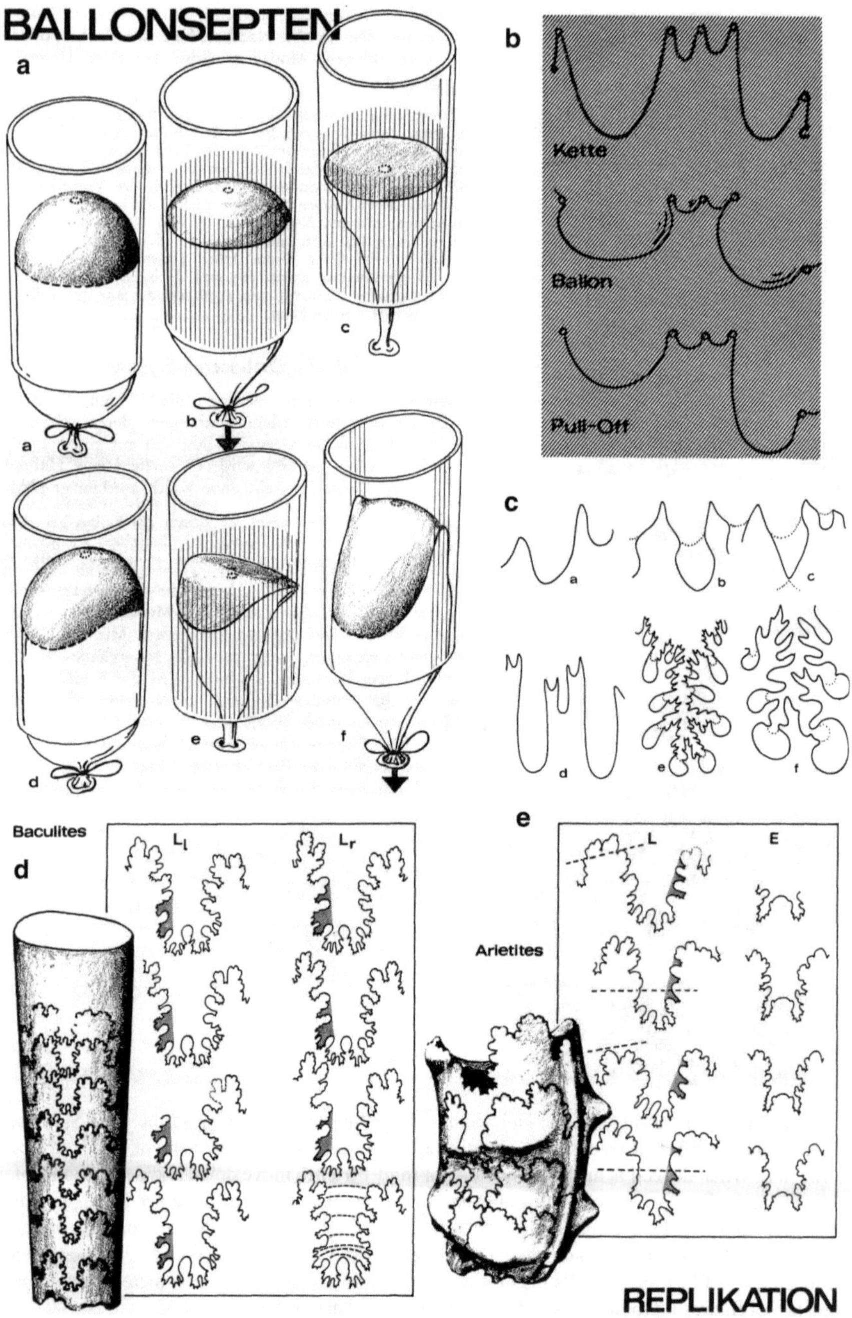

Fig. 3.8 Tie Point Model (images and captions from Seilacher 1975). **a** Balloon Model: "The spherical surface of a balloon pressed into a cylinder may be mechanically modified into flatter or anticlastic surfaces by completely or partially glueing and then retracting the balloon or by doing the same in a cylinder with elliptical cross section." **b** Simulations of mechanical pull-off lines to

towards the aperture due to positive hydrostatic pressure. Thereby the saddles and folioles were formed and attached to the inner surface of the wall subsequently. As a rule, the more tie points are present, the more complicated the septal margin appears. Tie points are not set contemporaneously but according to their hierarchic order (Seilacher 1988). After the fixation of all tie points, calcification of the membrane started. It remains unclear whether or not the posterior part of the mantle (septal mantle) was involved in shaping the septum and if so, to which degree, whether it was smooth or permanently fluted (Hewitt et al. 1991; Seilacher 1975; Westermann 1975; Seilacher 1988). Another option was discussed by Zaborski (1986), Lewy (2002, 2003) and Hewitt and Westermann (2003): They suggested that the mantle could control its form. For a further discussion, see Yacobucci and Manship (2004, 2011).

3.2.3 Reaction-Diffusion Model

Hammer (1999) explains the formation of septal sutures as the result of morphogen diffusion (Fig. 3.9). He used therefore the Reaction-Diffusion Model (Turing 1952; Meinhardt 1998) to explore the growth of the septal membrane by interactions between a slowly diffusing activator morphogen and a more rapidly diffusing inhibitor morphogen. This model was also applied to explain ammonoid ornamentation (Hammer and Bucher 1999) and Buckman's law of covariation (Guex et al. 2003). The Reaction-Diffusion Model serves to simulate complex structures such as the fluted septal margins that are reproducible without requiring direct genetic control or a complex developmental regulatory system. This model suggests that the formation of epithelial invaginations is controlled by morphogens and is partially consistent with Seilacher's Tie-Point Model (Seilacher 1975, 1988). It was inspired by developmental models for epithelial folding used to explain folding in kidneys, lungs, teeth, mammary glands and other organs. Mainly the morphogens or signal proteins regulate shape and/ or rate of proliferation, causing invagination and, in the case of the septal membrane, the formation of lobes. Similar invaginations caused by inhomogeneous lateral proliferation of cells was shown for teeth by Jernvall (1995) and suggested to explain the shape of giant single celled algae by Harrison and Kolar (1988). By contrast, the model was only proposed on theoretical grounds and has not yet been tested on ammonoid or other cephalopod septa (Checa 2003).

generate the smooth shape of saddles and folioles. Top: chain ("Kette"); middle: balloon; bottom: pull-off. **c** "bottle necks and pointed saddles (*b, c*) suggest a two-point fixation, while tongue-like saddles (*d*) may indicate the channelling effect of longitudinal fibres during the pull-off process. Balloon-shaped tips (*e, f*) are possibly due to final pressurization from the apical side. **d, e** "In contrast to the exact replication in the smooth shells of *Baculites*, successive suture lines of *Arietites* vary the pattern of established incisions in response to the corrugations of the shell wall"

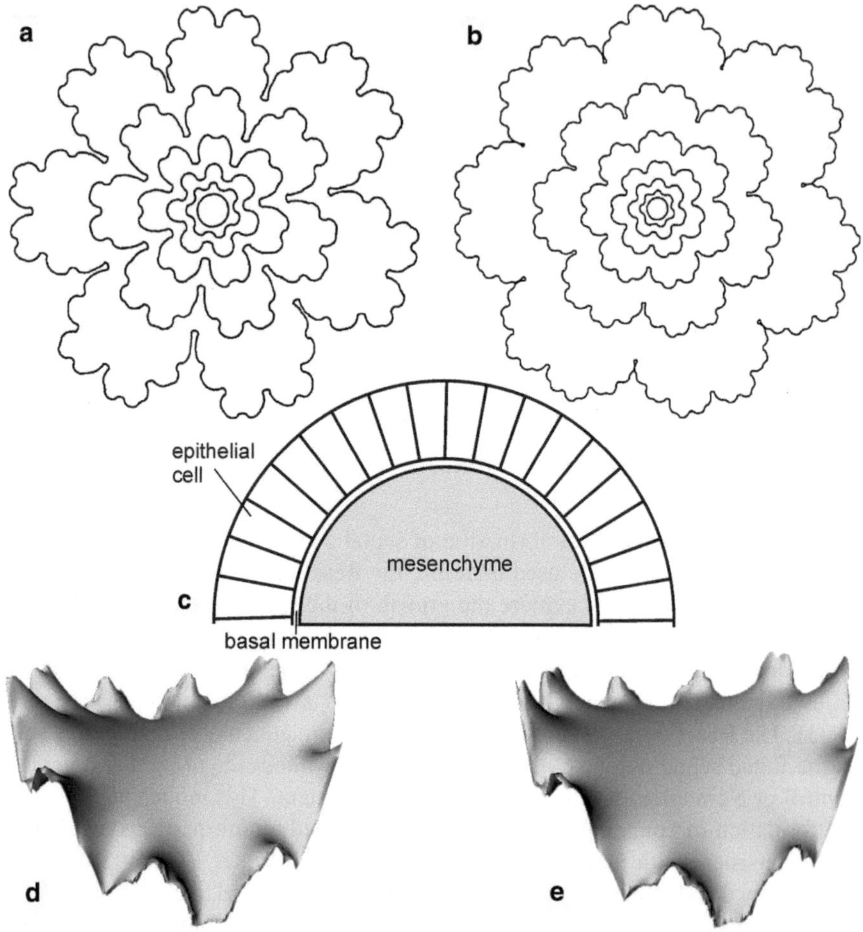

Fig. 3.9 Reaction-Diffusion Model (images and captions from Hammer 1999; courtesy by Ø. Hammer). **a** *"simulation of ammonoid septum formation using a coupled geometrical/chemical model. [...] Note the successive introduction of primary, secondary and tertiary lobes as the suture grows. Running time was 50 full traversals of all grid points."* **b** *"Iterated invagination produced by a uniform growth rate."* **c** *"Chemical signal substances induce changes in cell adhesion forces and/or the cytoskeleton. This results in a change in epithelial curvature, causing folding of the membrane."* **d** *"An explicit minimum curvature surface with the suture as given boundary condition"*. **e** *"An explicit Laplace surface with the suture as given boundary condition"*

3.2.4 Composite Model

Each of the models presented above contributed much to the understanding of the morphogenesis of ammonoid septa. Nevertheless, a comprehensive model, be it mechanical or numerical, is not available yet, although a numerical model appears to be an only moderately difficult task for a person familiar with programming and modelling.

To some degree, the shape of new septa was probably copied from preceding septa (Seilacher 1975). During the forward movement of the soft body and prior to the insertion of the new septum, the overall course of the suture line was maintained, as is recorded in drag bands and pseudosutures (Fig. 3.10; Seilacher 1975; Lominadze et al. 1993; Checa and García-Ruiz 1996). Although the septal mantle was fixed in the lobes and lobules first (Seilacher 1973, 1975), the septal mantle became eventually fixed to the shell tube along the entire suture line (Klug et al. 2008). The suture thus can be partially understood as a soft-tissue attachment site. Due to the presence of muscles in the septal mantle, parts of the suture could even be defined as muscle attachment sites.

Development of septal shape is controlled by the whorl cross-section and its ontogenetic modifications (allometry); apparently, the phylogenetic increase in coiling of the initial whorls (De Baets et al. 2012) contributed much to the increase in sutural complexity. Additionally, a whorl cross section with narrow and wider parts (compressed or depressed whorl section) from early stages in ontogeny onwards made sutural complexity rise faster than in forms with simple subcircular cross sections. Sutural complexity increased through ontogeny and in most cases also through phylogeny (Saunders 1995; Saunders and Work 1996; Saunders et al. 1999); exceptions are the so-called pseudoceratitic and pseudogoniatitic sutures, but there are other, less prominent cases (personal communication D. Korn, Berlin). Sutural complexity is thus an effect of heterochronic processes (Hammer 1999). Remarkably, the nautilid genus with the most complex sutures, *Aturia*, also has the smallest embryonic shell and a markedly compressed whorl cross section (Schlögl et al. 2011; Laptikhovsky et al. 2013). Possibly, the number of septa might also correlate with sutural complexity, but data to test this hypothesis are lacking at this point.

The Viscous Fingering Model helps to understand how septum-like shapes with their fractal aspect formed including the complexity of the suture and the shape of the septal surface in space. The Tie-Point Model postulates that muscles held the septal membrane in place, where the lobes formed and that the saddles are put in place by simple physical processes. It appears plausible that the preseptal membrane/septal mantle attached in the main lobules first and the control over the regular placing of lobules involved largely self-generating morphogenetic processes (Seilacher 1973, 1975; Westermann 1975; García-Ruiz et al. 1990; García-Ruiz and Checa 1993; Hammer 1999).

For the increase of sutural complexity, it is crucial that the whorl section is differentiated in terms of vaulting into the area around the siphuncle, a more or less broad central part of the whorl and more or less constricted parts in the concave whorl zone (imprint zone, whorl overlap zone). Of course, a concave whorl zone is not always present; both the earliest ammonoids and many heteromorphs lack a concave whorl zone. Normally, one should thus expect similar sutures in these two groups, but the sutures in Cretaceous heteromorphs and the earliest ammonoids are quite different. The differences in shell shape between some taxa of these two groups are largely limited to the embryonic shell, which is tightly coiled with whorl overlap in derived forms and loosely coiled with no or minor whorl overlap and with or without an umbilical window in early ammonoids. This somewhat supports

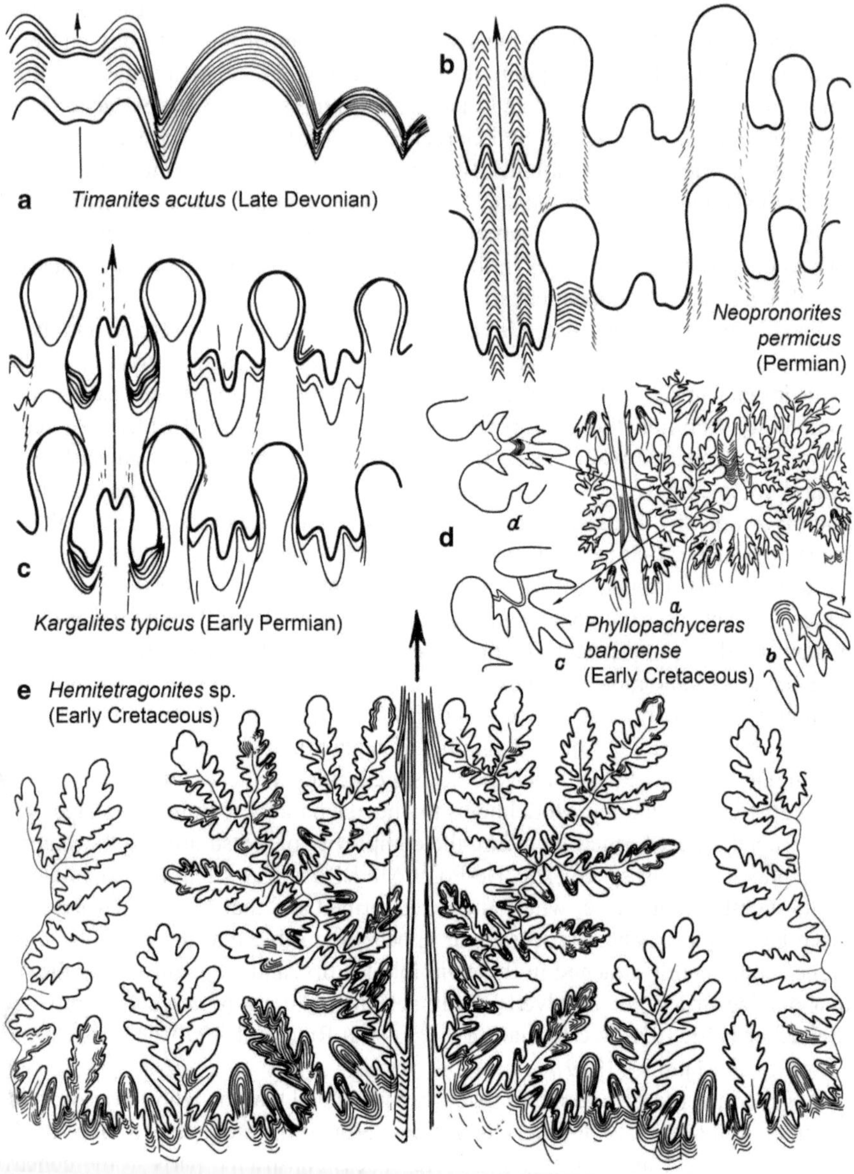

Fig. 3.10 Pseudosutures, reproduced from Lominadze et al. (1993). Note the slight interim simplification of pseudosutures compared to the sutures in **d** and **e**

the hypothesis that (i) the degree of coiling of the embryonal shell is linked with sutural complexity and (ii) the overall shape of the new septum is copied/preserved for the formation of the new septum.

As suggested by Klug et al. (2008), the musculature of the septal mantle was present in ammonoids and nautiloids. Additionally, it plays an important role in the septum formation cycle. Prior to the anterior movement of the soft body, the musculature of the septal mantle contracted, slightly reducing the area of the septal mantle and lifting it off the last septum and the shell wall, simultaneously pulling the mantle out of the lobules and folioles. Repeated contraction and relaxation with intermediate fixation is documented in pseudosutures (Fig. 3.10; Lominadze et al. 1993). Speculatively, part of the septum became stiffened in the middle by the contraction of the muscles before the anteriorward movement. This ensured that symmetry and thus balance was maintained and caused the harmonic increase in complexity as well as the great similarity between successive septa. Once the position for the new septum was reached, the septal mantle expanded slightly and was then probably attached more firmly. The differentiation in lobule and foliole shape might be simply the result of the slight forward pull, which had different effects on lobules and folioles. In that case, no true tie-points would have existed (compare Bayer 1978a, b) and the suture line course would be the result of the adapertural pull and the pull-and-release from the center of the septal mantle.

3.3 Differentiation and Mechanical Properties of the Septal Mantle: Form and Function

From pseudosutures (Fig. 3.10), Checa and García-Ruiz (1996) deduce that the septal mantle was not pressurized by the cameral fluid during most of the translocation process and that pressurization and differentiation into several-order flutes took place shortly before or after definitive emplacement of the rear body. More drastic are the changes in phylloceratids and lytoceratids with complexly fluted septa, where the septal mantle was totally invaginated before its final attachment (Lominadze et al. 1993). The septal lobe in *Pseudophyllites* with its septal tube, which is dorsally closed by the median part of the internal lobe, demonstrates that the septal recess had to contract and slide along the walls of the septal tube formed by the septal recess of the previous septum. An increase in complexity was possible only after the translocation occurred. Increasing lengths from the pseudosuture to the septal suture have been interpreted in terms of viscoelastic properties of the rear body. Highly viscoelastic mantles led to complex sutures of phylloceratids and lytoceratids, whereas progressively stiffer rear mantles produced simple ammonitic, ceratitic/pseudoceratitic, or goniatitic septal sutures.

Septal shape provided additional support to stiffen the rear of the soft body. This stiffness comes from the number of lobes in the primary suture and the type of subsequent sutural frilling. Wiedmann and Kullmann (1981) reported three lobes for goniatites, four lobes for ceratites, five lobes for ammonites as well as lytoceratids, and six lobes are present in the Late Cretaceous tetragonitids. According to Checa and Garcia-Ruiz (1996), the differentiation of the rear mantle into flutes takes place when the cameral liquid experiences increased pressure during the late stage of translocation. Although in Recent nautilids, the chamber liquid is slowly replaced by gas only after the new septum has reached about half of its final thickness (Ward

1979, 1987). Checa and Garcia-Ruiz (1996) attributed the increased pressure of the cameral liquid by hyperosmosis or muscular pull. They suggested that in early stages of translocation, pressure conditions were reverse. It remains unclear whether the rear mantle or the siphuncle was the chamber-filling organ in *Nautilus* (Ward 1987) and ammonites (e.g., Bayer 1975).

3.4 Septal Function

Palaeontologists have always been attracted by the complexly folded ammonite septa and the resulting suture line (Fig. 3.1; Allen 2007). Several attempts were made to explain the function of septa and their marginal corrugation. Westermann (1971, 1999), Kennedy and Cobban (1976), Hewitt and Westermann (1997, 2003) as well as Keupp (2000) presented an overview of all available hypotheses. These are briefly repeated here to elucidate the possible function or functions of complex ammonite septa and the septal lobe restricted to the Lytoceratoidea. Functional explanations for the frilled septal margin are twofold; physiological (1–7) and mechanical (8–12). These twelve explanations are listed and shortly discussed below:

1. Septum morphology reflects the shape of the posterior mantle, whose folding may serve for a better respiration (Newell 1949) or gas-production (Pia 1923). Accordingly, sutural complexity should be connected with ammonoid metabolism. Metabolic rates are often estimated as the rate of oxygen consumption but can also be considered as rate of buoyancy control or biomineralization; the hypothesis that sutural complexity reflects metabolic rates can be falsified by the facts (i) that the way the septum is folded is far from optimal to increase surface for maximal respiratory surface (it is a minimal-area surface as demonstrated by Hammer and Bucher 1999), (ii) that the septal mantle was probably not in direct contact with the sea water and all cephalopods developed gills for respiration and (iii) the gas infill of the chambers was provided through the siphuncle (Tanabe et al. 2015).
2. Septal folding increased the wettable conchiolin surface layer (pellicle) in the chambers to support faster liquid transport to the siphuncle for buoyancy control (Mutvei 1967); this explanation is supported by the discovery of organic intracameral sheets by Bayer (1975; see also Seuss et al. 2012 and Polizotto et al. 2015) and the presence of cameral membranes (Weitschat and Bandel 1991; Polizotto et al. 2015), but it is contradicted by: (i) the not optimal construction for area maximization of the septa and (ii) the low angles in folioles and lobules, potentially hampering (but accelerating) the complete emptying. Nevertheless, this hypothesis cannot be rejected entirely, but it appears insufficient as sole explanation for the evolution of sutural complexity.
3. Corrugation formed recesses where cameral liquid can be stored by surface tension to improve buoyancy adjustment (Kulicki 1979; Kulicki and Mutvei 1988; Weitschat and Bandel 1991; Saunders 1995); undoubtedly, the surface tension

would have held remaining chamber liquid in place, so lateral movements by shifts in chamber water would have been minimized.

4. The Cartesian Diver Model (Seilacher and Labarbera 1995; Seilacher & Gishlick 2015) assumes that the last septum remained uncalcified until the beginning of the next chamber cycle; in that interval, the space behind the septal mantle might have been filled with gas and functioned like a swim bladder. The enclosed gas volume could be adjusted by sutural muscles spanning across the septal mantle to change buoyancy. This would have allowed ammonoids to move vertically in the water column without high expenditure of energy; although it is a stimulating hypothesis, it lacks support (Jacobs 1996) and is potentially even contradicted by Seilacher's Tie-Point Model: Assuming that lobule formation is controlled by the insertion of muscles and that saddles are simply formed thereafter by the pull of the soft body in anterior direction, the insertion of the contractile fibers of the septal mantle would not be optimal to contract or expand the gas-filled preseptal space; a similar shape of lobules and folioles would be expected, if the muscles attached at the suture were functioning in opposite directions. In any case, it is not a parsimonious assumption knowing the function of the phragmocone from Recent nautilids.
5. Some suggested that the septal recesses facilitated the attachment of major adductor muscles (Seilacher 1975, 1988; Henderson 1984; Ebel 1992); the main muscle systems of ammonoids have been described for several groups of ammonites, and they are located anterior to the septa (Doguzhaeva and Mutvei 1991, 1996; Richter 2002; Klug et al. 2008; Richter and Fischer 2002). Therefore, the septal recesses were apparently not used to insert cephalic retractors or hyponome retractors.
6. The septa aided in emptying the chambers at almost any depth (Ward 1987); this appears likely for actualistic reasons because modern nautilids use the chambers like this. Only implosion depths of the various ammonoid taxa set limits to this. However, nautilid septa are only gently folded and therefore, this comparison is somewhat questionable.
7. The folded septa improved buoyancy control to escape predation (Daniel et al. 1997) and help to rapidly reflood chambers of regenerating ammonite shells to compensate shell loss (Kröger 2001, 2002); if the actualistic comparison with the phragmocone function of Recent nautilids (Ward 1979, 1987) is correct, chamber emptying was achieved by osmosis and thus was maybe too slow to aid in escaping predators. To regain buoyancy-control after sustaining buoyancy-altering injuries, the reflooding might have been fast enough, but then, does this suffice to explain the strong evolutionary trend in increasing complexity?
8. Septa helped to attach the soft body to the shell (von Buch 1829, 1830; Suess 1865; Tate and Blake 1876; Steinmann 1888; 1925; Diener 1912; Spath 1919; Reyment 1955; Hengsbach 1978; Henderson 1984); the suture was certainly a temporary soft-tissue attachment site, that was translocated regularly (Klug et al. 2008), but the connection was most likely not very strong.

9. Fluting strengthened the conch wall and the last septum against hydrostatic pressure (Buckland 1836, he also accepted von Buch's idea; Owen 1843; Tate and Blake 1876; Zittel 1884; Pfaff 1911; Nagao and Saito 1934; Reyment 1955; Westermann 1958, 1971, 1975; Seilacher 1975; Kennedy and Cobban 1976; Hewitt and Westermann 1997; Hassan et al. 2002; de Blasio 2008); Jacobs (1990, 1992, 1996) reviewed this so-called "*Buckland hypothesis*". Although there are limits to the functionality in terms of pressure resistance enhancement by the septal frilling, the presence of such walls in the partially gas-filled phragmocone most likely provided some hydrostatic support, which is reflected in the correlation between sutural approximation and shell curvature (Buckland 1836; Jacobs 1990, 1996; contra Saunders 1995). By contrast, the frilling itself probably did not contribute much to this function and was suggested not to be related to bathymetry and/or that there was no major differences in habitat depth for epicontinental and oceanic ammonites (Olóriz and Palmqvist 1995; Olóriz et al. 1997, 1999; Pérez-Claros 2005). In the previously listed papers, however, septum thickness was not taken into consideration. De Blasio (2008) describes a related function for suture frilling that was reinforcement against shell shrinkage due to hydrostatic pressure, which could potentially determine a loss of buoyancy.
10. Frilling helped the last septum withstand the pressure exerted through the soft body onto the shell wall (Pfaff 1911); there might be some truth in this hypothesis, because the non-mineralized or incompletely mineralized membrane of newly forming septa were potentially more tightly stayed by the frilled suture line, thus allowing greater pressure gradients to build up between phragmocone and soft parts (Klug et al. 2008).
11. The septa increased the overall weight of the shell for buoyancy control (Reyment 1958; Teichert 1967); there is no doubt that the septa added weight, but presumably, this was not the primary function of the septa because additional ballast could have been added by simply increasing shell thickness elsewhere. Additionally, when the low density chamber fillings are included in this consideration, this notion appears not very reasonable.
12. Lewy (2002, 2003) suggested that the complexly fluted septa are the result of a stronger connection between the soft body and the conch (temporary hold-anchorage system, no muscle attachment). This led Lewy (2002) to the assumptions that the greater the complexity of the marginal fluting of septa, the better the ammonoid could withstand the dragging force between the body and the buoyant conch, and hence the more aggressively the ammonoid predated and competed with other creatures; this hypothesis entirely lacks evidence from soft-tissues or muscle imprints (Doguzhaeva and Mapes 2015; Klug and Lehmann 2015).

Since the septal lobe (= I_{s1}—to avoid confusion with septal lobes, which are the apically oriented parts of a septum, dorsal lobe in Checa and García-Ruiz 1996) is a widely unrecognized structure and only present in Jurassic-Cretaceous Lytoceratoidea; only a few hypotheses are dealing with its possible function. Hoffmann (2010) suggested its main function was to improve the efficiency of the hydrostatic apparatus in both directions—emptying and flooding of phragmocone chambers. Westermann (1971) correctly stated that the septal lobe is not homologous to 'riding' sutures (septa, that partially attach to preceding septa), which occur irregularly due to adult septal approximation (Hölder 1952, p. 44). By contrast, Seilacher (1988) mentioned that the phenomenon described by Hölder (1952, p. 44) has become regular and a much more pronounced feature in the lytoceratid septal lobe. He suggested that the septal lobe may have served to increase septal vaulting to withstand apertural hydrostatic pressure.

Westermann (1971) argued that the septal lobe would have strengthened the last septum against hydrostatic pressure transmitted through the soft body and perhaps also reinforced the entire loosely coiled phragmocone. By contrast, heteromorph ammonoids did not develop a septal lobe or riding sutures. Likewise, Henderson (1984) hypothesized that, analogous to the fluted septal recesses, the septal lobe was related to facilitate the attachment of adductor muscles, an idea that was rejected by Keupp (2000), Hoffmann (2010) and herein (see point 5).

3.5 Sutures in Ammonoid Phylogeny

Reconstructing ammonoid phylogeny is hampered by several problems such as covariation between shell parameters as well as between shell shape changes and suture line morphology (Monnet et al. 2015). As far as suture line course is concerned, its value for phylogenetic reconstructions has been discussed repeatedly. There is no doubt that a phylogenetic signal can be extracted from suture lines, especially in the Paleozoic (e.g., Wedekind 1916; Schindewolf 1929, 1951, 1954, 1961; Wiedmann and Kullmann 1981; Saunders et al. 1999; Korn 2001; Korn et al. 2003). Limits of this use root in the sometimes high degree of variability of sutures. This variability has a set of origins; they vary (i) in covariation with shell shape (especially ornamentation), (ii) due to injuries (e.g., Rieber 1979; Keupp and Mitta 2004; Keupp 2012), (iii) due to asymmetries, which can also be induced by injuries (e.g., Lange 1929, 1941; Hölder 1956; Kemper 1961; Rein 2004; Longridge et al. 2009 and references therein) and (iv) throughout ontogeny (e.g., Schindewolf 1961–1968; Korn et al. 2003; Rein 2004). These possible sources of variation in suture lines have to be taken into account prior to their inclusion in phylogenetic analyses. In recent years, a number of researchers have experimented with methods to quantitatively evaluate suture line morphology. They have employed GIS (Geographic Information System; Manship 2004; Yacobucci and Manship 2011; Knauss and Yacobucci 2014), which can help comparing sutures quantitatively. Eigenshape analysis was employed by Ubukata et al. (2009) to quantitatively characterize ammonoid sutures. The same was attempted by Ubukata et al. (2014) using wavelet analysis. Undoubtedly, there is still room for research to facilitate the quantification of similarity between sutures, be it between or within taxa or between or within individuals (throughout ontogeny).

3.6 Phylogenetic Applicability of Sutures Case Study: Septal Lobe and Other Interpenetrating Septa

In the Ammonoidea, the general conch morphology is recognized as problematic for phylogenetic reconstructions due to the commonly occurring cases of convergent or parallel evolution (Monnet et al. 2011), speculated to be analogous adaptions to similar life habits by Westermann (1999), Ritterbush and Bottjer (2012) and Ritterbush et al. (2014). Therefore, conch characters appear to be often homoplastic and to contain no or little phylogenetic signals. By contrast, the suture line characters seemed to be more reliable (e.g., Schindewolf 1961–1968) due to their assumed genetically controlled formation (most of the septum formation models presented herein assume a low degree of genetic control and a high degree of self-organization); however, covariation of suture line characters with shell shape and intraspecific variability of suture lines should not be underestimated and the brightest solution will be in most cases to use the full set of available characters.

According to Guex (1987, 1995, 2006), Taylor (1998), Hillebrandt (2000), Guex et al. (2004), Bourillot et al. (2008), and Shigeta (2006), the Psiloceratida originated from the phylloceratids and gave rise to the Lytoceratoidea during the earliest Jurassic. The lytoceratid primary suture line is quinque- or sixlobate, thus representing the most complex stadium. Their adult suture is one of the most intricate structures (e.g., *Pseudophyllites*) compared to all other known ammonite sutures, and it displays a unique feature, the septal lobe (Fig. 3.11). All major lobes and saddles are trifid in early representatives but become bifid in later members of the group. The internal lobe of lytoceratids developed lateral corrugations, while the internal lobe of phylloceratids did not and appears tongue-shaped. In phylogenetically younger members (Middle Jurassic), the internal lobe of the Lytoceratoidea becomes cruciform by the horizontal position of lateral, elongated corrugations (Schindewolf 1961; Hoffmann 2010).

One of the most typical characters confined to lytoceratid ammonites is the septal lobe, which is also generated by the internal lobe. In contrast to the systematic significance of the septal shape and its ontogenetic and phylogenetic development, it is still remarkable that this feature lacked attention and was not seriously investigated for a long time (Salfeld 1924, Schindewolf 1961–1968). Owing to the lack of better knowledge, Arkell et al. (1957, p. L192) postulated for the lytoceratid ammonites that the "[…] *septal lobe* [is] *present in some*", indicating a polyphyletic origin of that morphological feature. Wiedenmayer (1979, p. 862) also suggested that the septal lobe is a non-diagnostic character. By contrast, Hoffmann (2010) demonstrated that the opposite is true and that the septal lobe represents the constituting character of the monophyletic group Lytoceratoidea. Thus, the septal lobe (term introduced by Uhlig 1883, p. 60, 62) is of particular interest here and is described in detail. Nagao and Saito (1934) recorded the presence of the septal lobe for *Gaudryceras*, *Kossmatella*, *Anagaudryceras*, and *Protetragonites* at that time and postulated that the septal lobe could be a synapomorphy for all lytoceratids.

With the verification of this phylogenetic hypothesis, they underlined the potential importance of traits related with the suture line or septal shape.

3.6.1 Morphology of the Septal Lobe

When viewed from the aperture, the surface of the central lytoceratid septum is planar or slightly convex but uniformly curved backwards at the mid-dorsal septal area forming a pronounced depression belonging to the internal lobe. This depression forms a short tube (= "pseudosiphuncular tube" of Nagao and Saito 1934; Fig. 3.11) with its end attached to the surface of the preceding septum generating the septal lobe (Fig. 3.11). The septal lobe therefore represents a modification of the internal lobe. As all lytoceratids show the tendency to develop bifid endings, the originally bifid internal lobe was attached to the preceding septum with two slightly curved main branches. Those attached two tips of the internal lobe surround but do not cover the median dorsal depression forming the tube of the septal lobe of the preceding septum. Between both endings of the internal lobe, at the median plane of the specimen, the internal lobe is retracted towards the aperture. Half of the septal tube of the preceding septum is now closed by the median part of the internal lobe. Obviously, the median internal lobe is not attached to the preceding septum but to the dorsal shell wall. Thereby, a small, rounded opening of the preceding septum still remains after precipitation of the latest septum. Through this opening, the larger ventral part of the chamber is connected with the smaller dorsal sack-like part (septal tube and septal lobe) created by the internal lobe (Schreiber and Hoffmann 2009).

Thus, in exact median section, the septal lobe is not visible (Fig. 3.11). The septal tube reaches about 50 % of the diameter of the septal lobe, i.e. the septal lobe is much broader than its corresponding tube, while the length of the structure covers the whole dorsal chamber length (Fig. 3.12). In several specimens, the terminal median part of the internal lobe does not simply close the septal tube but points to the aperture within the tube being still bifid (Fig. 3.11, 3.13) in *Lobolytoceras* (Schreiber and Hoffmann 2009), *Eogaudryceras*, *Anagaudryceras*, and *Gaudryceras* (Hoffmann 2010). Thereby, the septal tunnel itself becomes septate and does not connect the chambers of the phragmocone like the siphuncle. Thus, a funnel-shaped septate tunnel lies within the phragmocone on the dorsal part of the whorl ("*phragmocone in phragmocone structure*" of Henderson 1984). The short truncated subcircular septal tunnel measures approximately 40 % while the septal lobe reaches 66 % of the whorl diameter in late representatives of the Lytoceratoidea. The attached part of the septal lobe shows the same degree of complexity as the rest of the septum being attached to the conch wall (Henderson 1984). However, the septal lobe shows larger differences in the degree of incision and angle of attachment between saddles and lobes compared to other septal parts pronouncing the marked polarity.

Due to its attachment on septal surfaces, major parts of the internal lobe are not visible on lytoceratid internal moulds, which might be one of the reasons why the

Fig. 3.11 *Argonauticeras besairei*, Early Cretaceous (Albian) from Madagascar. **a** surface view of the conch with the suture line slightly shining through the shell wall, diameter = 11,5 cm.

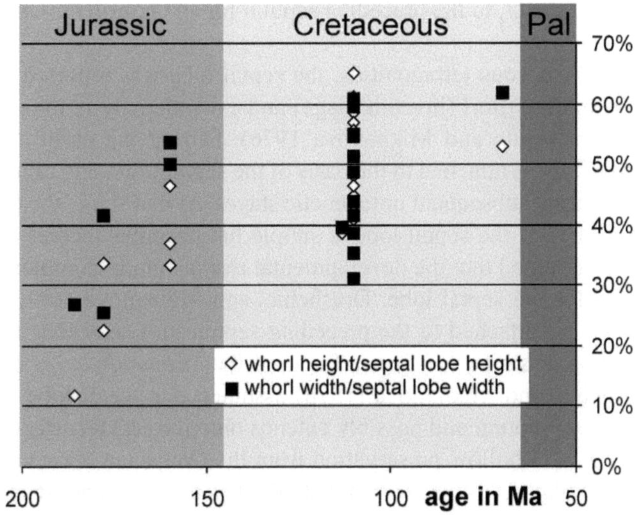

Fig. 3.12 Evolution of the dimensions of the septal lobe in relation to septum size

septal lobe was often overlooked by ammonite workers. Detection of a cruciform internal lobe on a mold allows for the assignment of the specimen to the lytoceratids and the statement that a septal lobe is present. In order to study this feature, the septal surface has to be prepared. Presence of the dorsal septal lobe accentuated the presence of a complete outer shell, with the dorsal part not being reduced.

3.6.2 Evolution of a Constituting Character

During the early phase of lytoceratid evolution, the internal lobe of adult to subadult specimens was attached only to the basis of the preceding septum, e.g., of the Lower Jurassic *Pleuroacanthites*. This evolutionary stage, with a low $Is_h:W_h$ ratio, is regarded as showing some characters that can be seen as a kind of predisposition for the evolution of a septal lobe (I_s) that climbs up the septal surface. During phylogeny, the septal lobe, like the lytoceratid suture in general, became more complex and the attached part enlarged (Fig. 3.12). The internal lobe was creeping over the tips of the saddles of the preceding septum throughout ontogeny and phylogeny. Schreiber and Hoffmann (2009) documented the oldest known case, which shows

b longitudinal cross section through the median plane of the phragmocone with the main part convex to the aperture, dorsal area with the septal tunnel, septal attachment (septal lobe) is not visible in the earliest chambers due to exact median cutting, septal lobe becomes partially visible after a half whorl in some translucent chamber fillings, diameter = 10 cm. **c** septal surface of the same specimen shown in 12A with a prominent septal lobe covering about 2/3 of the septal surface, Ww = 5,7 cm. **d1** *Gaudryceras* sp. (same specimen shown in Fig. 3.5) with the median closure of the septal tunnel, **d2** close up of the *Argonauticeras* septal lobe also with the median closure recognizable be weak remains of the septum, height of the septal lobe = 3,1 cm. **e–f** close ups of the frilled septal lobe margin as demonstration for how complex ammonite septa can be the attachment of the U_1 to the preceding septum for an Oxfordian (Upper Jurassic) giant *Lobolytoceras*.

For Lower Cretaceous tetragonitids, the septal lobe was reported to appear as early as at the fourth whorl (juvenile stage) and exceptionally in the middle of the third whorl (Drushchits and Mikhailova 1976). During the earliest ontogenetic stage, the septal lobe is attached to the basis of the septum first and climbs up on the septal surface during subsequent ontogenetic stages. At that stage, the margin of the two main branches of the septal lobe is simple but becomes increasingly incised. Therefore, it is assumed that the developmental rise in sutural complexity is apparently the same for the septal lobe. Drushchits and Mikhailova (1976) mentioned that parts of U_1 are attached to the preceding septum in Cretaceous tetragonitids. The same finding was reported from a Lower Albian *Eogaudryceras* by Hoffmann (2010). Drushchits and Mikhailova (1976) assumed that the E lobe is connected with the preceding septum and possibly extends onto the septal surface as well.

A *Gaudryceras* in hollow preservation from the Coniacian (Cretaceous) of Kamchatka (Coll. Shigeta, Tokyo) shows that all lobes are in contact with the saddles of the preceding septum (Fig. 3.5), thus representing true E_s, A_s, U_{2s}, U_{1s} and I_s; note that E_s is here the external lobe of the septal lobe and not the main external lobe *sensu* Checa and Garcia-Ruiz (1996). Maximum extension was reported for late Cretaceous representatives like *Pseudophyllites* with a Is_h:W_h-ratio of about 2:3.

A gradual heterochronous evolutionary pattern was recorded in several cases concerning shell ornamentation (e.g., Cecca and Rouget 2006), which is also assumed for the development of the septal lobe (I_s). For the oldest lytoceratid representatives, the septal lobe can only be recognized in the adult stage, while it appears approximately at about the fourth whorl in Early Cretaceous tetragonitids and about 8 mm in diameter in the Late Cretaceous *Pseudophyllites* (Henderson 1984). The same peramorphic trend was observed for the transition from the trifid (Early Jurassic) to the bifid (Middle Jurassic) stage of the lateral lobe (Hoffmann 2010).

3.6.3 Possible Function of the Septal Lobe

Due to the improved ratio of chamber volume to its inner surface, the chamber liquid can be pumped out in a shorter time. Therefore, the lytoceratids can grow faster and might have reached the stage of maturity and sexual reproduction much earlier compared to co-occurring ammonoids with simpler septa. Additionally, the refill of liquid should be improved by a larger inner surface being covered by the hydrophilic pellicle acting as a blotting paper (Ward 1987). Thereby, a larger volume can be flooded into the chambers in a shorter time interval, especially in the case of shell loss in the body chamber by a predation attempt, in order to compensate for the decreased total weight.

Presuming that a higher number of tie points facilitated the attachment of the soft body to the shell, the attachment of the septal mantle might have speculatively been one of the additional subordinate functions of the septal lobe. However, the idea that the septal lobe also improved the retraction of the soft body into the conch is rejected here due to a bad power transmission through the septal tunnel (Hoffmann 2010). Westermann (1971) and Seilacher (1988) suggested that the septal lobe may have served to strengthen the last septum against hydrostatic pressure transmitted through the soft body. At the beginning of a new chamber formation cycle, the hydrostatic pressure of the cameral liquid of derived ammonoids like the Lytoceratoidea was assumed to exceed that of the ambient seawater. If so, one would expect the septal lobe (I_s) to be curved in an apertural direction comparable to the centre of the septum of all other Mesozoic ammonoids, but the opposite can be seen (Fig. 3.5, 3.11, 3.13).

Another possible function for the septal lobe arises from the calculations made by Heptonstall (1970) and the experiments of Mutvei and Reyment (1973) and Kulicki and Mutvei (1988) that show that ammonoids with complex septa may have retained up to 25% of the chamber volume filled with liquid by surface tension. A high amount of cameral liquid has been interpreted as adaptive feature for vertical movements (diurnal vertical migrations) based on a rapid buoyancy regulation, which was supported by the documentation of pore canals in the siphuncular structure making a high rate of cameral liquid exchange possible (Mutvei et al. 2004). In this context, the septal lobe could have served as a water storage structure, too, to facilitate diurnal migrations and might have influenced the position of the centers of buoyancy and gravity, thus providing speculatively a stable swimming position. Nevertheless, these possible functions require further tests.

3.7 The Septal Mantle

3.7.1 The Septal Mantle and the Viscous Fingering Model

The septal mantle in *Nautilus* is a thin, translucent epithelium closing the rear body, which is attached along its periphery by the posterior aponeurosis (Ward and Westermann 1976). The septal mantle itself remains unknown from ammonoids. However, Klug et al. (2008) described soft-tissue imprints on the septa of early ammonoids, which are similar to those left by the spetal mantle on the septa of modern nautilids.

It was proposed by Westermann (1975) that the septal mantle in ammonoids was similarly attached to the inner surface of the conch wall, but only in the tie points. By contrast, Klug et al. (2008) suggested that the septal mantle was present in ammonoids and attached along the entire suture, comparable to nautilids. Contact of the septal mantle with the septum is corroborated by imprints in the mural parts of the septum. The idea of a membrane-like rear mantle has been well accepted for all different models (Bandel 1982; Seilacher 1988; Hewitt et al. 1991; Klug et al. 2008), except for the Viscous Fingering Model, although this model does not strictly contradict the presence of a thin and highly flexible membrane-like septal mantle.

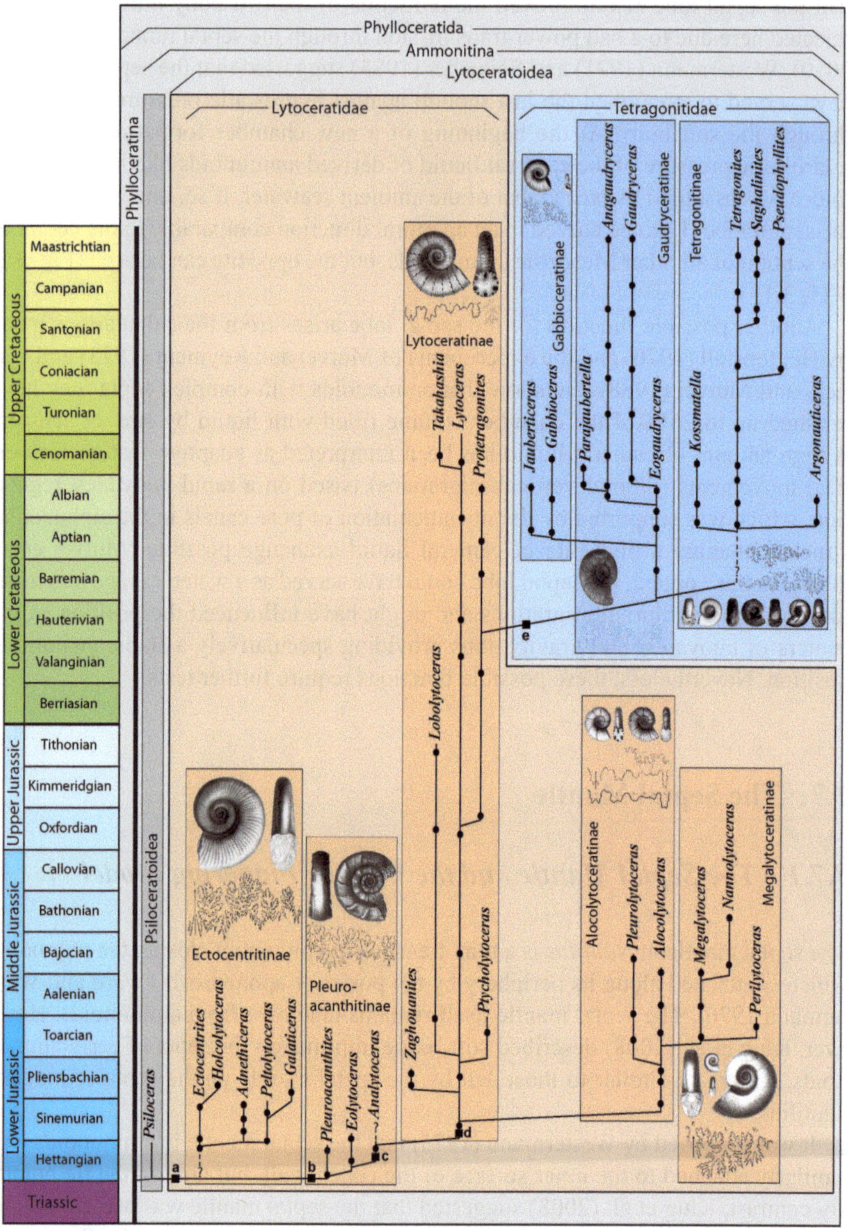

Fig. 3.13 Phylogeny of the Lytoceratoidea and its subgroups differentiated based on the following characters: **a** septal lobe, **b**—parabolic ribs (megastriae *sensu* Bucher et al. 1996), **c**—crinkled ribs, **d**—adult lateral lobe bifid, **e**—primary suture line six lobate, rhynchaptychus. (Modified after Hoffmann 2010)

For the explanation of septal morphology, Checa and García-Ruiz (1996) favor the Saffman-Taylor instability, in which a form develops spontaneously when two liquids of different viscosity (low = cameral liquid, high = viscoelastic rear soft body) are in direct contact without mixing, as explanation for ammonitic septal pattern. Accordingly, the pressure of the cameral liquid would have to exceed that of the soft body. Three questions arise here:

1. Did the chamber liquid or gas push or did musculature move the soft body to a more apertural position?
2. Does the Viscous Fingering Model apply at all and if yes, to the entire septum or only parts of it?
3. Could similar kinds of septal frilling occur in presence of a thin and flexible septal mantle?

For mineralization processes, mantle cells are able to secret a gel-like fluid, the extrapallial fluid (EPF), into the space between mantle and the membrane; this space or volume of the EPF, which is less than one micrometer wide in extant mollusks (Mutvei 1964), might have been too little to allow for septal formation following the Viscous Fingering Model. Checa (personal communication) argued that the required space would be produced by the forward movement of the soft body.

According to the Viscous Fingering Model, the rear septal mantle grows following the stress field (García-Ruiz, personal communication); therefore, the septal shape is retained between successive septa within one specimen although it would complicate progressively. In order to understand the similarity of sutures and septal shapes between specimens of the same species, the siphonal lobe (E) sets the position of the rest of the elements. If the viscosity of the rear mantle is the same across the species, the viscous fingers would be very similar. Experiments with Hele-Shaw cells show that this is the case (personal communication A. Checa). Maybe the septal membrane was first secreted and detached from the soft body after the translocation was completed. For a discussion see Checa and García-Ruiz (2000) and Hammer (1999).

In any case, the Viscous Fingering Model-experiments and the Saffman-Taylor instability theory were carried out with two fluids in direct contact. To our knowledge, nobody has produced such patterns in the presence of a highly flexible membrane (and it is likely that a membrane was there, namely the septal mantle). So the flexibility of this membrane is still unknown. A low flexibility is indicated by the high degree of similarity between two successive septa. Also, the membrane would have to grow in any case through ontogeny, requiring growth control, and then the Viscous Fingering Model would not be required (personal communication Ø. Hammer).

3.7.2 The Septal Mantle and the Tie-Point Model

Seilacher (1975, 1988) demonstrated that nearly all types of ammonite suture lines can be explained with his tie point model. If one accepts that tie points were set in a temporal order according to their hierarchy, this would allow the organic mantle membrane to creep slowly according to the direction of the highest hydrostatic pressure. With his model, Seilacher (1988) tried to explain the phenomenon of septal crowding first described by Hölder (1952). This phenomenon was interpreted to be a regular and much more pronounced feature in the lytoceratid septal lobe (I_s). Anyhow, we generally accept sequentially set tie points to fix the septal lobe within the conch and septal tube. Many ammonoid workers agree that septum formation starts with an organic, elastic preformed septal membrane, whose form is identical with the shape of the rear soft body. It was subjected to radial tension and mineralization after having reached its final position during the translocation process (see next chapter).

Nevertheless, both the complex structure of the lytoceratid septum, including the I_s (see above), and the documented changes of growth and pressure direction of at least five times over a quarter whorl (Fig. 3.11) cannot be explained by changes of pressure directions within the living animal and tie points alone. It appears difficult to envision that complexly fluted septa with a high degree of similarity among subsequent sutures were formed by soft parts with a high viscoelasticity being more or less completely contracted during translocation (Checa and García-Ruiz 1996). If the tips of all major lobes were fixed at first, this would imply a hydrostatic pressure of chamber liquid exceeding that of the ambient water. The soft body moved forwards adaperturally by muscular activity (or by the hydrostatic pressure of the newly filled chamber?) and then, the membrane became attached in the lobes. Compared to all other major lobes, the internal lobe became attached on the dorsal wall in a much more apertural direction. The transmitted ambient sea water pressure would have pushed the dorsal organic membrane backwards to the preceding septum. Then, the membrane became increasingly inflated till parts of it attached to the preceding septum. Due to increasing pressure of the cameral liquid, which is indicated by the shape of the lobules and their attachment sites being much narrower near the centre of the septum, the septal membrane was pushed to the aperture (Fig. 3.5, 3.11). Finally, mineralization began while the septal membrane was in contact with the rear body and its gel secreting mantle cells.

3.7.3 The Septal Mantle and the Reaction-Diffusion Model

According to Hammer (1999), the stiffness of the mantle sack membrane cannot be inferred from the degree of fluting across the surface of the septum and internal fluting cannot be explained by stiffness of the membrane alone. When the septal surface increases, liquid can be pumped out more rapidly and reflooding became faster. If the mantle attained a similar complexity, its surface was enlarged, too, i.e. more mantle cells would have been available for the secretion of extrapallial fluids to transport Ca^{2+} to the mineralization site of the new septum. This means that the ammonoid septa, which are thinner than nautiloid septa in general and become thinner towards their margin, could have been precipitated more rapidly. Nevertheless, all ammonoid septa have a minimal surface (minimal-curvature surface or minimal-area surface) constrained by the shape of their boundary, which represents the natural assumption for an ideal, elastic membrane under the given pressure distribution (Seilacher 1975; Bayer 1977a). This minimal surface requires the least possible amount of calcium carbonate. These conditions are seen as the major, possibly the most important driving forces of septal complexity. Additionally, the septa may also improve conch stability against hydrostatic pressure, once mineralized. Since fresh nacre is known to be elastic, it appears likely to allow for some compression without damaging the shell in greater water depths. With the implementation of GIS analyses as demonstrated by Yaccobucci and Manship (2011), a precise method for comparison of suture lines is available and can be applied to study spatial distribution of physical constraints and asymmetries and to clarify the formation of septa as well.

3.7.4 Function of the Septal Mantle

If the septal lobe functions as a structure to strengthen the phragmocone against implosion by ambient water pressure or by point-loading from predators (Hewitt and Westermann 1987), it would provide information that undermines some mechanical hypotheses. The septal lobe was not directly accessible to predators, and according to its dorsal position, it was in the pressure shadow of the preceding whorl and could hardly have any pressure absorbing function when mineralized. Also, suture complexity cannot be seen exclusively as adaptation to greater depths and pressures due to common findings of ammonites with complex sutures in rather shallow deposits, and isotopic data further corroborating a maximum diving depth shallower than that of Recent nautilids (Lukeneder et al. 2010). Again, the fluted shape of septa may be explained, bearing ammonites with the most complex sutures (e.g., lytoceratids) in mind, which inhabited the shelf, as having served to facilitate vertical movements (daily migration?) and/or to accelerate the growth process. If this holds true, an increased ratio of chamber volume to inner chamber surface area should be present. Nevertheless, *Nautilus* with simple but thick septa dives in great depths (up to 700 m; Dunstan et al. 2011), but only for short times due to potential flooding of the chambers under such high hydrostatic pressures. Finally, the Buckland hypothesis was rejected by Saunders (1995; but see Jacobs 1996); Saunders and Work (1996) as well as Daniel et al. (1997) highlighted that complex septa would have decreased the mechanical strength of the conch, which is in some way supported by the shallower maximum diving depth. By contrast, Hassan et al. (2002) demonstrated on numerical models using finite element analyses that more complex septa reduced the stresses in the center of the septa without weakening it elsewhere. Kiselev (2009) suggested the possibility of a multi-functional ammonitic septum, while Boiko (2012) favored the notion that the septa improved internal pneumocompensation. By contrast, Klug et al. (2008) suggested, in accordance with Seilacher (1975, p. 284), that the frilling served to enlarge the length of the attachment of the septal mantle prior to mineralization and helped the septum to withstand decreasing chamber pressure in the course of septum mineralization. It would thus have enabled faster growth and earlier functionality of the new formed chamber.

Presence and function of the septal mantle have been widely discussed. All mollusks (including cephalopods) today have a mantle surrounding their organs laterally, dorsally, ventrally, and posteriorly. Main differences between endo- and ectocochleates are the internal or external position of the shell and the number as well as thickness of the sheets of contractive tissue within the mantle. Therefore, we conclude that most likely, ammonoids also had a functional tissue, which contained fine layers of contractile fibers and blood vessels (Klug et al. 2008).

The function of the mantle is manifold in Recent cephalopods: (1) it contains muscles for locomotion, (2) it secretes the skeleton (except the mouth parts and arm hooks, if present), and (3) it holds internal organs in place. Therefore, it is likely that similar functions applied to the septal mantle in ammonoids. Its musculature most likely did not serve locomotion directly, but it was needed for the translocation of the soft body to form new chambers, and it was in charge of forming new septa (i.e. shell secretion). Additionally, during the forward movement of the soft body in the shell, the septal mantle probably preserved the rough shape of the septum because it appears highly unlikely that a membrane with muscles fits into a strongly folded surface (like that of the septum), and that all the little indentations would be completely or largely lost in a short instance (this would require a rather viscous behavior, which appears unlikely for a living tissue with muscle layers). Probably, the mantle shape displayed a slightly reduced complexity during its forward movement, which was regained after the fixation in the tie-points, and augmented thereafter (resembling viscous fingers).

3.8 Revised Chamber Formation Cycle

Klug et al. (2008) suggested a modified septum formation cycle, which will be repeated and shortly explained below in a modified form. To understand this model, it has to be taken into account that Klug et al. (2008) assume the presence of a septal mantle in ammonoids like in Recent nautilids. This assumption finds support in soft-tissue imprints found in septa of various cephalopods described in their article.

1. After the complete mineralization of the last formed septum, the muscle fibers, which left traces such as the structure known as the septal furrow (Teichert 1964; Shimanskij 1974; Chirat and Boletzky 2003, Kröger et al. 2005), might have held the soft part of the siphuncle in place when it grew anteriorly (compare Mutvei et al. 2004 and Landman et al. 2006).
2. More or less simultaneously, the contractile fibres of the septal mantle, which left radial traces on the mural part of the septum of bactritoids (*Devonobactrites*), some Early Devonian ammonoids (*Metabactrites*, *Gyroceratites*) and some nautilids (*Allonautilus, Cenoceras*, *Eutrephoceras*, *Germanonautilus*, *Nautilus*) contracted slightly, causing the septal mantle to separate from the last formed septum and the outer shell wall (Fig. 3.14a).

3. At the same time, the muscle fibers of the septal mantle, which are arranged perpendicular to those mentioned before (parallel to the suture line) became active, lifting the margin of the septal mantle slightly off the mural ridge of the last septum, the actual muscle-attachment site of the septal mantle musculature (Fig. 3.14b).
4. In that course, the septal mantle was retracted from the finest lobules, causing an interimistic simplification of the ‘suture’ as it was preserved in the mantle shape, which is documented in pseudosutures. A simplified general septum shape was preserved in the septal mantle, which still maintained some flexibility.
5. The muscle fibers, which were attached at the mantle myoadhesive ligament, contracted, thus pulling the septal mantle forward (Fig. 3.14c). At some point (probably more or less simultaneously), the insertion of all posteriorly inserted muscles had to be relocated anteriorly. This probably happened without a complete detachment of these muscles but rather with some kind of caterpillar-like or peristaltic creeping movement, enabled by the contraction of longitudinally arranged muscles in the mantle between the aperture and the muscle attachment. This slow forward movement of the cephalic retractors is documented in faint lines, which run subparallel to the mantle myoadhesive ligament attachment site in some fossil and Recent nautilids (Isaji et al. 2002; Klug and Lehmkuhl 2004) but also in ammonoids (Richter 2002).
6. The septal mantle was gently attached and detached several times, during the forward movement, which produced pseudosutures especially close to the position of the subsequent septum (Fig. 3.14c,d), but, more importantly, which contributed to the conservation of the overall septal shape in the septal mantle.
7. Possibly due to slowly rising chamber pressure, the septum was gently inflated anteriorly with liquid-filled bubbles (“*pneus*” *sensu* Seilacher 1975) forming in the saddles and its folioles between the tie points (lobes and lobules) until it became attached (probably slightly before). Afterwards, it was firmly attached in the tie points of the major tips of the lobes (Fig. 3.14e).
   Ontogenetic increase in sutural complexity was achieved by the marginal addition of folding (and thus increase in number of tie-points) by the increase in septum surface (marginal septal mantle growth) and suture length during the expansion of the whorl cross section (Seilacher (1975, p. 284). In this context, physical processes similar to processes as in a Hele-Shaw cell might have occurred (Fig. 3.14f), which would resemble such occurring in the Viscous Fingering Model of Checa and Garcia-Ruiz (1996), although this model does not include the presence of a membrane.
8. The septal mantle became attached along the entire new suture/mural ridge.
9. Subsequently, chamber pressure was probably reduced, thus partially emptying the bubbles in the folioles of the saddles. The combination of adapertural (anterior) and adapical (posterior) folding permitted the preseptum to become tightly stayed rapidly (and this might be one of the main functions of septal folding since it permitted an earlier onset of mineralization). Then, the formation and thus mineralization of the primarily organic septum began (Fig. 3.14f).

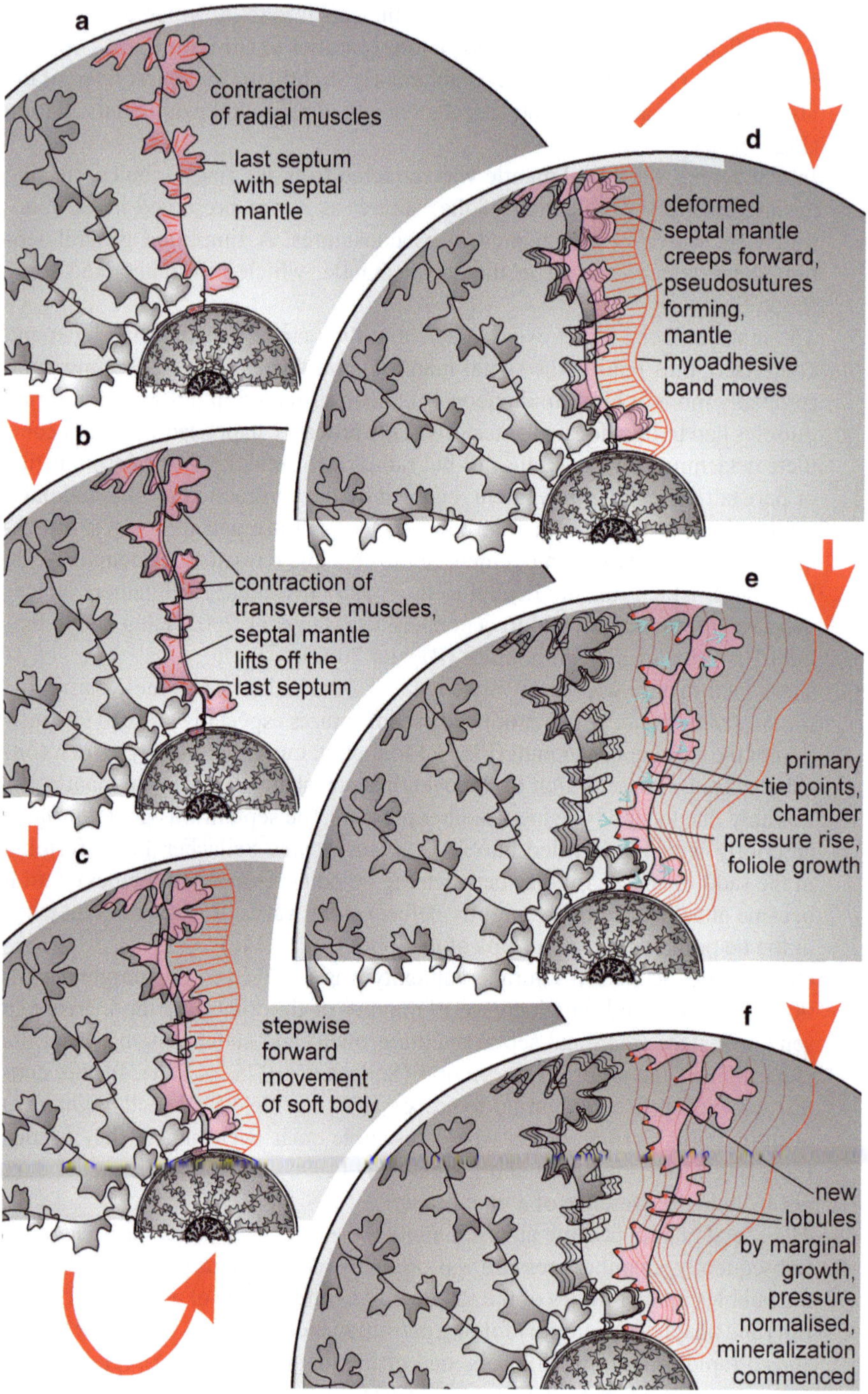

Fig. 3.14 The revised chamber formation cycle, incorporating principles from the Tie-Point Model of Seilacher (1975), the Viscous Fingering Model of Checa and Garcia-Ruiz (1996), as well as results from Klug et al. (2008)

10. At the point, when the new septum was partially mineralized and stable enough to withstand rising hydraulic pressure, chamber-emptying began.
11. When new shell was added at the aperture, the apertural mantle attachment was also translocated anteriorly. As documented in striations parallel to the mantle myoadhesive band in many cephalopods (e.g., Klug and Lehmkuhl 2004), these processes happened in small steps prior to the formation of a complete septum. Naturally, the cephalic retractor had to be moved anteriorly, too. This had to happen more or less synchronously with the forward movement of the mantle myoadhesive band and the palliovisceral band, which form a rather rigid unit in Recent *Nautilus pompilius*. For this action, possibly both the retractor muscles as well as (more certainly) the longitudinal muscle fibers in the mantle attached at the aperture became active (Stenzel 1964; Blind 1976; Doguzhaeva and Mutvei 1986; Mutvei et al. 1993; Mutvei and Doguzhaeva 1997).

3.9 Septal Precipitation Model

After the ammonoid soft body moved to its final position and became attached along the suture line to the conch wall via its muscular system, mineralization of the organic septum took place. This is documented by the tension furrows visible in well preserved specimens (Fig. 3.5c). According to Mutvei (1964), the mantle cells precipitated the periostracum, the three different cephalopod conch layers, and the septum and subsequently became part of the siphuncle, which impressively demonstrates the multiple functions of the mantle cells. It is assumed that between the shell secreting epithelium and the precipitated shell material a small gap, the extra pallial space (EPS), about 100–200 nanometers wide, existed (Nakahara 1991). Into this space, the mantle cells secreted the extra pallial fluid (EPF). The EPF is aqueous, gel-like and contains all constituents of nacre: polysaccharides (chitin), proteins (glycoproteins, silk fibroin and others), and minerals (e.g., aragonite; Nudelman et al. 2006). Mollusk shells are well known for their high amount of organic matter, which can be deduced from the reduced nacre density of about 2.62 g/cm^3 (Reyment 1958) compared to aragonite crystal density of 2.94 g/cm^3 (Trueman 1941). Also, direct thermogravimetric measurements estimate about 4.5% organic matter in nacre (personal communication A. Checa), while Olson et al. (2012) reported 2% organic matter. The model of ammonoid septal formation is based on what is known for *Nautilus* septal formation and general mechanisms of the mollusk nacre formation process. Precipitation of the septum occurred at the rear soft body, and no direct connection with the ambient sea water existed. The lack of direct connection with ambient sea water is also assumed for the precipitation of the conch wall because of the separation of the EPS from sea water by the periostracum. Therefore, the EPF can be considered as a body fluid secreted by the mantle cells (Weiner and Addadi 2011).

Nautilus pompilius secretes 33–38 septa, while *Nautilus belauensis* secreted 33–37 septa. Exact data for ammonoids are not available yet. During ontogeny, the time of successive chamber formation and septal formation increases (Westermann et al. 2004). Oba et al. (1992) demonstrated that mineralization of the septum was a continuous process for *Nautilus*, and they detected small scale cycles (15–40 μm) in the oxygen isotope composition of a single incomplete 1.53 mm thick *Nautilus* septum measured in analytical intervals of 50 μm and 5 μm. It was assumed that the septum records 40 to 100 cycles with a temperature difference of 1–2°C and was most likely caused by short term/ daily vertical movements of the animal. Compared to the conch, septa consist of relatively thin (about 500 nanometer thick) aragonitic platelets arranged in columns (nacre) and become precipitated simultaneously over its whole surface ("*Simultanskelett*" of Seilacher 1975). Septa show growth only in thickness, and therefore, show no growth lines like the conch wall. However, the precipitation of a septum takes about 105–180 days, according to Ward and Chamberlain (1983) as well as Ward (1985), both of which used radiographic examinations, and by Cochran et al. (1981), Landman and Cochran (1987), as well as Landman et al. (1989a, 1989b) by indirect radiometric dating of live-caught specimens. Their findings are summarized by Westermann et al. (2004). For comparison, the septum formation in *Sepia* takes only 2–3 days (Boletzky and Wiedmann 1978, Wiedmann and Boletzky 1983, Tanabe et al. 1985).

According to Bevelander and Nakahara (1969, 1980), the nacre formation starts with the formation of an organic compartment (interlamellar membrane) comparable to a brick wall without bricks before mineralization. Subsequently, precipitation of carbonatic nacre takes place. Mantle cells first secrete the polysaccharide β-chitin, a biopolymer, into the EPS. For *Nautilus*, Westermann et al. (2005) described the mantle edge gland, which produces numerous secretory vesicles responsible for shell formation. The outer epithelium of *Nautilus* shows many characteristics typical for transport-active epithelia comparable to the septal epithelium forming septa in *Sepia* (Wendling 1987). These characteristics are (i) the presence of Ca^{2+}-ATPase, an enzyme that transports Ca^{2+}-ions out of the cell into the EPS, (ii) large concentrations of mitochondria, (iii) a well-developed apical microvillar-border, (iv) a paucity of rough endoplasmatic reticulum, and (v) a Golgi complex within the outer epithelium (Westermann et al. 2005). Crystallization of the β-chitin polymer chains is due to contact with hydrogen, resulting in hundreds of nanometer-long needles. According to Levi-Kalisman et al. (2001), these crystallites are 20 nm thick and are arranged in such way that they form a felt-like core of the interlamellar membrane (ILM). The distance between different ILM sheets is initially 100 nm and later 300–500 nm (Imai et al. 2003, Cartwright and Checa 2007). Because the EPS in mollusks is narrow, no more than 1–3 sheets become aligned to form an ILM.

In some mollusks, e.g., gastropods and possibly *Nautilus*, a surface membrane (Nakahara 1991) exists and serves as a template for laying down new ILM. The ILMs detach regularly from the surface membrane and lay down at the precipitation site. All material found in the shell must pass through this surface membrane, which is impermeable for sea water.

The EPF surrounds all membranes. The mode of nacre growth implies that fresh liquid is constantly being secreted into the EPS because it is locked up in the earlier grown nacre. Mineralization of nacre proceeds by growth of calcium carbonate between the ILMs. Crystals first grow upwards and outwards from points on the lower membrane until they contact the upper membrane, then they grow outwards until they contact their neighboring crystals forming a random, closely packed structure within each layer; proteins promoting and inhibiting mineralization are also present (Marin and Luquet 2004, 2005). Nacre, i.e. irregular, polygonal aragonite tablets, grows at different levels simultaneously in gastropods, tablets of upper layers are centered on those in lower layers (Nakahara 1983). Accordingly, nacre growth in gastropods is best described as tower-like (columnar nacre). A similar growth mode occurs close to the aperture of *Nautilus* (personal communication A. Checa). Compared with the bivalve sheet nacre with two to three simultaneously formed tablet layers, in the gastropod and cephalopod columnar nacre between twenty and thirty layers of tablets can grow simultaneously. Tablets are stacked with the oldest at the bottom being the largest and the overlying ones decreasing in size forming an approximately conical column (Olson et al. 2012). The columns grow perpendicular to the organic membrane layers with its smallest part pointing to the gel-secreting mantle cells. The interlamellar membranes are not strongly bound together, but bonding becomes stronger by the addition of proteins present in the EPF on both sides of the ILM. The repulsion forces between the negatively charged protein sheets of the freshly laid down ILM and the previous ILM increases the space between them from 90 nm to the regular spacing of 400–500 nm, which equals the final thickness of the nacre tablets (compare Olson et al. 2012).

According to Mutvei and Dunca (2008), the organic ILMs determine the orientation of the c-axes to be right angles to the sheets. This can be seen as a pre-requisite for the crystallization of the tablets into "*single crystals*". By contrast, Checa et al. (2011) demonstrated that nacre tablets do not nucleate onto the ILMs. According to them, the consecutive tablets therefore acquire a structural continuity (homo- or heteroepitaxy) by "*organic-mineral bridges*", which have recently been detected in *Nautilus*, and are usually aligned along the nacre-tower axes (Checa et al. 2011). These bridges belong to primary pores present in the ILM. In gastropods, pores are invariably irregular-shaped but are regular in size and spacing. Note that there can only be one pore, i.e. nucleation site, per tablet. This prediction was verified by Nudelman et al. (2006). Further they demonstrated a zonation for each platelet with carboxylates, sulfates, and aragonite nucleation proteins in the center. This nucleation site is organic, thus it must not be continuous but porous to allow for propagation of crystal orientation from the underlying to the nucleating tablet above. Each tablet starts growing from this region without a real new nucleation event.

A fresh layer of nacre in mollusks is laid down every 1–24 h comprising a lamellar structure. Finally, nacre assemblage takes place outside the cell. The lack of faults coinciding with cell boundaries is a demonstration of the self-assembled nature of the material. It is assumed here that the final layers of the membrane forming the organic brick wall for septal and conch wall precipitation represents the wettable pellicle.

Acknowledgments Some of the insights grew in the course of the research projects with the numbers 200021-113956/1, 200020-25029, and 200020-132870 funded by the Swiss National Science Foundation SNF and HO 4674/2-1. RH acknowledges Dieter Buhl (Bochum) for the photographs shown in Fig. 3.12, Franziska Häuser (Bochum) for technical preparation of the Argonauticeras specimen shown in Fig. 3.12 and Yasunari Shigeta (Tokyo) for the loan of the hollow *Gaudryceras* shown in Fig. 3.5 as well as Helmut Keupp (Berlin) for fruitful discussions. Additionally, Øyvind Hammer (Oslo), Adolf Seilacher (Tübingen) and Antonio Checa (Granada) contributed illustrations and permissions for reproduction. Antonio Checa (Granada), Øyvind Hammer (Oslo) and Matthew Knauss (Riverside) kindly reviewed the chapter and provided constructive criticism.

References

Allen EG (2007) Understanding ammonoid sutures: new insight into the dynamic evolution of Paleozoic suture morphology. In: Landman NH, Davis RA, Mapes RH (eds) Cephalopods present and past: new insights and fresh perspectives. Springer, Dordrecht

Arkell WJ (1949) Jurassic ammonites. Sci Prog 147:401–417

Arkell WJ (1957) Sutures and septa in Jurassic ammonite systematic. Geol Mag 94:235–248

Arkell WJ, Furnish WM, Kummel B, Miller AK, Moore RC, Schindewolf OH, Sylvester-Bradley PC, Wright CW (1957) Treatise on invertebrate paleontology, Part L (Mollusca 4), Cephalopoda, Ammonoidea. Geological Society of America and University of Kansas Press, Lawrence

Bandel K (1982) Morphologie und Bildung der frühontogenetischen Gehäuse bei conchiferen Mollusken. Facies 7:1–198

Bayer U (1975) Organische Tapeten im Ammoniten-Phragmokon und ihr Einfluß auf die Fossilisation. Neues Jahrb Geol Paläontol, Monatshefte 1975(1):12–25

Bayer U (1977a) Cephalopoden-Septen. Teil 1: Konstruktionsmorphologie des Ammoniten-Septums. Neues Jahrb Geol Paläontol Abh 154:290–366

Bayer U (1977b) Cephalopoden-Septen. Teil 2: Regelmechanismen im Gehäuse und Septenbau der Ammoniten. Neues Jahrb Geol Paläontol Abh 155:162–215

Bayer U (1978a) The impossibility of inverted suture lines in ammonites. Lethaia 11:307–313

Bayer U (1978b) Constructional morphology of ammonite septa. Neues Jahrb Geol Paläontol Abh 157:150–155

Bayer U (1985) Pattern recognition problems in geology and paleontology. Springer, Berlin

Bevelander G, Nakahara H (1969) An electron microscope study of the formation of the nacreous layer in the shell of certain bivalve molluscs. Calc Tiss Res 3:84–92

Bevelander G, Nakahara H (1980) Compartment and envelope formation in the process of biological mineralization. In: Omori M, Watabe N (eds) The mechanisms of biomineralization in animals and plants. Tokai University, Tokyo

Blind W (1976) Die ontogenetische Entwicklung von *Nautilus pompilius* (Linné). Palaeontogr Abt A 153:117–160

Boiko MS (2012) Pneumocompensation in the cephalopod shells. In: Contributions to current cephalopod research: morphology, systematics, evolution, ecology and biostratigraphy. pp 119–122 [in Russian]

Boletzky Sv, Wiedmann J (1978) Schulp-Wachstum bei *Sepia officinalis* in Abhängigkeit von ökologischen Parametern. Neues Jahrb Geol Paläontol Abh 157:103–106

Bourillot R, Neige P, Pierre A, Durlet C (2008) Early-Middle Jurassic lytoceratid ammonites with constrictions from Morocco: palaeobiogeographical and evolutionary implications. Palaeontology 51:597–609

Boyajian GE, Lutz T (1992) Evolution of biological complexity and ist relation to taxonomic longevity in the Ammonoidea. Geology 20:983–986

Buch Lv (1829) Note sur les ammonites. Annales des sciences naturelles par Audouin. Ad Brogniart et Dumas 17:267–275

Buch Lv (1830) Note über Ammoniten. Jahrb Miner, Geognosie. Geol Petrefaktenk 1:397–398
Bucher H, Landman NH, Klofak SM, Guex J (1996) Mode and rate of growth in ammonoids. In: Landman NH, Tanabe K, Davis RA (eds) Ammonoid paleobiology. Plenum, New York
Buckland W (1836) Geology and mineralogy considered with reference to natural theology. Treatise VI. The bridgewater treatise on the power, wisdom, and goodness of god as manifested in the Creation. W. Pickering, London
Buckman SS (1909–1930) Yorkshire type ammonites, continued as Type ammonites. In: Cramer J, Swann HK (eds) 1972–1976. Reprint of vol 1–2. Yorkshire type ammonites. 1909–1912, vol 3–5. Type ammonites. 1919–1921, 6–7. Type ammonites. 1925–1927. 790 pls. London
Cartwright JHE, Checa AG (2007) The dynamics of nacre self-assembly. J R Soc Interface 4:491–504
Cecca F, Rouget I (2006) Anagenetic evolution of the early Tithonian ammonite genus *Semiformiceras* tested with cladistic analysis. Palaeontology 49:1069–1080
Checa AG (1991) Sectorial expansion and shell morphogenesis in molluscs. Lethaia 24:97–114
Checa AG (2003) Fabrication and function of ammonite septa—comment on Lewy. J Paleont 77:790–791
Checa AG, Cartwright JHE, Willinger M-G (2011) Mineral bridges in nacre. J Struct Biol 176:330–339
Checa AG, García-Ruiz JM (1996) Morphogenesis of the septum. In: Landman NH, Tanabe K, Davis RA (eds) Ammonoid paleobiology. Plenum, New York
Checa AG, García-Ruiz JM (2000) Discussion: the development of ammonoid septa: an epithelial invagination process controlled by morphogens or by viscous fingering? Hist Biol 14:299–301
Chirat R, Boletzky Sv (2003) Morphogenetic significance of the conchal furrow in nautiloids: evidence from early embryonic shell development of Jurassic Nautilida. Lethaia 36:161–170
Cochran JK, Rye DM, Landman NH (1981) Growth rate and habitat of *Nautilus pompilius* inferred from radioactive and stable isotope studies. Paleobiology 7:469–480
Damiani G (1986) Significato funzionalle dell'evoluzione dei setti e delle line di sutura dei nautiloidi e degli ammonoidi. In: Pallini G (ed) Atti I Convengo Internationale: Fossili, Evoluzione, Ambiente, Pergola 1984. Tecnoscienza, Roma
Damiani G (1990) Computer simulation of some ammonoid suture lines. In: Pallini G, Cecca F, Cresta S, Santantonio M (eds) Atti 2 Convengo Internationale: Fossili, Evoluzione, Ambiente, Pergola 1987. Tecnostampa, Ostra Vetere
Daniel TL, Helmuth BS, Saunders WB, Ward PD (1997) Septal complexity in ammonoid cephalopods increased mechanical risk and limited depth. Paleobiology 23:470–481
De Baets K, Klug C, Korn D, Landman NH (2012) Early evolutionary trends in ammonoid embryonic development. Evolution 66:1788–1806
De Blasio FV (2008) The role of suture complexity in diminishing strain and stress in ammonoid phragmocones. Lethaia 41:15–24
Deecke W (1913) Paläontologische Betrachtungen. I. Über Cephalopoden. Neues Jahrb Min, Geol Paläontol, Beilageband 35:241–276
Denton EJ, Gilpin-Brown JB (1966) On the buoyancy of the pearly *Nautilus*. J Mar Biol Assoc UK 46:723–759
Denton EJ, Gilpin-Brown JB (1973) Flotation mechanisms in modern and fossil cephalopods. Adv Mar Biol 11:197–268
Denton EJ (1974) On buoyancy and the lives of modern and fossil cephalopods. Proc R Soc London B. Biol Sci 185:273–299
Diener C (1912) Lebensweise und Verbreitung der Ammoniten. Neues Jahrb Min Geol Paläontol 192:67–89
Doguzhaeva LA, Mapes RH (2015) Muscle scars in ammonoid shells. This volume
Doguzhaeva LA, Mutvei H (1986) Functional interpretation of inner shell layers in Triassic ceratid ammonites. Lethaia 19:195–209
Doguzhaeva LA, Mutvei H (1991) Organization of the soft-body in *Aconeceras* (Ammonitina), interpreted on the basis of shell-morphology and muscle-scars. Palaeontol A 218:17–33
Doguzhaeva LA, Mutvei H (1996) Attachment of the body to the shell in ammonoids. In: Landman NH, Tanabe K, Davis R (eds) Ammonoid paleobiology. Plenum, New York

Drushchits VV, Mikhailova IA (1976) Ontogenetic development of Early Cretaceous tetragonitids (Ammonoidea). Paleontol J 10:159–169
Dunstan AJ, Ward PD, Marshall NJ (2011) Vertical distribution and migration patterns of *Nautilus pompilius*. PLoS ONE 6(2):e16311. doi:10.1371/journal.pone.0016311
Ebel K (1992) Mode of life and soft body shape of heteromorph ammonites. Lethaia 25:179–193
Erben HK (1960) Primitive Ammonoidea aus dem Unterdevon Frankreichs und Deutschlands. Neues Jahrb Geol Paläontol Abh 110:1–128
Erben HK (1962) Über den Prosipho, die Prosutur und die Ontogenie der Ammonoidea. Paläontol Z 36(1/2):99–108
Erben HK (1964) Die Evolution der ältesten Ammonoidea (Lieferung I). Neues Jahrb Geol Paläontol Abh 120(2):107–212
Ernst HU, Klug C (2011) Perlboote und Ammonshörner weltweit. Nautilids and Ammonites worldwide. Pfeil, München
García-Ruiz JM (1992) "Peacock" viscous fingers. Nature 356:113
García-Ruiz JM, Checa A (1993) A model for the morphogenesis of ammonoid septal sutures. GeoBios 26:157–162
García-Ruiz JM, Checa A, Rivas P (1990) On the origin of ammonite sutures. Paleobiology 16:349–354
Guex J (1981) Associations virtuelles et discontinuités dans la distribution des espèces fossils: Un exemple intéressant. Bull Soc Vaudoise Sci Nat 75:179–197
Guex J (1987) Sur la phylogenèse des ammonites du Lias inférieur. Bull Soc Vaudoise Sci Nat, Lausanne 78:455–469
Guex J (1992) Origine des sauts évolutifs chez les ammonites. Bull Soc Vaud Sci Nat 82:117–144
Guex J (1995) Ammonites hettangiennes de la Gabbs Valley Range (Nevada). Mém Géol Lausanne 27:1–130
Guex J (2006) Reinitialization of evolutionary clocks during sublethal environmental stress in some invertebrates. Earth Planet Sic Lett 242(2006):240–253
Guex J, Koch A, O'Dogherty L, Bucher H (2003) A morphogenetic explanation of Buckman's law of covariation. Bull de la Soc géol de Fr 174:603–606
Guex J, Bartolini A, Atudorei V, Taylor D (2004) High-resolution ammonite and carbon isotope stratigraphy across the Triassic Jurassic boundary at New York Canyon (Nevada). Earth Planet Sic Lett 225:29–41
Hammer Ø (1999) The development of ammonite septa: an epithelial invagination process controlled by morphogens? Hist Biol 13:153–171
Hammer Ø, Bucher H (1999) Reaction-diffusion processes: application to the morphogenesis of ammonoid ornamentation. Geobios 32:841–852
Harrison LG, Kolar M (1988) Coupling between reaction-diffusion prepattern and expressed morphogenesis, applied to desmids and dasyclads. J Theor Biol 130:493–515
Hassan MA, Westermann GEG, Hewitt RA, Dokainish MA (2002) Finite element analysis of simulated ammonoid septa (extinct Cephalopoda): septal and sutural complexities do not reduce strength. Paleobiology 28:113–126
Henderson RA (1984) A muscle attachment proposal for septal function in Mesozoic ammonites. Palaeontology 27:461–486
Hengsbach R (1978) Zur Sutur-Asymmetrie bei *Anahoplites* (Ammonoidea; Kreide). Senck leth 59:377–385
Heptonstall WB (1970) Buoyancy control in ammonoids. Lethaia 3:317–328
Hewitt RA (1985) Numerical aspects of sutural ontogeny in the Ammonitina and Lytoceratina. Neues Jahrb Geol Paläontol Abh 170:273–290
Hewitt RA, Westermann GEG (1986) Function of complexly fluted septa in ammonoid shells I. mechanical principles and functional models. Neues Jahrb Geol Paläontol Abh 172:47–69
Hewitt RA, Westermann GEG (1987) Function of complexly fluted septa in ammonoid shells II. Septal evolution and conclusions. Neues Jahrb Geol Paläontol Abh 174:135–169
Hewitt RA, Westermann GEG (1997) Mechanical significance of ammonoid septa with complex sutures. Lethaia 30:205–212

Hewitt RA, Westermann GEG (2003) Recurrences of hypotheses about ammonites and *Argonauta*. J Paleontol 77:792–795
Hewitt RA, Checa A, Westermann GEG, Zaborski PM (1991) Chamber growth in ammonites inferred from colour markings and naturally etched surfaces of Cretaceous vascoceratids from Nigeria. Lethaia 24:271–287
Hillebrandt Av (2000) Die Ammoniten-Fauna des südamerikanischen Hettangium (basaler Jura). Teil 3. Palaeontol A 258:65–116
Hoffmann R (2010) New insights on the phylogeny of the Lytoceratoidea (Ammonitina) from the septal lobe and its functional interpretation. Revue de Paléobiol 29:1–156
Hölder H (1952) Über den Gehäusebau, insbesondere den Hohlkiel jurassischer Ammoniten. Palaeontol A 102:18–48
Hölder H (1956) Über Anomalien an jurassischen Ammoniten. Paläontol Z 30:95–107
Imai T, Watanabe T, Yui T, Sugiyama J (2003) The directionality of chitin biosynthesis: a revisit. Biochem J 374:755–760. doi:10.1042/BJ20030145
Isaji S, Kase T, Tanabe K, Uchiyama K (2002) Ultrastructure of muscle-shell attachment in *Nautilus pompilius* Linnaeus (Mollusca: Cephalopoda). Veliger 45(4):316–330
Jacobs DK (1990) Sutural pattern and shell stress in *Baculites* with implications for other cephalopod shell morphologies. Paleobiology 16:336–348
Jacobs DK (1992) The support of hydrostatic load in cephalopod shells—adaptive and ontogenetic explanations of shell form and evolution from Hooke 1695 to the present. Evol Biol 26:287–349
Jacobs DK (1996) Chambered cephalopod shells, buoyancy, structure and decoupling: history and red herrings. Palaios 11:610–614
Jernvall J (1995) Mammalian molar cusp patterns: developmental mechanisms of diversity. Acta Zoologica Fennica 198:1–61
Kemper E (1961) Die Ammonitengattung *Platylenticeras* (= *Garnieria*). Beih Geol Jahrb 47:1–195
Kennedy WJ, Cobban WA (1976) Aspects of ammonite biology, biogeography, and biostratigraphy. Spec Pap Palaeontol 17:1–94
Keupp H (2000) Ammoniten. paläobiologische Erfolgsspiralen. Thorbecke, Stuttgart
Keupp H (2012) Atlas zur Paläopathologie der Cephalopoden. Berl Paläobiologische Abh 12:1–390
Keupp H, Mitta VV (2004) Septenbildung bei *Quenstedtoceras* (Ammonoidea) von Saratov (Russland) unter anomalen Kammerdruckbedingungen. Mitt Geol-Paläontol Inst Univ Hambg 88:51–62
Kiselev DN 2009. Evaluation of sutural complexity and functional aspects of septal shape in Ammonoids. In: Contributions to current Cephalopod Research: Morphology, Systematics, Evolution, Ecology and biostratigraphy. pp 131–136 [in Russian]
Klug C, Lehmann J (2015) Soft part anatomy of ammonoids: reconstructing the animal based on exceptionally preserved specimens and actualistic comparisons. This volume
Klug C, Lehmkuhl A (2004) Soft-tissue attachment structures and taphonomy of the Middle Triassic nautiloid *Germanonautilus*. Acta Palaeontol Pol 49:243–258
Klug C, Kröger B, Vinther J, Fuchs D, De Baets K (2015) Ancestry, origin and early evolution of ammonoids. Ammonoid paleobiology: from macroevolution to paleogeography. volume II
Klug C, Meyer E, Richter U, Korn D (2008) Soft-tissue imprints in fossil and recent cephalopod septa and septum formation. Lethaia 41:477–492
Knauss MJ, Yacobucci MM (2014) Geographic information systems technology as a morphometric tool for quantifying morphological variation in an ammonoid clade. Palaeontol Electron 17(1;19A):1–27 (palaeo-electronica.org/content/2014/721-gis-based-morphometrics)
Korn D (1997) A modified balloon model for septa of early ammonoids. Lethaia 30:39–40
Korn D (2001) Morphometric evolution and phylogeny of Palaeozoic ammonoids. Early and middle devonian. Acta Geol Pol 51:193–215
Korn D, Ebbighausen V, Bockwinkel J, Klug C (2003) The A-Mode sutural ontogeny in prolecanitid ammonoids. Palaeontology 46:1123–1132
Korn D, Klug C (2002) Ammoneae Devonicae. In: Riegraf W (ed) Fossilium Catalogus 1: animalia 138. Backhuys, Leiden, pp 1–375

Krivoshapkina VS (1978) The ontogeny of the suture in Late Cretaceous tetragonitids from Sakhalin. Paleontol J 12(1):64–70
Kröger B (2001) Discussion—comments on Ebel's benthic-crawler hypothesis for ammonoids and extinct nautiloids. Paläontol Z 75:123–125
Kröger B (2002) On the efficiency of the buoyancy apparatus in ammonoids: evidences from sublethal shell injuries. Lethaia 35:61–70
Kröger B, Mapes RH (2007) On the origin of bactritoids (Cephalopoda). Paläontol Z 81:316–327
Kröger B, Klug C, Mapes RH (2005) Soft-tissue attachment in Orthocerida and Bactritida of Emsian to Eifelian age (Devonian). Acta Palaeontol Pol 50:329–342
Kröger B, Vinther J, Fuchs D (2011) Cephalopod origin and evolution: a congruent picture emerging from fossils, development and molecules. BioEssays 1–12. doi:10.1002/bies.201100001
Kulicki C (1974) Remarks on the embryogeny and postembryonal development of ammonites. Acta Palaeontol Pol 19:201–224
Kulicki C (1979) The ammonite shell, its structure, development and biological significance. Palaeontol Pol 39:97–142
Kulicki C, Mutvei H (1988) Functional interpretation of ammonoids septa. In: Wiedmann J, Kullmann J (eds) Cephalopods-present and past. Schweitzerbart, Stuttgart
Kullmann J, Wiedmann J (1970) Significance of sutures in phylogeny of ammonoidea. Univ Kansas Paleontol Contrib, Kansas 47:1–32
Landman NH, Cochran JK (1987) Growth and longevity of *Nautilus*. In: Saunders WB, Landman NH (eds) *Nautilus*: the biology and paleobiology of a living fossil. Plenum, New York
Landman NH, Arnold JM, Mutvei H (1989a) Description of the embryonic shell of *Nautilus belauensis* (Cephalopoda). Am Mus Novit 2960:1–16
Landman NH, Cochran JK, Chamberlain JA Jr, Hirschberg JG (1989b) Timing of septal formation in two species of *Nautilus* based on radiometric and aquarium data. Mar Biol 102:65–72
Landman NH, Polizotto K, Mapes RH, Tanabe K (2006) Cameral membranes in prolecanitid and goniatitid ammonoids from the Permian Arcturus Formation, Nevada, USA. Lethaia 39:365–379
Lange W (1929) Zur Kenntnis des Oberdevons am Enkeberg und bei Balve (Sauerland). Abh Preuß Geol Landesanst N. F. 119:1–132
Lange W (1941) Die Ammonitenfauna der *Psiloceras*-Stufe Norddeutschlands. Paleontogr A 93:1–186
Laptikhovsky VL, Rogov MA, Nikolaeva SE, Arkhipkin AI (2013) Environmental impact on ectocochleate cephalopod reproductive strategies and the evolutionary significance of cephalopod egg size. Bull Geosci 88(1):83–93
Levi-Kalisman Y, Addadi L, Weiner S (2001) Structure of the nacreous organic matrix of a bivalve mollusc shell examined in the hydrated state using cryo-TEM. J Struct Biol 135:8–17
Lewy Z (2002) The function of the ammonite fluted septal margins. J Paleontol 76:63–69
Lewy Z (2003) Reply to Checa and to Hewitt and Westermann. J Paleontol 77:796–798
Lominadze T, Sharikadze M, Kvantaliani I (1993) On the mechanism of soft body movement within the body chamber in ammonites. Geobios 15:267–273
Longridge LM, Smith PL, Rawlings G, Klaptocz V (2009) The impact of asymmetries in the elements of the phragmocone of Early Jurassic ammonites. Paleontol Electron 12 (1):1–15. http://paleo-electronica.org/2009_1/160/index.html. Accessed 9 Feb 2015
Lukeneder A, Harzhauser M, Müllegger S, Piller WE (2010) Ontogeny and habitat change in Mesozoic cephalopods revealed by stable isotopes ($\delta^{18}O$, $\delta^{13}C$). Earth Planet Sic Lett 296:103–114
Manship LL (2004) Pattern matching: classification of ammonitic sutures using GIS. Palaeontol Electron 7(2);6A:p 15 doi:palaeo-electronica.org/2004_2/sutures/issue2_04.htm
Marin F, Luquet G (2004) Molluscan shell proteins. Comptes Rendus Palevol 3:469–492
Marin F, Luquet G (2005) Molluscan biomineralization: the proteinaceous shell constituents of *Pinna nobilis* L. Mater Sci Eng C25:105–111
Meinhardt H (1998) The algorithmic beauty of sea shells. Springer, Berlin
Miller AK, Furnish WM, Schindewolf OH (1957) Paleozoic Ammonoidea. In: Moore RC (ed) Treatise on invertebrate paleontology, Part L, Mollusca 4. GSA and University of Kansas Press, Lawrence

Monnet C, Klug C, De Baets K (2015) Evolutionary patterns of ammonoids: phenotypic trends, convergence, and parallel evolution. This volume
Monnet C, Klug C, De Baets K (2011) Parallel evolution controlled by adaptation and covariation in ammonoid cephalopods. BMC Evol Biol 11(115):1–21
Mutvei H (1964) Remarks on the anatomy of recent and fossil Cephalopoda. Stockh Contrib Geol 11(4):79–112
Mutvei H (1967) On the microscopic shell structure in some Jurassic ammonoids. Neues Jahrb Geol Paläontol Abh 129:157–166
Mutvei H, Arnold JM, Landman NH (1993) Muscles and attachment of the body to the shell in embryos and adults of *Nautilus belauensis* (Cephalopoda). Am Mus Novit 3059:1–15
Mutvei H, Doguzhaeva L (1997) Shell ultrastructure and ontogenetic growth in *Nautilus pompilius* L. (Mollusca, Cephalopoda). Palaeontol A 246:33–52
Mutvei H, Dunca E (2008) Structural relationship between interlamellar organic sheets and nacreous tablets in gastropods and the cephalopod *Nautilus*. Paläontol Z 82:84–95
Mutvei H, Reyment RA (1973) Buoyancy control and siphuncle function in ammonoids. Palaeontology 16:623–636
Mutvei H, Weitschat W, Doguzhaeva L, Dunca E (2004) Connecting ring with pore canals in two genera of Mesozoic ammonoids. Mitt Geol-Paläontol Inst Univ Hambg 88:135–144
Nagao T, Saito R (1934) Peculiar septal features observed in ammonites of certain lytoceratid genera. Proc Imp Acad 10:357–360
Nakahara H (1983) Calcification of gastropod nacre. In: Westbroek P, de Jong EW (eds) Biomineralization and biological metal accumulation. Reidel, Dordrecht
Nakahara H (1991) Nacre formation in bivalve and gastropod molluscs. In: Suga S, Nakahara H (eds) Mechanisms and phylogeny of mineralization in biological systems. Springer, Berlin
Newell ND (1949) Phyletic size increase, an important trend illustrated by fossil invertebrates. Evolution 3:103–124
Nudelman F, Gotliv BA, Addadi L, Weiner S (2006) Mollusk shell formation: mapping the distribution of organic matrix components underlying a single aragonitic tablet in nacre. J Struct Biol 153:176–187
Oba T, Kai M, Tanabe K (1992) Early life history and habitat of *Nautilus pompilius* inferred from oxygen isotope examinations. Mar Biol 113:211–217
Olóriz F, Palmqvist P (1995) Sutural complexity and bathymetry in ammonites: fact or artifact? Lethaia 28:167–170
Olóriz F, Palmqvist P, Pérez-Claros JA (1997) Shell features, main colonized environments, and fractal analysis of sutures in Late Jurassic ammonites. Lethaia 30:191–204
Olóriz F, Palmqvist P, Pérez-Claros JA (1999) Recent advances in morphometric approaches to covariation of shell features and the complexity of suture lines in Late Jurassic ammonites, with reference to the major environments colonized. In: Olóriz F, Rodríguez-Tovar FJ (eds) Advancing research on living and fossil cephalopods. Kluwer Academic, New York
Olson IC, Kozdon R, Valley JW, Gilbert PUPA (2012) Mollusk shell nacre ultrastructure correlates with environmental temperature and pressure. J Am Chem Soc 134:7351–7358
Owen R (1832) Memoir on the pearly *Nautilus*. Royal College of Surgeons, London
Owen R (1843) Lectures on the comparative anatomy and physiology of the invertebrate animals. p 392
Pérez-Claros JA (2005) Allometric and fractal exponents indicate a connection between metabolism and complex septa in ammonites. Paleobiology 31:221–232
Pfaff E (1911) Über Form und Bau der Ammonitensepten und ihre Beziehungen zur Suturlinie. Jahrb Niedersächs Geol Ver (Geol Abt Nat Ges Hann) 4:207–223
Pia J (1923) Über die ethologische Bedeutung einiger Hauptzüge in der Stammesgeschichte der Cephalopoden. Ann Nat Mus Wien 36:50–73
Polizzotto K, Landman NH, Klug C (2015) Cameral membranes, pseudosutures, and other soft tissue imprints in ammonoid shell. This volume
Reif WE, Thomas RDK, Fischer MS (1985) Constructional morphology: the analysis of constraints in evolution. Acta Biotheor 34:233–248

Rein S (2004) Zur Biologie der Ceratiten der spinosus-Zone—Ergebnisse einer Populationsanalyse, Teil II: Variationsbreite der Skulptur- und Suturbildungen. Veröff Naturkundemus Erfurt 23:33–50

Reyment RA (1955) Some examples of homeomorphy in Nigerian Cretaceous ammonites. Geologisca Foren i Stockh Forh 77:567–594

Reyment RA (1958) Some factors in the distribution of fossil Cephalopods. Acta Universitatis Stockholmiensis—Stockholm contributions. Geology 1:97–184

Richter U (2002) Gewebeansatz-Strukturen auf Steinkernen von Ammonoideen. Geol Beitr Hann 4:1–113

Richter U, Fischer R (2002) Soft tissue attachment structures on pyritized internal moulds of ammonoids. Abh Geol Bundesanst 57:139–149

Rieber H (1979) Eine abnorme, stark vereinfachte Lobenlinie bei *Brasilia decipiens* (Buckman). Paläontol Z 53:230–236

Ritterbush KA, Bottjer DJ (2012) Westermann Morphospace displays ammonoid shell shape and hypothetical paleoecology. Paleobiology 38:424–446

Ritterbush KA, Hoffmann R, Lukeneder A, De Baets K (2014) Pelagic palaeoecology: the importance of recent constraints on ammonoid palaeobiology and life history. J Zool 292:229–241

Ruzhencev VE (1960) Printsipy sistematiki, sistema i filogeniya paleozoyskikh ammonoidey. Trudy Paleontol Inst Akad Nauk SSSR 133:1–331

Ruzhencev VE (1962) Nadotryad Ammonoidea. Ammonoidei. Obshchaya chast'. In: Orlov YA, Ruzhencev VE (eds) Osnovy Paleontologii, 5, Mollyuski: Golovonogie 1; Akademiya Nauk SSSR, pp 243–334

Ruzhencev VE (1974) Superorder Ammonoidea. General section. In: Orlov YA, Ruzhencev VE (eds) Fundamentals of paleontology (Osnovy paleontologii). vol V. Mollusca—Cephalopoda I. pp 371–511

Salfeld H (1924) Die Bedeutung der Konservativstämme für die Stammesentwicklung der Ammonoideen. pp 1–16

Saunders WB (1995) The ammonoid suture problem: relationships between shell and septum thickness and suture complexity in Paleozoic ammonoids. Paleobiology 21:343–355

Saunders WB, Work DM (1996) Shell morphology and suture complexity in Upper Carboniferous ammonoids. Paleobiology 22:189–218

Saunders WB, Work DM, Nikolaeva SV (1999) Evolution of complexity in Paleozoic ammonoid sutures. Science 286:760–763

Schindewolf OH (1929) Vergleichende Studien zur Phylogenie, Morphologie und Terminologie der Ammoneen-Lobenlinie. Abh Preuß Geol Landesanst N. F. 115:1–102

Schindewolf OH (1951) Zur Morphogenie und Terminologie der Ammoneen-Lobenlinie. Paläontol Z 25(1/2):11–34

Schindewolf OH (1954) Über die Lobenlinie der Ammonoidea. Neues Jahrb Geol Paläontol Monatshefte 1954:123–140

Schindewolf OH (1961–1968) Studien zur Stammesgeschichte der Ammoniten; Lfg 1–7. Abh Math-Naturw Kl Akad Wiss Lit Mainz 1960(10)–1968(3):1–901

Schindewolf OH (1972) Über Clymenien und andere Cephalopoden. Abh Akad Wiss Lit Mainz, Math-Naturw Kl 1971(3):1–89

Schlögl J, Chirat R, Balter V, Joachimski M, Hudáčková N, Quillévéré F (2011) *Aturia* from the Miocene Paratethys: an exceptional window on nautilid habitat and lifestyle. Palaeogeogr Palaeoclimatol Palaeoecol 308:330–338. doi:10.1016/j.palaeo.2011.05.037

Schmidt M (1925) Ammonitenstudien. Fortschr Geol Paläontol 10:275–363

Schreiber G, Hoffmann R (2009) Der Septallobus als diagnostisches Merkmal für die Überfamilie Lytoceratoidea (Cephalopoda: Ammonitina). Berl Paläobiol Abh 10:307–310

Seilacher A (1973) Fabricational noise in adaptive morphology. Syst Zool 22:451–465

Seilacher A (1975) Mechanische simulation und funktionelle evolution des Ammoniten-Septums. Paläontol Z 49:268–286

Seilacher A (1988) Why are nautiloid and ammonite sutures so different? Neues Jahrb Geol Paläontol Abh 177:41–69

Seilacher A, Gishlick AD (2015) Morphodynamics. CRC Press, London

Seilacher A, Labarbera M (1995) Ammonites as Cartesian Divers. Palaios 10:493–506
Seuss B, Mapes RH, Klug C, Nützel A (2012) Exceptional cameral deposits in a sublethally injured carboniferous orthoconic nautiloid from the Buckhorn Asphalt Lagerstätte in Oklahoma, USA. Acta Palaeontol Pol 57:375–390
Shigeta Y (2006) Ammonoid evolution inferred for the early shell features. Iden 60:23–26
Shimanskij VN (1974) Superorder Actinoceratoidea. General section. In: Orlov YuA, Ruzhencev VE (eds) Fundamentals of paleontology, vol V. Mollusca—Cephalopoda I, Jerusalem
Solger F (1901) Die Lebensweise der Ammoniten. Naturw Wochenschr 17:89–94
Spath LF (1919) Notes on Ammonites. Geol Mag 56:27–35
Stanley EH (1987) Role of fluctuation in fluid mechanics and dendritic solidification. In: Guttinger W, Dangelmayr H (eds) The physics of structure formation
Steinmann G (1888) Vorläufige Mittheilung über die Organisation der Ammoniten. Ber Naturf Ges Freibg Breisgau 4:113–129
Steinmann G (1925) Beiträge zur Stammesgeschichte der Cephalopoden I. *Argonauta* und die Ammoniten. Z Indukt Abstamm- Vererb 36:350–416
Stenzel HB (1964) Living Nautilus. In: Teichert C et al (eds) Treatise on invertebrate paleontology, Part K, Mollusca 3, cephalopodageneral features—Endoceratoidea—Actinoceratoidea—Nautiloidea—Bactritoidea (1:59–93). GSA and The University of Kansas, Lawrence
Suess E (1865) Über Ammoniten. Sitzungsber Math-Naturw Kl Kaiserl Akad Wissensch, Abt 1(52):71–89
Swinnerton H, Trueman AE (1918) The morphology and development of the ammonite septum. Quat J Geol Soc 73(1917):26–58
Tanabe K, Fukuda Y, Ohtsuka Y (1985) New chamber formation in the Cuttlefish *Sepia esculenta* Hoyle. Venus (Jap Jour Malac) 44:55–67
Tanabe K, Sasaki T, Mapes RH (2015) Soft-part anatomy of the siphuncle in ammonoids. This volume
Tate R, Blake JF (1876) The Yorkshire lias. J. Van Vorst, London
Taylor DG (1998) Late Hettangian—early Sinemurian (Jurassic) ammonite biochronology of the Western Cordillera, United States. Geobios 31:467–497
Teichert C (1964) Morphology of hard parts. In: Teichert C et al (eds) Treatise on invertebrate paleontology, Part K, Mollusca 3, Cephalopoda general features—Endoceratoidea—Actinoceratoidea—Nautiloidea—Bactritoidea. GSA and The University of Kansas Press, Lawrence, pp 13–53
Teichert C (1967) Major features of cephalopod evolution. Essays in paleontology and stratigraphy. Spec Pub 2:162–210
Thompson DAW (1942) On growth and form. Cambridge University, Cambridge
Trueman AE (1941) The ammonite body chamber with special reference to the buoyancy and mode of life of the living ammonite. Quat J Geol Soc Lond 96:339–383
Turing A (1952) The chemical basis of morphogenesis. Phil Trans R Soc Lond B 237:37–72
Ubukata T, Tanabe K, Shigeta Y, Maeda H, Mapes RH (2009) Eigenshape analysis of ammonoid sutures. Lethaia 43:266–277. doi:10.1111/j.1502-3931.2009.00191.x
Ubukata T, Tanabe K, Shigeta Y, Maeda H, Mapes RH (2014) Wavelet analysis of ammonoid sutures. Palaeontol Electron 17(1;9A):1–17. palaeo-electronica.org/content/2014/678-wavelet-analysis-of-sutures
Uhlig V (1883) Die Cephalopodenfauna der Wernsdorfer Schichten. Denkschr Math-Naturwiss Cl Kaiserl Akad Wiss 46:127–290
Valentini FG, Finks RM (1974) The functional significance of the ammonoid suture pattern and the origin of the ammonoids. Abstracts GSA meeting Miami 994
Vicencio R (1973) Models for the morphology and morphogenesis of the ammonoid shell. Unpubl. PhD thesis
Ward PD (1979) Cameral liquid in *Nautilus* and ammonites. Paleobiology 5:40–49
Ward PD (1985) Periodicity of chamber formation in chambered cephalopods: evidence from *Nautilus macromphalus* and *Nautilus pompilius*. Paleobiology 11:438–450
Ward PD (1987) The natural history of *Nautilus*. Allen and Unwin, Boston

Ward PD, Chamberlain JA Jr (1983) Radiographic observations of chamber formation in *Nautilus pompilius*. Nature 304:57–59

Ward PD, Westermann GEG (1976) Sutural inversion in a heteromorph ammonite and its implication for septal formation. Lethaia 9:357–361

Wedekind R (1916) Über Lobus, Suturallobus und Inzision. Centralbl Miner, Geol Paläontol 8:185–195

Weiner S, Addadi L (2011) Crystalization pathways in biomineralization. Ann Rev Mat Res 41:21–40

Weitschat W (1986) Phosphatisierte Ammonoideen aus der Mittleren Trias von Central-Spitzbergen. Mitt Geol-Paläontol Inst Univ Hambg 61:249–279

Weitschat W, Bandel K (1991) Organic components in phragmocones of boreal Triassic ammonoids: implications for ammonoid biology. Paläontol Z 65:269–303

Wendling J (1987) On the buoyancy system of *Sepia officinalis* L. (Cephalopoda). PhD Thesis, University of Basel

Westermann B, Beck-Schildwächter I, Beuerlein K, Kaleta EF, Schipp R (2004) Shell growth and chamber formation of aquarium-reared *Nautilus pompilius* (Mollusca, Cephalopoda) by X-Ray analysis. J Exp Zool 301A:930–937

Westermann B, Schmidtberg H, Beurlein K (2005) Functional morphology of the mantle of *Nautilus pompilius* (Mollusca, Cephalopoda). J Morph 264:277–285

Westermann GEG (1956) Phylogenie der Stephanocerataceae und Perisphinctaceae des Dogger. Neues Jahrb Geol Paläontol Abh 103:233–279

Westermann GEG (1958) The significance of septa and sutures in Jurassic ammonite systematics. Geol Mag 95:441–455

Westermann GEG (1965) Septal and sutural patterns in evolution and taxonomy of Thamboceratide and Clydoniceratidea (M. Jurassic, Ammonitina). J Paleontol 39:864–874

Westermann GEG (1971) Form, structure and function of shell and siphuncle in coiled Mesozoic ammonoids. Life sciences contributions. R Ont Mus 78:1–39

Westermann GEG (1975) Model for origin, function and fabrication of fluted cephalopod septa. Paläontol Z 49:235–253

Westermann GEG (1999) Recent hypotheses on mechanical and metabolic functions of septal fluting and sutural complexity in post-carboniferous ammonoids. Ber Geol Bundesanst 46:120

Wiedenmayer F (1979) *Exomiloceras* (Analytoceratidae, Ammonoidea), eine neue Gattung aus dem unteren Lias der Tethys. Eclogae Geol Helv 72:859–870

Wiedmann J, Boletzky Sv (1983) Wachstum und Differenzierung des Schulps von *Sepia officinalis* unter künstlichen Aufzuchtbedingungen—Grenzen der Anwendung im palökologischen Modell. Neues Jahrb Geol Paläontol Abh 164:118–133

Wiedmann J, Kullmann J (1981) Ammonoid sutures in ontogeny and phylogeny. In: House MR, Senior JR (eds) The Ammonoidea. The systematics association special vol 18. Clarendon Press, Oxford, pp 215–255

Yacobucci MM, Manship LL (2004) GIS analysis of sutural constraints and asymmetry in ammonoids: testing the tie-point model for septal formation. University of Arkansas 2004:165–166

Yacobucci MM, Manship LL (2011) Ammonoid septal formation and suture asymmetry explored with a geographic information systems approach. Palaeontol Electron 14(3A):17 p. http://palaeo-electronica.org/2011_1/136/index.html

Zaborski PMP (1986) Internal mould markings in a Cretaceous ammonite from Nigeria. Palaeontology 29:725–738

Zittel KA (1884) Cephalopoda. In: Zittel KA (ed) Handbuch der Palaeontologie. 1. Abth, 2. Band. R. Oldenbourg, Munich

Chapter 4
Cameral Membranes, Pseudosutures, and Other Soft Tissue Imprints in Ammonoid Shells

Kristin Polizzotto, Neil H. Landman and Christian Klug

4.1 Introduction

An essential aim of ammonoid paleobiology is understanding the growth, locomotion, and mode of life of these animals. Evidence of these processes can be gathered from the remains of soft tissues or their traces preserved inside the shell such as cameral membranes, pseudosutures, and muscle scars, as well as other soft tissue imprints. Although these structures have been recognized for at least 100 years, in the last 15 years many important studies relating to their occurrence, ultrastructure, composition, and probable function have been published. This chapter summarizes previous research and focuses on recent advances in understanding cameral membranes, pseudosutures, drag lines, and soft tissue imprints unrelated to muscle attachment.

K. Polizzotto (✉)
Department of Biological Sciences, Kingsborough Community College,
2001 Oriental Boulevard, Brooklyn, NY 11235, USA
e-mail: Kristin.Polizzotto@kingsborough.edu

N. H. Landman
Division of Paleontology (Invertebrates), American Museum of Natural History,
Central Park West at 79th Street, New York, NY 10024, USA
e-mail: landman@amnh.org

C. Klug
Paläontologisches Institut und Museum, University of Zurich,
Karl Schmid-Strasse 6, 8006 Zurich, Switzerland
e-mail: chklug@pim.uzh.ch

C. Klug et al. (eds.), *Ammonoid Paleobiology: From Anatomy to Ecology*,
Topics in Geobiology 43, DOI 10.1007/978-94-017-9630-9_4

4.2 Cameral Membranes

Cameral membranes (or cameral sheets) are the remains of thin, originally organic structures within the chambers of ammonoids (Fig. 4.1). The term "cameral sheets" is preferred by some in order to avoid the implication of a physiological, cellular membrane, but either term is widely accepted. These structures can be divided into two general types: chamber linings, which coat the internal surfaces of the chambers, and suspended sheets, which are three-dimensional sheets internally attached to the shell at two or more different points (Landman et al. 2006, Polizzotto et al. 2007) or along longer lines. Suspended sheets may occur as siphuncular membranes, which extend between the siphuncle and the septum and/or ventral shell floor; transverse membranes, which extend between different points on the septum; and horizontal sheets, which divide the chamber into dorsal and ventral compartments (Weitschat and Bandel 1991).

4.2.1 Taxonomic Occurrence

Cameral membranes have been described in the ammonoid literature for over a century (John 1909; Grandjean 1910; Schoulga-Nesterenko 1926; Hölder 1952, 1954; Schindewolf 1968; Erben and Reid 1971; Westermann 1971; Bayer 1975; 1977; Bandel and Boletsky 1979; Kulicki 1979; Bandel 1981, 1982; Tanabe et al. 1982; Hagdorn 1983; Grégoire 1984; Henderson 1984; Weitschat 1986; Weitschat and Bandel 1991; Keupp 1992; Checa and Garcia-Ruiz 1996; Kulicki 1996; Tanabe and Landman 1996). The taxonomic occurrence of the membranes reported in these publications is wide, including phylloceratids, lytoceratids, ceratitids and ammonitids. More recently, such membranes have also been found in Paleozoic ammonoids such as goniatites (Polizzotto et al. 2007) and prolecanitids (Mapes et al. 2002, Landman et al. 2006) as well as in Cretaceous scaphitids (Polizzotto and Landman 2010). Schoulga-Nesterenko (1926) reported cameral membranes in the goniatite *Agathiceras uralicum* but may have misidentified the specimen (see Polizzotto et al. 2007).

Early work most often described cameral sheets associated with the siphuncle. Weitschat and Bandel (1991) described the most intricate and extensive cameral membranes reported up to that time (Fig. 4.1), including transverse and horizontal

Ladinian. Berlichingen (Germany), note the membranes on both sides of the siphuncle. **b** *C.* cf. *sublaevigatus*, Künzelsau, Garnberg (Germany), 0–7 m above Cycloidesbank gamma, Ladinian, membranes on both sides of the siphuncle. **c** *C.* cf. *sublaevigatus*, Nitzenhausen (Germany), 70 cm below Tonsteinhorizont 4 (delta), Ladinian, membranes on both sides of the siphuncle. **d** *C.* cf. *sublaevigatus*. Nitzenhausen (Germany), between Tonsteinhorizont 3 and 4 (γ and δ), Ladinian, phosphatized membranes on both sides of the siphuncle. **e** *C.* cf. *evolutus subspinosus*, Heming (France), evolutus Zone, Ladinian, phosphatized siphuncle. **f** *Bukkenites* sp., 104, Dienerian, Amb, Spiti valley, India, membranes on both sides of the siphuncle. **g** Gen. et sp. indet., 104, Amb, Spiti valley, India. **h** *Ambites* sp., Nam 53–20, Dienerian, Nammal (Pakistan), membranes crossing the saddles (separating fillings that differ in color). **i, j** horizontal lamellae in *Aristoptychites kolymensis*, Late Ladinian, Barentsøya Formation, Bertylryggen, Spitsbergen. **i** detail. **j** note the protoconch and the absence of horizontal sheets in the first whorls

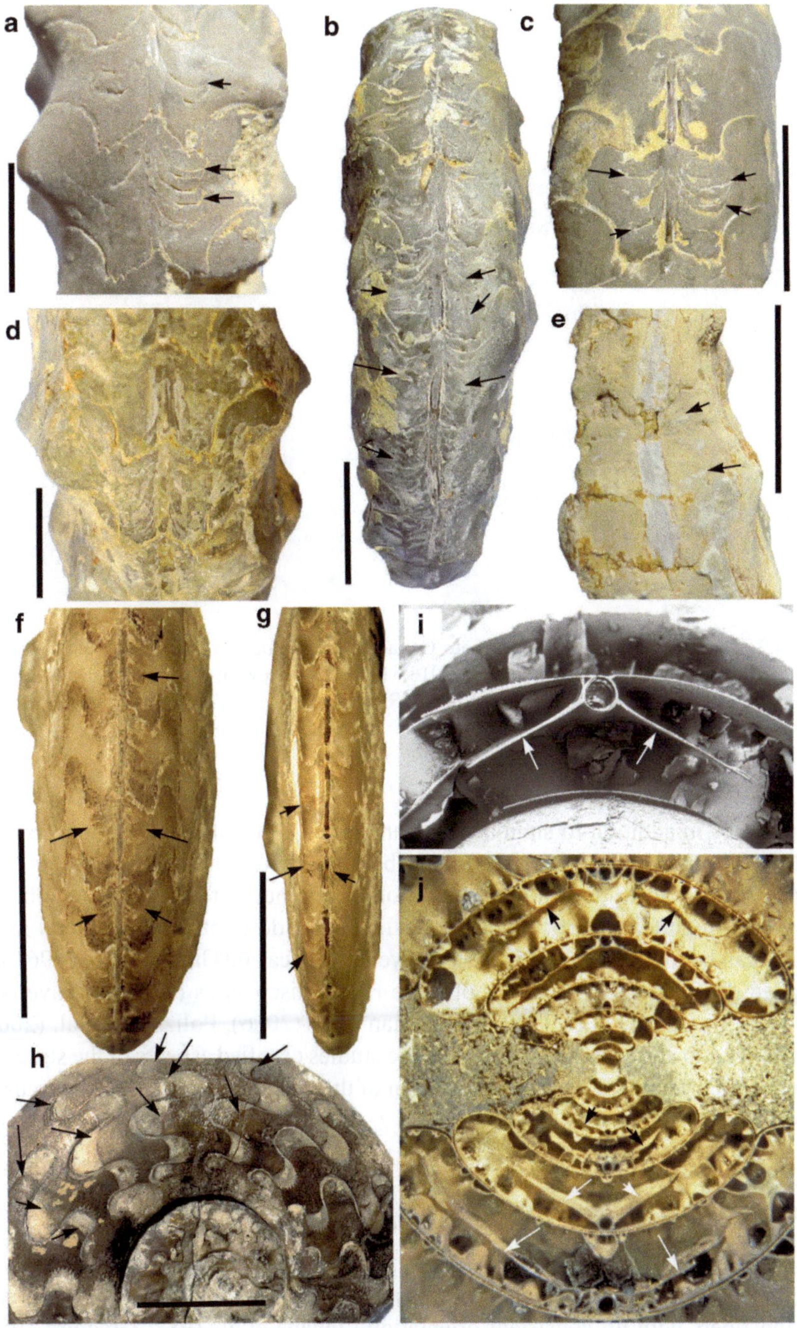

Fig. 4.1 Siphuncular sheets and cameral membranes/intracameral lamellae in various Triassic ammonites; images **a–e** courtesy of H. Hagdorn (Ingelfingen), **g–i** courtesy of D. Ware (Zürich), **j** and **k** courtesy of W. Weitschat (Hamburg). **a** *Ceratites* cf. *münsteri*, postspinosus to enodis-Zone,

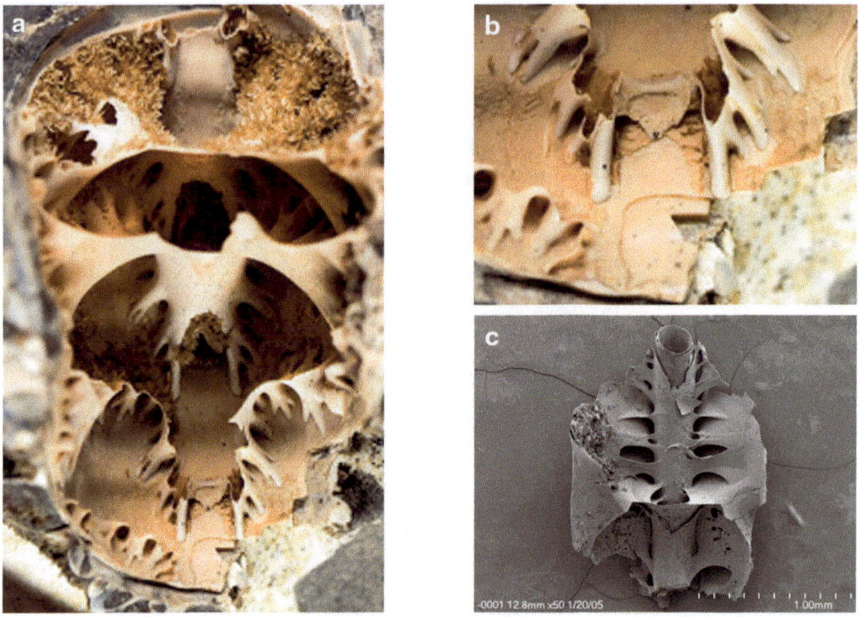

Fig. 4.2 **a** Hollow chambers from a Late Cretaceous *Rhaeboceras halli* from Montana, USA (modified from Polizzotto and Landman 2010). **b** A closer view of the siphuncular sheets and pseudosutures from the specimen shown in **a**. **c** Siphuncular sheets in a Permian *Akmilleria electraensis* from Nevada, USA

membranes in addition to siphuncular membranes. Since that time, transverse and siphuncular membranes have been described in several other ammonoids as mentioned above, but to our knowledge, no other instances of horizontal membranes have been reported. Brief summaries of the earlier descriptions of cameral membranes can be found in Kulicki (1996) as well as Checa and Garcia-Ruiz (1996), and detailed descriptions and images of more recent discoveries (Fig. 4.2) have been produced by Tanabe et al. (2005), Landman et al. (2006), Polizzotto et al. (2007), and Polizzotto and Landman (2010). These studies clarified aspects of the structure, composition, origin, and probable function of the cameral membranes, as discussed below.

4.2.2 Structure and Composition

The ultrastructure of the membranes has been described as thin (<0.2 μm) conchiolin fibers with no consistent orientation (Tanabe et al. 1982; Grégoire 1987). The chamber linings in particular have been compared to the pellicle in *Nautilus* and *Spirula*, but Kulicki (1996) pointed out that the fibers of ammonoid cameral membranes are considerably finer. Polizzotto and Landman (2010) described a well-preserved Late Cretaceous scaphite (*Rhaeboceras halli*) in which siphuncular membranes and chamber linings (as well as pseudosutures) were present in the same chamber. Membranes and chamber linings are 1–2 μm thick and composed of irregular globular particles. This is in contrast to the composition reported by Kulicki (1996), but may be more similar to the organic layer secreted on the inner chamber surface in *Nautilus* (Mutvei 1963). Study of the membranes in prolecanitids confirms phosphatic composition (Tanabe et al. 2000). Energy dispersive spectroscopy (EDS) on the *Rhaeboceras* specimen indicated high phosphorus content in the membranes and chamber linings, suggesting an organic origin (Polizzotto and Landman 2010). This hypothesis is accepted by most researchers; however, there is a difference of opinion as to the mode of formation of such membranes (see also the discussion on intracameral deposits in Seuss et al. 2012).

4.2.3 Formation

Two models have been proposed for the formation of cameral membranes. The first proposes that cameral membranes were secreted by the rear mantle as the animal moved forward during chamber formation, and that the shape of the membranes replicates the shape of the rear mantle (the secretion model; Weitschat and Bandel 1991). The second model contends that the membranes are simply the desiccated remains of a hydrogel formed by cameral fluid enriched with organic molecules and shaped by surface tension (the desiccation model; Hewitt et al. 1991; Westermann 1992; Checa 1996). Landman et al. (2006) argued that siphuncular membranes (Fig. 4.2 and 4.4) are not solely the result of cameral liquid dehydration, based on the absence of membranes from early whorls and a consistent first appearance at the end of the neanic stage of ontogeny, as well as the presence of membranes in body chambers in some ammonoids (Polizzotto et al. 2007). The formation of chamber linings is less clear, but the similarity in ultrastructure between chamber linings and siphuncular membranes suggests a similar origin, at least in scaphitids (Polizzotto and Landman 2010). Evidence for the morphogenesis of transverse membranes has not been investigated, and their overall morphology does not immediately rule out either the secretion or the desiccation hypothesis. An examination of the ultrastructure and ontogenetic pattern of occurrence of transverse membranes may shed light on this issue.

In summary, the secretion hypothesis is well supported for the formation of siphuncular membranes in at least some groups of ammonoids (prolecanitids, goniatites, phylloceratids, and scaphitids). There is also some evidence for the secretion hypothesis for chamber linings in scaphitids. Cameral membranes in other ammonoid groups, as well as transverse membranes in all ammonoids, may have been formed either by secretion, desiccation, or a combination of both processes (Checa and Garcia-Ruiz 1996).

4.2.4 Function

Researchers have long proposed that cameral membranes functioned in absorbing cameral fluid, either for fluid transport or for maintenance of a fluid reservoir (Mutvei 1967; Kulicki 1979; Kulicki and Mutvei 1988; Ward 1987; Weitschat and Bandel 1991; Kulicki 1996; Kröger 2002; Landman et al. 2006). This process may have helped in decoupling fluid reservoirs, which may have conferred some physiological benefit. Some researchers have tested a model demonstrating that the presence of cameral membranes may have maintained a reservoir that aided in buoyancy control by rapid fluid re-filling following sublethal shell loss from injury (Daniel et al. 1997; Kröger 2002). Kröger (2002) found that ammonoids survived shell loss up to four times greater than in *Nautilus*, suggesting some sort of buoyancy compensation mechanism. The evidence from Kröger (2002) clearly indicates that a high volume of cameral membranes (up to 14 % of the chamber volume; Hewitt and Westermann 1996) would have made a significant difference in rapidly compensating for shell loss due to injury. Whether or not most ammonoids possessed such a volume of cameral membranes is not yet known.

The capillary action of cameral membranes may also have aided in fluid transport, resulting in faster chamber emptying and thus faster growth rates in ammonoids that formed such membranes. Kröger (2002) suggests that this may be one explanation for increasingly more complex septa during the course of ammonoid evolution, which would have added to the volume of liquid reserved in correspondingly more complex and extensive cameral membranes.

Here, we suggest an additional possible function: the cameral membranes subdivided the chamber volume into smaller volumes. Taking the potentially large amount of chamber water (up to 30 % of phragmocone volume; Heptonstall 1970; Mutvei and Reyment 1973; Reyment 1973; Ward 1979; 1987; Tajika et al. 2014) in the phragmocone into account, water movement might have altered the orientation of the shell *syn vivo*. Cameral membranes would have limited the water movement within the phragmocone chambers and thus enhanced stability. However, the possible effect of moving chamber water needs to be modeled in order to test the potential physical effect of chamber water movements.

4.3 Pseudosutures and Drag Lines

Pseudosutures are incomplete replicas of the suture that are often preserved as raised ridges on the internal surface of the chamber or as lines or etched furrows on the surface of the steinkern between the sutures themselves (Fig. 4.3). Pseudosutures should not be confused with phantom sutures (Seilacher 1968, 1988), which formed when the ammonoid's surface was corroded by a pressurized solution and a phantom of the suture was copied on a lower level of the internal mould. Pseudosutures often occur in series and have sometimes been interpreted as the margins of

Fig. 4.3 Pseudosutures in various Jurassic and Cretaceous ammonites: **a** *Cadoceras* sp., Callovium, Kostroma Region, Russia, whorl height 12 mm, image courtesy of R. Hoffmann (Bochum). **b** *Baculites mariasensis* with multiple pseudosutures and drag lines associated with all four sutures in the image. Pseudosutures are found near the adapical side of the lobules of each suture. Adoral direction is to the right. **c–f**, *Craspedites* sp., Craspedites nodiger Zone, Cretaceous, Kostroma Region, Russia, images courtesy of R. Hoffmann (Bochum). **c** dm 24 mm. **d**, dm 34 mm. **e, f** image width at top 25 mm

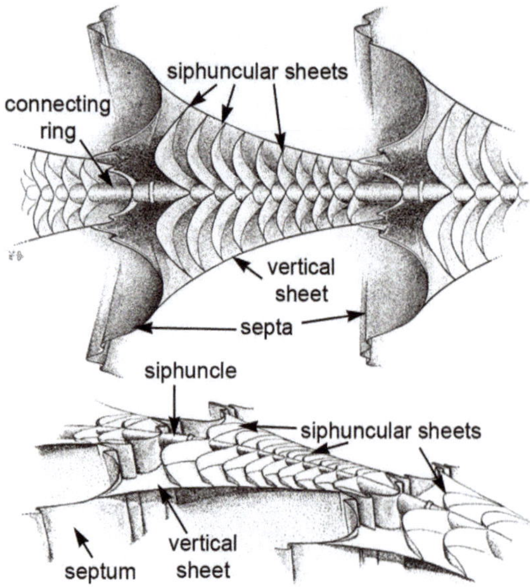

Fig. 4.4 Reconstruction of the siphuncular sheets and the vertical sheet in the Triassic ammonoid *Anagymnotoceras* from Spitsbergen, modified after Weitschat and Bandel (1991)

pseudosepta (Hewitt et al. 1991). The replicated portion may be lobes, saddles, or both, although pseudosutures mimicking lobes appear to be slightly more common. The pseudosutures may be evenly spaced throughout the chamber between sutures (Zaborski 1986), or they may occur singly or in a cluster on one side approaching the lobe or saddle. Pseudosutures most frequently appear on the flanks or ventrolateral portion of the chamber.

Drag lines (or drag bands), which are often associated with pseudosutures, are spiral markings that often (but not always) extend throughout the chamber from lobule to lobule. Like pseudosutures, drag lines form ridges on the internal chamber surface, or their imprints occur as grooves in the surface of the steinkern. In well-preserved specimens, the drag lines extend in pairs from the flanks of the lobe or lobules (Zaborski 1986; Hewitt et al. 1991; Polizzotto and Landman 2010).

4.3.1 Taxonomic Occurrence

Pseudosutures have been described and discussed in numerous ammonoid groups (John 1909; Hölder 1954; Vogel 1959; Schindewolf 1968; Bayer 1977; Hagadorn and Mundlos 1983; Zaborski 1986; Seilacher 1988; Hewitt et al. 1991; Weitschat and Bandel 1991, 1992; Westermann 1992; Landman et al. 1993; Lominadze et al. 1993; Bucher et al. 1996; Checa 1996; Checa and Garcis-Ruiz 1996; Doguzhaeva and Mutvei 1996; Tanabe et al. 1998; Keupp 2000; Richter 2002; Richter and Fischer 2002; Klug et al. 2007; Polizzotto et al. 2007; Klug et al. 2008; Polizzotto 2010; Polizzotto and Landman 2010). The groups in which pseudosutures have been most widely reported include ceratitids, lytoceratids, phylloceratids, perisphinctids, vascoceratids, scaphitids, and goniatitids, but they are found fairly often and probably

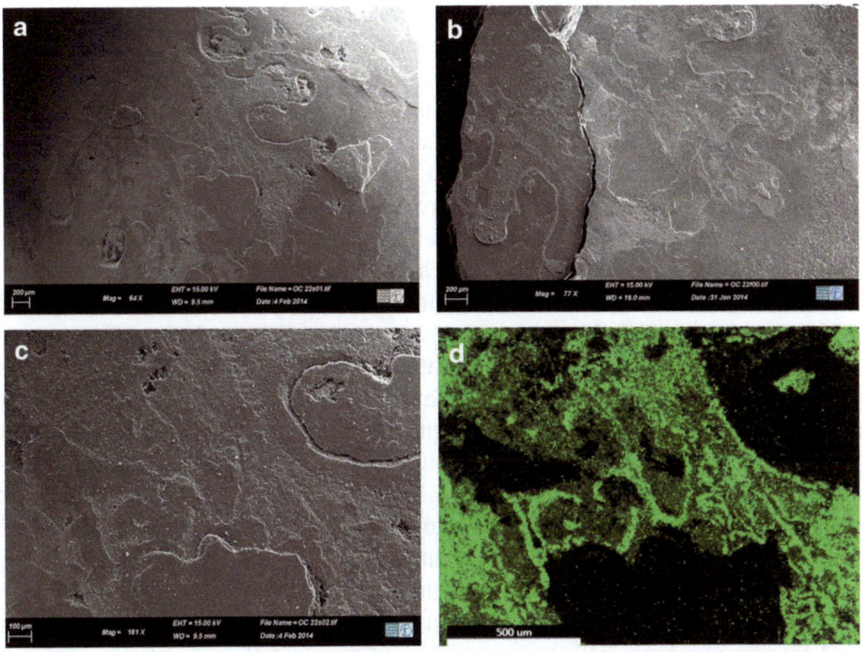

Fig. 4.5 *Eubaculites latecarinatus* from the Late Cretaceous Owl Creek Formation, Mississippi, USA. The upper left image (**a**) is the mould, and the upper right image (**b**) is the corresponding shell fragment. Each image shows the suture line, as well as associated pseudosutures. The lower left image (**c**) is a closer view of some of the suture and pseudosutures on the mould. EDS analysis indicates phosphorus enrichment in the pseudosutures and chamber surface (**d**), but no phosphorus is present on the inner surface of the shell (see text for explanation)

occur to one degree or another in most ammonoid groups. Pseudosutures have also been noted in baculitids (Fig. 4.3b and 4.5, W.A. Cobban unpublished observations, Polizzotto 2010). Drag lines are often reported together with pseudosutures (Fig. 4.1b, c, 4.2, Zaborski 1986; Lominadze et al. 1993; Richter 2002; Klug et al. 2007, 2008; Polizzotto and Landman 2010).

4.3.2 Structure and Composition

Pseudosutures form ridges on the internal surface of the shell, but the imprints (furrows or grooves) of these ridges on the surface of the steinkern are also called pseudosutures. The height and width of the ridges vary, with some pseudosutures quite prominent (20 μm wide and 25 μm high; see Polizzotto et al. 2007), and others less so (2 μm wide and 10 μm high; see Polizzotto and Landman 2010). Drag lines, while oriented differently than pseudosutures (parallel to the direction of growth rather than parallel to sutures; Richter 2002), appear to have a similar morphology in width and height (Polizzotto and Landman 2010).

The original composition of pseudosutures and drag lines is still unclear. Polizzotto et al. (2007) figured various goniatitid pseudosutures in the same specimen, some of which were made up of regular crystals arranged vertically along an asymmetrical slope (similar to the mural ridge in *Nautilus*), and others with a more random crystal arrangement. The pseudosutures in these goniatites sometimes dissolved when etched in acid, indicating an underlying carbonatic composition. At other times, however, the pseudosutures remained intact following etching, and were assumed in such cases to be coated by a now phosphatic and probably originally organic layer. In support of this idea, the thinner pseudosutures described by Polizzotto and Landman (2010) in a scaphitid were composed of (or perhaps coated with) an irregular, globular substance 1–2 μm thick. EDS analysis of these pseudosutures and of the drag lines in the same chamber revealed a high phosphorus content, corroborating an originally organic composition (at least for the surface of the pseudosutures and drag lines). Recent EDS analysis of well-preserved baculitids also indicated phosphorus enrichment (8–10 weight%) in pseudosutures (Polizzotto 2014). In that study, the shell was carefully removed from the mould, and pseudosutures on the surface of the mould were analyzed (Fig. 4.5). In addition, the imprints of pseudosutures on the inner surface of the corresponding shell fragment were analyzed. It is interesting to note that although phosphorus enrichment was found in the pseudosutures themselves, it was absent in the imprints. This implies that pseudosutures may have been composed of an originally mineralized substance (similar to the shell and septum), and then coated with an organic secretion that likewise coated the entire interior of the chamber (see also Polizzotto et al. 2007 and Polizzotto and Landman 2010). When the shell was removed from the mould, the originally organic coating adhered to the mould, explaining the presence of phosphorus on the mould and its absence on the inner shell surface. Though more evidence is needed in additional taxa, it seems likely that the original composition of pseudosutures and drag lines was carbonatic, with an overlying organic coating. It should be possible to verify this by performing EDS analysis on cross-sections of well-preserved pseudosutures. This hypothesis for the composition of pseudostutures leads to the question of how these structures were formed.

4.3.3 Formation

Pseudosutures likely formed as an accumulation of secretions from the rear mantle during pauses in forward movement (Weischat and Bandel 1991; Keupp 2000; Landman et al. 2006; Klug et al. 2007; Polizzotto et al. 2007 and references therein). It has also been proposed that siphuncular membranes and pseudosutures formed by a single process as parts of a continuous structure, and that both siphuncular membranes and pseudosutures are simply the remnants of pseudosepta (originally organic membranes that replicated the entire surface of the rear mantle; Hewitt et al. 1991; Westermann 1992; Checa 1996). These pseudosepta would have formed by either secretion or desiccation, as outlined earlier. We have described evidence above in favor of the secretion hypothesis for siphuncular membranes, and similar evidence suggests that pseudosutures formed by secretion as well (Polizzotto et al. 2007; Polizzotto and Landman 2010).

In addition, recent research has demonstrated that although siphuncular membranes and pseudosutures formed by a similar process (as accumulations of secreted material in the shape of the rear mantle), they are not parts of a single continuous structure (Polizzotto et al. 2007; Polizzotto and Landman 2010). This was confirmed by examining specimens in which siphuncular membranes and pseudosutures occurred in the same specimen or even in the same chamber, which revealed that differences in ultrastructure, position in the chamber, and spacing argue against the single-origin hypothesis.

If it is the case that pseudosutures are composed of an originally mineralized ridge overlain by an originally organic secretion, then two hypotheses of formation are possible. Either two separate but closely located populations of rear mantle cells produced two different secretions in sequence, or the same population of cells produced two different secretions at various points in the chamber formation cycle. It is difficult to test either hypothesis, but it may be fruitful to identify the specific cells that secrete organic and inorganic components of modern molluscan shells.

Drag lines have always been assumed to mark the progress of the rear mantle during translocation, and similarities to pseudosutures in structure and composition suggest that drag lines represent an accumulation of rear-mantle secretions. In contrast to the portion of the rear mantle that secreted pseudosutures, however, the parts of the mantle that formed drag lines must have remained in continuous contact with the shell wall. Alternatively, Klug et al. (2008) hypothesized that some spirally arranged drag lines might represent impressions of muscle fiber bundles in the posterior mantle (rather than secretions).

4.3.4 Implications for Growth

The shape of the pseudosutures and their probable origin as secretory products of the rear mantle corroborates the hypothesis that they formed during pauses in the forward movement of the animal. Many authors have proposed such an explanation for pseudosutures (e.g., Zaborski 1986; Seilacher 1988; Hewitt et al. 1991; Lominadze et al. 1993; Checa and Garcia-Ruiz 1996; Polizzotto et al. 2007; Polizzotto and Landman 2010). Interpretations differ, however, in what the pseudosutures reveal about the process of translocation. Some suggest that the temporary points of attachment served as critical, possibly genetically determined points that helped to maintain the shape of the rear mantle (and thus the consistent shape of the septum) between septa (Henderson et al. 2002; Polizzotto and Landman 2010). This line of reasoning gives rise to the hypothesis that the pseudosutures represent points of temporary attachment for the rear body during translocation (Klug and Hoffmann 2015). In any animal possessing a chambered shell, growth requires repeated detachment of the body from the shell, yet it is unlikely that the animal would have detached the entire body simultaneously. It is clear that *Nautilus* attaches to the mural ridge prior to septal formation, and it appears likely that extinct nautiloids, bactritoids, early coleoids, and many ammonoids did the same. Given the extremely similar morphology and ultrastructure of at least some ammonoid pseudosutures to the mural ridge (Polizzotto et al. 2007), it is possible that pseudosutures also represent points of temporary attachment, at least in some instances. As the occurrence of pseudosutures at particular points along the suture is remarkably consistent within species, this corresponds well to the tie-point hypothesis of septal morphogenesis, first proposed by Seilacher (1975, 1988).

Some have interpreted drag lines as candidates for these tie points (Zaborski 1986; Seilacher 1988); others, however, point out that the coincidence of drag lines with the flanks of the lobules rather than the tips, and their paired occurrence, suggests that drag lines are more likely fused, telescoped pseudosutures (Hewitt et al. 1991; Checa and Garcia-Ruiz 1996; Klug et al. 2007; Klug and Hoffmann 2015). Polizzotto and Landman (2010) reported several different drag lines in a single chamber, none of which were continuous with the pseudosutures in the same chamber. Additionally, some of the drag lines in this specimen were paired and apparently diverged from a single drag line adapically (Polizzotto and Landman 2010, Fig. 5, 6), while other, single drag lines continued nearly all the way to the lobule before ramifying into a short series of concentric ridges at the base of the lobule (Polizzotto and Landman 2010, Fig. 7). Based on observations from all these different specimens, it may be that drag lines formed in more than one way, but in every case they represent a point at which the rear mantle was in contact with (and possibly attached to) the inner shell wall.

An alternative interpretation of pseudosutures proposes that rather than acting as points of attachment, they may represent accumulations of secreted material at points determined by the interaction of the viscoelastic rear body and the varying pressure of cameral fluid and gas behind the body (Checa and Garcia-Ruiz 1996). The mantle did not necessarily attach at the location of the pseudosutures, but simply paused. This corresponds to the viscous fingering model of septal morphogenesis (Garcia-Ruiz et al. 1990; Garcia-Ruiz and Checa 1993; Checa and Garcia-Ruiz 1996). This model, however, would not explain the evidence of the attachment-like ultrastructure in at least some pseudosutures.

Klug et al. (2008) introduced a "*tension model*" of septal morphogenesis that incorporates elements of both the tie-point model and the viscous fingering model, in which muscle fibers at the edge of the rear mantle attached to the inner shell, and the more complex the shape of the septum, the more tension could develop in the rear mantle and in the organic pre-septum prior to mineralization. While Klug et al. (2008) did not elaborate specifically on the consequences of this model for the formation of pseudosutures, the model implies that the shape of temporary attachment points would have depended on components of translocation that were not so much genetically influenced, but mainly affected by changes in chamber pressurization. For more details on septum formation see Klug and Hoffmann (2015).

In addition to implications for septal morphogenesis, the number, placement, and spacing of pseudosutures may indicate the pace and timing of chamber formation. These interpretations, however, rely on the assumption that pseudosutures were secreted periodically, which is currently not possible to verify. These pauses, if they are such, might also correspond to growth rhythms recorded in growth lines or lirae formed at the aperture; however, this hypothesis has not yet been tested. There does not seem to be any analogous structure or process in *Nautilus* or *Sepia*, leaving us with little to substantiate the periodicity of pseudosuture formation one way or the other. The only similar structure seen in modern nautilids are spiral drag lines, possibly representing imprints of mantle muscle fibers and/or fine arteries in the mantle (Klug et al. 2008).

In summary, it appears plausible that pseudosutures record short pauses in forward movement during the chamber formation cycle, and it is likely that at least some pseudosutures (or pseudosutures in some taxa) represent the imprints of ephemeral soft-tissue attachment. Drag lines do not offer any evidence of the pace of translocation or of pauses, but they do show that at least some parts of the mantle were in continuous contact with the shell wall and may have formed attachment points. Such attachments may have helped to anchor parts of the animal in the body chamber during translocation, as well as helping to maintain the fundamental shape of the rear mantle in between formation of consecutive septa.

4.4 Other Soft Tissue Imprints

Other soft tissue imprints, particularly on the septum, have been recently described in ammonoids by Klug et al. (2008) (Fig. 4.6). These include the septal furrow and associated subparallel furrows (extending from the mid-dorsal suture to the siphuncular perforation); striations on the mural band and on the annular elevation; the conchal furrow on the venter (especially on the body chamber); and deformed septa (non-taphonomic deformities). These specific features had previously been reported only in nautiloids and bactritoids (Klug et al. 2008 and references therein), and were interpreted as the imprints of muscle fibers and blood vessels in the septal mantle. The presence of these features in nautiloids, bactritoids, and early ammonoids (Devonian), and their apparent absence in more derived ammonoids (with the exception of the conchal furrow, which appears in Cretaceous ammonites, Landman and Waage 1993), led Klug et al. (2008) to hypothesize that higher tension in the organic pre-septum due to the higher order septal folding may have prevented imprinting of soft tissue structures (see the summary of Klug et al.'s tension model of septal formation above).

Klug et al. (2007) found a black layer on the dorsal surface of some ammonoid shells (see also Keupp 2000). This layer is presumably originally organic and most likely similar to the black layer found in recent and fossil nautiloids.

Fig. 4.6 Soft-tissue imprints in the shell and septa of Early Devonian ammonoids, modified after Klug et al. (2008). **a–c**, *Erbenoceras advolvens* (Erben 1960), PIMUZ 7494, early Emsian, Hassi Chebbi, Tafilalt, Morocco, length 13 mm. **a** Septal view to show the septal furrow. **b** Lateral view.

4.4.1 Blood Vessel Imprints

Some septa of the Early and Middle Devonian ammonoids *Chebbites*, *Erbenoceras*, *Gracilites*, *Metabactrites*, *Rherisites*, and *Pinacites* show long wrinkles on the surface (Klug et al. 2008). Some of these wrinkles lie in the plain of symmetry (septal furrow; Stenzel 1964; Teichert 1964; Chirat and von Boletzky 2003) and some diverge from the plain of symmetry at a low angle, originating near the siphuncle (Fig. 4.6). Similar shallow imprints can be seen in modern nautilids: In some *Nautilus* conchs, Klug et al. (2008) found imprints of the left and right septal arteries, the siphuncular artery and the accessory siphuncular arteries (see also Deecke 1913, and Stenzel 1964, who figured a *Nautilus* septum showing soft tissue imprints). Corresponding to this fact, the imprints on the septa of the Devonian ammonoids were interpreted as imprints of arteries, providing the septal mantle with arterial blood. The absence of these structures in more derived ammonoids has been linked with the increase in septal frilling, which might be linked with a higher tension of the organic septum prior to mineralization. Accordingly, the more tightly stayed pre-septum would have prevented the formation of soft-tissue impressions in the septum.

On the inside of the shell wall, both nautiloids and ammonoids of various ages show the conchal furrow ("Fadenkiel" of von Bülow 1918; see also Teichert 1964; Shimanskij 1974; Landman and Waage 1993; Chirat and von Boletzky 2003; Keupp 2012). This furrow can be traced in nautilids from the cicatrix to the aperture, while in ammonoids, no corresponding structure has yet been found on or near the initial chamber. In ammonoids, it is rarely seen but throughout the entire phylogeny of ammonoids from the Early Devonian to the Late Cretaceous (Landman and Waage 1993; Klug et al. 2008). Doguzhaeva and Mutvei (1996) suggested that this midventral longitudinal elevation was the "attachment site for a ligament or a muscle used to maintain the shape and position of the circumsiphonal invagination during the growth and forward migration" of the soft body (Chirat and von Boletzky 2003, p. 168). Griffin (1900, fig. 11) illustrated the "lesser aorta and its branches", including the midventrally running pallial artery. Potentially, it may have been this artery that occasionally left a mid ventral imprint on the inside of the shell of various ectocochleates.

---

c *Ventral* view. **d** *Chebbites reisdorfi* Klug (2001), PIMUZ 27000, septal view, early Emsian, Hassi Chebbi, Tafilalt, Morocco; note the septal furrow and additional subparallel furrows on the right; length 15 mm. **e, f,** *Rherisites* sp., PIMUZ 27072, early Emsian, Hassi Chebbi, northern Tafilalt, Morocco; length 16 mm. **e** Ventral view, note the conchal furrow. **f** Lateral view, note the epizoans. G, H, The ammonoid *Gracilites* sp., PIMUZ 27070, early Emsian, Hassi Chebbi, northern Tafilalt, Morocco; length 30 mm. **g** Lateral view, note the pseudosutures. **h** Septal view, note the striations that run subparallel to the plane of symmetry. **i-l,** *Metabactrites ernsti* Klug et al. (2008), PIMUZ 7404, early Emsian, Ouidane Chebbi, Tafilalt, Morocco, length 25 mm. **i** detail of note the septal furrow, the striation on the mural part of the septum and the striation in the body chamber (probably within the attachment site of the dorsal muscle); height of detail 1.6 mm. **j** Ventral view. **k** Lateral view. **l** Dorsal view. **m** Agoniatitida gen. et sp. indet., MB.C.0782, late Emsian, Wissenbach Slate, Wissenbach, Germany

4.4.1.1 Muscle Imprints

Similar to the above mentioned arteries, bundles of muscle fibers in the posterior mantle might have left imprints on the septum as well as on the inside of the outer shell, as noted in some Devonian ammonoids (Klug et al. 2008). This can also be seen in nautilids (Deecke 1913; Stenzel 1964), where the orientation of some imprints coincides with the orientation of muscle fibers in the septal mantle (Klug et al. 2008). In agreement with the findings in modern nautilids, both the spirally arranged drag lines on the inside of the shell and the radially arranged impressions on the margin of the septum on its mural part (Fig. 4.6i) were interpreted as imprints of mantle musculature (Klug et al. 2008). Of course, attachment scars of various larger muscles are occasionally seen in well preserved ammonoid specimens. These structures are described in Doguzhaeva and Mapes (2015).

4.4.1.2 Other Imprints

From a few ammonoid taxa, tension wrinkles have been described (Checa and Garcia-Ruiz 1996; Klug et al. 2007). These wrinkles are one to several micrometers wide and ten to several tens of micrometers long. They are situated at the mural part of the septum. According to Checa and Garcia-Ruiz (1996), these wrinkles document the flexibility of the organic membrane prior to the mineralization of the septum. For an illustration, see Klug and Hoffmann (2015).

Acknowledgements The authors are grateful to Helmut Keupp (Berlin) and Anthea Lacchia (Dublin) for helpful reviews of the manuscript. Some of the results presented herein were obtained in the course of the research projects funded by the Swiss National Science Foundation SNF, #200021–113956/1, #200020–25029, and #200020–132870. Some of the images were kindly provided by Hans Hagdorn (Ingelfingen), David Ware (Zürich), René Hoffmann (Bochum) and Wolfgang Weitschat (Hamburg). We greatly acknowledge these contributions.

References

Bandel K (1981) The structure and formation of the siphuncular tube of *Quenstedtoceras* compared with that of *Nautilus* (Cephalopoda). N Jahrb Geol Paläont Ab 161:153–171

Bandel K (1982) Morphologie und Bildung der frühontogenetischen Gehäuse bei conchiferen Mollusken. Facies 7:1–198

Bandel K, Boletzky SV (1979) A comparative study of the structure, development, and morphological relationships of chambered cephalopod shells. Veliger 21:313–354

Bayer U (1975) Organische Tapeten im Ammoniten-Phragmocon und ihr Einfluß auf die Fossilisation. N Jahrb Geol Paläont Mh 1:12–25

Bayer U (1977) Cephalopoden-Septen. I. Konstruktionsmorphologie des Ammoniten-Septums. N Jahrb Geol Paläont Abh 154:290–366

Bucher H, Landman NH, Klofak SM, Guex J (1996) Mode and rate of growth in ammonoids. In: Landman NH, Tanabe K, Davis RA (eds) Ammonoid Paleobiology. Plenum, New York

Bülow Ev (1918) Über einige abnorme Formen bei den Ammoniten. Zeitschrift der deutschen geologischen Gesellschaft. Monatsber 69:132–139

Checa AG (1996) Origin of intracameral sheets in ammonoids. Lethaia 29:61–75

Checa AG, Garcia-Ruiz JM (1996) Morphogenesis of the septum in ammonoids. In: Landman NH, Tanabe K, Davis RA (eds) Ammonoid Paleobiology. Plenum, New York

Chirat R, Boletzky SV (2003) Morphogenetic significance of the conchal furrow in nautiloids: Evidence from early embryonic shell development of Jurassic Nautilida. Lethaia 36:161–170

Daniel TL, Helmuth BS, Saunders B, Ward PD (1997) Septal complexity in ammonoid cephalopods increased mechanical risk and limited depth. Paleobiology 23:470–481

Deecke W (1913) Paläontologische Betrachtungen. I. Über Cephalopoden. N Jahrb Min Geol Paläont Beilageband 35:241–276

Doguzhaeva LA, Mapes RH (2015) Muscle scars in ammonoid shells. This volume

Doguzhaeva LA, Mutvei H (1996) Attachment of the body to the shell in ammonoids. In: Landman NH, Tanabe K, Davis RA (eds) Ammonoid Paleobiology. Plenum, New York

Erben HK, Reid REH (1971) Ultrastructure of shell, origin of conellae, and siphuncular membranes in an ammonite. Biomineralization Res Rep 3:22–31

Erven HK (1960) Primitive Ammonoidea aus dem Unterdevon Frankreichs und Deutschlands. N Jahrb Geol Paläont Abh 110(1):1–128

Garcia-Ruiz JM, Checa AG (1993) A model for the morphogenesis of ammonoid septal sutures. Geobios. Mém Spéc 15:157–162

Garcia-Ruiz JM, Checa AG, Rivas P (1990) On the origin of ammonite sutures. Paleobiology 16:349–354

Grandjean F (1910) Le siphon des ammonites et des belemnites. Bull Soc Géol France 10:496–519

Grégoire C (1984) Remains of organic components in the siphonal tube and in the brown membrane of ammonoids and fossil nautiloids. Hydrothermal simulation of their diagenetic alterations. Akad Wiss Lit: Abh Math-Naturwiss Kl Mainz 5:5–56

Grégoire C (1987) Ultrastructure of the *Nautilus* shell. In: Saunders WB, Landman NH (eds) Nautilus-the biology and paleobiology of a living fossil. Plenum, New York

Griffin LE (1900) The anatomy of *Nautilus pompilius*. Memoirs of the. Natl Acad Sci 8:101–203

Hagdorn H, Mundlos R (1983) Aspekte der Taphonomie von Muschelkalk-Cephalopoden. Teil 1: Siphozerfall und Füllmechanismus. N Jahrb Geol Paläont Abh 16:369–403

Henderson RA (1984) A muscular attachment proposal for septal function in Mesozoic ammonites. Palaeontology 27:461–486

Henderson RA, Kennedy WJ, Cobban WA (2002) Perspectives of ammonite paleobiology from shell abnormalities in the genus *Baculties*. Lethaia 35:215–230

Heptonstall WB (1970) Buoyancy control in ammonoids. Lethaia 3:317–328

Hewitt RA, Westermann GEG (1996) Post-mortem behavior of early Palaeozoic nautiloids and paleobathymetry. Paläontol Z 70(3/4):405–424

Hewitt RA, Westermann GEG, Zaborski PMP (1991) Chamber growth in ammonites inferred from colour markings and naturally etched surfaces of Cretaceous vascoceratids from Nigeria. Lethaia 24:271–287

Hölder H (1952) Über Gehäusebau, insbesondere Hohlkiel jurassicher Ammoniten. Paläontogr A 102:18–48

Hölder H (1954) Über die Sipho-Anheftung bei Ammoniten. N Jahrb Geol Paläont Mh 8:372–379

John R (1909) Über die Lebenweise und Organisation des Ammoniten. Inaugural-Dissertation, Universität Tübingen, Stuttgart

Keupp H (1992) Organische Lamellen in einem Ammoniten-Gehäuse (*Craspedites*). Fossilien 3/1992:283–290

Keupp H (2000) Ammoniten, paläobiologische Erfolgsspiralen. Thorbecke, Stuttgart

Keupp H (2012) Atlas zur Paläopathologie der Cephalopoden. Berliner Paläobiologische Abh 12:1–390

Klug C, Hoffmann R (2015) Ammonoid septa and sutures. This volume

Klug C, Montenari M, Schulz H, Urlichs M (2007) Soft-tissue attachment of Middle Triassic Ceratitida from Germany. In: Landman NH, Davis RA, Tanabe K (eds) Cephalopods present and past: new insights and fresh perspectives. Springer, Netherlands

Klug C, Meyer EP, Richter U, Korn D (2008) Soft-tissue imprints in fossil and Recent cephalopod septa and septum formation. Lethaia 41:477–492

Kröger B (2002) On the efficiency of the buoyancy apparatus in ammonoids: evidences from sublethal shell injuries. Lethaia 35:61–70

Kulicki C (1979) The ammonite shell: its structure, development, and biological significance. Acta Palaeontol Polonica 39:97–142

Kulicki C (1996) Ammonoid shell microstructure. In: Landman NH, Tanabe K, Davis RA (eds) Ammonoid Paleobiology. Topics in Geobiology. Plenum, New York

Kulicki C, Mutvei H (1988) Functional interpretation of ammonoid septa. In: Wiedmann J, Kullmann J (eds) Cephalopods in present and past. Schweitzerbart, Stuttgart, pp. 215–228

Landman NH, Waage KL (1993) Scaphitid ammonoids of the Upper Cretaceous (Maastrichtian) Fox Hills Formation in South Dakota and Wyoming. Bull Am Mus Nat Hist 215:1–257

Landman NH, Tanabe K, Mapes RH, Klofak SM, Whitehill J (1993) Pseudosutures in Paleozoic ammonoids. Lethaia 26:99–100

Landman NH, Polizzotto K, Mapes RH, Tanabe K (2006) Cameral membranes in prolecanitid and goniatitid ammonoids from the Permian Arcturus Formation, Nevada, USA. Lethaia 39:365–379

Lominadze T, Sharikadze M, Kvantaliani I (1993) On mechanism of soft body movement within body chamber in ammonites. Geobios Mém spéc 15:267–273

Mapes RH, Landman NH, Tanabe K, Maeda H (2002) Intracameral membranes in Permian ammonoids from the Buck Montain, Nevada Lagerstätte. GSA Abstracts with program 34:354

Mutvei H (1963) Structure of siphonal tube in Recent and fossil cephalopods. Paläontol Z 37(1–2):16

Mutvei H (1967) On the microscopic shell structure in some Jurassic ammonoids. N Jahrb Geol Paläont Abh 129:157–166

Mutvei H, Reyment R (1973) Buoyancy control and siphuncle function in ammonites. Palaeontology 6:623–636

Polizzotto K (2010) Pseudosutures in *Baculites mariasensis*. 8th International Symposium on Cepahlopods-Present and Past Abstracts Volume, p. 86

Polizzotto K (2014) Organic origin of pseudosutures in Late Cretaceous ammonites. 10th North American Paleontological Convention abstracts book. Paleontological Soc Spec Pub 13:89

Polizzotto K, Landman NH (2010) Pseudosutures and siphuncular membranes in *Rhaeboceras* (Scaphitidae): implications for chamber formation and shell growth. In: Tanabe K, Shigeta Y, Sasaki T, Hirano H (eds) Cephalopods present and past. Tokai University, Tokyo

Polizzotto K, Landman NH, Mapes RH (2007) Cameral membranes in Carboniferous and Permian goniatites: description and relationship to pseudosutures. In: Landman NH, Davis RA, Tanabe K (eds) Cephalopods present and past: New insights and fresh perspectives. Springer, Netherlands

Reyment RA (1973) Factors in the distribution of fossil cephalopods. Part 3: experiments with exact models of certain shell types. Bull Geological Inst Univ Uppsala New Ser 4(2):7–41

Richter U (2002) Gewebeansatz-Strukturen auf Steinkernen von Ammonoideen. Geologische Beiträge Hannover 4:113

Richter U, Fischer R (2002) Soft tissue attachment structures on pyritized internal moulds of ammonoids. Abh der Geol BA 57:139–149

Schindewolf O (1968) Analyse eines Ammoniten-Gehäuses. Akad Wiss und der Lit. Abh Math-Naturwiss Kl Mainz 8:139–188

Schoulga-Nesterenko M (1926) Nouvelles données sur l'oranisation intérieure des conques des ammonites de l'étage d'Artinsk. Bulletin de la Societé des Naturalistes de Moscou Section Géologique 34:81–99

Seilacher A (1968) Sedimentationsprozesse in Ammonitengehäusen. Akad Wiss und der Lit Abh Math-Naturwiss Kl Mainz 9:191–203

Seilacher A (1975) Mechanische Simulation und funktionelle Evolution des Ammoniten-Septums. Palaontol Z 49:268–286

Seilacher A (1988) Why are nautiloid and ammonoid sutures so different? N Jahrb Geol Paläont Abh 177(l):41–69

Seuss B, Mapes RH, Klug C, Nützel A (2012) Exceptional cameral deposits in a sublethally injured Carboniferous orthoconic nautiloid from the Buckhorn Asphalt Lagerstatte in Oklahoma, USA. Acta Pal Pol 57:375–390

Shimanskij VN (1974) Superorder Actinoceratoidea. General section. In: Orlov YA, Ruzhencev VE (eds) Fundamentals of Paleontology (Osnovy paleontologii) 5:323–352 Mollusca-Cephalopoda I, Israel Program for Scientific Translation, Jerusalem (translated from Russian)

Stenzel HB (1964) Living nautilus. In: Teichert C (ed) Treatise on invertebrate paleontology, part K, Mollusca 3, Cephalopoda. General features—Endoceratoidea—Actinoceratoidea—Nautiloidea-Bactritoidea, vol. 1. GSA, Boulder, Colorado, and University of Kansas, Lawrence

Tajika A, Naglik C, Morimoto N, Pascual-Cebrian E, Hennhöfer DK, Klug C (2014) Empirical 3D-model of the conch of the Middle Jurassic ammonite microconch *Normannites*, its buoyancy, the physical effects of its mature modifications and speculations on their function. Historical Biol 20. doi:10.1080/08912963.2013.872097

Tanabe K, Landman NH (1996) Septal neck-siphuncular complex. In: Landman NH, Tanabe K, Davis RA (eds) Ammonoid Paleobiology. Plenum, New York

Tanabe K, Fukuda Y, Obata I (1982) Formation and function of the siphuncle-septal neck structures in two Mesozoic ammonites. Transactions Proc Palaeontological Soc Japan New Ser 128:433–443

Tanabe K, Landman NH, Mapes RH (1998) Muscle attachment scars in a Carboniferous goniatite. Paleont Res 2:130–136

Tanabe K, Mapes RH, Sasaki T, Landman NH (2000) Soft-part anatomy of the siphuncle in Permian prolecanitid ammonoids. Lethaia 33:83–91

Tanabe K, Kulicki C, Landman NH (2005) Precursory siphuncular membranes in the body chamber of *Phyllopachyceras* and comparisons with other ammonoids. Acta Pal Pol 50:9–18

Teichert C (1964) Morphology of hard parts. In: Teichert C (ed) Treatise on invertebrate paleontology, part K, Mollusca 3, Cephalopoda general features—Endoceratoidea—Actinoceratoidea—Nautiloidea—Bactritoidea. GSA, Boulder, Colorado, and University of Kansas Press, Lawrence

Vogel KP (1959) Zwergwuchs bei Polyptychiten (Ammonoidea). Geol Jahrb 76:469–540

Ward PD (1979) Cameral liquid in *Nautilus* and Ammonites. Paleobiology 5:40–49

Ward PD (1987) The natural history of Nautilus. Allen and Unwin, Boston

Weitschat W (1986) Phosphatisierte Ammonoideen aus der Mittleren Trias von Central-Spitsbergen. Mitt Geol Paläont Inst Univ Hamburg 61:249–279

Weitschat W, Bandel K (1991) Organic components in phragmocones of boreal Triassic ammonoids: implications for ammonoid biology. Paläontol Z 65:269–303

Weitschat W, Bandel K (1992) Formation and function of suspended organic cameral sheets in Triassic ammonoids: reply. Paläontol Z 66:443–444

Westermann GEG (1971) Form, structure, and function of shell and siphuncle in coiled Meozoic ammonites. Life Sciences Contributions. ROM 78:1–39

Westermann GEG (1992) Formation and function of suspended organic cameral sheets in Triassic ammonoids-discussion. Paläontol Z 66:437–441

Zaborski PMP (1986) Internal mould markings in a Cretaceous ammonite from Nigeria. Palaeontology 29:725–738

Part II
Ontogeny

Chapter 5
Ammonoid Embryonic Development

Kenneth De Baets, Neil H. Landman and Kazushige Tanabe

5.1 Introduction

Reconstruction of the early ontogeny of extinct organisms is an important subject in paleobiological research and developmental biology (e.g., Donoghue and Dong 2005; Sánchez-Villagra 2012; Urdy et al. 2013). Non-mineralized embryos and larval soft-tissues of invertebrates are only preserved in exceptional circumstances (e.g., Donoghue et al. 2006; Maas et al. 2006), so that the most abundant information about early ontogeny comes from groups with mineralized hard parts secreted at the embryonic or larval stage like ammonoids (Tanabe et al. 2008) and other shelled mollusks (Jablonski and Lutz 1983; Nützel et al. 2006; Manda and Frýda 2010). As in other mollusk shells, the aragonitic outer shell wall of ammonoids was formed by accretionary growth, so that the embryonic shell prior to hatching (ammonitella of Drushchits and Khiami 1970) is occasionally preserved in the apical shell portion of larger ammonoids or more rarely as accumulations or isolated finds of embryonic or juvenile shells.

The embryonic development of ammonoids has attracted the attention of ammonoid workers since at least two centuries (e.g., Branco 1879, 1880; Hyatt 1894; Smith 1897, 1898, 1899, 1900, 1901; Grandjean 1910; Schindewolf 1932, 1933,

K. De Baets (✉)
GeoZentrum Nordbayern, Fachgruppe PaläoUmwelt, Universität Erlangen, Loewenichstr. 28, D-91054 Erlangen, Germany
e-mail: kenneth.debaets@fau.de

N. H. Landman
Division of Paleontology (Invertebrates), American Museum of Natural History, Central Park West at 79th St., NY 10024–5192, New York, USA
e-mail: landman@amnh.org

K. Tanabe
Department of Historical Geology and Paleontology, The University Museum, The University of Tokyo, Hongo 7-3-1, Tokyo 113-0033, Japan
e-mail: tanabe@um.s.u-tokyo.ac.jp

C. Klug et al. (eds.), *Ammonoid Paleobiology: From Anatomy to Ecology*,
Topics in Geobiology 43, DOI 10.1007/978-94-017-9630-9_5

1934; Spath 1933), particularly for larger scale systematic questions and evolutionary studies on development. Early ontogenetic development played a central role in the discussion of the derivation of ammonoids from coiled nautiloids (e.g., Spath 1933, 1936; Böhmers 1936; Donovan 1964) or straight bactritoids (e.g., Schindewolf 1932, 1933, 1934). Mainly due to the work of Erben (1960, 1962a, b, 1964, 1965, 1966) and Bogoslovsky (1969), it is now well accepted that ammonoids can be traced back to a bactritoid ancestor (Dzik 1981, 1984; Klofak et al. 1999; Doguzhaeva 2002; Klug and Korn 2004; De Baets et al. 2009; Kröger et al. 2011; Ritterbush et al. 2014), which in turn can be traced back to an orthoceratid ancestor (Kröger and Mapes 2007; compare Ristedt 1971, Dzik 1984, Holland 2003; Klug et al. 2015). This is supported by the presence of a similar early shell with a small, elliptical initial chamber and a transitional series linking all intermediate forms from straight orthocerid nautiloids, through straight to slightly curved bactritoids, to coiled ammonoids, which is also consistent with their stratigraphic appearance (Kröger and Mapes 2007; De Baets et al. 2013b).

Based on microscopic observations on embryonic shells, some authors (e.g., Erben 1964; Erben et al. 1968, 1969) have postulated that ammonoids went through a post-hatching larval phase before metamorphosis like modern gastropods and bivalves. With the discovery of more well-preserved ammonitellae and increasing knowledge of the early ontogeny of modern shelled mollusks, particularly modern *Nautilus* (e.g., Uchiyama and Tanabe 1999) and gastropods (Bandel 1982, 1985; Hickman 1992, 2004), it is now a generally accepted theory that, like all modern cephalopods, ammonoids developed directly without a larval stage (e.g., Drushchits and Khiami 1970; Drushchits et al. 1977b; Drushchits and Doguzhaeva 1981; Birkelund and Hansen 1974; Kulicki 1974, 1979, 1996; Tanabe et al. 1980; Tanabe and Ohtsuka 1985; Tanabe 1989; Landman 1982, 1985, 1987; Bandel 1982; Bandel et al. 1982; Westermann 1996; Klug 2001b; Sprey 2002; Tanabe et al. 2008; De Baets et al. 2012; Ritterbush et al. 2014). This theory of direct development is supported by a number of morphological features (synchronous changes of ornament, shell microstructure, whorl growth at the primary constriction) and discoveries of ammonitellae (e.g., Bandel 1982, 1986; Landman 1982, 1985; Kulicki and Wierzbowski 1983; Kulicki and Doguzhaeva 1994; Tanabe et al. 1993, 1995). The exact sequence of embryonic development, however, is still debated as various models have been suggested for derived ammonoids, ranging from an accretionary growth model (as in *Nautilus*: compare Drushchits et al. 1977b; Drushchits and Doguzhaeva 1981; Kulicki 1979; Tanabe et al. 1980, 1993), a model with an originally organic (non-mineralized) shell (similar to "archaeogastropods": compare Bandel 1982, 1986; Kulicki and Doguzhaeva 1994), or an endocochleate embryo model (Tanabe 1989).

The embryonic development of ammonoids was recently extensively reviewed by Landman et al. (1996). We will therefore not repeat what has been written before, but will update accounts on the description of the embryonic shell (including terminology, shape, size, ornamentation, microstructure, septa, siphuncle and muscle scars), the sequence of embryonic development, reproductive strategy, and post-hatching mode of life followed by conclusions and possible future areas of research. We will focus on studies that have been done in the last two decades.

Landman et al. (1996) pointed out that only little was known about the embryonic development of Paleozoic ammonoids compared with Mesozoic ones. Numerous studies have recently focused on the embryonic shell structure and development of Paleozoic ammonoids (House 1996; Landman et al. 1999; Klofak et al. 1999, 2007; Shigeta et al. 2001; Tanabe et al. 2001; Kulicki et al. 2002; Klofak and Landman 2010, 2012; De Baets et al. 2012, 2013b). Additional studies on Mesozoic ammonoids, particularly on embryonic shell structure, have also come out since 1996 (Sprey 2001, 2002; Landman et al. 2001; Shigeta et al. 2001; Rouget and Neige 2001; Tanabe et al. 2003, 2008, 2010). A larger database on embryonic shell measurements is now available for both the Paleozoic and the Mesozoic (given as a table here; compare De Baets et al. 2012; Laptikhovsky et al. 2013).

Recently, increased attention has been paid to the large intraspecific variability in embryonic development in ammonoids, particularly in Jurassic and Cretaceous forms (e.g., Rouget and Neige 2001; Tanabe et al. 2003; Nishimura et al. 2010; Tajika and Wani 2011; compare Tanabe et al. 1995; Stephen and Stanton 2002 for Paleozoic ammonoids). This phenomenon was qualitatively demonstrated by Erben (1950, 1962b, 1964) in the earliest ammonoids several decades before and is also well known from post-embryonic stages (De Baets et al. 2015).

New studies have also been performed on paleoecological and paleobiological aspects such as reproductive and egg-laying strategies (Klug 2001b, 2007; Etches et al. 2009; Walton et al. 2010; De Baets et al. 2012, 2013b) and facies distribution of embryonic or juvenile shells (Tomašových and Schlögl 2008; Mapes and Nützel 2009; Stephen et al. 2012). Some studies have also investigated links between early ontogeny and biogeographic distribution (Tajika and Wani 2011) and evolutionary longevity (Stephen and Stanton 2002), although these are still rare.

5.2 Description of the Ammonitella

5.2.1 Terminology

The terms used to describe morphological features of the early whorls are illustrated on Fig. 5.1. The same terms are used for all ammonoids, although the size and shape of these features differ between taxa, particularly in Early to Middle Devonian forms (compare 5.2.2 and 5.2.3). The ammonitella is defined as the shell up to the end of the primary constriction (Drushchits and Khiami 1969, 1970; Drushchits et al. 1977a, b; Tanabe et al. 1980; Birkelund 1981; Landman 1987), and is considered by most authors to represent the embryonic shell.

The initial chamber ("Anfangskammer": Branco 1879, 1880; Schindewolf 1933; Erben 1960; "first whorl": Bandel 1982) refers to the portion of the ammonitella up to the proseptum (first septum). It is often known also as the protoconch (e.g., Owen 1878; Hyatt 1894; Miller 1938; House 1985), but the term should not be used to avoid confusion (Tanabe et al. 1994, 2008; De Baets et al. 2012) with the

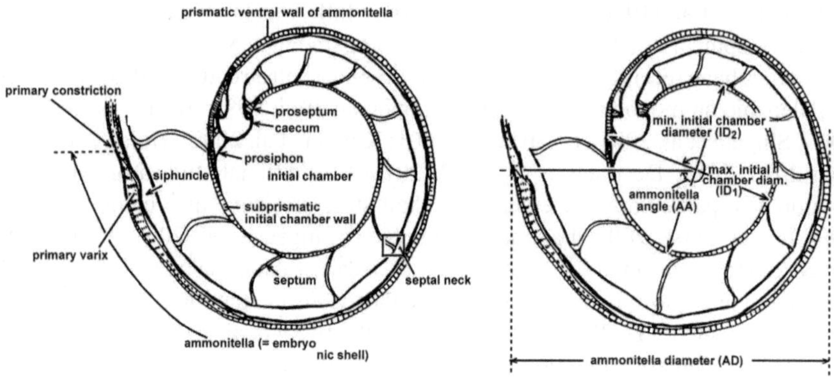

Fig. 5.1 Diagrams of the ammonoid early internal shell structures in median section, showing the terminology (left) and measurements (right). Abbreviations for measurements are indicated in *parentheses* (modified from Tanabe and Ohtsuka 1985, Fig. 1 and Tanabe et al. 2003, Fig. 1)

gastropod protoconch, which is not homologous as it can represent both embryonic and/ or larval shell (Frýda et al. 2008; Nützel 2014). Some authors have resolved this confusion by referring to the initial chamber as the "ammonoid protoconch" (House 1996). We herein will use the terms initial chamber and ammonitella. The embryonic shell without the initial chamber can be referred to as the ammonitella coil (in derived ammonoids) or as ammonitella shaft in early Emsian ammonoids like Anetoceratinae (Mimosphinctidae: Fig. 5.2a, d), where it is still straight or only slightly curved (e.g., De Baets et al. 2012, 2013b).

5.2.2 Shape

The initial chamber has a circular to lenticular outline in transverse cross-section and its shape ranges from globular (egg-shaped) to spindle-like. The earliest forms have an egg-shaped initial chamber and the succeeding whorls are loosely coiled or even straight, leaving an umbilical perforation. The most basal taxa such as *Metabactrites* (Fig. 5.2d), *Ivoites*, and *Erbenoceras* have an almost straight embryonic shell shaft, and loosely coiled post-embryonic whorls (De Baets et al. 2013b); the umbilical window is very large and not yet enclosed (open umbilical window sensu De Baets et al. 2012, where the initial chamber does not touch the subsequent whorl; Fig. 5.3). Early representatives of other slightly more derived lineages such as *Gyroceratites* (Mimoceratidae; Fig. 5.2f), *Irdanites* and *Convoluticeras* (Paleogoniatitinae: Fig. 5.2b, c), and *Teicherticeras* (Teicherticeratinae) have an open umbilical window (Klug 2001a; Korn and Klug 2002; De Baets et al. 2012, 2013b). In more derived taxa of these lineages, the initial chamber touches the subsequent whorl (enclosed umbilical window), although these forms occasionally still have a secondary "opening" between the embryonic shell and the subsequent whorl due to the less-curved embryonic shell (Bogoslovsky 1969; compare Fig. 5.3). All known em-

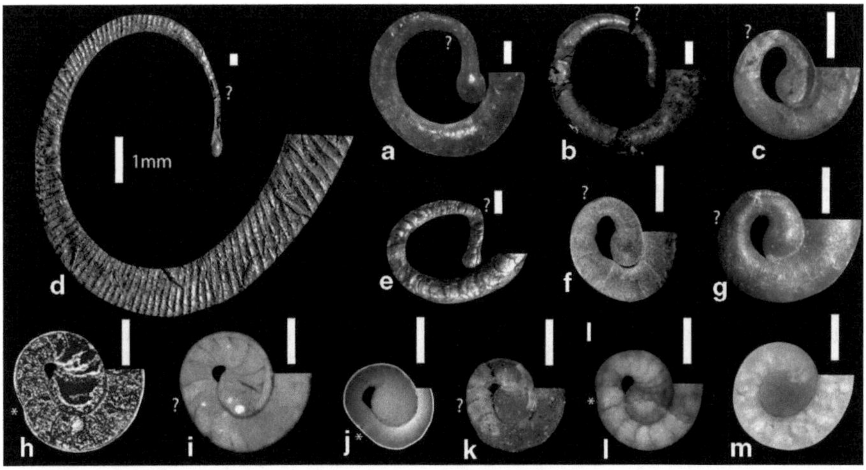

Fig. 5.2 Embryonic shells of the geologically youngest ammonoids of late Early and early Middle Devonian age (after De Baets et al. 2012). They are arranged more or less according to their stratigraphic sequence. Note the reduction of the embryonic shell size and the size of the umbilical window. **a** Mimosphinctidae, *Mimosphinctes khanakasuense*, PIN 3762/4. **b** Teicherticeratidae, *Irdanites kaufmanni*, GPIT 1849/2069. **c** Teicherticeratidae, *Convoluticeras lardeuxi*, GPIBo 10. **d** Mimosphinctidae, *Metabactrites fuchsi*, PWL2010/5251-LS. **e** Mimoceratidae, *Gyroceratites heinricherbeni*, PWL2010/1-LS. **f** Mimoceratidae, *Gyroceratites* cf. *laevis*, PIMUZ 29566. **g** Auguritidae, *Gaurites sperandus*, PIN 3981/27. **h** Latanarcestidae, *Latanarcestes* aff. *noeggerathi* auct., MB.C. 22202. **i** Anarcestidae, *Sellanarcestes cognatus*, L19729. **j** Anarcestidae, *Anarcestes lateseptatus*, AMNH 45372 (Klofak). **k** Anarcestidae, *Paranarcestes pictus*, L19440. **l** Werneroceratidae, *Praewerneroceras suchomastense*, L19443 **m** Werneroceratidae, *Fidelites fidelis*, L19824. The end of the embryonal shell is marked by an asterisk (clear case) or a question mark (ill-defined), where possible

bryonic shells of the Mimosphinctidae, Mimoceratidae, and Teicherticeratidae have an embryonic shell coil, which completes less than one-half whorl (Fig. 5.2; Erben 1964; Bogoslovsky 1969; House 1996). In more derived taxa such as *Mimagoniatites* and *Archanarcestes* (Mimagoniatitidae) as well as *Anarcestes* and *Sellanarcestes* (Anarcestidae; Fig. 5.2i, j), the embryonic shell coil is about one-half whorl, but the umbilical window is still partially surrounded by the post-embryonic whorls (Klofak et al. 1999, 2007; De Baets et al. 2012). In more derived taxa, the umbilical window becomes fully enclosed by the embryonic shell coil (e.g., *Praewernoceras*: Fig. 5.2l) and eventually disappears (e.g., *Fidelites*: Fig. 5.2m).

More derived taxa from the Middle Devonian to Late Cretaceous typically have a barrel- to spindle-shaped initial chamber, where the succeeding whorls are closely coiled around the initial chamber leaving a shallow to deep dorsal impression in the whorl profile as seen in transverse cross section. Even in heteromorph ammonoids from the Mesozoic, the embryonic shells are tightly coiled (without an umbilical perforation: Drushchits et al. 1977b; Dietl 1978; Landman 1982; House 1996; De Baets et al. 2012).

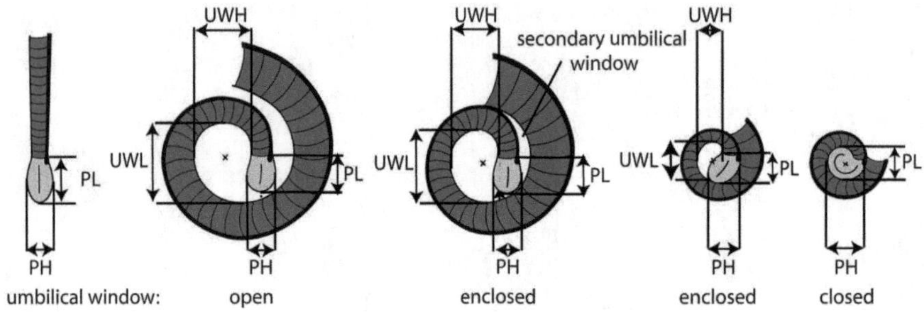

Fig. 5.3 Morphology, terminology and measurements of the initial chamber and umbilical window in bactritoids and Early and Middle Devonian ammonoids (*PL* "protoconch" length; *PH* "protoconch" height, *UWL* umbilical window length, *UWH* umbilical window height). From left to right: *Bactrites*, *Teicherticeras*, *Chebbites*, *Gyroceratites*, *Agoniatites* (modified after Bogoslovsky 1969; De Baets et al. 2012). The *x* marks the approximate position of the initial coiling axis

A second change in shell shape occurs at the primary constriction, which marks the end of the ammonitella (approximately one whorl from the start of the initial chamber in closely coiled ammonitellas). The primary or nepionic constriction typically appears as a groove in the shell wall, which is especially well expressed along the venter and coincides with abrupt changes in coiling, whorl shape, ornamentation (5.2.4), and microstructure (5.2.5) in derived ammonoids (e.g., Birkelund 1981; Landman et al. 1996). Landman and Waage (1982) noted the importance of distinguishing the actual constriction in the shell wall (primary constriction), from the shell thickening at this point (primary varix) and from the trace of this thickening on the internal mold or steinkern (varix trace). The presence of a primary varix in the earliest ammonoids (Anetoceratinae) has not yet been adequately demonstrated (House 1996; De Baets et al. 2012, 2013b). This might be related to the preservation of these fossils as internal molds so that the primary varix might not have left any trace on the inner surface of the shell tube or alternatively, these forms might have lacked this structure altogether such as bactritoids, where the nacre does not form an abrupt swelling (Doguzhaeva 2002; Kröger and Mapes 2007; Kröger 2008). In bactritids, this change in shape corresponds with a gradual narrowing of the shell, followed by a widening. In early ammonoids, the end of the embryonic shell is usually near the onset of distinct ornamentation (ribbing), changes in whorl profile, coiling and/ or cracks, which might preferentially occur at the transition from the embryonic to the post-embryonic shell (House 1996; De Baets et al. 2012, 2013b).

5.2.3 *Size*

Traditionally, three types of measurements are taken from the median cross section of ammonitellae (Fig. 5.1): the initial chamber diameter (ID = protoconch diameter PD), the ammonitella diameter (AD) and the ammonitella angle (AA). Note that differences in embryonic shell measurements within and between fossil assemblages can be related to various factors including intrinsic intraspecific variation, ecophenotypic or geographic variation, phylogenetic variation, taphonomic biases, taxonomic uncertainty (due to missing post-embryonic shell), and simple errors in measurement (Rouget and Neige 2001; Stephen and Stanton 2002; De Baets et al. 2012). Measurement errors can be relatively substantial due to the small dimensions of embryonic shells (particularly in transverse sections when the section is not in the plane of symmetry). Most Mesozoic data are from the Triassic (Zakharov 1974, 1978; Alekseev et al. 1984; Arkadiev and Vavilov 1984a, b; Vavilov 1992; Arkadiev et al. 1993) and Cretaceous (Tanabe et al. 1979, 2003; Drushchits and Doguzhaeva 1981; Shigeta 1993; Landman and Waage 1993), comparatively less is known from the Jurassic where most studies focused on the Oppeliidae (Palframan 1966, 1967, 1969; Neige 1997; Rouget and Neige 2001). A fair amount of data is available for the Permo-Carboniferous (Bogoslovskaya 1959; Zakharov 1974; Tanabe et al. 1994; Shigeta et al. 2001; Stephen and Stanton 2002), but comparatively less in the Devonian (Chlupáč and Turek 1983; House 1996; De Baets et al. 2012).

The diameter of the initial chamber (ID = ID_1) is measured from the ventral edge of the proseptum through the center of the initial chamber to the opposite side (Landman et al. 1996; Fig. 5.1). Various authors (Drushchits et al. 1977a; Zakharov 1978; Drushchits and Doguzhaeva 1981; Alekseev et al. 1984; Arkadiev and Vavilov 1984a, b; Vavilov 1992; Landman and Waage 1993; Rouget and Neige 2001; Shigeta et al. 2001; Sprey 2002; Tanabe et al. 2003) also distinguished a first or "maximum" diameter and a second or "minimum" diameter of the initial chamber. The former (ID_1) corresponds with the initial chamber (ID) or protoconch diameter (PD) and passes through the proseptum, the latter (ID_2) is the distance perpendicular to the former measured through the center of the initial chamber (Drushchits et al. 1977a; Landman and Waage 1993; Tanabe et al. 2003; Fig. 5.1). The use of the terms maximum and minimum is a bit confusing as sometimes the maximum diameter is of equal dimensions or even smaller than the minimum diameter (compare De Baets et al. 2012). The smallest initial chambers have been reported from the Jurassic (*Distichoceras bicostatum*: ID_1: 0.24–0.30 mm, *Taramelliceras richei*: ID_1: 0.25–0.30 mm: Palframan 1966, 1967) and the Cretaceous (*Scaphites whitfieldi*: ID_1: 0.25–0.40 mm: Landman 1987). Drushchits et al. (1977b) reported even smaller protoconchs from the Cretaceous (*Aconeceras*: ID_1: 0.20–0.35 mm; *Ptychophylloceras*: ID_1: 0.22–0.28 mm), although these accounts could not be verified and are absent from subsequent compendia by these authors (e.g., Drushchits and Doguzhaeva 1981). The largest protoconchs reported so far are known from the Middle Devonian (House 1985, 1996; De Baets et al. 2012). House (1985, 1996) reported an initial chamber with a diameter of 1.7 mm in *Agoniatites* cf. *costulatus* and De Baets et al. (2012) reported an even larger initial chamber with a diameter up to 2.2 mm in *Agoniatites costulatus* (based on an illustration by Petter 1959). Several Devonian taxa have initial chamber diameters over 1 mm, which are only rarely reported from younger periods including the Permian (e.g., *Perrinites*: Landman et al. 1996; *Pseudohalorites* and *Yinoceras*: Zhou et al. 2002) and Cretaceous (*Tetragonites terminus*: Shigeta 1989; *Boreophylloceras*: Repin 1998; Igolnikov

2007). The record holder in the Mesozoic is the genus *Boreophylloceras* with an initial chamber diameter ranging from 1.14 mm (*B. densicostatum*: Igolnikov 2007) to 1.54 mm (*B. praeinfundibulum*: Repin 1998).

The width of the initial chamber (IW = PW) is best studied in isolated embryonic shells or those broken out from larger specimens (Branco 1879; Palframan 1967), which might explain why it has been comparatively less studied (e.g., Erben 1964; Bogoslovsky 1969; Palframan 1966, 1967, 1969; Mikhailova 1974; Zakharov 1978; Drushchits and Doguzhaeva 1981). Some workers have used transversal cross sections to study the width of the initial chamber (e.g., Landman 1987) or the whorl width of the ammonitella (e.g., Sprey 2002), although this yields only approximate values as it is very difficult to cut through the middle of the initial chamber or ammonitella in the proper orientation, particularly in larger, tightly coiled specimens. Drushchits and Khiami (1969) distinguished between elliptical ($IW - ID_1 < 0.1$ mm), ridge-like ($0.1 < IW - ID_1 < 0.2$ mm) and fusiform initial chambers ($IW - ID_1 > 0.2$ mm) in Cretaceous ammonoids, although it is probably better to work with the ratio between IW and ID_1 to distinguish initial chamber shapes (compare Landman 1987). The latter ratio is close to 1 in early ammonoids with circular initial chambers (or even below one in elongate, ellipsoidal forms), while it can reach up to 1.5 or more in forms with spindle-shaped initial chambers (Palframan 1966, 1967). On the internal molds of the geologically oldest ammonoids (Fig. 5.3; De Baets et al. 2012), it is easier to measure the length and width of the initial chamber, but most commonly, the initial chamber diameter is measured from illustrations published in the literature.

The ammonitella diameter (AD) is defined as the distance from the adoral end of the primary constriction through the center of the initial chamber to the opposite side of the ammonitella. It varies from small (AD < 1.0), through medium (1.0 < AD < 1.5) via large (1.5 < AD < 2) to very large (AD > 2.0). Some of the largest embryonic shells are known from the Early Devonian (3.74 mm in *Erbenoceras advolvens*: Bogoslovsky 1969; 3.99 mm in *Mimosphinctes zlichovensis*; 3.1 mm in *Mimosphinctes discordans*: compare Bogoslovsky 1969; House 1996; De Baets et al. 2012). The ammonitellae might have been even larger in more primitive ammonoids with estimates up to 5 mm in *Ivoites schindewolfi* and 6 mm in *Metabactrites fuchsi* (De Baets et al. 2013b), when the onset of coarse ornamentation is used as a criterion for the end of the embryonic shell (compare House 1996). De Baets et al. (2013b) documented fractures in various internal molds of *Ivoites opitzi* around 3 mm (before the start of coarse ornamentation at about 3.5 mm), which could be explained by the transition from the ammonitella to the post-embryonic shell acting as a natural weak zone (Landman 1985; Sprey 2002) yielding more conservative estimates of embryonic shell size (De Baets et al. 2012).

The ammonitella angle (AA) is defined as the angle from the ventral edge of the proseptum (edge of the initial chamber) to the adoral end of the primary constriction (Landman et al. 1996; this definition slightly differs from Grandjean 1910: compare Drushchits and Khiami 1970). In loosely coiled embryonic shells, the shell angle depends on the size of the initial chamber/ embryonic shell coil and also coiling, i.e., the number of whorls the ammonitella encompasses (Fig. 5.4). The angle

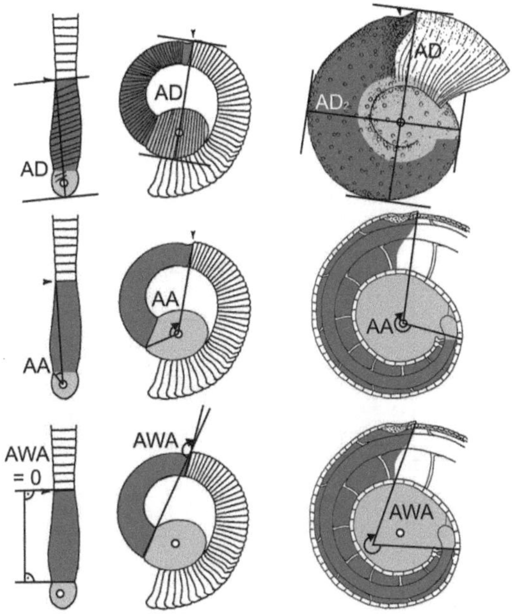

Fig. 5.4 Morphology of the ammonitella and commonly measured parameters of the ammonitella in the literature (*AD* ammonitella diameter, *AA* ammonitella angle, and *AWA* ammonitella whorl angle). From left to right: *Pseudobactrites*, *Mimagoniatites*, *Scaphites* (modified from Landman et al. 1996; De Baets et al. 2012). The initial chamber is marked in light gray and the ammonitella shaft or coil in dark gray. The white circles indicate the point where the two limbs delimiting the angle meet

between the initial chamber aperture and the minimum whorl height at the end of the embryonic shell (Amonitella Whorl Angle, AWA) might be a more appropriate measure for the number of whorls and would be independent of the size of the embryonic shell shaft/ whorl as well as the initial chamber size and shape as opposed to the AA. In bactritoids, AWA is zero, whereas AA is greater than zero and dependent on the length of the embryonic shell. However, AWA has not been used in the literature and to facilitate comparisons, we herein used the standard parameter (AA). It varies between small (<250°), through medium (250°<AA<350°) to large (AA>350°). AA ranges from small in early ammonoids to medium at the end of the Devonian in forms with coiled embryonic shells (De Baets et al. 2012). In loosely coiled forms, these changes are not only related to an increase of the length of the embryonic body chamber, but mostly to an increase in the degree of coiling and to a decrease in size of the embryonic shell. AA is mostly large in Permo-Carboniferous ammonoids (particularly Goniatitina: up to 410° in *Peritrochia erebus*: Tanabe et al. 1994) and mostly medium in the Mesozoic, although it can range from 223° (*Hypophylloceras ramosum*: Zakharov 1978) to 377° (*Phyllopachyceras ezoense*) in the Cretaceous Phylloceratina (Landman et al. 1996).

In the Devonian, there is a trend towards smaller, more tightly coiled embryonic shells from the Early to the Late Devonian (Fig. 5.5), which correlates with a decrease in AD and an increase in AA. This is also visible in a reduction of the umbilical window length and width (Fig. 5.3; De Baets et al. 2012). It is driven both by a preferential extinction of forms with larger embryonic shells during biotic events and by the progressive coiling of the embryonic shell within lineages (De Baets et al. 2012). A possible exception to this trend is a specimen of *Agoniatites* cf. *cos-*

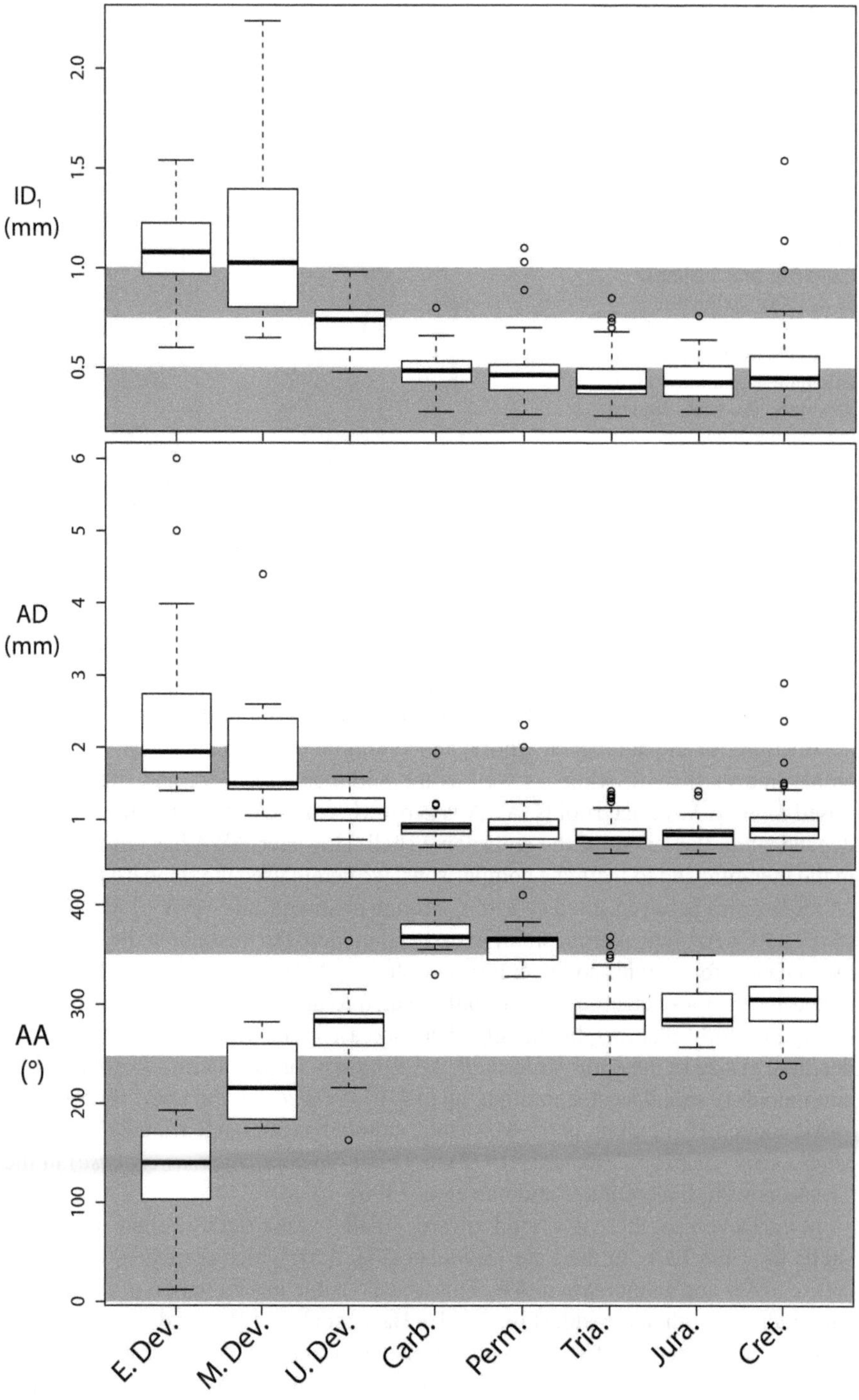

Fig. 5.5 Distribution and fluctuations of initial chamber (ID_1) as well as ammonitella size (*AD*) and coiling (*AA*) through time (updated from De Baets et al. 2012; whiskers = minima and maxima excluding outliers, boxes = middle two quartiles, horizontal stripe = median value, circles = outliers); *E. Dev*

tulatosus from the Middle Devonian, where House (1985) estimated an embryonic shell diameter of 4.4 mm, although other taxa of *Agoniatites* (where the ammonitella has been measured) have embryonic shell diameters between 2.3 and 2.6 mm (Erben 1964; Landman et al. 1996; De Baets et al. 2012). This trend is less clear in the evolution of initial chamber size, which reaches its largest diameter in the Middle Devonian (*Agoniatites*). In Post-Devonian times, the initial chamber and ammonitella diameter fluctuate much less; most ammonitellae measure between 0.5 and 1.5 mm (Landman et al. 1996), although it can range between 0.48 mm in *Distichoceras bauggeri* (compare Palframan 1967) to 2.9 mm in *Boreophylloceras praeinfundibulum* (Repin et al. 1998) in the Mesozoic. Forms with very large embryonic shells (AD>2 mm) are not restricted to the Devonian (Permian: 2.31 mm in *Perrinites* sp., Landman et al. 1996; Cretaceous: 2.37–2.9 mm in *Boreophylloceras*, Repin 1998; Igolnikov 2007). The ammonitella angle reached its maximum values in the Carboniferous-Permian (related with the high values of AA in Goniatitina and Prolecanitina: compare Landman et al. 1996) and decreases again in the Mesozoic (Fig. 5.5).

The diameter of the initial chamber diameter and ammonitella did, however, also fluctuate after the Devonian, particularly when looking at a higher stratigraphic resolution (e.g., Laptikhovsky et al. 2013; Table 5.1). Landman et al. (1996) reported an increase of the ammonitella diameter from the Middle Jurassic to the Late Cretaceous within the Lytoceratina. Tajika and Wani (2011) suggested that this might even have been the case within single lytoceratin taxa in the Cretaceous. Laptikhovsky et al. (2013) attributed the large fluctuations in embryonic shell and initial chamber sizes of fossil cephalopods in space and time throughout the Phanerozoic to changes in temperature, although it cannot explain the trend in the Devonian (compare De Baets et al. 2012). In extant cephalopods, hatchling size and embryonic development does not only depend on egg size, but also on developmental temperature and other individual hatching conditions (Boletzky 2003). Currently, the available measurements are heavily biased towards certain regions, taxa and timeframes (Table 5.1), so that this hypothesis needs to be further tested by quantitative analyses taking phylogenetic, biogeographic or taxonomic biases into account. Furthermore, if ammonoids migrated far during their life from hatching to adulthood, their final resting place may reflect, where the later ontogenetic stages died (if post-mortem drift can be ruled out) and not necessarily where they hatched; this hampers linking the paleobiogeographic distribution of the fossils with the habitat where they hatched.

In cases of more recently erected taxa, the lower order taxonomy follows the original publications, while higher order classification of the most closely related taxon was chosen. Paleozoic taxonomy was crosschecked with the Ammon Database, but no similar database is available for Mesozoic ammonoids. We are aware that some change has occurred in classification of some of these taxa, but revision

Early Devonian, *M. Dev.* Middle Devonian, *L. Dev.* Late Devonian, *Carb.* Carboniferous, *Perm.* Permian, *Tria* Triassic, *Jura* Jurassic, *Cret.* Cretaceous. Histograms were produced in the freely available statistical environment R (http://www.r-project.org/). The *grey/white* areas demarcate the subdivisions between small, medium, large to very large for ID_1 and *AD* as well as small, medium to large for the *AA* as defined by Landman et al. (1996)

Table 5.1 Compilation of initial chamber (ID_1) and ammonitella (AD, AA) measurements of Devonian—Cretaceous ammonoids (updated from Landman et al. 1996; De Baets et al. 2012). Minimum, mean and maximum are listed, when available; when only one value is available, this is listed under the mean. Taxa are grouped and ordered alphabetically per geological period. Species are tentatively assigned to families according to the latest comprehensive published ammonoid classification for each period: Korn and Klug (2002) for the Devonian; Furnish et al. (2009) for the Carboniferous-Permian; Tozer (1981) as modified by Hewitt et al. (1993) and Page (1996) for the Triassic; Donovan et al. (1981) for Jurassic Ammonitina as modified by Page (1996; except Psiloceratoidea, Eodoceratoidea, Hildoceratoidea: Howarth 2013); Wright (1996) for the Cretaceous. ID1 and AD in mm and AA in degrees

Taxon	Age	ID_1			AD			AA			Reference	Family
		min	mean	max	min	mean	max	Min	mean	max		(-t. = -tidae)
Achguigites tafilaltensis	Dev.		1.20								Klug 2002	Agoniatit.
Agoniatites cf. *costulatus*	Dev.		1.70			4.40					House 1985, 1996	Agoniatit.
Agoniatites costulatus	Dev.		2.24								Petter 1959; De Baets et al. 2012	Agoniatit.
Agoniatites fulguralis	Dev.		1.60			2.50			282		Erben 1964; Landman et al. 1996	Agoniatit.
Agoniatites holzapfeli	Dev.		1.40			2.30					Erben 1964; Landman et al. 1996	Agoniatit.
Agoniatites obliquus	Dev.		1.40			2.40					Wissner and Norris 1991; Landman et al. 1996	Agoniatit.
Agoniatites sp.	Dev.					2.60					Erben 1964; Landman et al. 1996	Agoniatit.
Amoenophyllites doeringi	Dev.		1.10								Klug 2002	Mimagoniatit.
Anarcestes densistriatus	Dev.	0.50	0.60	0.70							Chlupáč and Turek 1983	Anarcest.
Anarcestes lateseptatus	Dev.		1.02			1.40			170		Klofak et al. 1999; De Baets et al. 2012	Anarcest.
Anarcestes latissimus	Dev.		0.97			1.42			193		Chlupáč and Turek 1983; De Baets et al. 2012	Anarcest.
Anarcestes plebeius	Dev.		0.92			1.45			175		Chlupáč and Turek 1983; De Baets et al. 2012	Anarcest.
Archanarcestes boreus	Dev.		0.98								Bogoslovsky 1972; De Baets et al. 2012	Mimagoniatit.
Archanarcestes kakvensis	Dev.		1.33								Bogoslovsky 1969; De Baets et al. 2012	Mimagoniatit.

Table 5.1 (continued)

Taxon	Age	ID_1			AD			AA			Reference	Family
		min	mean	max	min	mean	max	Min	mean	max		(-t. = -tidae)
Archanarcestes obesus	Dev.		1.04								Bogoslovsky 1969; De Baets et al. 2012	Mimagoniatit.
Archanarcestes pronini	Dev.		0.99								Bogoslovsky 1969; De Baets et al. 2012	Mimagoniatit.
Archoceras paeckelmanni	Dev.					0.99					Bensaid 1974	Gephurocerat.
Atlantoceras tataense	Dev.					1.06					Bensaid 1974; De Baets et al. 2012	Atlantocerat.
Balacheiloceras balapanense	Dev.					1.12					Bogoslovsky 1971	Cheilocerat.
Cabrieroceras crispiforme	Dev.		1.74								Petter 1959	Cabrierocerat.
Carinoceras galeatum	Dev.	0.70	0.80	0.90							Clausen 1969; Landman et al. 1996	Gephurocerat.
Celaeceras mirandus	Dev.		1.07								Bogolovsky 1969; De Baets et al. 2012	Augurit.
Chebbites lissovi	Dev.		0.96								Bogoslovsky 1969	Mimosphinct.
Chebbites lissovi	Dev.		0.97								Bogoslovsky 1969	Mimosphinct.
Chebbites reisdorfi	Dev.		1.22								Klug 2001a; De Baets et al. 2012	Mimosphinct.
Chlupacites praeceps	Dev.	0.70	0.80	0.90							Chlupáč and Turek 1983	Latanarcest.
Convoluticeras flexuosum	Dev.		0.90								Bogoslovsky 1984; De Baets et al. 2012	Teicherticerat.
Convoluticeras lardeuxi	Dev.	0.80	0.90	1.02		1.75			103		De Baets et al. 2012	Teicherticerat.
Convoluticeras nikolaevi	Dev.		0.88								Bogolovsky 1969; De Baets et al. 2012	Teicherticerat.
Cyrtoclymenia frechi	Dev.		0.49			0.96			292		Bogoslovsky 1981	Cyrtoclymeniidae
Cyrtoclymenia frechi	Dev.					0.99			283		Bogoslovsky 1981	Cyrtoclymeniidae
Devonopronorites ruzhencevi	Dev.					1.30					Bogoslovsky 1969	Devonopronorit.

Table 5.1 (continued)

Taxon	Age	ID_1			AD			AA			Reference	Family
		min	mean	max	min	mean	max	Min	mean	max		(-t. = -tidae)
Diallagites lenticulifer	Dev.		0.80								Klug 2002	Wernerocerat.
Diallagites testatus	Dev.		1.09								Chlupáč and Turek 1983	Wernerocerat.
Erbenoceras advolvens	Dev.		0.90			3.74			44		Bogoslovsky 1969; De Baets et al. 2012	Mimosphinct.
Erbenoceras kimi	Dev.		0.95								Bogoslovsky 1984; De Baets et al. 2012	Mimosphinct.
Exopinacites singularis	Dev.		1.00								Klug 2002	Pinacit.
Fidelites bicanaliculatus	Dev.		1.49								Petter 1959; De Baets et al. 2012	Agoniatit.
Fidelites fidelis	Dev.		1.39								Chlupáč and Turek 1983; De Baets et al. 2012	Agoniatit.
Gaurites sperandus	Dev.	0.94	1.00	1.09							Bogoslovsky 1984; De Baets et al. 2012	Augurit.
Genuclymenia angelini	Dev.		0.60								Bogoslovsky 1981	Cymaclymeniidae
Genuclymenia frechi	Dev.		0.58			1.07			279		Bogoslovsky 1981	Cymaclymeniidae
Genuclymenia frechi	Dev.		0.60			1.13			289		Bogoslovsky 1981	Cymaclymeniidae
Gracilites svetlanae	Dev.		0.61								Bogoslovsky 1972	Teicherticerat.
Gyroceratites cf. *laevis*	Dev.		0.88								De Baets et al. 2012	Mimocerat.
Gyroceratites gracilis	Dev.		0.87								Bogoslovsky 1969	Mimocerat.
Gyroceratites gracilis	Dev.					1.49			80		Chlupáč and Turek 1983	Mimocerat.
Gyroceratites gracilis	Dev.		0.75			1.51			118		Erben 1964	Mimocerat.
Gyroceratites gracilis	Dev.		1.01								De Baets et al. 2012	Mimocerat.
Gyroceratites gracilis	Dev.		1.09								De Baets et al. 2012	Mimocerat.
Gyroceratites gracilis	Dev.		0.90								Schindewolf 1933	Mimocerat.
Gyroceratites gracilis	Dev.		0.85								Schindewolf 1933	Mimocerat.
Gyroceratites laevis	Dev.		0.98								De Baets et al. 2012	Mimocerat.
Irdanites kaufmanni	Dev.		0.76								Klug 2001a	Teicherticerat.

Table 5.1 (continued)

Taxon	Age	ID_1			AD			AA			Reference	Family
		min	mean	max	min	mean	max	Min	mean	max		(-t. = -tidae)
Irdanites korni	Dev.		1.42								De Baets et al. 2012	Teicherticerat.
Ivoites opitzi	Dev.		1.21			3.50			23		De Baets et al. 2013	Mimosphinct.
Ivoites schindewolfi	Dev.		1.29			5.00					De Baets et al. 2013b	Mimosphinct.
Ivoites sp.	Dev.		0.98								De Baets et al. 2012	Mimosphinct.
Koenenites styliophylus	Dev.	0.58	0.68	0.80							House and Kirchgasser 2008	Koenenit.
Koenenites uralensis	Dev.		0.87								Bogoslovsky 1969	Koenenit.
Latanarcestes aff.*noeg-gerathii* auct.	Dev.		1.16			1.79			164		De Baets et al. 2012	Latanarcest.
	Dev.		1.10								De Baets et al. 2012	Latanarcest.
	Dev.		1.21								De Baets et al. 2012	Latanarcest.
Lunupharciceras lunulicosta	Dev.		0.88								Petter 1959	Pharcicerat.
Maenioceras aff. *terebratum*	Dev.		0.65								House 1996	Maeniocerat.
Manticoceras adorfense	Dev.	0.65	0.74	0.83							Clausen 1969; Landman et al. 1996	Gephurocerat.
Manticoceras drevermanni	Dev.	0.68	0.73	0.78							Landman et al. 1996	Gephurocerat.
Manticoceras intumescens	Dev.	0.65	0.73	0.80							Landman et al. 1996	Gephurocerat.
Manticoceras sinuosum	Dev.					1.20					Bogoslovsky 1969	Gephurocerat.
Manticoceras sp.	Dev.					1.60			163		De Baets et al. 2012	Gephurocerat.
Maternoceras calculiformis	Dev.		0.74								Branco 1880	Gephurocerat.
Maternoceras calculiformis	Dev.		0.84								Branco 1880	Gephurocerat.

Table 5.1 (continued)

Taxon	Age	ID_1			AD			AA			Reference	Family
		min	mean	max	min	mean	max	Min	mean	max		(-t. = -tidae)
Metabactrites fuchsi	Dev.		1.54			6.00			12		De Baets et al. 2012	Mimosphinct.
Mimagoniatites aff. *fecundus*	Dev.		1.14			2.04			121		Erben 1964	Mimagoniatit.
Mimagoniatites aff. *fecundus*	Dev.		1.13			1.83			131		Erben 1964	Mimagoniatit.
Mimagoniatites cf. *fecundus*	Dev.		1.12								De Baets et al. 2012	Mimagoniatit.
Mimagoniatites cf. *fecundus*	Dev.		0.99								De Baets et al. 2012	Mimagoniatit.
Mimagoniatites cf. *zorgensis*	Dev.		1.20			1.95			159		Erben 1964	Mimagoniatit.
Mimagoniatites cf. *zorgensis*	Dev.		1.19			1.92			152		Erben 1964	Mimagoniatit.
Mimagoniatites fecundus	Dev.		1.08								De Baets et al. 2012	Mimagoniatit.
Mimagoniatites fecundus	Dev.		1.26								De Baets et al. 2012	Mimagoniatit.
Mimagoniatites zorgensis	Dev.		1.39								Schindewolf 1934	Mimagoniatit.
Mimosphinctes discordans	Dev.					3.10			103		Erben 1964	Mimosphinct.
Mimosphinctes khanakuense	Dev.		1.36								Bogoslovsky 1980	Mimosphinct.
Mimosphinctes tripartitus	Dev.		1.27								Bogoslovsky 1980	Mimosphinct.
Mimosphinctes zlichovensis	Dev.		1.27			3.99			80		Chlupáč and Turek 1983	Mimosphinct.
Mimotornoceras djemeli	Dev.		1.27								Petter 1959	Pinacit.

Table 5.1 (continued)

Taxon	Age	ID_1			AD			AA			Reference	Family
		min	mean	max	min	mean	max	Min	mean	max		(-t. = -tidae)
Pachyclymenia intermedia	Dev.		0.48			0.87			315		Bogoslovsky 1981	Pachyclymeniidae
Palaeogoniatites lituus	Dev.		1.32								Chlupáč and Turek 1983	Teicherticerat.
Palaeogoniatites lituus	Dev.		1.40								Erben 1960	Teicherticerat.
Paranarcestes chalix	Dev.		1.01								Klug 2002	Anarcest.
Paranarcestes chalix	Dev.		1.02								Klug 2002	Anarcest.
Paranarcestes pictus	Dev.		1.04			1.66			191		Chlupáč and Turek 1983	Anarcest.
Parentites praecursor	Dev.		1.05								Bogolovsky 1969; De Baets et al. 2012	Augurit.
Pinacites jugleri	Dev.		1.20								Schindewolf 1934; De Baets et al. 2012	Pinacit.
Platyclymenia richteri	Dev.		0.53			0.95			282		Bogoslovsky 1981	Cyrtoclymeniidae
Praewerneroceras suchomastense	Dev.		1.11			1.66					Chlupáč and Turek 1983	Wernerocerat.
Praewerneroceras Suchomastense	Dev.		1.09			1.64					Chlupáč and Turek 1983	Wernerocerat.
Probeloceras lutheri	Dev.					1.06					Landman et al. 1996	Belocerat.
Pseudofidelites bockwinkeli	Dev.		1.00								Klug 2002	Agoniatit.
Rherisites tuba	Dev.		1.05								Klug 2001b	Mimagoniatit.
Sellanarcestes cognatus	Dev.		1.25								Chlupáč and Turek 1983	Anarcest.
Sellanarcestes cognatus	Dev.		1.34								Chlupáč and Turek 1983	Anarcest.
Sellanarcestes eos	Dev.		1.20								Klug 2002	Anarcest.
Sellanarcestes neglectus	Dev.	1.28	1.33	1.38	1.89	2.01	2.12	170		172	Chlupáč and Turek 1983; De Baets et al. 2012	Anarcest.
Sellanarcestes perfectus	Dev.		1.52			2.38			140		Chlupáč and Turek 1983; De Baets et al. 2012	Anarcest.

Table 5.1 (continued)

Taxon	Age	ID_1			AD			AA			Reference	Family
		min	mean	max	min	mean	max	Min	mean	max		(-t. = -tidae)
Sellanarcestes sp.	Dev.		1.36			2.20			149		De Baets et al. 2012	Anarcest.
Sellanarcestes sp.	Dev.		1.31			2.20			177		De Baets et al. 2012	Anarcest.
Sellenarcestes certus	Dev.		1.10								Chlupac and Turek 1983	Anarcest.
Serramanticoceras serratum	Dev.		0.59								Branco 1880; De Baets et al. 2012	Gephurocerat.
Serramanticoceras serratum	Dev.	0.63	0.67	0.70							Clausen 1969; Landman et al. 1996	Gephurocerat.
Sphaeromanticoceras affine	Dev.	0.75	0.93	1.10							Clausen 1969; Landman et al. 1996	Gephurocerat.
Sphaeromantic. bullatum	Dev.	0.68	0.77	0.85		1.20			216		Clausen 1969; Landman et al. 1996; De Baets et al. 2012	Gephurocerat.
Sphaeromantic. cordatum	Dev.	0.68	0.76	0.83							Clausen 1969; Landman et al. 1996	Gephurocerat.
Sphaeromantic. crassum	Dev.	0.68	0.76	0.83							Clausen 1969; Landman et al. 1996	Gephurocerat.
Sphaeromantic. orbiculum	Dev.	0.68	0.78	0.88							Clausen 1969; Landman et al. 1996	Gephurocerat.
Sphaeromantico-ceras rhynchostoma	Dev.		0.80			1.10					House and Kirchgasser 2008	Gephurocerat.
Subanarcestes coronatus	Dev.		0.80			1.50					Klug 2002	Sobolewiidae
Subanarcestes macrocephalus	Dev.		0.80			1.50					Klug 2002	Sobolewiidae
Teicherticeras erbeni	Dev.		0.96								Bogoslovsky 1969; De Baets et al. 2012	Teicherticerat.
Timanites keyerlingi	Dev.		0.81			1.32			238		Bogoslovsky 1969; De Baets et al. 2012	Koenenit.

Table 5.1 (continued)

Taxon	Age	ID_1			AD			AA			Reference	Family
		min	mean	max	min	mean	max	Min	mean	max		(-t. = -tidae)
Tornoceras aldenense	Dev.	0.98	0.99	1.00		1.49					House 1965; Landman et al. 1996	Tornocerat.
Tornoceras arcuatum	Dev.		0.73								House 1965; Landman et al. 1996	Tornocerat.
Tornoceras arkonense	Dev.		0.80			1.50					Landman et al. 1996	Tornocerat.
Tornoceras concentricum	Dev.		0.55			1.40					House 1965; Landman et al. 1996	Tornocerat.
Tornoceras obesum	Dev.		0.98								House 1965; Landman et al. 1996	Tornocerat.
Tornoceras typum	Dev.					1.20					Bogoslovsky 1971; De Baets et al. 2012	Tornocerat.
Tornoceras uniangulare	Dev.		0.78								House 1965; Landman et al. 1996	Tornocerat.
Tornoceras widderi	Dev.					1.50					House 1965; Landman et al. 1996	Tornocerat.
Trevoneites foxi	Dev.		1.30								House 1963	Parodocerat.
Trimanticoceras bisulcatum	Dev.		0.77								Branco 1880; De Baets et al. 2012	Gephurocerat.
Truyolsoceras bicostatum	Dev.		0.50		1.40	1.45	1.50		364		Landman et al. 1996	Tornocerat.
Uraloclymenia kazakhstanica	Dev.		0.51			0.94			283		Bogoslovsky 1981	Pachyclymeniidae
Wendtia ougarta	Dev.		0.75								Klug 2002	Cabrierocerat.
Werneroceras ruppachense	Dev.		1.05								Schindewolf 1933; De Baets et al. 2012	Wernerocerat.
Werneroceras subnautilinum	Dev.		1.10								Branco 1880; De Baets et al. 2012	Wernerocerat.

Table 5.1 (continued)

Taxon	Age	ID_1			AD			AA			Reference	Family
		min	mean	max	min	mean	max	Min	mean	max		(-t. = -tidae)
Wocklumeria sphaeroides	Dev.					0.73					Schindewolf 1937	Wocklumeriidae
Anthracoceras missouriense	Carb.		0.28								Miller and Unklesbay 1943	Anthracocerat.
Aristoceras sp.	Carb.		0.36			0.75			368		Tanabe et al. 1994; Landman et al. 1996	Thalassocerat.
Arkanites relictus	Carb.		0.51			0.81			386		Tanabe et al. 1994; Landman et al. 1996; Shigeta et al. 2001	Reticulocerat.
Arkanites relictus	Carb.				0.71	0.78	0.87				Stephen and Stanton 2002	Reticulocerat.
Bisatoceras greenei	Carb.		0.28								Miller and Unklesbay 1943	Bisatocerat.
Bisatoceras sp.	Carb.		0.36			0.62			358		Tanabe et al. 1994; Landman et al. 1996; Shigeta et al. 2001	Bisatocerat.
Boesites texanus	Carb.		0.43								Miller and Unklesbay 1943	Daraelit.
Branneroceras branneri	Carb.				0.78	0.81	0.85				Stephen and Stanton 2002	Schistocerat.
Cancelloceras huntsvillense	Carb.				0.78	0.81	0.85				Stephen and Stanton 2002	Gastriocerat.
Cravenoceras lineolatum	Carb.		0.48			0.82			368		Shigeta et al. 2001	Cravenocerat.
Dombarites choctawensis	Carb.	0.54		0.56	1.00		1.06	345		383	Tanabe et al. 1994; Landman et al. 1996; Shigeta et al. 2001	Delepinocerat.
Edmooroceras plummeri	Carb.					1.03					Shigeta et al. 2001	Girtyocerat.
Eothalassoceras inexpectans	Carb.	0.37	0.38	0.39	0.66	0.67	0.68	356	361	365	Tanabe et al. 1994; Landman et al. 1996; Shigeta et al. 2001	Thalassocerat.
Epicanites loeblichi	Carb.		0.43			0.87			355		Shigeta et al. 2001	Daraelit.
Gaitherites morrowensis	Carb.		0.42			0.88			385		Shigeta et al. 2001	Reticulocerat.
Girtyoceras mesleranium	Carb.		0.45			0.96			400		Tanabe et al. 1994	Girtyocerat.

Table 5.1 (continued)

Taxon	Age	ID_1			AD			AA			Reference	Family
		min	mean	max	min	mean	max	Min	mean	max		(-t. = -tidae)
Girtyoceras mesleranium	Carb.		0.54			0.91			368		Shigeta et al. 2001	Girtyocerat.
Glaphyrites angulatus	Carb.					0.78					Stephen and Stanton 2002	Glaphyrit.
Glaphyrites hyattianus	Carb.		0.59			1.03			405		Tanabe et al. 1994; Landman et al. 1996	Glaphyrit.
Glaphyrites hyattianus	Carb.		0.59			1.05			370		Shigeta et al. 2001	Glaphyrit.
Glaphyrites jonesi	Carb.		0.54			0.92			372		Tanabe et al. 1994; Landman et al. 1996; Shigeta et al. 2001	Glaphyrit.
Glaphyrites warei	Carb.	0.45	0.48	0.53	0.86	0.89	0.92	369	375	385	Tanabe et al. 1994; Landman et al. 1996; Shigeta et al. 2001	Glaphyrit.
Glaphyrites welleri	Carb.		0.35			0.81			356		Tanabe et al. 1994; Landman et al. 1996	Glaphyrit.
Goniatites aff. *crenistria*	Carb.		0.48			0.95			382		Landman et al. 1996	Goniatit.
Goniatites aff. *crenistria*	Carb.		0.60			1.10			372		Landman et al. 1996; Shigeta et al. 2001	Goniatit.
Goniatites multiliratus	Carb.		0.53			0.94			384		Landman et al. 1996	Goniatit.
Goniatites multiliratus	Carb.		0.55			0.98			360		Shigeta et al. 2001	Goniatit.
Gonioloboceras welleri	Carb.		0.80			1.92			384		Miller and Unklesbay 1943; Tanabe et al. 1994; Landman et al. 1996	Goniatit.
Isohomoceras subglobosum	Carb.		0.53			0.91			385		Tanabe et al. 1994; Landman et al. 1996	Homocerat.
Isohomoceras subglobosum	Carb.		0.50			0.93			367		Shigeta et al. 2001	Homocerat.
Neoglyphioceras abramovi	Carb.	0.49	0.51	0.52	0.92	0.93	0.94	360	365	370	Zakharov 1978; Landman et al. 1996; Shigeta et al. 2001	Neoglyphiocerat.
Neogoniatites kentuckiensis	Carb.		0.45								Miller and Unklesbay 1943	Goniatit.

Table 5.1 (continued)

Taxon	Age	ID_1			AD			AA			Reference	Family
		min	mean	max	min	mean	max	Min	mean	max		(-t. = -tidae)
Neopronorites praepermicus	Carb.		0.61			1.22			330		Tanabe et al. 1994; Landman et al. 1996	Pronorit.
Owenoceras bellilineatum	Carb.					0.87			355		Tanabe et al. 1994; Landman et al. 1996	Gastriocerat.
Paraschistoceras sp.	Carb.					1.20					Ruzhencev and Shimansky 1954; Lap-tikovsky et al. 2013	Schistocerat.
Politoceras politum	Carb.		0.45			0.90					Miller and Unklesbay 1943; Tanabe et al. 1994	Neodimorphoc.
Pseudopronorites arkansiensis	Carb.				0.92	0.99	1.03				Stephen and Stanton 2002	Pronorit.
Retites semiretia	Carb.				0.73	0.80	0.87				Stephen and Stanton 2002	Reticulocerat.
Richardsonites richardsonianus	Carb.		0.49			0.81			367		Landman et al. 1996; Shigeta et al. 2001	Glaphyrit.
Stenoglaphyrites incisus	Carb.		0.53			0.95			380		Landman et al. 1996	Glaphyrit.
Stenoglaphyrites incisus	Carb.		0.49			0.91			360		Shigeta et al. 2001	Glaphyrit.
Stenoglaphyrites incisus	Carb.		0.46			0.90					Smith 1897; Laptikovsky	Glaphyrit.
Syngastrioceras clinei	Carb.		0.52			0.94			364		Tanabe et al. 1994; Landman et al. 1996	Glaphyrit.
Syngastrioceras clinei	Carb.		0.41			0.72			379		Tanabe et al. 1994; Landmane et al. 1996; Shigeta et al. 2001	Glaphyrit.
Syngastrioceras oblatum	Carb.				0.75	0.83	0.92				Stephen and Stanton 2002	Glaphyrit.
Vallites diadema	Carb.		0.66			0.90					Smith 1897; Laptikovsky	Homocerat.
Verneuilites pygmaeus	Carb.				0.75	0.78	0.82				Stephen and Stanton 2002	Reticulocerat.
Vidrioceras sp.	Carb.		0.44			0.80			361		Tanabe et al. 1994	Vidriocerat.
Vidrioceras sp.	Carb.				0.72	0.75	0.78				Tanabe et al. 2001	Vidriocerat.
Adrianites dunbari	Perm.		0.63			1.00			365		Tanabe et al. 1994	Adrianit.

Table 5.1 (continued)

Taxon	Age	ID_1			AD			AA			Reference	Family
		min	mean	max	min	mean	max	Min	mean	max		(-t. = -tidae)
Agathiceras applini	Perm.		0.48			1.03					Landman et al. 1996	Agathicerat.
Agathiceras applini	Perm.		0.52			1.01			365		Shigeta et al. 2001	Agathicerat.
Agathiceras uralicum	Perm.		0.51			0.95			369		Shigeta et al. 2001	Agathicerat.
Akmilleria electraensis	Perm.		0.48			1.01			340		Landman et al. 1996	Medlicottiidae
Akmilleria electraensis	Perm.		0.52			1.16			355		Landman et al. 1996	Medlicottiidae
Akmilleria electraensis	Perm.		0.63			1.25			338		Shigeta et al. 2001	Medlicottiidae
Almites invariabilis	Perm.		0.38			0.77			366		Shigeta et al. 2001	Marathonit.
Artinskia artiensis	Perm.	0.43	0.44	0.44		0.90			345		Bogoskovskaya 1959; Landman et al. 1996	Medlicottiidae
Artioceras rhipaeum	Perm.		0.36			0.70			334		Shigeta et al. 2001	Medlicottiidae
Crimites elkoensis	Perm.		0.37			0.65			345		Tanabe et al. 1994; Landman et al. 1996	Adrianit.
Crimites elkoensis	Perm.		0.38			0.73			365		Shigeta et al. 2001	Adrianit.
Crimites krotowi	Perm.	0.34	0.37	0.39	0.73	0.77	0.81		365		Bogoslovskaya 1959; Landman et al. 1996	Adrianit.
Crimites subkrotowi	Perm.		0.38			0.68			365		Shigeta et al. 2001	Adrianit.
Daraelites elegans	Perm.		0.47			0.91			350		Shigeta et al. 2001	Daraelit.
Daraelites elegans	Perm.		0.45			1.06					Bogoslovskaya 1959; Landman et al. 1996	Daraelit.
Eothinites kargalensis	Perm.		0.38			0.67			372		Shigeta et al. 2001	
Kargalites typicus	Perm.		0.47			0.91			360		Shigeta et al. 2001	Metalegocerat.
Kargalites typicus	Perm.	0.40	0.41	0.42		0.88			340		Bogoslovskaya; Landman et al. 1996	Metalegocerat.
Martoceras subinterruptum	Perm.	0.39	0.43	0.46		0.84			380		Bogoslovskaya 1959; Landman et al. 1996	Vidriocerat.
Medlicottia orbignyana	Perm.		0.34			0.72		326	328	330	Bogoslovskaya 1959; Landman et al. 1996	Medlicottiidae
Mesometalegoc. baylorense	Perm.		0.48			0.87			365		Tanabe et al. 1994; Landman et al. 1996; Shigeta et al. 2001	Metalegocerat.

Table 5.1 (continued)

Taxon	Age	ID_1			AD			AA			Reference	Family
		min	mean	max	min	mean	max	Min	mean	max		(-t. = -tidae)
Metalegoceras sp.	Perm.		0.47			0.83			365		Shigeta et al. 2001	Metalegocerat.
Mexicoceras guadalupense	Perm.		0.52			0.93			378		Tanabe et al. 1994; Landman et al. 1996	Cyclolobidae
Neogeoceras girtyi	Perm.		0.49								Miller and Unklesbay 1943	Medlicottiidae
Neopronorites permicus	Perm.	0.44	0.50	0.56	0.96	1.03	1.10		340		Bogoslovskaya 1959; Landman et al. 1996	Pronorit.
Neopronorites skvorzovi	Perm.		0.65			1.15			328		Shigeta et al. 2001	Pronorit.
Paraceltites elegans	Perm.		0.46			0.92			342		Shigeta et al. 2001	Paraceltit.
Paragastrioceras artolobatum	Perm.		0.41			0.74			365		Shigeta et al. 2001	Paragastriocerat.
Paragastrioceras kirghizorum	Perm.		0.40			0.69			365		Shigeta et al. 2001	Paragastriocerat.
Paragastrioceras sp.	Perm.		0.48			0.84			360		Bogoslovskaya 1959; Landman et al. 1996	Paragastriocerat.
Peritrochia erebus	Perm.		0.45			0.88			410		Tanabe et al. 1994; Landman et al. 1996	Vidriocerat.
Perrinites sp.	Perm.		1.03			2.31			383		Landman et al. 1996	Perrinit.
Perrinites sp.	Perm.		0.89			2.00			382		Landman et al. 1996	Perrinit.
Popanoceras annae	Perm.		0.36			0.66			352		Tanabe et al. 1994; Landman et al. 1996; Shigeta et al. 2001	Popanocerat.
Properrinites bakeri	Perm.		0.70			1.19					Tanabe et al. 1994; Landman et al. 1996	Perrinit.
Pseudohalorites sp.	Perm.		1.10								Zhou et al. 2002	Pseudohalorit.
Sakmarites vulgaris	Perm.	0.44	0.49	0.54	1.00	1.05	1.10		340		Bogoslovskaya 1959; Landman et al. 1996	Pronorit.
Stacheoceras toumanskyae	Perm.		0.27								Miller and Unklesbay 1943	Vidriocerat.
Stenolobulites simulator	Perm.		0.40			0.80			364		Tanabe et al. 1994; Landman et al. 1996	Paragastriocerat.

Table 5.1 (continued)

Taxon	Age	ID_1			AD			AA			Reference	Family
		min	mean	max	min	mean	max	Min	mean	max		(-t. = -tidae)
Texoceras sp.	Perm.					0.96					Tanabe et al. 1994; Landman et al. 1996; Shigeta et al. 2001	Adrianit.
Thalassoceras gemmellaroi	Perm.	0.30	0.32	0.34	0.58	0.62	0.66		380		Bogoslovskaya 1959; Landman et al. 1996	Thalassocerat.
Thalassoceras gemmellaroi	Perm.		0.37			0.69			356		Shigeta et al. 2001	Thalassocerat.
Uraloceras complanatum	Perm.		0.41			0.74			362		Shigeta et al. 2001	Paragastriocerat.
Uraloceras fedorowi	Perm.	0.38	0.39	0.40	0.74	0.75	0.76		370		Bogoslovskaya 1959; Landman et al. 1996	Paragastriocerat.
Uraloceras sp.	Perm.		0.52			0.85			371		Shigeta et al. 2001	Paragastriocerat.
Uraloceras suessi	Perm.		0.46								Miller and Unklesbay 1943	Paragastriocerat.
Yinoceras trifurcatus	Perm.		1.10								Zhou et al. 2002	Pseudohalorit.
Amphipopanoceras asseretoi	Trias.		0.38			0.69			335		Landman et al. 1996	Parapopanocerat.
Anagymnotoceras varium	Trias.		0.38			0.86			340		Landman et al. 1996	Ceratit.
Anasibirites nevolini	Trias.	0.46	0.48	0.49							Zakharov 1978	Prionit.
Arcestes sp.	Trias.		0.37			0.63					Zakharov 1978; Landman et al. 1996	Arcest.
Arctoceras septentrionale	Trias.		0.51			0.96					Zakharov 1974, 1978; Landman et al. 1996	Arctocerat.
Arctogymnites cf. *spectori*	Trias.		0.65			1.17			270		Vavilov 1992	Ceratit.
Arctogymnites sonini	Trias.	0.82	0.85	0.88	1.30	1.40	1.47		270		Vavilov 1992	Ceratit.
Arctogymnites spectori	Trias.		0.68			1.15			270		Vavilov 1992	Ceratit.
Arctohungarites evolutus	Trias.	0.36	0.39	0.42	0.69	0.74	0.78		270		Alekseev et al. 1984; Vavilov 1992	Danubit.

Table 5.1 (continued)

Taxon	Age	ID_1			AD			AA			Reference	Family
		min	mean	max	min	mean	max	Min	mean	max		(-t. = -tidae)
Arctohungarites involutus	Trias.	0.37	0.39	0.41	0.70	0.74	0.77		270		Alekseev et al. 1984; Vavilov 1992	Danubit.
Arctohungarites kharaulakhensis	Trias.		0.40		0.60	0.66	0.72		270		Vavilov 1992	Danubit.
Arctohungarites sp.	Trias.	0.39	0.41	0.42				229	230	230	Zakharov 1978	Danubit.
Arctohungarites triformis	Trias.	0.35	0.40	0.46	0.69	0.75	0.85		270		Alekseev et al. 1984; Vavilov 1992	Danubit.
Arctomeekoceras rotundatum	Trias.		0.37			0.82		280	283	285	Zakharov 1978; Landman et al. 1996	Meekocerat.
Arctoprionites prontchischevi	Trias.		0.39								Zakharov 1978	Prionit.
Arctoptychites euglyphus	Trias.		0.62			0.85			360		Vavilov 1992	Ptychit.
Arctoptychites kruzini	Trias.		0.65		0.90	0.91	0.92		350		Vavilov 1992	Ptychit.
Aristoptychites kolymensis	Trias.		0.54			0.96			368		Landman et al. 1996	Ptychit.
Aristoptychites kolymensis	Trias.	0.61	0.64	0.67	0.97	0.99	1.00		360		Vavilov 1992	Ptychit.
Boreomeekoceras keyserlingi	Trias.		0.35			0.73			325		Landman et al. 1996	Meekocerat.
Boreomeekoceras keyserlingi	Trias.		0.41			0.84			295		Zakharov 1978	Meekocerat.
Columbites ussuriensis	Trias.		0.46			0.73			240		Zakharov 1978; Landman et al. 1996	Columbit.
Czekanowskites decipiens	Trias.	0.35	0.36	0.37	0.60	0.66	0.70		270		Alekseev et al. 1984; Vavilov 1992	Danubit.
Czekanowskites rieberi	Trias.		0.39			0.80			305		Landman et al. 1996	Danubit.
Dieneroceras dieneri	Trias.		0.29								Zakharov 1978	Meekocerat.

Table 5.1 (continued)

Taxon	Age	ID_1			AD			AA			Reference	Family
		min	mean	max	min	mean	max	Min	mean	max		(-t. = -tidae)
Dieneroceras spathi	Trias.		0.29								Zakharov 1978	Meekocerat.
Discophyllites taimyrensis	Trias.		0.67			1.30			300		Vavilov 1992	Discophyllit.
Eophyllites sp.	Trias.	0.42	0.43	0.43							Zakharov 1971; Landman et al. 1996	Paleophyllit.
Frechites laqueatus	Trias.		0.46			0.71			288		Landman et al. 1996	Ceratit.
Frechites nevadanus	Trias.	0.41	0.49	0.59		0.77			270		Arkadiev and Vavilov 1984b; Landman et al. 1996; Vavilov 1992	Ceratit.
Frechites sp.	Trias.		0.60			1.32			270		Arkadiev and Vavilov 1984b; Vavilov 1992	Ceratit.
Grambergia olenekensis	Trias.	0.34	0.35	0.36		0.76		229	230	230	Zakharov 1978	Longobardit.
Grambergia taimyrensis	Trias.	0.45	0.49	0.53	0.85	0.88	0.92		290		Alekseev et al. 1984; Vavilov 1992	Longobardit.
Gymnotoceras falciforme	Trias.	0.36	0.37	0.38	0.66	0.68	0.69		270		Arkadiev and Vavilov 1984b; Vavilov 1992; Landman et al. 1996	Ceratit.
Gymnotoceras rotelliforme	Trias.	0.45	0.46	0.47		0.85			270		Arkadiev and Vavilov 1984b; Landman et al. 1996; Vavilov 1992	Ceratit.
Hedenstroemia hedenstroemi	Trias.		0.63			1.07		287	292	296	Zakharov 1974, 1978; Landman et al. 1996	Hedenstroemidae
Hedenstroemia mojsisovicsi	Trias.	0.45	0.63	0.53	0.90	0.93	0.95		296		Zakharov 1978; Landman et al. 1996	Hedenstroemidae
Hemiprionites contortus	Trias.		0.50								Zakharov 1978	Prionit.
Indigirites constantis	Trias.		0.45			0.77			290		Vavilov 1992	Nathorstit.
Indigirites krugi	Trias.	0.35	0.36	0.37	0.60	0.65	0.68	290	292	300	Arkadiev and Vavilov 1984a; Vavilov 1992	Nathorstit.
Indigirites tozeri	Trias.	0.32	0.33	0.33	0.59	0.64	0.68		330		Landman et al. 1996	Nathorstit.

Table 5.1 (continued)

Taxon	Age	ID$_1$			AD			AA			Reference	Family
		min	mean	max	min	mean	max	Min	mean	max		(-t. = -tidae)
Indigirophyllites bytschkovi	Trias.		0.70			1.25			300		Vavilov 1992	Discophyllit.
Indigirophyllites l.c. *Spitsbergensis*	Trias.		0.46			1.02			330		Landman et al. 1996	Discophyllit.
Karangatites evolutus	Trias.	0.26	0.27	0.28	0.75	0.76	0.77		290		Vavilov 1992	Aplococerat.
Kingites korostelevi	Trias.		0.42			0.68			347		Zakharov 1974, 1978; Landman et al. 1996	Meekocerat.
Lenotropites boschoensis	Trias.		0.38		0.73	0.74	0.74		270		Vavilov 1992	Longobardit.
Lenotropites cf. *tardus*	Trias.		0.32			0.57			270		Alekseev et al. 1984; Vavilov 1992	Longobardit.
Lenotropites solitarius	Trias.		0.32			0.62			270		Alekseev et al. 1984	Longobardit.
Lenotropites sp.	Trias.	0.37	0.38	0.39	0.69	0.72	0.74		270		Alekseev et al. 1984	Longobardit.
Lenotropites tardus	Trias.	0.34	0.36	0.37	0.64	0.66	0.67		270		Alekseev et al. 1984; Vavilov 1992	Longobardit.
Lepiskites kolymensis	Trias.		0.44			0.82			280		Zakharov 1978	Meekocerat.
Longobardites nevadanus	Trias.	0.40	0.44	0.47	0.80	0.84	0.88		290		Alekseev et al. 1984; Vavilov 1992	Longobardit.
Meekoceras subcristatum	Trias.	0.43	0.44	0.45							Zakharov 1978	Meekocerat.
Megaphyllites prometheus	Trias.	0.53	0.57	0.60	0.94	1.00	1.05	270	275	280	Drushchits and Doguzhaeva 1981; Landman et al. 1996	Megaphyllit.
Nathorstites argatassensis	Trias.	0.35		0.37		0.66			270		Arkadiev and Vavilov 1984a; Vavilov 1992	Nathorstit.
Nathorstites gibbosus	Trias.		0.37			0.62			270		Alekseev et al. 1984; Arkadiev and Vavilov 1984a; Vavilov 1992	Nathorstit.
Nathorstites intermedia	Trias.	0.30	0.34	0.37		0.63			270		Arkadiev and Vavilov 1984a; Vavilov 1992	Nathorstit.

Table 5.1 (continued)

Taxon	Age	ID_1			AD			AA			Reference	Family
		min	mean	max	min	mean	max	Min	mean	max		(-t. = -tidae)
Nathorstites lenticularis	Trias.		0.26			0.54			270		Arkadiev and Vavilov 1984a	Nathorstit.
Nathorstites mcconnelli	Trias.	0.36	0.39	0.42	0.59	0.68	0.76	270	280	290	Alekseev et al. 1984; Arkadiev and Vavilov 1984a; Vavilov 1992	Nathorstit.
Neoprotrachyceras sp.	Trias.		0.48			0.77			290		Vavilov 1992	Trachycerat.
Neosirenites irregularis	Trias.		0.37		0.56	0.57	0.57		290		Vavilov 1992	Trachycerat.
Nevadisculites smithi	Trias.		0.36			0.70			310		Arkadiev et al. 1993	Proarcest.
Nevadisculites taylori	Trias.		0.38			1.00			320		Arkadiev et al. 1993	Proarcest.
Nordophiceras schmidti	Trias.	0.37	0.38	0.39	0.77	0.78	0.79		260		Zakharov 1978; Landman et al. 1996	Arpadit.
Obrutchevites prodigialis	Trias.		0.51			0.82			290		Vavilov 1992	Trachycerat.
Olenekoceras middendorffi	Trias.		0.64		1.29	1.30	1.31		300		Zakharov 1978; Landman et al. 1996	Sibiridae
Olenikites spiniplicatus	Trias.		0.35			0.73			325		Landman et al. 1996	Sibiridae
Olenikites spiniplicatus	Trias.	0.34	0.37	0.40	0.62	0.67	0.72	300	308	315	Zakharov 1978	Sibiridae
Ophiceras sp.	Trias.		0.39			0.74					Zakharov 1974, 1978; Landman et al. 1996	Ophicerat.
Otoceras boreale	Trias.		0.73			1.11					Zakharov 1978	Otocerat.
Owenites koeneni	Trias.		0.37						270		Zakharov 1978	Paranannit.
Palaeokazachstanites Ussuriensis	Trias.		0.45			0.76			265		Zakharov 1978; Landman et al. 1996; Smyshlyaeva and Zakharov 2013	Meekocerat.
Paracladiscites sp.	Trias.		0.50			0.87			290		Vavilov 1992	Cladiscit.
Parafrechites meeki	Trias.	0.39	0.42	0.45	0.72	0.73	0.73		270		Arkadiev and Vavilov 1984b; Landman et al. 1996; Vavilov 1992	Ceratit.
Parahedenstroemia nevolini	Trias.		0.44								Zakharov 1978	Hedenstroemidae

Table 5.1 (continued)

Taxon	Age	ID_1			AD			AA			Reference	Family
		min	mean	max	min	mean	max	Min	mean	max		(-t. = -tidae)
Paranannites aspenensis	Trias.		0.37			0.66			238		Zakharov 1978; Landman et al. 1996	Paranannit.
Paranannites spathi	Trias.		0.35			0.67			330		Landman et al. 1996	Paranannit.
Parapopanoceras asseretoi	Trias.	0.30	0.36	0.40	0.60	0.65	0.68		270		Arkadiev and Vavilov 1984a; Vavilov 1992	Parapopanocerat.
Parapopanoceras janaensis	Trias.	0.33	0.42	0.50		0.64			270		Arkadiev and Vavilov 1984a; Vavilov 1992	Parapopanocerat.
Parapopanoceras medium	Trias.	0.35	0.38	0.42							Arkadiev and Vavilov 1984a; Vavilov 1992	Parapopanocerat.
Parapopanoc. paniculatum	Trias.		0.34			0.66			333		Landman et al. 1996	Parapopanocerat.
Parapopanoc. paniculatum	Trias.		0.39								Arkadiev and Vavilov 1984a	Parapopanocerat.
Parapopanoc. paniculatum	Trias.		0.39								Zakharov 1978	Parapopanocerat.
Parapopanoceras sp.	Trias.	0.26	0.33	0.40							Arkadiev and Vavilov 1984a	Parapopanocerat.
Parasibirites grambergi	Trias.		0.40			0.71			265		Zakharov 1978; Landman et al. 1996	Sibiridae
Parussuria semenovi	Trias.		0.43								Zakharov 1978	Ussurit.
Phyllocladiscites basarginensis	Trias.		0.59		0.76	0.82	0.96		265		Zakharov 1974, 1978; Landman et al. 1996	Cladiscit.
Placites polydactylus	Trias.		0.39								Zakharov 1978; Landman et al. 1996	Gymnit.
Proarcestes korchinskajae	Trias.	0.48	0.50	0.52	0.85	0.87	0.90		340		Vavilov 1992; Arkadiev et al. 1993	Arcest.
Proarcestes verchojanicus	Trias.		0.62			0.95			340		Vavilov 1992; Arkadiev et al. 1993	Arcest.

Table 5.1 (continued)

Taxon	Age	ID_1			AD			AA			Reference	Family
		min	mean	max	min	mean	max	Min	mean	max		(-t. = -tidae)
Prosphingites czekanowskii	Trias.		0.44			0.79			265		Zakharov 1978	Paranannit.
Prosphingites grambergi	Trias.	0.38	0.40	0.41	0.68	0.74	0.79		265		Zakharov 1971; Landman et al. 1996	Paranannit.
Prosphingites hexagonalis	Trias.		0.41			0.71					Zakharov 1978	Paranannit.
Prosphingites ovalis	Trias.	0.38	0.41	0.44		0.68					Zakharov 1978	Paranannit.
Pseudosageceras borealis	Trias.		0.57			0.96			295		Zakharov 1978; Landman et al. 1996	Hedenstroemidae
Pseudosageceras sp.	Trias.		0.57			0.96			295		Landman et al. 1996	Hedenstroemidae
Pterosirenites nelgechensis	Trias.		0.55			0.72			290		Vavilov 1992	Arpadit.
Sakhaites subleptodiscus	Trias.		0.38			0.73			280		Zakharov 1978	Ophicerat.
Shigetaceras dunajensis	Trias.		0.54								Zakharov 1978	Galfettit.
Sibirites eichwaldi	Trias.	0.35	0.36	0.37	0.60	0.64	0.67		260		Zakharov 1978; Landman et al. 1996	Sibiridae
Sphaerocladiscites omolonensis	Trias.		0.55			0.80			290		Vavilov 1992	Cladiscit.
Stenopopanoceras mirabile	Trias.		0.38			0.72			270		Landman et al. 1996	Parapopanocerat.
Stenopopanoceras mirabile	Trias.	0.43	0.45	0.46		0.75					Arkadiev and Vavilov 1984a; Vavilov 1992	Parapopanocerat.
Stolleyites tenuis	Trias.		0.36			0.62			320		Landman et al. 1996	Nathorstit.
Striatosirenites solonis	Trias.		0.45			0.70			290		Vavilov 1992	Trachycerat.
Subfengshanites multiformis	Trias.	0.34	0.38	0.41		0.63			240		Zakharov 1978; Landman et al. 1996	Columbit.
Subolenekites altus	Trias.		0.40			0.72			315		Landman et al. 1996	Olenikit.

Table 5.1 (continued)

Taxon	Age	ID_1			AD			AA			Reference	Family
		min	mean	max	min	mean	max	Min	mean	max		(-t. = -tidae)
Svalbardiceras sibiricum	Trias.		0.39		0.89	0.90	0.91		260		Zakharov 1978; Landman et al. 1996	Olenikit.
Svalbardiceras spitzbergensis	Trias.		0.37			0.70			290		Landman et al. 1996	Olenikit.
Tsvetkovites dolioliformis	Trias.	0.35	0.36	0.37		0.65			290		Vavilov 1992	Meekocerat.
Ussuriflemingites abrekensis	Trias.		0.49			0.93			287		Smyshlyaeva and Zakharov 2013	Melagathicerat.
Ussurijuvenites artyomensis	Trias.		0.50			0.75			275		Zakharov et al. 2012; Smyshlyaeva and Zakharov 2012	Melagathicerat.
Ussurites sp.	Trias.		0.75			1.35			300		Vavilov 1992	Ussurit.
Wangoceras berissense	Trias.	0.42	0.44	0.45	0.68	0.73	0.75	290	297	300	Vavilov 1992	Arpadit.
Wyomingites chaoi	Trias.		0.29								Zakharov 1974; Landman et al. 1996	Meekocerat.
Wyomingites spathi	Trias.		0.29								Zakharov 1974; Landman et al. 1996	Meekocerat.
Amaltheus margaritatus	Jura.		0.42						310		Grandjean 1910	Amaltheidae
Amauroceras ferrugineum	Jura.		0.56			1.00			300		Landman et al. 1996	Amaltheidae
Amoeboceras kitchini	Jura.		0.60								Knyazev 1975; Laptikovsky et al. 2013	Cardiocerat.
Binatisphinctes mosquensis	Jura.	0.35	0.41	0.48	0.63	0.73	0.83	274	283	308	Sprey 2002	Perisphinct.
Bredyia subinsignis [M/m]	Jura.	0.38	0.44	0.50	0.80	0.99	1.09				Senior 1977	Hammatocerat.
Cadoceras modiolare	Jura.		0.52								Drushchits et al. 1977	Cardiocerat.
Cadoceras sp.	Jura.		0.52		0.83	0.86	0.88	260		290	Drushchits et al. 1977	Cardiocerat.
Cadoceras tschefkini	Jura.	0.48	0.51	0.53							Drushchits et al. 1977	Cardiocerat.

Table 5.1 (continued)

Taxon	Age	ID_1			AD			AA			Reference	Family
		min	mean	max	min	mean	max	Min	mean	max		(-t. = -tidae)
Chondroceras allani [M]	Jura.		0.50			0.80					Hall and Westermann 1980; Laptikovsky et al. 2013	Sphaerocerat.
Coroniceras kridion	Jura.		0.37								Grandjean 1910; Landman et al. 1996	Arietit.
Coroniceras reynsei	Jura.		0.30			0.59			280		Landman et al. 1996	Arietit.
Coroniceras sp.	Jura.		0.35			0.64			284		Tanabe and Ohtsuka 1985; Landman et al. 1996	Arietit.
Craspedites cf. *ivanovi*	Jura.		0.53								Mikhailova 1974	Polyptychit.
Creniceras renggeri [M + m]	Jura.	0.28	0.29	0.30	0.53	0.54	0.56				Palframan 1966	Oppeliidae
Creniceras renggeri [M + m]	Jura.	0.27	0.33	0.37	0.53	0.57	0.60	250	264	281	Neige 1997	Oppeliidae
Dactylioceras sp.	Jura.	0.49	0.51	0.52	1.05	1.08	1.10				Kutygin and Knyazev 2000	Dactyliocerat.
Distichoceras bicostatum [M + m]	Jura.	0.24	0.28	0.30	0.48	0.51	0.57				Palframan 1967	Oppeliidae
Eleganticeras elegantulum	Jura.	0.40	0.41	0.42	0.80	0.84	0.88	277	285	292	Landman et al. 1996	Hildocerat.
Eurycephalites gottschei	Jura.		0.55								Parent 1997	Sphaerocerat.
Hecticoceras brightii [M + m]	Jura.	0.30	0.33	0.36	0.60	0.66	0.70				Palframan 1969	Oppeliidae
Hecticoceras canaliculatum	Jura.	0.36	0.40	0.45	0.59	0.66	0.74	234	257	288	Rouget and Neige 2001	Oppeliidae
Hecticoceras sp.	Jura.	0.34	0.35	0.36	0.55	0.60	0.70	277	285	294	Sprey 2002	Oppeliidae
Kolymophylloceras turomchense	Jura.		0.64			1.40			320		Repin et al. 1998	Yukagirit.
Kosmoceras sp.	Jura.	0.38	0.43	0.49	0.65	0.77	0.89	271	281	298	Sprey 2002	Kosmocerat.

Table 5.1 (continued)

Taxon	Age	ID_1			AD			AA			Reference	Family
		min	mean	max	min	mean	max	Min	mean	max		(-t. = -tidae)
Kosmoceras sp.	Jura.	0.39	0.44	0.49	0.76	0.81	0.85	310		315	Drushchits et al. 1977a	Kosmocerat.
Leioceras opalinum	Jura.		0.36								Landman et al. 1996	Graphocerat.
Nannolytoceras polyhelictum	Jura.	0.39	0.41	0.42	0.78	0.81	0.84	290	295	300	Landman et al. 1996	Nannolytocerat.
Osperleioceras lapparenti	Jura.	0.35	0.41	0.45	0.71	0.76	0.83				Morard and Guex 2003	Hildocerat.
Osperleioceras reynesi	Jura.	0.37	0.42	0.48	0.66	0.78	0.87				Morard and Guex 2003	Hildocerat.
Peronoceras fibulatum	Jura.		0.45			0.89			312		Landman et al. 1996	Dactyliocerat.
Phylloceras omkuchanicum	Jura.		0.76			1.08			350		Repin et al. 1998	Phyllocerat.
Pleuroceras sp.	Jura.		0.50			1.06			277		Landman et al. 1996	Amaltheidae
Promicroceras sp.	Jura.	0.42	0.43	0.43	0.71	0.73	0.75	280	283	286	Landman et al. 1996	Eoderocerat.
Pseudocadoceras sp.	Jura.	0.49	0.51	0.55	0.83	0.85	0.87	270		280	Drushchits et al. 1977a	Cardiocerat.
Sigaloceras enodatum	Jura.		0.46			0.85			315		Drushchits et al. 1977a	Kosmocerat.
Sphaeroceras brongniarti	Jura.		0.34								Landman et al. 1996	Sphaerocerat.
Spiroceras calloviense	Jura.		0.47			0.84			325		Landman et al. 1996	Stephanocerat.
Stephanoceras itinsae [m]	Jura.		0.43			0.80					Hall and Westermann 1980; Laptikovsky et al. 2013	Stephanocerat.
Sublunoloceras virguloides	Jura.	0.30	0.36	0.41	0.49	0.60	0.71	225	259	298	Rouget and Neige 2001	Oppeliidae
Taramelliceras richei	Jura.	0.25	0.28	0.30	0.52	0.53	0.56				Palframan 1966	Oppeliidae
Yukagirites kinasovi	Jura.		0.60								Repin et al. 1998	Yukagirit.
Zemistephanus crickmayi [M]	Jura.					0.85					Hall and Westermann 1980; Laptikovsky et al. 2013	Stephanocerat.

Table 5.1 (continued)

Taxon	Age	ID_1			AD			AA			Reference	Family
		min	mean	max	min	mean	max	Min	mean	max		(-t. = -tidae)
Zemistephanus richardsoni [M]	Jura.		0.35			0.80					Hall and Westermann 1980	Stephanocerat.
Acanthohoplites sp.	Cret.	0.33	0.39	0.45	0.62	0.75	0.88	270	275	280	Drushchits and Doguzhaeva 1981	Parahoplit.
Aconeceras (Sanmartinoceras) sp.	Cret.	0.42	0.44	0.45		0.84			300		Drushchits and Doguzhaeva 1981; Landman et al. 1996	Oppeliidae
Aconeceras trautscholdi	Cret.	0.30	0.35	0.38		0.63			295		Drushchits and Khiami 1970	Oppeliidae
Aconeceras trautscholdi	Cret.	0.34	0.36	0.38	0.64	0.77	0.90	275	318	360	Drushchits and Doguzhaeva 1981	Oppeliidae
Anagaudryceras limitatum	Cret.		0.62			1.28			365		Landman et al. 1996	Gaudrycerat.
Anagaudryceras matsumotoi	Cret.		0.59			1.32			338		Landman et al. 1996	Gaudrycerat.
Anagaudryceras nanum	Cret.		0.56			1.26			315		Landman et al. 1996	Gaudrycerat.
Anagaudryceras tetragonum	Cret.		0.58			1.26			323		Landman et al. 1996	Gaudrycerat.
Anagaudryceras yokoyamai	Cret.		0.69			1.41			365		Landman et al. 1996	Gaudrycerat.
Anahoplites intermedius	Cret.	0.60	0.62	0.63							Mikhailova 1974	Hoplit.
Baculites sp.	Cret.					0.78			331		Landman 1982; Landman et al. 1996	Baculit.
Berriasella jauberti	Cret.		0.50								Bogdanova and Arkadiev 2005	Neocomit.
Beudanticeras beudanti	Cret.		0.44			1.06					Landman et al. 1996	Desmocerat.
Beudanticeras laevigatum	Cret.	0.49	0.52	0.55	0.77	0.88	0.99	310	320	330	Landman et al. 1996	Desmocerat.
Boreophylloceras densicostatum	Cret.		1.14			2.37					Igolnikov 2007	Boreophyllocerat.

Table 5.1 (continued)

Taxon	Age	ID_1			AD			AA			Reference	Family
		min	mean	max	min	mean	max	Min	mean	max		(-t. = -tidae)
Boreophylloceras praeinfundibulum	Cret.		1.54			2.90			270		Repin et al. 1998	Boreophyllocerat.
Calliphylloceras subalpinum	Cret.		0.45			0.80			275		Landman et al. 1996	Phyllocerat.
Calliphylloceras velledae	Cret.	0.43	0.50	0.56	0.80	0.84	0.88	280	290	300	Landman et al. 1996	Phyllocerat.
Calycoceras orientale	Cret.		0.46			0.93			281		Landman et al. 1996	Acanthocerat.
Canadoceras kossmati	Cret.		0.41			0.89					Landman et al. 1996	Pachydiscidae
Canadoceras mystricum	Cret.		0.44			1.00			312		Landman et al. 1996	Pachydiscidae
Clioscaphites vermiformis	Cret.	0.29	0.37	0.44	0.60	0.71	0.82	259	283	306	Landman et al. 1996	Scaphit.
Collignoniceras woollgari	Cret.		0.45			0.82			294		Landman et al. 1996	Collignonicerat.
Colombiceras sp.	Cret.	0.35	0.38	0.41	0.60	0.62	0.63	280	290	300	Landman et al. 1996	Parahoplit.
Damesites ainuanus	Cret.	0.35	0.37	0.38	0.67	0.70	0.73	290	299	308	Landman et al. 1996	Desmocerat.
Damesites damesi	Cret.	0.36	0.41	0.46	0.83	0.89	0.95	307	334	360	Landman et al. 1996	Desmocerat.
Damesites latidorsatus	Cret.		0.41			0.85			320		Landman et al. 1996	Desmocerat.
Damesites semicostatus	Cret.	0.34	0.41	0.47	0.77	0.84	0.91	313	317	320	Landman et al. 1996	Desmocerat.
Damesites sugata	Cret.	0.46	0.47	0.48	0.82	0.83	0.84	307	308	309	Zakharov 1978	Desmocerat.
Deshayesites deshayesi	Cret.	0.42	0.49	0.56	0.91	1.00	1.09		300		Landman et al. 1996	Deshayesit.
Desmoceras dawsoni	Cret.	0.72	0.73	0.73	1.51	1.52	1.52		313		Kawabe and Haggart 2003	Desmocerat.
Desmoceras ezoanum	Cret.		0.48			1.22			305		Landman et al. 1996	Desmocerat.
Desmoceras japonicum	Cret.	0.45	0.47	0.48	0.95	0.97	0.98	317	327	337	Landman et al. 1996	Desmocerat.
Desmoceras kossmati	Cret.		0.47			0.85					Kawabe and Haggart 2003	Desmocerat.

Table 5.1 (continued)

Taxon	Age	ID_1			AD			AA			Reference	Family
		min	mean	max	min	mean	max	Min	mean	max		(-t. = -tidae)
Desmoceras kossmati	Cret.		0.40			0.90			305		Landman et al. 1996	Desmocerat.
Desmoceras poronaicum	Cret.	0.45	0.51	0.56		0.84					Kawabe and Haggart 2003	Desmocerat.
Desmophyllites diphylloides	Cret.	0.44	0.46	0.47	0.83	0.86	0.89		309		Landman et al. 1996	Desmocerat.
Desmophyllites diphylloides	Cret.	0.40	0.44	0.48	0.75	0.81	0.86		307		Zakharov 1978	Desmocerat.
Desmophyllites sp.	Cret.		0.43			0.84			317		Landman et al. 1996	Desmocerat.
Desmophyllites sp.	Cret.		0.60			0.89			322		Zakharov 1978	Desmocerat.
Diadochoceras nodosocotatiforme	Cret.		0.30			0.77			291		Landman et al. 1996	Douvilleicerat.
Diadochoceras sinuosocostatus	Cret.	0.38	0.40	0.41	0.71	0.74	0.76	250	260	270	Landman et al. 1996	Douvilleicerat.
Diadochoceras sp.	Cret.	0.38	0.42	0.45	0.70	0.75	0.80				Landman et al. 1996	Douvilleicerat.
Discoscaphites conradi	Cret.	0.37	0.40	0.43	0.72	0.76	0.80	299	310	320	Landman et al. 1996	Scaphit.
Discoscaphites gulosus	Cret.	0.34	0.36	0.38	0.67	0.72	0.77	294	304	313	Landman et al. 1996	Scaphit.
Discoscaphites rossi	Cret.	0.31	0.34	0.37	0.65	0.69	0.72	296	305	314	Landman et al. 1996	Scaphit.
Eogaudryceras (Eotetragonites) aurarium	Cret.		0.29			0.93					Landman et al. 1996	Gaudrycerat.
Eogaudryceras (Eotetragonites) balmensis	Cret.		0.50			0.98					Landman et al. 1996	Gaudrycerat.
Eogunnarites unicus	Cret.		0.35			0.76			319		Landman et al. 1996	Kossmaticerat.
Eupachydiscus haradai	Cret.	0.37	0.45	0.53	0.73	0.90	1.07	306	320	333	Landman et al. 1996	Pachydiscidae
Gabbioceras angulatum	Cret.		0.44			0.88					Landman et al. 1996	Tetragonit.

Table 5.1 (continued)

Taxon	Age	ID_1			AD			AA			Reference	Family
		min	mean	max	min	mean	max	Min	mean	max		(-t. = -tidae)
Gabbioceras latericarinatum	Cret.		0.49		0.88	0.93	0.98	280	290	300	Landman et al. 1996	Tetragonit.
Gabbioceras michelianum	Cret.		0.42			0.90			281		Landman et al. 1996	Tetragonit.
Gaudryceras cf. *denseplicatum*	Cret.	0.69	0.72	0.74	1.21	1.34	1.47				Zakharov 1978	Gaudrycerat.
Gaudryceras cf. *tenuiliratum*	Cret.	0.75	0.76	0.77	1.39	1.40	1.41		330		Zakharov 1978	Gaudrycerat.
Gaudryceras denseplicatum	Cret.	0.71	0.79	0.86	1.39	1.52	1.65	320	328	335	Landman et al. 1996	Gaudrycerat.
Gaudryceras stefaninii	Cret.		0.40			0.93			318		Landman et al. 1996	Gaudrycerat.
Gaudryceras striatum	Cret.	0.66	0.68	0.69	1.12	1.26	1.39	342	348	353	Landman et al. 1996	Gaudrycerat.
Gaudryceras tombetsense	Cret.		0.64			1.42			348		Landman et al. 1996	Gaudrycerat.
Hauericeras angustum	Cret.		0.33			0.70				312	Landman et al. 1996	Desmocerat.
Hauericeras gardeni	Cret.	0.38	0.44	0.50	0.70	0.72	0.73			315	Landman et al. 1996	Desmocerat.
Holcophylloceras guettardi	Cret.	0.45	0.50	0.55	0.70	0.81	0.91	270	275	280	Landman et al. 1996	Phyllocerat.
Holcophylloceras sp.	Cret.	0.46	0.53	0.59	0.70	0.81	0.91				Landman et al. 1996	Phyllocerat.
Hoploscaphites comprimus	Cret.				0.65	0.66	0.67				Landman et al. 1996	Scaphit.
Hoploscaphites nebrascensis	Cret.		0.41			0.68			310		Landman et al. 1996	Scaphit.
Hoploscaphites nicolletii	Cret.	0.39	0.43	0.46	0.72	0.77	0.81	288	302	316	Landman et al. 1996	Scaphit.
Hoploscaphites spedeni	Cret.	0.40	0.41	0.42	0.72	0.76	0.80	292	297	302	Landman et al. 1996	Scaphit.

Table 5.1 (continued)

Taxon	Age	ID_1			AD			AA			Reference	Family
		min	mean	max	min	mean	max	Min	mean	max		(-t. = -tidae)
Hypacanthohoplites sp.	Cret.	0.38	0.44	0.49	0.73	0.84	0.95	260	265	270	Landman et al. 1996	Parahoplit.
Hypacanthohoplites subcornuenianus	Cret.		0.40			0.93			290		Landman et al. 1996	Parahoplit.
Hypophylloceras hetonaiensis	Cret.		0.43			0.90			280		Landman et al. 1996	Phyllocerat.
Hypophylloceras ramosum	Cret.	0.58	0.63	0.67	0.94	1.02	1.09	223	229	235	Zakharov 1978	Phyllocerat.
Hypophylloceras Subramosum	Cret.	0.53	0.60	0.66	0.90	1.03	1.15	270	281	292	Landman et al. 1996	Phyllocerat.
Karsteniceras obatai	Cret.		0.27			0.75			305		Landman et al. 1996	Ancylocerat.
Kitchinites ishikawa	Cret.		0.42								Zakharov 1978	Desmocerat.
Kossmatella agassiziana	Cret.	0.52	0.60	0.67	0.94	1.11	1.27	300	308	315	Landman et al. 1996	Gaudrycerat.
Luppovia sp.	Cret.		0.37			0.70			290		Doguzhaeva and Mikhailova 1982; Land-man et al. 1996	Ancylocerat.
Lytoceras subsequens	Cret.	0.32	0.33	0.34							Landman et al. 1996	Lytocerat.
Mantelliceras japonicum	Cret.		0.40			0.89			284		Landman et al. 1996	Acanthocerat.
Marshallites compressus	Cret.		0.41			0.97			313		Landman et al. 1996	Kossmaticerat.
Melchiorites sp.	Cret.	0.32	0.34	0.36	0.67	0.69	0.70	270	280	290	Landman et al. 1996	Desmocerat.
Menuites pusilus	Cret.		0.50			0.87			328		Landman et al. 1996	Pachydiscidae
Menuites yezoensis	Cret.		0.35			0.76			315		Landman et al. 1996	Pachydiscidae
Metaplacentic. subtilistriatum	Cret.	0.49	0.52	0.54	1.09	1.14	1.19	315	336	356	Tanabe 1979; Landman et al. 1996	Placenticerat.
Microdesmoceras tetragonum	Cret.		0.43			0.94			305		Landman et al. 1996	Desmocerat.
Nolaniceras sp.	Cret.	0.38	0.42	0.45	0.73	0.79	0.85	270	275	280	Landman et al. 1996	Parahoplit.

Table 5.1 (continued)

Taxon	Age	ID_1			AD			AA			Reference	Family
		min	mean	max	min	mean	max	Min	mean	max		(-t. = -tidae)
Parahoplites melchioris	Cret.	0.55	0.63	0.70	1.13	1.21	1.29	315	323	330	Landman et al. 1996	Parahoplit.
Parajaubertella kawakitana	Cret.	0.60	0.63	0.66		1.12			270		Landman et al. 1996	Tetragonit.
Phylloceras japonicum	Cret.		0.46			0.92			272		Landman et al. 1996	Phyllocerat.
Phylloceras sp.	Cret.	0.41	0.45	0.49	0.63	0.74	0.85	260	270	280	Landman et al. 1996	Phyllocerat.
Phyllopachyceras ezoense	Cret.	0.47	0.53	0.58	0.85	1.08	1.30	284	331	377	Tanabe et al. 1979; Landman et al. 1996	Phyllocerat.
Phyllopachyceras ezoense	Cret.	0.56	0.58	0.60	0.86	0.91	0.96	234	241	248	Zakharov 1978	Phyllocerat.
Phyllopachyceras sp.	Cret.		0.44			0.76					Landman et al. 1996	Phyllocerat.
Placenticeras gaurdakense	Cret.		0.70								Mikhailova 1974	Placenticerat.
Placenticeras sp.	Cret.	0.63	0.64	0.65							Mikhailova 1974	Placenticerat.
Placenticeras sp.	Cret.		0.75								Mikhailova 1974	Placenticerat.
Protetragonites tauricus	Cret.		0.65								Landman et al. 1996	Lytocerat.
Protexanites minimus	Cret.		0.35			0.74			280		Shigeta 1993	Collignonicerat.
Pseudohaploceras nipponicus	Cret.		0.32			0.79			302		Landman et al. 1996	Desmocerat.
Pseudophyllites indra	Cret.		0.58			1.48					Landman et al. 1996	Tetragonit.
Ptychoceras renngarteni	Cret.		0.50			0.85			330		Landman et al. 1996	Ptychocerat.
Ptychophylloceras ptychoicum	Cret.	0.38	0.40	0.41		0.69			260		Landman et al. 1996	Phyllocerat.
Pusozia takahashii	Cret.		0.37			0.89			306		Landman et al. 1996	Desmocerat.
Puzosia orientale	Cret.		0.36			0.83			310		Landman et al. 1996	Desmocerat.

Table 5.1 (continued)

Taxon	Age	ID_1			AD			AA			Reference	Family
		min	mean	max	min	mean	max	Min	mean	max		(-t. = -tidae)
Puzosia pacifica	Cret.	0.37	0.40	0.43	0.83	0.84	0.84	302	306	310	Landman et al. 1996	Desmocerat.
Puzosia yubarensis	Cret.		0.28			0.61			302		Landman et al. 1996	Desmocerat.
Saghalinites teshioensis	Cret.		0.56			1.19					Landman et al. 1996	Tetragonit.
Scaphites carlilensis	Cret.				0.59	0.60	0.60		274		Landman et al. 1996	Scaphit.
Scaphites corvensis	Cret.		0.31			0.67			282		Landman et al. 1996	Scaphit.
Scaphites depressus	Cret.	0.32	0.36	0.40	0.68	0.76	0.83	282	287	292	Landman et al. 1996	Scaphit.
Scaphites larvaeformis	Cret.	0.28	0.32	0.36	0.52	0.59	0.65	266	283	300	Landman et al. 1996	Scaphit.
Scaphites nigricollensis	Cret.	0.34	0.37	0.40	0.60	0.68	0.76	253	281	308	Landman et al. 1996	Scaphit.
Scaphites planus	Cret.	0.40	0.48	0.55	0.73	0.84	0.95		280		Landman et al. 1996	Scaphit.
Scaphites preventitricosus	Cret.	0.29	0.36	0.42	0.58	0.65	0.71	257	275	292	Landman et al. 1996	Scaphit.
Scaphites pseudoaequalis	Cret.	0.35	0.41	0.46	0.64	0.72	0.80		295		Landman et al. 1996	Scaphit.
Scaphites warreni	Cret.	0.27	0.33	0.38	0.60	0.66	0.72	266	278	290	Landman et al. 1996	Scaphit.
Scaphites whitfieldi	Cret.	0.25	0.33	0.40	0.55	0.65	0.74	260	284	308	Landman et al. 1996	Scaphit.
Scaphites yonekurai	Cret.		0.44			0.87			295		Landman et al. 1996	Scaphit.
Schloenbachia sp.	Cret.		0.55								Mikhailova 1974	Schloenbachidae
Schloenbachia sp.	Cret.		0.50								Mikhailova 1974	Schloenbachidae
Schloenbachia varians	Cret.	0.50	0.52	0.53							Mikhailova 1974	Schloenbachidae
Simbirskites coronatiformis	Cret.	0.55	0.58	0.60	1.05	1.09	1.12	300	308	315	Landman et al. 1996	Polyptychit.
Simbirskites discofalcatus	Cret.	0.55	0.63	0.70	1.13	1.17	1.20	300	308	315	Landman et al. 1996	Polyptychit.
Simbirskites elatus	Cret.	0.57	0.59	0.60	1.05	1.06	1.06				Landman et al. 1996	Polyptychit.

Table 5.1 (continued)

Taxon	Age	ID_1			AD			AA			Reference	Family
		min	mean	max	min	mean	max	Min	mean	max		(-t. = -tidae)
Simbirskites sp.	Cret.		0.65			1.26			330		Landman et al. 1996	Polyptychit.
Simbirskites sp.	Cret.		0.48			0.98			308		Landman et al. 1996	Polyptychit.
Simbirskites versicolor	Cret.	0.53	0.58	0.63	1.02	1.09	1.15		300		Landman et al. 1996	Polyptychit.
Subprionocyclus bakeri	Cret.	0.36	0.39	0.42	0.74	0.75	0.75	270	292	313	Landman et al. 1996	Collignonicerat.
Subprionocyclus minimum	Cret.	0.35	0.43	0.50	0.59	0.74	0.89	250	299	348	Landman et al. 1996	Collignonicerat.
Subprionocyclus neptuni	Cret.	0.37	0.44	0.50	0.70	0.78	0.85	242	275	307	Landman et al. 1996	Collignonicerat.
Teshioites sp.	Cret.		0.46			0.92			305		Landman et al. 1996	Pachydiscidae
Tetragonites duvalianus	Cret.	0.50	0.60	0.70	0.94	1.06	1.18	300	315	330	Landman et al. 1996	Tetragonit.
Tetragonites glabrus	Cret.		0.56			1.08			331		Landman et al. 1996	Tetragonit.
Tetragonites hulensis	Cret.		0.50			1.04			312		Landman et al. 1996	Tetragonit.
Tetragonites minimus	Cret.	0.50	0.55	0.60	0.90	0.98	1.05	320	330	340	Landman et al. 1996	Tetragonit.
Tetragonites popetensis	Cret.		0.42			0.97			340		Landman et al. 1996	Tetragonit.
Tetragonites popetensis	Cret.	0.62	0.63	0.63	0.88	1.08	1.27	305	312	318	Zakharov 1978	Tetragonit.
Tetragonites terminus	Cret.	0.93	0.99	1.05	1.70	1.80	1.90	330	338	345	Landman et al. 1996	Tetragonit.
Texanites kawasakii	Cret.		0.47			0.93			280		Shigeta 1993	Texanit.
Tragodesmoceroides Subcostatus	Cret.	0.36	0.42	0.47	0.83	0.88	0.92	304	309	314	Landman et al. 1996	Desmocerat.
Valdedorsella akuschaensis	Cret.		0.28			0.68			308		Landman et al. 1996	Desmocerat.
Yezoites klamathensis	Cret.	0.37	0.43	0.48	0.67	0.75	0.83		285		Landman et al. 1996	Scaphit.
Yezoites matsumotoi	Cret.		0.43			0.80			285		Landman et al. 1996	Scaphit.
Yezoites puerculus	Cret.	0.43	0.51	0.58	0.71	0.82	0.92		285		Landman et al. 1996	Scaphit.

Table 5.1 (continued)

Taxon	Age	ID_1			AD			AA			Reference	Family
		min	mean	max	min	mean	max	Min	mean	max		(-t. = -tidae)
Yokoyamaoceras ishikawai	Cret.	0.39	0.50	0.61	0.80	0.89	0.97	297	321	344	Landman et al. 1996	Kossmaticerat.
Zelandites aff. *inflatus*	Cret.	0.67	0.68	0.68	1.19	1.22	1.24	320	328	336	Zakharov 1978; Landman et al. 1996	Gaudrycerat.
Zelandites kawanoii	Cret.		0.57			1.19			345		Landman et al. 1996	Gaudrycerat.
Zelandites mihoensis	Cret.		0.46			1.00			312		Landman et al. 1996	Gaudrycerat.
Zelandites varuna	Cret.		0.54			1.22			342		Landman et al. 1996	Gaudrycerat.
Zuercherella falcistriata	Cret.	0.38	0.40	0.42	0.66	0.75	0.84	270	280	290	Drushchits and Doguzhaeca 1981; Landman et al. 1996	Desmocerat.

of ammonoid classification is not the scope of this study. Where specimens were measured more than once, we used the values of the most recently published article. Some additional measurements can be found in Laptikhovsky et al. (2013), but we were unable to verify these values in the original publications and as these authors only give average values, we refrain from listing them here. Abbreviations: Dev = Devonian; Carb. = Carboniferous; Perm. = Permian; Trias. = Triassic; Jura. = Jurassic; Cret. = Cretaceous.

Additionally, various authors have demonstrated quite large intraspecific variation in embryonic shell and initial chamber diameters in samples of Jurassic (Palframan 1966, 1967, 1969; Neige 1997; Rouget and Neige 2001; Morard and Guex 2003) and Cretaceous ammonoids (Tanabe 1977a, b; Landman 1987; Kawabe and Haggart 2003; Tanabe et al. 2003; Nishimura et al. 2010; Tajika and Wani 2011) deriving from the same locality and stratigraphic unit (Table 5.2). Erben (1950, 1962b, 1964) qualitatively illustrated large intraspecific differences in shape and size of the initial chamber and embryonic shell in the early Devonian *Mimagoniatites fecundus*; ranging from forms with a smaller initial chamber and more elliptical first whorl ("*forme elliptique*" of Erben 1950, 1962b) to forms with a larger initial chamber and more circular first whorl ("*forme circulaire*" of Erben 1950, 1962b), which he also observed in other taxa (Erben 1962b).Tanabe et al. (1995) reported the intraspecific variation in the Carboniferous *Homoceras subglobosum* and Stephen and Stanton (2002) reported large ranges of intraspecific variation in ammonitella diameter of seven Carboniferous ammonoids. Arkadiev and Vavilov (1984a) as well as Vavilov (1992) listed measurements of various Triassic taxa, occasionally multiple specimens of the same species (up to five in *Parapopanoceras asseretoi*, *P. medium* and *Indirigites krugi*).

We herein use the coefficient of variation (CV) to quantify and compare the range of intraspecific variation in various measurements of the initial chamber and the ammonitella (Fig. 5.6; Table 5.2); the total range of the measurement should not be used for these purposes as it is very dependent on the sample size (Van Valen 2005; De Baets et al. 2015). The measurements of the initial chamber show the greatest range of intraspecific variation as well as the most extreme values, particularly the ID_1 and ID_2, while the measurements of the ammonitella (AD, AA) show a smaller range of intraspecific variation on average within the studied species (Fig. 5.6). It is unclear if these differences are due to biological reasons or if these could be related to measurement errors and sampling biases (e.g., comparatively little measurements are available for protoconch width). Comparisons between these values, even if they were derived from the same locality and time interval, are not straightforward due to time-averaging and other biases (De Baets et al. 2015). In some cases, extreme values could have originated from mixing specimens from different localities or time intervals (e.g., *Scaphites warreni*), measurement or preparation errors, or the low number of available specimens (e.g., *Parapopanoceras*).

Rouget and Neige (2001) also measured the area of the initial chamber. However, almost none of these approaches consider the width of the initial chamber and ammonitella and therefore, the 3D-structure or volume of the embryonic shell. House (1985) estimated the initial chamber volume to vary between about 0.01

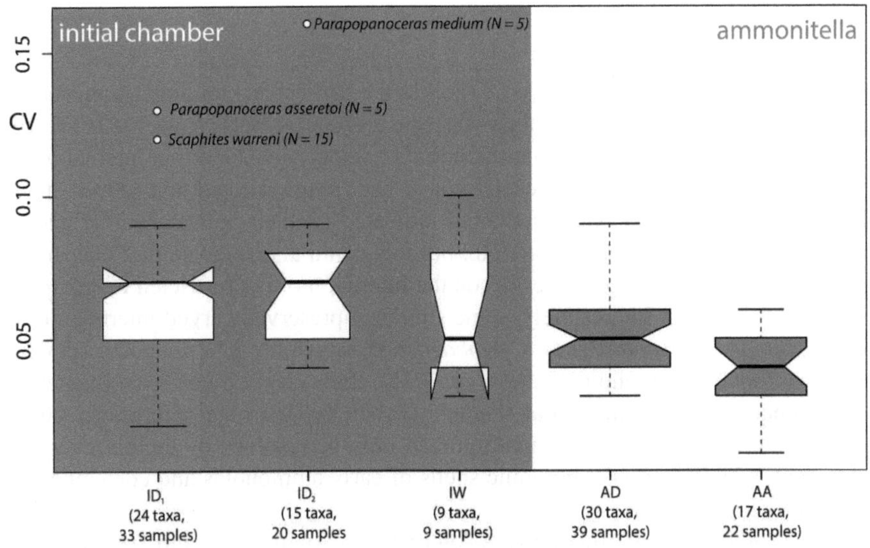

Fig. 5.6 Notched boxplots of the coefficient of variation within measurements of the initial chamber (*ID_1*, *ID_2*, *IW*) and the ammonitella (*AD*, *AA*) observed within samples of the same species. If two box notches do not overlap, there is strong evidence that their medians differ (compare Chambers et al. 1983, p. 62). The extremely large values in *Parapopanoceras* might be related to the low sample number and measurement errors (as sometimes multiple values are given for the same specimen: compare Arkadiev and Vavilov 1984a; Vavilov 1992), while the values might be artificially inflated the coefficient of variation in *Scaphites warreni* (compare Landman 1987). The width of the boxes is proportional to the amount of available samples. Histograms were produced in the freely available statistical environment R. (http://www.r-project.org/)

mm^3 in the spindle-shaped protoconch of *Distichoceras* to 1.5 mm^3 in the subspherical protoconch of *Agoniatites*. The lower limits of embryonic shell (0.4–0.5 mm) and initial chamber size (~0.2–0.3 mm) and their volumes might point to certain constructional and physiological constraints on egg and hatchling size (Vance 1973; Boletzky 1993, 1997; De Baets et al. 2012). Yolky eggs of a few hundred microns are considered to be a minimum size from which an actively jetting cephalopod with a muscular-funnel complex for jet propulsion and a muscular arm crown for seizure of prey can be formed (Boletzky 1989, 1993).

Several authors also measured the dimensions of the caecum and the prosiphon (compare 5.2.7; Landman 1987; Landman et al. 1996; Tanabe et al. 2003), particularly in the Russian literature (e.g., Zakharov 1974, 1978; Drushchits et al. 1977b; Drushchits and Doguzhaeva 1981; Arkadiev and Vavilov 1984a, b; Alekseev et al. 1984; Vavilov 1992; Arkadiev et al. 1993). These include the diameter of the caecum measured along the spiral (C1) and a second diameter measured perpendicular to the first (C2) as well as the length of the prosiphon (abbreviated as PL in Landman et al. 1996; sometimes also known as the length of the fixator: compare Drushchits et al. 1977a). Rouget and Neige (2001) also measured the area of the caecum (CS).

5.2.4 Ornamentation

The ornamentation of the embryonic shell in the earliest ammonoids (Anetoceratinae, Mimosphinctidae) is still unknown (House 1996; De Baets et al. 2012, 2013b). Their embryonic shells (including the initial chamber) are smooth in internal mold preservation as is the case in closely related bactritoids (Kröger and Mapes 2007; Kröger 2008; Fig. 5.2). This fact could indicate that their initial chamber and/or the embryonic shell were also smooth, but this could also indicate that their ornamentation is too fine to be preserved on the internal mold or restricted to the upper surface of the shell. Interestingly, some similarly-preserved curved internal molds of bactritoids (*Pseudobactrites*, *Cyrtobactrites*), which are interpreted by some to be transitional to ammonoids (Klug et al. 2015), show coarse transverse ornamentation immediately after the initial chamber (Erben 1964; Kröger and Mapes 2007; De Baets et al. 2013b).This conundrum can only be resolved by the discovery of abundant, well-preserved embryonic shells of early ammonoids and contemporary (Devonian) bactritoids.

Embryonic shells of more derived Agoniatitina, Anarcestina, and Tornoceratina are ornamented with transverse lirae (Fig. 5.7). The lirae are not uniformly distributed but are more closely spaced at the end of the initial chamber and at the end of the ammonitella (Fig. 5.7a, b; Klofak et al. 2007; compare House 1965). Erben (1962b, 1964, 1966) reported that the lirae develop a slight backward projection on the venter at the end of the initial chamber, which become more pronounced over the course of the first whorl. Wrinkle-like creases are also present on the shell and extend perpendicular to the lirae, giving the shell a reticulate appearance (Fig. 5.7e; Klofak et al. 1999). The lirae are symmetrical in cross section and represent periodic thickenings on the shell surface. As a result, they do not leave an impression on the steinkern. Adoral of the primary constriction, the lirae abruptly change in size, spacing, and shape (Fig. 5.7b, c). They become larger, more widely spaced, and asymmetrical in cross section. In addition, finer, more closely spaced striae (growth lines?) develop between the lirae (Klofak et al. 2007).

Kulicki et al. (2002) described the embryonic whorls of goniatites based on material with aragonitic preservation from the Late Carboniferous Buckhorn Asphalt of Oklahoma. The outer surface of the ammonitella is smooth without any trace of ornamentation or growth lines. This contrasts with earlier reports describing longitudinal lirae on the ammonitellae of goniatites from the Upper Carboniferous of Kansas (Tanabe et al. 1993). However, the specimens studied here are not as well preserved as those from the Buckhorn Asphalt, and reexamination of these and additional specimens revealed that these lirae represent micro-ornamentation on the inner side of the dorsal shell wall and that the exposed surface of the embryonic shell is smooth (Tanabe et al. 2001).

In contrast to Paleozoic ammonoids, the exposed portions of the ammonitella in Mesozoic ammonoids, including the Lytoceratina, Phylloceratina, Ancyloceratina, and Ammonitina, are covered with a tuberculate micro-ornamentation (e.g., Brown 1892; Kulicki 1974, 1979, 1996; Kulicki and Doguzhaeva 1994; Bandel

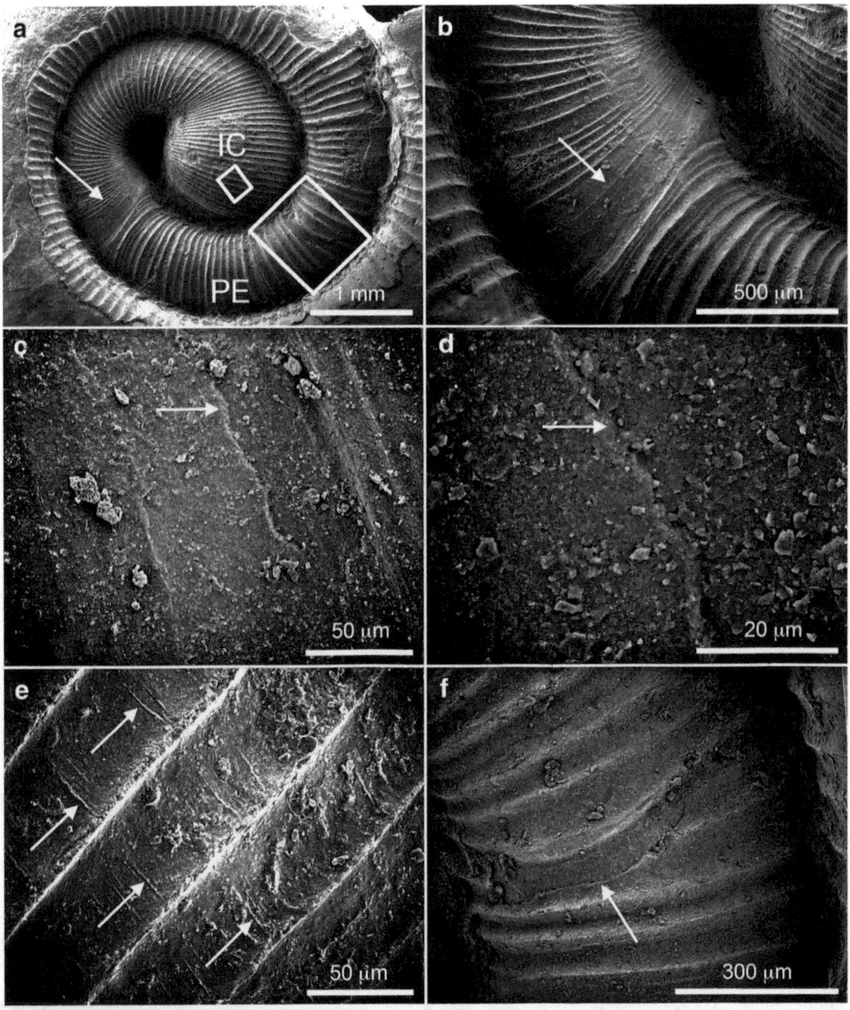

Fig. 5.7 Micro-ornamentation on the embryonic shell of the Agoniatitina. **a–f** *Mimagoniatites fecundus*, AMNH 46645, Devonian, Morocco. **a** Lateral view of the ammonitella showing the initial chamber, the apertural edge of the ammonitella (arrow), and approximately one-half whorl of the post-embryonic shell. **b** Close-up of aperture (arrow) showing the reduction in size and spacing of transverse lirae. Post-embryonic shell is to the right. **c** Close-up from **b**, rotated approximately 45° to the right, showing the apertural edge of the ammonitella (arrow). The post-embryonic shell is visible emerging from beneath the ammonitella edge on the right. **d** Close-up of the apertural edge of the ammonitella from C (arrow). **e** Close-up of the transverse lirae on the initial chamber showing the "wrinkle-like" creases stretched perpendicularly between them (arrows). Area of photograph is indicated by the small box on the initial chamber in 7A. **f** Close-up of the post-embryonic shell showing a healed break in the shell (arrow) which disrupted the production of the ornament. Area of close-up is indicated by box on the post-embryonic shell in 7A (from Klofak et al. 2007). Abbreviations: IC initial chamber; PE post-embryonic shell

1982; Bandel et al. 1982; Landman 1985, 1987, 1988, 1994; Landman and Waage 1993; Landman et al. 1996; Tanabe 1989; Sprey 2001, 2002). Based on his study of the early whorls of *Quenstedtoceras*, Kulicki (1979) documented that the tubercles represent extensions of the prisms of the outer prismatic layer. Tanabe et al. (2010) further noted that the tubercles are "*embedded*" in the outer prismatic layer, leaving "*holes*" where the tubercles have fallen off. Landman et al. (2001) documented this same micro-ornamentation on the embryonic shells of the Late Triassic ceratite *Trachyceras* from the San Cassiano (St. Cassian) Formation of the Italian Dolomites (Fig. 5.8). Thus, the presence of a tuberculate micro-ornamentation appears to represent a synapomorphy for all Mesozoic ammonoids.

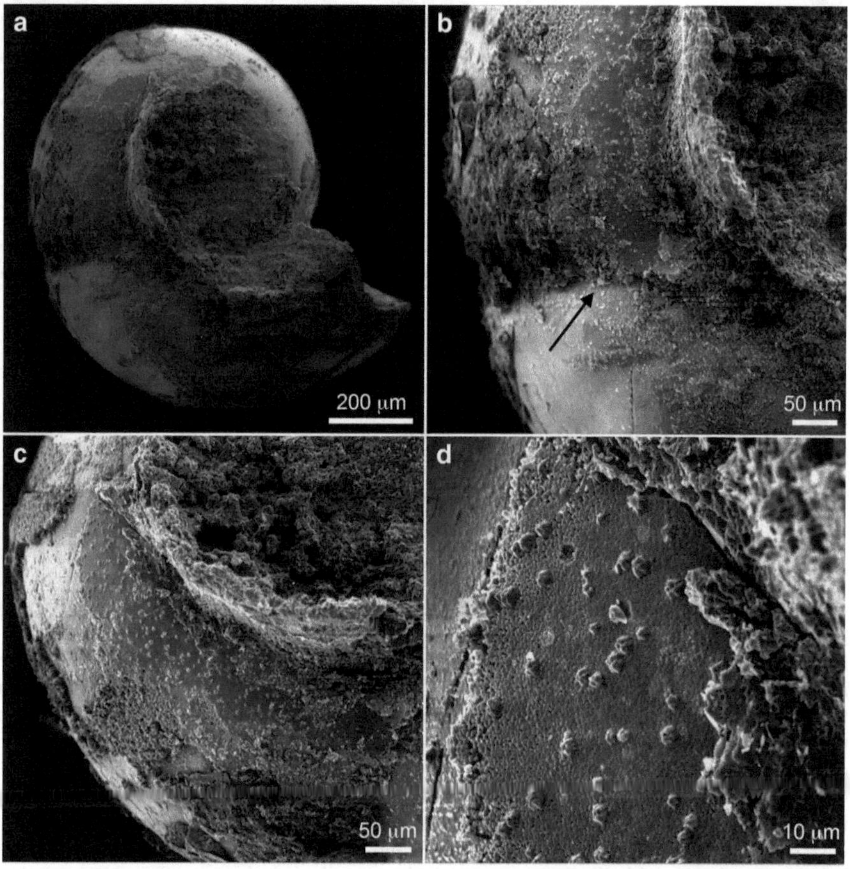

Fig. 5.8 Micro-ornamentation on the embryonic shell of the Ceratitina. **a–d** *Trachyceras* (*Trachyceras*) *aon* (Münster 1834), AMNH 46596, San Cassiano Formation, Dolomites, Italy. **a** Right lateral view of the ammonitella and part of the first whorl. **b** A patch of shell is visible on the adoral end of the ammonitella near the primary constriction (arrow). **c** The tubercles are irregularly distributed on the shell surface. The specimen has been rotated 30° clockwise relative to **a**. **d** The tubercles are composed of multiple sectors converging to a conical top. (from Landman et al. 2001)

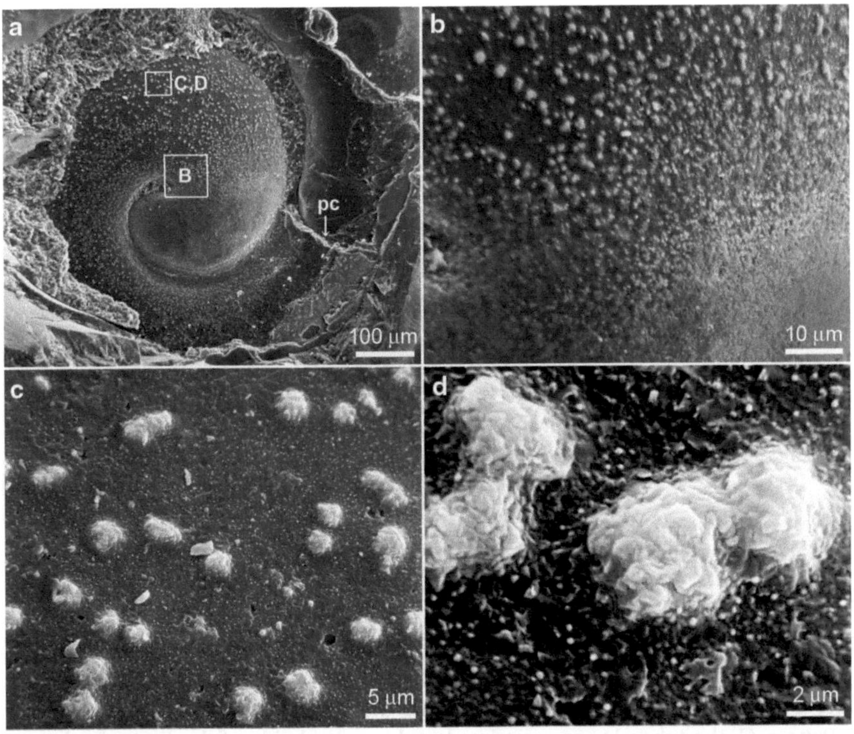

Fig. 5.9 Micro-ornamentation on the embryonic shell of the Ammonitina. **a–d** *Menuites* sp., UMUT MM 18327, lower Campanian, Nakagawa area, Hokkaido, Japan. **a** Lateral view of the embryonic shell. **b–d** Tuberculate micro-ornamentation on the initial chamber (**b**) and first whorl (**c**, **d**). Tubercles coalesce on the initial chamber (from Tanabe et al. 2010). Abbreviations: *pc* primary constriction

The size of the tubercles varies slightly among Mesozoic suborders and this variation could possibly be used for phylogenetic analysis. However, even within a single specimen, the size of the tubercles can vary, with smaller tubercles usually located nearer the ammonitella edge. In general, the tubercle size ranges from 3–4 μm in the Ceratitina (*Trachyceras*; Fig. 5.8), 2–4 μm in the Ammonitina (*Aconeceras, Anapachydiscus, Binatisphinctes, Desmophyllites, Kosmoceras, Menuites, Metaplacenticeras, Quenstedtoceras*: Fig. 5.9, 5.10, 5.11), 3–4 μm in the Phylloceratina (*Hypophylloceras* and *Phyllopachyceras*; Fig. 5.12a, b, c, d), 3.5–7 μm in the Lytoceratina (*Gaudryceras, Anagaudryceras*; Fig. 5.12e, f), and 4.5–7.5 μm in the Ancyloceratina (*Scaphites, Yezoites, Hoploscaphites, Clioscaphites, Discoscaphites*;Fig. 5.13; see references above).

The tubercles occur on the exposed portion of the ammonitella and abruptly disappear on the adoral side of the primary constriction. However, they never occur on the inner portion of the initial chamber covered by the first whorl or on the area immediately adjacent to the apertural edge. Tanabe et al. (2010) documented four patterns of distribution. In Pattern A, the tubercles are irregularly distributed on the

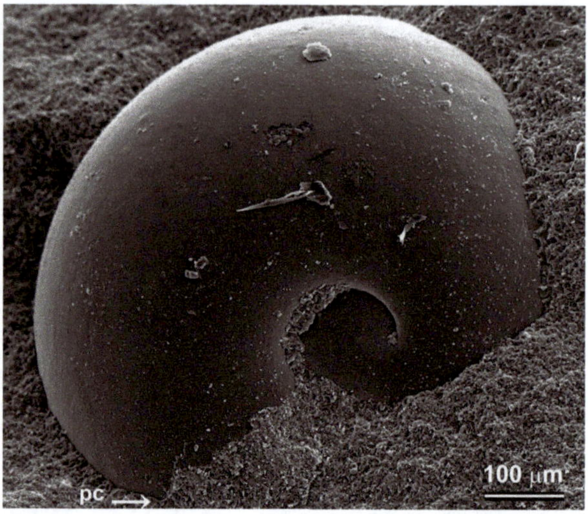

Fig. 5.10 Micro-ornamentation on the embryonic shell of the Ammonitina. *Aconeceras* cf. *trautscholdi*, UMUT MM 29439–4, lower Aptian, Symbirsk, Russia. Tubercles are absent on the lateral sides of the initial chamber (from Tanabe et al. 2010). Abbreviation: *pc* primary constriction

exposed surface. They sometimes become smaller and more closely spaced on the sides of the initial chamber and even fuse together in this region, forming a smooth layer (e.g., *Menuites*; Fig. 5.9a, b). In Pattern B, the tubercles are also regularly distributed on the exposed surface of the embryonic shell except for the lateral sides of the initial chamber where they are absent (e.g., *Aconeceras*; Fig. 5.10). In Pattern C, the tubercles are restricted to the ventral side of the first whorl (e.g., *Phyllopachyceras*; Fig. 5.12c, d) and in Pattern D, the tubercles are restricted to the initial chamber near the umbilical seam (Fig. 5.11e, f).

5.2.5 Microstructure of the Shell Wall

Studies of the microstructure of the embryonic shell have mostly been restricted to Jurassic and Cretaceous ammonoids because many of these forms retain their original mineralogy and microstructure (Erben et al. 1969; Kulicki 1974, 1979; Birkelund and Hansen 1974; Birkelund 1981; Drushchits and Khiami 1970; Drushchits and Doguzhaeva 1974, 1981; Drushchits et al. 1977a, b). Studies of Paleozoic forms are much rarer and include those by Kulicki et al. (2002) on the Goniatitina using well preserved material from the Carboniferous Buckhorn Asphalt of Oklahoma and by Doguzhaeva (2002) on the Bactritina using well preserved specimens from the lower Permian of the southern Urals.

The microstructure of the embryonic shell of Mesozoic ammonoids has been studied by many authors (for a review, see Landman et al. 1996; Kulicki 1996; Kulicki et al. 2015). One of the principle differences in their observations relates to the microstructure of the shell wall at the end of the initial chamber and the beginning of the first whorl. Erben et al. (1968) as well as Birkelund and Hansen (1968) argued that the wall of the initial chamber wedges out on the outer side before or

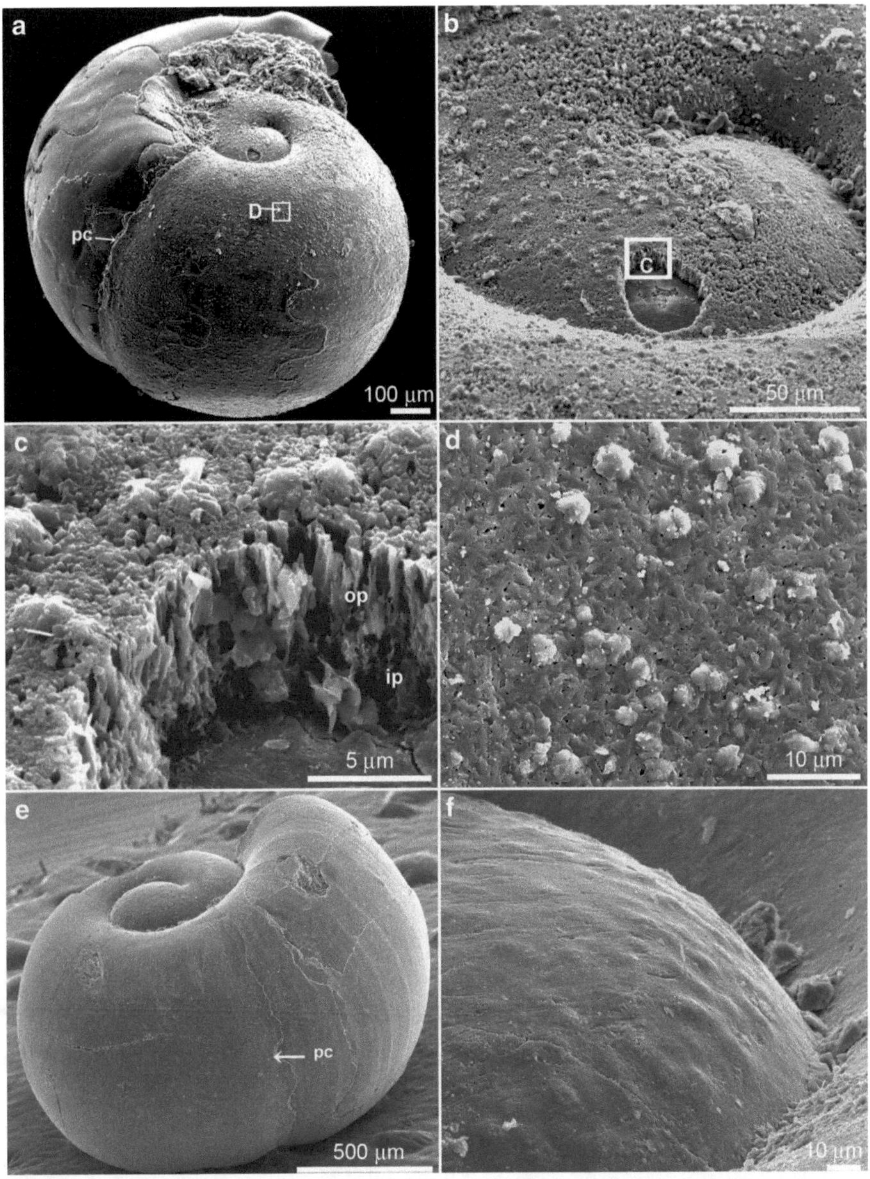

Fig. 5.11 Micro-ornamentation on the embryonic shell of the Ammonitina. **a–d** *Quenstedtoceras* sp., UMUT MM 29445, Callovian, Łuków, Poland. **a** Ventrolateral view of the specimen showing a clear primary constriction at the boundary between the embryonic and post-embryonic stages. **b** Close-up of the lateral shell surface of the initial chamber with numerous tubercles. **c** Part of b, showing the microstructure of the initial chamber wall consisting of outer prismatic and inner prismatic layers. **d** Ventrolateral shell surface of the first whorl, with tubercles. **e, f** Ammonitina, gen. et sp. indeterminate, morphotype B, UMUT MM 29446, middle Bathonian, Faustianka, Poland. **e** Ventrolateral view of the early post-embryonic shell showing a clear primary constriction at the boundary between embryonic and post-embryonic stages. **f** Ventrolateral view of a portion of the initial chamber. The embryonic shell surface portion is mostly smooth except for the initial chamber surface (from Tanabe et al. 2010). Abbreviations: *ip* inner prismatic layer; *op* outer prismatic layer; *pc* primary constriction

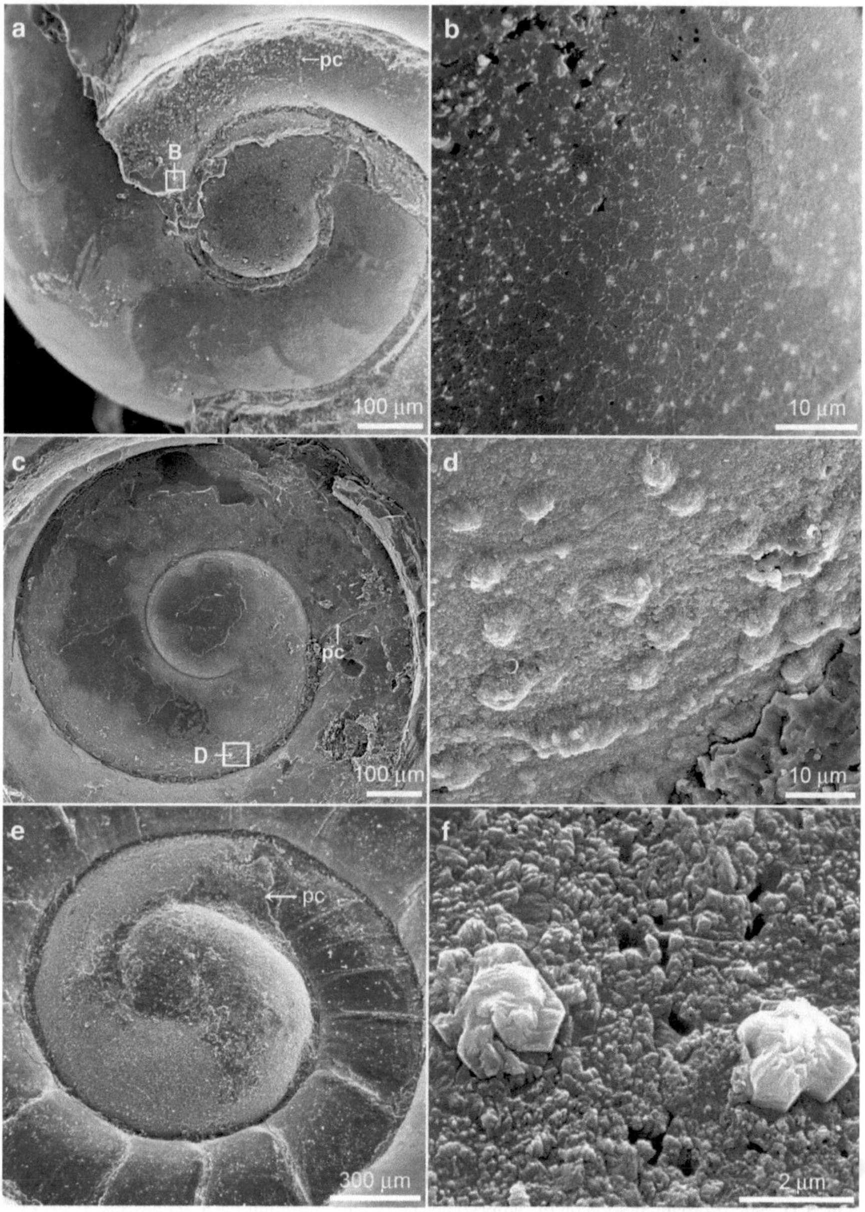

Fig. 5.12 Micro-ornamentation on the embryonic shell of the Phylloceratina and Lytoceratina. **a**, **b** *Hypophylloceras subramosum* (Shimizu 1934), UMUT MM 18320, lower Campanian, Nakagawa area, Hokkaido, Japan. **a** Lateral view of the embryonic shell. **b** Close-up of tubercles on the umbilical shoulder of the first whorl. **c**, **d** *Phyllopachyceras ezoense* (Yokoyama 1890), UMUT MM 18321, lower Campanian, Nakagawa area, Hokkaido, Japan. **c** Lateral view of the embryonic shell. **d** Close-up of tubercles on the ventrolateral side of the first whorl. **e**, **f** *Gaudryceras denseplicatum* Yabe 1903, UMUT MM 183222, Coniacian, Haboro area, Hokkaido, Japan. **e** Lateral view of the embryonic shell. **f** Close-up of tubercles on the lateral side of the first whorl (from Tanabe et al. 2010). Abbreviation: *pc* primary constriction

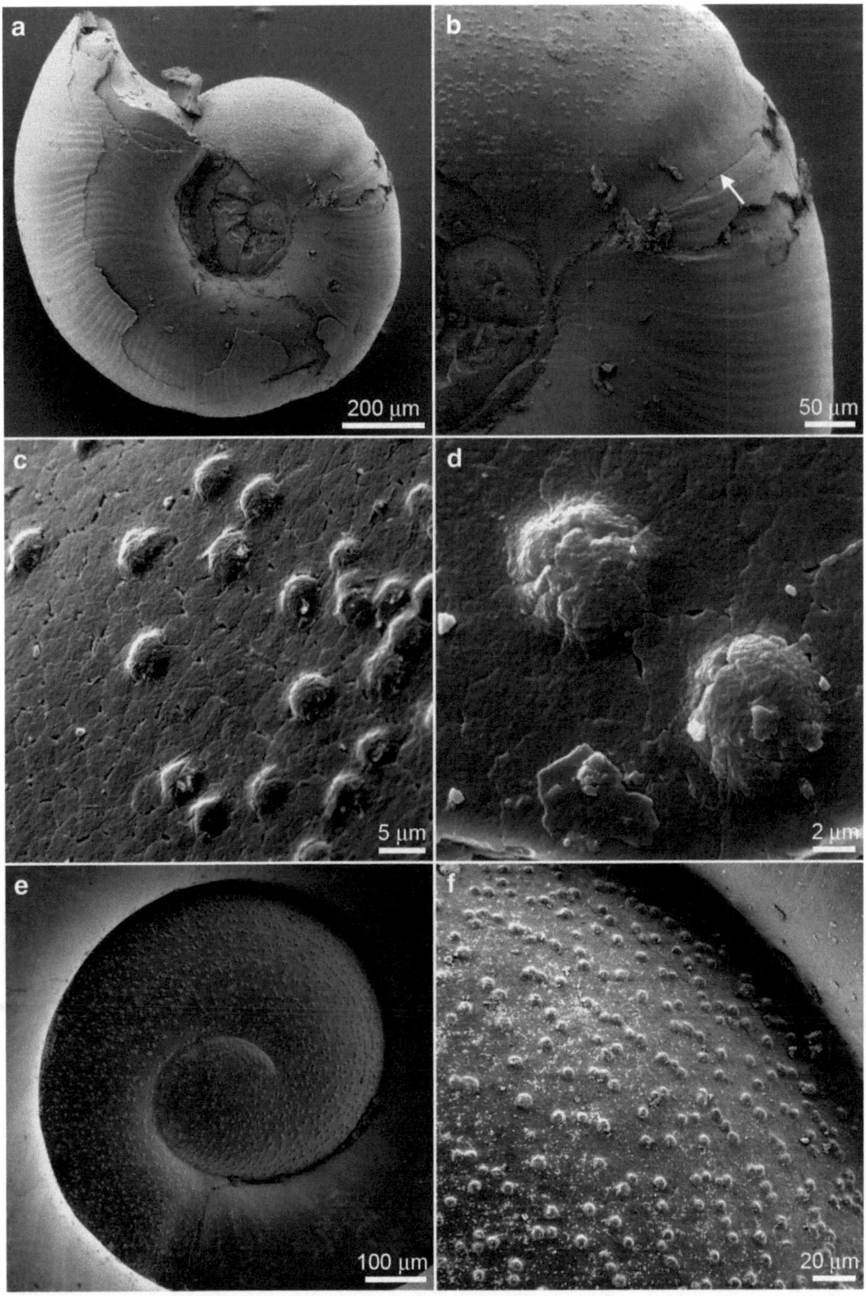

Fig. 5.13 Micro-ornamentation on the embryonic shell of the Ancyloceratina. **a–d** *Scaphites whitfieldi* Cobban, 1951, AMNH 44833, Turonian, South Dakota, U.S.A. **a** Lateral view of the early whorls of a juvenile. **b** The embryonic shell/post-embryonic shell boundary is marked by an arrow. Note that the tubercles abruptly disappear just adapical of the primary constriction, and are replaced by growth lines on the post-embryonic shell. **c, d** Close-ups of tubercles on the lateral side of the first whorl. **e, f** *Hoploscaphites* sp., YPM 34113, Upper Maastrichtian, South Dakota, U.S.A. **e** Lateral view of the embryonic shell. **f** Close-up of tubercles, some of which form discontinuous rows. (from Tanabe et al. 2010)

at the beginning of the first whorl. Kulicki (1979) argued that the wall of the initial chamber does not wedge out at all but forms the external layer of the wall of the first whorl. In contrast, Bandel (1982) and Tanabe (1989) argued that the wall of the initial chamber wedges out, but on the inner side, and that the external layer of the wall of the first whorl first appears on the outer side near the distal end of the initial chamber. Evidence from specimens actually preserved at early ontogenetic stages supports the views of Bandel (1982) and Tanabe (1989).

In all ammonoids whose microstructure has been examined, the most marked change in microstructure occurs at the primary constriction. In the Tornoceratina, Goniatitina, Ceratitina, Phylloceratina, Lytoceratina, Ammonitina, and Ancyloceratina, the outer prismatic layer of the first whorl decreases in thickness, and a large pad of nacre (the primary varix) develops on the inside of the shell (Grandjean 1910; House 1965; Dauphin 1975, 1977; Drushchits and Khiami 1970; Drushchits and Doguzhaeva 1974; Drushchits et al. 1977a, b, 1980; Kulicki 1979; Landman and Waage 1982). The varix parallels the primary constriction and lies close to its adapical end. Occasionally, the outer prismatic layer doubles back along the inside surface of the primary varix forming a short return section (Kulicki 1979). A primary varix has not been documented in the Agoniatitida (Klofak et al. 1999, 2007). In the Bactritina, a nacreous layer appears gradually on the straight part of the shell adapical of the primary constriction (Doguzhaeva 2002). A constriction on the steinkern, which is interpreted to correspond to the end of the embryonic shell (nepionic constriction), has been reported from various taxa of Agoniatitida, including Agoniatitina (*e.g., Mimagoniatites, Latanarcestes, Agoniatites*: Erben 1964; House 1996; Klofak et al. 1999, 2007), Anarcestina (*Anarcestes, Sellanarcestes, Praewernoceras*: Chlupáč and Turek 1983; Klofak et al. 1999, 2007; De Baets et al. 2012) and Gephuroceratina (e.g., *Manticoceras, Probeloceras*: Korn and Klug 2002; Landman et al. 1996; De Baets et al. 2012). Such constrictions have so far not been reported from less derived ammonoids (House 1996; De Baets et al. 2012, 2013b). The nacreous layer rapidly thickens but does not form a varix. Regardless of suborder, the post-ammonitella shell consists of both nacreous and prismatic layers and emerges from below the embryonic shell.

5.2.6 Septa

The first septum develops at the distal end of the initial chamber and is called the proseptum to distinguish it from all other septa (Schindewolf 1928, 1929, 1951, 1954). According to Branco (1879, 1880), the prosuture (the suture corresponding to the proseptum) can be divided into three character states: asellate, latisellate, and angustisellate, depending on the size of the dorsal and ventral saddles. However, as noted by Landman et al. (1996), these character states do not accommodate the full range of variation in the Ammonoidea (compare House 1965; Bensaid 1974). The proseptum is prismatic in microstructure. However, Landman and Bandel (1985, Fig. 33, 34) documented that, in *Euhoplites*, it consists of three layers: a middle irregularly prismatic layer sandwiched between two more homogeneous layers, which originally may have been organic. Wrinkles are visible on the surface of the proseptum in well-preserved specimens free of matrix (Landman and Bandel 1985, Fig. 18).

The second septum, sometimes called the primary septum, is the basis for the ontogenetic development of all later septa. The shape of the second septum is completely different from that of the proseptum. The suture corresponding to the second septum (primary suture) is characterized by a ventral, dorsal, and lateral lobe, and a maximum of three umbilical lobes, depending on the suborder. For example, according to Kullmann and Wiedmann (1970), the primary suture in the Tetragonitoidea (Lytoceratina) contains a total of six lobes (sixlobate). The microstructure of the second septum also differs from that of the proseptum. Like all subsequent septa, the second septum is mainly composed of nacre.

Drushchits and Doguzhaeva (1981) documented the variation in the distance between the proseptum and second septum in Mesozoic ammonoids. For example, in *Quenstedtoceras*, the second septum rides dorsally on the proseptum, although the two septa are distinct ventrally (Bandel 1982; Landman and Bandel 1985). In the Goniatitina, Landman et al. (1999) noted that the proseptum and second septum are very closely spaced. Because of this close spacing, Miller and Unklesbay (1943) identified both of these structures as prosepta. However, later studies have demonstrated that these structures represent the proseptum and second septum, which are joined together around the siphuncular opening on the dorsal side.

The relationship of the first few septa has also been examined in more primitive ammonoids. Klofak and Landman (2010) documented that the first three septa including the proseptum are joined together at a triple junction along the dorsal wall of the initial chamber in the Agoniatitina (Fig. 5.14). In some cases, authors have reported the subdivision of the initial chamber into two chambers (e.g., in the Devonian *Tornoceras*: Bogoslovsky 1971; Korn and Klug 2002; in the Permian *Pseudohalorites*: Zhou et al. 2002). Korn and Klug (2002) indicated that the first septum of *Tornoceras typum* (Tornoceratina) lacks a septal neck, is strongly concave and reaches the inner surface of the initial chamber, while the second septum is of normal form. They interpreted the latter to be connected with the outer surface of the initial chamber and to form a septal neck; the flange is situated inside the second gas-filled chamber (Bogoslovsky 1971). Zhou et al. (2002) reported that *Pseudohalorites* (Tornoceratina) has a subdivided initial chamber similar to *Tornoceras*, but this was only based on a transversal cross section. Klofak and Landman (2012) also investigated specimens of Tornoceratidae (compare House 1965, 1996) and concluded that this first septum might be the projection of the dorsal shell wall within the initial chamber. They suggested that the confusion might be related to the dorsal wall, which projected into the initial chamber, so that it might have extended far enough or bent in a way to appear to be two structurally similar attached septa; this might also explain the description of two prosepta by Miller and Unklesbay (1943) as well as Böhmers (1936). Additional studies and better preserved embryonic shells (with preserved shell structure) are necessary to further test the presence of two prosepta.

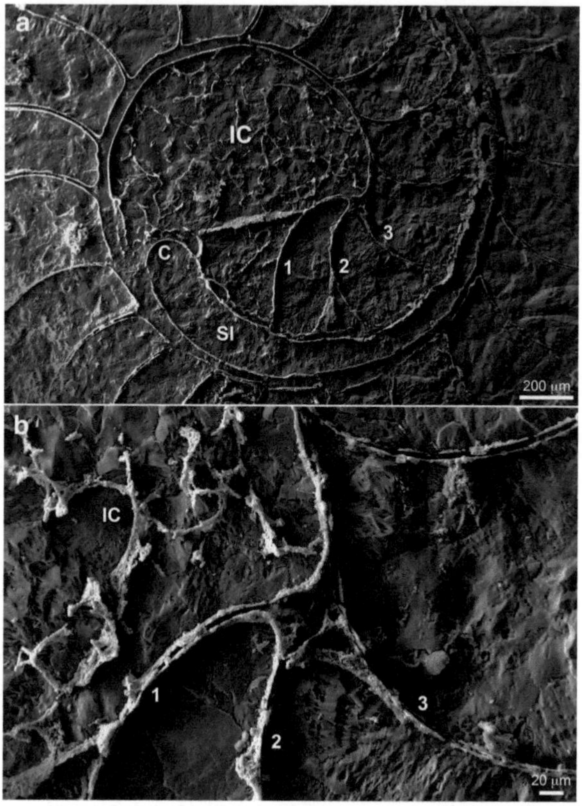

Fig. 5.14 Caecum, siphuncle, and first few septa in the Agoniatitina. **a, b** *Agoniatites vanuxemi*, AMNH 53357, middle Devonian, Cherry Valley Limestone, New York State, U.S.A., median cross section. **a** Overview of the ammonitella. **b** Close-up of initial chamber (from Klofak and Landman 2010). Abbreviations: *C* caecum; *IC* initial chamber, *SI* siphuncle; *1, 2* number of septum

5.2.7 *Siphuncle, Caecum, and Prosiphon*

The siphuncle originates in the initial chamber as a bulbous structure called the caecum (C in Fig. 5.14, 5.15 and 5.16). Like the siphuncle, the caecum was probably originally composed of organic membranes, which are usually preserved as phosphatic material. Indeed, Kulicki (1979) reported traces of fine wrinkles on the caecum of *Quenstedtoceras*, thus corroborating its original organic composition.

As described in Landman et al. (1996) and Klofak and Landman (2010) (1999), the shape of the caecum varies among ammonoids (Fig. 5.14, 5.15, 5.16). They reported three shapes: elliptical in the Bactritina, Agoniatitina, Goniatitina, Ceratitina, Phylloceratina, Ancyloceratina, and Ammonitina; rectangular in the Prolecanitina; and hemicircular (semicircular) in the Lytoceratina. Since then, Shigeta et al. (2001) redescribed the caecum of the Prolecanitina as bottle-shaped or gourd-shaped. Tanabe et al. (2003) argued that the shape is stable at the family or superfamily level in the Ammonitina but more variable even at the species level in the Lytoceratina. For example, Shigeta (1989) documented variation in the shape of the caecum from elliptical to hemicircular in *Tetragonites*.

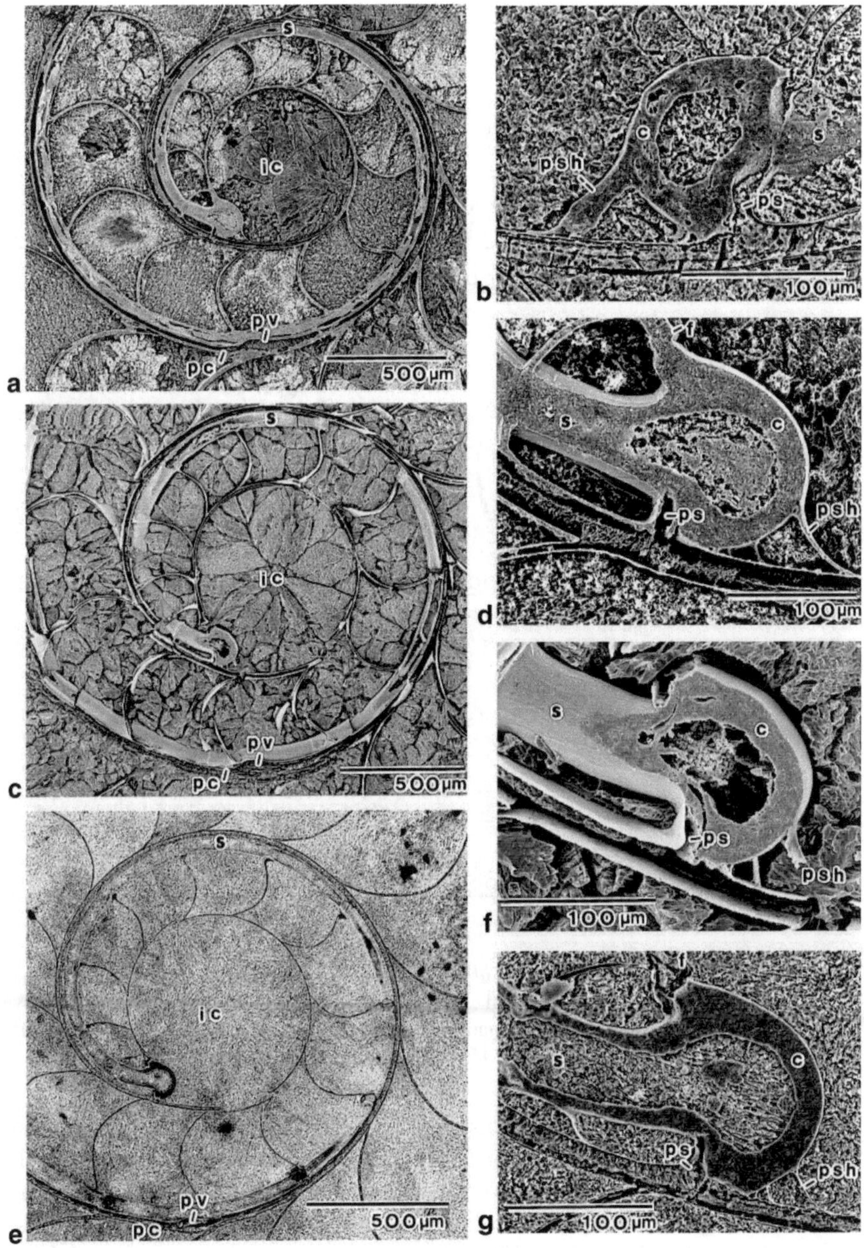

Fig. 5.15 Early internal shell structure in two species of *Gaudryceras* (Late Cretaceous Lytoceratina). Overall views of the ammonitella (**a**, **d**, **f**) and close-up views of the ventral side of the initial chamber (**b**, **c**, **e**, **g**) in median cross section. **a–c** *G. denseplicatum*, lower Santonian, Hokkaido. **a**, **c** UMUT MM 28224. **b** UMUT MM 28223. **d–g** *G. tenuiliratum*, lower Campanian, Hokkaido. **d**, **e** UMUT MM 28225. **f**, **g** UMUT MM 28226 (from Tanabe et al. 2003). Abbreviations: *c* caecum; *ic* initial chamber; *f* flange; *pc* primary constriction; *ps* proseptum; *psh* prosiphon; *pv* primary varix; *s* siphuncle

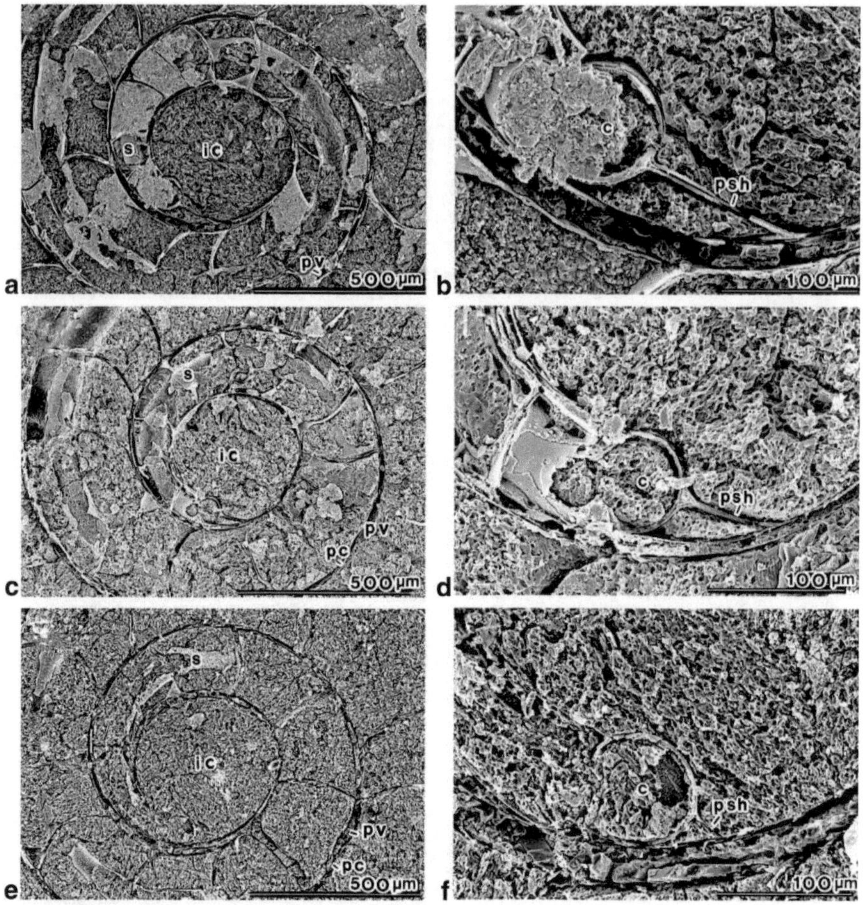

Fig. 5.16 Early internal shell structure in *Subprionocyclus minimus* (Late Cretaceous Ammonitina). Overall views of the ammonitella (**a**–**c**) and close-up views of the ventral side of the initial chamber (**d**–**f**) in median cross section, upper Turonian, Hokkaido. **a**, **b** UMUT MM 28230. **c**, **d** UMUT MM 28231. **e**, **f** UMUT MM 28232. See Fig. 5.15 for abbreviations. (from Tanabe et al. 2003)

The caecum is attached to the inside surface of the initial chamber by means of the prosiphon, which usually consists of a sheet with prominent folds along the sides and with smaller folds in between (psh in Fig. 5.14, 5.15, 5.16). Like the caecum, it was originally composed of organic tissues and commonly exhibits wrinkles along its length. Landman et al. (1996) distinguished two main types of prosiphons: (1) a long, nearly straight prosiphon in most Ammonitina except for the Amaltheidae, Collignoniceratidae, and Placenticeratidae; (2) a short, curved prosiphon in the Ancyloceratina, Bactritina, Goniatitina, Lytoceratina, Phylloceratina, and Prolecanitina. However, size and shape of the prosiphon commonly vary even between and within closely related species (e.g., *Subprionocyclus minimus*; Fig. 5.16b, d, f). Rouget and Neige (2001) documented both types of prosiphons in two closely related species of Oppeliidae (Ammonitina): *Sublunuloceras virguloides* and *Hecticoceras canaliculatum*.

Initially, the position of the siphuncle was marginal or central, but even if it was central, it became typically marginal by the end of the second whorl (with the exception of the Upper Devonian Clymeniida: Bogoslovsky 1976 or Upper Carboniferous-Permian Pseudohaloritoidea: Frest et al. 1981; Furnish et al. 2009). Landman et al. (1996, table 1) documented the initial position of the siphuncle in seven suborders of ammonoids, but noted wide variation within each suborder. Several recent studies have provided additional data. Shigeta et al. (2001) has confirmed that the initial position of the siphuncle is marginal in the Prolecanitina and central or subcentral in two genera of the Goniatitina (Landman et al. 1999). Klofak et al. (1999) documented that the initial position of the siphuncle is marginal in the Agoniatitina and Tanabe et al. (2003) noted the same in two species of *Gaudryceras* (Lytoceratina) (Fig. 5.15) and central in one species each of *Damesites*, *Subprionocyclus* (Fig. 5.16), and *Metaplacenticeras* (all Ammonitina).

5.2.8 Muscle Scars

Muscle scars in the embryonic shells of ammonoids have only been reported in a few species of Ammonitina, Ceratitina, and Goniatitina. In *Quenstedtoceras*, Bandel (1982) documented the ontogenetic sequence of muscle scar attachment. He noted an elongate muscle scar on the inside surface of the flange, a pair of muscle scars on the adoral face of the proseptum on either side of the proseptal opening, and another pair of muscle scars on the inside surface of the dorsal wall adoral of the second septum. These two muscle scars merge into a single muscle scar adoral of the third septum. A similar sequence of muscle scar attachment has been noted in *Euhoplites* by Landman and Bandel (1985) and in several genera of Ceratitina (*Amphipopanoceras, Anagymnotoceras, Czekanowskites, Nathorstites, Arctoptychites, Aristoptychites, Sphaerocladiscites, Paracladiscites, Stolleites, Indigirites*) by Weitschat and Bandel (1991). In the Goniatitina, Landman et al. (1999: *Glaphyrites*) and Kulicki et al. (2002; gen. et sp. indet.) documented a single elongate muscle scar immediately above the flange in the initial chamber. As in Mesozoic ammonoids, this scar is demarcated by a region of grainy texture.

5.3 Sequence of Embryonic Development

The development of the embryonic shell of Mesozoic ammonites has been reconstructed based on the study of the early whorls of larger specimens as well as the examination of specimens actually preserved at early ontogenetic stages. Reconstructions based on the early whorls of larger specimens have been reviewed in

Landman et al. (1996). Their summary is based on four lines of evidence, namely, the microstructure of the shell wall, the ornamentation, the presence of the primary varix, and the shape of the proseptum (Landman et al. 1996, Fig. 19). These observations, in analogy with studies of the early ontogeny of modern cephalopods, especially modern nautilids (Arnold et al. 1987; Landman 1988; Tanabe and Uchiyama 1997), suggest that ammonites developed directly without a larval stage, which is also in line with their larger size compared to other mollusks (more yolk?). According to this theory of direct development in ammonoids, the ammonitella, consisting of the initial chamber and the whorl terminating at the primary varix, formed within the egg capsule. The internal structures at this stage of development consist of the proseptum, caecum, and prosiphon.

Studies of specimens actually preserved at various stages during early ontogeny have further elucidated the sequence of embryonic shell development. These studies have been based on Late Cretaceous (Santonian) *Baculites* from North America (Landman 1982; Landman et al. 1996) and Early Cretaceous (Aptian) *Aconeceras* from Symbirsk, Russia (Fig. 5.17, 5.18, 5.19, 5.20; Kulicki and Doguzhaeva 1994; Tanabe et al. 2008). These studies suggest that in the initial stage of development, the embryo secreted an organic, unmineralized cap-shaped shell (Fig. 5.21a), although this stage has never actually been preserved as a fossil (such a stage has been observed in the early ontogentic development of extant gastropods; Iwata 1980; Bandel 1982; Hickman 1992). Subsequently, a thin, poorly mineralized shell was secreted approximately 1.5 mm in diameter (equal to the final size of the ammonitella) and 1.0 whorl in spiral length (Fig. 5.18, 5.21b). This is the first stage that is actually preserved as a fossil ('*Stage 1*'). The shell wall consists of a thin homogeneous layer (1–2 μm thick), probably composed of amorphous calcium carbonate (ACC), which passes into a slightly thicker prismatic layer at the adoral end of the shell (Fig. 5.18d). In closely coiled ammonitellae, only the exposed parts of the ammonitella including the lateral walls of the initial chamber are mineralized. The portion of the initial chamber covered by the first whorl is still organic (Fig. 5.18b). Similarly, the nacreous swelling (primary varix) has not yet developed. This area exhibits a constriction because it is the site of soft tissue (muscle) attachment.

In the next stage of development ('*Stage 2*'), an outer prismatic layer is secreted, so that the ventral shell wall of the ammonitella consists of three layers: an inner prismatic layer, a (now) middle homogeneous layer (ACC), and an outer prismatic layer, with a total thickness of 7–8 μm (Fig. 5.19c). The outer prismatic layer is covered with tubercles (*t* in Fig. 5.19c, d), which were secreted from the outside. The embryonic shell at this stage is approximately 1.0–1.5 whorls long, terminating at the primary constriction (arrow in Fig. 5.19a). The wall of the initial chamber covered by the first whorl is secreted along with the proseptum.

In the final stage of embryonic development ('*Stage 3*'), the shell is fully mineralized and consists of an initial chamber and whorl terminating at the constricted aperture (Fig. 5.20a, b, c). No growth lines are present on the surface of the shell, implying that the shell was secreted in direct and uninterrupted contact with the gland cells of the mantle. A thin prismatic layer is secreted on the dorsal side of the shell. The middle homogeneous layer, which was present at earlier stages of

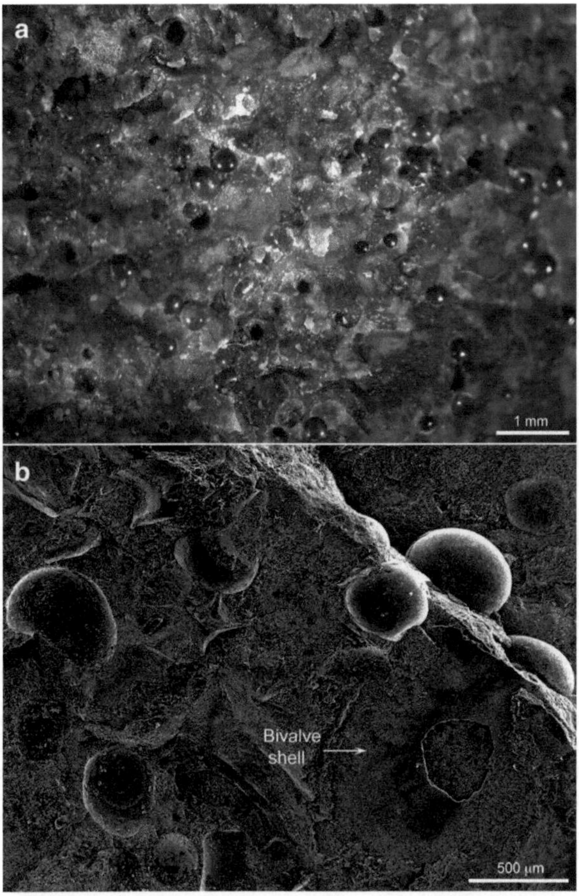

Fig. 5.17 Embryonic shells of *Aconeceras* cf. *trautscholdi*, lower Aptian, Symbirsk, Volga River Basin, Russia. **a** Optical micrograph of part of a calcareous concretion, UMUT MM 29439-1, showing the mode of occurrence of embryonic shells preserved as coprolite remains. **b** SEM of embryonic shells on the broken surface of the concretion, UMUT MM 29440-1. (from Tanabe et al. 2008)

development, is no longer visible and must have merged with the inner prismatic layer. Finally, a thick nacreous swelling forms on the inner side of the shell wall near the aperture, probably right before hatching during a temporary withdrawal of the mantle margin (Fig. 5.21d).

The sequence of embryonic development based on these specimens of *Aconeceras* closely matches that outlined by Bandel (1982) based on an examination of the early whorls of Jurassic *Quenstedtoceras*. The most important difference between the interpretations of Bandel (1982) and Tanabe et al. (2008) is in the timing of formation of the outer prismatic layer and tubercles. Bandel (1982) hypothesized that the outer prismatic layer was mineralized from the inside, preserving the original ornamentation of the organic shell. The observations of Tanabe et al. (2008) suggest

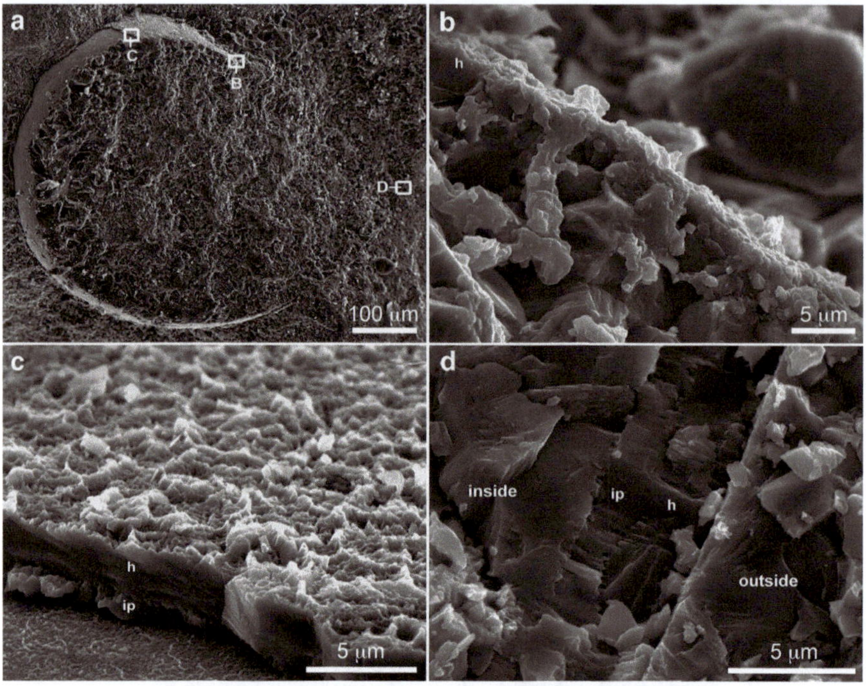

Fig. 5.18 Sequence of embryonic development (Stage 1) in *Aconeceras* cf. *trautscholdi*, UMUT MM 29439-2, lower Aptian, Symbirsk, Russia. **a** Lateral view of medially broken embryonic shell. **b–d** Microstructure of the wall of the embryonic shell. Note that the wall at the adapical end (**b**) consists of a very thin homogeneous, possibly amorphous calcium carbonate layer, whereas it is composed of an outer homogeneous and inner prismatic layer at the adoral end. The outer surface of the shell is bumpy but lacks tubercles (from Tanabe et al. 2008). Abbreviations: *h* homogeneous, possibly amorphous calcium carbonate layer, *ip* inner prismatic layer

that the outer prismatic layer and overlying tubercles were secreted simultaneously at a later stage of mineralization. According to them, the embryonic ammonoid was probably surrounded by fluid within the egg capsule. Under these circumstances, the outer prismatic layer and tubercles may have been deposited from this fluid, without any direct influence of the shell gland. This involves a process of secretion via non-epithelial mineralization with only weak biological control (Tanabe et al. 2010). Such a process has been postulated to account for the formation of micro-ornamentation on some gastropod larval shells (Hickman 2004).

As in Mesozoic ammonoids, the sequence of embryonic development in Paleozoic ammonoids is also based on an examination of the early whorls of larger specimens and the description of specimens actually preserved at early ontogenetic stages. However, with the exception of the Late Carboniferous goniatites from the Buckhorn Asphalt, none of the specimens is preserved with the original aragonitic mineralogy. Studies of the goniatites from the Buckhorn Asphalt demonstrated that the microstructure of the ammonitella, including the presence of a primary varix, and the relationship between the proseptum and the shell wall are identical to that

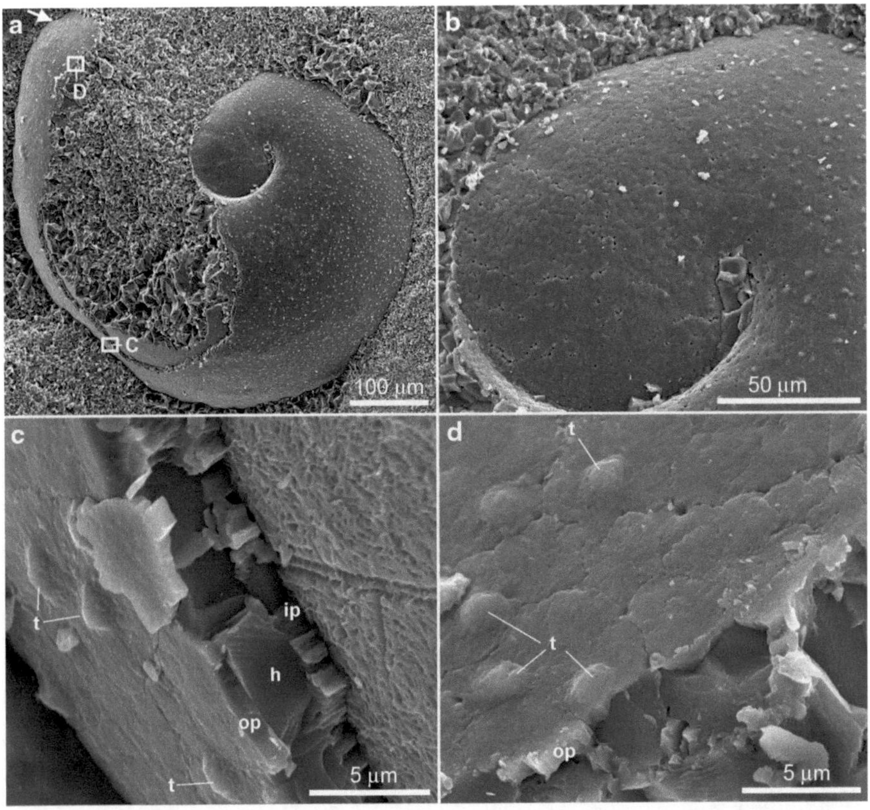

Fig. 5.19 Sequence of embryonic development (Stage 2) in *Aconeceras* cf. *trautscholdi*, UMUT MM 29439-3, lower Aptian, Symbirsk, Russia. **a** Lateral view of the embryonic shell partly exposed on the fractured surface of the carbonate concretion. The arrow points to the constricted aperture. **b** Close-up of the lateral side of the initial chamber without tubercles. **c** The microstructure of the wall consists of an inner prismatic, a middle homogeneous, and an outer prismatic layer on the ventral side of the first whorl. **d** The microstructure of the wall consists of a thin outer prismatic layer near the primary constriction (from Tanabe et al. 2008). Abbreviations: *h* homogeneous, possibly amorphous calcium carbonate layer; *ip* inner prismatic layer, *op* outer prismatic layer, *t* tubercle

in Mesozoic ammonoids (Kulicki et al. 2002). These similarities suggest that both groups shared a similar mode of embryonic development. This conclusion contrasts with earlier interpretations based on less well preserved Carboniferous material (Tanabe et al. 1993).

The sequence of embryonic development in more primitive ammonoids has not yet been satisfactorily worked out. However, in contrast to the hypothesis of Erben (1964, 1966) and Erben et al. (1968, 1969), these forms probably developed directly without a larval stage (House 1996; De Baets et al. 2012). Thus, the embryonic shell consists of both the initial chamber and subsequent whorl terminating at the primary constriction. Nevertheless, the exact sequence of embryonic development

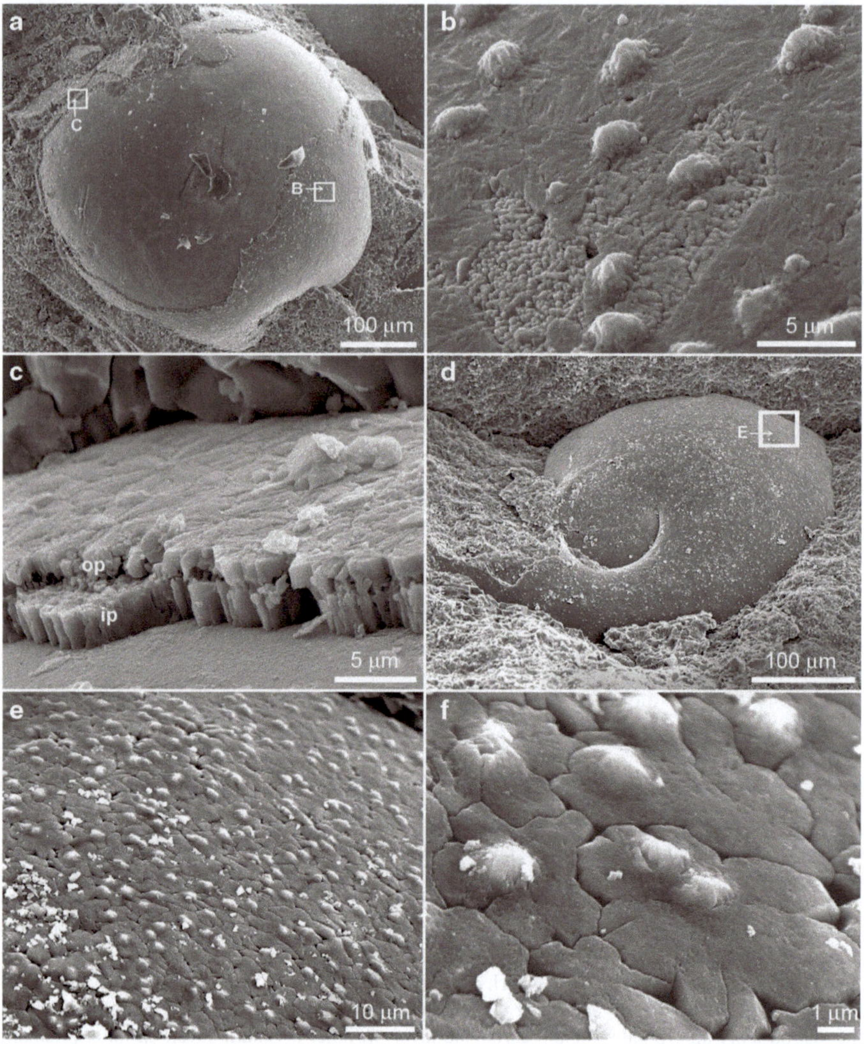

Fig. 5.20 Sequence of embryonic development (Stages 2 and 3) in *Aconeceras* cf. *trautscholdi*, lower Aptian, Symbirsk, Russia. **a–c** UMUT MM 29440-2. This specimen possibly belongs to Stage 3 although its aperture is not exposed. **a** Ventral view. **b** Close-up of the shell surface with tubercles. **c** The microstructure of the shell wall consists of an inner and an outer prismatic layer with tubercles on the outer layer. **d–f** UMUT MM29443. This specimen may belong to Stage 2. **d** Lateral view. **e, f** Close-ups of the outer surface with tubercles. Note that the tubercles rest on the underlying outer prismatic layer, and occasionally a couple of tubercles occur on a single prismatic crystal showing pseudo hexagonal trilling. (from Tanabe et al. 2008)

is still unknown. The presence of wrinkle-like creases on the embryonic shell of some primitive ammonoids like, e.g., the Agoniatitina (Klofak et al. 1999) suggests that the shell was secreted in close contact with an organic membrane (for an interpretation of the significance of such creases, see Checa 1994). The initial chamber

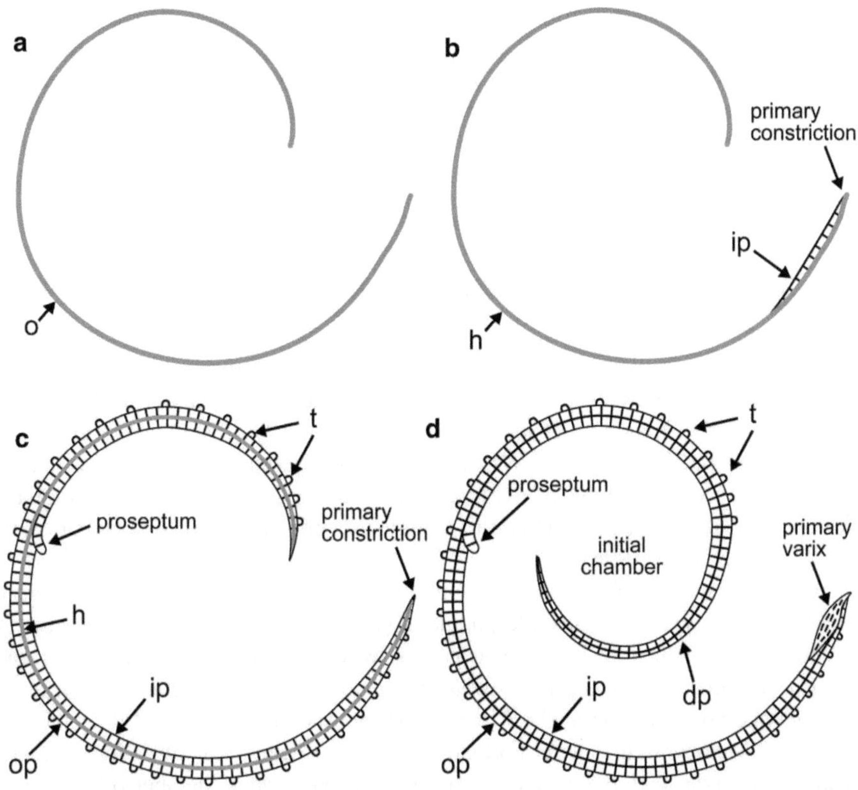

Fig. 5.21 Diagrams showing the sequence of embryonic shell development in Mesozoic ammonites. **a** Stage 0. Secretion of an unmineralized organic shell (*o*) (not observed). **b** Stage 1. Initial mineralization of a homogeneous layer (*h*), possibly consisting of amorphous calcium carbonate (ACC). Subsequent transformation of the inner portion of ACC into an inner prismatic layer (*ip*) on the adoral side. The aperture is distinctly constricted. **c** Stage 2. Secretion of an outer prismatic layer (*op*), followed by deposition of irregularly distributed tubercles (*t*) on top. **d** Stage 3. Secretion of a dorsal prismatic layer (*dp*) and a nacreous swelling (primary varix) on the inner side near the aperture. The middle homogeneous layer (ACC) that occurred in Stage 2 is mineralized and merged with the inner prismatic layer (*ip*). (from Tanabe et al. 2008)

of primitive ammonoids is similar in size and shape to the whole ammonitella of more advanced ammonoids, and in analogy with the embryonic development of these forms, could have originally been secreted as a thin organic membrane. The rest of the shell after the initial chamber may have formed by accretionary growth. The lirae on this part of the shell are not uniformly distributed but are more closely spaced near the end of the ammonitella, perhaps reflecting a deceleration in growth rate right before hatching. In addition, the appearance of a slight backward projection of the lirae on the venter at the end of the initial chamber, which gradually becomes more pronounced over the course of the first whorl, probably reflects the growth of an organ during embryogenesis (Bogoslovsky 1969). Thus, the lirae on the shell could not have been secreted simultaneously everywhere, but must have formed via accretionary growth. However, House (1965, p. 87–89) reported more or less equally spaced lirae in *Tornoceras* on the ammonitella (but see Klofak et al. 2007; 5.2.4) and the appearance of clear ventral (hyponomic) and lateral (ocular) sinuses only in post-embryonic growth at or slightly after the nepionic constriction (House 1996), which he interpreted to confirm their freedom from the egg capsule and that adult free-living life has begun at this stage of development.

5.4 Post-Hatching Mode of Life

It is now well accepted that ammonoid hatchlings, particularly those of derived forms with sizes below 2–3 mm, had a planktic mode of life (Zakharov 1972; Kulicki 1974, 1979; Landman et al. 1996; Westermann 1996; De Baets et al. 2012; Ritterbush et al. 2014; compare Wetzel 1959 for an alternative point of view). This is based on three main lines of evidence: facies of the host rock of ammonitellae-occurrences, density calculations and actualistic comparisons of embryonic shell size (Landman et al. 1996).

A pelagic mode of life is consistent with the occurrences of ammonitellae and very small juveniles with older juveniles, adult and/or other nektonic and planktic taxa in environments in which the bottom water (or at least the sediment-water interface) was dysoxic to anoxic with oxygenated water above and little or no benthic fauna (e.g., Late Mississippian Ruddle Shale, Arkansas: Mapes and Nützel 2009; Middle Triassic Fossil Hill Member of the Favret Formation, Nevada: Bucher et al. 1996; Jurassic: Kulicki and Wierzbowski 1983; Late Cretaceous Sharon Springs Member of the Pierre Shale, Wyoming: Landman 1988; Late Cretaceous Mancos Shale in Utah: Stephen et al. 2012). However, there are numerous examples of mixed assemblages of juvenile and adult ammonoids from well-oxygenated environments in which very small juveniles (<4 mm) are rare or absent, which suggest that newly hatched ammonoids might have lived in a different environment from that of the older juveniles and adults. This pattern could, however, be related to taphonomic processes, although isotope analyses through ontogeny are consistent with such a hypothesis for various Jurassic and Cretaceous ammonoids (Lukeneder et al. 2010; Ritterbush et al. 2014; Lukeneder 2015; Moriya 2015). The ecological differences between hatchlings and later ontogenetic stages might earn them the name of paralarvae (as it was introduced by Young and Harman 1988 for an early developmental stage of extant cephalopods that resides in the near-surface plankton and that differs from later stages in habit, habitat and, frequently, morphology). The term larvae cannot be used for direct developers like ammonoids (Kulicki 1979 used the term pseudolarvae). In the plankton, young ammonoids may have secreted up to two whorls reaching shell diameters between 3 and 5 mm (neanoconch of Westermann 1996).

In extant cephalopods, hatchlings deriving from eggs smaller than 3 mm in diameter are planktic, while hatchlings larger than about about 10 mm in diameter are typically nektobenthic. Between 3 and 7.5 mm, both planktic and benthic hatchlings are known in extant cephalopods (Fig. 5.22; De Baets et al. 2012), so that the large size of 3 mm or more of some Devonian and post-Devonian embryonic shells might indicate a nektobenthic mode of life. However, facies analysis indicate that their, bactritoid (Carboniferous: Mapes and Nützel 2009) and orthocerid ancestors (Ordovician: Kröger et al. 2009) with small initial chambers were also planktic. Size on its own is insufficient to postulate a nektobenthic mode of life for these larger hatchlings, and it is probably the most parsimonious to assume that early ammonoids also had a planktic mode of life like their ancestors and more derived forms (De Baets et al. 2012). This is supported by the co-occurrence of embryonic shells of bactritoids and ammonoids since the Early Devonian (e.g., Early Emsian of France, Morocco, Uzbekistan: compare Erben 1965; Klug et al. 2008; Kröger 2008; Becker et al. 2010; De Baets et al. 2012), just like in the Carboniferous (Mapes and Nützel 2009). Among octopods, a planktonic mode of life at hatching is considered to be the phylogenetically primitive condition, whereas a nektobenthic mode of life at this stage is considered derived (Boletzky 1978, 1987b, 1992).

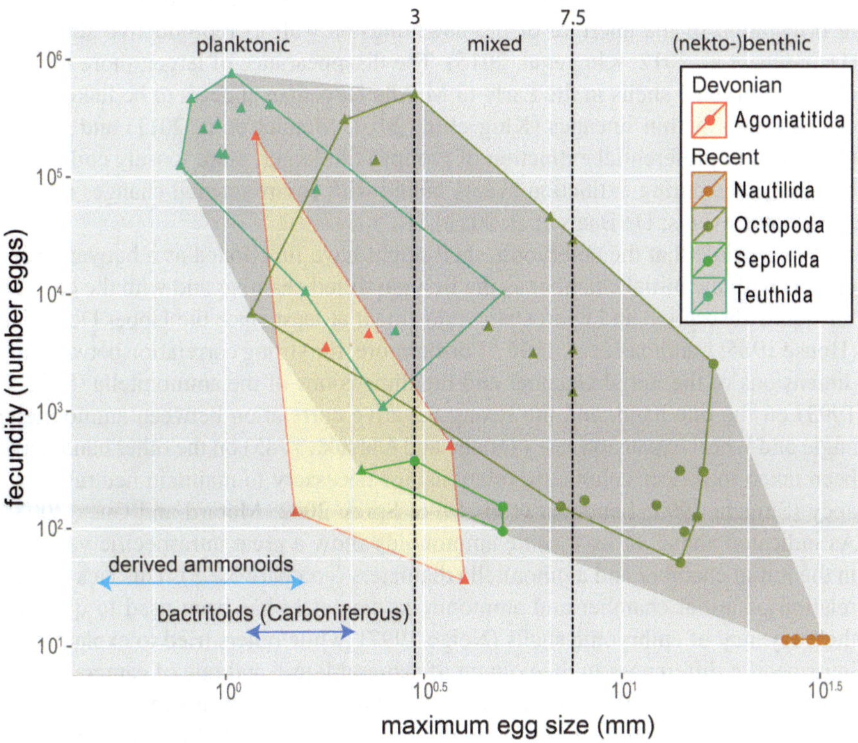

Fig. 5.22 Relationship between the average fecundity (number of spawned eggs) and maximum egg size in extant cephalopods and Devonian ammonoids (modified from De Baets et al. 2012). For the maximum embryonic shell diameter was used as a proxy for ammonoid egg size. The arrows mark the main ranges of embryonic shell size (AD) of derived ammonoids (0.5–1.5 mm) and Carboniferous bactritoids (1.2–2 mm). We also indicate which extant taxa have planktic (circles) and nektobenthic (triangles) hatchlings; between 3 and 10 mm, both planktic and non-planktic hatchlings are known

A pelagic mode of life is also supported by density calculations on ammonitellae of Paleozoic (Carboniferous Goniatitida: Tanabe et al. 1995) and Mesozoic (Cretaceous Ammonitida: Shigeta 1993), which have yielded values within the range of neutral buoyancy (Westermann 1993, 1996; Landman et al. 1996). Ammonoid hatchlings were probably rather planktonic than nektobenthic as they were too small to swim effectively against currents (Jacobs 1992; Jacobs and Chamberlain 1996, p. 213). Derived ammonoids probably already had a functional hyponome and functional eyesight at hatching as indicated by the presence of hyponomic and ocular sinuses at or slightly after the nepionic constriction in Late Devonian *Tornoceras* (House 1965, 1996). Not only the small size, but also its spherical shape in derived ammonoids might have been an adaptation to a planktonic mode of life in derived ammonoids (Kulicki 1979). At 1 mm, their spherical form is the optimal shape to minimize surface area and maximize volume (Jacobs 1992; Jacobs and Chamberlain 1996). The most basal ammonoids have straight, curved to loosely coiled embryonic shells (De Baets et al. 2012; De Baets et al. 2013b). The progressive coiling of the initial chamber and embryonic shell probably coincided with changes in orientation of the aperture of the hatchlings as well as reproductive strategies (De Baets et al. 2012; Klug et al. 2015). The disappearance of larger, more loosely coiled embryonic shells in the Early to Middle Devonian appears to be linked with driven trends within lineages (Klug et al. 2010, Monnet et al. 2011) and sorting trends due to preferential extinction of groups with larger, more loosely coiled embryonic shells during extinction events linked with environmental changes (Daleje and Choteč events; De Baets et al. 2012).

It is possible that the embryonic shell might have functioned as a buoyancy apparatus with the initial chamber as the first gas-filled chamber and with the caecum for liquid absorption and hydrostatic adjustment at least since the Upper Devonian (House 1985; Landman et al. 1996). Furthermore, the strong correlation between the dimensions of the initial chamber and the dimensions of the ammonitella (Shigeta 1993) on the one hand, and the strong negative correlation between ammonitella angle and whorl expansion rate (Tanabe and Ohtsuka 1985) on the other hand, have been taken to reflect volumetric relationships necessary to maintain neutral buoyancy (Shigeta 1993; Landman et al. 1996; Sprey 2002; Morard and Guex 2003). As indicated above, at least some ammonoids show a great intraspecific variation in the initial chamber and ammonitella diameters (compare 5.2.3). This lack of correlation of initial chamber and ammonitella diameters has been used to question the buoyancy of embryonic shells (Neige 1997), while others tried to explain it by intraspecific differences in the volume of cameral liquid and rate of cameral liquid discharge (Shigeta 1993; Rouget and Neige 2001), although this might be unnecesary. The correlation between the embryonic shell and initial chamber diameter has usually been interpreted in the light of neutral buoyancy, but the problem is that in most studies, the 3D dimensional form of the ammonitella and initial chamber is ignored or approximated (Shigeta 1993; Tanabe et al. 1995), which might be problematic, particularly in forms like Oppeliidae with a markedly barrel-shaped to spindle-shaped embryonic shell (Palframan 1966, 1967). Furthermore, these could also be influenced by the accuracy and precision of measurements derived from median sections.

The wide dispersal of planktonic hatchlings through oceanic currents might explain the global distribution of several heteromorph species, which are interpreted to be poor swimmers (Ward and Bandel 1987; De Baets et al. 2012), both in the Early Devonian (De Baets et al. 2013a, b) and the Mesozoic (Landman et al. 1996; Westermann 1996). The wide distribution of the extant giant squid *Architeuthis*, which probably belong to one single species, has also been attributed to drifting of paralarval stages in addition to the migration of larger individuals (Winkelmann et al. 2013). Tajika and Wani (2013) found no significant intraspecific differences in initial chamber and ammonitella diameters in the Late Cretaceous ammonoid populations (*Gaudryceras tenuiliratum* and *Hypophylloceras subramosum*) between the Haboro and Ikushumbetsu areas, which are 110 km apart. They interpreted that these ammonoids had been transported at least 110 km, which corresponds with a time of more than 5 days (if they assume a velocity of 0.25 m/s for the paleocurrent in the Cretaceous between these areas). The overlapping ranges could potentially also have other explanations (De Baets et al. 2015). Yahada and Wani (2013) found a significant difference in post-embryonic shell shape between samples of *Scaphites planus* (= macroconch of *Yezoites puerculus*; Davis et al. 1996) between Oyubari and Kotanbetsu areas of Hokkaido (Japan), although the samples do not show significant differences in embryonic shell size. They attributed these differences to limited migration between the two areas (currently ~ 130 km apart) related to a nektobenthic habitat after the post-hatching stages. This hypothesis is interesting but should be verified by quantitative analysis of large samples throughout the wide geographic range of this species in the North Pacific (see Tanabe 1977a) as well as the amount of time-averaging and other taphonomic biases (De Baets et al. 2015). The external shell of (adult) ammonoids might have impeded efficient long-distance migrations, although some authors have suggested that at least some ammonoids might have swum or drifted for considerable distances later in ontogeny (Westermann 1996; Brayard and Escarguel 2013; Ritterbush et al. 2014).

So, several lines of evidence speak for a planktic mode of life of ammonoid hatchlings, which might have contributed to their extinction at the end of the Cretaceous (Kennedy 1989; Tanabe 2011; De Baets et al. 2012) as juvenile ammonoids might have been particularly vulnerable to changes in the planktic food chain (Boletzky 2002; Kruta et al. 2011) or ocean acidification (Alegret et al. 2012; Arkhipkin and Laptikhovsky 2012). Interestingly, nautilids with larger, nektobenthic hatchlings survived this Cretaceous mass extinction (Wani et al. 2011), but it is still unclear if these can be linked with differences in habitat, feeding or reproductive strategy between the two groups (Nesis 1986; Kennedy 1989; Ward 1996; Kruta et al. 2011; Tanabe 2011; Arkhipkin and Laptikhovsky 2012; De Baets et al. 2012). Boletzky (2002) hypothesized that the yolk reserve might have been insufficient for smaller hatchlings to carry them through the juvenile period marked by food limitations in terms of available living zooplankton of appropriate size. He suggested that larger hatchings could have survived as scavengers or that they were capable of catching benthic or demersal prey immediately after hatching (or after some post-hatching growth based on yolk absorption).

5.5 Reproductive Strategy and Egg-Laying

In Recent cephalopods, there is a clear trade-off between egg size and fecundity (Calow 1987; Boyle and Rodhouse 2005; De Baets et al. 2012; Fig. 5.22), which becomes even clearer when the egg size relative to adult size is considered (Boletzky 1978, 1997; De Baets et al. 2012). Because the size of the ammonitella (embryonic shell) is considered to be a good proxy of egg size (as opposed to the initial chamber: House 1985, 1996), the progressive coiling and reduction in size of the embryonic shell in the early evolution of ammonoids might therefore have partially involved a change in reproductive strategy (House 1996; De Baets et al. 2012), especially in forms with larger adult sizes (e.g., *Manticoceras*: Korn and Klug 2007). This is not only supported by actualistic comparisons, but also by egg number estimates assuming a volume of the gonads between 8% (~gonadosomatic index of modern *Nautilus*: Tanabe and Tsukahara 1987) and 25% of the body chamber volume (Klug 2001b, 2007; Korn and Klug 2007; De Baets et al. 2012; Fig. 5.22). Most authors agree that derived ammonoids had a large fecundity compared with extant nautilids (House 1985; Ward and Bandel 1987; Landman et al. 1996). Fecundities higher than extant *Nautilus* (up to 10 to 20 eggs: Okubo et al. 1995) have even been estimated in some of the most primitive ammonoids (30–500 eggs in *Erbenoceras*: Klug 2001b, 2007; De Baets et al. 2012) or Devonian forms with very small adult sizes (e.g., 200–300 eggs in *Prolobites*: Walton et al. 2010). Nevertheless, such estimates are complicated by several factors (De Baets et al. 2012): lack of knowledge of the amount of yolk, size of the soft-tissues or growth of the embryos after egg-laying as well as the unknown volume of the gonads or spawning strategy of the adults (single spawning vs. repeated spawning: compare Rocha et al. 2001 for extant cephalopods). Furthermore, the embryonic shell (sometimes referred to as the nauta: Stephen and Stanton 2002, Laptikhovsky et al. 2013) of extant *Nautilus* falls within the upper range of the embryonic shell size spectrum of Cretaceous and Cenozoic nautilids (Wani et al. 2011). More importantly, the diameter of the embryonic shell of coiled 'nautiloids' varied greatly through the Phanerozoic from about 4 to 68 mm (Chirat 2001; Manda and Frýda 2010; Wani 2011; Laptikhovsky et al. 2013), so that some primitive nautiloids might have had a variety of reproductive strategies ranging from such similar to ammonoids (r-strategy) and others even more extreme than extant *Nautilus* with a low number of hatchlings (K-strategy).

Both a high fecundity of derived ammonoids and a high mortality of their hatchlings are indicated by large numbers of preserved ammonitellae and/or small juveniles in local accumulations (Dreyfuss 1933; Schindewolf 1959; Wetzel 1959; Landman 1982; Kulicki and Wierzbowski 1983; Landman and Bandel 1985; Landman et al. 1996; De Baets et al. 2012; Stephen et al. 2012). Occasionally, they have also been found in lenses or concretions, which have sometimes been interpreted as egg-laying assemblages (e.g., Tanabe et al. 1993, 1995) or coprolites (e.g., Tanabe et al. 2008; Fig. 5.17). That juvenile ammonoids were an abundant food source (Ritterbush et al. 2014) is also indicated by the presence of remains of juveniles (aptychi: Lehmann and Weitschat 1973; Nixon 1996; Jäger and Fraaye 1997; Keupp

2000) within larger ammonoids as well as embryonic shells in a coprolite of a yet unidentified predator (Fig. 5.17; Tanabe et al. 2008; Wetzel 1959 reported larger juveniles as small as 6 mm in putative coprolites he attributed to marine reptiles). At the end of the 19th century, such accumulations of small aptychi were wrongly interpreted as brood remains (Michael 1894). In some cases, the embryonic shells within accumulations are well-preserved and show no signs of prolonged transport (e.g., Tanabe et al. 1994). In other cases, they might have been transported, e.g., concentrations of juvenile ammonoids in turbidites or storm-deposits (Elmi and Rulleau 1991; House 1996; Suan et al. 2013).

Assemblages with adult specimens of both sexes (microconchs, macroconchs) and rich in juvenile specimens are thought to be typical for species that live and breed in the same area (Fernández-López 1995, 2007; Fernández-López and Meléndez 1996; Cecca 2002; Fernández-López 2013), i.e., which are "*eudemic taxa*" *sensu* Fernández-López (1991; the term *eudemic* was first introduced by Callomon 1985, who used it in a different context: Cecca 2002). If significant taphonomic alterations can be ruled out, a high proportion of hatchlings (neanoconchs) and juveniles in an assemblage (Kulicki and Wierzbowski 1983; Landman 1987; Landman and Waage 1993) speak not only for a high degree of mortality but also for a high proportion of early ontogenetic stages within the population (Kulicki 1979; Landman et al. 1996; Landman 1987; Tomašových and Schlögl 2008). The latter would imply "r-selected" taxa with a type-III survivorship curve (Landman et al. 1996; Klug 2001b; Stephen and Stanton 2002; Klug et al. 2012). Although the use of terms r-selected and k-selected for life history traits as introduced by Pianka(1970) are outdated (Stearns 1977, 1992; Reznick et al. 2002) and should be avoided, particularly in cephalopods which show a combination of characteristics of both strategies (Boletzky 1981). Embryonic shells and post-embryonic shells are, however, often broken and intermixed with small gastropods, bivalves, foraminifers and fish remains (e.g., Wetzel 1959; Birkelund 1979; Kulicki 1979; Landman 1982, 1985; Suan et al. 2013). Such associations were sometimes used to infer a benthic mode of life for young ammonoids (Wetzel 1959), but Landman (1985) suggested that these peculiar occurrences might be thanatocenoses of organic components from different environments, which might have been mixed by currents (compare Westermann 1996). Such assemblages are already known from the Devonian (e.g., *Manticoceras* juveniles in the Frasnian of Morocco associated with rare brachiopods, orthocerids and wood, own observation). In geologically older ammonoids, embryonic shells co-occur with larger juveniles and mature specimens (e.g., *Mimagoniatites fecundus* in the early Emsian of Morocco, own observations).

As already mentioned above, there are also multiple examples of mixed assemblages of larger juvenile and adult ammonoids from presumably well-oxygenated environments in which very small juveniles (<4 mm) are rare or absent from the Paleozoic (Manger et al. 1999; *Agoniatites*: Klofak 2002; *Prolobites*: Walton et al. 2010) to the Mesozoic (*Graphoceras*: Morton 1988; Shigeta 1993; *Scaphites*: Landman and Waage 1993; Bucher et al. 1996), which suggest that at least some newly hatched ammonoids might have lived in a different environment than older juveniles or adults (ontogenetic separation: Landman et al. 1996). Interestingly, Manger et al. (1999) reported Middle Carboniferous cephalopod assemblages dominated by a small number of ammonoid species of adult size that occur in unusual abundances in environments of both high energy (conglomeratic sandstone and oolite) and low energy (concretionary, dark shales). They interpreted these as possibly reflecting mass mortality following reproduction (semelparity) as they found no evidence for post-mortem sorting (both small and large species are represented). Ontogenetic separation for at least some Jurassic and Cretaceous taxa is also suggested by oxygen isotope studies (Lukeneder et al. 2010; Ritterbush et al. 2014; Lukeneder 2015; Moriya 2015). However, the lack of reports of embryonic shells within these assemblages might also be related with collection biases, transport or diagenetic destruction of minute, fragile aragonitic shells (Landman et al. 1996; Westermann 1996). Fernández-López (1991) introduced the term "*miodemic*" for species that were inferred to have lived in the area, but did not breed and were also not transported there (typical, but not exclusive to migratory species with migration capacity and ontogenetic segregation); "*parademic*" for species that reached the area by passive dispersal during life, and he used the termed "*ademic*" for specimens which were transported in the area after death.

Independent of the mode of life of the hatchlings, extant cephalopods have various strategies to lay their eggs. Most of these reproductive strategies have been suggested for ammonoids including egg laying on the seabottom (Tanabe et al. 1993, 1995; Etches et al. 2009), or attached to seaweed (Westermann 1996), floating egg masses (Mapes and Nützel 2009) or even brood care up to hatching (Ward and Bandel 1987; Walton et al. 2010). However, the egg-laying and brooding behavior in ammonoids is largely speculative and based on little supporting evidence (Tanabe et al. 1993; Etches et al. 2009; Nützel and Mapes 2009; Walton et al. 2010). Furthermore, the behavior might have differed between different taxa and time intervals (Ward and Bandel 1987; Westermann 1996; Ritterbush et al. 2014). As discussed above, some extant taxa with extreme sexual dimorphism (e.g., *Argonauta*: compare Boletzky 1992) brood their eggs until hatching and it has been suggested that ammonoids showing pronounced dimorphism and small embryonic shells might also have brooded their young (Ward and Bandel 1987; De Baets et al. 2012). However, ammonoid shells can not be directly compared with egg cases of *Argonauta* females as has been done by some authors (e.g., Naef 1922; Lewy 2002), because ammonoid shells are not just secreted only for attachment and brooding of eggs (Hewitt and Westermann 2003).

Apart from brooding, it is still debated if ammonoids laid their eggs on the seabottom or deposited them in floating egg-masses (Tanabe et al. 1993; House 1996; Landman et al. 1996; Westermann 1996; Etches et al. 2009; Mapes and Nützel 2009). Based on the fact that there was virtually no evidence of transport such a sorting and other current derived sedimentary features, Tanabe et al. (1993) suspected that at least some Paleozoic ammonoids might also have laid their eggs on the sea bottom. Etches et al. (2009) suggested that the swarming of embryonic shells (or eggs) might point to a preceding non-dispersal phase of development such as egg pouches that were anchored to a firm substrate (including algae) on a well-aerated seafloor below annual storm wave-based on the reproductive pattern of some extant neritic squid. However, as already discussed by Westermann (1996), these observations could also indicate planktic egg masses that sank to the seafloor before or after they were overcome by anoxia (compare Stephen et al. 2012 for a similar scenario proposed for juvenile mortality). All of these reproductive strategies might have been possible and a broad spectrum of strategies might have existed within ammonoids as is the case for extant cephalopods (Boyle 1983; Boletzky 1992; Boyle and Rodhouse 2005; Rosa et al. 2013a, b).

Mapes and Nützel (2009) found abundant embryonic and juvenile shells of bactritoids and ammonoids in various stages of development associated with minute planktic gastropod larvae in the Carboniferous Ruddle Shale. Both the facies analysis and the faunal composition (little to no benthic organisms) suggest oxygen depletion at the sediment/ water interface. Mapes and Nützel (2009), therefore, suggested that eggs of Carboniferous ammonoids were deposited as floating egg masses in mid-water (or attached to floating debris). Attachment to benthic algae is improbable in anoxic facies as eggs need more oxygen (Westermann 1996). So far, no direct evidence has been found that ammonoids attached their eggs to floating algae as suggested by Westermann (1996), but this does not falsify this hypothesis.

Invertebrate eggs can be quite resistant to decay (Martin et al. 2003, 2005) and various structures (mostly associated with ammonoid remains) interpreted as ammonoid eggs have been reported from the Carboniferous (e.g.,?*Rhadinites* or?*Anthracoceras*: Landman et al. 2010), Triassic (*Ceratites*: Müller 1969), Jurassic (Dreyfuss 1933; Wetzel 1959; *Eleganticeras*: Lehmann 1966, 1981, 1990; *Kachpurites*: Baranov 1985; Etches et al. 2009; Fig. 5.23) and Cretaceous (Baculitidae gen. et sp. indet.; Klug et al. 2012). If these interpretations are correct, they could shed some light on the egg-laying and brooding behavior of ammonoids. Unfortunately, these putative eggs do not provide information on the embryonic development of the shell as no embryonic shells or other mineralized structures have so far been found within these structures (Landman et al. 1996; Etches et al. 2009). The lack of embryonic shells inside these structures is usually interpreted to be related with an early stage of development (Kulicki 1979; Landman et al. 1996, 2010; Etches et al. 2009; possibly before the mineralization of the organic embryonic shell praecursor = stage 0: compare 5.3), which would also be supported by their small size compared to ammonoid embryonic shells (often in the size range of initial chambers); eggs of many modern cephalopods grow in size during embryogenesis (even after egg deposition: Boyle 1983, Boletzky 1987a, Boyle and Rodhouse 2005). However, the absence of mineralized structures within the eggs might also be due to taphonomic reasons (e.g., no internal mineralization was observed in laboratory decay experiments of invertebrate eggs: Martin et al. 2003, 2005). Furthermore, it cannot be entirely ruled out that they might be eggs of other mollusks including cephalopods and gastropods (Etches et al. 2009; Zaton et al. 2009) or other invertebrates; they might even potentially be food remains (Royal Mapes, personal communication 2014). The egg-like structures are mostly associated with ammonoid remains (Fig. 5.23), particularly within adult body chambers of macroconchs (*Eleganticeras*: Lehmann 1966; *Ceratites*: Müller 1969, Urlichs 2006; Landman et al. 2010), which might support their interpretation as females

Fig. 5.23 Subspherical structures interpreted as ammonite eggs by Etches et al. (2009) associated with *Aulacostephanus autissiodorensis* from the Lower Kimmeridge Clay. (K1486; Picture: Courtesy of Steve Etches)

and could indicate brooding behavior (Lehmann 1966; Landman et al. 1996). A report of such structures in a "Triassic" *Desmophyllites* by Lehmann (1981) based on a personal communication of Zakharov is rather dubious as no reference is made to it in works of the Zakharov himself and *Desmophyllites* is a Cretaceous taxon: see discussion in Etches et al. (2009). However, it cannot be ruled out that larger shells (after death) may been preferentially used as attachment or brood shelters for other taxa (Lehmann 1966; Etches et al. 2009). In some cases, smaller ammonoids might also be accumulated around larger ammonoids or within the body chamber of larger ammonoids post-mortem through currents (Maeda 1991; Maeda and Seilacher 1996). Maeda (1991) described one of the most extreme examples, the presence of more than 100 well-preserved juveniles of *Desmoceras* (together with plant remains) within the body chamber of a *Calycoceras* and introduced the term "sheltered preservation" for this phenomenon. Klug et al. (2012) speculated that certain ovoid structures with a diameter of about 1 mm, occurring within the body chamber of an exceptionally preserved baculitid, as eggs that might have been still present in the oviduct. Landman et al. (2010) found similar structures within the adoral part of the body chamber of a Carboniferous goniatite (?*Anthracoceras* or?*Rhadinites*) - about. 90° adapical of the aperture. These authors interpreted these structures as possible egg capsules and suggested this larger specimen might have been a macroconch. The large number of these structures interpreted as eggs, which are occassionally associated with a high number of juvenile jaw remains (Klug et al. 2012), might support the high fecundity suggested for derived ammonoids.

Table 5.2 Intraspecific variability in initial chamber (ID_1, ID_2, IW) and ammonitella (AD, AA) measurements (in mm for min. mean, max. and SD) reported in the literature for selected Carboniferous, Triassic, Jurassic and Cretaceous ammonoids where larger samples have been studied (N≥5 measurements). It includes: N (number of specimens measured), minimum, average and maximum values of ID_1, ID_2, IW, AD and AA, where available, the standard deviation and the coefficient of variation. Data derive from Arkadiev and Vavilov (1984a); Landman (1987); Vavilov (1992), Tanabe et al. (1995), Stephen and Stanton (2002), Tanabe et al. (2003), and Tajika and Wani (2011). Graphs with additional data on Jurassic and Cretaceous ammonoids can be found in Morard and Guex (2003) and Tanabe (1977a, b), respectively. The latter publications did not list the original measurements, so we could not calculate the CV and we do not list them here

Taxon	Reference	Period	sample	Meas.	N	min	mean	max	SD	CV
Arkanites relictus	Stephen and Stanton 2002	Carbonif.	multiple?	AD	47	0.71	0.78	0.87	0.03	0.04
Branneroceras branneri	Stephen and Stanton 2002	Carbonif.	multiple?	AD	6	0.78	0.81	0.85	0.03	0.04
Cancelloceras huntsvillense	Stephen and Stanton 2002	Carbonif.	multiple?	AD	6	0.78	0.81	0.85	0.03	0.04
Homoceras subglobosum	Tanabe et al. 1995	Carbonif.	Stonehead Beck	ID_1	21	0.43	0.49	0.52	0.02	0.05
				AD	21	0.85	0.92	0.98	0.04	0.04
Pronorites arkanasensis	Stephen and Stanton 2002	Carbonif.	multiple?	AD	9	0.92	0.99	1.03	0.06	0.06
Retites semiretia	Stephen and Stanton 2002	Carbonif.	multiple?	AD	22	0.73	0.80	0.87	0.04	0.05
Syngastrioceras oblatum	Stephen and Stanton 2002	Carbonif.	multiple?	AD	41	0.75	0.83	0.92	0.03	0.04
Verneuilites pygmaeus	Stephen and Stanton 2002	Carbonif.	multiple?	AD	6	0.75	0.78	0.82	0.03	0.04
Indirigites krugi	Arkadiev and Vavilov 1984a; Vavilov 1992	Triassic	multiple?	ID_1	5	0.35	0.36	0.37	0.01	0.02
				ID_2	5	0.32	0.33	0.35	0.01	0.04
				AD	5	0.60	0.65	0.68	0.03	0.05
Parapopanoceras asseretoi	Arkadiev and Vavilov 1984a; Vavilov 1992	Triassic	multiple?	ID_1	7	0.30	0.36	0.40	0.04	0.13
				ID_2	7	0.30	0.32	0.33	0.01	0.04
				AD	6	0.60	0.65	0.68	0.04	0.07
Parapopanoceras medium	Arkadiev and Vavilov 1984a; Vavilov 1992	Triassic	multiple?	ID_1	6	0.35	0.38	0.42	0.04	0.09
				ID_2	6	0.26	0.33	0.40	0.05	0.16

Table 5.2 (continued)

Taxon	Reference	Period	sample	Meas.	N	min	mean	max	SD	CV
Creniceras renggeri	Neige 1997	Jurassic	multiple?	ID_1	23	0.27	0.33	0.37	0.02	0.06
				ID_2	15	0.21	0.24	0.37	0.02	0.07
				AD	18	0.53	0.57	0.60	0.02	0.04
				AA	18	250	264	281	8	0.03
Creniceras renggeri [M + m]	Palframan 1966	Jurassic	Woodham	ID	14	0.25	0.28	0.30	0.01	0.05
				IW	9	0.38	0.41	0.42	0.01	0.03
				AD	13	0.52	0.54	0.56	0.01	0.03
Distichoceras bicostatum [M + m]	Palframan 1967	Jurassic	Woodham	ID	5	0.24	0.28	0.30	0.02	0.09
				IW	5	0.40	0.44	0.48	0.04	0.08
				AD	5	0.48	0.51	0.57	0.04	0.07
Hecticoceras (Brightia) canaliculatum	Bouget and Neige 2001	Jurassic	Athleta Zone, Burgundy	ID_1	18	0.36	0.39	0.45	0.03	0.07
				ID_2	18	0.27	0.30	0.35	0.02	0.07
				AD	18	0.59	0.66	0.74	0.05	0.08
				AA	18	234	257	288	15	0.06
Hecticoceras brightii [M + m]	Palframan 1969	Jurassic	Woodham	ID	19	0.30	0.33	0.36	0.01	0.03
				IW	22	0.41	0.46	0.50	0.02	0.03
				AD	19	0.60	0.66	0.70	0.04	0.05
Sublunuloceras virguloides	Rouget and Neige 2001	Jurassic	Athleta Zone, Burgundy	ID_1	34	0.30	0.36	0.41	0.03	0.07
				ID_2	34	0.24	0.28	0.33	0.02	0.08
				AD	34	0.49	0.60	0.71	0.04	0.07
				AA	34	225	259	298	14	0.05

Table 5.2 (continued)

Taxon	Reference	Period	sample	Meas.	N	min	mean	max	SD	CV
Clioscaphites vermiformis	Landman 1987; Tanabe et al. 2003	Cretac.	B235a	ID_1	43	0.29	0.33	0.44	0.03	0.08
				ID_2	44	0.23	0.27	0.37	0.02	0.09
				IW	18	0.43	0.44	0.46	0.02	0.04
				AD	43	0.60	0.66	0.80	0.04	0.05
				AA	44	262	289	310	11	0.04
Damesites semicostatus	Tanabe et al. 2003	Cretac.	R2672	ID_1	15	0.39	0.43	0.46	0.02	0.05
				ID_2	15	0.33	0.36	0.39	0.02	0.04
				AD	15	0.72	0.77	0.81	0.03	0.04
				AA	15	288	314	346	15	0.05
Damesites sugata	Tanabe et al. 2003	Cretac.	H60p	ID_1	6	0.39	0.40	0.42	0.01	0.03
				ID_2	6	0.34	0.35	0.38	0.02	0.05
				AD	6	0.73	0.76	0.79	0.02	0.03
				AA	6	308	315	324	6	0.02
Gaudryceras tenuiliratum	Tajika and Wani 2011	Cretac.	Haboro	ID_1	14	0.69	0.73	0.81	0.03	0.04
				AD	18	1.27	1.36	1.46	0.05	0.04
		Cretac.	Ikushum-betsu	ID_1	14	0.59	0.73	0.81	0.05	0.07
				AD	17	1.18	1.36	1.46	0.07	0.05
Hypophylloc. subramosum	Tajika and Wani 2011	Cretac.	Haboro	ID_1	11	0.57	0.64	0.71	0.04	0.06
				AD	11	0.99	1.03	1.11	0.03	0.03
		Cretac.	Ikushum-betsu	ID_1	10	0.56	0.63	0.70	0.04	0.07
				AD	14	0.91	1.03	1.13	0.07	0.07

Table 5.2 (continued)

Taxon	Reference	Period	sample	Meas.	N	min	mean	max	SD	CV
Subprionocyclus minimum	Tanabe et al. 2003	Cretac.	R4018	ID_1	8	0.46	0.49	0.53	0.03	0.05
				ID	8	0.37	0.40	0.44	0.02	0.05
				AD	8	0.83	0.86	0.89	0.03	0.03
				AA	8	260	260	272	4	0.01
		Cretac.	R4020	ID_1	8	0.45	0.53	0.58	0.05	0.09
				ID_2	8	0.37	0.44	0.48	0.03	0.08
				AD	8	0.72	0.89	0.97	0.08	0.09
				AA	8	263	268	276	4	0.01
	Tanabe et al. 2003	Cretac.	IK1003	ID_1	30	0.36	0.42	0.48	0.03	0.07
				ID_2	30	0.33	0.37	0.43	0.03	0.07
				AD	30	0.66	0.75	0.89	0.06	0.09
				AA	30	264	282	307	11	0.04
		Cretac.	IK946	ID_1	26	0.35	0.39	0.45	0.03	0.07
				ID_2	26	0.30	0.35	0.39	0.03	0.08
				AD	26	0.68	0.76	0.83	0.04	0.05
				AA	26	265	292	330	16	0.05
		Cretac.	M135a	ID_1	43	0.36	0.40	0.47	0.03	0.07
				ID_2	43	0.31	0.36	0.42	0.02	0.06
				AD	42	0.67	0.75	0.85	0.05	0.06
				AA	42	252	291	328	16	0.05

Table 5.2 (continued)

Taxon	Reference	Period	sample	Meas.	N	min	mean	max	SD	CV
		Cretac.	M136b	ID_1	21	0.35	0.40	0.47	0.03	0.08
				ID_2	21	0.35	0.37	0.38	0.03	0.09
				AD	21	0.66	0.76	0.88	0.06	0.08
				AA	21	260	293	320	17	0.06
		Cretac.	M163a	ID_1	23	0.39	0.42	0.46	0.02	0.05
				ID_2	23	0.32	0.37	0.40	0.02	0.06
				AD	22	0.78	0.80	0.85	0.03	0.03
				AA	22	270	294	348	16	0.05
Scaphites larvaeformis	Landman 1987	Cretac.	multiple	ID_1	36	0.28	0.32	0.36	0.02	0.06
				IW	16	0.36	0.40	0.46	0.03	0.08
				AD	38	0.53	0.58	0.65	0.03	0.06
				AA	36	266	280	300	9	0.03
Scaphites nigricollensis	Landman 1987	Cretac.	B240	ID_1	16	0.34	0.36	0.38	0.01	0.03
				IW	10	0.43	0.46	0.49	0.02	0.05
				AD	16	0.60	0.68	0.74	0.04	0.06
				AA	15	264	276	300	9	0.03
Scaphites preventricosus	Landman 1987; Tanabe et al. 2003	Cretac.	B228a	ID_1	39	0.30	0.34	0.39	0.02	0.05
				IW	21	0.39	0.46	0.49	0.02	0.05
				AD	39	0.58	0.65	0.69	0.03	0.04
				AA	39	257	273	292	9	0.03
Scaphites warreni	Landman 1987	Cretac.	multiple	ID_1	15	0.27	0.34	0.37	0.04	0.12
				IW	11	0.43	0.48	0.57	0.05	0.10
				AD	14	0.60	0.64	0.72	0.04	0.06
				AA	12	266	277	290	10	0.04

Table 5.2 (continued)

Taxon	Reference	Period	sample	Meas.	N	min	mean	max	SD	CV
Scaphites whitfeldi	Landman 1987; Tanabe et al. 2003	Cretac.	B176a	ID_1	108	0.25	0.31	0.27	0.02	0.07
				IW	28	0.39	0.44	0.50	0.03	0.06
				AD	110	0.55	0.61	0.68	0.03	0.05
				AA	108	308	287	375	10	0.04
Subprionocyclus neptuni	Tanabe et al. 2003	Cretac.	PM3007i	ID_1	66	0.36	0.42	0.50	0.03	0.07
				ID_2	66	0.32	0.37	0.44	0.03	0.07
				AD	66	0.67	0.76	0.85	0.05	0.06
				AA	66	247	284	307	13	0.05
Tetragonites popetensis	Tanabe et al. 2003	Cretac.	S3006	ID_1	19	0.53	0.58	0.67	0.04	0.08
				ID_2	19	0.46	0.52	0.60	0.05	0.09
				AD	19	0.96	1.05	1.14	0.05	0.05
				AA	18	305	330	348	11	0.03
Tragodesmoceroides subcostatus	Tanabe et al. 2003	Cretac.	R4001	ID_1	20	0.49	0.51	0.55	0.02	0.04
				ID_2	20	0.40	0.43	0.48	0.02	0.05
				AD	20	0.83	0.87	0.93	0.03	0.04
				AA	20	262	296	315	12	0.04
Yezoites puerculus	Tanabe et al. 2003	Cretac.	R2110a	ID_1	11	0.48	0.53	0.56	0.02	0.04
				ID_2	11	0.38	0.41	0.44	0.02	0.04
				AD	11	0.76	0.80	0.83	0.03	0.03
				AA	11	256	266	281	8	0.03

Wide interspecific variation in embryonic and adult sizes of ammonoids including the different development of sexual dimorphism suggests that a wide variety of reproductive and egg-laying strategies might have been present in different taxa or time frames as in extant coleoids (Ward and Bandel 1987; Mapes and Nützel 2009; De Baets et al. 2012; Ritterbush et al. 2014). Determination of these strategies will require analysis that combine facies analysis and isotopic studies through ontogeny (as it was attempted for the nautilid *Aturia*: Schlögl et al. 2011) as well as studies on size class distribution (e.g., Klofak 2002; Bucher et al. 1996; Walton et al. 2010).

5.6 Conclusions and Future Perspectives

Our knowledge of the embryonic development of Paleozoic ammonoids (House 1996; Klofak et al. 1999, 2007; Landman et al. 1999; Shigeta et al. 2001; Tanabe et al. 2001; Kulicki et al. 2002; Stephen and Stanton 2002; Klofak and Landman 2010, 2012; De Baets et al. 2012, 2013b) has considerably improved since the last review on ammonoid embryonic development (Landman et al. 1996). There are still many unresolved questions, particularly in Devonian ammonoids (e.g., the presence or absence of a nepionic constriction and the nature of embryonic ornamentation) and the sequence of embryonic development, which can only be tackled by finding better-preserved material (e.g., Klofak et al. 2010, 2012). Our database with embryonic shell measurements has considerably enlarged (e.g., De Baets et al. 2012; Laptikhovsky et al. 2013) with data for more than 500 taxa, but some stratigraphic intervals still have comparably fewer data than others (e.g., Devonian, Jurassic). Despite a considerable range of intraspecific variability in the dimensions and shape of the embryonic shell (Landman 1987; Tanabe et al. 1993; Rouget and Neige 2001; Stephen and Stanton 2002; Tanabe et al. 2003; Tajika and Wani 2011), such characters are still useful for phylogenetic purposes, particularly qualitative characters for distinguishing higher taxonomic relationships (e.g., Shigeta 2001; Tanabe et al. 2003). Furthermore, even in cases of extreme intraspecific variability, large enough samples can be used to statistically compare the mode and range of variation between antidimorphs, populations or taxa (e.g., Landman 1987; Neige 1997; Rouget and Neige 2001; Tanabe et al. 2003; Tajika and Wani 2011; De Baets et al. 2015).

Comparatively little direct evidence is available to constrain the ecology of newly hatched ammonoids as well as the reproductive strategies of the adults (e.g., egg-laying behavior). Facies analysis, buoyancy calculations and actualistic comparisons corroborate that derived ammonoids had planktic hatchlings; it is also the most parsimonious interpretation to assume that hatchlings of primitive ammonoids (Agoniatitina) already had a similar mode of life, too. More data on the width of initial chambers and embryonic shells are necessary to confidently calculate their volume and estimate buoyancy, egg size and fecundity, which could partially be achieved with novel tomographic methods (e.g., Hoffmann et al. 2013). The high concentration of embryonic shells in some assemblages speaks not only for a high mortality of young ammonoids, but also for a high fecundity of adult ammonoids (certainly when compared with extant nautilids). Nevertheless, a large range of interspecific variation in their reproductive strategies might have existed among different time frames and taxa. Our knowledge of post-hatching modes of life and reproductive strategies can only be expanded by combining studies on facies distribution (e.g., Westermann 1996), size class distribution (e.g., Bucher et al. 1996; Walton et al. 2010) and isotope analyses through ontogeny of various taxa and time frames (e.g., Lukeneder et al. 2010; Lukeneder 2015; Moriya 2015) as well as the further discoveries of exceptionally preserved ammonoid eggs and adults (Etches et al. 2009; Landman et al. 2010; Klug et al. 2012; Ritterbush et al. 2014).

Acknowledgments Some of the insights described in this chapter grew during the course of research projects 200021-113956/1, and 200020-125029 funded by the Swiss National Science Foundation SNF (to KDB). Mikhail A. Rogov (Moscow) and Carlo Romano (Zürich) helped with obtaining some of the literature. Steve Etches (Wareham) kindly put pictures at our disposal. We dedicate this chapter to Susan Klofak for her inspiring work on the embryonic development of Paleozoic ammonoids.

References

Alegret L, Thomas E, Lohmann KC (2012) End-Cretaceous marine mass extinction not caused by productivity collapse. Proc Natl Acad Sci U S A 109:728–732

Alekseev SN, Arkadiev VV, Vavilov MN (1984) Internal structure and ontogeny of certain Middle Triassic ceratites. Paleontol J 1984:51–64

Arkadiev VV, Vavilov MN (1984a) Middle Triassic Parapopanoceratidae and Nathorstitidae (ammonoidea) of Boreal region: internal structure, ontogeny and phylogenetic patterns. Geobios 17:397–425

Arkadiev VV, Vavilov MN (1984b) Inner structure and ontogeny of Late Anisian Beyrichitidae (Ammonoidea) of Middle Siberia. Paleontol J 17:63–72

Arkadiev VV, Bucher H, Vavilov MN (1993) Structure and systematic position of the middle Anisian genus *Nevadisculites* (Ammonoidea) from Nevada (USA). Paleontol Zh 1993(3): 30–36 [in Russian]

Arkhipkin AI, Laptikhovsky VV (2012) Impact of ocean acidification on plankton larvae as a cause of mass extinctions in ammonites and belemnites. Neues Jahrb Geol Palaeontol Abh 266:39–50

Arnold JM, Landman NH, Mutvei H (1987) Development of the embryonic shell of *Nautilus*. In Saunders WB, Landman NH (eds) Nautilus: the biology and paleobiology of a living fossil. Plenum Press, New York

Bandel K (1982) Morphologie und Bildung der frühontogenetischen Gehäuse bei conchiferen Mollusken. Facies 7:1–197

Bandel K (1986) The ammonitella: a model of formation with the aid of the embryonic shell of archaeogastropods. Lethaia 19:171–180

Bandel K, Landman NH, Waage KM (1982) Micro-ornament on early whorls of Mesozoic ammonites: implications for early ontogeny. J Paleontol 56:386–391

Baranov VN (1985) On the egg remains in the body chambers of Late Volgian. Bull Soc Nat Mosc Geol 60:89–91 [in Russian]

Becker RT, De Baets K, Nikolaeva S (2010) New ammonoid records from the Lower Emsian of the Kitab reserve (Uzbekistan)—preliminary results. SDS Newsl 25:20–28

Bensaïd M (1974) Étude sur des goniatites à la limite du Dévonien moyen et supérieur du Sud Marocain. Notes Serv Géol Maroc 36:81–140

Birkelund T (1979) The last Maastrichtian ammonites. In: Birkelund T, Bromley RG (eds) Cretaceous-tertiary boundary events. University of Copenhagen, Copenhagen, pp 51–57

Birkelund T (1981) Ammonoid shell structure. In: House MR, Senior JR (eds) The Ammonoidea: the evolution, classification, mode of life, and geological usefulness of a major fossil group. Academic Press, London, pp. 177–214

Birkelund T, Hansen HT (1968) Early shell growth and structures of the septa and the siphuncular tube in some Maastrichtian ammonites. Medd Dan Geol Foren 18:95–101

Birkelund T, Hansen HJ (1974) Shell ultrastructures of some Maastrichtian Ammonoidea and Coleoidea and their taxonomic implications. K Danske Vidensk Selsk Biol Skr 20:2–34

Bogdanova TN, Arkadiev VV (2005) Revision of species of the ammonite genus *Pseudosubplanites* from the Berriasian of the Crimean mountains. Cretac Res 26:488–506

Bogoslovskaya MF (1959) The internal structure of certain Artinskian ammonoid shells. Paleontol Zh 1:49–59 [in Russian]

Bogoslovsky BI (1969) Devonian ammonoids. I. Agoniatitids. Trudy Paleont Inst Akad Nauk SSSR 124:1–341 [in Russian]

Bogoslovsky BI (1971) Devonian ammonoids. II. Goniatitids. Trudy Paleont Inst Akad Nauk SSSR 127:1–228 [in Russian]

Bogoslovsky BI (1972) New Early Devonian cephalopods of Novaya Zemlya. Paleontol. Zh 4:44–51 [in Russian]

Bogoslovsky B (1976) Early ontogeny and origin of clymeniid ammonoids. Paleontol J 1976:150–158

Bogoslovsky BI (1981) Devonian ammonoids. III. Clymeniids. Trudy Paleont Inst Akad Nauk SSSR 191:1–123 [in Russian]

Bogoslovsky BI (1984) A new genus of the family Auguritidae and associated ammonoids from the Lower Devonian of the Zeravshan range. Paleontol Zh 1984(1):30–36 [in Russian]

Böhmers JCA (1936) Bau und Struktur von Schale und Sipho bei permischen Ammonoidea. Drukkerij Universitas, Appeldoorn

Boletzky Sv (1978) Nos connaissances actuelles sur le développement des octopodes. Vie Milieu 28:85–120

Boletzky Sv (1981) Reflexions sur les strategies de reproduction chez les céphalopodes. Bull Soc Zool Fr 106:293–304

Boletzky Sv (1987a) Embryonic phase. In: Boyle PR (ed) Cephalopod life cycles, Vol. II. Academic Press, London

Boletzky Sv (1987b) Juvenile behavior. In: Boyle PR (ed) Cephalopod life cycles, Vol. II. Academic Press, London

Boletzky Sv (1989) Early ontogeny and evolution: the cephalopod model viewed from the point of developmental morphology. Geobios 22 (Suppl 2):67–78

Boletzky Sv (1992) Evolutionary aspects of development, life style, and reproductive mode in incirrate octopods (Mollusca, Cephalopoda). Rev Suisse Zool 99:755–770

Boletzky Sv (1993) Development and reproduction in the evolutionary biology of Cephalopoda. Geobios 26(Suppl 1):33–38

Boletzky Sv (1997) Developmental constraints and heterochrony: a new look at offspring size in cephalopod molluscs. Geobios 30(Suppl 2):267–275

Boletzky Sv (2002) Yolk sac morphology in cephalopod embryo. Abh Geol Bundesanst (Vienna) 57:57–68

Boletzky Sv (2003) Biology of early life stages in cephalopod molluscs. Adv Mar Biol 44:143–203

Boyle PR (1983) Cephalopod Life Cycles, Vol. I. Academic Press, New York

Boyle P, Rodhouse P (2005) Cephalopods: ecology and Fisheries. Wiley, Oxford

Branco W (1879) Beiträge zur Entwicklungsgeschichte der fossilen Cephalopoden I. Palaeontographica 26:15–50

Branco W (1880) Beiträge zur Entwicklungsgeschichte der fossilen Cephalopoden II. Palaeontographica 27:12–81

Brayard A, Escarguel G (2013) Untangling phylogenetic, geometric and ornamental imprints on Early Triassic ammonoid biogeography: a similarity-distance decay study. Lethaia 46:19–33
Brown AP (1892) The Development of the Shell in the Coiled Stage of *Baculites compressus* Say. Proc Acad Nat Sci Phila 44:136–141
Bucher H, Landman NH, Klofak SM, Guex J (1996) Mode and rate of growth in ammonoids. In: Landman NH, Tanabe K, Davis RA (eds) Ammonoid paleobiology. Plenum, New York
Callomon J (1985) The evolution of the Jurassic ammonite Family Cardioceratidae. Spec Pap Palaeontol 33:49–90
Calow S (1987) Fact and theory-an overview. In: Boyle PR (ed) Cephalopod life cycles, volume ii: comparative reviews. Academic Press, London, pp. 351–365
Cecca F (2002) Palaeobiogeography of marine fossil invertebrates: concepts and methods. Taylor and Francis, London
Chambers JM, Cleveland WS, Kleiner B, Tukey PA Graphical methods for data analysis. Chapman and Hall, Belmont
Checa A (1994) A model for the morphogenesis of ribs in ammonites inferred from associated microsculptures. Palaeontology 37:863–863
Chirat R (2001) Anomalies of embryonic shell growth in post-Triassic Nautilida. Paleobiology 27:485–499
Chlupáč I, Turek V (1983) Devonian goniatites from the Barrandian area. Rozpr Ustred Ust Geol 46:1–159
Clausen C-D (1969) Oberdevonische Cephalopoden aus dem Rheinischen Schiefergebirge. II. Gephuroceratidae, Beloceratidae. Palaeontogr A 132: 95–178
Dauphin Y (1975) Anatomie de la protoconque et des tours initiaux de *Beudanticeras beudanti* (Brongniart) et *Desmoceras Latidorsatum* (Michelin), (Desmoceratidae, Ammonitina), Albien de Gourdon (Alpes-Maritimes). Ann Paléontol Invertébr 61:3–16
Dauphin Y (1977) Anatomie de la protoconque et des tours initiaux de *Uhligella walleranti* Jacob (Desmoceratidae, Ammonitina), Albien de Gourdon (Alpes–Maritimes). Ann Paléontol Invertébr 63:77–84
Davis RA, Landman NL, Dommergues J-L, Marchand D, Bucher H (1996) Mature modifications and dimorphism in ammonoid cephalopods. In: Landman NH, Tanabe K, Davis RA (eds) Ammnoid paleobiology. Plenum, New York, pp. 464–539
De Baets K, Klug C, Korn D (2009) Anetoceratinae (Ammonoidea, Early Devonian) from the Eifel and Harz Mountains (Germany), with a revision of their genera. Neues Jahrb Geol Palaeontol Abh 252:361–376
De Baets K, Klug C, Korn D, Landman NH (2012) Early evolutionary trends in ammonoid embryonic development. Evolution 66:1788–1806
De Baets K, Klug C, Monnet C (2013a) Intraspecific variability through ontogeny in early ammonoids. Paleobiology 39:75–94
De Baets K, Klug C, Korn D, Bartels C, Poschmann M (2013b) Emsian Ammonoidea and the age of the Hunsrück Slate (Rhenish Mountains, Western Germany). Palaeontogr Abt A 299:1–113
De Baets K, Bert D, Hofmann R, Monnet C, Yacobucci MM, Klug C (2015) Ammonoid intraspecific variability. This volume
Dietl G (1978) Die heteromorphen Ammoniten des Dogger. Stuttg Beitr Natkde B 33:1–97
Doguzhaeva L (2002) Adolescent bactritoid, orthocerold, ammonoid and coleoid shells from the Upper Carboniferous and Lower Permian of the South Urals. Abh Geol Bundesanst (Vienna) 57:9–55
Doguzhaeva L, Mikhailova I (1982) The genus *Luppovia* and the phylogeny of Cretaceous heteromorphic ammonoids. Lethaia 15:55–65
Donoghue PCJ, Dong X-P (2005) Embryos and ancestors. In: Briggs DEG (ed) Evolving form and function: fossils and development. Yale Peabody Museum of Natural History, Yale University, New Haven, pp. 81–99
Donoghue PCJ, Bengtson S, Dong X-p, Gostling NJ, Huldtgren T, Cunningham JA, Yin C, Yue Z, Peng F, Stampanoni M (2006) Synchrotron X-ray tomographic microscopy of fossil embryos. Nature 442:680–683

Donovan DT (1964) Cephalopod phylogeny and classification. Biol Rev 39:259–287
Donovan DT, Callomon JH, Howarth MK (1981) Classification of the Jurassic Ammonitina. In: House MR, Senior JR (eds) The Ammonoidea. Academic Press, London, pp. 101–155
Dreyfuss M (1933) Découverte de nodules phosphatés à jeunes ammonites dans le Toarcien de Créveney (Haute Saône). C R Somm Séances Soc Géol Fr 14:224–226
Drushchits VV, Doguzhaeva LA (1974) Some morphogenetic characteristics of phylloceratids and lytoceratids (Ammonoidea). Paleontol J 8:37–48
Drushchits VV, Doguzhaeva LA (1981) Ammonites under the electron microscope. Moscow University Press, Moscow. [in Russian]
Drushchits VV, Khiami N (1969) Characteristics of the early stages in the ontogeny of some Early Cretaceous ammonites. Moskov Obshch Ispytateley Prirody Byull Otd Geol 2:156–157
Drushchits VV, Khiami N (1970) Structure of the septa, initial chamber walls and initial whorls in Early Cretaceous ammonites. Paleontol J 1970:26–38
Drushchits VV, Doguzhaeva LA, Lominadze TA (1977a) Internal structural features of the shell of middle Callovian ammonites. Paleontol J 1977:16–29
Drushchits VV, Doguzhayeva LA, Mikhaylova IA (1977b) The structure of the ammonitella and the direct development of ammonites. Paleontol J 1977:188–199
Drushchits VV, Mikhailova IA, Kabanov GK, Knorina MV (1980) Morphogenesis of the *Simbirskites* group. Paleontol J 14:42–57
Dzik J (1981) Origin of the Cephalopoda. Acta Palaeontol Pol 26:161–19
Dzik J (1984) Phylogeny of the Nautiloidea. Palaeontol Pol 45:3–203
Elmi S, Rulleau L (1991) Le Toarcien des carrières Lafarge (Bas-Beaujolais, France): Cadre biostratigraphique de référence pour la région lyonnaise. Geobios 24:315–331
Erben HK (1950) Bemerkungen zu Anomalien mancher Anfangswindungen von *Mimagoniatites fecundus* (Barr.). Neues Jahrb Geol Paläontol Mh 25–32
Erben HK (1960) Primitive Ammonoidea aus dem Unterdevon Frankreichs und Deutschlands. Neues Jahrb Geol Paläontol Abh 110:1–128
Erben HK (1962a) Über böhmische und türkische Vertreter von *Anetoceras* (Ammon., Unterdevon). Paläontol Z 36:14–27
Erben HK (1962b) Über die "forme elliptique" der primitiven Ammonoidea. Paläontol Z 36:38–44
Erben HK (1964) Die Evolution der ältesten Ammonoidea (Lieferung I). Neues Jahrb Geol Paläont Abh 120:107–212
Erben HK (1965) Die Evolution der ältesten Ammonoidea (Lieferung II). Neues Jahrb Geol Paläont Abh 122:275–312
Erben HK (1966) Uber den Ursprung der Ammonoidea. Biol Rev 41:641–658
Erben HK, Flajs G, Siehl A (1968) Ammonoids: early ontogeny of ultra-microscopical shell structure. Nature 219:396–398
Erben HK, Flajs G, Siehl A (1969) Die frühontogenetische Entwicklung der Schalenstruktur ectocochleater Cephalopoden. Palaeontogr Abt A 132:1–54
Etches S, Clarke J, Callomon J (2009) Ammonite eggs and ammonitellae from the Kimmeridge Clay Formation (Upper Jurassic) of Dorset, England. Lethaia 42:204–217
Fernández-López SR (1991) Taphonomic concepts for a theoretical biochronology. Rev Esp Paleontol 6:37–49
Fernández-López S (1995) Taphonomie et interprétation des paléoenvironnements. Geobios 28(Suppl 1):137–154
Fernández-López SR (2007) Ammonoid taphonomy, palaeoenvironments and sequence stratigraphy at the Bajocian/Bathonian boundary on the Bas Auran area (Subalpine Basin, south-eastern France). Lethaia 40:377–391
Fernández-López SR (2013) Dimorphism and evolution of *Albarracinites* (Ammonoidea, Lower Bajocian) from the Iberian Range (Spain). J Syst Palaeontol 12:669–685
Fernández-López S, Meléndez G (1996) Phylloceratina ammonoids in the Iberian Basin during the Middle Jurassic: a model of biogeographical and taphonomic dispersal related to relative sea-level changes. Palaeogeogr Palaeoclimatol Palaeoecol 120:291–302

Frest TJ, Glenister BF, Furnish WM (1981) Pennsylvanian-Permian Cheiloceratacean ammonoid families Maximitidae and Pseudohaloritidae. Memoir (The Paleontological Society) 11:1–46
Frýda J, Nützel A, Wagner PJ (2008) Paleozoic Gastropoda. In: Ponder W, Lindberg DR (eds) Phylogeny and evolution of the Mollusca. University of California Press, Berkeley, pp. 239–270
Furnish WM, Glenister BF, Kullmann J, Zuren Z (2009) Carboniferous and Permian Ammonoidea. In: Selden PA (ed) Treatise on invertebrate paleontology. part L, Mollusca 4, Revised, vol. 2. The University of Kansas Paleontological Institute, Lawrence, Kansas
Grandjean F (1910) Le siphon des ammonites et des belémnites. Bull Soc Geol Fr 10:496–519
Hall BK, Westermann GEG (1980) Lower Bajocian (Jurassic) cephalopod faunas from western Canada and proposed assemblage zones for the Lower Bajocian of North America. Palaeontogr A 9:1–93
Hewitt RA, Westermann GEG (2003) Recurrences of hypotheses about ammonites and *Argonauta*. J Paleontol 77:792–795
Hewitt R, Kullmann J, House M, Glenister B, Yi-Gang W (1993) Mollusca: Cephalopoda (Pre-Jurassic Ammonoidea). In: Benton MJ (ed) Fossil record 2. Chapman & Hall, London, pp 189–212
Hickman CS (1992) Reproduction and development of trochacean gastropods. The Veliger 35:245–272
Hickman CS (2004) The problem of similarity: analysis of repeated patterns of microsculpture on gastropod larval shells. Invertebr Biol 123:198–211
Hoffmann R, Schultz JA, Schellhorn R, Rybacki E, Keupp H, Gerden SR, Lemanis R, Zachow S (2013) Non-invasive imaging methods applied to neo- and paleontological cephalopod research. Biogeosci Discuss 10:18803–18851
Holland CH (2003) Some observations on bactritid cephalopods. Bull Geosci 78:369–372
House MR (1963) Devonian ammonoid successions and facies in Devon and Cornwall. Quarterly Journal of the Geological Society 119:1–23
House MR (1965) A study in the Tornoceratidae: the succession of *Tornoceras* and related genera in the North American Devonian. Phil Trans R Soc Lond B, Biol Sci 250:79–130
House MR (1985) The ammonoid time-scale and ammonoid evolution. Geol Soc Lond Mem 10:273–283
House MR (1996) Juvenile goniatite survival strategies following Devonian extinction events. Geol Soc Lond Spec Publ 102:163–18
House MR, Kirchgasser WT (2008) Late Devonian goniatites (Cephalopoda, Ammonoidea) from New York State. Bull Am Paleontol 374:1–288
Howarth MK (2013) Part L, Revised, Volume 3B, Chapter 4: Psiloceratoidea, Eodoceratoidea, Hildoceratoidea. Treatise Online 57:1–139
Hyatt A (1894) Phylogeny of an acquired characteristic. Proc Am Philos Soc 32:349–647
Igolnikov AE (2007) A new species of the genus *Boreophylloceras* Alekseev et Repin, 1998 (Ammonitida) from the Berriasian kochi zone of north-central Siberia. Paleontol J 41:128–131
Iwata K (1980) Mineralization and architecture of the larval shell of *Haliotis discus hannai* Ino (Archaeogastropoda). J Fac Sci Hokkaido Univ Ser 4 Geol Miner 19:305–320
Jablonski D, Lutz RA (1983) Larval ecology of marine benthic invertebrates: paleobiological implications. Biol Rev 58:21–89
Jacobs DK (1992) Shape, drag, and power in ammonoid swimming. Paleobiology 18:203–220
Jacobs DK, Chamberlain JA (1996) Buoyancy and hydrodynamics in ammonoids. In: Landman NH, Tanabe K, Davis RA (eds) Ammonoid paleobiology. Plenum, New York, pp. 170–224
Jäger M, Fraaye R (1997) The diet of the Early Toarcian ammonite *Harpoceras falciferum*. Palaeontology 40:557–574
Kawabe F, Haggart JW (2003) The ammonoid *Desmoceras* in the Upper Albian (Lower Cretaceous) of Japan. J Paleontol 77:314–322
Kennedy WJ (1989) Thoughts on the evolution and extinction of Cretaceous ammonites. Proc Geol Assoc 100:251–279
Keupp H (2000) Ammoniten–Paläobiologische Erfolgsspiralen. Jan Thorbecke Verlag, Stuttgart

Klofak SM (2002) Size classes in ammonoids from the Middle Devonian Cherry Valley Limestone of New York state, USA. Abh Geol Bundesanst (Vienna) 57:443–457

Klofak SM, Landman NH, Mapes RH (1999) Embryonic development of primitive ammonoids and the monophyly of the Ammonoidea. In: Olóriz F, Rodríguez-Tovar FJ (eds) Advancing research on living and fossil cephalopods. Kluwer Academic/Plenum, New York, pp. 23–45

Klofak SM, Landman NH, Mapes RH (2007) Patterns of embryonic development in Early to Middle Devonian ammonoids. In: Landman NH, Davis RA, Mapes RH (eds) Cephalopods—present and past: new insights and fresh perspectives. Springer Netherlands, pp 15-56

Klofak SM, Landman NH (2010) Some exceptionally well preserved specimens of *Agoniatites vanuxemi* from the Middle Devonian Cherry Valley Limestone of New York State. In: Tanabe K, Shigeta Y, Sasaki T, Hirano H (eds) Cephalopods—present and past. Tokai University Press, Tokyo, pp. 93–103

Klofak SM, Landman NH (2012) Internal features of ammonitellas of tornoceratids from the Middle Devonian Cherry Valley Limestone, New York State, USA. Geobios 45:49–56

Klug C (2001a) Early Emsian ammonoids from the eastern Anti-Atlas (Morocco) and their succession. Palaeontol Z 74:479–515

Klug C (2001b) Life-cycles of some Devonian ammonoids. Lethaia 34:215–233

Klug C (2002) Quantitative stratigraphy and taxonomy of late Emsian and Eifelian ammonoids of the eastern Anti-Atlas (Morocco). Cour Forsch Senckenberg 238:1–109

Klug C (2007) Sublethal injuries in Early Devonian cephalopod shells from Morocco. Acta Palaeontol Pol 52:749–759

Klug C, Korn D (2004) The origin of ammonoid locomotion. Acta Palaeontol Pol 49:235–242

Klug C, Kröger B, Korn D, Rücklin M, Schemm-Gregory M, De Baets K, Mapes RH (2008) Ecological change during the early Emsian (Devonian) in the Tafilalt (Morocco), the origin of the Ammonoidea, and the first African pyrgocystid edrioasteroids, machaerids and phyllocarids. Palaeontogr A 283:83–176

Klug C, Kröger B, Kiessling W, Mullins GL, Servais T, Fryda J, Korn D, Turner S (2010) The Devonian nekton revolution. Lethaia 43:465–477

Klug C, Kröger B, Klug C, Riegraf W, Lehmann J (2012) Soft–part preservation in heteromorph ammonites from the Cenomanian–Turonian Boundary Event (OAE 2) in north–west Germany. Palaeontology 55:1307–1331

Klug C, Kröger B, Vinther J, Fuchs D, De Baets (2015) Ancestry, origin and early evolution of ammonoids. In: Klug C, Korn D, De Baets K, Kruta I, Mapes RH (eds) Ammonoid Paleobiology: From macroevolution to paleogeography. Springer, The Netherlands

Knyazev VG (1975). Ammonites and zonal stratigraphy of the Lower Oxfordian of North Siberia. Transactions of the Institute of Geology and Geophysics, Siberian Branch of Academy of Sciences of the USSR 275:1–139 [in Russian]

Korn D, Klug C (2002) Ammoneae Devonicae. Fossilium Catalogus 138. Backhuys, Leiden

Korn D, Klug C (2007) Conch form analysis, variability, morphological disparity, and mode of life of the Frasnian (Late Devonian) ammonoid *Manticoceras* from Coumiac (Montagne Noire, France). In: Landman NH, Davis RA, Mapes RH (eds) Cephalopods—present and past: new insights and fresh perspectives. Springer, Dordrecht, pp. 57–85

Kröger B (2008) Nautiloids before and during the origin of ammonoids in a Siluro-Devonian section in the Tafilalt, Anti-Atlas, Morocco. Spec Pap Palaeontol 79:5–110

Kröger B, Mapes RH (2007) On the origin of bactritoids (Cephalopoda). Paläontol Z 81:316–327

Kröger B, Servais T, Zhang Y (2009) The origin and initial rise of pelagic cephalopods in the Ordovician. PLoS ONE 4:e7262

Kröger B, Vinther J, Fuchs D (2011) Cephalopod origin and evolution: a congruent picture emerging from fossils, development and molecules. Bioessays 33:602–613

Kruta I, Landman N, Rouget I, Cecca F, Tafforeau P (2011) The role of ammonites in the Mesozoic marine food web revealed by jaw preservation. Science 331:70–72

Kulicki C (1974) Remarks on the embryogeny and postembryonal development of ammonites. Acta Palaeontol Pol 19:201–22

Kulicki C (1979) The ammonite shell: its structure, development and biological significance. Acta Palaeontol Pol 39:97–142

Kulicki C (1996) Ammonoid shell microstructure. In: Landman N, Tanabe K, Davis R (eds) Ammonoid paleobiology. Plenum, New York, pp. 65–101

Kulicki C, Doguzhaeva LA (1994) Development and calcification of the ammonitella shell. Acta Palaeontol Pol 39:17–44

Kulicki, C., Wierzbowski, H. (1983) The Jurassic juvenile ammonites of the Jagua Formation, Cuba. Acta Palaeontol Pol 28:369–384

Kulicki C, Landman NH, Heaney MJ, Mapes RH, Tanabe K (2002) Morphology of the early whorls of goniatites from the Carboniferous Buckhorn Asphalt (Oklahoma) with aragonitic preservation. Abh Geol Bundesanst (Vienna) 57:205–224

Kulicki C, Tanabe K, Landman NH, Kaim A (2015) Ammonoid shell microstructure. This volume

Kullmann J, Wiedmann J (1970) Significance of sutures in phylogeny of Ammonoidea. Kansas University Paleontological Contributions, Paper 47:1–32

Kutygin R, Knyazev V (2000) Ontogeny of the ammonoid genus *Dactylioceras* from northeastern Russia. Paleontol J 34:263–271

Landman NH (1982) Embryonic Shells of *Baculites*. J Paleontol 56:1235–1241

Landman NH (1985) Preserved ammonitellas of *Scaphites* (Ammonoidea, Ancyloceratina). Am Mus Novit 2815:1–10

Landman NH (1987) Ontogeny of Upper Cretaceous (Turonian-Santonian) scaphitid ammonites from the Western Interior of North America: systematics, developmental patterns, and life history. Bull Am Mus Nat Hist 185:117–241

Landman NH (1988) Early ontogeny of Mesozoic ammonites and nautilids. In: Wiedmann J, Kullmann J (eds) Cephalopods—present and past. Schweizerbart, Stuttgart, pp. 215–228

Landman NH (1994) Exceptionally well-preserved ammonites from the Upper Cretaceous (Turonian-Santonian) of North America: Implications for ammonite early ontogeny. Am Mus Novit 3086:1–15

Landman NH, Bandel K (1985) Internal structures in the early whorls of Mesozoic ammonites. Am Mus Novit 2823:1–21

Landman NH, Waage KM (1982) Terminology of structures in embryonic shells of Mesozoic ammonites. J Paleontol 56:1293–1295

Landman NH, Waage KM (1993) Scaphitid ammonites of the Upper Cretaceous (Maastrichtian) Fox Hills Formation in South Dakota and Wyoming. Bull Am Mus Nat Hist 215:1–257

Landman NH, Tanabe K, Shigeta Y (1996) Ammonoid embryonic development. In: Landman NH, Tanabe K, Davis A (eds) Ammonoid paleobiology. Plenum Pess, New York

Landman NH, Mapes RH, Tanabe K (1999) Internal features of the embryonic shells of Late Carboniferous Goniatitina. In: Olóriz F, Rodríguez-Tovar FJ (eds) Advancing research on living and fossil cephalopods. Kluwer Academic/Plenum, New York, pp. 243–254

Landman NH, Bizzarini F, Tanabe K, Mapes RH, Kulicki C (2001) Micro-ornamentation on the embryonic and post-embryonic shells of Triassic ceratites (Ammonoidea). Am Malacol Bull 16:1–12

Landman NH, Mapes RH, Cruz C (2010) Jaws and soft tissues in ammonoids from the Lower Carboniferous (Upper Mississippian) Bear Gulch Beds, Montana, USA. In: Tanabe K, Shigeta Y, Sasaki T, Hirano H (eds) Cephalopods—present and past. Tokai University Press, Tokyo, pp. 147–153

Laptikhovsky VL, Rogov MA, Nikolaeva SE, Arkhipkin AI (2013) Environmental impact on ectocochleate cephalopod reproductive strategies and the evolutionary significance of cephalopod egg size. Bull Geosci 88:83–94

Lehmann U (1966) Dimorphismus bei Ammoniten der Ahrensburger Lias-Geschiebe. Paläontol Z 40:26–55

Lehmann U (1981) The ammonites: their life and world. Cambridge University Press, New York

Lehmann U (1990) Ammonoideen: Leben zwischen Skylla und Charybdis. Enke, Stuttgart

Lehmann U, Weitschat W (1973) Zur Anatomie und Ökologie von Ammoniten: Funde von Kropf und Kiemen. Paläontol Z 47:69–76

Lewy Z (2002) New aspects in ammonoid mode of life and their distribution. Geobios 35(Suppl 1):130–139
Lukeneder A, Harzhauser M, Müllegger S, Piller WE (2010) Ontogeny and habitat change in Mesozoic cephalopods revealed by stable isotopes ($\delta^{18}O$, $\delta^{13}C$). Earth Planet Sci Lett 296:103–11
Lukeneder A (2015) Ammonoid habitats and life history. This volume
Maas A, Braun A, Dong X-P, Donoghue PCJ, Müller KJ, Olempska E, Repetski JE, Siveter DJ, Stein M, Waloszek D (2006) The 'Orsten'-more than a Cambrian Konservat-Lagerstätte yielding exceptional preservation. Palaeoworld 15:266–28
Maeda H (1991) Sheltered preservation: a peculiar mode of ammonite occurrence in the Cretaceous Yezo Group, Hokkaido, north Japan. Lethaia 24:69–82
Manda S, Frýda J (2010) Silurian-Devonian boundary events and their influence on cephalopod evolution: evolutionary significance of cephalopod egg size during mass extinctions. Bull Geosci 85:513–540
Maeda H, Seilacher A (1996) Ammonoid taphonomy. In: Landman NH, Tanabe K, Davis RA (eds) Ammonoid paleobiology. Plenum, New York, pp. 543–578
Manger W, Stephen D, Meeks L (1999) Possible cephalopod reproductive mass mortality reflected by middle Carboniferous assemblages, Arkansas, Southern United States. In: Olóriz F, Rodríguez-Tovar F (eds) Advancing research on living and fossil cephalopods. Kluwer Academic/Plenum, New York
Mapes RH, Nützel A (2009) Late Palaeozoic mollusc reproduction: cephalopod egg-laying behavior and gastropod larval palaeobiology. Lethaia 42:341–356
Martin D, Briggs DEG, Parkes RJ (2003) Experimental mineralization of invertebrate eggs and the preservation of Neoproterozoic embryos. Geology 31:39–42
Martin D, Briggs DEG, Parkes RJ (2005) Decay and mineralization of invertebrate eggs. Palaios 20:562–572
Michael R (1894) Ammoniten-Brut mit Aptychen in der Wohnkammer von *Oppelia steraspis* Oppel sp. Z Dtsch Geol Ges 46:697–702
Mikhailova IA (1974) The relationship between Early Cretaceous and Late Cretaceous Hoplitaceae. Rev Bulg Geol Soc 35:117–132
Miller AI (1938) Devonian ammonoids of America. GSA Special Papers 14:1–294
Miller AK, Unklesbay AG (1943) The siphuncle of Late Paleozoic ammonoids. J Paleontol 17:1–25
Monnet C, De Baets K, Klug C (2011) Parallel evolution controlled by adaptation and covariation in ammonoid cephalopods. BMC Evol Biol 11:115
Morard A, Guex J (2003) Ontogeny and covariation in the Toarcian genus *Osperleioceras* (Ammonoidea). Bull Soc Geol Fr 174:607–615
Moriya M (2015) Isotope signature of ammonoid shells. This volume
Morton N (1988) Segregation and migration patterns in some *Graphoceras* populations (Middle Jurassic). In: Wiedmann J, Kullmann J (eds) Cephalopods—present and past. Schweizerbart, Stuttgart, pp. 377–386
Müller A (1969) Ammoniten mit "Eierbeutel" und die Frage nach dem Sexual dimorphismus der Ceratiten (Cephalopoda). Monatsber Dtsch Akad Wiss Berl 11:411–420
Münster, GG von (1834) Über das Kalkmergel-Lager von St. Cassian in Tyrol und die darin vorkommenden Ceratiten. Neues Jahrbuch für Mineralogie, Geognosie und Petrefactenkunde [Stuttgart] 1834:1–15
Naef A (1922) Die fossilen Tintenfische: eine paläozoologische Monographie. Fischer, Jena
Neige P (1997) Ontogeny of the Oxfordian ammonite *Creniceras renggeri* from the Jura of France. Eclogae Geol Helv 90:605–616
Nesis K (1986) On the feeding habits and the causes of the extinction of some heteromorph ammonites. Paleontol Zh 1986 :8–15 [in Russian]
Nishimura T, Maeda H, Tanaka G, Ohno T (2010) Taxonomic evaluation of various morphological characters in the Late Cretaceous desmoceratine polyphyletic genus "*Damesites*" from the Yezo Group in Hokkaido and Sakhalin. Paleont Res 14: 33–55

Nixon M (1996) Morphology of the jaws and radula in Ammonoids. In Landman NH, Tanabe K,Davis A (eds) Ammonoid Paleobiology. Plenum Press, New York, pp. 23–42

Nützel A (2014) Larval ecology and morphology in fossil gastropods. Palaeontology 57:479–503

Nützel A, Lehnert O, Frýda J (2006) Origin of planktotrophy—evidence from early molluscs. Evol Dev 8:325–330

Okubo S, Tsujii T, Watabe N, Williams DF (1995) Hatching of *Nautilus belauensis* Saunders, 1981, in captivity: culture, growth and stable isotope compositions of shell, and histology and immunohistochemistry of the mantle epithelium of the juveniles. The Veliger 38:192–202

Owen C (1878) On the relative positions to their constructors of the chambered shells of cephalopods. Proc Zool Soc Lond 1878:955–977

Page KN (1996) Mesozoic Ammonoids in space and time. In: Landman NH, Tanabe K, Davis RA (eds) Ammonoid paleobiology. Plenum, New York, pp. 755–794

Palframan DFB (1966) Variation and ontogeny of some Oxfordian ammonites: *Taramelliceras richei* (de Loriol) and *Creniceras renggeri* (Oppel), from Woodham, Buckinghamshire. Palaeontology 9:290–311

Palframan DFB (1967) Variation and ontogeny of some oxford clay ammonites: *Distichoceras bicostatum* (Stahl) and *Horioceras baugieri* (D'Orbigny), from England. Palaeontology 10:60–94

Palframan D (1969) Taxonomy of sexual dimorphism in ammonites: morphogenetic evidence in *Hecticoceras brightii* (Pratt). In: Westermann GEG (ed) Sexual dimorphism in fossil Metazoa and taxonomic implications. Schweizerbart, Stuttgart, pp. 126–154

Parent H (1997) Ontogeny and sexual dimorphism of *Eurycephalites gottschei* (Tornquist) (Ammonoidea) of the Andean Lower Callovian (Argentine-Chile). Geobios 30:407–419

Petter G (1959) Goniatites dévoniennes du Sahara. Publications de la Carte géologique de l'Algérie, nouvelle série, Paléontologie, Mémoires 2:1–369

Pianka ER (1970) On r- and K-selection. Am Nat 104:592–597

Repin YS, Meledina SV, Alexeev SN (1998) Phylloceratids (Ammonoidea) from the Lower Jurassic of Northeastern Asia. Paleontol J 32:461–473

Reznick D, Bryant MJ, Bashey F (2002) r-and K-selection revisited: the role of population regulation in life-history evolution. Ecology 83:1509–1520

Ristedt H (1971) Zum Bau der Orthoceriden Cephalopoden. Palaeontogr Abt A 137:155–195

Ritterbush KA, Hoffmann R, Lukeneder A, De Baets K (2014) Pelagic palaeoecology: the importance of recent constraints on ammonoid palaeobiology and life history. J Zool 292:229–241

Rocha F, Guerra Á, Gonzalez ÁF (2001) A review of reproductive strategies in cephalopods. Biol Rev 76:291–304

Rosa R, Pierce GJ, O´Dor R (2013a) Advances in squid biology, ecology and fisheries. Part I—Myopsid Squids. Nova Science Publishers, New York

Rosa R, O´Dor R, Pierce GJ (2013b) Advances in squid biology, ecology and fisheries. Part II—Oegopsid Squids. Nova Science Publishers, New York

Rouget I, Neige P (2001) Embryonic ammonoid shell features: intraspecific variation revisited. Palaeontology 44:53–64

Ruzhentsev VE, Shimanskij VN (1954) Lower Permian coiled and curved nautiloids of the Southern Urals. Trans Paleontol Inst Akad Nauk SSSR 50:1–150 [in Russian]

Sánchez-Villagra MR (2012) Embryos in deep time: the rock record of biological development. University of California Press, Oakland

Schindewolf OH (1928) Zur Terminologie der Lobenlinie. Paläontol Z 9:181–186

Schindewolf OH (1929) Vergleichende Studien zur Phylogenie, Morphogenie und Terminologie der Ammoneenlobenlinie. Abh Preuss Geol Landesanst 115:1–102

Schindewolf OH (1932) Zur Stammesgeschichte der Ammoneen. Paläontol Z 14:164–180

Schindewolf OH (1933) Vergleichende Morphologie und Phylogenie der Anfangskammern tetrabranchiater Cephalopoden. Eine Studie über Herkunft, Stammesentwicklung und System der niederen Ammoneen. Abh Preuss Geol Landesanst 148:1–115

Schindewolf OH (1934) Zur Stammesgeschichte der Cephalopoden. Jahrb Preuss Geol Landesanst 55:258–283

Schindewolf, OH (1937) Zur Stratigraphie und Paläontologie der Wocklumer Schichten (Oberdevon). Abh Preuss Geol Landesanst 178:1–132
Schindewolf OH (1951) Zur Morphogenie und Terminologie der Ammoneen-Lobenlinie. Paläontol Z 25:11–34
Schindewolf O (1954) On development, evolution and terminology of ammonoid suture line. Bull Mus Comp Zool 112:217–237
Schindewolf OH (1959) Adolescent cephalopods from the exshaw formation of Alberta. J Paleontol 33:971–976
Schlögl J, Chirat R, Balter V, Joachimski M, Hudáčková N, Quillévéré F (2011) *Aturia* from the Miocene Paratethys: an exceptional window on nautilid habitat and lifestyle. Palaeogeogr Palaeoclimatol Palaeoecol 308:330–338
Senior JR (1977) The Jurassic ammonite *Bredyia* Buckman. Palaeontology 20:675–693
Shigeta Y (1989) Systematics of the ammonite genus *Tetragonites* from the Upper Cretaceous of Hokkaido. Trans Proc Palaeontol Soc Jpn 156:319–342
Shigeta Y (1993) Post-hatching early life history of Cretaceous Ammonoidea. Lethaia 26:133–145
Shigeta Y, Zakharov YD, Mapes RH (2001) Origin of the Ceratitida (Ammonoidea) inferred from the early internal shell features. Paleontol Res 5:201–213
Shimizu S (1934) Ammonites. In: Shimizu S, Obata T (eds) Ammonites: Iwanami's Lecture Series of Geology and Palaeontology, Tokyo, pp. 137 [in Japanese]
Smith JP (1897) The development of *Glyphioceras* and the phylogeny of the Glyphioceratidae. Proc Calif Acad Sci (Geol) 1:105–128
Smith JP (1898) The development of *Lytoceras* and *Phylloceras*. Proc Calif Acad Sci, 3rd Ser Geol 1:129–160
Smith JP (1899) Larval stages of *Schloenbachia*. J Morphol 16:237–268
Smith JP (1900) The development and phylogeny of *Placenticeras*. Proc Calif Acad Sci, 3rd Ser Geol 1:181–240
Smith JP (1901) The larval coil of *Baculites*. Am Nat 35:39–49
Smyshlyaeva OP, Zakharov YD (2012) New representatives of the family Melagathiceratidae (Ammonoidea) from the Lower Triassic of South Primorye. Paleontol J 46:142–147
Smyshlyaeva OP, Zakharov YD (2013) New members of the family Flemingitidae (Ammonoidea) from the Lower Triassic of South Primorye. Paleontol J 47:247–255
Spath LF (1933) The evolution of the Cephalopoda. Biol Rev 8:418–462
Spath LF (1936) The phylogeny of the Cephalopoda. Paläontol Z 18:156–181
Sprey AM (2001) Tuberculate micro-ornament on the juvenile shell of Middle Jurassic ammonoids. Lethaia 34:31–35
Sprey AM (2002) Early Ontogeny of three Callovian ammonite Genera (*Binatisphinctes*, *Kosmoceras* (*Spinikosmoceras*) and *Hecticoceras*) from Ryazan (Russia). Abh Geol Bundesanst (Vienna) 57:225–255
Stearns SC (1977) The evolution of life history traits: a critique of the theory and a review of the data. Annu Rev Ecol Syst 8:145–171
Stearns SC (1992) The evolution of life histories. Oxford University Press, Oxford
Stephen DA, Stanton. RJ (2002) Impact of reproductive strategy on cephalopod evolution. Abh Geol Bundesanst (Vienna) 57:151–155
Stephen DA, Bylund KG, Garcia P, McShinsky RD, Carter HJ (2012) Taphonomy of dense concentrations of juvenile ammonoids in the Upper Cretaceous Mancos Shale, east-central Utah, USA. Geobios 45:121–128
Suan G, Rulleau L, Mattioli E, Sucheras-Marx B, Rousselle B, Pittet B, Vincent P, Martin JE, Lena A, Spangenberg JE (2013) Palaeoenvironmental significance of Toarcian black shales and event deposits from southern Beaujolais, France. Geol Mag 150:728–742
Tajika A, Wani R (2011) Intraspecific variation of hatchling size in Late Cretaceous ammonoids from Hokkaido, Japan: implication for planktic duration at early ontogenetic stage. Lethaia 44:287–298
Tanabe K (1977a) Functional evolution of *Otoscaphites puerculus* (Jimbo) and *Scaphites planus* (Yabe), Upper Cretaceous ammonites. Mem Fac Sci Kyushu Univ Ser D (Geol) 23:367–407

Tanabe K (1977b) Mid-Cretaceous scaphitid ammonites from Hokkaido. Palaeontol Soc Jpn Spec Pap 21:11–22

Tanabe K (1989) Endocochliate embryo model in the Mesozoic Ammonitida. Hist Biol 2:183–196

Tanabe K (2011) The feeding habits of ammonites. Science 331:37–38

Tanabe K, Ohtsuka Y (1985) Ammonoid early internal shell structure: its bearing on early life history. Paleobiology 11:310–322

Tanabe K, Tsukahara J (1987) Biometric analysis of *Nautilus pompilius* from the Philippines and the Fiji Islands. In: Saunders WB, Landman NH (eds) *Nautilus*. Plenum Press, New York, pp. 105–113

Tanabe K, Uchiyama K (1997). Development of the embryonic shell structure in *Nautilus*. The Veliger 40:203–215

Tanabe K, Obata I, Fukuda Y, Futakami M (1979) Early shell growth in some Upper Cretaceous ammonites and its implications to major taxonomy. Bull Natl Sci Mus Ser C (Geol) 5:155–176

Tanabe K, Fukuda Y, Obata I (1980) Ontogenetic development and functional morphology in the early growth stages of three Cretaceous ammonites. Bull Natl Sci Mus Ser C (Geol) 6:9–26

Tanabe K, Landman NH, Mapes RH, Faulkner CJ (1993) Analysis of a Carboniferous embryonic ammonoid assemblage implications for ammonoid embryology. Lethaia 26:215–224

Tanabe K, Landman NH, Mapes RH (1994) Early shell features of some Late Paleozoic ammonoids and their systematic implications. Trans Proc Palaeontol Soc Jpn 173:384–400

Tanabe K, Shigeta Y, Mapes RH (1995) Early life history of Carboniferous ammonoids inferred from analysis of shell hydrostatics and fossil assemblages. Palaios 10:80–86

Tanabe K, Kulicki C, Landman NH, Mapes RH (2001) External features of embryonic and early post-embryonic shells of a Carboniferous goniatite *Vidrioceras* from Kansas. Paleontol Res 5:13–19

Tanabe K, Landman NH, Yoshioka Y (2003) Intra- and interspecific variation in the early internal shell features of some Cretaceous ammonoids. J Paleontol 77:876–887

Tanabe K, Kulicki C, Landman NH (2008) Development of the embryonic shell structure of mesozoic ammonoids. Am Mus Novit 3621:1–19

Tanabe K, Landman NH, Kaim A (2010) Tuberculate micro-ornamentation on embryonic shells of Mesozoic ammonoids: microstructure, taxonomic variation, and morphogenesis. In: Tanabe K, Shigeta Y, Sasaki T, Hirano H (eds) Cephalopods—present and past. Tokai University Press, Tokyo, pp. 105–121

Tomašových A, Schlögl J (2008) Analyzing variations in cephalopod abundances in shell concentrations: the combined effects of production and density-dependent cementation rates. Palaios 23:648–666

Tozer ET (1981) Triassic Ammonoidea: Classification, evolution and relationship with Permian and Jurassic Forms. In: House MR, Senior JR (eds.) The Ammonoidea: the evolution classification, mode of life and geological usefulness of a major fossil group. Academic Press, London, pp. 66–100

Uchiyama K, Tanabe K (1999) Hatching experiment of *Nautilus macromphalus* in the Toba Aquarium, Japan. In: Olóriz F, Rodríguez-Tovar FJ (eds) Advancing research on living and fossil cephalopods. Kluwer/Plenum, New York, pp. 13–16

Urdy S, Wilson LB, Haug J, Sánchez-Villagra M (2013) On the unique perspective of paleontology in the study of developmental evolution and biases. Biol Theory :1–19

Urlichs M (2006) Dimorphismus bei Ceratites aus dem Germanischen Oberen Muschelkalk (Ammonoidea, Mitteltrias) mit Revision einiger Arten. Stuttg Beitr Natkd Ser B 363:1–85

Vance RR (1973) On reproductive strategies in marine benthic invertebrates. Am Nat 107:339–352

Van Valen L (2005) The statistics of variation. In: Hallgrímsson B, Hall BK (eds) Variation: a central concept in biology. Academic Press, Burlington

Vavilov MN (1992) Stratigraphy and ammonoids of the Middle Triassic deposits of North-East Asia. Nedra, Moscow [in Russian]

Walton SA, Korn D, Klug C (2010) Size distribution of the Late Devonian ammonoid *Prolobites*: indication for possible mass spawning events. Swiss J Geosci 103:475–494

Wani R, Kurihara K, Ayyasami K (2011) Large hatchling size in Cretaceous nautiloids persists across the end-Cretaceous mass extinction: new data of Hercoglossidae hatchlings. Cretac Res 32:618–622

Ward P (1996) Ammonoid extinction. In: Landman NH, Tanabe K, Davis RA (eds) Ammonoid paleobiology. Plenum, New York, pp. 815–824

Ward PD, Bandel K (1987) Life history strategies in fossil cephalopods. In: Boyle PR (ed) Cephalopod life cycles, Vol. II. Academic Press, London, pp. 329–350

Weitschat W, Bändel K (1991) Organic components in phragmocones of boreal Triassic ammonoids: Implications for ammonoid biology. Paläontol Z 65:269–303

Westermann GEG (1993) On alleged negative buoyancy of ammonoids. Lethaia 26:246–246

Westermann GEG (1996) Ammonoid life and habitat. In: Landman NH, Tanabe K, Davis RA (eds) Ammonoid paleobiology. Plenum, New York, pp. 607–707

Wetzel W (1959) Über Ammoniten-Larven. N Jahrb Geol Paläont Abh 107:240–252

Winkelmann I, Campos PF, Strugnell J, Cherel Y, Smith PJ, Kubodera T, Allcock L, Kampmann M-L, Schroeder H, Guerra A, Norman M, Finn J, Ingrao D, Clarke M, Gilbert MTP (2013) Mitochondrial genome diversity and population structure of the giant squid *Architeuthis:* genetics sheds new light on one of the most enigmatic marine species. Proc R Soc Lond Ser B Biol Sci 280: 20130273

Wissner UFG, Norris AW (1991) Middle Devonian goniatites from the Dunedin and Besa River formations of northeastern British Columbia. Geol Surv Can Bull 412:45–79

Wright CW (1996) Treatise on Invertebrate Paleontology: Part L (Revised) Mollusca 4, vol. 4, Cretaceous Ammonoidea. University of Kansas Paleontological Institute, Lawrence

Yahada H, Wani R (2013) limited migration of scaphitid ammonoids: evidence from the analysis of shell whorls. J Paleont 87: 406–412

Yokoyama M (1890) Versteinerungen aus der japanischen Kreide. Palaeontogr 36:159–202

Young RE, Harman RF (1988) 'Larva', 'paralarva' and 'subadult' in cephalopod terminology. Malacologia 29:201–207

Zakharov YD (1971) Some features of the development of the hydrostatic apparatus in early Mesozoic ammonoids. Paleontol J 5(1):24–33

Zakharov YD (1972) Formation of the caecum and prosiphon in ammonoids. Paleontol J 6:201–206

Zakharov YD (1974) New data on internal shell structures in Carboniferous, Triassic and Cretaceous ammonoids. Paleontol J 8(1):25–36

Zakharov YD (1978) Lower Triassic ammonoids of the East USSR. NAUKA, Moscow [in Russian]

Zakharov YD, Smyshlyaeva OP, Simanenko LF (2012) Triassic ammonoid succession in South Primorye: 6. Melagathiceratid ammonoids (inner shell structure, phylogeny, stratigraphical and palaeobiogeographical importance). Albertiana 40:28–36

Zatoń M, Niedźwiedzki G, Pieńkowski G (2009) Gastropod egg capsules preserved on bivalve shells from the Lower Jurassic (Hettangian) of Poland. Palaios 24:568–577

Zhou Z, Glenister BF, Furnish WM (2002) Endemic Permian ammonoid genus *Yinoceras*, Central Hunan, South China. J Paleontol 76:424–430

Chapter 6
Theoretical Modelling of the Molluscan Shell: What has been Learned From the Comparison Among Molluscan Taxa?

Séverine Urdy

6.1 Introduction

Behind the aesthetic beauty of mollusc shells lies a still poorly understood developmental process. Yet, the mollusc shell is well suited to address the connections between development and evolution. Mollusc shells are abundant in the fossil record, making it possible to reconstruct evolutionary lineages. The mode of shell growth is accretionary, meaning that normally a complete record of ontogeny is preserved in each shell. Because of their mathematically appealing geometry, which often conforms to a logarithmic spiral, mollusc shells have also been instrumental in the emergence of the theoretical morphology research field.

However, the future role of the mollusc shell as a successful model for "*Evo-Devo*" depends on a better understanding of shell morphogenesis—in particular the covariation among the shell characters upon which most of the taxonomic studies rest upon. These taxonomic features include the shape of the aperture, the degree of coiling of the shell tube, the ornamentation (ribs, tubercles, spines, keels) and growth features (growth halts, constrictions, varices). Some of these taxonomic characters may vary extensively within even a single species and traditional taxonomic studies often note that this phenotypic plasticity is a hallmark of mollusks, in particular in ammonoids (Buckman 1987; Westermann 1966; Rieber 1972; Kennedy and Cobban 1976; Meister 1988; Reyment and Kennedy 1991; Seilacher 1991; Dagys and Weitschat 1993; Checa et al. 1996; Dagys et al. 1999; Korn and Klug 2003; Monnet

S. Urdy (✉)
Department of Anatomy, Department of Bioengineering and Therapeutic Sciences, University of California San Francisco, Genentech Hall, 600, 16th Street, San Francisco, CA 94143–2140, USA
e-mail: severine.urdy@ucsf.edu

Physics Institute, Disordered and Biological Soft Matter, University of Zurich, Winterthurerstrasse 190, 8057 Zurich, Switzerland
e-mail: severine.urdy@physik.uzh.ch

C. Klug et al. (eds.), *Ammonoid Paleobiology: From Anatomy to Ecology*, Topics in Geobiology 43, DOI 10.1007/978-94-017-9630-9_6

and Bucher 2005; De Baets et al. 2015; Klug et al. 2015; Monnet et al. 2015) and in gastropods (Vermeij 1980; Kemp and Bertness 1984; Trussell 1996; Yeap et al. 2001; Vermeij 2002). Moreover, the evolution of the molluscan shell is characterized by frequent convergences in form and ornamentation (Schander and Sundberg 2001; Chirat et al. 2013). As a consequence, the recognition of evolutionary lineages crucially depends on our understanding of the mechanisms generating these shapes.

The shell is secreted by the mantle, a soft sheet of connective tissue covered by an epithelium. During shell growth, the mantle edge extends slightly beyond the aperture and adds a shell increment to the margin. Thus, the shape and orientation of successive shell increments are "fossilizing" the growth field of the mantle edge as development proceeds.

The molluscan shell has been the subject of extensive theoretical work, emphasizing the regularity of accretionary growth (Moseley 1838; Thompson 1917; Raup 1961; Raup and Michelson 1965; Raup 1966), although drastic deviations have also been noted and extensively studied (Okomoto 1988a, b, c; Checa et al. 2002; Clements et al. 2008). There is also a rich diversity of theoretical models of mollusc shells: the different kinds of mathematical models of morphogenesis proposed in almost all developmental contexts so far have also been applied to molluscan shell morphogenesis. Sometimes, theoretical studies on molluscs have even set the foundations for important classes of developmental models applied later to the development of model organisms such as frogs, fruit flies and mice (Gierer and Meinhart 1972; Meinhardt and Gierer 1974; Meinhardt and Klinger 1988; Meinhardt 1995; Ermentrout et al. 1986, 1998; Meinhardt and Gierer 2000; Meinhardt et al. 2008). In most cases, the models discussed here would apply to all molluscs with an external shell, although the original formulation of these models may have been primarily motivated by the study of ammonoids (Okamoto 1988a, b, c; Checa et al. 2002; Hammer and Bucher 1999, 2005a, b), gastropods (Hutchinson 1990; Morita 1991a, b, 1993; Rice 1998; Urdy et al. 2010a, b), bivalves (Savazzi 1987, 1991; Ackerly 1989; Ubukata 1997, 2001, 2003, 2005), or even brachiopods (McGhee 1978, 1980; Ackerly 1992). Here, the discussion is not restricted to models motivated by studies exclusively on ammonoids. However, models, which would apply only to ammonoids—such as those related to septa formation (i.e. Bayer 1977; García-Ruiz et al. 1990; Checa and García-Ruiz 1996; Pérez-Claros et al. 2007; Klug and Hoffmann (2015))—will not be discussed here.

This chapter will focus on the comparison between different clades of molluscs, since comparative studies are often informative with regards to the evolutionary conserved rules underlying accretionary growth. In particular, it has been pointed out that common rules of accretionary growth underlie the morphogenesis of the shell and its evolution in ammonoids and gastropods (Bucher 1997; Checa et al. 1998). Evidences come from the comparison of intraspecific patterns of covariation between shell characters, from the description of changes occurring at maturity and from the analysis of teratological shells with malformations caused by injuries or changes in living conditions.

This chapter will first describe the current and past theories of morphogenesis of coiling and the ornamentation of molluscan shells. These two aspects are often viewed as geometrically independent, although from a developmental point of view, they are closely connected. The growth field in the mantle edge is responsible for determining the shape and position of the aperture during development (hence aperture shape and coiling), while local and dynamic perturbations of the same growth field is responsible for determining the ornamentation. In a second part, this chapter will discuss what insights have been gained from the comparison between ammonoids, gastropods and bivalves with regards to the morphogenesis of their shell and its ornamentation.

6.2 Shell Morphology

From geometrical models to mechano-chemical models, there exist a wide range of models of molluscan shell morphogenesis. Geometrical descriptions are particularly encouraged by the simple morphology of molluscan shells and such descriptions date back to the nineteenth century (Moseley 1838; Whitworth 1862). D'Arcy Thompson (1917) extensively used geometrical approaches to investigate the relationships between growth and form. His main thesis is that growth in mollusc shells follows rather rigid mathematical laws—in most cases, the curve traced in space by the shell outline is really close to a logarithmic spiral. D'Arcy Thompson recognized that most of the differences between species generally ascribed by taxonomists correspond to variations in four parameters describing this spiral. Importantly, his approach represents an attempt to diagnose shell form in terms of parameters related to growth instead of arbitrary parameters such as shell length or shell width. His work inspired a morphologically-oriented research program, which flourished after Raup's pioneer model (1961, 1966), whose practical applications were manifold.

A major theme in the so-called "*theoretical morphology field*" is the investigation of processes and constraints that determine the occupancy of theoretical morphospace—the spectrum of all possible morphologies that can be exhibited by a given taxonomic group (Raup and Michelson 1965). Since Raup set the foundations, mollusc shells have become a favorite in theoretical morphospace research (Dera et al. 2008). McGhee (1999) provides a definition of theoretical morphology as comprising the mathematical simulation of form and the construction of morphospaces where the empirical and theoretical distributions of forms are represented and compared. The simulation of form may not take morphogenesis explicitly into account, but modeling of growth and development has become the primary goal of more recent theoretical morphological studies (reviewed in Eble 1999). In this section, I will successively discuss the theoretical morphospace models ("*form models*") and the more recent generative mathematical models of shell form ("*morphogenetic models*"). When equations are provided, the constants are designated in bold, while the variables are in italics.

6.2.1 Geometric Models

Early qualitative theoretical morphology studies let to the general recognition that variation in species is bounded and has some structure or order. This field, inspired in most aspects by the famous book of D'Arcy Thompson "*On growth and form*", is motivated by the search for natural laws, which allow the inference of unity from diversity. In the specific case of mollusc shells, there is nowadays a general agreement that the bulk of coiled forms observed in this group represent variations of a rather simple model: the logarithmic spiral model. Over the years, this model has been more and more mathematically sophisticated, and studies have become more and more quantitative and detailed. To start with, I recall some basic properties of the planar and helicoidal logarithmic spirals.

6.2.1.1 Planar Logarithmic Spiral

The simple logarithmic spiral model states that the radius of a logarithmic spiral (Fig. 6.1) is an exponential function of the angle of revolution, whose equation in polar coordinates (d, theta) is given by:

$$d(theta(t)) = \mathbf{a} * exp(\mathbf{b} * theta(t))$$

where *d* is the distance from a point of the spiral to the origin (it is the radius of the spiral), *theta* is the angle from the x-axis, and **a** and **b** are arbitrary constants. Take **n** as the number of loops, *theta* is between [0, **n***2pi] and *t* is between [1, length(theta)]. The logarithmic spiral is also known as the equiangular spiral or *spira mirabilis*.

It can be expressed parametrically in Cartesian coordinates (x, y) as:

$$x(t) = \ d * cos(theta(t)) = \mathbf{a} * cos(theta(t)) * exp(\mathbf{b} * theta(t))$$
$$y(t) = \ d * sin(theta(t)) = \mathbf{a} * sin(theta(t)) * exp(\mathbf{b} * theta(t))$$

The *rate of change of the distance from the origin d with respect to theta* is:

$$(d/theta)' = \mathbf{a} * \mathbf{b} * exp(\mathbf{b} * theta) = \mathbf{b} * d$$

The angle between the tangent and a radial line is a constant, called the **equiangular angle** (Thompson 1917, Fig. 6.1):

$$\mathbf{phi} = acot(\mathbf{b}) \ \ [inv\ cot]$$

So, as **b** tends towards 0, **phi** tends toward pi/2 and the spiral approaches a circle.

The arc length (as measured from the origin) is:

$$s(theta) = (\mathbf{a} * \mathrm{exp}(\mathbf{b} * \mathrm{theta}) * \mathrm{sqrt}(1 + \mathbf{b}^2))/\mathbf{b}$$

The *curvature* is:

$$Kappa(theta) = \|de1/dS\| \text{ where e1 is the unit tangent vector}$$
$$= exp(-\mathbf{b} * theta)/(\mathbf{a} * sqrt)(1 + \mathbf{b}^2))$$

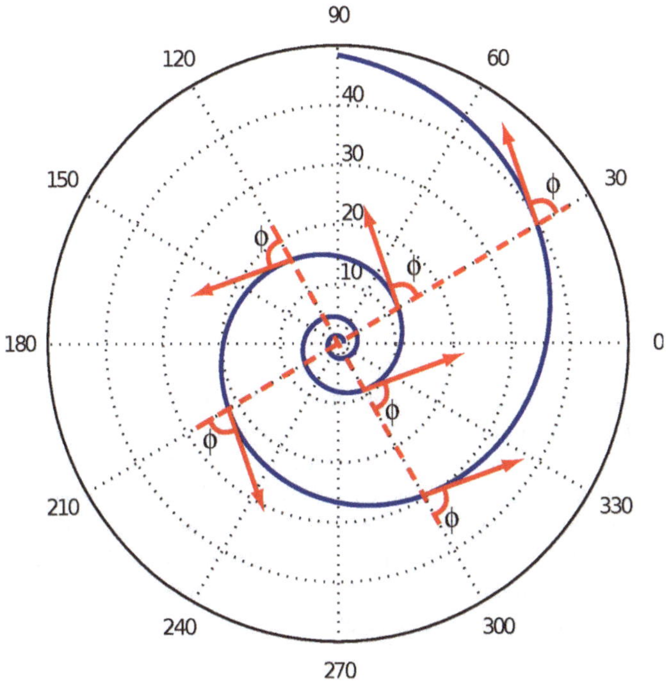

Fig. 6.1 Planar logarithmic spiral. The two radii marked with the broken *red line* measure 30 and 40 units, while the rotation angle theta are 300° and 390° respectively. Raup's W is given by: **W**=(r2/r1)2π/Δθ, so here **W**=(40/30) 2π/(π/2) =3.1605. The equi-angular angle phi is given by: **phi**=cot -1 **(b)**≈1.3897 radians or 80°

6.2.1.2 Helicoidal Logarithmic Spiral

A helicoidal logarithmic spiral is a space curve whose parametric equation in Cartesian coordinates (x, y, z) is given by:

$$x(t) = d * cos(theta(t)) = \mathbf{a} * cos(theta(t)) * exp(\mathbf{b} * theta(t))$$
$$y(t) = d * sin(theta(t)) = \mathbf{a} * sin(theta(t)) * exp(\mathbf{b} * theta(t))$$
$$z(t) = \mathbf{c} * theta(t)$$

where *d* is the distance from a point of the helix to the origin (it is the radius of the helix), *theta* is the angle from the x-axis, and **a, b** and **c** are arbitrary constants. Take **n** as the number of loops, *theta* is between [0, **n***2pi] and *t* is between [1, length(theta)].

The **vertical separation of the helix's loops** is a constant, which equals to 2pi***c**. The *arc length* of a helix is given by:

$$s(theta) = theta * sqrt(d^2 + \mathbf{c}^2) \text{ with } d(theta(t)) = \mathbf{a} * exp(\mathbf{b} * theta(t))$$
$$= theta * sqrt(\mathbf{a}^2 * exp(2 * \mathbf{b} * theta) + \mathbf{c}^2)$$

The *curvature* of a helix is:

$$Kappa(theta) = \| de1/dS \| \text{ where e1is the unit tangent vector}$$
$$= d/(d^2 + \mathbf{c}^2) \text{ with } d(theta(t)) = \mathbf{a} * exp(\mathbf{b} * theta(\mathrm{t}))$$

The *torsion* of the helix is:

$$Tau(theta) = c/(d^2 + c^2) with\ d(theta(t)) = a * exp(b * theta(t))$$

A shell can be graphically constructed by revolving a generating curve (e.g., a circle) about an axis passing through the spiral pole (coiling axis) along a logarithmic trajectory. In a logarithmically coiled shell, the diameter of the generating curve increases exponentially at each equal interval of rotation, so that each new growth increment is shape invariant. Thus, each new growth increment is simply an enlarged version of a previous increment. The angle between the tangent of a spiral trajectory at one point of the generating curve and the radius line at this point is a constant (equiangular angle). The apical angle (Thompson 1917) is defined as the angle of the cone enveloping the spire of the shell.

6.2.1.3 Raup's Model

In a single shell, the following parameters are usually approximately constant:

- the shape of the aperture
- the equiangular angle as defined by Thompson
- the relative amount of whorl overlap
- the ratio of the size of the aperture to its distance from the coiling axis

Based on these observations, Raup (1961) has shown that the basic form of mollusc shells can be defined by four parameters (not the same four parameters that D'Arcy Thompson used). Raup's model (1966) represents a shell as the successive positions of a generating curve rotating about a coiling axis in a cylindrical coordinate system (Fig. 6.2). Biologically, the generating curve represents a cross-sectional outline of the body whorl when a shell is sectioned in a plane passing through the presumptive coiling axis. All four parameters are defined with respect to one whorl revolution of the generating curve about this coiling axis. They are constants for a given logarithmically coiled form. This means that successive shell segments—defined by any sections of two planes passing through the coiling axis separated by a given angle of revolution *theta*—are simple enlarged or reduced versions of other shell segments corresponding to the same revolution angle *theta*. Thus, these segments are homothetic to each other and growth is self-similar (isometric).

The first parameter (S) describes the shape of the generating curve. It is assumed that, with each revolution, the generating curve retains a constant shape—described

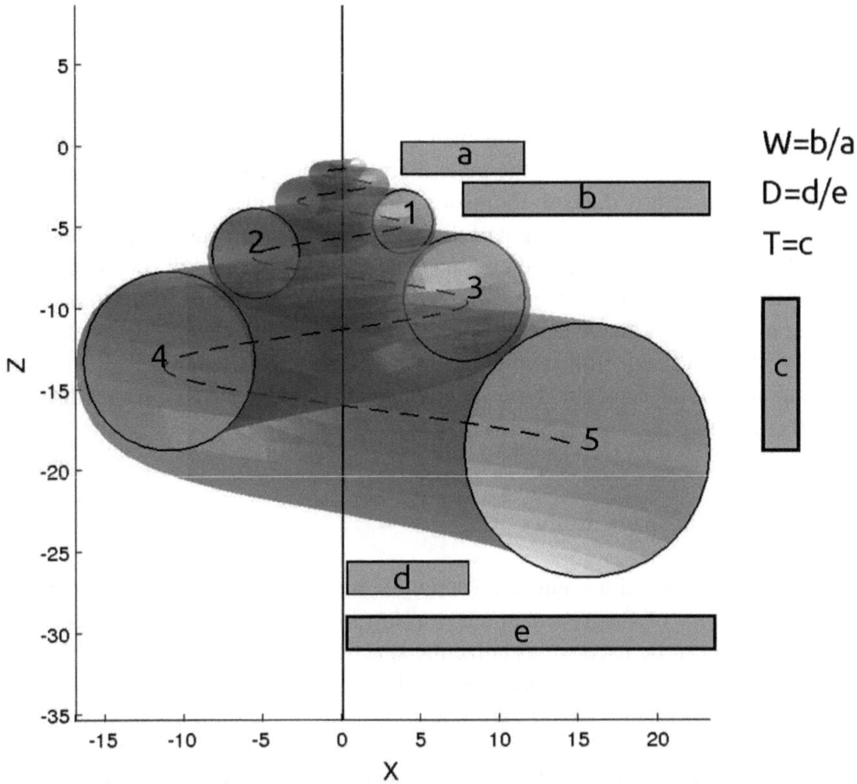

Fig. 6.2 Raup's parameters. **a** is the diameter of the whorl cross-section number 3. **b** is the diameter of the whorl cross-section number 5. **c** is the distance between the centroids of whorl cross-sections 3 and 5, parallel to the coiling axis (z-*axis*). **d** is the distance of the inner most point of whorl cross-section number 5 from the coiling axis. **e** is the distance of the outer most point of the whorl cross-section number 5 from the coiling axis

heuristically by the parameter S. However, the generating curve expands with each revolution about the coiling axis, and this rate of increase in size per revolution is described by a second parameter (W). A third parameter (D) defines the position of the generating curve relative to the coiling axis. A fourth parameter (T) describes the rate of translation of the generating curve along the coiling axis.

Figure 6.2 represents a theoretical shell as cross-sectioned by a plane passing through the coiling axis. Therefore, we see replicas of the circular generating curve at 180° intervals. The shell outline in 3D is added as an aid for visualization. If one assumes that the coiling axis coincides with the z-axis and that the first generating curve is positioned in the (x, y) plane as in Fig. 6.2, the W, D and T parameters are defined numerically and are obtained geometrically as follows:

- W is the factor by which any linear dimension of the generating curve is enlarged by each full revolution: it is the whorl expansion rate measured as the ratio of one

linear measurement of two whorl cross-sections one full revolution apart (i.e. whorl diameters, whorl width, whorl height…).
- D is the distance of the generating curve from the coiling axis relative to the size of the generating curve (outward migration of the generating curve): it is defined as the ratio of the position of the inner margin of the generating curve to the position of its outer margin. If the inner margin of the generating curve lies on the coiling axis, as it is the case for most gastropods, D equals 0.
- T is the proportion of the height of the generating curve (measured parallel to the coiling axis), which is covered by the generating curve one full revolution apart (downward migration of the generating curve): it is the rate of whorl translation. It is measured as the distance along the coiling axis traveled by two generating curve centroids one full revolution apart. In planispiral shells, T equals zero, meaning that the centroid of the generating curve has a constant z-value throughout ontogeny.

6.2.1.4 Morphospace Studies

Raup (1962) proposed a computer version of his earlier model (1961), where he produced images of theoretical snails varying in W, T, and D. Raup's geometric model (1966) can be used to visualize the spectrum of theoretically possible shell forms in a three dimensional space defined by W, D and T.

One important result of Raup's morphospace study is that the portion of the W-D-T space occupied by actual species known from the fossil and living records looks smaller than one would expect. The bulk of species represented by four invertebrate groups—brachiopods, bivalves, gastropods and ammonoids—are confined to relatively small and only slightly overlapping areas of the morphospace (Raup 1966). This morphospace shows that the distribution of species is not random. Typically, brachiopods and bivalves have higher values of W than ammonoids and gastropods. Brachiopods and bivalves appear to have been constrained by the functional necessity of articulating their two valves together: this function is only possible when there is no whorl overlap. The "*no overlap region*" corresponds to high values of W, high values of T, and high values of D.

Gastropods and bivalves usually explore a wide range of the trochispiral plane ($T>0$), whereas ammonoids and brachiopods are mostly confined to planispiral regions ($T=0$). The reasons underlying such a taxonomic splitting between the planispiral and trochospiral regions are still unclear. However, Raup (1967) argued that the confinement of most ammonoids to the planispiral region contributes to their mode of life as swimming organisms (requirements imposed by their orientation in the sea water and their streamlining capabilities).

Raup's ideas have been well received in the ammonoid research community. Consequently, there exist numerous morphospace studies based on ammonoids. Raup's model has been extensively applied to the evolution of ammonoid families to assess the role of geometry on various factors such as the orientation and stability of life position and the efficiency in shell strength, streamlining and swimming (Raup 1967; Raup and Chamberlain 1967; Ward 1980; Saunders and Swan 1984; Saunders and Shapiro 1986; Klug 2001; Korn and Klug 2003; Korn and Klug 2007). Many studies showed that some ammonoid geometries have evolved more often than others (evolutionary convergence) and that some families were extremely variable while some other families were rather restricted to small regions of the morphospace (Raup 1967; Dommergues et al. 1996; Gerber et al. 2007, 2008; Neige et al. 2009).

The reasons for such a success are manifold. Notably, morphospace studies in ammonoids are facilitated by the fact that the W-D plane resumes much of the variation observed among species/ genera, covering the transition from evolute to involute forms. Moreover, in ammonoids, the whorl cross-section can reasonably be approximated by simple variations about a circle. If the parameter S is said to represent the ratio of whorl width to whorl height (lateral and dorso-ventral respectively), variations in S represent variation from depressed ellipses ($S>1$) to compressed ellipses ($S<1$). This fits relatively well with the kind of apertures shapes generally exhibited by ammonoids.

To the contrary, much of the variation among gastropod species is not reasonably captured by any 2D morphospace (W-T, W-D, T-D). Additionally, Raup's model assumes that the apertures are radial but almost all gastropod species exhibit strongly tangential apertures, which are also tilted relative to the coiling axis (Linsley 1977, 1978). Consequently, most gastropod researchers had the feeling that Raup's model was missing two critical components of the gastropod morphology: the degree of aperture retardation (which was one of D'Arcy Thompson's parameters) as well as the degree of aperture inclination (mu in Cortie 1989). Please note that radial apertures seem to be a rarity in ammonoids too, however, contrary to gastropods, this parameter has not been strongly linked to other important aspects of ammonoids physiology or mode of life. Additionally, the determination of W and T in gastropods requires pictures of the shell in the apical and apertural views, whereas the determination W and D in ammonoids only requires a single photograph in lateral view. Generally, in gastropods, aperture shapes are asymmetrical and their allometries are often non-linear (e.g., Urdy et al. 2010a, b, but see also Korn 2012 for multi-phasic allometries in ammonoids). As a consequence, it is not easy to define the parameter S in a way that would reach a consensus acceptance or general significance.

Regarding other invertebrate groups, the impact of Raup's model has been rather limited. Although McGhee pursued extensive morphospace studies on brachiopods (1978, 1980) and bivalves (1999), he noted additional difficulties arising with shapes, which do not coil over several whorls (McGhee 2001). Most bivalves and brachiopods also have tangential apertures (not passing through the coiling axis) and many bivalves evolved towards the unequivalve geometry, turning morphospace studies on bivalves less straightforward than those on ammonoids.

However, these are only additional difficulties and they cannot be claimed as inherent limitations of Raup's model, which in its elegance and simplicity represents the quintessence of theoretical morphology. Nevertheless, there is one important limitation in Raup's model and I will discuss it below.

6.2.1.5 Limitations of Raup's Model

Raup assumed that his parameters were geometrically independent and that the variation along the defined axes could be represented on orthogonal axes; all combinations of W, D, and T are theoretically possible. The reality is unfortunately not that straightforward, and a more recent study has shown that D and T were both dependent on W (Schindel, 1990). In other words, the axial margin of two shells may move away or down from the coiling axis at the same rate (in mm/whorl), but if they differ in W, their values of D and T will differ as well. The direct consequence is that the apparent proportion of unoccupied morphospace results more from Raup's algebra than from the biology of the studied taxa (Schindel 1990). Proposing a revised morphospace, with the parameters W-U-M, Schindel (1990) compared the morphospace occupation of 63 genera of gastropods in his revised morphospace and in the original Raup's morphospace. He concluded that Raup's scaling of D and T with respect to W leads to a biased morphospace occupation that appears smaller than in his morphospace occupation. Furthermore, he noted that some shells seem fairly uniform in external aspects like spire index or equiangular angle, but they do so through a great variety of compensatory changes in coiling parameters as well as in aperture shape, orientation and inclination (the last 2 parameters being left out in Raup's morphospace).

Another limitation of Raup's model lies in its core assumption that growth follows the rules of the logarithmic spiral. Although this assumption is what provides this model with its elegance and generality at the clade level in the first place, it is also one of its pitfalls when applied below the family level. For instance, ammonoids often show considerable ontogenetic change in geometry, particularly in whorl shape and involution (Klug 2001; Korn 2012). Thus, it is commonly impossible in practice to find a single logarithmic spiral that fits the shell throughout ontogeny as recognized by Raup (1966). Moreover, for shells coiled over few whorls—like brachiopods and bivalves—it is also challenging to find a single logarithmic spiral that fits the shell throughout ontogeny especially if one does not carefully record the position of real apertures via growth lines (McGhee 2001). It should be noted that variation in average parameters among major taxonomic groups can be less than the variation observed in a single individual. Some modifications of Raup's model have been undertaken to simulate more varied aperture shapes (Savazzi 1985, 1987), some allometries (Bayer 1978; Checa 1991; Korn 2012) as well as non-radial apertures (McGhee 1978). Some of these models did not assume a generating curve, but rather focused on individual helicospirals (McGhee 1978; Savazzi 1985; Checa 1991; Checa and Aguado 1992). Shell shape was then represented by the set of spiral trajectories of given landmarks (e.g. longitudinal ornamentation).

So although the logarithmic spiral case is fully justified as a null model since it implies that growth occurs without change in shape, followers of Raup were driven toward new mathematical formulations that would allow ontogenetic changes to be taken into account, while relaxing the coiling axis assumption. Models using a coiling axis (fixed reference frames) have been extensively criticized: first, the coiling axis has no biological meaning as in real shells an approximated coiling axis emerges only *a posteriori* as a result of accretionary growth (Ackerly 1989; Schindel 1990; McGhee 1999); second, this approximated axis is not easily located on real shells and small deviations from the "real axis" (provided that one exists at all) may have large consequences for theoretical and empirical data interpretation; third, many shells do not have a coiling axis at all, or may not have a single coiling axis. Thus, assuming a coiling axis in theoretical or empirical studies may force the interpretations in predefined directions (and perhaps lead to contradictory conclusions, e.g., McGhee 1980, 2001; Aldridge 1998). It could also bias the observed morphospace occupation if one is unable to take into account the spiral limit cases (e.g. conical, heteromorph or irregularly coiled shells) or if parameters are not algebraically independent, as discussed above. These new moving reference frame models facilitate the application of theoretical models to intraspecific and ontogenetic variation, which—as argued by Raup (1967) himself—often provides the key to many of the functional problems that he has raised with his theoretical framework.

6.2.2 Space Curve Models

As Savazzi (1990) pointed out, theoretical models of mollusc shells can be divided into two broad categories, (1) those using fixed-frame methods (and thus referring to a coiling axis) and (2) those using moving-frame methods (and thus simulating growth locally at the aperture without reference to a coiling axis). The models developed around the same time by Okamoto (1988a, b, c), Ackerly (1989) and Illert (1987, 1989, 1990) belong to the second category. These models appear as a necessary step towards the simulation of the processes underlying morphogenesis ("*morphogenetic models*") and depart significantly from the previously discussed models ("*form models*").

In order to determine the general mode of coiling of ammonoids, Okamoto (1988a, b, c) focused mainly on heteromorph ammonites, because some of the coiling anomalies observed in these peculiar ammonites provide important clues for understanding the processes underlying shell morphogenesis in general. Okamoto described the coiling of heteromorph ammonites using a space curve model based on a moving reference frame, the so-called Frenet Frame. This frame is composed of three orthogonal unit vectors (e1, e2, e3) representing the tangent, normal and binormal vectors of a curve parametrized by its arc length. Okamoto's "*growing tube model*" simulates the growth of the shell by using this frame locally to define, by infinitesimal steps, the growth trajectory of the generative curve centroid throughout ontogeny. The tangent vector e1 indicates the growth direction, while the normal vector e2 links the generating curve centroid to the point of maximum growth (MGP), and the binormal vector e3 is orthogonal to e1 and e2. So if the growth patterns are described accurately, the resulting shape is fully specified. Reciprocally, if shell form is described completely, the growth patterns can be inferred *a posteriori*. Moving frame analyzes, such as those undertaken by Okamoto

(1988a, b, c), enable the description, analysis and comparison of highly allometric and distorted coiling in heteromorph shells. If ontogenetic similarity indicates close phylogenetic relationships, the moving frame analysis can be helpful to reconstruct them. If a shell is successfully simulated, several other physical quantities—such as center of gravity, center of buoyancy—can be determined to make inferences on the life position. Below, I describe Okamoto's model in more details.

6.2.2.1 Frenet Frame

Given a space curve F parametrized by its *arc length s,* F(s)

$$
\begin{aligned}
e1 &= F^{'}(s)\\
e2 &= e1^{'}/norm(e1^{'})\\
e3 &= cross(e1,e2)
\end{aligned}
$$

with *e1*, *e2*, *e3* being the tangent, normal and binormal unit vectors.

The Frenet equations define the following relationships between *e1*, *e2*, and *e3* and the *curvature (Kappa)* and *torsion (Tau)* of the curve F:

$$
\begin{aligned}
e1 &= Kappa(\mathrm{s}) * e2;\\
e2 &= -Kappa(s) * e1 + Tau(\mathrm{s}) * e3\\
e3 &= -Tau(s) * e2
\end{aligned}
$$

The curvature *Kappa*(s) represents the revolution rate of the tangent vector with the change in *arc length s*. The torsion *Tau*(s) represents the revolution rate of the tangent plane with the change in *arc length s*. The utility of the Frenet frame is that, given *Kappa*(s) and *Tau*(s), it is possible to find a single space curve F.

6.2.2.2 Okamoto's Model Aka the Growing Tube Model

Let's define *r* as the radius of the shell tube at the time *t*. Define **r0** as the initial radius of the shell tube. Okamoto (1988a) defines three parameters, **E**, **C**, and **TO**.

- **E** is the **radius enlarging ratio** (a constant for a given isometric shell) defined as: $r(t) = ro * E^t$
- **C** is the **standardized curvature** (a constant for a given isometric coiling mode) defined as: $\mathbf{C} = r(t) * Kappa(t)$
- **TO** is the **standardized torsion** (a constant for a given isometric coiling mode) defined as: $\mathbf{TO} = r(t) * Tau(t)$

To the extent that a shell grows proportionally, the parameters **E**, **C**, and **TO** are constants. **E** is the ratio of enlargement of the tube radius. **C** (between 0 and 1) represents the degree of tube bending. **TO** represents the revolution rate of the point of maximum growth (MGP) in the generating curve. Typical ammonoids are characterized by having low **E**, constant **C**, and zero **TO**, whereas in heteromorph ammonites like *Nipponites*, **C** and **TO** change with growth.

6.2.2.3 Application of the Growing tube Model to *Nipponites*

Well-known heteromorph ammonites like the Late Cretaceous *Nipponites* were first viewed as a challenge to the simulation of their ontogeny and their evolution (Illert 1990). However, Okamoto (1988a, b, c) and Illert (1989, 1990) derived mathematical models of the peculiar meandering whorls of *Nipponites* independently (Fig. 6.3). In particular, Okamoto argued that in any simulation of ammonoid morphogenesis, the hydrostatics of the phragmocone must be considered, as the growth direction may be affected by the living position of the organism (defined as the angle between the aperture and the sea floor). Indeed, the simulation of the meandering whorls of *Nipponites* relies on a simple assumption: the angle between the growth direction and the sea floor is kept between a lower and upper value throughout ontogeny.

The posture of a neutrally buoyant animal in sea water is determined by the distribution of its center of buoyancy and gravity. The center of buoyancy coincides with the center of volume, while the center of gravity coincides with the center of mass (center of body chamber). The distance between these two centers provides a measure of the stability of the organism: the closer they are, the less hydrodynamically

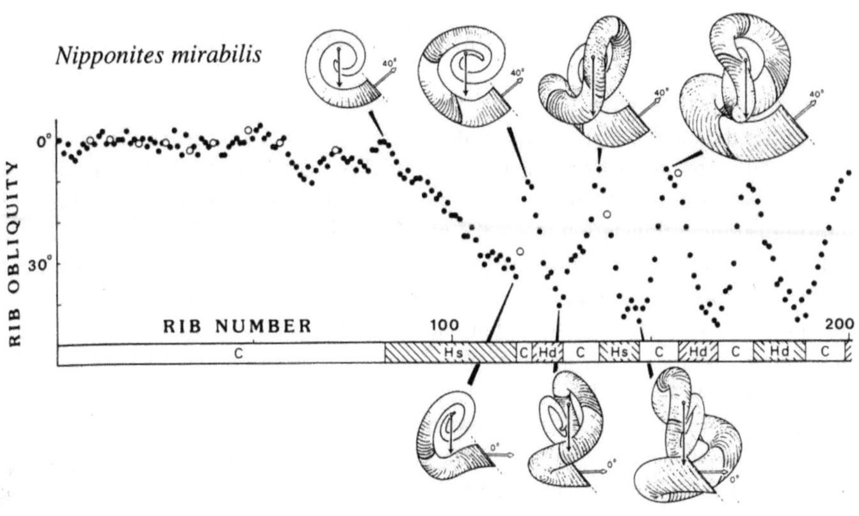

Fig. 6.3 Okamoto's model used to simulate the meandering whorls of *Nipponites mirabilis*. There are three modes of coiling (Hs, Hd, C), which represent sinistral helicoid, dextral helicoid and cryoceratoid (planispiral) coiling. This figure illustrates how rib obliquity and life orientation change with changes in coiling. The growth direction changes between 0 ° and 40 ° with respect to the sea floor (*horizontal*)

stable. Truman (1941) estimated the life orientation of ammonoids under the assumption of neutral buoyancy. The ratio of length of the body chamber to the length of the shell is approximately constant in most ammonoids and ranges from 0.55 to 0.60.

Okamoto (1988b, c; 1996) observed that *Nipponites* regularly changes between three modes of coiling during ontogeny - where dextral (1) and sinistral coiling (2) occur in alternation with planispiral coiling (3). Change in coiling mode occurs when the rib obliquity reaches a minimum value. Starting from the planispiral coiling mode, the hypothetical ammonite is simulated by assuming that growth continues in this mode as long as the growth direction is within a certain given range of variation. When the life orientation exceeds some value, the coiling is changed to dextral or sinistral, which acts to decrease the elevation angle of the growth direction. If the life orientation drops below a certain value, the mode of coiling changes again to planispiral. A meandering shell is automatically created under these conditions. So, assuming neutral buoyancy and an approximately constant aperture angle relative to sea bottom, Okamoto's model shows that the meandering whorls morphology is controlled, under hydrostatic constraints, by the permitted range of variation in the growth direction relative to the sea bottom (Fig. 6.3). If the lower limit of the growth direction is set extremely low, the helicoid coiling does not return to the planispiral coiling, producing shapes reminiscent of *Eubostrychoceras*, a possible ancestor of *Nipponites*. The simulations suggest that the morphological transition from a simple helicoidal form to a meandering form occurs abruptly without any intermediary form as a result of a minor change in the upper and lower limits of variation in the growth direction.

6.2.2.4 Ackerly's Model Aka the Moving Reference Model

Using also a Frenet frame, Ackerly (1989) focused on the motion of an actual aperture rather than a hypothetical generating curve that is always perpendicular to the growth direction as in the growing tube model of Okamoto. With the introduction of a rotation parameter **alpha** (around a rotation axis) and a translation parameter **gamma** (along a translation axis), the movement of the centroid of the aperture is defined. The rate of enlargement of the aperture is defined by a third parameter **delta**. Two more parameters are necessary to define the aperture orientation; **phi**, the angle between the aperture plane and the rotation axis; and **sigma** the angle between the normal of the aperture plane and the translation axis. Ackerly's model is well suited to the analysis of conical and loosely coiled shells in which a consistent direction of the coiling axis can be defined throughout growth. This model facilitates the empirical analysis of the shells in which the position of the coiling axis is difficult to define, as exemplified in many representatives of inarticulate and articulate brachiopods, monoplacophorans, bivalves, and gastropods.

Recently, a model akin to Ackerly's model has been proposed by Moulton et al. (2012): the CDRT model, where the shell aperture undergoes coiling (**C**), dilation (**D**), rotation (**R**), and translation (**T**). One difference between this model and that of Ackerly is that the aperture is not described by the position of its centroid but by a set of points attached to it. However, given that there is almost a one-to-one correspondence between the parameters of these two models (C = alpha; D = delta, T = gamma, R corresponds to an angle not considered as relevant by Ackerly), Moulton et al.'s (2012) model won't be discussed in details. Also, a number of other mathematical formulations exist that are described as generalized models of molluscan growth (Fowler et al. 1992; Pappas and Miller 2013; Moulton and Goriely 2014). These formulations are stemming from the study of 3D surface kinematics (Skalak et al. 1997) and can provide analytical solutions when the generating curve is planar and maintains a fixed shape throughout growth (Moulton & Goriely 2014). Although mathematically convenient, these formulations do not bring further insight with regards to the study of allometry and phenotypic variation at lower phylogenetic levels or with regards to morphospace studies at high taxonomic levels.

6.2.2.5 Limitations of the Space-Curve Models

D'Arcy Thompson's coordinate grid transformations already clearly emphasized the necessity for analyzing shape changes in their spatio-temporal dimensions as "*[i]t is obvious that the form of an organism is determined by its rate of growth in various directions…organic form itself is found, mathematically speaking, to be a function of time*" (Thompson 1917, p. 76). Although all models discussed so far address the spatial component of shell growth, none explicitly addresses its temporal component. The mathematical formulation found in Ackerly's model, however, not directly assuming logarithmic spiral growth, was instrumental in transitioning towards the kinematic models explicitly considering timing of growth processes. These models are discussed in the next section.

6.2.3 Kinematic models

Since the turn of the twenty-first century, a number of studies have connected shell shape to relative and absolute growth rate (Rice 1998), notably through growth vector models (Hammer and Bucher 2005a; Urdy et al. 2010a, b). These new models are typically high dimensional and do not allow the straightforward construction of theoretical morphospaces any more. However, these kinematic models—if time is an explicit parameter as in the models discussed in this section—are well suited to answer two questions in particular: (1) What kind of morphological variation or pattern of covariation is expected given some basic rules of growth? (2) What kind of rules could underlie the observed patterns of variation?

To address these issues, one has to simulate *in silico* plausible processes of growth, representing simple properties of developmental processes such as timing or interactions among parts. Morphometric variables can be subsequently derived from the simulated morphologies. Then, one can investigate theoretically how changes in hypothetical processes are recorded in the covariation between morphometric variables. Although the processes are only hypothetical with respect to real processes, the interest of this approach is that it can draw a causal link between variation in processes and variation in the resulting patterns of covariation. A large part of phenotypic variation and covariation between characters may be explained by the fact that morphogenesis is governed by a small set of basic (sometimes generic) processes that themselves may account for the robustness of the developmental outcome. This kind of models deeply roots in the concept of allometry, coined by Huxley and Teissier (1936) and which refers to the problem of relative growth: what are the "laws" of growth, which underlie the correlations between the changes in the relative dimensions of parts of the body and the changes in overall size?

6.2.3.1 Rice's Model Aka the Biogeometric Model

Building on previous studies raising the issue of growth timing in mollusc shells (Løvtrup and Løvtrup 1988; Hutchinson 1990), Rice (1998) framed his hypotheses with reference to relative and absolute shell growth rates. In this model, the shell is simulated in a similar manner as fixed-frame generating curve models. However, several biologically relevant parameters are derived from the model, namely the pattern of relative growth rates called aperture map (scalar field of relative growth rates of the aperture's homologous landmarks), an absolute growth rate (scaling of the aperture map), an aperture growth rate ("*body growth*") and various spatial parameters defining the initial conditions and the helicospiral geometry. Although the relationships between shell form and growth rates have been analyzed earlier, Rice's article marks the first consideration of time as an explicit parameter in shell shape simulation and allows the separation of the magnitude of growth from the direction of growth.

For instance, Rice (1998) generated a *Cerion*-like morphology by assuming that the absolute rate of calcification was proportional to aperture size (rather than aperture growth rate) and was a logistic function of time. From these hypotheses, it follows that shell shape is a function of the growth curve of aperture size over time (logistic in that case). The resulting shell morphology looks like the beehive-shaped shell of *Cerion*: shell growth first follows a (isometric) helicospiral in the exponential phase of the aperture growth curve, then, during the asymptotic phase of aperture growth curve, the shell continues coiling while maintaining aperture size nearly constant, thus naturally simulating to the "*barrel*" phase of *Cerion* growth (Gould 1984, 1989; Stone 1995, 1996a). The allometry in Rice's example is thus biphasic: the first phase is isometric, whereas the second phase leads to a decrease in shell apical angle, meaning that the shell becomes relatively narrower with increasing number of whorls.

6.2.3.2 Urdy et al's Model Aka the Growth Vector Model

Hammer and Bucher (2005a) developed a free-form vector model, subsequently modified by Urdy et al. (2010a) so that it can simulate apertural shape changes and nonlinear allometries (Fig. 6.4). The growth vector model assumes simple addition of growth vectors, which represent mantle growth during arbitrary (but constant) time steps. Shell morphology is generated by iteratively adding a growth increment onto the last computed aperture position. The first growth increment defines the so-called growth vectors, which are assumed to be constant in direction (relative to the last computed aperture position) during the simulation of a shell (ontogeny). These growth vectors are uniformly scaled at each time step according to various growth rate curves that simulate mantle growth over time.

Instead of starting with the logarithmic spiral model, as in models discussed previously, Urdy et al. (2010a) postulated first simple rules related to the scaling and relative arrangement of successive growth increments and then derived the conditions satisfying the simple logarithmic spiral model. It is shown that, given an initial increment, there is a single growth curve that can satisfy the conditions for isometry. In all other cases, non-linear allometries are generated. The resulting allometry increases quadratically with the rotation angle, as defined in the first increment. Because this model focuses the issue on time, it highlights a plausible effect of growth rate on shell shape and illustrates some fundamental geometrical properties of the logarithmic spiral, in particular the close relationship between the size and the geometry of growth increments. This null hypothesis model generates various correlations between morphometric variables and provides insights into potential growth-dependent shape changes.

For instance, basing model parameters on a study documenting the results of transplants experiments of three gastropods ecomorphs (Johnson and Black 2008), Urdy et al. (2010b) reproduced the main aspects of the variation in size, shape, and growth rates within and among populations when bred in their own habitat or transplanted to another ecotype habitat (Fig. 6.4). In agreement with empirical results, the simulation of ontogenetic, intra- and inter-population variation shows that a flatter growth profile corresponds to conditions of rapid growth (**parameter r**). The model also allows the comparison of allometric slopes using different subdata-sets that correspond to static and ontogenetic allometry (Fig. 6.4). This model highlights that depending on subdata-sets extracted from the simulations, the main statistical effects could be attributed to source population or environment. In addition, convergence or divergence of allometric slopes is observed depending on the subdata-sets. Although there is evidence that shell shape in gastropods is to some extent growth rate dependent, gaining a general overview of the issue is still challenging, in particular because of the scarcity of studies referring to allometry. Urdy et al. (2010b) argue that the dynamics of development at the phenotypic level constitute a non-reducible level of investigation if one seeks to relate the observed amount of phenotypic variation to variability in the underlying factors. Keeping this in mind, the study of fossil lineages like ammonoids can greatly benefit from the study of living counterparts like gastropods.

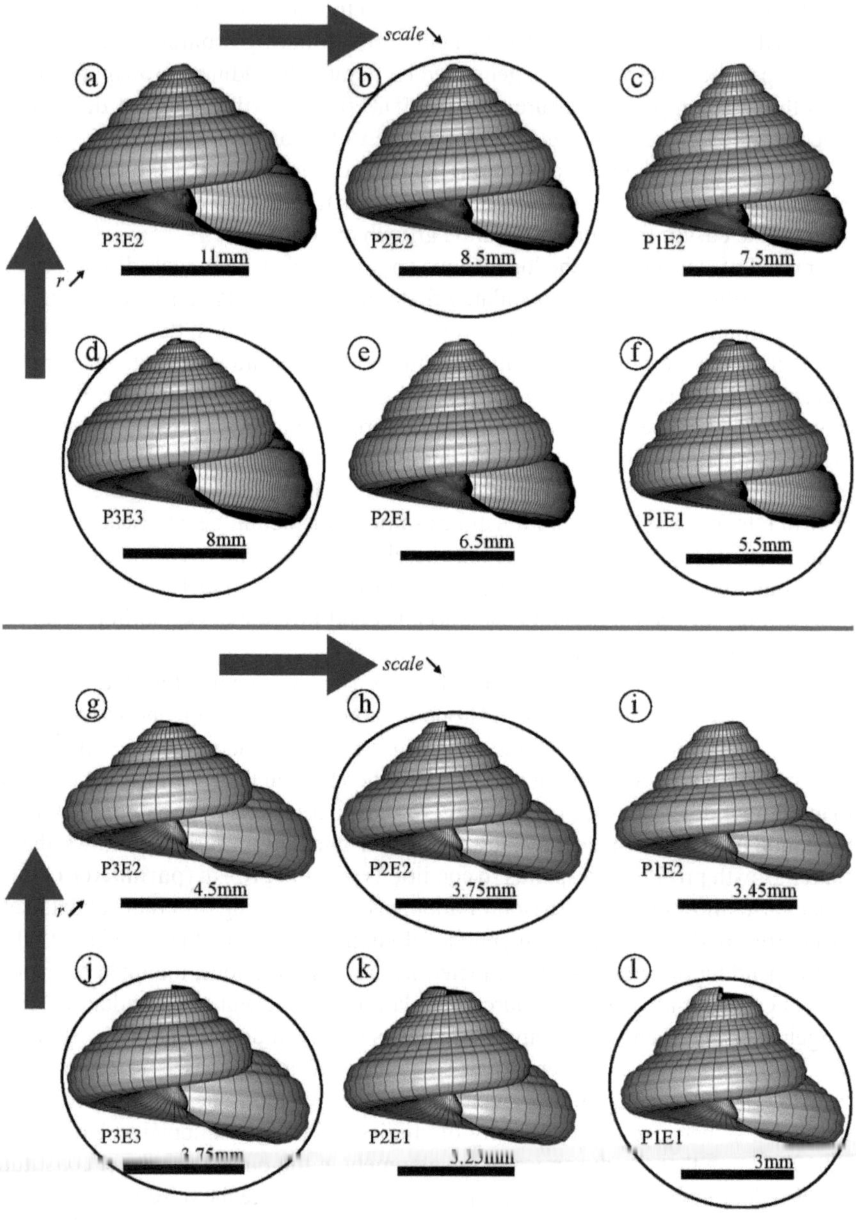

Fig. 6.4 Urdy et al.'s (2010a) model used to simulate variation in 3 populations of *Bembicium vittatum* (P1, P2, P3) when grown in 3 different environments (E1, E2, E3). *a–f*, Mean shell shape in each of the six groups at adulthood (300 time steps), redrawn at the same shell length. *g–l*, Mean shell shape in each of the six groups at a juvenile "*stage*" (150 time steps) redrawn at the same shell length. a/g, P3E2. b/h, P2E2. c/i, P1E2. d/j, P3E3. e/k, P2E1. f/l, P1E1. In each panel, from *bottom to top*: the growth rate **parameter *r*** increases; from *left to right*: the **parameter *scale*** *decreases.* The **parameter scale** is equivalent to the **parameter delta** of Ackerly (1989).

6.2.3.3 Limitations of the Kinematic Models

Kinematic models can create fairly general shell evolutions and explicitly include the aspect of time in shell growth. These models illustrate the role of growth rates and timing in the generation of allometries and phenotypic plasticity. The growth vector model in particular provides a theoretical link between variation at the ontogenetic, population, and species levels (Urdy et al. 2010a, b). Also, growth vectors can be changed through ontogeny, in an *a priori* way, or in response to shell shape, thus allowing the testing of the "*road-holding hypothesis*" (Hutchinson 1989; Morita 2003; Hammer and Bucher 2005a). The regulation of the magnitude and direction of growth is deeply related to the mechanics of the mantle over time, and it is expected that the effect of genes and environmental factors on growth are mediated by these physical factors. Unfortunately, no kinematic model directly addresses these underlying factors. In order to explain the variability of shell morphogenesis in relation to genetically-regulated tissue properties, kinematic models are thus insufficient and one must look in the direction of the more explicit "*morphogenetic models*" discussed in the next section.

6.2.4 *Mechanical Models*

Relating shell shape to genetic and environmental parameters require the investigation of the mechano-chemical aspects of shell morphogenesis, which provide the 'causal' factors determining the shape and size of growth increments at each growth step upon the constraints of previous built shell. Such analyses are notoriously challenging, especially because of the non-linearity of soft tissue growth and gene interactions, and because most of molluscan development is unknown.

6.2.4.1 Morita's Model Aka The Dead Spiral Model

One notable attempt is provided by the "*dead spiral model*" of Morita (1991a, b, 1993, 2003). This model has been used to draw conclusions on coiling and aperture shape, as well as on the morphogenesis of the ornamentation. Although it does not make sense to discuss these results separately, the fact that most other models of the same type have been proposed either for shape or ornamentation only, I will discuss Morita's model in two separate sections for matters of comparison between different studies.

Morita (1991a) defined the whole mantle as a hydro-skeleton, which is usually in a state of expansion resulting from internal haemolymph pressure. Consequently, the mantle is simulated as a double elastic membrane connected by internal springs. Its physical state is supposed to be in balance between its internal stress and the forces acting on it, such as the pressure of the haemolymph, the pressure induced by the foot and the boundary of the shell. The deformation of the mantle is then deduced from its stress field using a finite element analysis.

In a first step, Morita (1991a) proposed to answer the following (puzzling) question: How does the mantle edge usually make contact with the inner surface of the shell wall so as to ensure continuous accretion of shell materials to the aperture margin? His model shows that above an internal pressure level, the tube will expand outward rather than shrink inward. Where the shell wall is present, the mantle outward expansion is prevented, but the mantle can slide along the shell instead, allowing accretionary growth. Note that if some portion of the shell at the aperture is removed while the specimen is alive, the mantle edge curves backward, demonstrating that it is under high internal pressure, as illustrated in Fig. 6.5b,c.

In a second step, Morita (1991a) asked the following question: Why do openly coiled or minimally overlapping shells have generally circular apertures, while shells with apertures overlapped by whorls have non-uniformly curved outer apertural lips? Morita investigated the effect of a zone where the mantle cannot deform—presumably because of the foot/soft parts pressing on the mantle edge. He showed that initially circular walls change into elliptically elongated ones with pressure rising. In other words, the existence of a fixed zone (large or small) breaks the initial symmetry in a specific manner: the direction of elongation is perpendicular to the fixed zone. On the contrary, all tube shapes tend to converge to circular outlines when no fixed zone exists.

Morita (1991b, 1993, 2003) argues that this fixed zone represents whorl overlap (=imprint zone, concave whorl zone) and may explain why most openly coiled or minimally overlapping whorls of gastropods have circular apertures. By contrast, outer apertural lips accompanied by a distinct whorl overlap zone are either extended perpendicularly to the overlap zone or are abapically inflated. Tight coiling in gastropods (defined as large whorl overlap) is then viewed as the consequence of the contact of the body parts with the mantle edge at the insertion of the collumelar muscle: because the mantle is unable to deform in this contact zone, the shell coils in the opposite direction. To the contrary, loosely coiled specimens usually exhibit a wide muscle insertion all around the aperture. This makes the mantle edge free to deform uniformly in any direction and apertures are generally close to circular.

Several studies have pointed out a general correspondence between the life position and the shell morphology in recent gastropods (Linsley 1977, 1978), indicating that the life position of gastropods is almost equal to the gravitationally stable position of their empty shells. As an exception, Linsley (1978) stated that some trochiform gastropods (the Xenophoridae or some Astraeinae) live with the basal part of the aperture margin directly resting on the substrate. This "*unconventional*" life position provides a boundary condition different from those assumed by Morita's model, and as such, their aperture shape can depart from the morphological rules he has described. Since the free expansion of the mantle edge is prevented by the substrate, the mantle may be forced to elongate parallel to the substrate, if the animal maintains this peculiar position at the time of shell growth. In fact, the apertures of these trochiforms are shaped as though formed in such a way: they are not elongated perpendicularly to the overlap zone but rather in parallel to their base, as if the apertural margins had been compressed by the substrate (Morita 1993).

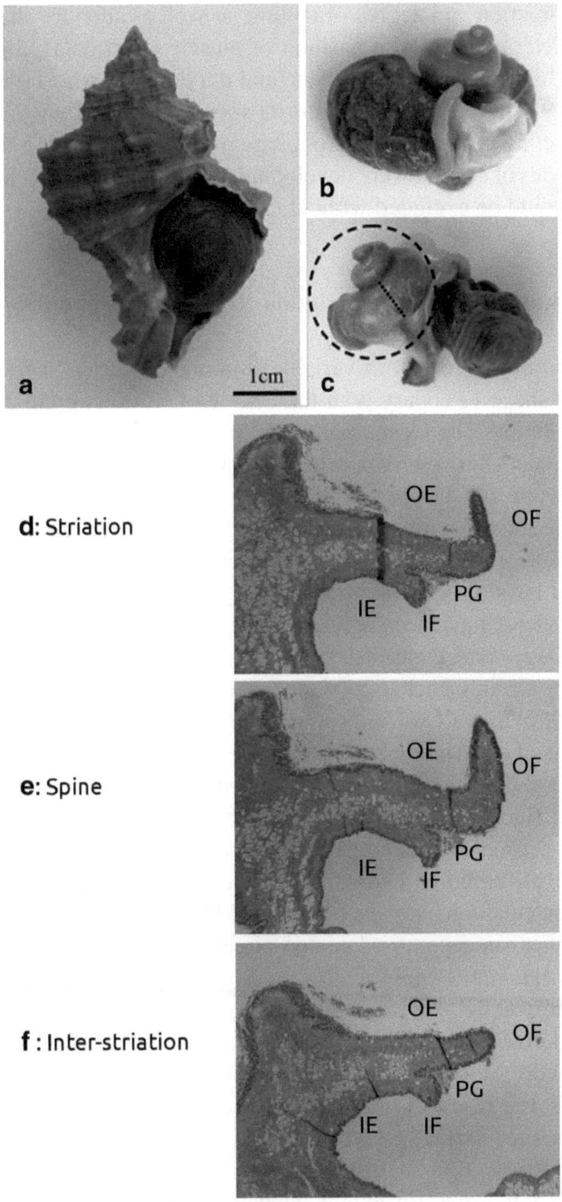

Fig. 6.5 Local longitudinal protrusions in length of the outer fold of the mantle edge of juveniles of *Hexaplex trunculus*. *a–c,* A living juvenile specimen of *Hexaplex trunculus*. *b, c,* If the shell is removed on a living specimen, the mantle edge curves backward (towards the digestive gland) demonstrating that the mantle edge is under internal (hydrostatic) stress. Broken line shows the approximate position of the coiling axis. *d–f,* Histological sections of the mantle edge on the same specimen at different locations along the aperture, corresponding to a spiral striation (*d*), a spine (*e*) and in between two spiral striations (*f*). Growth direction is toward the right. The shell was on the *top*. OE: Outer Epithelium responsible for shell secretion. OF—Outer Fold responsible for generating ornamentation. PG—Periostracal groove. IF—inner fold. IE—inner epithelium. Reproduced with permission from Vasconcelos et al. (2006), copyright Oxford University Press (license number 3511891006437)

Hence, the direction and degree of coiling, as well as aperture shape seem to be at least partly determined by the position of muscle insertion(s), the animal's living position at the time of shell secretion, and the previous whorl ("*road-holding*", Hutchinson 1989; Checa et al. 1998). Morita's work suggested several developmental constraints acting on the shell morphology of molluscs, arguing that the muscle patterns and modes of coiling were developmentally coupled so that malfunctioning combinations would be excluded without natural selection.

6.2.4.2 Buckman's Laws of Covariation: Hydrostatics and Proportionality?

It has been often observed that in variable ammonoid species, there is a tendency for the aperture shape to covary with the degree of whorl overlap and the robustness of ornamentation. The most ornamented specimens tend to exhibit a depressed aperture and small whorl overlap, whereas smooth specimens exhibit a laterally compressed aperture and a large whorl overlap. In the same species, the septal spacing as well as the sutural complexity also covary with the strength of ornamentation and aperture shape. These patterns of intraspecific variation, known as Buckman's first and second laws of covariation (coined by Westermann 1966; see Monnet et al. 2015 for a review), have been repeatedly observed in Triassic boreal ammonoids (Rieber 1972; Dagys and Weitschat 1993; Checa et al. 1996; Bucher 1997; Dagys et al. 1999; Hammer and Bucher 2005a; Monnet and Bucher 2005), and in Jurassic (Westermann 1966) as well as Cretaceous ammonites (Kennedy and Cobban 1976). As these patterns of covariation have been observed in phylogenetically distant ammonoids at several different time periods, they are evolutionary recurrent.

Hammer and Bucher (2006) tentatively explained the developmental origin of the Buckman's second law of covariation between ribbing and sutural complexity using arguments based on hydrostatic calculations. They proposed a numerical procedure for calculating buoyancy, apertural orientation and rotational stability of ammonoids that can accommodate non-circular apertures and allometries. They showed that in *Amaltheus margaritatus*, hydrostatic properties and body chamber length vary significantly as a result of intraspecific variation in shell shape as described by Buckman's first law of covariation. Assuming that new septa are laid down as the hydrostatic properties (orientation and buoyancy) change beyond threshold values as a result of growth, the covariation between septal spacing and whorl shape observed in this species can be accounted for

Hammer and Bucher (2005a) also proposed that the negative correlation between the compression of the aperture and the intensity of ornamentation (Buckman's first law) can be satisfactorily accounted for by assuming that lateral rib heights increase isometrically with aperture width, whereas ventral rib heights increase isometrically with aperture height. Simple scaling relationships lead to produce proportionally stronger lateral ribs on depressed specimens than on compressed specimens, which only exhibit strong ribs on venter. Hammer and Bucher (2005a) argue that, in a shell that is strongly compressed laterally, the diameter of the aperture, and hence the soft parts within, is small in the lateral direction. For such a shell, coarse lateral ribbing would imply impossibly strong relative compression and expansion of the soft body through the formation of a rib. They conclude that any departure from this covariation would imply strong deformations of the soft body as ornamentation was formed. Although this scaling model is compelling by its simplicity, it does not directly address the morphogenetic mechanisms underlying ornamentation patterns, for which other models are necessary. This is the topic of the next section.

6.2.4.3 Limitations of Mechanical Models

Although mechanical models can yield insight in developmental constraints on a large phylogenetic scale as exemplified above, they are generally really qualitative in nature. The testing of these models in a developmental context is something that is left open for future studies, once more of the developmental biology of molluscs will be understood. As these mechanical models are constrained to remain really generic (non-specific), they provide little insight on the mechanical factors that govern variation in shell shape at low taxonomic levels, especially in relation to variation in environmental factors. Such a crude understanding of morphogenesis then hinders the comparison between experimental and theoretical data on variation in populations or species.

6.3 Shell Ornamentation and Pigmentation

During shell growth, the mantle moves forward slightly beyond the calcified shell edge while secreting the periostracum, which isolates the extrapallial fluid from which the calcified shell is precipitated from the external environment. The periostracum is secreted in the periostracal groove, between the outer and middle mantle lobes in cephalopods and bivalves and between the outer and inner mantle lobes in gastropods (Fig. 6.5). When calcification in the extrapallial cavity occurs, the periostracum becomes fixed on the outer shell surface (it is often abraded later).

One can distinguish a variety of pigmentation patterns and ornamental features in molluscs. However, they can usually be split into three broad categories—antimarginal undulations, commarginal undulations, and traveling waves. Antimarginal pigmentation patterns and spiral ribs usually correspond to undulations perpendicular to the aperture: from a developmental point of view, these are spatial patterns, which do not vary in time. Commarginal pigmentation patterns and ribs are parallel to the aperture and thus correspond to temporal undulations of the mantle activity and growth direction respectively. Spines have a spatial and a temporal component. They are hollow and they first emerge as a bulge of the shell edge that curves both longitudinally and transversely as growth proceeds, resulting in a fold that eventually closes in on itself as the lateral edges converge toward the fold axis (spine constriction). Traveling waves correspond to a periodic function of both space and

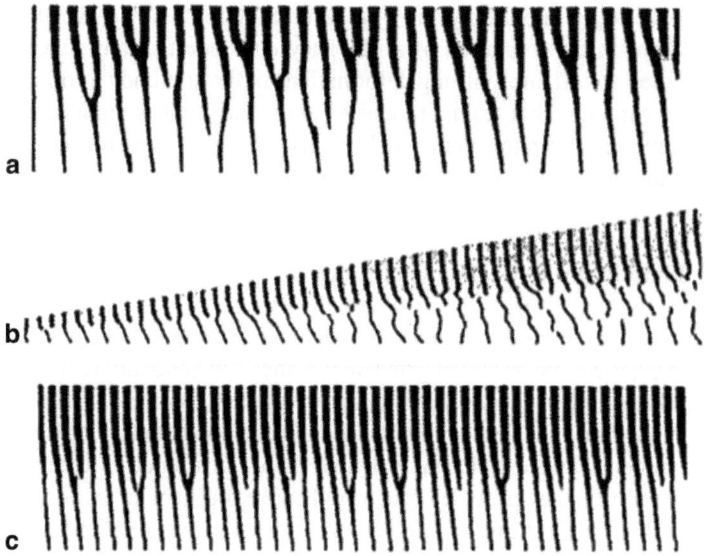

Fig. 6.6 Different patterns generated by reaction-diffusion models. **a** Bifurcations can form in an oscillating activator-inhibitor model when the basic activator production increases from *bottom to top* (smooth gradient). **b** Branching and insertion of ribs in a growing geometry. Ribs are continuous in early stages, but break up as the structure grows. **c** A sharp gradient in the basic activator production gradient causes the secondary ribs to form at a constant position

time, and would correspond to patterns of divaricate ribs for instance (e.g., Hayami and Okamoto 1986; Ubukata and Nakagawa 2000). Basically, two types of dynamic models have been proposed to account for the morphogenesis of these structures and pigmentation patterns. The first type belongs to the class of lateral inhibition models and the second type belongs to the class of mechanical models. The similarities and differences between these two classes of models—as well as their historical roots—are discussed at length in Urdy (2012), using various examples from developmental biology.

6.3.1 Lateral Inhibition Models

6.3.1.1 Reaction-Diffusion and Neuronal Models

Reaction-diffusion models (Fig. 6.6) have been extensively applied to problems of pattern formation—strictly defined as the dynamical process by which spatio-temporal arrangements of repeated features arise. The reaction-diffusion models root in the pioneer work of Turing (1952). The original Turing model (1952) was concerned with the emergence of spatial inhomogeneities (the so-called Turing structures) by means of two interacting substances, which he called morphogens, diffusing in the external milieu and exhibiting antagonistic behavior: one molecule is called activator and is assumed to exhibit autocatalytic kinetics and to stimulate the synthesis of another molecule called inhibitor that counter-acts activator synthesis by consuming it. Diffusion is generally thought of as a homogenizing process. However, in combination with two coupled antagonists reactions (auto-catalysis and inhibition), a stable homogeneous concentration of molecules can be destabilized by diffusion in response to small random perturbations. This phenomenon is called "*diffusion-driven instability*". According to the relative diffusion range of the activator and inhibitor, the initial perturbations of the stable homogeneous solution can be either smoothed out or amplified. If so, the whole system stabilizes to another predictable equilibrium state (e.g., spatial patterns, temporal oscillations; or traveling waves if one considers the interaction between three morphogens) independently of the initial perturbations.

Indeed, only opposing reacting kinetics of two molecules are necessary for their concentration to oscillate over time. This is exactly the same generic process that has been proposed to be involved in the well known Lotka-Volterra predator-prey models. One has to assume that the more prey, the more predators (predator number increases when plenty of food is available) but the more predators, the less prey animals (since predators also consume more prey). Such a mechanism can give rise to temporal patterns of varying number of prey and predators: at one moment, there are more prey than predators, than the situation reverses. To the contrary, the emergence of "*Turing structures*" (spatial patterns such as ribs perpendicular to growth lines) generally requires that inhibitor diffuses more rapidly than activator, so that the activator concentration does not smooth out all over the domain. This hypothesis is more generally known as "*short range activation and long range inhibition*" (Gierer and Meinhardt 1972) or "*local self-activation and lateral inhibition*" (Meinhardt and Gierer 2000).

The signaling molecules (morphogens) are assumed to control the subsequent differentiation of the cells during morphogenesis. The spatio-temporal distribution of these morphogens is often named "*pre-pattern*". This implies the idea that the emerging distribution of the gene products will control the differentiation of the cells to finally create the associated pattern. Reaction-diffusion models have been largely developed by Meinhardt and his coworkers (e.g. Meinhardt and Gierer 1974) and one-dimensional models have been applied to the pigmentation of sea shells to reproduce a vast diversity of coloration patterns including stripes, dots and chevrons (Meinhardt and Klinger 1988; Meinhardt 1995). Several authors have proposed that an oscillating biochemical system of the reaction-diffusion type could also underlie the formation of commarginal and divaricate ribs (Meinhardt and Klingler 1988; Savazzi 1990; Meinhardt 1995; Bucher 1997). Hammer and Bucher (1999) used similar models to study ornamental patterns in ammonoids, but regarded the reaction-diffusion model as an abstraction, without insisting on the chemical diffusion mechanism that is generally assumed. In Fig. 6.6, some patterns simulated by Hammer and Bucher (1999) are represented. One can note that new elements can be inserted, or form as branches, as the distances between adjacent peaks exceed some threshold because of growth or if there exists a gradient in activator production.

Many theoretical systems exhibit similar properties of 'local self-activation and lateral inhibition' (Meinhardt and Gierer 2000) but they do not necessarily imply chemical diffusion. For instance, the neuronal model of Ermentrout et al (1986) simulates pigmentation of sea shells by computing the positive and negative influxes of neurons that are supposed to trigger or prevent the secretion of pigment by the mantle cells. In general, these models highlighted the emerging capabilities of molecular networks through local self-enhancement and lateral inhibition, capturing some of the dynamics of developmental systems at the molecular level.

6.3.1.2 Limitations of Lateral Inhibition Models

Authors generally conclude that lateral inhibition models capture some of the essence of biological pattern formation, although these models should only be regarded as formal conceptual frameworks for understanding development. Indeed, a general drawback of lateral inhibition models is the difficulty of linking these microscopic descriptions of processes (which echo the genetic level of description of development) to the three-dimensional macroscopic shape changes that have to be explained, especially when biochemical components, cell movements and tissues continuously interact to generate structures of specific shape (Murray 1989). Goodwin (1988, p. 636) captured the essence of this limitation by saying that "*Turing's achievement was remarkable, but it does not provide the solution to the problem of morphogenesis. The reason is that a spatial pattern in the concentration of metabolites within a developing organism does not itself explain the actual geometry of say, the tentacles on a hydroid, the leaves on a plant, or the limbs of an amphibian. Morphogenesis, as the name implies, is the generation of structures of specific shape, whereas spatial patterns of chemical concentration arise within some predefined geometry. In order to get morphology, work has to be done in deforming cells or cell sheets into specific shapes, and growth must be localized to generate specific structures*". This issue can be overcome if one takes conjointly into account the mechanical aspects of morphogenesis (e.g., Oster et al. 1988) or simulates some kind of feedback between the interacting biochemical components (genes products) and the growth of the developing structure, an approach that has been called 'morphodynamic' in opposition to pure reaction-diffusion models that are qualified as 'morphostatic' (Salazar-Ciudad et al. 2003, see also Urdy 2012). In the next section, I discuss models that focus on the mechanical aspects underlying the morphogenesis of shell ornamentation.

6.3.2 *Mechanical Models*

6.3.2.1 Morita's Model Aka the Dead Spiral Model

As discussed earlier, Morita's model (1991a, b, 1993, 2003) includes several aspects on the mantle deformation such as the influence of the whorl overlap or the effect of the previously secreted shell (Hutchinson 1989). Morita's interpretation thus implies that the life position of a gastropod is involved directly in the morphogenetic process of forming the shell aperture. This model has other implications, however. Besides the general outward expansion and the shift from circular to elliptical outward expansion, Morita (1991a) defined two other behaviors of the mantle edge: antimarginal ridge formation and spine formation. His main concerns are: What is the cause of antimarginal ridges? What is the cause of long open spines and why do they always appear on antimarginal ridges?

Morita (1991a) showed that if the initial tube exhibits local protrusions in its length, antimarginal ridges will occur. Such local longitudinal protrusions in length have been observed in the outer fold of the mantle edge of *Hexaplex trunculus*, while the size of the inner fold is almost constant along the aperture (Fig. 6.5). Similar structures have been mentioned (but not really investigated) in other gastropods as well (Jackson et al. 2006, 2007). Morita (1991a) further demonstrated that if the shell wall already presents antimarginal ridges and if internal pressure increases, this folded protrusion can gradually become prominent in size and at the same time, its basal parts conspicuously constrict inward, forming an open spine. Morita concluded that the formation of a spine is probably associated with an event, which leads up to an abrupt increase of haemolymph pressure (or growth) higher than that at previous stages, because prominent constriction inward never appears without concomitant high mechanical expansion. If the rate of the mantle expansion is at the same level as in previous growth stages, the constriction would not appear but the antimarginal ridge would grow continuously, scaling with the size of the mantle edge.

6.3.2.2 Hammer's Model Aka the Regulative Oscillation Model

What about the formation of commarginal ribs? Hammer (2000) developed a model to explain their emergence and ubiquity in molluscs. He assumed that there should be a way for the mantle to sense—through its compression or tension—if it is a little too wide or narrow compared to the previous aperture. Then, this sensing mechanism could give feedback on mantle growth to adjust its size. This system can easily fall into oscillatory behavior, producing waves akin to ribs. This can happen because the regulative feedback mechanism is not fast enough to stop its compensation behavior at the right moment in time. The system will therefore compensate too much, and increase or decrease the size of the aperture too far before turning back again. Then it will grow too far in the opposite direction, and the cycle repeats.

Hammer (2000) assumed that the rate of change in angle of growth to the central axis (dphi/dt) is proportional to the distance from the current position of the shell edge (r) to an ideal mean R, divided by R:

$$d\ phi/dt = \mathbf{c1}[(R - r)/R]$$

The value R represents the radius of the relaxed soft parts within the shell. The value **c1** represents the stiffness of the mantle/periostracum system. With this equation, it is thus assumed that the rate of change in the angle *phi* (curvature) is then proportional to the strain caused by stretching or compression of the soft parts. Another differential equation links the rate of change in the tube radius (dr/dt) to *phi*:

$$dr/dt = \mathbf{c2} sin\ (phi)$$

where **c2** represents the growth rate of the tube radius in units of distance per time (assuming a constant growth rate).

This set of equations appears sufficient to explain the occurrence of commarginal ribs where the amplitude and the wavelength of the ribs increase during growth, as in accordance with observations in ammonites. Using a non-linear form for the feedback mechanism, it is possible to produce waves, which are sharper and more widely spaced than in the sinusoidal case, as is often observed in ammonites (i.e. *Promicroceras*). If the parameter **c1** is low, the stiffness of the mantle/ periostracum system is increased, and this can lead to the disappearance of oscillations, thus mimicking a smooth shell. Hammer (2000) concludes that if his simple model reflects reality to some extent, though perhaps in an abstract way, it is clear that commarginal ribs can evolve very naturally in a mollusc shell. In fact, it may even be more difficult to produce a smooth shell than a ribbed one, because this needs a more stringent/ rapid regulative capability. This is in contrast with the perhaps intuitive idea that a ribbed shell is morphologically more complex than a smooth one, and therefore needs a more complex developmental system, possibly with additional genes. If true, this model implies that commarginal ribs will evolve almost unavoidably in any shelled mollusc group, as long as they are not selectively disadvantageous.

6.3.2.3 The Elastic Rod Model and its Coupling with the CRDT Model

Moulton et al. (2012) recently developed a mathematical model for the morphogenesis of commarginal ribs, which is similar in essence to that of Hammer (2000). This model is based on the mechanical interaction between the secreting mantle edge and the calcified shell edge to which the mantle “adheres” during shell growth. They assume that the periostracum—attached at both extremities along the calcified shell edge and inside the periostracal groove—establishes a close physical elastic link between the calcified shell and the mantle edge. The mantle edge is simulated as a one-dimensional elastic rod, connected to a rigid foundation representing the shell edge by elastic springs, supposedly representing the periostracum.

As in Hammer (2000), commarginal ribs are simulated by assuming that the rate of change in aperture radius is a sinusoidal function of the growth angle, which itself depends on the stress in the mantle. However, this linear feedback produces fairly uniform oscillations in aperture radius over time. Assuming that there is a non-linear feedback response when the mantle is in tension, sharp ribs separated by wide valleys are simulated. However, the wavelength remains nearly constant throughout development, leading to ribs, which attenuate with growth. Decreasing this feedback during development leads to an increase in wavelength and in the amplitude of the ridges with time, which is then consistent with most observations in ammonites and Hammer's model. Discrepancies between these two models are unfortunately not discussed by Moulton et al. (2012). One can then speculate that these differences stem from the fact that Moulton et al. represent the mantle edge as a one-dimensional rod. Indeed, Moulton et al. (2012) combine this mechanical model with the CDRT model mentioned earlier. They define first an aperture shape and 4 CDRT parameters to describe the basic evolution of aperture shape without ornamentation. The ornamentation is then added as a second feature. As a consequence, the deformed aperture is independent of its coiling, translation and rotation parameters. This may explain why there is no scaling between the wavelength/ amplitude of commarginal ribs and the aperture dimensions, when there is no temporal variation in feedback strength.

Building of this model, Chirat et al. (2013) simulated the formation of spines in Muricidae. Assuming that there is an excess length in the mantle edge compared to the previously secreted shell margin, the elastic rod is expected to buckle to accommodate its extra length. Starting from an initially flat rod under clamped-clamped conditions, they solve the equations of mechanical equilibrium to deduce the form taken by the rod while it grows. The number of folds taken by this rod is governed by the elastic properties of the rod, the strength of the attachment to the rigid foundation, and the excess of rod length compared with the previous rod length. There can be several different modes of buckling, which are interpreted as the number of the spines that can emerge along the aperture. In the regime where only a single fold will form—corresponding to a fixed value for strength of attachment—Chirat et al. (2013) investigated the effects of the two remaining parameters affecting buckling outcome. Above a certain length threshold, the rod folds into a tall structure. Increasing this length parameter produces a smaller structure that folds and closes on itself (spine constriction), as predicted by Morita (1991a). A larger length parameter produces a more highly curved and shorter structure. Long straight spines, as seen in the murex *Bolinus cornutus* for instance, are simulated by assuming that the mantle has a heterogeneous bending stiffness. Then, the rod will increase its curvature in regions where its stiffness is decreased. Large stiffness variation along the rod leads to narrow spines reminiscent of the Muricidae genus *Hexaplex* and *Bolinus*, while decreased stiffness variation leads to wide spines reminiscent of the Muricidae genus *Pterynotus*. This model shows that a large diversity of spine structures can then be accounted for through variations in control parameters of a simple mechanical model. The authors recall that hollow spines evolved independently in at least 55 genera and 21 families of current gastropods; 10 genera and families of current bivalves; 11 genera and 8 families of ammonoids; and 6 fossil nautiloid genera. This physical mechanism suggests that convergent evolution of spines can be understood through a simple generic morphogenetic process.

The elastic rod model is compelling on its conclusions and provides much progress from the analytical point of view—shedding light in particular on how the rod length affects the shape taken by the aperture/ spines. However, this model holds some shortcomings that were absent in previous efforts aimed at studying the mechanical principles underlying shell morphogenesis. For instance, this model relies on the distinction between the overall aperture shape and the ornamentation, a distinction that is technically unfeasible on real shells. For that reason, this model appears hardly testable in practice. Additionally, the CRDT parameters are independent of the mechanical stress building in the rod. As the deformation of the rod does not affect its growth in return, this model can be considered morphostatic and it shares some shortcomings with the lateral inhibition models discussed earlier. The shortcomings implied by the decomposition of growth into a size and a shape component with regards to the underlying mechanisms of morphogenesis are discussed at length in Urdy et al (2010a, b) in the case of molluscs and Urdy (2012) in general. To summarize this argument, I can quote Klingenberg (1998, p. 84): "*although the separation of growth as isometric size increase from all shape changes agrees with our intuitive concept of size and shape based on geometric similarity, it does not reflect a corresponding dichotomy of underlying biological processes.*"

6.3.2.4 Limitations of Mechanical Models

In conclusion, the development of mechanical models for 3D ornamental features is still in its infancy. Morita has outlined the major elements of a mechanical model of shell growth, although most of his numerical simulations are not calculated in 3D, but on cross-sections. This model is more than 20 years old, but the broad range and generality of its insights—with its hydrostatic basis and its coupling of aperture shape and coiling with life position—has not been superseded by more recent models. Analytical treatment has been made possible recently by the reduction of the mantle to a one-dimensional rod. This comes at the expense of hindering the comparison between experimental and theoretical data, especially with regards to the relationships between growth rates and ornamentation. In the last section, comparative empirical studies are discussed.

6.4 Empirical Data and Comparison with other Molluscs

The comparison of shell shape between and within different clades of molluscs is informative with regards to the basic rules of accretionary growth. For instance, many authors argued that common rules of growth underlie the morphogenesis of the shell and its evolution in ammonoids and gastropods (Bucher 1997; Bucher and Guex 1990; Bucher et al. 1996; Checa and Jiménez-Jiménez 1997; Checa et al. 1998, 2002). Evidences come from the comparison of intraspecific and/ or interspecific patterns of covariation among shell characters (Westermann 1966; Morita 1991a, b, 2003; Dagys and Weitschat 1993; Checa et al. 1996; Hammer and Bucher 2005a), from the description of changes occurring at maturity in different species or clades (Thompson 1917; Burnaby 1966; Bucher 1997; Chirat et al. 2008), and from the analysis of teratological shells in response to injuries (Thompson 1917; Guex 1967, 1968; Bayer 1970; Landman and Waage 1986; Bond and Saunders 1989; Hammer and Bucher 2005b) or to change in living conditions (Linsley 1977; Checa and Jimenez-Jimenez 1987; Checa et al. 2002).

6.4.1 *Similar Principles Underlie Growth Features*

6.4.1.1 Post-Embryonic Stages

Changes in shell shape, ornamentation and septal spacing in the ontogeny of most ammonoids is used to define three distinctive post-embryonic stages: a neanic stage starting just after hatching (ammonitella), a juvenile stage and a mature stage (reviewed in Bucher et al. 1996). The neanic shell commonly displays little or no ornamentation with more and more widely spaced septa. During the neanic stage, ammonoids may have lived in the plankton and the transition to the juvenile stage may correspond to a change in mode of life in a different habitat. In general, the juvenile shell is more compressed and involute than the neanic shell and ornamentation is almost always present. The septa are usually more closely and more constantly spaced than in the neanic shell, although occasional short-term fluctuations can occur (discussed below). There is usually important variation in the number of septa formed during this stage among individuals of the same species, as well as extensive variation in shell shape and suture outline. Maturity is defined by the approximation of several septa and is sometimes associated with a change in coiling, increased whorl width, strengthening or disappearance of ornamentation and increase in thickness of the apertural lip. These changes are interpreted as the onset of maturation of sexual organs and indicate that growth was determinate in most ammonoids. Similar as in gastropods, maturity is often marked by changes in coiling and thickening of the apertural lip as well as apertural expansions (e.g. in strombids).

6.4.1.2 General Shape of Growth Curves

Bucher et al. (1996) established a generalized growth curve for ammonoids, where the overall mode of growth of ammonoids is derived from comparison with *Nautilus*, the only externally shelled cephalopod with chambers that is still extant. Ammonoids are, in fact, phylogenetically more closely related to coleoids than to *Nautilus* (Engeser 1990; Jacobs and Landman 1993). However, the retention of an external shell in ammonoids implies that they shared with *Nautilus* basic similarities in mode of growth and developmental constraints, although not necessarily a similarity in their absolute rate of growth (i.e. mm/day) or age at maturity. In ammonoids, as in *Nautilus*, the timing and volumetric growth of chambers were, no doubt, coordinated with the growth of the shell at the aperture and the growth of the soft body parts, insuring a near-neutral buoyancy throughout ontogeny. In *Nautilus*, the

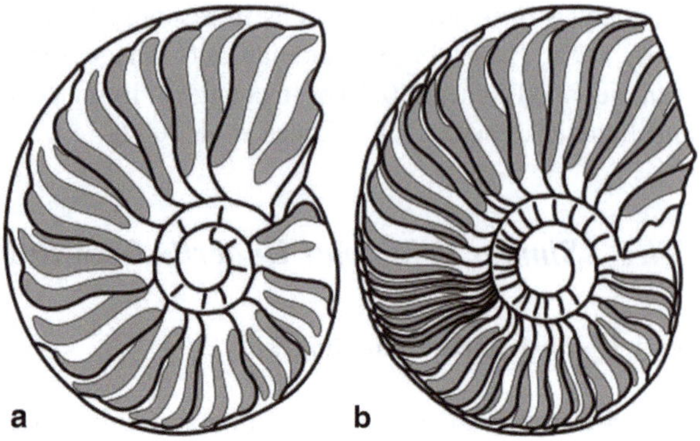

Fig. 6.7 Lateral view of two variants of *Gymnotoceras rotelliformis*. **a** depressed variant. **b** compressed variant. The *shaded* areas represent ribs; the *dark lines* represent growth halts. Growth halts, ribs and septa (not shown) are all more closely spaced in the more compressed specimen (*b*). Redrawn from Bucher (1997)

time required for chamber formation increases exponentially throughout ontogeny, while successive septa are spaced at constant angular intervals and progressively become thicker (Ward 1985). The shell diameter is increasing as a linear function of time. In consequence, variation in the spacing between septa may represent a response to variation in growth rate at the aperture, variation in shell shape and variation in the density of the shell and soft parts. Ammonoids generally follow a pattern similar to *Nautilus* during their juvenile stage, but the neanic and mature stages significantly depart from this pattern with their often abrupt changes in septal spacing. As the neanic stage is characterized by an interval of increasingly wider septal spacing, this suggests that the rate of growth in shell diameter over time accelerated during this stage. To the contrary, at the approach of maturity, the thickening of apertural margin and the more and more closely spaced septa indicate that the rate of growth in diameter decelerated. By combining these observations, one can estimate that the temporal evolution of most ammonoids shell diameter would follow a sigmoid curve throughout ontogenetic time as illustrated in Bucher et al. 1996. Empirical evidence in living molluscs also suggests that growth rates in linear dimensions often decrease with increasing size, indicating that sigmoid curves are the rule rather than the exception (e.g., Bretos 1980; Picken 1980; Guzman and Rios 1987; Black et al. 1994; Iijima 2001; Schöne et al. 2002, 2007).

6.4.1.3 Short-term Fluctuations in Growth Rate

In ammonoids, short-term fluctuations in growth rate (Fig. 6.7) superimposed on the generalized growth curve are evidenced by the presence of morphological features like constrictions and megastriae. Besides the usual growth lines, which correspond to intermittent shell secretion of a short temporal scale of the order of days (spacing between 1 and 100 µm, Bucher et al. 1996), other discontinuities in shell growth correspond to growth rate variation on a higher temporal scale (of the order of weeks and perhaps months, Spight 1973; Spight and Lyons 1974; Spight et al. 1974). Megastriae correspond to distinctive thick lines, different from growth lines.

Microstructural studies (Bucher et al. 1996) revealed that megastriae correspond to major shell discontinuities in the outer prismatic and nacreous layers, implying a retreat of the mantle edge and a relatively long pause in shell secretion (growth halts). In contrast, the inner prismatic layer is continuous, indicating that it was laid subsequently to the mantle edge retreat after growth at the aperture resumed. Megastriae occur in a wide variety of ammonoid taxa and also in many bivalves and gastropods (i.e. Spight 1973; Spight and Lyons 1974; Spight et al. 1974). In ammonoids, megastriae are more frequent during the juvenile stage than during the mature stage. For instance, in the Middle Triassic *Parafrechites meeki*, the juvenile whorls display numerous, variably spaced megastriae. However, these features are generally absent on the adult body chamber. In only a few ammonoid groups, e.g. some lytoceratids, did megastriae persist to maturity. The presence of numerous closely spaced megastriae can sometimes result in composite ribs, consisting of a number of juxtaposed rib elements. This indicates that some parts of the secreted shell have been dissolved at the time of megastriae formation (see also Fig. 6.7). Also in some lytoceratids, megastriae are at an angle with growth lines. Moreover, hollow spines and parabolic tubercles are typically associated with megastriae (Bucher et al. 1996; Bucher 1997).

Some patterns of variation of shell shape and its associated growth features, particularly growth halts, in ammonoids and gastropods point out that similar rules of accretionary growth underlie the morphogenesis of the shell and its evolution in both clades (Bucher 1997). At maturity, the spacing between successive growth halts tends to decrease in ammonoids and gastropods. Growth halts approximation is generally accompanied by a change in aperture shape and/ or coiling. For instance, in the gastropod *Epitonium scalare*, the shell is isometric (or nearly so) until maturity, which is recorded in the shell by a more elliptic aperture and a few (about 5) approximated varices (growth halts). In Muricidae, growth halts are more closely spaced at maturity. For instance, in *Bolinus brandaris*, a decrease in the length of spines is coinciding with two approximated growth halts (personal observation). Bucher (1997) noted also that in *Murex haustellum*, the first growth halt, at about 3.5 whorls after the embryonic constriction, is coinciding with an apertural shape change. Closely spaced growth halts at maturity are also quite frequent in ammonoids, for instance in *Parafrechites* and in *Gymnotoceras*.

More generally, Bucher (1997) suggests that a regular/irregular spacing between growth halts (more or less constant angle or not) during ontogeny could be related to isometric/ allometric growth of the aperture. Throughout ontogeny and within populations, aperture shape and ornamentation tend to covary with the spacing between growth halts in ammonoids (Bucher 1997). For instance, in *Gymnotoceras rotelliformis*, growth halts and ribs are more closely spaced in the compressed variants than in the depressed variants (Fig. 6.7). Specimens tend to become more compressed during ontogeny while the number of growth halts per whorl tends to increase. Also, during ontogeny, the spacing between growth halts is reflected in the spacing between septa (Bucher 1997; see also Bucher and Guex 1990; Bucher et al. 1996 for documentation in *Parafrechites* and *Eotetragonites*). These patterns of variation between growth halts, ornamentation, aperture shape and septa are quite frequent in the Triassic subfamilies Berichitinidae and Paraceratitinae (Monnet and Bucher 2005). It remains to be investigated if similar patterns occur in gastropods and bivalves.

6.4.2 Teratology as Natural Experiments

6.4.2.1 Coiling

Teratological shells, including fossil ones, often provide an additional useful source of information about the way development generally proceeds. For instance, planispiral ammonites that were infested by epizoans during their lifetime exhibit alterations of their coiling geometry (Checa et al. 2002). These authors pointed out that, most commonly, the epizoans settled on the venter of ammonoids and constituted an obstacle to the subsequent growth. This disturbance probably initiated changes in the hydrostatic conditions of the ammonite and caused a lateral shifting of the growth direction compared to the previous whorl in attempts to avoid the obstacle. Using a hydrostatic model based on earlier papers of Okamoto (1988a, b, c), Checa et al. (2002) showed that the shell tube should periodically cross the venter, thus leading to zigzag coiling, if the ammonite tried to maintain the growth direction perpendicular to the substrate (sea bottom). If the epizoan was positioned on the midventer, the whorl could be detached from the previous whorl. Under constant growth direction relative to the substrate, a lateral placement of the epizoan would rather result in trochospiral coiling as illustrated in Fig. 6.8, especially if the epizoan had a certain non-negligible weight, which could cause the tilting of the ammonite.

A similar role for life orientation in determining the growth direction has been experimentally tested in gastropods. In the benthic freshwater Planorbidae (Gastropoda), specimens experimentally altered by extra weights on one side of the shell revealed that the growth direction remained perpendicular to the substrate (Checa and Jiménez-Jiménez 1997). Similarly, the benthic prosobranch gastropods exhibiting a tangential aperture with regards to the coiling axis have been shown to live with the aperture parallel to the substrate (Linsley 1977). These gastropods have the ability to regulate the amount of torsion/ detorsion of the foot to place the center of gravity of the shell and body over the midline of the cephalopodial mass, thus allowing the maintenance of a constant life orientation. A well-known example of the influence of change of mode of life on shell morphology is provided by the gastropod of the genus *Distorsio*, which, once settled on the substrate, displays distorted coiling (Checa and Aguado 1992). These comparative studies put a strong emphasis on the role of life orientation in the determination of growth direction in both gastropods and ammonoids (Linsley 1977, 1978; Checa and Jiménez-Jiménez 1997; Checa et al. 2002).

Fig. 6.8 Example of teratological shells found in an assemblage of *Aplococeras vogdesi* from Triassic, Nevada. First row: a regular planispiral specimen. Second and third rows: two specimens exhibiting trochospiral coiling. No scars are visible and the trochospiral coiling appears in the early growth stages. The settlement of an epizoan on the left side of these specimens could have induced the tilting of the ammonite because of changes in the hydrostatic condition, especially if the weight of the epizoan was comparable to the weight of the ammonite. A trochospiral coiling is subsequently produced, if a constant growth direction with respect to the sea floor is maintained. Photographs by Rosi Roth (Paläontologisches Institut und Museum der Universität Zürich, Switzerland). Specimen courtesy of H. Bucher (Universität Zürich), redrawn from Urdy et al. (2013)

6.4.2.2 Ornamentation

In ammonoids, shells with regenerated damage are commonly found (Landman and Waage 1986; Bond and Saunders 1989; Kröger 2002; Hoffmann and Keupp 2015), Particularly, some changes in the ornamental features have been described in response to the location of injuries reaching the mantle (Guex 1967, 1968; Bayer 1970; Hammer and Bucher 2005b). For example, some shells with a ventral keel associated with ribs on the flanks can lose their keel in response to a wound located on the venter. Then, the ribs in the post-damaged shell cross the venter, whereas before they were interrupted by the keel. Some other shells bearing bifurcating ribs on the venter rather construct simple ribs after being damaged on one side (Hammer and Bucher 2005b). These examples are described in terms of "*ornamental compensation*" (Guex 1967, 1968).

In the Muricidae, ornamental features may also be greatly modified in some specimens, sometimes as a consequence of damages of the mantle (Fig. 6.9). In *Bolinus brandaris*, the teratological specimens are particularly remarkable with respect to the extremely limited extent of the intraspecific variation. In this species, the number of rows of spines, as well as the number of spines per whorl is highly stable. But there exists a small proportion of specimens with a higher number of rows of spines. When supernumerary rows of spines are present, spines may look typical or may be larger, curved backward, subdivided and/ or more opened. For instance, specimen 1 (Fig. 6.9) exhibits three rows of spines on the body whorl and the second one from the umbilical line is subdivided. Specimen 2 exhibits four rows of spines on the body whorl and the second one from the umbilical line is subdivided, too. Note that the subdivision of these spines mimics the spine morphology of related species ('foliated spines'). The presence of injuries in these two specimens cannot be ascertained, since no marks of injuries are visible on the last whorl. But in some specimens, the addition and subdivision of spine rows in response to injuries is clear (Fig. 6.9, specimen 3). For instance, after a large breakage of the aperture and the siphon (black arrow), specimen 3 exhibits a subdivision of the first row of spines from the umbilical line and the spatial periodicity of growth halts is lost. In this specimen, the region of the mantle, which was related to the normal second row of spines from the umbilical line slightly shifted to the posterior part of the aperture (white arrow) immediately after damage. In the subsequent growth increments, this row of spines tends to move more anteriorly, probably as a consequence of the regrowing of the siphon. This illustrates that the mantle withdrew into the shell just after the damage and became progressively stretched during the following growth steps. Shell damages can also cause the spines to be lost. For instance, after a breakage of the aperture (Fig. 6.9, arrow), specimen 4 is almost completely smooth. This change is accompanied by a slight increase in whorl overlap and the loss of the spatial periodicity of growth halts, indicating that the wound reached the mantle and probably affected its visco-elastic properties. A moderate modification of the shell in response to a shell breakage away from the aperture is illustrated in specimen 5 (Fig. 6.9). Note that the antimarginal ridges of the repaired shell are not strictly concordant with the remains of the shells built before damage. A growth halt is built just next to the remains of the growth halt built before damage. The next increment looks normal. This documents that the mantle edge has not been (seriously) damaged and that the repaired shell has not been secreted by the mantle edge.

In conclusion, these teratological changes in ornamentation can be seen as a generic outcome of modes of shell growth, whether one interprets such results in terms of reaction-diffusion (Guex et al. 2003; Hammer and Bucher 2005b), or mechanical effects (Hammer and Bucher 2005a), both in ammonoids and gastropods.

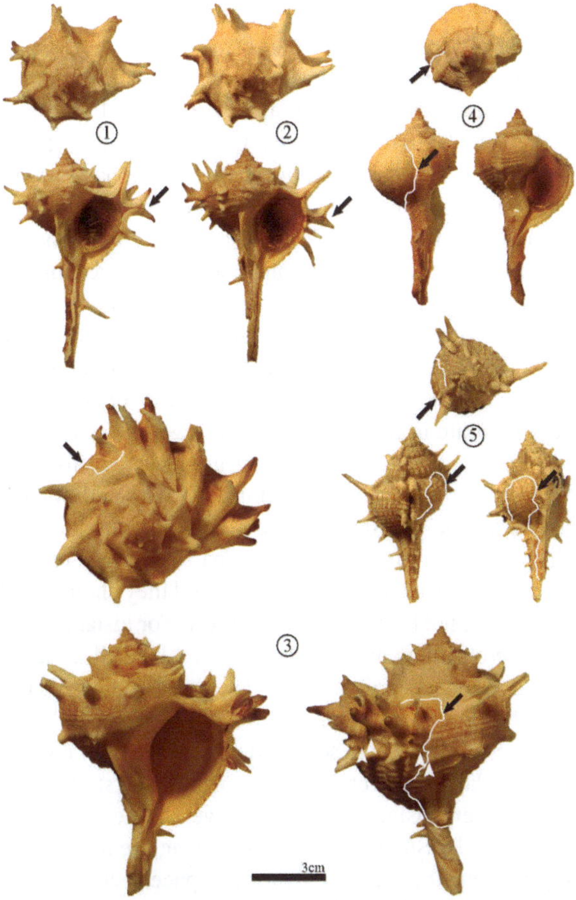

Fig. 6.9 Example of teratological specimens of *Bolinus brandaris*. Specimens 1–4 have been trawled by fishermen at 30–40 m depth in Malaga (Spain). 1, a variant with three rows of spines on the body whorl. 2, a variant with four rows of spines on the body whorl. The presence of a precocious wound in 1 and 2 is not ascertained. *Arrows*: subdivided spines. 3, shell anomalies in response to a breakage of the aperture and siphon (*black arrow*, contour breakage in *white*). Note the two step displacement of the mantle after the breakage and the closely spaced growth halts (*white arrow* heads). 4, shell anomalies in response to a breakage of the aperture (*arrow*, breakage in *white*). Note the change in whorl overlap and approximated growth halts. 5, shell repair in response to a damage not affecting the aperture (breakage in *white*). Note that a growth halt is built just next the previous one. Undetermined species. Photographs by Rosi Roth (Paläontologisches Institut und Museum der Universität Zürich, Switzerland)

6.5 Conclusions

The morphology of organisms is central to different fields of biological research such as taxonomy, evolutionary biology, ecology, and functional biology. About half of the models in this field have been developed for organisms with accretionary growth—a majority of which is devoted to mollusk shells (Dera et al. 2008).

As such, the theoretical morphospace of mollusk shells is subjected to a greater scrutiny than the morphospace of any other taxa.

A variety of forces act to constraint evolutionary paths among well-structured lines, most notably development, external environmental factors and interaction among species. Seilacher (1970, 1991) has qualitatively defined the kind of constraints, which are at work in separating the forms we have seen from those we have not seen and will probably never see. However, the study of evolutionary constraints needs more than a qualitative description: it requires a quantitative metric to define a morphospace and then visualize and compare the occupied and unoccupied regions of that morphospace. Alternatively, it requires a morphogenetic model that can link the timing of growth processes and/ or the mechano-chemical basis of morphogenesis to specific patterns of variation at some taxonomic level.

Simple models are generally focusing on geometry only. The simplest models are generally less accurate in representing a given geometry but they are generally applicable across many taxa. Any subsequent improvement for accuracy often requires a completely new mathematical formulation with additional parameters and it sometimes decreases the generality of the model. Because any theoretical model is created for the purpose of interpreting some particular data or phenomenon, they cannot be necessarily exported to other contexts and they should not be. Several examples have been discussed throughout this review. For instance, Okomoto's model is ideal to study heteromorph ammonoids, but reveals itself unnecessarily cumbersome while applied to loosely coiled shells. The exact contrary is true for Ackerly's model. With their few parameters, Raup's model and Schindel's model (as well as Okamoto's and Ackerly's models to some extant) are ideal for morphospace studies, but Urdy et al.'s (2010a) model render any morphospace study impracticable. This is the cost to pay for new insights on quantitative allometric growth. As the "*phylogeny of shell models*" by Stone (1996b) made clear, the models using a fixed axis ("*form models*") facilitate the analysis of morphospace occupation, whereas moving reference models ("*growth-like models*") conform better to the accretionary growth but do not usually provide convenient shape descriptors. Similarly, models tending more toward the morphogenetic side—although they can lead to broad insight on developmental constraints on a large phylogenetic scale—tend to render any quantitative study of shape or growth impracticable at a lower taxonomic level. Insights from morphogenetic models are virtually qualitative; such a crude understanding hinders the comparison between experimental and theoretical data on variation in populations or species. So there exists basically no all-purpose model. In any cases, the best models are those that do not sacrifice important unresolved biological questions for simple mathematical convenience or practical issues. So it is pushing Raup's model too far to apply it below the family level. It is pushing lateral inhibition models too far to interpret changes in 3D ornamentation structure. It is pushing the linear elasticity theory too far to interpret how ornamentation and growth rates are related.

As a general conclusion, one can note that the macroevolutionary patterns observed in the fossil record provide independent tests of the validity of morphogenetic models proposed on living species (Urdy et al. 2013). Besides the briefly discussed Buckman's laws of covaration, well-known examples of macro-evolutionary patterns come from the studies of the tetrapod limb (Oster et al. 1988; Shubin et al. 2009; Zhu et al. 2010) and the molar proportions in mammals (Polly 2007; Kavanagh et al. 2007; Renvoisé et al. 2009; Salazar-Ciudad and Jernvall 2010; Wilson et al. 2012). Reciprocally, the observation of these macroevolutionary patterns often act as a source of inspiration to investigate the underlying rules of development, because, at the end, they are the patterns that the Neo-Darwinian theory was unable to account for.

Acknowledgments I thank Hugo Bucher (PIMUZ) for courteously providing the specimens illustrated in Fig. 6.8 and allowing them to be published here. I thank Nicolas Goudemand for fruitful discussions, as well as Takao Ubukata and an anonymous reviewer for providing useful comments during the reviewing process. This work would not have been possible without the support of the Swiss National Science Foundation (200021_124784/1 and PA00P3–136478) and the University of Zurich.

References

Ackerly SC (1989) Kinematics of accretionary shell growth, with examples from brachiopods and molluscs. Paleobiology 15:147–164

Ackerly SC (1992) Morphogenetic regulation in the shells of bivalves and brachiopods—Evidence from the geometry of the spiral. Lethaia 25:249–256

Aldridge AE (1998) Brachipod outline and the importance of the logarithmic spiral. Paleobiology 24/2:215–226

Bayer U (1970) Anomalien bei Ammoniten des Aaleniums und Bajociums und ihre Beziehung zur Lebensweise. N Jahrb Geol Paläont Abh 135:19–41

Bayer U (1977) Cephalopod septa I. Constructional morphology of the ammonite septum. N Jahrb Geol Paläont Abh 154:19–41

Bayer U (1978) Morphogenetic programs, instabilities and evolution: a theoretical study. N Jahrb Geol Paläont Abh 156:226–261

Black R, Turner SJ, Johnson MS (1994) The early life history of *Bembicium vittatum* Philippi, 1846 (Gastropoda, Littorinidae). Veliger 37:393–399

Bond PN, Saunders WB (1989) Sublethal injury and shell repair in upper Mississippian ammonoids. Paleobiology 15:414–428

Bretos M (1980) Age-determination in the keyhole Limpet *Fissurella crassa* Lamarck (Archaeogastropoda, Fissurellidae) based on shell growth rings. Biol Bull 159:606–612

Bucher H (1997) Caractères périodiques et modes de croissance des ammonites: comparaison avec les gastéropodes. Geobios 20:85–99

Bucher H, Guex J (1990) Rythmes de croissance chez les ammonites triasiques. Bulletin des Laboratoires de Geologie, Mineralogie, Geophysique et du Musee géologique de l'Université de Lausanne 308:191–209

Bucher H, Landman NH, Klofak SM, Guex J (1996) Mode and rate of growth in ammonoids. In: Landman NH, Tanabe K, Davis RA (eds) Ammonoid paleobiology. Plenum, New York

Buckman SS (1887–1907) Monograph of the ammonites of the inferior Oolite series. Palaeontographical Society, London, p. 456

Burnaby TP (1966) Allometric growth of ammonoid shells: a generalization of logarithmic spiral. Nature 209:904–906

Checa A (1991) Sectorial expansion and shell morphogenesis in mollusks. Lethaia 24:97–114

Checa A, Aguado R (1992) Sectorial expansion analysis of irregularly coiled shells: application to the recent gastropod *Distorsio*. Palaeontology 35:913–925

Checa AG, García-Ruiz JM (1996) Morphogenesis of the septum in ammonoids. In: Landman NH, Tanabe K, Davis RA (eds) Ammonoid Paleobiology. Topics in Geobiology 13. Plenum, New York

Checa AG, Jiménez-Jiménez AP (1997) Regulation of spiral growth in planorbid gastropods. Lethaia 30:257–269

Checa AG, Company M, Sandoval J, Weitschat W (1996) Covariation of morphological characters in the Triassic ammonoid *Czekanowskites rieberi*. Lethaia 29:225–235

Checa AG, Jiménez-Jiménez AP, Rivas P (1998) Regulation of spiral coiling in the terrestrial gastropod *Sphincterochila*: an experimental test of the road-holding model. J Morph 235:249–257

Checa AG, Okamoto T, Keupp H (2002) Abnormalities as natural experiments: a morphogenetic model for coiling regulation in planispiral ammonites. Paleobiology 28:127–138

Chirat R, Enay R, Hantzpergue P, Mangold C (2008) Developmental integration related to buoyancy control in nautiloids: evidence from unusual septal approximation and ontogenetic allometries in a Jurassic species. Palaeontology 51:251–261

Chirat R, Moulton DE, Goriely A (2013) Mechanical basis of morphogenesis and convergent evolution of spiny seashells. PNAS USA 110:6015–6020

Clements R, Liew TS, Vermeulen JJ, Schilthuizen M (2008) Further twists in gastropod shell evolution. Biol Lett 4:179–182

Cortie MB (1989) Models for mollusk shell shape. S Afr J Sci 85:454–460

Dagys AS, Weitschat W (1993) Extensive intraspecific variation in a Triassic ammonoid from Siberia. Lethaia 26:113–121

Dagys A, Bucher H, Weitschat W (1999) Intraspecific variation of *Parasibirites kolymensis* Bychkov (Ammonoidea) from the lower Triassic (Spathian) of Arctic Asia. Mitt Geol-Paläont Inst Univ Hamb 83:163–178

De Baets K, Bert D, Hoffmann R, Monnet C, Yacobucci MM, Klug C (2015) Ammonoid intraspecific variability. This volume

Dera G, Eble GH, Neige P, David B (2008) The flourishing diversity of models in theoretical morphology: from current practices to future macroevolutionary and bioenvironmental challenges. Paleobiology 34/3:301–317

Dommergues JL, Laurin B, Meister C (1996) Evolution of ammonoid morphospace during the early Jurassic radiation. Paleobiology 22:219–240

Eble GJ (1999) Developmental and non-developmental morphospaces in evolutionary biology. In: Chapman RE, Rasskin-Gutman D, Wills M (eds) Morphospace concepts and applications. Cambridge University Press, Cambridge

Engeser TS (1990) Major events in cephalopod evolution. Major evolutionary radiations. Syst Assoc Spec 42:119–138

Ermentrout B (1998) Neural networks as spatio-temporal pattern-forming systems. Rep Prog Phys 61:353–430

Ermentrout B, Campbell J, Oster G (1986) A model for shell patterns based on neural activity. Veliger 28:369–388

Fowler DR, Meinhardt H, Prusinkiewicz P (1992) Modeling Seashells. Comput Graphics 26:379–387

García-Ruiz JM, Checa A, Rivas P (1990) On the origin of ammonoid sutures. Paleobiology 16:349–354

Gerber S, Neige P, Eble GJ (2007) Combining ontogenetic and evolutionary scales of morphological disparity: a study of early Jurassic ammonites. Evol Dev 9:472–482

Gerber S, Eble GJ, Neige P (2008) Allometric space and allometric disparity: a developmental perspective in the macroevolutionary analysis of morphological disparity. Evolution Int J Org Evolution 62:1450–1457

Gierer A, Meinhardt H (1972) Theory of biological pattern formation. Kybernetik 12:30–39

Goodwin BC (1988) Problems and prospects in morphogenesis. Experientia 44:633–637

Gould SJ (1984) Morphological channeling by structural constraint: convergence in styles of dwarfing and gigantism in *Cerion*, with a description of two new fossil species and a report on the discovery of the largest *Cerion*. Paleobiology 10:172–194

Gould SJ (1989) A developmental constraint in *Cerion*, with comments of the definition and interpretation of constraint in evolution. Evolution 43:516–539

Guex J (1967) Contribution à l'étude des blessures chez les ammonites. Bull géologique de Lausanne 165:1–23

Guex J (1968) Sur deux conséquences particulières des traumatismes du manteau des ammonites. Bull de la Societé vaudoise des sciences naturelles 70:121–126

Guex J, Koch A, O'Dogherty L, Bucher H (2003) A morphogenetic explanation of Buckman's law of covariation. Bulletin de la Société géologique de France 174:603–606

Guzman LF, Rios CF (1987) Age and growth of the sub-antarctic limpet *Nacella (Patinigera) magellanica magellanica* (Gmelin, 1791) from the Strait of Magellan, Chile. Veliger 30:159–166

Hammer Ø (2000) A theory for the formation of commarginal ribs in mollusc shells by regulative oscillation. J Molluscan Stud 66(3):383–392

Hammer Ø, Bucher H (1999) Reaction-diffusion processes: application to the morphogenesis of ammonoid ornamentation. Geobios 32:841–852

Hammer Ø, Bucher H (2005a) Models for the morphogenesis of the molluscan shell. Lethaia 38:111–122

Hammer Ø, Bucher H (2005b) Buckman's first law of covariation: a case of proportionality. Lethaia 38:67–72

Hammer Ø, Bucher H (2006) Generalized ammonoids hydrostatics modelling, with application to Intornites and intraspecific variation in *Amaltheus*. Paleontol Res 10:91–96

Hayami I, Okamoto T (1986) Geometric regularity of some oblique sculptures in pectinid and other bivalves: recognition by computer simulations. Paleobiology 12:433–449

Hoffmann R, Keupp H (2015) Ammonoid Paleopathology. In: Klug C (ed) Ammonoid Paleobiology, Chapter 21

Hutchinson JMC (1989) Control of gastropod shell shape: the role of the preceding whorl. J Theor Biol 140:431–444

Hutchinson JMC (1990) Control of gastropod shell form via apertural growth rates. J Morph 206:259–264

Huxley JS, Teissier G (1936) Terminology of relative growth. Nature 137:780–781

Iijima A (2001) Growth of the intertidal snail, *Monodonta labio* (Gastropoda, Prosobranchia) on the Pacific coast of central Japan. Bull Mar Sci 68:27–36

Illert C (1987) Formulation and solution of the classical problem. I. Shell geometry. Nuovo Cimento 9:791–813

Illert C (1989) Formulation and solution of the classical problem. II. Tubular three-dimensional seashell surfaces. Nuovo Cimento 11:761–780

Illert C (1990) *Nipponites mirabilis*: a challenge to seashell theory. Nuovo Cimento D 12:1405–1421

Jackson DJ et al (2006) A rapidly evolving secretome builds and patterns a sea shell. BMC Biol 4:40

Jackson DJ, Worheide G, Degnan BM (2007) Dynamic expression of ancient and novel molluscan shell genes during ecological transitions. BMC Evol Biol 7:160

Jacobs DK, Landman NH (1993) *Nautilus*—a poor model for the function and behavior of ammonoids? Lethaia 26:101–111

Johnson MS, Black R (2008) Adaptive responses of independent traits to the same environmental gradient in the intertidal snail *Bembicium vittatum*. Heredity 101:83–91

Kavanagh KD, Evans AR, Jernvall J (2007) Predicting evolutionary patterns of mammalian teeth from development. Nature 449:427–432

Kemp P, Bertness MD (1984) Snail shape and growth rates: evidence for plastic shell allometry in *Littorina littorea*. PNAS USA Biol Sci 81:811–813

Kennedy WJ, Cobban WA (1976) Aspects of ammonite biology, biogeography, and biostratigraphy. Spec Pap in. Palaeontology 17:1–94

Klingenberg CP (1998) Heterochrony and allometry: the analysis of evolutionary change in ontogeny. Biol Rev 73:79–123

Klug C (2001) Life-cycles of some Devonian ammonoids. Lethaia 34:215–233

Klug C, Hoffmann R (2015) Ammonoid Septa and Sutures. This volume

Klug C, Korn D, Landman NH, Tanabe K, De Baets K, Bert D, Naglik C (2015) Describing ammonoid conchs. This volume

Korn D (2012) Quantification of ontogenetic allometry in ammonoids: ontogenetic allometry in ammonoids. Evol Dev 14:501–514

Korn D, Klug C (2003) Morphological pathways in the evolution of early and middle Devonian ammonoids. Paleobiology 29:329–348

Korn D, Klug C (2007) Conch form analysis, variability, morphological disparity, and mode of life of the Frasnian (Late Devonian) ammonoid *Manticoceras* from Coumiac (Montagne Noire, France). In: Landman NH, Davis RA, Mapes RH (eds) Cephalopods present and past: new insights and fresh perspectives. Springer Science & Business Media, New York

Kröger B (2002) On the efficiency of the buoyancy apparatus in ammonoids: evidences from sublethal shell injuries. Lethaia 35:61–70

Landman NH, Waage KM (1986) Shell abnormalities in scaphitid ammonites. Lethaia 19:211–224

Linsley RM (1977) Some "laws" of gastropod shell form. Paleobiology 3:196–206

Linsley RM (1978) Shell form and the evolution of gastropods. Am Sci 66:432–441

Løvtrup S, Løvtrup M (1988) The morphogenesis of molluscan shells: a mathematical account using biological parameters. J Morph 197:53–62

McGhee GR (1978) Analysis of shell torsion phenomenon in Bivalvia. Lethaia 11:315–329

McGhee GR (1980) Shell form in the biconvex articulate Brachiopoda: a geometric analysis. Paleobiology 6:57–76

McGhee GR (1999) Theoretical morphology. Columbia University, New York

McGhee GR (2001) The question of spiral axes and brachiopod shell growth: a comparison of morphometric techniques. Paleobiology 27:716–723

Meinhardt H (1995) The algorithmic beauty of sea shells. Springer, Berlin

Meinhardt H (2008) Models of biological pattern formation: from elementary steps to the organization of embryonic axes. Curr Top Dev Biol 81:1–63

Meinhardt H, Gierer A (1974) Applications of a theory of biological pattern formation based on lateral inhibition. J Cell Sci 15:321–346

Meinhardt H, Gierer A (2000) Pattern formation by local self-activation and lateral inhibition. Bioessays 22:753–760

Meinhardt H, Klingler H (1988) A model for pattern formation of shells of mollusks. J Theor Biol 126:63–89

Meister C (1988) Ontogeny and evolution of the Amaltheidae (Ammonoidea). Eclogae Geol Helv 81:763–841

Monnet C, Bucher H (2005) New middle and late Anisian (Middle Triassic) ammonoid faunas from northwestern Nevada (USA): taxonomy and biochronology. Fossils Strata 52:1–60

Monnet C, De Baets K, Yacobucci MM (2015) Buckman's rules of covariation. In: Klug C, Korn D, De Baets K, Kruta I, Mapes RH (eds) Ammonoid paleobiology: from macroevolution to paleogeography. Springer, Dordrecht

Morita R (1991a) Finite element analysis of a double membrane tube (DMS-tube) and its implication for gastropod shell morphology. J Morph 207:81–92

Morita R (1991b) Mechanical constraints on aperture form in gastropods. J Morph 207:93–102

Morita R (1993) Development mechanics of retractor muscles and the "Dead Spiral Model" in gastropod shell morphogenesis. N Jahrb Geol Paläont Abh 190:191–217

Morita R (2003) Why do univalve shells of gastropods coil so tightly? A head-foot guidance model of shell growth and its implication on developmental constraints. In: Sekimura T, Noji S, Ueno N, Maini PK (eds) Morphogenesis and pattern formation in biological systems: experiments and models. Springer, Tokyo

Moseley H (1838) On the geometrical forms of turbinated and discoid shells. Philos Trans R Soc Lond 128:351–370

Moulton DE, Goriely A (2014) Surface growth kinematics via local curve evolution. J Math Biol 68:81–108

Moulton DE, Goriely A, Chirat R (2012) Mechanical growth and morphogenesis of seashells. J Theor Biol 311:69–79

Murray JD (1989) Mathematical biology. Springer, Berlin

Neige P, Brayard A, Gerber S, Rouget I (2009) Ammonoids (Mollusca, Cephalopoda): recent advances and contributions to evolutionary paleobiology. Comptes Rendus Pale 8:167–178

Okamoto T (1988a) Analysis of heteromorph ammonoids by differential geometry. Palaeontology 31:37–51

Okamoto T (1988b) Changes in life orientation during the ontogeny of some heteromorph ammonoids. Paleontology 31:281–294

Okamoto T (1988c) Developmental regulation and morphological saltation in the heteromorph ammonite *Nipponites*. Paleobiology 14/3:272–286

Okamoto T (1996) Theoretical modeling of ammonoid morphology. In: Landman NH, Tanabe K, Davis RA (eds) Ammonoid Paleobiology. Topics in Geobiology, 13. Plenum, New York

Oster GF, Shubin N, Murray JD, Alberch P (1988) Evolution and morphogenetic rules: the shape of the vertebrate limb in ontogeny and phylogeny. Evolution 42(5):862–884

Pappas JL, Miller DJ (2013) A Generalized approach to the modeling and analysis of 3D surface morphology in organisms. PLoS ONE 8(10):e77551. doi:10.1371/journal.pone.0077551

Pérez-Claros JA, Olóriz F, Palmqvist P (2007) Sutural complexity in Late Jurassic ammonites and its relationship with phragmocone size and shape: a multidimensional approach using fractal analysis. Lethaia 40:253–272

Picken GB (1980) Distribution, growth, and reproduction of the antarctic limpet *Nacella (Patinigera) Concinna* (Strebel, 1908). J Exp Mar Biol Ecol 42:71–85

Polly PD (2007) Development with a bite. Nature 449:413–415

Raup DM (1961) The geometry of coiling in gastropods. Proc Natl Acad Sci USA 47:602–609

Raup DM (1962) Computer as aid in describing gastropod shells. Science 138:150–152

Raup DM (1966) Geometric analysis of shell coiling: general problems. J Paleont 40:1178–1190

Raup DM (1967) Geometric analysis of shell coiling: coiling in ammonoids. J Paleont 41:43–65

Raup DM, Chamberlain JA Jr (1967) Equations for volume and center of gravity in amnonoid shells. J Paleont 41:566–574

Raup DM, Michelson A (1965) Theoretical morphology of the coiled shell. Science 147:1294–1295

Renvoisé E, Evans AR, Jebrane A, Labruère C, Laffont R, Montuire S (2009) Evolution of mammal tooth patterns: new insights from a developmental prediction model. Evolution 63:1327–1340

Reyment RA, Kennedy WJ (1991) Phenotypic plasticity in a Cretaceous ammonite analyzed by multivariate statistical methods—a methodological study. Evol Biol 25:411–426

Rice SH (1998) The bio-geometry of mollusc shells. Paleobiology 24:133–149

Rieber H (1972) Die Triasfauna der Tessiner Kalkalpen. XXII. Cephalopoden aus der Grenzbitumenzone (Mittelere Trias) des Monte San Giorgio (Kanton Tessin, Schweiz). Schweizerische Paläont Abh 93:1–95

Salazar-Ciudad I, Jernvall J (2010) A computational model of teeth and the developmental origins of morphological variation. Nature 464:583–586

Salazar-Ciudad I, Jernvall J, Newman S (2003) Mechanisms of pattern formation in development and evolution. Development 130:2027–2037

Saunders WB, Shapiro EA (1986) Calculation and simulation of Ammonoid hydrostatics. Paleobiology 12:64–79

Saunders WB, Swan ARH (1984) Morphology and morphologic diversity of mid-Carboniferous (Namurian) ammonoids in time and space. Paleobiology 10:195–228

Savazzi E (1985) Shellgen: a basic program for the modeling of molluscan shell ontogeny and morphogenesis. Comput Geosci 11:521–530

Savazzi E (1987) Geometric and functional constraints on bivalve shell morphology. Lethaia 20:293–306

Savazzi E (1990) Biological aspects of theoretical shell morphology. Lethaia 23:195–212

Savazzi E (1991) Constructional Morphology of strombid gastropods. Lethaia 24:311–331

Schander C, Sundberg P (2001) Useful characters in gastropod phylogeny: soft information or hard facts? Syst Biol 50:136–141
Schindel DE (1990) Unoccupied morphospace and the coiled geometry of gastropods: architectural constraint or geometrical covariation. In: Ross RM, Allmon WD (eds) Causes of evolution: a paleontological perspectives. University of Chicago, Chicago
Schöne BR, Lega J, Flessa KW, Goodwin DH, Dettman DL (2002) Reconstructing daily temperatures from growth rates of the intertidal bivalve mollusk *Chione cortezi* (northern Gulf of California, Mexico). Palaeogeogr Palaeoclimat Palaeoeco 184:131–146
Schöne BR, Rodland DL, Wehrmann A, Heidel B, Oschmann W, Zhang ZJ, Fiebig J, Beck L (2007) Combined sclerochronologic and oxygen isotope analysis of gastropod shells (*Gibbula cineraria*, North Sea): life-history traits and utility as a high-resolution environmental archive for Kelp forests. Marine Biol 150:1237–1252
Seilacher A (1970) Arbeitskonzept zur Konstruktionmorphologie. Lethaia 3:393–396
Seilacher A (1991) Self-organizing mechanisms in morphogenesis and evolution. In: Schmidt-Kittler N, Vogel K (eds) Constructional Morphology and Evolution. Springer Berlin Heidelberg, pp. 251–271
Shubin N, Tabin C, Carroll S (2009) Deep homology and the origins of evolutionary novelty. Nature 457:818–823
Skalak R, Farrow DA, Hoger A (1997) Kinematics of surface growth. J Math Biol 35:869–907
Spight T (1973) Ontogeny environment and shape of a marine snail *Thais lamellosa*. J Exp Mar Biol Ecol 13:215–228
Spight TM, Lyons A (1974) Development and functions of the shell sculpture of the marine snail *Ceratostoma foliatum*. Marine Biol 24:77–83
Spight TM, Birkelan C, Lyons A (1974) Life histories of large and small murexes (Prosobranchia-Muricidae). Marine Biol 24:229–242
Stone JR (1995) Cerioshell: a computer program designed to simulate variation in shell form. Paleobiology 21:509–519
Stone JR (1996a) Computer-simulated shell size and shape variation in the caribbean land snail genus *Cerion*: a test of geometrical constraints. Evolution 50:341–347
Stone JR (1996b) The evolution of ideas: a phylogeny of shell models. Am Nat 148:904–929
Thompson D'AW (1917) On growth and form. Cambridge University, Cambridge
Truman AE (1941) The ammonite body chamber with special reference the buoyancy and mode of life of the living ammonite. Q J Geol Soc Lond 96:339–383
Trussell GC (1996) Phenotypic plasticity in an intertidal snail: the role of a common crab predator. Evolution 50:448–454
Turing AM (1952) The chemical basis of morphogenesis. Philos Trans R Soc Lond B Biol Sci 237:37–72
Ubukata T (1997) Microscopic growth of bivalve shells and its computer simulation. Veliger 40:165–177
Ubukata T (2001) Morphological significance of the orientation of shell coiling and the outline of "the aperture" in bivalve molluscs. N Jahrb Geol Paläont Abhandlungen 221:249–270
Ubukata T (2003) Computer modeling of microscopic features of molluscan shells. In: Sekimura T, Noji S, Ueno N, Maini PK (eds) Morphogenesis and pattern formation in biological systems: experiments and models. Springer, Tokyo
Ubukata T (2005) Theoretical morphology of bivalve shell sculptures. Paleobiology 31:643–655
Ubukata T, Nakagawa Y (2000) Modelling various sculptures in the Cretaceous bivalve *Inoceramus hobetsensis*. Lethaia 33:313–329
Urdy S (2012) On the evolution of morphogenetic models: mechano-chemical interactions and an integrated view of cell differentiation, growth, pattern formation and morphogenesis. Biol Rev Camb Philos Soc 87:786–803
Urdy S, Goudemand N, Bucher H, Chirat R (2010a) Allometries and the morphogenesis of the molluscan shell: a quantitative and theoretical model. J exp zool B 314:280–302
Urdy S, Goudemand N, Bucher H, Chirat R (2010b) Growth-dependent phenotypic variation of molluscan shells: implications for allometric data interpretation. J exp zool B 314:303–326

Urdy S, Wilson LAB, Haug JT, Sánchez-Villagra MR (2013) On the unique perspective of paleontology in the study of developmental evolution and biases. Biol Theory 8:293–311

Vasconcelos P, Gaspar MB, Castro M (2006) Development of indices for nonsacrificial sexing of imposex-affected Hexaplex (Truncularioppsis) Trunculus (Gastropoda: Muricidae). J Molluscan Stud 72(3):285–294

Vermeij GJ (1980) Gastropod growth rate, allometry, and adult size: environmental implications. In: Rhoads DC, Lutz RA (eds) Skeletal growth of aquatic organisms: biological records of environmental change. Plenum, New York

Vermeij GJ (2002) Characters in context: molluscan shells and the forces that mold them. Paleobiology 28:41–54

Ward PD (1980) Comparative shell shape distributions in Jurassic-Cretaceous ammonites and Jurassic-Tertiary nautiloids. Paleobiology 6:32–43

Ward PD (1985) Periodicity of chamber formation in chambered cephalopods: evidences from Nautilus macromphalus and Nautilus pompilius. Paleobiology 11:438–450

Westermann GEG (1966) Covariation and taxonomy of the Jurassic ammonite Sonninia adrica (Waagen). N Jahrb Geol Paläont Abh 124:289–312

Whitworth WA (1862) The equi-angular spiral, its chief properties proved geometrically. Messenger Math 1:5–13

Wilson LAB, Madden RH, Kay RF, Sanchez-Villagra MR (2012) Testing a developmental model in the fossil record: molar proportions in South American ungulates. Paleobiology 38:308–321

Yeap KL, Black R, Johnson MS (2001) The complexity of phenotypic plasticity in the intertidal snail *Nodilittorina australis*. Biol J Linn Soc 72:63–76

Zhu J, Zhang YT, Alber MS, Newman SA (2010) Bare bones pattern formation: a core regulatory network in varying geometries reproduces major features of vertebrate limb development and evolution. PLoS ONE 5:e10892

Chapter 7
Mature Modifications and Sexual Dimorphism

Christian Klug, Michał Zatoń, Horacio Parent, Bernhard Hostettler and Amane Tajika

7.1 Introduction

Allometric growth between different parts of the shell often hampers the identification of mollusk shells, particularly in such cases where preadult shell growth varies strongly. Especially in gastropods, the terminal aperture is often less variable and yields morphological information essential for species determination (e.g. Vermeij 1993; Urdy et al. 2010a, b). In fossil mollusk shells, the adult aperture (peristome) is often missing, partially due to an early death, and partially due to destructive processes, which occurred *post mortem* (taphonomy). Therefore, the entire shell ontogeny is known only from a small fraction of all ammonoid taxa (e.g., Landman et al. 2012). Nevertheless, knowledge of the adult shell of ammonoids is very important since it can yield morphological information essential for systematics and for the reconstruction of various aspects of their paleobiology.

C. Klug (✉) · A. Tajika
Paläontologisches Institut und Museum, University of Zurich,
Karl Schmid-Strasse 6, 8006 Zurich, Switzerland
e-mail: chklug@pim.uzh.ch

A. Tajika
e-mail: amane.tajika@pim.uzh.ch

M. Zatoń
Faculty of Earth Sciences, University of Silesia,
Będzińska 60, 41-200 Sosnowiec, Poland
e-mail: mzaton@wnoz.us.edu.pl

H. Parent
Laboratorio de Paleontología, IFG–FCEIA, Universidad Nacional de Rosario,
Pellegrini 250, 2000 Rosario, Argentina
e-mail: parent@fceia.unr.edu.ar

B. Hostettler
Naturhistorisches Museum, Bernastrasse 15,
3005 Bern, Switzerland
e-mail: bernhard.hostettler@nmbe.ch

C. Klug et al. (eds.), *Ammonoid Paleobiology: From Anatomy to Ecology*,
Topics in Geobiology 43, DOI 10.1007/978-94-017-9630-9_7

In the past five decades, numerous researchers have worked on documenting mature modifications and it can be said that the maturity of an ammonoid shell can be determined with some confidence (e.g. Makowski 1962, 1971, 1991; Callomon 1963; Brochwicz-Lewiński & Różak 1976; Bucher and Guex 1990; Brooks 1991; Bucher et al. 1996; Davis et al. 1996; Schweigert and Dietze 1998; Parent 1997; Klug 2004; Parent et al. 2008a; Zatoń 2008; Landman et al. 2012). The reliable identification of mature shells is the logical prerequisite to determine sexual dimorphism. Both mature modifications and sexual dimorphism are discussed in this chapter, since these are intimately linked with each other. Much of the information contained herein comes from the original work of Davis et al. (1996).

7.2 Mature Modifications

7.2.1 Modifications in Recent Nautilida

Modern Nautilida have been studied for over a century (e.g., Griffin 1900). Much of this research was summarized in Ward (1987). Therein, he listed the mature modifications that have been seen in shells of Recent nautilids (see also Collins and Ward 1987). This list was summarized by Klug (2004) and is repeated here:

1. Shell growth band (shell thickening at the apertural edge, 25 mm wide and up to 1 mm thick).
2. Black band (evenly distributed around the aperture, 1 to 5 mm wide).
3. Deepening of the ocular sinuses.
4. Reduction of relative whorl height by a decrease in whorl expansion rate.
5. Reduction of whorl width by a decrease in whorl width expansion rate; this is accompanied by a more rounded venter.
6. Septal thickening (the terminal septum is up to 30 % thicker than the preceding ones).
7. Septal crowding.
8. Maximum shell diameter (unreliable character because of variability).
9. White ventral area.
10. Increase in body chamber length.
11. Reduction of cameral liquid (probably to compensate for the additional shell material at the aperture and the longer body chamber).

7.2.2 Modifications in Ammonoidea

Among the mature modifications known from nautilids listed above, the majority has also been documented from ammonoids, except the shell growth band, the septal thickness, the white venter, and the reduction of the cameral liquid. Some of these mature modifications that are unknown in ammonoids potentially are unknown because they are not or only poorly preserved or expressed in a different way. For example, the shell growth band could be homologized with a (sub-)terminal shell thickening (a constriction), the white venter might be unknown because of the poor knowledge of color patterns in ammonoids (Mapes and Larson 2015), and the mature reduction of cameral liquid could be tested in the future using volume models of ammonoid shells (Hoffmann et al. 2013; Tajika et al. 2014).

Some of these structures, however, may occur in similar forms in earlier growth stages, either as consequence of an injury, adverse living conditions, and illnesses, or as a recurrent growth feature such as megastriae (growth halts; Bucher and Guex 1990; Bucher et al. 1996). These similar structures may be misinterpreted, what represents a general problem that occurs in research related to mature modifications. Therefore, to ascertain the quality of any such structure as a mature modification, it is helpful to look for other modifications supporting the hypothesis of adulthood for the material under consideration. For instance, a specimen may show septal crowding, which is insufficient as an isolated character to prove adulthood. If, however, it is additionally associated with, e.g., a crowding of growth lines and a change in shell geometry, it is more likely that the given specimen had actually reached maturity.

Another difficulty is linked with the questions of sexual maturity, semelparity, and iteroparity. Is the formation of mature modifications linked with sexual maturity in ammonoids as it is in modern nautilids? Do recurrent structures such as late ontogenetic pre-terminal growth halts coincide with phases of reproduction and would thus indicate iteroparity? These questions are currently difficult to test scientifically, because the soft-part evidence needed to do it is missing. Nevertheless, it appears likely that the ammonoids were sexually mature at the time when growth had terminated and mature modifications of the shell had formed because this is the case in Recent Nautilida.

It might appear trivial, but we still want to point out that in most cases, only a couple of the criteria for maturity listed below will be fulfilled or visible in one specimen. It is also highly unlikely that all criteria will be met in a single specimen. This is due to the fact that in some species, some of these modifications were never realized and certain modes of preservation allow the recording of some characters while others are lost (e.g., Ruzhencev 1962, 1974; Davis et al. 1996).

In the following, we will briefly discuss the most important mature modifications that have become known. Naturally, this list will be incomplete, since many taxa may have formed their own unique adult shell morphology.

7.2.2.1 Septal Crowding

Septal crowding is potentially one of the most widely recognized and published mature modifications in ammonoids, which is reflected in an overwhelming number of publications in which this prominent feature is mentioned (e.g., Westermann

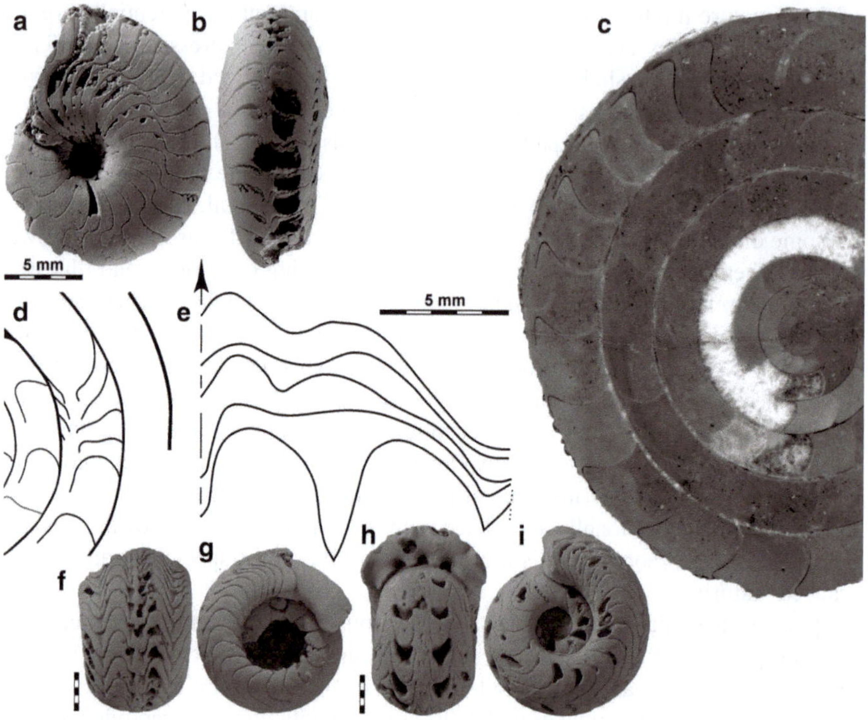

Fig. 7.1 Examples of septal crowding from the Devonian and Carboniferous. **a, b** *Pernoceras crebriseptum*, MB.C.9140.1, Milligan Canyon, Montana, US, lateral and ventral views; dm 24 mm. **c** *Sellanarcestes* sp., PIMUZ 28586, late Emsian, Oufrane, Morocco; dm 77 mm. **d, e** *Wocklumeria sphaeroides*, adult specimen, MB.C.9306.1, Bou Tlidat, Maïder, Morocco, from Ebbighausen and Korn (2007). **d** Septal section, note the change in septal angle and siphuncle position. **e** suture lines, note the extreme simplification (dm = 25 mm). **f–i**, *Ouaoufilalites creber*, S of Oued Temertasset, Algeria (from Korn et al. 2010). **f, g** ventral and lateral view of MB.C.18733.3. **h, i** ventral and lateral view of MB.C.18733.2

1971; Kulicki 1974; Zakharov 1977; Blind and Jordan 1979; Doguzhaeva 1982; Weitschat and Bandel 1991; Klug 2001, 2004; Ebbighausen and Korn 2007; Kraft et al. 2008). Septal crowding affects the distance of at least the last two septa (for nautilids, see, e.g., Willey 1902). This term applies to cases in which the distance between septa (best measured in angles) is reduced (Fig. 7.1). Such a reduction in septal spacing is, however, not only found in adult specimens but sometimes also in preadult ones (Korn and Titus 2006; Kraft et al. 2008 and references therein). Premature septal crowding can be caused by various factors, which can only rarely be identified. More than twenty septa might be more tightly arranged than the preceding ones (e.g., *Pernoceras crebriseptum* in Korn and Titus 2006), documenting a prolonged reduction of the growth rate near the termination of growth (Fig. 7.1). Nevertheless, septal crowding is a good indicator for adulthood when combined with other mature modifications.

7.2.2.2 Thickness of Septa and Sutural Complexity

In several ammonoids, septal thickness increases towards adulthood, mainly the last adult, crowded septa (Westermann 1971, p. 15, fig. 7.8), as in modern nautilids (Collins and Ward 1987). Furnish and Knapp (1966) reported a case of simplification of the terminal suture in Paleozoic forms. Davis et al. (1996) illustrated a *Texoceras* from the Permian of Texas, where the last sutures were not only approximated but the last suture also displays shallower lobes, which are less parabolic than the preceding ones. An impressive example has been illustrated by Ebbighausen and Korn (2007). In their fig. 7, they show the last few septa of a Late Devonian *Wocklumeria* (Fig. 7.1). In this genus, the normal septum displays some deep parabolic pointed lobes. These lobes are completely reduced in the last four septa, which are very strongly approximated and also show a strong change in inclination. This reduction (Fig. 7.1) in sutural frilling might be a consequence of the reduced space between two successive septa due to the limited forward movement of the soft body, which did not produce sufficient space to create lobes of similar length as in the preceding suture. Alternatively, the smaller chamber volume might have required a lower surface to remove the lesser amount of cameral fluid from the new chamber.

7.2.2.3 Change in Coiling and Whorl Cross Section

Many Paleozoic and Mesozoic ammonoids display a more or less strong change in coiling near the termination of growth (e.g., Trueman 1941; Parent 1997; Klug 2001; Klug and Korn 2003). In the earliest ammonoids such as *Metabactrites*, *Anetoceras* and *Erbenoceras*, the last whorl is usually more openly coiled than the preceding ones (e.g., De Baets et al. 2013a, b). One of the most common changes in coiling in planispiral ammonoids is the umbilical egression, i.e. the increase in the relative umbilical width close to adulthood. Lehmann (1981) dubbed this phenomenon "*retraction*". Conspicuous examples are found within Late Devonian Wocklumeriidae (Ebbighausen and Korn 2007), Middle Triassic Ceratitidae (e.g. Wenger 1957), Late Triassic Haloritidae (Mojsisovics 1882), Middle Jurassic Tulitidae (e.g., Hahn 1971; Zatoń 2008) and Late Cretaceous Acanthoceratidae (Kennedy and Cobban 1976).

The changes in coiling in the terminal whorl of Cretaceous heteromorphs range among the most conspicuous and thus most famous mature modifications. Many taxa formed a U-shaped terminal demi-whorl, which sometimes deviates from the coiling plane of preceding whorls. In the Late Cretaceous *Didymoceras*, the U-shaped part is separated from a helicospirally coiled preadult shell, whose coiling axis forms an angle of 60–90° to that of the terminal demi-whorl (e.g. Kennedy et al. 2000). In the Late Cretaceous *Pravitoceras*, the coiling direction changes in the opposite direction from the penultimate to the terminal demi-whorl (Matsunaga et al. 2008). In the Early Cretaceous *Hamulina* and *Heteroceras*, the U-shaped hook represents the largest part of the shell (Orbigny 1850).

Several evolutionary lineages independently produced small to medium sized forms, in which the terminal whorl is strongly elliptical or even forms a kink. For instance, the last whorl of the Late Devonian Prolobitidae is slightly elliptical and ends in a nearly straight shaft. Simultaneously, the umbilical wall closes the umbilicus (Walton et al. 2010 and references therein). The Permian *Hyattoceras* produced a similar shell form with the main difference being that the whorl forms a subtriangular cross section about 180° behind the terminal aperture (marked by a constriction), preceded and followed by a much more rounded cross section (Gemmelaro 1887; Davis 1972; Davis et al. 1969, 1996). A similar morphology evolved convergently in the Triassic families Haloritidae and Lobitidae (e.g., Mojsisovics 1882). In fully grown specimens of both groups, the last whorl is elliptical. Where the whorl height is largest, the whorl width is reduced and the whorl tapers towards the venter, while both before and after this short whorl segment, the venter is more or less broadly rounded. In the Jurassic, a couple of genera evolved comparable morphologies, but in these cases, they represent microconchs of less than 5 cm diameter and with strong apertural modifications (lappets). In the Middle Jurassic, all representatives of *Oecoptychius* display a strongly elliptical terminal whorl and some even form a distinct kink a demi-whorl posterior of the terminal aperture (Schweigert and Dietze 1998; Schweigert et al. 2003). *Cadomoceras* (Middle Jurassic; Schweigert et al. 2007), *Sutneria* (Late Jurassic; Parent et al. 2008a), and *Protophites* (Bert 2003) evolved quite similar changes in coiling in the terminal whorl.

Especially in Paleozoic forms, such a change in coiling is not always obvious. In such cases, adulthood/maturity is often reflected in more or less distinct changes in whorl expansion rate. For example, Devonian Anarcestidae commonly have a whorl expansion rate around 1.5. In the terminal whorl, the whorl expansion rate (Raup and Michelson 1965) increases to values around 2 (Klug 2001; Korn 2012). In the Devonian agoniatitids, the whorl expansion rate rises in the preadult whorls. When the specimen approached maturity, this increasing trend is inverted. At least for these Devonian ammonoids, the rule applies that forms with high whorl expansion rates show a terminal decrease while those with low whorl expansion rates display a terminal increase.

7.2.2.4 Changes in Ornament

A change in ornament near the terminal aperture is very common in ammonoids (e.g., Davis et al. 1996). Many show a decrease in ornament strength, especially as far as ribbing is concerned. This applies to such genera as Triassic *Ceratites*, Jurassic *Dactylioceras*, and Cretaceous *Acanthoceras* among many others. In some ammonoid taxa, the ornament became initially stronger and then smoothed directly behind the terminal aperture. In macroconchiate Jurassic perisphinctids, the preadult whorls sometimes carry rather closely spaced fine and sharp ribs, which more or less abruptly change into coarse and broad ribs on the last whorl (variocostation; e.g., *Crussoliceras, Lithacoceras, Perisphinctes*). Usually, however, the last 10–20° behind the terminal aperture are devoid of strong ribs and commonly display tightly spaced growth lines and/or lirae.

7.2.2.5 Terminal Apertural Constriction or Shell Thickening

A sudden reduction in the whorl cross section at the terminal peristome is very common in the Ammonoidea (e.g., Devonian: *Parawocklumeria*, *Wocklumeria*; Permian: *Agathiceras*, *Hyattoceras*; Triassic: *Arcestes*, *Lobites*; Jurassic: *Bullatimorphites*, *Cadoceras*; Cretaceous: *Baculites*, *Saynoceras*, *Scaphites*, *Valanginites*; e.g., Davitashvili and Khimshiashvili 1954; see Davis et al. 1996 for further examples). In some genera, this constriction is combined with a shell thickening or the terminal shell thickening may appear like a constriction in the internal mould (e.g., Devonian *Agoniatites*, *Manticoceras*; e.g., Klug 2001; De Baets et al. 2012).

7.2.2.6 Formation of Adult Apertural Modifications

Changes in the shape of the aperture (Fig. 7.2) are the most conspicuous mature modification. In some taxa, the undulation of the apertural margin with its projections and sinuses increased only slightly, while in others, this undulation became so extreme that long projections formed adjacent to the supposed ocular sinus. In microconchs of *Kosmoceras phaeinum*, these projections or lappets approached the diameter of the adult shell in length in some specimens (Arkell et al. 1957; Krimholc et al. 1958b; Makowski 1962; Callomon 1963). These extensions of the terminal peristome developed various shapes.

From the Paleozoic, only a few examples have become known. Davis (1972) and Davis et al. (1969, 1996) published Permian examples of *Adrianites* and *Hyattoceras* with strong projections at the terminal aperture. Zhao and Zheng (1977) introduced the Permian genus *Elephantoceras*, which is a small, globular form with strong ornament and long apertural lappets. Some Triassic Arcestidae carry strong ventrolateral or ventral projections (Mojsisovics 1882), while the ceratitids often lack strong lappets (e.g. Sun 1928).

Prominent lateral apertural lappets became common among Middle and Late Jurassic microconchs (Keupp and Riedel 2010). In the Haploceratoidea, several microconchs carry drop-shaped lappets, while in many Stephanoceratoidea and Perisphinctoidea, the lappets are rather straight and tongue-shaped (e.g., Zatoń 2008, 2010; Tajika et al. 2014). In *Oecoptychius*, the lateral lappets are hammer-shaped and combined with a ventral hemispherical projection (Schweigert and Dietze 1998; Schweigert et al. 2003, 2007). Another example has been discussed by Keupp and Riedel (2010): in the Middle Jurassic microconch *Ebrayiceras*, the lateral lappets are very large (half the size of the last whorl) and nearly fused with smaller ventrolateral lappets, thus forming oval ventrolateral openings.

Several groups produced more or less long ventral projections. For example, all species of the Early Jurassic Amaltheidae formed ventral projections when mature. In the Cretaceous, the genus *Mortoniceras* produced a more or less strongly curved midventral spine (Marcinowski and Wiedmann 1990; Amedro 1992).

Fig. 7.2 Middle and Late Jurassic microconchs. **a** *Cadomoceras cadomense*, Bajocian. **b** *Paralingulaticeras lithographicum*, early Tithonian, Mörnsheim, Germany, dm ca. 50 mm. **c** *Kosmoceras* "*compressum*", Callovian. **d** *Neomorphoceras*, sp. Oxfordian. **e** *Normannites orbignyi*, Bajocian, Thorigné. **f** *Grossouvria* sp., Callovian. **g** *Indosphinctes* sp., Callovian, Pamproux, with three successive growth halts that all contain lappets. **h** *Cleistosphinctes* sp., Bajocian. **i** *Indosphinctes* sp.,

7.2.2.7 Muscle Scars

An increasing number of ammonoid species have become known for the preservation of muscle scars (e.g., Doguzhaeva 1981; Doguzhaeva and Kabanov 1988; Doguzhaeva and Mikhailova 1991, 2002; Doguzhaeva and Mutvei 1991, 1993, 1996; Tanabe et al. 1998; Kennedy et al. 2002; Richter 2002; Klug et al. 2008 Chap. 2.4). In most cases, the muscle scars became visible in specimens that were adult. This can be explained by the fact that in mollusks, the secretion of carbonate is commonly linked with muscle attachment, be it at the aperture or at muscle attachment sites. The longer the muscles stayed at the same place, the more carbonate was secreted, thus increasing the likelihood of its preservation. In preadult growth stages, the interim attachment sites apparently existed too briefly in one place to allow the deposition of a sufficient amount of aragonite to become visibly preserved. An additional bias might be the size of the specimen, although some small (probably adult) cheiloceratids (<30 mm) have been reported (Richter 2002) that nicely show muscle attachment structures.

An illustrative example of sexual dimorphism in muscle scars, with connotations in soft-body organization, was described by Palframan (1969: text-fig. 11) from adult macro- and microconchs of *Hecticoceras brightii*. Besides the usual ventrolateral muscle scars in both dimorphs (Doguzhaeva and Mutvei 1991), the macroconchs have an additional ventrolateral scar behind the peristome. The microconchs also bear these additional scars but extended ventro-laterally and projected on the flanks until, at least, the umbilical shoulder.

7.2.2.8 Colour Pattern

Colour patterns are rarely preserved in ammonoids (e.g., Mapes and Davis 1996; Mapes and Larson 2015). Adult modifications of these patterns are even rarer. We are aware of only the one record already reported by Mapes and Davis (1996), namely Mapes and Sneck (1987), who described an *Owenites* in which the transverse color bands were more tightly spaced near the terminal aperture.

7.2.2.9 The Black Layer

The black layer is well-known from modern nautilids (Ward 1987). In shells of adult nautilids, a black chitinous layer less than 0.5 mm thick in the dorsal part of the shell extends beyond the apertural edge. It covers a tongue-shaped surface with an adult thickening, which is formed at the termination of growth. A similar black layer has been found in various ammonoids (Fig. 7.3), including e.g., Devo-

---

Callovian, Pamproux, with bent lappet. **j** *Bigotites* sp., Bajocian. **k** *Parataxioceras latifasciculatum*, middle Kimmeridgian, Gräfenberg, Germany, dm ca. 145 mm. **l** *Parataxioceras* cf. *lothari*, middle Kimmeridgian, Geisingen, Germany, dm ca. 100 mm. **a, d, h, j** Ste. Honorine Des Pertes, France, col. C. Obrist. B, K, L, col. V. Schlampp. C, F, Aichelberg, Germany, from Dietl (2013). E, G, I, col. P. Branger

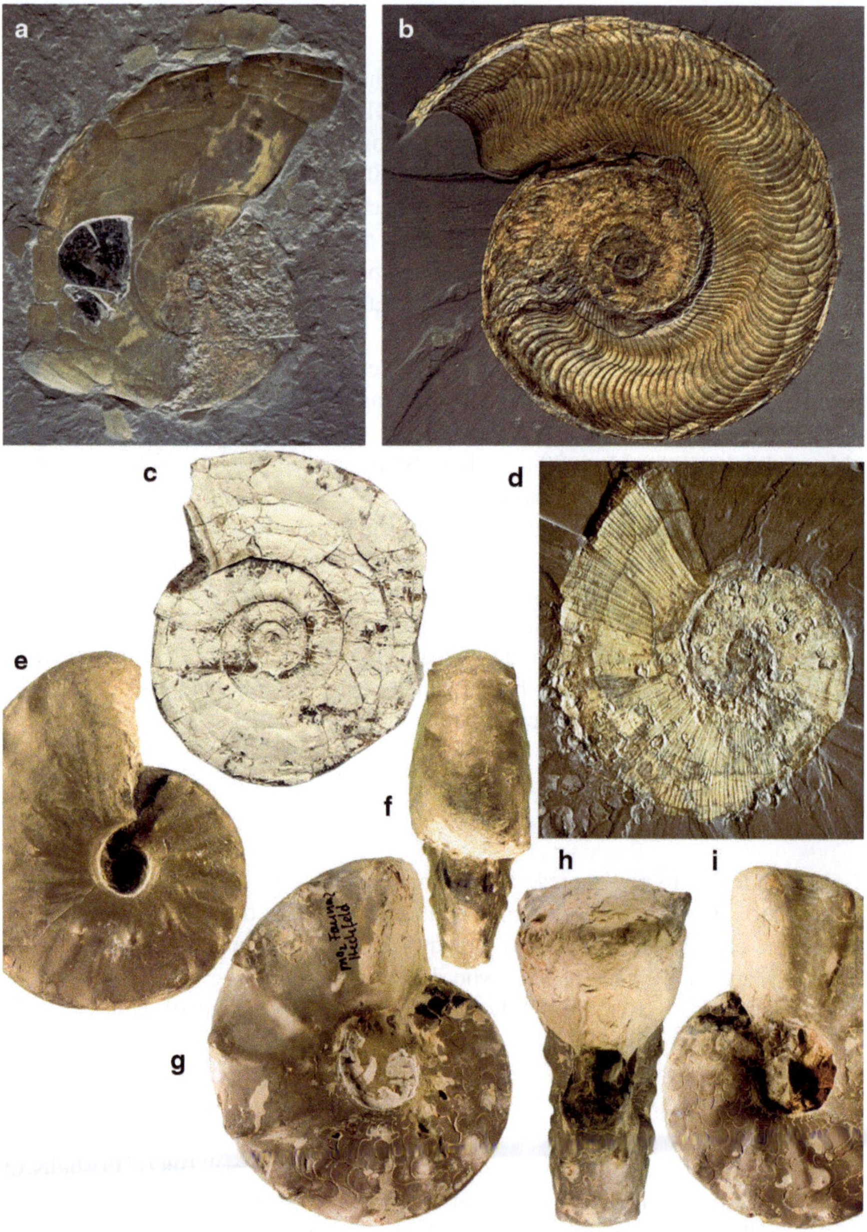

Fig. 7.3 Black layer and black band in Mesozoic ammonoids (**a–d** from Klug et al. 2007; **e–i** from Klug et al. 2004). **a** *Phylloceras heterophyllum*, SMNS 26462, tenuicostatum Zone, Ohmden, Germany; dm 87 cm. Note the jaws and the black band. **b** *Harpoceras falciferum,* falciferum-bifrons Zone, Holzmaden, dm 24 cm. **c** *Psiloceras planorbis*, PIMUZ 6519; planorbis Zone, Hettangian, Blue Anchor, Somerset, UK, dm 45 mm. Note the black band and black layer. **d** *Lytoceras ceratophagum,* SMNS 26465; falciferum Zone, Toarcian, Ohmden, Germany; dm 41 cm. **e, f** *Paraceratites atavus*, lateral and dorsal views, SMNS 24503, atavus Zone, Neckarrems, col. M. Warth; dm 61 mm. **g–i**, *Ceratites spinosus*, lateral and dorsal views, SMNS 25255–33, spinosus Zone, Heckfeld; dm 102 mm. Images: **a, b** Staatliches Museum für Naturkunde Stuttgart. **d** Urweltmuseum Hauff. **c** T. Galfetti. **e–i** W. Gerber

nian gephuroceratids (Keupp 2000), Triassic ceratitids (Klug et al. 2004), as well as Jurassic ammonites (Klug et al. 2007) and is considered as either homologous or convergent with the structure in nautilids. Of the criteria of homology, only those of position and specific structure are fulfilled, since transitional states are missing (Klug et al. 2004). Nevertheless, it is likely that the anterior edge of the black layer served for the attachment of the dorsal mantle.

7.2.2.10 The Black Band

The black band is a thin organic coating on the shell, which forms a narrow band surrounding the adult aperture in some cephalopods; it is sometimes found in modern nautilids (Ward 1987) and rarely in ammonoids (Klug 2004; Klug et al. 2007). Like the black layer, it is black due to its melanin content. So far, it has been found in *Psiloceras*, *Phylloceras*, *Lytoceras*, and *Harpoceras* from the Early Jurassic (Fig. 7.3) and two questionable specimens from the Triassic (Klug 2004; Klug et al. 2007). It is apparently linked with the adult cessation of growth and in modern nautilids, the black band has been recorded from adult females, although not all individuals appear to develop this structure. Like the black layer, the black band is probably also linked with mantle attachment at the aperture.

Davis (1972), summarized in Davis et al. (1996, p. 469), found indications for "*an actual change in the nature of shell deposition late in ontogeny*". Accordingly, the shells of adult *Adrianites* and other Permian ammonoids displayed small "*pits in the internal mold*". Davis et al. (1996) suggested that these structures are possibly homologous to the apertural attachment of the mantle at the black band in mature modern nautilids.

7.2.2.11 The Wrinkle Layer

The wrinkle layer is a structure of uncertain function that occurs in a number of ammonoids in the form of irregular shell wrinkles in the dorsal part of the shell, usually near maturity (Barrande 1877; House 1970; Walliser 1970; Senior 1971; Doguzhaeva 1981; Kulicki et al. 2001). Strength of the wrinkle layer is a character that is rarely preserved and thus of limited use. The wrinkle layer is predominantly found in nearly adult or fully mature specimens (Korn et al. 2014). Kulicki et al. (2001) already pointed out that the wrinkle layer might be comparable or even homologous to the black layer of nautilids.

7.2.3 Constructional and Functional Morphology

Some of the mature modifications of ammonoid shells are so profound that it is hard to imagine that the altered adult morphology did not affect the life style of the ammonoid and thus their evolution. Recurrent morphologies (such as convergent evolution of apertural lappets) support this hypothesis. Several suggestions have been made with regard to functional as well as constructional interpretations of the modified adult morphology (Tajika et al. 2014): (1) change of habitat (Davis et al. 1996); (2) defense against predators (Keupp and Riedel 2010); (3) sexual display (Keupp and Riedel 2010); (4) attachment of reproductive organs (a modified arm; Landman et al. 2012); (5) change in locomotion/behavior (Klug 2001); (6) fast and metabolically economic construction of the terminal shell segment; (7) fabricational noise with a lack of function (Seilacher 1974).

(1) The change of habitat did possibly occur since the relative abundance of macroconchs or microconchs varies between localities. Especially in such cases with a large difference in adult size, it might have been important that the sexes stayed separate until the time of mating in order to reduce the time of exposure to potential "*sexual*" cannibalism (Hanlon and Forsythe 2008; Keupp and Riedel 2010). Nevertheless, this hypothesis is difficult to test.

(2) Many ammonoid species reinforced their terminal apertures by shell thickenings (e.g., *Agoniatites*, *Arcestes*, *Manticoceras*). Even constrictions without shell thickenings might have increased the resistance of the aperture against breakage by predators (Landman and Waage 1986; Keupp and Riedel 2010; Keupp 2012). It is, however, not possible at this point to test whether these modifications are effects of the terminal deceleration of growth or whether they represented antipredatory adaptations. Seilacher (1974) argued against such a function in the microconchs because in his opinion, such defensive structures would be more meaningful in the females.

(3) and (4) are nice ad hoc hypotheses and are difficult to test, especially with the lack of knowledge of the soft parts in general and the reproductive organs in particular. The extreme differences between some antidimorphs (especially *Phlycticeras* and *Oecoptychius*) suggest a comparison to the modern octobrachian *Argonauta*, in which the male measures only 2 cm in length, while the female may reach over 40 cm, when the shell is included. The male of *Argonauta* has a modified arm (hectocotylus), which is stored in a ventral sac prior to mating. This structure is somewhat reminiscent of some of the apertural modifications. Landman et al. (2012) hypothesized that ammonites with a high aperture angle had this type of arm to improve mating efficiency. Nevertheless, direct evidence for such a convergence is still missing.

(5) Klug (2001) showed that the whorl expansion rate in Middle Devonian ammonoids changed close to the cessation of growth. He argued that this change in whorl expansion rate was linked to a change in body chamber length, which, in turn, caused a change in the *syn vivo* orientation of the shell. In the main lineages of Devonian ammonoids, the adult aperture would have moved to a more horizontal position than in preceding growth stages (see also Korn and Klug 2002; Klug and Korn 2004). The latter authors concluded that this change in shell orientation improved mobility and manoeuvrability, both valuable traits to find a mating partner and good spawning grounds.

Tajika et al. (2014) empirically tested the effect of apertural lappets in the Middle Jurassic *Normannites* and found that absence or presence of these lappets would not have altered the shell orientation significantly. Therefore, these lappets most likely did not serve the function of altering the shell orientation.

(6) Microconchs of the Haploceratoidea and Perisphinctoidea produced a pair of lateral projections or lappets in the peristome (Fig. 7.2), and most frequently their body chambers are shorter than those of macroconchs. Thus, the pair of lappets could be interpreted as the terminal shell segment for accommodating the cephalic portion of the animal. This "*shell segment*" could have been secreted rapidly and economically considering the low amount of aragonite necessary compared to a complete "*tubular*" shell segment. The muscle scars around the peristome of the haploceratoid *Hecticoceras brigthii* (see above; Palframan 1969) would indicate additional muscle development providing for support and mobility of the cephalic portion of the body.

7.3 Dimorphism

As far as the history of research on ammonoid dimorphism is concerned, we only want to mention briefly that de Blainville (1840) and d'Orbigny (1847, p. 441) were probably the first to discuss sexual dimorphism in ammonoids (see also Foord and Crick 1897; Haug 1897). In any case, broader interest in the topic grew in with the important monographs of Makowski (1962) and Callomon (1963).

Commonly, there is a smaller and a larger form in those taxa in which dimorphism is more apparent due to clear differences combined with equally visible shared juvenile characters. For these, Callomon (1955) introduced the widely used terms microconch and macroconch, respectively. These are called antidimorphs.

7.3.1 Monomorphism, Dimorphism, and Polymorphism

Adults of the two sexes of any given animal species may have similar or different shapes. If they are monomorphic, there is no significant difference in adult shape. In the case of dimorphism, two different adult morphologies can develop from morphologically similar juveniles (Davis 1972). In ammonoids, these are traditionally called microconchs (for the smaller variant) and macroconchs for the larger variant. The corresponding pairs are named antidimorphs. In the case of polymorphism, there are more than two (usually three or more) different adult morphologies. Polymorphism in modern biology refers to natural genetic variation (with phenotypic expression or not), undetectable in fossils; here, we use the term to refer to morphological differences between supposed conspecific phenotypes, which could have either a genetic or an environmental cause (compare De Baets et al. 2015a).

As did our forerunners (Davis et al. 1996), we will not repeat all details of the research on dimorphism from its beginnings in the nineteenth century (de Blainville

1840; Orbigny 1847). Instead, we recommend looking up these details in the excellent monographs by Makowski (1962) and Callomon (1963).

There are numerous articles dealing with polymorphism (McCaleb and Furnish 1964; McCaleb et al. 1964; Ivanov 1971, 1975, 1985; Kant 1973; Hirano 1978, 1979; Matyja 1986, 1994; Makowski 1991; Melendez and Fontana 1993). According to these authors, some Jurassic and Cretaceous ammonoids produced more than two forms. After Ivanov (1971, 1975, 1985) had dubbed exceptionally large forms "*megaconchs*", Matyja (1986, 1994) introduced the term "*miniconchs*" for exceptionally small specimens. He suggested that certain environmental parameters controlled the point of maturation, inducing monomorphism, dimorphism or polymorphism. In his work on modern coleoids, Mangold-Wirz (1963), Mangold-Wirz et al. (1969), as well as Mangold (1987) demonstrated how hormones produced by the optic gland can control the timing of maturation and thus size depending on the developmental state of the gonads. It was also demonstrated for Recent coleoids in captivity that environmental factors such as light intensity, temperature or food availability can have an effect on maturation and therefore adult size (e.g., Gabr et al. 1998; Moltschaniwskyj and Martínez 1998; Tafur et al. 2001; Jackson and Moltschaniwskyj 2002). Callomon (1988) criticized Matyja's ideas about polymorphism, suggesting that a larger database would be needed to test some of his hypotheses (see also De Baets et al. 2015a). Later, Dzik (1990a), analysing a rich collection of Callovian *Quenstedtoceras* ammonites from the classic locality at Łuków in eastern Poland, did not find any evidence for polymorphism.

7.3.2 Classification of Dimorphism

Bearing in mind the vast diversity and impressive variability of ammonoids, it is not surprising that dimorphism is far from uniform within this group. Consequently, various authors have attempted to meaningfully classify ammonoid dimorphism. In his pioneer monograph, Makowski (1962) introduced two kinds of dimorphs. In Type A, the microconch has five (four) to six whorls and the macroconch has seven (six) to nine whorls. Type B microconchs have seven (six) to nine whorls and the macroconchs have one additional whorl. Guex (1968) added Type O, where the microconch has three to four whorls.

Westermann (1964a) and Houša (1965) also differentiated between two types of dimorphism, where one type differs only in size, while the other differs in size and other characters, especially in the shape of the peristome. Zeiss (1969) added a third group to these two, in which dimorphism was not recognized. This leads to the question, whether dimorphism is the rule and that it only can sometimes not be identified due to taphonomic processes (loss of soft-tissues and subtle conch characters). In that case, a lot of work would await ammonoid researchers, because many more cases of dimorphism would await their detection.

7.3.3 Criteria for Dimorphism

In order to verify the hypothesis of conspecificity of two or more different adult forms, the following criteria (Makowski 1962; Callomon 1963, 1981; Westermann 1964a; Davis 1972; Davis et al. 1996) should be fulfilled:

1. The antidimorphs should differ in adult morphology;
2. They should have more or less identical early developmental stages;
3. They should occur in strata of the same stratigraphic range;
4. They should have overlapping geographic occurences;
5. They should have the same ancestors;
6. The ratio of numbers of micro- to macroconchs should be about the same through time and throughout the evolution of their clade.

Most ammonoid workers would agree on points (1) and (2). However, a few exceptions exist. McCaleb (1968) stated that in the Late Carboniferous *Syngastrioceras oblatum*, the morphological differences between macro- and microconch are larger in juvenile/preadult growth stages than in the mature stage/the last whorl. Similarly, Rein (2001, 2003) suggested that species of the Middle Triassic genus *Ceratites* show a similar morphological separation of the antidimorphs. He introduced the terms E- (referring to the smooth species *C. enodis*) and P-morph (referring to the strongly ornamented species *C. posseckeri*) for forms with smooth or strongly ornamented preadult whorls. Although further work on this issue would be welcome, we would like to point out that the coiled shells of many mollusks display the highest degree of intraspecific variability in preadult whorls (e.g., Urdy et al. 2010a, b; De Baets et al. 2015). Therefore, the question arises whether these two exceptions, where supposedly the middle whorls differ in antidimorphs instead of the adult morphology, are artifacts from normal intraspecific variability (Urlichs 2009).

Davis et al. (1996) pointed out that differences in geographical occurrences (point 4) of the antidimorphs would not be surprising since their differing morphologies might reflect differing ecological requirements. This might hold true for parts of their life but at least at some point, males and females had to meet in order to reproduce. It is still conceivable that the intersexual differences in behaviour and habitat varied between species, when the extreme differences in dimorphism throughout the ammonoid clade are taken into account.

For various reasons, the morphologic evolutionary rates among the microconch part of a lineage may seem (1) higher (Lehmann 1981; Davis et al. 1996) or (2) lower (e.g., Callomon 1969, Westermann and Riccardi 1979: p. 134) than those of the macroconchs. In the first case, this might be a primary signal, i.e. the microconchs evolved morphologically faster because the mature modifications were directly prone to sexual selection or of great importance for reproduction. Alternatively, this seeming difference in evolutionary rates might be an artifact because the microconchs might attract more attention due to their peculiar morphology, or because evolutionary change is easier to track in microconchs since they display more distinct morphological character states. In any case, these differences in morphological evolution between antidimorphs may hamper evolutionary studies. In studies of dimorphism, it is important to know the phylogenetic framework of the ammonoid lineage under consideration (point 5), because this knowledge optimally contains information on the development of ancestors and other members of the clade, as well as plesiomorphies and degrees of conservativeness of traits. Finally, this phylogenetic test is needed to falsify the hypothesis that the antidimorphs under consideration indeed belong to two separate species.

In the second case, the slower morphologic evolutionary change of the microconchs with respect to the macroconchs produced the opposite pattern, like a morphological stasis of the males. This pattern is produced in lineages where the main morphologic changes developed in the subadult and/or adult ontogeny of the macroconchs. The microconchs typically stop their growth in the early ontogeny of the species, thus not reflecting the changes seen in the macroconchs.

Of course, there may be traits that are not preserved or only rarely or poorly preserved, which could potentially be used to discriminate between antidimorphs, where shell characters alone do not suffice. Till (1909, 1910) searched for dimorphic characters in the jaws, while Parent et al. (2013, p. 32) found evidence of sexual dimorphism in the aptychus (*Praestriaptychus*) of *Lithacoceras* [M]/*Silicisphinctes* [m]. Mapes and Sneck (1987) found two kinds of colour patterns in *Owenites*. Nothing is known about differences in the soft part anatomy between the antidimorphs and we can only hope that one day, exceptionally preserved soft-tissue ammonoids will be discovered, shedding more light on the internal organisation of ammonoids.

7.3.4 Sexing of Ammonoid Antidimorphs

For some, sexing of ammonoid shells seems a trivial task and it happens quite commonly that the microconch is automatically considered the male. This confidence is surprising because the ultimate evidence, namely soft-tissue preservation of reproductive organs in the antidimorphs, is still missing. The background for this slightly premature conclusion is probably the actualistic comparison with some Recent octobrachians. In *Argonauta*, which was already mentioned above, the size differences are just as striking as in *Ocythoe tuberculata*, where the female is 1 m long and so ten times as long as the male (Makowski 1962; Wells 1962, 1966; Mangold-Wirz 1963; Westermann 1969a; Mangold-Wirz et al. 1969; Roper and Sweeney 1975). By contrast, the male is slightly larger in Recent Nautilida (Willey 1895, 1902; Saunders and Spinosa 1978; Saunders and Ward 1987; Hayasaka et al. 1987), but they have a different reproductive strategy. Remarkably, aptychi have been interpreted as protecting the nidamental glands (Keferstein 1866) and Siebold (1848) even suggested that the aptychi were the micromorphic males. Nowadays, there is not much doubt that the aptychi were part of the buccal apparatus and had nothing to do with reproduction.

Numerical ratios between the antidimorphs were another line of evidence that has been explored to assign sexes to each of them. Davis et al. (1996) gave an overview of the contradicting results of various authors who worked on ammonoids or on Recent cephalopods (Willey 1902; Coëmme 1917; Pelseneer 1926; Mangold-Wirz 1963; Makowski 1962; Mangold-Wirz et al. 1969; Westermann 1969a; Saunders and Ward 1987; Hayasaka et al. 1987). In ammonoids, the results are biased by the fact that the numerical ratios are influenced by facies and taphonomy (e.g., Callomon 1981, 1985). The most plausible line of reasoning appears to be that of Lehmann (1981), who inferred that the macroconchs were the females because the maturation of eggs takes longer than that of spermatophores, implying a longer lifespan and thus a larger adult size. Moreover, reproductive organs of females (e.g., ovaries, nidamentary glands) are often larger than the simpler reproductive organs of males.

Ammonoid eggs have been reported by several authors (Lehmann 1966; Müller 1969; Zakharov 1969; Maeda 1991; Urlichs 2009; Etches et al. 2009; Landman et al. 2010; Klug et al. 2012). In spite of these findings, there is as yet no report of a discovery of eggs within an ammonoid shell that is free of doubt (De Baets et al. 2015b). Either the preservation is insufficient to detect whether they are truly ammonoid eggs or it is unclear if such an egg mass is really *in situ* within the ammonoid.

7.3.5 Development and Dimorphism

To detect dimorphism in ammonoids, knowledge of their ontogeny is needed (see the criteria for dimorphism). Developmental heterochronies have been suggested as processes generating the differences between the antidimorphs (Gould 1977; Shea 1986). Davis et al. (1996) discussed whether microconchs were accordingly progenetic and/or hypomorphic (Landman et al. 1991; Neige 1992).

In any case, many antidimorphs display a congruent pattern of development of various shell parameters, which diverge at some point with the microconchs maturing and stopping growth at a size-wise earlier point (e.g., Makowski 1962; Guex 1973; Parent 1997; Parent et al. 2008b, 2009).

Plotting certain shell parameters throughout ontogeny (versus diameter) is an important and powerful method to demonstrate dimorphism. The graphs in Fig. 7.4 illustrate some of the patterns and features discussed above from a moderately large sample of the Early Callovian sphaeroceratid *Eurycephalites gottschei*. The growth rate measured by the relative whorl ventral height is dimorphic (Fig. 7.4a upper) with an increase in both dimorphs up to about 10 mm diameter, after which both dimorphs inverted the trend; from a diameter of about 18 mm onward, the microconch diverged by increasing the rate of growth towards the peristome (Parent 1997). The shape of the whorl cross section changed strongly during growth, but the ontogenetic trajectory is the same in micro- and macroconchs, i.e., monomorphic (Fig. 7.4a bottom). The ventral ribbing is another feature that is dimorphic with a similar trend to that found in the growth rate (Fig. 7.4b). The microconch trend diverged from a diameter of 18 mm compared to that of the macroconch, which is taken as the standard.

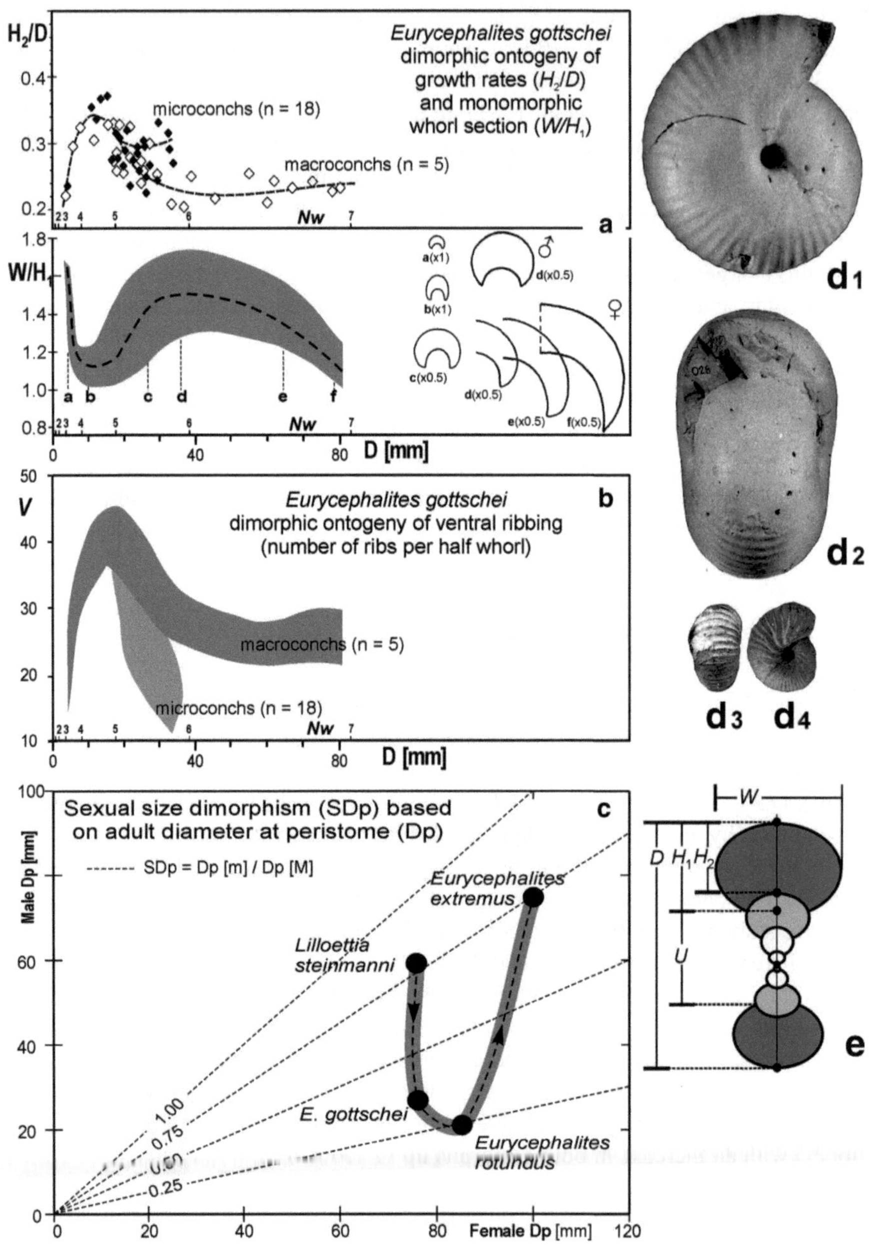

Fig. 7.4 Dimorphism of *Eurycephalites gottschei* (Tornquist 1898) [Sphaeroceratidae]. **a** (*upper*): dimorphic ontogeny of the growth rate measured as H_2/D versus diameter (D) and whorl number (Nw); **a** (*lower*), monomorphic ontogeny of the whorl section measured by W/H_1 versus D and Nw (*grey area*) and selected cross-sections (**a–f**). **b** dimorphic ontogeny of ventral rib number per half whorl (V) versus D and Nw. **c** evolution of the sexual size dimorphism in the lineage *Lilloettia* (late Bathonian)—*Eurycephalites* (Early Callovian). **d** dimorphic pair, a complete adult macroconch ($\mathbf{d_1}$–$\mathbf{d_2}$) and an adult microconch ($\mathbf{d_3}$–$\mathbf{d_4}$), both x0.4. **e** shell size dimensions. **a–b, d–e** modified from Parent (1997). **c** from data in Parent (1998)

7.3.6 *Evolution of Dimorphism*

Evolution of sexual dimorphism is a complex and interesting aspect in ammonoid paleobiology. However, it is also very demanding with respect to material and raw data. One of the most detailed studies is that of Schweigert and Dietze (1998) for the evolution of *Phlycticeras/Oecoptychius*. Sexual size dimorphism, the feature more readily captured by the observer, shows important changes during the evolution of many ammonoid lineages. One of them is the lineage *Lilloettia* (Late Bathonian)-*Eurycephalites* (Early Callovian), which shows important changes in the ratio of the adult micro- versus macroconch size (Fig. 7.4c), ranging from microconchs of three-quarters (*L. steinmanni* and *E. extremus*) to one third (*E. gottschei* and *E. rotundus*) the size of the macroconch.

7.3.7 *Occurrences of Dimorphism*

From the Hettangian onwards, Guex (1981) stated that, except for the phylloceratids, the majority of ammonites do show dimorphism. It thus appears that at least in the Jurassic and Cretaceous, whether dimorphism is detected or not largely depends on quality and quantity of the material plus the motivation of a researcher to quantitatively analyze an ammonite lineage. Presuming that the dimorphism of ammonoids really represents the different sexes, a more or less omnipresent dimorphism appears not so surprising, especially because most modern cephalopods also show more or less strong dimorphism.

In contrast to the work of Davis et al. (1996), we will not list mature modifications in great taxonomic detail in the text below, because it appears that the majority of ammonoids actually did undergo some kind of terminal growth and thus produced mature modifications. It is striking that strong apertural modifications became common only in the Jurassic, although a few Permian species (Zhao and Zheng 1977), as well as some Paleozoic nautilids (e.g., Dzik 1984) did produce strong apertural appendages.

7.3.7.1 Palaeozoic Dimorphism

Devonian: For the earliest, loosely coiled ammonoids, De Baets et al. (2013a) tested for sexual morphism in Moroccan *Erbenoceras* and *Anetoceras* and found no clear indication of it. It could be argued that dimorphism might be camouflaged in the strong intraspecific variability (Kakabadze 2004 discussed dimorphism in relation to variability) of this group, but among these two genera, the intrageneric and probably also intraspecific variability is markedly reduced toward the end of growth. Not much has been published on Devonian dimorphism after Makowski (1962) had listed several cases (Table 7.1). Walliser (1963) only shortly mentioned its existence without any detail. Later, Makowski (1991) determined the relative abundance of

Table 7.1 Sexual dimorphism in Devonian to Triassic ammonoids

Superfamily	Family	Genus	Age	Source
Tornoceratoidea	Tornoceratidae	*Tornoceras*	Famennian	Makowski 1962
Cheiloceratoidea	Cheiloceratidae	*Cheiloceras*	Famennian	Makowski 1962
	Gephuroceratidae	*Manticoceras*	Frasnian	Makowski 1962
Ceratitoidea	Acrochordiceratidae	*Acrochordiceras*	Anisian	Dzik 1990b
	Ceratitidae	*Ceratites*	Anisian, Ladinian	Müller 1969; Rein 2001, 2003; Urlichs 2009

micro- versus macrococonchs in various species of *Tornoceras*. In *T. frechi parvum,* he found 28 % macroconchs ($n=95$), in *T. subacutum* there were 47 % macroconchs ($n=65$), and in *T. sublentiforme*, macroconchs varied between 40 % ($n=133$) and 45 % ($n=95$).

Most authors have focused on descriptions of adult modifications (Table 7.2; e.g., Ruzhencev 1962; Korn 1992). Septal crowding has been mentioned commonly (e.g., Korn and Titus 2006; Ebbighausen and Korn 2007; Kraft et al. 2008), elliptical coiling is common in some clymeniids (e.g., Ebbighausen and Korn 2007), increasing umbilical width is characteristic for the earliest, loosely coiled ammonoids (e.g., De Baets et al. 2013a, b), and, of course, changes in ornament spacing have been documented (Fig. 7.5).

Carboniferous: Remarkably, we have not found an unequivocal report on ammonoid dimorphism in the Carboniferous (Nettleship and Mapes 1993). Davis et al. (1996) mentioned the work of Trewin (1970), who suggested that *Eumorphoceras* produced antidimorphs, but he used poorly preserved materials. Frest et al. (1981) examined Late Pennsylvanian *Maximites oklahomensis* and found that 40 % of the examined specimens belonged to "*form a*", which might be the macroconch according to its less strong adult modifications (Davis et al. 1996).

In some taxa, a strong wrinkle layer was secreted when the specimen approached maturity (Korn et al. 2014 and references therein). Septal crowding is also not rare (e.g. Korn et al. 2010), although we have to repeat that its value to determine maturity is limited (Kraft et al. 2008). Umbilical egression (Fig. 7.5; e.g., Frest et al. 1981) and other changes in coiling (e.g., Ruzhencev 1962) also occur in Carboniferous forms, which are visible in some of the cross sections figured in Korn et al. (2010). Ruzhencev (1962) illustrated *Dombarites*, *Homoceras*, and *Praedarelites*, which formed ventral keels near maturity.

Permian: Although many Permian ammonoids are known to have formed distinct mature modifications (Table 7.3, Fig. 7.5; Miller and Furnish 1940; Miller 1944; Ruzhencev 1962; Davis 1972; Zhao and Zheng 1977; Frest et al. 1981; Zhou 1985; Schiappa et al. 1995), not much has been published, suggesting the presence of dimorphism of the shell. Davis et al. (1969, 1996) and Davis (1972) counted the specimens of Permian *Agathiceras uralicum* and found 75 % macroconchs ($n=110$). Table 7.3 lists mature modifications (modified from Davis et al. 1996).

Table 7.2 Mature modifications in Devonian and Carboniferous ammonoids

Taxon	Coiling	Constrictions	Ornament	Whorl section	Wrinkle layer	Septal crowding	Suture simplified	Reference
Anetoceratidae								
Anetoceras	•		•			•		De Baets et al. 2013a, b
Erbenoceras	•		•			•		De Baets et al. 2013a, b
Anarcestidae								
Sellanarcestes	•			•		•		Klug 2001
Agoniatitidae								
Agoniatites	•	•	•			•		Klug 2001
Gephuroceratidae								
Manticoceras	•	•						Korn and Klug 2007
Prolobitidae								
Prolobites	•	•	•	•		•		Bogoslovsky 1969
Wocklumerioidea								
Wocklumeria	•	•				•	•	Ebbighausen and Korn 2007
Kamptoclymenia	•	•	•					Schindewolf 1937
Pericyclidae								
Oaoufilalites	•					•		Korn and Ebbighausen 2008
Maxigoniatitidae								
Maxigoniatites	•				•			Korn et al. 1999
Girtyoceratidae								
Calygirtyoceras	•							Korn et al. 1999

Since the account of Davis et al. (1996) appeared, not a lot of new data have been published and it appears that these are still insufficient to analyze evolutionary trends in Paleozoic dimorphism. Accordingly, McCaleb's (1968, p. 29) statement that "*dimorphism is a predominant feature at the inception of an evolutionary lineage and decreases through phylogeny*" appears premature at best, if not wrong.

7.3.7.2 Triassic Dimorphism

Davis et al. (1996) listed mature modifications in Triassic ammonoids in relation to their shell shape (Table 7.4, Fig. 7.6). This information is summarized in Table 7.4. In spite of the sometimes quite conspicuous adult modifications and the incredible diversity as well as morphological disparity of Triassic ammonoids, reliably dem-

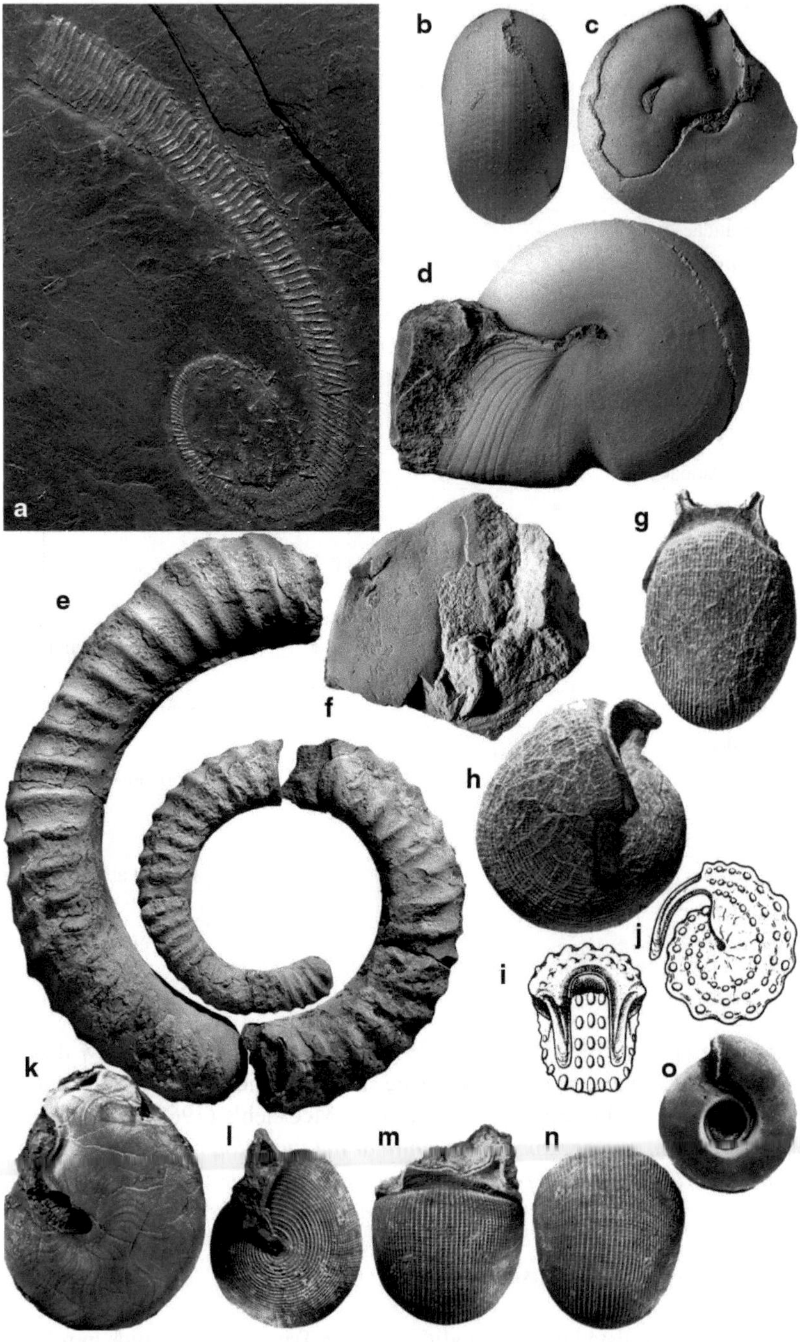

Fig. 7.5 Paleozoic mature modifications. **a** *Metabactrites fuchsi*, PWL2010/5251-LS, middle Kaub Formation, Hunsrück, Bundenbach, from De Baets et al. (2013b), dm 80 mm. Note the changes in coiling and ribbing. **b–d** *Prolobites aktubensis*, col. Ademmer, Kattensiepen, Germany,

Table 7.3 Mature modifications in Permian ammonoids. (modified after Davis et al. 1996)

Taxon	Coiling	Constrictions	Ornament	Lappets	Whorl section	Wrinkle layer	Thick septum	Septal crowding	Suture simplified
Adrianitidae									
Adrianites	●	●		●	●	●			
Crimites	●	●			●				
Epadrianites	●	●							
Hoffmannia	●	●	●		●				
Neocrimites	●	●	●	●					
Palermites	●	●		●	●				
Pseudagathic	●	●	●						
Sizilites	●	●	●		●				
Texoceras	●	●				●			●
Cyclolobidae									
Cyclolobus	●	●		●	●				
Mexicoceras	●	●			●		●		●
Waagenoceras	●	●			●	●		●	●
Hyattoceratidae									
Hyattoceras	●	●	●		●				
Marathonitidae									
Marathonites	●	●			●				
Pseudovidrioc		●	●		●				●
Vidrioceratidae									
Peritrochia		●		●					
Stacheoceras	●	●		●	●				
Agathiceratidae									
Agathiceras	●	●	●	●	●				
Pseudohaloritidae									
Elephanticeras		●	●	●					
Sangzhites	●	●	●	●	●				
Shangraoceras	●	●		●	●				

from Korn and Klug (2002). **b, c** SMF 34694. lateral and ventral view; note the elliptical constriction (internal shell projection) and the strong subterminal constriction; dm 21 mm. **d** SMF 34691; note the change in coiling, lirae spacing and the constriction; dm 28 mm. **e** *Erbenoceras advolvens*, early Esmian, Tafilalt, Morocco, from De Baets et al. (2013b); dm 156 mm; note the change in coiling. **f** *Fidelites* sp., GPIT 1862-133, *costatus* conodont Zone, Eifelian, Tafilalt, Morocco; with broad terminal constriction, from Klug (2001). **g–l** Permian ammonoids from Davis et al. (1996). **g, h** *Adrianites* sp., lateral and ventral views, GIUA Drawer 55, T328, Maoen Mollo, Timor, Indonesia, dm 26 mm. **i, j** Elephantoceras sp., Permina, from Zhao and Zheng (1977). **k** *Cyclolobus walkeri*, MNHN B 7520, Ankitohazo, Madagascar; dm 93 mm. **l-n** *Adrianites* cf. *insignis*, BMNH C37654, Sosio Limestone, Province of Palermo, Italy; dm 21 mm. **o**, *Wocklumeria sphaeroides*, nr. 572, Famennian, Kowala, Poland, from Czarnocki (1989)

Table 7.4 Mature modifications in Triassic ammonoids. (modified after Davis et al. 1996)

Taxon	Shell shape	Coil-ing	Elliptical coiling	Constric-tions	Orna-ment	Apertural extensions	Whorl section	Thick shell	Septal crowding	Occluded umbilicus	Reference
Longobarditidae											
Intornites	Oxycone	•			•		•	•	•	•	Hammer and Bucher 2006
Beyrichitidae											
Eogymnotoceras	Discocone				•		•		•		Tozer 1994
Favreticeras	Discocone				•		•		•		Bucher 1992
Balatonitidae.											
Platycuccoceras	Platycone				•				•		Davis et al. 1996
Kashmiritidae											
Kashmirites	Serpenticone				•				•		Davis et al. 1996
Ussuritidae											
Palaeophyllites	Serpenticone				•				•		Davis et al. 1996
Haloritidae											
Halorites	Sphaerocone	•	•	•	•		•		•		Mojsisovics 1893
Homerites	Sphaerocone	•	•	•	•	•			•		
Arcestidae											
Amphipopanoceras	Sphaerocone	•		•			•		•		Davis et al. 1996
Arcestes	Sphaerocone	•		•		•	•	•	•		Davis et al. 1996
Tropitidae											
Tropites	Cadicone	•		•	•		•		•		Mojsisovics 1893
Ceratitidae											
Ceratites	Discocone	•			•		•		•		Wenger 1957
Discoceratites	Oxycone	•			•		•		•		Wenger 1957

Fig. 7.6 Triassic mature modifications. **a**, **b** *Halorites macer*, PIMUZ 31068, Norian, Bihati?, Timor, Indonesia, dm 103 mm. Note the change in whorl section. **c**, **d** *Orestites* cf. *frechi*, PIMUZ 31069, Norian, Timor, Indonesia, dm 75 mm. Elliptical coiling and constricted aperture. **e**, **f** *Lobites ellipticus*, PIMUZ SQL43399, Norian, Timor, Indonesia, dm 62 mm. **g** *Arcestes* sp., PIMUZ 31067, Norian, Timor, Indonesia, dm 152 mm. **h** *Halorites* cf. *phaonis*, PIMUZ 31066, Norian, Timor, Indonesia, dm 124 mm. **i**, **j** *Ceratites* cf. *compressus*, MHI 688732, Anisian?, Garnberg, Germany, dm 75 mm. **k**, **l** *Protrachyceras archelaus*, from Mojsisovics (1882), Ladinian, Agordo, Italy, dm 132 mm

onstrated cases of dimorphism are exceedingly rare. A first report was published by Müller (1969), who found a questionable egg mass in the shell of *Ceratites*. This specimen is undoubtedly interesting, but it has neither been proven that the preserved globules are eggs nor that they are part of this ceratitid. It is also questionable because no additional soft parts are preserved in this specimen.

Similarly, the report of Dzik (1990b) of acrochordiceratid antidimorphs is doubtful, because the species of this family have been shown to be highly variable (Monnet et al. 2010). By contrast, the account of dimorphism in Middle Triassic *Ceratites* by Urlichs (2009) appears to represent one of the first profound accounts of Triassic dimorphism. He collected adult specimens and measured their sizes in populations in combination with morphometric data from juvenile to adult whorls. Thereby, he could show that the antidimorph's juvenile whorls are quite similar and they begin to diverge morphologically late in ontogeny, with clearly separated adult sizes and conch parameters. Probably, many more cases of dimorphism will be detected among Triassic ammonoids when well-preserved materials are carefully analyzed for this aspect.

7.3.7.3 Jurassic Dimorphism

Jurassic ammonoids probably contain some of the most convincing and most impressive, as well as famous, examples of sexual dimorphism (Table 7.5, Fig. 7.7). For example, Makowski's (1962) kosmoceratids became the icon of the journal *Acta Palaeontologica Polonica* and the impressive combination of extreme size difference, as well as the exotically modified aperture in the dimorphic pair *Phlycticeras* and *Oecoptychius* made them well-known among collectors (Fig. 7.8, 7.9).

Several excellent monographs on Jurassic dimorphism are readily available (Makowski 1962; Callomon 1963, 1981; Tintant 1963; Westermann 1964a, b; Elmi 1967; Davis et al. 1996). It is thus not necessary to repeat all their results. At this occasion, however, we want to summarize the categories (Fig. 7.10), which were introduced by Davis et al. (1996):

Category A: Sexual dimorphism has been shown convincingly, applying the criteria of different adult morphologies, close phylogenetic relationship reflected in their early ontogenies, shared geographic as well as stratigraphic occurrences and shared habitats.

Category B: Davis et al. (1996) grouped forms with dubious sexual dimorphism here. In many of the species included in this category, their phylogenetic relationship may be unclear, geographic ranges may differ, and preadult ontogenetic stages are either poorly known, do not match perfectly, or lack diagnostic characters (Ziegler 1974, 1987).

Category C: In this category, species are included that either appear as monomorphic (i.e., minute or no morphological differences in the shells of the sexes) or where the preservation or other factors make the assignment to an antidimorphic pair impossible. Davis et al. (1996) included several genera with strong apertural modifications such as *Gemmellaroceras* and *Cymbites*.

Table 7.5 Sexual dimorphism in Jurassic ammonoids (incomplete list)

Superfamily	Family	Genus	Age	Source
Psiloceratoidea	Psiloceratidae	*Neophyllites*	Hettangian	Guex 1981
		Psiloceras	Hettangian	Guex 1981; Hillebrandt and Krystyn 2009
		Discamphiceras	Hettangian	Guex 1981
		Badouxia	Hettangian—Sinemurian	Longridge et al. 2006
	Schlotheimiidae	*Kammerkarites*	Hettangian	Guex 1981
		Saxoceras	Hettangian	Guex 1981
		Schlotheimia	Hettangian	Guex 1981
Eoderoceratoidea	Liparoceratidae	*Liparoceras*	Pliensbachian	Callomon 1963; Callomon and Gradinaru 2005
	Dactylioceratidae	*Coeloceras*	Toarcian	Dommergues 1993
		Gabillytes/Zugodactylites	Toarcian	Guex 1973
		Mucrodactylites/Catacoeloceras	Toarcian	Guex 1973
		Collina/Porpoceras	Toarcian	Guex 1973
Hildoceratoidea	Hammatoceratidae	*Hammatoceras/Onychoceras*	Toarcian	Guex 1967
		Erycitoides	Aalenian	Howarth 2013
		Spinammatoceras	Aalenian	Howarth 2013
		Podagrosiceras	Bajocian	Howarth 2013
	Phymatoceratidae	*Haugia*	Toarcian	Keupp and Riedel 2010
	Hildoceratidae	*Tmetoceras/Tmetoites*	Aalenian	Westermann 1964b
		Leioceras	Aalenian	Makowski 1962; Keupp and Riedel 2010; Howarth 2013
		Grammoceras	Toarcian	Callomon 1963; Howarth 2013
	Graphoceratidae	*Ludwigia*	Aalenian	Howarth 2013
		Graphoceras	Aalenian	Callomon 1963; Keupp and Riedel 2010; Howarth 2013
	Sonniniidae	*Witchellia/Pelekodites*	Bajocian	Sandoval and Chandler 2000; Dietze et al. 2005

Table 7.5 (continued)

Superfamily	Family	Genus	Age	Source
		Fontannesia/Nannoceras	Bajocian	Dietze et al. 2005
Haploceratoidea	Strigoceratidae	*Strigoceras/Cadomoceras*	Bajocian	Dietze et al. 2007
		Phlycticeras/Oecoptychius	Bajocian-Callovian	Basse 1952; Arkell et al. 1957; Krimholc et al. 1958b; Makowski 1962; Schweigert and Dietze 1998; Schweigert et al. 2003
		Protophites	Oxfordian	Bert 2003
	Oppeliidae	*Oxycerites*	Bathonian	Elmi and Mangold 1966
		Ochetoceras/Glochiceras	Oxfordian	Makowski 1962; Keupp and Riedel 2010
		Neochetoceras/Lingulaticeras	Tithonian	Makowski 1962; Keupp and Riedel 2010; Schweigert 1998
		Hecticoceras	Bathonian-Callovian	Makowski 1962; Callomon 1963
		Taramelliceras/Creniceras	Oxfordian	Palframan 1966
		Trimarginites	Oxfordian	Keupp and Riedel 2010
		Cymaceras	Kimmeridgian	Keupp and Riedel 2010
		Semiformiceras/Cyrtosiceras	Tithonian	Ziegler 1974; Cecca and Rouget 2006; Keupp and Riedel 2010
		Oppelia/Oecotraustes	Bajocian	Callomon 1963
		Distichoceras/Horioceras	Callovian	Palframan 1967; Baloge and Cariou 2001
	Lissoceratidae	*Lissoceratoides*	Oxfordian	Makowski 1962
Stephanocera-toidea	Otoitidae	*Otoites/Emileia*	Bajocian	Makowski 1962; Callomon 1963; Westermann 1964a; Westermann and Riccardi 1979
		Docidoceras/Trilobiticeras	Bajocian	Keupp and Riedel 2010; Westermann 1964a
	Stephanoceratidae	*Stephanoceras/Itinsaites*	Bajocian	Dietze et al. 2013
		Cadomites/Polyplectites	Bajocian	Makowski 1962; Keupp and Riedel 2010
		Orthogarantiana/Strenoceras	Bajocian	Gauthier et al. 2002

Table 7.5 (continued)

Superfamily	Family	Genus	Age	Source
	Sphaeroceratidae	*Sphaeroceras*	Bajocian	Krimholc et al. 1958a; Makowski 1962; Callomon 1985
		Chondroceras	Bajocian	Makowski 1962; Westermann and Riccardi 1979
		Macrocephalites	Bathonian-Callovian	Krimholc et al. 1958a; Thierry 1978
		Eurycephaites	Callovian	Parent 1997
	Cardioceratidae	*Arctocephalites*	Bajocian	Callomon 1963, 1985
		Cranocephalites	Bathonian	Callomon 1963, 1985
		Longaeviceras	Callovian	Callomon 1963, 1985
		Cadoceras	Callovian	Basse 1952; Arkell et al. 1957; Callomon 1963, 1985
		Cardioceras	Oxfordian	Krimholc et al. 1958a; Makowski 1962; Callomon 1963, 1985
	Kosmoceratidae	*Kosmoceras*	Callovian	Brinkmann 1929; Raup and Crick 1981
		Sigaloceras	Callovian	Callomon 1963
Perisphinctoidea	Parkinsonidae	*Parkinsonia*	Bajocian	Makowski 1962
	Morphoceratidae	*Morphoceras/Ebrayiceras*	Bathonian	Krimholc et al. 1958b; Makowski 1962; Mangold 1970
		Asphinctites/Polysphinctites	Bathonian	Matyja and Wierzbowski 2001; Zatoń 2010
	Pachyceratidae	*Pachyceras*	Callovian	Makowski 1962; Charpy and Thierry 1976
		Tornquistes	Oxfordian	Thierry and Charpy 1982
	Tulitidae	*Morrisiceras/Holzbergia*	Bathonian	Zatoń 2008; Keupp and Riedel 2010
		Bullatimorphites	Bathonian	Krimholc et al. 1958a; Makowski 1962
		Tulites	Bathonian	Makowski 1962
	Reineckeiidae	*Rehmannia*	Callovian	Cariou 1984
		Reineckeia	Callovian	Cariou 1984; Krimholc et al. 1958b

Table 7.5 (continued)

Superfamily	Family	Genus	Age	Source
	Perisphinctidae	*Prorsisphinctes/Vermisphinctes*	Bajocian	Keupp and Riedel 2010
		Cleistosphinctes/Leptosphinctes	Bajocian	Chimšiašvili et al. 1958; Krimholc et al. 1958b; Pavia and Zunino 2012
		Procerites/Siemieradzkia	Bathonian	Keupp and Riedel 2010
		Procerozigzag/Zigzagiceras	Bathonian	Sturani 1966
		Indosphinctes/Elatmites	Callovian	Keupp and Riedel 2010; Cox 1988; Mangold 1971
		Grossouvria	Callovian	Mangold 1971
		Choffatia	Callovian	Cox 1988; Bonnot et al. 2008
		Parachoffatia/Homoeoplanulites	Callovian	Mangold 1971
		Arisphinctes/Dichotomosphinctes	Oxfordian	Callomon 1963; Enay 1966
		Perisphinctes/Dichotomoceras	Oxfordian	Enay 1966; Chimšiašvili et al. 1958; Makowski 1962; Callomon 1963
		Pseudorthosphinctes/ Orthosphinctes	Kimmeridgian	Atrops 1982; Keupp and Riedel 2010
		Proplanulites	Callovian	Chimšiašvili et al. 1958; Makowski 1962
	Ataxioceratidae	*Danubisphinctes/ Parapallasiceras*	Tithonian	Keupp and Riedel 2010
		Lithacoceras/Silicisphinctes	Tithonian	Schweigert 1998
		Choicensisphinctes	Tithonian	Parent et al. 2011, 2013
		Catutosphinctes	Tithonian	Leanza and Zeiss 1992; Parent et al. 2011
		Parataxioceras/Ataxioceras	Oxfordian	Chimšiašvili et al. 1958; Atrops 1982
	Aulacostephanidae	*Microbiplices/Ringsteadia*	Oxfordian	Wright 2010
		Aulacostephanus	Kimmeridgian	Scherzinger and Mitta 2006
	Virgatitidae	*Virgatites*	Kimmeridgian-Tithonian	Chimšiašvili et al. 1958; Makowski 1962

Table 7.5 (continued)

Superfamily	Family	Genus	Age	Source
	Aspidoceratidae	*Aspidoceras*	Kimmeridgian-Tithonian	Basse 1952; Chimšiašvili et al. 1958; Makowski 1962
		Clambites/Epipeltoceras	Oxfordian	Schweigert 1997
		Hybonoticeras/Hybonotella	Tithonian	Schweigert 1997
		Euaspidoceras/Mirosphinctes	Oxfordian	Makowski 1962; Schweigert 1997, Bonnot et al. 1994
		Peltoceras	Callovian	Chimšiašvili et al. 1958; Makowski 1962
		Peltoceratoides	Oxfordian	Arkell et al. 1957; Makowski 1962; Matyja 1994
		Physodoceras/Sutneria	Kimmeridgian-Tithonian	Geyer 1969; Schweigert 1997; Parent et al. 2008a; Keupp and Riedel 2010
		Simocosmoceras/ Pseudhimalayites	Tithonian	Schweigert 1997; Keupp and Riedel 2010
	Polyptychitidae	*Craspedites*	Tithonian	Chimšiašvili et al. 1958; Makowski 1962
		Kachpurites	Tithonian	Chimšiašvili et al. 1958; Makowski 1962; Mitta 2010

Fig. 7.7 Jurassic antidimorphs. Examples from the Jurassic of Switzerland. The *grey* triangles connect the antidimorphs, with the narrow end at the microconch. This is an example, how antidimorphic pairs could be given one species name

Fig. 7.8 Adult specimens of the antidimorphic couple *Oecoptychius refractus* (Reinecke 1818) (*top left*; microconch) and *Phlycticeras* pustulatum (Reinecke 118) (*bottom*, macroconch). Herznach-Member, Callovian, Swiss Jura Mountains. Note the morphologic and size difference

Category D: Nonsexual dimorphism and polymorphism is contained here. Such polymorphic structures have been shown by various authors (Marchand 1976; Charpy and Thierry 1976; Tintant 1976; Thierry 1978; Contini et al. 1984). According to Davis et al. (1996), this variation has nothing to do with sexual dimorphism. The problem here could be the multitude of factors controlling shape and size of antidimorphs. As discussed above, polymorphism may be caused by a variety of processes such as ecological factors (phenotypic plasticity), hormonal processes, and potentially by diseases and parasites. All these factors contribute to intraspecific variability that may have affected microconchs and macroconchs in different ways. Therefore, some of the cases included in Category D may actually represent blurred cases of sexual dimorphism.

Taking the seeming absence of dimorphism or at least its weak expression in the Triassic into account, it is remarkable how common sexual dimorphism already was in the Early Jurassic (Callomon 1963; Lehmann 1966; Guex 1967, 1968, 1973, 1981; Howarth 2013). Depending on the opinion of authors that focused on Jurassic dimorphism, i.e. which case of dimorphism to include or exclude in sexual dimorphism, the abundance of dimorphism became high to very high in the Middle to Late Jurassic (e.g. Ziegler 1974, 1987; Parent 1997; Schweigert 1997; Schweigert and Dietze 1998, Schweigert et al. 2003, 2007; Matyja and Wierzbowski 2001; Parent et al. 2008a, b, 2009; Zatoń 2008, 2010; Keupp and Riedel 2010; Bardhan et al.

Fig. 7.9 Model of an antidimorphic pair of the Middle Jurassic *Phlycticeras*/*Oecoptychius* by B. Scheffold (Zurich). Note that *Oecoptychius* might have used a modified arm to transmit a spermatophore. Naturally, many aspects of this reconstruction are based on speculations

2012). The great success of this reproductive strategy probably originated partially in the fact that many Middle and Late Jurassic ammonoids phylogenetically root in the Hildoceratoidea (Donovan et al. 1981; Davis et al. 1996), which gave rise to many of the younger clades except for the lytoceratids and phylloceratids, which are not known to have produced clear cases of dimorphism (Fig. 7.11), except few cases such as *Juraphyllites* studied by Cope (1992). This also shows that dimorphism may well be of use as a character for phylogenetic reconstructions. Another factor for the Jurassic success of sexual dimorphism in ammonoids is certainly ecological. The energetic cost of reproduction (the energy that is available for the ovaries and thus eggs) can be significantly reduced when the size of the males, and thus their energy intake, is reduced.

The monophyletic nature of Jurassic sexual dimorphism has another important implication. Genera such as *Taramelliceras* and *Creniceras* in the Late Jurassic become more likely to have been antidimorphs (in contrast to the doubts of Davis et al. 1996) because there is a strong phylogenetic component in dimorphism; nevertheless, the situation is complicated by difficulties in taxonomic assignments, and other, similar genera such as *Proscaphites*, which also show dimorphism. It appears that in the Haploceratoidea, the majority of species produced a pronounced dimorphism. Additionally, the majority of the haploceratoidean microconchs had subcircular to ear-shaped lappets, while those in the stephanoceratids are rather elongate spatulate (*Kosmoceras*) or subovally elliptical (*Normannites*).

Interestingly, the peak in well documented sexual dimorphism in conjunction with extreme adult modifications is in the Middle Jurassic, followed by an increasing number of dubious cases in the Callovian to Kimmeridgian and decreasing abundance of mature modifications in the Tithonian (Davis et al. 1996). From the Tithonian onwards, dimorphism continued to exist.

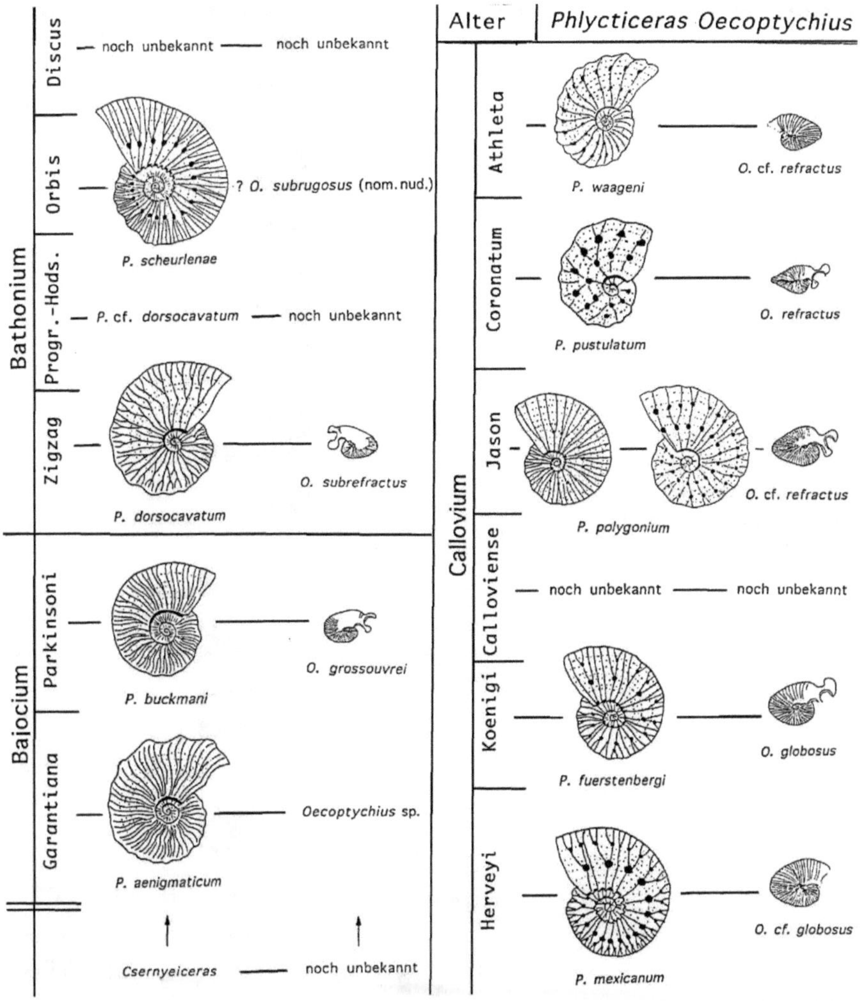

Fig. 7.10 Evolutionary change in the lineage *Phlycticeras/Oecoptychius*. (slightly modified from Schweigert and Dietze 1998)

7.3.7.4 Cretaceous Dimorphism

As in the Jurassic, each ammonoid clade produced different mature modifications, except for the phylloceratids. The seeming absence of dimorphism in phylloceratids might be due to the scarcity of adult specimens (Davis et al. 1996) and/or to the commonly simple ornament and more or less straight apertures.

In contrast to the phylloceratids, several Cretaceous lytoceratids do display mature modifications (Fig. 7.13). For example, some gaudryceratids changed the whorl cross section and a slight umbilical egression may occur, as well as changes in ornament (Wiedmann 1973; Cooper and Kennedy 1979; for a review of the

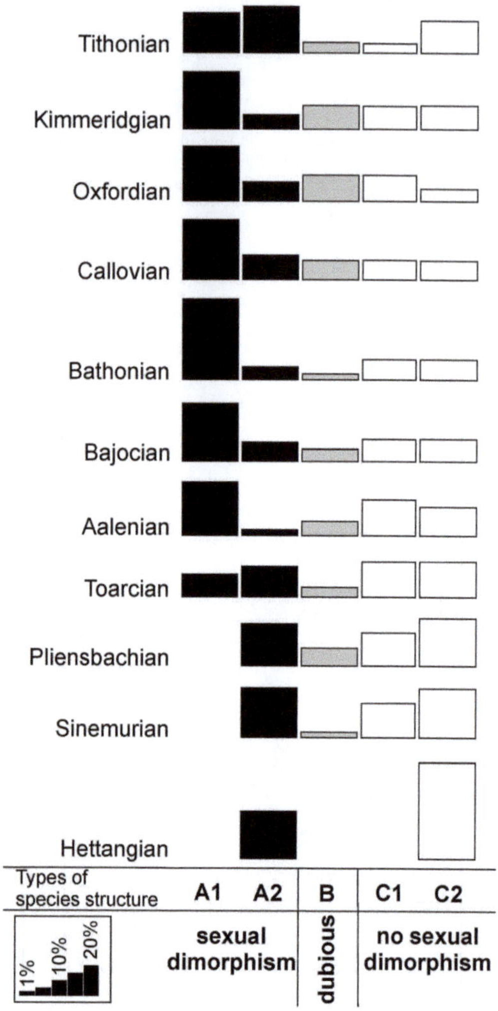

Fig. 7.11 History of dimorphism in Jurassic ammonoids, taking into account the types of species structure, i.e. the predominant type of adults depending on the amount of dimorphic forms (redrawn from Davis et al. 1996). *A1* microconchs with strong mature modifications (e.g., rostrum, lappets); *A2* microconchs with weak mature modifications; *B* groups that look like sexually dimorphic taxa but lack some features to determine dimorphism; *C1* monomorphic, small forms with adult modifications; *C2* monomorphic, moderately large to large forms without adult modifications

group see Hoffmann 2010). In haploceratoideans such as *Aconeceras*, the apertural lappets are less distinct than in the Jurassic, but also present, and combined with a short triangular ventral rostrum (Doguzhaeva and Mutvei 1991). Davis et al. (1996) mention an interesting mode of terminal countdown in *Menuites*, which is remotely reminiscent of the white venter in nautilid shells: in the microconchs of this genus, the ventral tubercles and spines vanish about a demi-whorl behind the terminal aperture (Cobban and Kennedy 1993). The venter and the ventrolateral part stay smooth until shortly behind the terminal aperture, where they re-appear. *Hauericeras*, by contrast, resembles in its terminal aperture the haploceratoideans (Obata et al. 1978). According to Klinger and Kennedy (1989), the hoplitoidean *Placenticeras kaffarium* displays a rather strong umbilical egression, which gives it a scaphitoid adult morphology. Additionally, the venter became rounded and the ornament changed. The possibly most conspicuous adult modification in the Cretaceous among the regularly coiled ammonites is that of the Albian acanthoceratoidean *Mortoniceras*, which forms a long ventral spine, which may be directed ventrally (*M. equidistans*) or even curved posteriorly (*M. rostratum*), forming almost a complete loop (e.g., Cooper and Kennedy 1979). Additionally, the whorl cross section and ornament changed.

The most famous kinds of mature modifications are undoubtedly those of the Cretaceous heteromorphs of the superfamilies Ancyloceratoidea and Turrilitoidea. A good example is the ancyloceratoidean scaphitids (Fig. 7.13, 7.14), which have a wide geographic distribution and a rather impressive diversity. In most species of this group, a more or less straight shaft with a terminal hook follows the normally planispirally coiled phragmocone (Cobban 1951; Landman 1987). Additionally, the aperture is constricted (e.g., Landman et al. 2012). The degree of uncoiling and the length of the straight shaft in relation to the terminal diameter vary as well as the changes in ornament (e.g., in *Hoploscaphites* or *Scaphites*). The microconchs of both *Worthoceras* and *Yezoites* carry broad lappets (Fig. 7.13) with strong convex growth lines (Tanabe 1977; Kennedy 1988). In the Santonian *Scaphites (Pteroscaphites) coloradensis*, the lateral lappets occur in both antidimorphs and have a peculiar hollow spine-like morphology (Kennedy 1988; Landman 1989).

The mature modifications of some Turrilitoidea appear even more unusual. In the baculitids, the adult modifications are usually limited to changes in ornament (stronger ribs on the venter), sometimes a slight dorsal turn of the aperture and dorsal as well as ventral lappets (Kennedy 1988; Cobban and Kennedy 1991c; Klug et al. 2012). The ventral projection or rostrum can be rather long, clearly exceeding the shell diameter. Davis et al. (1996) figured a *Baculites* in which this projection is very long. They assumed that the rostrum might have attained this long size due to an injury or infection. By contrast, we have seen other specimens with similarly shaped adult apertures, thus indicating that this might be a normal adult aperture of this species (Fig. 7.13).

In many genera, such as *Nostoceras*, *Didymoceras*, *Allocrioceras* or *Emericiceras*, the terminal demi-whorl is characterized by a U-shaped part (e.g., Stephenson 1941; Kennedy 1988; Cobban and Kennedy 1994a). In most cases, the coiling direction differs more or less strongly from the preceding whorls. Sometimes, the plane of coiling stayed the same (like in the turrlilitoideans *Allocrioceras*, *Emericiceras* and *Labeceras*, and the lytoceratid *Macroscaphites*), sometimes the plane of coiling changed: in *Eubostrychoceras*, *Hyphantoceras*, *Nostoceras* and *Didymoceras*, for example, the coiling axis turned for 50–90 °. In *Didymoceras nebrascense*, this change in coiling axis is merely a continuation of a similar change in the preceding whorls; in this species, the coiling axis appears to be coiled in itself (Meek and Hayden 1856). By contrast, the Japanese *Pravitoceras* might be the only genus in which the coiling axis switches rapidly for 90 °. In all these cases, this terminal countdown of heteromorphs (Seilacher and Gunji 1993) is linked with changes in ornament.

The abundance of such a U-shaped terminal demi-whorl raises the question of the selective force behind it. Although the ultimate evidence is lost due to the ex-

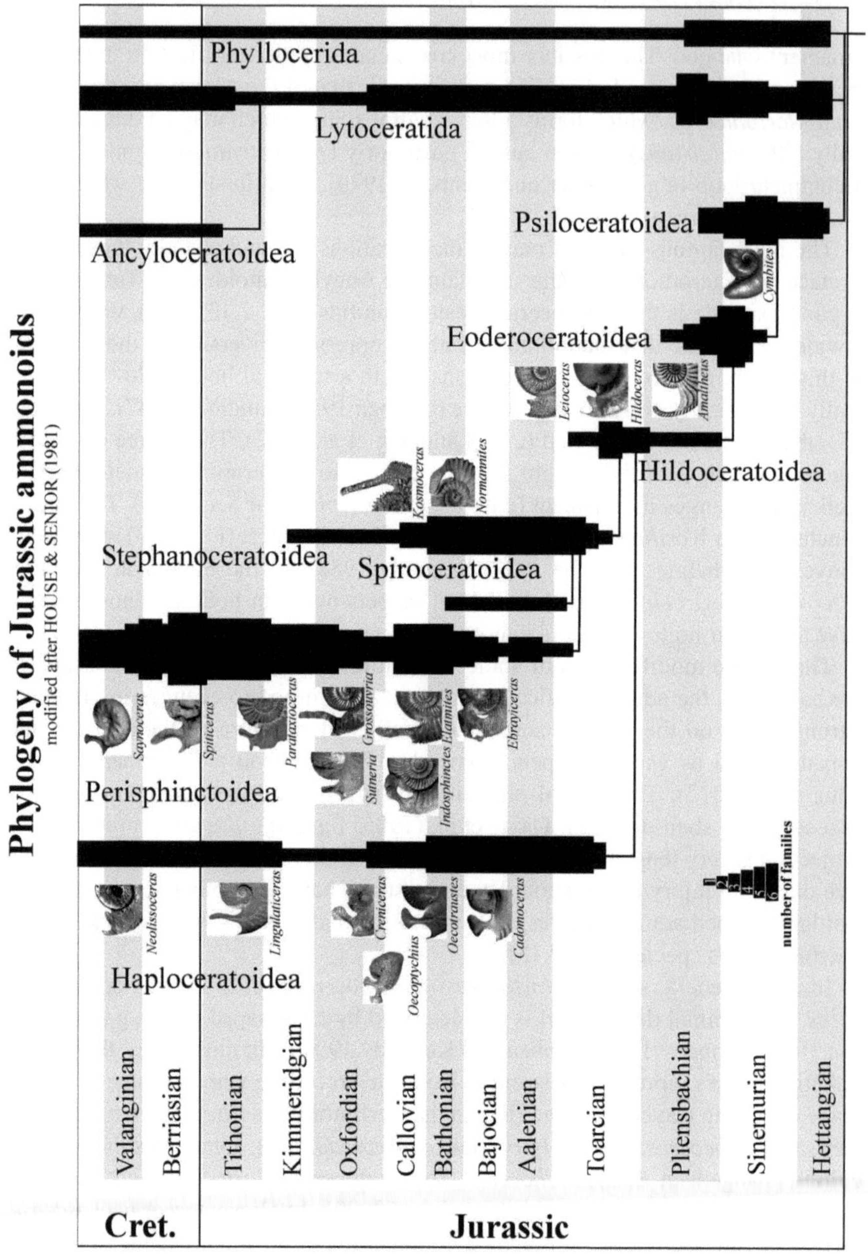

Fig. 7.12 Modified peristomes of adult microconchs of the Jurassic and Early Cretaceous. Note the rather uniform lappets in the haploceratoids, the rather straight lappets in the perisphinctoids, and the disparity in the lappets among the stephanoceratoids. Image sources: V. Schlamp (*Elatmites*, *Paralingulaticeras*, *Parataxioceras lothari*), J.-S. David (*Cadomoceras cadomense*; *Cymbites laevigatus*; *Ebrayiceras pseudoanceps*; *Hildoceras lusitanicum*; *Morrisiceras schwandorfense*; *Oecotraustes bomfordi*); D. Bert (*Kosmoceras phaeinum*); H. Chatelier (*Saynoceras*); Quenstedt (1885: *Amaltheus margaritatus*; *Leioceras opalinum*); R. Roth (*Creniceras crenatum*; *Sutneria platynota*); P. Branger (*Indosphinctes*; *Normannites orbignyi*); Dietl (2013: *Grossouvria*). Djanelidzé (1922: *Spiticeras kiliani*, modified or reduced); Atrops and Reboulet (1995: *Neolissoceras grasianum*, modified or reduced); Ernst and Klug (2011: *Oecoptychius refractus*)

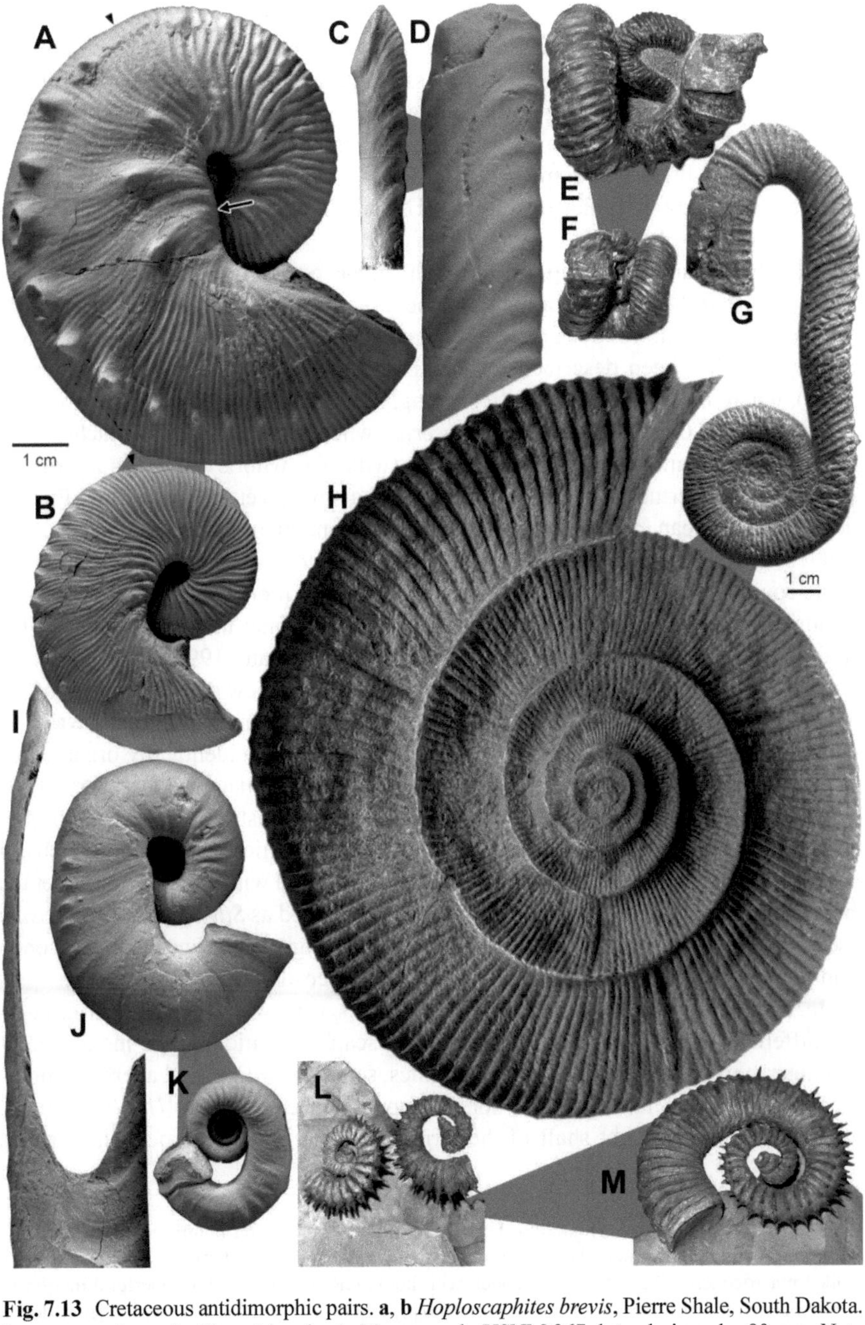

Fig. 7.13 Cretaceous antidimorphic pairs. **a**, **b** *Hoploscaphites brevis*, Pierre Shale, South Dakota. Last septum is marked by a triangle. A, Macroconch, USNM 367, lateral view, dm 90 mm. Note the bulge along the umbilical margin (*arrow*). B, Microconch, lateral view, AMNH 58514, dm 55 mm. **c**, **d** *Sciponoceras gracile*, Cenomanian, Texas. C, microconch, USNM 411539, wh 9 mm. D, Macroconch (USNM 411537, wh 17 mm. The antidimorphs differ mainly in size. **e**, **f** *Nippo-*

tinction of the Ammonoidea, one line of reasoning shall be mentioned here: it is conceivable that this upward turn of the terminal aperture enabled the heteromorphs to approach the level of the center of mass with their hyponome, thus significantly improving their swimming abilities in a horizontal direction. Such a horizontal alignment of aperture and centre of mass was achieved in most ammonoid clades and apparently there was some selection for that trait (e.g., Korn and Klug 2003; Klug and Korn 2004; Tajika et al. 2014). This might have been of particular importance at the time of mating.

Like the mature modifications, dimorphism has been reported from most Cretaceous superfamilies except for the phylloceratines (Table 7.6; Kennedy and Wright 1985a; Davis et al. 1996). Concerning members of the Desmoceratoidea, sexual dimorphism has been described for Campanian *Menuites* (Cobban and Kennedy 1993), where the antidimorphs differ strongly in size. Maeda (1993) examined the dimorphism of Campanian *Yokoyamaoceras*, where the microconch reaches only a third of the diameter of the macroconch (with two whorls less), has a stronger ornament, an aperture with lateral lappets, and a strong ventral projection. For the acanthoceratoidean *Metoicoceras* from the Cenomanian, Cobban and Kennedy (1991b) described a size difference where the macroconch is more than twice as big as its counterpart (Cobban 1953). Its microconchs are more robust and thus have a stronger ornament. Except for the ornament, which is finer in the last whorl of the microconch, the same applies to *Subprionocyclus* (Futakami 1990).

Interestingly, albeit controversial in our opinion, one view on dimorphism of the Valanginian ammonite *Valanginites nucleus* from Wąwał in central Poland was presented by Ploch (2003, 2007). Here, the size differences in identically ornamented and identically coiled specimens with purely macroconchiate modification of the terminal part of the shell (prominent lip preceded by a constriction and without lappets) have been used as the only criterion in separating micro- and macroconchs. It is all the more strange as these ammonites are associated with similarly ornamented but much smaller shells having lateral lappets, classified as *Saynoceras verrucosum* (see Dzik 1990a). Thus, it appears that *V. nucleus* might represent the macroconch and *S. verrucosum* its antidimorphic microconch (see also Bulot et al. 1990).

In scaphitids (Ancyloceratina), dimorphism is very well known. Morphological differences between the terminal whorl of scaphitid antidimorphs include adult diameter, coiling, ribbing and nodes or spines, septal crowding, and apertural modifications including lateral and/or dorsal lappets. For example, in *Hoploscaphites*, the more or less straight shaft of the terminal whorl of the macroconch carries a

---

nites mirabilis, Campanian, Hokkaido, Japan. The antidimorphs differ mainly in size. **g, h** *Macroscaphites yvani*, Barremian, Angles, France. **g** microconch with large terminal hook. **h** regularly coiled macroconch, G12/336. **i** microconch with hook, GRY/903b. **i** mature apertural margin of a microconch (?) of the Santonian *Baculites thomi*, USGS 21419, Montana, with a short dorsal and a long, ventral rostrum. **j, k** *Yezoites puerculus*, Turonian, Hokkaido, Japan. **j** Macroconch, AMNH 45280. **k** microconch, AMNH 45281, note the lateral lappets and the absence of nodes. **l, m** *Imerites dichotomum*, Barremian, Alpes de Haute Provence, France. **l** notice that one of the microconchs is sinistral and the other dextral; dm 45 mm. **m** macroconch, dm 75 mm. (Images: **a–d, i–k** (N. Landman). **e, f, l, m** (W. Grulke). **h, i** (D. Bert))

Table 7.6 Sexual dimorphism in Cretaceous ammonoids. (incomplete list, largely taken from Davis et al. 1996)

Superfamily	Family	Genus	Source
Haploceratoidea	Aconeceratidae	*Gyaloceras*	Kennedy and Klinger 1979
		Protaconeceras	Riccardi et al. 1987
		Sanmartino-ceras	Kennedy and Klinger 1979; Riccardi et al. 1987; Parent 1991
		Sinzovia	Riccardi et al. 1987
	Binneyitidae	*Borrisiakoceras*	Kennedy and Cobban 1976
	Haploceratidae	*Haploceras*	Bujtor 1993
Perisphinctoidea	Neocomitidae	*Berriasella*	Howarth 1992; Wright et al. 1996
		Neocomites	Fatmi 1969; Reboulet 1995
		Thurmanniceras	Bujtor 1993; Wright et al. 1996
	Olcostephanidae	*Groebericeras*	Howarth 1992
		Olcostephanus	Fatmi 1969; Bulot et al. 1990; Bujtor 1993; Wright et al. 1996
		Saynoceras	Bulot et al. 1990; Dzik 1990a
Desmoceratoidea	Desmoceratidae	*Anapuzosia*	Matsumoto 1988
		Achilleoceras	Matsumoto 1988
		Austiniceras	Matsumoto 1988
		Bhimaites	Matsumoto 1988
		Epipuzosia	Matsumoto 1988
		Grandidiericeras	Matsumoto and Saito 1987; Matsumoto 1988
		Hauericeras	Matsumoto et al. 1990b; Wright et al. 1996
		Jimboiceras	Matsumoto 1988
		Kitchinites	Matsumoto 1988; Matsumoto et al. 1990b
		Matsumotoceras	Matsumoto 1988; Matsumoto et al. 1990a
		Mesopuzosia	Matsumoto 1988; Matsumoto et al. 1990a, b

Table 7.6 (continued)

Superfamily	Family	Genus	Source
		Neopuzosia	Matsumoto 1988; Matsumoto et al. 1990b
		Pachydesmoceras	Matsumoto 1987b, 1988
		Parapuzosia	Matsumoto 1988
		Puzosia	Marcinowski 1980; Wright and Kennedy 1984; Cooper and Kennedy 1987; Matsumoto 1988; Matsumoto et al. 1990a, b; Matsumoto and Skwarko 1993
	Kossmaticeratidae	*Grossouvrites*	Olivero and Medina 1989
		Gunnarites	Kennedy and Klinger 1985
		Kossmaticeras	Matsumoto 1991b
		Maorites	Macellari 1986
		Yokoyamaoceras	Matsumoto 1991b; Maeda 1993; Keupp and Riedel 2010
	Pachydiscidae	*Anapachydiscus*	Kennedy 1986b, d, 1989; Kennedy et al. 1986; Kennedy and Henderson 1992; Kennedy and Klinger 1993
		Canadoceras	Kennedy 1986d
		Eupachydiscus	Kennedy 1986d
		Lewesiceras	Wright and Kennedy 1984; Kennedy 1986d
		Menuites	Kennedy 1986d, 1989; Kennedy and Henderson 1992; Kennedy and Klinger 1993; Cobban and Kennedy 1993
		Pachydiscoides	Kennedy 1984
		Pachydiscus	Kennedy and Summesberger 1984; Kennedy 1986d; Jagt 1989
		Teshioites	Kennedy 1986d
	Silesitidae	*Silesites*	Cecca and Landra 1994
Hoplitoidea	Hoplitidae	*Anahoplites*	Marcinowski and Wiedmann 1990
		Callihoplites	Marcinowski and Wiedmann 1990
		Euhoplites	Amedro 1992

Table 7.6 (continued)

Superfamily	Family	Genus	Source
		Hyphoplites	Wright and Kennedy 1984
	Placenticeratidae	*Hoplitoplacenticeras*	Kennedy 1986d
		Hypengonoceras	Klinger and Kennedy 1989
		Karamaites	Kennedy and Wright. 1981, 1983; Klinger and Kennedy 1989
		Placenticeras	Kennedy and Wright 1981, 1983; Kennedy 1984, 1986d, 1988; Kennedy and Cobban 1991a; Cobban et al. 1989; Klinger and Kennedy 1989; Matsumoto and Skwarko 1991; Ganguly and Bardhan 1993
Acanthoceratoidea	Acanthoceratidae	*Acanthoceras*	Moreau et al. 1983; Wright and Kennedy 1987; Kennedy and Cobban 1990; Matsumoto and Skwarko 1991
		Acompsoceras	Wright and Kennedy 1987
		Benueites	Reyment 1971, 1988
		Calycoceras	Wright and Kennedy 1987, 1990; Cobban et al. 1989; Cobban and Kennedy 1990; Matsumoto and Skwarko 1991; Kennedy and Juignet 1994
		Conlinoceras	Kennedy and Cobban 1990
		Cunningtoniceras	Wright and Kennedy 1987; Matsumoto et al. 1989; Kennedy and Cobban 1990
		Eucalycoceras	Cobban 1988a; Wright and Kennedy 1990
		Euomphaloceras	Kennedy 1988; Wright and Kennedy 1990
		Mantelliceras	Wright and Kennedy 1984; Kennedy 1989; Matsumoto and Takahashi 1992
		Metoicoceras	Kennedy et al. 1981a, 1989; Kennedy 1988, 1989; Forster et al. 1983; Cobban and Kennedy 1991b
		Mrhiliceras	Kennedy and Wright 1985b; Wright and Kennedy 1987
		Nannometoioceras	Kennedy 1988, 1989; Cobban et al. 1989
		Neocardioceras	Cobban 1988a
		Paraburoceras	Cobban et al. 1989

Table 7.6 (continued)

Superfamily	Family	Genus	Source
		Paraconlinoceras	Kennedy and Cobban 1990
		Plesiacanthoceras	Kennedy and Cobban 1990
		Plesiacanthoceratoides	Kennedy and Cobban 1990
		Protacanthoceras	Wright and Kennedy 1980, 1987; Kennedy and Wright 1985a
		Pseudaspidoceras	Matsumoto 1991a
		Pseudocalycoceras	Kennedy 1988
		Spathites	Kennedy and Cobban 1988b
		Sumitomoceras	Cobban 1988a
		Tarrantoceras	Kennedy 1988; Kennedy and Cobban 1990; Cobban 1988a
		Thomelites	Wright and Kennedy 1990
		Watinoceras	Zaborski 1987; Cobban 1988b
	Brancoceratidae	*Euhystrichoceras*	Kennedy and Wright 1981; Wright and Kennedy 1984
		Hysteroceras	Amedro 1992
		Mortoniceras	Marcinowski and Wiedmann 1990; Amedro 1992
	Coilopoceratidae	*Coilopoceras*	Kennedy and Wright 1984b; Kennedy 1988; Reyment 1988; Luger and Groschke 1989; Meister et al. 1992
		Hoplitoides	Kennedy and Cobban 1988b; Reyment 1988
	Collignoniceratidae	*Barroisiceras*	Immel 1987
		Collignoniceras	Kennedy et al. 1980, 1989
		Forresteria	Kennedy et al. 1983; Kennedy 1984; Kennedy and Cobban 1991a
		Gauthiericeras	Kennedy 1984
		Paratexanites	Kennedy 1984
		Peroniceras	Kennedy 1984
		Prionocyclus	Kennedy 1988

Table 7.6 (continued)

Superfamily	Family	Genus	Source
		Protexanites	Kennedy 1984; Kennedy and Cobban 1991a
		Submortoni-ceras	Kennedy et al. 1981b
		Subpriono-cyclus	Reyment 1982; Futakami 1990
		Yabeiceras	Kennedy et al. 1983
	Flickiidae	*Litophragmatoceras*	Kennedy and Cobban 1988a
		Salaziceras	Kennedy and Wright 1984a; Keupp and Riedel 2010
	Forbesiceratidae	*Forbesiceras*	Kennedy and Juignet 1984; Kennedy and Cobban 1990; Wright and Kennedy 1984; Matsumoto 1987a
	Lyelliceratidae	*Neophlycticeras*	Kennedy and Delamette 1994; Wright and Kennedy 1994
		Stoliczkaia	Wright and Kennedy 1984
	Sphenodiscidae	*Libycoceras*	Kassab and Hamama 1991
		Manambolites	Luger and Grosehke 1989
		Sphenodiscus	Kennedy 1986d
	Tissotiidae	*Metatissotia*	Kennedy 1984
	Vascoceratidae	*Fagesia*	Kennedy and Wright 1985a; Kennedy et al. 1987
		Hourcquia	Matsumoto and Toshimitsu 1984
		Microdiphaso-ceras	Cobban et al. 1989
		Neoptychites	Kennedy and Wright 1979; Kennedy and Cobban 1988b; Cobban and Hook 1983; Zaborski 1987
		Pseudobarroisiceras	Matsumoto and Toshimitsu 1984
		Rubroceras	Cobban et al. 1989
		Thomasites	Zaborski 1987
		Vascoceras	Cobban et al. 1989; Meister et al. 1992

Table 7.6 (continued)

Superfamily	Family	Genus	Source
Ancyloceratoidea	Ancyloceratidae	*Ancyloceras*	Klinger and Kennedy 1977; Forster and Weier 1983
		Acantholytoceras	Delanoy et al. 1995
		Acrioceras	Klinger and Kennedy 1992
		Crioceratites	Klinger and Kennedy 1992
		Emericiceras	Delanoy et al. 1995
		Hemihoplites	Delanoy et al. 1995
		Lytocrioceras	Delanoy et al. 1995
		Pseudomoutoniceras	Delanoy et al. 1995
	Heteroceratidae	*Colchidites*	Aguirre-Urreta and Klinger 1986
		Heteroceras	Aguirre-Urreta and Klinger 1986, Delanoy et al. 1995
Douvilleiceratoidea	Astiericeratidae	*Astiericeras*	Kennedy 1986a
	Douvilleiceratidae	*Douvilleiceras*	Amedro 1992
		Paraspiticeras	Aguirre-Urreta and Rawson 1993
Deshayesitoidea	Parahoplitidae	*Hypacanthoplites*	Kemper 1982
Scaphitoidea	Scaphitidae	*Acanthoscaphites*	Birkelund 1982; Kennedy 1986c; Kennedy and Summesberger 1987; Jagt and Kennedy 1989; Jagt et al. 1992
		Clioscaphites	Landman 1987
		Discoscaphites	Jeletzky and Waage 1978; Kennedy and Cobban 1993c; Landman and Waage 1993
		Hoploscaphites	Makowski 1962; Birkelund 1982; Kennedy 1986b, c, 1989, 1993; Kennedy et al. 1986; Kennedy and Summesberger 1987; Kennedy and Cobban 1993a; Landman and Waage 1993; Machalski 2005
		Jeletzkytes	Landman and Waage 1993; Kennedy and Cobban 1993a; Cobban and Kennedy 1994a; Jagt and Kennedy 1994; Kennedy and Cobban 1993c
		Rhaeboceras	Cobban 1987

Table 7.6 (continued)

Superfamily	Family	Genus	Source
		Scaphites	Cobban 1969, 1984; Cobban and Kennedy 1991a; Kennedy 1984, 1986d, 1988, 1989; Kennedy et al. 1989, 1992; Kennedy and Christensen 1991; Kennedy and Cobban 1991a, b; Marcinowski 1980, 1983; Immel 1987; Kaplan et al. 1987; Landman 1987, 1989; Jagt 1989
		Trachyscaphites	Kennedy and Summesberger 1984; Kennedy 1986d; Cobban and Kennedy 1992, 1994b
		Worthoceras	Forster et al. 1983; Kennedy 1988; Kennedy and Cobban 1988a; Kennedy et al. 1989; Cobban et al. 1989; Bujtor 1991; Keupp and Riedel 2010
		Yezoites	Tanabe 1977; Kennedy 1984, 1988; Kennedy and Christensen 1991; Kaplan et al. 1987; Keupp and Riedel 2010
Turrilitoidea	Anisoceratidae	*Allocrioceras*	Kennedy 1988
	Baculitidae	*Baculites*	Kennedy 1984, 1986b, c; Jagt 1989; Cobban and Kennedy 1991b
		Boehmoceras	Kennedy and Cobban 1991b
		Eubaculites	Klinger and Kennedy 1993
		Lechites	Cooper and Kennedy 1977; Kennedy and Wright 1985a
		Sciponoceras	Marcinowski 1980; Kennedy and Juignet 1983; Kennedy 1988
	Diplomoceratidae	*Oxybeloceras*	Cobban et al. 1989
		Solenoceras	Cobban and Kennedy 1994a
	Hamitidae	*Metaptychoceras*	Cobban et al. 1989
	Labeceratidae	*Labeceras*	Klinger 1989
		Myloceras	Klinger 1989

Table 7.6 (continued)

Superfamily	Family	Genus	Source
	Nostoceratidae	*Anaklinoceras*	Davis et al. 1996
		Axonoceras	Davis et al. 1996
		Bostrychoceras	Kennedy 1986d
		Didymoceras	Cobban and Kennedy 1994a
		Eubostrychoceras	Kennedy 1986d
		Exiteloceras	Davis et al. 1996
		Hyphantoceras	Kennedy and Wright 1985a
		Nostoceras	Luger and Gröschke 1989; Kennedy and Cobban 1993a, b; Cobban and Kennedy 1994a
	Ptychoceratidae	*Lytocrioceras*	Delanoy and Poupon 1992
	Turrilitidae	*Mariella*	Kennedy and Wright 1985a
Lytoceratoidea	Macroscaphitidae	*Costidiscus*	Cecca and Landra 1994
		Macroscaphites	Cecca and Landra 1994
Tetragonitoidea	Gaudryceratidae	*Gaudryceras*	Hirano 1978, 1979

thick dorsal swelling, which gives it a pregnant appearance (e.g., Morton 1834; Kennedy and Cobban 1993a; Landman and Waage 1993; Machalski 2005). Speculatively, this bump provided extra space for the ovaries. In the corresponding microconch, the dorsal wall of the shaft is subparallel to the venter. There are also some differences in ornament (Davis et al. 1996). Landman and Waage (1993) examined the size-differences between the antidimorphs of the Maastrichtian species *Hoploscaphites (Jeletzkytes) spedeni.* Although the macroconchs are in average almost twice as large as their counterparts, the size distribution of both antidimorphs does overlap (Fig. 7.14). This size overlap varies between the species (Landman and Waage 1993), but it is not entirely clear whether the presence or absence of an overlap and its quality are predominantly controlled by (1) difference in adult diameter of the antidimorphs, (2) difference in intrasexual variability, (3) sample size or (4) ecology.

It is also remarkable that in the Maastrichtian *Hoploscaphites comprimus*, some morphological differences between the antidimorphs already occur in the normally coiled juvenile part (Landman and Waage 1993), making sexing of juvenile specimens possible.

Among the Turrilitoidea, there are also many cases of likely dimorphism, although often the main difference between the antidimorphs is size (e.g., *Didymoceras*, *Bostrychoceras*, *Nipponites*, *Oxybeloceras, Sciponoceras*; Kennedy 1988; Cobban and Kennedy 1994a). A nice example for dimorphism in the Lytoceratina is the genus *Macroscaphites*, in which the microconch develops a long straight shaft with a U-shaped hook at the end, while the macroconch consists only of a regularly coiled shell with a terminal constriction (Fig. 7.13). Most other lytoceratines display more normal kinds of dimorphism, i.e. mainly differences in size (*Gaudryceras*, *Costidiscus*, *Tetragonites*; Wiedmann 1973).

Davis et al. (1996) reported a couple of possible cases of Cretaceous trimorphism. One case was published by Hirano (1978, 1979) and concerns the lytoceratin *Gaudryceras*. A second case comprises scaphitids of the genera *Scaphites, Clioscaphites*, and *Scaphites (Pteroscaphites)*. Wiedmann (1965) thought that the species of the latter genus were the microconchs of those of the former two genera, based on the same morphology of juvenile shells, same stratigraphic occurrences and the adult size and morphological differences. This interpretation appears to be incorrect because dimorphism was demonstrated for the macroconchs as proposed by Wiedmann (1965) by both Cobban (1951) and Landman (1987), and later also in *Scaphites (Pteroscaphites)* by Landman (1989). In many cases, the mix of intraspecific variability within both sexes, evolutionary changes, phenotypic plasticity (Wilmsen and Mosavinia 2011) and dimorphism blurs the patterns of disparity in ammonoid populations to such extent that the various phenomena can hardly be distinguished (Kennedy and Wright 1979; Reyment 1988; Kassab and Hamama 1991).

The ratios of numbers of macroconchs (M) to microconchs (m) have also been determined for various species (Davis et al. 1996):

Hoploscaphites constrictus, France: 1.9 M: 1 m (Kennedy 1986b),

H. constrictus, Poland: 2.2 M: 1 m (Makowski 1962),

H. nicolletii, South Dakota: 20 M: 1 m (Landman and Waage, 1993),

H. comprimus, South Dakota: 1.5 M: 1 m (Landman and Waage 1993),
Menuites oralensis, Colorado: 2 M: 1 m (Cobban and Kennedy 1993),
M. portlocki complexus, Wyoming: 3.2 M: 1 m (Cobban and Kennedy 1993),
Scaphites hippocrepis, Wyoming: 0.5 M: 1 m (Cobban 1969),
S. hippocrepis III, Montana: 0.8 M: 1 m (Cobban 1969),
Scaphites leei III, New Mexico: 0.7 M: 1 m (Cobban 1969).

Another interesting aspect of dimorphism is the geographically varying ratio of the antidimorphs, for example in *Metoicoceras* (Kennedy 1988; Cobban et al. 1989) and *Hoploscaphites* (Landman and Waage 1993; Machalski 2005).

Some evolutionary trends in sexual dimorphism in Cretaceous ammonites have been described. Klinger and Kennedy (1989) examined *Placenticeras* from the Albian to the Maastrichtian and discovered that the early antidimorphs of this genus differed mainly in size, while younger, more derived forms differed also in ornament strength. Landman (1987) studied a population of Turonian *Scaphites whitfieldi* in which some of the specimens can be assigned to macroconchs or microconchs, while many forms display intermediate sizes and morphologies. In more derived scaphitids from the Maastrichtian, the assignment of antidimorphs can be done more easily because the dimorphism is more strongly expressed (Landman and Waage 1993).

7.4 Open Questions

7.4.1 Intraspecific Variability of Antidimorphs

Only a few studies are available dealing with the intraspecific variability of dimorphic species (compare De Baets et al. 2015a). This is understandable, because often it is difficult or impossible to get hold of a sufficiently large collection of mature specimens that are suitably preserved. Nevertheless, we are convinced that such populations of various ages are available in several museums worldwide, awaiting examination. Potential outcomes of such studies are a better understanding of the biological background of polymorphism, more confident separation of consecutive dimorphic species in evolutionary lineages, additional support for (or falsification of) dimorphism in cases of dubious dimorphism, an enhanced knowledge of the differences in variability between the antidimorphs, and raw data for further evolutionary studies.

7.4.2 Macroevolution of Mature Modifications and Dimorphism

Similar to the preceding topic, evolutionary aspects of dimorphism have only rarely or indirectly been addressed (one example is the work by Schweigert and Dietze 1998 on *Oecoptychius* and *Phlycticeras*; Fig. 7.9). It appears that the Haploceratoidea

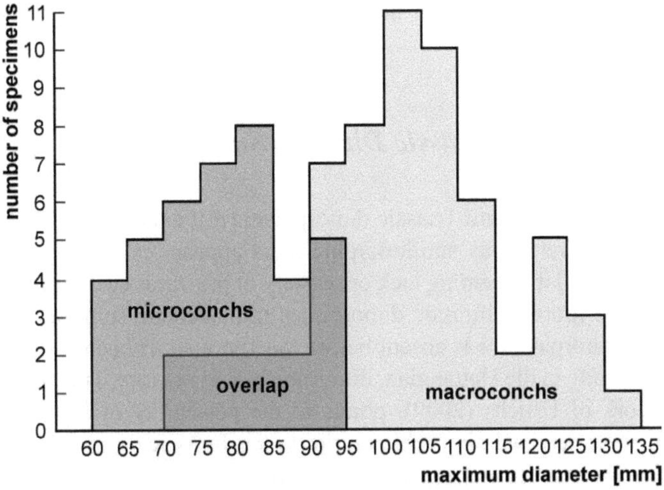

Fig. 7.14 Histogram of adult size in a collection of Maastrichtian *Hoploscaphites spedeni* from South Dakota. Macroconchs are usually larger than microconchs, but the size ranges overlap. (Redrawn from Davis et al. 1996 and Landman and Waage 1993)

could be an especially rewarding group in which to examine evolutionary changes in dimorphs (Fig. 7.11).

It has been suggested that evolutionary rates differed between antidimorphs; these rates may be difficult to quantify. By contrast, differences in variability of the antidimorphs through phylogeny could be studied in some small lineages (Fig. 7.14)

7.4.3 Taxonomic Treatment of Antidimorphs

Classically, partially because of the lack of knowledge, most antidimorphic pairs have been assigned to different taxa, occasionally reaching family level. Normally, members of one biological or morphological species should carry the same name according to the International Code of Zoological Nomenclature. As in trace fossils, the certainty of identity with respect to systematic nomenclature is often not given (for discussions see Callomon 1969; Lehmann 1981; Westermann 1969b). This is probably the reason why Demanet (1943) as well as Furnish and Knapp (1966) added various terms to the species name in order to state that they assume that two forms belong to the same species and at the same time, mark which form belongs to which sex. Although it appears reasonable to assign such antidimorphs, where it has been convincingly shown that they are conspecific, to the same species, this would imply applying different nomenclatorial rules depending on the state of knowledge (the problem could be solved by using partial names for cases that are not clear). An additional problem arises when diversity counts are carried out. If one species is knowingly subdivided into two, namely the antidimorphs, this would increase diversity artificially. Ammonoid researchers need to agree on a unified treatment that addresses this problem.

7.4.4 Devonian to Triassic Dimorphism

Compared to Paleozoic and Triassic dimorphism (if it existed at all), identification of Jurassic and Cretaceous antidimorphic pairs appears easy. Davis et al. (1996) have already found the seeming lack or scarcity of pre-Jurassic dimorphism intriguing. Taking the roots of Jurassic dimorphism into account, some researchers considered that dimorphism was absent before the Toarcian. By contrast, Guex (1981) stated that already in the Hettangian, dimorphism was not rare. This, in combination with the work of Urlichs (2009), points to the possibility of a reasonably common but not yet detected dimorphism prior to the Jurassic. Further support for this hypothesis comes from the repeated occurrences of morphologies that resemble Jurassic microconchs in various respects, such as e.g., Devonian *Prolobites* and *Wocklumeria*, Permian *Elephantoceras* and *Adrianites*, Triassic *Coroceras* and dwarf *Arcestes* or *Lobites*.

These are just some out of many open questions. Davis et al. (1996, p. 521) actually listed many more such questions at the end of their article. We do not repeat this here but recommend it to those further interested.

Acknowledgments Some of the insights in this chapter grew in the course of research projects with the numbers 200021-113956/1, 200020-25029, and 200020-132870 funded by the Swiss National Science Foundation SNF. Images were kindly provided by Neil Landman (New York, USA), Günter Schweigert and Gerd Dietl (both Stuttgart, Germany), Jean-Stéphane David (Saint Dolay, France), Victor Schlamp (Lappersdorf, Germany), Christian Obrist (Rickenbach, Switzerland), Didier Bert (La Mure-Argens, France), Patrick Branger (Poitiers cedex, France), Wolfgang Grulke (Oborne, UK), Andreas E. Richter (Augsburg, Germany), Pierre-Yves Boursicot (Hauts-de-Seine, France). Margaret Yacobucci (Bowling Green), Izabela Ploch (Warsaw), and Kristin Polizzotto (Brooklyn) reviewed the manuscript and provided valuable constructive suggestions to improve it.

References

Aguirre-Urreta MB, Klinger HC (1986) Upper Barremian Heteroceratinae (Cephalopod, Ammonoidea) from Patagonia and Zululand, with comments on the systematics of the subfamily. Ann S Afr Mus 96(8):315–358

Aguirre-Urreta MB, Rawson PF (1993) The Lower Cretaceous ammonite *Paraspiticeras* from the Neuquen Basin, west-central Argentina. Neues Jahrb Geol Paläontol 188:187–233

Amedro F (1992) L'Albien du bassin anglo-Parisien: ammonites, zonation phylétique, séquences. Bull Cent Rech Explor Prod Elf-Aquitaine 16:187–233

Arkell WJ, Furnish WM, Kummel B, Miller AK, Moore RC, Schindewolf OH, Sylvester-Bradley PC, Wright CW (1957) Part L. Mollusca 4: Ammonoidea. In: Kaesler RL (ed) Treatise on invertebrate paleontology, Part L, Mollusca 4 (revised). GSA and University of Kansas Press, Lawrence

Atrops F (1982) La sous-famille des Ataxioceratinae dans le Kimmeridgien inférieur du Sud-Est de la France. Doc Lab Géol Lyon 83:1–371

Atrops F, Reboulet S (1995) *Neolissoceras* (*Carinites*), nouveau sous-genre d'ammonites du Valanginien du bassin vocontien (SE de la France). CR Acad Sci Paris 321(IIa):1203–1210

Baloge P-A, Cariou E (2001) Les Distichoceratinae (Ammonitina) du Centre-Ouest de la France. Palaeontogr A 261:125–159

Bardhan S, Dutta R, Chanda P, Mallick S (2012) Systematic revision and sexual dimorphism in *Choffatia* (Ammonoidea: Perisphinctoidea) from the Callovian of Kutch, India. Palaeoworld 21:29–49

Barrande J (1877) Systême Silurien du centre de la Bohème, Première Partie: Récherches paléontologiques. 2, Classes des Mollusques, Ordre des Céphalopodes. By the author, Paris

Basse, E (1952) Céphalopodes, Nautiloidea, Ammonoidea. In: Pivéteau J (ed) Traité de Paléontologie, vol 2. J. B. Baillière, Paris, pp 522–688

Bert D (2003) Etude de *Protophites vannii* sp. nov. (Ammonoidea), sous-zone a *Cardioceras vertebrale*, Oxfordien moyen, et evolution du genre *Protophites* Ebray, 1860. Riviéra Sci 87:69–84

Birkelund T (1982) Maastrichtian ammonites from Hemmoor, Niederelbe (NW-Germany). Geol J A 61:13–33

de Blainville MHD (1840) Prodrome d'une monographie des ammonites. In: Supplément du Dictionnaire des Sciences Naturelles, Bertrand. Paris, pp 1–31

Blind W, Jordan R (1979) "Septen-Gabelung" an einer *Dorsetensia romani* (Oppel) aus dem nordwestdeutschen Dogger. Paläontol Z 53(3/4):137–141

Bogoslovsky BI (1969) Devonskie ammonoidei. I. Agoniatity, vol 124 (Trudy Paleontologicheskogo Instituta Akademiya Nauk SSSR) Nauka, Moskva, pp 1–341

Bonnot A, Neige P, Tarkowski R, Marchand D (1994) *Mirosphinctes* Schindewolf et *Euaspidoceras* Spath du Niveau Vert de Zalas [Pologne] (Oxfordien Inférieur, Zone a Cordatum): Dimorphes sexuels? Bull Pol Acad Sci (Earth Sciences) 42:181–205

Bonnot A, Boursicot P-Y, Ferchaud P, Marchand D (2008) Les Pseudoperisphinctinae (Ammonitina, Perisphinctidae) de l'horizon à Leckenbyi (Callovien supérieur, zone à Athleta) de Montreuil-Bellay (Maine-et-Loire, France) et description d'une nouvelle espèce, *Choffatia isabellae*. Carnets Géol 5:1–27

Brinkmann R (1929) Monographie der Gattung *Kosmoceras*. Abh Ges Wiss Göttingen, Math-Phys Kl, NF 13(4):1–123

Brochwicz-Lewiński W, Różak Z (1976) Some difficulties in recognition of sexual dimorphism in Jurassic perisphinctids (Ammonoidea). Acta Palaeontol Pol 21:115–124

Brooks MJ (1991) The ontogeny of sexual dimorphism: Quantitative models and a case study in labrisomid blennies (Teleostei: *Paraclinus*). Syst Zool 40(3):71–283

Bucher H (1992) Ammonoids of the Shoshonensis Zone (Middle Anisian, Middle Triassic) from NW Nevada. Jahrb Geol Bundesanst 135(2):423–466

Bucher H, Guex J (1990) Rythmes de croissance chez les ammonites triasiques. Bull Geol Lausanne 308:191–209

Bucher H, Landman NH, Klofak, SM, Guex J (1996) Mode and rate of shell growth. In: Landman NH, Tanabe K, Davis RA (eds) Ammonoid paleobiology. Plenum, New York

Bujtor I (1991) A new *Worthoceras* (Ammonoidea, Cretaceous) from Hungary, and remarks on the distribution of *Worthoceras* species. Geol Mag 128:537–542

Bujtor L (1993) Valanginian ammonite fauna from the Kisújbánya Basin (Mecsek Mts., South Hungary) and its palaeobiogeographical significance. Neues Jahrb Geol Paläontol Abh 188:103–131

Bulot L, Company M, Thieuloy J-P (1990) Origine, évolution et systematique du genre Valanginien *Saynoceras* (Ammonitina, Olcostephaninae). Geobios 23:399–413

Callomon JH (1955) The ammonite succession in the Lower Oxford Clay and Kellaway beds at Kidlington, Oxfordshire, and the zones of the Callovian Stage. Phil Trans R Soc Lond (Biol) 239:215–264

Callomon JH (1963) Sexual dimorphism in Jurassic ammonites. Trans Leic Lit Philos Soc 57:21–56

Callomon JH (1969) Dimorphism in Jurassic Ammonoidea. Some reflections. In: Westermann GEG (ed) Sexual dimorphism in fossil Metazoa and taxonomic implications. International union of geological sciences. A 1 Schweizerbart, Stuttgart, pp 111–125

Callomon JH (1981) Dimorphism in ammonoids. In: House MR, Senior JR (eds) The Ammonoidea, vol 18. Systematics Association by Academic Press, London pp 257–273

Callomon JH (1985) The evolution of the Jurassic ammonite family Cardioceratidae. Spec Pap Palaeontol 3349–90

Callomon JH (1988) [Review of] Matyja BA (1986) Developmental polymorphism in Oxfordian ammonites. Acta Geol Pol 36:37–68 (Cephalopod Newsl 9:14–16)

Callomon JH, Gradinaru E (2005) From the thesaurus of the museum collections. I. Liassic ammonites from Munteana (Svinita Zone, Southern Carpathians, Romania). Acta Palaeontol Rom 5:49–65

Cariou E (1984). Les Reineckeiidae (Ammonitina, Callovien) de la Tethys occidentales. Dimorphisme et evolution. Etude a partir des gisements du centre-ouest de la France. Doc Lab Géol Lyon HS 8:1–599

Cecca E, Landra G (1994) Late Barremian-early Aptian ammonites from the Maiolica formation near Cesana Brianza (Lombardy Basin, northern Italy). Riv It Paleont Strat 100(3):395–422

Cecca F, Rouget I (2006) Anagenetic evolution of the Early Tithonian ammonite genus *Semiformiceras* tested with cladistic analysis. Palaeontology 49:1069–1080

Charpy N, Thierry (1976) Dimorphisme et polymorphisme chez *Pachyceras* Bayle (Ammonitina, Stephanocerataceae) du Callovien supérieur (Jurassique moyen). Haliotis 6:185–218

Chimšiašvili NG, Kamsyeva-Elpatevskaja VG, Bodylevskij VI et al. (1958) Nadsemejstvo Perisphinctaceae. In: Orlov JA (ed) Osnovy paleontologii, molluski-golovonogie, II. Moskva, pp 85–96

Cobban WA (1951) Scaphitoid cephalopods of the Colorado Group. US Geol Surv Prof Pap 239:1–42

Cobban WA (1953) Cenomanian ammonite fauna from the Mosby Sandstone of central Montana, US Geol Surv Prof Pap 243D:45–55

Cobban WA (1969) The Late Cretaceous ammonites *Scaphites leei* Reeside and *Scaphites hippocrepis* (DeKay) in the Western Interior of the United States. US Geol Surv Prof Pap 619:1–29

Cobban WA (1984) Molluscan record from a mid-Cretaceous borehole in Weston County, Wyoming. US Geol Surv Prof Pap 1271:1–24

Cobban WA (1987) The Upper Cretaceous ammonite *Rhaeboceras* Meek in the Western Interior of the United States. US Geol Surv Prof Pap 1477:1–15

Cobban WA (1988a) *Tarrantoceras* Stephenson and related ammonoid genera from Cenomanian (Upper Cretaceous) rocks in Texas and the Western Interior of the United States. US Geol Surv Prof Pap 1473:1–30

Cobban WA (1988b) The Upper Cretaceous ammonite *Watinoceras* Warren in the Western Interior of the United States. US Geol Surv Bull 1788:1–15

Cobban WA, Hook SC (1983) Mid-Cretaceous (Turonian) ammonite fauna from Fence Lake area of west-central New Mexico, NM. Bur Mines Miner Resour Mem 41:1–50

Cobban WA, Kennedy WJ (1990) Variation and ontogeny of *Calycoceras (Proeucalycoceras) canitaurinum* (Haas, 1949) from the Upper Cretaceous (Cenomanian) of the Western Interior of the United States. US Geol Surv Bull 1881:BI–B7

Cobban WA, Kennedy WJ (1991a) A giant scaphite from the Turonian (Upper Cretaceous) of the Western Interior of the United States. US Geol Surv Bull 1934:A1–A2

Cobban WA, Kennedy WJ (1991b) Evolution and biogeography of the Cenomanian (Upper Cretaceous) ammonite *Metoicoceras* Hyatt. 1903, with a revision of *Metoicoceras praecox* Haas, 1949. US Geol Surv Bull 1934:BI–B11

Cobban WA, Kennedy WJ (1991c) *Baculites thomi* Reeside 1927, an Upper Cretaceous ammonite in the Western Interior of the United States. US Geol Surv Bull 1934:C

Cobban WA, Kennedy WJ (1992) Campanian *Trachyscaphites spiniger* ammonite fauna in northeast Texas. Palaeontology 35(1):63–93

Cobban WA, Kennedy WJ (1993) The Upper Cretaceous dimorphic pachydiscid ammonite *Menuites* in the Western Interior of the United States. US Geol Surv Prof Pap 1533:1–14

Cobban WA, Kennedy WJ (1994a) Upper Cretaceous ammonites from the Coon Creek Tongue of the Ripley Formation at its type locality in McNairy County, Tennessee. US Geol Surv Bull 2073:BI–B12

Cobban WA, Kennedy WJ (1994b) Middle Campanian (Upper Cretaceous) ammonites from the Pecan Gap Chalk of central and northeastern Texas. US Geol Surv Bull 2073:D1-D9

Cobban WA, Hook SC, Kennedy WJ (1989) Upper Cretaceous rocks and ammonite faunas of southwestern New Mexico. NM Bur Mines Miner Resour Mem 45:1–137

Coemme S (1917) Note critique sur le genre *Cadomoceras*. Bull Soc Géol Fr 417:44–54

Collins D, Ward PD (1987) Adolescent growth and maturity in Nautilus. In: Saunders WB, Landman NH (eds) Nautilus. The biology and paleobiology of a living fossil, vol 6. Plenum Press, New York 421–432 (Topics in Geobiology)

Contini D, Marchand D, Thierry J (1984) Réflexion sur la notion de genre et de sous-genre chez les ammonites: Exemples pris essentiellement dans le Jurassique moyen. Bull Soc Géol Fr 26(4):653–661

Cooper MR, Kennedy WJ (1977) A revision of the Baculitidae of the Cambridge Greensand. Neues Jahrb Geol Paläontol Mh 11:641–658

Cooper MR, Kennedy WJ (1979) Uppermost Albian (Stoliczkaia dispar zone) ammonites from the Angolan littoral. Ann S Afr Mus 77(10):175–308

Cooper MR, Kennedy WJ (1987) A revision of the Puzosiinae (Cretaceous ammonites) of the Cambridge Greensand. Neues Jahrb Geol Paläontol Abh 174(1):105–121

Cope JCW (1992) Dimorphism in a Tethyan Early Jurassic *Juraphyllites*. Lethaia 25:439–441

Cox B.M. (1988) English Callovian (Middle Jurassic) perisphinctid ammonites. Part 1. Monogr Palaeontogr Soc Lond 140:1–54

Czarnocki J (1989) Klimenie Gór Świętokrzyskich (Prace Państwowego Instytutu Geologicznego), vol 127 Wydawnictwa Geologiczne, Warszawa, pp 1–91

Davis RA (1972) Mature modification and dimorphism in selected late Paleozoic ammonoids. Bull Am Paleontol 62(272):23–130

Davis RA, Furnish WM, Glenister BF (1969) Mature modification and dimorphism in late Paleozoic ammonoids. In: Westermann GEG (ed) Sexual dimorphism in fossil Metazoa and taxonomic implications (IUGS, Series A1). Schweizerbart, Stuttgart, pp 101–110

Davis RA, Landman NH, Dommergues J-L, Marchand D, Bucher H (1996) Mature modifications and dimorphism in ammonoid cephalopods. In: Landman NH, Tanabe K, Davis RA (eds) Ammonoid paleobiology. Plenum, New York

Davitashvili LSh, Khimshiashvili NG (1954) On the question of the biological significance of the apertural formation of ammonites. Works Paleobio Sect Acad Sci Georgian Soviet Soc Re 2:44–76 [in Russian]

De Baets K Klug C Korn D Landman NH (2012) Evolutionary trends in ammonoid embryonal development. Evolution 66:1788–1806

De Baets K Klug C Monnet C (2013a) Intraspecific variability through ontogeny in early ammonoids. Paleobiology 39:75–94

De Baets K Klug C Korn D Bartels C Poschmann M (2013b) Emsian Ammonoidea and the age of the Hunsrück Slate (Rhenish Mountains, Western Germany). Palaeontogr A 299(1–6):1–113

De Baets K, Bert D, Hoffmann R, Monnet C, Monnet C, Yacobucci MM, Klug C (2015a) Ammonoid Intraspecific Variability. This volume

De Baets K, Landman NH, Tanabe K (2015) Ammonoid embryonic development. This volume

Delanoy G, Poupon A (1992) Sur le genre *Lytocrioceras* Spath, 1924 (Ammonoidea, Ancyloceratina). Geobios 25(3):367–382

Delanoy G, Ropolo P, Magnin A, Autran G, Poupon A, Gonnet R (1995) Sur le dimorphisme chez les Ancyloceratina (Ammonoidea) du Crétacé inférieur. Comptes Rendus Acad Sci Sér IIa 321:537–543

Demanet F (1943) Les horizons marins du Westphalien de la Belgique et leurs faunes. Mém Mus R Hist Nat Belg 101:1–166

Dietl G (2013) Der Braune Jura ober-ε und ζ. Fossilien. Sonderheft “Der Braunjura am Fuß der Schwäbischen Alb”, pp 30–46

Dietze V, Callomon JH, Schweigert G, Chandler RB (2005) The ammonite fauna and biostratigraphy of the Lower Bajocian (Ovale and Laeviscula zones) of E Swabia (S Germany). Suttg Beitr Naturk B 353:1–82

Dietze V, Chandler RB, Callomon JH (2007) The ovale zone (Lower Bajocian, Middle Jurassic) at Little Down Wood (Dundry Hill, Somerset, SW England). Stuttg Beitr Naturk B 368:1–45

Dietze V, Bosch K, Franz M, Kutz M, Schweigert G, Wannenmacher N, Studer S (2013) Die Humphriesianum-Zone (Unter-Bajocium, Mitteljura) am Kahlenberg bei Ringsheim (Oberrheingraben, SW Deutschland). Palaeodiversity 6:29–61

Djanélidzé A (1922) Les *Spiticeras* du sud-est de la France. Mém Explic Carte Géol détaill Fr VI: 1–255

Doguzhaeva L (1982) Rhythms of ammonoid shell secretion. Lethaia 15:385–394

Doguzhaeva LA (1981) The wrinkle layer in the shell of ammonoids. Paleontol Z 1:38–48 [In Russian]

Doguzhaeva LA, Kabanov GK (1988) Muscle scars in ammonoids. Dokl Akad Nauk USSR 301:210–212

Doguzhaeva LA, Mikhailova IA (1991) New data on muscle system of heteromorphic ammonites. Dokl Akad Nauk USSR 318(4):981–984

Doguzhaeva LA, Mikhailova IA (2002) The jaw apparatus of the heteromorphic ammonite *Australiceras whitehouse*, 1926 (Mollusca: Cephalopoda) from the Aptian of the Volga Region. Dokl Akad Nauk USSR 382:38–40

Doguzhaeva LA, Mutvei H (1991) Organization of the soft body in *Aconeceras* (Ammonitina), interpreted on the basis of shell morphology and muscle scars. Palaeontogr A 218:17–33

Doguzhaeva LA, Mutvei H (1993) Shell ultrastructure, muscle scars, and buccal apparatus in ammonoids. Geobios 15:111–119

Doguzhaeva LA, Mutvei H (1996) Attachment of the body to the shell in ammonoids. In: Landman NH, Tanabe K, Davis RA (eds) Ammonoid paleobiology. Plenum, New York

Dommergues J-L (1993) The Jurassic ammonite *Coeloceras*: An atypical example of dimorphic progenesis elucidated by cladistics. Lethaia 27(2):143–152

Donovan DT, Callomon JH, Howarth MK (1981). Classification of the Jurassic Ammonitina. In: House MR, Senior JR (eds) The Ammonoidea, vol 18. Systematics Association by Academic Press, pp 101–155

Dzik J (1984) Phylogeny of the Nautiloidea. Palaeont Pol 45:1–220

Dzik J (1990a) The concept of chronospecies in ammonites. In: Pallini G, Cecca F, Cresta S, Santantonio M (eds) Atti del secondo convegno internazionale Fossili Evoluzione Ambiente. Pergola

Dzik J (1990b) The ammonite *Acrochordiceras* in the Triassic of Silesia. Acta Palaeontol Pol 35(1–2):49–65

Ebbighausen V, Korn D (2007) Conch geometry and ontogenetic trajectories in the triangularly coiled Late Devonian ammonoid *Wocklumeria* and related genera. Neues Jahrb Geol Paläontol Abh 244:9–41

Elmi S (1967) Le Lias supérieur et le Jurassique moyen de l'Ardeche. Docum Lab. Géol Lyon 19(1–3):1–845

Elmi S, Mangold C (1966) Eude de quelques *Oxycerites* du Bathonien Inférieur. Trav Lab Géol Fac Sc Lyon, NS 13:143–181

Enay R (1966) L'Oxfordien dans la moitié sud du Jura français. Nouv Arch Mus Hist Nat Lyon 8:1–624

Ernst HU, Klug C (2011) Perlboote und Ammonshörner weltweit. Nautilids and Ammonites worldwide. Pfeil Verlag, München

Etches S, Clarke J, Callomon J (2009) Ammonite eggs and ammonitellae from the Kimmeridge clay formation (Upper Jurassic) of Dorset, England. Lethaia 42:204–217

Fatmi AN (1969) Dimorphism in some Jurassic and Lower Cretaceous ammonites from West Pakistan. Geonews (Geol. Surv. Pakistan) 1(2):6–13

Foord AH, Crick GC (1897) Catalogue of the fossil Cephalopoda in the British Museum (Natural History). Part III. Containing the Bactritidae and Part of the suborder Ammonoidea. British Museum (Natural History), London

Förster R, Weier H (1983) Ammoniten und Alter der Niongala-Schichten (Unterapt, Slid-Tanzania). Mitt Bayer Staatsslg Paläontol Hist Geol 23:51–76

Förster R, Meyer R, Risch H (1983) Ammoniten und planktonische Foraminiferen aus den Eibrunner Mergeln (Regensburger Kreide, Nordostbayern). Zitteliana 10:123–141

Frest TJ, Glenister BF, Furnish WM (1981) Pennsylvanian-Permian Cheiloceratacean ammonoid families Maximitidae and Pseudohaloritidae. Paleontol Soc Mem 11 (Paleontol suppl to) 55(3):1–46

Furnish WM, Knapp WD (1966) Lower Pennsylvanian fauna from eastern Kentucky; Part 1, Ammonoids. J Paleontol 40(2):296–308

Futakami M 1990. Turonian collignoniceratid ammonites from Hokkaido, Japan. Stratigraphy and paleontology of the Cretaceous in the Ishikari province, central Hokkaido. (Part 3, l, vol 1). Kawamura Gakuen Women's University, Japan, pp 235–260

Gabr HR, Hanlon RT, Hanafy MH, El-Etreby SG (1998) Maturation, fecundity and seasonality of reproduction of two commercially valuable cuttlefish, *Sepia pharaonis* and *S. dollfusi*, in the Suez Canal. Fisch Res 36:99–115

Ganguly T, Bardhan S (1993) Dimorphism in *Placenticeras mintoi* from the Upper Cretaceous Bagh Beds, central India. Cretac Res 14:747–756

Gauthier H, Branger P, Boursicot P-Y, Trévisan M, Marchand D (2002) La faune *d'Orthogarantiana* Bentz (Garantianinae, Stephanoceratidae) de la sous-zone à Polygyralis (z. à Niortense, Bajocien sup.) nouvellement découverte au nord de Niort (Deux-Sèvres, France). Une preuve du dimorphisme *Orthogarantiana/Strenoceras*. Géol Fr 1:81–86

Gemmellaro GG (1887) La fauna dei calcari con Fusulina della valle del Fiume Sosio nella Provincia di Palermo. Fascio I-Cephalopoda, Ammonoidea. G Sci Nat Econ 19:1–106

Geyer OF (1969) The ammonite genus *Sutneria* in the Upper Jurassic of Europe. Lethaia 2:63–72

Gould SJ (1977) Ontogeny and phylogeny. Harvard University Press, Cambridge

Griffin LE (1900) The anatomy of *Nautilus pompilius*. Mem Natl Acad Sci 8:101–203

Guex J (1967) Dimorphisme sexuel d'un groupe d'*Hammatoceras* et position systématique du genre *Onychoceras*. Bull Soc Vaud Sci Nat 69:423–434

Guex J (1968) Note préliminaire sur le dimorphisme sexuel des Hildocerataceae du Toarcien moyen et supérieur de l'Aveyron (France). Soc Vaud Sci Nat Lausanne Bull 70(327):57–84

Guex J (1973) Dimorphisme des Dactylioceratidae du Toarcien. Eclogae Geol Helv 66:545–583

Guex J (1981) Quelques cas de dimorphisme chez les ammonïodés du Lias inférieur. Bull Soc Vaud Sci Nat 75:239–248

Hahn W (1971) Die Tulitidae S. Buckman, Sphaeroceratidae S. Buckman und Clydoniceratidae S. Buckman (Ammonoidea) des Bathoniums (Brauner Jura ε) im südwestdeutschen Jura. Jahresh Geol Landesamtes Baden-Württ 13:55–122

Hammer Ø, Bucher H (2006) Generalized ammonoid hydrostatics modelling, with application to *Intornites* and intraspecific variation in *Amaltheus*. Paleontol Res 10:91–96

Hanlon RT, Forsythe JW (2008) Sexual cannibalism by *Octopus cyanea* on a Pacific coral reef. Mar Freshw Behav Physiol 41:19–28

Haug E (1897) Observations a la suite d'une note de Ph. Glangeaud sur las forme de l'ouverture de quelques ammonites. Bull Soc Géol Fr Ser 325:107

Hayasaka S, Oki K, Tanabe K, Saisho T, Shinomiya A (1987) On the habitat of *Nautilus pompilius* in Tafton Strait (Philippines) and the Fiji Islands. In: Saunders WB, Landman NH (eds) *Nautilus*. The biology and paleobiology of a living fossil. Plenum, New York

Hillebrandt A, Krystyn L (2009) On the oldest Jurassic ammonites of Europe (Northern Calcareous Alps, Austria) and their global significance. Neues Jahrb Geol Paläontol Abh 253:163–195

Hirano H (1978) Phenotypic substitution of *Gaudryceras* (a Cretaceous ammonite). Trans Proc Palaeontol Soc Jpn NS 109:235–258

Hirano H (1979) Importance of transient polymorphism in systematics of Ammonoidea. The Gakujutsu Kenkyu Sch Educ Waseda Univ Ser Biol Geol 28:35–43

Hoffmann R (2010) New insights on the phylogeny of the Lytoceratoidea (Ammonitina) from the septal lobe and its functional interpretation, vol 29. Revue de Paléobiologie, Genève, pp 1–156

Hoffmann R, Schultz JA, Schellhorn R, Rybacki E, Keupp H, Gerden SR, Lemanis R, Zachow S (2013) Non-invasive imaging methods applied to neo- and paleontological research. Biogeosci Discuss 10:18803–18851. doi:10.5194/bgd-10-18803-2013

Housa V (1965) Sexual dimorphism and the system of Jurassic and Cretaceous Ammonoidea (Preliminary note). Casas Nar Muz 134:33–35

House MR (1970) The goniatite wrinkle layer. Smithson Contrib Paleontol 3:23–32

Howarth MK (1992) Tithonian and Berriasian ammonites from the Chia Gara Formation innorthern Iraq. Palaeontology 35(3):597–655

Howarth MK (2013) Part L, Revised, Volume 3B, Chapter 4: Psiloceratoidea, Eoderoceratoidea, Hildoceratoidea. Treatise 57:1–139

Immel H (1987) Die Kreideammoniten der nordlichen Kalkalpen. Zitteliana 15:3–163

Ivanov AN (1971) About some growth alterations in ammonite shells. Bull Mosc Soc Nat Hist Geol Sect 46:155 [in Russian]

Ivanov AN (1975) Late ontogeny of ammonites and, in particular, of the micro-, macro-, and megaconchs. Coll Stud Notes Sci W Yarosl St Ped Inst 142:5–57 [in Russian]

Ivanov AN (1985) Were micro- and macroconchs of ammonites sexual dimorphs? In: Taxonomy and ecology of Cephalopoda. Scientific papers. Academy of Sciences of the USSR, Zoological Institute, Scientific Council on the Problem of "Biological Bases of Utilization, Remaking, and Protection of the Animal World", Malacological Committee, Leningrad, 32–34 [in Russian]

Jackson GD, Moltschaniwskyj NA (2002) Spatial and temporal variation in growth rates and maturity in the Indo-Pacific squid *Sepioteuthis lessoniana* (Cephalopoda: Loliginidae). Mar Biol 140:747–754

Jagt JWM (1989) Ammonites from the early Campanian Vaals Formation at the CPL Quarry (Haccourt, Liege, Belgium) and their stratigraphic implications. Meded. Rijks Geol Dienst 43(1):1–18

Jagt JWM, Kennedy WJ (1989) *Acanthoscaphites varians* (Lopuski, 1911) (Ammonoidea) from the Upper Maastrichtian of Haccourt, NE Belgium. Geol Mijnb 68:237–240

Jagt JWM, Kennedy WJ (1994) *Jeletzkytes dorfi* Landman and Waage 1993, a North American ammonoid marker from the lower Upper Maastrichtian of Belgium, and the numerical age of the Lower/Upper Maastrichtian boundary. Neues Jahrb Geol Paläontol Mh 4:239–245

Jagt JWM, Kennedy WJ, Burnett J (1992) *Acanthoscaphites tridens* (Kner, 1848) (Ammonoidea) from the Vijlen Member (Lower Maastrichtian) of Gulpen, Limburg, The Netherlands. Geol Mijnb 71:15–21

Jeletzky JA, Waage KM (1978) Revision of *Ammonites conradi* Morton 1834, and the concept of *Discoscaphites* Meek 1870. J Paleontol 52:1119–1132

Kakabadze MV (2004) Intraspecific and intrageneric variabilities and their implication for the systematics of Cretaceous heteromorph ammonites; a review. Scr Geol 128:17–37

Kant R (1973) Allometrisches Wachstum paläozoischer Ammonoideen: Variabilität und Korrelation einiger Merkmale. Neues Jahrb Geol Paläontol Abh 143:153–192

Kaplan U, Kennedy WJ, Wright CW (1987) Turonian and Coniacian Scaphitidae from England and North-Western Germany. Geol J 103:5–39

Kassab AS, Hamama HH (1991) Polymorphism in the Upper Cretaceous ammonite *Libycoceras ismaeli* (Zittel). J Afr Earth Sci 12(3):437–448

Keferstein W (1866) Cephalopoden. In: Die Klassen und Ordnungen des Thierreichs wissenschaftlich dargestellt in Wort und Bild von Hans Georg Bronn. Fortgesetzt von Wilhelm Keferstein, (vol 3, part 2). Verlag Winter, Leipzig, pp 1337–1406

Kemper E (1982) Die Ammoniten des späten Apt und frühen Alb Nordwestdeutschlands. Geol Jahrb A 65:553–557

Kennedy WJ (1984) Systematic paleontology and stratigraphic distribution of the ammonite faunas of the French Coniacian. Spec Pap Palaeontol 31:1–160

Kennedy WJ (1986a) Observations on *Astiericeras astierianum* (d'Orbigny, 1842) (Cretaceous Ammonoidea). Geol Mag 123(5):507–513

Kennedy WJ (1986b) The ammonite fauna of the Calcaire a *Baculites* (Upper Maastrichtian) of the Cotentin Peninsula (Manche, France). Palaeontology 29:25–83

Kennedy WJ (1986c) The ammonite fauna of the type Maastrichtian with a revision of *Ammonites colligatus* Binkhorst 1861. Bull Inst R Sci Nat Belg (Sciences de la Terre) 56:151–267

Kennedy WJ (1986d) Campanian and Maastrichtian ammonites from northern Aquitaine, France. Spec Pap Palaeontol 36:1–145

Kennedy WJ (1988) Late Cenomanian and Turonian ammonite faunas from north-east and central Texas. Spec Pap Palaeontol 39:1–131

Kennedy WJ (1989) Thoughts on the evolution and extinction of Cretaceous ammonites. Proc Geol Assoc 100(3):251–279

Kennedy WJ (1993) Campanian and Maastrichtian ammonites from the Mons Basin and adjacent areas. vol 63. Bull lnst R Sci Nat Belg (Sciences de la Terre) 63:99–131 (Belgium)

Kennedy WJ, Christensen WK (1991) Coniacian and Santonian ammonites from Bornholm, Denmark. Bull Geol Soc Den 38:203–226

Kennedy WJ, Cobban WA (1976) Aspects of ammonite biology, biogeography, and biostratigraphy. Spec Pap Palaeontol 17:1–93

Kennedy WJ, Cobban WA (1988a) *Litophragmatoceras incomptum* gen. et sp. nov. (Cretaceous Ammonoidea), a cryptic micromorph from the Upper Cenomanian of Arizona. Geol Mag 125(5):535–539

Kennedy WJ, Cobban WA (1988b) Mid-Turonian ammonite faunas from northern Mexico. Geol Mag 125:593–612

Kennedy WJ, Cobban WA (1990) Cenomanian ammonite faunas from the Woodbine Formation and lower part of the Eagle Ford Group, Texas. Palaeontology 33:75–154

Kennedy WJ, Cobban WA (1991a) Coniacian ammonite faunas from the United States Western Interior. Spec Pap Palaeontol 45:1–96

Kennedy WJ, Cobban WA (1991b) Upper Cretaceous (upper Santonian) *Boehmoceras* fauna from the Gulf Coast region of the United States. Geol Mag 128(2):167–189

Kennedy WJ, Cobban WA (1993a) Ammonites from the Saratoga Chalk (Upper Cretaceous), Arkansas. J Paleontol 67:404–434

Kennedy WJ, Cobban WA (1993b) Campanian ammonites from the Annona Chalk near Yancy, Arkansas. J Paleontol 67:83–97

Kennedy WJ, Cobban WA (1993c) Maastrichtian ammonites from the Corsicana Formation in northeast Texas. Geol Mag 130(1):57–67

Kennedy WJ, Delamette M (1994) *Neophlycticeras* Spath, 1922 (Ammonoidea) from the Upper Albian of Ain, France. Neues Jahrb Geol Palaontol Abh 191(1):1–24

Kennedy WJ, Henderson RA (1992) Non-heteromorph ammonites from the Upper Maastrichtian of Pondicherry, South India. Palaeontology 35:381–442

Kennedy WJ, Juignet P (1983) A revision of the ammonite faunas of the type Cenomanian. 1. Introduction, Ancyloceratina. Cretac Res 4:3–83

Kennedy WJ, Juignet P (1984) A revision of the ammonite faunas of the type Cenomanian. 2. The families Binneyitidae, Desmoceratidae, Engonoceratidae, Placenticeratidae, Hoplitidae, Schloenbachiidae, Lyelliceratidae and Forbesiceratidae. Cretac Res 5:93–161

Kennedy WJ, Juignet P (1994) A revision ofthe ammonite faunas of the type Cenomanian, 5. Acanthoceratinae *Calycoceras* (*Calycoceras*), *C.* (*Gentoniceras*) and *C.* (*Newboldiceras*). Cretac Res 15:17–57

Kennedy WJ, Klinger HC (1979) Cretaceous faunas from Zululand and Natal, South Africa. The ammonite superfamily Haplocerataceae Zittel, 1884. Ann S Afr Mus 77(6):85–121

Kennedy WJ, Klinger HC (1985) Cretaceous faunas from Zululand and Natal, South Africa. The ammonite family Kossmaticeratidae Spath. 1922. Ann S Afr Mus 95(5):165–231

Kennedy WJ, Klinger HC (1993) On the affinities of *Cobbanoscaphites* Collignon, 1969 (Cretaceous Ammonoidea). Ann S Afr Mus 102(7):265–271

Kennedy WJ, Summesberger H (1984) Upper Campanian ammonites from the Gschliefgraben (Ultrahelvetic, Upper Austria). Beitr Paläontol Österr 11:149–206

Kennedy WJ, Summesberger H (1987) Lower Maastrichtian ammonites from Nagoryany (Ukrainian SSR). Beitr Paläontol Österr 13:25–78

Kennedy WJ, Wright CW (1979) Vascoceratid ammonites from the type Turonian. Palaeontology 22:665–683

Kennedy WJ, Wright CW (1981) *Euhystrichoceras* and *Algericeras*, the last mortoniceratine ammonites. Palaeontology 24:417–435

Kennedy WJ, Wright CW (1983) *Ammonites polyopsis* Dujardin, 1837, and the Cretaceous ammonite family Placenticeratidae Hyatt, 1900. Palaeontology 26:855–873

Kennedy WJ, Wright CW (1984a) The Cretaceous ammonite *Ammonites requienianus* d'Orbigny, 1841. Palaeontology 27:281–293

Kennedy WJ, Wright CW (1984b) The affinities of the Cretaceous ammonite *Neosaynoceras* Breistroffer, 1947. Palaeontology 27:159–167

Kennedy WJ, Wright CW (1985a) Evolutionary patterns in Late Cretaceous ammonites. Spec Pap Palaeontol 33:131–143

Kennedy WJ, Wright CW (1985b) *Mrhiliceras* n.g. (Cretaceous Ammonoidea), a new Cenomanian mantelliceratine. Neues Jahrb Geol Paläontol Mh 9:513–526

Kennedy WJ, Wright CW, Hancock JM (1980) Collignoniceratid ammonites from the mid-Turonian of England and northern France. Palaeontology 23:557–603

Kennedy WJ, Juignet P, Hancock JM (1981a) Upper Cenomanian ammonites from Anjou and the Vendee, western France. Palaeontology 24:25–84

Kennedy WJ, Klinger HC, Summesberger H (1981b) Cretaceous faunas from Zululand and Natal, South Africa. Additional observations on the ammonite subfamily Texanitinae Collignon, 1948. Ann S Afr Mus 86(4):115–155

Kennedy WJ, Wright CW, Klinger HC (1983) Cretaceous faunas from Zululand and Natal, South Africa. The ammonite subfamily Barroisiceratinae Basse, 1947. Ann S Afr Mus 90(6):241–324

Kennedy WJ, Bilotte M, Lepicard B, Segura F (1986) Upper Campanian and Maastrichtian ammonites from the Petites-Pyrenees, southern France. Eclogae Geol Helv 79:1001–1037

Kennedy WJ, Wright CW, Hancock JM (1987) Basal Turonian ammonites from west Texas. Palaeontology 30:27–74

Kennedy WJ, Cobban WA, Hancock JM, Hook SC (1989) Biostratigraphy of the Chispa Summit Formation at its type locality: A Cenomanian through Turonian reference section for trans-Pecos Texas. Bull Geol Inst Univ Upps NS 15:39–119

Kennedy WJ, Hansotte M, Bilotte M, Burnett J (1992) Ammonites and nannofossils from the Campanian of Nalzen (Ariege, France). Geobios 25(2):263–278

Kennedy WJ, Landman NH, Cobban WA, Scott GR (2000) Late Campanian (Cretaceous) heteromorph ammonites from the western interior of the United States. Bull Am Mus Nat Hist 251:1–88

Kennedy WJ, Cobban WA, Klinger HC (2002) Muscle attachment and mantle-related features in Upper Cretaceous *Baculites* from the United States Western Interior. Abh Geol Bund-Anst 57:89–112

Keupp H (2000) Ammoniten—paläobiologische Erfolgsspiralen. Thorbecke, Stuttgart

Keupp H (2012) Atlas zur Paläopathologie der Cephalopoden. Berl Paläobiol Abh 12:1–390

Keupp H, Riedel F (2010) Remarks on the possible function of the apophyses of the Middle Jurassic microconch ammonite *Ebrayiceras sulcatum* (Zieten 1830), with a discussion on the palaeobiology of Aptychophora in general. Neues Jahrb Geol Paläontol Abh 255:301–314

Klinger HC (1989) The ammonite subfamily Labeceratinae Spath, 1925. Systematics, phylogeny, dimorphism and distribution (with a description of a new species). Ann S Afr Mus 98(7):189–219

Klinger HC, Kennedy WJ (1977) Cretaceous faunas from Zululand, South Africa and southern Mozambique. The Aptian Ancyloceratidae (Ammonoidea). Ann S Afr Mus 73(9):215–359

Klinger HC, Kennedy WJ (1989) Cretaceous faunas from Zululand and Natal. South Africa. The ammonite family Placenticeratidae Hyatt, 1900, with comments on the systematic position of the genus *Hypengonoceras* Spath, 1924. Ann S Afr Mus 98(9):241–408

Klinger HC, Kennedy WJ (1992) Cretaceous faunas from Zululand and Natal, South Africa. Barremian representatives of the ammonite family Ancyloceratidae Gill, 1871. Ann S Afr Mus 101(5):71–138

Klinger HC, Kennedy WJ (1993) Cretaceous faunas from Zululand and Natal, South Africa. The heteromorph ammonite genus *Eubaculites* Spath, 1926. Ann S Afr Mus 102(6):185–264

Klug C (2001) Life-cycles of Emsian and Eifelian ammonoids (Devonian). Lethaia 34:215–233

Klug C (2004) Mature modifications, the black band, the black aperture, the black stripe, and the periostracum in cephalopods from the Upper Muschelkalk, vol (Middle Triassic, Germany). Mitt Geol-Paläont Inst Univ Hamburg 88:63–78

Klug C, Korn D (2003) Morphological pathways in the evolution of Early and Middle Devonian ammonoids. Paleobiology 29:329–348

Klug C, Korn D (2004) The origin of ammonoid locomotion. Acta Palaeontol Pol 49:235–242

Klug C, Korn D, Richter U, Urlichs M (2004) The black layer in cephalopods from the German Muschelkalk (Middle Triassic). Palaeontology 47:1407–1425

Klug C, Brühwiler T, Korn D, Schweigert G, Brayard A, Tilsley J (2007) Ammonoid shell structures of primary organic composition. Palaeontology 50:1463–1478

Klug C, Meyer E, Richter U, Korn D (2008) Soft-tissue imprints in fossil and Recent cephalopod septa and septum formation. Lethaia 41:477–492

Klug C, Riegraf W, Lehmann J (2012) Soft-part preservation in heteromorph ammonites from the Cenomanian-Turonian Boundary Event (OAE 2) in the Teutoburger Wald (Germany). Palaeontology 55:1307–1331

Korn D (1992) Heterochrony in the evolution of Late Devonian ammonoids. Acta Palaeontol Pol 37(1):21–36

Korn D (2012) Quantification of ontogenetic allometry in ammonoids. Evol Dev 14(6):501–514

Korn D, Ebbighausen V (2008) The Early Carboniferous (Mississippian) ammonoids from the Chebket el Hamra (Jerada Basin, Morocco). Foss Rec 11:83–156. doi:10.1002/mmng.200800004

Korn D, Klug C (2002). Ammoneae Devonicae. In: Riegraf W (ed) Fossilium Catalogus 1: Animalia, vol 138.Backhuys, Leiden, pp 1–375

Korn D, Klug C (2003) Morphological pathways in the evolution of early and middle Devonian ammonoids. Paleobiology 29:329–348

Korn D, Klug C (2007) Conch form analysis, variability, and morphological disparity of a Frasnian (Late Devonian) ammonoid assemblage from Coumiac (Montagne Noire, France). In: Landman NH, Davis RA, Manger W, Mapes RH (eds) Cephalopods—present and past. Springer, New York

Korn D, Titus A (2006) The ammonoids from the Three Forks Shale (Late Devonian) of Montana. Foss Rec 9:198–212

Korn D, Klug C, Mapes RH (1999) Viséan and Early Namurian Ammonoids from the Tafilalt (Eastern Anti-Atlas, Morocco). Abh Geol Bundesanst 54:345–375

Korn D, Bockwinkel J, Ebbighausen V (2010) The ammonoids from the Argiles de Teguentour of Oued Temertasset (early Late Tournaisian; Mouydir, Algeria). Foss Rec 13:35–152

Korn D, Mapes RH, Klug C (2014) The coarse wrinkle layer of Palaeozoic ammonoids: new evidence from the Early Carboniferous of Morocco. Palaeontology 57:771–781. doi:10.1111/pala.12087

Kraft S, Korn D, Klug C (2008) Ontogenetic patterns of septal spacing in Carboniferous ammonoids. Neues Jahrb Geol Miner Abh 250(1):31–44

Krimholc GJ, Sazonov NT, Kamsyeva-Elpatevskaja VG (1958a) Nadsemejstvo Stephanocerataceae. In: Orlov JA (ed) Osnovy Paleontologii, Molluski-Golovonogie, II. Izdatel'stvo Akademii Nauk SSSR, Moskva, pp 75–79 [in Russian]

Krimholc GJ, Kamsyeva-Elpatevskaja VG, Kachadze IP (1958b) Nadsemejstvo Kosmocerataceae. In: Orlov JA (ed) Osnovy Paleontologii, Molluski-Golovonogie, II. Izdatel'stvo Akademii Nauk SSSR, Moskva, pp 79–82 [in Russian]

Kulicki C (1974) Remarks on the embryogeny and postembryonal development of ammonites. Acta Palaeontol Pol 19:201–224

Kulicki C, Tanabe K, Landman NH, Mapes RH (2001) Dorsal shell wall in ammonoids. Acta Palaeontol Pol 46:23–42

Landman NH (1987) Ontogeny of Upper Cretaceous (Turonian-Santonian) scaphitid ammonites from the Western Interior of North America: Systematics, developmental patterns, and life history. Bull Am Mus Nat Hist 185(2):117–241

Landman NH (1989) Iterative progenesis in Upper Cretaceous ammonites. Paleobiology 15:95–117

Landman NH, Waage KM (1986) Shell abnormalities in scaphitid ammonites. Lethaia 19:211–224
Landman NH, Waage KM (1993) Scaphitid ammonites of the Upper Cretaceous (Maastrichtian) Fox Hills Formation in South Dakota and Wyoming. Bull Am Mus Nat Hist 215:1–257
Landman NH, Dommergues J-L, Marchand D (1991) The complex nature of progenetic species—examples from Mesozoic ammonites. Lethaia 24:409–421
Landman NH, Mapes RH, Cruz C (2010) Jaws and soft tissues in ammonoids from the Lower Carboniferous (Upper Mississippian) Bear Gulch Beds, Montana, USA. In: Tanabe K, Shigeta Y, Sasaki T, Hirano H (eds) Cephalopods—present and past. Tokai University Press, Tokyo
Landman NH, Cobban WA, Larson NL (2012) Mode of life and habitat of scaphitid ammonites. Geobios 45:87–98
Leanza H, Zeiss A (1992) On the ammonite fauna of the lithographic limestones from the Zapala region (Neuquén province, Argentina), with the description of a new genus. Zbl Geol Paläontol I 1991(6):1841–1850
Lehmann U (1966) Dimorphismus bei Ammoniten der Ahrensburger Lias-Geschiebe. Paläontol Z 40(1–2):26–55
Lehmann U (1981) The ammonites: their life and their world. Cambridge University Press, New York
Longbridge LM, Smith PL, Tipper H (2006) The Early Jurassic ammonite *Badouxia* from British Columbia, Canada. Palaeontology 49:795–816
Luger P, Groschke M (1989) Late Cretaceous ammonites from the Wadi Qena area in the Egyptian Eastern Desert. Palaeontology 32(2):355–407
Macellari CE (1986) Late Campanian-Maaastrichtian ammonite fauna from Seymour Island (Antarctic Peninsula). Paleontol Soc Mem 18:1–55
Machalski M (2005) Late Maastrichtian and earliest Danian scaphitid ammonites from central Europe: Taxonomy, evolution, and extinction. Acta Palaeontol Pol 50:653–696
Maeda H (1991) Sheltered preservation: a peculiar mode of ammonite occurrence in the Cretaceous Yezo Group, Hokkaido, north Japan. Lethaia 24:69–82
Maeda H (1993) Dimorphism of Late Cretaceous false-puzosiine ammonites, *Yokoyamaoceras* Wright and Matsumoto, 1954 and *Neopuzosia* Matsumoto, 1954. Trans Proc Palaeontol Soc Jpn NS 169:97–128
Makowski H (1962) Problem of sexual dimorphism in ammonites. Palaeontol Pol 12:1–92
Makowski H (1971) Some remarks on the ontogenetic development and sexual dimorphism in the Ammonoidea. Acta Geol Pol 21:321–340
Makowski H (1991) Dimorphism and evolution of the goniatite *Tornoceras* in the Famennian of the Holy Cross Mountains. Acta Palaeontol Pol 36:241–254
Mangold C (1970) Morphoceratidae (Ammonitina-Perisphinctoidea) Bathoniens du Jura Méridional, de la Niévre et du Portugal. Geobios 3:43–130
Mangold C (1971) Les Perisphinctidae (Ammonitina) du Jura meridional au Bathonien et au Callovien. Doc Lab Géol Fac Sci Lyon 41:1–246
Mangold K (1987) Reproduction. In: Boyle PR (ed) Cephalopod life cycles. Comparative reviews, vol 2. Academic Press, London, pp 157–200
Mangold-Wirz K (1963) Biologie des Cephalopodes benthiques et nectoniques de la Mer Catalane. Vie Milieu (Suppl) 13:1–285
Mangold-Wirz K, Lu CC, Aldrich EA (1969) A reconsideration of forms of squid of the genus *Illex* (Illicinae, Ommastrephidae). II. Sexual dimorphism. Can J Zool 47:1153–1156
Mapes RH, Davis RA (1996) Color patterns in ammonoids. In: Landman NH, Tanabe K, Davis RA (eds) Ammonoid paleobiology. Plenum, New York
Mapes RH, Larson NL (2015) Ammonoid Color Patterns. This volume
Mapes RH, Sneck DA (1987) The oldest ammonoid "colour" patterns: description, comparison with *Nautilus,* and implications. Palaeontology 30:299–309
Marchand D (1976) Quelques précisions sur le polymorphisme dans la famille des Cardioceratidae Douville (Ammonoides). Haliotis 6:119–140
Marcinowski R (1980) Cenomanian ammonites from German Democratic Republic, Poland, and the Soviet Union. Acta Geol Pol 30(3):215–325

Marcinowski R (1983) Upper Albian and Cenomanian ammonites from some sections of the Mangyshlak and Tuarkyr regions, Transcaspia, Soviet Union. Neues Jahrb Geol Paläontol Mh 3:156–180
Marcinowski R, Wiedmann J (1990) The Albian ammonites of Poland. Palaeontol Pol 50:1–94
Matsumoto T (1987a) Notes on *Forbesiceras* (Ammonoidea) from Hokkaido (Studies of Cretaceous ammonites from Hokkaido-LX). Trans Proc Palaeontol Soc Jpn NS 145:16–31
Matsumoto T (1987b) Notes on *Pachydesmoceras*, a Cretaceous ammonite genus. Proc Jpn Acad 63B:5–8
Matsumoto T (1988) A monograph of the Puzosiidae (Ammonoidea) from the Cretaceous of Hokkaido). Palaeont. Soc Jpn Spec Pap 30:1–131
Matsumoto T (1991a) On some acanthoceratid ammonites from the Turonian of Hokkaido (Studies of the Cretaceous ammonites from Hokkaido-LXIX). Trans Proc Palaeontol Soc Jpn NS 164:910–927
Matsumoto T (compiler, 1991b) The mid-Cretaceous ammonites of the family Kossmaticeratidae from Japan. Palaeontol Soc Jpn Spec Pap 33:1–143
Matsumoto T, Saito R (1987) Little known ammonite *Crandidiericeras* from Hokkaido (Studies of Cretaceous ammonites from Hokkaido-LVIII). Trans Proc Palaeontol Soc Jpn NS 145:1–9
Matsumoto T, Skwarko SK (1991) Ammonites of the Cretaceous Ieru Formation, western Papua New Guinea. BMR J Aust Geol Geophys 12(3):245–262
Matsumoto T, Skwarko SK (1993) Cretaceous ammonites from south-central Papua New Guinea. AGSO J Aust Geol Geophys 14(4):411–433
Matsumoto T, Takahashi T (1992) Ammonites of the genus *Acompsoceras* and some other acanthoceratid species from the Ikushunbetsu Valley, central Hokkaido. Trans Proc Palaeontol Soc Jpn NS 166:1144–1156
Matsumoto T, Toshimitsu S (1984) On the systematic positions of the two ammonite genera *Hourcquia* Collignon, 1965 and *Pseudobarroisiceras* Shimizu, 1932. Mem Fac Sci Kyushu Univ Ser D Geol 25(2):229–246
Matsumoto T, Suekane T, Kawashita Y (1989) Some acanthoceratid ammonites from the Yubari Mountains, Hokkaido-Part 2. Sci Rep Yokosuka City Mus 37:29–44
Matsumoto T, Nemoto M, Suzuki C (1990a) Gigantic ammonites from the Cretaceous Futaba Group of Fukushima Prefecture. Trans Proc Palaeontol Soc Jpn NS 157:366–381
Matsumoto T, Toshimitsu S, Kawashita Y (1990b) On *Hauericeras* de Grossouvre, 1894, a Cretaceous ammonite genus. Trans Proc Palaeontol Soc Jpn NS 158:439–458
Matsunaga T, Maeda H, Shigeta Y, Hasegawa K, Nomura S-I, Nishimura T, Misaki A, Tanaka (2008) First discovery of *Pravitoceras sigmoidale* Yabe from the Yezo Supergroup in Hokkaido, Japan. Paleontol Res 12:309–319. doi:10.2517/prpsj.12.309
Matyja BA (1986) Developmental polymorphism in Oxfordian ammonites. Acta Geol Pol 36(1–3):37–68
Matyja BA (1994) Developmental polymorphism in the Oxfordian ammonite subfamily Peltoceratinae. In: Palaeopelagos Special Publication 1. Proceedings of the 3rd Pergola International Symposium, Rome, pp 277–286
Matyja BA, Wierzbowski A (2001) Palaeogeographical distribution of early Bathonian ammonites of the *Asphinctites-Polysphinctites* group. Hantkeniana 3:89–103
McCaleb JA (1968) Lower Pennsylvanian ammonoids from the Bloyd Formation of Arkansas and Oklahoma. Geol Soc Am Spec Pap 96:1–123
McCaleb JA, Furnish WM (1964) The Lower Pennsylvanian ammonoid genus *Axinolobus* in the southern midcontinent. J Paleontol 38(2):249–255
McCaleb JA, Quinn JH, Furnish WM (1964) Girtyoceratidae in the southern midcontinent. Okla Geol Surv Circ 67:1–41
Meek FB, Hayden FV (1856) Descriptions of new fossil species of Mollusca collected by Dr. F. V. Hayden, in Nebraska Territory; together with a complete catalogue of all the remains of Invertebrata hitherto described and identified from the Cretaceous and Tertiary formations of that region. Proc Acad Nat Sci Phila 8:265–286

Meister C, Alzouma K, Lang J, Mathey B (1992) Les ammonites du Niger (Afrique occidentale) et la transgression transsaharienne au cours du Cénomanien-Turonien. Geobios 25(1):55–100

Melendez G, Fontana B (1993) Intraspecific variability, sexual dimorphism, and non-sexual polymorphism in the ammonite genus *Larcheria* Tintant (Perisphinctidae) from the middle Oxfordian of western Europe. In: House MR (ed) The ammonoidea: environment, ecology, and evolutionary change (Systematics Association Special), vol 47. Clarendon Press, Oxford

Miller AK (1944) Permian cephalopods. In: King RE, Dunbar CO, Cloud PE Jr, Miller AK (eds) Geology and paleontology of the Permian area northwest of Las Delicias, southwestern Coahuila, Mexico, vol 52. Geological Society of America, Washington, DC, pp 71–127 (Geologial Society of America Special Papers)

Miller AK, Furnish WM (1940) Permian ammonoids of the Guadalupe Mountain region and adjacent areas. Geol Soc Am Spec Pap 26:1–242

Mitta VV (2010) Late Volgian *Kachpurites* Spath (Craspeditinae, Ammonoidea) of the Russian Platform. Paleontol J 44:622–631

Mojsisovics von Mojsvar E (1893) Das Gebirge um Hallstatt, Theil I, Die Cephalopoden der Hallstätter Kalke. K-K Geol Reichsanst Wien Abh 6(2):1–835

Mojsisovics von Mojsvar JAE (1882) Die Cephalopoden der mediterranen Triasprovinz. Abh K-K Geol Reichsanst Wien Abh 10:1–322

Moltschaniwskyj NA, Martínez P (1998) Effect of temperature and food levels on the growth and condition of juvenile *Sepia elliptica* (Hoyle, 1885): an experimental approach. J Exp Mar Biol Ecol 229:289–302

Monnet C, Bucher H, Wasmer M, Guex J (2010) Revision of the genus *Acrochordiceras* Hyatt, 1887 (Ammonoidea, Middle Triassic): Morphology, biometry, biostratigraphy and intraspecific variability. Palaeontology 53:961–996

Moreau P, Francis IH, Kennedy WJ (1983) Cenomanian ammonites from northern Aquitaine. Cretac Res 4:317–339

Morton SG (1834) Synopsis of the organic remains of the Cretaceous groups of the United States. Illustrated by nineteen plates, to which is added an appendix containing a tabular view of the Tertiary fossils discovered in America. Key & Biddle, Philadelphia, pp. 1–88

Müller AH (1969) Ammoniten mit "Eierbeutel" und die Frage nach dem Sexualdimorphismus der Ceratiten (Cephalopoda). Monatsber Dtsch Akad Wiss Berl 11:411–420

Neige P (1992) Mise en place du dimorphisme (sexuel) chez les Ammonoides. Approche ontogénétique et interprétation hétérochronique. DEA University Bourgogne, pp 1–50

Nettleship MT, Mapes RH (1993) Morphologic variation, maturity, and sexual dimorphism in an Upper Carboniferous ammonoid from the Midcontinent. GSA, Abstracts with Program 25(2):67

Obata I, Futakami M, Kawashita Y, Takahashi T (1978) Apertural features in some Cretaceous ammonites from Hokkaido. Bull Natl Sci Mus (Tokyo) Ser C (Geol) 4(3):139–155

Olivero EB, Medina FA (1989) Dimorfismo en *Grossouvrites gemmatus* (Huppe) (Ammonoidea) del Cretacico superior de Antartica. Actas Cuarto Congreso Argentino de Paleontología y Bioestratigrafía (Mendoza), pp 65–74

Orbigny A d' (1847) Paléontologie Française. Terraines jurassiques. Part I: Céphalopodes. Masson, Paris

Orbigny A d' (1850) Prodrôme de paléontologie stratigraphique universelle des animaux mollusques & rayonnés faisant suite au cours élémentaire de paléontologie et de géologie stratigraphiques. Tom, vol 2. Masson, Paris, pp 1–427

Palframan DFB (1966) Variation and ontogeny of some Oxfordian ammonites. *Taramelliceras richei* (de Loriol) and *Creniceras renggeri* (Oppel) from Woodham Buckinghamshire. Palaeontology 9:290–311

Palframan DFB (1969) Taxonomy of sexual dimorphism in ammonites: mor-phogenetic evidence in *Hecticoceras brightii* (Pratt). In: Westermann GEG (ed) Sexual dimorphism in fossil Metazoa and taxonomic implications, vol 1. Stuttgart, Schweizerbart, pp 125–154 (IUGS A)

Parent H (1991) Ammonites Cretácicos de la Formación Rio Mayer (Patagonia austral) *Hatcheriiceras patagonense* Stanton (Barremiano) y *Sanmartinoceras patagonicum* Bonarelli (Albiano). Inst Fisiogr Univ Nac Rosario Notas A 15:1–8

Parent H (1997) Ontogeny and sexual dimorphism of *Eurycephalites gottschei* (Tornquist) (Ammonoidea) of the Andean Lower Callovian (Argentine-Chile). Geobios 30:407–419

Parent H (1998) Upper Bathonian and lower Callovian ammonites from Chacay Melehué (Argentina). Acta Palaeontol Pol 43:69–130

Parent H, Scherzinger H, Schweigert G (2008a) Sexual phenomena in Late Jurassic Aspidoceratidae (Ammonoidea). Dimorphic correspondence between *Physodoceras hermanni* (Berckhemer) and *Sutneria subeumela* Schneid, and first record of possible hermaphroditism. Palaeodiversity 1:181–187

Parent H, Schweigert G, Scherzinger A, Enay R (2008b) *Pasottia*, a new genus of Tithonian oppeliid ammonites (Late Jurassic, Ammonoidea: Haploceratoidea). Bol Inst Fisiogr Geol 78:23–30

Parent H, Greco AF, Bejas M (2009) Size-Shape Relationships in the Mesozoic planispiral ammonites. Acta Palaeontol Pol 55:85–98

Parent H, Garrido AC, Schweigert G, Scherzinger A (2011) The Tithonian ammontie fauna and stratigraphy of Picún Leufú, southern Neuquén Basin, Argentina. Rev Paléobiol 30:45–104

Parent H, Garrido AC, Schweigert G, Scherzinger A (2013). Andean Lower Tithonian (Picunleufuense Zone) ammonites and aptychus from Estancia Maria Juana, southern Neuquén Basin, Argentina. Boletín del Instituto de Fisiografía y Geología 83:27–34

Pavia G, Zunino M (2012) Ammonite assemblages and biostratigraphy at the Lower to Upper Bajocian boundary in the Digne area (SE France). Implications for the definition of the Lower Bajocian GSSP. Rev Paléobiol Vol Spéc 11:205–227

Pelseneer P (1926) La proportion relative des sexes chez les animaux et particulièrement chez les Mollusques. Acad R Belg Cl Sci (Mem 2ieme Ser) 8(11):1–258

Ploch I (2003) Taxonomic interpretation and sexual dimorphism in the Early Cretaceous (Valanginian) ammonite *Valanginites nucleus* (Roemer, 1841). Acta Geol Pol 53:201–208

Ploch I (2007) Intraspecific variability and problematic dimorphism in the Early Cretaceous (Valanginian) ammonite *Saynoceras verrucosum* (d'Orbigny, 1841). Acta Geol Sin 81:877–882

Quenstedt FA (1885) Die Ammoniten des Schwäbischen Jura. Schweizerbart, Stuttgart

Raup DM, Crick RE (1981) Evolution of single characters in the Jurassic ammonite *Kosmoceras*. Paleobiology 7:200–215

Raup DM, Michelson A (1965) Theoretical morphology of the coiled shell. Science 147:1294–1295

Reboulet S (1995) L'évolution des ammonites du Valanginien-Hauterivien inférieur du Bassin Vocontien et de la Plate-forme Provencale (Sud-Est de la France). Doc Lab Géol Lyon 137:1–371

Rein S (2001) Neue Erkenntnisse zur Evolutionsbiologie der germanischen Ceratiten – Ontogenese, Phylogenese und Dimorphismusverhalten. Freib Forsch C 492:99–120

Rein S (2003) Zur Biologie der Ceratiten der *spinosus*-Zone – Ergebnisse einer Populationsanalyse, Teil I: Populationsstatistik, Sexual-Dimorphismus und Artkonzept. Veröff Naturkundemus Erf 22:29–50

Reyment RA (1971) Vermuteter Dimorphismus bei der Ammonitengattung *Benueites*. Bull Geol Inst Univ Uppsl NS 3(1):1–18

Reyment RA (1982) Size and shape variation in some Japanese upper Turonian (Cretaceous) ammonites. Stock Contrib Geol 37(16):201–214

Reyment RA (1988) Does sexual dimorphism occur in Upper Cretaceous ammonites? Senckenb leth 69(1/2):109–119

Riccardi AC, Aguirre Urreta MB, Medina FA (1987) Aconeceratidae (Ammonitina) from the Hauterivian-Albian of southern Patagonia. Palaeontogr A 196:105–185

Richter U (2002) Gewebeansatz-Strukturen auf pyritisierten Steinkernen von Ammonoideen. Geol Beitr Hann 4:1–113

Roper CFE, Sweeney MJ (1975) The pelagic octopod *Ocythoe tuberculata* Rafinesque, 1814. Bull Am Malacol Union 1975:21–28

Ruzhencev VE (1962) Superorder Ammonoidea. The ammonoids—general part. In: Ruzhencev VE (ed) Molluscs Cephalopods. I. Publishing House of the Academy of Science of the USSR, Moscow [in Russian]

Ruzhencev VE (1974) Superorder Ammonoidea. General section. In: Orlov YA, Ruzhencev VE (ed) Fundamentals of paleontology. V. Mollusca: Cephalopoda I. Jerusalem

Sandoval J, Chandler RB (2000) The sonniniid ammonite *Euhoploceras* from the Middle Jurassic of South-West England and southern Spain. Palaeontology 43:495–532

Saunders WB, Spinosa C (1978) Sexual dimosphism in *Nautilus* from Palau. Paleobiology 4:349–358

Saunders WB, Ward PD (1987) Ecology, distribution and population characteristics of *Nautilus*. In: Saunders WB, Landman NH (eds) Nautilus. The biology and paleobiology of a living fossil. Plenum Press, New York

Scherzinger A., Mitta V. (2006) New data on ammonites and stratigraphy of the Upper Kimmeridgian and Lower Volgian (Upper Jurassic) of the middle Volga Region (Russia). Neues Jahrb Geol Paläontol Abh 241:225–251

Schiappa TA, Spinosa C, Snyder WS (1995) *Nevadoceras*, a new Early Permian adrianitid (Ammonoidea) from Nevada. J Paleontol 69:1073–1079

Schindewolf OH (1937) Zur Stratigraphie und Paläontologie der Wocklumer Schichten (Oberdevon). Abh Preuß Geol Landesanst, NF 178:1–132

Schweigert G (1997) Die Ammonitengattungen *Simocosmoceras* Spath und *Pseudhimalayites* Spath (Aspidoceratidae) im süddeutschen Oberjura. Stuttg Beitr Naturk B 246:1–29

Schweigert G (1998) Die Ammonitenfauna des Nusplinger Plattenkalks (Ober-Kimmeridgium, Beckeri-Zone, Ulmense-Subzone, Württemberg). Stuttg Beitr Naturk B 267:1–61

Schweigert G, Dietze V (1998) Revision der dimorphen Ammonitengattungen *Phlycticeras* Hyatt—*Oecoptychius* Neumayr (Strigoceratidae, Mitteljura). Stuttg Beitr Naturk B 269:1–59

Schweigert G, Dietl G, Dietze V (2003) Neue Nachweise von *Phlycticeras* und *Oecoptychius* (Ammonitina: Strigoceratidae: Phlycticeratinae). Stuttg Beitr Naturk B 335:1–21

Schweigert G, Dietze V, Chandler RB, Mitta V (2007) Revision of the Middle Jurassic dimorphic ammonite genera *Strigoceras*/*Cadomoceras* (Strigoceratidae) and related forms. Stuttg Beitr Naturk B 373:1–74

Seilacher A (1974) Fabricational noise in adaptive morphology. Syst Biol 22:451–465

Seilacher A, Gunji YP (1993) Morphogenetic countdown: another view on heteromorph shells in gastropods and ammonites. Neues Jahrb Geol Paläontol Abh 190:237–265

Senior JR (1971) Wrinkle-layer structures in Jurassic ammonites. Palaeontology 14:107–113

Shea BT (1986) Ontogenetic approaches to sexual dimorphism in anthropods. Hum Evol 1(2):97–110

Siebold CT (1848) Lehrbuch der vergleichenden Anatomie, vol 1. Veit, Berlin

Stephenson LW (1941) The larger invertebrates of the Navarro Group of Texas (exclusive of corals and crustaceans and exclusive of the fauna of the Escondido Formation). (University of Texas) Bulletin 4101:1–641

Sturani C (1966) Ammonites and stratigraphy of the Bathonian in the Digne- Barreme area (southeastern France, Dept. Basses- Alpes). Boll Soc Paleontol Ital 5:3–57

Sun YC (1928) Mundsaum und Wohnkammer der Ceratiten des Oberen deutschen Muschelkalks. Weg, Leipzig

Tafur R, Villegas P, Rabí M, Yamashiro C (2001) Dynamics of maturation, seasonality of reproduction and spawning grounds of the jumbo squid *Dosidiscus gigas* (Cephalopoda: Ommastrephidae) in Peruvian waters. Fish Res 54:33–50

Tajika A, Naglik C, Morimoto N, Pascual-Cebrian E, Hennhöfer DK, Klug C (2014) Empirical 3D-model of the conch of the Middle Jurassic ammonite microconch *Normannites,* its buoyancy, the physical effects of its mature modifications and speculations on their function. Historical Biology, 11 pp. doi: 10.1080/08912963.2013.872097

Tanabe K (1977) Functional evolution of *Otoscaphites puerculus* (Jimbo) and *Scaphites planus* (Yabe), Upper Cretaceous ammonites. Mem Fac Sci Kyushu Univ Ser D Geol 23(3):367–407

Tanabe K, Landman NH, Mapes RH (1998) Muscle attachment scars in a Carboniferous goniatite. Paleontol Res 2(2):130–136

Thierry J (1978) Le genre *Macrocephalites* au Callovien inférieur (Ammonites, Jurassique moyen). Mém Géol Univ Dijon 4:1–490

Thierry J, Charpy N (1982) Le genre *Tornquistes* (Ammonitina, Pachyceratidae) a l'Oxfordien inférieur et moyen en Europe occidentale. Geobios 15:619–677
Till A (1909) Die fossilen Cephalopodengebisse. K-K Geol Reichsanst Jahrb 58(4):573–608
Till A (1910) Die fossilen Cephalopodengebisse. Folge 3. K-K Geol Reichsanst Jahrb 59:407–426
Tintant H (1963) Les Kosmoceratides du Callovien inférieur et moyen d'Europe occidentale. University of Dijon, France
Tintant H (1976) Le polymorphisme intraspécifique en paléontologie. Exemple pris chez les ammonites. Haliotis 6:49–69
Tornquist A (1898) Der Dogger am Espinazito Pass. Paläontol Abh NF 3(2):3(135)–69(201)
Tozer KT (1994) Canadian Triassic ammonoid faunas. Geol Surv Can Bull 467:1–663
Trewin NH (1970) A dimorphic goniatite from the Namurian of Cheshire. Palaeontology 13:40–46
Trueman AE (1941) The ammnonite body chamber, with special reference to the buoyancy and mode of life of the living ammonite. Q J Geol Soc Lond 96:339–383
Urdy S, Goudemand N, Bucher H, Chirat R (2010a) Allometries and the morphogenesis of the molluscan shell: a quantitative and theoretical model. J Exp Zool B 314:280–302
Urdy S, Goudemand N, Bucher H, Chirat R (2010b) Growth dependent phenotypic variation of molluscan shell shape: implications for allometric data interpretation. J Exp Zool B 314:303–326
Urlichs M (2009) Weiteres über Dimorphismus bei *Ceratites* (Ammonoidea) aus dem Germanischen Oberen Muschelkalk (Mitteltrias) mit Revision einiger Arten. Neues Jahrb Geol Paläontol Abh 251:199–223
Vermeij GJ (1993) A natural history of shells. Princeton University Press, Princeton
Walliser OH (1963) Dimorphismus bei Goniatiten. Paläontol Z 37(1–2):21
Walliser OH (1970) Über die Runzelschicht bei Ammonoidea. Gött Arb Geol Paläontol 5:115–126
Walton SA, Korn D, Klug C (2010) Size distribution of the late devonian ammonoid *Prolobites*: indication for possible mass spawning events. Swiss J Geosci 103:475–494
Ward PD (1987) The natural history of *Nautilus*. Allen and Unwin, Boston
Weitschat W, Bandel K (1991) Organic components in phragmocones of boreal Triassic ammonoids; implications for ammonoid biology. Paläontol Z 65:269–303
Wells MJ (1962) Brain and behavior in cephalopods. Stanford University Press, Stanford
Wells MJ (1966) The brain and behavior of cephalopods In: Wilbur KM, Younge CM (eds) Physiology of mollusca. Academic Press, New York
Wenger R (1957) Die Germanischen Ceratiten. Paleontogr A 108:57–129
Westermann GEG (1964a) Sexual-Dimorphismus bei Ammonoideen und seine Bedeutung für Taxionomie der Otoitidae (einschliesslich Sphaeroceratinae; Ammonitina, M. Jura). Palaeontogr A 124(1–3):33–73
Westermann GEG (1964b) The ammonite fauna of the Kialagvik formation at Wide Bay, Alaska Peninsula. Part I. Lower Bajocian (Aalenian). Bull Am Palacontol 47:327–503
Westermann GEG (1969a) Supplement: sexual dimorphism, migration, and segregation in living cephalopods. In: Westermann GEG (ed) Sexual dimorphism in fossil Metazoa and taxonomic implications (IUGS, Series A1). Schweizerbart, Stuttgart
Westermann GEG (1969b) Proposal: classification and nomenclature of dimorphs at the genus-group level [with discussion]. In: Westermann GEG (ed) Sexual dimorphism in fossil Metazoa and taxonomic implications (IUGS, Series A1). Schweizerbart, Stuttgart
Westermann GEG (1971) Form, structure, and function of shell and siphuncle in coiled Mesozoic ammonoids. Life Sci Contrib R Ont Mus 78:1–3
Westermann GEG, Riccardi AC (1979) Middle Jurassic ammonoid fauna and biochronology of the Argentine-Chilean Andes. Part II: Bajocian Stephanocerataceae. Palaeontogr A 164:85–188
Wiedmann J (1965) Origins, limits, and systematic position of *Scaphites*. Palaeontology 8:397–453
Wiedmann J (1973) The Albian and Cenomanian Tetragonitiae (Cretaceous Ammonoidea), with special reference to the Circum-Indic species. Eclogae Geol Helv 66:585–616
Willey A (1895) In the home of the *Nautilus*. Nat Sci Lond 6(40):405–414

Willey A (1902) Contribution to the natural history of the pearly nautilus. In: Zoological results based on material from New Britain, New Guinea, Loyalty Islands, and Elsewhere, collected during the years 1895, 1896, and 1897. Cambridge University Press, Cambridge

Wilmsen M, Mosavinia A (2011) Phenotypic plasticity and taxonomy of *Schloenbachia varians* (J. Sowerby, 1817) (Cretaceous Ammonoidea). Paläontol Z 85:169–184

Wright CW, Kennedy WJ (1980) Origin, evolution and systematics of the dwarf acanthoceratid *Protacanthoceras* Spath, 1923 (Cretaceous Ammonoidea). Bull Br Mus (Nat Hist) Geol 34(2):65–107

Wright CW, Kennedy WJ (1984) The Ammonoidea of the Lower Chalk. Part I. (Monograph Palaeontographical Society London 567, part of vol 137 for 1983). Palaeontographical Society, London, pp 1–126

Wright CW, Kennedy WJ (1987) The Ammonoidea of the Lower Chalk. Part 2. (Monograph Palaeontographical Society London 573, part of vol 139 for 1985). Palaeontographical Society, London, pp 127–218

Wright CW, Kennedy WJ (1990) The Ammonoidea of the Lower Chalk. Part 3. (Monograph Palaeontographical Society London 585, part of vol 144 for 1990). Palaeontographical Society, London, pp 219–294

Wright CW, Kennedy WJ (1994) Evolutionary relationships among Stoliczkaiinae (Cretaceous ammonites) with an account of some species from the English Stoliczkaia dispar Zone. Cretac Res 15:547–582

Wright CW, Callomon JH, Howarth MK (1996) Cretaceous Ammonoidea. In: Kaesler RL (ed) Treatise on invertebrate paleontology, Part L, Mollusca 4 (revised). GSA and University of Kansas Press, Lawrence

Wright JK (2010) The Aulacostephanidae (Ammonoidea) of the Oxfordian/Kimmeridgian boundary beds (Upper Jurassic) of Southern England. Palaeontology 53:11–52

Zaborski PMP (1987) Lower Turonian (Cretaceous) ammonites from south-east Nigeria. Bull Br Mus (Nat Hist) Geol 41(2):31–66

Zakharov YD (1969) Problems of sexual dimorphism in fossil cephalopods, an impotant subject in modern systematics. In Gramm N, Krassilov VA (eds) Problems of phylogeny and systematics. Acad Sci USSR, Far Eastern Geological Institute, All Union Paleontology Society of Vladivostok [in Russian]

Zakharov YD (1977) Ontogeny of ceratites of the genus *Pinacoceras* and developmental features of the suborder Pinacoceratina. Paleontol J 4:445–451

Zatoń M (2008) Taxonomy and palaeobiology of the Bathonian (Middle Jurassic) tulitid ammonite *Morrisiceras*. Geobios 41:699–717

Zatoń M (2010) Bajocian-Bathonian (Middle Jurassic) ammonites from the Polish Jura. Part 2: families Stephanoceratidae, Perisphinctidae, Parkinsoniidae, Morphoceratidae and Tulitidae. Palaeontogr A 292:115–213

Zeiss A (1969) Dimorphismus bei Ammoniten des Unter-Tithon. Mit einigen allgemeinen Bemerkungen zum Dimorphismus-Problem. In: Westermann GEG (ed) Sexual dimorphism in fossil Metazoa and taxonomic implications. International Union Geological Science, A1. Schweizerbart, Stuttgart

Zhao J, Zheng Z-G (1977) The Permian ammonoids from Zhejiang and Jiangxi. Acta Palaeontol Sin 16(2).217–254

Zhou Z (1985) Several problems on the Early Permian ammonoids from south China. Palaeontol. Cathayana 2:179–210

Ziegler B (1974) Über Dimorphismus und Verwandtschaftsbeziehungen bei 'Oppelien' des oberen Juras (Ammonoidea: Haplocerataceae). Stuttg Beitr Naturk B 11:1–42

Ziegler B (1987) Der weiße Jura der Schwäbischen Alb. Stuttg Beitr Naturk C 23:1–71

Chapter 8
Ammonoid Shell Microstructure

Cyprian Kulicki, Kazushige Tanabe, Neil H. Landman and Andrzej Kaim

8.1 Introduction

This chapter is not devoted to shell microstructure alone. In addition to presenting a description of the structure of the individual layers that compose the ammonoid shell, we also discuss the distribution and relationships of these layers to one another as well as their ultrastructure whenever possible. Because aragonite, the chief mineral that makes up the ammonoid shell, is metastable and transforms into calcite as a function of time, pressure, and temperature (Dullo and Bandel 1988), it is difficult to obtain specimens for study, which are preserved well enough to observe fine details of their microstructure. The oldest known occurrence of shells with pristine aragonite preserved derives from the Pennsylvanian Buckhorn Asphalt, USA. This explains why nearly all micro- and ultrastructural studies of ammonoids have been conducted on materials collected from Mesozoic platform deposits.

C. Kulicki (✉)
Institute of Paleobiology, Polish Academy of Sciences, ul. Twarda 51/55,
PL 00-818 Warszawa, Poland
e-mail: kulicki@twarda.pan.pl

K. Tanabe
Department of Historical Geology and Paleontology, The University Museum,
The University of Tokyo, Hongo 7-3-1, Tokyo 113-0033, Japan
e-mail: tanabe@um.u-tokyo.ac.jp

N. H. Landman
Division of Paleontology (Invertebrates), American Museum of Natural History,
Central Park West at 79th St. New York 10024-5192, USA
e-mail: landman@amnh.org

A. Kaim
Institute of Paleobiology, Polish Academy of Sciences,
ul. Twarda 51/55, Warszawa PL 00-818, Poland
e-mail: kaim@twarda.pan.pl

C. Klug et al. (eds.), *Ammonoid Paleobiology: From Anatomy to Ecology*,
Topics in Geobiology 43, DOI 10.1007/978-94-017-9630-9_8

8.2 Embryonic Stage

8.2.1 Existing Structural Models

The term "*ammonitella*" was proposed by Drushchits and Khiami (1969) to denote the initial chamber plus the first shell whorl up to the primary constriction. Additional internal structure elements of ammonitella are proseptum, prosiphon, and cecum. Therefore, the term ammonitella is generally understood as an ammonite embryonic shell (compare De Baets et al. 2015). Prior to Drushchits and Khiami (1969), the same structure was known as "*protoconch*" (Ruzhentsev and Shimanskij 1954; Makowski 1962, 1971) though the later term should rather be refereed to initial chamber than to entire ammonitella. Because of controversies on the structure and relationships of the layers in the wall of the ammonitella, and because of diverse opinions on the embryogenesis of ammonoids, the problem of the microstructure of the ammonitella is of special interest.

The distinct morphological and microstructural features of the ammonitella had already been observed in the nineteenth century (Hyatt 1872; Branco 1880), but knowledge on the subject rapidly improved with the use of electron microscopy. Birkelund (1967) and Birkelund and Hansen (1968) were the first to apply transmission electron microscopy (TEM) in studies of the microstructure of ammonitellae of Late Cretaceous *Saghalinites* and *Scaphites*. The preparation method was described by Hansen (1967) and depended generally on slight EDTA etching of the surfaces of polished cross-sections and then removing the colloidal replicas and sputtering them with carbon. By means of this method, the following was determined: (1) The wall of the initial chamber is built of two layers without a distinct boundary in-between; (2) The inner layer consists of crystals perpendicular to the inner shell surface, whereas crystals in the external layer are distributed without a distinct orientation; (3) Both layers of the initial chamber wedge out in the vicinity of the proseptum base.

The wall of the first whorl appears as a prismatic layer on the inner surface of the initial chamber. After the wall of the initial chamber wedges out, the wall of the first whorl continues without much change to the primary constriction. It is similar in construction to the wall of the initial chamber; i.e., it consists of two sublayers, an inner sublayer having more regular crystals perpendicular to the inner shell surface and an external, thinner sublayer with less regularly oriented crystals. The terminal part of the ammonitella aperture is delimited by a structure known as a primary constriction. The prismatic and subprismatic sublayers of the first whorl become much thinner, and the nacreous primary varix develops beneath them. The proseptum is constructed of the same crystalline matter as that of the internal prismatic layer of the initial chamber and the wall of the first whorl.

Erben et al. (1968, 1969) first introduced scanning electron microscopy (SEM) to study the shell microstructure of ammonoids. This technique is much easier in specimen preparation and generally more appropriate than TEM. Based on SEM observations, Erben et al. (1968, 1969) proposed a model for the structure and development of the ammonitella. According to this model, the wall of the initial chamber initially consists of two subprismatic layers. These wedge out, and only somewhat later the fully prismatic sublayers appear on the inner surface of the initial chamber. Two of these sublayers also wedge out, and only the sublayer beyond the base of the proseptum continues until the end of the primary constriction. The dorsal part of the proseptum consists of an additional layer on the inner surface of the initial chamber, whereas the ventral part of the proseptum is a continuation of the innermost layer of the initial chamber and the first whorl. The flange is composed of a prismatic layer and is separated from the proseptum and the wall of the initial chamber by a discontinuity surface, and according to this model it is suggested that the flange was formed later in ontogeny.

Another model of ammonitella formation has been presented by Kulicki (1979) based on excellently preserved material of *Quenstedtoceras* and *Kosmoceras* from Łuków (Callovian, Poland). In this model, the wall of the initial chamber has two layers, best seen in its dorsal and apical parts. The inner layer has a regular prismatic structure and represents the wall proper of the initial chamber, continuing into the outer prismatic layer of the first whorl up to the ammonitella edge. The outer layer of the wall of the initial chamber is a continuation of the mural part of the proseptum and represents the dorsal wall of the first whorl of the ammonitella. This layer is thickest opposite the primary varix and is subprismatic. Crystallites in this layer are oriented in parallel, diagonally, or perpendicularly to the shell surface. The boundary between this layer and the wall proper of the initial chamber is not distinct. A distinct boundary between the dorsal wall and the outer prismatic layer of the first whorl develops at the appearance of the tuberculate sculpture characteristic of the ammonitella in Mesozoic ammonoids.

In medial and paramedial cross sections through the outer saddle of the proseptum on the inner surface of the venter of the initial chamber, there is an inner prismatic layer of regular structure linked to the base of the proseptum. This layer is separated from the wall proper of the initial chamber by a thin layer of microcrystalline structure. This thin layer is thickest in the middle part of the base of the proseptum. The prismatic layer of the proseptum commonly continues as one of the main components of the wall of the first whorl ("medial prismatic layer" of Kulicki 1979).

All of the above models assume simultaneous secretion of the organic phase of the shell together with mineralization. This is what occurs in the formation of the postembryonic shell in Recent mollusks. In contrast, Bandel (1982, 1986) presented a model based on shell development in some Archaeogastropoda in which the larval shell is formed in two phases. In the first phase, the shell consists only of elastic, organic matter, and in the second phase, the organic primary shell is calcified. The direction of calcification may not have been consistent with the direction of secretion of the organic shell. In the case of the ammonitella, the wall of the first whorl and umbilical walls of the initial chamber would have been calcified first. Only later would the remaining wall of the initial chamber and proseptum have been mineralized.

Bandel's (1982, 1986) interpretation has been confirmed by well-preserved ammonitellae of *Aconeceras* (Albian) representing different calcification stages (Kulicki 1989; Kulicki and Doguzhaeva 1994). Four stages of calcification of the ammonitella have been recognized.

Stage 1 is represented by specimens in which the wall of the first whorl, including the primary constriction and the lateral walls of the initial chamber are calcified.

Stage 2 is represented by specimens in which, in addition, the part of the wall of the initial chamber that separates the interior of the initial chamber from the lumen of the first whorl is calcified.

Stage 3 is represented by specimens that have a calcified first whorl, initial chamber, proseptum, and nacreous primary varix. In *Quenstedtoceras* ammonitellae from Łuków, there is another septum adapertural of the proseptum (the first nacroseptum).

Stage 4 is represented by ammonitellae of larger, postembryonic specimens. This stage is characterized by a distinct thickening from the inside of the inner prismatic layer and commonly by the addition of extra prismatic layers from the inside.

Another model has been proposed by Tanabe (1989), which assumed that the embryo of Mesozoic ammonoids might have temporarily had an endocochliate body plan late in embryonic development during which the outer prismatic layer with tubercles was secreted from the outer reflected mantle. Finally, Tanabe et al. (2008), after investigation of exceptionally well preserved embryonic shells of *Aconoceras*, returned to the model of Bandel (1982) and Kulicki and Doguzhaeva (1994) but refined the timing of formation of the outer prismatic layer and tubercles. Tanabe et al. (2008) demonstrated for the first time the presence of amorphous calcium carbonate (ACC) in the wall of the embryonic shell.

8.2.2 Structure of the Ammonitella Walls

A cross section through the wall of the first whorl in Paleozoic and Mesozoic ammonoids is shown in Fig. 8.1–8.3. The outer layer has a regular prismatic structure and a thickness of about 1–3 μm in Mesozoic ammonoids while in the Pennsylvanian Buckhorn Asphalt goniatite it is indistinct. Prisms consist of aragonitic needles 0.1–0.2 μm in diameter oriented perpendicular to the outer surface. The diameter of the crystallites is 0.2 μm .

The medial layer, sandwiched between the outer and inner prismatic layers, has an irregular, grainy structure. The irregular, grainy structure of the medial layer results most likely from the limiting effect of organic matter on the growth of aragonitic crystals. In transverse cross sections through the initial chamber and first whorl, the outer prismatic layer and the medial layer wedge out on the umbilical seam, and only the inner prismatic layer continues across the dorsum forming the lateral walls of the initial chamber, which are two or three times thicker than the walls covered by the following whorl. This confirms earlier assumptions of Bandel (1982, 1986) and Kulicki (1979). In Mesozoic ammonoids, the crystallites of the outer layer in the outer wall of the first whorl of the ammonitella and lateral parts of the initial

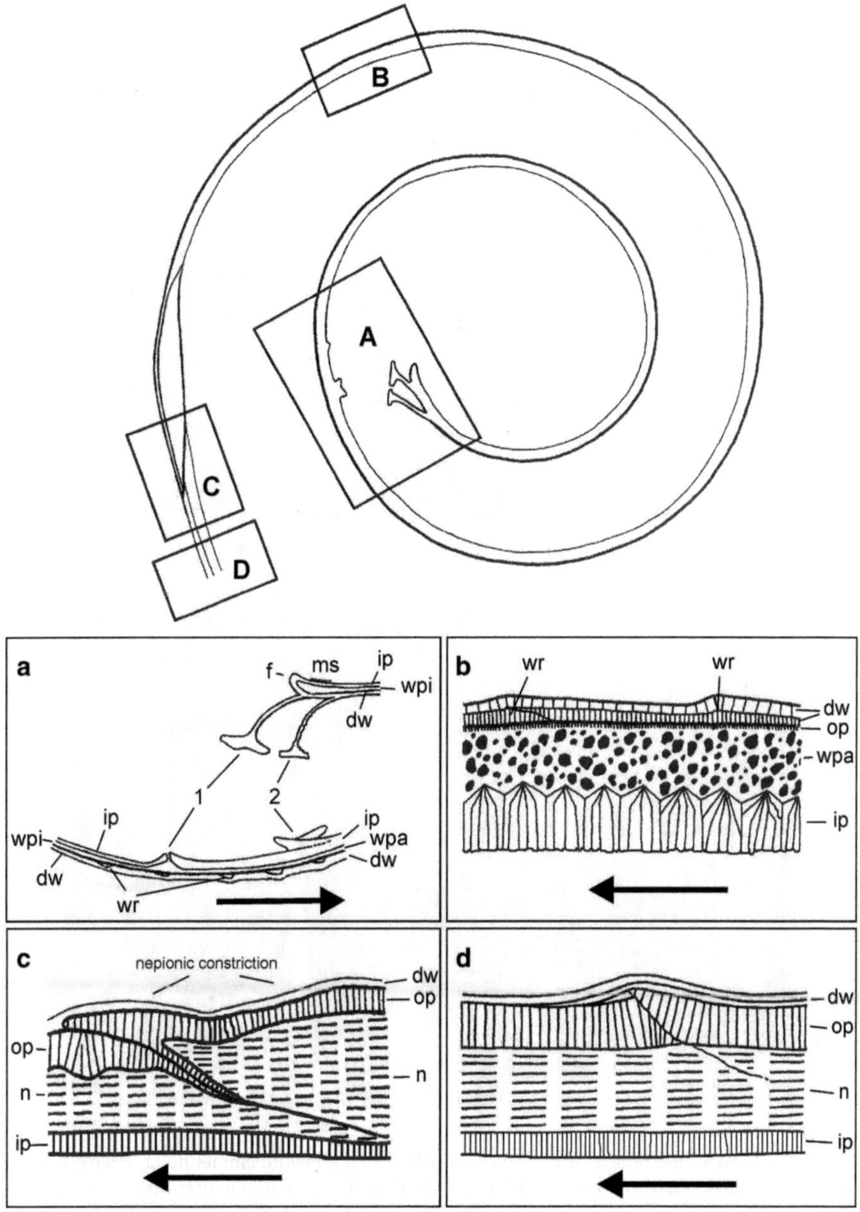

Fig. 8.1 Schematic drawing of median cross-section through ammonoid ammonitella and early postembryonic shell (modified after Kulicki et al. 2002). The arrows indicate adoral direction. **a** Overview figure with the places of close-ups indicated. **b** Close-up of the initial chamber in the vicinity of the first two septa. **c** Close-up of the wall of the first whorl. **d** Close-up of the ammonitella edge. **e** Close-up of the wall of the post-embryonic shell showing a lira in the outer prismatic layer (open arrowhead). Abbreviations: *dw*, dorsal wall; *f*, flange; *ip*, inner prismatic layer; *ms*, muscle scar; *n*, nacreous layer; *op*, outer prismatic layer; *wpa*, wall proper of the ammonitella; *wpi*, wall proper of the initial chamber; *wr*, wrinkle layer; *1*, first septum (proseptum); *2*, second septum

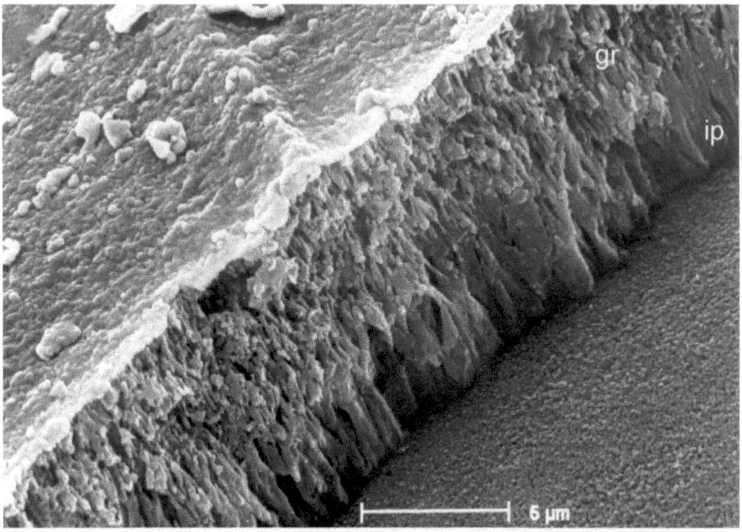

a

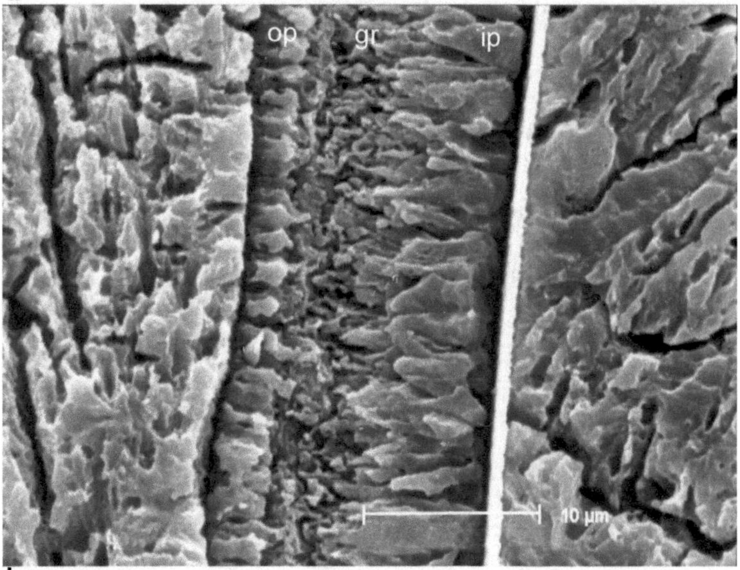

b

Fig. 8.2 **a** View of the broken wall of the first whorl of a goniatite ammonitella, Pennsylvanian, Buckhorn Asphalt quarry, Oklahoma. **b** Longitudinal section through the first whorl of the ammonitella of *Damesites sugata* Forbes, 1846, Coniacian, Cretaceous, Nakafutamata Rivulet, Haboro area northwestern Hokkaido. In case of Mesozoic ammonitellae the outer prismatic layer is more distinct and better developed. Abbreviations: *gr*, granular layer; *ip*, inner prismatic layer; *op*, outer prismatic layer

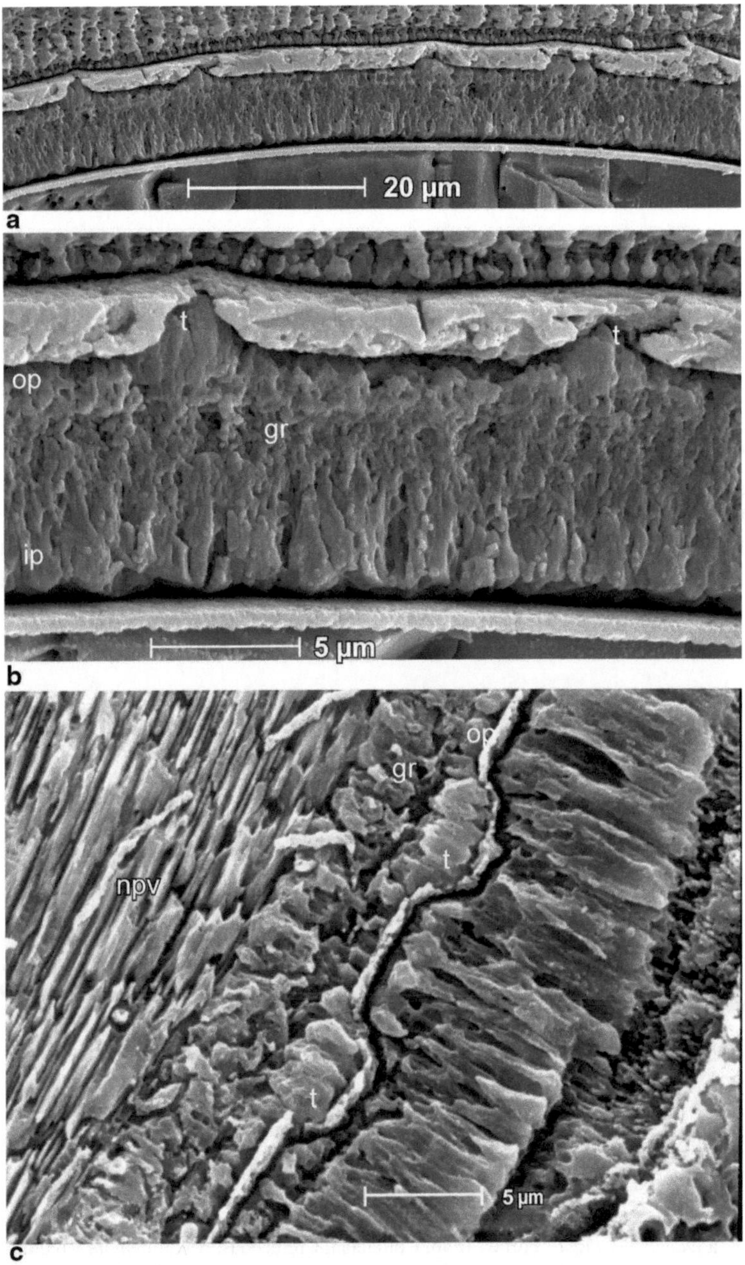

Fig. 8.3 **a**, **b** *Ptychoceras* sp. Aptian, Cretaceous, Khocods Rivier, NW Caucasus. Longitudinal section through tuberculate ventral wall of ammonitella whorl. **c** *Damesites sugata* Forbes, 1846, Coniacian, Cretaceous, Nakafutamata Rivulet, Haboro area northwestern Hokkaido. Longitudinal section through tuberculate ventral wall of ammonitella whorl above primary varix. Abbreviations: *gr*, granular layer; *ip*, inner prismatic layer; *npv*, nacreous layer of the primary varix; *op*, outer prismatic layer; *t*, tubercule

chamber become aggregated into structural elements of a higher order, the so-called pseudohexagonal trillings (Kulicki and Doguzhaeva 1994; Tanabe et al. 2008, 2010).

8.2.3 Apertural Zone of the Ammonitella

The terminology used here to describe different structural and morphological elements of ammonoid embryonic shells is derived from Drushchits et al. (1977) and Landman and Waage (1982). The morphological-structural distinctness of the apertural zone of ammonitellae was first recorded by Hyatt (1872). Please note, however, that the significance of this zone in terms of the ontogeny of ammonoids is beyond the scope of this chapter (see De Baets et al. 2015 for a review).

Fully developed ammonitellae terminate in a structure called the nepionic constriction, where they form, for the first time in ontogeny, the nacreous layer comprising the so-called nepionic swelling (primary varix; Fig. 8.4a). The outermost thin prismatic layer continues until the ammonitella edge, where it forms a short return section directed toward the shell interior underlying internally the nacreous layer of the nepionic swelling or primary varix (Fig. 8.4a). The middle, granular, or subprismatic layer decreases in thickness towards the aperture and wedges out in

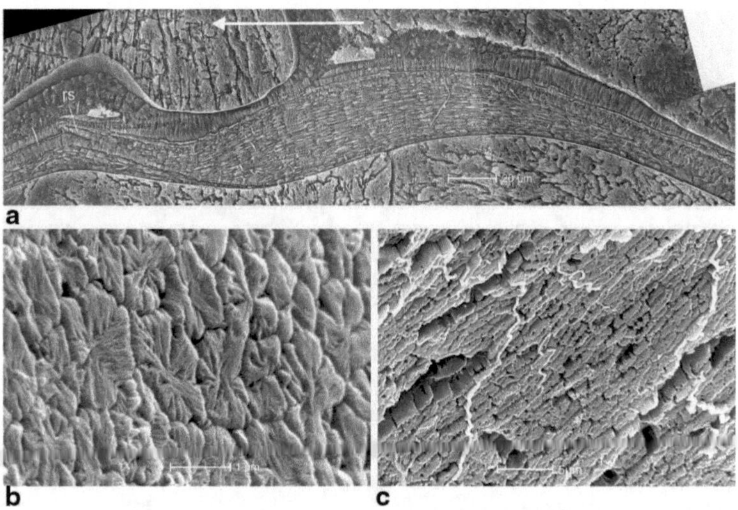

Fig. 8.4 **a** *Damesites sugata* Forbes 1846, Coniacian, Cretaceous, Nakafutamata Rivulet, Haboro area northwestern Hokkaido, primary varix (nepionic constriction), **b** *Gaudryceras tenuiliratum* Yabe, 1903, Campanian, Abeshinai River, Nakagawa Town, north Hokkaido, view of outer prismatic layer from the outside. Aggregation in form of pseugohexagonal trillings is clearly visible. **c** *Phyllopachyceras ezoense* (Yokoyama 1890), Middle Campanian, Cretaceous, Osoushunai Rivulet, Nakagawa Town, Hokkaido, view of outer prismatic layer from the outside. Fuselar aggregation is visible

the middle of the primary varix zone. The inner prismatic layer of the ammonitella terminates at the appearance of the nacreous layer. The border between the granular layer and the inner prismatic layer is indistinct and the structure of the prismatic layer itself is characterized in this zone by a slightly spherulitic distribution of the crystals with the center located at the interface with the granular layer. The lamellae of the nacreous layer form a characteristic arrangement, described for the first time by Kulicki (1974). The most external lamellae of the nacreous layer are relatively short and end on the inner surface beneath the outer prismatic layer on the posterior side of the shell while on the anterior side they reach the return section of the outer prismatic layer. The inner lamellae reach further backward and are generally longer. The plates of the nacreous layer of the primary varix are arranged in vertical stacks, similar in structure to those of the nacreous layer of the postembryonic shell. The vertical stacks in the nacreous layer of the primary varix have been illustrated by Birkelund and Hansen (1974); Drushchits et al. (1977); Drushchits and Doguzhaeva (1981), and Ohtsuka (1986).

8.2.4 Structure of the Initial Chamber Wall and Proseptum

The umbilical walls of the initial chamber in ammonitellae of Mesozoic and Paleozoic ammonoids are relatively thick, being comparable in thickness to that of the first whorl wall. With respect to their structure, they are also alike; i.e. they are composed of the same three layers. Transverse cross sections through the first whorl of ammonitellae of *Aconeceras* (see Kulicki and Doguzhaeva 1994) and *Quenstedtoceras* (see Kulicki 1979; Bandel 1982, 1986) show that the most external components of the umbilical wall of the initial chamber and of the lateral wall of the first whorl wedge out at the umbilical seam. Only the internal regular prismatic layer of the umbilical wall goes under the umbilical seam as the primary wall of the initial chamber (Fig. 8.4). In the apical part of the initial chamber of the ammonitella of *Quenstedtoceras,* the 3.0–3.5-μm-thick wall made of organic matter is penetrated through its entire thickness by needle-like aragonitic crystallites. In longitudinal cross section, this layer appears to reach the end of the flange.

In medial and paramedial cross sections through the ventral wall of the initial chamber, one or two prismatic layers are visible under the wall proper of the initial chamber (Fig. 8.5b; Kulicki 1979). Kulicki (1979) and Kulicki and Doguzhaeva (1994) claimed that the wall proper of the initial chamber continues as the middle granular or subprismatic layer of the first whorl. Birkelund (1967, 1981); Birkelund and Hansen (1968); Erben et al. (1969); Tanabe et al. (1980, 1993b), and Tanabe and Ohtsuka (1985) concluded that the wall of the initial chamber wedges out in the vicinity of the base of the proseptum. According to Bandel (1982, 1990), in median section, the inner prismatic layer of the initial chamber extends from the base of the proseptum and continues as the primary wall of the initial chamber, ending in the flange. These observations of Bandel are contradictory to those of Kulicki (1979) conducted on the same material, namely, *Quenstedtoceras* from Łuków.

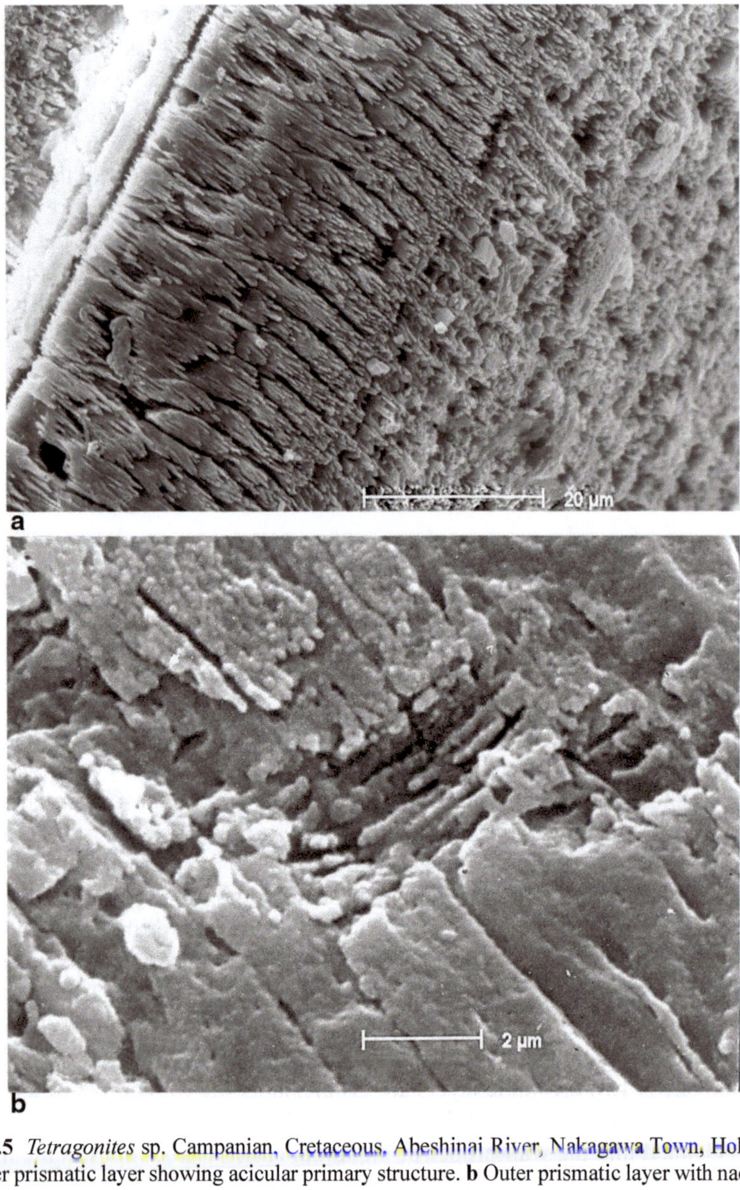

Fig. 8.5 *Tetragonites* sp. Campanian, Cretaceous, Abeshinai River, Nakagawa Town, Hokkaido. **a** Outer prismatic layer showing acicular primary structure. **b** Outer prismatic layer with nacre like intercalation

Opinions concerning the relationship between the proseptum and the shell wall are not consistent either. Some authors, such as Grandjean (1910); Miller and Unklesbay (1943); Arkell (1957); Birkelund and Hansen (1968); Erben et al. (1969); Kulicki (1975, 1979), Drushchits and Doguzhaeva (1981); Bandel (1982), and Landman and Bandel (1985) have considered the prismatic proseptum to be continuous with the internal layers of the whorl and initial chamber of ammonitella while Hyatt (1872); House (1965); Erben (1962, 1966), Drushchits and Khiami (1970), and Drushchits and Doguzhaeva (1981) have suggested that the relationship of the proseptum to the shell wall is the same as that of all other septa.

In the dorsal region, the proseptum has a very long prismatic mural part that covers a large area of the wall proper of the initial chamber. Bandel (1982, 1990) interpreted the mural part of the proseptum as continuing to the ammonitella aperture; this is consistent with the occurrence of a dorsal wall in the ammonitella whorl. As shown in paramedial cross sections distant from the plane of bilateral symmetry, the ventral part of the proseptum continues as the inner prismatic layer of the first whorl (Birkelund and Hansen 1974, Fig. 2; Kulicki 1979, Fig. 10C). Most authors consider the proseptum to be a one-or two-layered structure made of aragonitic prisms oriented perpendicular to the outer surface. Only Landman and Bandel (1985, Fig. 33) illustrated the three-layered structure of the proseptum in *Euhoplites* sp., in which the irregularly prismatic proseptal layer is sandwiched between two layers of a more homogeneous prismatic material that originally may have been organic. These authors (1985, Fig. 21) also noted distinct wrinkles on the surface of the proseptum in *Baculites* sp.

8.2.5 Dorsal Wall of the Ammonitella

Birkelund (1967, 1981), Birkelund and Hansen (1968, 1974), and Erben et al. (1969) have denied the existence of a dorsal wall in the ammonitella. However, Kulicki (1979) has shown that the outermost layer overlying the wall proper of the initial chamber as seen in medial and paramedial cross sections is the dorsal wall of the first whorl. Its structure is subprismatic, and it is not separated from the wall proper of the initial chamber by a sharp boundary (Fig. 8.1a, 8.1b, 8.3a, 8.3b, 8.3c). The dorsal wall is reduced in thickness in the apertural region of the ammonitella and the distinct boundary between the dorsal wall and the ventral wall of the previous whorl appears with the advent of tuberculate sculpture.

In the terminal stage of the ammonitella in *Quenstedtoceras* there are two septa and a nacreous primary varix. Kulicki and Doguzhaeva (1994) have shown that in this developmental stage the outer surface of the initial chamber in the apertural region is covered by distinct structures resembling those of the wrinkle layer. Inside the living chamber these structures reach about 90° from the aperture. In longitudinal cross section, these structures are triangular, with the gentle slope pointing adapically and the much steeper slope facing adaperturally. These triangular elements have a prismatic structure with the long axes of the prisms perpendicular to the overall surface of the wall of the ammonitella rather than to the surfaces of the gentle adapical slopes of the triangles.

Kulicki (1979) and Doguzhaeva and Mutvei (1986b) investigated the complex structure of the dorsal wall. Kulicki (1979) distinguished two components, outer and inner ones in the dorsal wall. The wrinkle-like layer of the ammonitella is the outer component. It was produced by the anterior part of the mantle and, later in ontogeny, was covered by more internal components that were produced by the posterior part of the body. Generally, the ammonitellae available for studies are in the final calcification stage; i.e., they are the apical parts of large specimens. Thus, the dorsal wall at what had been the edge of the ammonitella has the thickness and structure characteristic of the posterior part of the living chamber. On the basis of such specimens, Kulicki (1979) stated that the dorsal wall of the ammonitella covers the outer surface of the wall proper of the initial chamber, that the dorsal wall has a subprismatic structure, and that the boundary between these two layers is not too distinct. The outermost component of the dorsal wall, i.e., a wrinkle-like layer of Paleozoic ammonitellae is present in the apertural region and it is characterized by a specific ornamentation described by Tanabe et al. (2010).

8.2.6 *Ornamentation of the Ammonitella*

The presence of a tuberculate micro-ornamentation in Mesozoic ammonitellae has been already documented in the early works of Brown (1892); Smith (1901), and Smith (1905) . More detailed investigations were possible due to use of SEM techniques (Kulicki 1974, 1979; Bandel 1982; Bandel et al. 1982; Landman 1985, 1987, 1988; Tanabe 1989; Tanabe et al. 2001; Tanabe et al. 2008, 2010). All investigated Mesozoic ammonitellae possess this characteristic tuberculate micro-ornamentation (De Baets et al. 2015), which is limited to the outer surface and is not observed on the initial chamber covered by the whorl of the ammonitella. However, there is wide variation in the density of tubercles, their size, and their distribution on the shell surface (Tanabe et al. 2010).

Bandel et al. (1982, p. 387) found that "*some smaller tubercles appear to be emergent ends of single large prisms, but the larger ones show the complex spherulitic structure.*" On the basis of cross sections through the walls of ammonitellae of *Quenstedtoceras*, Kulicki (1979) determined that tubercles are a continuation of the prisms of the outer prismatic layer and that they are not separated from the outer shell surface by any discontinuity (Fig. 8.2b, 8.3a, b, c).

Tanabe (1989, Fig. 5) presented the tubercles as separate elements on the surface of the outer prismatic layer or, as in the umbilical wall of *Anapachydiscus*, sometimes fused into a continuous layer, forming a smooth outer surface. On this basis, Tanabe (1989) formulated an endocochliate embryo model in which the tuberculate micro-ornamentation, or the layer corresponding to it, was formed by a fold of the mantle covering the outside of the ammonitella.

8.3 Postembryonic Stage

8.3.1 *Products of the Anterior Mantle Edge*

Ammonoids resemble Recent *Nautilus* and other Recent mollusks (e.g., bivalves, gastropods, and monoplacophorans), both in the products of secretion and in the morphology of the apertural edge of the shell. This similarity makes it possible to argue that these groups are almost identical with respect to the processes of biocalcification taking place at the mantle edge. Biocalcification in *Nautilus* has been described by Crick and Mann (1987) and Gregoire (1987) and is briefly summarized below. In *Nautilus* the epithelium of the mantle edge is typically folded. The outer fold is separated from the rest of the mantle by a periostracal groove, where the outermost layer of the mollusk shell, the periostracum, is secreted. The inner wall of the periostracal groove adheres strongly to the outer surface of the periostracum, thus protecting the space filled with extrapallial fluid from the external environment. In this way, all the products of secretion participating in the construction of the outer shell pass through the underlying epithelium before being exuded into the extrapallial fluid. The periostracum insures the isolation of the secretory environment and is, at the same time, a substrate for the nucleation of mineral components. The first zone lying immediately behind the periostracal groove is where the outer prismatic layer is produced, and the deposition of the nacreous layer is associated with another zone, located farther away. Production of $CaCO_3$ is controlled by the disequilibrium of the extrapallial fluid as the CO_2 content varies. The organic matrix accounts for the crystal orientation. The deposition of the inner prismatic layer is confined to the zone of the myoadhesive epithelium (Mutvei 1964).

The anterior mantle edge is responsible for the production of sculpture in ammonoids. Some sculptural elements, e.g., ribs and nodes, are formed from both the outer prismatic and middle nacreous layers, so that convex shapes on the outer surface correspond to concave ones on the inner surface. Both layers generally retain the same thickness throughout the extent of the rib or node. In the posterior part of the body chamber and in the phragmocone, where the inner prismatic layer is present, a certain smoothening out of the interior surface of the sculpture may occur. Occasionally, as in *Hypophylloceras,* sculptural elements are composed mostly of the outer prismatic layer so that the inner surface of the shell wall remains flat (Birkelund and Hansen 1974, 1975; Birkelund 1981). Protruding sculptural elements-spines or a sharp keel-may be totally cut off from the cavity of the body chamber by the inner prismatic layer; as a result, the space within such elements remains free (Hölder 1952a, 1952b; Erben 1972b). Numerous Paleozoic and Mesozoic ammonoids produce regularly spaced varices, which are thickenings of the nacreous layer. Such varices appear as constrictions on inner molds. Constrictions as outer sculptural elements may occur alone or can be accompanied by varices. Mature growth stages of micro- and macroconchs may show modifications related to the thickenings of the apertural edge; these thickenings are composed chiefly of the nacreous layer.

Periostracum This is the most external layer of the shell wall, which is made of an organic substance in the majority of mollusks. In the case of *Yokoyamaoceras* from Hokkaido, the periostracum is thick (5 μm), multilayered and made of stabilized ACC matrix (Fig. 8.6b) similar to the black film of Recent *Nautilus* (CK unpublished data). Periostracum is produced in the periostracal groove of the mantle. In the ontogeny of ammonoids, the periostracal groove would have already been formed in the final stage of embryonic development. The periostracum in extant

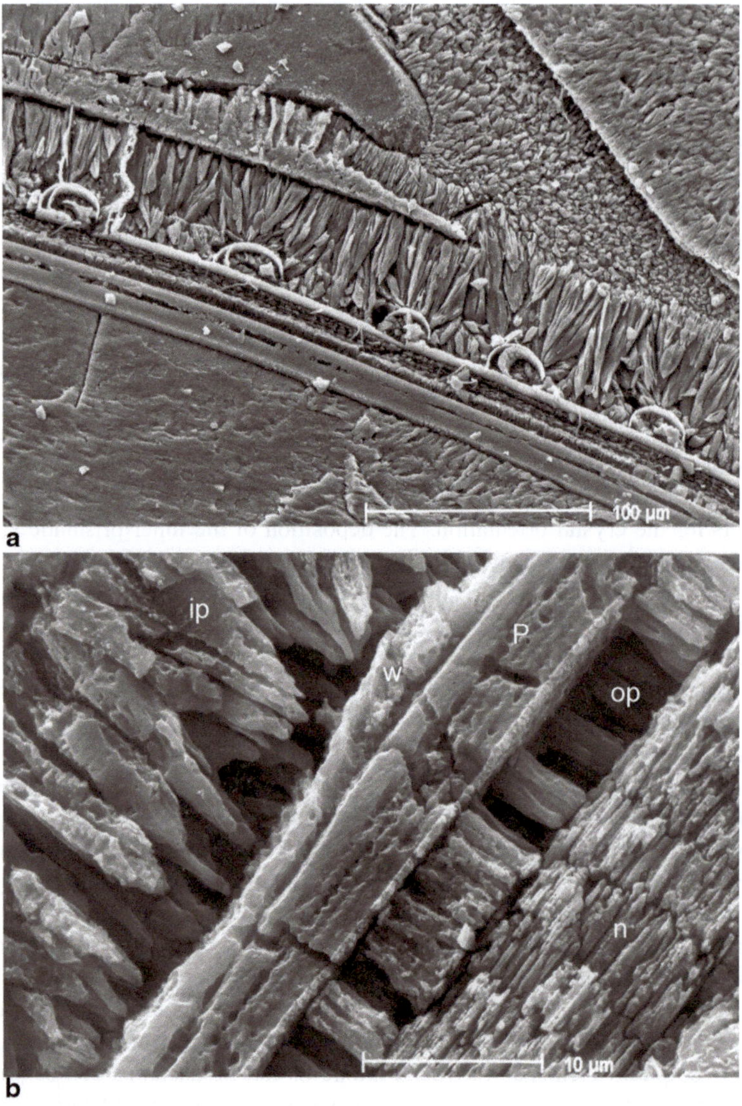

Fig. 8.6 **a** *Phyllopachyceras ezoense* (Yokoyama 1890), Middle Campanian, Cretaceous, Osoushunai Rivulet, Nakagawa Town, Hokkaido. Longitudinal section through the ventral wall covered by distinct periostracum showing periodically free periostracal edges directed adapically. **b** *Yokoyamaoceras jimboi* Matsumoto 1955, Middle Campanian, Cretaceous, Abeshinai River, Nakagawa Town, Hokkaido. Thick multilayered periostracum composed with stabilized amorphous calcium carbonate. Abbreviations: *ip*, inner prismatic layer of the next whorl; *n*, nacreous layer; *op*, outer prismatic layer; *p*, periostracum; *w*, wrinkle layer

mollusks may have a complex, multilayered structure as in, e.g., *Mytilus edulis* (see Dunachie 1963), be double-layered as in *Nautilus macromphalus* or very thick as on the lower half of the shell in *N. scrobiculatus* (see Ward 1987). In phylloceratids *Phyllopachyceras* and *Hypophylloceras* the periostracum does not form a continuous layer (as in many other ammonoids) but displays periodic bending outwards of the outer layers with small collars inclined posteriorly (*Phyllopachyceras*) or anteriorly (*Hypophylloceras*) (Kulicki et al. 2001).

Outer Prismatic Layer The mantle zone directly behind the periostracal groove is responsible for formation of the outer prismatic layer. It is deposited directly on the inner surface of the periostracum. In all normally coiled ammonoids (Erben et al. 1969; Kulicki 1979), this layer occurs in the ventral, lateral, and umbilical portions of the wall but wedges out at the umbilical seam. Earlier reports of the total lack of an outer prismatic layer in some normally coiled ammonoids (e.g., Bøggild 1930; Mutvei 1967; Birkelund and Hansen 1974) are based on incompletely preserved material (Erben et al. 1969; Birkelund and Hansen 1975). In heteromorphs in which a portion of the shell has no contact with the shell of earlier developmental stages, the outer prismatic layer occurs not only ventrally and laterally but also in the dorsal part (Doguzhaeva and Mikhailova 1982; Doguzhaeva and Mutvei 1989). In early postembryonic stages, the outer prismatic layer comprises a considerable percentage of the total thickness of the shell wall: over 75 % in *Aconeceras,* and about 50 % in *Quenstedtoceras.* During later development, the thickness of the outer prismatic layer decreases considerably to only a small fraction of the total shell wall thickness. In the Cretaceous heteromorph *Ptychoceras,* the thickness of the outer prismatic layer is about 1 μm. From the microstructural point of view, the outer prismatic layer consists of regular prisms, each with needle-like crystallites 0.2–0.5 μm in diameter; these are perpendicular to the outer shell surface (Fig. 8.5a). The relatively thick outer prismatic layer like the one in *Tetragonites*, indicates some irregularities in development of prisms, i.e. empty crystals or inclusions of nacre-like inserts (Fig. 8.5b).

The monolayered structure of the outer prismatic layer of ammonoids distinguishes this layer from the outer prismatic layer of the present-day *Nautilus,* which consists of two sublayers with different ultrastructure (Mutvei 1964). Namely, the outer sublayer has a grainy structure; each grain is made of crystallites that are mutually parallel in arrangement or slightly spherulitic.

Nacreous Layer As in *Nautilus,* the nacreous layer of ammonoids was produced in two secretory zones located in the shell wall and in the septa respectively. The first of these has a belt-like shape, encircling the inside of the aperture except for its dorsal part in the normally coiled ammonoids. The second zone of nacreous secretion is a large area at the posterior part of the body. Although the secretory product is the same in both zones, the relationship of the surface of secretion to the elements of the nacreous layer (lamellae) varies. Generally, the growth direction in the zone of septal secretion is perpendicular to the surface of the septum and to the interlamellar membranes. On the other hand, in the shell wall, the interlamellar membranes occur in very long sections parallel to the inner surface of the outer prismatic layer, and the growth zone cuts them diagonally (see Erben 1972a, Fig. 1). In fossil and modern gastropods and nautilids, as well as in Paleozoic ammonoids, the mineral component of the nacreous layer is characterized by a uniform orientation of the crystallographic axes (crystallographic texture) while in Mesozoic ammonoids the aragonite crystals display variable orientations of the crystallographic axes (Frýda et al. 2007, 2009).

According to these authors the molecular mechanism driving the origin and the development of gastropod and nautilus nacre are thus extremely old and remained unchanged for at least 220 million years. A quite different crystallographic texture of the nacre is known from each of the Mesozoic ammonoid suborders Ammonitina, Phylloceratina, and Lytoceratina. The presence of different nacre textural patterns in nautiloid and ammonoid lineages as well as the extreme stability of those patterns supports the conclusion that this patterns can be used as a tool to resolve cephalopod phylogenetic relationships (Frýda et al. 2007).

The basic structural elements of the nacreous layer are aragonitic hexagonal plates, which are arranged in sheets separated by interlamellar membranes (Fig. 8.7a). These plates are placed one on top of another, forming vertical stacks (Fig. 8.7c). Such a columnar arrangement of nacre is characteristic of ectocochliate cephalopods and of gastropods. The growth surfaces of columnar nacre described by Wise (1970) and Erben (1972a) show that the nuclei of newly formed plates are always deposited in the central part of subordinate plates and afterwards accreted. The lateral walls of neighboring hexagonal plates are separated from each other by intercrystalline membranes of conchiolin. According to Mutvei (1980), the hexagonal tablets of stack nacre are composed of a variable number of crystalline sectors, which are separated from one another by vertical radial organic membranes. These sectors represent contact and interpenetrant twins. The central portion of each tablet in the stack is occupied and connected vertically by a central organic accumulation (Mutvei 1980, 1983); these accumulations commonly occur in ammonoid nacre (Fig. 8.7b).

Ultrastructural studies of the conchiolin of the mollusk nacreous layer have been conducted by many authors, but only Gregoire (1966, 1980) included ammonoids in his investigations. Experimental studies on the conchiolin of the mollusk nacreous layer show that it is thermoresistant. Even after having been heated to 900°C for 5 h, it is still biuret-positive (Gregoire 1966, 1968, 1972, 1980; Voss-Foucart and Gregoire 1971). During diagenesis, the chemical composition of mollusk conchiolin changes rapidly but afterwards can remain stable and unchanged over millions of years, even under metamorphic conditions (Voss-Foucart and Gregoire 1971; Weiner et al. 1979; Gregoire 1980). The mineral component of the hexagonal plates of nacre has been examined since the nineteenth century, but even present-day authors offer differing opinions on its structure.

In addition to its presence in ectocochliate cephalopods, stack nacre is characteristic of gastropods, monoplacophorans, and primitive bivalves such as *Nucula*

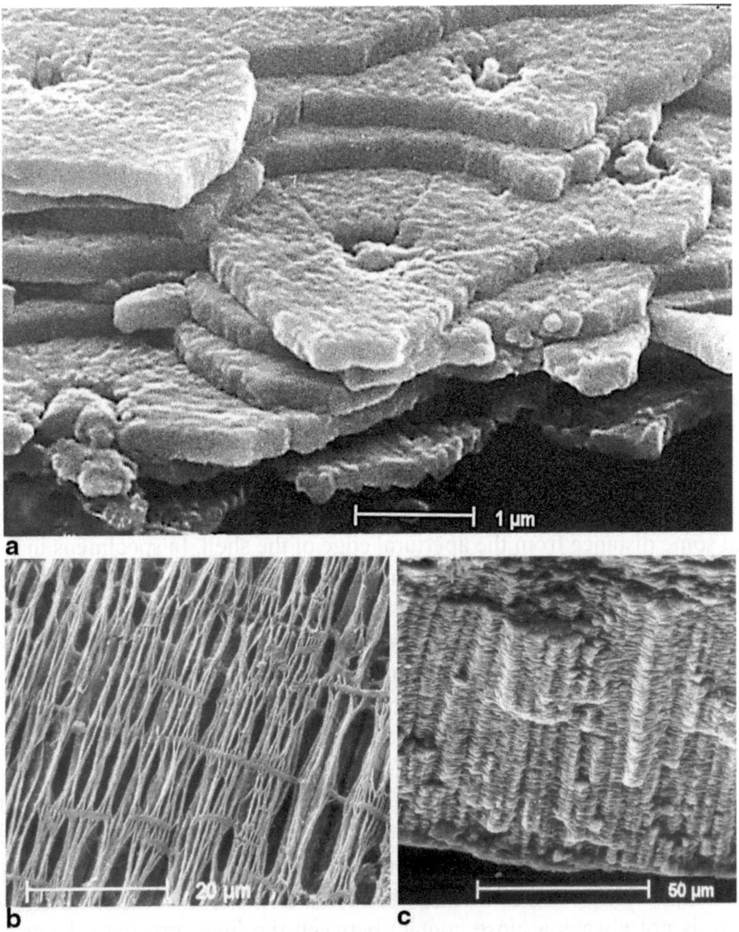

Fig. 8.7 **a**, **b** *Aconeceras trautscholdi* (Sinzov 1870), Aptian, Cretaceous, vicinity of Simbirsk, Russia. **c** *Bochianites neocomiensis* (d'Orbigny 1840), Valanginian, Cretaceous, Wąwał, Poland. **a**, Tablets of the nacre with central organic accumulations and sectors of tablets. **b**, Interlamellar organic membranes and central organic accumulations of stack nacre. **c**, Stacks of the nacre visible in broken specimen

and *Trigonacea* (see Erben et al. 1968). Orientation of axes *a* and *b*, respectively, in different plates is parallel in cephalopods (Gregoire 1962; Wise 1970) but random in gastropods (Wise 1970). Studies of the nacre in gastropods by Erben (1974) concluded that the hexagonal plates represent compact crystals rather than polycrystalline aggregates, as suggested by Mutvei (1969, 1970, 1979). From a study of chromium-sulfate-treated shells, Erben (1974) also observed, apart from interlamellar and intercrystalline membranes, intracrystalline organic elements such as diagonal sheets or vertical pillars. Mutvei (1969, 1970, 1979), on the other hand, interpreted these elements as acicular crystallites. Chromium sulfate preparation of the nacreous layer of *Quenstedtoceras* reveals detailed structural characters as seen in *Nautilus* (see Mutvei 1972a) and *Haliotis* (see Erben 1974, pl. 4, Fig. 1–5). The same *Quenstedtoceras* specimens prepared in H_2O_2 (the method used by Kulicki and Doguzhaeva 1994) display a great similarity in their nacre to that of modern *Mytilus edulis* treated with sodium hypochlorite solution and etched with 25 % glutaraldehyde solution for 2 days (Mutvei 1979). The lamellae of the nacreous layer of ammonoids are generally 0.25 µm thick in different parts of the shell, but are several times thicker in the transition zone to the outer prismatic layer. Doguzhaeva and Mutvei (1989) described pores in the nacreous layer of the heteromorph *Ptychoceras.* They considered them to be an inherent characteristic of these ammonoids. In the nacreous layer of the Early Cretaceous heteromorph *Bochianites,* similar pores are also observed, but it is difficult to dismiss the possibility that they were produced by boring organisms even though they display a "rowlike distribution" (Doguzhaeva and Mutvei 1989).

Inner Prismatic Layer This layer is situated on the inner surface of the nacreous layer at some distance from the apertural edge of the shell. In specimens in the final stages of ontogeny, the inner prismatic layer comes closer to the apertural zone (Doguzhaeva and Mutvei 1986b). This layer results from secretory activity of the myoadhesive epithelium (Mutvei 1964). Although the literature provides various data on the occurrence of the inner prismatic layer in the ontogeny of different ammonoids, in all ammonoids examined by us it starts to develop as far adapically as the second or third septum. In transverse cross sections of normally coiled ammonoids, the inner prismatic layer is the only shell layer that extends across the venter and dorsum; this is in contrast to the outer prismatic and nacreous layers, both of which wedge out at the umbilical seam and do not occur in some Phylloceratina; on the dorsum, this layer is more than twice as thick as the whole ventral wall of the previous whorl (Birkelund and Hansen 1974, 1975; Birkelund 1981).

There is not always a close contact between the inner prismatic layer and the overlying layer. The general tendency is to smooth out all unevenness or hollows, e.g., as in the floored hollow spines of *Kosmoceras* (see Erben 1972b), or to smooth out the fine ribbing on the exterior of the previous whorl (Birkelund and Hansen 1974, 1975: Howarth 1975). The inner prismatic layer generally consists of regular prisms with visible needle-like crystallites. In the heteromorph *Ptychoceras* it shows irregular nacre like inclusions (Fig. 8.8a, b). In the Triassic *Phyllocladiscites,* well-developed pores filled with crystalline matrix occur in the inner prismatic layer (Doguzhaeva and Mutvei 1986b). These pores are not visible in cross sections of the wall but can be seen only on the surface.

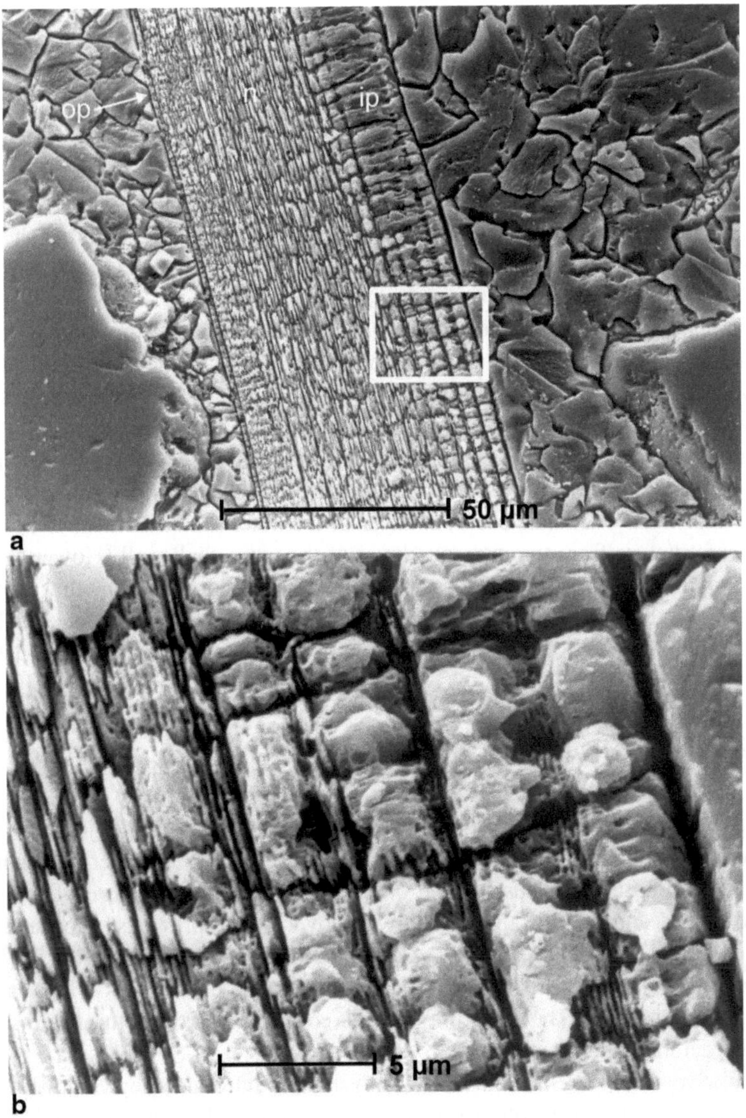

Fig. 8.8 *Ptychoceras* sp. Aptian, Cretaceous, Khocods Rivier, NW Caucasus. **a** Longitudinal section through ventral wall. Outer prismatic layer extremely thin and inner prismatic very thick. **b** Magnification of a indicating inclusions of nacre between prisms. Abbreviations: *ip*, inner prismatic layer; *n*, nacreous layer; *op*, outer prismatic layer

8.4 Modifications of the Shell Wall

8.4.1 *Inner Shell Wall of the Dactylioceratidae*

Howarth (1975) described the so-called inner shell wall in the Dactylioceratidae, a modification unknown in other groups of ammonoids. The main shell wall in representatives of the Dactylioceratidae consists of layers with a normal sequence: outer prismatic layer, nacreous layer, and inner prismatic layer. The inner shell wall consists of an outer prismatic layer and an inner nacreous layer. On the inside, the inner shell wall is covered by a septal prismatic layer.

The appearance of the inner shell wall coincides with the appearance of ornamentation. It is not closely attached to the inner surface of the main shell wall but smoothes out this wall from the inside, leaving empty spaces between it and the concave surfaces of ribs. The outer prismatic layer of the inner shell wall is not continuous—it is missing in some places. In ontogenetic development, the nacreous layer of the inner shell wall first appears as an insert in the inner prismatic layer of the main shell wall. The nacreous layer of the inner shell wall does not differ in its microstructure from the nacreous layer of the main shell wall (Howarth 1975). Although the inner prismatic layer is not split by a nacreous layer in other groups of ammonoids, there is an analogy to the splitting of the inner prismatic layer by the conspicuous organic membranes in *Nautilus* (see Doguzhaeva and Mutvei 1986b).

8.4.2 *Dorsal Wall*

In heteromorphs, where the shell walls of adjacent "whorls" do not join, the dorsal wall of the "free whorls" usually has the same structure as do other shell parts. In normally coiled ammonoids, on the other hand, the microstructure of the dorsal wall changes, and the dorsal wall is a wall covering the previous whorl and laterally limited by the umbilical seams. According to Kulicki (1979), the dorsal wall of ammonoids generally consists of two components: the outer wrinkle layer and the inner prismatic layer. The latter may have a complex microstructure, as discussed in the previous section.

The term *"Runzelschicht"* (wrinkle layer) was introduced by Sandberger and Sandberger (1850) to denote a thin layer superimposed on the test of Devonian goniatites. This layer had been recognized earlier by Keyserling (1846). The descriptions of these authors leave no doubt that the layer described is connected with the dorsal wall. Sandberger and Sandberger (1850) also used another expression, "*Ritzstreifen,*" to denote the markings preserved on internal molds of the inside of the lateral and ventral parts of the ammonoid whorl. Subsequent authors have used the term *Runzelschicht* for both categories of elements. The problem of the wrinkle layer in Paleozoic goniatites was discussed by House (1971), who insisted that the terms *"Runzelschicht"* and *"Ritzstreifen"* denote the same structure and suggested that this structure be referred to as the dorsal or ventral wrinkle layer, respectively. According to House (1971), all Paleozoic ammonoids possessing such structures are smooth and ribless: the Anarcestoidea and Pharciceratoidea display both kinds of wrinkle layers while the Cheiloceratoidea only the dorsal one, and the Clymeniina has just a narrow belt on the dorsal side. Tozer (1972) verified and precisely defined the meaning of these expressions.

The wrinkle layer occurs on the dorsal side of normally coiled ammonoids in the vicinity of the aperture; in some instances, it has been observed beyond the aperture (Senior 1971; Tozer 1972; Tanabe et al. 2001). The secretory zone of this layer is connected with the mantle edge, and, therefore, it is comparable to the black film of *Nautilus* with which it shares a similar microornamentation. *Ritzstreifen,* as defined by Sandberger and Sandberger (1850), are situated at the back of the body chamber, and, therefore, the zone of their formation lies behind the secretory zone of the nacreous layer.

Kulicki (1979) described the development of the wrinkle layer in the body chamber of *Quenstedtoceras.* A similar wrinkle layer and its developmental sequence have been presented in Triassic *Proarcestes* by Doguzhaeva and Mutvei (1986b). Walliser (1970) described the variability in ornamentation and occurrence of the *Runzelschicht* and *Ritzstreifen* without differentiating these two categories.

The main elements of the wrinkle layer of *Quenstedtoceras* are triangular in longitudinal cross section and are composed of prisms perpendicular to the surfaces of the steep adapertural sides of the triangles. Each triangle is filled with smaller prisms, arranged more chaotically as those known from perpendicular sections in *Damesites* and *Tetragonites* (Fig. 8.9). In addition to the prisms, the triangular elements contain abundant organic matter. Nassichuk (1967) assumed that the wrinkle layer of *Clistoceras* may have been organic. Zakharov and Grabovskaya (1984), however, assigned the wrinkle layer in *Zelandites japonicus* to the nacreous layer, but this feature is not clearly shown in their Figure (Zakharov and Grabovskaya 1984, Fig. 3). Bayer (1974) described the wrinkle layer of Mesozoic ammonoids as occurring only on the dorsal side. He argued that the wrinkle layer did not constitute a separate layer but was merely part of the inner prismatic layer. Kulicki et al. (2001) described the dorsal shell wall in many genera of Goniatitida, Ammonitida, Phylloceratida, and Lytoceratida. Apart from the external component of the dorsal wall expressed as wrinkle layer, there are also other forms of the external component as e.g., in *Aconeceras* illustrated herein (Fig. 8.10).

8.4.3 Umbilical Plugs and Encrusting Layers

Umbilical plugs comparable to those in *Nautilus pompilius* are observed in some Paleozoic and Triassic ammonoids, especially in strongly involute forms such as *Clistoceras* (see Nassichuk 1967), *Nathorstites* (see Tozer 1972), and a number of Triassic ammonoids and nautilids (Klug et al. 2004). In the present-day *Nautilus pompilius,* the surface of the umbilicus is partly covered by a black film on which

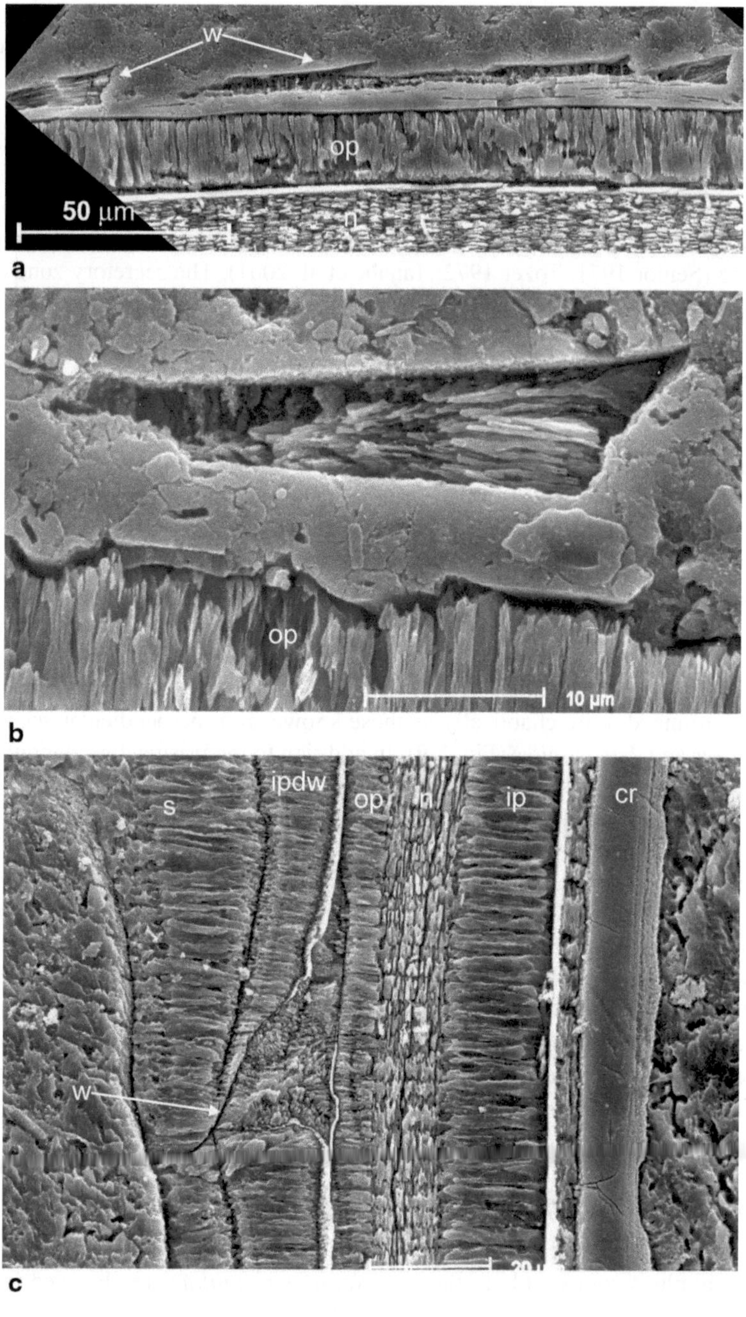

Fig. 8.9 **a**, **b**, *Tetragonites glabrus* (Jimbo 1894), Early Campanian, Cretaceous, Abeshinai River, Nakagawa Town, Hokkaido. **c**, *Damesites sugata* Forbes, 1846. Coniacian, Cretaceous Nakafutamata Rivulet, Haboro area northwestern Hokkaido. **a**, Longitudinal section through the ventral

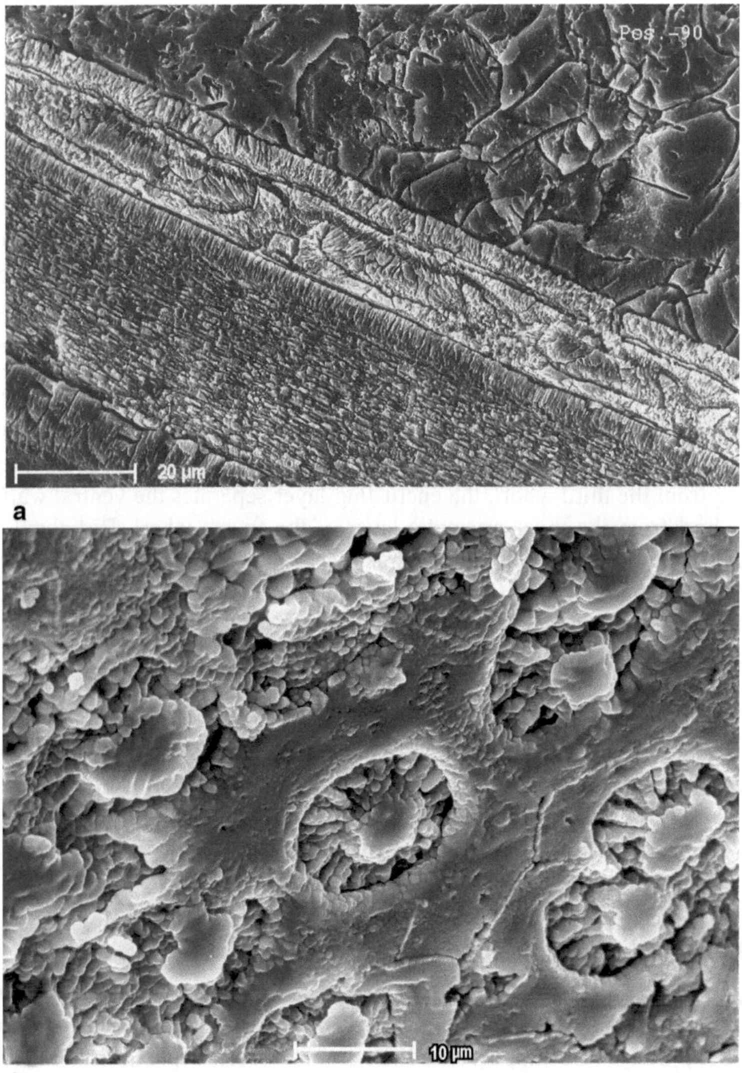

Fig. 8.10 *Aconeceras trautscholdi* (Sinzov 1870), Aptian, Cretaceous, vicinity of Simbirsk, Russia. **a**, Longitudinal section through ventral wall covered by dorsal one of the next whorl. **b**, View on exfoliated dorsal wall from the basal part

---

wall covered by triangular elements of the dorsal wall of the next whorl. **b**, Triangular element of the same specimen as **a**. **c**, Longitudinal section through the ventral wall covered by dorsal one, consisting of three constituents: triangular elements, middle prismatic layer and mural part of the septum. Abbreviations: *cr*, ventral wall of the connecting ring; *ip*, inner prismatic layer; *ipdw*, inner prismatic layer of the dorsal wall of the next whorl; *n*, nacreous layer; *op*, outer prismatic layer; *w*, wrinkle the material of the umbilical plug lies. This umbilical callus consists of spherulitic prismatic deposits and elements of the nacreous layer (Gregoire 1966; CK, unpublished observations). In adult shells the umbilical plug does not fill the whole umbilical space but only the outer part. Nassichuk (1967) called the formations described in *Clistoceras* as “helicolateral deposits” but did not confirm their homology to the umbilical plugs in *Nathorstites* and *Nautilus pompilius.* Tozer (1972), in contrast, confirmed the close homology between the helicolateral deposits of *Clistoceras* and the umbilical plugs of *Nathorstites* and *N. pompilius.* There is no detailed description of the structure of the above mentioned deposits apart from the statement by Tozer (1972) that, in the formation of the umbilical callus in *Nathorstites,* a secondary nacreous material is included. Bogoslovsky (1969) illustrated specimens of *Prolobites* in which the umbilicus is covered by a calcium carbonate wall. Drushchits et al. (1978) described multilayered structures encrusting the umbilical surface in the Late Cretaceous lytoceratid *Gaudryceras.* These structures do not cover the entire shell surface in that some spaces are left out, especially at the umbilical seams. Starting from the third whorl, the encrusting layer separates the ventral wall of the third whorl from the inner prismatic layer of the fourth whorl. Birkelund (1981) confirmed the occurrence of such encrusting layers in specimens of *Gaudryceras* from Japan. The thickness of the encrusting layers may be several times thicker than that of the shell wall itself. Our reinvestigation of well-preserved material from Hokkaido suggests that the coating layers of *Gaudryceras* are homologous to the black film of *Nautilus*, being very similar both in color and the elemental composition (Fig. 8.11). There are significant amounts of Ca and variable amounts of C in these structural elements which may indicate that the primary material has been formed from more or less stabilized ACC (amorphous calcium carbonate). In case of *Gaudryceras* illustrated herein (Fig. 8.11b) some crystallization could occur as a result of diagenesis.

8.4.4 Septa

According to the commonly accepted view, the first two septa are called prosepta (Fig. 8.12). They were supposed to differ from all other septa in their relationship to the wall of the outer shell (Grandjean 1910; Böhmers 1936; Voorthuysen 1940; Miller and Unklesbay 1943; Miller et al. 1957; Arkell 1957; Drushchits and Doguzhaeva 1974; Tanabe et al. 1980). Miller and Unklesbay (1943) pointed out that prosepta are adaperturally concave in medial cross section, whereas all remaining septa are convex toward the aperture.

On the basis of SEM observations, Erben et al. (1969) confirmed the distinct microstructural character of prosepta in relation to all other septa. They also observed that the second septum (primary septum, “*Primarseptum*”), like all subsequent septa, is separated from the shell wall by a conchiolin layer.

A nacreous layer in the second septum has been identified by Birkelund and Hansen (1974); Kulicki (1979); Bandel (1982); Landman and Bandel (1985), and

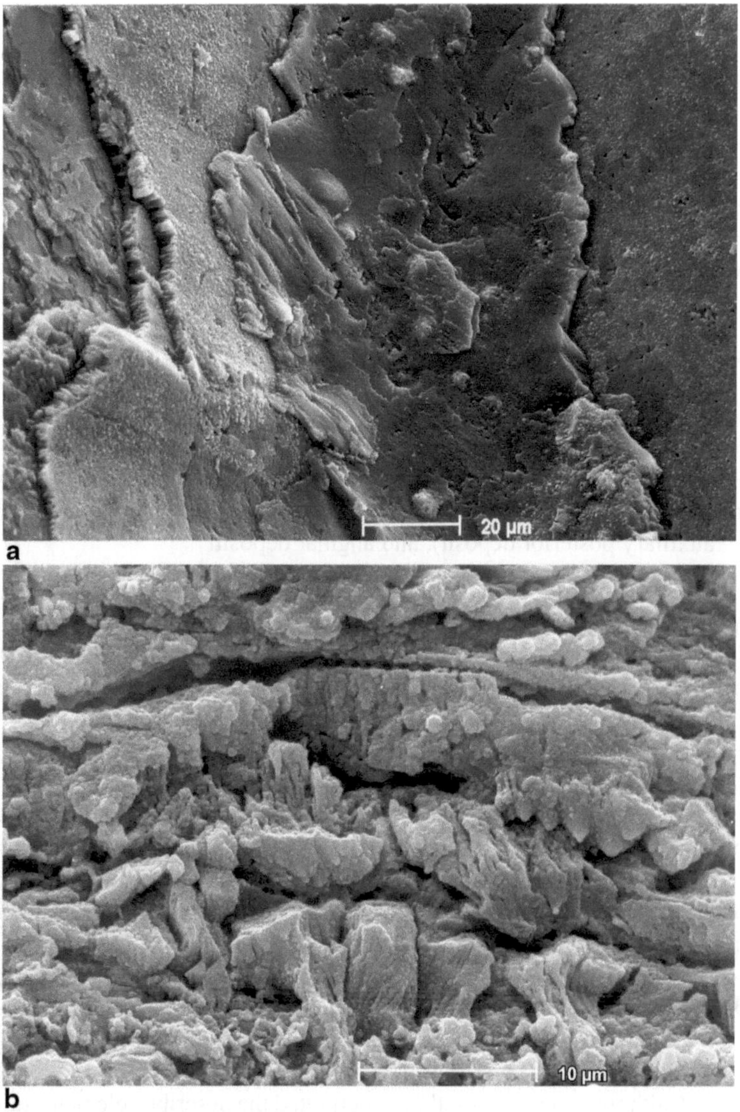

Fig. 8.11 *Gaudryceras tenuiliratum* Yabe, 1903, Campanian, Cretaceous, Abeshinai River, Nakagawa Town, north Hokkaido. **a**, Coating layers covering seam between two whorls. **b**, Broken view of several coating layers

Landman (1987). Erben et al. (1969), using strictly medial cross sections, may not have recorded the occurrence of a nacreous layer in the structure of the second septum. The main component in the construction of septa, with the exception of the proseptum, is the nacreous layer. Sporadically, the adapertural surface and, less commonly, the adapical one may be covered by a prismatic layer and, more rarely, a spherulitic prismatic one (Birkelund and Hansen 1974; Howarth 1975; Kulicki 1979; Kulicki and Mutvei 1982). The connection between the septum and septal neck on the adapertural surface in *Quenstedtoceras* illustrates the prismatic layer passing collaterally and continuously into the lamellae of the nacreous layer (Kulicki and Mutvei 1982, text-Fig. 3, Pl. 5, Fig. 1). At the connection points of the septa and the shell wall a distinct boundary separating the two is visible. Lamellae of the nacreous layer bend strongly adaperturally and pass continuously into the prismatic layer, the so-called septal prismatic layer of Howarth (1975), which may be developed to different degrees.

8.4.5 Septal Neck-Siphuncular Complex

The expression "siphuncular complex," introduced by Tanabe et al. (1993a), includes the following elements: cecum and prosiphon, siphuncular tube (= connecting ring), siphuncular membranes, auxiliary deposit (= auxiliary anterior deposit), cuff (= auxiliary posterior deposit), and angular deposit.

Cecum and Siphuncular Tube The cecum (caecum) is the spherical ending of the siphuncular tube in the initial chamber. Originally composed of organic conchiolin, the cecum generally is phosphatized in the fossil state. The problem of preservation of the cecum applies equally well to the other organic elements of the siphuncular complex, i.e., the siphuncular tube and the siphuncular membranes. The main mineral in these elements is francolite (Andalib 1972; Hewitt and Westermann 1983). The siphuncular tube is very rarely calcified (Joly 1976; Kulicki et al. 2007; Tanabe et al. 2005). Nautiloids of the same age have non-phosphatized siphuncular tubes (Fukuda in Obata et al. 1980; CK, unpublished observations). Hewitt and Westermann (1983) presented a tentative interpretation of the ammonoid siphuncular tube as having been originally chitinous, but stiffened internally with phosphate crystals.

In its anterior part, the cecum in Mesozoic forms, e.g., *Kosmoceras* and *Quenstedtoceras,* is connected directly or by a calcareous deposit to the achoanitic proseptum (Kulicki 1979). In *Tornoceras* the relationship between the proseptum and the cecum was illustrated by House (1965, Fig. 2) and also by Bogoslovsky (1971, Fig. 6), and, although these two authors interpreted the described elements quite differently, they both indicated a retrochoanitic septal neck in the proseptum. Thus, it can be expected that the relationship between the conchiolin wall of the cecum and the proseptum is the same as that between the siphuncular tube and the retrochoanitic septal neck in *Nautilus.* The outer part of the cecum in many genera is covered by a calcareous deposit (Drushchits and Doguzhaeva 1974, 1981; Drushchits et al. 1983; Tanabe et al. 1979, 1980).

The cecum is connected to the inner surface of the initial chamber by means of the so-called prosiphon, which is an organic structure of complex construction. Prosiphons, even in the same species, may display considerable morphological variability. Apart from tube-like elements, which vary as to the degree of flatten-

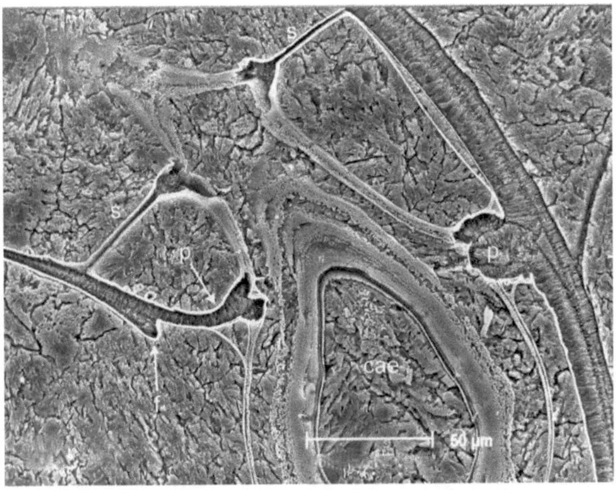

Fig. 8.12 *Damesites sugata* Forbes, 1846. Coniacian, Cretaceous, Nakafutamata Rivulet, Haboro area northwestern Hokkaido, Prosepta: prismatic proseptum and nacreous primary septum. Abbreviations: *cae*, caecum; *f*, flange; *p*, proseptum; *s*, first nacroseptun

ing and bending, there is another very variable element: membranes that join the tubular elements to the inner surface of the initial chamber (Kulicki 1979; Bandel 1982; Ohtsuka 1986; and citations above). The cecum in cross section may show a multilayered concentric structure (Ohtsuka 1986 and Fig. 8.12 herein). The cecum and prosiphon commonly exhibit wrinkles or tension lines, proving their former elasticity.

The siphuncular tube in ammonoids, as opposed to that in *Nautilus,* is constructed exclusively of organic matter, i.e., multilayered concentric membranes of conchiolin (Fig. 8.13a, c, d), each of which consists of lace-like, tuberculate microfibrils 1–2 µm in diameter with elongate pores (Obata et al. 1980; Gregoire 1984).

In the retrochoanitic condition, comparable to that in present-day *Nautilus,* the adapertural end of each segment of the siphuncular tube is a continuation of the organic membranes of the nacreous layer of the septal neck. The adapical end of each segment of the siphuncular tube terminates in an auxiliary deposit attached to the inner surface of the previous septal neck. In the prochoanitic condition, the only difference is the relationship between the siphuncular tube and the septal neck at the adapertural end of the tube segment. In this case, the organic segment of the siphuncular tube is connected with the cuff and not with the distal part of the septal neck.

According to Mutvei (1972a), in present-day *Nautilus* there are no organic layers connecting the adjacent segments of the siphuncular tube. Therefore, Gregoire's (1987, Fig. 12) schematic drawing, in which the auxiliary ridge is connected to both the adapertural and adapical segments of the siphuncular tube is misleading. Ammonoids in the retrochoanitic condition do not exhibit such connections either while they were reported from the ammonoids with the prochoanitic condition (Kulicki and Mutvei 1982) as being present between adjacent segments of the siphuncular

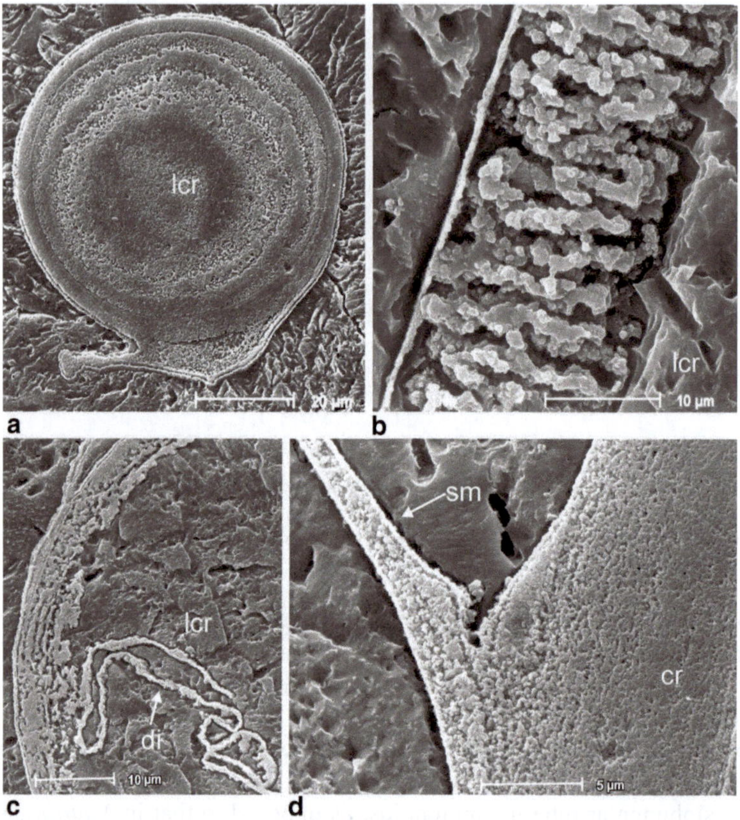

Fig. 8.13 **a**, **c**, **d**, *Damesites sugata* Forbes, 1846. Coniacian, Cretaceous, Nakafutamata Rivulet, Haboro area northwestern Hokkaido. **b**, *Aconeceras trautscholdi* (Sinzov 1870), Aptian, Cretaceous, vicinity of Simbirsk, Russia. **a**, **c**, Phosphatized siphuncular tube showing multilayered structure of the siphuncular wall. **b**, Cross section through partly phosphatized and partly calcified wall of the siphuncular tube. **d**, The outermost siphuncular layer passing in to siphuncular membrane. Abbreviations: *cr*, connecting ring; *di*, delaminated inner layer of the connecting ring; *lcr*, lumen of the connecting ring; *sm*, siphuncular membrane

tube via the basal lamella of the cuff. This phenomenon is easy to explain because the siphuncular tube protruding into the body chamber was longer than the distance between septa. In this case, after the soft body moved forward into a position appropriate to form a new septum, the basal lamella undergoing calcification became connected to both the adapical and adapertural segments of the siphuncular tube, which was already partially secreted. An additional connection of the outer layer of the siphuncular tube to the adapertural tip of the prochoanitic septal neck has also been reported by Kulicki and Mutvei (1982). According to Kulicki (1979), the siphuncular tube in the prochoanitic condition may have been secreted both from the inside, by the epithelium of the siphuncular cord, and from the outside, by the epithelium of the circumsiphonal invagination. The outermost layer of the tube segment would have been secreted by the epithelium of the circumsiphonal invagination.

Birkelund and Hansen (1974) have described a double-layered siphuncular tube in a specimen of the Maastrichtian ammonoid *Saghalinites wrighti.* The inner, initially organic layer of this siphuncular tube displays a fibrous structure, whereas the outer layer is composed of calcium carbonate with a prismatic structure. The state of preservation of the specimen prevented these authors from determining whether the mineral of the outer layer was calcite or aragonite. But the fact that the strontium content of this layer differed from that in the surrounding rock was, in their opinion, a confirmation of the primary nature of this calcium carbonate layer and, hence, evidence of a similarity between the siphuncular structure in ammonoids and that in present-day *Nautilus.*

Septal Necks Siphuncular deposits in the Ammonitina are best known in *Quenstedtoceras* (see Kulicki 1979; Kulicki and Mutvei 1982) and *Gaudryceras* (Tanabe et al. 1993a; Tanabe and Landman 1996, 2010). Generally, the inner surface of the septal neck is covered by a cuff, which is the adapertural termination of the siphuncular tube segment. An auxiliary deposit, which is the adapical termination of the next segment of the siphuncular tube, frequently covers the cuff from the inside. In middle and late ontogenetic stages of *Quenstedtoceras* the outermost part of the cuff, called the basal lamella, is distinguished as a separate element. It remains in direct contact with the nacreous layer of the septal neck. Kulicki and Mutvei (1982) distinguished it because of the difference in its microstructure with respect to the rest of the septal neck and because it is connected to the wall of the siphuncular tube both adaperturally and adapically. They have also distinguished the basal lamella in the auxiliary deposit, which is connected at its distal end to the wall of the siphuncular tube.

The cuff and auxiliary deposit are identical in microstructure. They both show fine lamellae (about 1 μm thick) and a spherulitic-prismatic system of mineral components similar to the modified nacreous layer in the septal neck of *Nautilus* (see Kulicki and Mutvei 1982). The basal lamellae have a stronger spherulitic-prismatic calcification than the rest of the deposit, and, thus, the lamellar structure is less visible there (Kulicki and Mutvei 1982). These studies did not confirm the interpretation of Bandel and Boletzky (1979) and Bandel (1981, 1982) that the siphuncular deposits in the Ammonitina were originally porous.

In a newly formed septum, one commonly can observe that the basal lamella of the cuff is not in contact with the inner surface of the proximal part of the septal neck and that there is an empty space between these two elements (H. Mutvei and C. Kulicki, unpublished data). In more adapical septa, this space is filled by sparry calcareous material and forms so-called "angular deposit" (Kulicki 1979). Angular deposits are formed as a result of inorganic calcification; they are almost always present between the cuff and the adapical part of the prochoanitic septal neck, with the exception of the last septum. This suggests rapid formation of angular deposits during the lifetime of the animal (Kulicki 1979).

Intermediate Condition Septal necks characteristic of this condition represent an intermediate form between typical retro-and prochoanitic necks. Such necks have been described by Doguzhaeva (1973), Doguzhaeva and Mutvei (1986a), Drushchits and Doguzhaeva (1974), Kulicki (1979), Kulicki (1994), Tanabe et al. (1993a), and Tanabe and Landman (1996).

During the transition from retrochoanitic septal necks to fully prochoanitic necks, the secretion of the siphuncular tube became more independent of the secretion of the septum and septal neck than in the retrochoanitic condition. In the retrochoanitic condition, the siphuncular tube, septal neck, and septum must have been secreted simultaneously. In the prochoanitic condition, secretion of the siphuncular tube was independent of the formation of the septum. According to Kulicki and Mutvei (1982), the organic layers of siphuncular tube are formed prior to the appearance of the septum and siphuncular neck.

8.4.6 Intracameral Membranes

The inner surface of chamber walls and the adapical and adoral surfaces of septa are covered with phosphatized membranes that were originally organic. These are called, in general, intracameral, conchiolin, or cameral membranes. Those that remain in contact with the siphuncular tube are specifically known as siphuncular membranes. Siphuncular membranes have been described by Grandjean (1910); Schulga-Nesterenko (1926); Erben and Reid (1971); Westermann (1971); Bayer (1975); Kulicki (1979); Bandel and Boletzky (1979); Bandel (1981, 1982); Tanabe et al. (1982); Weitschat (1986); Weitschat and Bandel (1991), and Landman et al. (2006).

The ultrastructure of intracameral membranes was described by Tanabe et al. (1982) and Gregoire (1984). According to these authors, these membranes are composed of thin (0.015–0.15 μm) fibers that lack a distinct orientation. These membranes are comparable to the pellicle ("*brown membrane*") in *Nautilus* and *Spirula.* However, they differ considerably in having finer fibers that have no preferred orientation.

Siphuncular membranes display great variability among different ammonoid groups. Although Permian and Triassic genera seem to have especially well-developed and numerous siphuncular membranes (Schulga-Nesterenko 1926; Weitschat and Bandel 1991), they are less numerous and not-so-well-developed in Jurassic and Cretaceous forms. These membranes connect the siphuncular tube to both the adapical and adapertural surfaces of the septa. They may also form connections to the ventral or ventrolateral shell wall. Especially complex systems of siphuncular membranes have been described by Weitschat and Bandel (1991) in the Triassic genera *Anagymnotoceras* and *Czekanowskites.*

Horizontal membranes, as noted by Weitschat and Bandel (1991), represent a case of extremely well-developed siphuncular membranes. These horizontal membranes divide the phragmocone chamber into two main compartments of nearly equal size, a ventral and a dorsal compartment. These membranes are connected to the siphuncular tube, to the septa at their mid-height, and to the inner surface of the outer shell wall flanks. These membranes are commonly perforated. There is another kind of intracameral membrane known as a transverse membrane (Weitschat and Bandel 1991). These membranes are not always connected to the siphuncle but generally separate the chamber into compartments. Transverse membranes also display perforations.

Zaborski (1986) and Hewitt et al. (1991) have described membranous structures, so-called "pseudosepta", in vascoceratids from Nigeria. Pseudosutures, i.e., the lines of attachment of pseudosepta to the shell walls, have been reported from Mesozoic ammonoids (John 1909; Hölder 1954; Vogel 1959; Schindewolf 1968; Bayer 1977) as well as from Paleozoic goniatites (Landman et al. 1993; Tanabe et al. 1998) and prolecanitids (Landman et al. 2006).

There are two interpretations for the formation of intracameral membranes. The first, accepted by the majority of researchers, posits that organic membranes are secreted in close contact with the mantle surface, like the periostracum or other mineral-organic layers of the shell, e.g., the nacreous layer. Weitschat and Bandel (1991) have critically assessed this point of view and they proposed an alternative interpretation which agrees with the former one that the organic matter of the intracameral membranes is secreted by the mantle surface but the membranes themselves are produced by dehydration of a jelly-like cameral fluid (Bandel and Boletzky 1979; Hewitt et al. 1991; Westermann 1992).

Functionally, hydrophilic intracameral membranes could have served either as pathways for transport of cameral fluid or as structures helping to maintain the fluid within certain reservoirs and preventing any overflow. This may have made it possible to keep cameral and circumsiphonal fluids in a state of decoupling, which may have been beneficial from the physiological point of view (Mutvei 1967; Kulicki 1979; Kulicki and Mutvei 1988; Ward 1987; Weitschat and Bandel 1991). For further discussion of intracameral membranes, see Polizzotto et al. (2015).

8.5 Conclusions

The major limitation to micro-and ultrastructural studies of ammonoids is diagenesis, i.e., the state of preservation of the specimens used. This explains why Jurassic and Cretaceous ammonoids are among the best studied, while the data on Paleozoic and Triassic forms are generally rare. It is only under exceptionally favorable conditions of preservation, such as those in the Buckhorn asphalts of Pennsylvanian age, that the study of the ultrastructure of orthoconic cephalopods was possible, revealing a great similarity between these forms and present-day *Nautilus* (see Mutvei 1972b). At the ultrastructural level, the three main layers of the postembryonic shell of ammonoids do not differ significantly from those known from the shell of present-day *Nautilus*. The same is also true for the septa. However, a clearly distinctive element is the ammonitella, which is now the focus of attention of many ammonitologists. The new model of calcification of the ammonitellae, based on well-preserved specimens of *Aconeceras* (see Kulicki and Doguzhaeva 1994, Tanabe et al. 2008, 2010) and the importance of more or less stabilized ACC (amorphous calcium carbonate) as well as the remotely controlled development of the tubercles, requires further investigations. Knowledge of the ultrastructure of both the mineral and organic components in ammonoids is meager compared with what is known about Recent mollusks. Some results obtained in the course of ultrastructural studies by Gregoire (1984) and Obata et al. (1980) seem to show that such research is feasible.

References

Andalib F (1972) Mineralogy and preservation of siphuncles in Jurassic cephalopods. N Jb Geol Paläont Abh 140:33–48

Arkell WJ (1957) Introduction to Mesozoic Ammonoidea. In: Moore RC (ed) Treatise on invertebrate paleontology, Part L, Mollusca 4. GSA and University Kansas Press, Lawrence, pp. 81–129

Bandel K (1981) The structure and formation of the siphuncular tube of *Quenstedtoceras* compared with that of *Nautilus* (Cephalopoda). N Jb Geol Paläont Abh 161:153–171

Bandel K (1982) Morphologie und Bildung der frühontogenetischen Gehäuse bei conchiferen Mollusken. Facies 7:1–198

Bandel K (1986) The ammonitella: a model of formation with the aid of the embryonic shell of archaeogastropods. Lethaia 19:171–180

Bandel K (1990) Cephalopod shell structure and general mechanisms of shell formation. In: Carter G (ed) Skeletal biomineralization: patterns, processes and evolutionary trends, vol I. Van Nostrand Reinhold, New York

Bandel K, Boletzky SV (1979) A comparative study of the structure, development, and morphological relationships of chambered cephalopod shells. Veliger 21:313–354

Bandel K, Landman NH, Waage KM (1982) Micro-ornament on early whorls of Mesozoic ammonites: implications for early ontogeny. J Paleontol 56:386–391

Bayer U (1974) Die Runzelschicht—ein Leichtbauelement der Ammonitenschale. Paläontol Z 48(1–2):6–15

Bayer U (1975) Organische Tapeten im Ammoniten-Phragmocon und ihr Einfluß auf die Fossilisation. N Jb Geol Paläont Mh 1975(1):12–25

Bayer U (1977) Cephalopoden Septen. Teil I. Konstruktionsmorphologie des Ammoniten-Septums. N Jb Geol Paläont Abh 154:290–366

Birkelund T (1967) Submicroscopic shell structures in early growth stage of Maastrichtian ammonites (*Saghalinites* and *Scaphites*). Medd Dan Geol Foren 17(1):95–101

Birkelund T (1981) Ammonoid shell structure. In: House MR, Senior JR (eds) The Ammonoidea (Systematics Association Special vol 10). Academic Press, London, pp. 177–214

Birkelund T, Hansen HT (1968) Early shell growth and structures of the septa and the siphuncular tube in some Maastrichtian ammonites. Medd Dan Geol Foren 18:95–101

Birkelund T, Hansen HJ (1974) Shell ultrastructures of some Maastrichtian Ammonoidea and Coleoidea and their taxonomic implications. K Dan Vidensk Selsk Biol Skr 20(6):2–34

Birkelund T, Hansen HJ (1975) Further remarks on the post-embryonic *Hypophylloceras* shell. Bull Geol Soc Den 24:87–92

Bøggild OB (1930) The shell structure of the molluscs. K Dan Vidensk Selsk Skr Raekke 92(2):233–326

Bogoslovsky BI (1969) Devonskie Ammonoidei. I. Agoniatity. Trans Paleont Inst Akad Nauk SSSR 124:1-341 [in Russian]

Bogoslovsky BI (1971) Devonskie Ammonoidei. II. Goniatity. Trans Paleont Inst Akad Nauk SSSR 127:1-216 [in Russian]
Böhmers JCA (1936) Bau und Struktur von Schale und Sipho bei permischen Ammonoidea. Dissertation. Drukkerij University, Amsterdam, Apeldoorn
Branco W (1880) Beiträge zur Entwicklungsgeschichte der fossilen Cephalopoden. Palaeontographica 27:17–81
Brown A (1892) The development of the shell the coiled stage of *Baculites compressus* Say. Proc Acad Nat Sci Phila 44:136–142
Crick RE, Mann KO (1987) Biomineralization and systematic implications. In: Saunders WB, Landman NH (eds) *Nautilus*—The biology and paleobiology of a living fossil. Plenum, New York, pp. 115–134
De Baets K, Landman NH, Tanabe K (2015) Ammonoid embryonic development.This volume
Doguzhaeva LA (1973) Vnutriennoe stroienie rakoviny roda *Megaphyllites*. Byull Mosk Ova Ispyt Prir Otd Geol 48(6):161
Doguzhaeva LA, Mikhailova IA (1982) The genus *Luppovia* and the phylogeny of Cretaceous heteromorph ammonoids. Lethaia 15:55–65
Doguzhaeva LA, Mutvei H (1986a) Retro- and prochoanitic septal necks in ammonoids, and transition between them. Palaeontogr A 195:1–18
Doguzhaeva LA, Mutvei H (1986b) Functional interpretation of inner shell layers in Triassic ceratid ammonites. Lethaia 19:195–209
Doguzhaeva LA, Mutvei H (1989) *Ptychoceras,* a heteromorphic lytoceratid with truncated shell and modified ultrastructure. Palaeontogr A 208:91–121
Drushchits VV, Doguzhaeva LA (1974) Some morphogenetic characteristics of phylloceratids and lytoceratids (Ammonoidea). Paleontol J 8(1):37–48
Drushchits VV, Doguzhaeva LA (1981) Ammonity pod elektronnym mikroskopom. Moskva University Press [in Russian], p. 240
Drushchits VV, Khiami N (1969) O niekotorykh voprosakh izuchenia rannikh stadia ontogeneza ammonitov. In Tez Dokl na sveshch po probl Puti i zakonomiernosti istoritscheskogo rozvitia zivotnykh i rostitelnykh organizmov. Moscow, pp. 26–30 [in Russian]
Drushchits VV, Khiami N (1970) Stroienie sept, stenki protokonkha i natchalnykh oborotov rakoviny nekotorykh ranniemelovykh ammonitov. Paleontol Zh 1970(1):35–47 [in Russian]
Drushchits VV, Doguzhaeva LA, Mikhailova IA (1977) The structure of the ammonitella and the direct development of antmonites. Paleontol J 11(2):188–199
Drushchits VV, Doguzhaeva LA, Mikhailova IA (1978) Neobytchnye oblekayusche sloi ammonitov. Paleontol Zh 1978(2):36–44 [in Russian]
Drushchits VV, Muravin ES, Baranov VN (1983) Morfogenez rakovin srednevolzhskikh ammonitov roda *Virgatites*, *Lomonosovella*, *Epivirgatites*. Vestn Mosk Univ Ser 4 Geol 1983:35–44 [in Russian]
Dullo WC, Bandel K (1988) Diagenesis of molluscan shells: a case study. In: Wiedmann J, Kullmann J (eds) Cephalopods-Present and past. Schweizerbart, Stuttgart, pp 719–729
Dunachie JF (1963) The periostracum of *Mytilus edulis*. R Soc Edinb 65(15):383–411
Erben HK (1962) Über den Prosipho, die Prosutur und die Ontogenie der Ammonoidea. Paläontol Z 36:99–108
Erben HK (1966) Über den Ursprung der Ammonoidea. Biol Rev 41:6–19
Erben HK (1972a) Über die Bildung und das Wachstum von Perlmutt. Biomineralisation 4:15–46
Erben HK (1972b) Die Mikro- und Ultrastruktur abgedeckter Hohlelemente und die Conellen des Ammonitem-Gehäuses. Paläontol Z 46:6–19
Erben HK (1974) On the structure and growth of the nacreous tablets in gastropods. Biomineralisation 4:14–22
Erben HK, Reid REH (1971) Ultrastructure of shell, origin of conellae and siphuncular membranes in an ammonite. Biomineralisation 3:22–31
Erben HK, Flajs G, Siehl A (1968) Ammonoids: early ontogeny of ultramicroscopical shell structure. Nature 219:396–398
Erben HK, Flajs G, Siehl A (1969) Die frühontogenetische Entwicklung der Schalenstruktur ectocochleater Cephalopoden. Palaeontogr A 132:1–54

Frýda J, Weitschat W, Tycova P, Haloda J, Mapes RH (2007) Crystallographic textures of cephalopod nacre: its evolution, time stability, and phylogenetic significance. Seventh International Symposium, Cephalopods—Present and past, Sapporo Japan, Abstracts Volume, pp. 56–57

Frýda J, Bandel K, Frýdova B (2009) Crystallographic texture of late Triassic gastropod nacre: evidence of long-term stability of the mechanism controlling its formation. Bull Geosci 84:745–754

Grandjean F (1910) Le siphon des ammonites et des belemnites. Bull Soc Géol Fr Sér 4(10):496–519

Grégoire C (1962) On submicroscopic structure of the *Nautilus* shell. Bull Inst R Sci Nat Belg 38(49):1–71

Grégoire C (1966) On organic remains of Paleozoic and Mesozoic cephalopods (nautiloids and ammonoids). Bull Inst R Sci Nat Belg 42(39):1–36

Grégoire C (1968) Experimental alteration of the *Nautilus* shell by factors involved in diagenesis and metamorphism, Part I. Thermal changes in conchiolin matrix of mother-of-pearl. Bull Inst R Sci Nat Belg 44(25):1–69

Grégoire C (1972) Experimental alteration of the *Nautilus* shell by factors involved in diagenesis and in metamorphism. Part III, Thermal and hydrothermal changes in the mineral and organic components of the mural mother-of-pearl. Bull Inst R Sci Nat Belg 48(6):1–85

Grégoire C (1980) The conchiolin matrices in nacreous layers of ammonoids and fossil nautiloids: a survey. Akad Wiss Lit Abh Math Naturwiss Kl Mainz 1980(2):1–128

Grégoire C (1984) Remains of organic components in the siphonal tube and in the brown membrane of ammonoids and fossil nautiloids. Hydrothermal simulation of their diagenetic alterations. Akad Wiss Lit Abh Math Naturwiss Kl Mainz 1984(5):5–56

Grégoire C (1987) Ultrastructure of the *Nautilus* shell. In: Saunders WB, Landman NH (eds) *Nautilus*—The biology and paleobiology of a living fossil. Plenum, New York, pp. 463–486

Hansen HJ (1967) A technique for depiction of grind sections of foraminifera by aid of compiled electronmicrographs. Medd Dan Geol Foren 17:128

Hewitt RA, Westermann GEG (1983) Mineralogy, structure and homology of ammonoid siphuncles. N Jb Geol Paläont Abh 165(3):378–396

Hewitt RA, Checa A, Westermann GEG, Zaborski PM (1991) Chamber growth in ammonites inferred from colour markings and naturally etched surfaces of Cretaceous vascoceratids from Nigeria. Lethaia 24:271–287

Hölder HH (1952a) Über Gehäusebau, insbesondere den Hohlkiel jurassischer Ammoniten. Palaeontogr A 102:18–48

Hölder HH (1952b) Der Hohlkiel der Ammoniten und seine Entdeckung durch F. A. Quenstedt. Jb Vaterl Naturk Württ 1952:37–50

Hölder HH (1954) Über die Sipho-Anheftung bei Ammoniten. N Jb Geol Paläont Mh 1954(8):372–379

House MR (1965) A study in the Tornoceratidae: the succession of *Tornoceras* and related genera in the North American Devonian. Phil R Soc Lond B 250(763):79–130

House MR (1971) The goniatite wrinkle-layer. Smithson Contrib Paleobiol 3:23–32

Howarth MK (1975) The shell structure of the Liassic ammonite family Dactylioceratidae. Bull Br Mus (Nat Hist) Geol 26:45–67

Hyatt A (1872) Fossil cephalopods of the museum of comparative zoology: embryology. Bull Mus Comp Zool Harv 3:59–111

John R (1909) Über die Lebensweise und Organisation des Ammoniten. Inaugural Dissertation, University of Tübingen, Stuttgart

Joly B (1976) Les Phylloceratidae malgaches au Jurassique. Généralités sur la Phylloceratidae et quelques Juraphyllitidae. Doc Lab Géol Fac Sci Lyon 67:1–471

Keyserling A (1846) Wissenschajtliche Beobachtungen auf einer Reise in das Petschora-Land im Jahre 1843. St. Petersburg

Klug C, Korn D, Richter U, Urlichs M (2004) The black layer in cephalopods from the German Muschelkalk (Triassic). Palaeontology 47:1407–1425

Kulicki C (1974) Remarks on the embryogeny and postembryonal development of ammonites. Acta Palaeontol Pol 19:201–224

Kulicki C (1975) Structure and mode of origin of the ammonite proseptum. Acta Palaeontol Pol 20(4):535–542

Kulicki C (1979) The ammonite shell: its structure, development and biological significance. Palaeontol Pol 39:97–142

Kulicki C (1989) Archaeogastropod model of mineralization of ammonitella shell. In: Carter JG (ed) Skeletal biomineralization: patterns, processes and evolutionary trends. Short course in geology, vol 5, Part 2. American Geophysical Union, Washington

Kulicki C (1994) Septal neck-siphuncular complex in *Stolleyites* (Ammonoidea), Triassic, Svalbard. Polish Polar Research 15(1–2):37–49

Kulicki C, Doguzhaeva LA (1994) Development and calcification of the ammonitella shell. Acta Palaeontol Pol 39:17–44

Kulicki C, Mutvei H (1982) Ultrastructure of the siphonal tube in *Quenstedtoceras* (Ammonitina). Stockholm Contrib Geol 37:129–138

Kulicki C, Mutvei H (1988) Functional interpretation of ammonoid septa. In: Wiedmann J, Kullmann J (eds) Cephalopods-Present and past. Schweizerbart, Stuttgart

Kulicki C, Tanabe K, Landman NH, Mapes RH (2001) Dorsal shell wall in ammonoids. Acta Palaeontol Pol 46:23–42

Kulicki C, Landman NH, Heaney MJ, Mapes RH, Tanabe K (2002) Morphology of the early whorls of goniatites from the Carboniferous Buckhorn Asphalt (Oklahoma) with aragonitic preservation. Abh Geol BA Wien 205:205–224

Kulicki C, Tanabe K, Landman NH (2007) Primary structure of the connecting ring of ammonoids and its preservation. Acta Palaeontol Pol 53:823–827

Landman NH (1985) Preserved ammonitellas of *Scaphites* (Ammonoidea. Ancyloceratina). Am Mus Novit 2815:1–21

Landman NH (1987) Ontogeny of Upper Cretaceous (Turonian-Santonian) scaphitid ammonites from the Western interior of North America: systematics, developmental patterns, and life history. Bull Amer Mus Nat Rist 185:117–241

Landman NH (1988) Early ontogeny of Mesozoic ammonites and nautilids. In: Wiedmann J, Kullman J (eds) Cephalopods—Present and past. Schweizerbart, Stuttgart, pp. 215–228

Landman NH, Bandel K (1985) Internal structures in the early whorls of Mesozoic ammonites. Am Mus Novit 2823:1–21

Landman NH, Waage K (1982) Terminology of structures in embryonic shells of Mesozoic ammonites. J Paleontol 56:1293–1295

Landman NH, Tanabe K, Mapes RH, Klofak SM, Whitehill J (1993) Pseudosutures in Paleozoic ammonoids. Lethaia 26:99–100

Landman NH, Polizzotto K, Mapes RH, Tanabe K (2006) Cameral membranes in prolecanitid ammonoids from the Permian Arcturus Formation, Nevada. Lethaia 39:365–379

Makowski H (1962) Problem of sexual dimorphism in ammonites. Palaeontol Pol 12:1–92

Makowski II (1971) Some remarks on the ontogenetic development and sexual dimorphism in the Ammonoidea. Acta Geol Pol 21:321–340

Miller AK, Unklesbay AG (1943) The siphuncle of late Paleozoic ammonoids. J Paleontol 17:1–25

Miller AK, Furnish WM, Schindewolf OH (1957) Paleozoic Ammonoidea. In: Moore RC (ed) Treatise on invertebrate paleontology. Part L, Mollusca 4. GSA and University Kansas Press, Lawrence, pp. 11–79

Mutvei H (1964) On the shells of *Nautilus* and *Spirula* with notes on the shell secretion in non-cephalopod molluscs. Ark Zool 16(14):221–278

Mutvei H (1967) On the microscopic shell structure in some Jurassic ammonoids. N Jb Geol Paläont Abh 129(2):157–166

Mutvei H (1969) On the micro and ultrastructure of the conchiolinin the nacreous layer of some recent and fossil Molluscs. Stockholm Contrib Geol 20(1):1–17

Mutvei H (1970) Ultrastructure of the mineral and organic components of molluscan nacreous layers. Biomineralisation 2:48–61

Mutvei H (1972a) Ultrastructural studies on cephalopod shells, Part 1, The septa and siphonal tube in *Nautilus*. Bull Geol Inst Univ Upps 3:237–261

Mutvei H (1972b) Ultrastructural studies on cephalopod shells, Part 2. Orthoconic cephalopods from the Pennsylvanian Buckhorn Asphalt. Bull Geol Inst Univ Upps 3:263–272

Mutvei H (1979) On the internal structures of the nacreous tablets in molluscan shells. Scanning Electron Microsc 1979(II):451–462
Mutvei H (1980) The nacreous layer in molluscan shells. In: Omori M, Watabe N (eds) The mechanisms of biomineralization in animals and plants: proceedings 3rd International Biomineralization Symposium. Tokai University Press, Tokyo, pp. 49–56
Mutvei H (1983) Flexible nacre in the nautiloid *Isorthoceras*, with remarks on the evolution of cephalopod nacre. Lethaia 16:233–240
Nassichuk WW (1967) A morphological character new to ammonoids portrayed by *Clistoceras* gen. nov. from Pennsylvanian of Arctic Canada. J Paleontol 41:237–242
Obata I, Tanabe K, Fukuda Y (1980) The ammonite siphuncular wall: its microstructure and functional significance. Bull Natl Sci Mus (Tokyo) Ser C (Geol) 6(2):59–72
Ohtsuka Y (1986) Early internal shell microstructure of some Mesozoic Ammonoidea: implications for higher taxonomy. Trans Proc Palaeontol Soc Jpn New Ser 141:275–288
Polizzotto K, Landman NH, Klug C (2015) Cameral membranes, Pseudosutures, and other Soft-Tissue Imprints in Ammonoid Shells. This volume
Ruzhentsev VE, Shimanskij VN (1954) Nizhnepermskye svernutye i sognutye nautiloidei juzhnogo Urala. Trans Paleontol Inst Akad Nauk SSSR 50:1-150 [in Russian]
Sandberger G, Sandberger F (1850) Die Versteinerungen des rheinischen Schichtensystems in Nassau. Kreidel & Nieder, Wiesbaden, p. 564
Schindewolf OH (1968) Analyse eines Ammoniten-Gehäuses. Akad Wiss Lit Abh Math Naturwiss Kl Mainz 1968(8):139–188
Schulga-Nesterenko MJ (1926) Internal structure of the shell in Artinskian ammonites. Byull Mosk Ova Ispyt Prir Otd Geol 4(1–2):81–100 [in Russian]
Senior JR (1971) Wrinkle-layer structures in Jurassic ammonites. Palaeontology 14:107–113
Smith JP (1901) The larval coil of *Baculites*. Am Nat 35(409):39–49
Smith WD (1905) The development of *Scaphites*. J Geol 13:635–654
Tanabe K (1989) Endocochliate embryo model in the Mesozoic Ammonoidea. Hist Biol 2:183–196
Tanabe K, Ohtsuka Y (1985) Ammonoid early internal shell structure: Its bearing on early life history. Paleobiology 11:310–322
Tanabe K, Landman NH (1996) Septal neck-siphuncular complex of ammonoids. In: Landman NH, Tanabe K, Davis RA (eds) Ammonoid paleobiology. Plenum, New York, pp. 129–165
Tanabe K, Obata I, Fukuda Y, Futakami M (1979) Early shell growth in some Upper Cretaceous ammonites and its implications to major taxonomy. Bull Natl Sci Mus (Tokyo). Ser C (Geol) 5(4):153–176
Tanabe K, Fukuda Y, Obata I (1980) Ontogenetic development and functional morphology in the early growth stages of three Cretaceous ammonites. Bull Natl Sci Mus (Tokyo), Ser C (Geol) 6(1):9–26
Tanabe K, Fukuda Y, Obata I (1982) Formation and function of the siphuncle-septal neck structures in two Mesozoic ammonites. Trans Proc Palaeontol Soc Jpn New Ser 128:433–443
Tanabe K, Landman NH, Weitschat W (1993a) Septal necks in Mesozoic Ammonoidea: structure, ontogenetic development, and evolution. In: House MR (ed) The Ammonoidea: environment, ecology, and evolutionary change. Clarendon, Oxford, pp. 57–84
Tanabe K, Landman NH, Mapes RH, Faulkner C (1993b) Analysis of a Carboniferous embryonic ammonoid assemblage from Kansas. USA-Implications for ammonoid embryology. Lethaia 26:215–224
Tanabe K, Landman NH, Mapes RH (1998) Muscle attachment scars in a Carboniferous goniatite. Paleontological Research 2:130–136
Tanabe K, Kulicki C, Landman NH, Mapes RH (2001) External features of embryonic and early postembryonic shells of a Carboniferous goniatite *Vidrioceras* from Kansas. Paleontological Research 5:13–19
Tanabe K, Kulicki C, Landman NH (2005) Precursory siphuncular membranes in the body chamber of *Phyllopachyceras* and comparisons with other ammonoids. Acta Palaeontol Pol 50:9–18
Tanabe K, Kulicki C, Landman NH (2008) Development of the embryonic shell structure of mesozoic Ammonoidea. Novitates of American Museum of. Natural History 3621:1–19

Tanabe K, Kulicki C, Landman NH, Kaim A (2010) Tuberculate micro-ornamentation on embryonic shells of mesozoic ammonoids: microstructure, taxonomic variation, and morphogenesis. In: Tanabe K, Shigeta Y, Sasaki T, Hirano H (eds) Cephalopods—Present and past. Tokai University, Tokyo, pp. 105–121

Tozer ET (1972) Observations on the shell structure of Triassic ammonoids. Palaeontology 15:637–654

Vogel KP (1959) Zwergwuchs bei Polyptychiten (Ammonoidea). Geol Jahrb 76:469–540

Voorthuysen JH (1940) Beitrag zur Kenntnis des inneren Baus von Schale und Sipho bei Triadischen Ammoniten. Dissertation, Amsterdam University. Van Gorcum & Co., Assen

Voss-Foucart MF, Grégoire C (1971) Biochemical composition and submicroscopic structure of matrices of nacreous conchiolin in fossil cephalopods (nautiloids and ammonoids). Bull Inst R Sci Nat Belg 47(41):1–42

Walliser OH (1970) Über die Runzelschicht bei Ammonoidea. Göttinger Arb Geol Paläont 5:115–126

Ward PD (1987) The Natural History of *Nautilus*. Allen & Unwin, Boston, p. 267

Weiner S, Lowenstam HA, Taborek B, Hood I (1979) Fossil mollusk shell organic matrix components preserved for 80 million years. Paleobiology 5:144–150

Weitschat W (1986) Phosphatisierte Ammonoideen aus der Mittleren Trias von Central-Spitzbergen. Mitt Geol Paläont Inst Univ Hamburg 61:249–279

Weitschat W, Bandel K (1991) Organic components in phragmocones of Boreal Triassic ammonoids: implications for ammonoid biology. Paläontol Z 65(3–4):269–303

Westermann GEG (1971) Form, structure and function of shell and siphuncle in coiled Mesozoic ammonoids. Life Sci Contrib R Ont Mus 78:1–39

Westermann GEG (1992) Formation and function of suspended organic cameral sheets in Triassic ammonoids-discussion. Paläontol Z 66(3–4):437–441

Wise SW (1970) Microarchitecture and mode of formation of nacre (mother-of-pearl) in pelecypods, gastropods and cephalopods. Eclogae Geol Helv 63:775–797

Zaborski PME (1986) Internal mould markings in a Cretaceous ammonite from Nigeria. Palaeontology 29:725–738

Zakharov YuD, Grabovskaia BS (1984) Stroienie rakoviny roda *Zelandites* (Lytoceratida). Paleontol Zh 1984(1):19–29

Chapter 9
Ammonoid Intraspecific Variability

Kenneth De Baets, Didier Bert, René Hoffmann, Claude Monnet, Margaret M. Yacobucci and Christian Klug

9.1 Introduction

Individual organisms within extant species, including cephalopods (Boyle and Boletzky 1996), vary morphologically (size, shape, colour), physiologically, behaviorally, and demographically (Wagner 2000). This was no different for ammonoids,

K. De Baets (✉)
GeoZentrum Nordbayern, Fachgruppe PaläoUmwelt, Universität Erlangen, Loewenichstr 28, 91054 Erlangen, Germany
e-mail: kenneth.debaets@fau.de

D. Bert
UMR-CNRS 6118 Géosciences, Université de Rennes 1, campus Beaulieu, bâtiment 15, 35042 Rennes cedex, France
e-mail: paleo-db@orange.fr

Laboratoire du Groupe de recherche en paléobiologie et biostratigraphie des ammonites (GPA), Bois-Mésanges, quartier St-Joseph, 04170 La Mure-Argens, France

R. Hoffmann
Department of Earth Sciences, Institute of Geology, Mineralogy, and Geophysics, Ruhr-Universität Bochum, 44801 Bochum, Germany
e-mail: Rene.Hoffmann@rub.de

C. Monnet
UMR CNRS 8198 Evo-Eco-Paleo, Université de Lille, UFR Sciences de la Terre (SN5), Avenue Paul Langevin, 59655 Villeneuve d'Ascq, France
e-mail: claude.monnet@univ-lille1.fr

M. M. Yacobucci
Department of Geology, Bowling Green State University, 190 Overman Hall, Bowling Green, Ohio 43403-0218, USA
e-mail: mmyacob@bgsu.edu

C. Klug
Paläontologisches Institut und Museum, University of Zurich, Karl Schmid-Strasse 6, 8006 Zurich, Switzerland
e-mail: chklug@pim.uzh.ch

C. Klug et al. (eds.), *Ammonoid Paleobiology: From Anatomy to Ecology*, Topics in Geobiology 43, DOI 10.1007/978-94-017-9630-9_9

which are well-known for intraspecific variation in conch shape, ornamentation, ontogeny, size, as well as the morphology of the suture line (e.g., Westermann 1966; Kennedy and Cobban 1976; Tintant 1980; Dagys and Weitschat 1993a, b; Kakabadze 2004; Bersac and Bert 2012a, b; De Baets et al. 2013a; Bert 2013). Some specimens of a species were large, others were smaller at maturity; some specimens of a species were more involutely coiled and less densely ribbed, while others were more loosely coiled and more coarsely ribbed. Intraspecific variation also occurs in the shape or position of the suture line (e.g., Yacobucci and Manship 2011) or in dextral or sinistral coiling of the conch (e.g., Matsunaga et al. 2008) as seen in extant gastropods. Ammonoids might also have differed intraspecifically in colour patterns (e.g., Mapes and Sneck 1987; Bardhan et al. 1993; see also Mapes and Larson 2015), buccal mass (Davis et al. 1996; Keupp 2000; Keupp and Mitta 2013; compare Kruta et al. 2015) or other characteristics which are rarely or not preserved at all such as soft-tissues (Klug et al. 2012) or other peculiar structures (Landman et al. 2012). We will herein focus on intraspecific variation in shell shape, ornamentation and size, as well as spacing and shape of the septa (suture lines), for which more data are available.

Mollusks in general and ammonoids in particular are known to display a sometimes profound morphological intraspecific variability of their shell. Although this phenomenon is of greatest importance, it has rarely been investigated and quantified in large samples adequately. Studies of intraspecific variability in ammonoids have focused on coiled Mesozoic ammonoids, while Mesozoic heteromorphs (Kakabadze 2004; Bert 2013) and Paleozoic ammonoids (Kaplan 1999; Korn and Klug 2007; De Baets et al. 2013a) have been comparatively less investigated. Not properly taking intraspecific variability into account mostly leads to taxonomic oversplitting (or lumping) and thus not only significantly biases taxonomy and diversity counts, but also biostratigraphic, evolutionary and paleobiogeographic studies (e.g., Kennedy and Cobban 1976; Tintant 1980; Dzik 1985, 1990a; Hughes and Labandeira 1995; Nardin et al. 2005; Korn and Klug 2007; De Baets et al. 2013a, b; Bert 2013; compare Sect. 9.5), particularly if the authors have very different principles when defining species between certain timeframes or regions. Geographic variation might also lead to specimens of a single biological species being erroneously assigned to different morphospecies based on differences in shell morphology, ornamentation and/or size (e.g., Kennedy and Cobban 1976; Courville and Thierry 1993).

More importantly, heritable (genetic) variation is believed to be the raw material for evolution and natural selection (e.g., essays by Charles Darwin and Arthur Wallace compiled in De Beer 1958; Mayr 1963; Hallgrimmson and Hall 2005; Hunt 2007). This makes the mode and range of intraspecific differences interesting with respect to their genetic heritability. They are of ecological and evolutionary interest in terms of the environmental influences that shape them, both non-genetically in the present and genetically over an evolutionary time scale (Wagner 2000). It is, however, hard to separate heritable phenotypic variation from variation resulting from a plastic response to the environment (Urdy et al. 2010a), especially in extinct groups. For instance, a large part of the intraspecific variability in shelled mollusks could be caused by differences in growth rates (Urdy et al. 2010a) and development

(Courville and Crônier 2003, 2005). This could also explain certain recurrent patterns in intraspecific variation in the shells of ammonoids and other mollusks with coiled shells (e.g., Dommergues et al. 1989; Urdy et al. 2010b, 2013; Urdy 2015). Extant cephalopods can comprise a high intraspecific variability, particularly in their variable size-at-age, which can be related to intrinsic as well as extrinsic (environmental) factors (Boyle and Boletzky 1996; compare De Baets et al. 2015a; Keupp and Hoffmann 2015 for pathologies affecting growth).

The main goal of this chapter is to review the main types of intraspecific variation reported in shell shape, ornamentation, suture line and adult size within and between ammonoid populations and how they might have been shaped by development and the environment. Additionally, we briefly review the main methods that can be used to quantitatively study intraspecific variation. For this purpose, we focus on studies that have specifically dealt with intraspecific variation as well as more general studies that have discussed particular patterns of intraspecific variation, including case studies from the literature and our own studies ranging from Devonian to Cretaceous ammonoids. Before doing this, we will set up the main terminology used to study intraspecific variation and possible sources of variation between and within fossil populations, including those not related to intraspecific variation, which might bias the results of studies on intraspecific variation in fossil samples.

9.2 Definitions

Here, we define some commonly used terms related to variation within and between populations of the same species, which are ubiquitous in extant species. This is generally referred to as intraspecific or phenotypic variation, sometimes as "*Individual variability*" (Darwin 1859) or occasionally somewhat confusingly as phenotypic "*polymorphism*" (e.g., Fusco and Minelli 2010; see below for a stricter definition of polymorphism). Phenotypic variation results from both genetic and environmental factors. Traditionally, evolution is assumed to consist of (genetic) changes in populations over time (Tintant 1980), making it the central goal of biology to understand the complex interactions that mediate the translation from genotype (the genetic make-up or precise genetic constitution of an organism: Lawrence 2000) to phenotype (the visible or otherwise measurable physical and biochemical characteristics of an organism, resulting from the interaction of the genotype and the environment: Lawrence 2000). Phenotypic variability is closely related to phenotypic variation and is defined as the potential or tendency of an organism (e.g., a species) to vary (Wagner and Altenberg 1996). This means that variation can be documented as a series of static observations within a sample—each observation representing a single instance of the many phenotypic expressions resulting from interactions of genetic and environmental factors—while variability can be seen as a more abstract view of the range or distribution of potential variation, which comprises all possible outcomes, realized or not (Willmore et al. 2007). Note, that analyzing variation in fossil samples or "populations" is even more complex than in extant populations, because differences between specimens can relate to other factors than intraspecific variation (Tintant 1980; discussed in Sect. 9.3).

Intraspecific variation in certain characters of a species can be continuous (e.g., following a unimodal Gaussian distribution) and/or discontinuous such as polymorphism. Polymorphism is traditionally defined as the occurrence together in the same habitat of two or more distinct forms of a species in such proportions that the rarest of them cannot be maintained merely by recurrent mutation (Ford 1955, 1965). According to the definition of Ford (1940, 1945, 1955, 1965), this excludes geographic and seasonal forms as well as continuous variation falling within a curve of normal distribution. Mayr (1963) introduced the term polyphenism to distinguish environmentally induced phenotypic variation ("*the occurrence of several phenotypes in a population, the differences between which are not the result of genetic differences*"; Mayr 1963, p. 670) from genetically controlled phenotypic variation or genetic polymorphism. Although Mayr (1963) specifically included both continuous and discontinuous variation, the term polyphenism is often restricted to refer to two or more distinct phenotypes produced by the same genotype (e.g., Simpson et al. 2011), which would make polyphenism a particular case of phenotypic plasticity (West-Eberhard 2003). The term polyphenism has occasionally also been used for ammonoids (e.g., Reyment 2003, 2004), sometimes interchangeably with polymorphism (Parent 1998; Parent et al. 2008). The switch between forms is believed to be environmental in polyphenism (e.g., Fusco and Minelli 2010), while the switch is believed to be "almost always" genetic in (genetic) polymorphism (Ford 1966; this should not be confused with the use of the same terminology by molecular biologists for certain point mutations in the genotype, which do not necessarily correlate with recognizable phenotypic effects: Fusco and Minelli 2010). In ammonoids, polymorphism has been traditionally used to refer to two or more discrete coexisting forms within the same fossil population (Tintant 1980; Davis et al. 1996 and references therein; Klug et al. 2015), although others have used it more generally to include also continuous variation (e.g., Beznosov and Mitta 1995). We suggest using the term polymorphism only to refer to discontinuous variation in ammonoids to avoid confusion and to be in line with its most common use. We will therefore use polymorphism here to refer to discontinuous intraspecies variation without interpreting a potential genetic or environmental switch between these forms or variants, although in some cases (like sexual dimorphism) a genetic mechanism is obvious (at least in cephalopods).

Polymorphism or polyphenism should not be confused with polytypism. The latter term refers to the presence of geographically or ecologically isolated populations within a species, which differ morphologically (Tintant 1980). It is not uncommon that the mode and range of intraspecific variation varies between different samples or populations depending on the environment (ecophenotypic variation) or region (geographic variation). Phenotypic variation that is attributable to environmental variation is referred to as ecophenotypic variation (Foote and Miller 2007). The tendency of a single genotype to produce different phenotypes depending on environment gradients is known as phenotypic plasticity (Lawrence 2000). West-Eberhard (1989) defined it differently as the ability of a single genotype to produce more than one alternative form of morphology, physiological state, and/or behavior in response to environmental conditions; both definitions are hard to verify in the fossil record. All these types of variation might also have occurred in ammonoids, but their study is hampered by the difference between biological populations and fossil populations, which are affected by various taphonomic and collection biases, as well as various other factors, discussed in more detail in the Sect. 9.3.

9.3 Sources of Variation within and between Fossil Populations

Measurements of individuals of the same species within and between fossil populations (separated in time and/or space) can show variations that can not only be associated with intraspecific or phenotypic variation (which includes ecophenotypic variation and geographic variation), but also with ontogenetic variation, phylogenetic variation, taphonomic biases (including post-mortem transport and distortion, time-averaging and differences in preservation), taxonomic uncertainty, and simple measurement errors, particularly in the case of small size (compare Tintant 1980; Stephen and Stanton 2002; Foote and Miller 2007; Bert 2013; De Baets et al. 2012, 2013a, 2015b).

Individual organisms of the same species can vary in their phenotype, resulting from the interaction of its genotype and the environment. More precisely, the features of individual organisms result from developmental processes, which are influenced by environmental conditions as well as its genetic make-up. Intraspecific variation refers to the variation within a species at a comparable ontogenetic stage, age and/or size (Foote and Miller 2007). Changes in ontogeny and differences between sexes might also contribute to the variation of the overall population. Ontogenetic variation is therefore factored out by studying specimens only at comparable ontogenetic stages or sizes (De Baets et al. 2013a). Traditionally, only a single set of measurements from "mature" specimens are used (so-called cross-sectional data by opposition to longitudinal data based on measurements of the same individuals at several developmental stages: compare Klingenberg 1996; Foote and Miller 2007), which are recognized by adult modifications. Studying the entire ontogeny, in the form of ontogenetic trajectories or changes in these measured characters through development might be more meaningful, particularly in taxa where the earlier ontogeny is more variable than the later ontogeny (e.g., De Baets et al. 2013a). This means that for each ontogenetic stage, a statistically significant number of measurements should ideally be available (>30: compare Bert 2013; De Baets et al. 2013a; Sect. 9.9).

Sexual variation is sometimes factored out too by studying only specimens of the same sex (Foote and Miller 2007) or antidimorphs (Sect. 9.4.2). This might not always be advisable, e.g., when subjectively sorting out specimens based on size and subsequently testing for significant differences between them: compare Tintant (1980). Populations of a species can also vary in features between different localities or regions (interpopulational variation), although it might be hard to attribute this purely to geographic variation in the fossil population. Phylogenetic variation related to changes through time might also play a role, although it is hard to separate such variation from geographic variation without proper time constraints (Kennedy and Cobban 1976; Tintant 1980). Furthermore, fossil ammonoid populations might include specimens from different paleoenvironments, water depths and seasons depending on the environment as well as the degree of transport and time-averaging; these are important factors, which should be considered when studying ecophenotypic and geographic variation.

These complications are related to the fact that fossil populations have passed through various taphonomic filters such as post-mortem transport, post-mortem distortion and time-averaging (Foote and Miller 2007). To avoid time-averaging (Fig. 9.1) as much as possible, populations are usually studied from a restricted stratigraphic interval, such as a single layer, horizon or preferentially a single concretion or nodule (Reeside and Cobban 1960; Dzik 1990a; Dagys and Weitschat 1993a, b; Fig. 9.2, 9.23); nevertheless, opinions vary in this respect and sometimes some compromises (pooling of samples resulting in "analytical time averaging": compare Fürsich and Aberhan 1990) have to be made to have a sufficiently large sample for statistical analysis (Dzik 1990a; Bert 2013). Any estimate of population variability is potentially falsifiable by further studies on material collected from a narrower stratigraphic interval or different localities (Dzik 1990a). Note that even for fossils deriving from a single bed, there is little control over the time range of the specimens contained within this bed (Foote and Miller 2007). Ammonoid assemblages like all other shell assemblages (Olóriz 2000; Wani 2001)—even those contained within a single nodule—have typically undergone various amounts of taphonomic filters and time-averaging (Kidwell 2002), which, in modern shelf environments, might range between 100 and 10,000 years (Powell and Davies 1990; Kidwell and Bosence 1991; Flessa and Kowalewski 1994; Kidwell 1998; Wani 2001; Kowalewski 2009). Short-term time-averaging (on the order of up to several thousand years) prevails in nearshore shallow environments, whilst long-term time-averaging (in the order of 10^4 to 10^5 years) becomes more important towards lower shelf and deep sea environments (Fürsich and Aberhan 1990). However, some evidence suggests that variances within fossil samples are not necessarily dominated by time-averaging (Tintant 1980; Hunt 2004a, b). Well-preserved specimens, particularly those with *in situ* buccal masses, as well as the lack of preferred orientation, size distribution and other biostratinomic data have been used to support lack of (strong) condensation or (long) transport (compare Wani and Gupta 2015). Not only animals that lived during different times, but also organisms coming from various depths (i.e., living in different parts of the water column or at different seafloor depths) might be mixed within these assemblages, particularly in pelagic shell assemblages, which can also be related to post-mortem floating or transport of their shells. Taphonomic studies can be important to disentangle different post-mortem histories of shells within the same fossil sample and can also give information on the faunal succession (Fernández-López 1995, 2000; Wani and Gupta 2015). However, it is not generally true that less well-preserved shells are older—their preservation mainly depends on how much time they spent on the sea bottom (or afloat) and the degree to which they were subjected to diagenesis during burial as well as

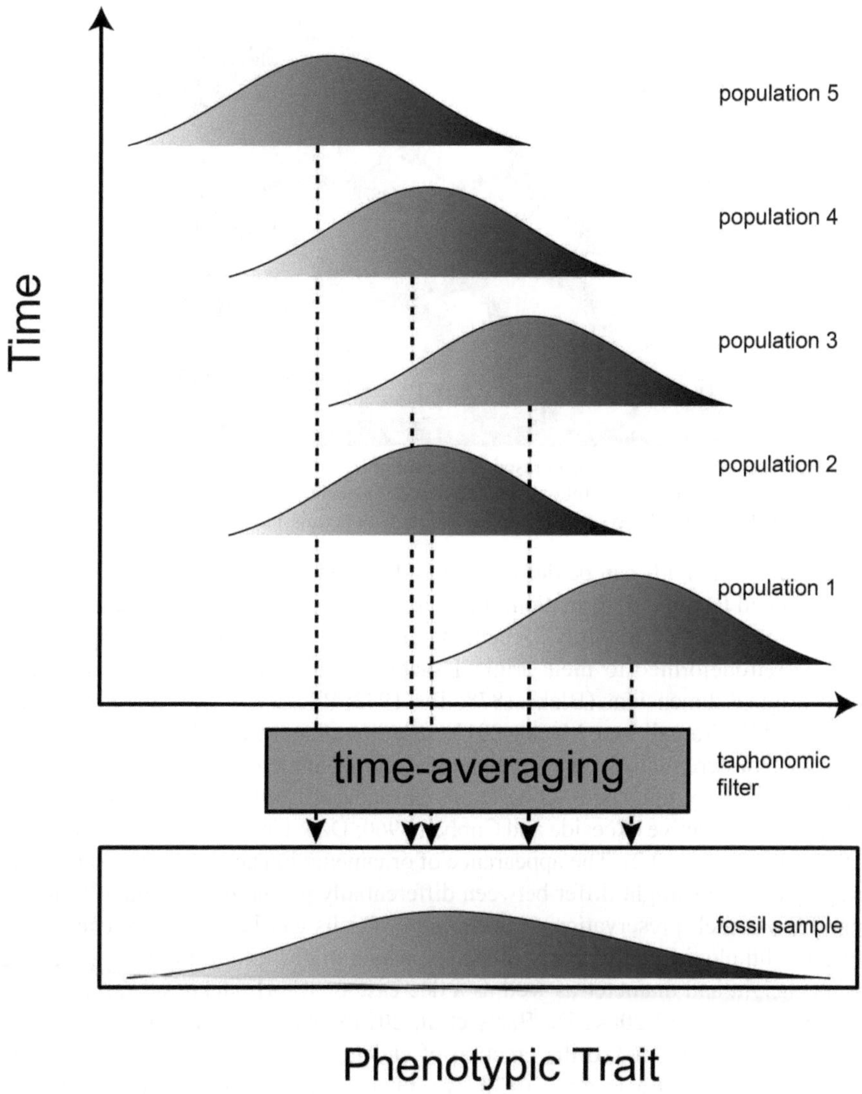

Fig. 9.1 Schematic illustration of the effect of fossil mixing and/or analytical lumping on the observed phenotypic variance (modified from Hunt 2004a; with permission from the author). The variance of the time-averaged (fossil) sample is greater than the variance of the individual populations representing five stages in the evolutionary sequence of an evolving lineage, which shows an almost steady decrease in a phenotypic trait over time. The density distribution of the fossil sample is also flatter (more platykurtic) than is ever observed in a single time slice

in which paleoenvironments they resided (e.g., Flessa et al. 1993). According to Tintant (1980), long-term time-averaging or condensation as well as reworking can even be seen in some quantitative analyses, which might reveal a distribution flatter than a normal distribution (e.g., platykurtic distribution due to considerable time-averaging: Fig. 9.1) or a polymodal distribution (due to reworking).

Fig. 9.2 The extensive range of intraspecific variation observed within *Czekanowskites rieberi* within a single carbonatic concretion from the Lower Anisian of Mount Tuaray-Khayata in Arctic Siberia (modified from Dagys and Weitschat 1993b; with permission from the author)

Fossil ammonoids can be deformed and distorted by abrasion, compaction, dissolution and tectonic deformation, these processes represent an important obstacle to research on their variability. Depending on the degree of deformation, specimens can be retrodeformed to their original shape in many cases, but not necessarily their original dimensions (Blake 1878; Tan 1973; Rocha and Dias 2005; De Baets et al. 2013b; Yamaji and Maeda 2013). Studies of intraspecific variability have therefore preferentially used largely undeformed, three-dimensionally preserved specimens. Early diagenetic concretions are ideal to study intraspecific variation from this perspective (Reeside and Cobban 1960; Dzik 1990a; Dagys and Weitschat 1993a, 1993b; Fig. 9.2). The appearance of ornamentation and measurements of the same parameters might differ between differentially preserved specimens (internal moulds vs. shell preservation). Furthermore, fossils can be extremely compacted in some lithologies, particularly in shales, which might lead to the increase of the whorl height and diameter as well as a decrease of whorl thickness and umbilical width (e.g., Morard 2004; De Baets et al. 2013b; Wani and Gupta 2015). These effects could even lead to the erection of endemic "species" restricted to certain lithologies (see De Baets et al. 2013b for such a case in the early Emsian ammonoid *Ivoites*). In such cases, it probably makes sense to correct for taphonomic processes in the most conservative way to avoid artificially inflating diversity (De Baets et al. 2013b). Specimens from the same sample (with similar preservation) are generally deformed in the same way, so that the introduced systematic error might be less significant (Dzik 1985). More importantly, certain parameters such as rib count per half-whorl and diameter at mid-whorl height can be affected by differential compaction in different lithologies and thus, the material should be examined for such deformation prior to the beginning of data collection (De Baets et al. 2013b). Differential compaction might contribute to trends in increased whorl compression from shallower environments with coarser sedimentation to deeper environments with finer sedimentation (Wilmsen and Mosavinia 2011; Bert 2013).

Additionally, variation might be related to false assignment of specimens to the same species, which depends on the objectivity and opinion of the scientists involved. Authors explicitly or implicitly include a range of intraspecific variation in their definitions of taxa, which might artificially inflate (oversplitting, often the case in strict typological approaches: compare Sect. 9.5) or deflate diversity (lumping). Oversplitting might occur when only a little well-preserved material is available with a precise age assignment and/or locality/region, while lumping occurs particularly when considering specimens from a wide range of stratigraphic ages and localities/regions to belong to the same taxon. If the entire range of variation occurs at the same age and place, this might be a good indicator that they belong to one and the same species (Tintant 1980). The effects of lumping or oversplitting might be partially counteracted by randomly distributed new discoveries, revalidations and/or invalidations of species over time (compare Nardin et al. 2005). The variable interpretation of the range of intraspecific variation also affects the disparity (morphological richness) recognized within a species (e.g., Courville and Crônier 2005). A study by Nardin et al. (2005) on Jurassic ammonoids demonstrated that extreme forms are often identified and named before intermediate forms (particularly for ornamentation, while shell geometry is often underused to define species). Such problems can only be resolved by quantitatively studying as many characters as possible in large samples, which can make it easier to recognize species (by finding significant differences in these characters) and their range of intraspecific variation. Each measurement or count carries with it a possibility of error (Van Valen 2005). Variation in measurements within a single sample might also be related to these measurements errors, which are usually estimated by repeated and independent measurements of the same specimens, or a randomly chosen appropriate subset of them (e.g., Bailey and Byrnes 1990; Van Valen 2005). In some cases, errors might be small enough to be neglected (Van Valen 2005), while in other cases, when the magnitude of the variable of interest is close to the measuring precision, they can blur (De Baets et al. 2013a) or even erase the original signal.

9.4 Types of Intraspecific Variation in Ammonoids

9.4.1 Continuous Variation

Most authors agree that continuous variation is recognized by a series of interconnected morphologies in a restricted interval in time and space (Reeside and Cobban 1960; Kennedy and Cobban 1976; Silberling and Nichols 1982; Dzik 1985, 1990a). Typically, all intermediate forms should be present and more common than extreme morphologies leading to a unimodal distribution. The best evidence for continuous intraspecific variation is often believed to be a unimodal, normal (Gaussian) distribution (Tintant 1980; Silberling and Nichols 1982; Dagys and Weitschat 1993b;

Weitschat 2008; Monnet et al. 2010). However, even in such cases, it cannot be entirely ruled out that such distributions contain various sympatric species (inhabiting the same or overlapping geographic areas), which are inseparable based on their hard part anatomy alone (Tintant 1980; Dzik 1990a) and therefore cannot be picked up in the fossil record. For example, Dommergues et al. (2006) showed that in the extant gastropod *Trivia* the differences of the hard part anatomy (excluding the colour patterns) between such closely related species is insufficient to infer the existence of two separate sympatric species, masking the true underlying biodiversity. On the other hand, when the distribution is not normal or unimodal, it does not necessarily mean that the specimens belong to different species either (Tintant 1980). Such a distribution could originate from environmental influences, taphonomic biases, sampling biases or the fact that the distribution is not of a Gaussian kind (for example in the case of discrete variation within a species such as dimorphism or non-sexual polymorphism: Klug et al. 2015).

Continuous variation has typically been analyzed from the perspectives of covariation among traits and development. Studies have focused particularly on strongly ornamented, coiled Mesozoic ammonoids, including taxa deriving from:

- the Triassic (e.g., Silberling 1956; Silberling and Nichols 1982; Dagys and Weitschat 1993a, b; Checa et al. 1996; Dagys et al. 1999; Dagys 2001; Monnet and Bucher 2005; Weitschat 2008; Monnet et al. 2010),
- the Jurassic (e.g., Tintant 1963, 1980; Westermann 1966; Sturani 1971; Howarth 1973; Dzik 1985, 1990a; Westermann and Callomon 1988; Mitta 1990; Bhaumik et al. 1993; Beznosov and Mitta 1995; Guex et al. 2003; Courville and Crônier 2005; Morard and Guex 2003; Bert 2004, 2009; Morard 2004, 2006; Zatoń 2008; Chandler and Callomon 2009; Baudouin et al. 2011, 2012; Bersac and Bert 2012a, b),
- and the Cretaceous (e.g., Haas 1946; Reeside and Cobban 1960; Kennedy and Hancock 1970; Kennedy and Cobban 1976; Kennedy and Wright 1985; Meister 1989; Kassab and Hamama 1991; Reyment and Kennedy 1991, 1998; Courville and Thierry 1993; Tanabe 1993; Aguirre-Urreta 1998; Courville and Crônier 2005; Yacobucci 2004b; Wiese and Schulze 2005; Ploch 2007; Wilmsen and Mosavinia 2011; Bersac and Bert 2012a, b; Knauss and Yacobucci 2014).

These studies have demonstrated strong variations in shell shape (whorl cross section, coiling) and ornamentation (strength, spacing). Many authors discussed a marked covariation between shell shape and strength of ornamentation and more rarely also with shape, frilling and spacing of the suture line (compare Sect. 9.3). One peculiar case of such covariation is often coined as Buckman's rules of covariation (Westermann 1966; for further details, see Monnet et al. 2015a). Such covariations have also been reported above the species level between different taxa, both in the Paleozoic (Swan and Saunders 1987; Kaplan 1999) and Mesozoic (e.g., Yacobucci 2004a; Brayard and Escarguel 2013). It is, however, not obvious that this rule can be extended beyond intraspecific variation. Yacobucci (2004a), for example, measured the variance of ornamentation and whorl shape within a number of ammonite genera and found that they do not correlate. Hammer and Bucher (2005) attributed this to varying ratios of proportionality of Buckman's law across species, which could potentially weaken the interspecific correlation between ornamentation and whorl shape (e.g., some species have stronger lateral ribs relative to shell width than others). Such exceptions might form a problem for studies that interpret such continuous Buckman's type intraspecific variation within taxa based on limited material (compare Monnet et al. 2008) or without properly quantitatively analyzing this intraspecific variation (e.g., Howarth 1973, 1978). One should remain cautious in such cases as discussed by Tintant (1976, 1980). Howarth (1973) studied *Dactylioceras* from four distinct levels in the Lower Toarcian of Yorkshire and interpreted a large continuous variation (compare Morard 2004, 2006) between forms (classically attributed to *Orthodactylites*) with more evolute inner whorls, a compressed whorl section and weak ornamentation (thin ribs, often bifurcated and non-tuberculated) to forms (traditionally attributed to *Kedonoceras* and *Nodicoeloceras*) with more involute inner whorls, a depressed whorl section and strong ornamentation (thick, more widely spaced ribs with tubercles) in earlier ontogeny. Tintant (1976, 1980) investigated a French sample of *Dactylioceras* from the first level and reported both a marked dimorphism and possible non-sexual polymorphism in the form of the coexistence of forms with compressed inner whorls without lateral tubercles (morphotype "*Orthodactylites clevelandicum*") and forms with a depressed whorl section and lateral tubercles (morphotype "*Nodicoeloceras acanthum*"). Interestingly, Tintant (1976, 1980) reported that the whorl width index (whorl width/ whorl height) is strongly bimodal below 50 mm, but in later growth stages, the forms become progressively more similar, resulting in remarkably similar final body chambers. All intermediates are available between these forms, but the extreme forms appear to be most abundant and the intermediate forms the least abundant. Tintant (1980) suggested that this might indicate polymorphism or even the presence of two species with similar evolutionary trends and convergence in their adult body chambers (compare Monnet et al. 2015b). However, Tintant's (1976, 1980) analyses were preliminary and more detailed analyses of the evolutionary history and intraspecific variability of these groups are necessary to corroborate such hypotheses. Furthermore, the influence of potential environmental differences (e.g., Wilmsen and Mosavinia 2011) as well as taphonomic and sampling biases also needs to be considered (compare Sect. 9.3).

Continuous variation has been studied less in larger samples of Paleozoic ammonoids (e.g., Kaplan 1999; Korn and Vöhringer 2004; Ebbighausen and Korn 2007; Korn and Klug 2007; De Baets et al. 2013a). Korn and Klug (2007) reported a large variation in several conch parameters in *Manticoceras* throughout the ontogeny of a single specimen (ontogenetic variation) as well as at the same size between different specimens (e.g., intraspecific variation). This variation in *Manticoceras* had already been noticed by Clarke (1899) but it was largely ignored by subsequent authors, resulting in a plethora of species and genera with (small) differences in conch shape, but comparable suture lines and ornamentation, thus making the genus a kind of waste basket taxon (Korn and Klug 2007). In some cases, intraspecific variation consistent with Buckman's first law of covariation might also be present in Paleozoic ammonoids (e.g., Kaplan 1999; De Baets et al. 2013a).

In Mesozoic taxa showing this covariation, a remarkable range of intraspecific variation in ornamentation still remains in forms with the same shell morphology and size (Morard and Guex 2003). Wiese and Schulze (2005) reported a marked variation in the umbilical width of *Neolobites vibrayeanus* from a funnel-like deepening to a well-developed umbilicus reaching 18 % of the diameter, which did not show a covariation with either the ribbing strength or the degree of inflation. The range and mode of intraspecific variation might also depend on shell morphology, particularly the degree of coiling (De Baets et al. 2013a). Such hypotheses are best tested by comparing closely related and/or contemporary species with different shell morphologies. Dagys (2001, p. 546) stated that the range and degree of covariation decreased towards taxa with very involute subcadiconic shells on the one hand and increased with most evolute platyconic shells on the other hand. This observation illustrates that the mode and range of interspecific correlation between ornamentation and whorl shape might depend on shell morphology (cf. Ubukata et al. 2008). Tanabe and Shigeta (1987) studied the intraspecific variation of whorl thickness (S) and distance of the venter from the coiling axis (D) at the same growth stage in cross-sections of Cretaceous ammonoids. This variation was the highest in heavily ornamented (e.g., Acanthocerataceae) and heteromorph forms (e.g. Scaphitaceae), intermediate in finely ribbed platycones (Lytocerataceae) and the smallest in weakly ribbed, heavily streamlined involute-compressed morphotypes (*Hypophylloceras*, *Placenticeras* and most Desmocerataceae) in the small samples of Cretaceous ammonoids they investigated. Further studies on larger samples are necessary to further corroborate these results and rule out the potential interference of ornamentation on measurements of these parameters in cross sections (which could introduce apparent variation which is not actually there).

Heteromorph ammonoids might be particularly useful for testing such hypotheses. Many authors have acknowledged high intraspecific variability in heteromorph ammonoids (e.g., Egojan 1969; Rawson 1975a, b; Dietl 1978; Ropolo 1995; Delanoy 1997; Wiedmann and Dieni 1968; Wiedmann 1969; Kennedy 1972; reviewed in Kakabadze 2004; De Baets et al. 2013a, b; Bert 2013), maybe even more than in normally coiled planispiral ammonoids (Dietl 1978; Kakabadze 2004). However, besides studies on *Scaphites* (reviewed in Landman et al. 2010; Knauss and Yacobucci 2014), which only uncoils at the end of ontogeny, intraspecific variation has been only rarely studied in numerically large samples and/or quantitatively in heteromorphs with openly coiled and/or trochospirally coiled whorls (Aguirre-Urreta and Riccardi 1988; Dietl 1978; Ropolo 1995; Delanoy 1997; De Baets et al. 2013a; Bert 2013). This lack of research might be due to their fragmentary preservation (De Baets et al. 2013a) and difficulties in quantifying some of their shell characters using traditional morphometrics and classical "Raupian" parameters (Tsujino et al. 2003; Parent et al. 2009, 2011; Bookstein and Ward 2013).

Dimorphism is also known from several Mesozoic (Jurassic and Cretaceous) heteromorphs (see reviews in Delanoy et al. 1995; Davis et al. 1996). In some cases, however, continuous variation in shell and/or ornamentation is present, which could

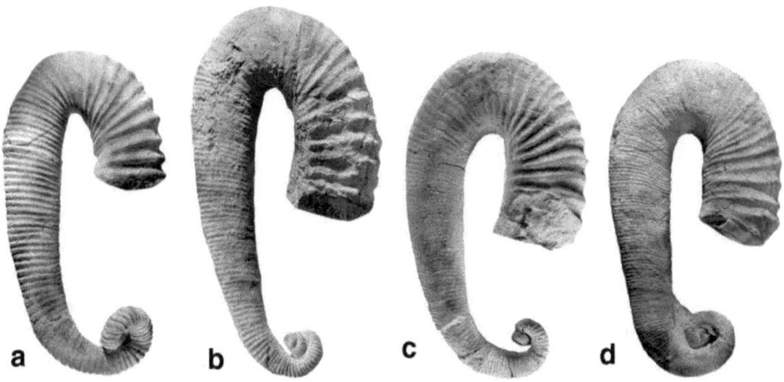

Fig. 9.3 Range of intraspecific variation observed in the coiling and ornamentation of the Late Barremian heteromorph *Heteroceras emerici* (after Delanoy 1997; modified from Bert 2013)

potentially be confused with dimorphism in studies using small samples and/or lacking quantitative analyses (compare Ropolo 1995). Dietl (1978) reported intraspecific variation in planispiral to trochospiral coiling in the Jurassic genus *Spiroceras* without a clear correlation with strength of ornamentation and any ornamental or thickness influence of the whorl section. Delanoy (1997) reported continuous variation from the Cretaceous ammonoid *Heteroceras emerici* (Fig. 9.3) between a pole with heterocone coiling (*imericum* morphology: large turricone and no planispiral part of the shell before the shaft) and a pole with colchicone coiling (*leenhardtii* morphology: small turricone preceding a substantial planispiral portion before the shaft) interconnected by all intermediates (e.g., the *tardieui* and *emerici* morphologies). Similar variation has also been reported from *Imerites* (Bert et al. 2011).

De Baets et al. (2013a) reported that in the Early Devonian, loosely coiled *Erbenoceras*, the more coarsely ornamented specimens are also those with the thickest whorl section. This fits with the redefinition by Hammer and Bucher (2005) of Buckman's First Law of Covariation (Monnet et al. 2015a). However, the coiling shows an opposite covariation with ornamentation to that seen in coiled Mesozoic morphs and some heteromorph forms, as the more tightly coiled conchs are the most heavily ornamented forms instead of the most loosely coiled forms. The correlation of ribbing strength with coiling might be indirect because covariation between whorl shape and coiling geometry are also known from weakly ornamented to smooth or unornamented coiled taxa such as from some Lytoceratina and Phylloceratina (Joly 2003; Morard 2004). Joly (2003) reported that "less-thick shells have an elliptic section and the thickest shells have an oval section". Ubukata et al. (2008) also attributed part of the covariation with constructional linkage between whorl shape and coiling geometry. The rule specifically refers to strength of ornamentation, but usually the spacing of ribs is used as this is more readily available and less affected by preservation and preparation (e.g., Yacobucci 2004a; De Baets et al. 2013a; Monnet et al. 2015a). Bert (2013) reported that ornament strength did not correlate perfectly with rib density in *Gassendiceras*, which he attributed to the large distance between ribs in this taxon, thus leaving more room for strength varia-

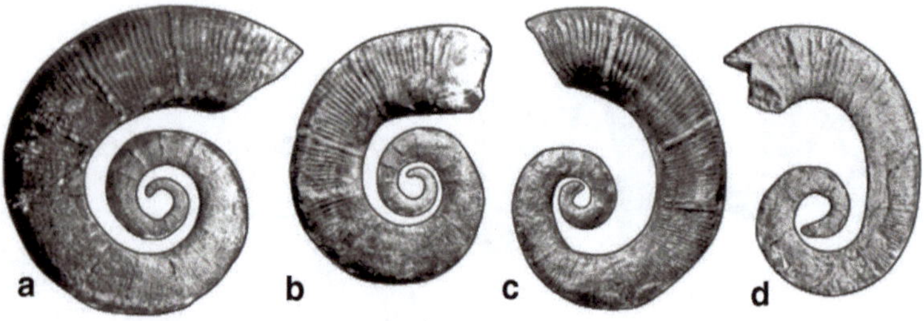

Fig. 9.4 Range of intraspecific variation observed in the coiling of the Hauterivian heteromorph *Crioceratites*. (after Ropolo 1995; modified from Bert 2013)

tion without changing their spacing (compare Bert 2013). The lack of correlation of ornamentation with coiling in some species of *Crioceratites* (Ropolo 1995; Bert 2013; Fig. 9.4) could potentially also be related to differences in growth between *Crioceratites*, characterized by a "discontinuous" mode of growth as documented in their megastriae as well as constrictions on the one hand, and other ammonoids with a "differential" mode of growth on the other hand (e.g., Bucher 1997, p. 98). Things are further complicated by the fact that such a correlation might be present in more primitive species like *C. loryi* (Bert 2013), although this still needs to be quantitatively investigated.

Continuous intraspecific variation usually ranges between two extreme morphologies, but some authors have reported more complex patterns of intraspecific variation between three or more morphological poles in shell shape and/or ornamentation (Rieber 1973; Vermeulen 2002; Bert 2009, 2013: review in the latter; Courville 2011). Bert (2013) quantitatively studied the intraspecific variation in Cretaceous *Gassendiceras alpinum* and reported continuous intraspecific variation between three poles: (1) robust specimens characterized by a depressed whorl section and strong ornamentation, (2) more traditional gracile or slender specimens characterized by a finer ornamentation and compressed whorl section and (3) specimens characterized by a depressed whorl section and non-robust ornamentation (Fig. 9.5a). All poles are connected by intermediates. Interestingly, Bert (2009) reported an inversed pattern in *Tornquistes* (Jurassic; Fig. 9.5c), which is characterized by morphological poles with a thin whorl section and respectively thin and robust ornamentation, and a morphological pole with a thick whorl section and weak ornamentation. The whorl section can be differently affected by compaction, which could blur the relationship between the whorl section thickness and the strength of ornamentation, but it cannot explain all aspects of tripolar patterns of intraspecific variation in these taxa (Bert 2013). Similar patterns were reported from the Pulchellidae (Vermeulen 2002; Fig. 9.5b) and the Kosmoceratidae (Courville 2011).

Rieber (1973) reported continuous variation between four extreme poles ranging from unornamented or smooth forms to forms with ribs and/or tubercles in *Repossia acutenodosa* (Triassic) from a single bed (Fig. 9.6), although the relationship between shell shape and ornamentation was not discussed and no quantitative analysis was performed.

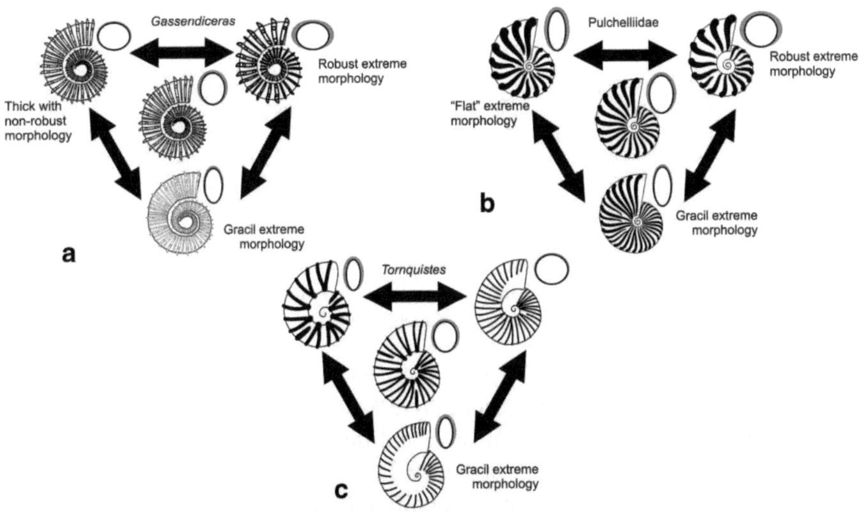

Fig. 9.5 Tripolar intraspecific variation observed in *Gassendiceras* (Late Barremian), Pulchelliidae (Barremian) and *Tornquistes* (Middle Oxfordian) (modified from Bert 2013)

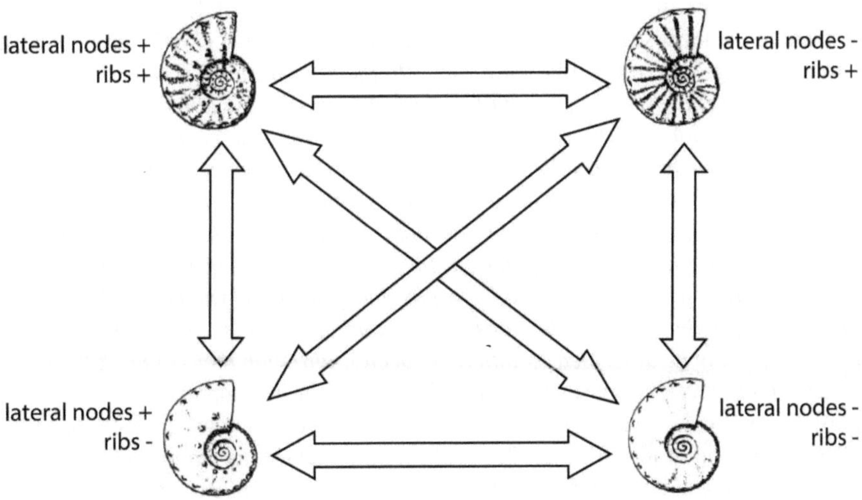

Fig. 9.6 Intraspecific variation reported in *Repossia acutenodosa* (Anisian) between four extreme poles of ornamentation (modified from Rieber 1973)

Other authors have used relative shifts in development (i.e., heterochronies of the development), which have often been used in the context of ontogeny/phylogeny relationships, to describe intraspecific morphological variations (Schmidt 1926; Dommergues et al. 1986; Meister 1989; Mitta 1990; Beznosov and Mitta 1995; Courville and Crônier 2003; Bersac and Bert 2012a, b; Bert 2013; also dimorphism: Neige et al. 1997a; Fig. 7, 8a). In some cases, specimens might even omit entire

Fig. 9.7 Range of intraspecific variation in timing of ontogenetic development observed in *Virgatites pusillus* (Tithonian; modified from Mitta 1990; with permission from the author); from left to right: a bradymorphic, a normomorphic and a tachymorphic individual

ontogenetic stages during development (e.g., in *Placenticeras*: compare Klinger and Kennedy 1989; Bert 2013). According to Mitta (1990), Michalsky (1890) was one of the first to notice this phenomenon of different rates of shell morphogenesis in individuals of the same species within Volgian ammonoids. Schmidt (1926) interpreted a similar phenomenon in Carboniferous ammonoids and introduced the terms bradymorphic (in terms of heterochronies: paedomorphic) and tachymorphic (in terms of heterchronies: peramorphic) to refer to the extreme end members of these series within a species, which possess characteristics of earlier whorls later in development or which possess characteristics of later stages of development earlier, respectively (Beznosov and Mitta 1995). Beznosov and Mitta (1995) defined these terms (see Fig. 9.7) in the following way: "*In the tachymorphic forms, the shell or separate elements of it (the sculpture, crosssectional shape, width of the umbilicus, angle of inclination of the umbilical wall) even at small diameters have already taken on an appearance usually typical of a later stage of development. In the bradymorphic individuals the shell for a long time retains features typical of a younger individual. Brady- and tachymorphy are most clearly manifested in the duration of one or another stage of development of the sculpture, the extreme representatives of the variation series (typical bradymorphs and typical tachymorphs) often differing so strongly that, if the collections do not contain "normal" (or normomorphic) forms, they may be described as different species, although they occur at the same stratigraphic level.*" Such intraspecific differences in development have been reported in particular from large samples of Jurassic (Dommergues et al. 1986; Mitta 1990; Baudouin et al. 2011, 2012; Fig. 9.7) and Cretaceous ammonoids (Meister 1989; Courville and Crônier 2003; Bersac and Bert 2012a, b; Bert 2013; Fig. 9.8a). It was interpreted to be a dominant factor in the variability of *Nigericeras gadeni* (Courville and Crônier 2003) and *Vascoceras* (Meister 1989) of the Cenomanian of Nigeria, while in other taxa like the Middle Liassic *Aegoceras capricornus*, it was only a residual factor (Dommergues et al. 1986, Fig. 6). In some taxa such as Deshayesitidae (Bersac and Bert 2012a, b; Fig. 9.8a), *Gassendiceras* (Bert 2013), as well as *Streblites* and *Taramelliceras* (Baudouin et al. 2011), this type of variation was also reported to be combined with variation following Buckman's first law of covariation between whorl section and ornamentation.

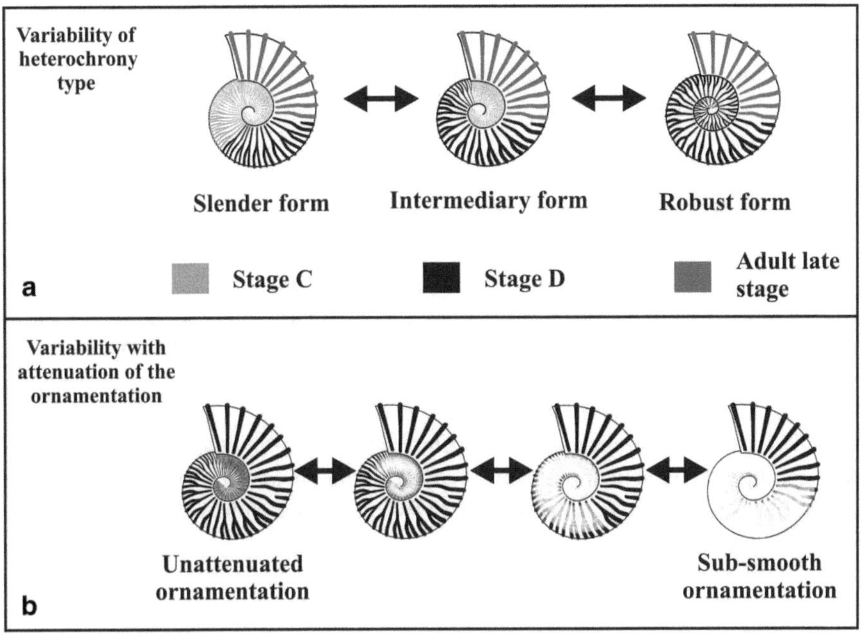

Fig. 9.8 Types of intraspecific variation differing from the first Buckman's Law of covariation observed within the Deshayesitidae (Early Cretaceous; after Bersac and Bert 2012a, b; modified from Bert 2013)

Not all types of intraspecific variation relate to covariation between shell shape and ornamentation or relative shifts in development. Bersac and Bert (2012a, b; Fig. 9.8b) reported intraspecific variation in the relative timing of ornamentation attenuation in the Aptian Deshayesitidae (another classical example of oversplitting), which was independent of Buckman's type covariation between shell shape and ornamentation as well as a heterochronic shift in development (which were also present), as they cut across the entire range of morphologies associated with these intraspecific patterns. This type of variation might also play a role in other taxa (Bert 2013) such as the Douvilleiceratidae, particularly the genus *Douvilleiceras* (Courville and Lebrun 2010), or in the genus *Vascoceras* (Courville 1993). In the former genus, disappearance of ornamentation may occur from medium diameters irrespective of the type of morphology (ranging from slender forms with weak ornamentation to the hyper-ornamented robust forms).The two approaches to studying intraspecific variation might also be unifiable as several authors have reported links between differences in rates and shifts in development on the one hand and shell morphology on the other hand. Some authors have interpreted the presence of gracile "peramorphic" forms (thin whorl section, almost smooth) to robust "paedomorphic" forms (thick whorl section, strong ornament) linked by all intermediates in the same species (Courville and Crônier 2003; Baudouin et al. 2011, 2012; Bersac and Bert 2012; Bert 2013). In other taxa such as *Streblites weinlandi*, the relationship might however be reversed, with the most slender specimens being the most paedomorphic (Baudouin et al. 2011, 2012). Others have tried to explain the covariation of the suture line and shell shape from a developmental perspective (Hammer and Bucher 2006).

9.4.2 Discontinuous Variation

The most accepted pattern of discontinuous intraspecific variation in ammonoids is dimorphism (Makowski 1962; Callomon 1963; Tintant 1963; Westermann 1964), which is often interpreted to represent the two sexes (Lehmann 1981; Delanoy et al. 1995; Davis et al. 1996). Such dimorphism is supported by having overlapping stratigraphic and geographic distributions as well as distinct ratios within a population. This dimorphism is thought to result typically in a bimodal signal in adult size and/or morphology in later ontogeny (e.g., Palframan 1966, 1967; Ploch 2003; Zatoń 2008; Fig. 9.9). However, the presence of two morphs within an ammonoid species

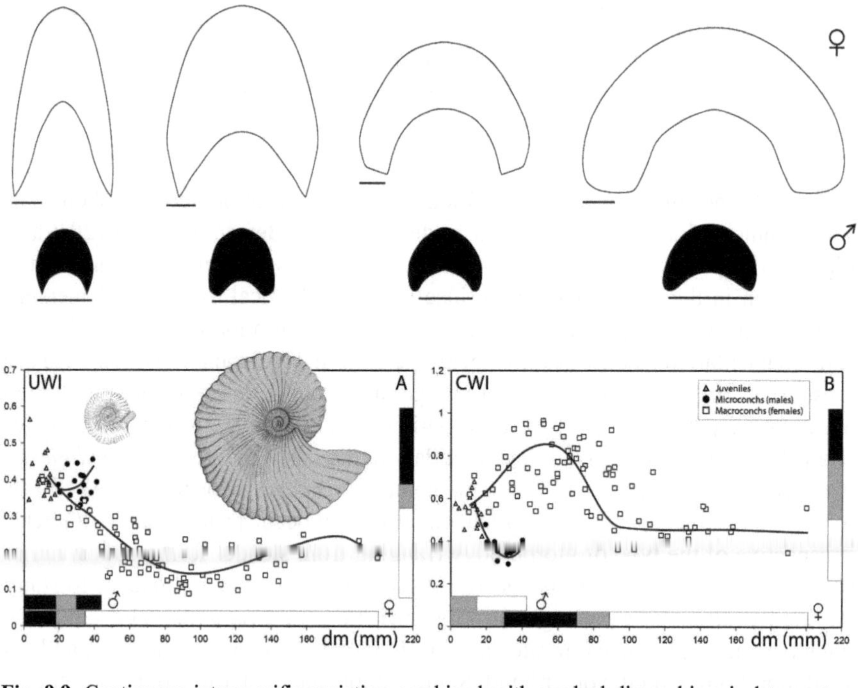

Fig. 9.9 Continuous intraspecific variation combined with marked dimorphism in later ontogeny within *Morrisiceras morrisi* (Bathonian; modified from Zatoń 2008; with permission of the author). Note the similar early ontogeny of both antidimorphs, but marked bimodal differences in adult morphology and size between the microconch and macroconch. Both the microconch and macroconch show a continuous variation in whorl cross section at similar sizes

does not necessarily reflect sexual dimorphism (e.g., Reyment 1988), particularly when no clear differences can be found in later ontogeny or pre-adult forms are more dissimilar than adult forms (e.g., Tintant 1980; Davis et al. 1996). Additionally, interpretations of dimorphism within taxa might differ between authors (e.g., Brochwicz-Lewinski and Rózak 1976). Dimorphism is discussed in further detail in Klug et al. 2015. In some cases, more than two discrete forms in adult size within a species have been reported (e.g., Ivanov 1971a, 1975; Matyja 1986, 1994; compare (Sect. 9.7)), which might be related to differences in rate and length of development. Furthermore, sometimes multiple morphs might be present in one or both of the sexes (Sonny Walton 2014, personal communication). Several authors studying homogenous and synchronous populations of Jurassic ammonoids have noted the presence of morphologically similar groups only distinguishable with the presence or absence of certain characters (Tintant 1963, 1976; Tintant 1980; Contini et al. 1984; Atrops and Mélendez 1993; Meléndez and Fontana 1993; Davis et al. 1996; Bardhan et al. 2010), particularly in ornamentation (presence of one or two rows of lateral tubercles, trifurcation vs. single and biplicate ribs, presence or absence of parabolic ribs, etc.). Interestingly, this variation can occur independently of sexual dimorphism as it affects both macroconchs and microconchs of these taxa in the same way (Tintant 1963, 1976, 1980; Charpy and Thierry 1976). A classic example is *Kosmoceras* (Tintant 1963, 1976, 1980), which possesses "*a clear sexual dimorphism with a microconch bearing mature modifications*" (Davis et al. 1996, p. 501). For a long time, two genera or subgenera were distinguished only differing in the presence of one ("*Zugokosmoceras*") or two rows of lateral tubercles (*Kosmoceras*) in contemporaneous populations. If no intermediate morphologies are found, it would be more conservative to interpret these as separate taxa, but some authors have argued that such forms should be interpreted as cases of intraspecific polymorphism, when these two groups display parallel, evolutionary changes or trends (Tintant 1980; Atrops and Meléndez 1993) in other characters. Such assertions of evolutionary trends (Monnet et al. 2015b) in these characters still have to hold to novel statistical methods which can analytically support the presence of evolutionary trends (Hunt 2006; Monnet et al. 2011a) and test how parallel evolutionary (or ontogenetic) trajectories really are (Adams and Collyer 2009, Collyer and Adams 2013; applied to ammonoids in Monnet et al. 2011a; De Baets et al. 2013a). Tintant (1963, 1976) has supported his claims not only by the study of numerous populations, but also with the discovery of a pathological macroconch displaying a "*Zugokosmoceras*" pattern on one side and a "*Kosmoceras*" pattern on the opposite side. Reports of polymorphism are not restricted to ornamentation, but might also occur in shell morphology, particularly shell thickness (Fig. 9.10).

Several authors have reported the presence of discontinuous variation in whorl section thickness in some Jurassic ammonoid taxa (Charpy and Thierry 1976; Marchand 1976; Tintant 1980; Contini et al. 1984; Fig. 9.10). In some cases, the two forms are more similar in morphology at the end of the ontogeny, as discussed by Tintant (1976, 1980). Nevertheless, such discontinuous distributions might also be related to differences in paleoenvironments, taphonomic and collection biases (temporal and spatial mixing), and sample sizes (Tintant 1980; Wilmsen and Mo-

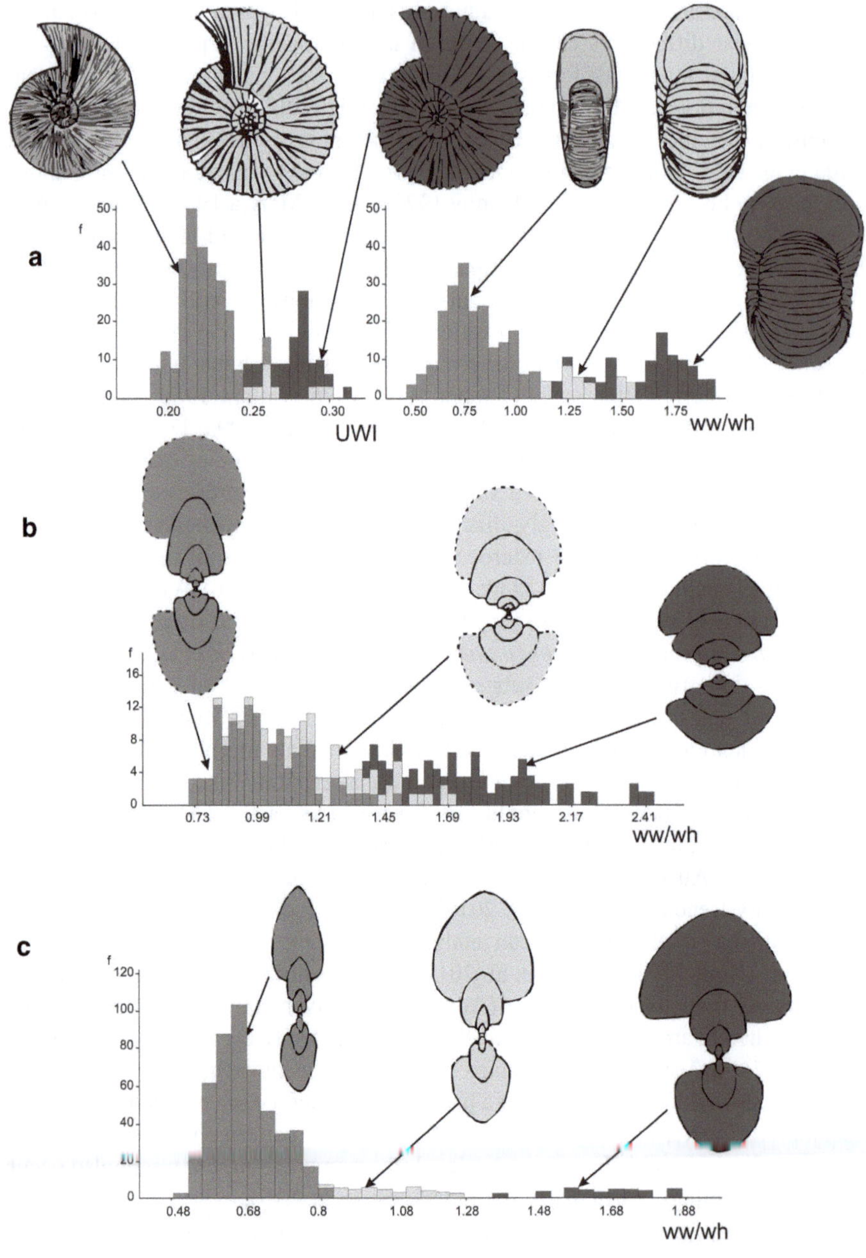

Fig. 9.10 Possible cases of discontinuous intraspecific variation interpreted as non-polymorphism in Jurassic ammonoids (redrawn from Contini et al. 1984): A sample of microconch specimens of *Macrocephalites* (Upper Callovian, Gracilis Zone, Michalskii sub-zone of Arino, Spain) interpreted to be three different morphs of a single species, which were previously interpreted to belong to three different subgenera and species: *Dolikephalites gracilis* (compressed shell with narrow umbilicus and fine ribbing), *Kamptokephalites herveyi* (round section with intermediate umbilicus

savinia 2011; De Baets et al. 2013a). As evidenced by Wilmsen and Mosavinia (2011), variation in a trait might be discontinuous even in taxa at some localities/ paleoenvironments showing otherwise continuous variation. This might depend on paleoenvironmental conditions, further stressing the need to study intra- and inter-population variation in ammonoids to further elaborate these patterns (Tintant 1980; Callomon 1988; Sect. 9.8).

Polymorphism in ornamentation and/or whorl thickness has also been reported from Cretaceous ammonoids (e.g., Hirano 1978, 1979; Reyment 1988; Kassab and Hamama 1991; Gangopadhyay and Bardhan 2007), although it is not always clear whether it reflects sexual dimorphism or could even be part of a more continuous variation. In Upper Cretaceous ammonoids, dimorphism is recognized in most families, especially based on differences in size and only occasionally on mature modifications, and intraspecific variation is important though of unequal proportion between some groups (Kennedy and Wright 1985). Various authors have described the presence of polymorphism (two or more variants) recognizable in whorl thickness, coiling and/or ornamentation in Carboniferous ammonoids (McCaleb et al. 1964; McCaleb and Furnish 1964; Furnish and Knapp 1966; McCaleb 1968), which is reminiscent of continuous variation in Mesozoic coiled ammonoids, particularly the covariation of ornamentation with coiling and whorl shape. Davis et al. (1996) raised doubts about these Carboniferous accounts because of the presence of intermediates between these morphs, particularly as in some cases the forms seem to differ more in juvenile stages than in the adult stage (Davis et al. 1996; but they might still represent cases of non-sexual polymorphism according to Tintant 1980). Tintant (1980) has also discussed the possibility of non-sexual polymorphism in whorl width index and ornamentation in *Dactylioceras* (Jurassic), which varies less in early ontogeny than in late ontogeny. At least in some cases (e.g., *Arkanites*), as evidenced by Stephen et al. (2002), these Paleozoic morphs might be indistinguishable in juvenile stages and then later on during ontogeny show a bimodal distribution in shell parameters, which might indicate sexual dimorphism (compare Davis et al. 1996; Sarti 1999; Fig. 9.9). Nevertheless, large differences in adult size and mature modifications of the aperture used to recognize sexual dimorphism in Jurassic ammonoids seem to be absent in Paleozoic forms (e.g., Davis et al. 1996; compare Makowski 1962, 1991). The different nature of this dimorphism does not necessarily speak against it being of a sexual nature, as in some cases, there is little differences in size (or morphology) between sexes in extant cephalopods (e.g., for

---

and strong ribbing) and *Pleurocephalites folliformis* (depressed shell with large umbilicus and very strong ribbing); B sample with distinct morphologies of *Pachyceras lalandeanum* (Upper Callovian, Lamberti Zone of Villers-sur-Mer, Calvados, France; ordered from compressed to depressed section) interpreted as intraspecific polymorphism, which were previously considered to belong to three distinct species and two different genera (*P. lalandeanum*, *P. crassum*, *Pachyerymnoceras jarryi*); C sample with distinct morphologies of Quenstedtoceras (Upper Callovian Lamberti zone, Lamberti subzone of Magny-les-Villers, Champs Mollous, Côte-dÓr, France), which were previously described as three different (sub)genera and species: *Q. (Lamberticeras) lamberti* with a compressed section, *Quenstedtoceras (Eboraciceras) ordinarium* with an intermediate section and *Quenstedtoceras (Sutherlandiceras) carinatum* with a depressed section (modified from Contini et al. 1984)

Nautilus: Ward 1987; for squids: Zuev 1971). Other possible reports of polymorphism in Paleozoic ammonoids or intraspecific variation (Kant 1973a, b; 1975; Davis et al. 1996, p. 490–491) are dubious because of the low sample size and the fact that specimens described as one species are now known to belong to multiple taxa (Dieter Korn 2013, personal communication).

It is not uncommon to see discontinuous shape and ornament within a species, including mollusks, alongside more continuously expressed variations in size, shape or ornament (Reyment and Kennedy 1991). Continuous variation in shell shape, ornamentation and/or suture line can be combined with dimorphism and/or non-sexual polymorphism within the same species (e.g., Jurassic: Tintant 1963; Westermann 1966; Zatoń 2008; Chandler and Callomon 2009; Cretaceous: Ploch 2003; Landman et al. 2010). Sexual dimorphism might also be associated with non-sexual polymorphism as discussed by Charpy and Thierry (1976), Tintant (1963, 1976, 1980) and reviewed by Contini et al. (1984) for several Middle Jurassic ammonoids, although it is unclear if all these cases represent discontinuous variation. In other cases, no evidence is found for the presence of dimorphism (or polymorphism) associated with continuous variation (Reeside and Cobban 1960; Dagys and Weitschat 1993b; Monnet et al. 2010; De Baets et al. 2013a; but compare Sarti 1999), which might speak for the absence of polymorphism (at least in shell morphology) in these ammonoids showing more continuous variation.

In the Triassic, low sample size has often led to confusion of dimorphism with continuous variation (e.g., Dzik 1990b vs. Monnet et al. 2010 for *Acrochordiceras;* Lehmann 1990). De Baets et al. (2013a) suspected dimorphism in Early Devonian *Erbenoceras*, but could not find evidence for it in a quantitative analysis of a larger sample (82 specimens). They found a bimodal size distribution, but the lower peak was not associated with adult modifications. Furthermore, morphs of both modes overlapped in several characters and already differed throughout earlier ontogeny, suggesting the presence of continuous variation rather than (sexual) dimorphism. Clearly more studies are necessary on intraspecific variability in several time intervals to fully understand the relative contribution of different types of continuous and discontinuous intraspecific variation (including dimorphism and polymorphism) in ammonoids. Such knowledge can only be achieved by quantitative studies on numerically large populations of precisely known geological ages derived from a wide variety of ages, paleoenvironmental or geographic areas, taxa and shell morphologies (cf. Tintant 1980; Davis et al. 1996). At the moment, it appears therefore most reasonable to assign co-occurring specimens to different species when no evidence is available for continuous variation with unimodal distribution or discontinuous variation in the form of morphs that evolve in parallel and/or co-occur with similar early ontogeny and/or later ontogeny, particularly in the Paleozoic. This interpretation might, however, change when additional material becomes available.

9.4.3 Variation in the Suture Line

While numerous studies have focused on intraspecific variability of shell shape and ornamentation, variation in the suture line has been less studied, particularly from

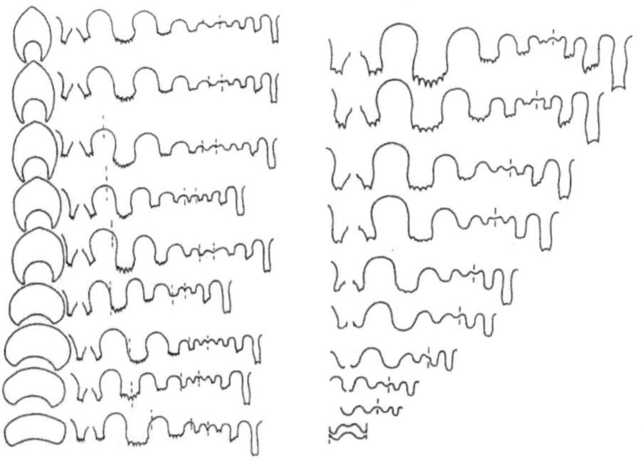

Fig. 9.11 Intrapopulation variation in the last whorl and sutures (left) and ontogenetic variation in the development of the suture line through ontogeny (right) in *Tuaroceras rieberi* from the Lower Anisian (modified form Dagys 2001)

a quantitative point of view. We herein discuss some intraspecific factors of suture line and septal variability.

The development of the suture line not only varies throughout ontogeny (Klug and Hoffmann 2015; Fig. 9.11) or on both sides of the plane of symmetry (asymmetry) within the same specimen (compare Klug and Hoffmann 2015; De Baets et al. 2015a), but can also vary between specimens of the same species at similar diameters (intrapopulational or intraspecific variation: e.g., Seilacher 1973; Dagys 2001; Fig. 9.11). Despite the common, perhaps over-emphasized use of the suture for taxonomic purposes (Arkell 1957), intraspecific variation in the suture line is only rarely studied quantitatively (Manship 2004, 2008; Waggoner 2006; Yacobucci and Manship 2011). Suture intraspecific variability appears to be particularly large in Jurassic and Cretaceous pseudoceratites (Arkell 1957) and heteromorphs (Kakabadze 2004). According to Arkell (1957), variation is the greatest among regressive types in which the suture line is secondarily simplified (e.g., in Jurassic and Cretaceous "pseudoceratites"). The suture line is also quite variable in several Mesozoic heteromorphs and might therefore be of little help for taxonomy (Kakabadze 2004), particularly at lower taxonomic levels (e.g., Hoffmann et al. 2009). Differences in suture line between specimens of the same species at a similar diameter have been related to differences in whorl shape (e.g., Arkell 1957; Reeside and Cobban 1960), ornamentation (e.g., Casey 1961; Westermann 1966) and/or ontogenetic development (e.g., Hammer and Bucher 2006), but they might also be more random (e.g., no clear correlation with other properties of the shell or ontogeny).

A marked variation of the suture line with whorl shape has been long known (e.g., Pictet 1854; Zittel 1885; Pfaff 1911; Spath 1919; Arkell 1957; Seilacher 1988; Klug and Hoffmann 2015), which manifests itself through the ontogeny of the same specimen or the evolution of taxa through time (reviewed by Monnet et al. 2011a). The phenomenon has been particularly discussed on large taxonomic scales, when

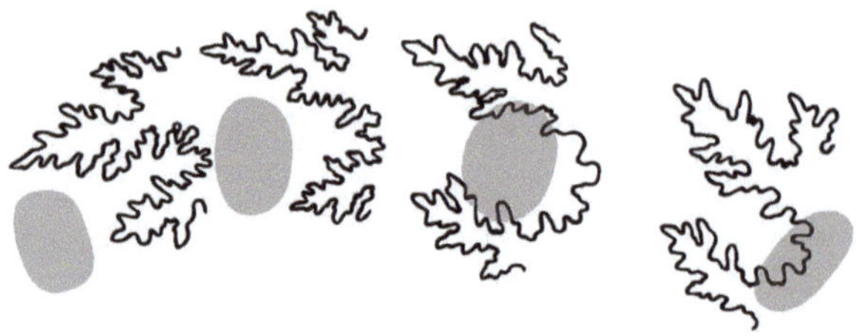

Fig. 9.12 Variation in four consecutive suture lines of the holotype of *Cheloniceras disparile* (Early Aptian) and its relationship with the position of the lateral tubercle (grey) (modified from Casey 1961, p. 216)

comparing taxa with different shell shapes (e.g., Westermann 1971; Ward 1980; Seilacher 1988; Jacobs 1990; Olóriz et al. 1997, 1999). Several authors have discussed complex covariations between shell morphology, ornamentation and/or suture line above the species level (e.g., Ward and Westermann 1985; Olóriz et al. 1997, 1999). Differences in the shape of the suture line between specimens of the same species at similar diameters have been less discussed. Pictet (1854, p. 669) already pointed out that the inflated varieties of a species often differed from the compressed ones in the number of accessory lobes (and that modification of the umbilicus can cause the same alteration of sutural element number). Reeside and Cobban (1960) found that more inflated and heavily ornamented forms within individual *Neogastroplites* species tended to have taller lateral saddles (see also Yacobucci and Manship 2011). Buckman (1892, p. 313) reported that in *Sonninia* and *Amaltheus*, the complexity of the suture varies with the ornament (and through the first law of covariation also with whorl cross section). Westermann (1966) dubbed the covariation of whorl cross section and number of lobes/ saddles 'Buckman's second law of covariation' (see Monnet et al. 2015a). He also stated that this might explain the statements of Oechsle (1958) on the moderately incised suture line of "*S. adicra*" and intensively incised suture line of "*S. modesta*", which he considered to be extreme variants of the same species. He suggested that the covariation between septal suture and shell plication could be explained by functional requirements ("the stiffening of the phragmocone against shear, a strongly incised suture line furnishing a better fixture of the septum against shear and more even distribution of stresses from the septum onto the outer shell vice versa"), particularly when this covariation could be demonstrated in multiple, not closely related genera. By contrast, Morard and Guex (2003) stated that the sculpture probably does not influence suture complexity directly, but that the sculptural and sutural elements both depend on a common third factor, the whorl shape (Guex 2001, 2003). Casey (1961) illustrated that the shape of the suture line at more or less the same diameter can also vary with ornamentation (tubercles) within the same specimen (Fig. 9.12). Some additional differences in the suture line within and between specimens can also be associated with pathologies (compare De Baets et al. 2015b; Keupp and Hoffmann 2015)

Hammer and Bucher (2006) tried to explain the correlation between whorl shape and suture line complexity partially in developmental terms in the following way: "*Most ammonoids with compressed shells have more circular whorl sections early in ontogeny. The intraspecific variation in whorl shape can therefore be explained as heterochronic: the more rounded forms are retaining their juvenile shape and can be regarded as paedomorphic. In such forms, where development is delayed, it would not be surprising if sutural development was similarly delayed, retaining the simple suture line of the juvenile into more mature stages.*" They acknowledged, however, that "*other physical mechanisms may also influence the fine shape of the suture*". They also reported intraspecific variation in septal spacing within *Amaltheus* that correlated with whorl shape, which they explained functionally in terms of hydrostatic properties. The smaller interseptal spacing in the compressed form has a positive impact on hydrostatic consistency through chamber formation (i.e., the smaller septal spacing leads to a smaller relative loss of buoyancy and smaller rotations of the aperture between consecutive septae). Similar covariation was also reported in *Dactylioceras* by Morard (2004), where the septal distance (septal angle) was larger in evolute forms with a depressed whorl section than in involute forms with a compressed whorl section.

One of the prime examples of high continuous intraspecific variation in shell shape and ornamentation are Triassic faunas from Siberia (e.g., Dagys and Weitschat 1993a, b; Checa et al. 1996; Dagys et al. 1999; Dagys 2001; Weitschat 2008; Fig. 9.2, 9.11). Several of these authors report an absence or no straightforward relationship between shell shape and suture line within these species. An exception is Dagys (2001, p. 548), who reported that more compressed forms had the highest number of umbilical lobes, although in the systematic descriptions, he stated that the covariation of shell shape with the suture line was not straightforward (compare Fig. 9.13). Dagys and Weitschat (1993b) only reported that the position of the first saddle changes with the morphology of the conchs in *Czekanowskites rieberi*. Dagys et al. (1999) found that the outline of the saddles is highly variable in *Parasibirites kolymensis*.

Manship (2008) specifically investigated variation of the suture line in the Late Cretaceous acanthoceratoid *Coilopoceras springeri*, which has a marked intraspecific variation from robust, strongly ornamented to gracile, weakly ornamented shells. She found a subtle, gradational variation in suture forms, which was only weakly tied to shell morphology. Interestingly, Yacobucci and Manship (2011) reported a higher degree of constraint in the suture line pattern of the Cretaceous hoplitid *Neogastroplites muelleri* (known for its wide range of intraspecific variability in shell shape and ornamentation: Reeside and Cobban 1960; see Fig. 9.13) than in *C. springeri*. While overall there is much less variation in suture line than in shell shape and ornamentation in *Neogastroplites*, it is true that the "subglobose spinose" forms tended to have taller suture elements (compare Reeside and Cobban 1960).

Kassab and Hamama (1991) also figured the intraspecific variation in suture line in morphs with a depressed and compressed whorl shape of *Libycoceras ismaeli* showing no straightforward relationship with whorl shape. Morard and Guex (2003) reported, based on qualitative observations, that suture elements in involute morphotypes of the Early Jurassic ammonoid *Osperleioceras* tend to be more finely fringed and that the lateral saddle lies proportionally lower on the flanks in involute morphotypes. Similar qualitative differences were reported by Bersac and Bert (2012a, b) from Cretaceous Deshayesitidae. More quantitative analyses on large samples are necessary to further investigate these patterns of intraspecific variation in the suture line.

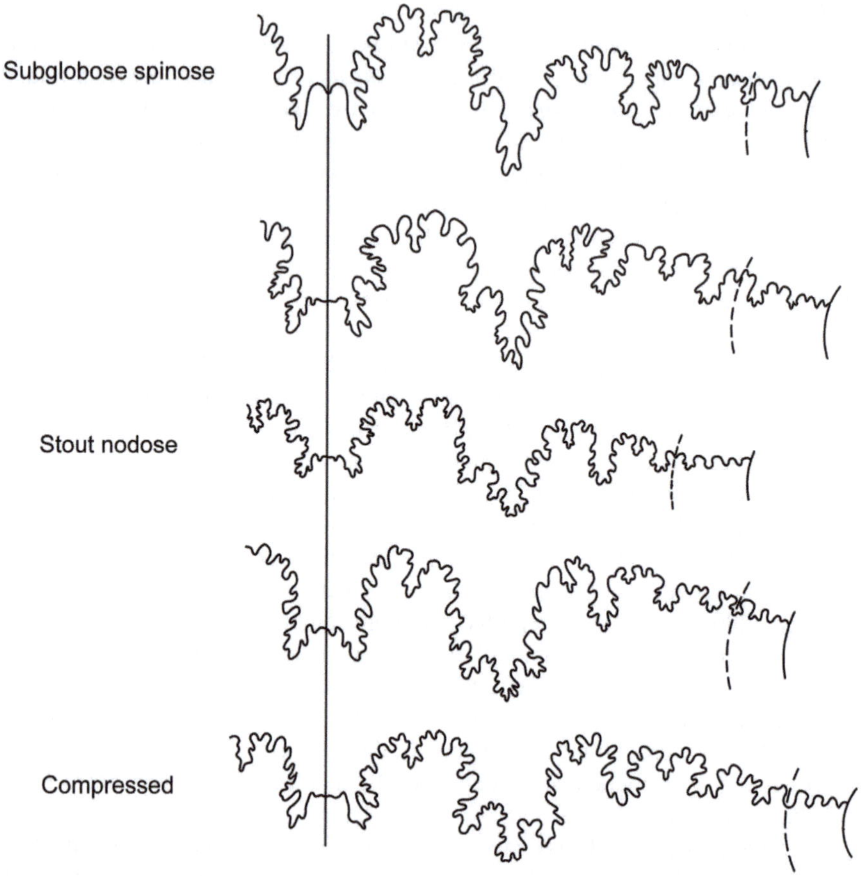

Fig. 9.13 Suture variation in morphs of *Neogastroplites muelleri* (Early Cenomanian). All sutures taken at shell diameters between 30 and 35 mm. While some variations in the suture line appear correlated with shell shape (e.g., taller lateral saddles in more globose shells), other variation is uncorrelated with shell shape; the overall path of the suture line is similar in all forms. Suture patterns redrawn from Reeside and Cobban (1960). From top to bottom: Variant G, uncatalogued specimen; Variant E-F, USNM 129496; Variant C-D, USNM 129468; Variant B, USNM 129435; Variant A, USNM 129418

A special kind of intraspecific variation is the suture asymmetry (Kakabadze 2004; Paul 2011; Keupp 2012) in some taxa, which is reflected in the different position and development of the suture line elements between the left and right sides of the whorl. In at least some cases, this asymmetry could be related to asymmetrical development of the soft tissues (Yacobucci and Manship 2011). In some taxa, the symmetrical development of the suture line is poorly constrained and is very variable between different specimens (e.g., the labile position of the external lobe in some taxa: Lange 1929, 1941; Hölder 1956; Schindewolf 1961; Hengsbach 1976, 1980, 1986; Landman and Waage 1986; Keupp 2012 for a review), while in others, the development and direction of asymmetry seems to be genetically fixed (constant excentrical position of the lateral lobe in Platylenticeratidae and *Anahoplites*: Hölder 1956; Keupp 2012). The development of asymmetry in the suture can also be due to pathologies when it only appears in a small percentage of the population (some authors have linked this to parasitic infestations: see discussion in De Baets et al. 2015a).

9.5 Influence of Intraspecific Variation on Ammonoid Studies

Ammonoids have often suffered extreme taxonomic oversplitting (Kennedy and Cobban 1976; Kennedy and Wright 1985; Dagys and Weitschat 1993b; Donovan 1994), but lumping is also not uncommon (e.g., Westermann 1966; Howarth 1973; Callomon 1985). Underestimating (or overestimating) intraspecific variation can bias taxonomy and diversity counts, as well as biostratigraphic, evolutionary and paleobiogeographic analyses (e.g., Dzik 1985; Kennedy and Wright 1985; Hughes and Labandeira 1995; Nardin et al. 2005; Korn and Klug 2007; Monnet et al. 2010; De Baets et al. 2013a; Bert 2013).

Traditionally, many ammonoid workers have used a strict typological approach, erecting narrowly defined morphospecies, which has led to an artificial inflation of paleodiversity. Some authors like Buckman (1887) had already realized the problem early on, but still kept using this typological approach resulting in the oversplitting of species. A typical example of this approach is the Jurassic ammonite *Sonninia*, for which Buckman alone erected over 60 species. Westermann (1966) subsequently lumped 69 species of *Sonninia* (including the ones erected by Buckman) together with *Sonninia adicra* as they all form part of a continuum in morphology (as well as dimorphism) and based on this work, defined the Buckman laws of covariation following observations that had already been reported by the former author in 1892. Although a large degree of intraspecific variability in *Sonninia* is still acccpted, it is now well established that specimens of *Sonninia* lumped together by Westermann (1966) come from multiple (bio)stratigraphic levels (e.g., Callomon 1985; Westermann 1996; Sandoval and Chandler 2000; Dietze et al. 2005). When better preserved or better stratigraphically controlled material becomes available, this can still lead to the erection of additional species or the re-establishment of older ones based on previously overlooked differences in ontogeny or morphology. Hence, we frequently see in the history of ammonoid taxonomy an initial rapid increase in taxonomic diversity as a result of taxonomic oversplitting related to a strict typological approach, followed by a decline and then potentially a slight rise in diversity again, when a better numerical grasp on intraspecific variability and even finer stratigraphic resolution is achieved, as illustrated by Buckman's *Sonninia* (Fig. 9.14).

In the middle of the last century, various authors realized the problems related to a strict typological or morphospecies approach, which resulted in the introduction of multiple co-occurring species at the same stratigraphic interval and region (e.g.,

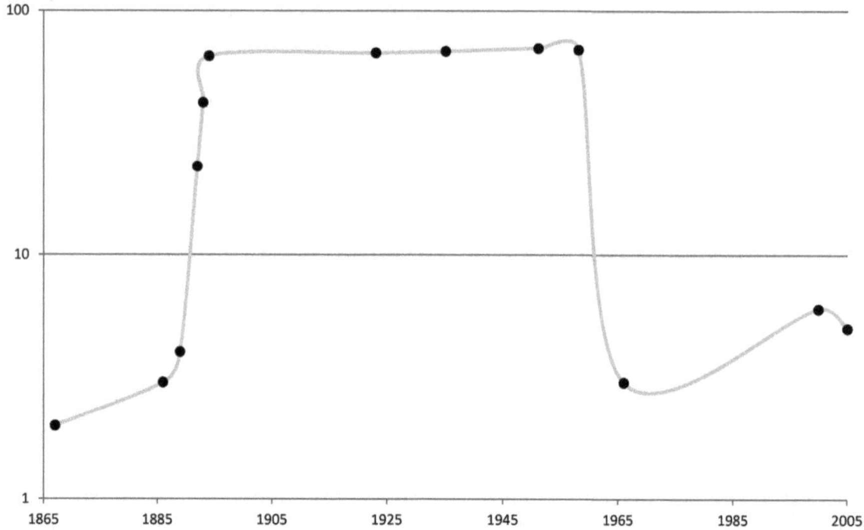

Fig. 9.14 Number of species of *Sonninia* (Bajocian) in use through time. Note the steep growth in the beginning during the era of a strict typological approach (mostly erected by Buckman), followed by a steep decline due to the biospecies approach of Westermann (1966) and a slight increase in modern times as a result of a more highly resolved stratigraphic framework and the discovery of new material (Sandoval and Chandler 2000; Dietze et al. 2005)

Haas 1946; Barber 1957; Callomon 1963; Tintant 1963; Westermann 1966), and have rallied for a more 'biological' species concept. Multiple authors have promoted the merits of using a horizontal (population or biospecies) approach as opposed to a vertical (index or morphospecies) approach (e.g., Callomon 1963, 1985; Tintant 1963, 1976, 1980; Tozer 1971; Kennedy and Cobban 1976; Silberling and Nichols 1982; Dzik 1985, 1990a, 1994; Westermann and Callomon 1988; Atrops and Meléndez 1993; Chandler and Callomon 2009; Fig. 9.15), which does not only include a wide range of continuous variation, but potentially also discontinuous variation (sexual dimorphism and non-sexual polymorphism). These authors may explicitly or implicitly claim that a fossil assemblage from a single stratigraphic horizon is more likely to represent a true biospecies (i.e., a reproductively isolated population) than assemblages from different horizons. However, it is obviously difficult to test claims about reproductive isolation in fossil samples, and we suggest great caution in applying the term biospecies to extinct taxa (see Yacobucci et al. 2015).

Oversplitting might also artificially create two or more lineages evolving seemingly in parallel (Tintant 1980; Atrops and Meléndez 1993). Many authors keep using morphological species or variants to refer to different morphologies or morphs of these species for practical purposes (e.g., Dietze et al. 2005; Chandler and Callomon 2009), but it is incorrect to give these forms the rank of subspecies from a biological point of view (where species need to be reproductively isolated by geography or other factors), following the International Code of Zoological Nomenclature. Similar views were expressed for the subgeneric ranking of dimorphs (e.g.,

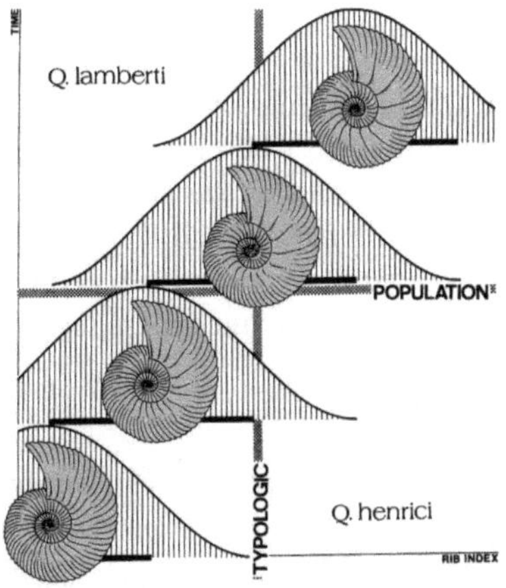

Fig. 9.15 Differences between typological and population concepts of an ammonoid species as demonstrated by Late Callovian *Quenstedtoceras*. According to a typological approach all specimens having more than three secondary ribs per primary rib belong to *Q. lamberti* (irrespective of their age), while according to a population approach all populations showing modal values of the rib index exceeding 3 at size class 50-60 are to be included in that species (modified from Dzik 1990a; with permission from the author)

Schweigert et al. 2007), which should not be used because a subgeneric placement should express (paleo-) biogeographic or habitat differentiation within a genus rather than pure morphological differences. Nomenclatorial reasons also speak against the use of a subgeneric ranking (e.g., Pavia 2006).

Not properly taking intraspecific variation into account can cause obvious problems for taxonomy and systematics, but it can also significantly influence biostratigraphic studies (Dzik 1985, 1990a). As explained by Dzik (1985), the probability of finding a particular morphotype in a sample is not only related to the sample size but also to the horizon being sampled, as different morphotypes are more common in different horizons (Fig. 9.15). Thus, definitions of time correlation units based on the first known occurrence (FAD) of a morphotype does not provide a completely reliable basis for a study, which has led to the common use of assemblage zones. Morphological variation in contemporary populations might greatly exceed evolutionary changes between successive faunas (e.g., *Schloenbachia*: Kennedy and Cobban 1976, Wilmsen and Mosavinia 2011; Kennedy 2013 or *Neogastroplites*: Reeside and Cobban 1960, Reyment and Minaka 2000), so that in many cases, successive faunas can only be separated on the basis of the mode of variation of the population, as individual morphotypes have relatively long stratigraphic ranges. This might also make it difficult to compare specimens from different localities or regions, when only limited material is available (e.g., De Baets et al. 2013a). Properly taking into account intraspecific variability is therefore a very important prerequisite for studying temporal and spatial patterns of diversity through time and their relation with environmental changes and extinction events (e.g., Kennedy and Wright 1985; Monnet et al. 2011b; De Baets et al. 2013a), for which ammonoids are often used (e.g., Brayard et al. 2009; Dera et al. 2011). This problem is illustrated by a study of Korn and Klug (2007) on the intraspecific variability of *Manticoceras*, which indicates that the effect of the Frasnian–Famennian extinction on ammonoids might be significantly overestimated when ignoring intraspecific variability. Frasnian diversity is based mainly on manticoceratid diversity, which was artificially inflated by taxonomic oversplitting (Korn and Klug 2007). Clearly, analysis of intraspecific variability is a prerequisite for many paleobiological and evolutionary studies (e.g., Monnet et al. 2015b), and much more research in this field is needed.

9.6 Intraspecific Variation through Ontogeny

The mode and range of intraspecific variation might change through ontogeny. Several authors have reported the largest range of continuous intraspecific variation from the middle whorls (e.g., Dagys and Weitschat 1993b; Morard 2004, 2006; Korn and Klug 2007; De Baets et al. 2013a). In these cases, specimens of the same species (and even different species and genera) are morphologically more similar to each other in the last whorl (recognized by adult modifications discussed in Klug et al. 2015) and early whorls than during intermediate growth stages. Others have reported higher variation in juvenile and adult forms (e.g., Korn and Vöhringer 2004). Tanabe and Shigeta (1987) reported a higher variation in early whorls than in later whorls in several Cretaceous ammonoid taxa, although they only investigated a limited number of specimens per species. Monnet et al. (2012) reported also the same pattern of decreasing intraspecific variation through ontogeny (i.e. high juvenile plasticity) in some Triassic ammonoids that is also independent of evolutionary trends through time and also may have a very different variance (range) in different morphological characters. However, this pattern must be cautiously treated because it may be biased by the usual higher abundance of intermediate-sized specimens within preserved "populations" of species. The most extreme example of large differences in adult form is the dimorphism in late ontogeny typical of sexual dimorphism (Klug et al. 2015). There are also examples of non-sexual polymorphism, where the forms are at their most dissimilar in earlier ontogeny and become more similar in later ontogeny (e.g., in *Dactylioceras* as discussed by Tintant 1976, 1980) or where they have similar ontogenies differing only in discrete characters (compare Sect. 9.4.2) Furthermore, a large degree of intraspecific variation in ammonoids might be related to differences in development, more specifically growth rates and the length and shape of ontogenetic trajectories through development. For these reasons, intraspecific variability should be studied throughout ontogeny or development from early to late growth stages (e.g., Neige 1997; Morard and Guex 2003; Urdy et al. 2010a, b, 2013; De Baets et al. 2013a; Bert 2013: compare Urdy 2015). Only limited studies have focused on quantitatively analyzing changes in intraspecific variation throughout ontogeny (e.g., De Baets et al. 2013a), but they might be particularly important to understand the mode of growth as well as paleobiology and paleoecology of ammonoids.

9.7 Size-At-Age Variation in Ammonoids

The growth of extant cephalopods is notoriously variable and is influenced by various biotic and abiotic factors. Populations of extant cephalopod species can show large intra-annual (seasonal) and inter-annual variations in growth rate and adult size (e.g., Boyle and Ngoile 1993; Arkhipkin and Laptikhovsky 1994; Nigmatullin et al. 2001; Jackson and Moltschaniwskyi 2002; Arkhipkin 2004; Keyl et al. 2011; Hoving et al. 2013). This typically results in a high intraspecific variation in size-at-age data (e.g., Boyle and Boletzky 1996), which has been corroborated both by laboratory rearing experiments and population studies in the field (Pecl et al. 2004 and references therein). In particular, ambient temperature as well as food quality and quantity have been implicated in growth rates (e.g., Pecl et al. 2004), but various other environmental factors such as light intensity, pressure, dissolved oxygen, and the abundance and kind of predators might have stressed or otherwise influenced the growth rates of marine organisms (see Bucher et al. 1996 for a review). Food and temperature are also partially interrelated because feeding rate is a function of temperature (Mangold 1983; Hewitt and Stait 1988). Such ecologically driven differences in size are exemplified by a study by Hoving et al. (2013), who recorded a spectacular decrease in the adult size of the jumbo squid (*Dosidicus gigas*) from more than 55 cm to less than 30 cm mantle length after an El Niño event, which is a periodic development of anomalously warm surface water temperatures (and can coincide with a reduction in primary production associated with wind-driven upwelling as well as a variety of effects on higher trophic levels) off the western coast of South America. This variation might also be partially related to differences in temperature or other environmental conditions at hatching. Climate change might also affect the life history of squids as warmer temperatures are expected to reduce embryonic duration and hatchling size, increase growth rates and shorten the overall life-span, resulting in maturation at smaller sizes and younger ages. Individual squids will require more food per unit body size, need more oxygen for faster metabolism and have a reduced capacity to cope with food scarcity (Pecl and Jackson 2008). Climate change might well have affected ammonoid populations in a similar way (e.g., Matyja and Wierzbowski 2000), although this needs be further investigation.

Similar size-at-age variation might have existed in ammonoids, and could be reflected in differences in adult size of fossil ammonoid populations at the same locality and age (e.g., Ivanov 1971a, 1975: microconchs, macroconchs and megaconchs; Matyja 1986, 1994: miniconchs, microconchs and macroconchs: compare Klug et al. 2015), over evolutionary time (Hewitt and Hurst 1977; Dzik 1990a; Landman et al. 2008) as well as interpopulational differences in adult size between different regions and/or paleoenvironments (Kummel 1948; Vogel 1959; Wendt 1971; Mancini 1978; Elmi and Benshili 1987; Stevens 1988; Courville and Thierry 1993; Mignot 1993; Mignot et al. 1993; Matyja and Wierzbowski 2000; Reboulet 2001; Urlichs 2004, 2012; Ploch 2007; Landman et al. 2008). It should be noted, however, that it can be hard to disentangle intra- and inter-annual environmental factors due to the inherent time-averaging each shell assemblage (fossil population) has gone through (compare Sect. 9.3). Also, to document size-at-age variation requires a method for determining the age of individuals. Even in extant cephalopods, estimating age (often indirectly using growth increments in statoliths: Jackson 1994) and maturity (e.g., Arkhipkin 1992) is not always straightforward. In fossil cephalopods, such studies are further complicated. Diameter is most commonly used as a proxy for size, although other parameters like body chamber (or shell) volume or its square root might be more suitable, as they better reflect the growth of the soft tissues contained within the body chamber, particularly when comparing taxa or ontogenetic stages with different shell shapes (e.g., Bucher et al. 1996; Dommergues et al. 2002; De Baets et al. 2013a). The volumetric parameters are more difficult to obtain and not commonly used in the literature, making the diameter still the most commonly used proxy for size. The diameter is also not directly comparable with mantle length, which is the most widely used measure of size in extant coleoids. The body chamber length would be more directly comparable with mantle length. The diameter has been used more rarely as a proxy for age, but this probably makes even less sense. Other proxies for age have been employed like septal spacing (e.g., Stephen et al. 2002) or rib spacing counts (e.g., Dommergues 1988), although these also have their problems as they are based on several assumptions and do not necessarily stay constant through ontogeny or between taxa (compare Bucher et al. 1996). Ideally, we would need to know growth rates through ontogeny, which are, however, hard to estimate in fossil organisms like ammonoids. Several methods making various assumptions have been suggested, which can often only be used in particular cases and show various degrees of success and reliability (comprehensively reviewed in Bucher et al. 1996; compare Urdy 2015). These methods are:

- Assumptions about the periodicity of shell secretion in septa, constrictions or pseudosutures (e.g., Ivanov 1971b; Hirano 1981; Doguzhaeva 1982; Weitschat and Bandel 1991; Hewitt et al. 1991),
- Detection of seasonal signals in morphology (e.g., shell volume: Trueman 1940; rates of whorl expansion and septal spacing: Westermann 1971; septal spacing: Kulicki 1974; Zakharov 1977; jaw increments: Hewitt et al. 1993) and isotopic data (e.g., Jordan and Stahl 1971; Lécuyer and Bucher 2006),
- Presence of distinct age classes or cohorts in ammonoid species, which are assumed to have bred (and spawned) periodically at regular times (e.g., Trueman 1940; Bucher et al. 1996),
- Using growth rates of epizoans with modern counterparts growing on ammonoids during their lifetime (e.g., Schindewolf 1934; Merkt 1966; Meischner 1968; Hirano 1981; Andrew et al. 2011).

Some of the most pronounced size differences in ammonoids might be associated with sexual dimorphism (classically expressed as microconchs and macroconchs), although the sexual nature or presence of dimorphism is not always clear, even during the Jurassic where the phenomenon is most accepted (compare Sect. 9.4.2). Some authors have reported only one morphology, resembling microconchs and/or macroconchs of other taxa, while others have reported the presence of three morphologies (e.g., Ivanov 1971a, 1975; Brochwicz-Lewinski and Rózak 1976; Matyja 1986, 1994) within ammonoid species (compare Klug et al. 2015). The latter phenomenon might reflect the presence of morphologically distinct "populations", which might be largely geographically isolated or represent cohorts hatched at different times (and which were therefore subjected to different environmental conditions at different times in their development). Ivanov (1971a, 1975) reported the presence of megaconchs, which do not show signs of maturation (although this might potentially be a taphonomic bias) and Matyja (1986, 1994) reported the presence of miniconchs in addition to microconchs and macroconchs within some ammonoid taxa. The presence of more than two morphologies does not necessarily rule out the presence of sexual dimorphism as there might be more than one morph in one or both sexes. Interestingly, the presence of three adult size-classes has also been reported from extant cephalopods (e.g., *Sthenoteuthis pteropus*: Zuev 1976; *Dosidicus gigas*: Nigmatullin et al. 2001), although it probably does not reflect sexual dimorphism, as the three size classes occur in both sexes (Nigmatullin et al. 2001). Matyja (1986, 1994) has dubbed this phenomenon developmental polymorphism, but this does not necessarily correspond to the biological definition of polymorphism, as these might have lived in different times or places and might therefore not be polymorphic in a biological sense (as such forms could represent seasonal or geographic variants: compare Sect. 9.3). Matyja (1986) attributed the expression of "developmental polymorphism" as mono-, di- or trimorphism in a given population to environmental factors, more specifically geography and lithofacies. Nevertheless, one should remain cautious in interpreting size structure in fossil populations as various factors could also contribute to a non-unimodal size distribution of fossil assemblages, including taphonomic and collection biases (size sorting, spatial and temporal mixing) as well as paleoecological factors (different ratios or separation of size classes or ontogenetic stages during the lifetime). The studies of Matyja (1986, (1994) and Ivanov (1971a, 1975) are therefore open to scrutiny as they did not use the proper quantitative methods and have a low sample size (compare Callomon 1988: Davis et al. 1996).

Several authors have reported intraspecific differences in adult size of populations deriving from different regions and/or paleoenvironments (Elmi and Benshili 1987; Mignot 1993; Mignot et al. 1993; Reboulet 2001; Urlichs 2004; Ploch 2007; Landman et al. 2008). Many of these studies have suggested a correlation between miniaturization and a locally confined paleogeographic and/or paleoenvironmental context (e.g., Elmi and Benshili 1987; Mignot 1993; Mignot et al. 1993; Olóriz et al. 2000; Urlichs 2004; Ploch 2007). Such a relationship has been particularly supported by comparative studies on Tethyan populations of Jurassic ammonites from small basins in the northwest European platform (e.g., Elmi and Benshili 1987; Mignot 1993; Mignot et al. 1993). Stunting ("*Kümmerwuchs*" sensu Ager 1963; see Keupp and Hoffmann 2015) preferentially affected populations in small isolated basins, which belong to species that are widespread and have larger individuals outside of the confined basins. Dommergues et al. (2002) attributed this size decrease to the scarcity of nutrients because such stunting is found experimentally in miniaturized (undernourished) cuttlefish (Boletzky 1974), although such populations have so far not been reported from natural environments. Reboulet (2001) reported that representatives of several Valanginian ammonoid species in the deeper water facies of the Vocontian basin were smaller than their counterparts in the shallow water facies in the Provence Platform, which he attributed to a decrease in growth rate under conditions of higher hydrostatic pressure as well as lower water temperature and nutrient supply in the basinal areas. Ploch (2007) reported that specimens of *Saynoceras verrucosum* from the deeper Vocontian basin were smaller than those from the shallower epicratonic Polish basin, which she attributed to their isolation and differences in depth or temperature. Landman et al. (2008) investigated differences in body size of the Late Cretaceous *Hoploscaphites nicolletii* over time and between geographically separated contemporary populations. They reported that macroconchs from the northwestern portion of their study area were smaller than those from the southwestern portion, which they explained by unfavorable conditions (lower oxygen levels, less than normal marine salinity) also reflected in the low diversity and abundance of nektic and benthic organisms in general. Landman et al. (2008) also reported that body size of the same species correlated with environmental changes over time. Urlichs (2004) discovered that smaller Triassic *Lobites nautilinus/pisum* specimens occurred in distinct clay to marly clay beds rich in pyrite while larger specimens were found in marl and limestone beds poor in pyrite. He interpreted the larger specimens as normal-sized when compared with contemporary specimens from the Hallstätt Limestone and the smaller specimens as stunted adults based on septal crowding (compare Kraft et al. 2008 for an alternative interpretation). Interestingly, the environment did not seem to have affected several other species (e.g., *Lecanites glaucus*, *Megaphyllites jarbas*, *Proarcestes klipsteini*, etc.), which are represented by normal-sized juveniles. Often, other shelled organisms from the St. Cassian Formation such as bivalves and brachiopods (Urlichs 2012) or in other, similar faunas are miniaturized (e.g., Kummel 1948; Mancini 1978). Urlichs interpreted this as a possible indication for a nektobenthic or demersal mode of life of *Lobites* and the confined paleogeographic condition of this basin. The close correlation between body size and paleoenvironment has often been used to indicate a deep nektonic or demersal/nektobenthic mode of life (Reboulet 2001; Urlichs 2004), although it might similarly affect forms with a pelagic mode of life as both facies and body size might be influenced by a second factor that controls both.

Stunting ("*Kümmerwuchs*"), which has ecological reasons, should not be confused with dwarfism ("*Zwergwuchs*"), which is genetically fixed (compare Ager 1963; Hallam 1965; Mancini 1978; Urlichs 2004; Keupp 2012). Paedomorphic processes that might lead to evolutionary trends towards smaller representatives are not uncommon in ammonoids, although they are often discussed above the species level (e.g., Dommergues et al. 1986; Landman and Geyssant 1993; Korn 1995). Logically, the first thing one needs to rule out in miniaturized faunas is that their small size does not relate to an early stage of development (e.g., juveniles) or taphonomic bias (e.g., only inner whorls preserved). Some authors have reported smaller representatives of species from fissure-fillings compared to 'normal' individuals from other sediments in the Triassic (e.g., Krystyn et al. 1971; Urlichs 2004) and the Jurassic (e.g., Wendt 1971; Aubrecht and Schlögl 2011). Some have attributed this to taphonomic biases (preferential preservation of smaller specimens in these fissures: Krystyn et al. 1971), while others attributed these to stunting (Wendt 1971; Urlichs 2004). Aubrecht and Schlögl (2011) could not establish whether these ammonites were juvenile, dwarfed specimens adapted to limited cave space or size-sorted adult specimens. According to Stevens (1988), environmental constraints could, however, also lead to gigantism in adults; using recent squids as a model, he predicted that large specimens might have preferred colder environments and should therefore be more numerous in the fossil record during transgressive episodes. Rare abnormally large or "giant" specimens within an ammonoid species could potentially be the outcome of pathological conditions such as parasitic castration, in analogy with cases seen in extant gastropods, or hormonal disorders (Dommergues et al. 1986; Stevens 1988; Manger et al. 1999; this is discussed in more detail in De Baets et al. 2015a).

Environmental factors do not explain all variation in size-at-age, as individuals reared under identical conditions still display a wide range of sizes (Pecl et al. 2004). Many studies have attributed size-at-age variation to intrinsic factors or environmental factors affecting cephalopods after hatching, overlooking the effect of seasonal temperature variation and individual hatching size heterogeneity. Such differences in hatching size might amplify throughout the lifespan of the cephalopods. In Recent cephalopods, hatchling size not only depends on egg size, but also on developmental temperature (varying with seasons, depth or latitude) and individual hatching conditions (Boletzky 2003). A negative relationship between egg size and environmental temperature is known from extant cephalopods and has been reported both within species and between species (e.g., Laptikhovsky 2006). Latitudinal temperature-related differences in embryonic shell size might also have been present in ammonoids (Laptikhovsky et al. 2013), but so far, the fossil record is spatially too patchy to test geographic differences in contemporary populations of the same species over large latitudinal distances, where this could be relevant (Tajika and Wani 2011). In extant cephalopods, there is also a large intrinsic variation at hatching at the same locality (De Baets et al. 2015b), which might also contribute to the large intraspecific variation reported in various ammonoids sampled from restricted intervals in time and space from the Paleozoic (Erben 1950, 1964; Tanabe et al. 1995; Stephen and Stanton 2002) to the Mesozoic (Tanabe 1977a, b; Landman 1987; Rouget and Neige 2001; Tanabe et al. 2003; Tajika and Wani 2011). It has been hypothesized that both seasonal temperature variation and individual hatchling size heterogeneity might influence subsequent growth and contribute to the (adult) size-at-age variation of cephalopods (Pecl et al. 2004; Leporati et al. 2007). Pecl et al. (2004) hypothesized that small changes in temperature might be particularly relevant during the early exponential growth phase and then amplify throughout the lifespan. An exception might be longer-lived species that exhibit asymptotic growth, where small initial differences between individuals might be minimized as the organisms grow. Individual hatchling variation might, however, play a subordinate role, at least in some ammonoids such as *Scaphites* (Landman 1987) or *Creniceras renggeri* (Neige 1997), where authors reported a lack of correlation between size at hatching and adult size. Nevertheless, such correlations might still be present earlier in ontogeny or in other taxa, as the influence of range of variation in early ontogeny seems to depend on the growth pattern of the cephalopods.

All these studies demonstrate that size is a highly variable parameter among populations within ammonoid species when subjected to specific environmental constraints. This indicates that size can be a useful tool to recognize spatially separated fossil populations (e.g., Courville and Thierry 1993). However, size-at-age may not be an effective criterion to separate ammonoid species. For example, the only character separating "*Mimagoniatites falcistria*" (restricted to the Hunsrück Slate) from *M. fecundus* (more widely distributed) was its size, combined with differences in preservation between clayey and carbonate sediments (Chlupáč and Turek 1983; Göddertz 1989). More complete specimens of "*M. falcistria*" with preserved inner whorls show that they are conspecific with *M. fecundus* (De Baets et al. 2013b). De Baets et al. (2013b) could also demonstrate that the larger specimens from the Hunsrück Slate show indications of adulthood (Chlupáč and Turek 1983), while larger, complete specimens are rare or absent in its type region in the Czech Republic.

9.8 Ecophenotypic Variation

Intraspecific morphological variation has been well-documented, particularly in coiled Mesozoic ammonoids, but it still remains poorly understood from an ecological point of view (e.g., Westermann 1996; Ritterbush et al. 2014). Strong intraspecific variation without apparent facies association has often been used to question a close correlation between shell shape and ecology in ammonoids (Dagys and Weitschat 1993a, b; Dagys et al. 1999; Weitschat 2008; Kennedy and Cobban 1976; compare Naglik et al. 2015). Others have suggested it might be related to a lack of selection on shell shape in these taxa and/or certain environments (Kennedy and Cobban 1976; Westermann 1966, 1996; Keupp 2000). The occurrence of broadly varying species together with narrowly varying species in the same family (e.g., *Czekanowskites* with *Arctohungarites*: Dagys and Weitschat 1993a) or co-occurring in the same strata (*Sonninia* with *Fissilobiceras*: Dietze et al. 2005) has been used by some authors (Westermann 1996; Keupp 2000) to suggest differences in selection pressure on streamlining and/or "defensive" sculpture depending on the paleoenvironment. Westermann (1996) hypothesized that streamlining varied much more in populations of planktic drifters and vertical migrants compared with nektic and demersal swimmers that "*depended on speed, acceleration and/or steerage for catching prey*". Keupp (2000) suspected that forms showing high variation might have lived in shallower water with higher water energy, where they would have been exposed to stronger selection for forms with better "streamlining" (Naglik et al. 2015). Interestingly, other ammonoid workers have related shell shape strictly to environmental influences as fossil taxa that show an array of forms are often assumed *a priori* to be ecophenotypically plastic (e.g., Crick, 1978; Kassab and Hamama 1991; Reyment 1988; Reyment and Kennedy 1991; Kin 2010), even if the specimens are more or less contemporary and derive from the same locality or region (Reyment and Kennedy 1991; Kin 2010). On the other hand, shell morphology is not necessarily strongly controlled by environmental conditions. Callomon (1985), for instance, noted that a high range of morphological variation in the Jurassic cardioceratids persisted for several million years despite migrations and changes in habitat, suggesting that internal controls on shell shape also existed. Some authors (Reyment 1988; Reyment and Kennedy 1991; Reyment 2003) have attributed the pronounced variation observed in some ammonoids to genetically based, "*multiple niche polymorphism*" (Bulmer 1980) in stressful environments, which is common in modern gastropods (Goodfriend 1986), although this is hard to prove.

Differences in intraspecific variation between different paleoenvironments and regions have rarely been studied in ammonoids, although they are the only way to test hypotheses of phenotypic plasticity and ecophenotypic variation in ammonoids. Some authors have not only reported intraspecific differences in size (discussed in more detail above in Sect. 9.7), but also differences in shell morphology and/or ornamentation between different paleoenvironments (Dietl 1978; Tintant 1980; Courville and Thierry 1993; Jacobs et al. 1994; Diedrich 2000; Ploch 2007; Wilmsen and Mosavinia 2011). Potentially, this link between shell form and environment might also be related to differences in growth rates and development between localities (Mignot et al. 1993; Reboulet 2001). Phenotypic plasticity in shell shape and ornamentation has mainly been reported in several Cretaceous ammonoids (e.g., Jacobs et al. 1994; Reyment and Kennedy 1991; Diedrich 2000; Kin 2010, 2011; Machalski 2010; Wilmsen and Mosavinia 2011; Fig. 9.16). Unfortunately, most of these studies only semi-quantitatively investigated these patters (i.e., in categories), having access only to rather small sample sizes, or they did not list the original measurements of their specimens, which would make it possible to assess these patterns statistically. In any case, phenotypic plasticity can never be fully demonstrated for ammonoids, since it would require proof that morphologically differing individuals shared the same genome (which is obviously unknown from ammonoids) and varied due to differing ecological conditions.

Jacobs et al. (1994) reported the presence of more compressed morphs of *Scaphites whitfieldi* in nearshore sandy facies, while more depressed morphs of this species occurred in offshore muds. The phenomenon of shell compression varying with lithofacies has also been recognized above the species level (Jurassic: Bayer and McGhee 1984, 1985; Cretaceous: Landman and Waage 1993; Kawabe 2003). This is seemingly consistent with a hydrodynamic explanation, where thinner, more compressed morphs would be able to swim more efficiently at higher velocities (typical for nearshore environments) and depressed morphs more efficiently at low velocities (see Naglik et al. 2015). A similar phenomenon might be present in *Thomasites gongilensis*, for which Courville and Thierry (1993) reported a continuous intraspecific variation from extreme platycone (*compressum*), over intermediate morphologies (*tectiforme*), to subsphaeroconic or subcadiconic morphologies (*gongilense*) linked by intermediate forms. The main morphology also depended on the paleoenvironments, with more compressed platycones dominating in shallower areas, while more inflated morphs dominated in deeper, subsiding areas (*tectiforme*

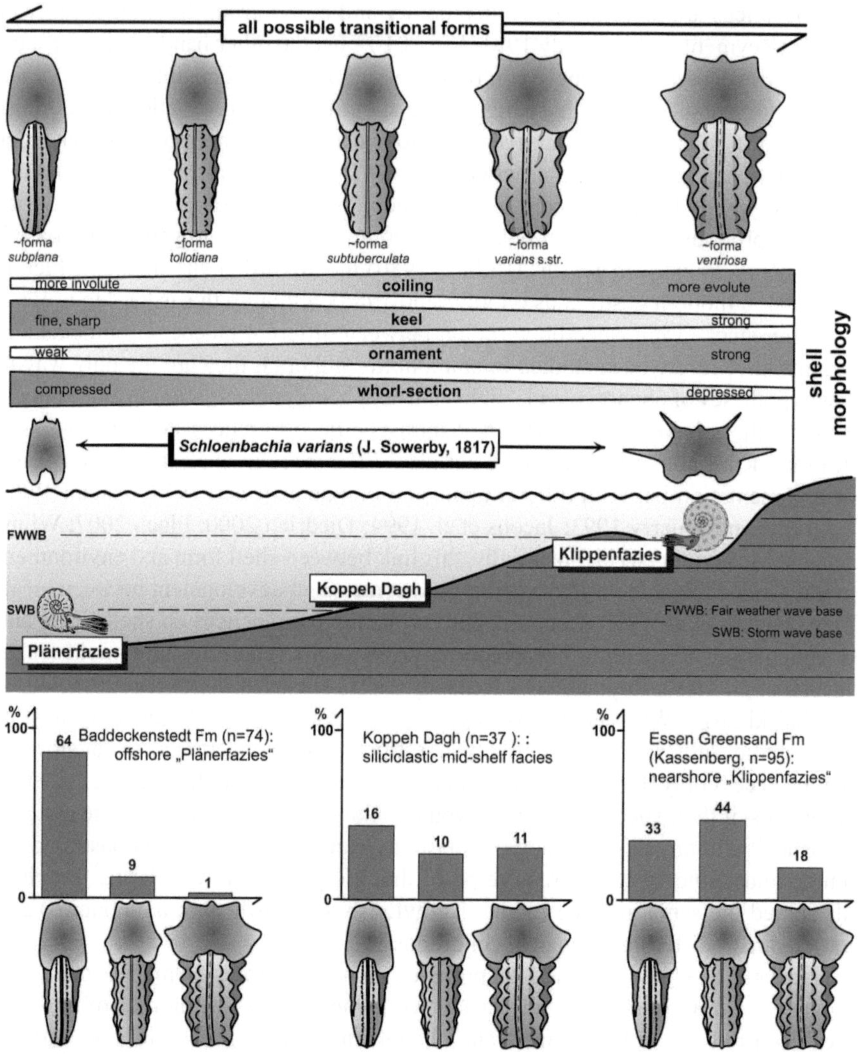

Fig. 9.16 Schematized distribution and semi-quantitative analysis of the relative proportion of morphological variants within contemporary samples of *Schloenbachia varians* (Cenomanian) from different paleoenvironments (modified from Wilmsen and Mosavinia 2011, with permission from the author)

and *gongilensis*). The situation might have been even more complicated as they also found a group (*inflatum*), which differed in ornamentation (branching of ribs and presence of furrows) and which they assigned to the same species; this species did not show this ecophenotypic variation (compare Reyment 2004 for a different point of view). Courville and Thierry (1993) speculated that the *inflatum* group might have been a geographic subspecies or a more complex case of extreme intraspecific variation as both groups were only occasionally found together in a transitional area between both regions, where these morphs are typical (compare also Reyment 2003). There might also be a link with development as the representatives of the species from different localities showed markedly different body sizes. The authors explained this by internal (genetic) factors as well as external constraints (water depth and energy, interactions with other organisms including predation and competition for food).

Complicating the situation, Wilmsen and Mosavinia (2011) reported the opposite pattern in *Schloenbachia varians*, with compressed, weakly ornamented morphs being more common in open (and deeper) waters and depressed, strongly tuberculate forms in shallower environments (Fig. 9.16). They attributed the dominance of strongly tuberculate, depressed forms in shallow, nearshore environments to higher water energy and predation pressure. They only studied this semi-quantitatively, but their histograms suggest non-unimodal and strongly skewed distributions in some paleoenvironments. A similar type of ecophenotypic variation was reported by Diedrich (2000) for *Pusozia*, where finely ribbed ecotypes are typical for ramp facies in the deeper shelf, more coarsely ribbed varieties for slope environments of the middle shelf and heavily ornamented forms for shallower submarine swells. He attributed the facies-dependence of shell morphology to the nektobenthic or demersal ecology of the ammonoids in question (similar claims have been made for facies-dependence of size: Reboulet 2001; Urlichs 2004). Even if a relationship between shell morphology and an environmental factor can be shown, such a link does not by itself prove that morphology is controlled by that factor (Yacobucci 2008). For example, given a correlation between shell shape and facies, the underlying cause of the differing shell shapes is certainly not the facies *per se*, but rather one or more associated factors, such as the water energy, predatory pressure, hydrostatic pressure, light, nutrient availability, geographical region, or stage of a sea-level cycle in which the ammonoids preferentially lived. These components might be difficult to disentangle in fossil "populations" as they are typically mixed in time (different seasons, years or decades to hundreds and thousands of years; natural and empiral time-averaging of shell assemblages) and space (bathymetrically, geographically, paleoenvironmentally; due to their pelagic lifestyle as well as potential post-mortem transport: Kennedy and Cobban 1976; Ritterbush et al. 2014; Naglik et al. 2015). Reyment and Kennedy (1991) and Kin (2010, 2011) discussed phenotypic plasticity as a possible reason for the large intraspecific variation in *Knemiceras* and *Acanthoscaphites*, respectively. They could not, however, provide evidence that particular phenotypic classes or morphologies were linked to different environmental conditions. An external influence on phenotype might also be present in Hemihoplitidae as Bert (2012) reported that more robust individuals are usually found on the edges of the platform, while these morphotypes are rare in the pelagic environment. Similar environmental factors might also have played a role for the evolution of *Gassendiceras* towards more slender morphologies over time (Bert and Bersac 2013); both patterns have still have to be studied quantitatively.

We can extend environmental variation to geographic variation (Bert 2013), which is a scaled-up version of local and regional environmental differences that often remain undetected (Westermann 1996) given the limits of stratigraphic and

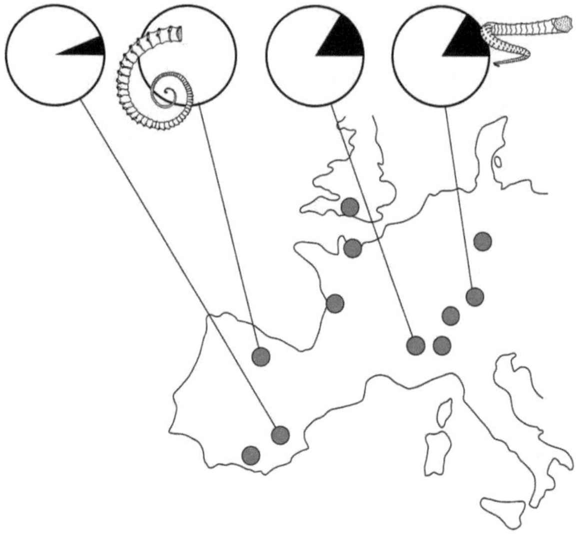

Fig. 9.17 Differences in coiling (planispiral: white; trochospiral: black) within contemporary populations of *Spiroceras orbignyi* (Late Bajocian) depending on the locality/paleoenvironment (modified from Dietl 1978)

spatial resolution in temporally and/or spatially mixed fossil populations. Besides geographic differences in adult size (Sect. 9.7), authors have also reported geographic differences in shell shape (planispiral vs. trochospiral coiling: Dietl 1978; thickness ratio of shell whorls: Ikeda and Wani 2012; Yahada and Wani 2013) and ornamentation (e.g., differences in ribbing and furrows: Courville and Thierry 1993; Olóriz et al. 2000). Other shell shape parameters and the shape of the suture line might also be involved, but they have to our knowledge not yet been adequately studied. Some of these patterns might also be related to differences in growth rate and length of development, which have also been reported (e.g., Mignot et al. 1993; Reboulet 2001; Ploch 2007). Dietl (1978) noted a reduced variation in coiling (e.g., lack of three-dimensionally coiled or trochospiral forms) with increasing geographic distance from the "optimal biotope" (Fig. 9.17). He interpreted the high variation in shell shape in *Spiroceras* as an indication of a benthic mode of life on algal meadows (compare also Westermann 1996). Other authors have reported differences in whorl thickness ratios of shells (whorl breadth/shell diameter) between different areas and/or paleoenvironments, both in the Jurassic (e.g., Tintant 1963, 1980) and Cretaceous (Ploch 2007; Ikeda and Wani 2012; Yahada and Wani 2013). Tintant (1963) described differences in whorl height and umbilical width at the same diameters between different samples of *Kepplerites gowerianus* deriving from different localities, but the same subzone. Tintant (1980) stated this pattern could potentially be related to geographic variation, but might equally reflect a slightly different age of these samples, as *Kepplerites* shows similar changes in umbilical width and whorl height over evolutionary time. Ploch (2007) wrote that the specimens of *Saynoceras verrucosum* from the Vocontian Basin are more inflated (e.g., showed relatively larger whorl height than whorl width) than specimens from the Polish Basin. Some authors have attributed differences in certain characters such as whorl thickness ratios of shells between areas and/or paleoenvironments (Ikeda and Wani 2012; Yahada and Wani 2013) as an indication that members of different populations did not frequently migrate between these two areas in later ontogeny. Interestingly, the early ontogeny was indistinguishable in these specimens as these differences only developed later in ontogeny (compare Ikeda and Wani 2012; Yahada and Wani 2013). Inherent problems with temporal resolution in fossil populations (even within taxa deriving from the same biostratigraphic unit) hamper the separation of evolutionary changes (microevolution: compare Bert 2013) over time from geographic variation at the same time within species and lineages, particularly in the case of high intraspecific variability (compare Kennedy and Cobban 1976; Tintant 1980; Landman et al. 2008).

Not only the mode of intraspecific variation might change, but also the range of intraspecific variation can vary through time and/or space. Bert (2004) reported, for example, that the range of variation within species of *Gregoryceras* increases from the Cordatum Subzone (Lower Oxfordian, Jurassic) to reach its maximum in the Luciaeformis Subzone (Middle Oxfordian) and decreases thereafter. In the Cretaceous ammonoid *Deshayesites fissicostatus,* the proportion of smoother shells was reported by Bersac and Bert (2012a) to vary between the Fissicostatus and Obsoletus subzones (Lower Aptian) in southern England (Casey 1963), probably for local ecological reasons. Monnet et al. (2012) also reported similar irregular fluctuations through time in the range of intraspecific variation for the Triassic *Acrochordiceras* that are not clearly related to sample size or facies changes. Wiese and Schulze (2005) reported that local/regional populations of *Neolobites vibrayeanus* were apparently morphologically stable, but that little morphological overlap occurred between individuals of geographically separated assemblages.

High levels of morphological variability within a species have been attributed to various ecological and developmental mechanisms, including selection for ecological generalists in an unstable environment (Simpson 1944; Levins 1968), inherent developmental plasticity (West-Eberhard 2003, 2005), and the absence of competitors in an empty ecosystem (Westermann 1966, 1996). Yacobucci (2004b) investigated the response of Cretaceous *Neogastroplites*, known for its notorious range of intraspecific variation in shell shape and ornamentation (Reeside and Cobban 1960), to the invasion of a potential competitor, *Metengonoceras*, in the Mowry Sea. A competitive interaction model would predict that a variable species would contract its variation when encountering a new competitor. However, *Neogastroplites* responded not by decreasing its morphological variation but by expanding into a previously unoccupied region of its morphospace. Yacobucci (2004b) attributed the variability of *Neogastroplites* to environmental instability or developmental flexibility. Tanabe and Shigeta (1987) explained differences in the range of variation of shell shape, from high in heavily ornamented and heteromorph forms, to intermediate in finely ribbed platycone forms, to small in weakly ornamented forms, to possible differences in the mode of life of these ammonoids (see also Westermann 1996; Naglik et al. 2015). In some time intervals like the Triassic (e.g., Dagys and Weitschat 1993a, b; Dagys et al. 1999; Dagys 2001; Weitschat 2008), extreme intraspecific variation appears to have been particularly concentrated in high-latitude “polar” regions (a phenomenon also seen in extant gastropods: Clarke 1978).

However, similar cases are known from more or less contemporary faunas from lower latitudes (Monnet et al. 2010) suggesting this interpretation might represent a preservational or collection bias. Geographic differences in size within species have often been attributed to the differences in environmental or geographic context (Mignot 1993; Mignot et al. 1993; Reboulet 2001; Ploch 2007). There also seems to be no clear link to endemism as a high range of intraspecific morphological variation is known not only from more endemic taxa (*Thomasites*: Courville and Thierry 1993) but also from more widespread, globally distributed taxa (*Acrochordiceras*: Monnet et al. 2010). Both phenomena still need to be studied more extensively and consistently.

9.9 Quantification, Analysis and Comparison of Intraspecific Variation

One must first quantify the degree and nature of the variation (by making measurements, calculating variance/standard deviation/coefficient of variation, making histograms that show the spread of the data, etc.) and then one can analyze the variation (e.g., looking for correlations among traits, tracking changes in variation through time and across space, etc.). Fundamentally, three types of quantitative methods are available to study intraspecific variation: univariate, bivariate and multivariate methods (e.g., Charpy and Thierry 1976; Bert 2013). Univariate methods (e.g., histograms, descriptive statistics) are typically employed to visualize and test the homogeneity of a sample restricted in time and/or space (Tintant 1980), but can also be used to test or plot intraspecific variation through ontogeny for various ontogenetic stages or size classes. Bivariate methods can be applied to test correlations between parameters and see changes in particular characters through ontogeny (De Baets et al. 2013a). Multivariate methods are commonly used to investigate the relative contribution of each measured character to the total variation of the sample and to group individuals by morphological or ontogenetic similarities considering all measured characters simultaneously (Charpy and Thierry 1976; Bert 2013).

9.9.1 Univariate and Bivariate Methods

Intraspecific variation of a single measurable character is often simply visualized by using a box-and-whisker plot for certain size classes or ontogenetic stages (e.g., Korn and Klug 2007; Monnet et al. 2010; Fig. 9.10, 9.18). However, such graphs do not show whether the specimens show a uni-, bi-, or multimodal distribution. As explained above, one might intuitively expect a unimodal distribution if the specimens belong to the same taxon with continuous variation (e.g., Monnet et al. 2010), whereas discontinuous variation or polymorphism might result in multimodality for certain characters. A particular case might be sexual dimorphism, where bimodality at the end of ontogeny can be expected (e.g., Palframan 1966, 1967).

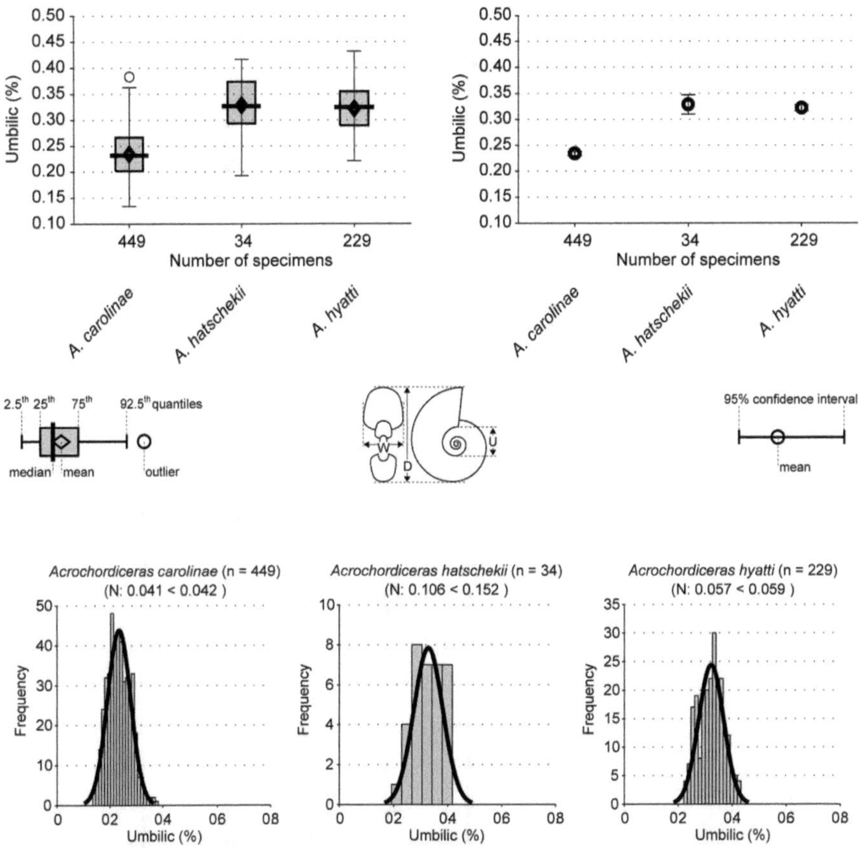

Fig. 9.18 Univariate tools used to study the range and distribution of intraspecific variation of the umbilical width with three species of *Acrochordiceras* (Middle Triassic; modified from Monnet et al. 2010): boxplots (upper left) and confidence intervals on the mean (upper right), which give only limited information on the density distribution of the data; histograms (bottom) illustrating the density distribution of intraspecific variation; note that not all specimens pass the normality test

Frequency distributions and (multi)modality of single traits are usually depicted with histograms, that is, a graphical representation of the data distribution that bins data into discrete intervals (Pearson 1895; Hammer and Harper 2006; Foote and Miller 2007; Fig. 9.18, 9.19). A histogram therefore only approximates the distribution of a variable (Pearson 1895). Histograms have certain disadvantages because they are discontinuous, have a fixed bin width, and are dependent on bin size and origin (Salgado-Ugarte et al. 2000). An alternative to the histogram is kernel density estimation, which uses a kernel to smooth samples. This approach will construct a smooth probability density function, which will in general more accurately reflect the underlying variable. In contrast to histograms, kernel density estimators are smoother and continuous, and allow for easier recognition of outliers, skewness,

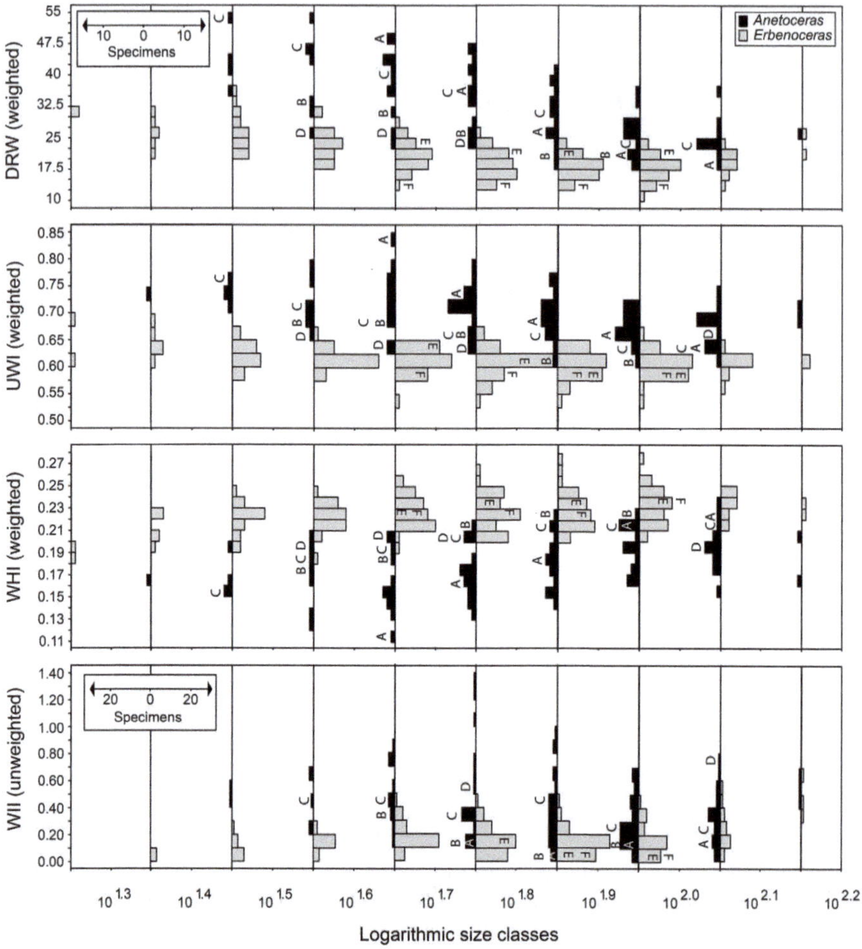

Fig. 9.19 Comparison of density distributions between two taxa using histograms over various size classes (modified from De Baets et al. 2013a)

and multimodality (Sanvicente-Añorve et al. 2003). A beanplot (Fig. 9.20, 9.21) is a combination of a 1-D scatterplot and a density trace (Kampstra 2008). In the 1-D scatterplot, each measurement is represented by a line, which makes it easy to spot outliers. If multiple measurements have the same value, the individual lines are added together increasing the length of the line. Beanplots also clearly show whether values are rounded or discontinuous. An alternative to the beanplot is the violin plot (Fig. 9.21), which is a combination of a box plot and a kernel density plot. It is probably best to investigate the frequency distribution and multimodality of ammonoid shell characters by combining several methods (Fig. 9.21).

Testing for a normal distribution is not only important to see if the sample is homogenous and that specimens might belong to the same species, but also for further

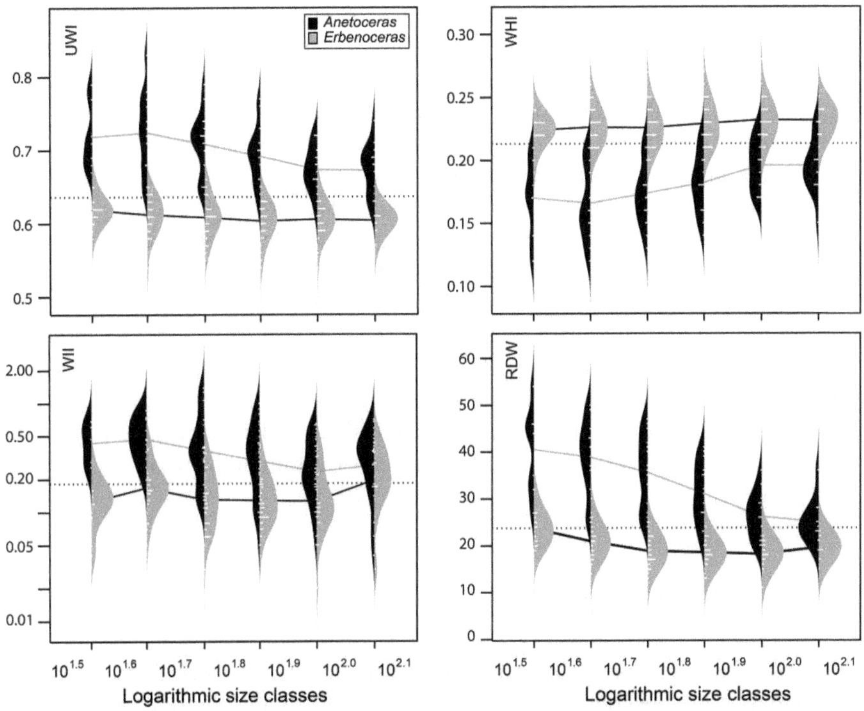

Fig. 9.20 Comparison of density distributions between two taxa using beanplots over various size classes. Note the unimodal distribution of the group in grey, while the second group shows a polymodal distribution, which might be due to a low sample size (modified from De Baets et al. 2013a)

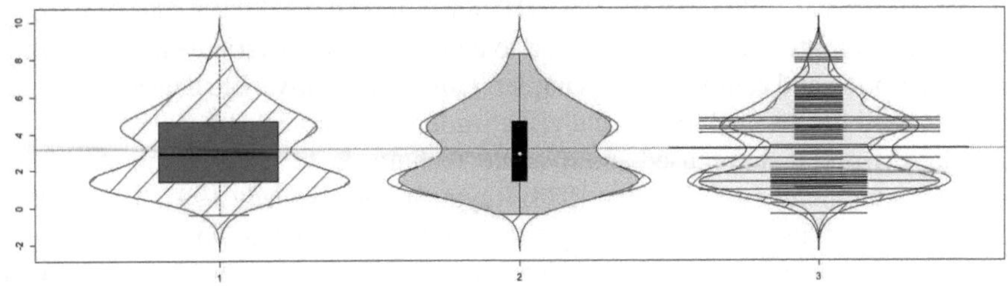

Fig. 9.21 Comparisons of a bimodal distribution using a boxplot (density distribution in the back for comparison), violin plot and beanplots (all graphs produced with R using packages UsingR, Vioplot and Beanplot)

statistical analyses as many (parametric) statistical tests assume a normal distribution (e.g., Hammer and Harper 2006; Monnet et al. 2010). Distribution-free or non-parametric statistical methods exist, but they often have a lower statistical power than parametric ones (Hammer and Harper 2006). For the purposes of analysis, a normal distribution can also be achieved by conducting various transformations of the raw data (e.g., Bert 2013). Although normal distribution determines the types of statistical methods that can be applied to the data and is expected for "biological" species, ammonoid species often deviate from normality "simply" by allometric changes and/or adult morphological changes (see e.g., Monnet and Bucher 2005; Fig. 9.18) and it is therefore not a strict prerequisite.

Bivariate methods are useful to test correlations between two parameters or to see changes in particular characters during growth by plotting the character against the diameter or another measure (e.g., De Baets et al. 2013a; Bert 2013; Fig. 9.9). Linear relationships between two parameters can be tested with a Pearson correlation coefficient (Pearson 1901) and non-linear monotonic correlations with a Spearman rank correlation coefficient (Spearman 1904; compare Sprent 1989). The relationship between two parameters can also be evaluated by fitting a model to the data. The model is often linear (e.g., Type I regression, reduced major axis regression) but need not be; for instance, exponential and logarithmic models may be more biologically appropriate. Analysis of residuals (that is, deviations from the fitted model) can reveal changes in the degree of variation that exists at different shell sizes.

9.9.2 Multivariate Methods

Often with ammonoids several different measurements are taken from one specimen and to study the collected data simultaneously requires multivariate methods. With the increasing power of computer processors that allows analysis of very large datasets, multivariate statistics have become indispensable analytical tools in paleontology. Multivariate techniques can be considered as an extension of univariate and bivariate techniques into multidimensional space, and many univariate and bivariate tests have multivariate analogs (Hammer and Harper 2006). The purpose of many multivariate approaches is similar to that of uni- and bivariate ones, involving the description and comparison of samples. However, multivariate data analysis can also be used for the exploration and visualization of complex data.

Many workers have used multivariate techniques to analyze morphometric data in ammonoids (e.g., Thierry 1978; Reyment and Kennedy 1991, 1998, 2000; Kassab and Hamama 1991; Hohenegger and Tatzreiter 1992; Reyment and Minaka 2000; Reyment 2003, 2004, 2011; Kennedy et al. 2009; Bert 2013; see Hammer and Harper 2006 for a more general review). Some of these methods have helped to discriminate species in ammonoids, to investigate overlap of morphological variants and intraspecific variation (including dimorphism) in specimens from a single locality and stratigraphic interval, or to understand how the morphological diversity of a sample is structured and how it is located with respect to other samples (e.g. factorial planes; the concept of morphological space of Neige et al. 1997b). It is also possible to avoid bias due to taxonomic classifications and analyze shape disparity among specimens directly (morphodiversity), through the use of shape parameters or landmarks (geometric morphometric analysis: see e.g., Neige and Dommergues 1995; Reyment and Kennedy 1991, 1998). For ammonoids, Saunders and Swan 1984 as well as Swan and Saunders 1987 performed some of the earliest morphospace analyses using ordination techniques (although these have sometimes been criticized: see Reyment and Kennedy 1991). Multivariate analysis can be a powerful tool for interpreting shape changes in terms of variability, evolution or paleoecology.

Using multivariate methods does require selecting appropriate data transformations and standardizations (Kenkel 2006). An appropriate multivariate analytical strategy should take into account the statistical relevance, data structure and the objectives of the study. Therefore, before proceeding with a formal multivariate analysis, it is important to complete a detailed exploratory analysis of the data (e.g. univariate and bivariate analysis).

Many multivariate analyses are rooted in ordination techniques. Ordination reduces a multivariate dataset to fewer, uncorrelated axes that capture most of the variation contained in the original data (Hammer and Harper 2006; Kenkel 2006). The ordination results can then be used to assess the relative contribution of each variable to each axis and to the overall variance structure of the data. In other terms, ordination gives an image (a morphological map) of the total measured variation of a sample, depending on the characters selected, and gives a summative model of the underlying data structure, by means of a projection of a point cloud of n-dimensional space onto a biplot (e.g., two-dimensional space defined by the first two ordination axes). Ordination methods actually used in biology are derived from Pearson's Factorial Analysis (Pearson 1901). The classical Principal Components Analysis (PCA, see Jolliffe 2002 for a review; compare Bert 2013) is one of the multivariate ordination methods most often applied to ammonoids (e.g., Reyment and Kennedy 1991; Kassab and Hamama 1991; Reyment and Minaka 2000; Reyment 2004; Bert 2013). Its purpose is to describe the total variance of a sample with the smallest possible number of factors, taking into account all the variables. Therefore, the PCA identifies the axes of maximum variance (the principal components, PC) in order to preserve as much variance as possible through the data compression process. With measurement data, the first axis (PC1), which contains the largest portion of the overall variation in the data, is typically interpreted as capturing variation in size, while the other axes express variation in shape (Hammer and Harper 2006). The specimens can then be projected onto a biplot, typically of PC1 vs. PC2, PC1 vs. PC3 or PC2 vs. PC3, with a minimum of distortion, so that two individuals that are similar morphologically will fall close to each other in the projected point cloud. The convex hulls (i.e., the limits of the point cloud) produced correspond to the occupation of the morphological space of the sample.

Disparity between several samples (or individuals) can be tested using analysis of similarity methods (one-way ANOSIM, MANOVA, etc: compare Hammer and Harper 2006; Bert 2013), which are based on comparing a measure of distance between groups with distances within groups. Here the homoscedasticity of the variable and the normal Gaussian distribution of the data (depending on the method) have to be tested. The results are given in a matrix of probability and graphically investigated also by ordination methods such as Canonical Variate Analysis (CVA: Ter Braak 1986; see Reyment 1998, 2003 for examples) or Discriminant Analysis

(Fisher 1936; see Hohenegger and Tatzreiter 1992 for an example), which closely resembles PCA but produces factors (axes) that capture the maximum difference (instead of maximum variance for PCA) between predefined groups (e.g., species) and can help identify variables (characters), which contribute to these differences. These methods can be used, for example, to test for the presence of significant morphological differences between samples. Other methods are also available to isolate deviating specimens (ecomorphs) in a genetically homogeneous sample (compare Reyment 2004).

Despite their effectiveness, one of the main problems with multivariate analyses is that, as with most other statistical methods, they loose robustness with decreasing sample size. Hence, large samples are required as much as possible. This critical threshold is usually set to at least 32 specimens (see Tintant 1963; Bert 2013), which could be problematic when abundant and well-preserved material is scarce. When the sample is small, the risk is also high that the specimens studied do not represent the full range of variation of the whole population. This issue is however less problematic when the statistics are used only for a comparative and/or graphical purpose (Reyment and Kennedy 1991). Multivariate ordination methods also involve some loss of information, which in practice will likely not hamper the analysis. However, if the percentage of total variance explained by the first few ordination axes is low (<70–90 % depending on the analysis and the number of first "few" ordination axes retained for the analysis; see Joliffe 2002; Bert 2013), special caution is needed when analyzing the results. Another source of error lies with imperfections in the data, such as the inclusion of aberrant specimens or specimens with measurement errors that can produce misleading results. Such outliers have to be detected with other methods (e.g., a bivariate analysis is usually sufficient) and removed before the analysis. Problems with the data can also occur with the choice of the variables included into the analysis. Of particular concern is the inclusion of redundant characters. While it is impossible for all morphometric variables to be independent of each other (i.e., uncorrelated), if only because most measurements increase with size, efforts should be taken to minimize the redundancy of the variables so that certain aspects of shell form are not overemphasized in the resulting data set.

9.9.3 Comparing the Range of Intraspecific Variation

In order to understand and interpret morphological variation within ammonoid species, it is useful to quantify and compare the range of intraspecific variation among different samples or species. We will focus here on continuous variation (alternative methods are available for discrete variation that cannot be approximated by continuous distributions: Van Valen 2005). Extreme values (e.g., minima and maxima) are very sensitive to sample size, as is the total range of variation, so these metrics should be avoided in most cases unless sample sizes are high. The robustness of extreme values can be assessed by bootstrapping under conditions of different sampling densities (Monnet et al. 2010). The variance and its square root, the standard deviation, are most suitable for comparing intraspecific variation in single variables.

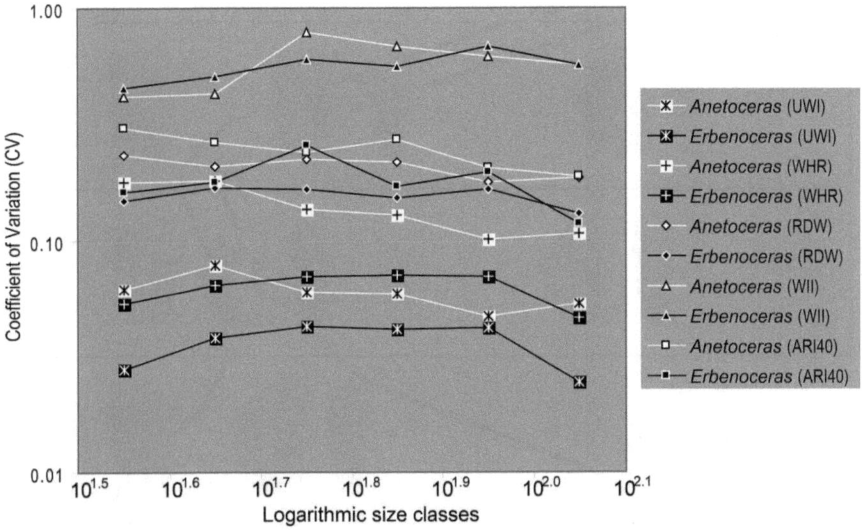

Fig. 9.22 Comparisons of the coefficient of variation of various characters between six ontogenetic (size) classes and two taxa (*Anetoceras obliquecostatum* and *Erbenoceras solitarium*, Early Devonian (modified from De Baets et al. 2013a)

The standard deviation is usually divided by the mean or "normalized" to obtain the coefficient of variation (CV). The coefficient of variation is a relative measure and should always be used "*when the mean differs enough to matter*" (Van Valen 2005). The CV is commonly used to compare the range of variation in ammonoids (e.g., Neige 1997; Parent 1998; Rouget and Neige 2001; Joly 2003; Tanabe et al. 2003; Yacobucci 2004a; De Baets et al. 2013a). It has the advantage that the range of variation of different characters and taxa can be directly compared with one another (e.g., De Baets et al. 2013a; Fig. 9.22). It can also be generalized for multivariate cases (Van Valen 1978, 2005). Van Valen (2005) reviewed the statistics of variation and suggested Levene's test (Levene 1960), Smith's Test (published in Grüneberg et al. 1966) and jackknifing (Arvesen and Schmitz 1970; Miller 1974; Bissell and Ferguson 1975), which can be used to compare absolute and relative variation, to be most suitable, depending on the situation (Van Valen 1978, 2005). He advised against using the classical F-test for the equality of variances as it is very sensitive to non-normality (Van Valen 2005).

A problem related to the pronounced variation seen in ammonoids is that successive faunas separated in time or contemporary faunas separated in space can only be compared on the basis of the mode and range of intraspecific variation within populations (Reeside and Cobban 1960; Kennedy and Cobban 1976; Dagys 2001; De Baets et al. 2013a; Fig. 9.23). This might also lead to small samples of intergrading populations, which can be considered to belong to different taxa, obscuring the synchronicity and identity of faunas (Kennedy and Cobban 1976). When only one specimen is available, it makes no sense to test if the mean of this population differs significantly from that of another population. Some approximate this by testing if

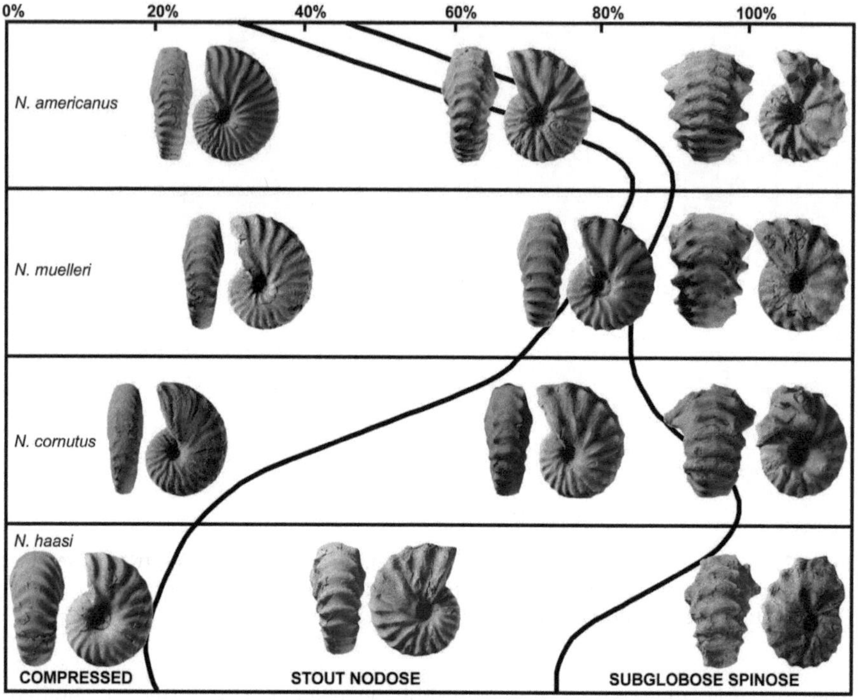

Fig. 9.23 Intraspecific variation and relative proportion of forms within four successive species of the Early Cenomanian hoplitid ammonoid *Neogastroplites* (modified from Kennedy and Cobban 1976: with permission from the authors). As first documented by Reeside and Cobban (1960), specimens recovered from single concretions within the Mowry Shale of the American Western Interior show extreme variation in shell shape and ornamentation. Reeside and Cobban (1960) subdivided this continuum of variation into three morphs, whose relative proportions vary from species to species. Sample sizes per concretion used to calculate percentages: *N. haasi*, *N*=333; *N. cornutus*, *N*=2471; *N. muelleri*, *N*=3708; *N. americanus*, *N*=1286. Specimen photos from Reeside and Cobban (1960): N. haasi from USGS Mesozoic Locality 24566: compressed, USNM 129308 (Plate 11, Fig. 4, 6); stout nodose, AMNH 28098:11 (Plate 13, Fig. 14, 15); subglobose spinose, AMNH 28095:25 (Plate 15, Fig. 12, 15). *N. cornutus* from USGS Mesozoic Locality 23021: compressed, USNM 129320a (Plate 5, Fig. 1, 2); stout nodose, USNM 129320f (Plate 5, Fig. 16, 17); subglobose spinose, USNM 129320n (Plate 5, Fig. 40, 41). *N. muelleri* from USGS Mesozoic Locality 24065: com-pressed, USNM 129416f (Plate 6, Fig. 16, 17); stout nodose, USNM 129416j (Plate 6, Fig. 28, 29); subglobose spinose, USNM 129416q (Plate 6, Fig. 49, 50). *N. americanus* from USGS Mesozoic Locality 23042: compressed, USNM 129528a (Plate 7, Fig. 1, 2); stout nodose, USNM 129528f (Plate 7, Fig. 16, 17); subglobose spinose, USNM 129528l (Plate 7, Fig. 34, 35)

the measurement falls in the 95 % range of that of the other population, which is actually a different question, or by assuming that the variance of one population is similar to that of the other, in which case the problem is reduced to comparing the means of both populations (Van Valen 2005). To accurately capture this variation, a sufficient number of specimens (>32: Tintant 1963, Bert 2013) needs to be collected from an interval restricted in time and space, which might also depend on the amount of variation observed within the population. If intraspecific variation is studied through ontogeny, a sufficient number of specimens needs to be available for each ontogenetic stage or size class that shall be used (De Baets et al. 2013a; Fig. 9.20, 9.22), which might mean a prohibitively large sample is required.

9.10 Conclusions and Future Perspectives

Intraspecific variation in shell shape, ornamentation, size and suture line is quite common in ammonoids and can in some cases be larger within a single fossil assemblage than differences in morphology over space and/or time within these lineages. Some authors, like Lehmann (1990, p. 23), have stated that intraspecific variation in ammonoids does not follow general laws or rules, but it is evident from our review that many ammonoid workers agree that at least some generalizations or recurrent patterns can be recognized. Nevertheless, the type and degree of intraspecific variation can be highly different from taxon to taxon, which is not unexpected considering the large differences in shell shape of ammonoids and intraspecific variation in extant cephalopods. Two main types of intraspecific variation are commonly recognized, which are not necessarily mutually exclusive:

- Continuous intraspecific variation in shell shape, ornamentation, suture line and size and
- Discontinuous intraspecific variation in the above mentioned aspects (dimorphism and polymorphisms).

Continuous intraspecific variation has mainly been studied from the perspective of covariation between shell shape, ornamentation and more rarely studied more often lines as well as the perspective of variation in ontogenetic development. The few studies using a large amount of material and employing quantitative methods have particularly focused on Mesozoic coiled ammonoids showing extensive ranges of intraspecific variation (e.g., Reeside and Cobban 1960, Dagys and Weitschat 1993b; Checa et al. 1996; Morard and Guex 2003; Weitschat 2008; Monnet et al. 2010), but a large range of intraspecific variation also seems to be present in at least some Paleozoic taxa (e.g., Klug and Korn 2007; De Baets et al. 2013a) and Mesozoic heteromorph ammonoids (e.g., Bert 2013; Knauss and Yacobucci 2014).

The presence of discrete morphologies within a species in the form of intraspecific sexual dimorphism and non-sexual polymorphism has been suggested based on polymodal distributions in shell shape, ornamentation and/or size in late and/or early ontogeny in contemporary specimens. Nevertheless, studies show that the paleoenvironment (Wilmsen and Mosavinia 2011), taphonomic biases (Tintant 1980) and undersampling (De Baets et al. 2013a) can also lead to polymodal density distributions in forms of the same age and region with continuous intraspecific variation.

Heritable (genetic) variation is the raw material for evolution, but a larger part of the intraspecific variation seen in ammonoids might be related to differences in growth rates and ontogenetic trajectories (Urdy et al. 2010a, b). The large intraspecific variation documented in many ammonoids might have had a partially intrinsic component, but a large part might be linked with the interplay between developmental and environmental parameters. This is corroborated by the range of intraspecific variation in size and shell morphology observed at the same localities and between different paleoenvironments and regions. The large range of intraspecific variation within and between ammonoid populations has typically led to an artificial inflation of diversity (oversplitting) by using strict typological approaches. But mixing together specimens from different regions and ages (analytical time-averaging) can also lead to an artificial deflation of diversity (lumping). Intraspecific variability is often not properly taken into account or quantitatively analyzed, which is detrimental not only to systematic and evolutionary studies, but also to biostratigraphic, biogeographic and diversity studies.

Intraspecific variability in Paleozoic and heteromorph taxa as well as Mesozoic coiled taxa showing little intraspecific variation should be quantitatively studied more often to better understand the type and range of intraspecific variation as well as possible relations with shell morphology. Previous studies can also be further refined when more material, more ontogenetic stages or a finer stratigraphic framework becomes available. Various statistical methods are available and easily accessible through free software (PAST: Hammer et al. 2001; The R Project for Statistical Computing: http://www.r-project.org/), which makes it possible to analyze interpopulation and intrapopulation variation within ammonoid species quantitatively. New approaches to quantifying morphological variation that are rooted in spatial statistics can also be applied to many ammonoid groups (Manship 2004, 2008; Yacobucci and Manship 2011; Knauss and Yacobucci 2014). A better understanding of the types and drivers of intraspecific variation in ammonoids can only be achieved by quantitative analyses of numerically large samples from a wide variety of ages, paleoenvironments, geographic areas, taxa and shell morphologies.

Acknowledgments Some of the insights described in this chapter grew during the course of research projects 200021-113956/1, 200020-25029, and 200020-132870 funded by the Swiss National Science Foundation SNF. David Ware (Zürich) and Isabelle Rouget (Paris) helped with obtaining some of the literature. Markus Wilmsen (Senckenberg Natural History Collections, Dresden), Gene Hunt (Smithsonian Institution, Washington, USA) and Michał Zatoń (University of Silesia, Sosnowiec) kindly put figures at our disposal. Jerzy Dzik (Institute of Paleobiology, Polish Academy of Sciences, Warsaw), Helga Weitschat on behalf of Wolfgang Weitschat (Geological–Paleontological Institute and Museum, University of Hamburg, retired), Vasily Mitta (Paleontological Institute, Russian Academy of Sciences, Moscow), Lionel Cavin (Natural History Museum of Geneva) on behalf of the journal Revue de Paléobiologie, and Jim Kennedy (University of Oxford, retired) gave permission to use their figures. We thank the reviewers Michał Zatoń (University of Silesia, Sosnowiec) and Sonny A. Walton (Naturkunde Museum, Berlin) for the constructive comments and suggestions. We would like to dedicate this chapter to the inspiring work of Algirdas Dagys and Wolfgang Weitschat on intraspecific variation in Triassic ammonoids.

References

Adams DC, Collyer ML (2009) A general framework for the analysis of phenotypic trajectories in evolutionary studies. Evolution Int J org Evolution 63:1143–1154

Ager DV (1963) Principles of paleoecology. McGraw Hill, New York

Aguirre-Urreta MB (1998) The ammonites *Karakaschiceras* and *Neohoploceras* (Valanginian Neocomitidae) from the Neuquen basin, west-central Argentina. J Paleontol 72:39–59

Aguirre-Urreta MB, Riccardi AC (1988) Albian heteromorph ammonoids from southern Patagonia, Argentina. J Paleontol 62:598–614

Andrew C, Howe P, Paul CRC, Donovan SK (2011) Epifaunal worm tubes on Lower Jurassic (Lower Lias) ammonites from Dorset. Proc Geol Assoc 122:34–46

Arkell WJ (1957) Introduction to Mesozoic Ammonoidea. In: Moore RC (ed) Treatise on invertebrate paleontology, Part L, Mollusca 4, Cephalopoda-Ammonoidea. GSA and University of Kansas Press, L80–L100

Arkhipkin A (1992) Reproductive system structure, development and function in cephalopods with a new general scale for maturity stages. J Northw Atl Fish Sci 12:63–74

Arkhipkin AI (2004) Diversity in growth and longevity in short-lived animals: squid of the suborder Oegopsina. Mar Freshw Res 55:341–355

Arkhipkin A, Laptikhovsky V (1994) Seasonal and interannual variability in growth and maturation of winter-spawning *Illex argentinus* (Cephalopoda, Ommastrephidae) in the Southwest Atlantic. Aquat Living Resour 7:221–232

Arvesen JN, Schmitz TH (1970) Robust procedures for variance component problems using the jackknife. Biometrics 26:677–686

Atrops F, Mélendez G (1993) Current trends in systematics of Jurassic Ammonoidea: the case of Oxfordian-Kimmeridgian perisphinctids from southern Europe. Geobios 26(Suppl 1):19–31. doi:http://dx.doi.org/10.1016/S0016-6995(06)80357-8

Aubrecht R, Schlögl J (2011) Jurassic submarine troglobites: is there any link to the recent submarine cave fauna? Hydrobiologia 677:3–14

Bailey RC, Byrnes J (1990) A new, old method for assessing measurement error in both univariate and multivariate morphometric studies. Syst Biol 39:124–130

Barber WM (1957) The Lower Turonian ammonites of northeastern Nigeria. Bull Geol Surv Nigeria 26:1–86

Bardhan S, Jana SK, Datta K (1993) Preserved color pattern of a phylloceratid ammonoid from the Jurassic Chari Formation, Kutch, India, and its functional significance. J Paleontol 67:140–143

Bardhan S, Jana SK, Roy P (2010) Sexual dimorphism and polymorphism in a Callovian *Phlycticeras* (Ammonoidea) assemblage of Kutch, India. Geobios 43:269–281

Baudouin C, Boselli P, Bert D (2011) The Oppeliidae of the Acanthicum zone (Upper Kimmeridgian) from Mount Crussol (Ardèche, France): ontogeny, variability and dimorphism of the genera *Taramelliceras* and *Streblites* (Ammonoidea). Rev Paleobiol 30:619–684

Baudouin C, Bert D, Boselli P (2012) Preview on the ontogeny, variability and dimorphism of the genera *Taramelliceras* and *Streblites* (Ammonoidea) of the Acanthicum zone (Upper Kimmeridgian) from Mount Crussol (Ardèche, France). Bol Inst Fisiog Geol 82:19–21

Bayer U, McGhee GR Jr (1984) Iterative evolution of middle Jurassic ammonite faunas. Lethaia 17:1–16

Bayer U, McGhee GR Jr (1985) Evolution in marginal epicontinental basins: the role of phylogenetic and ecological factors. Ammonite replacements in the German Lower and Middle Jurassic. In: Bayer U, Seilacher A (eds) Sedimentary and evolutionary cycles. Springer, Berlin

Bert D (2004) Révision, etude systématique et evolution du genre *Gregoryceras* Spath, 1924 (Ammonoidea, Oxfordien). Ann Mus Hist Nat Nice 19:1–184

Bert D (2009) Discussion, evolution and new interpretation of the *Tornquistes* Lemoine, 1910 (Pachyceratidae, Ammonitina) with the exemple of the verte-brale subzone sample (Middle Oxfordian) of southeastern France. Rev Paleobiol 28:471–489

Bert D (2012) Phylogenetic relationships among the Hemihoplitidae Spath 1924 (Ammonoidea, Upper Barremian). Boletin del Instituto de Fisiografia y. Geologia 82:17–18

Bert D (2013). Factors of intraspecific variability in ammonites, the example of *Gassendiceras alpinum* (d'Orbigny, 1850) (Hemihoplitidae, Upper Barremian). Annales de Paléontologie doi:10.1016/j.annpal.2013.11.007

Bersac S, Bert D (2012a) Ontogenesis, variability and evolution of the Lower Greensand Deshayesitidae (Ammonoidea, Lower Cretaceous, Southern England): reinterpretation of literature data; taxonomic and biostratigraphic implications. Ann Mus Hist Nat Nice 27:197–270

Bersac S, Bert D (2012b) Variability and evolution of the Deshayesitidae (Ammonoidea, Lower Aptian, Lower Cretaceous) from southern England. Bol Inst Fisiog Geol 82:27–30

Bert D, Bersac S (2013) Evolutionary patterns-tested with cladistics-and pro-cesses in relation to palaeoenvironments of the Upper Barremian genus *Gassendiceras* (Ammonitina, Lower Cretaceous). Palaeontology 56:631–646

Bert D, Delanoy G, Bersac S (2011) The dichotomus horizon: a new biochronologic unit of the Giraudi zone of the Upper Barremian of southeastern France, and considerations regarding the genus *Imerites* Rouchadze (Ammonoidea, Gassendiceratinae). Carnets Geol 2011/01: http://paleopolis.rediris.es/cg/CG2011_A01/

Beznosov NV, Mitta VV (1995) Polymorphism in the Jurassic ammonoids. Paleontol J 29:46–57

Bhaumik D, Datta K, Jana-Sudipta K, Bardhan S (1993) Taxonomy and intraspecific variation of *Macrocephalites formosus* (Sowerby) from the Jurassic Chari Formation, Kutch, western India. J Geol Soc India 42:163–179

Bissell A, Ferguson R (1975) The jackknife-toy, tool or two-edged weapon? The Statistician:79–100

Blake JF (1878) On the measurements of curves formed by cephalopods and other mollusks. Philsoph Mag 5:241–262

Boletzky Sv (1974) Effets de la sous-nutrition prolongée sur le développement de la coquille de *Sepia officinalis* L. (Mollusca, Cephalopoda). Bull Soc Zool Fr 99:667–673

Boletzky Sv (2003) Biology of early life stages in cephalopod molluscs. Adv Mar Biol 44:143–203

Bonnot A, Marchand D, Neige P (1999) Les Oppeliidae (Ammonitina) de l'horizon à Collotiformis (Callovien supérieur, zone à Athleta) de la région Dijonnaise (Côte-d'Or, France). Annales de Paléontologie 85:241–263

Bookstein FL, Ward PD (2013) A modified procrustes analysis for bilaterally symmetrical outlines, with an application to microevolution in *Baculites*. Paleobiology 39:214–234

Boyle P, Ngoile M (1993) Population variation and growth in *Loligo forbesi* (Cephalopoda: Loliginidae) from Scottish waters. In: Okutani T, O'Dor RK, Kubodera T (eds) Recent advances in cephalopod fisheries biology. Tokai University Press, Tokyo

Boyle PR, von Boletzky S (1996) Cephalopod populations: definition and dynamics. Philos Trans R Soc B-Biol Sci 351(1343):985–1002. doi:10.2307/56291

Brayard A, Escarguel G (2013) Untangling phylogenetic, geometric and ornamental imprints on Early Triassic ammonoid biogeography: a similarity-distance decay study. Lethaia 46:19–33

Brayard A, Escarguel G, Bucher H, Monnet C, Brühwiler T, Goudemand N, Galfetti T, Guex J (2009) Good genes and good luck: ammonoid diversity and the end-permian mass extinction. Science 325:1118–1121

Brochwicz-Lewinski W, Rozak Z (1976) Some difficulties in recognition of sexual dimorphism in Jurassic perisphinctids (Ammonoidea). Acta Palaeontol Polonica 21:115–124

Bucher H (1997) Caractères périodiques etmode de croissance des ammonites: Comparaison avec les gastéropodes. Geobios 30(Suppl 1):85–99

Bucher H, Landman NH, Klofak SM, Guex J (1996) Mode and rate of growth in ammonoids. In: Landman NH, Tanabe K, Davis RA (eds) Ammonoid paleobiology. Plenum, New York

Buckman SS (1887-1907) A monograph of the ammonites of the inferior oolite series. Palaeontogr Soc 40-61:1–456

Bulmer MG (1980) The mathematical theory of quantitative genetics. Oxford Science Publications, Oxford

Callomon JH (1963) Sexual dimorphism in Jurassic ammonites. Trans Leicester Lit Philos Soc 57:21–56
Callomon J (1985) The evolution of the Jurassic ammonite family Cardioceratidae. Spec Pap Palaeontol 33:49–90
Callomon JH (1988) Review of Matyja 1986. Cephalopod Newsletter 9:14–16
Casey R (1961) A monograph of the Ammonoidea of the Lower Greensand, part II. Palaeontogr Soc Lond 493:45–118
Casey R (1963) A monograph of the Ammonoidea of the Lower Greensand, part V. Palaeontogr Soc Lond 502:289–398
Chandler R, Callomon J (2009) The inferior oolite at Coombe quarry, near Mapperton, Dorset, and a new Middle Jurassic ammonite faunal horizon, Aa-3b, *Leioceras comptocostosum* n. biosp. in the Scissum zone of the Lower Aalenian. Proc Dorset Nat Hist Archaeol Soc 130:99–132
Charpy N, Thierry J (1976) Dimorphisme et polymorphisme chez *Pachyceras* Bayle (Ammonitina, Stephanocerataceae) du Callovien Supérieur (Jurassique Moyen). Haliotis 6:185–218
Checa A, Company M, Sandoval J, Weitschat W (1996) Covariation of morpho-logical characters in the Triassic ammonoid *Czekanowskites rieberi*. Lethaia 29:225–235
Chlupáč I, Turek V (1983) Devonian goniatites from the Barrandian area. Rozpr Ustred Ust Geol 46:1–15
Clarke JM (1899) The Naples fauna (fauna with *Manticoceras intumescens*) in western New York. New York State Museum. Annu Rep Regents 50:31–161
Clarke AH (1978) Polymorphism in marine mollusks and biome development. Smithson Contrib Zool 274:1–14
Collyer ML, Adams DC (2013) Phenotypic trajectory analysis: comparison of shape change patterns in evolution and ecology. Hystrix 24:75–83
Contini D, Marchand D, Thierry J (1984) Reflexions sur la notion de genre et de sous-genre chez les Ammonites: exemples pris essentiellement dans le Jurassique moyen. Bull Soc Geol Fr 26:653–666
Courville P (1993) Les formations marines et les faunes d'ammonites cénomaniennes et turoniennes (Crétacé supérieur) dans le Fosséde la Bénoué (Nigéria). Impacts des facteurs locaux et globaux sur les échanges fauniques à l'interface Téthys/ Atlantique Sud. Unpubl PhD Thesis, Univ de Dijon, p. 360
Courville P (2011) Caractères ornementaux, disparité et diversité chez les Ammonitina: exemple des Kosmoceratinae (Stephanoceratoidea), Callovien moyen et supérieur (Jurassique moyen, Bassin parisien). C R Palevol 10:155–170
Courville P, Crônier C (2003) Les hétérochronies du développement: un outil pour l'étude de la variabilité et des relations phylétiques: Exemple de *Nigericeras*, Ammonitina du Crétacé supérieur africain. C R Palevol 2:535–546
Courville P, Crônier C (2005) Diversity or disparity in the Jurassic (Upper Callovian) Genus *Kosmoceras* (Ammonitina): a morphometric approach. J Paleontol 79:944–953
Courville P, Lebrun P (2010) L'Albien (Crétacé) de la region de Troyes (Aube) et ses ammonites: Hoplitidae et Douvilleiceratidae. Fossiles 4:4–30
Courville P, Thierry J (1993) Sous-espèces géographiques et/ou contrôle environnemental de la variabilité morphologique chez "*Thomasites*" *gongilensis* (Woods, 1911), (Ammonitina, Acanthocerataceae, Vascoceratinae) du Turonien inférieur de la Haute Bénoué (Nigéria). Geobios 26(Suppl 1):73–89
Crick RR (1978) Morphological variations in the ammonite *Scaphites* of the Blue Hill member, Crlile Shale, Upper Cretaceous. Univ Kans Paleontol Contrib 88:1–30
Dagys AS (2001) The ammonoid family Arctohungaritidae from the boreal Lower-Middle Anisian (Triassic) of arctic Asia. Rev Paleobiol 20:543–546
Dagys AS, Weitschat W (1993a) Intraspecific variation in Boreal Triassic ammonoids. Geobios 26:107–109
Dagys AS, Weitschat W (1993b) Extensive intraspecific variation in a Triassic ammonoid from Siberia. Lethaia 26:113–121

Dagys AS, Bucher H, Weitschat W (1999) Intraspecific variation of *Parasibirites kolymensis* Bychkov (Ammonoidea) from the Lower Triassic (Spathian) of arctic Asia. Mitt aus dem Geol-Paläont Inst Universität Hamburg 83:163–178

Darwin CR (1859) On the origin of species by means of natural selection, or the preservation of favoured races in the struggle for life. John Murray, London

Davis RA, Landman NH, Dommergues J-L, Marchand D, Bucher H (1996) Mature modifications and dimorphism in ammonoid cephalopods. In: Landman NH, Tanabe K, Davis RA (eds) Ammonoid paleobiology. Plenum, New York

De Baets K, Klug C, Monnet C (2013a) Intraspecific variability through ontogeny in early ammonoids. Paleobiology 39:75–94

De Baets K, Klug C, Korn D, Bartels C, Poschmann M (2013b) Emsian Ammonoidea and the age of the Hunsrück Slate (Rhenish Mountains, Western Germany). Palaeontogr A 299:1–113

De Baets K, Keupp H, Klug C (2015a) Parasites of ammonoids. This volume

De Baets K, Landman NH, Tanabe K (2015b) Ammonoid embryonic development. This volume

De Beer G (1958) Evolution by natural selection: a centenary commemorative volume. Papers by Charles Darwin and Alfred Wallace. Cambridge University, Cambridge

Delanoy G (1997) Biostratigraphie des faunes d'Ammonites à la limite Barrémien-Aptien dans la région d'Angles-Barrême-Castellane. Étude particulière de la Famille des Heteroceratidae Spath 1922 (Ancyloceratina, Ammonoidea). Ann Mus Hist Nat Nice 12:1–270

Delanoy G, Ropolo P, Magnin A, Autran G, Poupon A, Gonnet R (1995) Sur le dimorphisme chez les Ancyloceratina (Ammonoidea) du Crétacé Inférieur. C R Acad Sci Ser IIa 321:537–543

Dera G, Neige P, Dommergues J-L, Brayard A (2011) Ammonite paleobiogeography during the Pliensbachian-Toarcian crisis (Early Jurassic) reflecting paleoclimate, eustasy, and extinctions. Glob Planet Change 78:92–105

Diedrich C (2000) Faziesabhängige Schalenmorphologie des Großammoniten *Puzosia dibleyi* (Spath 1922) aus dem Puzosia-Event I (Ober-Cenoman) von Europa. Senckenb Lethaea 80:463–483

Dietl G (1978) Die heteromorphen Ammoniten des Dogger. Stuttg Beitr Natur B 33:1–97

Dietze V, Callomon JH, Schweigert G, Chandler RB (2005) The ammonite fauna and biostratigraphy of the Lower Bajocian (Ovale and Laeviuscula zones) of E Swabia (S Germany). Stuttg Beitr Natur B353:1–82

Doguzhaeva L (1982) Rhythms of ammonoid shell secretion. Lethaia 15:385–394

Dommergues J-L (1988) Can ribs and septa provide an alternative standard for age in ammonite ontogenetic studies? Lethaia 21:243–256

Dommergues J-L, David B, Marchand D (1986) Les rélations ontogenèse-phylogenèse: applications paléontologiques. Geobios 19:335–356

Dommergues J-L, Cariou E, Contini D, Hantzpergue P, Marchand D, Meister C, Thierry J (1989) Homéomorphies et canalisations évolutives: Le rôle de l'ontogenèse. Quelques exemples pris chez les Ammonites du Jurassique. Geobios 22:5–48

Dommergues J-L, Montuire S, Neige P (2002) Size patterns through time: the case of the early Jurassic ammonite radiation. Paleobiology 28:423–434

Dommergues E, Dommergues J-L, Dommergues C-H (2006) Deux espèces sous un même masque. Le point de vue paléontologique piégé par les coquilles de deux espèces européennes de *Trivia* (Mollusca, Gastropoda). Rev Paleobiol 25:775–790

Donovan DT (1994) History of classification of Mesozoic ammonites. J Geol Soc 151:1035–1040

Dzik J (1985) Typologic versus population concepts of chronospecies: implications for ammonite biostratigraphy. Acta Palaeontol Pol 30:71–92

Dzik J (1990a). The concept of chronospecies in ammonites. In: Cecca F, Cresta S, Pallini G, Santantonio M (eds) Atti del Secondo Convegno Inter-nazionale Fossili, Evoluzione, Ambiente, Pergola 25-30 ottobre 1987 estratto, Pergola, Comitato Centenario Raffaele Piccinini

Dzik J (1990b) The ammonite *Acrochordiceras* in the Triassic of Silesia. Acta Palaeontol Pol 35:49–65

Dzik J (1994) Sexual dimorphism in the virgatitid ammonites. Palaeopelagos Spec Publ 1:129–141

Ebbighausen V, Korn D (2007) Conch geometry and ontogenetic trajectories in the triangularly coiled Late Devonian ammonoid *Wocklumeria* and related genera. Neues Jahrb Geol Paläontol Abh 244:9–41

Egojan VL (1969) Ammonites from the Clanseysian beds of the western Caucasus. Trud Krasnodar Fil Vses Neftegazov Nauchnoissledovatel'sk Inst 19:126–188 [in Russian]

Elmi S, Benshili K (1987) Relations entre la structuration tectonique, la composition des peuplements et l'évolution; exemple du Toarcien du Moyen-Atlas méridional (Maroc). Boll Soc Paleontol Ital 26:47–62

Erben HK (1950) Bemerkungen zu Anomalien mancher Anfangswindungen von *Mimagoniatites* fecundus (Barr.). Neues Jahrb Geol Paläont Mh:25–32

Erben HK (1964) Die Evolution der ältesten Ammonoidea (Lieferung I). Neues Jahrb Geol Paläontol Abh 120:107–212

Flessa KW, Kowalewski M (1994) Shell survival and time-averaging in nearshore and shelf environments: estimates from the radiocarbon literature. Lethaia 27:153–165

Fernández-López S (1995) Taphonomie et interpretation des paléoenvironements. Géobios 18:137–154

Fernández-López S (2000) Temas de Tafonomía. Departamento de Paleontología. Universidad Complutense de Madrid, Madrid

Fisher RA (1936) The use of multiple measurements in taxonomic problems. Ann Eugen 7:179–188

Flessa KW, Cutler AH, Meldahl KH (1993) Time and taphonomy: quantitative estimates of time-averaging and stratigraphic disorder in a shallow marine habitat. Paleobiology 19:266–286

Foote M, Miller AI (2007) Principles of paleontology. Freeman, New York

Ford EB (1940) Polymorphism and taxonomy. In: Huxley JS (ed) The new systematics. Oxford University, Oxford

Ford EB (1945) Polymorphism. Biol Rev 20:73–88

Ford EB (1955) Polymorphism and taxonomy. Heredity 9:255–264

Ford EB (1965) Genetic polymorphism. Faber and Faber, London

Ford EB (1966) Genetic polymorphism. Proc R Soc B-Biol Sci 164:350–361

Fürsich FT, Aberhan M (1990) Significance of time-averaging for palaeocommunity analysis. Lethaia 23:143–152

Furnish WM, Knapp WD (1966) Lower Pennsylvanian fauna from eastern Kentucky; Part 1, Ammonoids. J Paleontol 40:296–308

Fusco G, Minelli A (2010) Phenotypic plasticity in development and evolution: facts and concepts. Philos Trans R Soc B-Biol Sci 365:547–556

Gangopadhyay TK, Bardhan S (2007) Ornamental polymorphism in *Placenticeras kaffrarium* (Ammonoidea; Upper Cretaceous of India): evolutionary implications. In: Landman N, Davis R, Mapes R (eds) Cephalopods present and past: new insights and fresh perspectives. Springer, Netherlands

Göddertz B (1989) Unterdevonische hercynische Goniatiten aus Deutschland, Frankreich und der Türkei. Palaeontogr A 208:61–89

Goodfriend GA (1986) Variation in land-snail shell form and size and its causes: a review. Syst Biol 35:204–223

Grüneberg H, Bains GS, Berry BJ, Riles L, Smith C, Weiss R (1966) A search for genetic effects of high natural radioactivity in south India. Spec Rep Ser Med Res Counc 307:1–59

Guex J (2001) Environmental stress and atavism in ammonoid evolution. Eclogae Geol Helv 94:321–328

Guex J (2003) A generalization of Cope's rule. Bull Soc Geol Fr 174:449–452

Guex J, Koch A, O'Dogherty L, Bucher H (2003) A morphogenetic explanation of Buckman's law of covariation. Bull Soc Geol Fr 174:603–606

Haas O (1946) Intraspecific variation in, and ontogeny of, *Prionotropis woollgari* and *Prionocyclus wyomingensis*. Bull Am Mus Nat Hist 86(4):141–224

Hallam A (1965) Environmental causes of stunting in living and fossil marine benthonic invertebrates. Palaeontology 8:132–155

Hallgrímsson B, Hall BK (2005) Variation: a central concept in biology. Elsevier, Amsterdam
Hammer Ø, Bucher H (2005) Buckman's first law covariation-a case of proportionality. Lethaia 38:67–72
Hammer Ø, Harper DAT (2006) Paleontological data analysis. Wiley-Blackwell, United Kingdom
Hammer Ø, Bucher H (2006) Generalized ammonoid hydrostatics modelling, with application to *Intornites* and intraspecific variation in *Amaltheus*. Paleontol Res 10:91–96
Hammer Ø, Harper DAT, Ryan PD (2001) PAST: paleontological statistics software package for education and data analysis. Palaeontol Electron 4:A4
Hengsbach R (1976) Über die Sutur-Assymetrie bei *Cymbites laevigatus* (Ammonoidea; Jura). Senckenb Lethaea 56:463–468
Hengsbach R (1980) Über die Sutur-Asymmetrie bei *Hecticoceras* (Ammonoidea; Jura). Senckenb Lethaea 60:463–473
Hengsbach R (1986) Zur Kenntnis der Asymmetrie der Sutur-Asymmetrie bei Ammoniten. Senckenb Lethaea 67:119–149
Hewitt RA, Hurst JM (1977) Size changes in Jurassic liparoceratid ammonites and their stratigraphical and ecological significance. Lethaia 10:287–301
Hewitt RA, Stait B (1988) Seasonal variation in septal spacing of *Sepia officinalis* and some Ordovician actinocerid nautiloids. Lethaia 21:383–394
Hewitt RA, Checa A, Westermann GEG, Zaborski PM (1991) Chamber growth in ammonites inferred from colour markings and naturally etched surfaces of Cretaceous vascoceratids from Nigeria. Lethaia 24:271–287
Hewitt RA, Westermann GEG, Checa A (1993) Growth rates of ammonites estimated from aptychi. Geobios 26(Suppl 1):203–208
Hirano H (1978) Phenotypic substitution of *Gaudryceras* (a Cretaceous ammonite). Trans Proc Palaeontol Soc Jpn New Ser 109:235–258
Hirano H (1979) Importance of transient polymorphism in systematics of Ammonoidea. Gakujutsu Kenkyu Sch Educ Waseda Univ Ser Biol Geol 28:35–43
Hirano H (1981) Growth rates in *Nautilus macromphalus* and ammonoids: its implications. In: Martinell J (ed) International symposium on conceptions and methods in paleontology. University of Barcelona, Barcelona
Hoffmann R, Keupp H, Wiese F (2009) The systematic position of the Lower Cretaceous heteromorphic ammonite *Pictetia* Uhlig, 1883. Paläontol Z 83:521–531
Hohenegger J, Tatzreiter F (1992) Morphometric methods in determination of ammonite species, exemplified through *Balatonites* shells (Middle Triassic). J Paleontology 66:801–816
Hölder H (1956) Über Anomalien an jurassischen Ammoniten. Paläontol Z 30:95–107
Hoving H-JT, Gilly WF, Markaida U, Benoit-Bird KJ, Brown ZW, Daniel P, Field JC, Parassenti L, Liu B, Campos B (2013) Extreme plasticity in life-history strategy allows a migratory predator (jumbo squid) to cope with a changing climate. Glob Change Biol 19:2089–2103
Howarth MK (1973) The stratigraphy and ammonite fauna of the Upper Liassic grey shales of the Yorkshire coast. Bull Br Mus (Nat Hist) Geol 24:235–277
Howarth MK (1978) The stratigraphy and ammonite fauna of the Upper Lias of Northamptonshire. Bull Br Mus (Nat Hist) Geol 29:235–288
Hughes NC, Labandeira CC (1995) The stability of species in taxonomy. Paleobiology 21:401–403
Hunt G (2004a) Phenotypic variation in fossil samples: modeling the consequences of time-averaging. Paleobiology 30:426–443
Hunt G (2004b) Phenotypic variance inflation in fossil samples: an empirical assessment. Paleobiology 30:487–506
Hunt G (2006) Fitting and comparing models of phyletic evolution: random walks and beyond. Paleobiology 32:578–601
Hunt G (2007) Variation and early evolution. Science 317:459–460
Ikeda Y, Wani R (2012) Different modes of migration among late cretaceous ammonoids in northwestern Hokkaido, Japan: evidence from the analyses of shell whorls. J Paleontol 86:605–615
Ivanov AN (1971a) Problems of the periodization of ontogeny in ammonites. Yarosl Ped Inst Uch Zap geol i paleont 87:76–119

Ivanov AN (1971b) On the problem of periodicity of the formation of septa in ammonoid shells and in that of other cephalopods. Yarosl Ped Inst Uch Zap geol i paleont 87:127–130

Ivanov AN (1975) Late ontogeny in ammonites and its characteristics in micro-, macro- and megaconchs. Yarosl Ped Inst Sb Nauchn Trudy 142:5–57

Jackson GD (1994) Application and future potential of statolith increment analysis in squids and sepioids. Can J Fish Aquat Sci 51:2612–2625

Jackson G, Moltschaniwskyj N (2002) Spatial and temporal variation in growth rates and maturity in the Indo-Pacific squid *Sepioteuthis lessoniana* (Cephalopoda: Loliginidae). Mar Biol 140:747–754

Jacobs DK (1990) Sutural pattern and shell stress in *Baculites* with implications for other cephalopod shell morphologies. Paleobiology 16:336–348

Jacobs DK, Landman NH, Chamberlain JA (1994) Ammonite shell shape co-varies with facies and hydrodynamics: iterative evolution as a response to changes in basinal environment. Geology 22:905–908

Jolliffe IT (2002) Principal component analysis. Springer, Netherlands

Joly B (2003) L'évolution chez les Phylloceratacea, la variabilité des paramètres dimensionnels et relatifs. Variabilité de la complexité de la ligne cloisonnaire: Variabilité et paedomorphose. C R Pale 2:231–240

Joly B, Fonters B (2007) Morphotypes, polymorphism and peristome in the species of the genus *Holcophylloceras* Spath, 1927. Hypothesis of the dimorphism in the species *Holcophylloceras zignodianum* (d'Orbigny, 1848). Bull Soc Geol Fr 178:217–229

Jordan R, Stahl W (1971) Isotopische Paläotemperatur-Bestimmungen an Jurassischen Ammoniten und grundsätzliche Voraussetzungen für diese Methode. Geol Jahrb 89:33–62

Kakabadze MV (2004) Intraspecific and intrageneric variabilities and their implication for the systematics of Cretaceous heteromorph ammonites; a review. Scr Geol 128:17–37

Kampstra P (2008) Beanplot: a boxplot alternative for visual comparison of distributions. J Stat Softw 28, Code Snippet 1.

Kant R (1973a) Allometrisches Wachstum paläozoischer Ammonoideen: Variabilität und Korrelation einiger Merkmale. Neues Jahrb Geol Paläontol Abh 143:153–192

Kant R (1973b) Untersuchungen des allometrischen Gehäusewachstums paläozoischer Ammonoideen unter besonderer Berücksichtigung einzelner "Populationen.". N Jahrb Geol Paläont Abh 144:206–251

Kant R (1975) Biometrische Untersuchungen an Ammonoideen-Gehäusen. Paläontol Z 49:203–220

Kaplan P (1999) Buckman's rule of covariation and other trends in Paleozoic Ammonoidea: morphological integration as key innovation. GSA 31:172

Kassab AS, Hamama HH (1991) Polymorphism in the upper Cretaceous ammonite *Libycoceras ismaeli* (Zittel). J Afr Earth Sci (Middle East) 12:437–448

Kawabe F (2003) Relationship between Mid-Cretaceous (upper Albian-Cenomanian) ammonoid facies and lithofacies in the Yezo forearc basin, Hokkaido, Japan. Cretac Res 24:751–763

Kenkel NC (2006) On selecting an appropriate multivariate analysis. Can J Plant Sci 86:663–676

Kennedy WJ (1972) The affinities of *Idiohamites ellipticoides* Spath (Cretaceous Ammonoidea). Palaeontology 15:400–404

Kennedy WJ (2013) On variation in *Schloenbachia varians* (J. Sowerby, 1817) from the lower Cenomanian of western Kazakhstan. Acta Geol Pol 63:443–446

Kennedy WJ, Cobban WA (1976) Aspects of ammonite biology, biogeography, and biostratigraphy. Spec Pap Palaeontol 17:1–94

Kennedy WJ, Hancock JM (1970) Ammonites of the genus *Acanthoceras* from the Cenomanian of Rouen, France. Palaeontology 13:462–490

Kennedy WJ, Wright CW (1985) Evolutionary patterns in Late Cretaceous ammonites. Spec Pap Palaeont 33:131–143

Kennedy WJ, Reyment RA, MacLeod N, Krieger J (2009) Species discrimination in the Lower Cretaceous (Albian) ammonite genus Knemiceras Von Buch 1848. Palaeontogr A 290:1–63

Keupp H (2000). Ammoniten: paläobiologische Erfolgsspiralen. Thorbecke, Stuttgart

Keupp H (2012) Atlas zur Paläopathologie der Cephalopoden. Berl Palaeobiol Abh 12:1–392

Keupp H, Hoffmann R (2015) Ammonoid paleopathology. This volume

Keupp H, Mitta V (2013) Cephalopod jaws from the Middle Jurassic of Central Russia. Neues Jahrb Geol Paläontol Abh 270:23–54

Keyl F, Argüelles J, Tafur R (2011) Interannual variability in size structure, age, and growth of jumbo squid (*Dosidicus gigas*) assessed by modal progression analysis. ICES J Mar Sci J Cons 68:507–518

Kidwell SM (1998) Time-averaging in the marine fossil record: overview of strategies and uncertainties. Geobios 30:977–995

Kidwell SM (2002) Time-averaged molluscan death assemblages: palimpsests of richness, snapshots of abundance. Geology 30:803–806

Kidwell SM, Bosence DWJ (1991) Taphonomy and time-averaging of marine shelly faunas. In: Allison PA, Briggs DEG (eds) Taphonomy: releasing the data locked in the fossil record. Plenum, New York

Kin A (2010) Early Maastrichtian ammonites and nautiloids from Hrebenne, southeast Poland, and phenotypic plasticity of *Acanthoscaphites tridens* (Kner, 1848). Cretaceous Res 31:27–60

Kin A (2011) Phenotypic plasticity of *Acanthoscaphites tridens* (Late Cretaceous ammonites): additional data. Cretaceous Res 32:131–134

Klingenberg CP (1996) Multivariate allometry. In: Marcus LF et al (eds) Advances in morphometrics. Plenum, New York

Klinger HC, Kennedy WJ (1989) Cretaceous faunas from Zululand and Natal, South Africa. The ammonite family Placenticeratidae hyatt, 1900; with comments on the systematic position of the genus *Hypengonoceras* Spath, 1924. Ann S Afr Mus 98:241–408

Klug C, Riegraf W, Lehmann J (2012) Soft-part preservation in heteromorph ammonites from the Cenomanian-Turonian boundary event (OAE 2) in north-west Germany. Palaeontology 55:1307–1331

Klug C, Zatoń M, Parent H, Hostettler B, Tajika A (2015) Mature modifications and sexual dimorphism. This volume

Knauss MJ, Yacobucci MM (2014) Geographic information systems technology as a morphometric tool for quantifying morphological variation in an ammonoid clade. Palaeontol Electron 17:19A

Korn D (1995) Impact of environmental perturbations on heterochronic development in Palaeozoic ammonoids. In: McNamara KJ (ed) Evolutionary change and heterochrony. Wiley, Chichester

Korn D, Klug C (2007) Conch form analysis, variability, morphological disparity, and mode of life of the Frasnian (Late Devonian) ammonoid *Manticoceras* from Coumiac (Montagne Noire, France). In: Landman NH, Davis RA, Mapes RH (eds) Cephalopods-present and past: new insights and fresh perspectives. Springer, Dordrecht

Korn D, Vöhringer E (2004) Allometric growth and intraspecific variability in the Basal Carboniferous ammonoid *Gattendorfia crassa* Schmidt, 1924. Paläontol Z 78:425–432

Kowalewski M (2009) The youngest fossil record and conservation biology: holocene shells as eco-environmental recorders. In: Dietl GP, Flessa KW (eds) Conservation paleobiology: using the past to manage for the future. Paleontological Society, New Haven

Kraft S, Korn D, Klug C (2008) Patterns of ontogenetic septal spacing in Carboniferous ammonoids. Neues Jahrb Geol Paläontol Abh 250:31–44

Kruta I, Landman NH, Tanabe K (2015) Ammonoid radulae. This volume

Krystyn L, Schäffer G, Schlager W (1971) Über die Fossil-Lagerstätten in den triadischen Hallstätter Kalken der Ostalpen. Neues Jahrb Geol Paläontol Abh 137:284–304

Kulicki C (1974) Remarks on the embryogeny and postembryonal development of ammonites. Acta Palaeontol Pol 19:201–224

Kummel B (1948) Environmental significance of dwarfed cephalopods. J Sediment Res 18:61–64

Landman NH (1987) Ontogeny of Upper Cretaceous (Turonian-Santonian) scaphitid ammonites from the western interior of North America: systematics, developmental patterns, and life history. Bull Am Mus Nat Hist 185:117–241

Landman NH, Geyssant JR (1993) Heterochrony and ecology in Jurassic and Cretaceous ammonites. Geobios 26(Suppl 1):247–255

Landman NH, Waage KM (1986) Shell abnormalities in scaphitid ammonites. Lethaia 19:211–224

Landman NH, Waage KM (1993) Scaphitid ammonites of the Upper Cretaceous (Maastrichtian) Fox Hills formation in South Dakota and Wyoming. Bull Am Mus Nat Hist 215:1–257

Landman N, Klofak SM, Sarg KB (2008) Variation in adult size of scaphitid ammonites from the Upper Cretaceous Pierre Shale and Fox Hills formation. In: Harries PJ (ed) High-resolution approaches in stratigraphic paleontology. Springer, Netherlands

Landman NH, Kennedy WJ, Cobban WA, Larson NL (2010) Scaphites of the "Nodosus Group" from the Upper Cretaceous (Campanian) of the Western Interior of North America. Bull Am Mus Nat Hist 342:1–242

Landman NH, Cobban WA, Larson NL (2012) Mode of life and habitat of scaphitid ammonites. Geobios 45:87–98

Lange W (1929) Zur Kenntnis des Oberdevons am Enkeberg und bei Balve (Sauerland). Abh Preuss Geol Landesanst NF 119:1–132

Lange W (1941) Die Ammonitenfauna der Psiloceras-Stufe Norddeutschlands. Palaeontogr A 93:1–186

Laptikhovsky V (2006) Latitudinal and bathymetric trends in egg size variation: a new look at Thorson's and Rass's rules. Mar Ecol 27:7–14

Laptikhovsky VL, Rogov MA, Nikolaeva SE, Arkhipkin AI (2013) Environmental impact on ectocochleate cephalopod reproductive strategies and the evolutionary significance of cephalopod egg size. Bull Geosci 88:83–94

Lawrence E (2000) Henderson's dictionary of biological terms. Pearson, Essex

Lécuyer C, Bucher H (2006) Stable isotope compositions of a late Jurassic ammonite shell: a record of seasonal surface water temperatures in the southern hemisphere? eEarth 1:1–7

Lehmann U (1981) The ammonites: their life and their world. Cambridge University, New York

Lehmann U (1990) Ammonoideen. Enke, Stuttgart

Leporati S, Pecl G, Semmens J (2007) Cephalopod hatchling growth: the effects of initial size and seasonal temperatures. Mar Biol 151:1375–1383

Levene H (1960) Robust test for the equality of variances. In: Olkin I, Ghurye SG, Hoeffding W, Madow WG, Mann HB (eds) Contributions to probability and statistics: essays in honor of harold hotelling. Stanford University, Stanford

Levins R (1968) Evolution in changing environments. Princeton University, Princeton, p 120

Machalski M (2010) Early Maastrichtian ammonites and nautiloids from Hrebenne, southeast Poland, and phenotypic plasticity of *Acanthoscaphites tridens* (Kner, 1848): a commentary. Cretaceous Res 31:593–595

Makowski H (1962) Problem of sexual dimorphism in ammonites. Palaeontol Pol 12:1–92

Makowski H (1991) Dimorphism and evolution of the goniatite *Tornoceras* in the Famennian of the holy cross mountains. Acta Palaeontol Pol 36:241–254

Mancini EA (1978) Origin of micromorph faunas in the geologic record. J Paleontol 52:311–322

Manger WL, Meeks LK, Stephen DA (1999) Pathologic gigantism in middle Carboniferous cephalopods, southern midcontinent, United States. In: Olóriz F, Rodríguez-Tovar FJ (eds) Advancing research on living and fossil cephalopods. Kluwer Academic/Plenum, New York

Mangold K (1983) Food, feeding and growth in cephalopods. Mem Natl Mus Vic 44:81–93

Manship LL (2004) Pattern matching: classification of ammonitic sutures using GIS. Palaeontol Electron 7(6A):1–15

Manship LL (2008) Variation analysis of ammonites and conodonts (implementing Geographic Information Systems): a qualitative and quantitative method. Texas Tech University, Lubbock

Mapes RH, Larson NL (2015) Colour patterns. This volume

Mapes RH, Sneck DA (1987) The oldest 'colour' patterns: description, comparison with *Nautilus*, and implications. Palaeontology 30:299–309

Marchand D (1976) Quelques précisions sur le polymorphisme dans la famille des Cardioceratidae Douville (Ammonoidea). Haliotis 6:119–140

Matsunaga T, Maeda H, Shigeta Y, Hasegawa K, Nomura S-I, Nishimura T, Misaki A, Tanaka G (2008) First discovery of *Pravitoceras sigmoidale* Yabe from the Yezo supergroup in Hokkaido, Japan. Paleontol Res 12:309–319
Matyja BA (1986) Developmental polymorphism in Oxfordian ammonites. Acta Geol Pol 36:37–67
Matyja BA (1994) Developmental polymorphism in the Oxfordian ammonite subfamily Peltoceratinae. Palaeopelagos Spec Publ 1:277–286
Matyja BA, Wierzbowski A (2000) Biological response of ammonites to changing environmental conditions: an example of Boreal *Amoeboceras* invasions into Submediterranean province during Late Oxfordian. Acta Geol Pol 50:45–54
Mayr E (1963) Animal species and evolution. Belknap of Harvard University , Cambridge
McCaleb JA (1968) Lower Pennsylvanian ammonoids from the Bloyd formation of Arkansas and Oklahoma. GSA Special Papers 96:1–118
McCaleb JA, Furnish WM (1964) The Lower Pennsylvanian ammonoid genus *Axinolobus* in the southern Midcontinent. J Paleontol 38:249–255
McCaleb JA, Quinn JH, Furnish WM (1964) Girtyoceratidae in the southern Midcontinent. Okla Geol Surv Circ 67:1–41
Meischner D (1968) Perniciöse Epökie von *Placunopsis* auf *Ceratites*. Lethaia 1:156–174
Meister C (1989) Les ammonites du Crétacé supérieur d'Ashaka (Nigéria). Bull Centres Rech Explor-Prod Elf-Aquitaine 13(Suppl):1–84
Meléndez G, Fontana B (1993) Intraspecific variability, sexual dimorphism, and non-sexual polymorphism in the ammonite genus *Larcheria* Tintant (Perisphinctidae) from the middle Oxfordian of western Europe. In: House MR (ed) The Ammonoidea: environment, ecology, and evolutionary change. Clarendon, Oxford
Merkt J (1966) Über Austern und Serpeln als Epöken auf Ammonitengehäusen. Neues Jahrb Geol Paläontol Abh 125:467–479
Michalsky AO (1890) Ammonites of the Lower Volgian stage. Tr Geol kom-ta St. Petersburg 8:361–369 [in Russian]
Mignot Y (1993) Un problème de paléobiologie chez les ammonoides (Cephalopoda): croissance et miniaturisation en liaison avec les environnements. Doc Lab Geol Lyon 124:1–113
Mignot Y, Elmi S, Dommergues J-L (1993) Croissance et miniaturisation de quelques *Hildoceras* (Cephalopoda) en liaison avec des environnements contraignants de la Téthys Toarcienne. Geobios 26(Suppl 1):305–312
Miller RG (1974) The jackknife-a review. Biometrika 61:1–15
Mitta VV (1990) Intraspecific variability in the Volgian ammonites. Paleontol J 1990:10–15
Monnet C, Bucher H (2005) New middle and late Anisian (Middle Triassic) ammonoid faunas from northwestern Nevada (USA): taxonomy and biochronology. Fossils Strata 52:1–121
Monnet C, Brack P, Bucher H, Rieber H (2008) Ammonoids of the middle/late Anisian boundary (Middle Triassic) and the transgression of the prezzo limestone in eastern Lombardy-Giudicarie (Italy). Swiss J Geosci 101:61–84
Monnet C, Bucher H, Wasmer M, Guex J (2010) Revision of the genus *Acrochordiceras* Hyatt, 1877 (Ammonoidea, Middle Triassic): morphology, biometry, biostratigraphy and intraspecific variability. Palaeontology 53:961–996
Monnet C, De Baets K, Klug C (2011a) Parallel evolution controlled by adaptation and covariation in ammonoid cephalopods. BMC Evol Biol 11:115
Monnet C, Klug C, Goudemand N, De Baets K, Bucher H (2011b) Quantitative biochronology of Devonian ammonoids from Morocco and proposals for a refined unitary association method. Lethaia 44:469–489
Monnet C, Bucher H, Guex J, Wasmer M (2012) Large-scale evolutionary trends of Acrochordiceratidae Arthaber, 1911 (Ammonoidea, Middle Triassic) and Cope's rule. Palaeontology 55:87–107
Monnet C, De Baets K, Yacobucci MM (2015a) Buckman's rules of covariation. In: Klug C, Korn D, De Baets K, Kruta I, Mapes RH (eds) Ammonoid paleobiology: from macroevolution to paleogeography. Springer, Dordrecht

Monnet C, Klug C, De Baets K (2015b) Evolutionary patterns of ammonoids: phenotypic trends, convergence, and parallel evolution. In: Klug C, Korn D, De Baets K, Kruta I, Mapes RH (eds) Ammonoid paleobiology: from macroevolution to paleogeography. Springer, Dordrecht

Morard A (2004) Les événements du passage Domérien-Toarcien entre Thétys occidentale et Europe du Nord-Ouest. 1–338. Thèse de Doctorat, Université de Lausanne, Lausanne

Morard A (2006) Covariation patterns in ammonoids: observations, models, and open questions. Proceedings of the 4th Swiss Geoscience Meeting, Bern

Morard A, Guex J (2003) Ontogeny and covariation in the Toarcian genus *Osperleioceras* (Ammonoidea). Bull Soc Geol Fr 174:607–615

Naglik C, Tajika A, Chamberlain JA, Klug C (2015) Ammonoid locomotion. This volume

Nardin E, Rouget I, Neige P (2005) Tendencies in paleontological practice when defining species, and consequences on biodiversity studies. Geology 33:969–972. doi:10.1130/g21838.1

Neige P (1997) Ontogeny of the Oxfordian ammonite *Creniceras renggeri* from the Jura of France. Eclogae Geol Helv 90:605–616

Neige P, Dommergues J-L (1995) Morphometric and phenetics versus cladistic analysis of the early Harpoceratinae (Pliensbachian ammonites). Neues Jahrb Geol Paläontol Abh 196:411–438

Neige P, Marchand D, Laurin B (1997a) Heterochronic differentiation of sexual dimorphs among Jurassic ammonite species. Lethaia 30:145–155

Neige P, Chaline J, Chone T, Courant F, David B, Dommergues J-L, Laurin B, Madon C, Magniez-Jannin F, Marchand D, Thierry J (1997b) La notion d'espace morphologique, outil d'analyse de la morphodiversité des organismes. Geobios 30(Suppl 1):415–422

Nigmatullin CM, Nesis KN, Arkhipkin AI (2001) A review of the biology of the jumbo squid *Dosidicus gigas* (Cephalopoda: Ommastrephidae). Fish Res 54:9–19

Oechsle E (1958) Stratigraphie und Ammonitenfauna der Sonninien-Schichten des Filsgebiets unter besonderer Berucksichtgung der Sowerbyi Zone (Mittlerer Doggers, Wuttemberg). Palaeontogr A 111:47–129

Olóriz F (2000) Time-averaging and long-term palaeoecology in macroinvertebrate fossil assemblages with ammonites (Upper Jurassic). Rev Paleobiol 19:123–140

Olóriz F, Palmqvist P, Pérez-Claros JA (1997) Shell features, main colonized environments, and fractal analysis of sutures in Late Jurassic ammonites. Lethaia 30:191–204

Oloriz F, Palmqvist P, Perez-Claros JA (1999) Recent advances in morphometric approaches to covariation of shell features and the complexity of suture lines in Late Jurassic ammonites, with reference to the major environments colonized. In: Oloriz F, Rodriguez-Tovar FJ (eds) Advancing research on living and fossil cephalopods. Kluwer Academic/Plenum, New York

Olóriz F, Villaseñor AB, González-Arreola C (2000) Geographic control on phenotype expression. The case of *Hybonoticeras mundulum* (Oppel) from the Mexican Altiplano. Lethaia 33:157–174

Palframan DFB (1966) Variation and ontogeny of some Oxfordian ammonites: *Taramelliceras richei* (de Loriol) and *Creniceras renggeri* (Oppel), from Woodham, Buckinghamshire. Palaeontology 9:290–311

Palframan DFB (1967) Variation and ontogeny of some oxford clay ammonites: *Distichoceras bicostatum* (Stahl) and *Horioceras baugieri* (D'Orbigny), from England. Palaeontology 10:60–94

Parent H (1998) Upper Bathonian and lower Callovian ammonites from Chacay Melehué (Argentina). Acta Palaeontol Pol 43:69–130

Parent H, Scherzinger A, Schweigert G (2008) Sexual phenomena in late Jurassic Aspidoceratidae (Ammonoidea). Dimorphic correspondence between *Physodoceras hermanni* (Berckhemer) and *Sutneria subeumela* Schneid, and first record of possible hermaphroditism. Palaeodiversity 1:181–187

Parent H, Greco AF, Bejas M (2009) Size-shape relationships in the Mesozoic planispiral ammonites. Acta Palaeontol Pol 55:85–98

Parent H, Bejas M, Greco A, Hammer O (2011) Relationships between dimensionless models of ammonoid shell morphology. Acta Palaeontol Pol 57:445–447

Paul CRC (2011) Sutural variation in the ammonites *Oxynoticeras* and *Cheltonia* from the Lower Jurassic of Bishop's Cleeve, Gloucestershire, England and its significance for ammonite growth. Palaeogeogr. Palaeoclimatol Palaeoecol 309:201–214

Pavia G (2006) Nomenclatural suitability in ammonoid classification: generic versus subgeneric status of dimorphic pairs. Volumina Jurassica 7:254

Pearson K (1895) Contributions to the mathematical theory of evolution II. Skew variation in homogeneous material. Philos Trans Roy Soc Lond A 186:343–414

Pearson K (1901) On lines and planes of closest fit to systems of points in space. Philos Phenomenol 2:559–572

Pecl G, Jackson G (2008) The potential impacts of climate change on inshore squid: biology, ecology and fisheries. Rev Fish Biol Fish 18:373–385

Pecl GT, Steer MA, Hodgson KE (2004) The role of hatchling size in generating the intrinsic size-at-age variability of cephalopods: extending the Forsythe hypothesis. Mar Freshw Res 55:387–394

Pfaff E (1911) Über Form und Bau der Ammonitensepten und ihre Beziehungen zur Suturlinie. Jb Niedersächs Geol 4:208–223

Pictet FJ (1854) Traité de paléontologie, Céphalopodes, 2. B. Baillière, Paris

Ploch I (2003) Taxonomic interpretation and sexual dimorphism in the Early Cretaceous (Valanginian) ammonite *Valanginites nucleus* (ROEMER, 1841). Acta Geol Pol 53:201–208

Ploch I (2007) Intraspecific variability and problematic dimorphism in the Early Cretaceous (Valanginian) ammonite *Saynoceras verrucosum* (d'Orbigny, 1841). Acta Geol Sin 81:877–882

Powell EN, Davies DJ (1990) When Is an "Old" shell really old? J Geol 98:823–844

Rawson PF (1975a) The interpretation of the Lower Cretaceous heteromorph ammonite genera *Paracrioceras* and *Hoplocrioceras* Spath, 1924. Palaeontology 18:275–283

Rawson PF (1975b) Lower Cretcaeous ammonites from north-east England: the Hauterivian heteromorph *Aegocrioceras*. Bull Br Mus (Nat Hist) Geol 26:139–159

Reboulet S (2001) Limiting factors on shell growth, mode of life and segregation of Valanginian ammonoid populations: evidence from adult-size variations. Geobios 34:423–435

Reeside JB, Cobban WA (1960) Studies of the Mowry shale (Cretaceous) and contemporary formations in the United States and Canada. US Geol Surv Prof Pap 355:1–126

Reyment RA (1988) Does sexual dimorphism occur in cretaceous ammonoids? Senckenb Lethaea 69:109–119

Reyment RA (2003) Morphometric analysis of variability in the shell of some Nigerian Turonian (Cretaceous) ammonites. Cretaceous Res 24:789–803

Reyment R (2004) Instability in principal component analysis and the quantification of polyphenism in palaeontological data. Math Geol 36:629–638

Reyment RA (2011) Morphometric analysis of polyphenism in Lower Cretaceous ammonite genus *Knemiceras*. In: Elewa AMT (ed) Computational paleontology. Springer, Berlin

Reyment RA, Kennedy WJ (1991) Phenotypic plasticity in a cretaceous ammonite analyzed by multivariate statistical methods. Methodol Study Evol Biol 25:411–426

Reyment RA, Kennedy WJ (1998) Taxonomic recognition of species of *Neogastroplites* (Ammonoidea, Cenomanian) by geometric morphometric methods. Cretaceous Res 19:25–42

Reyment RA, Kennedy WJ (2000) Morphological links in an evolutionary sequence of the cretaceous ammonite genus *Metoicoceras* Hyatt. Cretaceous Res 21:845–849

Reyment RA, Minaka N (2000) A note on reiterated phenotypes in species of *Neogastroplites* (Ammonoidea, Cenomanian, Cretaceous). Cretaceous Res 21:173–175

Rieber H (1973) Ergebnisse paläontologisch-stratigraphischer Untersuchungen in der Grenzbitumenzone (Mittlere Trias) des Monte San Giorgio (Kanton Tessin, Schweiz). Eclog Geol Helv 66:667–685

Ritterbush KA, Hoffmann R, Lukeneder A, De Baets K (2014) Pelagic palaeoecology: the importance of recent constraints on ammonoid palaeobiology and life history. J Zool 292:229–241

Rocha R, Dias R (2005) Finite strain analysis using ammonoids: an interactive approach. J Struct Geol 27:475–479

Ropolo P (1995) Implications of variation in coiling in some Hauterivian (Lower Cretaceous) heteromorph ammonites from the Vocontian basin, France. Mem Descr Cart Geol Ital 51:137–165
Rouget I, Neige P (2001) Embryonic ammonoid shell features: intraspecific variation revisited. Palaeontology 44:53–64
Salgado-Ugarte IH, Shimizu M, Taniuchi T, Matsushita K (2000) Size frequency analysis by averaged shifted histograms and kernel density estimators. Asian Fish Sci 13:1–12
Sandoval J, Chandler RB (2000) The sonniniid ammonite *Euhoploceras* from the Middle Jurassic of south-west England and southern Spain. Palaeontology 43:495–532
Sanvicente-Añorve L, Salgado-Ugarte I, Castillo-Rivera M (2003) The use of kernel density estimators to analyse length-frequency distributions of fish larvae. In: Browman HI, Skiftesvik AB (eds) The big fish bang. Proceedings of the 26th annual larval fish conference. Institute of Marine Research, Bergen
Sarti C (1999) Whorl width in the body chamber of the ammonites as a sign of dimorphism. In: Olóriz F, Rodríguez-Tovar F(eds) Advancing research on living and fossil cephalopods. Kluwer Academic, Plenum, New York
Saunders WB, Swan ARH (1984) Morphology and morphologic diversity of Mid-Carboniferous (Namurian) ammonoids in time and space. Paleobiology 10:195–228
Schindewolf OH (1934) Über Epöken auf Cephalopoden-Gehäusen. Paläontol Z 16:15–31
Schindewolf OH (1961) Die Ammoniten-Gattung *Cymbites* im deutschen Lias. alaeontogr Abt A Palaeozool-Stratigr 117:193–232
Schmidt H (1926) Neotenie und beschleunigte Entwicklung bei Ammoneen. Paläontol Z 7:197–205
Schweigert G, Dietze V, Chandler RB, Mitta VV (2007) Revision of the Middle Jurassic dimorphic ammonite genera *Strigoceras/Cadomoceras* (Strigoceratidae) and related forms. Stuttg Beitr Nat Ser B (Geol Paläont) 373:1–74
Seilacher A (1973) Fabricational noise in adaptive morphology. Syst Zool 22:451–465
Seilacher A (1988) Why are nautiloid and ammonite sutures so different? Neues Jahrb Geol Paläontol Abh 177:41–69
Silberling NJ (1956) "Trachyceras Zone" in the Upper Triassic of the western United States. J Paleontol 30:1147–1153
Silberling NJ, Nichols KM (1982) Middle Triassic molluscan fossils of biostratigraphic significance from the Humboldt Range, northwestern Nevada. U.S. Geological Survey Professional Paper 1207:1–77
Simpson GG (1944) Tempo and mode in evolution. Columbia University Press, New York, p. 237
Simpson Stephen J, Sword Gregory A, Lo N (2011) Polyphenism in insects. Curr Biol 21:R738–R749
Spath LF (1919) V.-Notes on Ammonites: I. Geol Mag 6:27–35
Spearman C (1904) The proof and measurement of association between two things. Am J Psychol 15:72–10
Sprent P (1989) Applied nonparametric statistical methods. Chapman & Hall, London
Stephen DA, Stanton RJ (2002) Impact of reproductive strategy on cephalopod evolution. Abh Geol Bundesanst 57:151–155
Stephen DA, Manger WL, Baker C (2002) Ontogeny and heterochrony in the middle Carboniferous ammonoid *Arkanites relictus* (Quinn, McCaleb, and Webb) from northern Arkansas. J Paleontol 76:810–821
Stevens GR (1988) Giant ammonites: a review. In: Wiedmann J, Kullmann J (eds) Cephalopods-present and past. Schweizerbart, Stuttgart
Sturani C (1971) Ammonites and stratigraphy of the "Posidonia Alpina" beds of the Venetian Alps (Middle Jurassic, Mainly Bajocian). Mem Ist Geol Mineral Univ Padova 28:1–190
Swan ARH, Saunders WB (1987) Function and shape in late Paleozoic (Mid-Carboniferous) ammonoids. Paleobiology 13:297–311
Tajika A, Wani R (2011) Intraspecific variation of hatchling size in late Cretaceous ammonoids from Hokkaido, Japan: implication for planktic duration at early ontogenetic stage. Lethaia 44:287–298

Tan BK (1973) Determination of strain ellipses from deformed ammonoids. Tectonophys 16:89–101

Tanabe K (1977a) Mid-Cretaceous scaphitid ammonites from Hokkaido. Palaeontological Society of Japan. Special Papers 21:11–22

Tanabe K (1977b) Functional evolution of *Otoscaphites puerculus* (Jimbo) and *Scaphites planus* (Yabe), Upper Cretaceous ammonites, Series D (Geology) 23. Memoirs of the Faculty of Science, Kyushu University, pp 367–407

Tanabe K (1993) Variability and mode of evolution of the Middle Cretaceous ammonite *Subprionocyclus* (Ammonitina: Collignoniceratidae) from Japan. Geobios 26(Suppl 1):347–357

Tanabe K, Shigeta Y (1987) Ontogenetic shell variation and streamlining of some Cretaceous ammonites. Transactions and Proceedings of the Palaeontological Society of Japan, New Series 147:165–179

Tanabe K, Shigeta Y, Mapes RH (1995) Early life history of Carboniferous ammonoids inferred from analysis of shell hydrostatics and fossil assemblages. Palaios 10:80–86

Tanabe K, Landman NH, Yoshioka Y (2003) Intra- and interspecific variation in the early internal shell features of some Cretaceous ammonoids. J Paleontol 77:876–887

Thierry J (1978) Le genre *Macrocephalites* au Callovien Inférieur (Ammonites, Jurassique Moyen). Mémoires Géologiques de l'Université de Dijon 4:1-490Tintant H (1963) Les kosmocératidés du Callovien inférieur et moyen d'Europe occidentale: essai de paléontologie quantitative. 29. Presses universitaires de France. Publications de l'Université de Dijon 29:1–491

Tintant H (1963) Les kosmocératidés du Callovien inférieur et moyen d'Europe occidentale: essai de paléontologie quantitative, vol 29. Publications de l'Universite de Dijon, Paris, pp. 1–491

Tintant H (1976) Le polymorphisme intraspécifique en paléontologie. Haliotis 6:49–69

Tintant H (1980) Problématique de l'espèce en Paléozoologie. Mem Soc Zool Fr 40:321–372

Tozer ET (1971) Triassic time and ammonoids: problems and proposals. Canad J Earth Sci 8:989–1031

Trueman AE (1940) The ammonite body-chamber, with special reference to the buoyancy and mode of life of the living ammonite. Q J Geol Soc 96:339–383

Tsujino Y, Naruse H, Maeda H (2003) Estimation of allometric shell growth by fragmentary specimens of *Baculites tanakae* Matsumoto and Obata (a Late Cretaceous heteromorph ammonoid). Paleontol Res 7:245–255

Ubukata T, Tanabe K, Shigeta Y, Maeda H, Mapes RH (2008) Piggyback whorls: a new theoretical morphologic model reveals constructional linkages among morphological characters in ammonoids. Acta Palaeontol Pol 53:113–128

Urdy S (2015) Theoretical modelling of the molluscan shell: what has been learned from the comparison among molluscan taxa? This volume

Urdy S, Goudemand N, Bucher H, Chirat R (2010a) Allometries and the morphogenesis of the molluscan shell: a quantitative and theoretical model. J Exp Zool B 314:280–302

Urdy S, Goudemand N, Bucher H, Chirat R (2010b) Growth dependent phenotypic variation of molluscan shell shape: implications for allometric data interpretation. J Exp Zool B 314:303–326

Urdy S, Wilson LAB, Haug JT, Sánchez-Villagra MR (2013) On the unique perspective of paleontology in the study of developmental evolution and biases. Biol Theory 8:1–10. doi:10.1007/s13752-013-0115-1

Urlichs M (2004) Kümmerwuchs bei *Lobites* Mosjsisovics, 1902 (Ammonoidea) aus dem Unter-Karnium der Dolomiten (Ober-Trias, Italien) mit Revision der unterkarnischen Arten. Stuttg Beitr Nat Ser B (Geol Palaont) 344:1–37

Urlichs M (2012) Stunting in some invertebrates from the Cassian formation (Late Triassic, Carnian) of the Dolomites (Italy). Neues Jahrb Geol Paläontol Abh 265:1–25

Van Valen L (1978) The statistics of variation. Evolut Theory 4:33–43

Van Valen L (2005) The statistics of variation. In: Hallgrímsson B, Hall BK (eds) Variation: a central concept in biology. Academic, Burlington

Vermeulen J (2002) Étude stratigraphique et paléontologique de la famille des Pulchelliidae (Ammonoidea, Ammonitina, Endemocerataceae). Geol Alp Hors Ser 42:331–333

Vogel K-P (1959) Zwergwuchs bei Polyptychiten (Ammonoidea). Geol Jahrb 76:469–540
Waggoner KJ (2006) Sutural form and shell morphology of *Placenticeras* and systematic descriptions of late cretaceous ammonites from the big bend region, Texas. 1-398. Texas Tech University, Lubbock
Wagner FH (2000) Intraspecific variation. McGraw-Hill Yearbook of Science and Technology
Wagner GP, Altenberg L (1996) Perspective: complex adaptations and the evolution of evolvability. Evolution Int J org Evolution 50:967–976
Wani R (2001) Reworked ammonoids and their taphonomic implications in the upper cretaceous of northwestern Hokkaido, Japan. Cretaceous Res 22:615–625
Wani R, Gupta NS (2015) Ammonoid taphonomy. In: Klug C, Korn D, De Baets K, Kruta I, Mapes RH (eds) Ammonoid paleobiology: from macroevolution to paleogeography. Springer, Dordrecht
Ward P (1980) Comparative shell shape distributions in Jurassic-Cretaceous ammonites and Jurassic-Tertiary nautilids. Paleobiology 6:32–43
Ward PD (1987) The natural history of *Nautilus*. Allen & Unwin, Boston
Ward PD, Westerman GEG (1985) Cephalopod paleoecology. In: Broadhead TW (ed) Mollusks, notes for a short cource. University of Tennessee, Knoxville
Weitschat W (2008) Intraspecific variation of *Svalbardiceras spitzbergensis* (Frebold) from the early Triassic (Spathian) of Spitsbergen. Polar Res 27(3):292–297
Weitschat W, Bandel K (1991) Organic components in phragmocones of Boreal Triassic ammonoids: implications for ammonoid biology. Paläontol Z 65:269–303
Wendt J (1971) Genese und Fauna submariner sedimentärer Spaltenfüllungen im mediterranen Jura. Palaeontogr A 136:121–192
West-Eberhard MJ (1989) Phenotypic plasticity and the origins of diversity. Annu Rev Ecol Syst 20:249–278
West-Eberhard MJ (2003) Developmental plasticity and evolution. Oxford University, Oxford
West-Eberhard MJ (2005) Developmental plasticity and the origin of species differences. Proc Natl Acad Sci USA 102:6543–6549
Westermann G (1964) Sexual-Dimorphismus bei Ammonoideen und seine Bedeutung für die Taxonomie der Otoitidae (einschließlich Sphaeroceratinae; Ammonitina, M. Jura). Palaeontogr A:33–73
Westermann GEG (1966) Covariation and taxonomy of the Jurassic ammonite *Sonninia adicra* (Waagen). Neues Jahrb Geol Paläontol Abh 124:289–312
Westermann GEG (1971) Form, structure and function of shell and siphuncle in coiled Mesozoic ammonoids. Life Sci Contrib, Royal Ontario Museum 78:1–39
Westermann GEG (1996) Ammonoid life and habitat. In: Landman NH, Tanabe K, Davis RA (eds) Ammonoid paleobiology. Plenum, New York
Westermann GEG, Callomon J (1988) The Macrocephalitinae and associated Bathonian and early Callovian (Jurassic) ammonoids of the Sula Islands and New Guinea. Palaeontogr A 203:1–90
Wiedmann J (1969) The heteromorphs and ammonoid extinction. Biol Rev 44:563–602
Wiedmann J, Dieni I (1968) Die Kreide Sardiniens und ihre Cephalopoden. Palaeontogr Italica 64:1–171
Wiese F, Schulze F (2005) The Upper Cenomanian (Cretaceous) ammonite *Neolobites vibrayeanus* (d'Orbigny, 1841) in the Middle East: taxonomic and palaeoecologic remarks. Cretaceous Res 26:930–946
Willmore KE, Young NM, Richtsmeier JT (2007) Phenotypic variability: its components, measurement and underlying developmental processes. Revolut Biol 34(3–4):99–120
Wilmsen M, Mosavinia A (2011) Phenotypic plasticity and taxonomy of *Schloenbachia varians* (J. Sowerby, 1817) (Cretaceous Ammonoidea). Paläontol Z 85:169–184
Yacobucci MM (2004a) Buckman's paradox: variability and constraints on ammonoid ornament and shell shape. Lethaia 37:57–69
Yacobucci MM (2004) *Neogastroplites* meets *Metengonoceras*: morphological response of an endemic hoplitid ammonite to a new invader in the mid-cretaceous mowry sea of North America. Cretaceous Res 25:927–944

Yacobucci MM (2008) Controls on shell shape in acanthoceratid ammonites from the Cenomanian-Turonian Western Interior seaway. In: Harries PJ (ed) High-resolution approaches in stratigraphic paleontology. Springer, Netherlands

Yacobucci MM, Manship LL (2011) Ammonoid septal formation and suture asymmetry explored with a geographic information systems approach. Palaeontol Electron 14:3A:17

Yahada H, Wani R (2013) Limited migration of scaphitid ammonoids: evidence from the analyses of shell whorls. J Paleontol 87:406–412

Yamaji A, Maeda H (2013) Determination of 2D strain from a fragmented single ammonoid. Isl Arc 22:126–132

Zakharov YD (1977) Ontogeny of ceratites of the genus *Pinacoceras* and developmental features of the suborder Pinacoceratina. Paleontol J 4:445–445

Zatoń M (2008) Taxonomy and palaeobiology of the Bathonian (Middle Jurassic) tulitid ammonite *Morrisiceras*. Geobios 41:699–717.

Zittel KA von (1885) Handbuch der Paläontologie, Abt. 1, Bd. 2. R. Oldenbourg, München

Zuev GV (1971) Cephalopods from the north-western part of the Indian Ocean. Naukova Dumka, Kiev in Russian

Zuev GV (1976) Physiological variability of the females of the squid *Symplectoteuthis pteropus* (Steenstrup). Biol Sea 38:55–62 [in Russian]

Part III
Anatomy

Chapter 10
Ammonoid Buccal Mass and Jaw Apparatus

Kazushige Tanabe, Isabelle Kruta and Neil H. Landman

10.1 Introduction

Virtually all modern cephalopod mollusks possess a well-developed jaw apparatus consisting of upper and lower elements (synonymous with beaks or mandibles) and a radula as primary feeding organs. These feeding apparatuses are housed in a globular-shaped muscular organ called the buccal mass, in the proximal portion of the digestive tract (Clarke 1962; Nixon 1988a, b; Tanabe and Fukuda 1999). Because of their essential role for feeding, it is safely considered that extinct cephalopods including ammonoids also possessed a jaw apparatus and a radula.

Since the early nineteenth century, calcified and horny remains of cephalopod lower jaws have been discovered from the middle Paleozoic and younger marine deposits. In most cases, they occur as isolated specimens, but are rarely preserved in place, either closing the aperture or within the body chamber of ammonoids. Two general types have been recognized among them, i.e., aptychus and anaptychus. The aptychus (Greek άπτυχοξ—a body folded into two parts) named by Meyer (1829) is represented by an open bivalve shell-like disk consisting of an inner dark-colored horny lamella (rarely preserved as fossils) and outer symmetrical paired calcitic plates, either in juxtaposition or coalesced along a median dorsoventral line. Aptychi are known to occur from the Jurassic and Cretaceous rocks and have

K. Tanabe (✉)
Department of Historical Geology and Paleontology, The University Museum,
The University of Tokyo, 113-0033 Tokyo, Japan
e-mail: tanabe@um.u-tokyo.ac.jp

I. Kruta · N. H. Landman
Division of Paleontology (Invertebrates), American Museum of Natural History,
New York, NY 10024, USA
e-mail: ikruta@amnh.org

N. H. Landman
e-mail: landman@amnh.org

C. Klug et al. (eds.), *Ammonoid Paleobiology: From Anatomy to Ecology*,
Topics in Geobiology 43, DOI 10.1007/978-94-017-9630-9_10

been classified into several form genera by Trauth (1927–1936, 1938) based on the overall shape and external sculpture. The anaptychus (ανάπτυχοξ—unfolding), a morphological term introduced by Oppel (1856), is a simple arched structure consisting wholly of a black, horny substance. Anaptychi have been found from the Devonian to the Cretaceous (Moore and Sylvester-Bradley 1957).

Meek and Hayden (1864) were the first to describe the co-occurrence of an aptychus and a jaw-like appendage in the body chamber of the Late Cretaceous scaphitid *Hoploscaphites nebrascensis* and explained them as the lower and upper jaws. On the other hand, subsequent authors (e.g., Schmidt 1925, 1928; Trauth 1927–1936; Nagao 1931a, b, c; 1932; Matern 1931; Schindewolf 1958) interpreted the aptychi and anaptychi as ammonoid opercula, because the general outlines of aptychi and anaptychi more or less fit with the corresponding ammonoid apertures and are also similar to the hood of modern *Nautilus* and opercula of some gastropods.

Meek and Hayden's (1864) view was later justified by Closs (1967) and Lehmann (1967, 1970, 1971, 1972), who reported the anaptychus or aptychus-upper jaw-radula association within the body chambers of some Late Paleozoic and Jurassic ammonoids. As a result of these and subsequent works, anaptychi and aptychi are now conclusively regarded anatomically as ammonoid lower jaws (Lehmann 1976, 1981, 1990; Nixon 1988a, b, 1996; Tanabe and Fukuda 1999).

Remains of jaws and radulae that are referred to the Ammonoidea have been reported from Devonian to Cretaceous marine deposits (e.g., Woodward 1885; Closs 1967; Lehmann 1967, 1971; Mapes 1987; Dagys and Weitschat 1988; Tanabe and Landman 2002; Kruta et al. 2011). Most are found individually, although they are rarely preserved *in situ* within the body chambers of ammonoid conchs whose taxonomic relationships are known (e.g., Lehmann 1967, 1971, 1979; Doguzhaeva and Mutvei 1992, 1993; Tanabe and Landman 2002; Kruta et al. 2010, 2011; Landman et al. 2010; Tanabe et al. 2012, 2013; Klug and Jerjen 2012) and/or within the mouthpart portion of exceptionally well-preserved ammonoids with soft-tissue remains from Konservat Lagerstätten (Klug et al. 2012). Based on comparative morphologic examinations of these *in situ* materials with modern cephalopod counterparts, morphological features of the ammonoid jaw apparatus and radula and their taxonomic and ecological implications have been synthesized by previous authors (Lehmann 1976, 1990; Nixon 1988a, 1996; Tanabe and Fukuda 1999; Tanabe and Landman 2002). In addition, recent studies of well-preserved materials by means of new techniques such as X-ray computer tomographic (CT) scans and X-ray synchrotron analyses (e.g., Kruta et al. 2011, 2013; Tanabe et al. 2013) made it possible to restore the three-dimensional architecture, microstructure, and mineralogy of the ammonoid buccal apparatuses. This chapter synthesizes the current knowledge on ammonoid jaw apparatuses and discusses it from taxonomic and paleoecological points of view. The descriptive terms of cephalopod jaw apparatuses utilized in this article follow Clarke (1962, 1986) for modern coleoids and nautilids, and Tanabe et al. (1980b, fig. 1) and Tanabe (1983, fig. 1) for ammonoids. Suprageneric taxonomy of the species treated in this chapter is based on House (1980) for Devonian ammonoids, Furnish et al. (2009) for Carboniferous and Permian ammonoids, Tozer (1980) for Triassic ammonoids, Howarth (2013) for Jurassic Psiloceratoidea, Eoderoceratoidea and Hildoceratoidea (Ammonitina), Donovan et al. (1980) for other Jurassic Ammonitina, and Wright et al. (1996) for Cretaceous ammonoids.

Institutional abbreviations: GPIMH: Geologisch-Paläontologisches Institut und Museum, Universität Hamburg, Germany, SMNS: Staatliches Museum für Naturkunde, Stuttgart, Germany, YPM: Yale Peabody Museum, New Haven, CT, USA, USNM: National Museum of Natural History, Smithsonian Institution, Washington, DC., USA, AMNH: American Museum of Natural History, New York, NY, USA, BHMNH: Black Hills Museum of Natural History, Hill City, SD, USA, UH: Hokkaido University Museum, Sapporo, Japan, GK: Kyushu University Museum, Fukuoka, Japan, KMNH: Kitakyushu Museum of Natural History and Human History, Kitakyushu, Japan, UMUT: University Museum, University of Tokyo, Tokyo, Japan.

10.2 Buccal Mass Structures of Modern Cephalopods

Before documenting the current knowledge of ammonoid jaw apparatuses, comparative anatomical features of the buccal organs in modern cephalopods are summarized here, because they are essential to restore the corresponding structures in ammonoids adequately. The buccal mass of living cephalopods largely consists of well-developed muscular systems that support active movement of the jaws and radula. Inside the buccal cavity is the radular complex, which comprises the radular sac, salivary glands, salivary papilla, and a pair of lateral buccal palps (Tanabe and Fukuda 1987a; Nixon 1988a; Sasaki et al. 2010; Fig. 10.1a). The inner side of the lower jaw and outer side of the upper jaw are connected by jaw muscles, with a thin layer of tall columnar cells named beccublasts by Dilly and Nixon (1976) situated between them (Tanabe and Fukuda 1987a; Tanabe 2012; Fig. 10.1b–c). The outer lamellae of the upper and lower jaws are mostly covered by a thin connective tissue and are free of muscles.

The buccal mass musculature of coleoid cephalopods is classified into four major jaw muscles: the anterior, posterior, superior, and lateral mandibular muscles (Ueno and Kier 2005, 2007), among which the first three muscles connect the upper jaw, and the superior mandibular muscles connect the upper and lower jaws. The lateral mandibular muscles originating on the upper jaw do not connect to the lower jaw and instead insert on a connective tissue surrounding the buccal mass. The anterior and superior mandibular muscles are interpreted as responsible for jaw closing and shearing movements, whereas the posterior mandibular muscles may act synergistically with the lateral mandibular muscles to open the jaws in addition to jaw closing and shearing (Ueno and Kier 2005).

The mode of attachment of beccublasts on jaw lamellae differs between nautilids and coleoids. In nautilids, the distal ends of beccublasts are deeply inserted within

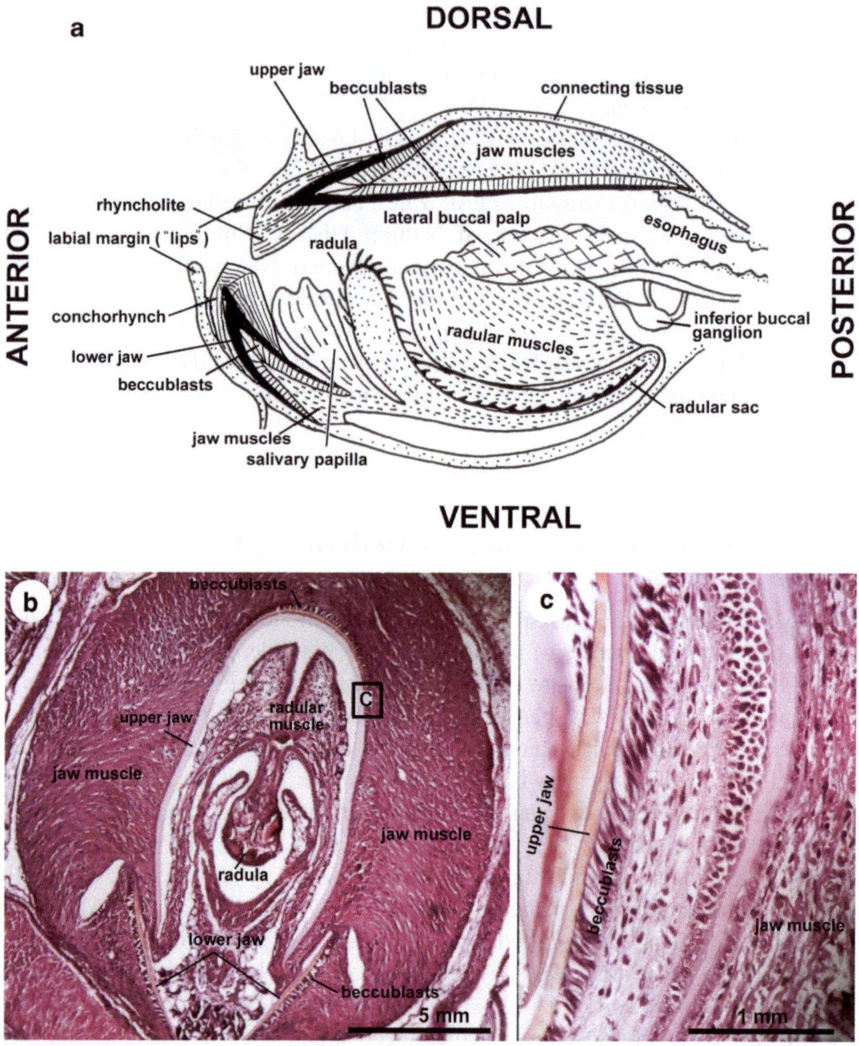

Fig. 10.1 Modern cephalopod buccal mass structure. **a** Diagram of the buccal mass structure of modern *Nautilus* (median section). Modified from Tanabe and Fukuda (1987a). **b** Photomicrograph of a cross-sectioned buccal mass of a cuttlefish, *Sepia esculenta*, captured in the Sea of Japan, west Japan. Stained with haematoxylin-eosin solution. **c** Close-up of the portion shown in **b**, showing beccublasts between upper jaw plate and jaw muscle. **b** and **c** from Tanabe (2012, fig. 1B, C; reproduced by permission of Schweizerbart, Stuttgart)

the hard chitinous plate, forming numerous micropores on it (Tanabe and Fukuda 1983, 1987a; Fig. 10.2a). In coleoids, the attachment scars of beccublasts on the jaw plates lack micropores and instead consist of anchor-type aligned polygonal imprints (Dilly and Nixon 1976; Tanabe and Fukuda 1983) (Fig. 10.2b). Therefore, in addition to the primary function of secreting the chitin-protein complex of the

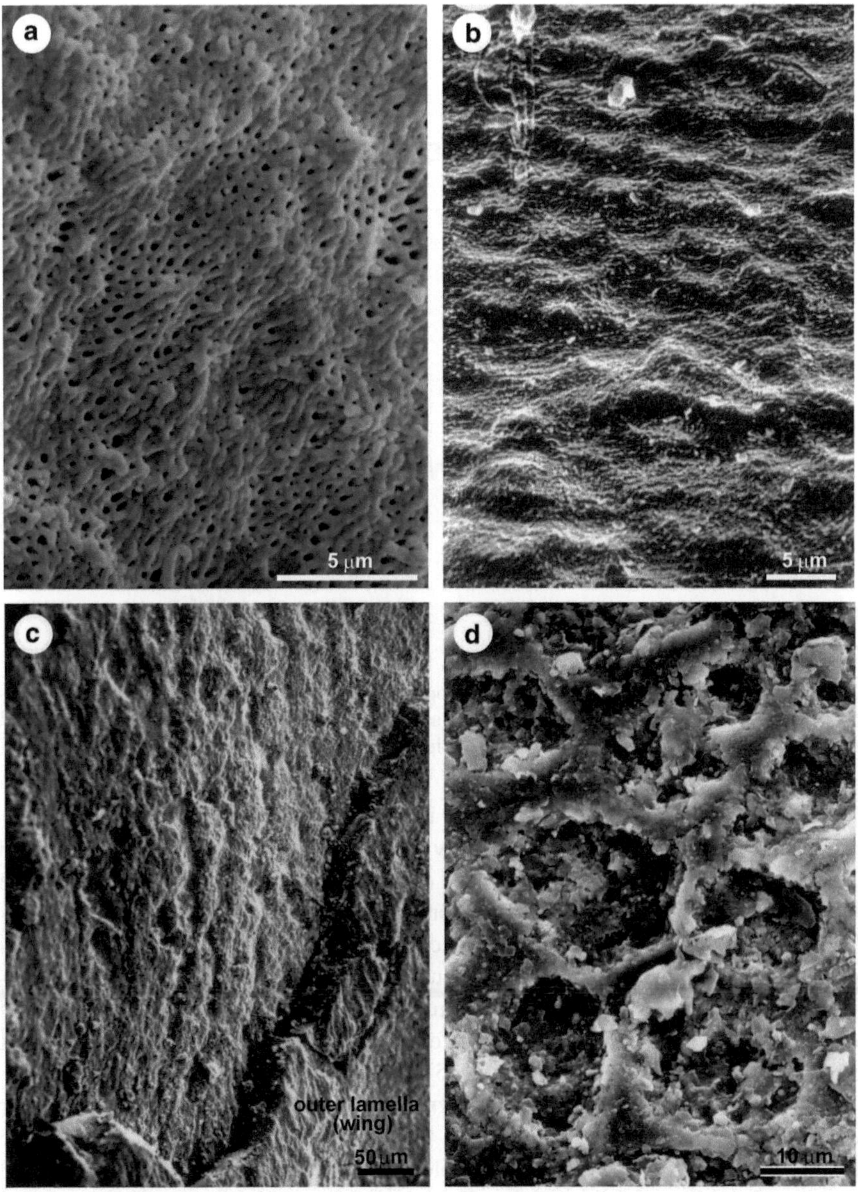

Fig. 10.2 Scanning electron micrographs of the attachment scars of beccublasts on the jaw surface. **a** *Nautilus pompilius* Linnaeus (Nautilida). Inner surface of the lower jaw, showing the micropores into which the branching ends of the beccublasts are deeply inserted. UMUT RM 199552, from Tañon Strait, Philippines. **b** *Sepia esculenta* (Hoyle) (Sepiida). Outer surface of the upper jaw showing the enlarged polygonal imprints of the beccublasts. UMUT RM 19953. Sea of Japan. **c** Isolated rhynchaptychus-type lower jaw attribuited to *Gaudryceras* sp. (Cretaceous Lytoceratina). Inner mold of the outer lamella (wing), showing the impression of aligned beccublasts covering the inner side. UMUT MM 18219, from the upper Santonian of Hokkaido, Japan. **d** Isolated lower jaw with a bivalved calcitic plate (laevaptychus) attributed to an aspidoceratid ammonoid. Inner surface of the inner chitinous lamella shows the aligned elliptical imprints of the beccublasts. UMUT MM 19945, from the Oxfordian (Jurassic) of Nusplingen, Germany

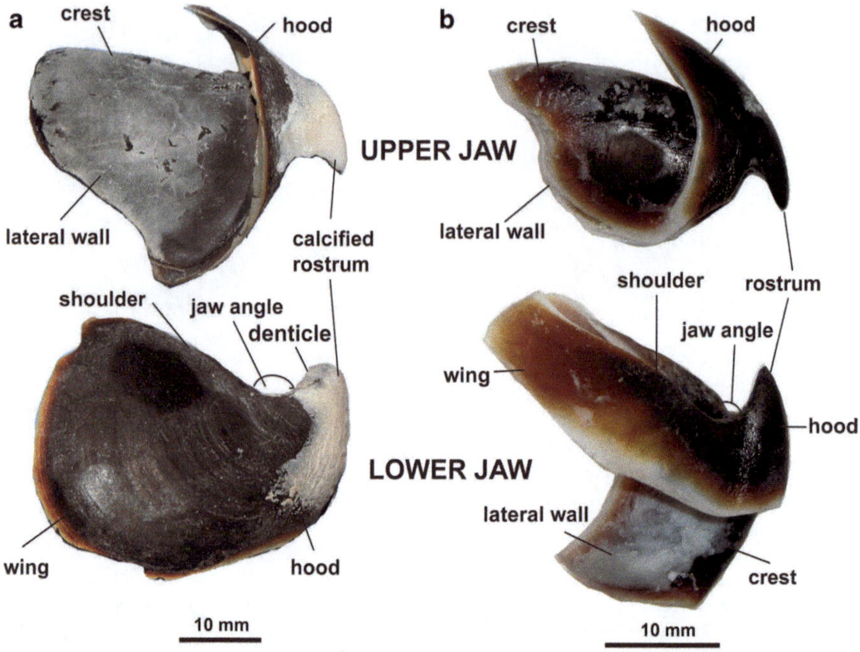

Fig. 10.3 Jaw apparatuses of modern cephalopods (right lateral views). **a** *Nautilus belauensis* Saunders (Nautilida). UMUT RM 27882-1, Palau. **b** *Sepia officinalis* Linnaeus (Sepiida). AMNH unregistered specimen from the Atlantic. Descriptive terms after Clarke (1962, 1986)

jaw lamellae, the branching ends of the beccublasts may serve to provide firm attachment of the jaw muscles onto the jaw plates, especially in nautilids (Tanabe and Fukuda 1987a, 1999). The outer surface of the buccal mass is wholly covered by a thin layer of connective tissue, except for the oral opening.

The labial margin ('lips') encircles the oral opening and consists of several rows of dense, triangular projections that include tall columnar epithelial, mucus-secreting, and sensory cells. The sharply pointed rostral tips of the upper and lower jaws of extant cephalopods appear to serve for biting and cutting up prey by means of strong jaw muscles (Ueno and Kier 2005).

The upper and lower jaws of modern cephalopods are both built up of outer and inner lamellae, which are joined in the anterior portion (Fig. 10.3). They are largely darkly tinted, more or less flexible when the animal is alive, and composed mainly of a chitin-protein complex (Saunders et al. 1978; Hunt and Nixon 1981; Lowenstam et al. 1984; Gupta et al. 2008). In nautilids (*Nautilus* and *Allonautilus*), thick calcified deposits cover the chitinous lamellae of the upper and lower jaws in the anterior rostral portion (Fig. 10.3); they were secreted by the tall columnar epithelial cells of the labial margin (Fukuda 1980; Tanabe and Fukuda 1987a). The calcified deposit in the upper jaw is arrowhead-shaped, whereas that in the lower jaw is scallop-shaped, with distinct denticles on the dorsal side; they are called the rhyncholite and conchorhynch, respectively (Saunders et al. 1978).

The gross morphology of the upper jaws is essentially similar among modern cephalopods, in having a posteriorly elongated, large inner lamella (crest and lateral wall) and a short reduced outer lamella (hood), both of which are prominently arched dorsally (Fig. 10.4e, f, g, h). The structure of the lower jaw differs markedly at higher taxonomic levels; namely, in the Nautilida, the inner lamella (crest and lateral wall) is short and reduced (Fig. 10.4e), whereas in the Coleoidea, it is well developed and projects posteroventrally (Clarke 1986). The degree of posteroventral projection of the inner lamella is most prominent in the Octopoda, intermediate in the Teuthida and Sepiida, and relatively weak in the Vampyromorpha and Cirroctopodida (Clarke 1986; Kubodera 2005; Tanabe 2012; Fig. 10.4f–h).

The radula of cephalopods is composed of a series of tooth rows. The radular teeth are made of a chitin-protein complex (Hunt and Nixon 1981) and are secreted by a thick epithelium consisting of columnar cells (odontoblasts) that line the pos-

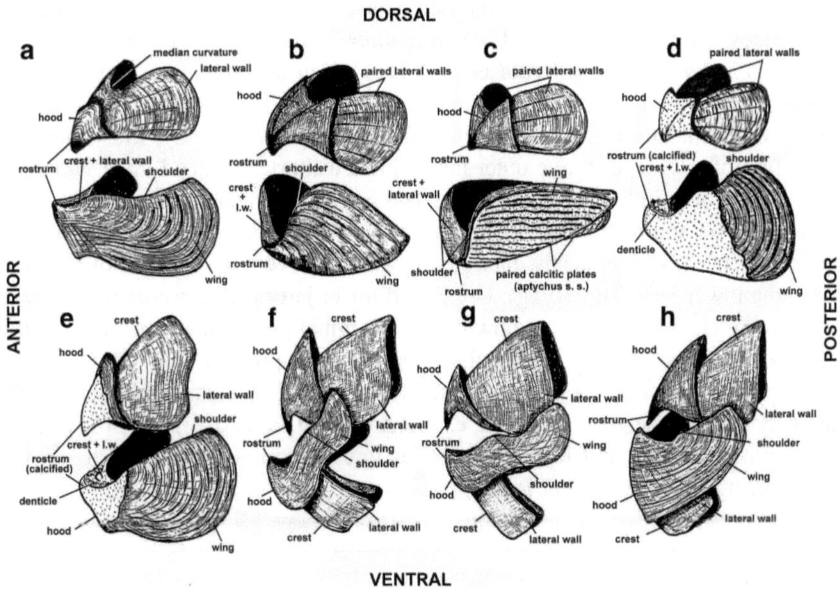

Fig. 10.4 Three-dimensional reconstruction of the jaw apparatuses of ammonoids (**a–d**) and extant cephalopods (**e–h**) (viewed from the anterolateral side for **a–e** and lateral side for **f–h**). Upper and lower jaws are shown on upper side and underside respectively for each figure. Intermediate-type jaw apparatus of the Desmoceratoidea (Cretaceous Ammonitina) is not shown in this figure. **a** Normal type. *Girtyoceras* (Early Carboniferous Goniatitina; after Doguzhaeva et al. 1997). **b** Anaptychus type. *Psiloceras* (Early Jurassic Ammonitina; after Lehmann 1975, fig. 3). **c** Aptychus type. *Hildoceras* (Early Jurassic Ammonitina; after Lehmann 1975, fig. 4). **d** Rhynchaptychus type. *Hypophylloceras* (Late Cretaceous Phylloceratina; modified from Tanabe et al. 2013, fig. 5). Outer calcareous layer on the outer chitinous lamella of lower jaw is partly taken off in this figure. **e** *Nautilus* (Nautilida, Nautiloidea). **f** *Sepia* (Sepiida, Coleoidea). **g** *Octopus* (Octopodida, Coleoidea). **h** *Vampyroteuthis* (Vampyromorpha, Coleoidea). Modified from Tanabe and Fukuda (1999, Fig. 19.3). Descriptive terms after Clarke (1962, 1968) for modern cephalopods and Tanabe et al. (1980b, fig. 1) and Tanabe (1983, fig. 1) for ammonoids

terior portion of the radular sac (Raven 1958) (Fig. 10.1a). Comparative anatomical and morphological features of radulae of extant cephalopods and ammonoids are discussed in Kruta et al. (2015).

10.3 Restoration of Ammonoid Jaw Apparatus

10.3.1 Recognition of in Situ *Jaw Apparatus*

Upper and lower jaws with radula are rarely found together *in situ* retaining their original life orientation in the body chambers of some ammonoids of Late Paleozoic (e.g., *Glaphyrites*; Closs 1967; *Cravenoceras*; Tanabe and Mapes 1995; Doguzhaeva et al. 1997) and Mesozoic ages (e.g., *Eleganticeras, Arnioceras, Hildoceras*, Lehmann 1967, 1971; Lehmann and Weitschat 1973; *Aconeceras*, Doguzhaeva and Mutvei 1992, 1993; *Baculites*, Kruta et al. 2011; baculitid, Klug et al. 2012; *Rhaeboceras*, Kruta et al. 2013). Such exceptionally well-preserved specimens make it possible to recognize upper and lower jaws through comparison with the buccal mass structure of modern cephalopods. Namely, the larger form consisting of a wide, concave outer lamella and a shorter reduced inner lamella on the ventral side is judged as a lower jaw. The smaller form consisting of paired wide inner lamellae and a short reduced outer one on the dorsal side is identified as an upper jaw, because it is partly encircled by the concave outer lamella of the larger form, the lower jaw (Fig. 10.5g). Upper and lower jaws are occasionally preserved in the body chamber, without showing their original position, suggesting that they have been moved slightly during the biostratinomic process (Fig. 10.5b, c, d, f; Kruta et al. 2010, Fig. 6). More frequently, a single jaw element is found within the body chamber (e.g., Figs. 10.5a, e, 10.6a, b, c, 10.7a, b, d, e), and in such cases, recognition of upper and lower jaws relies upon morphological comparison with co-occurring upper and lower jaws.

Fig. 10.5 Mode of occurrence of the jaw apparatus in the body chambers of selected ammonoid specimens. Abbreviations. *lj* lower jaw, *uj* upper jaw, *mg* median groove. **a** *Harpoceras falciferum* Sowerby (Hildoceratidae, Ammonitina) with an aptychus-type lower jaw. UMUT MM 31054 from the Lower Toarcian (Jurassic) of Dotternhausen, Germany. **b** *Physodoceras nattheimense* Schweigert (Aspidoceratidae, Ammonitina), with an aptychus-type jaw apparatus. SMNS Inv. Nr. 70072 from the Kimmeridgian (Jurassic) of Nurspring, Germany. **c** *Metahaploceras* sp. (Oppeliidae, Ammonitina) with an aptychus-type jaw apparatus. SMNS Inv. Nr. 63998 from the Kimmeridgian (Jurassic) of Nurspring, Germany. Same specimen as that figured by Schweigert (2009, fig. 3). **d** *Hypophylloceras subramosum* (Shimizu) (Phylloceartidae, Phylloceratina) with a rhychaptychus-type jaw apparatus. KMNH IvP 902011 from the Santonian (Cretaceous) of Haboro area, Hokkaido, Japan. **e** *Anagaudryceras limatum* (Yabe) (Gaudryceratidae, Lytoceratina) with a rhynchaptychus-type lower jaw. UMUT MM 30877 from the Coniacian (Cretaceous) of Haboro area, Hokkaido. **f** *Menuites naumanni* (Yokoyama) (Desmoceratidae, Ammonitina) with an intermediate-type jaw apparatus. UMUT MM 27835 from the Campanian (Cretaceous) of Naiba area, South Sakhalin, Far East Russia. **g** *Polyptychoceras* sp. (Diplomoceratidae, Ancyloceratina) with an aptychus-type jaw apparatus. UMUT MM 30878 from the Santonian (Cretaceous) of Furenai area, Hokkaido

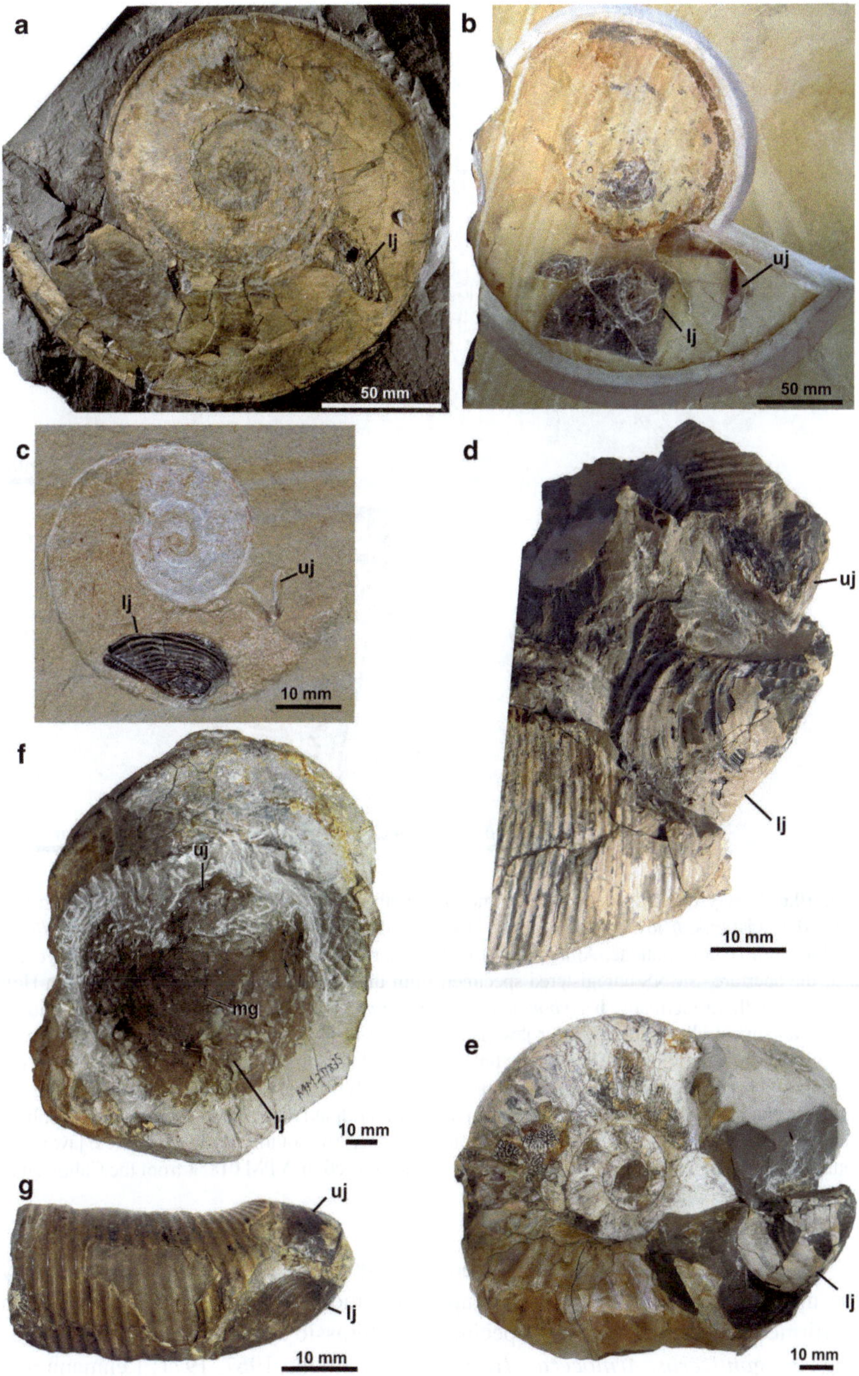
a
lj
50 mm
b
uj
lj
50 mm
c
uj
lj
10 mm
d
uj
lj
10 mm
f
uj
mg
lj
10 mm
e
lj
10 mm
g
uj
lj
10 mm

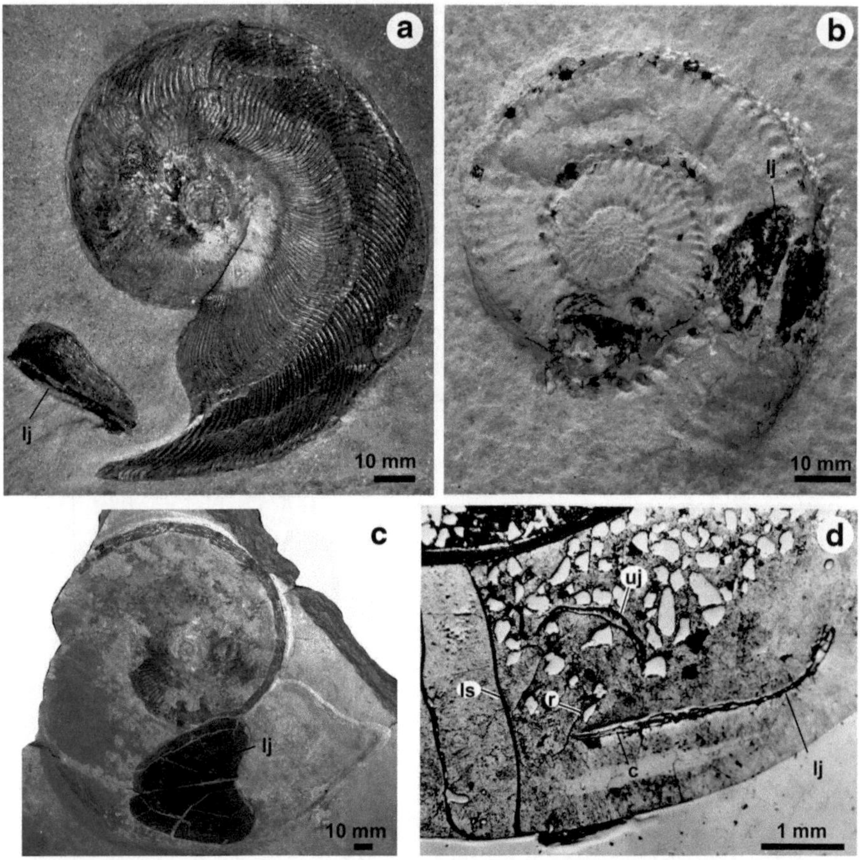

Fig. 10.6 Newly recognized Jurassic ammonoids with preserved aptychus-type jaw apparatus *in situ*. Abbreviations. *lj* lower jaw, *uj* upper jaw, *r* radula, *ls* last septum. **a** *Harpoceras capellinum* Schlotheium (Hildoceratidae, Ammonitina) with a laterally compressed lower jaw (cornaptychus) near the aperture. SMNS unregistered specimen from the Lower Toarcian Posidonia Shale in Holzmaden, southern Germany. **b** *Hybonoticeras hybonotum* (Oppel) (Aspidoceratidae, Ammonitina) with a dorsoventrally flattened lower jaw (laevaptychus) in the body chamber. SMNS unregistered specimen from the Tithonian lithographic limestone in Solnhofen, Germany. **c** *Streblites* cf. *zlatarskii* (Sapunov) (Oppeliidae, Ammonitina) with a dorsoventrally flattened lower jaw (lamellaptychus) in the body chamber. SMNS Inv. Nr. 67886 from the Kimmeridgian lithographic limestone in Nuspling, Germany. **d** *Sphaeroceras callovianus* Morris (Sphaeroceratidae, Ammonitina) with a complete jaw apparatus and a radula near the base of body chamber. Median section. YPM 01854 from the Callovian in Sutton, Wiltshire, England

The radula is occasionally preserved in the space (buccal cavity) surrounded by the upper and lower jaws. The jaw apparatus-radula-conch association has been confirmed in median-sectioned specimens of Jurassic and Cretaceous ammonoids (e.g., *Eleganticeras, Arnioceras, Hildoceras*: Lehmann 1967, 1971; Lehmann and Weitschat 1973; *Aconeceras*: Doguzhaeva and Mutvei 1992, 1993) removed from

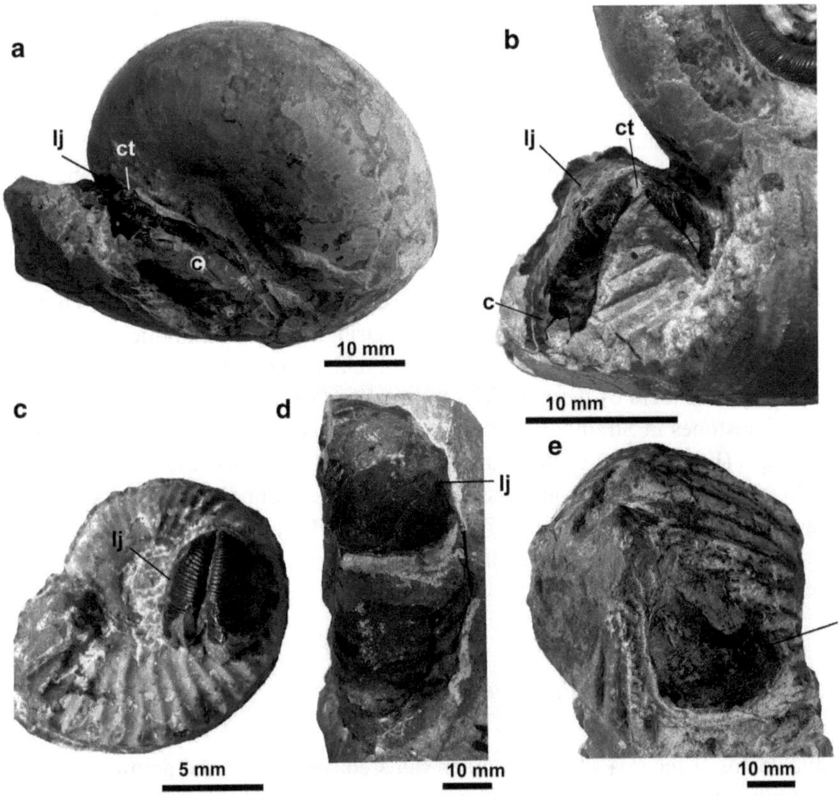

Fig. 10.7 Newly recognized Cretaceous ammonoids with preserved aptychus-type lower jaw *in situ*. Abbreviations. *lj* lower jaw, *c* calcareous layer on the outer chitinous layer, *ct* calcified tip. **a** *Tetragonites glabrus* (Jimbo) (Tetragonitidae, Lytoceratina) with a dorsoventrally compressed rhynchaptychus-type lower jaw within the body chamber. UMUT MM 19873 from the ?Santonian of Hokkaido (detailed locality unknown). **b** *Gaudryceras denseplicatum* (Jimbo) (Gaudryceratidae, Lytoceratina) with a rhynchaptychus-type lower jaw (laevaptychus) in the body chamber. UMUT MM 19875 from the Santonian in the Minoru-zawa Creek, Haboro area, northwest Hokkaido. **c** *Prohysteroceras* sp. (Brancoceratidae, Ammonitina) in association with an aptychus-type lower jaw on the lateral side of a conch. UMUT MM 31056 from the upper Albian in the Sounnai River, Seoushinai area, northwest Hokkaido. This specimen was found alone in a concretion, suggesting that the lower jaw can be attributed to this ammonite. **d** *Subptychoceras* sp. (Diplomoceratidae, Ancyloceratina) with an aptychus-type lower jaw in the body chamber. UMUT MM 31058 from the Santonian in the Pomporomui Creek, Manji area, central Hokkaido. **e** *Nostoceras* sp. (Nostoceratidae, Ancyloceratina) with an aptychus-type lower jaw in the body chamber. UMUT MM 31057 from the lower Maastrichitian in the Alfred Creek, Talkeetna Mountains, southern Alaska

calcareous concretions, in the phosphatized specimens of Late Paleozoic Gonatitina whose jaw lamellae are partly exfoliated (e.g., *Glaphyrites*: Closs 1967; *Cravenoceras*: Tanabe and Mapes 1995), and in the specimens of Late Cretaceous Ancyloceratina (*Baculites* and *Rhaeboceras*) by means of synchrotron X-ray examinations (Kruta et al. 2011, 2013).

An autochthonous nature of the jaw remains in body chambers has been judged on the basis of the following two conditions: (1) co-occurrence of upper and lower jaws within the ammonoid body chamber, and (2) a single jaw element-bearing ammonoid conch with a complete aperture that was deposited under a low-energy bottom environment. Typical examples satisfying the second condition are known from the Lower Jurassic bituminous black shales called Posidonienschiefer or Posidonia Shale (Figs. 10.5a, 10.6a; Hauff 1953) and Upper Jurassic lithographic limestones (Fig. 10.6b, c; Schweigert 1998; Schweigert and Dietl 1999) in southern Germany.

Under high-energy depositional conditions, jaw remains might have been trapped into the body chamber of other ammonoids during the biostratinomic process. For example, lamellaptychi, one form genus of aptychi, described by Bachmayer (1963) in the body chambers of the phylloceratid *Calliphylloceras* from the Late Jurassic oolitic limestones in Stramberg, Austria, are now interpreted to have an allochthonous origin (Engeser and Keupp 2002; Vašíček 2010), because the lamellaptychus-type lower jaw-upper jaw-conch association has been found in the Jurassic Haploceratoidea of the suborder Ammonitina (see Table 10.1). Exact taxonomic relationships cannot be determined for individually found jaw remains, even if they could be referred to the Ammonoidea by comparison with *in situ* jaws.

10.3.2 Taphonomic Problems

Available fossil records of paired outer calcareous plates of the ammonoid lower jaws (aptychi *sensu stricto*) from Jurassic and Cretaceous marine deposits indicate that they had a higher preservation potential than the chitinous jaw portions. Chitinous jaw portions of ammonoids are more rarely preserved than the aptychi, but are found as fossils, when they have been diagenetically replaced by phosphatic material in calcareous concretions (Kanie 1982; Tanabe and Fukuda 1983; Tanabe et al. 2012) or by carbonaceous remnants in lithographic limestones (Schweigert 1998; Klug et al. 2012).

Jaw apparatus bearing ammonoid shells that are horizontally or vertically embedded in bituminous black shales (Fig. 10.5a, 10.6a) and lithographic limestones (Figs. 10.5b, c, 10.6b, c) have suffered from an intensive lateral deformation during sediment compaction, in association with differential dissolution of the aragonitic shell during early diagenesis. This type of ammonoid preservation makes it difficult to restore the original shape and composition of the jaw apparatus, leading to misidentification of the jaw morphotypes. On the other hand, jaw fossils preserved in calcareous and/ or phosphate concretions generally retain their primary three-dimensional architecture and microstructure (e.g., Figs. 10.5d, e, f, g, 10.7b, c). Previous reconstructions of the ammonoid jaw apparatuses have been done on the basis of such well-preserved *in situ* material whose taxonomic relationships are known (e.g., Lehmann 1976, 1981, 1990; Doguzhaeva et al. 1997; Tanabe and Fukuda 1999; Tanabe and Landman 2002).

Table 10.1 List of ammonoids with preserved jaw apparatus in the body chamber *in situ*. Abbreviations: Uj upper jaw, Lj lower jaw, (○) known,? unknown

Suborder	Superfamily	Family	Species	Uj	Lj	Jaw morphotype	Age	Sources
Goniatitina	Gastriocera-toidea	Glaphyritidae	*Glaphyrites* sp.	○	○	Normal type	Carboniferous-Permian	Closs 1967; Bandel 1988
	Goniolobocceratoidea	Wiedeyocera-tidae	*Wiedeyoceras* sp.	?	○	Normal type	L. Carbon. (Pennsylva-nian)	Saunders and Richardson 1979
	Dimorpho-ceartoidea	Anthracocera-tidae	*Anthracoceras?* sp.	?	○	Normal type?	E. Carbon. (Mississippian)	Mapes 1987
		Girtyoceratidae	*Girtyoceras lima-tum* (Miller and Faber)	○	○	Normal type	E. Carbon. (Mississippian)	Mapes 1987; Doguzhaeva et al. 1997
	Thalassoc-eratoidea	Thalassocera-tidae	*Thalassoceras* sp.	○	○	Normal type	Late Carboniferous	Doguzhaeva 1999
			Prothalassoceras bashkiricum Ruzhencev	?	○	Normal type?	Late Carboniferous	Doguzhaeva 1999
	Neoglyphio-ceratoidea	Cravenocera-tidae	*Cravenoceras fay-ettevillae* Gordon	?	○	Normal type?[a]	E. Carbon. (Mississippian)	Mapes 1987; Tanabe and Mapes 1995
	Gastriocera-toidea	Homoceratidae	*Vallites* sp.	○	○	Normal type?	E. Carbon. (Mississippian)	Bandel 1988
Gephuroceratina	Gephurocera-toidea	Gephurocera-tidae	*Manticoceras intumescens* Sowerby	?	○	Normal type?	Late Devonian	Woodward 1885
			Crickites koeneri (Clarke)	?	○	Normal type?	Late Devonian	Matern 1931
Prolecanitina	Medlicot-tioidea	Medlicottiidae	*Uddenites* sp.	?	○	Normal type?	Late Carboniferous	Doguzhaeva 1999

Table 10.1 (continued)

Suborder	Superfamily	Family	Species	Uj	Lj	Jaw morphotype	Age	Sources
Ceratitina	Notitoidea	Olenekitidae	*Olenekites spiniplicatus* (Mojsisovics)	?	○	Normal type?[b]	Early Triassic (Olenekian)	Zakharov 1974
		Meekoceratidae	*Bajarunia euomphala* (Keyserling)	?	○	Normal type?	Early Triassic (Spathian)	Dagys and Weitschat 1988
			Epiboreoceras lenaense (Dagys and Konst)	?	○	Normal type?	Early Triassic	Keupp 2000
		Ophiceratidae	*Nordophiceras karpinskii* (Mojsisovics)	?	○	Normal type?	Early Triassic (Olenekian)	Keupp 2000
			Ophiceras commune Spath	○	○	Normal type?	Early Triassic (Griesbachian)	Lehmann 1985
	Ceratitoidea	Ceratitidae	*Ceratites nodosus* Philippi	○	○	Normal type	Middle Triassic (Anisian)	Lehmann 1990
			Ceratites penndorfi Rothe	○	○	Normal type	Middle Triassic (Anisian)	Klug and Jerjen 2011
			Anagymnotoceras cf. *varium* McLearn	○	○	Normal type	Middle Triassic (Anisian)	Dagys and Weitschat 1988
			Gymnotoceras rotelliformis (Meek)	?	○	Normal type?	Middle Triassic (Anisian)	Dagys and Weitschat 1988
		Keyserlingitidae	*Keyserlingites subrobustus* (Mojisisovics)	○	○	Normal type	Early Triassic (Spathian)	Dagys and Weitschat 1988; Lehmann 1988

Table 10.1 (continued)

Suborder	Superfamily	Family	Species	Uj	Lj	Jaw morphotype	Age	Sources
Ceratitina	Xenodiscoidea	Paraceltitidae	*Paraceltites rectangularis* Miller	?	○	Normal type?	Permian	Spinosa 1968
	Arcestoidea	Arcestidae	*Arcestes* sp.	?	○	Normal type?	Middle-Late Triassic	Mojsosovics (1873-1902)
	Pinacoceratoidea	Ptychitidae	*Aristoptychites kolymensis* (Kiparisova)	?	○	Normal type?	Middle Triassic (Ladinian)	Köhler-Lopez and Lehmann 1984; Dagys and Weitschat 1988
			Aristoptychites? sp.	○	○	Normal type	Middle Triassic (Ladinian)	Dagys and Weitschat 1988
	Trachyceratoidea	Trachyceratidae	*Austrotrachyceras* sp.	○	○	Normal type?	Late Triassic (Carnian)	Doguzhaeva et al. 2007
Phylloceratina	Phylloceratoidea	Phylloceartidae	*Phylloceras heterophyllum* (Sowerby)	?	○	Anaptychus?	E. Jurassic (Toarcian)	Hauff 1953; Lehmann 1990
			Hypophylloceras subramosum (Shimizu)	○	○	Rhynchaptychus	Turonian-Campanian	Tanabe et al. 2013; this study (KMNH IvP 902011)
			Phyllopachyceras yessoense (Yokoyama)	○	○	Rhynchaptychus	Turonian-Campanian	Tanabe and Landman 2002; Tanabe et al. 2013
Lytoceratina	Lytoceratoidea	Lytoceartidae	*Lytoceras cornucopiae* Young and Bird	?	○	Anaptychus?	E. Jurassic (Toarcian)	Schmidt 1928
	Tetragonitoidea	Gaudryceartidae	*Gaudryceras denseplicatum* (Jimbo)	?	○	Rhynchaptychus	L. Cretaceous (Turon.-Camp.)	Tanabe et al. 1980a; Kanie 1982; this study (UMUT MM 19875)

Table 10.1 (continued)

Suborder	Superfamily	Family	Species	Uj	Lj	Jaw morphotype	Age	Sources
Lytoceratina	Tetragoni-toidea	Gaudryceratidae	*Gaudryceras tenuiliratum* Yabe	?	○	Rhynchaptychus	L. Cretaceous (Con.-Camp.)	Nagao 1931a, c; Tanabe et al. 1980a; Kanie 1982
			Anagaudryceras limatum (Yabe)	?	○	Rhynchaptychus	L. Cretaceous (Turon.-Con.)	Tanabe 2012; Tanabe et al. 2012; this study (UMUT MM 30877)
		Tetragonitidae	*Tetragonites glabrus* (Jimbo)	?	○	Rhynchaptychus	L. Cretaceous (Turon.-Camp.)	Tanabe et al. 1980a; Kanie 1982; this study (UMUT MM 19873)
			Zelandites inflatus Matsumoto	?	○	Rhynchaptychus	L. Cretaceous (Cenomanian)	Tanabe et al. 1980a
Ammonitina	Psiloocera-toidea	Psiloceratidae	*Psiloceras planorbis* Sowerby	?	○	Anaptychus	E. Jurassic (Hettangian)	Schmidt 1928
			Psiloceras sp.	○	○	Anaptychus	E. Jurassic (Hettangian)	Lehmann 1970, 1975
		Arietitidae	*Arnioceras* sp.	○	○	Anaptychus	E. Jurassic (Sinemurian)	Lehmann 1970, 1971
			Arnioceras cf. *semicostatoides* (Buckman)	?	○	Anaptychus	E. Jurassic (Sinemurian)	Cope and Sole 2000
			Asteroceras stellare (J. Sowerby)	○	○	Anaptychus	E. Jurassic (Sinemurian)	Cope and Sole 2000
			Asteroceras obtusum (Sowerby)	○	○	Anaptychus	E. Jurassic (Sinemurian)	Keupp 2000
			Caenisites cf. *turneri* (Sowerby)	?	○	Anaptychus	E. Jurassic (Sinemurian)	Schmidt 1928
			Alsatites laqueus (Quenstedt)	?	○	Anaptychus	E. Jurassic (Hett.-Sinem.)	Schmidt 1928

Table 10.1 (continued)

Suborder	Superfamily	Family	Species	Uj	Lj	Jaw morphotype	Age	Sources
Ammonitina	Psilocera-toidea	Arietitidae	*Coroniceras conyberari* (Sowerby)	?	○	Anaptychus	E. Jurassic (Hett.-Sinem.)	Schmidt 1928
	Eoderocera-toidea	Dactyliocera-tidae	*Dactylioceras tenuicostatum* (Young and Bird)	○	○	Anaptychus	E. Jurassic (Toarcian)	Lehmann 1979
			Dactylioceras semicelatum (Simpson)	○	○	Anaptychus	E. Jurassic (Toarcian)	Lehmann 1979
		Eoderoceratidae	*Promicroceras* sp.	?	○	Anaptychus	E. Jurassic (Sinemurian)	Cope 1994; Cope and Sole 2000
		Amaltheidae	*Amaltheus costatus* Quenstedt	?	○	Anaptychus	E. Jurassic (Pliensbachian)	Schmidt 1928
			Amaltheus margaritatus Montfort	?	○	Anaptychus	E. Jurassic (Pliensbachian)	Schmidt 1928
			Pleuroceras sp.	○	○	Anaptychus	E. Jurassic (Pliensbachian)	Lehmann 1970
	Hildocera-toidea	Hildoceratidae	*Hildoceras levisoni* (Simpson)	○	○	Aptychus	E. Jurassic (Toarcian)	Lehmann 1972, 1975
			Phlyseogrammoceras sp.	?	○	Aptychus (Cornaptychus)	E. Jurassic (Toarcian)	Keupp (2000)
			Eleganticeras elegantulum (Young and Bird)	○	○	Aptychus	E. Jurassic (Toarcian)	Lehmann 1972
			Harpoceras capellinum Schlotheim	?	○	Aptychus (Cornaptychus)	E. Jurassic (Toarcian)	SMNS unregistered specimen (this study)

Table 10.1 (continued)

Suborder	Superfamily	Family	Species	Uj	Lj	Jaw morphotype	Age	Sources
Ammonitina	Hildoceratoidea	Hildoceratidae	*Harpoceras falciferum* Sowerby	?	○	Aptychus (Cornaptychus)	E. Jurassic (Toarcian)	This study (UMUT MM 31054, 31055)
			Oxyparoniceras buckmani (Bonarelli)	?	○	Aptychus	E. Jurassic (Toarcian)	Martinez Escauriaza 2007
		Graphoceratidae	*Leioceras opalinum* (Reinecke)	?	○	Aptychus (Cornaptychus)	M. Jurassic (Bajocian)	Trauth 1930
		Hammatoceratidae	*Sonninia espinazitensis* Tornquist	?	○	Aptychus (Cornaptychus)	M. Jurassic (Bajocian)	Dietze et al. 2012
			Sonninia arenata (Quenstedt)	?	○	Aptychus	M. Jurassic (Bajocian)	Morton 1973
			Sonninia mesacantha (Waagen)	?	○	Aptychus	M. Jurassic (Bajocian)	Morton 1973
			Shirbuirnia trigonalis Buckman	?	○	Aptychus	M. Jurassic (Bajocian)	Morton 1975
			Euhoploceras fissilobatum (Waagen)	?	○	Aptychus	M. Jurassic (Bajocian)	Morton 1973, 1975
	Stephanoceratoidea	Otoitidae	*Normanites* sp.	○	○	Aptychus (Praestriaptychus)	M. Jurassic (Bajocian)	Lehmann 1972
		Cardioceratidae	*Quenstedtoceras* sp.	?	○	Aptychus	M. Jurassic (Callovian)	Lehmann 1972
		Kosmoceratidae	*Kosmoceras* cf. *galilaeii* (Oppel)	?	○	Aptychus (Praestriaptychus)	M. Jurassic (Callovian)	Trauth 1930
	Perisphinctoidea	Parkinsoniidae	*Parkinsonia* sp.	○	○	Aptychus (Praestriaptychus)	M. Jurassic (Baj.-Bathon.)	Lehmann 1978

Table 10.1 (continued)

Suborder	Superfamily	Family	Species	Uj	Lj	Jaw morphotype	Age	Sources
Ammonitina	Perisphinc-toidea	Aspidoceratidae	*Physodoceras nattheimense* Schweigert	○	○	Aptychus (Laevaptychus)	L. Jurassic (Kimmeridg-ian)	Schweigert and Dietl 1999; this study (SMNS Inv. Nr. 70072)
			Physodoceras cf. *altenense* (d'Orbigny)	○	○	Aptychus (Laevaptychus)	L. Jurassic (Kimmeridg-ian)	Schindewolf 1958; Morton 1973, 1981
			Physodoceras sp.	○	○	Aptychus (Laevaptychus)	L. Jurassic (Kimmeridg-ian)	Morton 1981; Keupp 2007
			Hybonoticeras hybonotum (Oppel)	?	○	Aptychus (Laevaptychus)	L. Jurassic (Tithonian)	SMNS unregistered specimen (this study)
			Sutneria cf. *reb-holzi* Berckhemer	?	○	Aptychus (Laevaptychus)	L. Jurassic (Kimmeridg-ian)	Schweigert 1998
			Sutneria? sp.	?	○	Aptychus (Laevap-tychus)[c]	L. Jurassic (Kimmeridg-ian)	Zakharov and Lominadze 1983
		Perisphinctidae	*Subplanites?* sp.	?	○	Aptychus (Granulaptychus)	L. Jurassic (Kimm.-Tithon.)	Keupp 2007
			Proplanulites koenigi (Sowerby)	?	○	Aptychus (Praestriaptychus)	L. Jurassic (Kimm.-Tithon.)	Rogov and Gulyaev 2003
			Perisphinctes desertorum Stehn	?	○	Aptychus (? Granulaptychus)	L. Jurassic (Oxfordian)	Trauth 1930

Table 10.1 (continued)

Suborder	Superfamily	Family	Species	Uj	Lj	Jaw morphotype	Age	Sources
Ammonitina	Perisphinc-toidea	Perisphinctidae	*Perisphinctes rueppelianus* Schneidt	?	○	Aptychus (Granulaptychus)	L. Jurassic (Oxfordian)	Closs 1960
			Perisphinctes sp.	?	○	Aptychus (Granulaptychus)	L. Jurassic (Oxfordian)	Closs 1960
		Ataxioceratidae	*Silicisphinctes keratinitiformis* Schweigert	?	○	Aptychus (Praestriaptychus)	L. Jurassic (Kimmeridg-ian)	Schweigert 1998
			Lithacoceras fasciferum (Neumayr)	?	○	Aptychus (Praestriaptychus)	L. Jurassic (Kimmeridg-ian)	Schweigert 1998; Schweigert and Dietl 1999
		Polyptychitidae	*Simbirskites dech-eni* (Roemer)	?	○	Aptychus (Striaptychus?)	E.Cretaceous (Hauterivian)	Baraboshkin and Shumiilkin 2010
			Simbirskites staffi Wedekind	?	○	Aptychus (Striaptychus)	E.Cretaceous (Hauterivian)	Engeser and Keupp 2002
	Haplocera-toidea	Oppeliidae	*Fontannesiella prolithographica* (Fontannes)	○	○	Aptychus (Lamellaptychus)	L. Jurassic (Kimmeridg-ian)	Schweigert 2009; this study (SMNS Inv.-Nr. 66075)
			Fontannesiella? sp.	○	○	Aptychus (Lamellaptychus)	L. Jurassic (Kimmeridg-ian)	Schindewolf 1958; Keupp 2007
			? *Neochetoceras steraspis* (Oppel)	?	○	Aptychus (Lamellaptychus)	L. Jurassic (Kimmeridg-ian)	Closs 1960
			Neochetoceras subnudatum (Fontannes)	?	○	Aptychus (Lamellaptychus)	L. Jurassic (Kimmeridg-ian)	Schweigert 1998; Schweigert and Dietl 1999

Table 10.1 (continued)

Suborder	Superfamily	Family	Species	Uj	Lj	Jaw morphotype	Age	Sources
Ammonitina	Haplocera-toidea	Oppeliidae	*Neochetoceras* cf. *steraspis* (Oppel)	?	○	Aptychus (Lamellaptychus)	L. Jurassic (Kimmeridg-ian)	Trauth 1938
			Neochetoceras sp.	?	○	Aptychus (Lamellaptychus)	L. Jurassic (Kimmeridg-ian)	Michael 1894
			Metahaploceras sp.	○	○	Aptychus (Lamellaptychus)	L. Jurassic (Kimmeridg-ian)	Schweigert 1998; 2009
			Streblites cf. *zlatarskii* (Sapunov)	?	○	Aptychus (Lamellaptychus)	L. Jurassic (Kimmeridg-ian)	This study (SMNS Inv.-Nr.67886)
			Lingulaticeras planulatum Berckhemer	○	○	Aptychus (Lamellaptychus)	L. Jurassic (Kimmeridg-ian)	Schweigert 1998, 2009
			Lingulaticeras sp.	?	○	Aptychus (Lamellaptychus)	L. Jurassic (Kimmeridg-ian)	Schweigert 1998
			Cieneguiticeras perlaevis (Steuer)	○	○	Aptychus (Lamellaptychus)	L. Jurassic (Tithonian)	Parent et al. 2011
			Bradfordia liom-pha Buckman	?	○	Aptychus (Lamellaptychus)	M. Jurassic (Bajocian)	Arkell 1957; Morton 1981
			Taramelliceras trachynotus (Oppel)	?	○	Aptychus (Lamellaptychus)	L. Jurassic (Oxford.-Kimm.)	Trauth 1938

Table 10.1 (continued)

Suborder	Superfamily	Family	Species	Uj	Lj	Jaw morphotype	Age	Sources
Ammonitina	Haplocera-toidea	Oppeliidae	*Taramelliceras holbeini* (Oppel)	?	○	Aptychus (Lamellaptychus)	L. Jurassic (Kimmeridg-ian)	Trauth 1938
			Taramelliceras sp.	?	○	Aptychus (Lamellaptychus)	Late Jurassic	Schindewolf 1958; Closs 1960
			Oppelia litho-graphica (Oppel)	?	○	Aptychus (Lamellaptychus)	M. Jurassic (Bajocian)	Trauth 1938
			Aconeceras trautscholdi Sinzow	○	○	Aptychus	E. Cretaceous (Aptian)	Doguzhaeva and Mutvei 1992, 1993
			Oppeliidae, genus & species indeterminate	?	○	Aptychus (Lamellaptychus)	Late Jurassic	Closs 1960
		Haploceratidae	*Glochiceras thoro* (Oppel)	?	○	Aptychus (Lamellaptychus)	L. Jurassic (Kimmeridg-ian)	Trauth 1938
			Glochiceras lithographicum (Oppel)	?	○	Aptychus (Lamellaptychus)	L. Jurassic (Kimmeridg-ian)	Zakharov and Lominadze 1983
			Glochiceras sp.	?	○	Aptychus (Lamellaptychus)	Late Jurassic (Tithonian)	Schindewolf 1958
			Neolissoceras grasianum (d'Orbigny)	?	○	Aptychus	E. Cretaceous (Hauterivian)	Stephanov 1961; Reboulet 1996
			Haploceras eli-matum (Oppel)	?	○	Aptychus	E. Cretaceous (Berriasian)	Retowski 1891

Table 10.1 (continued)

Suborder	Superfamily	Family	Species	Uj	Lj	Jaw morphotype	Age	Sources
Ammonitina	Spiroceratoidea	Spiroceratidae	*Spiroceras calloviense* Morris	○	○	Aptychus	M. Jurassic (Callovian)	This study (YPM 01854)
	Desmoceratoidea	Desmoceratidae	*Pseudohaploceras matheroni* (d'Orbigny)	?	○	Intermediate[d]	E. Cretaceous (Barremian-Aptian)	Vašíček 2010
			Tragodesmoceroides subcostatus Matsumoto	?	○	Intermediate	L. Cretaceous (Turonian)	Tanabe 1983
			Damesites semicostatus Matsumoto	○	○	Intermediate	L. Cretaceous (Con.-Sant.)	Nagao 1932; Tanabe 1983
			Damesites ainuanus Matsumoto	○	?	Intermediate?	L. Cretaceous (Turonian)	Tanabe 1983
			Damesites aff. *sugata* Forbes	?	○	Intermediate	L. Cretaceous (Con.-Camp.)	Tanabe and Landman 2002; Tanabe et al. 2012
			Menuites naumanni (Yokoyama)	○	○	Intermediate	L. Cretaceous (Campanian)	Tanabe and Landman 2002; this study (UMUT MM 27835)
			Pachydiscus kamishakensis Anderson	?	○	Intermediate	L. Cretaceous (Maastrichtian)	Tanabe et al. 2012
	Acanthoceratoidea	Collignoniceratidae	*Subprionocyclus minimus* (Hayasaka and Fukada)	○	○	Aptychus	L. Cretaceous (Turonian)	Tanabe and Fukuda 1987a
			Texanites soutoni (Baily)	?	○	Aptychus (Spinaptychus)	L. Cretaceous (Turonian)	Kennedy and Klinger 1972

Table 10.1 (continued)

Suborder	Superfamily	Family	Species	Uj	Lj	Jaw morphotype	Age	Sources
Ammonitina	Acanthocera-toidea	Acanthocera-tidae	*Pseudaspidoceras flexuosum* Powell	?	○	Aptychus	L. Cretaceous (Turonian)	Ifrim 2013
		Brancoceratidae	*Prohysteroceras* sp.	?	○	Aptychus[e]	E. Cretaceous (Albian)	This study (UMUT MM 31056)
	Hoplitoidea	Placenticeratidae	*Placenticeras costatum* Hyatt	?	○	Aptychus	L. Cretaceous (Campanian)	Landman et al. 2006
			Placenticeras sp. (*P. costatum* or *P. meeki* Böhm)	○	○	Aptychus[f]	L. Cretaceous (Campanian)	Landman et al. 2006
			Metaplacen-ticeras subtilis-triatum (Jimbo)	?	○	Aptychus[f]	L. Cretaceous (Campanian)	Landman et al. 2006
Ancyloceratina	Ancylocera-toidea	Ancyloceratidae	*Ancyloceras* cf. *matheronianum* d'Orbigny	?	○	Aptychus (Praestriaptychus)	E. Cretaceous (Barremian)	Casey 1980
			Australiceras sp.	○	○	Aptychus (Praestriaptychus)	E. Cretaceous (Aptian)	Doguzhaeva and Mikhailova 2002
			Karsteniceras ternbergense Lukeneder and Tanabe	○	○	Aptychus (Praestri-aptychus)[g]	E. Cretceous (Barremian)	Lukeneder and Tanabe 2002
		Crioceratidae	*Crioceratites* cf. *nolani* (Kilian)	?	○	Aptychus[h]	E. Cretaceous (Hauterivian)	Keupp 2000; Engeser and Keupp 2002; Frerichs 2004
			Aegocrioceras compressum Rawson	○	○	Aptychus[h]	E. Cretaceous (Hauterivian)	Engeser and Keupp 2002; Frerichs 2004

Table 10.1 (continued)

Suborder	Superfamily	Family	Species	Uj	Lj	Jaw morphotype	Age	Sources
Ancyloceratina	Turrilitoidea	Anisoceratidae	*Allocrioceras* cf. *annulatum* (Shumard)	?	○	Aptychus (Praestriaptychus)	L. Cretaceous (Turonian)	Wippich and Lehmann 2004
		Nostoceratidae	*Pravitoceras sigmoidale* Yabe	?	○	Aptychus (Striaptychus)[e]	L. Cretaceous (Maastrichtian)	Matsunaga et al. 2008
			Didymoceras nebrascense (Meek and Hayden)	○	○	Aptychus (? Striaptychus)	L. Cretaceous (Campanian)	Kruta et al. 2010
			Nostoceras sp.	?	○	Aptychus (Striaptychus)	L. Cretaceous (Maastrichtian)	This study (UMUT MM 31057)
		Diplomoceratidae	*Scalarites mihoensis* Wright and Matsumoto	○	○	Aptychus (Striaptychus)	L. Cretaceous (Coniacian)	Kanie et al. 1978; Tanabe et al. 1980b
			Polyptychoceras pseudogaultinum (Yokoyama)	?	○	Aptychus (Striaptychus)	L. Cretaceous (Santonian)	Nagao 1931b; c; Tanabe and Landman 2002; Kruta et al. 2009
			Polyptychoceras sp.	○	○	Aptychus (Striaptychus)	L. Cretaceous (Santonian)	Tanabe 2011; this study (UMUT MM 30878)
			Subptychoceras sp.	○	○	Aptychus (Striaptychus)	L. Cretaceous (Santonian)	Tanabe and Landman 2002; this study (UMUT MM 31058)
		Baculitidae	*Sciponoceras kossmati* (Nowak)	?	○	Aptychus (Striaptychus)	L. Cretaceous (Turonian)	Tanabe and Landman 2002
			Sciponoceras sp.	?	○	Aptychus (Striaptychus)	L. Cretaceous (Cenomanian)	Breitkreutz et al. 1991; Kaplan et al. 1998

Table 10.1 (continued)

Suborder	Superfamily	Family	Species	Uj	Lj	Jaw morphotype	Age	Sources
Ancyloceratina	Turrilitoidea	Baculitidae	*Baculites* cf. *princeps* Matsumoto and Obata	?	○	Aptychus (Rugaptychus)	L. Cretaceous (Campanian)	Tanabe and Landman 2002
			Baculites knorrianus Desmarest	?	○	Aptychus (Rugaptychus)	L. Cretaceous (Maastrichtian)	Schlüter 1875, 1876
			Baculites leopoldiensis Nowak	?	○	Aptychus (Rugaptychus)	L. Cretaceous (Campanian)	Nowak 1908
			Baculites sp.	○	○	Aptychus (Rugaptychus)	L. Cretaceous (Campanian)	Landman et al. 2007; Kruta et al. 2009, 2011
			Baculitidae, genus & species indeterminate	○	○	Aptychus	L. Cretaceous (Cenomanian)	Klug et al. 2012
	Scaphitoidea	Scaphitidae	*Yezoites puerculus* (Jimbo)	?	○	Aptychus (Striaptychus)	L. Cretaceous (Turonian)	Nagao 1931b, c), Tanabe and Landman 2002
			Yezoites bladensis (Schlüter)	?	○	Aptychus (Striaptychus)	L. Cretaceous (Turonian)	Fritsch 1893
			Scaphites geinitzi d'Orbigny	?	○	Aptychus (Striaptychus)	L. Cretaceous (Turonian)	Fritsch and Schlönbach 1872; Keupp 2000
			Scaphites fischeri Riedel	?	○	Aptychus (Striaptychus)	L. Cretaceous (Campanian)	Kennedy 1986; Kennedy and Kaplan 1995
			Scaphites cobbani Birkelund	?	○	Aptychus (Striaptychus)	L. Cretaceous (Campanian)	Birkelund 1965
			Hoploscaphites greenlandicus (Donovan)	?	○	Aptychus (Striaptychus)	L. Cretaceous (Campanian)	Schlüter 1876; Kennedy and Kaplan 1997

Table 10.1 (continued)

Suborder	Superfamily	Family	Species	Uj	Lj	Jaw morphotype	Age	Sources
Ancyloceratina	Scaphitoidea	Scaphitidae	*Hoploscaphites nicolletii* (Morton)	?	○	Aptychus (Striaptychus)	L. Cretaceous (Maastrichtian)	Tanabe and Landman 2002
			Hoploscaphites nebrascensis (Owen)	○	○	Aptychus (Striaptychus)	L. Cretaceous (Maastrichtian)	Meek and Hayden 1864; Kennedy and Cobban 1976; Landman and Waage 1993; Kennedy et al. 2002
			Hoploscaphites spedeni (Landman and Waage)	?	○	Aptychus (Praestriaptychus)	L. Cretaceous (Maastrichtian)	Landman and Waage 1993
			Hoploscaphites dorfi (Landman and Waage)	○	○	Aptychus (? Praestriaptychus)	L. Cretaceous (Maastrichtian)	Landman and Waage 1993
			Discoscaphites gulosus (Morton)	?	○	Aptychus (Praestriaptychus)	L. Cretaceous (Maastrichtian)	Landman and Waage 1993
			Trachyscaphites spiniger (Schlüter)	?	○	Aptychus (Striaptychus)	L. Cretaceous (Campanian)	Schlüter 1876; Kennedy 1986; Kennedy and Kaplan 1997
			Rhaeboceras halli (Meek and Hayden)	○	○	Aptychus (? Praestriaptychus)	L. Cretaceous (Campanian)	Kennedy et al. 2002; Kruta et al. 2013

[a] Described as anaptychus-type by Tanabe and Mapes (1995)
[b] Described as "anaptychus" by Zakharov (1974)
[c] Described as Lamellaptychus by Zakhrov and Lominadze (1983)
[d] Originally described as rhynchaptychus by Vašíček (2010)
[e] The lower jaw is preserved near the shell aperture
[f] Jaws were found individually, but their taxonomic relationships were assumed from their mode of occurrence
[g] Mistakenly identified as Lamellaptychus by Lukeneder and Tanabe (2002)
[h] Originally described as anaptychus by the listed authors, but reinterpreted here as the inner chitinous lamella of aptychus-type lower jaw

10.3.3 Ammonoid Jaw Morphotypes

In situ jaw apparatuses of the Ammonoidea are known from 152 species (including two indeterminate species) of 109 genera that are distributed in the 30 superfamilies of 8 suborders (Goniatitina, Gephuroceratina, Prolecanitina, Ceratitina, Phylloceratina, Lytoceratina, Ammonitina, and Ancyloceratina; Table 10.1). Measurements of the jaw apparatuses of 40 species of Mesozoic ammonoids belonging to 35 genera of 13 superfamilies and of 3 extant *Nautilus* species are shown in Table 10.2. They indicate that ammonoids generally have a larger jaw apparatus in relation to shell diameter or whorl height than in modern *Nautilus*. Co-occurrences of upper and lower jaws are known from 41 genera, and only lower jaws are reported for the rest of the genera. The reason for the rarer occurrence of upper jaws than lower jaws presumably owes to their smaller size versus lower jaws and the absence of a calcareous upper jaw element in most taxa. The jaw apparatuses of the Ammonoidea described previously can be classified into the following five morphotypes on the basis of their overall morphology and composition, especially of the presence or absence of calcareous jaw elements (Fig. 10.4a, b, c, d): (1) normal type (Lehmann et al. 1980; Lehmann 1980, 1988), (2) anaptychus type (Lehmann 1990), (3) aptychus type (Lehmann 1990), (4) rhynchaptychus type (Lehmann et al. 1980; Lehmann 1988, 1990; Tanabe et al. 2013; = neoanaptychus type of Dagys et al. 1989), and 5) intermediate type (this paper). The normal-type jaw apparatus is known from a number of pre-Jurassic ammonoid suborders (Table 10.1), namely, Gephuroceratina (Woodward 1885), Prolecanitina (Doguzhaeva 1999), Goniaitina (Figs. 10.4a, 10.8.1, 10.8.2, 10.8.3; Closs 1967; Saunders and Richardson 1979; Mapes 1987; Bandel 1988; Tanabe and Mapes 1995; Doguzhaeva et al. 1997), and Ceratitina (Fig. 10.8.4, 10.8.5; Zakharov 1974; Dagys and Weitschat 1988; Klug and Jerjen 2012). The remaining four jaw morphotypes are known from the order Ammonitida of Jurassic and Cretaceous age (Table 10.1, Fig. 10.4b, c, d). The morphotypic classification of ammonoid jaw apparatuses has been based mainly on the overall shape and structure of the lower jaws (Lehmann et al. 1980; Lehmann 1988, 1990).

10.3.3.1 Normal Type

The upper jaw is equally sized to and/ or slightly smaller than the lower jaw (Figs. 10.4a, 10.8). Both jaws consist of a black, possibly originally chitinous material. Doguzhaeva et al. (1997, fig. 10.2C, D) reported a calcareous layer that exists in the connecting space between the outer and inner chitinous lower jaw lamellae of a Carboniferous *Girtyoceras* (Dimorphoceratoidea, Goniatitina). However, there is a possibility that the calcareous layer was formed secondarily during diagenesis, because the anterior calcified tips of modern nautilid and rhynchaptychus-type ammonoid lower jaws rest on the underlying chitinous layer (see Tanabe et al. 1980a, figs. 3–6). The upper jaw consists of a short, reduced outer lamella (hood) and a large posteriorly elongated inner lamella (lateral wall and crest) with a distinct U- or

Table 10.2 Measurement data of *in situ* jaw apparatuses of Mesozoic ammonoids and extant *Nautilus*. Abbreviations: *SD* shell diameter, *WH* whorl height, *MLU* maximum lateral wall length of upper jaw, *MWL* maximum wing length of lower jaw, [M] macroconch, [m] microconch

Suborder	Superfamily	Species	Specimen and source	*SD* (mm)	*WH* (mm)	*MLU* (mm)	*MWL* (mm)	*MWL*/*MLU*	*MWL*/*SD*	*MWL*/*WH*
Ceratitina	Notitoidea	*Bajarunia euomphala* (Keyserling)	GPIMH 3567 (Dagys and Weitschat 1988)	c. 48	c. 25	?	c. 14	?	c. 0.29	c. 0.56
	Ceratitoidea	*Ceratites penndorfi* Rothe	SMNS 67494 (Klug and Jerjen 2012)	c. 194	?	c. 28	c. 35	c. 1.25	c. 0.18	?
		Gymnotoceras rotelliformis (Meek)	GPIMH 3566 (Dagys and Weitschat 1988)	> 46	> 24	?	> 16	?	c. 0.35	c. 0.67
	Ptychitoidea	*Aristoptychites kolymensis* (Kiparisova)	GPIMH 3564 (Dagys and Weitschat 1988)	c. 37	c. 18	?	c. 13	?	c. 0.35	c. 0.72
Phylloceratina	Phylloceratoidea	*Phyllopachyceras ezoense* (Yokoyama)	UMUT MM 27831	> 39.4	18.3	16.4	18.6	1.13	< 0.47	1.02
		Hypophylloceras subramosum (Shimizu)	KMNH IvP 902011	?	> 31.5	20.6	30.1	1.46	?	< 0.96
Lytoceratina	Tetragonitoidea	*Tetragonites glabrus* (Jimbo)	GK H 2073	81.1	38.6	?	28.4	?	0.35	0.74
		Anagaudryceras limatum (Yabe)	UMUT MM 30877	140.0	44.0	?	34.5	?	0.25	0.78
		Gaudryceras denseplicatum (Jimbo)	UMUT MM 19875	41.6	17.1	?	11.9	?	0.29	0.70
Ammonitina	Hildoceratoidea	*Harpoceras capellinum* Schlotheim	SMNS unregistered specimen	119.4	45.0	?	40.0	?	0.34	0.89
		Harpoceras falciferum Sowerby	UMUT MM 31054	197.5	66.9	?	44.1	?	0.22	0.66
			UMUT MM 31055	168.2	52.0	?	39.6	?	0.24	0.76
		Eleganticeras? sp.	SMNS unregistered specimen	122.6	39.7	?	32.7	?	0.27	0.82
		Sonninia espinazitensis Tornquist, [M]	SMNS 33738 (Dietze et al. 2012)	171.4	98.5	?	74.2	?	0.43	0.75

Table 10.2 (continued)

Suborder	Superfamily	Species	Specimen and source	*SD* (mm)	*WH* (mm)	*MLU* (mm)	*MWL* (mm)	*MWL/ MLU*	*MWL/ SD*	*MWL/WH*
Ammonitina	Perisphinctoidea	*Physodoceras nattheimense* Schweigert	SMNS 63232	32.6	10.9	11.8	12.3	1.04	0.38	1.13
			SMBS 70072	194.0	80.0	42.0	60.5	1.44	0.31	0.76
			SMNS 63288	61.5	27.3	?	24.0	?	0.39	0.88
			SMNS 63830	81.2	?	?	32.9	?	0.41	?
			SMNS unregistered specimen	37.1	16.6	?	14.8	?	0.40	0.89
		Subplanites? sp.	SMNS unregistered specimen	112.6	42.8	?	30.5	?	0.27	0.71
		Simbirskites decheni (Roemer)	Baraboshkin & Shumilkin (2010)	81.1	47.0	?	27.2	?	0.34	0.58
		Hybonoticeras hybonotum (Oppel)	SMNS unregistered specimen	60.3	20.4	?	20.2	?	0.33	0.99

Table 10.2 (continued)

Suborder	Superfamily	Species	Specimen and source	*SD* (mm)	*WH* (mm)	*MLU* (mm)	*MWL* (mm)	*MWL/MLU*	*MWL/SD*	*MWL/WH*
Ammonitina	Haploceratoidea	*Lingulaticeras* sp.	SMNS 67658	37.0	15.3	?	12.8	?	0.35	0.84
			SMNS unregistered specimen	30.6	11.2	?	11.8	?	0.39	1.05
		Fontannesiella prolithographica (Fontannes)	SMNS 66075	76.0	32.0	13.0	25.6	1.97	0.34	0.80
		Neochetoceras subnudatum (Fontannes)	SMNS 62784	112.7	45.4	?	42.0	?	0.37	0.93
		Metahaploceras sp.	SMNS 67274	47.8	18.1	9.7	18.3	1.89	0.38	1.01
			SMNS 63998	56.3	20.1	8.8	18.8	2.14	0.33	0.94
			SMNS unregistered specimen	58.5	23.4	?	20.3	?	0.35	0.87
			SMNS unregistered specimen	61.0	24.2	?	22.6	?	0.37	0.93
			SMNS unregistered specimen	52.1	19.9	?	17.9	?	0.34	0.90
			SMNS unregistered specimen	60.2	23.1	?	20.6	?	0.34	0.89
			SMNS unregistered specimen	57.4	22.7	?	20.0	?	0.35	0.88
		Streblites cf. *zlatarskii* (Sapunov)	SMNS 67886	177.8	63.0	?	56.0	?	0.31	0.89
	Acanthoceratoidea	*Prohysteroceras* sp.	UMUT MM 31056	13.2	6.0	?	5.3	?	0.40	0.88

Table 10.2 (continued)

Suborder	Superfamily	Species	Specimen and source	*SD* (mm)	*WH* (mm)	*MLU* (mm)	*MWL* (mm)	*MWL/MLU*	*MWL/SD*	*MWL/WH*
Ammonitina	Desmoceratoidea	*Damesites* aff. *sugata* Forbes	UMUT MM 27833	114.8	58.9	?	31.0	?	0.27	0.53
		Damesites semicostatus Matsumoto	UH 4545	29.7	14.8	11.0	11.2	1.02	0.38	0.76
			UMUT MM 30879	34.8	19.0	?	14.6	?	0.42	0.77
		Menuites naumanni (Yokoyama)	UMUT MM 27835	131.2	61.6	> 23.5	49.0	< 2.09	0.37	0.80
		Pachydiscus kamishakensis Jones	UMUT MM 30876	305.0	175.7	?	102.9	?	0.34	0.59
	Hopliceratoidea	*Placenticeras costatum* Hyatt	USNM 529075	> 170	> 86	?	29.3	?	< 0.17	< 0.34
Ancyloceratina	Scaphitoidea	*Hoploscaphites nicolletii* (Morton), [m]	YPM 32304	69.7	30.5	?	15.2	?	0.22	0.50
			YPM 32303	62.2	23.8	?	18.9	?	0.30	0.79
			YPM 32309	59.5	28.6	?	16.5	?	0.28	0.58
		Hoploscaphites nebrascenis (Owen), [m]	USNM 386	?	37.0	18.6	27.0	1.45	?	0.73
		Rhaeboceras halli (Meek & Hayden)	BHMNH 4334	> 130.8	56.9	25.7	34.6	1.35	< 0.26	0.61

Table 10.2 (continued)

Suborder	Superfamily	Species	Specimen and source	*SD* (mm)	*WH* (mm)	*MLU* (mm)	*MWL* (mm)	*MWL/ MLU*	*MWL/ SD*	*MWL/WH*
Ancyloceratina	Turritoidea	*Nostoceras* sp.	UMUT MM 31057		38.5	?	21.2	?	?	0.55
		Didymoceras nebrascense (Meek & Hayden)	AMNH 53811 (Kruta et al. 2010)		43.0	?	25.7	?	?	0.60
		Scalarites mihoensis Wright & Matsumoto	GK H 1451		40.0	22.1	24.4	1.10	?	0.61
		Polyptychoceras cf. *pseudogualtinum* Yokoyama	UH 4592		20.4	?	15.8	?	?	0.77
		Polyptychoceras sp.	UMUT MM 30878		16.0	11.3	16.7	1.48	?	1.04
		Subptychoceras sp.	UMUT MM 27834		32.4	23.1	26.0	1.13	?	0.80
			UMUT MM 31058		23.9	?	19.1	?	?	0.80
		Sciponoceras kossmati (Nowak)	GK H 4335		24.7	?	20.1	?	?	0.81
		Baculites cf. *princeps* Matsumoto & Obata	GK H 4909		7.6	?	7.1	?	?	0.93
		Baculites sp. (smooth)	BHMNH 5495		28.7	?	33.8	?	?	1.18
			BHMNH 47109		26.5	?	32.2	?	?	1.22
			BHMNH 5491		30.0	?	23.9	?	?	0.80
			BHMNH 5147		22.3	?	25.2	?	?	1.13

Table 10.2 (continued)

Order	Family	Species	Specimen and source	*SD* (mm)	*WH* (mm)	*MLU* (mm)	*MWL* (mm)	*MWL*/*MLU*	*MWL*/*SD*	*MWL*/*WH*
Nautilida	Nautilidae	*Nautilus belauensis* Saunders	Palau (T-3-1, male), UMUT RM 27882-1	206.5	132.8	42.8	40.5	0.94	0.20	0.30
		Nautilus belauensis Saunders	Palau (T-2-14, male), UMUT RM 27881-2	148.2	97.8	32.2	32.4	1.01	0.22	0.33
		Nautilus belauensis Saunders	Palau (T-9-4, male), UMUT RM 27883-1	197.4	127.5	41.5	42.1	1.01	0.21	0.33
		Nautilus pompilius Linnaeus	Suva, Fiji (SV28-3-2, male), UMUT unreg. Specimen	144.8	94.8	32.8	31.0	0.95	0.21	0.33
		Nautilus pompilius Linnaeus	Suva, Fiji (SV 28-4-1, male), UMUT RM 31011	139.8	91.1	30.8	26.8	0.87	0.19	0.29
		Nautilus macrompha-lus Sowerby	Toba Aquarium, hatchling, UMUT unreg. Speimen	27.8	16.0	6.1	6.1	1.00	0.22	0.38

V-shaped median curvature in the posterior portion (Fig. 10.8). Both lamellae are joined in the anterior portion.

The lower jaw is characterized by having a posteriorly elongated outer lamella (wing), a shortly reduced inner lamella (crest and lateral wall), and a sharply pointed rostrum, and in these features, it is fairly similar to the lower jaws of modern nautilids (Fig. 10.4e), except for the absence of an anterior calcareous covering.

10.3.3.2 Anaptychus Type

The upper jaw consists of short reduced outer and widely open longer inner chitinous lamellae, without any trace of a calcareous element. Unlike that of the normal type, the inner lamella is distinctly divided posteriorly into paired lateral walls. These lamellae are joined together in the anterior portion forming a sharply pointed rostrum. The lower jaw of this morphotype (= anaptychus *sensu stricto*, Figs. 10.4b, 10.9.2) was defined by having a widely open outer lamella (wing), a short and reduced inner lamella (crest and lateral wall), and a more or less pointed rostrum (Lehmann 1970, 1980). The anaptychus-type jaw apparatuses are known from the Jurassic Psiloceratidae and Arietitidae (Psiloceratoidea) (Schmidt 1928; Lehmann 1970, 1975; Cope and Sole 2000; Keupp 2000), Dactylioceratidae, Eoderoceratidae, and Amaltheidae (Eoderoceratoidea; see Schmidt 1928; Lehmann 1979; Cope 1994; Cope and Sole 2000; Keupp 2000), all of which belong to the suborder Ammonitina (Table 10.1). In addition, Schmidt (1928, pl. 15, figs. 1–2) and Hauff (1953, pl. 75, fig. F.) respectively figured as an anaptychus, a lower jaw preserved *in situ* in the body chambers of *Lytoceras cornucopiae* (Lytoceratina) and *Phylloceras heterophyllum* (Phylloceratina) from the Toarcian (Lower Jurassic) Posidonia Shale of Holzmaden, southern Germany. The lytoceratid and phylloceratid specimens with a lower jaw are, however, flattened on a plane, and their aragonitic shells have been dissolved away during early diagenesis. The mode of preservation of these specimens prevents exact morphotypic assignment of the lower jaws of the two species.

According to the classical definition by Arkell et al. (1957), anaptychi consist wholly of a chitinous material. This definition is applied to the lower jaws of the psiloceratids (Fig. 10.4b), dactylioceratids (Fig. 10.9.2), and amaltheids, but not to those of the arietitids (*Asteroceras* and *Arnioceras*), and eoderoceratid (*Promicroceras*), whose outer chitinous lamella is wholly covered with a thin univalved calcareous layer (Lehmann 1971; Cope 1994; Cope and Sole 2000; Keupp 2000). Based on this evidence, Tanabe et al. (2012) suggested that the anaptychus-type lower jaws might have originally had a thin calcareous layer on the ventral side of the outer chitinous lamella. According to their interpretation, the absence of the calcareous layer in some anaptychus-type lower jaws is due to the fact that it dissolved away during diagenesis and/or was secondarily exfoliated during preparation.

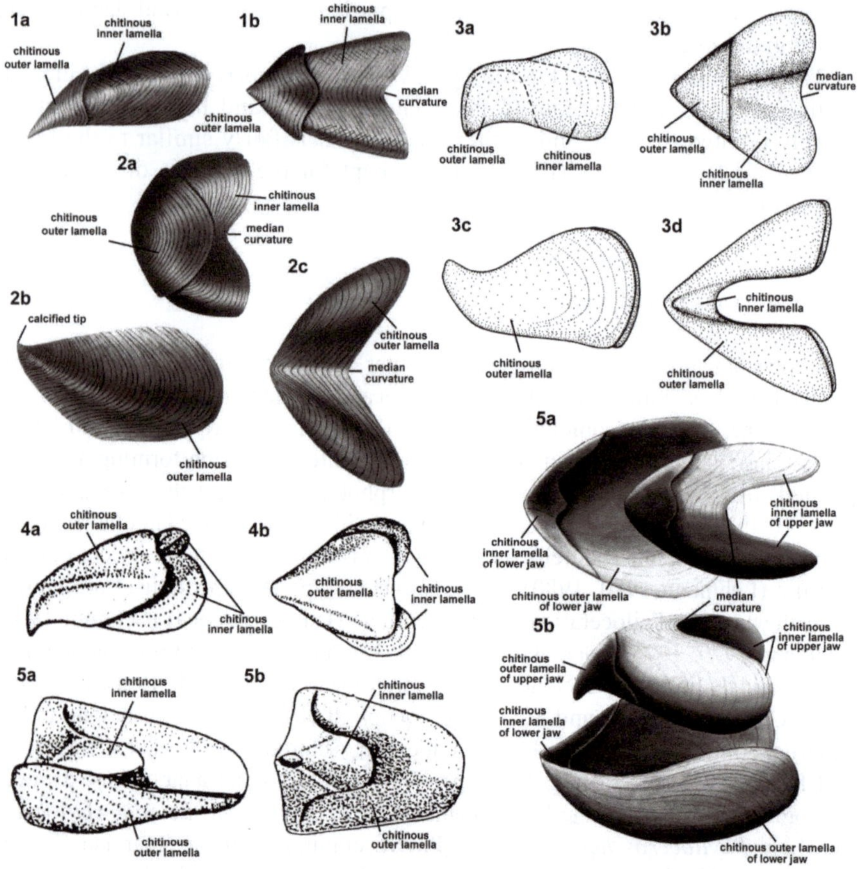

Fig. 10.8 Drawings of the normal-type jaw apparatuses of Paleozoic and Triassic ammonoids. **1** Right lateral (*1a*) and dorsal (*1b*) views of upper jaw of Goniatitina, genus and species inderminate (Late Mississippian Goniatitina). After Doguzhaeva et al. (1997, fig. 4). **2** Upper (*2a*; dorsal view) and lower (*2b, c*; left lateral and ventral views respectively) jaws of *Girtyoceras limatum* (Miller and Faber) (Late Mississippian Goniatitina). After Doguzhaeva et al. (1997, fig. 4). **3** Left lateral (*3a*) and dorsal (*3b*) views of upper jaw and left lateral (*3c*) and dorsal (*3d*) views of lower jaws of *Glaphyrites* sp. (Late Paleozoic Goniatitina). After Bandel (1988, fig. 6). **4** Left lateral (*4a*) and dorsal (*4b*) views of upper jaw and left lateral (*4c*) and dorsal (*4d*) views of lower jaw of a Middle Triassic ceratitid ammonoid. After Dagys and Weitschat (1988, fig. 1). **5** Dorsal (*5a*) and left lateral (*5b*) views of the jaw apparatus of *Ceratites penndorfi* Rothe (Middle Triassic Ceratitina), after Klug and Jerjen (2012, fig. 5A–C)

10.3.3.3 Aptychus Type

The jaw apparatus of this morphotype is characterized by the presence of a pair of calcareous plates (aptychus) in the lower jaws (Figs. 10.4c, 10.9.4b, 10.9.5c, d, 10.9.6b). The upper jaw is composed of a black chitinous material, and its architecture closely resembles that of the anaptychus-type upper jaw. The lower jaw comprises thinner inner chitinous and thicker outer calcareous elements. The inner chitinous element consists of a short inner lamella (crest and lateral wall) and a widely open, outer lamella with a deep median depression forming two wings. The short inner lamella appears to be vestigial and/ or absent in the Late Jurassic *Physodoceras* (Aspidoceratidae) and ?*Fontannesiella* (Oppeliidae) (Keupp 2007). The outer chitinous lamella is sculptured by dense, concentric growth lines and has an indentation at the midline joint in the posterior margin. The outer calcified element of the lower jaw, i.e., aptychus, is made of calcite (Schindewolf 1958; Landman et al. 2007; Kruta et al. 2009). It is distinctly partitioned into paired plates along the harmonic midline joint termed "*Symphysenrand*" by Trauth (1927), whose English equivalent, symphysis, was introduced by Arkell (1957). The aptychus covers the wide area of the underlying chitinous membrane.

The aptychus-type jaw apparatus is widely distributed in the Jurassic Ammonitina (Hildoceratoidea, Stephanoceratoidea, Perisphinctoidea, Haploceratoidea, Spiroceratoidea; Figs. 10.5a, b, c; 10.6a, b, c, d), the Cretaceous Ammonitina (Acanthoceratoidea, Holplitoidea; Fig. 10.7c), and the Cretaceous Ancyloceratina (Ancyloceratoidea, Turrilitoidea; Figs. 10.5g, 10.7d, e; Table 10.1). In most genera, the lower jaw is larger than the upper jaw (Fig. 10.5c), sometimes attaining 200% or more of the length of the upper jaw (e.g., *Fontannesiella prolithographica* and *Metahaploceras* sp. of the Haploceratoidea, see Table 10.2; Lehmann 1980; see also Kruta et al. 2011, fig. 1c). The Late Jurassic perisphinctoid *Physodoceras* and Late Cretaceous turrilitoids *Scalarites* and *Subptychoceras*, however, have almost equal-sized upper and lower jaws (Table 10.2; Fig. 10.5b; see also Schweigert and Dietl 1999, pl. 4, fig. 1; Tanabe et al. 1980b, pl. 20; Tanabe and Landman 2002, pl. 1, fig. 7). The Late Jurassic aptychi of the Aspidoceratidae consist of three layers with different microstructure, i.e., a basal lamellar layer (Fig. 10.10f), a middle 'honey-comb like' tubular layer (Fig. 10.10e), and an upper lamellar layer (Schindewolf 1958; Closs 1960; Farinacci et al. 1976; Hewitt et al. 1993). The basal layer occurs on the convex side of the aptychus and forms the co-marginal lirae that are visible on the inner (dorsal) side of the aptychus and are interpreted as growth lines (Kruta et al. 2009; Fig. 10.10f). The aptychi of Cretaceous Ancyloceratina (e.g., *Baculites, Hoploscaphites,* and *Polyptychoceras*) differ from those of Jurassic Ammonitina in having one or two layers without a sponge-like structure (Kruta et al. 2009; Fig. 10.10d). The surface sculpture and lamellar microstructure of aptychi exhibit wide taxonomic variation; hence, they have been classified into several form genera such as cornaptychus, striaptychus, granulaptychus, lamellaptychus, laevaptychus and synaptychus by Trauth (1927–1936).

In the Jurassic Ammonitina such as *Hildoceras* (Fig. 10.4c), *Streblites* (Fig. 10.6c) and *Dactylioceras* (Fig. 10.9.2), the anterior margin of the lower jaw is gently arched toward the ventral side without a beak-like projection (Lehmann 1972, 1975, 1979). By contrast, this portion is sharply pointed in some Cretaceous Ammonitina such as *Subprionocyclus* (Fig. 10.9.4b; Tanabe and Fukuda 1987b) and *Metaplacenticeras* (see Landman et al. 2006), and Ancyloceratina (e.g., *Scalarites*; Tanabe et al. 1980b). The lower jaws of *Subprionocyclus* and *Placenticeras* also respectively possess rows of serrated ridges and grooves and a V-shaped slit in the anterior portion (Fig. 10.9.4b, 10.9.5c; Tanabe and Fukuda 1987b; Landman et al. 2006).

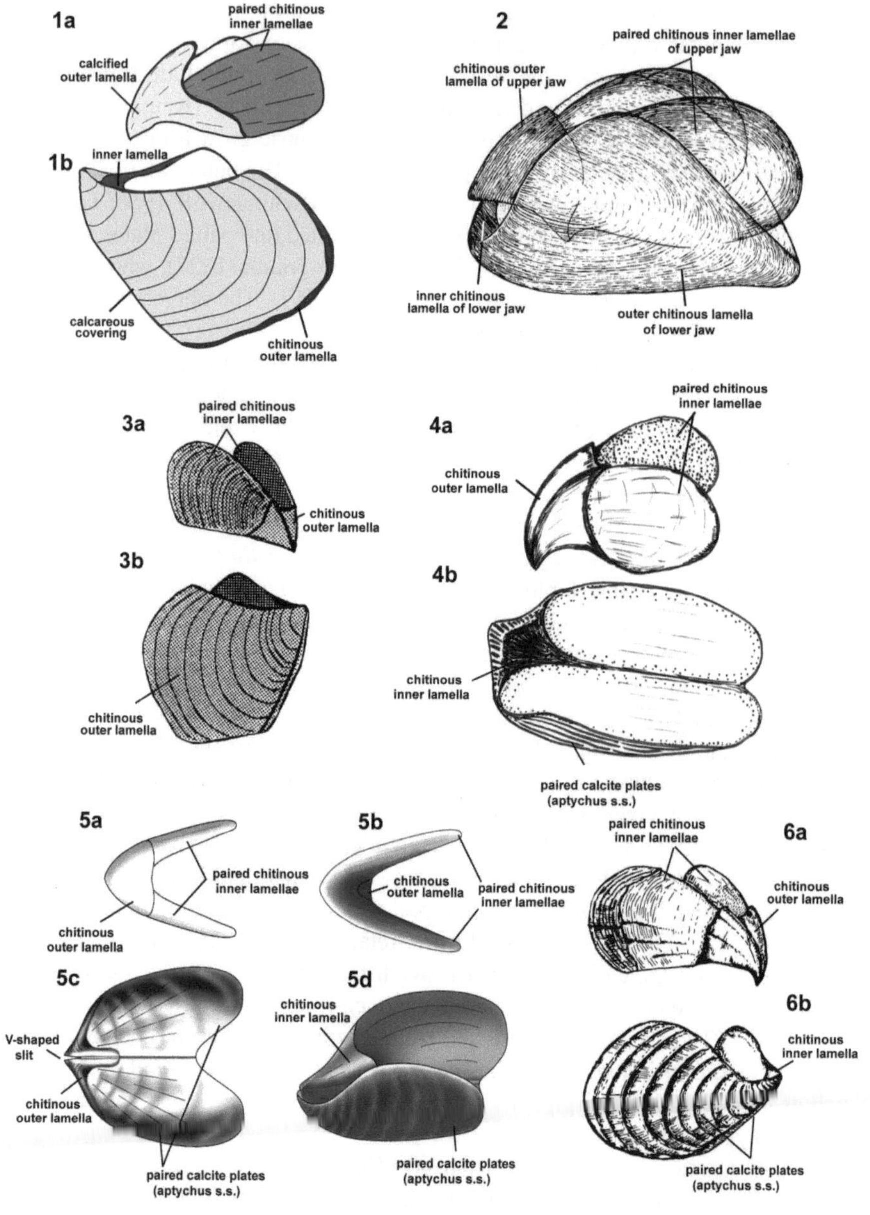

Fig. 10.9 Drawings of the jaw apparatuses of Mesozoic Ammonitida. **1** Left lateral view of upper (*1a*) and lower (*1b*) jaws of *Hypophylloceras subramosum* (Shimizu) (Late Cretaceous Phylloceratidae, Phylloceratina). **2** Jaw apparatus of *Dactylioceras semicelatum* (Simpson) (Early Jurassic Dactylioceratidae, Ammonitina). After Lehmann (1979, fig. 2). **3** Anterolateral views of upper (*3a*) and lower (*3b*) jaws of *Damesites semicostatus* Matsumoto (Late Cretaceous Desmoceratidae, Ammonitina). After Tanabe (1983, fig. 3A, C). **4** Dorsolateral views of upper (*4a*) and lower (*4b*) jaws of *Subprionocyclus minimus* (Hayasaka and Fukada) (Late Cretaceous Collignoniceratidae, Ammonitina). Modified from Tanabe and Fukuda (1987, fig. 3). **5** Dorsal (*5a*) and ventral (*5b*)

10.3.3.4 Rhynchaptychus Type

The rhynchaptychus-type jaw apparatus is characterized by the development of a calcareous rostral tip on both upper and lower jaws (Fig. 10.4d, 5d). The upper jaw consists of a shorter outer lamella (hood) and a pair of larger inner lamellae (lateral walls) that become narrower and join together in the anterior portion, as in those of anaptychus- and aptychus-types (Fig. 10.5d). The lower jaw is built up of a short reduced inner lamella (crest and lateral wall) and a large, gently convex outer lamella (wing) without a median groove that is wholly covered with a thin calcareous layer (Figs. 10.5d, e, 10.7a, b). The jaw apparatus of this morphotype is distributed in the Cretaceous Phylloceratina (*Hypophylloceras*: Fig. 10.5d; *Phyllopachyceras*: Tanabe and Landman 2002; Tanabe et al. 2013) and Lytoceratina (*Tetragonites*, *Gaudryceras*, *Aanagaudryceras*; Fig. 10.5e, 10.7a, b; Table 10.1; Tanabe et al. 1980a, 2012; Lehmann et al. 1980; Kanie 1982; Tanabe and Landman 2002). In the two phylloceratids, the lower jaw is slightly larger than the upper jaw (Table 10.2). Meanwhile, the 'upper' jaws of *Tetragonites* and *Gaudryceras* described by Tanabe et al. (1980a, fig. 9c, d) and Kanie (1982, figs. 4, 7) are judged as deformed lower jaws, because they show no morphological resemblance to the upper jaws of any other known ammonoids and instead exhibit similar shape and structure to the lower jaws of these and other genera of the Phylloceratina and Lytoceratina (Tanabe and Landman 2002). The upper jaws of the Lytoceratina are therefore still unknown. In addition to these *in situ* jaw apparatuses, three isolated lower jaws described by Lehmann et al. (1980, figs. 3b, c, e) as lytoceratoid anaptychi from the Jurassic of northern Europe may also belong to the rhynchaptychus-type in regard to the presence of a notch in the anterior portion that may have originally been occupied by a calcified tip. The anterior calcified tip of an isolated lower jaw referred to *Gaudryceras* has distinct denticles around the oral margin (Kanie et al. 1978; Tanabe et al. 1980a; Kanie 1982), as in modern nautilids (Okutani and Mikami 1977; Saunders et al. 1978).

The outer calcareous layer on the lower jaw of *Anagaudryceras* is made of aragonite and exhibits granular microstructure (Tanabe et al. 2012; Fig. 10.10a), whereas the calcified tip of the upper jaw and the outer calcareous layer of the lower jaw in *Hypophylloceras* are both made of calcite (Tanabe et al. 2013). The anterior calcareous tips of the rhynchaptychus-type jaws fairly resemble in shape and internal microstructure not only those of modern and fossil nautilid jaws but also of the isolated arrowhead- and scallop-shaped calcareous remains called rhyncholites and conchorhynchs that have been found from the Late Paleozoic and younger marine deposits (Teichert et al. 1964; Teichert and Spinosa 1971; Riegraf and Luterbacher 1989). This fact suggests that at least some previously known Jurassic and Creta-

---

views of upper jaw and dorsal (*5c*) and dorsolateral (*5d*) views of lower jaw of a placenticeratid ammonoid (Late Cretaceous Placenticeratidae, Ammonitina). After Landman et al. (2006, fig. 30). **6** Anterolateral views of upper (*6a*) and lower (*6b*) of *Scalarites mihoensis* Wright and Matsumoto (Late Cretaceous Diplomoceratidae, Ancyloceratina). After Tanabe et al. (1980b, fig. 1) ceous counterparts belonged to either Lytoceratina or Phylloceratina (Tanabe et al. 1980a, 2013; Tanabe and Fukuda 1999).

10.3.3.5 Intermediate Type

The lower jaws of the Desmoceratoidea (Cretaceous Ammonitina) have been summarized in the anaptychus type (Nagao 1931a, c; Moore and Sylvester-Bradley 1957), because the fossil materials examined by previous authors consist wholly of a chitinous element. The lower jaws of *Damesites*, *Tragodesmoceroides* and *Pachydiscus* described by subsequent workers, however, exhibit transitional features from the anaptychus-type to the aptychus-type, by the presence of a distinct median groove on the outer horny lamella (Tanabe 1983, pl. 71, figs. 1d, 3b; see also Figs. 10.5f, 10.9.3b) and a very thin univalved calcareous layer, which covers the horny lamella (Tanabe et al. 2012, Figs. 5, 6). The mineralogy and microstructure of the calcareous layer exhibit some taxonomic variation; i.e., the spherulitic prismatic structure of aragonite in *Damesites* (Fig. 10.10c) and the polygonal prismatic structure of calcite in *Pachydiscus* (Fig. 10.10b) (Tanabe et al. 2012; Figs. 10.10b, c). The upper jaws of desmoceratids *Damesites* and *Menuites* consist of short reduced outer and widely open longer inner chitinous lamellae, without any trace of a calcareous element, as in those of anaptychus and aptychus types (Tanabe 1983; Tanabe and Landman 2002). We previously reported that the outer lamella of the lower jaw of *Menuites* lacks the median groove (Tanabe and Landman 2002), but our reexamination of the same material confirmed the presence of a weak median groove (see mg in Fig. 10.5f). Based on these observations, we exclude the jaw apparatuses of the Desmoceratoidea from the anaptychus-type jaws and treat them herein as an intermediate type.

10.4 Restoration of Ammonoid Buccal Mass Structure

Attachments scars of beccublasts have been found on the inner surface of the outer chitinous lamella (wing) of the isolated rhynchaptychus- and aptychus (laevaptychus)-type lower jaws that are attributed to the Cretaceous lytoceratid *Gaudryceras* and a Late Jurassic aspidoceratid, respectively (Fig. 10.2c, d; Tanabe and Fukuda 1983, fig. 2A-C; Tanabe and Fukuda 1999, fig. 19.5c, d), and on the outer surface of the inner lamella of the upper jaws of the Early Carboniferous goniatitid *Girtyoceras* (Doguzhaeva et al. 1997, fig. 5B) and of an unknown ammonoid (Tanabe et al. 2001, fig. 4.5, 4.6). The scars on these ammonoid jaws all consist of equally-sized circular to hexagonal pits. Each pit, measuring about 10 μm in diameter in *Girtyoceras* and in an aspidoceratid and 20–30 μm in diameter in *Gaudryceras*, is surrounded by a distinct ridge. They are similar in shape to the anchor-type attachment scars of beccublasts on the jaws of modern coleoids (Dil-

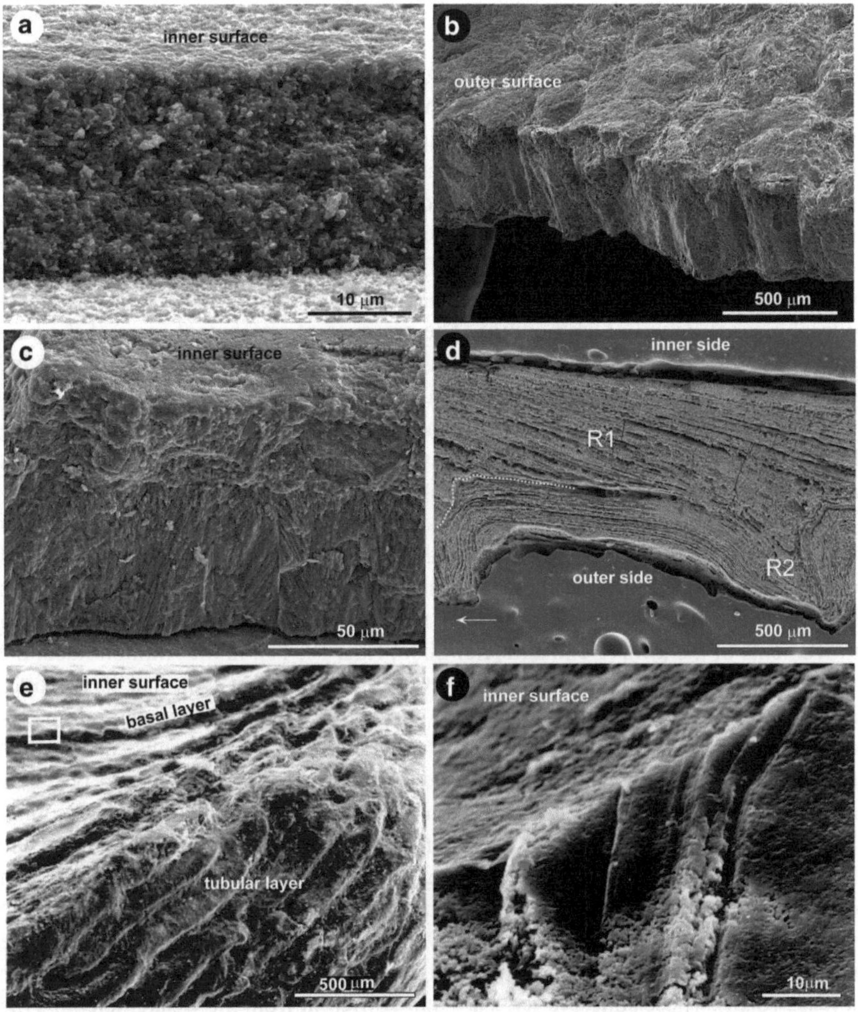

Fig. 10.10 Scanning electron micrographs of the outer calcareous element of lower jaws of Jurassic and Cretaceous ammonoids. **a** *Anagaudryceras limatum* (Yabe) (Gaudryceratidae, Lytoceratina). Rhynchaptychus-type. Cross section of the outer aragonitic layer, showing the granular microstructure. UMUT MM 30877 from the Coniacian of Haboro area, Hokkaido, Japan. **b** *Pachydiscus kamishakensis* Jones (Desmoceratidae, Ammonitina). Anaptychus-type. Fracture of the outer calcitic layer showing the columnar prismatic structure. UMUT MM 30876 from the Maastrichtian of southern Alaska. **c** *Damesites* aff. *sugata* Forbes (Desmoceratidae, Ammonitina). Anaptychus-type. Fracture of the outer aragonitic layer showing the spherulitic prismatic structure. UMUT MM 27833 from the Coniaciain of Haboro area, Hokkaido. **d** *Baculites* sp. (smooth to weak flank ribs) (Baculitidae, Ancyloceratina). Aptychus-type (rugaptychus). Cross section of the calcitic aptychus, showing the main lamellar layer (*R1*) and the outer layer (*R2*). BHMNH 5801 from the Campanian of Alabama. (**a–c** from Tanabe et al. 2012, Fig. 4B, D, F. **d** from Kruta et al. 2009, Fig. 3A; reproduced by permission of John Wiley & Sons Ltd., UK). **e–f** Bivalved calcitic plate (laevaptychus) of isolated lower jaw attributed to an aspidoceratid ammonoid. Cross-section of the laevaptychus consisting of thin basal layer and thick tubular layer (**e**) and close-up of the basal layer (**f**; squared portion in **e**) consisting of thin laminae inclined and overlapping one another. The upper lamellar layer is not shown in this figure. UMUT MM 19945, from the Oxfordian (Upper Jurassic) of Nusplingen, Germany

ley and Nixon 1976), but are clearly distinguished from those of modern nautilids, which are characterized by numerous micropores of 0.2–0.5 μm in diameter (Fig. 10.2a; Tanabe and Fukuda 1983, figs. 3D, 6). In modern cephalopods, a layer of tall beccublast cells, each about 5 μm in diameter, is intercalated in the space between the jaw muscles and the chitinous jaw lamellae, and the other sides of the jaw lamellae, i.e., the outer sides of the outer lamellae of the upper and lower jaws, in contrast, are free from jaw muscles, and are covered directly with a thin connecting tissue (Figs. 10.1a, 10.11a, b; Dilley and Nixon 1976; Tanabe and Fukuda 1983, 1999; Sasaki et al. 2010; Tanabe 2012). These anatomical relationships among jaw muscles, beccublasts and jaw lamellae in extant cephalopods help us to restore the buccal mass structure of ammonoids. The presence of attachment scars of the beccublasts on the inner side of the outer chitinous lamella (wing) of the lower jaw and on the outer surface of the inner lamella (lateral wall) of the upper jaw indicates that these sides were connected by jaw muscles, with a layer of beccublasts lying between them, and that the opposite sides of the jaws were free from jaw muscles and were covered directly with a thin connecting tissue (Figs. 10.11c, d).

In modern coleoids, the anterior and posterior mandibular muscles widely cover the outer surface of the posteriorly expanded inner lamella (crest and lateral wall) of the lower jaw and are connected to the lateral and dorsal sides of the inner lamella (lateral wall and crest) of the upper jaw; they likely serve for closing and shearing movements of the jaws (Ueno and Kier 2005, figs. 10, 11). In the Ammonoidea, the inner lamella of the lower jaw is, by contrast, much shorter than that in coleoids, and furthermore, the inner lamella of the upper jaw is distinctly divided into a pair of lateral walls without a crest region for the Jurassic and Cretaceous Ammonitida (Figs. 10.4b, c, d), suggesting a weaker development of the anterior and posterior mandibular muscles in Ammonoidea than in the Coleoidea. Reduction of the inner lamella of the lower jaw is also observed in modern and fossil nautilids (Okutani and Mikami 1977; Saunders et al. 1978; Klug 2001). To compensate for the reduction of the muscle attachment area, the beccublasts of extant *Nautilus* are branched into many fine trabeculae on the side of the jaw lamella, and the ends of the trabeculae are deeply inserted within the chitinous jaw lamella, forming numerous micropores on it (Fig. 10.2a). These adhesive scars are quite different from the weaker anchor-type attachment scars of beccublasts in modern coleoids and ammonoids. These observations indicate a firmer attachment of jaw muscles on the jaw lamella in *Nautilus* than in coleoids and ammonoids, thus underlining their phylogenetic relationship.

All of the calcified jaw elements in the Ammonoidea, i.e., anterior calcified tips of the upper and lower jaws, and the paired outer calcareous plates (aptychus), and a thin calcareous layer on the outer lamella of the lower jaw, were presumably secreted from the outside by the overlying epithelial tissue, as in the case of the calcareous tips of the present-day nautilids (Farinacci et al. 1976; Seilacher 1993; Tanabe and Fukuda 1999; Kruta et al. 2009).

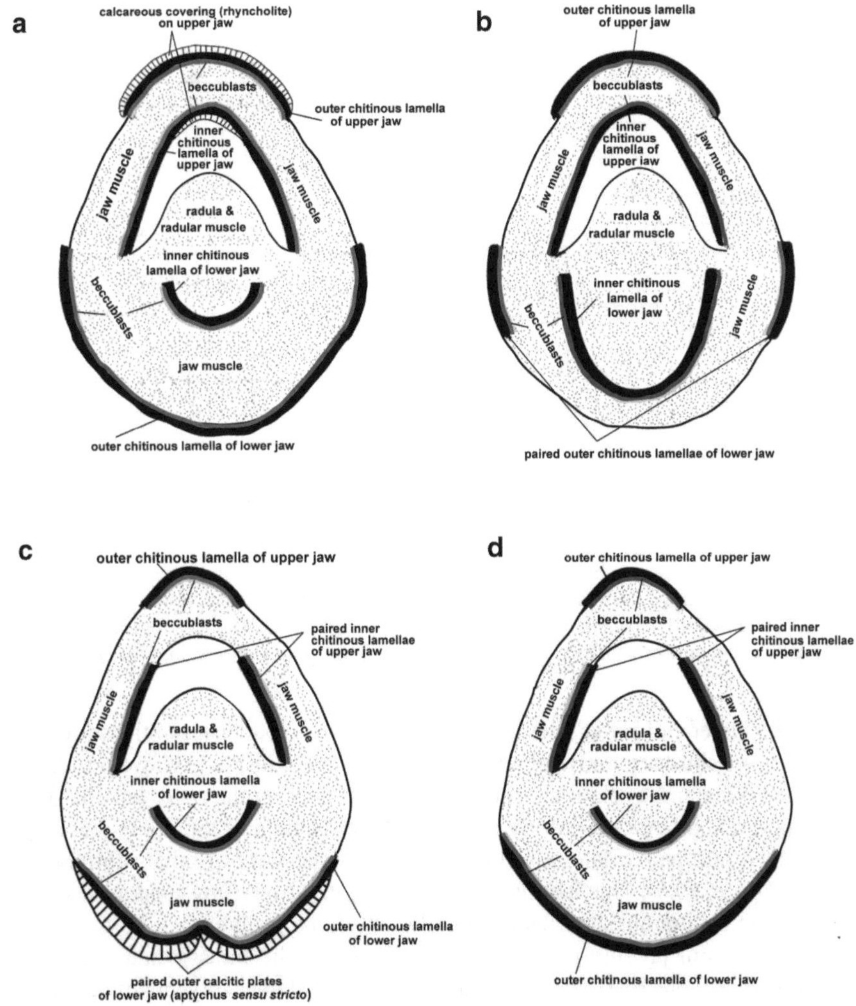

Fig. 10.11 Reconstruction of buccal mass structure of ammonoids in cross section with the aptychus- and anaptychus-type jaw apparatuses compared with those of modern cephalopods (*Nautilus* and *Sepia*). **a** *Nautilus* (Nautilida, Nautiloidea), **b** *Sepia* (Sepiida, Coleoidea), **c** Ammonoidea with aptychus-type jaw apparatus, **d** Ammonoidea with anaptychus-type jaw apparatus

10.5 Discussion

10.5.1 Taxonomic Evaluation of Jaw Morphology

Of the five recognized jaw morphotypes, the normal-type jaw apparatus is shared by the pre-Jurassic suborders, i.e., Gephuroceratina, Goniatitina, Prolecanitina and Ceratitina (Table 10.1; Fig. 10.8). Except for the absence of a thick calcified rostral

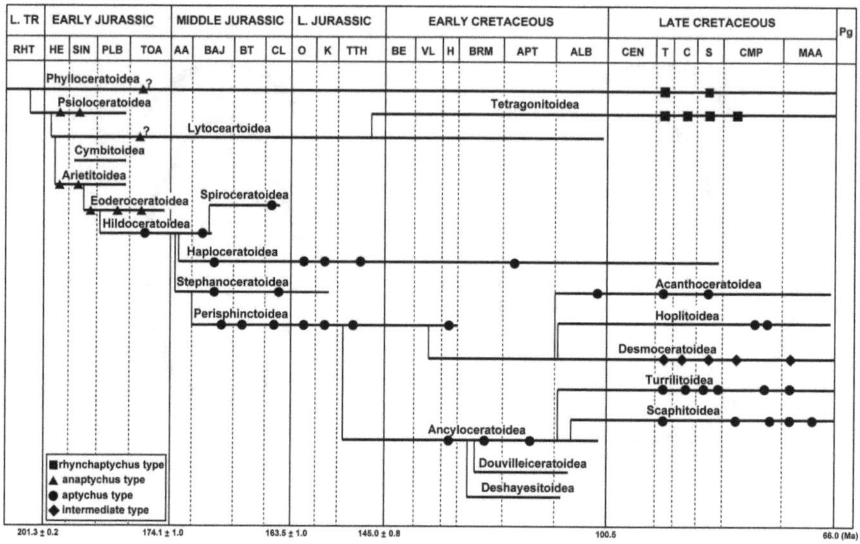

Fig. 10.12 Taxonomic distribution of jaw morphotypes in the Jurassic and Cretaceous ammonoid superfamilies. Modified from Page (1996, Fig. 2) for the phylogenetic tree, in which the Arietitoidea and Cymbioidea (*sensu* Page 1996) are treated as families in the Psiloceratoidea, on the basis of Jurassic ammonoid systematics by Howarth (2013). Dates from Cohen et al. (2012)

tip in both upper and lower jaws, the normal-type jaw apparatuses of the above four suborders are similar to those of modern and fossil nautilids, in having a posteriorly expanded wing and a relatively short and straight hood in the outer lamella of the lower jaw, and a large posteriorly elongated inner lamella of the upper jaw that is not divided into paired lateral walls, unlike the upper jaws of other morphotypes (Figs. 10.4a, e, 10.8). This fact suggests that the normal-type jaw apparatuses of the pre-Jurassic ammonoid suborders retain plesiomorphic features.

The other four jaw morphotypes are distributed in the Jurassic and Cretaceous Ammonitida (Table 10.1; Figs. 10.9 and 10.12). The rhynchaptychus-type jaw apparatuses are known from the Late Cretaceous Phylloceratina (Tanabe and Landman 2002; Tanabe et al. 2013) and Lytoceratina (Tanabe et al. 1980a, 2012; Kanie 1982; Lehmann et al. 1980; Tanabe and Landman 2002). In addition, wholly chitinous lower jaws of the Early Jurassic *Phylloceras* and *Lytoceras* were described as anaptychi (Schmidt 1928; Hauff 1953; Lehmann 1990). These sporadic jaw fossil records of the Phylloceratina and Lytocertina seem to suggest that the rhynchaptychus-type jaws had been derived from the anaptychus-type jaws independently in the two suborders, but this hypothesis should be verified based on better fossil material retaining original jaw morphology and mineralogy.

The anaptychus-type jaw apparatuses appeared earlier than the aptychus-type jaws. They are shared by the Early Jurassic members of the Ammonitina (Psiloceratoidea and Eoderoceratoidea) (Fig. 10.12).

The aptychus-type jaw apparatus was first developed in the Hildoceratoidea in the Toarcian (Early Jurassic) and became dominant in the Ammonitina and

Ancyloceratina during the Middle Jurassic to the end of the Cretaceous (Fig. 10.12; Engeser and Keupp 2002). This evidence suggests their apomorphic nature in the evolution of the ammonoid jaw apparatuses. Forms of the aptychus-type lower jaws defined by the surface sculpture appear to be common at the family level, e.g., Cornaptychus-type in the Hildoceratidae (Hildoceratoidea), Laevaptychus-type in the Aspidoceratidae (Perisphinctoidea), and Lamellaptychus-type in the Oppeliidae and Haploceratidae (Haploceratoidea) (Table 10.1; Trauth 1927–1936, 1938). Based on this fact, Engeser and Keupp (2002) proposed the new taxon Aptychophora for Jurassic and Cretaceous ammonoids with an aptychus-type lower jaw. However, the aptychus-type lower jaws observed in the Ammonitina and Ancyloceratina show marked variation in their shape with and without a sharply pointed rostral tip (Tanabe and Landman 2002) and internal microstructure (Kruta et al. 2009). This evidence suggests that the aptychus-type lower jaws might have been developed in a different way in the Ammonitina and Ancyloceratina, requiring careful testing of the hypothesis of monophyly of the Aptychophora.

The intermediate-type lower jaws of the Cretaceous Desmoceratoidea (Ammonitina) exhibit transitional features from the anaptychus-type to aptychus-type jaws, such as a well-developed, gently convex outer chitinous lamella with a shallow median groove, a more or less pointed rostral portion, and an univalved calcareous layer, which covers the outer chitinous lamella. The Desmoceratoidea are now regarded to have been derived from the Perisphinctoidea with an aptychus-type jaw apparatus in the Valanginian (Early Cretaceous) (Page 1996). Judging from these lines of evidence, it is postulated that the intermediate-type jaws of the Desmoceratoidea might have been developed from the aptychus-type jaws of the Perisphinctoidea, and that they had no phylogenetic relationships with the anaptychus-type jaws in Early Jurassic members of the Ammonitina (Psiloceratoidea and Eoderoceratoidea).

10.5.2 Feeding and Dietary Habits Inferred from Jaw Apparatuses and Gut Contents

Most extant cephalopods are carnivores that generally capture relatively large prey, but the microphagous mode of feeding has also been reported in some pelagic coleoids living in deeper waters (~900 m), including *Vampyroteuthis infernalis* (Vampyromorpha) and *Spirula spirula* (Spirulida) on the basis of analyses of gut contents and specialized morphological features of the feeding apparatus (Scott 1910; Robson 1930; Young 1977; Hoving and Robison 2012) as well as amino acid nitrogen isotopic analysis of soft tissues (Ohkouchi et al. 2013).

Since the jaw apparatus is a primary feeding organ of modern and extinct cephalopods, its morphological and structural features appear to reflect the feeding and dietary habits of living animals. Based on this idea, previous authors considered these aspects on the basis of functional morphologic comparison of the jaw apparatuses between Ammonoidea and modern cephalopods whose feeding and dietary habits are known (Lehmann 1975, 1980, 1988; Tanabe et al. 1980a, 2013) and analyses of organismic remains preserved in the body chambers of ammonoids, under the assumption that they represent the diet preserved in the crop and/or stomach (see Jäger and Fraaye 1997 for the data on the gut contents in Triassic and Jurassic ammonoids) and buccal cavity (Kruta et al. 2011).

Of the five recognized jaw morphotypes, the normal-type jaw apparatuses shared by pre-Jurassic ammonoids have a beak-like rostral tip on the chitinous upper and lower jaws as in modern coleoids, suggesting their primary role in feeding. The rhynchaptychus-type jaw apparatuses of the Cretaceous Phylloceratina and Lytoceratina are similar to those of modern and fossil nautilids (Saunders et al. 1978; Klug 2001), in having a thick and sharply pointed calcified tip in both upper and lower elements, with distinct denticles on the oral margin of the lower one. Saunders et al. (1978) interpreted that in modern nautilids, the sharp, arrowhead-shaped calcareous tip of the upper jaw (rhyncholite) is used as an incisor, while the distinctly denticulated, scallop-shaped calcareous tip of the lower jaw (conchorhynch) has a special function for grasping and shearing food. This interpretation is confirmed by aquarium-based observations of the feeding behavior (e.g., Mikami et al. 1980) and analyses of chewed pieces of food found in the crop of an animal captured in the wild (Tanabe et al. 1980a). Investigation of stomach and/ or crop contents in freshly captured animals has demonstrated that modern nautilids feed mainly on small fish, crustaceans (lobsters and shrimps), nematodes, echinoids, and tentacles of other nautilids in their natural habitat (Ward and Wicksten 1980; Saisho and Tanabe 1985; Ward 1987; Saunders and Ward 1987). Judging from these lines of evidence, the development of thick calcified tips in the jaws of the Cretaceous Phylloceratina and Lytoceratina as well as modern and fossil nautilids is interpreted as convergent adaptation to a scavenging-predatory mode of feeding (Tanabe et al. 1980a, 2013; Tanabe and Fukuda 1999).

The other three jaw morphotypes, the anapychus, aptychus and intermediate types exhibit a remarkable morphologic diversity in their relative size, shape and structure. The diversity is especially conspicuous in the lower jaws (Figs. 10.4, 10.5, 10.6, 10.7, 10.9), and the lower jaw became much larger than the upper jaw and lost the anterior rostral projection in some Jurassic ammonoids such as *Hildoceras* of the aptychus type (Fig. 10.4c; Lehmann 1975) and *Dactylioceras* of the anaptychus-type (Fig. 10.9.2; Lehmann 1979). Most of the Jurassic ammonoid specimens retaining organismic remains in their body chambers belong to such genera as *Hildoceras, Harpoceras, Arnioceras, Physodoceras,* and *Neochetoceras* of the Ammonitina. They are occasionally found with a preserved aptychus-and/ or anaptychus-type lower jaw *in situ* within the body chamber (Lehmann 1971, 1985; Lehmann and Weitschat 1973; Riegraf et al. 1984; Jäger and Fraaye 1997; Schweigert and Dietl 1999). The diet identified in the ‘crop/stomach’ remains includes small decapod crustaceans, ostracods, foraminifers, fragmented arms and calices of stalkless crinoids (*Saccocoma*), and calcified jaw remains (aptychi) of small ammonoids. Based on these facts, Lehmann (1975, 1980) and Morton and Nixon (1987) postulated that the jaws of these Jurassic ammonoids did not have the ability to bite and cut up prey, and instead their large shovel-like aptychus- and anaptychus-type lower jaws were likely used as a scoop to feed mainly on small benthic organisms on the seafloor. Meanwhile, Jäger and Fraaye (1997) suggested a weak biting ability in the Early Jurassic *Harpoceras* with an aptychus-type lower jaw (see Figs. 10.5a, 10.6a) on the basis of careful observations of food remains (mostly small crustaceans) preserved as stomach/ crop contents in adult macroconchs.

The mode of life of *Saccocoma* was previously interpreted as benthic (Milsom 1994). Recent functional morphological and biostratinomical studies, however, suggested that this stalkless crinoid had a planktonic mode of life (Seilacher and Hauff 2004; Hess and Etter 2011). Following this interpretation, Keupp (2007) considered that Jurassic aptychophoran (= aptychus-type lower jaw bearing) ammonoids were more or less passive demersal organisms and fed on small zooplankton including *Saccocoma* under more or less shallow water conditions, while maintaining neutral buoyancy in the water column.

Westermann (1996) proposed an interesting hypothesis that many Cretaceous heteromorph ammonoids of the Ancyloceratina were vertical migrants in the water column and some fed on mesopelagic organisms including sluggish juvenile ammonoids, while others caught zooplankton in tentacles modified into umbrella-nets. Consistent with this hypothesis, Kruta et al. (2011) suggested that *Baculites* (Late Cretaceous Ancyloceratina) with an aptychus-type jaw apparatus fed on zooplankton in the water column on the basis of the discovery of isopods and larval shells of gastropods in the buccal portion of a *Baculites* examined by means of synchrotron X-ray microtomography.

To sum up these available data, the fairly large variation of the jaw morphology and the variety of food remains in the crop, stomach and buccal cavities known in Mesozoic ammonoids probably reflects diversity of feeding and dietary habits among them.

10.5.3 Secondary Function of Aptychi?

Some Jurassic (e.g., *Physodoceras* of the Aspidoceratidae; Fig. 10.5b) and Cretaceous (e.g., *Texanites* of the Acanthoceratoidea; Kennedy and Klinger 1972, pl. 37) ammonoids developed a thick, but porous bivalved calcitic plate (aptychus) in their lower jaws. The flattened outline of the aptychus without a projected rostral tip in these ammonoids occasionally fits tightly against the corresponding ammonoid aperture (see Schmidt 1928, fig. 6; Schindewolf 1958, pls. 1, 8; remarkably, the same fit was documented for lower jaws in Late Devonian *Manticoceras* by Clausen 1969). For these peculiar features, some workers speculated that such highly specialized aptychi may have acquired a secondary function as opercula (Lehmann and Kulicki 1990; Seilacher 1993), as a response to increased predation pressure in shallow-water environments. Indeed, analysis of the large museum collections of Devonian to Late Cretaceous ammonoids has shown that fatally bitten ventral shell breakage possibly made by nautilids and some ammonoids with a sharp calcified tip in their jaws, 'teuthid' coleoids (most of them are now regarded as vampyropod coleoids; see Fuchs 2006), and teleost fishes are relatively abundant in the Jurassic and Cretaceous, whereas they are essentially absent in the Paleozoic (Klompmaker et al. 2009; Andrew et al. 2010). In addition, sublethal injuries occur abundantly on the shells of many Jurassic ammonoids. Durophagous fishes, crustaceans and coleoid cephalopods were probably responsible for some of these sublethal shell breaks (Kröger 2000). Thick calcitic aptychi in the lower jaws of some ammonoids could have been effective against a predatory attack from the apertural side under increasing predation pressure in shallow-water environments of Jurassic and Cretaceous time, if ammonoids were able to seal the shell aperture.

Morphological features of the aptychus-type lower jaws suggest that the semi-flexible outer chitinous lamella with a median depression ('hinge') was possibly connected with jaw muscles via beccublasts on the dorsal side. This may have permitted the living ammonite to fold the aptychus during foraging activity (Fig. 10.11). According to the hypothetical model by Lehmann and Kulicki (1990), the lower jaw could be tilted upwards to seal the aperture, when the head was drawn back by means of the head retractor muscles. As a consequence, the outermost connective tissue could be retracted backward, and the exposed calcitic aptychus served as a protective shield against predators. In addition to the protective function, Morton and Nixon (1986) suggested that calcified aptychi would have acted to weight the buccal mass for nektobenthic feeding and to make it more rigid, while Schweigert (2009) interpreted them as an added ballast weight to stabilize the conch of a living ammonoid for swimming in the water column.

In the Cretaceous Ammonitina (e.g., *Aconeceras*: Doguzhaeva and Mutvei 1992; *Subprionocyclus*: Tanabe and Fukuda 1987b; Fig. 10.9.4) and Ancyloceratina (e.g., *Scalarites*, Fig. 10.9.6; Tanabe et al. 1980b), their aptychus-type lower jaws possess a sharply pointed rostral tip. The lower jaws of these ammonoids presumably served for feeding, although an operculum-like secondary function cannot be ruled out because of its shape and the presence of a distinct median 'hinge' in the chitinous outer lamella.

10.6 Summary

Virtually all modern cephalopod mollusks possess a well-developed jaw apparatus consisting of upper and lower elements and a radula in the globular-shaped buccal mass as primary feeding apparatuses. Ammonoid jaw apparatuses that were preserved *in situ* within the body chambers are currently known to occur from 109 genera belonging to 30 superfamilies of 8 suborders that spanned from the Devonian to the Cretaceous. These jaw apparatuses can be classified into the normal, anaptychus, aptychus, rhynchaptychus and intermediate types by the differences in overall morphology and composition, especially in the presence or absence of calcareous jaw elements. The upper and lower jaws of the former two morphotypes are made mainly of a chitinous material, whereas the lower jaws of the aptychus and intermediate types respectively have a bivalved calcitic plate and a thin univalved calcareous layer on the outer chitinous lamella sculptured by a median groove. The upper and lower jaws of the rhynchaptychus type are characterized by the development of a thick calcified rostral tip.

The jaw apparatuses of the normal and rhynchaptychus types are known from the four pre-Jurassic ammonoid suborders (Gephuroceratina, Goniaitina, Prolecanitina, and Ceratitina) and the Cretaceous Phylloceratina and Lytoceratina. The anaptychus-type jaws are distributed in the Psiloceratoidea and Eoderoceratoidea of the Jurassic Ammonitina, whereas the apytchus-type jaws are widespread in the other superfamilies of Jurassic and Cretaceous Ammonitina and in the Cretaceous Ancyloceratina, showing marked variation in their overall shape with and without a sharply pointed rostral tip and internal microstructure. The intermediate-type jaws are known from the Desmocearatoidea of the Cretaceous Ammonitina. The aptychus-type lower jaws of the Ammonitina and Ancyloceratina show marked variation in their overall shape with and without a sharply pointed rostral tip (Tanabe and Landman 2002) and internal microstructure (Kruta et al. 2009), suggesting that the aptychus-type lower jaws might have developed in a different way in the Ammonitina and Ancyloceratina, requiring careful examination for the monophyletic nature of the aptychus-bearing taxa defined as the Aptychophora. Thick calcitic aptychi developed in the lower jaws of some Jurassic and Cretaceous ammonoids inhabiting shallow-water environments would have been effective against predatory attack from the apertural side, if they were able to seal the shell aperture.

The fairly large variation of the jaw morphology and the variety of food remains in the crop/ stomach and buccal cavities known in Mesozoic ammonoids may reflect diversity of feeding and dietary habits ranging from predatory-scavenging to microphagous (zooplankton-feeding) habits.

Acknowledgments

We thank Günter Schweigert (Staatliches Museum für Naturkunde, Stuttgart), Neal L. Larson (Larson Paleontology Unlimited), Tim White (Yale Peabody Museum), Makoto Kato (Hokkaido University) and Akihiro Misaki (Kitakyushu Museum of Natural History and Human History) for facilitating our study on ammonoid specimens with preserved jaws in their care, Manfred Jäger (Holcim GmbH, Germany) for help in collecting jaw-bearing ammonoids from the Lower Jurassic section in Dotternhausen, Germany, Yasunari Shigeta (National Museum of Nature and Science, Tsukuba) and Takashi Okamoto (Ehime University) for providing well-preserved jaw-bearing ammonoids from the Cretaceous of Hokkaido (Japan) for this study. KT and NHL thank Yoshio Fukuda (Chiba Prefectural Institute of Public Health), the late Tatsuro Matsumoto (Kyushu University), the late Karl M. Waage (Yale University), Isabelle Rouget and the late Fabrizio Cecca (both Université Pierre et Marie Curie) and the late Urlich Lehmann (Hamburg Universität) for helpful suggestions and encouragement during the course of this study. We acknowledge Christian Klug (Universität Zürich) and Horacio Parent (Universidad Nacional de Rosario) for critical comments to improve this manuscript.

References

Andrew C, Howe P, Paul CRC, Donovan SK (2010) Fatally bitten ammonites from the lower Lias Group (Lower Jurassic) of Lyme Regis, Dorset. Proc Yorks Geol Soc 58:81–94

Arkell WJ (1957) Aptychi. In: Moore RC (ed) Treatise on invertebrate paleontology, Part L, Mollusca 4, Cephalopoda, Ammonoidea. GSA & Univ Kansas Press, New York, pp L437–L441

Arkell WJ, Kummel B, Wright CW (1957) Mesozoic Ammonoidea. In: Moore RC (ed) Treatise on invertebrate paleontology, part L, Mollusca 4, Cephalopoda, Ammonoidea. Geol Soc Am & Univ Kansas Press, New York, pp L81–L464

Bachmayer F (1963) Beiträge zur Palaeontologie oberjurassischer Riffe. I. Die Aptychen (Ammonoidea) des Oberjura von Stramberg (ČSR). II. Die Aptychen der Klrntnitzer Serie in Österreich. Ann Naturhist Mus Wien 66:125–138

Bandel K (1988) Operculum and buccal mass of ammonites. In: Wiedmann J, Kullmann J (eds) Cephalopods—present and past. Schweizerbart, Stuttgart, pp 653–678

Baraboshkin EJ, Shumilkin IA (2010) Unique discovery of aptychi in ammonite subfamily Simbriskitinae Spath, 1924. In: Sadovnichy VA, Smurov A (eds) Life of the Earth. Geology, geodynamics, environment, museology. Earth Science at Museum Moscow State University, Moscow, pp 132–136 [in Russian]

Birkelund T (1965) Ammonites from the Upper Cretaceous of West Greenland. Medd Grønland 179:1–192

Breitkreutz, H, Diedrich R, Metzdorf R (1991) Fossilfunde aus der Schwarz-Bunten Weschselfolge (Ob. Cenoman bis Unter Turon) des Ostwestfalendammes bei Bielefeld. Ber Naturwiss Ver Bielef Umgeg 32:37–48

Casey R (1980) A monograph of the Ammonoidea of the Lower Greensand, part 9. Palaeontogr Soc Monogr 133:633–660

Clarke MR (1962) The identification of cephalopod 'beaks' and the relationship between beak size and total body weight. Bull Brit Mus (Nat Hist), [Zool] 8:419–480

Clarke MR (ed) (1986) A handbook for the identification of cephalopod beaks. Clarendon Press, Oxford

Clausen C-D (1969) Oberdevonische Cephalopoden aus dem Rheinischen Schiefergebirge. II.: Gephuroceratidae, Beloceratidae. Palaeontogr A 132:95–178

Closs D (1960) Contribuição ao estudo dos Aptychi (Cephalopoda-Ammonoidea) do Jurássico. Esc Geol Pórto Alegre, Publ Espec 2:1–67

Closs D (1967) Gonatiten mit Radula und Kieferapparat in der Itararé-Formation von Uruguay. Paläontol Z 41:19–37

Cohen KM, Finney S, Gibbard PL (2012) International chronostratigraphic chart. International Subcommision on Stratigraphy (July, 2012)

Cope JW (1994) Preservation, sexual dimorphism, and mode of life of Sinemurian eoderoceratid ammonites. In: Pallini G (ed) Proceedings of the 3rd Pergola International Symposium "Fossili, Evoluzione, Ambiente". Palaeopelagos, Spec Issue 1:57–66

Cope JW, Sole DTC (2000) Ammonite jaw apparatus from the Sinemurian (Lower Jurassic) of Dorset and their taphonomic relevance. J Geol Soc Lond 157:201–205

Dagys AS, Weitschat W (1988) Ammonoid jaws from the Boreal Triassic realm (Svalbard and Siberia). Mitt Geol-Paläont Inst Univ Hambg 67:53–71

Dagys AS, Lehmann U, Bandel K, Tanabe K, Weitscaht W (1989) The jaw apparati of ectocochleate cephalopods. Paläontol Z 63:41–53

Dietze V, von Hillebrandt A, Riccardi A, Schweigert G (2012) Ammonites and stratigraphy of a Lower Bajocian (Middle Jurassic) section in Sierra Chacaico (Neuquén Basinm Argentina). Zitteliana A 52:119–139

Dilly PN, Nixon M (1976) The cells that secrete the beaks in octopods and squids (Mollusca, Cephalopoda). Cell Tissue Res 167:229–241

Doguzhaeva LA (1999) Beaks of the Late Carboniferous ammonoids from the Southern Urals. In: Rozanov AY, Shevyrev AA (eds) Fossil cephalopods: recent advances in their study. Russian Academy of Science, Paleontological Institute, Moscow, pp 68–87 [in Russian]

Doguzhaeva LA, Mikhailova LA (2002) The jaw apparatus of the heteromorphic ammonite *Australiceras whitehouse,* 1926 (Mollusca: Cephalopoda) from the Aptian of the Volga region. Doklady Biol Sci 382:38–40

Doguzhaeva LA, Mutvei H (1992) Radula of the Early Cretaceous ammonite *Aconeceras* (Mollusca: Cephalopoda). Palaeontogr A 223:167–177

Doguzhaeva LA, Mutvei H (1993) Shell ultrastructure, muscle-scars, and buccal apparatus in ammonoids. Geobios Mém Spec 15:111–119

Doguzhaeva L, Mapes RH, Mutvei H (1997) Beaks and radulae of Early Carboniferous goniatites. Lethaia 30:305–313

Doguzhaeva LA, Mapes RH, Summesberger H, Mutevi H (2007) The preservation of body tissues, shell, and mandibles in the ceratitid ammonoids *Austrotrachyceras* (Late Triassic), Austria. In: Landman NH, Davis RA, Mapes RH (eds) Cephalopods present and past: new insights and fresh perspective, Springer, Dordrecht

Donovan DT, Callomon JH, Howart MK (1980) Classification of the Jurassic Ammonitina. In: House MR, Senior JR (eds) The Ammonoidea, Academic Press, London, pp 101–155

Engeser T, Keupp H (2002) Phylogeny of aptychi-possessing Neoammonoidea (Aptychophora nov., Cephalopoda). Lethaia 24:79–96

Farinacci A, Mariotti N, Matteucci R, Nicosia U, Pallini G (1976) Structural features of some Jurassic and early Cretaceous aptychi. Boll Soc Paleontol Ital 15:111–143

Frerichs U (2004) Anaptychen und Aptychen-Kieferapparate order Deckel? Allgemeine Einführung und Beschreibung von Funden aus der Kreide im Raum Hannover. Arb Paläontol Hann 32:1–15

Fritsch A (1893) Studien im Gebiete der böhmischen Kreideformation V. Die Priesener Schichten. Archiv Naturwiss Landesdurchforsch Böhm 9:1–135

Fritsch A, Schlönbach U (1872) Cephalopoden der böhmischen Kreideformation. F. Rivnác, Prague

Fuchs D (2006) Fossil erhaltungsfähige Merkmalskomplexe der Coleoidea (Cephalopoda) und ihre phylogenetische Bedeutung. Berl Paläobiol Abh 8:1–115

Fukuda Y (1980) Observations by SEM. In: Hamada T, Obata I, Okutani T (eds) *Nautilus macromphalus* in captivity. Tokai University Press, Tokyo, pp 23–33

Furnish WM, Glenister BF, Kullumann J, Zuren Z (2009) Treatise on invertebrate paleontology, Part L, Mollusca 4 Revised. Volume 2: Carboniferous and Permian Ammonoidea (Goniatitida and Prolecanitida). Univ Kansas Paleont Inst Lawrence, Kansas, pp 1–253

Gupta NS, Briggs DEG, Landman NH, Tanabe K, Summons RE (2008) Molecular structure of organic components in cephalopods: Evidence for oxidative cross linking in fossil marine invertebrates. Org Geochem 39:1405–1414

Hauff B (1953) Das Holzmadenbuch. Hohenlohesche Buchhandlung, Öhringen

Hess H, Etter W (2011) Life and death of *Saccocoma tenella* (GOLDFUSS). Swiss J Geosci 104(Suppl. 1):99–106

Hewitt, RA, Westermann GEG, Checa A (1993) Growth rates of ammonites estimated from aptychi. Geobios Mém Spec 15:203–208

House MR (1980) On the origin, classification and evolution of the early Ammonoidea. In: House MR, Senior JR (eds) The Ammonoidea. Academic Press, London, pp 3–36

Hoving HJT, Robison BH (2012) Vampire squid: detritivores in the oxygen minimum zone. Proc R Soc B 279:4559–4567

Howarth MK (2013) Treatise online number 57. Part L, revised, volume 3B, chapter 4: Psiloceratoidea, Eoderoceratoidea, Hildoceratoidea. Univ Kansas, Paleont Inst Laurence, Kansas, pp 1–139

Hunt S, Nixon M (1981) A comparative study of protein composition in the chitin-protein complexes of the beak, pen, sucker disc, radula and oesophageal cuticle of cephalopods. Comp Biochem Phys B 68:535–546

Ifrim C (2013) Paleobiology and paleoecology of the Early Turonian (Late Cretaceous) ammonite *Pseudaspidoceras flexuosum*. Palaios 28:9–22

Jäger M, Fraaye R (1997) The diet of the early Toarcian ammonite *Harpoceras falciferum*. Palaeontology 40:557–574

Kanie Y (1982) Cretaceous tetragonitid ammonite jaws: a comparison with modern *Nautilus* jaws. Trans Proc Palaeontol Soc Jpn N S 125:239–258

Kanie Y, Tanabe K, Fukuda Y, Hirano H, Obata I (1978) Preliminary study of jaw apparatus in some Late Cretaceous ammonites from Japan and Sakhalin. J Geol Soc Jpn 8: 629–631 [in Japanese]

Kaplan U, Kennedy WJ, Lehmann J, Marcinowski R (1998) Stratigraphie und Ammonitenfaunen des westfälischen Cenoman. Geol Paläontol Westfal 51:5–237

Kennedy WJ (1986) Campanian and Maastrichtian ammonites from northern Aquitaine, France. Spec Pap Palaeontol 36:1–145

Kennedy WJ, Cobban WA (1976) Aspects of ammonite biology, biogeography, and biostratigraphy. Spec Pap Palaeontol 17:1–94

Kennedy WJ, Kaplan U (1995) *Parapuzosia (Parapuzosia) seppenradensis* (Landois) und die Ammonitenfauna der Dülmener Schichten unteres Unter-Campan, Westfalen. Geol Paläontol Westfal 33:5–127

Kennedy WJ, Kaplan U (1997) Ammoniten aus dem Campan des Stemweder Berges, Dammer Oberkreidemulde, NW-Deutschland. Geol Paläontol Westfal 50:31–245

Kennedy WJ, Klinger HC (1972) A *Texanites-Spinaptychus* association from the Upper Cretaceous of Zululand. Palaeontology 15:394–399

Kennedy WJ, Landman NH, Cobban WA, Larson NL (2002) Jaws and radulae in *Rhaeboceras*, a Late Cretaceous ammonite. Abh Geol Bund 57:113–132

Keupp H (2000) Ammoniten. Paläobiologische Erfolgsspiralen. Thorbecke, Stuttgart

Keupp H (2007) Complete ammonoid jaw apparatuses from the Solnhofen plattenkalks: implications for aptychi function and microphagous feeding of ammonoids. Neues Jahrb Geol Paläontol Abh 245:93–101

Klompmaker AA, Waljaard NA, Fraaije RHB (2009) Ventral bite marks in Mesozoic ammonoids. Palaeogeogr Palaeoclimatol Palaeoecol 280:245–257

Klug C (2001) Functional morphology and taphonomy of nautiloid beaks from the Middle Triassic of southern Germany. Acta Palaeont Pol 46:43–68

Klug C, Jerjen I (2012) The buccal apparatus with radula of a ceratitic ammonoid from the German Middle Triassic. Geobios 45:57–65

Klug C, Riegraf W, Lehmann J (2012) Soft-part preservation in heteromorph ammonites from the Cenomanian-Turonian boundary event (OAE 2) in north-west Germany. Palaeontology 55:1307–1331

Köhler-Lopez M, Lehmann U (1984) The Triassic ammonite *Aristoptychites kolymensis* (Kiparisova) from Botneheia, Spitsbergen. Polar Res 2:61–75

Kröger B (2000) Schalenverletzungen an jurassischen Ammonitenï¼ihre paläobiologische und paläoökologische Aussagefähigkeit. Berl Geowiss Abh E 33:1–97

Kruta I, Rouget I, Landman NH, Tanabe K, Cecca F (2009) Aptychi microstructure in Late Cretaceous Ancyloceratina (Ammonoidea). Lethaia 42:312–321

Kruta I, Landman NH, Rouget I, Cecca F, Larson NL (2010) The Jaw apparatus of the Late Cretaceous ammonite *Didymoceras*. J Paleontol 84:556–560

Kruta I, Landman N, Rouget I, Cecca F, Tafforeau P (2011) The role of ammonites in the Mesozoic marine food web revealed by jaw preservation. Science 331:70–72

Kruta I, Landman NH, Rouget I, Cecca F, Tafforeau P (2013) The radula of the Late Cretaceous scaphitid ammonite *Rhaeboceras halli* (Meek and Hayden, 1856). Palaeontology 56:9–14

Kruta I, Landman NH, Tanabe K (2015) Ammonoid radula. This volume

Kubodera T (2005) Manual for the identification of cephalopod beaks in the Northwest Pacific. Natn Mus Sci Nature, Tokyo, http://research.kahaku.go.jp/zoology/Beak-E/index.htm

Landman NH, Waage KM (1993) Scaphitid ammonites of the Upper Cretaceous (Maastrichtian) Fox Hills Formation in South Dakota and Wyoming. Bull Am Mus Nat Hist 215:1–257

Landman NH, Tsujita CJ, Cobban WJ, Larson NL, Tanabe K (2006) Jaws of Late Cretaceous placenticeratid ammonites: how preservation affects the interpretation of morphology. Am Mus Novit 3500:1–48

Landman NH, Larson NL, Cobban WA (2007) Jaws and radula of *Baculites* from the Upper Cretaceous (Campanian) of North America. In: Landman NH, Davis RA, Mapes RH (eds) Cephalopods—present and past, new insights and fresh perspectives. Springer, Dordrecht, pp 257–298

Landman NH, Mapes RH, Cruz C (2010) Jaws and soft tissues in ammonoids from the Lower Carboniferous (Upper Mississippian) Bear Gulch Beds, Montana, USA. In: Tanabe K, Sasaki T, Shigeta Y, Hirano H (eds) Cephalopods—present and past. Tokai University Press, Tokyo, pp 147–153

Lehmann U (1967) Ammoniten mit Kieferapparat und Radula aus Lias-Geschieben. Paläontol Z 41:25–31

Lehmann U (1970) Lias-Anaptychen als Kieferelemente (Ammonoidea). Paläontol Z 44:38–45

Lehmann U (1971) Jaws, radula, and crop of *Arnioceras*. Palaeontology 14:338–341

Lehmann U (1972) Aptychen als Kieferelemente der Ammoniten. Paläontol Z 46:34–48

Lehmann U (1975) Über Nahrung und Ernährungsweise von Ammoniten. Paläontol Z 49:187–195

Lehmann U (1976) Ammoniten: ihr Leben und ihre Umwelt. Enke, Stuttgart

Lehmann U (1978) Über den Kieferapparat von Ammoniten der Gattung *Parkinsonia*. Mitt Geol-Paläontol Inst Univ Hambg 48:79–84

Lehmann U (1979) The jaws and radula of the Jurassic ammonite *Dactylioceras*. Palaeontology 22:265–271

Lehmann U (1980) Ammonite jaw apparatus and soft parts. In: House MR, Senior JR (eds) The Ammonoidea. Academic Press, London, pp 275–287

Lehmann U (1981) The Ammonites. Their life and their world. Cambridge University Press, Cambridge

Lehmann U (1985) Zur Anatomie der Ammoniten: Tintenbeutel, Kiemen, Augen. Paläontol Z 59:99–108

Lehmann U (1988) On the dietary habits and locomotion of fossil cephalopods. In: Wiedmann J, Kullmann J (eds) Cephalopods—present and past. Schweizerbart, Stuttgart, pp 633–644

Lehmann U (1990) Ammonoideen. Enke, Stuttgart

Lehmann U, Kulicki C (1990) Double function of aptychi (Ammonoidea) as jaw elements and opercula. Lethaia 23:325–331

Lehmann U, Weitschat W (1973) Zur Anatomie und Ökologie von Ammoniten: Funde von Kropf und Kiemen. Paläontol Z 47:69–76

Lehmann U, Tanabe K, Kanie Y, Fukuda Y (1980) Über den Kieferapparat der Lytoceratacea (Ammonoidea). Paläontol Z 54:319–329

Lowenstam HA, Traub W, Weiner S (1984) *Nautilus* hard parts: a study of the mineral and organic constitutions. Paleobiology 10:269–279

Lukeneder A, Tanabe K (2002) *In situ* finds of aptychi in the Barremian of the Alpine Lower Cretaceous (Northern Calcareous Alps, Upper Austria). Cret Res 23:15–24

Mapes RH (1987) Upper Paleozoic cephalopod mandibles: frequency of occurrence, modes of preservation, and paleoecological implications. J Paleont 61:521–538

Martinez Escauriaza G (2007) The aptychus of "*Oxyparoniceras* (*Oxyparoniceras*) *buckmani*"(Bonarelli, 1895) (Paroniceratinae Ammonoidea) from the Toracian of Ariño (Iberian Range, Spain). Coloq Paleontol 57:15–21

Mastunaga T, Maeda H, Shigeta Y, Hasegawa K, Nomura S, Nishimura T, Misaki A, Tanaka G (2008) First discovery of *Pravitoceras sigmoidale* Yabe from the Yezo Supergroup in Hokkaido, Japan. Paleontol Res 12:309–319

Matern H (1931) Oberdevon Anaptychen in situ und über die Erhaltung von Chitin-Substanzen. Senckenbergiana 13:160–167

Meek FB, Hayden FV (1864) Paleontology of the Upper Missouri. Smithon Contr Knowl 172:1–135

Meyer H (1829) Das Genus Aptychus. Verh Kaiserl Leopold Carolin Akad Naturfor 15(2):125–170

Michael R (1894) Ammoniten-Brut mit Aptychen in der Wohnkammer von *Oppelia steraspis* Oppel sp. Zeit Dtsch Geol Ges 46:697–702

Mikami S, Okutani T, Hirano H, Kanie Y, Hamada T (1980) Behavior in captivity. In: Hamada T, Obata I, Okutani T (eds) *Nautilus macromphalus* in captivity. Tokai University Press, Tokyo, pp 11–22

Milsom C (1994) *Saccocoma*: a benthic crinoid from the Jurassic Solnhofen Limestone, Germany. Palaeontology 37:121–129

Mojsisovics EV (1873–1902) Die Cephalopoden der Hallstätter Kalke (Teil I, Textteil+Atlas). Abh K-K Geol Reichsanst Wien 6(1,2):1–835

Moore RC, Sylvester-Bradley PC (1957) Taxonomy and nomenclature of aptychi. In: Moore RC (ed) Treatise on invertebrate paleontology, Part L, Mollusca 4, Cephalopoda, Ammonoidea. Geol Soc Am & Univ Kansas Press, Boulder, pp L465–L471

Morton N (1973) The aptychi of *Sonninia* (Ammonitina) from the Bajocian of Scotland. Palaeontology 16:195–203

Morton N (1975) Bajocian Sonniniidae and other ammonites from Western Scottland. Palaeontology 18:41–91

Morton N (1981) Aptychi—The myth of the ammonite operculum. Lethaia 14:57–61

Morton N, Nixon M (1987) Size and function of ammonite aptychi in comparison with buccal masses of modern cephalopods. Lethaia 20:231–238

Nagao T (1931a) The occurrence of anaptychus-like bodies in the Upper Cretaceous of Japan. Proc Imp Acad Tokyo 7:106–109

Nagao T (1931b) New discovery of aptychus in two species of ammonites from the Upper Cretaceous of Japan. Proc Imp Acad Tokyo 7:165–168

Nagao T (1931c) Anaptychus and aptychus lately acquired from the Upper Cretaceous of Hokkaido, Japan. J Fac Sci Hokkaido Imp Univ Ser 4: 215–222

Nagao T (1932) Discovery of a *Desmoceras*-operculum. Proc Imp Acad Jpn 8:175–178

Nixon M (1988a) The buccal mass of fossil and recent Cephalopoda. In: Clarke MR, Trueman ER (eds) The Mollusca, vol. 12, paleontology and neontology of cephalopods. Academic Press, San Diego, pp 103–122

Nixon M (1988b) The feeding mechanisms and diets of cephalopods-living and fossil, In: Wiedmann J, Kullmann J (eds) Cephalopods—present and past. Schweizerbart, Stuttgart, pp 641–652

Nixon M (1996) Morphology of the jaws and radula in ammonoids. In: Landman NH, Tanabe K, Davis RA (eds) Ammonoid paleobiology. Plenum, New York, pp 23–42

Nowak J (1908) Untersuchungen über die Cephalopoden der oberen Kreide in Polen. I. Genus *Baculites* Lamarck. Bull Intern Acad Sci Cracovie, Class Sci Math Nat 1911:326–353

Ohkouchi N, Tsuda R, Chikaraishi Y, Tanabe K (2013) A preliminary estimate of trophic position of a deep-water ram's horn squid *Spirula spirula* based on nitrogen isotopic composition of amino acids. Mar Biol 160:773–779

Okutnai T, Mikami S (1977) Description on beaks of *Nautilus macromphalus* Sowerby. Venus (Jpn J Malacol) 36:115–121

Oppel A (1856) Ueber einige Cephalopoden der Juraformation Wüttenbergs, 2. *Ammonites planorbis* Sow. (*psilonotus* Quenst.) mit erhaltenen Aptychus. Jahrresh Ver Vaterl Nat Württ Jahrg 12:107–108

Page KN (1996) Mesozoic ammonoids in space and time. In: Landman NH, Tanabe K, Davis RA (eds) Ammonoid paleobiology. Plenum, New York, pp 755–794

Parent H, Garrido C, Schweigert G, Scherzinger A (2011) The Tithonian ammonite fauna and stratigraphy of Picún Leufú, Neuquén Basin, Argentina. Rev Paleobiol 30:45–104

Raven CP (1958) Morphogenesis: the analysis of molluscan development. Pergamon Press, Oxford

Reboulet S (1996) L'evolution des ammonites du Valanginien–Hauterivien inferieur du bassin vocontien et de la plate–forme provencale (sud–est de la France): relations avec la stratigraphie sequentielle et implications biostratigraphiques. Doc Lab Geol Lyon 137:1–371

Retowski O (1891) Die Aptychen sind echte Ammonitendeckel. Neues Jahrb Miner 2:220–221

Riegraf W, Luterbacher H (1989) Jurassic and Cretaceous rhyncholites (cephalopod jaws) from the North Atlantic Ocean (Deep Sea Drilling Project Leg 1–79) and their European counterparts. Evidence for the uniformity of the Western Tethys. Geol Rundsch 78:1141–1163

Riegraf W, Werner G, Lörcher F (1984) Der Posidonienschiefer. Biostratigraphie, Fauna und Fazies des südwestdeutschen Untertoarciums (Lias epsilon). Enke, Stuttgart

Robson GC (1930) Cephalopoda, I. Octopoda. Discov Rep 2:373–401

Rogov MA, Gulyaev DB (2003) On the first find of aptychi in representatives of the subfamily Proplanulitinae Buckman (Perisphinctidae, Ammonitida). Paleontol J 37:382–385

Saisho T, Tanabe K (1985) Notes on the oesophagus and stomach-contents of *Nautilus pompilius* in Fiji. Kagoshima Univ Res Center S Pacific, Occ Pap 4:62–64

Sasaki T, Shigeno S, Tanabe K (2010) Anatomy of living *Nautilus*: Reevaluation of primitiveness and comparison with coleoids. In: Tanabe K, Shigeta Y, Sasaki T, Hirano H (eds) Cephalopods—present and past. Tokai University Press, Tokyo, pp 35–66

Saunders WB, Richardson ES Jr (1979) Middle Pennsylvanian (Desmoinesean) Cephalopoda of the Mazon Creek fauna, Northeastern Illinois. In: Nitecki MH (ed) Mazon Creek fossils. Academic Press, New York, pp 333–359

Saunders WB, Ward PD (1987) Ecology, distribution, and population characteristics of *Nautilus*. In Saunders WB, Landman NH (eds) *Nautilus*, the biology and paleobiology of a living fossil. Plenum, New York, pp 137–162

Saunders WB, Spinosa C, Teichert C, Banks RC (1978) The jaw apparatus of Recent *Nautilus* and its palaeontological implications. Palaeontology 21:129–141

Schlülter CA (1875) Ein *Baculites Knorrianus* von Lüneburg, in dessen Wohnkammer noch die beiden zugehörigen Aptychen stecken. Sitzungsber Niederrh Gesel Nat Heilkd 1875:31–32

Schlülter CA (1876) Cephalopoden der oberen deutschen Kreide, 2. Abtheilung. Palaeontogr 24:123–263

Schmidt M (1925) Ammonitenstudien. Fortschr Geol Paläontol 3(10): 275–363

Schmidt M (1928) Anaptychen von *Lytoceras cornucopiae* Young a. Bird. Neues Jb Miner etc. Beil 61B:399–432

Schweigert G (1998) Die Ammonitenfauna des Nusplinger Plattenkalks (Ober-Kimmeridgium, Beckeri-Zone, Ulmense-Subzone, Baden-Württemberg). Stuttg Beitr Nat Ser B 267:1–61

Schweigert G (2009) First three-dimensionally preserved *in situ* record of an aptychophoran ammonite jaw apparatus in the Jurassic and discussion of the function of aptychi. Berl paläobiol Abh 10:321–330

Schweigert G, Dietl G (1999) Zur Erhaltung und Einbettung von Ammoniten im Nusplinger Plattenkalk (Oberjura, Südwestdeutschland). Stuttg Beitr Nat Ser B 272:1–31

Scott T (1910) Notes on Crustacea found in the gizzard of a deep-sea cephalopod. Ann Mag Nat Hist 5:51–54

Seilacher A (1993) Ammonite aptychi: how to transform a jaw into an operculum? Am J Sci 293A:20–32

Seilacher A, Hauff RB (2004) Constructional morphology of pelagic crinoids. Palaios 19:3–16

Shindewolf OH (1958) Über Aptychen (Ammonoidea). Paleontogr A 3:1–46

Spinosa C (1968) The Xenodiscidae, Permian otoceratacean ammonoids. Unpubl. Ph.D. dissertation, Univ Iowa

Stephanov J (1961) Ammonoid operculums (aptychi) from the Lower Cretaceous of Bulgaria. Trav Géol BulgSér Paléontol 3:209–235 [in Bulgarian]

Tanabe K (1983) The jaw apparatuses of Cretaceous desmoceratid ammonites. Palaeontology 26:677–686

Tanabe K (2011) The feeding habits of ammonites. Science 331:37–38

Tanabe K (2012) Comparative morphology of modern and fossil coleoid jaw apparatuses. Neues Jahrb Geol Paläontol Abh 266:9–18

Tanabe K, Fukuda Y (1983) Buccal mass structure of the Cretaceous ammonite *Gaudryceras*. Lethaia 16:249–256

Tanabe K, Fukuda Y (1987a) Mouth part histology and morphology. In: Saunders WB, Landman NH (eds) *Nautilus*, the biology and paleobiology of a living fossil. Plenum, New York, pp 313–322

Tanabe K, Fukuda Y (1987b) The jaw apparatus of the Cretaceous ammonite *Reesidites*. Lethaia 20:41–48

Tanabe K, Fukuda Y (1999) Morphology and function of cephalopod buccal mass. In: Savazzi E (ed) Functional morphology of the invertebrate skeleton. Wiley, London, pp 245–262

Tanabe K, Landman NH (2002) Morphological diversity of the jaws of Cretaceous Ammonoidea. Abh Geol Bund 57:157–165

Tanabe K, Mapes RH (1995) Jaws and radula of the Carboniferous ammonoid *Cravenoceras*. J Paleontol 69:703–707

Tanabe K, Fukuda Y, Kanie Y, Lehmann U (1980a) Rhyncholites and conchorhynchs as calcified jaw elements in some late Cretaceous ammonites. Lethaia 13:157–168

Tanabe K, Hirano H, Kanie Y (1980b) The jaw apparatus of *Scalarites mihoensis*, a Late Cretaceous ammonite. In: Igo H, Noda H (eds) Professor Saburo Kanno memorial volume. University Tsukuba, Tsukuba, pp 159–165

Tanabe K, Mapes RH, Kidder DL (2001) A phosphatized cephalopod mouthpart from the Upper Pennsylvanian of Oklahoma, U.S.A. Paleontol Res 5:311–318

Tanabe K, Landman NH, Kruta I (2012) Microstructure and mineralogy of the outer calcareous layer in the lower jaws of Cretaceous Tetragonitoidea and Desmoceratoidea (Ammonoidea). Lethaia 45:191–199

Tanabe K, Misaki A, Landman NH, Kato T (2013) The jaw apparatuses of Cretaceous Phylloceratina (Ammonoidea). Lethaia 46:399–408

Teichert C, Spinosa C (1971) Cretaceous and Tertiary rhyncholites from the western Atlantic Ocean and from Mississippi. Univ Kansas, Paleontol Contr Pap 58:1–10

Teichert C, Moore RC, Nodine Zeller DE (1964) Rhyncholites. In: Moore RC (ed) Treatise on invertebrate paleontology, Part K, Cephaloda Nautiloidea, Geological Society am & University Kansas Press, New York, pp K467–K484

Tozer ET (1980) Triassic Ammonoidea: classification, evolution and relationship with Permian and Jurassic forms. In: House MR, Senior JR (eds) The Ammonoidea, Academic Press, London, pp 66–100

Trauth F (1927–1936) Aptychenstudien, I-VII. Ann Nat Mus Wien 41 (1927):171–259, 42 (1928):171–259, 44(1930):329–411, 45(1931):17–136, 47(1936):127–145

Trauth F (1938) Die Lamellaptychi des Oberjura und der Unterkreide. Palaeontogr A 88: 115–229

Uyeno TA, Kier WM (2005) Functional morphology of the cephalopod buccal mass: a novel joint type. J Morphol 264:211–222

Uyeno TA, Kier WM (2007) Electromyography of the buccal musculature of octopus (*Octopus bimaculoides*): a test of the function of the muscle articulation in support and movement. J Exp Biol 210:118–128

Vašíček Z (2010) Aptychi and their significance for taxonomy of Lower Cretaceous ammonites. J Nat Mus (Prague), Nat Hist Ser 179 (18):183–188

Ward PD (1987) The natural history of *Nautilus*. Allen and Unwin, Boston

Ward P, Wicksten MK (1980) Food sources and feeding-behavior of *Nautilus macromphalus*. Veliger 23:119–124

Westermann GEG (1996) Ammonoid life and habitat. In: Landman NH, Tanabe K, Davis RA. (eds) Ammonoid paleobiology. Plenum Press, New York, pp 607–707

Wippich MGE, Lehmann J (2004) *Allocrioceras* from the Cenomanian (Mid-Cretaceous) of the Lebanon and its bearing on the palaeobiological interpretation of heteromorphic ammonites. Palaeontology 47:1093–1107

Woodward H (1885) II.—On some Palaeozoic phyllopod-shields, and on *Nebalia* and its allies. Geol Mag (Decade III) 2:345–352.

Wright CW, Callomon JH, Howarth MK (1996) Treatise on invertebrate paleontology, Part L, Mollusca 4 Revised, vol. 4: Cretaceous Ammonoidea. Geological Society America & University Kansas Press, Boulder, pp 1–362

Young JZ (1977) Brain, behavior and evolution of cephalopods. Symp Zool Soc Lond 38:377–434

Zakarov YD (1974) A new find of an ammonoid jaw apparatus. Paleontol J 1974(4):554–556

Zakharov YD, Lominadze TA (1983) New data on the jaw apparatus of fossil cephalopods. Lethaia 16:67–78

Chapter 11
Ammonoid Radula

Isabelle Kruta, Neil H. Landman and Kazushige Tanabe

11.1 Introduction

The molluscan feeding device called the radula is present in virtually all Recent cephalopods (except if reduced or absent as a secondary loss in *Spirula* and a few cirrate octopods; Nixon 2011). The radula itself is a tongue-like ribbon covered by tiny teeth and plates arranged in a series of transverse rows one behind the other. Ammonoids possessed a radula as well, but the preservation of this delicate structure in the fossil record is considered exceptional. Very few specimens are well enough preserved to provide anatomical details about the shape, dimensions, and position in the buccal mass. Radulae are only known from twelve genera of ammonoids and commonly occur between or in close association with the upper and lower jaws. Although information is only available for a limited number of taxa, radulae are considered examples of exceptional preservation that could provide information on the paleobiology (e.g., diet) and phylogeny of the Ammonoidea.

Institutional abbreviations: American Museum of Natural History (AMNH), University Museum University of Tokyo (UMUT); Geological Survey collection,

I. Kruta (✉) · N. H. Landman
Division of Paleontology (Invertebrates), American Museum of Natural History, New York, NY 10024, USA
e-mail: ikruta@amnh.org

I. Kruta
Sorbonne Universités, UPMC Université Paris 06, CR2P-UMR 7207 CNRS, MNHN, 4 Place Jussieu, F-75005, Paris, France

N. H. Landman
e-mail: landman@amnh.org

K. Tanabe
Department of Historical Geology and Paleontology, The University Museum, The University of Tokyo, Tokyo 113-0033, Japan
e-mail: tanabe@um.u-tokyo.ac.jp

C. Klug et al. (eds.), *Ammonoid Paleobiology: From Anatomy to Ecology*, Topics in Geobiology 43, DOI 10.1007/978-94-017-9630-9_11

London (GSCL); Geologisch-Paläntologisches Intitut Universität Hamburg (GPIH); Fachbereich Geowissenschaften Paläontologische Sammlung, Universität Tübingen (GPIT); Staatliches Museum für Naturkunde in Stuttgart (SMNS); Geol. Staatsinst. Hamburg (GSH); Paleontological Institute of the Russian Academy of Sciences (PIN); Ohio University Zoological Collection (OUZC), this collection will be moved to AMNH.

11.2 Formation of the Radula in Cephalopods

The general organization of the radular apparatus is similar in all cephalopods. Differences exist regarding the number and shape of teeth, the morphology of the buccal palps, and the development of accessory organs. The radula of Recent cephalopods is here used as comparative material for the function and the morphology of non-preserved tissues in fossil material.

The radular apparatus is a complex structure found between the upper and lower jaws (Fig. 11.1a). It is composed of a toothed ribbon with the associated muscles (the odontophore), the hyaline shield, and accessory structures called radular organs (Messenger and Young 1999). The radula is already well formed at hatching (Messenger and Young 1999) and continues to grow throughout the entire life of the animal (Aldrich 1969). The teeth on the radular ribbon are secreted posteriorly in the radular sac by elongate cells with microvilli (the odontoblasts). The odontoblasts

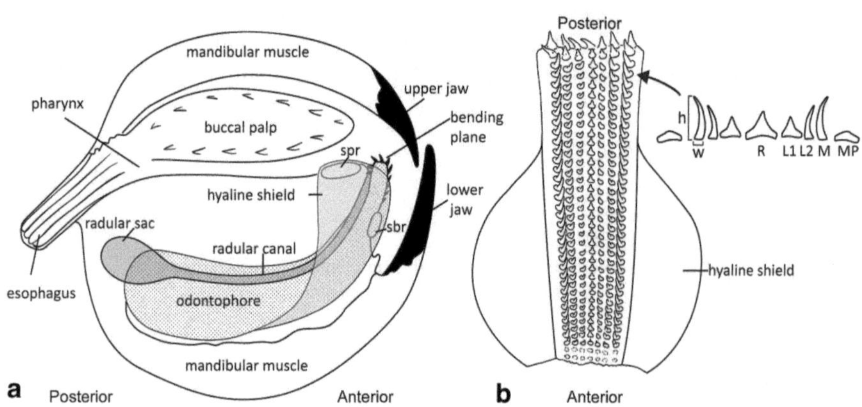

Fig. 11.1 Buccal mass structure and radular apparatus in Recent cephalopods. **a** Diagram of the buccal mass structure and radular apparatus in *Octopus*. Abbreviations. Spr: supraradular organ, sbr: subradular organ. Modified from Messenger and Young (1999). **b** Diagram of a radular ribbon showing the series of rows of teeth and the nomenclature for the teeth in a single row as well as the measurements of the teeth. The row of teeth has a central rachidian tooth (*R*), a first and second lateral tooth (*L1* and *L2* respectively) on each side of the rachidian tooth, marginal teeth (*M*) and marginal plate (*MP*). Measurements for the teeth provided in Table 11.1 have been done on the width (*w*) and the height (*h*) of the teeth. (Modified from Naef 1923)

are organized in two layers: the outer layer secretes the radular membrane, the base of the teeth, and the radular ribbon, while the internal layer secretes the cusps of the teeth (Messenger and Young 1999). The newly formed teeth move forward along the ribbon during ontogeny and increase in size during the growth of the animal (Nixon 1969). Therefore, the most recently formed teeth are the largest. Only one portion of the radula is exposed on the anterior part of the beak, the rest of the radula is folded on itself in a canal that originates in the radular sac.

Approaching the anterior part of the buccal mass, the ribbon unfolds and turns sharply downward (i.e. the bending plane) and then backward. The teeth that were flat lying in the U-shaped canal become erect as they approach the area where the ribbon bends at the anterior end of the beak, and become functional. The flexible ribbon with the radular teeth pointing towards the esophagus can be moved forward and backward in order to facilitate the transfer of food towards the esophagus. Such usage causes wear of the cusps and occasional loss of teeth, necessitating their continual replacement (Aldrich 1969). After the bending plane, the old teeth are dissolved by accessory organs (Nixon and Young 2003).

11.3 Terminology

The organization of the teeth in one row is reproduced in all the rows along the ribbon (Fig. 11.1b). Different types of teeth are present on a transverse row. Various terms have been used for describing the teeth and we follow the terminology of Nixon (1995). Except in cases of secondary loss (e.g., some cirrate octopods), each row has a central tooth called a rachidian tooth with a central cusp called the mesocone while those flanking are called ectocones (Robson 1925). The other teeth are symmetric on each side of the rachidian tooth and are numbered starting from the center of the ribbon (Fig. 11.1b). The rachidian tooth (R) is flanked by the first lateral tooth (L1), followed by the second lateral tooth (L2), then by the marginal tooth (M), and finally the marginal plates (MP). The number of teeth in the radula is expressed by the number of elements on each transverse row. The number of teeth per row varies slightly in cephalopods compared to other mollusks. The numbers of rows is defined as the number of transverse rows that are represented by at least one tooth (Kruta et al. 2014). Measurements of radular elements are presented in table 11.1. Several cusps can be present on a single tooth. The height of a tooth is the measure between the highest cusp and the base of the tooth. The width of a tooth is measure of the transverse width at the base of the tooth (Fig. 11.1b).

11.4 Composition

The teeth in Recent cephalopods are composed of a chitin-protein complex (α-chitin and protein; Hunt and Nixon 1981). Mineral elements can be present in the teeth, for example, the teeth in *Nautilus* are enriched in Ca, Mg, Si, P, K, Mn, Fe, and S

Table 11.1 Measurements in mm. of radular teeth and radular ribbons in ammonoids. The numbers with asterisk (*) indicate that the measures were done on published illustrations. The underlined number indicates the value corresponds to the diameter of the ammonite, no data on body chamber whorl height was available. Institutional abbreviations see introduction

Taxa	Suborder	Superfamily	Period/Stage	Specimen number	Type of jaw	Rachidian tooth	
						width	height
Cravenoceras fayettevillae Gordon	Goniatitina	Neoglphioceratoidea Plummer & Scott	Carboniferous (Chesterian)	AMNH 66407 (= OUZC 1001 by Tanabe and Mapes 1995)	Normal type	0.35	0.27
Girtyoceras limatum (Miller & Faber)	Goniatitina	Dimorphoceratoidea Hyatt	Carboniferous (Chesterian)	OUZC	Normal type	0.34	0.34
Isolated nodule-goniatite	Goniatitina	–	Carboniferous (Lower Missourian/Upper Pennsylvanian)	AMNH 66393 (=OUZC 4001by Tanabe et al. 2001)	Normal type	0.53	0.41
Glaphyrites	Goniatitina	Gastrioceratoidea Hyatt	Carboniferous (Pennsylvanian)	GPIT 1320/2	Normal type	0.8	1
Ceratites penndorfi Roth	Ceratitina	Ceratitoidea Mojsisovics	Triassic (Anisian)	SMNS 67494	Normal type	0.30/0.40	0.10/0.15
Arnioceras	Ammonitina	Psiloceratoidea Hyatt	Jurassic (Lower Lias)	GSCL 22703	Anaptychus	0.24*	0.19*
Dactylioceras tenuicostatum (Young & Bird)	Ammonitina	Eoderoceratoidea Spath	Jurassic (Lower Lias)	–	Anaptychus	–	0.2/0.3
Eleganticeras elegantulum (Young & Bird)	Ammonitina	Hildoceratoidea Hyatt	Jurassic (Toarcian)	GSH 1132	Aptychus	–	–
Aconeceras trautscholdi (Sinzow)	Ammonitina	Haploceratoidea Zittel	Cretaceous (Aptian)	PIN	Aptychus	0.08*	0.06
Baculites	Ancyloceratina	Turrilitoidea Meek	Cretaceous (Campanian)	AMNH 55901	Aptychus	0.62	0.4
Rhaeboceras halli (Meek & Hayden)	Ancyloceratina	Scaphitoidea Meek	Cretaceous (Campanian)	AMNH 51334	Aptychus	–	–

Table 11.1 (continued)

First Lateral tooth		Second Lateral tooth		Marginal tooth		Marginal plate		Radular ribbon	Body chamber whorl height
width	height	width	height	width	height	width	length	width	
0.21	0.26	0.23	0.46	0.11	0.8	–	–	2.5	12
–	–	–	–	–	0.5	–	–	1.5	–
0.45	0.75	0.42	1	0.21	1	0.09	0.2	–	–
–	–	0.58	1.9	–	–	–	–	5.5	45
0.30/0.40	0.10/0.15	0.30/0.40	0.10/0.15	0.3/0.4	0.1/0.15	–	–	–	70*
0.19	0.19*	–	–	–	–	–	–	0.78*	20
–	–	–	–	–	1.2	–	–	2.7	82
–	–	–	–	–	–	–	–	0.8	31
0.06*	0.06	0.06	0.06	0.08*	0.4	0.06*	0.04	0.45	15
0.5	0.9	1.2	1	0.4	1.5	0.3	0.3	–	33
1.6	0.8	3.7	1.1	0.5	4.5	–	–	–	37

(Fukuda 1980). Hardening of the teeth occurs during ontogeny (Messenger and Young 1999) due to the tanning of the chitin. Studies on experimental decay of Recent cephalopods show that the untanned portion of the radula (i.e., the younger portion nearer the radular sac) decays faster than the tanned portions and that teeth survive after the disaggregation of the ribbon (Kear et al. 1995). This implies that in fossil material the older part of the radula is more likely to be preserved and that teeth preserved on the younger portion of the ribbon indicate early fossilization. In fossils, the original, supposedly chitinous structure is altered and radular teeth can be preserved in several ways: (1) the teeth are preserved as external molds, (2) the teeth consist of altered organic matter, (3) the teeth are partially or fully mineralized, (4) the teeth are filled with matrix, and (5) any combination of the other three. When the teeth are filled with matrix, they have sometimes been interpreted as having been hollow or having had a pulp cavity (as explained in Kruta et al. 2014), a condition very different from that in Recent cephalopods, which have solid teeth. Whether the hollows in the fossil teeth reflect a biological reality or are due to a preservation artifact has yet to be established. Techniques using computerized tomography (μCT scan, PPC SRμCT scan) have been successfully applied in recent studies (e.g. Kruta et al. 2011; Klug and Jerjen 2012; Kruta et al. 2013, 2014) to describe the anatomy of fossil radular structures in ammonoids exploiting the differences in composition (absorption contrast) between the teeth and the matrix or gaps between them.

11.5 The Radula in the Nautilidae

The radular apparatus of Recent *Nautilus* is quite different from those of other cephalopods (Fig. 11.2a). Very few radulae of fossil Nautilidae have been discovered so far and the structures were discovered as isolated occurrences. Nonetheless, the strong resemblance between the fossil radulae and the radula in Recent *Nautilus* suggests that there is a strong evolutionary stability of the radular morphology in the Nautilidae (Doguzhaeva et al. 1997).

The radula of extant *Nautilus* is composed of 13 elements (R, L1, L2, M1, MP1, M2, MP2) with marginal plates inserted between marginal teeth (Fig. 11.2a, 3a). The morphology of the radula is very conservative in the genus (Tanabe et al. 1990). The radula is wide and is approximately 10 mm wide and 30 mm long (Saunders et al. 1978). Only one relatively short cusp is present on each tooth with the base as wide as the tooth. The cusps of the rachidian and lateral teeth point backward towards the esophagus. The marginal teeth are short and robust. The inner marginal plate (MP1) is a uniquely nautiloid feature that functions as a trigger for the inner marginal tooth as well as folding support for the outer marginal tooth (M2) (Solem and Roper 1975).

Two isolated radulae from the Late Carboniferous of Mazon Creek were discovered with a number of elements and tooth morphology similar to that in extant *Nautilus*. They are attributed to the genus *Paleocadmus* (*Paleocadmus herdinae* in Solem and Richardson 1975, and *Paleocadmus pohli* in Saunders and Richardson

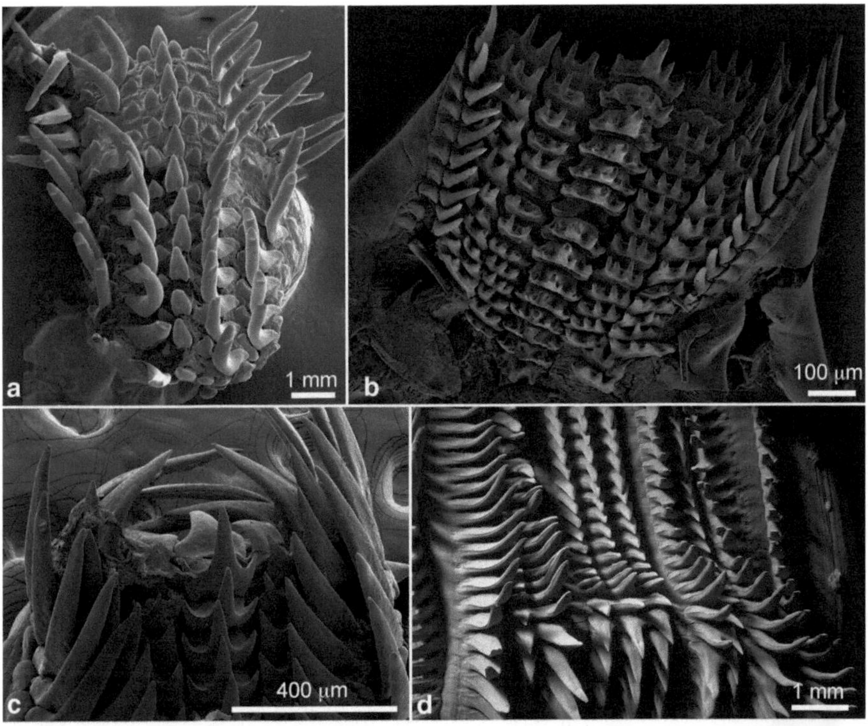

Fig. 11.2 Micrographs of radulae of Recent cephalopods. **a** Radula of *Nautilus pompilius* (Nautilidae, Nautiloidea). UMUT unregistered specimen. From off Suva, Fiji. **b** Radula of *Japetella diaphana* (Octopoda, Amphitretidae). **c** Radula of *Loligo pealei* (Loliginidae, Coleoidea). **d** Radula of *Dosidicus gigas* (Humboldt squid, Ommastrephinae, Coleoidea)

1979). The radular formula is R, L1, L2, M1, MP1, M2, MP2; the radulae are preserved as external molds in concretions (Fig. 11.3b, c). 25 rows of teeth are preserved in *P. herdinae,* the radula is 8 mm long and 3 mm wide while 18 rows are preserved in *P. pohli* and the radula is 5 mm long and 2 mm wide (Saunders

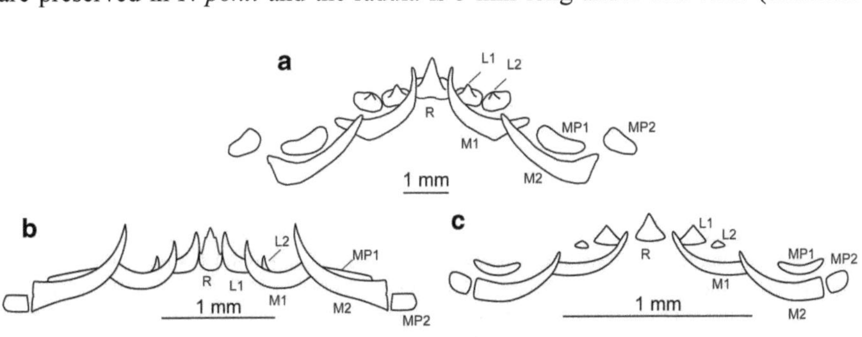

Fig. 11.3 Drawings of the radular rows in Nautilidae. Abbreviations. R: rachidian tooth, L1: first lateral tooth, L2: second lateral tooth, M1: first marginal tooth, M2: second marginal tooth, MP1: first marginal plate, MP2: second marginal plate **a** *Nautilus pompilius*. **b** *Paleocadmus herdinae,* Late Carboniferous. **c** *Paleocadmus pohli*, Late Carboniferous (modified from Saunders and Richardson 1979)

and Richardson 1979). The two species are distinguished by several differences in the teeth morphology. In *P. herdinae*, the teeth are elongated, the rachidian tooth is described as slightly tricuspid, and the marginal teeth have a deep narrow groove on the outer surface (Fig. 11.3b). In *P. pohli*, the rachidian tooth is triangular shape, both the rachidian and first lateral teeth relatively short and unicuspidate (Fig. 11.3c) and the marginal tooth is smooth.

11.6 The Radula in Other Cephalopods

Several radulae have been recorded from Paleozoic cephalopods including the oldest record of a cephalopod radula from the Ordovician. The Late Silurian orthoceratid *Michelinoceras* sp. is described as having seven elements per row (Mehl 1984). The rachidian tooth supposedly is tricuspid (Fig. 11.4a), but considering the state of preservation (only the marginal teeth are well exposed), the reconstruction might be somewhat speculative. Radulae preserved as external molds were also found in four unidentified Late Ordovician orthoconic cephalopods (Gabbot 1999), but the preservation does not allow to distinguish the shape of the teeth (Fig. 11.4b).

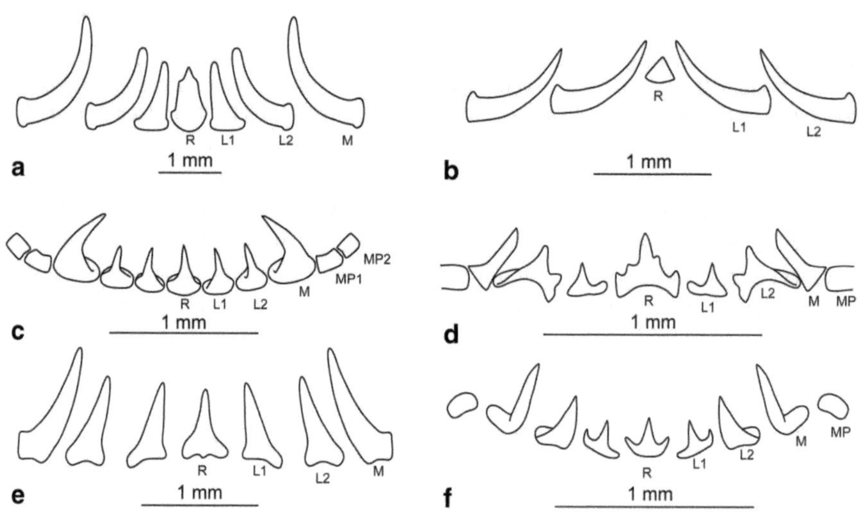

Fig. 11.4 Drawings of the radular row in fossil cephalopods (orthoceratid, unidentified orthocone and coleoids) and Recent Coleoidea. Abbreviations. R: rachidian tooth, L1: first lateral tooth, L2: second lateral tooth, M: marginal tooth, MP: marginal plate (first MP1, second MP2) **a** *Michelinoceras* sp., Late Silurian. Modified from Mehl (1984). **b** Unidentified Late Ordovician orthoconic cephalopod. Modified from Gabbot (1999). **c** *Saundersites ilinoissensis*, coleoid, Late Carboniferous. Modified from Doguzhaeva et al. (2007). **d** *Octopus vulgaris* (Octopodidae, Coleoidea). Modified from Naef (1923). **e** *Sepia officinalis* (Sepiida, Coleoidea). Modified from Naef (1923). **f** *Loligo vulgaris* (Loliginidae, Coleoidea)

The coleoid *Jeletzkya douglassae* from the Mazon Creek Lagerstätte (Pennsylvanian, Desmoinesian) is described as having a radula with 9 elements (Saunders and Richardson 1979). The two specimens were studied by Doguzhaeva et al. (2007) and redescribed as *Saundersites illinoissensis* (Fig. 11.4c). Two marginal plates are visible on the external side of the ribbon. Therefore, eleven elements are present on each transverse row (R, L1, L2, M, MP1, MP2). Radular remains were discovered in *Pohlsepia mazonensis*, another cephalopod from Mazon Creek, but the radula is obscured by matrix and unidentifiable (Kluessendorf and Doyle 2000).

The radula of extant cephalopods generally consists of 7 teeth with few exceptions due to secondary loss. A pair of marginal plates can be present as well, extending the number of elements to 9. Reduction or loss of the marginal plate is frequent in the teuthoids and sepiids. Unicuspid, bicuspid, and tricuspid rachidian teeth and unicuspid or bicuspid lateral teeth are reported in many groups (Solem and Roper 1975). For example, octopods (Fig. 11.4d) have a multicuspidate rachidian tooth with ectocones that vary in position (Robson 1925) and sometimes multicuspidate lateral teeth as for example in the Amphitredidae (Fig. 11.2b). In *Sepia officinalis* (Fig. 11.4e) as well as in other Sepiidae, the seven teeth are unicuspidate, moderately acute, and almost equal in height, the marginal tooth being always the longest (Aldrich, 1969). Intraspecific variability and similarities between different species were observed for the radula of *Loligo* (Aldrich et al. 1971). In *Loligo*, the rachidian tooth is tricuspidate (Figs. 11.2c, 11.4f), the first lateral tooth can be bicuspidate and the bases of the teeth are well developed (Aldrich 1969). *Spirula* lacks a radula but has well developed buccal palps replacing the radular function (Nixon and Young 2003).

11.7 The Radula in Ammonoids

Because of the rarity of the structure, radulae are unknown for many groups of ammonoids. We will describe briefly the major finds in the Ammonoidea. All the radulae in the Ammonoidea include 7 teeth; the presence of marginal plates is variable. Dimensions for teeth are given in table 11.1. However these values are just estimates as wear can decrease the size of the cusps and the radula grows during ontogeny. For systematic purposes, teeth present in the radular canal and not yet exposed to wear would be the most appropriate to measure (Aldrich 1969). Unfortunately, the exact position of teeth on the ribbon is very hard to identify in fossil material, except for a few cases of early fossilization where the ribbon is preserved three dimensionally.

11.7.1 Goniatitina

The radula is known in four genera of Carboniferous Goniatitina. Other radulae found as isolated occurrences in phosphatic nodules were attributed to this group

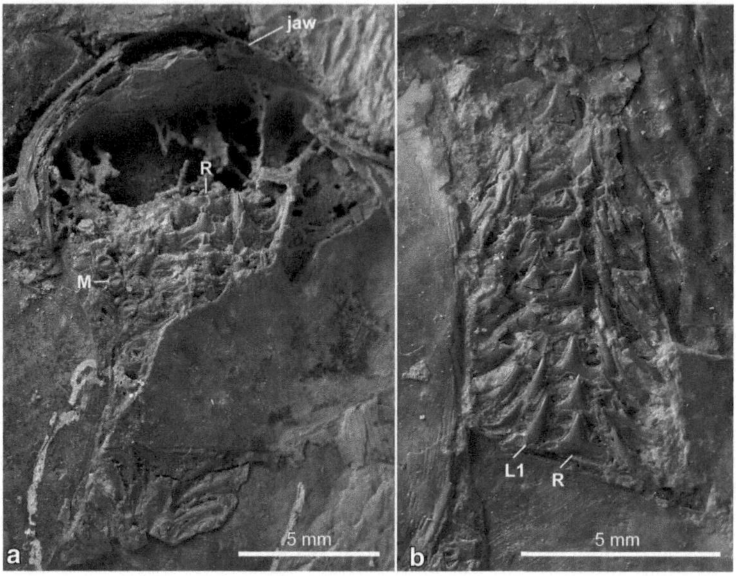

Fig. 11.5 The three dimensionally preserved radula of *Glaphyrites* (Goniatitina). Abbreviations. *R* rachidian tooth, *L1* first lateral tooth, *M* marginal tooth **a** Close up of the radula preserved three dimensionally between the upper and lower jaw in the body chamber, GPIT 1320/1b. **b** Counterpart of the radula GPIT 1320/1b

as well. The first record for a radula preserved in an ammonoid is from an exceptionally preserved goniatite of the genus *Glaphyrites* (described at the time as *Eosianites*) by Closs and Gordon (1966) and Closs (1967). This sensational discovery occurs in Carboniferous phosphatic nodules (Upper Carboniferous? Lower Permian, Itatare Formation, Uruguay). The study was done on eight specimens of *Glaphyrites* with the buccal mass preserved almost three dimensionally in the body chamber (Figs. 11.5, 11.6a). The specimens were restudied by Lehmann (1976), Bandel (1988), Tanabe and Mapes (1995), Lehmann et al. (2014), and Lehmann and Klug (2015). The radula was discovered between the upper and lower jaw and has 7 elements and 2 marginal plates identified by Tanabe and Mapes (1995). The teeth were dissolved and replaced by secondary calcite (Closs 1967).

Another example of good preservation is the radula of *Cravenoceras fayettevillae* (Upper Mississippian of Arkansas) found *in situ* between the upper and lower jaws (Fig. 11.6b). The radula is three dimensional and partially exposed on the surface; however, the presence of a mineral crust makes detailed observations difficult (Tanabe and Mapes 1995). Teeth embedded in the matrix were studied by Synchrotron propagation phase contrast microtomography and allowed the precise reconstruction of the radular row (Kruta et al. 2014). Radula fragments were found as well in five isolated beaks from the Upper Mississippian Fayetteville Shale, which were attributed to *Girtyoceras* (Doguzhaeva et al. 1997). Only one sectioned specimen allows documentation of some details on the radula morphology. Radulae were identified in five specimens of *Vallites* (Namurian, Arnsberger Schichten) by Bandel (1988) and studied through thin sections. The complete radula contained

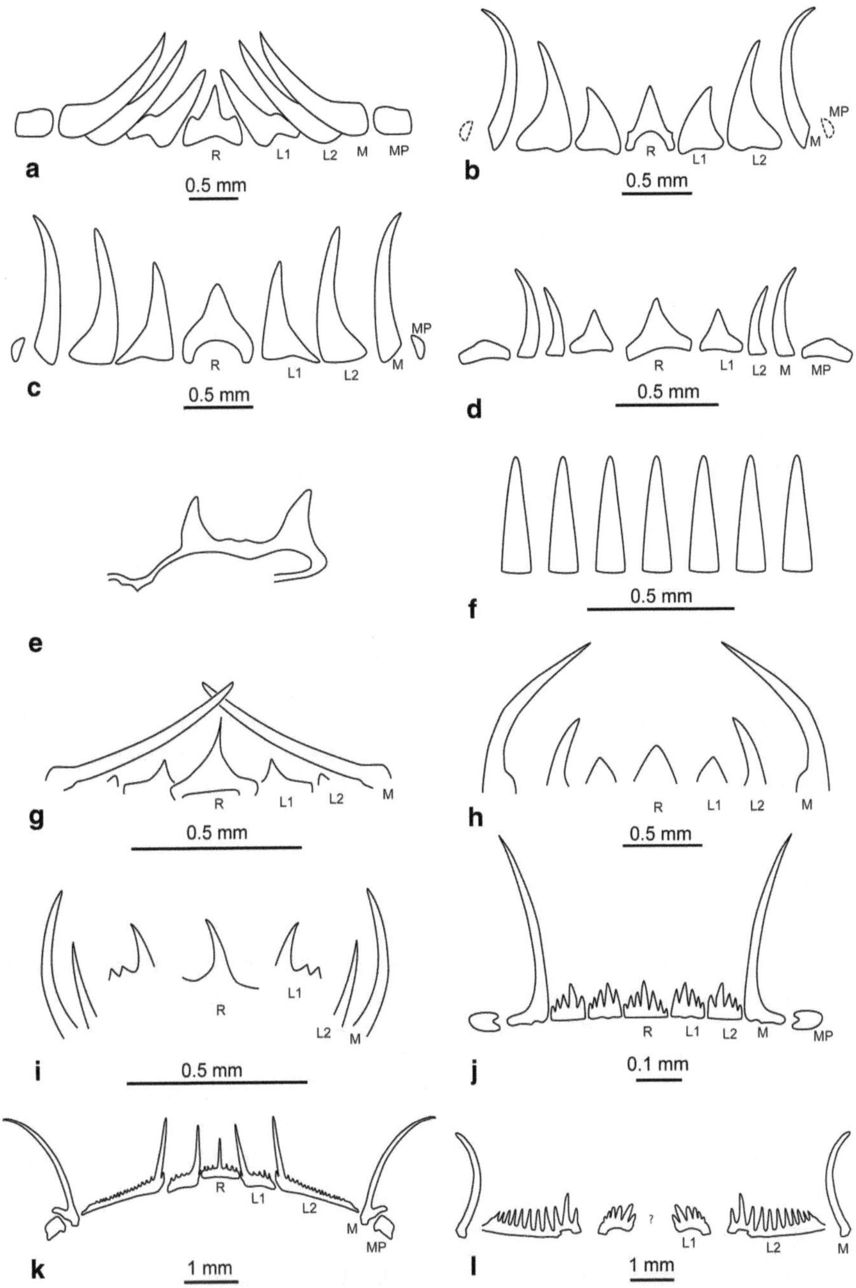

Fig. 11.6 Drawings of the radular row in ammonoids. **a–d** Goniatitina. **e–f** Ceratitina. **g–j** Ammonitina. **k–l** Ancyloceratina. Abbreviations. *R* rachidian tooth, *L1* first lateral tooth, *L2* second lateral tooth, *M* marginal tooth, *MP* marginal plate. **a** *Glaphyrites*, Late Carboniferous, modified from Tanabe and Mapes (1995). **b** *Cravenoceras fayettevillae*, Early Carboniferous. Modified from Kruta et al. (2014). **c** isolated Late Carboniferous radula, Goniatitina. Modified from Kruta et al. (2014). **d** *Girtyoceras*, Early Carboniferous. Modified from Doguzhaeva et al. (1997).

25–30 rows (Bandel 1988). Teeth are described as minute and hook-like in shape and no precise morphological description is given. Tanabe et al. (2001) described an extremely well-preserved phosphatized cephalopod buccal mass from the Upper Pennsylvanian of Oklahoma. The radula is preserved in the anterior portion of the buccal cavity; it is made of more than ten rows of teeth, each consisting of seven tooth elements with a pair of marginal plates. The overall features of the radula and jaws are essentially similar to those described in association with ammonoids rather than nautiloids and coleoids, suggesting that this buccal mass can be referred to the Ammonoidea, possibly to the Goniatitina. The radula consists of 23 rows of teeth preserved in their original position (Kruta et al. 2014); 6 rows can be observed anterior to the bending plane and were analyzed in order to find traces of wear. No diminution in the height of the cusp was observed (Kruta et al. 2014).

In all these specimens the radula consists of nine elements per row (Fig. 11.6a, b, c, d). The overall morphology of the teeth is similar to radulae present in Recent teuthoids and sepiids (Closs 1967; Bandel 1988; Kruta et al. 2014). Two very similar radular morphologies seem to be present in the Carboniferous, the only differences so far established consist in the dimensions of the first lateral tooth and slight variation in the shape of the base and central cusp of the rachidian tooth (Kruta et al. 2014). The rachidian tooth has generally a robust central cusp with a large more or less angular base. The first lateral tooth is pyramid shaped with the cusp inclined towards the outer margin of the radular ribbon. The cusp is higher in *Cravenoceras* (Neoglyphioceratoidea) and *Girtyoceras* (Dimorphoceratoidea) compared to the Pennsylvanian specimens. The second lateral tooth is similar in shape to the first lateral tooth but has a higher cusp. The marginal tooth is long and slender and the marginal plate oval-shaped.

11.7.2 *Ceratitina*

The fossil record for radulae in the Ceratitina is extremely poor and the preservation does not allow a precise description of the anatomy. Two teeth from the Triassic of Siberia were attributed to *Nordophiceras schmidti* by Zakharov (1979). They were found in two isolated anaptychi discovered in a concretion containing ceratitids (Spathian, Early Triassic). The shape of the teeth is illustrated (Fig. 11.6e) but not described in detail. The author suggests similarities with the teeth found in *Eosianites*. A radula *in situ* was discovered in the Anisian ceratitid *Ceratites*

---

e *Nordophiceras schmidti,* Triassic, Spathian. Modified from Zakharov (1979). **f** *Ceratites penndorfi*, Triassic, Anisian, **g** *Arnioceras* (Arietitidae, Psiloceratoidea), Sinemurian. Modified from Lehmann (1971). **h** *Dactylioceras* (Dactylioceratidae, Eoderoceratoidea), Lower Toarcian. Modified from Lehmann (1979). **i** *Eleganticeras elegantulum* (Early Toarcian Hildoceratidae, Hildoceratoidea). Modified from Nixon (1996). **j** *Aconeceras trautscholdi* (Aptian, Oppellidae, Haploceratoidea). Modified from Doguzhaeva and Mutvei (1992). **k** *Baculites* sp, (Campanian, Baculitidae, Turrilitoidea). Modified from Kruta et al. (2011). **l** *Rhaeboceras halli* (Campanian, Scaphitoidea). (Modified from Kruta et al. 2013) *penndorfi* from the German Muschelkalk (Klug and Jerjen 2012). The radula remains was found between the beak elements and in the buccal mass area, behind the two mandibles. Eighteen teeth were identified. The brownish color of the teeth is interpreted as a result of phosphatization. The teeth are uniform in shape and dimension (Fig. 11.6f) and are arranged in vaulted rows. Transverse rows contain seven elements. Each tooth is approximately 0.3–0.4 mm high and approximately 0.1–0.15 mm thick with an oval to circular cross section. The wall of the teeth is thin and the inside of the tooth is filled with sediment.

11.7.3 Ammonitina

Radulae were discovered in five genera of Ammonitina but described only in four of them. Most of the radulae belong to Jurassic (Lias) taxa. Only one radula was found in the Cretaceous taxa *Aconeceras* (Aptian Oppeliidae, Haploceratoidea). The radulae are found in association with two types of jaw morphology: the anaptychus type and the aptychus type (Tab. 11.1). The radular row is composed of seven teeth, marginal plates were only recorded in *Aconeceras*. The morphology is slightly different and will therefore be described for each taxon.

The radula of *Arnioceras* (Sinemurian Arietitidae, Psiloceratoidea) was found nestled in the jaws (anaptychus type) of three specimens (Fig. 11.7). The radula measures 0.72 mm in length. The marginal teeth are twice the height of the rachidian teeth and the lateral teeth are unicuspidate (Fig. 11.6g). The mesocone of the rachidian tooth is sharp (Lehmann 1971, 1981). The radula of *Dactylioceras* (Dactylioceratidae, Eoderoceratoidea) from the Lower Toarcian (Yorkshire Coast, Port Mulgrave) was found between the jaws (anaptychus-type; Lehmann 1979). The radula is incomplete but few radular teeth could be identified (Fig. 11.6h).

The radula discovered in two specimens of *Eleganticeras elegantulum* (Toarcian, Hildoceratidae, Hildoceratoidea) is 0.5 mm in length (Lehmann 1967). The marginal teeth are long and slender (Figs. 11.6i, 11.8). The rachidian tooth has a thin mesocone with smaller lateral cusps. The first lateral tooth has three cusps, and the most internal cusp is as high as the mesocone of the rachidian tooth. The study of the radula in the two specimens indicates that the size of the teeth increases with growth (the smaller specimens being considered younger: 17.5 mm compared to 31 mm for the size of the shell). Such an increase in size during growth is also observed in Recent cephalopods (Nixon 1969). Lehmann and Weitschat (1973) reported the a radula *in situ* (housed in the space surrounded by the jaws) of *Hildoceras levisoni* (Toarcian), comprising seven teeth per row. The morphology of the teeth is neither detailed nor is the radula illustrated.

Twenty juveniles of *Aconeceras trautscholdi* (Aptian Oppelliidae, Haploceratoidea) with the radula and the jaws present in the body chamber were described by Doguzhaeva and Mutvei (1990, 1992). Some buccal masses were found outside the body chamber. The radulae preserved their bending point. Forty transverse rows were counted on the most complete specimen. The radula is 0.45 mm wide and has seven teeth and two marginal teeth. The marginal tooth is slender and very long,

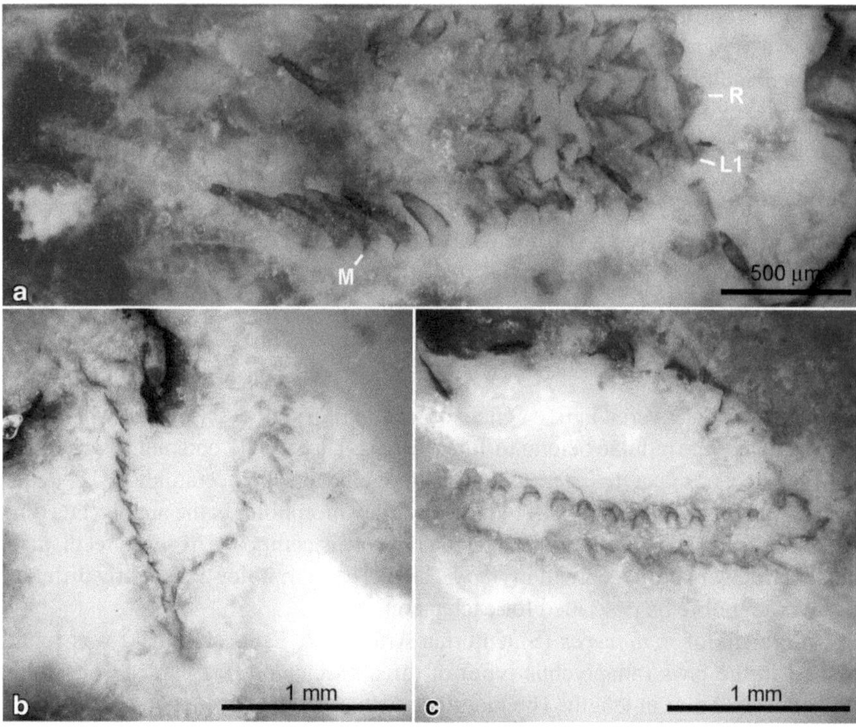

Fig. 11.7 Radula of *Arnioceras* (Goniatitina), thin sections described by Lehmann (1971). **a** Close up of the radula showing the rachidian tooth (*R*), first lateral tooth (*L1*), marginal tooth (*M*). **b** Overview of the radula preserved in GSCL 22703. **c** Overview of the radula preserved in GSCL 22703 (pictures S. Thurston AMNH)

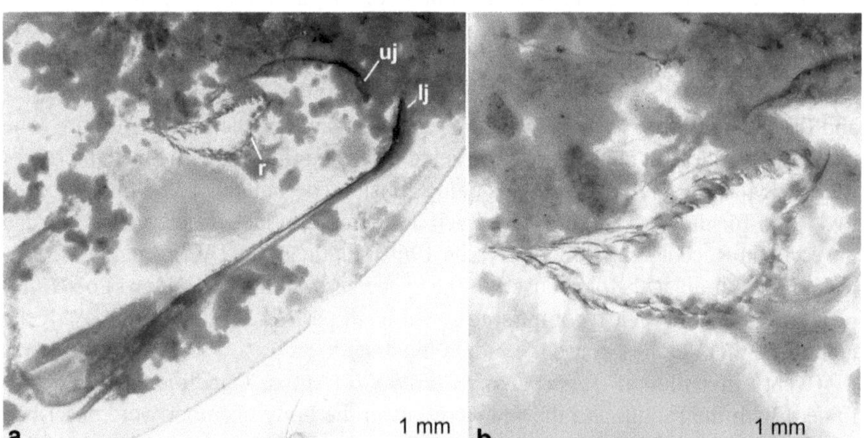

Fig. 11.8 Radula of *Eleganticeras elegantulum* described by Lehmann (1967) **a** Overview of the radula (*r*), between the upper (*uj*) and lower jaw (*lj*) GPIH CE003. **b** Close up of the radula between the upper and lower jaw GPIH CE003 (pictures S. Thurston AMNH)

all teeth except the marginal tooth are multicuspidate (Fig. 11.6j). The second cusp from the internal side of the radula is the highest on the tooth in L1 and L2. The teeth are described as having a central cavity and originally hollow.

11.7.4 Ancyloceratina

Radular structures were found in three specimens of *Baculites* sp. (smooth) (Baculitidae, Turrilitoidea) from the Campanian of the U.S. Western Interior (Gammon Ferruginous Member, Pierre Shale, Butte County, South Dakota). The radulae were identified and described using propagation phase contrast Synchrotron microtomography (i.e. PPC SRμCT scan, see method in Kruta et al. 2011). The teeth are preserved close to their original position in two specimens and in one specimen, they are preserved in a U-shaped structure possibly reflecting its position in the radular canal. If unfolded, dimensions of the radula are 6 mm wide and 7 mm long. Each radular row consists of nine elements including seven teeth and two marginal plates. Four elements appear on each side of the central rachidian tooth. The teeth vary in shape, are multicuspidate, and comb-like (Fig. 11.6k). The cusps of the rachidian tooth are narrow and sharp, and decrease in height with increasing distance from the central cone. Each lateral tooth (L1 and L2) bears a single small cusp on the inner side of a prominent cone approximately 1 mm high, and multiple cusps on the outer side, decreasing in height toward the margin, four on the first lateral tooth (L1) and 17 on the second lateral tooth (L2). The marginal tooth is unicuspidate, 1.6 mm high, curved, and sabre-like. The marginal plates are small and flat. High resolution scans with voxel (volumetric pixel) size of 1.4 μm revealed a carena running from the tip of the highest cone to its base on the second lateral tooth. These radulae are similar to a previously reported structure in the same species of *Baculites*, but most of the teeth in this previously reported specimen were not visible (Landman et al. 2007). The structures described in Landman et al. (2007) are here reinterpreted as a row of five lateral teeth (L2) preserved as external molds (Fig. 11.9). Only the base of the tooth can be observed on the surface of the sediment as well as imprints of the multiple little cusps, while the main cusp is still buried in the sediment.

Klug et al. (2012) described the radula in five heteromorph ammonoid specimens from the Cenomanian/Turonian boundary (Germany). The radulae are preserved as phosphatized or carbonized material in association with jaws. Shell material is not present and does not allow a closer identification: specimens were identified as a baculitid because of the radula morphology and the fact that baculitids occasionally occur in these strata. Although the preservation does not allow reconstruction of the morphology of the teeth, the overall shape seems to be very similar to the teeth described in *Baculites* sp. (Kruta et al. 2011). The smaller dimension of the teeth elements is related to the rather small size of the ammonite that is interpreted as a small sized species or as juvenile.

Several radular teeth were identified between the jaws of two specimens of *Rhaeboceras halli*, a scaphitid ammonite from the Campanian of the U.S. Western

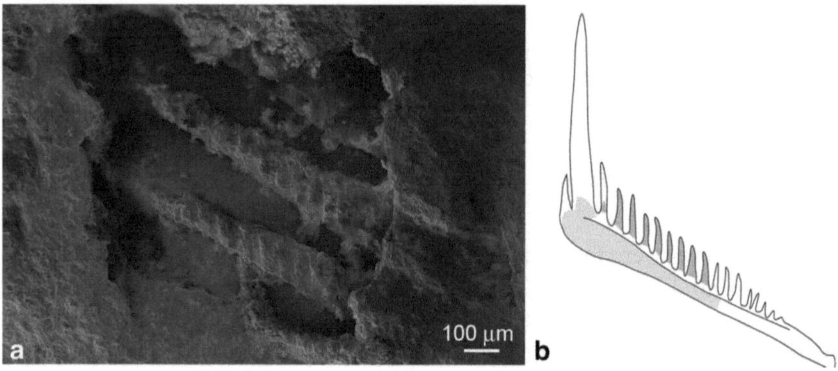

Fig. 11.9 Radula of *Baculites* sp. (smooth) BHMNH 5496 described in Landman et al. (2007). The structure is here reinterpreted as the external mold of a row of four/five second lateral teeth (L2). **a** Micrograph showing the series of lateral teeth. **b** Sketch of the second lateral tooth (L2), the grey areas indicate the surface of the tooth that left the impression on the sediment

Interior. The teeth were found between the upper and lower jaw and identified thanks to PPC SRµCT scan (Kruta et al. 2013). The teeth are disarticulated and only a few teeth are preserved in rows. The row of radular teeth is reconstructed with seven elements although the presence of a rachidian tooth could not be confirmed (Fig. 11.6l). Dissolution occurred in the specimens, which is why only a few teeth could be reconstructed. Some rows of teeth are present with the teeth aligned one below the other. The rachidian and marginal plates are absent probably because of dissolution. The first lateral tooth is semi-lunate with seven cusps, the second cusp from the inside being the highest. The other cusps decrease in height towards the external margin of the radula. The second lateral tooth (L2) is long with 11 cusps, and the second cusp higher than the other. The marginal tooth is saber-shaped and very high (4.6 mm).

Hook-shaped structures in body chambers of *Rhaeboceras halli* were previously interpreted as radular elements by Kennedy et al. (2002). However, a number of morphological features makes this interpretation unlikely. The supposed radular teeth end in two sharp points and present a arm hook-like morphology. They vary in shape and some are extremely long, a single tooth representing approximately 50% of the length of the upper jaw (Kruta et al. 2013). The dimensions of the teeth and the morphology is inconsistent with a radular function in modern and fossil cephalopods and to what is recognized as radulae in other ammonoids (Kruta et al. 2013). Landman and Klofak (2004) suggested that these hooks could be remnants of a hectocotylus structure used during mating.

11.8 Discussion

11.8.1 Phylogenetic Implications

Although very few data are available, fossil radulae highlight the differences in the feeding structures between the major groups of cephalopods. The radula in nautilids is different from the radula in other cephalopods by the larger number of elements and the morphology of the teeth, while similarities are present between the radulae of Recent coleoids and the radulae of ammonoids. Following these observations, Lehmann (1967, 1981) suggested regrouping ammonoids and coleoids in the same subclass: the Angusteradulata, while the Lateradulata included nautiloids. In the Angusteradulata, 7 elements are present per transverse row, with the addition of two marginal plates. The Lateradulata have 9 teeth per row and 4 marginal plates. Engeser (1996) used the character of reduction of tooth number (7) as a synapomorphy for ammonoids, orthoceratids, bactritids, belemnoids, and coleoids. The morphology of the Lateradulata is considered to be the plesiomorphic state because the radula of an undetermined cephalopod (*Campitius titanicus*) discovered in the Lower Cambrian by Firby and Durham (1974) presents several elements (9 elements reconstructed). This radula, although not attributed to a precise taxon, may belong to a cephalopod stem group (Firby and Durham 1974; Engeser 1996). Recent discoveries of fossil radulae in ammonoids (Klug and Jerjen 2012; Klug et al. 2012; Kruta et al. 2011, 2013, 2014) seem to reinforce this hypothesis, as all the specimens have 7 teeth per row. The number of marginal elements (absent, 2 or 4) seem to be more variable compared to what Lehmann (1967) predicted (cf. *Saundersites illinoisensis*).

Discussions on the variability within the Ammonoidea are limited by the few records available. The radula in the Goniatitina shows little variability and is similar to the one present in Recent coleoids with respect to both the number of elements and the shape of the teeth. Very few data are available for the Ceratitina but the number of tooth elements is consistent with the discoveries in other ammonoids. Multicuspidate radulae with extremely elongated marginal teeth (Tab. 11.1) are found both in the Ammonitina and in the Ancyloceratina (Fig. 11.6). Radulae with multicuspidate teeth seem to be associated with the aptychus-type jaw in the Ancyloceratina and some Ammonitina. The teeth reconstructed from the well preserved radula of *Baculites* are extremely slender and with sharper and more delicate cusps compared to all other ammonoids (Fig. 11.10) and Recent coleoids. Further discoveries will confirm if the slender multicuspidate teeth are related to the aptychus type of jaw and bear a phylogenetic signal.

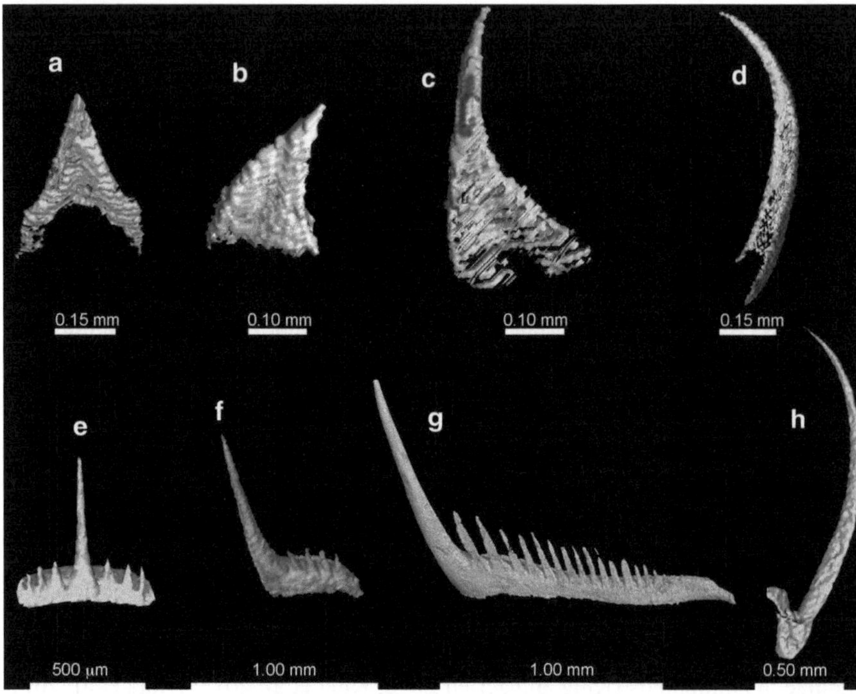

Fig. 11.10 Comparison between representatives of the radular morphology in the Carboniferous Goniatitina (*Cravenoceras fayettevillae* from Kruta et al. (2013) and the Cretaceous Ancyloceratina (*Baculites* from Kruta et al. 2011). **a–d** Teeth in *Cravenoceras fayettevillae* (Neoglyphioceratoidea, Cravenoceratidae. **a** Rachidian tooth. **b** First lateral tooth. **c** Second lateral tooth. **d** marginal tooth. **e–f** Teeth in *Baculites* (Baculitidae, Turrilitoidea). **e** Rachidian tooth. **f** First lateral. **g** Second lateral tooth. **h** Marginal tooth

11.8.2 Function of the Radula

Although if it is often used for systematic purposes (Aldrich 1969; Aldrich et al. 1971), the morphology of the radula in Recent cephalopods is relatively uniform. This lack of variability is believed to be related to the carnivorous feeding habit of cephalopods and the minor role of the radula in feeding habits compared to the important role of the brachial crown and jaws (Boucher-Rodoni et al. 1987). In Recent cephalopods, the radula is used to convey food towards the esophagus and only in a few octopods (e.g., *Eledone cirrhosa* or *Octopus vulgaris*; Runham et al. 1997; Nixon 1979) as a drilling device. For Solem and Richardson (1975), the slight variations in the radular morphology are related to the evolution of a variety of feeding characteristics throughout cephalopods. Recent studies indicate that the feeding habits in present-day cephalopods are more diversified than previously assumed (Hoving and Robison 2012), but the link between the feeding habits and the morphology of the feeding structures is hard to establish. In ammonoids, multicuspidate radulae (Fig. 11.11) seem to be related to the aptychus type of jaw. Several

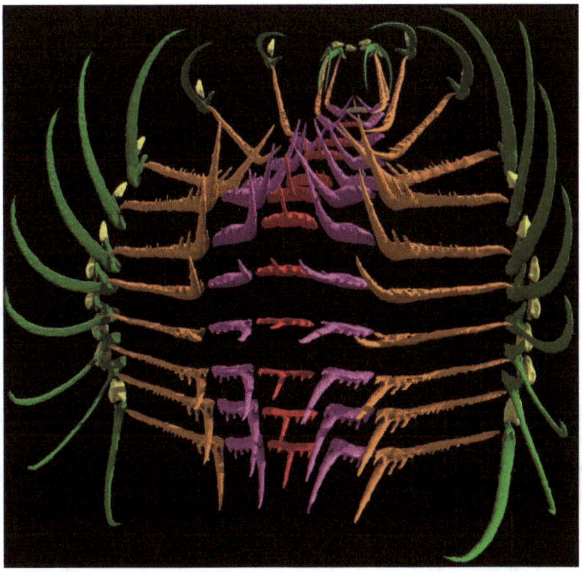

Fig. 11.11 3D reconstruction of the radula of *Baculites* with an anterior view of the bending plane. The different types of teeth are outlined in different colors. *Yellow* marginal plate, *Green* marginal teeth, *Orange* second lateral teeth, *Purple* first lateral teeth, *Red* rachidian teeth. Reconstruction I. Kruta/A. Lethiers, CR2P, MNHN, CNRS, UPMC-Paris 6

hypotheses were made on the feeding habits related to this particular jaw apparatus (see also Tanabe et al. 2014), most of them include a microphagous diet (Morton and Nixon 1987; Kennedy et al. 2002; Keupp 2007; Seilacher 1993; Tanabe and Landman 2002) and zooplankton (Kruta et al. 2011).

References

Aldrich MM (1969) The teuthoid radula as a taxonomic criterion, with special reference to families Architeuthidae and Omnastrephidae (Cephalopoda: Coleoidea). MS Thesis; Memorial University of Newfoundland, Canada

Aldrich MM, Barber VC, Emerson CJ (1971) Scanning electron microscopical studies of some cephalopoda radulae. Can J Zool 12:1569–1594

Bandel K (1988) Operculum and buccal mass of ammonites. In: Wiedmann J, Kullmann J (eds) Cephalopods—present and past, Schweizerbart Stuttgart, pp 653–678

Boucher-Rodoni R, Boucaud-Camou E, Mangold K (1987) Feeding and digestion. In: Boyle PR (ed.). Cephalopod life cycles: comparative reviews II. Academic Press, London, pp 85–108

Closs D (1967) Goniatiten mit Radula und Kieferapparat in der Itararé Formation von Uruguay. Paläontol Z 41:19–37

Closs D, Gordon M (1966) An upper Paleozoic goniatite radula. Esc Geol Porto Alegre, Notas Estud 1:73–75

Doguzhaeva LA, Mutvei H (1990) Radulae, aptychi and counteraptychi in Cretaceous ammonites (Mollusca: Cephalopoda), Dokl Akad Nauk SSSR 313:192–195

Doguzhaeva LA, Mutvei H (1992) Radula of the Early Cretaceous ammonite *Aconeceras* (Mollusca: Cephalopoda). Palaeontogr A 223:167–177

Doguzhaeva LA, Mapes RH, Mutvei H (1997) Beaks and radulae of Early Carboniferous goniatites. Lethaia 30:305–313

Doguzhaeva LA, Mapes RH, Mutvei H (2007) Late Carboniferous coleoid cephalopod from the Mazon Creek Lagerstätte (USA), with a radula, arm hooks, mantle tissues, and ink. In: Landman NH, Davis RA, Mapes RH (eds) Cephalopods-present and past: new insights and fresh presperctives. Topics in geobiology. Springer, Netherland pp 121–143

Engeser T (1996) The position of the Ammonoidea within the Cephalopoda. In: Landman NH, Tanabe K, Davis RA (eds) Ammonoid Paleobiology. Topics in geobiology 13. Plenum Press, New York

Firby JB, Durham JW (1974) Molluscan radula from earliest Cambrian. J Paleontol 48:1109–1119

Fukuda Y (1980) Observations by SEM. In: Hamada T, Obata I, Okutani T (eds) *Nautilus macromphalus* in captivity. Japanese expert consultation on living nautilus, Tokai University Press, Tokyo pp 23–33

Gabbot SE (1999) Orthoconic cephalopods and associated fauna from the late Ordovician Soom Shale Lagerstätte, South Africa. Palaeontology 42:123–148

Hoving HJT, Robison BH (2012) Vampire squid: detritivores in the oxygen minimum zone. Proc R Soc B 279(1747):4559–4567

Hunt S, Nixon M (1981) A comparative study of protein composition in the chitin-protein complexes of the beak, pen, sucker disc, radula and oesophageal cuticule of cephalopods. Comp Biochem Physiol 68:535–546

Kear AJ, Briggs DEG, Donovan DT (1995) Decay and fossilisation of non-mineralised tissue in coleoid cephalopods. Palaeontology 38:105–131

Kennedy WJ, Landman NH, Cobban WA, Larson NL (2002) Jaws and radulae in *Rhaeboceras*, a Late Cretaceous ammonite. Abh Geol B-A 57:394–399

Keupp H (2007) Complete ammonoid jaw apparatuses from the Solnhofen plattenkalks: implications for aptychi function and microphagous feeding of ammonoids. Neues Jahrb Geol Paläontol Abh 245:3–101

Kluessendorf J, Doyle P (2000) *Pohlsepia mazonensis*, an early 'octopus' from the Carboniferous of Illinois, USA. Palaeontology 43:919–926

Klug C, Jerjen I (2012) The buccal apparatus with radula of a ceratitic ammonoid from the German middle triassic. Geobios 45:57–65

Klug C, Riegraf W, Lehmann J (2012) Soft-part preservation in heteromorph ammonites from the Cenomanian-Turonian boundary event (OAE 2) in north-west Germany. Palaeontology 55:1307–1331

Kruta I, Landman NH, Rouget I, Cecca F, Tafforeau P (2011) The role of ammonites in the Mesozoic marine food web revealed by exceptional jaw preservation. Science 331(70):70–72

Kruta I, Landman NH, Rouget I, Cecca F, Tafforeau P (2013) The radula of the Late Cretaceous ammonite *Rhaeboceras halli*. Palaeontology 56:9–14

Kruta I, Mapes R, Pradel A, Tafforeau P, Landman NH (2014) New insights into the buccal apparatus of the Goniatitina; paleobiological and phylogenetic implications. Lethaia 47:38–48

Landman NH, Klofak SM (2004) Radulae and red herrings. 6th international symposium cephalop—present and past. Abstract vol 37:99

Landman NH, Larson NL, Cobban WA (2007) Jaws and radula of *Baculites* from the Upper Cretaceous (Campanian) of North America. In: Landman NH, Davis RA, Mapes RH (eds) Cephalopods-present and past, new insights and fresh perspectives. Topics in geobiology series. Springer, Netherlands pp 257–298

Lehmann U (1967) Ammoniten mit Kieferapparat und Radula aus Lias-Geschieben. Paläontol Z 41:38–45

Lehmann U (1971) Jaws, radula, and crop of *Arnioceras*. Palaeontology 14:338–341

Lehmann U (1976) Ammoniten, ihr Leben und ihr Umwelt. Enke, Stuttgart

Lehmann U (1979) The jaws and radula of the Jurassic ammonite *Dactylioceras*. Palaeontology 22:265–271

Lehmann U (1981) The ammonites. Their life and their world. Cambridge University Press, Cambridge
Lehmann J, Klug C (2015) Soft part anatomy of ammonoids: reconstructing the animal based on exceptionally preserved specimens and actualistic comparisons. In: Klug C, Korn D, De Baets K, Kruta I, Mapes RH (eds) Ammonoid paleobiology: from anatomy to ecology. Springer, Dordrecht
Lehmann U, Weitschat W (1973) Zur Anatomie und Ökologie von Ammoniten. Funde von Kropf und Kiemen. Paläontol Z 47:69–76
Lehmann J, Klug C, Wild F (2014) Did ammonoids possess opercula? Reassessment of phosphatised soft tissues in *Eoasianites* from Uruguay. Paläontol Z:1–10. doi:10.1007/s12542-013-0219-8
Mehl J (1984) Radula und Fangarme bei *Michelinoceras* sp. aus den Silur von Bolivien. Paläontol Z 58:211–228
Messenger JB, Young JZ (1999) The radular apparatus of cephalopods. Phil Trans R Soc B 354:161–182
Morton N, Nixon M (1987) Size and function of ammonite aptychi in comparison with buccal masses of modern cephalopods. Lethaia 20:231–238
Naef A (1923) Die Cephalopoden. Fauna e flora del Golfo di Napoli, Monographie 35, 1, Systematik, part I, pp. 863 (English translation by A. Mercado, 1972, Israel Program for Scientific Translations, Jerusalem)
Nixon M (1969) Growth of the beak and radula of *Octopus vulgaris*. J Zool 159:363–379
Nixon M (1979) Has *Octopus vulgaris* a second radula. J Zool 187:291–296
Nixon M (1995) A nomenclature for the radula of the Cephalopoda (Mollusca)-living and fossil. J Zool 236:73–81
Nixon M (1996) Morphology of the jaws and radula in ammonoids. In: Landman, NH, Tanabe K, Davis RA (eds) Ammonoid Paleobiology. Topics in Geobiology 13, Plenum Press, New York, pp 23–42
Nixon M (2011) Anatomy of recent forms. Treatise online, no. 17, part M, chapter 3:15–17
Nixon M, Young JZ (2003) The brains and lives of cephalopods. Oxford University Press Inc., New York
Robson GC (1925) On seriation and asymetry in the cephalopod radula. Zool J Linn Soc 36:99–106
Runham NW, Bailey CJ, Carr M, Evans CA, Malham S (1997) Hole-drilling in crab and gastropod shells by *Eledone cirrhosa* (Lamarck, 1789). Sci Mar 61:67–76
Saunders WB, Richardson ES Jr (1979) Middle Pennsylvanian (Desmoinesean) Cephalopoda of the Mazon Creek fauna, northeastern Illinois. In: Nitecki MH (ed) Mazon Creek fossils. Academic Press, New York pp 333–359
Saunders WB, Spinosa C, Teichert C, Banks RC (1978) The jaw apparatus of Recent *Nautilus* and its palaeontological implications. Palaeontology 21:129–141
Seilacher A (1993) Ammonite aptychi: how to transform a jaw into an operculum? Am J Sci 293A:20–32
Solem A, Richardson ES (1975) *Paleocadmus*, a nautiloid cephalopod radula from the Pennsylvanian Francis Creek Shale of Illinois. Veliger 17:233–242
Solem A, Roper CFE (1975) Structures of recent cephalopod radulae. Veliger 18:127–133
Tanabe K, Landman NH (2002) Morphological diversity of the jaws of Cretaceous Ammonoidea. Abh Geol B-A 57:157–165
Tanabe K, Mapes RH (1995) Jaws and radula of the Carboniferous ammonoid *Cravenoceras*. J Paleontol 69:703–707
Tanabe K, Tsukahara J, Hayasaka S (1990) Comparative morphology of living *Nautilus* (Cephalopoda) from the Philippines, Fiji and Palau. Malacologia 31:297–312
Tanabe K, Mapes RH, Kidder DL (2001) A phosphatized cephalopod mouthpart from the Upper Pennsylvanian of Oklahoma, USA. Paleontol Res 5: 311–318
Tanabe K, Kruta I, Landman NH (2015) Ammonoid buccal mass and jaw apparatus. (This volume)
Zakharov YD (1979) New finds of rhyncholites, anaptychi, aptychi, and remains of cephalopod radula at USSR territory. In: Gramm MI (ed) Fossil invertebrates of the Far East (Data on new discoveries), pp 80–91 [in Russian]

Chapter 12
Soft Part Anatomy of Ammonoids: Reconstructing the Animal Based on Exceptionally Preserved Specimens and Actualistic Comparisons

Christian Klug and Jens Lehmann

12.1 Introduction

Soft-tissue preservation, i.e. remains of altered organic structures from the animal living inside the body chamber, is generally rare in fossil Mollusca. In ammonoids, the original soft tissue is never preserved and the parts that are commonly preserved are remains of chitinous structures (mostly part of the buccal mass; Tanabe et al. 2015). Organic remains are generally poorly preserved making the reconstruction of the internal anatomy very difficult. Looking at the rich ammonoid collections worldwide, the low number of ammonoid specimens displaying remains of their soft parts appears particular.

Stratigraphically seen, the oldest cases of soft part preservation were reported from the Lower Devonian Hunsrück Slate and the Middle Devonian Wissenbach Slate (Stürmer 1968, 1969, 1970; Zeiss 1968, 1969). In an X-ray of a pyritized *Gyroceratites*, Stürmer (1968, 1969, 1970) thought to have discovered pyritized arms, mantle remains and some organs. The nature of almost all of these soft tissue specimens are called into question today (Otto 1994; Maeda and Seilacher 1996; De Baets et al. Otto 2013). Further cases or fossils with well-assignable soft-tissue preservation from the Devonian have not become known yet.

The picture changes in the Carboniferous. There still are not a lot of cases of known soft-tissue preservation, but a couple of localities did yield some fossilized ammonoid soft parts. One of the most famous cases is the specimens of *Glaphyrites* (earlier also referred to as *Eoasianites*; Doguzhaeva et al. 1997, 2007; Kruta et al. 2013; Lehmann et al. 2014) described by Closs and Gordon (1966) as well as Closs

C. Klug (✉)
Paläontologisches Institut und Museum, University of Zurich, Karl Schmid-Strasse 6, 8006 Zürich, Switzerland
e-mail: chklug@pim.uzh.ch

J. Lehmann
Fachbereich Geowissenschaften, University of Bremen, Klagenfurter Strasse, 28357 Bremen, Germany
e-mail: jens.lehmann@uni-bremen.de

C. Klug et al. (eds.), *Ammonoid Paleobiology: From Anatomy to Ecology*, Topics in Geobiology 43, DOI 10.1007/978-94-017-9630-9_12

(1967a, b) which display complete buccal masses including jaws and radula as well as extensive phosphatized soft-tissues, possibly the mantle (Bandel 1988; Tanabe and Mapes 1998; see Lehmann et al. 2014 for a review). Landman et al. (2010) described Carboniferous ammonoids from Bear Gulch with various soft parts. Further cases of special preservation in Carboniferous ammonoids comprises materials that contain the buccal mass (Miller et al. 1957; Kullmann 1981; Mapes 1987; Tanabe and Mapes 1998; Doguzhaeva et al. 1997; Kruta et al. 2013). From Permian strata, the only case of soft-tissue preservation to our knowledge is a specimen of *Akmilleria* which shows fine details of the soft parts inside the siphuncle (Tanabe et al. 2000).

Several occurrences of soft-tissue preservation have been described from the Mesozoic. In the Triassic, the preservation of organic structures (buccal masses excluded) has been published from various localities. The most spectacular and most widely discussed example probably is an *Ophiceras*, which stems from the Early Triassic of Spitsbergen (Lehmann 1981, 1985, 1987, 1990). An additional case from the Middle Triassic was reported by Klug and Jerjen (2012). In a ceratitid, the body chamber contained both mandibles, remains of the radula and the oesophagus.

Especially the Fossillagerstätten of Jurassic age yielded several interesting cases of soft-tissue preservation (Table 12.1). At least in Europe, it appears like the clay-dominated sedimentary systems of the Early Jurassic promoted the preservation of soft tissues. In addition to a Sinemurian occurrence (Lehmann 1967a, b), various Toarcian ammonoids from the Posidonia slate show poorly discernible remains of soft tissues (Riegraf et al. 1984). In a series of papers, Lehmann (1967a, b, 1985) described three-dimensionaly preserved materials with some unusual detail, also of Toarcian age. But also younger Jurassic ammonites became known with soft part remains (Lehmann 1971). One poorly known Middle Jurassic specimen of *Sigaloceras* (Hollingworth and Hilton 1999) shows extensive organic remains, which are, however, difficult to interpret. The Kimmeridgian and Tithonian Fossillagerstätten of the Solnhofen/Eichstätt region as well as Nusplingen yielded several ammonite specimens, displaying remains of the digestive tract with stomach contents and more or less complete buccal masses (Schweigert 1998, 2009; Frickhinger 1994, 1999; Schweigert and Dietl 1999, 2001; Keupp 2000, 2007, 2011a).

A surprisingly low number of cases of soft-tissue preservation in ammonites has been reported from the Cretaceous (Table 12.1). Alleged imprints of chitin-secreting cells (beccublast cells) are unique among ammonoids. Such imprints were, discovered on a rhynchaptychus referrable to *Gaudryceras* (Tanabe and Fukuda 1983). While *Aconeceras* and *Baculites* mainly became famous for the excellent preservation of their buccal masses including radulae (Doguzhaeva and Mutvei 1991; Kruta et al. 2011), two other examples were published in the past decade which do show some interesting soft-tissue remains. Wippich and Lehmann (2004) reported of *Allocrioceras* from the Fossillagerstätten Hâqel and Hjoula of Lebanon. This locality occasionally yields specimens of this small heteromorph with the buccal mass and phosphatized tissues *in situ*. The second example also concerns a heteromorph, namely a baculitid, possibly belonging to the genus "*Sciponoceras*", (taxonomic

Table 12.1 Published cases of preservation of soft-tissues

Genus	Age	Arms	Mantle	Buccal mass	Oesophagus	Crop, stomach	Muscles (retractors)	Gills	Siphuncle	Eyes	Ovaries, eggs	References
Girtyoceras	Visean			•								Miller et al. 1957; Kullmann 1981; Doguzhaeva et al. 1997
Glaphyrites	Gzhelian (?)		•	•			?					Closs and Gordon 1966; Closs 1967a, b; Lehmann et al. 2014
Rhadinites *Anthracoceras*	Serpukhovian	?		•		•	•				?	Landman et al. 2010
Cravenoceras	Serpukhovian			•								Mapes 1987; Tanabe and Mapes 1998; Kruta et al. 2013
Akmilleria	Artinskian										•	Tanabe et al. 2000
Ophiceras	Griesbachian			•		•		•				Lehmann 1985
Ceratites	Ladinian			•	•							Klug and Jerjen 2012
Arnioceras	Sinemurian			•								Lehmann 1967a
Eleganticeras	Toarcian			•				•				Lehmann 1967b, 1985
Hildoceras	Toarcian			•				•				Lehmann 1985
Harpoceras *Hildoceras* *Lytoceras* *Phylloceras*	Toarcian			•	•	•						Riegraf et al. 1984; Jäger and Fraaye 1997
Sigaloceras	Callovian	?	?	•			?					Hollingworth and Hilton 1999
Lithacoceras *Neochetoceras* *Physodoceras*	Kimmeridgian			•		•						Dietl et al. 1996; Schweigert 1998; Schweigert and Dietl 1999
Lingulaticeras	Kimmeridgian			•	?					?		Schweigert 2009
Hybonoticeras	Tithonian			•								Frickhinger 1999
Aconeceras	Albian			•								Doguzhaeva and Mutvei 1985

Table 12.1 (continued)

Genus	Age	Arms	Mantle	Buccal mass	Oesophagus	Crop, stomach	Muscles (retractors)	Gills	Siphuncle	Eyes	Ovaries, eggs	References
Allocrioceras	Cenomanian			•		•						Wippich and Lehmann 2004
? Sciponoceras	Cenomanian/ Turonian	?		•	•	•				?	?	Klug et al. 2012
Pseudaspido-ceras	Turonian			•		•						Ifrim 2013
Baculites	Maastrichtian			•								Kruta et al. 2011

assignment hampered by the flattened preservation) from the OAE2 of Germany (Klug et al. 2012). These baculitid remains will be further discussed below.

Indirect hints on soft tissues are muscle scars, repeatedly recorded from the Devonian to the Cretaceous which was reviewed by Doguzhaeva and Mapes (2015; see also Doguzhaeva and Mutvei 1993; Richter and Fischer 2002). In the Cretaceous heteromorph *Ptychoceras*, an unusual shell structure probably reflects an organization of the soft body different from all other ammonoids (Doguzhaeva and Mutvei 1989).

In only two cases, ichnofossils of more or less questionable imprints of ammonite soft parts have been described. Some Late Jurassic ammonites from Germany are associated with traces which are difficult to interpret, but which may represent imprints of soft-parts or of scavengers (Frickhinger 1999; Keupp 2007). Summesberger et al. (1999) found roll marks and supposedly soft-part imprints of Late Cretaceous ammonites in Slovenia.

In this chapter, we will provide a review of the state-of-the-art of the knowledge on ammonoid soft-part anatomy, hypotheses for soft part reconstructions as well as implications from applications of the extant phylogenetic bracket (Witmer 1995).

The acronym GPIT refers to the Palaeontological Collection of Tübingen University, Germany (after the former designation "Geologisch-Paläontologisches Institut Tübingen"). GSUB stands for the Geosciences Collection of the University of Bremen.

12.2 Digestive Tract

12.2.1 Oesophagus

In some extant cephalopods, like *Sepia officinalis*, the oesophagus can be rather rigid and may contain a chitinous layer according to Nixon and Young (2003). With this knowledge, it becomes understandable, why the likelihood of preservation of the oesophagus is actually reasonably high (Table 12.2). Nevertheless, the rare reports of oesophagus preservation finds its explanation in the fact that the chitin is untanned and thus disintegrated much faster than the tanned chitin of other parts (Table 12.1; Kear et al. 1995)

The oldest record of an oesophagus might be that of a Triassic (Ladinian) *Ceratites* (Klug and Jerjen 2012). In the specimen described therein, remains of the radula are associated with elongate carbonized structures that were interpreted as remains of the oesophagus. According to its dimension, position and overall shape, the criteria for a homologization with the oesophagus are largely fulfilled, although the criterion of the specific quality lacks support due to the lack of fossilized detail.

The only other example of oesophagus preservation is of Late Cretaceous age (Fig. 12.1). Klug et al. (2012) found extensive parts of the digestive tracts in several specimens of baculitids (questionably assignable to *Sciponoceras*). In these specimens, the buccal mass is well discernible including jaws, radula and other structures. Within the buccal mass, a long, drop-shaped strongly carbonized structure is preserved. These structures are linked with the mouth and with what was interpreted as crop and stomach. In combination with the consideration discussed below, we interpret these black structures as the oesophagus. The ca. 5–10 mm long duct behind the buccal mass most likely belongs to the oesophagus. Interestingly, it is carbonized in most specimens but phosphatized in another one. The assignment to the oesophagus is further corroborated by two specimens with poorly preserved and thus questionable remains of the cephalic cartilage associated. Because of its localization and shape, it appears also reasonably plausible that the most adapically situated sac-like structure represents the stomach-caecum-complex.

12.2.2 Crop and Stomach

Crop and stomach including stomach contents (Fig. 12.2) have been described repeatedly, but so far only from Mesozoic taxa (Lehmann and Weitschat 1973; Riegraf et al. 1984; Lehmann 1975, 1985, 1988, 1990; Nixon 1996; Jäger and Fraaye 1997; Schweigert 1998; Schweigert and Dietl 1999; Keupp 2000, 2007; Wippich and Lehmann 2004; Landman et al. 2010; Klug et al. 2012). Remarkably, the stomach- and crop-remains are preserved either in phosphate (e.g., Schweigert 1998; Schweigert and Dietl 1999; Wippich and Lehmann 2004) or in carbon (e.g., Lehmann 1985; Riegraf et al. 1984; Klug et al. 2012). All of these specimens lack anatomical detail, thus hampering the exact assignment of the preserved body parts to distinct organs.

To our knowledge, the Late Cretaceous material of Klug et al. (2012) shows probably the most macroscopic detail (Figs. 12.1 to 12.4), although the materials published by Lehmann (e.g., 1985, 1988, 1990) might reveal more when investigated by modern methods (e.g., synchrotron). In some of the questionable *Sciponoceras*, the digestive tract is subdivided in an elongate brownish to black spot (ca. 2–3 mm long) that originates in the buccal mass. This continues with a narrow duct (ca. 5–10 mm long), which becomes thicker adapically. This thickening has

Table 12.2 Consistency of organic parts in coleoids, which probably also occurred in the ammonoid animal and their sequence of decay (data from Kear et al. 1995). For the decay time, some of the times given below were not explicitly stated by Kear et al. (1995); these are labelled by a question mark

Tissue/organ	Composition	Reference	Approximate time until decayed
Chitinous beak	tanned α chitin	Kear et al. 1995	>50 weeks
Radula	tanned α chitin	Kear et al. 1995	>50 weeks
Buccal muscle	musculature	Kear et al. 1995	2 weeks
Lip lining	chitin	Kear 1990; Kear et al. 1995	ca. 2 weeks
Hyaline shield	chitin	Kear et al. 1995	2 weeks?
Buccal palp teeth	α chitin	Hunt and Nixon 1981; Kear et al. 1995	2 weeks?
Oesophagus lining	chitin	Kear 1990; Kear et al. 1995	2 weeks?
Stomach lining	γ+α chitin	Rudall and Kenchington 1973; Hunt and Nixon 1981; Kear et al. 1995	2 weeks?
Cephalic cartilage	Collagen	Nesis 1987; Kear et al. 1995	ca. 20 weeks
Eye lenses	Crystalline	Kear et al. 1995	6 weeks
Mantle locking cartilage	Collagen	Nesis 1987; Kear et al. 1995	20 weeks?

been interpreted as possibly being the crop, followed by another narrow duct (ca. 1–4 mm long). It connects the questionable crop with the stomach (ca. 3–5 mm long), which also is dark brown to black. One specimen shows a little pouch at the adoral end of the supposed stomach, which has been interpreted tentatively as the caecum (Klug et al. 2012). One specimen also displays the base of a further duct which might represent the base of the intestine. Most of these interpretations lack evidence from additional morphological detail.

Several other cases of digestive tracts that were assigned to stomach and/or crop display less morphological detail but contain discernible stomach contents (Table 12.3) although in some cases, alternative interpretations are at hand (in the case of small ammonoids as supposed stomach contents: ovary or brood care; in the case of other skeletal fragments: fossil trap). This is of great importance since it sheds light on the diet of ammonoids, on their position in the food web, and indirectly also on their habitat. Accordingly, ammonoids fed on foraminifers, planktonic crinoids, crustaceans, gastropods, ostracods, and sponges (Lehmann 1971, 1985; Riegraf et al. 1984; Nixon 1996; Jäger and Fraaye 1997; Schweigert and Dietl 1999; Keupp 2000, 2012; Wippich and Lehmann 2004; Kruta et al. 2011). All these food items were small, thus documenting that many Mesozoic ammonoids were microphagous (Fig. 12.2). This small diet fits well with the jaws of those Mesozoic forms whose morphologies do not appear useful for catching and cutting up large prey. A microphagous diet is also evident for some ammonites by prominent peristomal apophyses, simply not allowing to insert larger food particles through

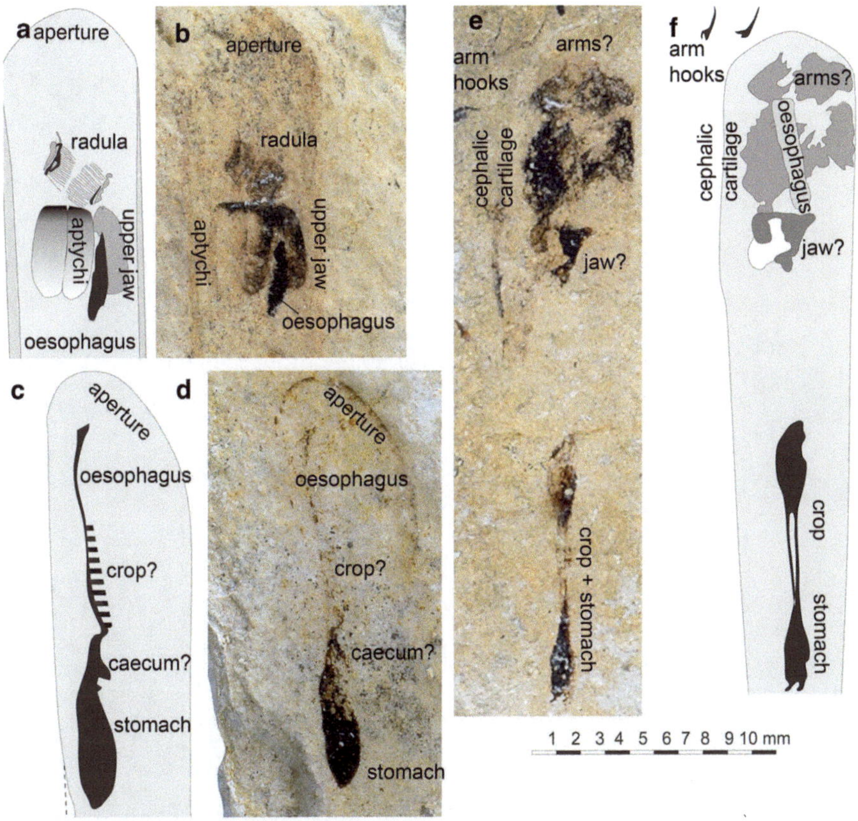

Fig. 12.1 Preservation of the digestive tract in ? *Sciponoceras* sp., OAE2, Late Cretaceous, Teutoburger Wald, Germany (stored at the University of Bremen). **a**, **c**, **f** Explanatory sketches. **a**, **b** GSUB C5833, showing remains of both beaks, radula and oesophagus. **c**, **d**, GSUB C5834, obliquely deposited shell fragment with aperture (dark band); oesophagus, stomach and maybe caecum are preserved and perhaps the crop (or a digestive gland?). **e**, **f** GSUB C5836, near-complete specimen, showing remains of digestive tract, probably parts of buccal mass, cephalic cartilage and questionable arm crown; associated pair of coleoid arm hooks

the mouth opening of the peristome (Keupp and Riedel 2010). Additionally, at least some of the prey animals were planktonic such as the crinoid *Saccocoma*. Schweigert and Dietl (1999) convincingly explained why a nonbenthonic mode of life in the water column of both *Saccocoma* and Late Jurassic ammonites is likely. This mode of life did not deter the ammonoids to pick up food from the sediment-surface like nautilids do today. Nevertheless, the great disparity in ammonoid mouthparts points at a similar diversity in diets (Tanabe et al. 2015).

Because of the repeated discovery of sclerites of *Saccocoma* in ammonite digestive tracts, the tracefossil *Lumbricaria* has been interpreted as ammonite droppings (Schweigert and Dietl 1999). This ichnogenus is one of the most common coprolites in the locally great Late Jurassic of the Solnhofen/Eichstätt region, which is in accordance with the locally great abundance of ammonites.

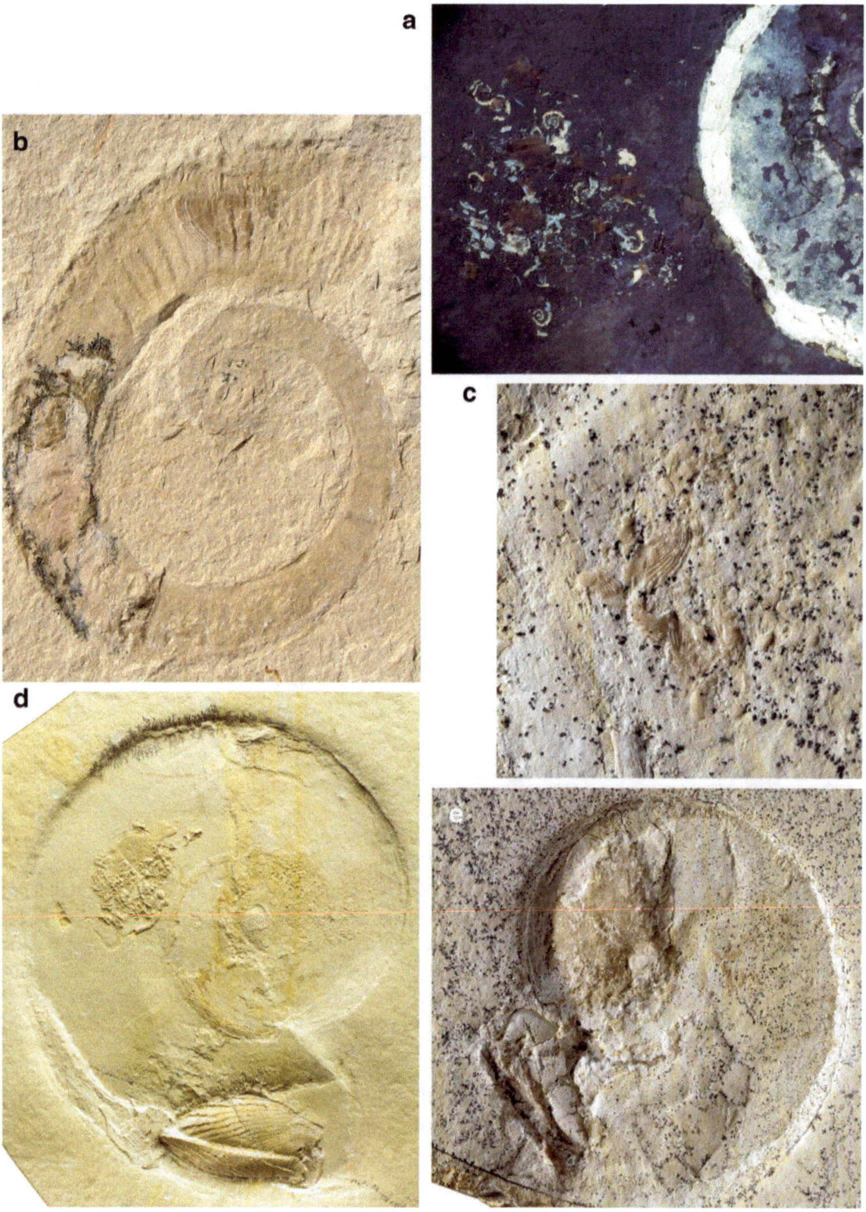

Fig. 12.2 Preservation of stomach contents in various Mesozoic ammonoids. Figures **a**, **c–e** courtesy H. Keupp Berlin (figured in Keupp 2000, p. 118 and in Keupp 2012: fig. 22–24; coll. H. Keupp, Berlin, stored at the Freie Universität Berlin). **a** Oppeliid, Tithonian, Solnhofen, Germany, image width ca. 30 mm; image taken under UV-light by H. Tischlinger; note the stomach content with small ammonites (siphuncles and jaws are visible). The rather complete preservation of the siphuncle casts doubts on the stomach-interpretation and some believe this might be a case of brood care. **b** *Allocrioceras* cf. *annulatum*, BSPG 1978 I 72 (Bayerische Staatssammlung für Paläontologie und Historische Geologie, Munich), Late Cenomanian, Hâqel, Lebanon, dm = 49 mm;

Table 12.3 Published cases of the possible diet of ammonoids as seen from oesophagus-, crop- or stomach-contents

Genus	Age	Content	Reference
Svalbardiceras	Griesbachian	Ostracods	Lehmann 1985
Arnioceras	Sinemurian	Ostracods, foraminifers	Lehmann 1971
Harpoceras	Toarcian	Crustaceans (Coleiida), aptychi, debris of ?*Pseudomytiloides*	Riegraf et al. 1984; Jäger and Fraaye 1997
Hildoceras	Toarcian	Small ammonite jaw apparatus	Lehmann and Weitschat 1973; Jäger and Fraaye 1997
Hildaites	Toarcian	Small aptychus, echinoderms? aragonitic shell debris	Riegraf et al. 1984; Jäger and Fraaye 1997
Phylloceras	Toarcian	Debris of *Pseudomytiloides*, echinoderm remains	Riegraf et al. 1984; Jäger and Fraaye 1997
Lithacoceras	Kimmeridgian	Foraminifers, sponges	Schweigert and Dietl 1999
Neochetoceras	Kimmeridgian, Tithonian	*Lamellaptychus* (conspecific?)	Michael 1894; Gürich 1924; Wetzel 1966; Lehmann 1976; Dietl et al. 1996; Keupp and Veit 1996; Schweigert and Dietl 1999; Keupp 2000, 2012
Physodoceras	Tithonian	*Saccocoma* (planktonic crinoid), echinoid spines	Lehmann 1972, 1975; Lehmann and Weitschat 1973; Riegraf et al. 1984; Schweigert and Dietl 1999; Keupp 2000, 2013
Baculites	Maastrichtian	Isopods, planktonic gastropods	Kruta et al. 2011
Allocrioceras	Cenomanian	Ophiuroids, comatulid crinoids	Wippich and Lehmann 2004

12.3 Cephalic Cartilage and Sensory Organs

Hardly any findings of structures that could be part of the cephalic cartilage or eyes have been found in ammonoids so far. Fossilized cephalic cartilages are known from the coleoid fossil record and are linked with the preservation of the eye capsules (Fuchs and Larson 2011). In ammonoids, however, like the *Lingulaticeras planulatum* described by Schweigert (2009, fig. 4) from Schamhaupten, Germany,

with jaws and phosphatized stomach content with questionable ophiuroid and comatulid ossicles (specimen was previously figured in Wippich and Lehmann 2004). **c**, **e** *Neochetoceras* sp., SHK MAa-7, Tithonian, Eichstätt, dm = 80 mm, leg. P. Ernst, with aptychus and stomach content (detail in **c**); the prey aptychi are possibly from the same species. **d** *Neochetoceras* sp., SHK MAa-7, Tithonian, Daiting (Germany), dm = 130 mm; stomach content with the nektoplanktic crinoid *Saccocoma* and aptychi.

a phosphatic patch lies lateral of the jaw apparatus. Speculatively, this could either be a part of the buccal mass musculature or parts of an eye. Currently, support for either hypothesis is missing.

The baculitids described by Klug et al. (2012) do show more detail, although it lacks morphological details to thoroughly test interpretations. Three of the specimens show structures that presumably belong to the cephalic cartilage (Fig. 12.2, 12.4). In order to understand these interpretations, it has to be understood that in coleoids, the cephalic cartilage surrounds the oesophagus behind the buccal mass. Since the oesophagus apparently has a reasonably high fossilization potential (see the preceding paragraphs), remains of this organ can serve as orientation to detect possible remains of the cephalic cartilage. In coleoids, the cephalic cartilage is linked with the eye capsules, which are preserved in Cretaceous coleoids from Lebanon (Fuchs and Larson 2011).

In specimen GSUB C5836, the light-colored elongate structure has been interpreted as the oesophagus. separating the two parts of the cephalic cartilage with the eye capsules attached, although this interpretation is questioned by the presence of the beak-like structure. Specimen GSUB C5839 shows the spatial relationships slightly more clearly. In the third specimen (GSUB C5835), the presence of eye-capsules is questionable. In contrast to Klug et al (2012), we present two hypotheses to explain the anatomical origin of the two oval patches present in the adapical position of the specimen. In accordance to their localization adjacent to an elongate structure, which might be the oesophagus, these patches might represent the eye capsules or alternatively demineralized aptychi or digestive glands. At this point, we lack evidence supporting either interpretation.

To sum up, based on comparisons with recent cephalopods, ammonoids probably possessed a cephalic cartilage (Fig. 12.3). Although no morphological details are revealed by the Cretaceous material, its overall shape appears to be reminiscent to that of coleoids. No good evidence is available yet for the preservation of ammonoid eyes. Taking the phylogenetic relationship with coleoids into account, we suggest that ammonoids had some kind of lens eye (compare Jacobs and Landman 1993). This hypothesis stands in contrast to the assumption of Lehmann (1985), who suggested that ammonoid eyes were of the pinhole camera eye type as in Recent nautilids.

12.4 Arms

No fossil specimens have been published that unequivocally display fossilized arm remains. To our knowledge, preservation of arms has been discussed for the genera *Gyroceratites* (Stürmer 1969; Bandel 1988), *Rhadinites* or *Anthracoceras* (Landman et al. 2010), *Sigaloceras* (Hollingworth and Hilton 1999), *Hybonoticeras* (Frickhinger 1999) and ? *Sciponoceras* (Klug et al. 2012). In the case of the Early Devonian *Gyroceratites* (Stürmer 1969), we agree with Bandel (1988) who already

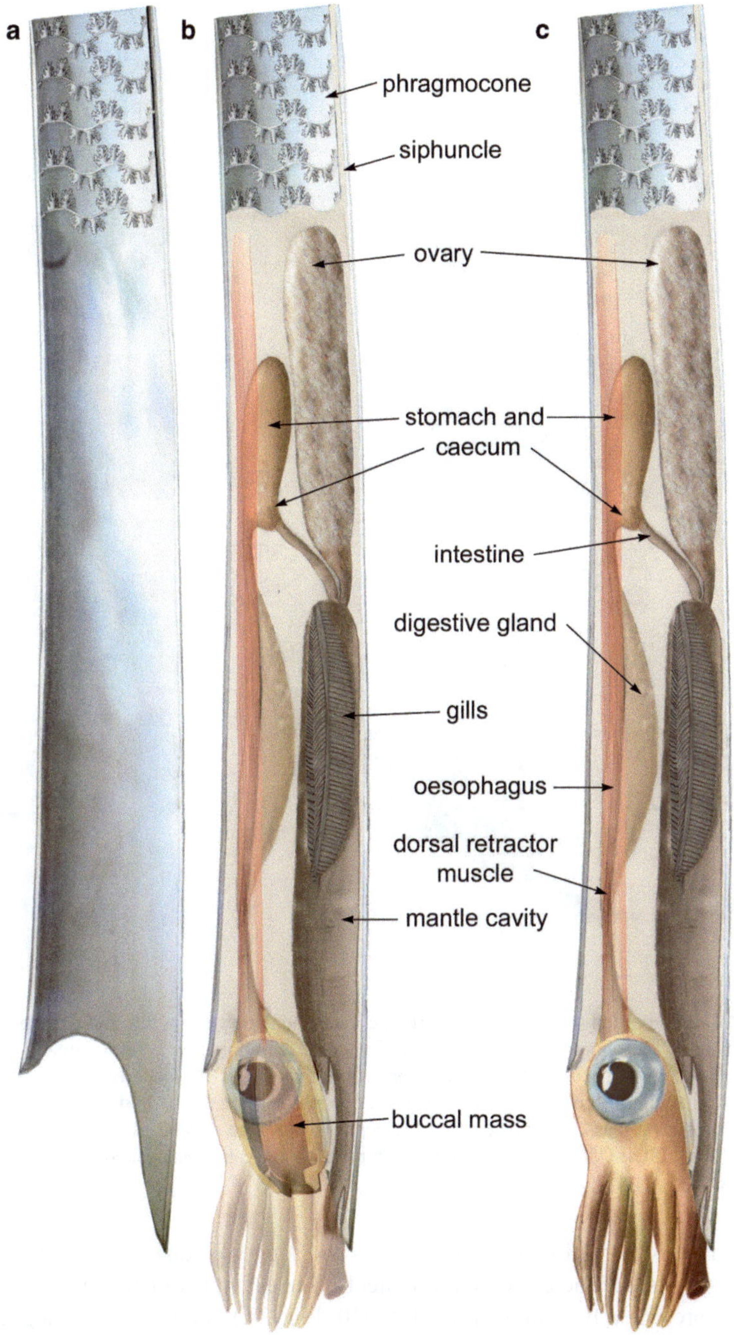

Fig. 12.3 Reconstruction of a baculitid, based on the findings of ? *Sciponoceras* sp., Late Cretaceous, Teutoburger Wald, Germany (Klug et al. 2012). Vertical position at rest (compare Hauschke et al. 2011; Westermann 2013). **a** inside of the shell. Muscle attachment according to Kennedy et al. (2002). **b**, **d** Digestive tract, buccal mass and eyes according to Klug et al. (2012); dorsal muscle following Kennedy et al. (2002)

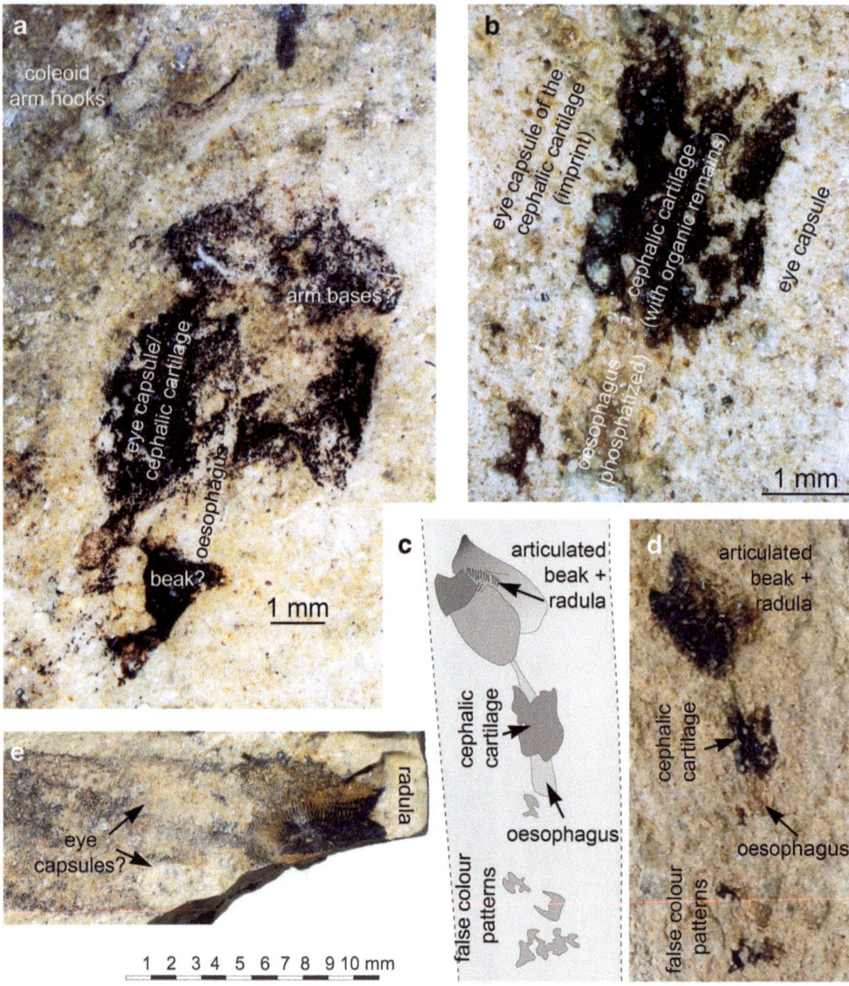

Fig. 12.4 Preservation of the cephalic cartilage, questionable eye capsules and perhaps a partial arm crown in? *Sciponoceras* sp., OAE2, Late Cretaceous, Teutoburger Wald, Germany (modified after from Klug et al. 2012). Specimens stored at University of Bremen. **a** GSUB C5836, with the questionable arm crown, cephalic cartilage, oesophagus and a jaw. **b–d** GSUB C5839, with articulated buccal mass, oesophagus and the cephalic cartilage as well as false colour bands. **b** detail of **d**. **e** GSUB C5835, body chamber with putative eye capsules. Scale bar at the bottom for images **c–e**

doubted the preservation of pyritized arms and suggested that the structures visible in radiographs were actually shell fragments. The example of *Rhadinites* or *Anthracoceras* presented by Landman et al. (2010, Fig. 2A) is doubtful; we suggest that the supposed arm is actually a trace fossil. In the Jurassic *Sigaloceras* of Hollingworth and Hilton (1999), a lot of organic matter appears to be preserved (Fig. 12.5). It thus represents an example, which deserves more attention and high resolution tomographic investigation, which hopefully will reveal more anatomical details of this ammonite. A completely different kind of preservation displays the Tithonian

Fig. 12.5 *Sigaloceras* (*Catasigaloceras*) *enodatum,* NMW.2002.69G.1 (National Museum and Galleries of Wales), adult microconch, Middle Jurassic, Callovian (Calloviense Biozone) Kellaways Sand Member of Colne Gravel Claydon Pike pit (SU 182 998), Fairford, Gloucestershire, UK, dm = 48 mm. Image courtesy J. Hilton (Edinburgh; previously figured in Hollingworth and Hilton 1999). Note the dark organic remains in the partially calcitic body chamber filling, perhaps containing muscles such as the dorsal muscle (which appears to be still attached at the dorsal attachment mark visible in the image), digestive tract and possibly arms?

Hybonoticeras of Frickhinger (1999, Fig. 32). In this case, the ammonite is largely flattened and shows radially arranged ridges in the sediment in front of the aperture. These also might represent trace fossils or arm imprints; the latter hypothesis is difficult to test since the imprints lack morphological detail.

The last example is also questionable (Klug et al. 2012). In this specimen (GSUB C5836), undoubted remains of the digestive tract as well as questionable remains of the cephalic cartilage and perhaps the arm bases or remains of the arm crown are seen. The latter part is also the non-mineralized part of the carbonized organs of this specimen that lies the closest to the aperture. In this irregular grey patch that narrows onion-shaped towards the peristome, a couple of thin projections point aperturally. These projections are reminiscent of a retracted arm-crown. It appears likely that, when in danger or when dying, the ammonite animal withdrew its arm-crown into the shell. On the one side, such a behavior would explain the difficulties of finding fossil ammonite arm crowns and on the other side, it would support our interpretation of this structure as arm crown. Although far from proven, some additional lines of reasoning support this hypothesis: In dead nautilids, the arms are often retracted in such way that the tips converge anteriorly. This is also seen in young *Nautilus* (Shigeno et al. 2008: Fig. 2). In the structure preserved in specimen GSUB C5836, the projections also converge towards the midline. Additionally, the scarcity of arm preservation suggests that the arms were rather fine and poorly muscular (e.g., Landman et al. 2010). In fact, the fossilized remains of this ? *Sciponoceras* resemble flattened remains of the arm crown as it was reconstructed for the scaphitid in Landman et al. (2010, Fig. 10). To sum up, we see the following lines of reasoning supporting the interpretation of the structure under consideration as arm crown: (1) it is the soft part that lies the closest to the aperture; (2) it displays a low number of projections; (3) it is slightly withdrawn into the aperture, possibly due to perimortal distress; (4) the putative fossil arm crown resembles the ventral aspect of the arm crown of Recent nautilid carcasses; (5) the shortness of the putative arms coincides with previous assumptions on the rather gracile morphology of ammonite arms to explain the scarcity of their fossils.

Presuming this interpretation is correct, we could state the following: (1) at least some ammonites had short arms; (2) the number of arms was low, probably ten; (3) they could be withdrawn into the body chamber.

Independent of the question whether the interpretations above are correct or not, we cannot extract much morphological knowledge out of it. Some essential questions remain open such as that for the number of arms of ammonoids. The following line of reasoning yields a plausible answer to that question. Most cephalopod workers agree that ammonites derived from bactritoids. Bactritoids also gave rise to coleoids, although later (Kröger et al. 2011, Fig. 5). Bactritoids probably derived from orthocerids, probably not very long (in a geological time scale) after the Nautilida split off the same clade, namely the Orthocerida (Kröger et al. 2011, Fig. 5). Thus, all these forms belong to one monophyletic group. Phylogenetically seen, the clade Ammonoidea is surrounded by Recent forms, on the one side the Nautilida and on the other the Coleoidea. Witmer (1995) termed such a position an "*extant phylogenetic bracket*". This implies that a character state that is shared by clades in an evolutionary lineage on both sides of the clade under consideration, it is very likely that this character state is similarly developed in the middle clade, where it might be unknown either due to the fact that the clade is extinct or poorly known for other reasons.

In early stemgroup coleoids such as phragmoteuthids, ten arms are present (e.g. Rieber 1970; Riegraf et al. 1984; Donovan 2006; Fuchs 2006). For nautilids, the question for the number of arms requires a look at its developmental biology. As adults, Recent nautilids have about 90 arms distributed over two circles, the inner and the outer arm crown. These 90 arms originate from ten anlagen (buds), a feature that is probably shared by all cephalopod groups (Shigeno et al. 2008, 2010). Therefore, ten arms appears to be a bauplan-trait of cephalopods and it also occurs in embryos of modern nautilids. In consequence, the criteria for an extant phylogenetic bracket are fulfilled and the ten arm-character state can be assumed for the Ammonoidea. Keupp (2007) speculated that aptychophoran ammonites might have reduced two arm pairs, in order to enable to use the aptychi as operculum. This hypothesis has found no empirical support yet.

12.5 Gills

Except for in the publications of Lehmann (1967b, 1973, 1985, 1987, 1990), no cases of gill preservation have been reported. Doubts that the structures described by Lehmann are gills arose since even the strongly phosphatized specimens of the Carboniferous *Glaphyrites* (Closs and Gordon 1966; Lehmann et al. 2014) do not show traces of gills on the surface. In depth tomographic examinations might shed

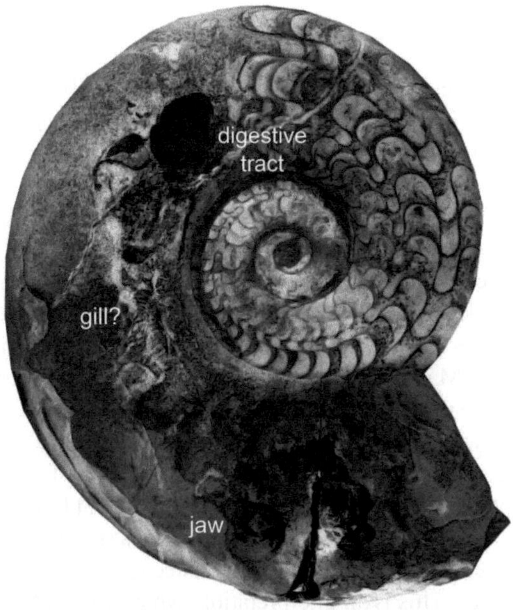

Fig. 12.6 *Ophiceras commune*, earliest Triassic, Griesbachian, Kap Stosch, Greenland, dm=75 mm (modified after Lehmann 1985, 1990). Repository unknown (perhaps Kopenhagen). This is one of the few specimens of ammonoids displaying remains of gills, jaws and parts of the digestive tract

light on the question about the nature of these structures and thus might reveal anatomical details.

The oldest reported case of gill preservation is probably the Triassic *Ophiceras* (Fig. 12.6) described by Lehmann and Weitschat (1973) as well as Lehmann (1985, 1990). In this specimen, parts of a supposed gill are visible in the body chamber between remains of the stomach and the buccal mass. This discovery is outclassed only by specimens from the Early Jurassic. According to Lehmann (1985), the best preserved gill is the one he discovered in a macroconch of *Eleganticeras* in northern Germany. In all three cases, mainly one branch with discernible filaments is preserved.

Lehmann reported that these ammonite gills are rather short and stout. He has also got the impression, that they might have had two branches. According to the shape of the preserved branches, he supposed that these ammonoid gills rather resemble those of nautilids than those of coleoids. It has to be taken into account that in this time, an actualistic comparison of ammonoids with nautilids was fashionable. Potentially, he was slightly biased by this view and would judge differently today, bearing the ideas of Jacobs and Landman (1993) of a closer affinity of ammonoids to coleoids in mind. This latter view has found support recently from developmental biology (Shigeno et al. 2008, 2010) and phylogeny (Kröger et al. 2011). Nevertheless, tomographic analyses of Lehmann's material might reveal further anatomical detail to add knowledge about the ammonoid bauplan.

12.6 Ink Sac or Not?

The presence or absence of an ink sac in ammonoids has been debated by several authors. Among the first to suggest the presence of ink sacs in ammonoids was Lehmann (1967a), interpreting a bag-shaped black structure in a specimen of *Eleganticeras* from the Toarcian of northern Germany as an ink sac. Others followed this idea, namely Wetzel (1979), who published on what he believed was an ink sac of the Early Cretaceous ammonite *Bochianites*, also from northern Germany. In the end, it was Lehmann himself, who started doubting his own interpretation of these structures as ink sacs. In his "*Epilog über den Tintenbeutel*" (Lehmann 1985), he reports of his observation that most ammonoids, even exceptionally preserved ones, lacked ink sacs. Mathur (1977) proved the presence of melanin in ammonoids, which seemed like the ultimate proof for ink sacs in some genera. For his *Ophiceras* from Greenland, however, Lehmann (1985) did not manage to proof the same. His doubts found support in the work of Riegraf et al. (1984), who concluded that ammonoids did not possess an ink sac, an opinion, Lehmann (1985) then adopted.

Doguzhaeva et al. (2004, 2010) reported bituminous structures in the body chamber of *Austrotrachyceras*. Due to their resemblance to melanin, the structures were interpreted as ink sacs. This is an interpretation, which we doubt since unequivocal evidence for ink sacs in ectocochleates is so far missing and does not make much sense. There are several other structures in ammonoids that contain melanin (Pezzella et al. 1997; Glass et al. 2012). It could well be, however, that the organic material is part of the black layer, the beaks or the oesophagus, which are all known to contain melanin in cephalopods. Additional comparable structures were published by Klug et al. (2007). Therein, several cases of black band, black layer and false color patterns were illustrated (see also Klug 2004; Klug et al. 2004).

12.7 Hyponome

To our knowledge, there is not a single article describing a fossilized ammonoid hyponome. The only exception is described from ammonite rollmarks with supposed hyponome imprints (Summesberger et al. 1999; Landman and Cobban 2007). Therefore, we have no good alternative to applying a similar approach as in the case of ammonoid arms. With respect to the extant phylogenetic bracket (Witmer 1995), the presence of a hyponome in ammonoids is highly likely, because it is present and strongly developed on both sides of the bracket, i.e. in the Nautilida and in all Coleoidea. Its presence in early ammonoids is corroborated by the persistence of a ventral hyponomic sinus in all Devonian ammonoids.

By contrast, many Mesozoic ammonoids have a ventral projection or even a rostrum (e.g., Pliensbachian *Amaltheus*, Toarcian *Harpoceras* or Albian *Mortoniceras*). This change in shell morphology led various ammonoid workers to contrasting opinions, ranging from the suggestion that the hyponome was more or less reduced to that the hyponome became bipartite or directed backwards, i.e.

dorsally (e.g., Schmidt 1930, Geczy 1960; Landman et al. 2010; *"twin-nozzle hypothesis"* of Westermann 2013). One of the involved questions is also the direction of swimming. From *syn vivo* epizoans, Seilacher (1960, 1982, 2004), Hauschke et al. (2011) as well as Keupp et al. (1999; 2011b) concluded that, for example, the regularly coiled *Buchiceras* swam like *Nautilus* and baculitid ammonites like *Sciponoceras* swam with their shell in a horizontal position, like orthoconic nautiloids and rather forwards than backwards (see also Westermann 1977, 2013).

12.8 *Glaphyrites:* Its Buccal Mass and Other Organic Remains

The most classical finding of phosphatized buccal masses are probably those described by Closs and Gordon (1966): *Glaphyrites* from the Late Carboniferous of Uruguay became famous for being the first ammonite described with the complete buccal mass *in situ*. Here, we discuss only the soft tissue aspect of the buccal mass, since beaks and radulae have been described in the preceding chapters (Kruta et al. 2015; Tanabe et al. 2015) and the alleged opercula elsewhere recently (Lehmann et al. 2014). The buccal mass shows great details, which had never been documented in such quality before, namely both the upper and the lower jaws, the radula, and phosphatized soft tissue remains (Closs 1967a, 1967b). These specimens have been occasionally referred to later (e.g., Kennedy and Cobban 1976; Lehmann 1987, 1990; Tanabe and Mapes 1995) and restudied by Bandel (1988), who also interpreted the taphonomic history besides focusing on the visceral mass-mantle compound.

All of the eight specimens known of *Glaphyrites* with soft tissue preservation (Closs 1967a) show an outline deviating from the logarithmic ammonite spiral and actually most show a rather ovoid outline, except for the largest shell with a diameter of about 120 mm (GPIT Ce1323/2), thus exceeding the size of the others (diameter range 90–120 mm). This irregular outline, in combination with the inner whorls tilted in relation to the outer whorls, probably encouraged Bandel (1988) to assume a vertical embedding of the ammonites. He suggests that after the ammonites settled onto the sediment surface, it has been covered by sediment nearly in living position until the onset of compaction. Subsequently, a large part of the shell of the living chamber was dissolved before further diagenetic processes began and thus, the shell twisted from a vertical to a nearly horizontal position without crushing, because the carbonate of the living chamber had vanished earlier. This is evidenced by wrinkle structures covering the visceral mass-mantle compound (Fig. 12.7), representing the inner layer of the demineralized shell. Shells of today's *Nautilus* and *Spirula* get similarly flexible when dissolved in weak acid and produce a comparable shrinking surface. In contrast to the living chamber, the inner whorls remain 3-dimensional during the taphonomic process with unfilled chambers.

The visceral mass-mantle compound is almost completely covered by the inner layer of the demineralized shell that is clearly distinguishable by the wrinkle structure (Fig. 12.7). Only minute portions of soft body remains are visible, interpreted

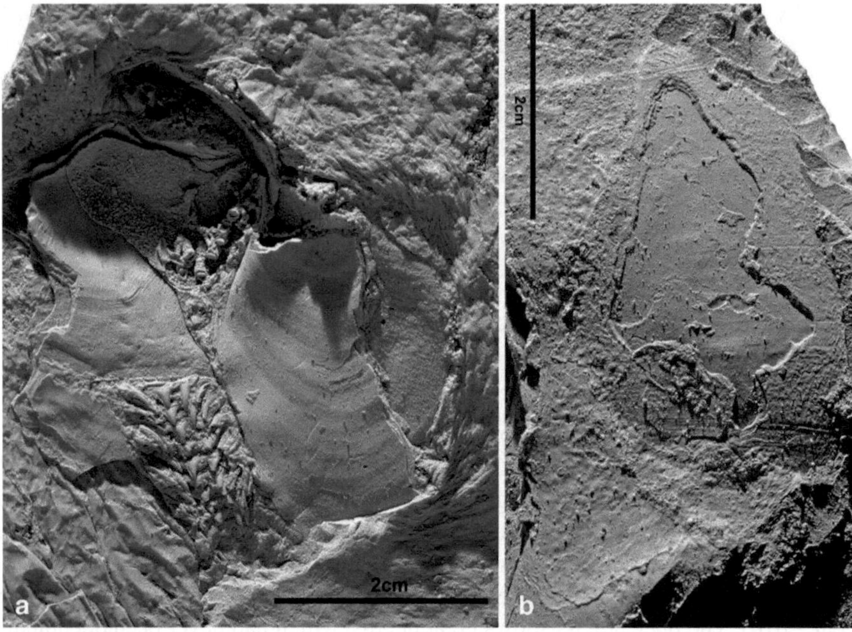

Fig. 12.7 Detail of *Glaphyrites* from the Late Carboniferous of Uruguay, preserved with soft parts. Coated with NH_4Cl. **a** GPIT Ce1320/1b, Institut für Geowissenschaften (Tübingen, Germany). Jaw-radula complex. Note the wrinkle structures of the demineralized shell, particularly well-visible in the left lower corner of the picture. **b** GPIT Ce1323/2a, growth lines on shell fragment from the aperture. The extremely fine lirae were probably not noted and thus the shell fragment has been interpreted as an operculum by Closs (1967b) and Bandel (1988)

as occurring at fissures in the wrinkle structure. The soft body is preserved as phosphatic mass, beige-brown in color; it lacks structural detail and thus can be well-distinguished from the wrinkle structured shell remains.

The jaws in *Glaphyrites* are distorted but three-dimensionally preserved (Fig. 12.7a). Most researchers agree that the position of the jaws in *Glaphyrites* is very proximal within the living chamber—in the anterior one-third or one-fourth of the body chamber in all four individuals showing jaws (Bandel 1988).

We differ in the interpretation of the key specimen of Bandel (1988; GPIT Ce 1323/2a) and Closs (1967b: pl. 2, Fig. 3) as possessing an operculum and assume it is the external rim of an apertural shell fragment (Lehmann et al. 2014). Bandel assumes a less solid operculum that has been folded along the median line, but apparently did not note the fine lines (Fig. 12.7). The lines are characteristic growth lines whose shape is not in accordance with concentric growth lines potentially characterizing an operculum. Additionally, with their straight shape vertical to the "*hinge*" of the "*operculum*", they revealed that they represent growth lines on the body chamber. Most other supposed anaptychi in *Glaphyrites* are also shell fragments. The one indicated by Closs (1967b: pl. 1, Fig. 1) is evidently an irregularly shaped shell fragment, since it retained the iridescent colours of the original shell (pers. comm. P. Havlik, Tübingen). It derives probably from the apertural region and is visible from the inner side. The supposed anaptychus focused on by Bandel (1988, Fig. 4) is merely a structure caused by shrinking during demineralization and possibly only partial dissolution of the outer shell at the phragmocone-body chamber transition.

Acknowledgments We thank Philipe Havlik (Tübingen) for the loan of the bulk of the *Glaphyrites* specimens and additional information about specimens we did not borrow. Technical support by Martin Krogmann (Bremen) with respect to the photographic works is highly appreciated. We greatly appreciate the constructive reviews of Isabelle Kruta (New York, Paris) and Didier Bert (La Mure-Argens).

References

Bandel K (1988) Operculum and buccal mass of ammonites. In:Wiedmann J, Kullmann J (eds) Cephalopods—present and past. Schweizerbart, Stuttgart

Closs D (1967a) Goniatiten mit Radula und Kieferapparat in der Itararé Formation von Uruguay. Paläontol Z 41:19–37

Closs D (1967b) Upper Carboniferous anaptychi from Uruguay. Ameghiniana 5:145–148

Closs D, Gordon M Jr. (1966) An upper Paleozoic radula. Escola de Geologia. Notas e Estudos 1:73–75

De Baets Klug C, Korn D, Bartels C, Poschmann M (2013) Emsian Ammonoidea and the age of the Hunsrück Slate (Rhenish Mountains, Western Germany). Palaeontographica A 299:1–114

Dietl G, Kapitzke M, Rieter M, Schweigert G, Hugger R (1996) Der Nusplinger Plattenkalk (Weißer Jura zeta)—Grabungskampagne 1995. Jahreshefte der Gesellschaft für Naturkunde in Württemberg 152:25–40

Doguzhaeva LA, Mapes RH (2015) Muscle scars in ammonoid shells. This volume

Doguzhaeva LA, Mutvei H (1989) *Ptychoceras*—a heteromorphic lytoceratid with truncated shell and modified ultrastructure (Mollusca: Ammonoidea). Palaeontographica A 208:91–121

Doguzhaeva LA, Mutvei H (1991) Organization of the soft body in *Aconeceras* (Ammonitina), interpreted on the basis of shell morphology and muscle scars. Palaeontographica A 218:17–33

Doguzhaeva LA, Mutvei H (1993) Shell ultrastructure, muscle-scars, and buccal apparatus in ammonoids. In: Elmi S, Mangold C, Alméras Y (eds) 3ème Symposium International: Céphalopodes actuels et fossiles. Symposium F. Roman. Géobios, Mém spéc 15:111–119

Doguzhaeva LA, Mapes RH, Mutvei H (1997) Beaks and radulae of Early Carboniferous goniatites. Lethaia 30:305–313

Doguzhaeva LA, Mapes RH, Mutvei H (2007) Late Carboniferous coleoid cephalopod from the Mazon Creek Lagerstätte (USA), with a radula, arm hooks, mantle tissues, and ink. In: Landman NH, Davis RA, Mapes RH (eds) Cephalopods—present and past: new insights and fresh perspectives. topics in geobiology. Springer, Dordrecht

Doguzhaeva LA, Mutvei H, Summesberger H, Dunca E (2004) Bituminous soft body tissues in the body chamber of the Late Triassic ceratitid *Austrotrachyceras* from the Austrian Alps. Mitt Geol-Paläont Inst Univ Hamburg 88:37–50

Doguzhaeva LA, Mapes RH, Bengtson S, Mutvei H (2010) A radula and associated cephalic remains of a Late Carboniferous coleoid from Oklahoma, USA. Ferrantia 59:37–50

Donovan DT (2006) Phragmoteuthida (Cephalopoda: Coleoidea) from the Jurassic of Dorset, England. Palaeontology 49:673–684

Frickhinger KA (1994) Die Fossilien von Solnhofen. Goldschneck, Korb

Frickhinger KA (1999) Die Fossilien von Solnhofen, 2. Goldschneck, Korb

Fuchs D (2006) Fossil erhaltungsfähige Merkmalskomplexe der Coleoidea (Cephalopoda) und ihre phylogenetische Bedeutung. Berliner paläobiol Abh 8:1–165

Fuchs D, Larson N (2011) Diversity, morphology and phylogeny of coleoid cephalopods from the Upper Cretaceous Plattenkalks of Lebanon—part I: Prototeuthidina. J Paleont 85:234–249

Geczy B (1960) A Neoammoideak eletmodjarol. Foldtani Kozlony Magyar Foldtani Tarsulat 90:200–203

Glass K, Ito S, Wilby PR, Sota T, Nakamura A, Bowers CR, Vinther J, Dutta S, Summons R, Briggs DEG, Wakamatsu K, Simon JD (2012) Direct chemical evidence for eumelanin pigment from the Jurassic period. PNAS 1–6. doi:10.1073/pnas.1118448109

Gürich G (1924) "Ammonitenbrut" von *Oppelia steraspis* nach Michael. Centralbl Miner Geol Paläont 1924:700–704

Hauschke N, Schöllmann L, Keupp H (2011) Oriented attachment of a stalked cirripede on an orthoconic heteromorph ammonite—implications for the swimming position of the latter. N Jahrb Geol Paläont Abh 202:199–212

Hollingworth NT, Hilton J (1999) Fossil sheds new light on ammonites. New Scientist, 4th Sept. 1999: 25

Hunt S, Nixon M (1981) A comparative study of protein composition in the chitin-protein complexes of the beak, pen, sucker, disc, radula and oesophageal cuticle of cephalopods. Comp Biochem Physiol 68B:535–546

Ifrim C (2013) Paleobiology and paleoecology of the early Turonian (Late Cretaceous) ammonite *Pseudaspidoceras flexuosum*. Palaios 28:9–22

Jacobs DK, Landman NH (1993) *Nautilus*—a poor model for the function and behavior of ammonoids? Lethaia 26:101–111

Jäger M, Fraaye RHB (1997) The diet of the early Toarcian ammonite *Harpoceras falciferum*. Palaeontology 40:557–574

Kear AJ (1990) Feeding mechanisms and diet in cephalopods: special reference to Antarctic mesopelagic squid. Unpubl PhD thesis, University of Aberdeen

Kear AJ, Briggs DEG, Donovan DT (1995) Decay and fossilization of nonmineralized tissue in coleoid cephalopods. Palaeontology 38:105–131

Kennedy WJ, Cobban WA (1976) Aspects of ammonite biology, biogeography, and biostratigraphy. Spec Pap in Palaeont 17:1–94

Kennedy WJ, Cobban WA, Klinger HC (2002) Muscle attachment and mantle-related features in upper Cretaceous *Baculites* from the United States Western Interior. Abh Geol BA Wien 57:89–112

Keupp H (2000) Ammoniten. Paläobiologische Erfolgsspiralen. Thorbecke, Sigmaringen

Keupp H (2007) Complete ammonoid jaw apparatuses from the Solnhofen plattenkalks: implications for aptychi function and microphagous feeding of ammonoids. N Jahrb Geol Paläont Abh 245:93–101

Keupp H (2012) Atlas zur Paläopathologie der Cephalopoden. Berliner paläobiologische Abhandlungen 12:1–390

Keupp H, Riedel F (2010) Remarks on the possible function of the apophyses of the Middle Jurassic microconch ammonite *Ebrayiceras sulcatum* (Zieten, 1830), with a discussion on the palaeobiology of Aptychophora in general. N Jahrb Geol Paläont, Abh 255:301–314

Keupp H, Veit B (1996) Ein *Phylloceras* mit Anaptychus. Fossilien 136:343–351

Keupp H, Röper M, Seilacher A (1999) Paläobiologische Aspekte von *syn vivo*-besiedelten Ammonoideen im Plattenkalk des Ober-Kimmeridgiums von Brunn in Ostbayern. Berliner geowiss Abh E 30:121–145

Keupp H, Engeser T, Fuchs D, Haeckel W (2011a) Fossile Spermatophoren von *Trachyteuthis hastiformis* (Cephalopoda, Coleoidea) aus dem Ober-Kimmeridgium von Painten/Bayern. Archaeopteryx 28:23–30

Keupp H, Röper M, Rothgaenger M (2011b) Serpuliden-Epökie auf Ammoniten aus dem Brunner Plattenkalk (Ober-Kimmeridgium) in Ostbayern. Archaeopteryx 29:1–12

Klug C (2004) Mature modifications, the black band, the black aperture, the black stripe, and the periostracum in cephalopods from the Upper Muschelkalk (Middle Triassic, Germany). Mitt Geol-Paläont Inst Univ Hamburg 88:63–78

Klug C, Jerjen I (2012) The buccal apparatus with radula of a ceratitic ammonoid from the German Middle Triassic. Geobios 45:57–65

Klug C, Korn D, Richter U, Urlichs M (2004) The black layer in cephalopods from the German Muschelkalk (Middle Triassic). Palaeontology 47:1407–1425

Klug C, Brühwiler T, Korn D, Schweigert G, Brayard A, Tilsley J (2007) Ammonoid shell structures of primary organic composition. Palaeontology 50:1463–1478

Klug C, Riegraf W, Lehmann J (2012) Soft-part preservation in heteromorph ammonites from the Cenomanian-Turonian Boundary Event (OAE 2) in the Teutoburger Wald (Germany). Palaeontology 55:1307–1331

Kröger B, Vinther J, Fuchs D (2011) Cephalopod origin and evolution: a congruent picture emerging from fossils, development and molecules. Bioessays:12. doi:10.1002/bies.201100001

Kruta I, Landman NH, Rouget I, Cecca F, Tafforeau P (2011) The role of ammonites in the Mesozoic marine food web revealed by jaw preservation. Science 331:70–72

Kruta I, Landman NH, Mapes RH, Pradel A (2013) New insights into the buccal apparatus of the Goniatitina; palaeobiological and phylogenetic implications. Lethaia 47:38–48. doi:10.1111/let.12036

Kruta I, Landman NH, Tanabe K (2015) Ammonoid radulae. In: Klug C, Korn D, De Baets K, Kruta I, Mapes RH (eds) Ammonoid paleobiology: from anatomy to ecology. Springer, Dordrecht

Kullmann J (1981) Carboniferous goniatites. In: House MR, Senior JR (eds) The Ammonoidea. Syst Ass Spec Pub 18:37–48

Landman NH, Cobban WA (2007) Ammonite touch marks in upper Cretaceous (Cenomanian-Santonian) deposits of the western interior sea. In: Landman NH, Davis RA, Mapes RH (eds) Cephalopods present and past: new insights and fresh perspectives. Springer, Dordrecht

Landman NH, Mapes RH, Cruz C (2010) Jaws and soft tissues in ammonoids from the lower Carboniferous (upper Mississippian) Bear Gulch Beds, Montana, USA. In: Tanabe K, Shigeta Y, Sasaki T, Hirano H (eds) Cephalopods—present and past. Tokai University, Tokyo

Lehmann U (1967a) Ammoniten mit Tintenbeutel. Paläontol Z 41:132–136

Lehmann U (1967b) Ammoniten mit Kieferapparat und Radula aus Lias-Geschieben. Paläontol Z 41:38–45

Lehmann U (1971) New aspects in ammonite biology. Proceedings of the North American Paleontological Convention 1:1251–1269

Lehmann U (1972) Aptychen als Kieferelemente der Ammoniten. Paläontol Z 46:34–48

Lehmann U (1973) Zur Anatomie und Ökologie der Ammoniten: Funde von Kropf und Kiemen. Paläontol Z 47:69–76

Lehmann U (1975) Über Nahrung und Ernährungsweise von Ammoniten. Paläontol Z 49:187–195

Lehmann U (1981) Ammonite jaw apparatus and soft parts. In: House MR, Senior JR (eds) The Ammonoidea. Syst Ass Spec 18:275–287

Lehmann U (1985) Zur Anatomie der Ammoniten: Tintenbeutel, Kiemen, Augen. Paläontol Z 59:99–108

Lehmann U (1987) Ammoniten. Ihr Leben und ihre Umwelt. Enke, Stuttgart

Lehmann U (1988) On the dietary habits and locomotion of fossil cephalopods. In: Wiedmann J, Kullmann J (eds) Cephalopods—present and past. Schweizerbart, Stuttgart

Lehmann U (1990) Ammonoideen—Leben zwischen Skylla und Charybdis. In: Erben HK, Hillmer G, Ristedt H (eds) Haeckel-Bücherei 2. Enke, Stuttgart

Lehmann U, Weitschat W (1973) Zur Anatomie und Ökologie von Ammoniten: Funde von Kropf und Kiemen. Paläontol Z 47:69–76

Lehmann J, Klug C, Wild F (2014) Did ammonoids possess opercula? Reassessment of phosphatised soft tissues in *Glaphyrites* from the Carboniferous of Uruguay. Paläontol Z 15. doi:10.1007/s12542-013-0219-8

Maeda H, Seilacher A (1996) Ammonoid Taphonomy. In: Landman NH, Tanabe K, Davis RA (eds) Ammonoid paleobiology. Topics in geobiology 13. Plenum, New York
Mapes RH (1987) Upper Paleozoic cephalopod mandibles: frequency of occurrence, modes of preservation, and paleoecological implications. J of Paleont 61:521–538
Mathur AC (1977) Über Ammoniten der Kössener Schichten und den Nachweis der Tintenbeutel-Substanz Melanin. PhD-thesis München
Michael R (1894) Ammoniten-Brut mit Aptychen in der Wohnkammer von *Oppelia steraspis* Oppel sp. Zeitschrift der Deut Geol Ges 46:697–702
Miller AK, Furnish WM, Schindewolf OH (1957) Paleozoic Ammonoidea (Anarcestina, Goniatitina, Prolecanitina). In: Moore RC (ed.) Treatise on invertebrate paleontology, Part L. University of Kansas, Lawrence
Nesis KN (1987) Cephalopods of the world. TFH Publications, New Jersey
Nixon M (1996) Morphology of the jaws and radula in ammonoids. In: Landman NH, Tanabe K, Davis RA (eds) Ammonoid paleobiology. Topics in geobiology 13. Plenum, New York
Nixon M, Young (2003) The brains and lives of cephalopods. Oxford University, Oxford
Otto M (1994) Zur Frage der "Weichteilerhaltung" im Hunsrückschiefer. Geol Palaeontol 28:45–63
Pezzella A, d'Ischia M, Napolitano A, Palumbo A, Prota G (1997) An integrated approach to the structure of *Sepia* melanin. Evidence for a high proportion of degraded 5,6-dihydroxyindole-2-carboxylic acid units in the pigment backbone. Tetrahedron 53:8281–8286
Richter U, Fischer R (2002) Soft tissue attachment structures on pyritized internal moulds of ammonoids. In: Summesberger H, Histon K, Daurer A (eds) Cephalopods—Present and past. Abh Geol BA 57:139–149
Rieber H (1970) *Phragmoteuthis? ticinensis* n. sp., ein Coleoidea-Rest aus der Grenzbitumenzone (Mittlere Trias) des Monte San Giorgio (Kt. Tessin, Schweiz). Paläontol Z 44:32–40
Riegraf W, Werner G, Lörcher F (1984) Der Posidonienschiefer. Biostratigraphie, Fauna und Fazies des südwestdeutschen Untertoarciums (Lias epsilon). Enke, Stuttgart
Rudall KM, Kenchington (1973) The chitin system. Biol Rev 48:597–636
Schmidt H (1930) Ueber die Bewegungsweise der Schalencephalopoden. Paläontol Z 12:194–208
Schweigert G (1998) Die Ammonitenfauna des Nusplinger Plattenkalkes (Oberes Kimmeridgium, Beckeri Zone, Ulmense Subzone, Baden Württemberg). Stutt Beitr Natkde B 267:1–61
Schweigert G (2009) First three-dimensionally preserved in situ record of an aptychophoran ammonite jaw apparatus in the Jurassic and discussion of the function of aptychi. Berliner paläobiol Abh E 10:321–330
Schweigert G, Dietl G (1999) Zur Erhaltung und Einbettung von Ammoniten im Nusplinger Plattenkalk (Oberjura, Südwestdeutschland). Stutt Beitr Natkde B 272:1–31
Schweigert G, Dietl G (2001) Die Kieferelemente von *Physodoceras* (Ammonitina, Aspidoceratidae) im Nusplinger Plattenkalk (Oberjura, Schwäbische Alb). Berliner geowiss Abh E 36:131–143
Seilacher A (1960) Epizoans as a key to ammonoid ecology. J Paleontol 34:189–193
Seilacher A (1982) Ammonite shells as habitats in the Posidonia Shales of Holzmaden—floats or benthic islands? N Jahrb Geol Paläont Abh 159:98–114
Seilacher A (2004) Trittbrettfahrer im Muschelkalkmeer. Fossilien 3:157–160
Shigeno S, Sasaki T, Moritaki T, Kasugai T, Kasugai T, Vecchione M, Agata K (2008) Evolution of the cephalopod head complex by assembly of multiple molluscan body parts: evidence from *Nautilus* embryonic development. J of Morph 269:1–17
Shigeno S, Takenori S, Boletzky SV (2010) The origins of cephalopod body plans: a geometrical and developmental basis for the evolution of vertebrate-like organ systems. In: Tanabe K, Shigeta Y, Sasaki T, Hirano H (eds) Cephalopods—present and past. Tokai University, Tokyo
Stürmer W (1968) Einige Beobachtungen an devonischen Fossilien mit Röntgenstrahlen. Nat Mus 98:413–417
Stürmer W (1969) Pyrit-Erhaltung von Weichteilen bei devonischen Cephalopoden. Paläontol Z 43:10–12

Stürmer W (1970) Soft parts of cephalopods and trilobites: some surprising results of X-ray examinations of Devonian Slate. Science 170:1300–1302

Summesberger H, Jurkivsek B, Kolar-Jurkovsek T (1999) Rollmarks of soft parts and a possible crop content of Late Cretaceous ammonites from the Slovenian karst. 335-344. In: Olóriz F, Rodríguez-Tovar FJ (eds) Advancing research on living and fossil cephalopods. Kluwer Academic, New York

Tanabe K, Fukuda Y (1983) Buccal mass structure of the Cretaceous ammonite *Gaudryceras*. Lethaia 16:249–256

Tanabe K, Mapes RH (1998) Jaws and radula of the Carboniferous ammonoid *Cravenoceras*. J of Paleont 69:703–707

Tanabe K, Mapes RH, Sasaki T, Landman NH (2000) Soft–part anatomy of the siphuncle in Permian prolecanitid ammonoids. Lethaia 33:83–91

Tanabe K, Kruta I, Landman NH (2015) Ammonoid buccal mass and jaw apparatus. This volume

Westermann GEG (1977) Form and Function of orthocone cephalopod shells with concave septa. Paleobiology 3:300–321

Westermann GEG (2013) Hydrostatics, propulsion and life-habits of the Cretaceous ammonoid *Baculites*. Revue de Paléobiologie 32:249–265

Wetzel W (1966) Über einige umstrittene Bath-Ammoniten nebst paläobiologischen Bemerkungen über die Neoammoniten. N Jahrb Geol Paläont Abh 124:84–102

Wetzel W (1979) Seltene Wohnkammerinhalte von Neoammoniten. N Jahrb Geol Paläont Mh 1979(1):46–53

Wippich MGE, Lehmann J (2004) *Allocrioceras* from the Cenomanian (mid-Cretaceous) of the Lebanon and its bearing on the palaeobiological interpretation of heteromorphic ammonites. Palaeontology 47:1093–1107

Witmer LM (1995) The extant phylogenetic bracket and the importance of reconstructing soft tissues in fossils. In: Thomason JJ (ed) Functional morphology in vertebrate paleontology. Cambridge University, Cambridge

Zeiss A (1968) Fossile Cephalopoden mit Weichteilen. Nat Mus 98:418–424

Zeiss A (1969) Weichteile ectocochleater paläozoischer Cephalopoden in Röntgenaufnahmen und ihre paläontologische Bedeutung. Paläontol Z 43:13–27

Chapter 13
Soft-Part Anatomy of the Siphuncle in Ammonoids

Kazushige Tanabe, Takenori Sasaki and Royal H. Mapes

13.1 Introduction

The fossil record of soft-parts of the Ammonoidea is extremely scarce despite the abundance of their external chambered shells in Devonian to Cretaceous marine deposits of various geological settings. Previous authors described several kinds of peculiar 'organismic' structures within the body chambers of some ammonoids as gills (Lehmann and Weitschat 1973), ink sacs and/or ink (Lehmann 1967; Wetzel 1968; Doguzhaeva et al. 2004, 2007), muscular tissues (Doguzhaeva et al. 2007; Landman et al. 2010), arms (Zeiss 1968; Stürmer 1969), and egg capsules (Landman et al. 2010). Unfortunately, detailed anatomical features of these organs are not preserved in the available fossil materials, so that their biological attribution is still controversial for the structures described as gills, arms and ink sacs. For these reasons, our knowledge on ammonoid soft-parts has long been virtually restricted to organic hard tissues of mouthparts (jaws and a radula) (Lehmann 1990; Nixon 1996; Tanabe and Fukuda 1999; see also Chaps. 10 and 11 in this volume) and attachment scars of the soft-body that are preserved on the inner surfaces of the

K. Tanabe (✉) · T. Sasaki
Department of Historical Geology and Paleontology,
The University Museum, The University of Tokyo,
Hongo 7-3-1, Tokyo 113-0033, Japan
e-mail: tanabe@um.u-tokyo.ac.jp

T. Sasaki
e-mail: sasaki@um.u-tokyo.ac.jp

R. H. Mapes
North Carolina Museum of Natural Sciences, West Jones St., Raleigh, NC, USA
e-mail: mapes@ohio.edu

North Carolina Museum of Natural Sciences,
11 West Jones St., Raleigh, NC 27601, USA

C. Klug et al. (eds.), *Ammonoid Paleobiology: From Anatomy to Ecology*,
Topics in Geobiology 43, DOI 10.1007/978-94-017-9630-9_13

shell and/or on the outer surface of its steinkerns (e.g., Crick 1898; Jordan 1968; Doguzhaeva and Mutvei 1996, among others).

The only exception of soft-parts whose anatomical features are known in ammonoids is the siphuncle that existed in the siphuncular tube within the phragmocone. Among the present-day cephalopods, chambered shells have been completely lost or vestigially reduced, for example, to chitinous gladii in most coleoids, and only three taxonomically different groups preserve chambered shells; namely *Nautilus* and *Allonautilus* of the Nautilida (Nautiloidea), *Sepia* of the Sepiida (Coleoidea), and *Spirula* of the Spirulida (Coleoidea) (Denton and Gilpin-Brown 1961, 1966, 1973).

The shells of these genera are characterized by the development of numerous chambers (phragmocone) filled with low-pressure gas and small amounts of cameral liquid in association with the siphuncle. The combination of the gas and liquids in the chambers was developed as hydrostatic floats to adjust the density of a living animal to that of seawater (neutral buoyancy) by means of low-pressure gas within the chambers, and the animal could ascend or descend in the water column by changing the amount of cameral liquid within the air chambers osmotically (Denton and Gilpin-Brown 1961, 1966, 1973; Greenwald et al. 1980, 1982; Ward et al. 1980, 1981; Ward and Chamberlain 1983).The siphuncle is a long and narrow segmented soft-tissue consisting mainly of blood vessels and surrounding epithelium (Denton and Gilpin-Brown 1973). In *Nautilus* and *Spirula*, it is housed in the siphuncular tube within the phragmocone and is connected to the rear part of the body at the base of the body chamber, while in *Sepia*, it is flattened underneath the cuttlebone without a mineralized wall (Appellöf 1893; Bandel and Boletzky 1979; Denton and Gilpin-Brown 1961, 1966, 1973; Ward et al. 1980).

The siphuncular tube whose wall consists mainly of an organic hard tissue (conchiolin) and a porous calcified element (Bandel and Boletzky 1979) is known to occur in virtually all modern and fossil chambered cephalopods, including ammonoids. In ammonoids, it passes through an opening in each septum and extends from the initial portion of the shell to the base of the body chamber, forming the septal neck-siphuncular complex (Tanabe and Landman 1996). Many authors speculated that, as in extant nautilids and *Spirula*, a soft-part element of the siphuncle must also have been present in ammonoids. This assumption has been verified by the discovery of fossilized siphuncular remains in the early Permian prolecanitids from South Urals (Zakharov 1996, pl. 6), Triassic ceratitids from Spitsbergen (Weitschat 1986, pl. 5, Fig. 2; Weitschat and Bandel 1991, Fig. 18) and the Jurassic ammonite *Virgatites* from Russia (Drushchits and Doguzhaeva 1981, pl. 22, Fig. 1e-f; Drushchits et al. 1982, pl. 6, Fig. 1e; Barskov 1996, Fig. 2–8, 11), but the detailed anatomical features are not clearly preserved in these materials. Better preserved remains of the siphuncle were later discovered within a phosphatized siphuncular tube in specimens of the early Permian prolecanitid *Akmilleria* from Nevada (Tanabe et al. 2000).

In this chapter, we describe anatomical characteristics of the siphuncle in four ammonoids including *Akmilleria* based on available fossil evidence and compare them with those of *Nautilus* and *Spirula*, discussing their taxonomic and functional morphologic implications.

13.2 Materials and Methods

The following four Late Paleozoic and Cretaceous ammonoid specimens with preserved soft tissue remains in their siphuncular tubes are examined in this study:

1. *Glaphyrites clinei* (Miller and Owen) (Glaphyritidae, Goniatitina). UMUT PM 19028 from the Desmoinesian (Pennsylvanian) in Collinsville, near Tulsa, Oklahoma. Same specimen as that studied by Tanabe and Landman (1996).
2. *Akmilleria electraensis* (Plummer and Scott) (Medliocottiidae, Prolecanitina). UMUT PM 27800 from the Wolfcampian (Lower Permian) Arcturus Formation, Buck Mountain, White Pine County, Nevada. The same specimen was described by Tanabe et al. (2000, Fig. 6).
3. *Paracanthoplites* sp. (Deshayesitidae, Ammonitina). From the upper Aptian in Pshekha River basin, northwestern Caucasus, Russia.
4. *Gaudryceras tenuiliratum* Yabe. UMUT MM 28225 (same specimen as that figured by Tanabe et al. 2003, Fig. 5.4, 5.5) from the lower Campanian in Nakagawa area, northern Hokkaido, Japan.

Each specimen was ground and polished with a graded series of carborundum and diamond pastes to just above the median plane, and the polished surface was etched with 5 % acetic solution for 1 min. The etched surface was washed in distilled water, dried, coated with platinum or carbon, and then observed with a Hitachi Model S 4500 scanning electron microscope. Also, energy dispersion X-ray (EDX) analysis was made for the specimen UMUT PM 19028 coated with carbon to determine elemental and mineralogical compositions of the fossilized siphuncular remains and tube wall.

For comparison, optical and SEM observations were made of a siphuncle in a specimen of *Nautilus pompilius* Linnaeus captured alive from a point off the Bohol Straits (Philippines). Details of preparation prior to SEM observations are described in Tanabe et al. (2000). This article follows Greenwald et al. (1982) for terminology of the microstructural elements of the siphuncle.

The specimen of *Paracanthoplites* sp. is housed in the Institute of Paleobiology, Georgian Academy of Science, Tbilisi, Georgia (GAS). The remaining ones of ammonoids and *N. pompilius* are deposited in the University Museum, University of Tokyo (UMUT).

13.3 Anatomy of Siphuncles of Extant *Nautilus* and Extinct Ammonoids

13.3.1 Nautilus

The siphuncle in *Nautilus* spp. is a segmented tubular structure consisting of a wide central vein, two pairs of arteries, porous connective tissue, and a siphuncular

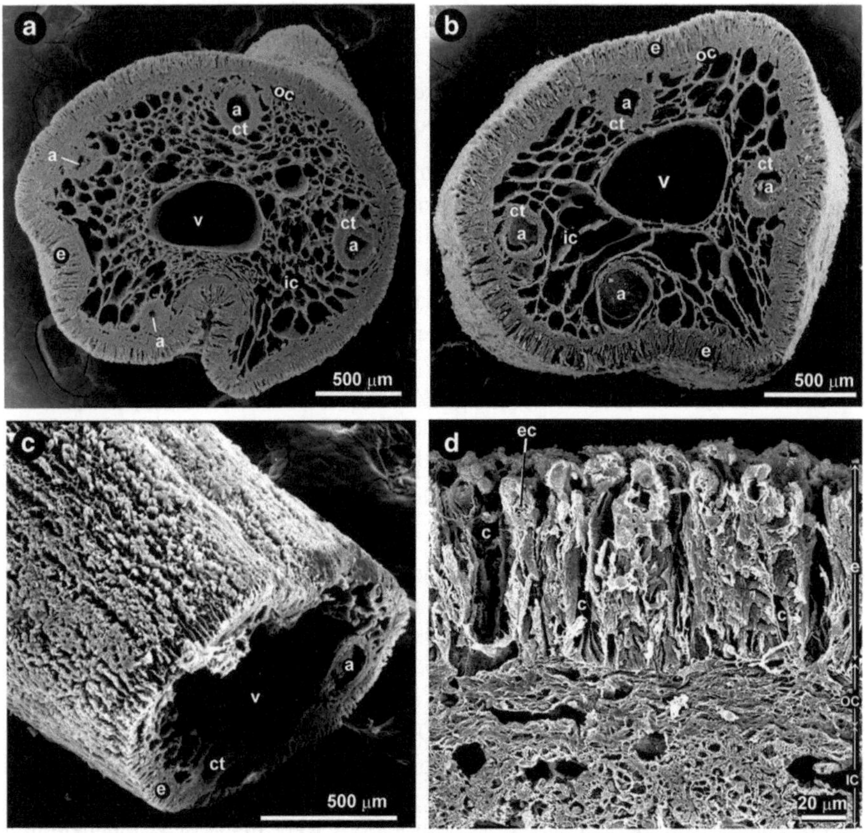

Fig. 13.1 SEM images of the siphuncle of *Nautilus pompilius*. UMUT RM 27811 (same specimen as that figured by Tanabe et al. 2000, Fig. 2). **a, b** Cross-sections of the adapical (**a**) and middle (**b**) portions of the siphuncular cord. **c** Oblique view, showing equally spaced, numerous longitudinal ridges of canaliculi. **d** Close-up of the marginal portion of b, showing the siphuncular epithelium and underlying outer connective tissue. *e* siphuncular epithelium, *c* canaliculus, *oc* outer connective tissue, *ic* inner connective tissue, *ct* connective tissue, *a* artery, *v* vein (hemocoele)

epithelium on the outside (Denton and Gilpin-Brown 1966, 1973; Bassot and Mortoja 1966; Greenwald et al. 1980, 1982; Fukuda et al. 1981; Bandel and Spaeth 1983, Fig. 13.1a, b). Of these anatomical elements, the siphuncular epithelium has been extensively investigated by Denton and Gilpin-Brown (1966) and Greenwald et al. (1980, 1982) for its ultrastructure in relation to its function in buoyancy regulation (see discussion for details). The structures of various anatomical elements of the *Nautilus* siphuncle are summarized below on the basis of the description by Tanabe et al. (2000).

The siphuncle has a three-layered structure with a central vein (Figs. 13.2, 13.3, 13.4, abbreviations used in these figures are shown in parentheses).

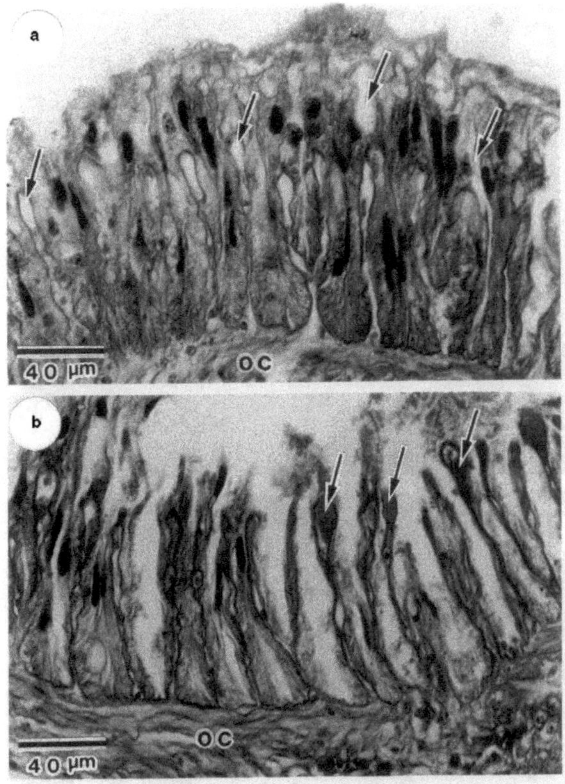

Fig. 13.2 Optical micrographs of siphuncular epithelium of *N. pompilius*. **a** Intact part of the siphuncular epithelium. **b** Damaged part of the siphuncular epithelium due to insufficient fixation. Most of the cytoplasm is detached and the ridge of basement membrane surrounding canaliculi (pointed by arrows) is exposed. The outer connective tissue (*oc*) has sporadic haemocoelic spaces. After Tanabe et al. (2000, Fig. 3); reproduced by permission of John Wiley & Sons, Inc

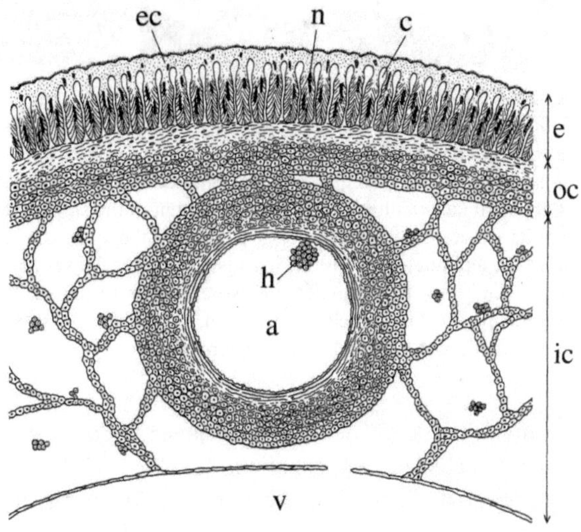

Fig. 13.3 Schematic illustration of the cross-section of the siphuncle in *N. pompilius*. *ec* siphuncular epithelial cell, *h* haemocyte, *n* nucleus of siphuncular epithelium. See the explanation of Fig. 13.1 for other abbreviations. After Tanabe et al. (2000, Fig. 4); reproduced by permission of John Wiley & Sons, Inc

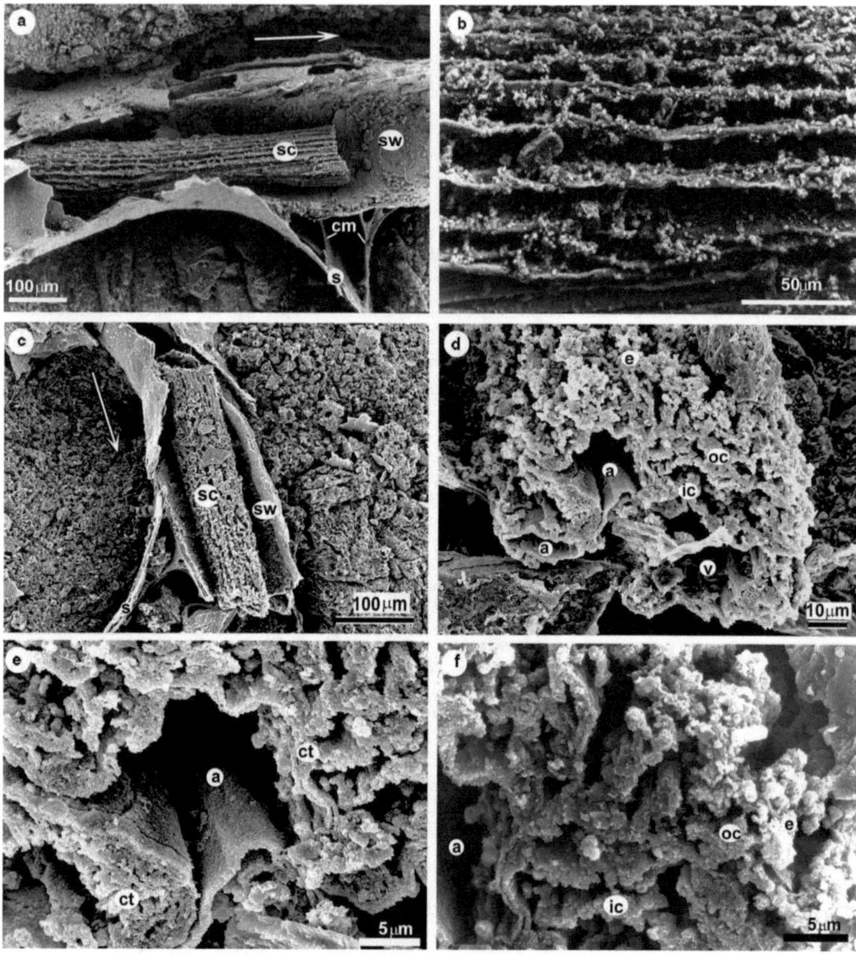

Fig. 13.4 SEM images of the phosphatized siphuncle of *Akmilleria electraensis*. UMUT PM 27800 (same specimen as that figured by Tanabe et al. 2000, Fig. 6). The arrows in **a** and **c** indicate the adoral direction. **a, b** Exposed segment of the siphuncle in the third whorl (**a**) and its close-up (**b**), showing numerous equal-spaced, longitudinal ridges and grooves. Each ridge represents the distal end of canaliculus. **c, d** Exposed segment of the siphuncle in the second whorl (**c**) and close-up of its structure in cross-section (**d**), showing the arteries, vein and connective tissue. The lower half of the siphuncle was lost during fossilization. **e** Close-up of d, showing a portion of the artery surrounded by the thick connective tissue. **f** Porous, sponge-like connective tissue surrounded by the thin siphuncular epithelium. *s* septum, *sw* siphuncular tube wall, *cm* cameral membranes. See the explanation of Fig. 13.1 for other abbreviations

1. Epithelium (*e*): The outermost layer is composed of a layer of columnar epithelial cells (*ec*) and canaliculi (*c*) (Figs. 13.1a, b, d, 13.2). The outer surface of the epithelial cells is wavy corresponding to the longitudinal arrangement of the equally spaced canaliculi (Figs. 13.1c, 13.2a).

2. Outer connective tissue layer (*oc*): The subepithelial basement is formed by firm sheets of connective tissue (*ct*) with sporadic haemocoelic spaces. In the outer sublayer, fibers are stratified almost parallel to the bases of epithelial cells, while in the inner sublayer they occur irregularly among connective tissue cells (Figs. 13.1d, 13.3).
3. Inner connective tissue layer (*ic*): The inner section contains the siphuncular arteries (*a*) and a large hemocoelic space partitioned reticulately by membranous network of connective tissue (Figs. 13.1a, b, 13.3). Two pairs of arteries surrounded by a thick wall of connective tissue exist throughout the entire length of the siphuncular cord (Fig. 13.1a, b). Occasionally the arteries possess small branches of arterioles in the outer connective tissue layer.
4. Siphuncular vein (*v*): This structure is delimited by a very thin layer of connective tissue (Fig. 13.1a, b) in contrast to the arteries, which are surrounded by a thick circular wall of connective tissue (Figs. 13.1a, b, 13.3). The vein increases its diameter from the proximal to the distal end of the siphuncular cord (Fig. 13.1a, b). The vein has connections with haemocoelic space of the inner connective tissue layer through small pores.
5. Canaliculi (*c*): The most specialized structures in the siphuncle of *Nautilus* are the canaliculi, which are formed by deep infoldings of the basement membrane of the siphuncular epithelial cells (Figs. 13.1d, 13.2a, 13.3). The apical end of each canaliculus is completely closed without connecting directly with the inner surface of the siphuncle, while the basal part is in communication with the subepithelial hemocoelic space.

The number of canaliculi around the siphuncle is extremely large, ranging from 240 near the distal end to 280 near the proximal end of the siphuncular cord in the specimen examined. Because epithelial cells seem almost paired on either side of a canaliculus (Fig. 13.2a), the number of epithelial cells is estimated to be approximately 480–560 around the whole siphuncle. The basement membrane forming the canaliculus is less susceptible to dissolution than the cytoplasm of the siphuncular cells. In the damaged portion whose cytoplasm is mostly removed due to insufficient fixation, the canaliculi remain intact and are exposed on the surface (Fig. 13.2b). The canaliculi in this portion exhibit a rough sculpture consisting of alternating tall ridges and deep furrows. This observation appears to provide a special implication for reconstruction of the fossilized siphuncular epithelium in ammonoids.

13.3.2 Ammonoids

13.3.2.1 Notes on Taphonomy

EDX-analyses of the carbon-coated specimen of *Akmilleria electraensis* (UMUT PM 27800) revealed that the fossilized siphuncular remains (Fig. 13.4) are rich in phosphorous and calcium with lesser amounts of fluorine and sulfur (Tanabe

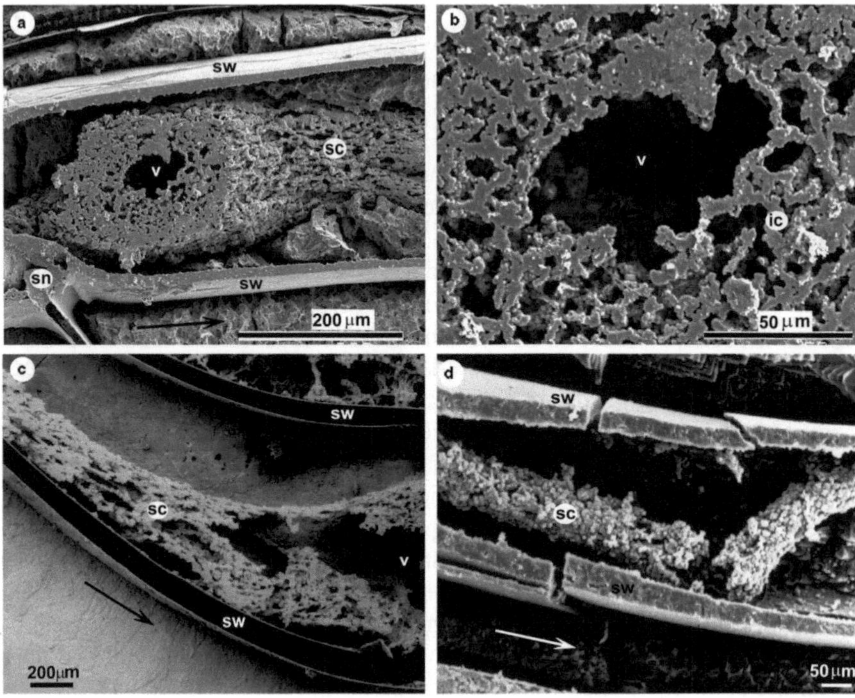

Fig. 13.5 SEM images of the phosphatized siphuncles of *Paracanthoplites* sp. (**a**, **b**), *Glaphyrites clinei* (**c**) and *Gaudryceras tenuiliratum* (**d**). The arrows in a, c and d indicate the adoral direction. **a** Exposed segment of the siphuncle, showing numerous longitudinal ridges and grooves on the outer surface and the central vein and surrounding connective tissue in the oblique section. **b** Close-up of a, showing the central vein and surrounding porous, sponge-like inner connective tissue. Unregistered specimen of GAS. **c** Exposed segment of the siphuncle that was inflated partially during fossilization. UMUT PM 19028. **d** Remnant of the siphuncular segment whose original structure is not preserved. UMUT MM 28225. See the explanation of Fig. 13.1 and 13.3 for abbreviations

et al. 2000, Fig. 5); the amounts of P_2O_5 and CaO are 40.31 wt% and 54.57 wt%, respectively. A similar chemical composition was detected for the siphuncular tube wall that appears to be primarily made of conchiolin. These data suggest that the fossilized siphuncle and surrounding tube wall in the specimen both consist of fluorapatite. In all probability, the siphuncular tissues might exist within the siphuncular tube even after the post-mortem detachment of the main soft body from the body chamber. They might have been phosphatized prior to decay by the activity of bacteria (Briggs et al. 1993).

Although we did not analyze chemical and mineral compositions of the siphuncular remains in the specimens of *Glaphyrites clinei*, *Paracanthoplites* sp. and *Gaudryceras tenuiliratum*, their microscopic features suggest that their siphuncles are preserved by phosphate, and that this mineralization of the tissue occurred early in the taphonomic pathway of fossil preservation (Fig. 13.5).

13.3.2.2 Microanatomy

The microanatomy of the siphuncle is best preserved in UMUT PM 27800 of *Akmilleria electraensis*. The siphuncle occurs within the siphuncular tube as truncated segments (Fig. 13.4a, c). Each segment is circular in cross-section with a diameter of approximately 60–70 % of the inner diameter of the siphuncular tube, indicating shrinkage of 30–40 % during fossilization. In life, the segments would have been connected to form a long continuous cord-shaped structure.

Comparative SEM observations of the siphuncles of *Nautilus pompilius* and *A. electraensis* help elucidate the similarity and dissimilarity of their microanatomy. The outer surface of the siphuncle of *A. electraensis* is sculptured by equally spaced, longitudinal ridges and furrows (Fig. 13.4a, b, c), as in the case of *N. pompilius* (Figs. 13.1a, b, 13.2, 13.3). The ridges correspond to the distal ends of individual canaliculi. The number of ridges around the siphuncle of *A. electraensis* is approximately 30 in the second and third whorls, whose siphuncle diameters are 60 µm (Fig. 13.4c) and 100 µm (Fig. 13.4a). These numbers are much lower than that observed on the siphuncle of *N. pompilius* (N=240–280), which measures 1–2 mm in diameter (Fig. 13.1c). Individual canaliculi of the siphuncular epithelium of *A. electraensis*, presuming no postmortem shrinkage, are 10 µm wide in the second whorl (Fig. 13.4c, d) and 15–20 µm wide in the third whorl (Fig. 13.4a, b), whose dimensions are comparable with those of the canaliculi of *N. pompilius*, even though the shell of the latter is much larger (ca. 180 mm in diameter).

In cross-section, the siphuncle of *A. electraensis* consists of a large central vein, a pair of arteries, a reticulate network of connective tissue, and a thin epithelium on the outside, though the lower half has shrunken toward the inside (Fig. 13.4d). A thick, smooth connective tissue surrounds each artery (Fig. 13.4e). The boundary between the outermost epithelium and the underlying connective tissue is unclear in the specimen examined, owing to destruction of the internal epithelial structure during fossilization (Fig. 13.4f).

The siphuncular remains of the specimens of *Glaphyrites clinei* (Fig. 13.5c), *Paracanthoplites* sp. (Fig. 13.5a, b) and *Gaudryceras tenuiliratum* (Fig. 13.5d) are more poorly preserved than that of the specimen of *A. electraensis*, but a large central vein and surrounding porous connective tissues, and longitudinally arranged infolded basement membranes of epithelial cells could be distinguished in the siphuncles of the former two species.

13.4 Discussion

13.4.1 Comparative Anatomy

The siphuncle of *Akmilleria electraensis* is essentially similar to those of *Nautilus pompilius* and *Spirula spirula* (see Chun 1915, pl. 73, Fig. 2) in the following features: (1) the epithelial layer consists of elongate epithelial cells and canaliculi, (2) the outer connective tissue layer is made up of a subepithelial basement, (3) the inner connective tissue layer comprises siphuncular arteries and hemocoelic space, and (4) the siphuncular vein exists in or near the center of the siphuncle (Fig. 13.5). These anatomical similarities support the homology of the siphuncle among *Akmilleria, Nautilus* and *Spirula*, on the basis of not only positional but also structural criteria.

In spite of these similarities, there are the following three major microstructural differences among the siphuncles of these genera. (1) **Number of canaliculi and epithelial cells around the siphuncle.** *Akmilleria* has about 30 canaliculi with epithelial cells based on the presence of approximately 15 ridges that are visible on the half of the siphuncle preserved (Fig. 13.4a). The number in *Nautilus* is much larger than in *Akmilleria,* attaining as many as 500. The siphuncular epithelium of *Spirula* has 60 or more longitudinal lines, which may represent canaliculi; accordingly approximately 120 epithelial cells around the siphuncle. (2) **Number of arteries.** *Akmilleria* probably had four arteries, equal to those in *Nautilus*. Chun (1915, pl. 73, Fig. 2) illustrated the siphuncular structure of *Spirula* in which there are as many as nine arteries surrounded by a thick connective tissue. (3) **Space of inner connective tissue layer.** *Akmilleria* and *Nautilus* share reticulate hemocoelic space. In contrast, the layer is filled mostly with connective tissue in *Spirula.*

The similarity between *Nautilus* and *Akmilleria,* i.e., four arteries within the sponge-like inner connective tissue layer, may represent the plesiomorphic state inherited from a cephalopod ancestor. However, it remains uncertain whether or not the above three microanatomical differences among the three distantly related genera are constrained phylogenetically or functionally, because of insufficient data from other fossil taxa.

In addition to the four ammonoid species described herein, phosphatized remains of the siphuncle were found in other ammonoids such as the early Permian prolecanitids (*Neopronorites* and *Artioceras*) from South Urals (Zakharov 1996, pl. 6), Triassic ceratitids from Spitsbergen (Weitschat 1986, pl. 5, Fig. 2; Weitschat and Bandel 1991, Fig. 18) and Late Jurassic (Volgian) *Virgatites virgatus* (Perisphinctidae, Ammonitina) from the bank of the Moskva River, Russia (Drushchits and Doguzhaeva 1981, pl. 22, Fig. 1e-f; Drushchits et al. 1982, pl. 6, Fig. 1e; Barskov 1996, Fig. 2–8, 11).

The siphuncular remains preserved in the specimen of *V. virgatus* consist of larger and smaller tubules in cross-section, each surrounded by a thin prismatic layer. Drushchits and Doguzhaeva (1981), Drushchits et al. (1982) and Barskov (1996) interpreted the tubules as remains of blood vessels. The number of tubules increases from two to six from the first to the second whorl, but neither a central vein nor a marginal epithelial layer is preserved in the specimen (Barskov 1996). As already described, the number of arteries in the siphuncle of *Nautilus* is invariable (two pairs) throughout the entire length of the siphuncular cord. These anomalies and the absence of any detailed tissue structure seem to suggest that the 'blood vessels' of *V. virgatus* described by the above authors were formed secondarily during the phosphatization of the siphuncle. The siphuncles of the specimens of the early Permian prolecanitids and Triassic ceratitids respectively figured by Zakharov (1996, pl. 6, Fig. 5, 6) and Weitschat (1986, pl. 5, Fig. 2) do not preserve detailed microstructure, retaining only a relatively large central vein.

13.4.2 Functional Morphology

In extant chambered cephalopods represented by chambered nautilids (*Nautilus* spp. and *Allonautilus scrobiculatus*), cuttlefish (*Sepia* spp.) and ram's horn squid (*Spirula spirula*), the siphuncular epithelium exhibits certain common characteristic features (Denton and Gilpin-Brown 1961, 1966, 1973; Bassot and Mortoja 1966; Greenwald et al. 1980; Ward et al. 1980; Fukuda et al. 1981; Bandel and Spaeth 1983). Namely, the epithelia of these cephalopods all consist of high columnar cells that change adorally to the cubic cells of the rear mantle. Each cell comprises a canaliculus, which is connected with the underlying drainage channel. Cytoplasm around the canaliculus contains numerous mitochondria and microcanals, the latter of which open into the canaliculus and drainage channel. Because these histological features are commonly observed in excretory organs of various animals, Denton and Gilpin-Brown (1973) proposed a model of buoyancy regulation in chambered cephalopods, explaining that the siphuncular epithelium serves to concentrate salt, permitting osmotic transfer of liquid between chambers and blood vessels within the siphuncle. According to this model, during active pumping, cameral liquid is temporally stored within the intracellular duct (= canaliculus) and is subsequently transported osmotically to the blood vessels via a drainage channel. Because osmotic emptying by simple osmotic pressure cannot be applied to nautilus and *Spirula* living in waters deeper than 400 m, Greenwald et al. (1980) modified the Denton and Gilpin-Brown's (1973) model and proposed another model ('local osmosis' model) for cameral liquid transport in nautilids that emphasizes enhanced solute concentrations within the intracellular space.

In summary, the essential similarity of the microanatomy of the siphuncle among *Akmilleria*, *Nautilus* and *Spirula* strongly suggests that ammonoids could control their buoyancy by transferring liquid osmotically between the air chambers and blood vessels of the siphuncle.

13.5 Conclusion

Fossilized siphuncular remains discovered in specimens of four ammonoid species are described, compared with the counterparts of the modern chambered cephalopods (*Nautilus pompilius* and *Spirula spirula*), and their functional morphologic implications are discussed. The four ammonoid species examined are as follows: *Glaphyrites clinei* (Middle Pennsylvanian Goniatitina) from Oklahoma, *Akmilleria electraensis* (Permian Prolecanitina) from Nevada, *Paracanthoplites* sp. (Early Cretaceous Ammonitina) from NW Caucasus, and *Gaudryceras tenuiliratum* (Late Cretaceous Lytoceratina) from Hokkaido. These four taxa represent four of the major suborders of the Ammonoidea.

In these specimens, the siphuncle is preserved as truncated segments within the siphuncular tube, whose outer surfaces are sculptured by longitudinal ridges and furrows in the former three species. The best preserved siphuncle is found in the specimen of *A. electraensis*. It is similar to those of *N. pompilius* and *S. spirula*

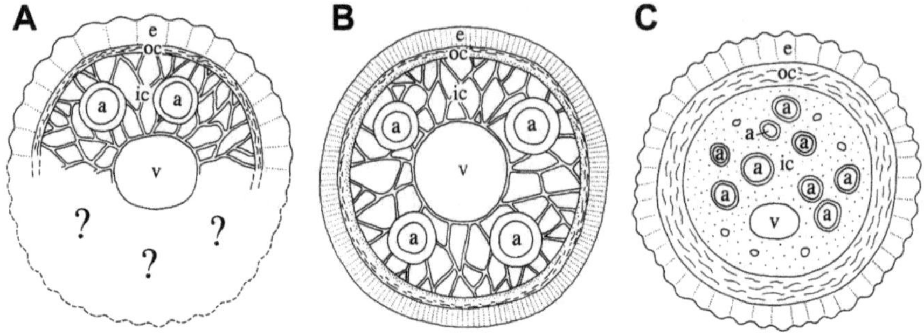

Fig. 13.6 Schematic diagram of the cross-section of the siphuncle in *Akmilleria* (**a**), *Nautilus* (**b**), and *Spirula* (**c**). Modified from Chun (1915) for *Spirula*. See the explanation of Fig. 13.1 for abbreviations. Modified from Tanabe et al. (2000, Fig. 8); reproduced by permission of John Wiley & Sons, Inc

in having the following common anatomical features; (1) elongate epithelial cells and canaliculi, (2) the outer connective tissue layer consisting of a subepithelial basement, (3) the inner connective tissue layer comprising siphuncular arteries and hemocoelic space, and (4) the siphuncular vein in or near the center of the siphuncle (Fig. 13.6). These anatomical similarities support the homology of the siphuncle among ammonoids and modern *Nautilus* and *Spirula*, on the basis of not only positional but also structural criteria. The essential resemblance of the microanatomy of the siphuncle of *Akmilleria* with modern *Nautilus* and *Spirula* strongly suggests that ammonoids controlled their buoyancy in the water column by transferring cameral liquid osmotically between air chambers and blood vessels of the siphuncle.

Acknowledgments We thank Tamaz A. Lominadze (Institute of Paleobiology, Georgian Academy of Science, Tbilisi) for providing an interesting specimen of *Paracanthoplites* sp. with preserved siphuncular tissue remains for this study, and Cyprian Kulicki (Institute of Paleobiology, Polish Academy of Science; Warsaw), Neil H. Landman (American Museum of Natural History, New York City), Yuri D. Zakharov (Far Eastern Geological Institute, Russian Academy of Science, Vladivostok), and an anonymous reviewer for helpful comments and discussion.

References

Appellöf A (1893) Die Schalen von *Sepia, Spirula* und *Nautilus*. Kongliga Svenska Vertenskaps-Akad Handl 25:1–106

Bandel K, Boletzky Sv. (1979) A comparative study of the structure, development and morphological relationships of chambered cephalopod shells. Veliger 21:313–354

Bandel K, Spaeth C (1983) Beobachtungen am rezenten *Nautilus*. Mitt Geol-Paläont Inst Univ Hamburg 54:9–26

Barskov IS (1996) Phosphatized blood vessels in the siphuncle of Jurassic ammonites. Bull Inst Oceanogr Monaco N S 14:335–341

Bassot J-M, Martoja M (1966) Histologie et fonction du siphon chez le Nautile. Comp Rend Akad Sci Paris 263:980–982

Briggs DFG, Kear AJ, Martill DM, Wilby PR (1993) Phosphatization of soft-tissue in experiments and fossils. J Geol Soc Lond 150:1035–1038

Chun C (1915) Die Cephalopoden. II: Myopsida, Octopoda. Wissenschaftliche Ergebnisse der Deutschen Tiefsee-Expedition 'Valdiva' 1898–1899 18:311–552

Crick GC (1898) On the muscular attachment of the animal to its shell in some Cephalopoda (Ammonoidea). Trans Linn Soc Lond 2nd Ser Zool 7:71–113

Denton EJ, Gilpin-Brown JB (1961) The buoyancy of the cuttlefish, *Sepia officinalis* (L.). J Mar Biol Ass UK 41:319–342

Denton EJ, Gilpin-Brown JB (1966) On the buoyancy of the pearly *Nautilus*. J Mar Biol Ass UK 46:723–759

Denton EJ, Gilpin-Brown JB (1973) Floatation mechanisms in modern and fossil cephalopods. In: Russel FS, Younge M (eds) Advances in marine biology 11. Academic, London, pp 197–268

Doguzhaeva L, Mutvei H (1996) Attachment of the body to the shell in ammonoids. In: Landman NH, Tanabe K, Davis RA (eds) Ammonoid paleobiology. Plenum, New York, pp 43–63

Doguzhaeva LA, Mutvei H, Summesberger H, Dunca E (2004) Bituminous soft body tissues in Late Triassic ceratitid *Austrotrachyceras*. Mitt Geol-Paläont Inst Univ Hamburg 88:37–50

Doguzhaeva LA, Mapes RH, Summesberger H, Mutvei H (2007) The preservation of body tissues, shell, and mandibles in the ceratitid ammonoid *Austrotrachyceras* (Late Triassic), Austria. In: Landman NH, Davis RA, Mapes RH (eds) Cephalopods present and past: new insights and fresh perspectives. Springer, New York, pp 221–238

Drushchits VV, Doguzhaeva LA (1981) Ammonity pod elektronnym mikroskopom (Ammonites under the electron microscope). Izdatelstvo Moskovskogo Universiteta, Moscow [in Russian]

Drushchits VV, Mesezhnikov MS, Alekseyev SB (1982) Structural characteristics of the siphonal system in Volgian ammonites. Paleont Zhur 1982–4:49–57 [in Russian]

Fukuda Y, Tanabe K, Obata I (1981) Histology of the siphuncular epithelium in *Nautilus* and its functional significance. J Fossil Res 14:29–40 [in Japanese]

Greenwald L, Ward PD, Greenwald OE (1980) Cameral liquid transport and buoyancy control in chambered nautilus (*Nautilus macromphalus*). Nature 286:55–56

Greenwald L, Cook CB, Ward PD (1982) The structure of the chambered *Nautilus* siphuncle: the siphuncular epithelium. J Morph 172:5–22

Jordan R (1968) Zur Anatomie mesozoischer Ammoniten nach den Strukturelementen der Gehäuse-Innenwand. Beih Geol Jahrb 77:1–64

Landman NH, Mapes RH, Cruz C (2010) Jaws and soft tissues in ammonoids from the Lower Carboniferous (Upper Mississippian) Bear Gulch Beds, Montana, USA. In: Tanabe K et al. (eds) Cephalopods–present and past. Tokai University Press, Tokyo, pp 147–153

Lehmann U (1967) Ammoniten mit Tintenbeutel. Paläontol Z 41:132–136

Lehmann U (1990) Ammonoideen. Enke, Stuttgart

Lehmann U, Weitschat W (1973) Zur Anatomie und Ökologie von Ammoniten: Funde von Kropf und Kiemen. Paläontol Z 47:69–76

Nixon M (1996) Morphology of the jaws and radula in ammonoids. In: Landman NH, Tanabe K, Davis RA (eds) Ammonoid Paleobiology. Plenum, New York, pp 23–42

Stürmer W (1969) Pyrit-Erhaltung von Weichteilen bei devonischen Cephalopoden. Paläontol Z 43:10–12

Tanabe K, Fukuda Y (1999) Morphology and function of cephalopod buccal mass. In: Savazzi E (ed) Functional morphology of the invertebrate skeleton. Wiley, London, pp 245–262

Tanabe K, Landman NH (1996) Septal neck-siphuncular complex of ammonites. In Landman NH, Tanabe K, Davis RA (eds) Ammonoid paleobiology. Plenum, New York, pp 129–165

Tanabe K, Mapes, RH, Sasaki T, Landman NH (2000) Soft-part anatomy of the siphuncle in Permian prolecanitid ammonoids. Lethaia 33:83–91

Tanabe K, Landman NH, Yoshioka Y (2003) Intra- and interspecific variation in the early internal shell features of some Cretaceous ammonoids. J Paleont 77:876–887

Ward PD, Chamberlain JA Jr (1983) Radiometric observation of chamber formation in *Nautilus pompilius*. Nature 304:57–59

Ward P, Greenwald L, Greenwald OE (1980) The buoyancy of the chambered Nautilus. Sci Amer 243:190–203
Ward PD, Greenwald L, Magnier Y (1981) The chamber formation cycle in *Nautilus macromphalus*. Paleobiology 7:481–493
Weitschat W (1986) Phosphatisierte Ammonoideen aus der Mittleren Trias von Central Spitzbergen. Mitt Geol-Paläont Inst Univ Hamburg 61:249–279
Weitschat W, Bandel K (1991) Organic components in phragmocones of Boreal Trassic ammonoids: implications for ammonoid biology. Paläontol Z 65:269–303
Wetzel W (1968) Seltene Wohnkammerinhalte von Neoammoniten. N Jb Geol Paläont Mh 1968(1):46–53
Zakharov YD (1996) Orthocerid and ammonoid shell structure: its bearing on cephalopod classification. Bull Ntn Sci Mus Tokyo Ser C 22:11–35
Zeiss A (1968) Fossile Cephalopoden mit Weichteilen. Natur Mus 98:418–424

Chapter 14
The Body Chamber Length Variations and Muscle and Mantle Attachments in Ammonoids

Larisa A. Doguzhaeva and Royal H. Mapes

14.1 Introduction

The body chamber with the preserved mantle and muscle attachment marks provide some insight into the poorly known soft body organization and locomotion of extinct cephalopods (Trueman 1941; Mutvei 1957, 1964, 2013; Ivanov 1975; Kennedy and Cobban 1976; Chamberlain 1980, 1993; Saunders and Shapiro 1986; Zaborski 1986; Hewitt et al. 1991; Jacobs and Chamberlain 1996; Landman et al. 1996; Westermann 1996; Westermann and Tsujita 1999; Doguzhaeva et al. 2002, 2003a, b, 2010; Richter 2002; Richter and Fischer 2002; Mutvei and Dunca 2007; Klug 2004; Klug et al. 2007, 2008; Korn and Klug 2007). One of the ways to achieve a better understanding of these paleobiological questions is to continue recollecting localities famous for preservation of non-mineralized parts to recover ammonoid shells with unbroken body chambers and shell apertures, weakly recrystallized original shell material, traces of mantle and muscle attachments in the body chamber, and fossilized mantle tissue. Using this approach, previously unknown "unusual" large ventrolateral marks of the muscle and body chamber attachment structures were discovered in complete body chambers showing shell apertures and also containing buccal apparatuses (Doguzhaeva and Kabanov 1988; Doguzhaeva and Mutvei 1990, 1991, 1992, 1993a, b; Doguzhaeva 2014). In several specimens, the additional small spot near the foremost part of the large mark is separated (Doguzhaeva and Mutvei 1991, pl. 2, fig. 4; pl. 4, fig. 1, pl. 6, fig. 2). This suggests, by comparison with the squids (Wells 1988), that in some ammonoids, the cephalic retractor and funnel retractor muscles might have been well-developed and permitted

L. A. Doguzhaeva (✉)
Department of Palaeobiology, Swedish Museum of Natural History, 104 05 Stockholm, Sweden
e-mail: larisa.doguzhaeva@nrm.se

R. H. Mapes
North Carolina Museum of Natural Sciences, 11 West Jones St., Raleigh, NC 27601, USA
e-mail: mapes@ohio.edu

C. Klug et al. (eds.), *Ammonoid Paleobiology: From Anatomy to Ecology*,
Topics in Geobiology 43, DOI 10.1007/978-94-017-9630-9_14

jet-powered swimming (Doguzhaeva and Mutvei 1991, 1996; Doguzhaeva 2014). The two side by side ventrolateral marks support the assumption of higher-energy coleoid-like mode of locomotion in some ammonoids (Wells and O'Dor 1991; Jacobs and Chamberlain 1996), but conflicts with the widely spread opinion (based on small attachment marks) of a weakly developed muscular system and insufficient capacity of the ammonoids for jet-powered swimming (see Mutvei and Reyment 1973; Jacobs and Chamberlain 1996; Monks and Young 1998). In *Nautilus,* the funnel retractor muscles are part of branches of the cephalic retractor muscles, which originate from the inside part of the head and funnel and are inserted on the shell laterally. The cephalic retractor muscles drive lower speed jet-powered swimming (see Griffin 1900; Mutvei 1957, 1964; Mutvei et al. 1993; O'Dor et al 1990; Sasaki et al. 2010).

The mantle, so far known in the Late Triassic ceratitid *Austrotrachyceras*, was also found due to investigation of cephalopod remains from the paleoenvironment providing non-biomineralized structures (Doguzhaeva et al. 2004, 2007a, b; Doguzhaeva and Summesberger 2012). Fine lamination of the fossilized mantle is thought to have resulted from the original lamination of the muscular tissue. In coleoids, the mantle is thickened with layers of muscles and connective tissue but it is thin in *Nautilus* (Kier 1988; Kier and Thompson 2003; Sasaki et al. 2010).

Herein, the ontogenetic and evolutionary variations of the body chamber length and traces of mantle and body chamber attachment in Paleozoic and Mesozoic ammonoids (Appendices 1, 2) are analyzed. The mantle and the body chamber attachment marks of the Early Cretaceous heteromorph ammonoid *Audouliceras* from Ulyanovsk, Russia, are worth to be herein demonstrated in detail. They were observed in two large mature shells with aragonitic preservation and nearly complete body chambers (Doguzhaeva 2014). The possibilities of jet-powered swimming in some ammonoids together with the fossilization of the mantle and the mantle and body chamber attachment marks are discussed.

The illustrated heteromorphs (Figs. 14.1–14.4) are from the private collection of G. K. Kabanov (Moscow) donated to the first author. Prefix DPC preceding specimen numbers denote repository: Doguzhaeva's private collection.

14.2 Previous Studies

Münster (1839) and Suess (1865) were apparently the first researchers who demonstrated that the body and body chamber attachment marks and the variability of the body chamber length are potentially important aspects of the paleobiological study of ammonoids. After them, Crick (1898) described small scars in 13 genera of Paleozoic and Mesozoic ammonoids, suggesting that they served for shell and soft tissue attachment. Following Suess's idea (1865) on the taxonomic value of the body chamber length, Haug (1898) subdivided the Devonian ammonoids into two groups: Brevidoma, including the Agoniatitidae, characterized by a short body chamber ranging from half a whorl to a complete whorl, and Longidoma, comprising the Anarcestidae, in which the body chamber length ranges from a complete whorl to 1.75 of a whorl. Arthaber (1911) accepted the erection of two groups in ammonoids based on the body chamber length and suggested the new names Microdoma and Macrodoma for them. Diener (1916a) established the intermediate group Metridoma for the ammonoids with body chamber lengths from 0.75 to one whorl. The essential variations of the body chamber length in taxa that were considered at that time to be closely related, were documented by Diener (1916a); he rejected the taxonomic value of the body chamber length (Diener 1916b). Trueman (1941) measured the body chamber length in about 80 ammonoid species and revealed two frequency peaks corresponding to 260 and 380 degrees. Ruzhencev (1962) suggested that a long ammonoid body chamber provided protection of the phragmocone; this conclusion was based on his observation of body chamber lengths that exceeded one whorl on specimens with complete apertures from the Late Paleozoic from southern Urals. Ruzhencev (1962) found also that the body chambers are usually shorter in prolecanitids than in goniatitids. Because of this, he confirmed a potential taxonomic value for body chamber length. Jones (1961) reported small umbilical and mid-ventral scars in a heteromorph ammonite. Jordan (1968) recorded the attachment scars in 22 Mesozoic genera, and in addition to small mid-dorsal scars, reported large lateral U-shaped marks interpreted as pallial line. In the end of the 20th and at the beginning of 21st century, the mid-dorsal and mid-ventral scars were revealed on the internal moulds of the phragmocones in a large number of Mesozoic genera of different systematic positions (Bandel 1982; Landman and Bandel 1985; Weitschat 1986; Weitschat and Bandel 1991; Sharikadze et al 1990; Richter 2002; Richter and Fischer 2002). Large ventrolateral marks extending from the last septum to about the midpoint of about 180 degrees long body chambers were documented in two Early Cretaceous genera (Doguzhaeva and Kabanov 1988). These unusually large and robust muscle scars co-occurred with the small mid-ventral, mid-dorsal and umbilical scars. The excellent preservation allowed the reconstruction of the mantle and muscle configuration inside the body chamber of *Aconeceras* (Doguzhaeva and Mutvei 1990, 1991, 1993a, b). Doguzhaeva and Mikhailova (1991) described the muscle scars in the Early Cretaceous heteromorph ammonite *Ancyloceras*. Sharikadze et al. (1990) documented mid-dorsal, mid-ventral and the large ventrolateral scars in a large number of ammonites from the Late Jurassic Volga Region and from Early Cretaceous sediments collected in the Adygeya Republic, north-west Caucasus, Russia. Doguzhaeva and Mutvei (1991) suggested principal anatomical differences between ammonoids and *Nautilus* in the muscular systems, shape of head-funnel region and its relationship to the shell aperture, and certain similarities between ammonoids and recent coleoids in the possible development of strong funnel retractors. Kyuma and Nishida (1987) recorded that the longest body chamber belonged to the Late Carboniferous goniatite *Akiyoshiceras* from Japan. While none of the specimens available to them had a complete body chamber with an aperture, the incomplete body chamber lengths of some of the Japanese specimens exceeded two complete whorls. A specimen of this genus from Oklahoma has a body chamber length of approximately two and a half whorls (RHM, personal observation) with a mature modification of the aperture giving a body chamber length for the genus of approximately 900°. Kyuma and Nishida (1987) also noted that there were paired raised ridges and depressions about midway on the body chambers on some Japanese specimens. These ridges and depressions are almost certainly related to muscle attachments that possibly allowed extension and retraction of the body mass of the animal. However, anatomical details have not been studied at this time. Saunders and Work (1997) carried out experimental study on shell buoyancy and calculated the body chamber lengths of some Paleozoic ammonoids. They came to the conclusion that the two stocks Goniatitida and Prolecanitida differ in body chamber length. Their experimental data were supported by their observations on the body chamber lengths in three prolecanitid genera. Thus, the hypothesis of Ruzhencev (1962) on the body chamber length as an additional trait distinguishing prolecanitids, which gave rise to the Mesozoic ammonoids, was supported. In the Middle Carboniferous *Goniatites,* Tanabe et al. (1998) recorded mid-ventral marks and traces of pseudosutures. Dagys and Keupp (1998) described a pair of ventral internal keels in the two Middle Triassic ceratitids *Czekanowskites* and *Arctohungarites* from Arctic Siberia. The keels extend in the body chamber for 80° from the last septum. The authors suggested that the ventral internal keels might be the attachment structures of the funnel retractors. Monks and Young (1998) suggested that to maintain hydrodynamic stability, the heteromorph ammonoids might have changed their centers of gravity by the forward-backward shifting of a comparatively short body within the long body chamber. Kröger (2002) used this model for the explanation of the healing of an extensive damage seen on the monomorph shells of Jurassic ammonoids with a long body chamber; this damage extends over about two thirds of the body chamber length. He also concluded that a withdrawal of more than 35 degrees into the body chamber was only a strategy of emergency, and that longidome ammonoids were able to withdraw deeper into the body chamber. He suggested that the dactyliocatids characterized by a long body chamber possibly had a normal variability of soft body position between the aperture and 35° behind it. Kröger (2002) published new data on the body chamber length in five Jurassic genera and came to the conclusion that long (more than one whorl) body chamber lengths protected the phragmocone against lifetime damage, an idea previously presented by Ruzhencev (1962). Richter (2002) as well as Richter and Fischer (2002) reported small mid-ventral, mid-dorsal, umbilical and large ventrolateral marks in a number of Paleozoic and Mesozoic ammonoids. Klug (2004) recorded the shortest known body chamber (about 120 degrees long) on the Middle Triassic genus *Ceratites*. Later, Klug et al. (2007) observed on a Middle Triassic specimen of *Ceratites* a set of structures on the internal mould of the body chamber and phragmocone that were interpreted by them as the attachment marks of the cephalic retractors, the mantle myoadhesive band, the paired dorsal muscles, the unpaired ventral muscle, and the black layer. In the Early Carboniferous goniatite (?*Rhadinites* or ?*Anthracoceras*), Landman et al. (2010) documented "*a spiral red band that extends from midway in the body chamber to the apical end of the jaw (Figs 3A, B, F). The position of this feature matches that of the cephalic retractor muscle illustrated in the reconstruction of Cretaceous ammonoids by Doguzhaeva and Mutvei* (1996, *Fig. 7*)". In the same specimens,

Landman et al. (2010) revealed the mid-dorsal band extending for about 200 degrees in the long body chamber, with the estimated length of more than one whorl.

14.3 Terminology and Abbreviations

The following terms are used in the present paper for characterizing the body chamber lengths and the body and body chamber attachment marks. Depending on their position in the body chamber, the marks are arranged in six categories: dorsal, umbilical, dorsolateral, lateral, ventrolateral and ventral.

Annular elevation (ae)—dorsolateral posterior structure in body chamber; formed by a swollen inner prismatic layer of shell wall;

Dorsal band (db)—dorsal dark color band-like attachment structure along the median plane of body chamber and phragmocone;

Dorsal ridge (dr)—dorsal thread-like ridge-shaped attachment structure along the median plane of body chamber and phragmocone; known in clymeniids;

Dorsal discrete mark (d)—dorsal small discrete attachment scars along the median plane of the body chamber and phragmocone;

Lateral mantle V-shaped sinus (lms)—this is a delicate mantle structure here recorded for the first time; it is evidenced by repetition of fine uneven lateral V-shaped lines on the body chamber in *Audouliceras* (Fig. 14.3). This structure is clearly different from the sinus-like attachment mark (Figs. 14.3, 14.5);

Lateral sinus (ls)—sinus-like attachment mark on sides of body chamber;

Long body chamber—length range is about 370°–900°;

Medium–long body chamber—length range is about 270°–370°;

Myoadhasive elevation (mae)—the elevation in front of the last septum;

Myoadhasive gutter (mg)—thin ridge surrounding the annular elevation;

Ventral band (vb)—ventral band-like structure along the median plane of body chamber and phragmocone visible due to its colour surface, lack a relief on the shell wall or on the internal mould of the body chamber;

Short body chamber—length range is about or shorter than 270 degrees;

Ventral discrete mark (v)—ventral small rounded or oval discrete marks along the median plane of body chamber and phragmocone;

Ventral ridge (vr)—ventral thread-like ridge along the median plane of body chamber and phragmocone;

Ventrolateral elevations (ve)—(= internal ventral keels by Dagys and Keupp 1998) two spiral elevations on the inner surface of the body chamber extending for about 80 degrees from the last septum;

Ventrolateral lobe-like marks (vl)—ventrolateral large lobe-like attachment marks; extend from the last septum approximately to the midpoint of the body chamber;

Umbilical small discrete mark (us)—discrete small, less than the interval between two neighbouring septa of the phragmocone, marks located at the umbilical corners of body chamber;

Umbilical tongue-like marks (ut)—elongated attachment marks in the umbilical corners of the body chamber; comparatively broad posteriorly and narrowing anteriorly, extend from the last septum for about 1/5-1/3 of the body chamber length.

14.4 Ontogenetic and Evolutionary Variations of Body Chamber Length

The data on body chamber length of 120 genera from the orders Goniatitida, Prolecanitida, Clymeniida, Ceratitida, Phylloceratida, Lytoceratida and Ammonitida are represented in Appendix 1. The body chamber lengths of ammonitellae of 73 genera and of the adult shells of 42 genera is shown as well; 18 genera show the body chamber length, both in the ammonitellae and in adult shells. These data are summarized as follows:

1. The range of body chamber length varies between 223°–410° in ammonitellae and between 180°–900° in adults. This shows that in ammonitellae, the body chamber length varies in smaller range than in the adults.
2. The minimum length of the body chamber of the ammonitella is 300° in goniatitids (*Glaphyrites*), 290° in prolecanitids (*Hedenstroemia*), 230° in ceratitids (*Arctohungarites* and *Crumbergia*) and phylloceratids (*Holcophylloceras*), 260° in lytoceratids (*Eurystomiceras*), and 242° in ammonitids (*Subprionocyclus*) (Appendix 1). The ammonitellae possessing the longest body chamber (410°) are recorded in the order Goniatitida; the ammonitellae with the shortest body chamber (223°) are from the order Phylloceratida. Thus, the evolutionary development of the angular length of the body chamber shows a long-term decreasing trend of the ammonitella body chamber length in the goniatitids–prolecanitids–ceratitids–phylloceratids branch. This long-term trend supposedly indicates the common early ontogenetic evolutionary direction of the goniatitids, prolecanitids, ceratitids and phylloceratids. The reversal of this trend in the lytoceratids and ammonitids presumably indicates a different and innovative early ontogenetic strategy of these two groups in the Jurassic.
3. The ontogenetic decrease of the body chamber length is documented in the following genera (Appendix 1): Prolecanitida, *Metapronorites*—from 340° in ammonitella to 250° in the adult shell; *Neopronorites*—from 310°–340° in the ammonitella to 230° in the adult shell; Ceratitida, *Sakmarites*—from 340° in the ammonitella to 225° in the adult shell; Lytoceratida, *Eurystomiceras*—from 290°–300° in the ammonitella to 270° in the adult shell; Perisphinctida *Indosphinctes*—from ca. 330° in the ammonitella to ca. 180°–220° in the adult shell; Ammonitida *Deshayesites*—from 307°–330° in the ammonitella to 180° in the adult shell; ammonitid *Aconeceras*—from 270°–360° in the ammonitella to 180° at the 7th to 8th whorls. The ontogenetic increase of the angular body chamber length is documented in the goniatite *Neoglyphioceras* (from 360° in the ammonitella to >450° at 32 mm diameter adult shell) and in the ceratitids *Proarcestes* (from c. 360° in the ammonitella to >450° in the 8th whorl), *Phyllocladiscites* (from c. 330° in the ammonitella to 360° at the 9th whorl), and *Prosphingites*

(from 265° in the ammonitella to c. 360° in the adult shell). Thus, the ontogenetic development of angular length of the body chamber in these ammonoid groups shows either a decrease in body chamber length (Miller et al. 1957) or an increase in body chamber length (Doguzhaeva 1990). In no known case does the ammonitella body chamber length remain the same throughout ontogeny.

14.5 Muscle and Mantle Attachment Marks, and their Interpretation

The data on the mantle and body chamber attachment marks or scars in mature shells of 102 genera from the orders Goniatitida, Prolecanitida, Clymeniida, Ceratitida, Phylloceratida, Lytoceratida and Ammonitida are given in Appendix 2. The marks are interpreted as traces of the mantle and the body chamber attachment because of their regular arrangement in different ammonoid taxa in various lineages from the Devonian to the Cretaceous. The marks originated on the inside of the body chamber wall, but during ontogeny, the marks became part of the phragmocone wall where some marks remained visible and others are probably hidden under the shell of the dorsal wall. The muscle attachments are best observed in mature shells because deposition of shell material around the mark sites probably took a long time. The castings of these marks by internal moulds formed by Fe-containing minerals have yielded extensive information on the mantle and body chamber attachment. The marks are arranged in five categories depending on their position in the body chamber: dorsal, dorsolateral, lateral, ventrolateral and ventral; these are briefly characterized and discussed in the following interpretations.

14.5.1 Dorsal Marks

The dorsal discrete scars, dorsal band, and dorsal internal ridge are the marks arranged along the median plane on the dorsal side of the body chamber (Appendix 2). The dorsal marks produced earlier in the ontogeny are often visible between the septa in the phragmocone. These marks were, in all probability, the attachment areas of the dorsal muscles. To judge from the small size of the dorsal marks and their remoteness from the anterior of the body chamber, the dorsal muscles must have been comparatively weakly developed and restricted in extension to the posterior part of the mantle.

14.5.1.1 Dorsal Small Discrete Marks

Relatively small, rounded, oval or cruciform marks with a typically rugose texture in the dorsal position are located in the body chamber immediately in front of the last septum inside the internal (dorsal) lobe of the last suture. These discrete dorsal marks are interpreted as evidence of the periodic interruption of the attachment of the body tissue during its periodic shifting forward with shell growth. The marks are often expressed on the interior of the body chamber shell surface by a positive relief rugose structure that is repeated at regular intervals in the juvenile portions of the phragmocone and at the posterior end of the mature body chamber. These marks seem to be a universal structure in the ammonoids or have higher potential for preservation as they are observed in the shells lacking any other attachment marks or in association with the other attachment marks (Appendix 2). Those produced earlier in ontogeny often visible between the septa in the phragmocone; every chamber has a solo mark.

14.5.1.2 Dorsal Band

The dorsal band is a brown or black stripe, which forms neither a relief, rugosity nor porous surface on the dorsal side of the body chamber (Figs. 14.1, 14.4). In long body chambers with lengths of more than one whorl, the middorsal band extends for about 200° (Landman et al. 2010). The intensive color difference between the middorsal band and the matrix apparently indicates the originally high content of non-biomineralized "soft" material in the body chamber wall along its middorsal side. During shell growth, the middorsal band remains visible in the chambers of the phragmocone. Contrary to the discrete dorsal marks, the dorsal band indicates continuous attachment of the body during its shifting within the growing shell.

14.5.1.3 Dorsal Internal Ridge

This is a thread-like elevation in the body chamber, which is located in the dorsal median plane. This dorsal internal ridge bears a weak transverse striation. The dorsal internal ridge extends for about one-third to one-fourth of the body chamber length. It is sometimes preserved on the external surface of the phragmocone and its length can indicate the length of a missing portion of the body chamber (B. I. Bogoslovky, personal communication, 1980's). To our knowledge, this feature is only known in clymeniids.

14.5.2 Umbilical Marks

These marks are situated in the shoulders of the body chamber (Appendix 2). There are two kinds of marks: umbilical small discrete rounded or transverse kidney-shaped marks and large tongue-like marks.

14.5.2.1 Umbilical Small Discrete Marks

These marks, which are shorter than the distance between two adjacent septa, are rounded imprints located in the umbilical region posterior to the terminal septum. A

Fig. 14.1 *Audouliceras renauxianum* (d'Orb., 1841), DPC/1. Aptian, Lower Cretaceous; Ulyanovsk Region, Central Russia. Fully-grown shell showing umbilical tongue-like marks in the posterior portion of the body chamber; **a** × 0.5, **b** × 0.8, **c** × 0.6. *db* dorsal band, *ut* umbilical tongue-like mark

set of these marks remains visible on the phragmocone so that every chamber of the phragmocone has a single mark. The imprints of these marks are formed by minute raised tubercles, which are the hole fillings in the inner surface of the inner shell surface. The discrete character of these scars indicates the periodically interrupted attachment between the mantle and the body chamber during the forward movement of the body in the growing shell. These marks are observed in shells with a short, medium and long body chamber (Appendices 1, 2).

Based on the relatively small size and location in the posterior portion of the body chamber, these small umbilical marks must have been the attachment sites of the branches of the muscles in the posterior portion of the muscular mantle to the inner shell wall.

14.5.2.2 Umbilical Large Tongue-Like Marks

The tongue-like umbilical marks have only been observed in the terminal body chambers of Cretaceous heteromorphs including *Audouliceras* (Figs. 14.1, 14.2, 14.3, 14.4; Doguzhaeva 2014), *Ancyloceras* (Doguzhaeva and Mikhailova 1991), *Ptychoceras* (Doguzhaeva and Mutvei 2015), *Pseudocrioceratites* (DLA, personal observation), and possibly in *Baculites* (Kennedy et al. 2002, pl. 6, Figs. 1, 2). These marks are relatively broad posteriorly, narrow anteriorly, and extend for about 20–30 % of the body chamber length along the umbilical corners (Figs. 14.1, 14.2, 14.3, 14.4). In *Pseudocrioceratites*, the mark is a tongue-like depression on the internal mould of an incomplete body chamber. In *Audouliceras, Ancyloceras* and *Ptychoceras*, the umbilical tongue-like marks are recognized by their nacreous incrustation retained on dark grey siderite internal moulds of body chambers. In these three genera, the tongue-like marks comprise small rounded brownish marks placed in the umbilical corners (Fig. 14.2). In *Audouliceras,* there are two equal sized marks situated close to each other located near the body chamber midpoint (Fig. 14.2). In *Ptychoceras*, the small rounded marks within the tongue-like umbilical marks are arranged in two pairs, one of which is situated in front of the last septum, and the next pair is situated near the midpoint of the body chamber (Doguzhaeva and Mutvei 2015).

The tongue-like umbilical marks show that in the heteromorphs listed above, the mantle and the body chamber attachment sites moved forward to the middle part of the body chamber and are probably the attachment sites of the powerful umbilical retractor muscles. In heteromorph ammonoids with a planispiral irregularly coiled or hook-shaped body chamber, the development of these muscles might have speculatively occurred in order to change the mantle cavity volume to improve filter feeding on small planktonic organisms (Doguzhaeva 2014). This mode of feeding is supported by the three-dimensionally preserved jaw apparatus extracted from the body chamber of the Early Cretaceous heteromorph *Australiceras* (Doguzhaeva and Mikhailova 2002). The jaw apparatus lacks a rostrum and has a large cavity between the lower and upper jaws, which is formed by the deep and wide lower jaw with a solid flat ventral side, and the long wings of the upper

Fig. 14.2 *Audouliceras renauxianum*, DPC/2. Aptian, Lower Cretaceous; Ulyanovsk Region, Central Russia. Fully-grown shell showing rounded umbilical scars within the umbilical tongue-like marks in the middle part of the body chamber; **a** Lateral view, × 0.6. **b** Close-up of (a) to show two side by side rounded marks, × 0.7. *c* umbilical rounded scar within the umbilical tongue-like mark

jaw. The absence of the rostrum suggests that *Australiceras* did not hunt for large prey; hence, there was no need to hold and crush large prey. Large surfaces of the wings could serve as sites for the attachment of well-developed buccal muscles, suggesting that the large cavity between the lower and upper jaws could be used for sucking and filtering seawater, which suggests that these organisms fed on minute plankton or suspended organic particles. This hypothesis is supported by the heavily ribbed shell of *Australiceras*, which has the shape of irregular coiled

Fig. 14.3 *Audouliceras renauxianum* (d'Orbigny, 1841), DPC/2. Aptian, Lower Cretaceous; Ulyanovsk Region, Central Russia. **a** Lateral view of fully-grown shell showing lateral V-shaped sinus of the mantle, × 0.6. **b** Close-up of (**a**) to show the repetition of muscle scar traces in the lateral sinus that were apparently left during the movement of the mantle; × 1. *ms* muscle scar, *ut* tongue-like mark

spiral, i.e., the form best suited to floating and perhaps vertical (diurnal) migrations (Doguzhaeva and Mikhailova 2002).

14.5.3 Lateral Marks

14.5.3.1 Lateral Sinus-Like Attachment Marks

The lateral sinus-like marks are adaperturally open, relatively large in comparison with the body chamber length, with sinus- or bay-shaped outlines on the body

Fig. 14.4 *Audouliceras renauxianum* (d'Orbigny, 1841) : A, C, DPC/1; B, DPC/2. Aptian, Lower Cretaceous; Ulyanovsk Region, Central Russia. **a** Ventral view of the body chamber (aperture is down) showing a colour ventral band with a rounded ventral mark ending the ventral band and the ventral ridge in front of the ventral mark; × 1.2. **b** Dorsal view of the posterior portion of the body chamber to show a colour dorsal band between the paired umbilical marks; × 1.5. **c** Two umbilical rounded scars within the umbilical tongue-like marks; × 1.3. *db* dorsal band, *ut* umbilical tongue-like mark, *c* umbilical rounded scar within the umbilical tongue-like mark, *v* ventral rounded scar, *vb* ventral band, *vr* ventral ridge

chamber sides. The lateral sinusoidal marks extend from the posterior portion of the body chamber to about its midpoint. This mark is narrowly rounded behind and gets broader in the adapertural direction. It may co-occur with ventrolateral and other marks (Appendix 2). The morphology of the sinus-like attachment marks is here illustrated by those of *Deshayesites* (Figs. 14.5, 14.7b). Their function is unknown.

Fig. 14.5 Shape variations of ventrolateral attachment marks in the body chamber of three specimens of *Deshayesites deshayesi*; DPC/3, 4, 5. Aptian, Lower Cretaceous; Ulyanovsk Region, Central Russia. **a** ×3.5. **b** ×3.8. **c** ×3. *bc* body chamber; *vl* ventrolateral mark

14.5.3.2 Lateral Mantle V-Shaped Marks

The lateral mantle V-shaped marks represent a delicate structure here illustrated for the first time; it is evidenced by repetition of fine uneven V-shaped lines on sides of the body chamber in *Audouliceras* (Fig. 14.3). This structure is clearly different from the sinus-like attachment mark by its extremely fine structure (compare Figs. 14.3, 14.5, 14.7b). It is highly likely that the lateral V-shaped lines reflect the similar pattern of the periphery of the mantle.

14.5.4 Ventrolateral Marks

These are marks situated ventrolaterally of the body chamber; there are lobe-like and sinus-like marks, which can have ridge-like elevations (Figs. 14.6, 14.7a, 14.8, 14.9, 14.10).

14.5.4.1 Ventrolateral Lobe-Like Marks

These are large, adaperturally closed, lobe-like marks on the ventrolateral portions of body chamber. The ventrolateral marks extend from the last septum for about the midpoint of the body chamber. Posteriorly, a pair of these marks is separated on the ventral side with a slit-like interspace, which gets broader towards the anterior portion of the marks (Figs. 14.6, 14.7a, 14.8, 14.9, 14.10). The outline of the ventrolateral marks is usually strengthened with a ridge-like elevation forming a thin rim. Within the marks, additional nacreous and prismatic layers thicken the shell wall.

This type of attachment marks has only been observed in the body chambers, which are shorter than a whorl (Appendices 1, 2). Their adapertural side shows two separated spots situated next to each other. They are considered as potential attachment sites for the cephalic retractor and funnel retractor muscles.

Fig. 14.6 Lateral sinus-like attachment mark in the posterior portion of body chamber in *Deshayesites deshayesi;* Doguzhaeva Private Collection DPC/6. Aptian, *Lower* Cretaceous; Ulyanovsk Region, Central Russia; ×2.6. *bc*, body chamber; *ls*, lateral sinus, *ph* phragmocone

Fig. 14.7 Soft-tissue attachment marks in the body chamber of *Deshayesites deshayesi*. DPC/7, 8. Aptian, Lower Cretaceous; Ulyanovsk Region, Central Russia; **a** ventrolateral muscle mark is getting narrower in adoral direction contrary to lateral sinus-like mark on **b** which is getting broader in this direction; × 2.5 and 2.7, respectively, *bc* body chamber; ls ventrolateral mark

Fig. 14.8 Ventrolateral lobe-like attachment mark in the body chamber of *Aconeceras trautscholdi* (Sinzow, 1870) Doguzhaeva Private Collection DPC/9. Aptian, Lower Cretaceous; Ulyanovsk Region, Central Russia; × 3. *bc* body chamber, *vl* ventrolateral lobe-like mark

Fig. 14.9 Ventrolateral lobe-like attachment mark in the body chamber of *Aconeceras trautsholdi* (Sinzow, 1870); DPC/10. Aptian, Lower Cretaceous; Ulyanovsk Region, Central Russia; × 2.6. *bc* body chamber; *vl* ventrolateral mark

Fig. 14.10 Lateral attachment mark on the phragmocone surface of *Kashpurites* sp. DPC/11. Upper Jurassic, Moscow Region, Russia; x 1.8, *ph* phragmocone; *vl* ventrolateral mark

14.5.4.2 Ventrolateral Elevations

These are ridge-like paired structures described as internal ventral keels by Dagys and Keupp (1998). They represent two elevations on the inner surface of the body chamber along the ventral keel and extend from the last septum for about a quarter of the body chamber length.

Fig. 14.11 The ventrolateral view on the last chambers of the phragmocone and posterior part of the body chamber of the gerontic shell of *Indosphinctes* sp. DPC/12; Upper Jurassic, Volga River, Central Russia; × 2. *mae* myoadhesive elevation; *v* ventral discrete scars

14.5.5 Ventral Marks

The ventral marks include the ventral small discrete scars, the ventral band, and the ventral ridge. During shell growth, the marks remained distinguishable in the chambers of the phragmocone and were also impressed into the internal mould. The midventral marks are observed in shells with short, moderately long, and long body chambers.

14.5.5.1 Ventral Discrete Marks

The rounded, oval or crescent-shaped mark is situated on the ventral side of the body chamber in front of the ventral lobe of the last suture at the maximum distance about one to three chamber lengths from the last septum, in rare cases in a greater distance (Figs. 14.4a, 14.11). This distance is shorter or equal to the length of the circumsiphonal invagination of the posterior portion of the body where the formation of the new segment of the siphonal tube began (Doguzhaeva 1988; Doguzhaeva and Mutvei 1996). Ventral marks produced earlier in the ontogeny are often visible between the septa in the phragmocone; every chamber has a solo mark (Fig. 14.11). In some shells (Jordan 1968), the ventral marks are partially black or brown; the color apparently originated from the pyritization of organic material within the attachment mark. The discrete ventral is considered evidence of the periodically interrupted attachment of the body during its shifting forward with shell growth. The marks can be observed in the shells lacking any other attachment marks or in association with the other marks. They are apparently a unifying morphologic element seen in the growth of all ammonoids (Appendix 2).

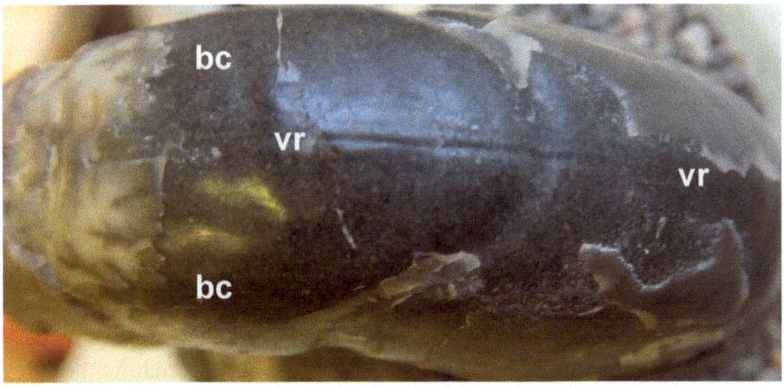

Fig. 14.12 Thin median incision on the internal mould of body chamber of *Damesites* that evidences a fine thread-like ventral ridge on the inner surface of body chamber. Aptian, Early Cretaceous, R. Hokodz, Adygeya Republic, NW Caucasus, Russia; DPC/13; x. *bc* body chamber, *vr* ventral ridge

14.5.5.2 Ventral Band

This is a brown or black band, not expressed as a feature with high relief, or a rugose or porous texture, which is located on the ventral side of the body chamber (Fig. 14.4a). The length of the midventral band in the body chamber is uncertain; the preserved portion that has been observed is usually short, equal to the length of the few last chambers of the phragmocone. The intensive color difference between the midventral band and the matrix apparently indicates the original structure had a higher concentration of non-biomineralized organic material in the body chamber wall along its ventral position. The ventral band is sometimes visible in the chambers of the phragmocone. Contrary to the discrete ventral marks, the ventral band indicates there was non-stop attachment of the body tissues during the forward movement of the growing shell. These marks have been observed in fewer genera than the discrete ventral marks (Appendix 2).

14.5.5.3 Ventral Ridge

The ventral ridge is present as a thin median incision in front of the ventral scar of the internal moulds of body chambers. This incision indicates that there was a fine thread-like ventral ridge on the inner surface of the body chamber. The ventral ridge is here illustrated in *Audouliceras* (Fig. 14.4a) and *Damesites* (Fig. 14.12). It extends to about half of the body chamber length in both of them. It was also observed in *Ptychoceras*. The ventral internal ridge is associated with a ventral discrete mark.

14.5.6 Pore Canals and Assumed Porous Apertural Band

Regularly spaced pore canals that penetrate the ventral keel and open from the outside have only been observed in *Aconeceras* so far. The canals are thought to have housed the mantle extensions and served for the body and body chamber attachment (Doguzhaeva and Mutvei 2015). In ammonoids, the mantle was probably attached along its apertural edge to the shell aperture in a manner similar to that of *Nautilus*. In the latter genus, the pore canals are developed on the inner surface of the body chamber wall along the apertural edge; they are arranged in a narrow apertural band that becomes hidden under the prismatic layers during shell growth Mutvei, Doguzhaeva 1997. By analogy with *Nautilus,* the apertural attachment of the mantle edge is suggested for the ammonoids. Coincident with this attachment at the aperture in *Nautilus* is a black layer (Stenzel 1964; Klug et al. 2004). A partial proof of this type of attachment in ammonoids is the presence of a black band at the shell aperture that was recently found in the mature shell of Triassic *Ceratites* and some other ammonoids (Klug 2004; Klug et al. 2004, 2007).

14.5.7 Myoadhesive Elevation

This is the elevation in front of the last septum, observed in association with enlarged ventral scars in gerontic shells (Fig. 14.11). The elevation is commonly formed by a thickening of the shell wall.

14.5.8 Septal Attachment

Henderson (1984) suggested that the muscles were attached to the periphery of the septum proper and the inner septum within the septal lobes in the Cretaceous lytoceratid *Pseudophyllites*. The septal lobe extends to the surface of the previous septum (Fig. 14.13); the successive septal recesses link one septum to the next, forming a septal tunnel. This is a unique pattern of muscle attachment in the posterior of this genus that is not present elsewhere in the ammonoid phylogeny but in lytoceratids (Hoffman 2010).

14.5.9 Cameral Membranes

Landman et al. (2006) suggested that in the Permian prolecanitid and goniatitid ammonoids, the cameral membranes which are continuations of septal organic layer strengthened the attachment of the siphuncle to the shell wall.

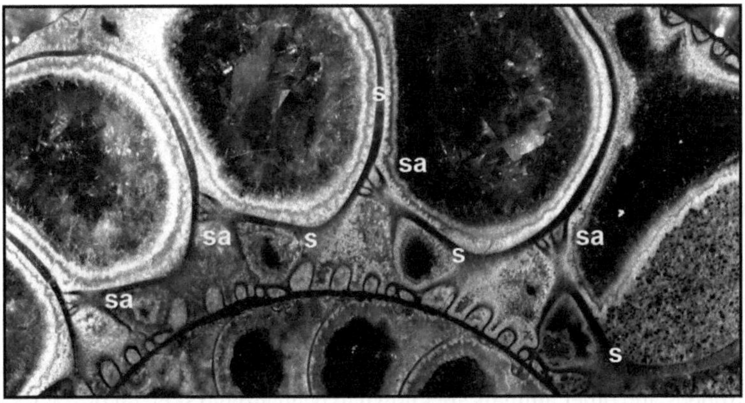

Fig. 14.13 The septal attachments in paramedial shell section of *Gaudryceras tenuiliratum*. Turonian-Campanian, Upper Cretaceous; island of Sakhalin; Russia; DPC/14.; ×3. *s* septum; *sa* septal attachment

14.6 Mantle Structure of the Late Triassic Ceratitid Austrotrachyceras

Remains of the fossilized mantle were detected in compacted body chambers of the Late Triassic ceratitid *Austrotrachyceras*, which co-occurs with the coleoids *Phragmoteuthis* and *Lunzoteuthis*. These coleoids show soft tissue preservation, which shows that the taphonomic conditions were suitable for soft tissue preservation of the co-occurring ammonoids as well (Doguzhaeva et al. 2004, 2006, 2007a, b; Doguzhaeva and Summesberger 2012). The mantle of *Austrotrachyceras* exhibits a thin laminate ultrastructure and in this respect, it is similar to the muscular mantle of the Jurassic coleoids *Belemnoteuthis* (Kar et al. 1995) and *Loligosepia* (Doguzhaeva et al. 2004). The mantle of *Austrotrachyceras* indicates that the external shell did not impede the development of a muscular mantle. Together with the cephalic retractor and funnel retractor muscles, this enabled ammonoids to jet-powered swimming.

14.7 Morphological Indications of Jet-Powered Swimming in Ammonoids

In adult shells of Paleozoic and Mesozoic ammonoids, the range of the body chamber length is about 180°–900°. From this range, the body chambers with lengths less than one whorl are associated with the large attachment marks located on the sides of body chambers extending to the midpoint of the body chamber. These situations can be seen in goniatitids (*Aulatornoceras, Cheiloceras*, *Linguatornoceras*,

Tornoceras) and ammonitids (*Aconeceras, Amaltheus, Amauroceras, Androgynoceras, Arieticeras, Armatites, Deshayesites, Hecticoceras, Kashpurites, Quenstedtoceras, Stolleyites*; see Appendices 1, 2). No large ventrolateral marks have been observed in the body chambers longer than one whorl, but this size-relationship still needs to be quantified in order to test its significance. Among the diverse sets of attachment marks that are seen on ammonoids, the large ventrolateral marks are the only ones that can be connected by a straight line with the rear of the cephalic region of the body. The remoteness of the dorsal, umbilical and ventral marks from the anterior ventrolateral part of the body chamber in planispiral shells makes it impossible to put a straight line between them and the cephalic area of the body. Therefore, the latter marks are here considered as less likely attachment sites for the cephalic retractor and funnel retractor muscles. Nevertheless, muscles in gastropods and other animals are known to bend around skeletal obstacles (shell, bone), so we cannot entirely exclude an insertion of such muscles at these remote posterior sites. By contrast, the large ventrolateral attachment marks are thought to be potential sites for the insertion of more or less strong retractor muscles. In several exceptionally well-preserved specimens of *Aconeceras*, the anterior part of the ventrolateral scar is subdivided into two minor lobes (Doguzhaeva and Mutvei 1991, pl. 2, Fig. 2, 4; pl. 3, Fig. 1; pl. 4, Fig. 1, 2). The functional interpretation of these marks arises from their comparison with the attachment of the cephalic retractor and funnel retractor muscles to the gladius in squids (Wells 1988; see Doguzhaeva and Mutvei 1996, Fig. 7a, b). The attachment sites of the powerful cephalic retractor and funnel retractor muscles in squids are located close to each other on the inside surface of the gladius (Wells 1988). In *Nautilus,* the attachment marks of the cephalic retractor muscles are relatively large, but those of the funnel retractor muscles are indistinctly separated branches of the cephalic retractor muscles (Mutvei et al. 1993). Doguzhaeva and Mutvei (1991), Fig. 8) postulated that the cephalic retractor muscles originated from the dorsal minor lobe and the hyponome retractors from the ventral minor lobe within the ventrolateral marks. This conclusion is applied to other genera showing large ventrolateral marks listed above.

In coleoids, the muscle contractions of the thick mantle due to contractions of the cephalic retractor muscle generate the water current into the pallial cavity and provide ventilation of the pallial cavity. In *Nautilus*, the ventilation of the pallial cavity is supported by undulating contractions of the funnel wings (Wells 1988; Wells and O'Dor 1991; Sasaki et al. 2010).

Thus, we assume that the preconditions for jet-powered swimming of ammonoids is a short body chamber with the appropriate location of the attachment of the cephalic and funnel retractor muscles, which should have been able to extend straight across to the head and funnel area (Doguzhaeva 2014). None of the small dorsal, umbilical and ventral marks are here considered to be the attachment sites of the cephalic and funnel retractor muscles. We suggest that the variations of the body chamber length (Appendix 1) in combination with the mantle and muscle attachment marks (Appendix 2) indicate that the ventrolateral attachment marks occur exclusively in such body chambers, which are maximally one whorl long, where the ventrolateral marks extend from the last septum to the midpoint of the body chamber. In *Aconeceras* and similar genera (Appendix 2) the ventrolateral marks probably represented the attachment site of the cephalic retractor muscles that extended to the cephalic region and of the funnel retractor muscle that extended to the funnel.

Assuming that a muscular mantle was present in these brevidome to mesodome ammonoids like in the Late Triassic ceratitid *Austrotrachyceras*, we suggest that they may have had strong cephalic retractor and funnel retractor muscles and a well developed funnel and muscular mantle that was utilized for propulsion. Such a muscular system would have made these ammonoids suitable for a nektonic mode of life.

The soft body and the body chamber are traditionally assumed to be of approximately equal lengths in ammonoids. This presumption has been used for the experimental estimation of the orientation of the aperture in the shells of different geometry, ornamentation and body chamber lengths (Trueman 1941; Saunders and Work 1997; Saunders and Shapiro 1986; Jacob and Chamberlain 1996).

The hypothesis of the soft body size exceeding the body chamber volume arose from the additional external shell wall layers in some taxa, which caused or required a mantle extension over the external shell surface for the extra shell secretion; this was proposed by Drushchits et al. (1978) and supported by Doguzhaeva and Mutvei (1989, 1991, 1996, 2015). This hypothesis received further support from the discovery of the thin fragile apertural margin in *Ptychoceras* (Doguzhaeva and Mutvei 1993) and parabolic structures on the external surface of the body chamber in *Indosphinctes* (Doguzhaeva 2012).

A different hypothesis was proposed by Monks and Young (1998). They proposed that the body was much smaller than the analogy with *Nautilus* suggests, and the ammonites, especially heteromorphs, were more like a gastropod. Until now, this hypothesis has not yet been confirmed or rejected by direct observations or other indirect assumptions based on morphological data. Their hypothesis apparently arose from the so far poorly understood mode of life of heteromorph ammonoids. The previously mentioned thin fragile apertural margin of *Ptychoceras* and the large tongue-like umbilical marks in the heteromoph shells of *Ancyloceras, Audouliceras, Baculites*, and *Ptychoceras* question the hypothesis of Monks and Young (1998), because a small gastropod-like soft body in a big shell appears unlikely to require big muscles. For perisphinctids and *Dactylioceras*, it was also suggested that the body was significantly smaller than suggested by the body chamber volume (Kröger 2002). This assumption arose from the observation of over half a whorl deep sublethal injuries of the body chamber wall that supposedly would have required substantial withdrawing of the body into the body chamber (Kröger 2002).

14.8 Fossilization of Mantle and Muscle Attachment Marks

Ammonoid shells showing both mantle and body chamber attachment marks are very rare. There are several causes of post mortem loss of such shell features. First of all, there is the relatively likely destruction of body chambers because of the lack of septal support and because the aperture is prone to mechanical damage during transport. Secondly, recrystallization of the shell material can result in a fusion of the shell wall with the matrix, which makes the inner surface of the body chamber invisible. Thirdly, the attachment marks, which were expressed in a relief of the body chamber wall, might have been cast by the internal moulds but the other marks were not. Fourthly, the marks, which did not form a relief, might have been preserved due to higher concentration of organic (non-biomineralized) material, but in that case, a burial environment that allowed fossilization of such materials (e.g., melanin, chitin) would be required. Such environments are rare in the fossil record.

14.8.1 Mantle Fossilization

The soft tissue of the mantle in the body chamber was detected in seven comparatively large (diameter = 50–75 mm) specimens of Late Triassic ceratitid *Austrotrachyceras.* The conchs were compacted by lithostatic pressure, and their preserved shells retain their original aragonitic mineralogy with only little alteration (Doguzhaeva et al. 2004). One of the shells was associated with its mandibles (Doguzhaeva et al. 2007a). The mantle is a black glossy pitch-like mass restricted in distribution to the body chambers. Under SEM observation, the mantle revealed fine laminations and a fibrous ultrastructure. The fibers consisted of numerous micro-grains of microbial size. The fossilized mantle showed a high content of C but lacked Ca and P. The elemental composition is here given in weight percents: C (60–65 %); O (30 %); S (2–6 %); Si (1–2 %); Cd (0.5–1.8 %); Fe (1 %) and K, Al, Zn (each < 1 %) (Doguzhaeva et al. 2004). A similar chemical composition was detected in the mandibles: C (53–58 %), O (30 %), Si (1–5 %), Fe (1–2 %), and Ca (< 1 %) (Doguzhaeva et al. 2007a). The replacement of the muscular mantle tissues by carbon apparently resulted from the metabolism of anaerobic carbon-accumulating bacteria (Doguzhaeva et al. 2004, 2007a, b). Lack of phosphorus in the mantle matter and in the mandibles reveals that the preservation of the non-biomineralized soft tissue was not controlled by a phosphorus-rich medium. Contrary to the mantle and mandibles, the shell wall in *Austrotrachyceras* is calcareous; it is composed of O (65–80 %), C (25 %), Ca (8–16 %), Sb (2 %), Mo (2 %), Mg, K (each 1 %), and S, Si, Mn, Fe (each < 1 %) (Doguzhaeva et al. 2004, 2007b). The low oxygen depositional environment (Griffith 1977) responsible for carbonization of the mantle and the mandibles of *Austrotrachyceras* also favored the preservation of the muscular mantle, ink sacs and arm-hooks of the coleoid *Phragmoteuthis bisinuata* (Doguzhaeva et al. 2007b, Doguzhaeva and Summesberger 2012).

14.8.2 Fossilization of Mantle and Muscle Attachment Marks

Traces of the soft tissue attachment sites are often distinct on the inner surfaces of the body chamber wall in shells in aragonite preservation in such cases when the shell wall can be split from the internal mould (Doguzhaeva and Kabanov 1988; Doguzhaeva and Mutvei 1991). More commonly, these marks are preserved as imprints on pyritized internal moulds completely lacking shell wall (Kennedy et al. 2002; Richter 2002; Richter and Fischer 2002) or in aragonitic shells extracted from siderite concretions with a low content of Fe- minerals of the phragmocones and in posterior portions of body chambers. On internal moulds, the attachment marks can occur as small tubercles, shallow depressions or spots. These show often dark brownish or reddish colors due to pyrite or manganese hydroxides (Jordan 1968; Richter 2002; Richter and Fisher 2002; Doguzhaeva and Mutvei 2015). These differences corroborate that the inside of the body chamber wall originally carried attachment marks, which were either shallow pits, elevations or spots rich in organic material. The color differences probably resulted from the higher content of organic material in the attachment sites than in the rest of the shell wall. The surfaces of the attachment marks are often rough and bear pyritized micro-pits.

Currently, the best preservation of attachment marks is known from the body chambers of shells recovered from ankerite-siderite concretions in black clays (Doguzhaeva and Kabanov 1988; Doguzhaeva and Mutvei 1991, 1992, 1993; Doguzhaeva and Mikhailova 1991). For example, exceptionally well-preserved attachment marks were found in more than 50 shells from such concretions from the Lower Aptian of the Uljanovsk Region in Russia (Doguzhaeva and Kabanov 1988). These ammonoids display different ontogenetic growth stages. Multiple attachment marks were observed on phragmocone surfaces, interior surfaces of shells when the matrix was split from the ammonite, and especially on the internal mould of the body chambers of the conchs.

Attachment marks are typically observed in the shells or internal moulds from the environments favoring the genesis of rapidly deposited Fe-, S- or Mn-minerals. Additionally, the rapid post-mortem burial of the recently deceased ammonoid animal in the oxygen-deficient fine-grained sediments promoted the substitution of the organic-rich shell components with Fe- or Mn-minerals by anaerobic microbes. Attachment marks lacking relief but showing black, brownish or yellow coatings on the inside of the shell wall or on internal molds possibly appeared post mortem at the attachment sites; these can be distinguished from the rest of the shell wall by a higher organic content. The iridescent nacreous spots on the dark internal moulds of body chambers were formed due to secretion of additional portions of shell matter inside the weak impressions on the inside body chamber wall. The attachment marks forming low tubercles on the internal mould surfaces of the phragmocones, which rarely occur in body chambers, were formed by casting shallow pit-like attachment sites on the inside of the shell.

14.9 Conclusions

The data on variations of body chamber lengths (Appendix 1), traces of mantle and body chamber attachment (Appendix 2), muscular mantle preservation, and mode of preservation lead us to the following conclusions.

1. In the ontogeny of Paleozoic and Mesozoic ammonoids, the angular body chamber length may decrease (Miller et al. 1957) or increase (Doguzhaeva 1990); the first trend was documented in six and the second trend in seven genera from a total of 14 genera showing the body chamber length from the ammonitella to the adult stages; in one genus, the adequacy of measurements is insufficient (Appendix 1).
2. The gradual decrease of body chamber length at the ammonitella stage (Appendix 1) is an evolutionary trend characterizing the development of goniatitid–prolecanitid–ceratitid–phylloceratid lineage. It might indicate that the common early ontogenetic condition aimed at the achievement of a higher shell stability during early post-hatching stages; the replacement of this tendency with the contrary one in lytoceratids and ammonitids supposedly indicates an innovative early ontogenetic strategy of these two groups in the Jurassic.
3. A set of discrete dorsal, small umbilical and ventral marks along the phragmocone indicates that the mantle was regularly detached from the body chamber wall in each of these sites before it moved forward during growth to a new place of fixation. The discrete marks show a stepwise forward movement of the body within the growing body chamber and the band-shaped dorsal and ventral marks indicate continuing attachment and non-detached forward movement of the mantle. The overlap of discrete and band-shaped middorsal and midventral marks (Kennedy et al. 2002, pl. 5, Fig. 9; pl. 7, Fig. 5) possibly reflects the alternating weakening and reinforcement of the mantle and body chamber attachment along the middorsal and midventral areas of the body chamber wall. It is most likely that during the forward movement, the growing mantle was locally detached from the body chamber but remained attached in other sites. This partial detachment of the mantle presumably would produce differential extension of the mantle.
4. The dorsal internal ridge of clymeniids and the ventral internal ridge of all other ammonoids may have been the attachment structure between mantle and the siphuncular side of the body chamber wall.
5. In the ammonoids with a monomorph shell, there might have been forms combining a relatively short body (approximately equal to the body chamber length, or more) with well developed cephalic retractor and funnel retractor muscles. Those forms with a long body commonly had weakly developed cephalic retractor and funnel retractor muscles. With respect to higher- or lower-energy mode of locomotion (O'Dor et al 1990; Wells and O'Dor 1991), the former might have been more similar to the coleoids.

6. In the Early Cretaceous *Ancyloceras*, *Audouliceras*, *Pseudocrioceras*, *Ptychoceras,* and possibly also in the Late Cretaceous *Baculites,* in which the shell axis lies in the same plane throughout ontogeny, the body chamber bears enlarged tongue-like umbilical marks. In the first three genera, the terminal body chamber occupies a hook-like portion of shell with an estimated length of about 270°. The ventrolateral muscle scars are missing and the umbilical marks are the largest ones. The anterior end of these marks is located closer to the apertural portion of the body chamber. Therefore, the umbilical muscles were possibly the most powerful in these ammonoids and might have been designed for changing the mantle cavity volume to improve filter feeding on small plankton (Doguzhaeva 2014). This feeding mode is supported by the morphology of the buccal apparatus of the heteromorph ammonite *Australiceras*, which co-existed with *Audouliceras* and *Ancyloceras* (Doguzhaeva and Mikhailova 2002). In *Australiceras*, the jaw apparatus lacked a rostrum. Instead, it had a large cavity between the lower and upper jaws, which is formed by the deep and wide lower jaw with a solid flat ventral side, and the long wings of the upper jaw. Being the largest muscles, the umbilical muscles might have served for a regular water exchange of the mantle cavity that would be needed for respiration and filtering plankton. This suggests that these heteromorph ammonites perhaps fed on small plankton and were able to float as part of the plankton or performed vertical migrations.
7. In *Ptychoceras,* the terminal body chamber occupies the third shaft and approximately equally long portion of the second shaft. The extremely thin apertural edge of the terminal body chamber must have been protected against destruction by the mantle fold. This shows that the soft body must have been longer than the body chamber (Doguzhaeva and Mutvei 1989, 1993a, b, 2015).
8. The mantle tissue of some ammonoids was probably elastic and muscular as it was shown by exceptionally preserved specimens of the Late Triassic ceratitid *Austrotrachyceras* (Doguzhaeva et al 2004). Also, some ammonoid genera possessed powerful cephalic retractor and funnel retractor muscles while others did not. The variety of different body chamber lengths and mantle and muscle attachment marks probably indicates highly diverse life modes among ammonoids (Doguzhaeva 2014). A similar pattern is expressed in the diverse life modes of Recent shell-bearing and shell-less coleoids.

Further search for cephalopod shells preserving body chambers with weakly recrystallized shell material showing details of soft tissue attachment marks is a way towards better understanding of the paleobiology of cephalopods, including ammonoids.

Acknowledgements We would like to thank Christian Klug (University of Zurich) and Dieter Korn (Museum of Natural History, Berlin) for giving us an impetus to prepare this article and Neil H. Landman (New York) for his review.

Appendix 1

Table A.1 The angular body chamber length of Paleozoic and Mesozoic ammonoid genera (abbreviations of ages: D-Devonian; C-Carboniferous; P-Permian; T-Triassic; J-Jurassic; K-Cretaceous; the subscript numbers 1 to 3 in combination with D, T and J mean early, middle and late; the subscript numbers 1 and 2 in combination with C, P and K mean early and late)

Genus	Age	Body chamber length (angle in degrees)		Reference
		Ammonitella	Adult	
Goniatitida				
Tornoceras	D_3	360°		Ruzhencev 1962
Aulatornoceras	D_3		< 360°	Richter and Fischer 2002
Linguatornoceras	D_3		< 360°	Richter and Fischer 2002
Cheiloceras	D_3		< 360°	Richter and Fischer 2002
Maeneceras	D_3		ca. 360°	Bockwinkel et al. 2002
Neoglyphioceras	C_1	360°	>450°	Zakharov 1974, 1978
Glaphyrites	C_1	300–405°		Landman et al. 1996, 1999a
?*Rhadinites*	C_1		360°	Landman et al. 2010
?*Anthracoceras*	C_1		360°	Landman et al. 2010
Gyrtioceras	C_2	400°		Landman et al. 1996
Cravenoceras	C_2	370–380°		Landman et al. 1996
Gatherites	C_2	385°		Landman et al. 1996
Pseudogastrioceras	C_2	370°		Landman et al. 1996
Paragastrioceras	C_2	360°		Landman et al. 1996
Goniatites	C_2	345–385°		Landman et al. 1996; Kulicki et al. 2002
Akiyoshiceras	C_2		900°	Kyuma and Nishida 1987
Agathiceras	C_2–P_1	360°	>370°	Shulga-Nesterenko 1925 Arthaber 1928; Miller et al. 1957; Ruzhencev 1962
Peritrochia	C_2–P_1	340–410°		Bogoslovskaya 1959; Landman et al. 1996
Thalassoceras	C_2–P_1	380°		Bogoslovskaya 1959
Hyattoceras	P		>450°	Miller et al. 1957
undet. juv. goniatite	C_2–P_1	360°		Doguzhaeva 2002
Uraloceras	P_1		ca. 360°	Ruzhencev 1956
Waagenina	P_1		?>360°	Ruzhencev 1956
Marathonites	P_1		ca. 360°	Ruzhencev 1956
Kargalites	P_1		?>360°	Ruzhencev 1956
Clymeniida				
Genuclymenia	D_3		275–280°	Bogoslovsky 1976
Cyrtoclymenia	D_3		275°–280°	Bogoslovsky 1976
Platyclymenia	D_3			Bogoslovsky 1976
Prolecanitida				

Table A.1 (continued)

Genus	Age	Body chamber length (angle in degrees)		Reference
		Ammo-nitella	Adult	
Neopronorites	C_2–P_1	310–340°	230°	Bogoslovskaya 1959; Saunders and Work 1997
Metapronorites	C_2–P_1	340°	250°	Ruzhencev 1962; Saunders and Work 1997
Pronorites	P	330°		Landman et al. 1996
Akmilleria	P_1		218°	Saunders and Work 1997
Medlicottia	P_2	326–340°		Bogoslovskaya 1959; Ruzhencev 1962
Sakmarites	P_1	340°	225°	Bogoslovskaya 1959; Saunders and Work 1997
Pseudosageceras	T_1	295°		Zakharov 1978
Hedenstroemia	T_1	287–296°		Zakharov 1978
Ceratitida				
Xenodiscus	T_1	280°		Zakharov 1978
Kingites	T_1	347°		Zakharov 1978
?*Paranorites*	T_1	280°		Zakharov 1978
Prosphingites	T_1	265°	>360°	Zakharov 1978
Paranannites	T_1	238°		Zakharov 1978
Owenites	T_1	270°		Zakharov 1978
Svalbardiceras	T_1	260°		Zakharov 1978
Nordophiceras	T_1	260°		Zakharov 1978
Boreomeekoceras	T_1	295°		Zakharov 1978
Arctomeekoceras	T_1	280–285°		Zakharov 1978
Columbites	T_1	240°		Zakharov 1978
Sucolumbites	T_1	240°		Zakharov 1978
Palaeokazachstanites	T_1	?265°		Zakharov 1978
Olenekites	T_1	300–315°		Zakharov 1978
Parasibirites	T_1	265°		Zakharov 1978
Sibirites	T_1	260°		Zakharov 1978
Keyserlingites	T_1	300°		Zakharov 1978
Arctohungarites	T_1	229–330°		Zakharov 1978
Crumbergia	T_1	229–330°		Zakharov 1978
Phyllocladiscites	T_1	265°		Zakharov 1978
Phyllocladiscites	T_2	330°	360°	Doguzhaeva 1990
Megaphyllites	T_2	270–280°		Drushchits and Doguzhaeva 1981
Ceratites	T_2		?120°	Klug 2004
Proarcestes	T_2	?350°	>450°	Doguzhaeva 1990

Table A.1 (continued)

Genus	Age	Body chamber length (angle in degrees)		Reference
		Ammo-nitella	Adult	
Placites	T_3	268°		Shigeta and Weitschat 2004
Choristoceras	T_3	295°		Shigeta and Weitschat 2004
Phylloceratida				
Rhacophyllites	T_3	286°		Shigeta and Weitschat 2004
Partschiceras	J_2	260–280°		Drushchits and Doguzhaeva 1981
Holcophylloceras	J_2	270°		Drushchits and Doguzhaeva 1981
Phyllopachyceras	K_1	234–300°	ca. 360°	Zakharov 1978; Drushchits and Doguzhaeva 1981
Holcophylloceras	J_2	223–270°		Zakharov 1978; Drushchits and Doguzhaeva 1981
Salfeldiella	K_1	280°		Drushchits and Doguzhaeva 1981
Hypophylloceras	K_2	280°		Drushchits and Doguzhaeva 1981
Lytoceratida				
Eurystomiceras	J_2	290–300°	270°	Drushchits and Doguzhaeva 1981; Doguzhaeva 1990
Tetragonites	K_1	300–330°		Zakharov 1978; Drushchits and Doguzhaeva 1981
Gabbioceras	K_1	280–300°		Drushchits and Doguzhaeva 1981
Eogaudryceras	K_1	330°		Doguzhaeva et al. 2010
Parajaubertella	K_2	270°		Drushchits and Doguzhaeva 1981
Gaudryceras	K_2	330°		Zakharov 1978
Zelandites	K_2	320°–336°		Zakharov 1978
Tanabeceras	K_2		300°	Shigeta and Izukuda 2013
Ammonitida				
Psiloceras	J_1	285°		Shigeta and Weitschat 2004
Peronoceras	J_1		>360°	Keupp 2000
Asteroceras	J_1		ca. 180°	Keupp 2000
Hecticoceras	J_1		ca. 270°	Keupp 2000
Cleviceras	J_1		ca. 270°	Keupp 2000
Hildoceras	J_1		260°	Kröger 2002
Kepplerites	J_2		180°	Mitta and Starodubceva 2000
Pseudoneuqueniceras	J_2		330°	Matsuoka et al. 2010
Pelekodites	J_3		ca. 270°	Keupp 2000

Table A.1 (continued)

Genus	Age	Body chamber length (angle in degrees)		Reference
		Ammo-nitella	Adult	
Perisphinctes	J_3		370°	Kröger 2002
Dactylioceras	J_3		390°	Kröger 2002
Harpoceras	J_3		260°	Kröger 2002
Pseudolioceras	J_3		220°	Kröger 2002
Indosphinctes	J_3	330°	?330°	Ruzhencev 1962; Doguzhaeva 2012
Valanginites	K_1		>360°	Keupp 2000
Speetoniceras	K_1	300°		Drushchits and Doguzhaeva 1981
Craspedodiscus	K_1	300–315°		Drushchits and Doguzhaeva 1981
Parahoplites	K_1	305–330°		Drushchits and Doguzhaeva 1981
Aconeceras	K_1	270–365°	180°	Drushchits and Doguzhaeva 1981; Kulicki and Doguzhaeva 1994
Deshayesites	K_1	300–330°	180°	DLA, personal observation
Desmophyllites	K_1	307–332°		Zakharov 1978
Melchiorites	K_1	270–290°		Drushchits and Doguzhaeva 1981
Zurcherella	K_1	270–285°		Drushchits and Doguzhaeva 1981
Beudanticeras	K_1	330°		Drushchits and Doguzhaeva 1981
Damesites	K_1	307–309°		Zakharov 1978
Ancyloceras	K_1		180–270°	Doguzhaeva and Mikhailova 1991
Audouliceras	K_1		180–270°	Figs 14.1–14.3
Subpriomocyclus	K_2	242–348°		Landman et al. 1996
Discoscaphites	K_2		180°	Landman et al. 2007

Apendix 2

Table A.2 Record of mantle and body chamber attachment marks Abbreviations of ages: D-Devonian; C-Carboniferous; P-Permian; T-Triassic; J-Jurassic; K-Cretaceous; the subscript numbers 1 to 3 in combination with D, T and J mean early, middle and late; the subscript numbers 1 and 2 in combination with C, P and K mean early and late (for further abbreviations see Terminology and abbreviations)

Genera	Age	Type of attachment marks	References
Goniatitida			
Tornoceras	D	vl	Richter and Fischer 2002
Aulatornoceras	D	vl	Richter and Fischer 2002
Linguatornoceras	D	vl	Richter and Fischer 2002
Cheiloceras	D	vl	Richter and Fischer 2002
?*Rhadinites*	C_1	db	Landman et al. 2010
?*Anthracoceras*	C_1	db	Landman et al. 2010
Goniatites	C	us	Crick 1898; Tanabe et al. 1998
Muensteroceras	C	us	Crick 1898; Jordan 1968
Clymeniida			
gen. indet.	D_3	dr	(B.I. Bogoslovsky, pers. comm.)
Ceratitida			
Anagymnotoceras	T_2	d	Weitschat 1986
Amphipopanoceras	T_2	d	Weitschat 1986; Weitschat and Bandel 1991; Keupp 2000
Czekanowskites	T_2	d, ve	Weitschat et Bandel 1991; Dagys and Keupp 1998; Keupp 2000
Indigirites	T_2	d	Weitschat et Bandel 1991
Nathorstites	T_2	d	Weitschat et Bandel 1991
Aristoptychites	T_2	d	Weitschat et Bandel 1991
Arctoptychites	T_2	d	Weitschat et Bandel 1991
Sphaerocladiscites	T_2	d	Weitschat et Bandel 1991
Stolleyites	T_2	d	Weitschat et Bandel 1991
Paracladiscites	T_2	d	Weitschat et Bandel 1991
Ceratites	T_2	ae, us, v	Klug 2004; Klug et al.2007
Arctohungarites	T_2	ve	Dagys and Keupp 1998
Phylloceratida			
Tragophylloceras	J_1	us	Jordan 1968
Holcophylloceras	J_2	v	DLA, personal observation
Phyllopachyceras	K_1	v	Sharikadze et al. 1990
Euphylloceras	K_1	v	Sharikadze et al. 1990
Salfeldiella	K_1	v	Sharikadze et al. 1990
Lytoceratida			
Lytoceras	J_{1-2}	us, ls	Crick 1898; Jordan 1968; Rakús 1978

Table A.2 (continued)

Genera	Age	Type of attachment marks	References
Derolytoceras	J_1	us	Rakús 1978
Pachylytoceras	J_1	us	Jordan 1968
Pleurolytoceras	J_1	us	Jordan 1968
Pictetia	K_1	d	Sharikadze et al. 1990
Tetragonites	K_1	v	Sharikadze et al. 1990
Hemitetragonites	K_1	us	Sharikadze et al. 1990
Ptychoceras	K_1	ut, d, vr	Sharikadze et al. 1990; Kakabadze and Sharikadze 1993; Doguzhaeva and Mutvei 2015
Hamites	K_1	us	Crick 1898; Jordan 1968; Doguzhaeva and Mutvei 1996
Diplomoceras	K_2	us	Crick 1898; Jordan 1968
Turillites	K_2	us	Crick 1898; Jordan 1968
Holcoscaphites	K_2	us	Landman and Waage 1993
Ammonitida			
Alsatites	J_1	us	Crick 1898; Jordan 1968
Arietites	J_1	us,?v,?ae	Crick 1898; Jordan 1968
Asteroceras	J_1	us	Jordan 1968
Paroxynoticeras	J_1	us	Rakús 1978
Eoderoceras	J_1	d	Rakús 1978
Amaltheus	J_1	us, d, v, ls, vl, ae	Crick 1898; Jordan 1968; Richter and Fischer 2002
Androgynoceras	J_1	us,?v	Crick 1898
Amauroceras	J_1	us, d, vl	Landman et al. 1990b; Richter and Fischer 2002
Armatites	J_1	vl	Münster 1839; Richter and Fischer 2002
Arieticeras	J_1	us, d, vl	Jordan 1968; Richter and Fischer 2002
Oedania	J_1	us, v	Richter and Fischer 2002
Polymorphites	J_1	us, v	Richter and Fischer 2002
Stolleyites	J_1	d, vl	Richter and Fischer 2002
Grammoceras	J_1	ls	Jordan 1968
Dorsetensia	J_2	us, v	Jordan 1968; Doguzhaeva and Mutvei 1996
Ludwigia	J_2	us	Jordan 1968
Sonninia	J_2	us, v	Crick 1898; Jordan 1968
Staufenia	J_2	us, ae	Jordan 1968
Leioceras	J_2	us	Jordan 1968
Distichoceras	$J_{2\text{-}3}$	ls	Crick 1898; Jordan 1968

Table A.2 (continued)

Genera	Age	Type of attachment marks	References
Paroecotraustes	J_2	d	Jordan 1968
Hecticoceras	J_2	us, ls, vl, mg, ve	Crick 1898; Jordan 1968; Dagys and Keupp 1998, 2000; Richter and Fischer 2002
Creniceras	J_3	us, ae	Keupp 2000; Crick 1898; Jordan 1968
Clydoniceras	J_2	us,?v, ae	Crick 1898; Jordan 1968
Bullatimorphites	J_2	v	Jordan 1968
Quenstedtoceras	J_2	d, v	Jordan 1968; Landman and Bandel 1985; Sharikadze et al. 1990; Doguzhaeva and Mutvei 1991, 1996; Tanabe et al. 1998
Cardioceras	J_3	us,?v, ae	Crick 1898; Jordan 1968
Grossouvria	J_1	v	Jordan 1968
Siemiradzkia	J_2	v	Jordan 1968
Peltoceras	J_2	us	Crick 1898; Jordan 1968
Aspidoceras	J_3	us	Crick 1898; Jordan 1968
Virgatites	J_3	v	Doguzhaeva and Mutvei 1996
Indosphinctes	J_2	us, v, vb	Sharikadze et al. 1990; Doguzhaeva 1986, 2012
Dorsoplanites	J_3	us, d	DLA, personal observation
Kashpurites	J_3	v, l	Fig. 14.10
Polyptychites	K_1	v	Vogel 1959
Olcostephanus	K_1	v	Doguzhaeva and Mutvei 1996
Beudanticeras	K_1	v	Doguzhaeva and Mutvei 1996
Melchiorites	K_1	vr, us, ls	Sharikadze et al. 1990
Zurcherella	K_1	ls	Sharikadze et al. 1990
Pseudosilesites	K_1	us, v	Sharikadze et al. 1990
Amoeboceras	K_1	us, ae	Crick 1898; Jordan 1968
Aconeceras	K_1	d, us, ls, vl, v	Figs 14.8, 9; Doguzhaeva and Kabanov 1988; Sharikadze et al. 1990; Landman et al. 1999c
Deshayesites	K_1	us, ls, vl, v	Figs 14.5-7; Doguzhaeva and Kabanov, 1988; Sharikadze et al. 1990; Doguzhaeva and Mutvei 1991
Ancyloceras	K_1	ut, v	Crick 1898; Jordan 1968; Doguzhaeva and Mikhailova 1991; Richter and Fischer 2002
Audouliceras	K_1	ut, db, v, vb, vr. lms	Fig. 14.1–4
Epicheloniceras	K_1	us	Sharikadze et al. 1990
Parahoplites	K_1	us, v	Landman and Bandel 1985

Table A.2 (continued)

Genera	Age	Type of attachment marks	References
Colombiceras	K_1	us, v	Sharikadze et al. 1990
Acanthohoplites	K_1	us, v	Sharikadze et al. 1990
Hypacanthoplites	K_1	d, v	Landman and Bandel 1985; Sharikadze et al. 1990
Diadochoceras	K_1	v	DLA, personal observation
Nodosohoplites	K_1	v	DLA, personal unpubl. observation
Euhoplites	K_1	d	Landman and Bandel 1985
Aegocrioceras	K_1	us	Crick 1898; Jordan 1968
Pseudocrioceratite	K_1	ut, d	Sharikadze et al. 1990; DLA, personal observation
Crioceratites	K_1	d	Richter and Fischer 2002
Bochianites	K_1	us, d	Richter and Fischer 2002
Hauericeras	K_2	d	Obata et al. 1978
Sciphoceras	K_2	ae	Henderson 1984
Tissotia	K_2	?d	Crick 1898; Jordan 1968
Scaphites	K_2	us, d	Landman and Waage 1993
Baculites	K_2	ae, d, us, ut, vb	Crick 1898; Kennedy et al. 2002

References

Arthaber GV (1911) Die Trias von Albanian. Beitr Paläont Geol Öster-Ungarns und des Orients 24:169–277

Arthaber GV (1928) Ammonoidea Leiostraca aus der oberen Trias von Timor. 2. Nederl Timor Expeditie 1916 onder Leiding von Dr. H. A. Brouwer. Jarb Mijnw Nederl Oost-Indiës 55(2):1–174

Bandel K (1982) Morphologie und Bildung der frühontogenetischen Gehäuse bei conchiferen Mollusken. Facies 7:1–198

Bockwinkel J, Becker RT, Ebbighausen V (2002) Morphometry and taxonomy of lower Famennian Sporadoceratidae (Goniatitida). Abh Geol BA 57:279–297

Bogoslovskaya MF (1959) Internal shell structure of some Artinskian ammonoids. Paleontol Zh 1:49–57

Bogoslovsky BI (1976) Early ontogeny and origin of clymeniids. Palaeont Zh 2:41–51

Chamberlain JA (1980) The role of body extension in cephalopod locomotion. Palaeontology 23:445–461

Chamberlain JA (1993) Locomotion in ancient seas: constraint and opportunity in cephalopod adaptive design. Geobios Mem Spéc 15:49–61

Crick GC (1898) On the muscular attachment of the animal to its shell in some Cephalopoda (Ammonoidea). Trans Linn Soc London (2) Zool 7:71–113

Dagys AS, Keupp H (1998) Internal ventral keels in Triassic ceratid ammonoids: description and functional interpretation as muscle scars. [Interne Ventralleisten bei ceratitischen Ammonoiden der Trias: Beschreibung und funktionale Interpretation als Muskelansätze.]. Z Dt Geol Ges 149:81–89

Diener C (1916a) Einiges über Terminologie und Entwicklung der Lobenelemente in der Ammonitensutur. Centralbl Min Geol Paläont 23(8):553–568; 24:578–592

Diener C (1916b) Bemerkungen über die Inzisionen der Suturlinie als Grundlage einer natürlichen Klassifikation der Ammoniten. Centralbl Min Geol Paläont 15:374–381
Doguzhaeva LA (1981) The wrinkle layer in the shell of ammonoids. Paleont Zh 1:38-48 [In Russian]
Doguzhaeva LA (1986) How long did ammonites live? Science USSR 3:113–117
Doguzhaeva LA (1988) Siphuncular tube and septal necks in ammonoid evolution. In: Wiedmann J, Kullmann J (eds) Cephalopods—present and past. Schweizerbart, Stuttgart
Doguzhaeva LA (1990) Analysis of shell growth of ammonoids. Trudy Paleont. Inst. USSR 243:15-28 [In Russian]
Doguzhaeva LA (2002) Adolescent bactritoid, orthoceroid, ammonoid and coleoid shells from the upper Carboniferous and lower Permian of South Urals. Abh Geol B-A 57:9–55
Doguzhaeva LA (2012) Functional significance of parabolae, interpreted on the basis of shell morphology, ultrastructure and chemical analyses of the Callovian ammonite *Indosphinctes* (Ammonoidea: Perisphinctidae), Central Russia. Rev Paléobiol, Genéve, spec vol 11:89–101.
Doguzhaeva LA (2014) Muscle and mantle attachment marks as well as body chamber lengths indicative of diverse life styles of coexisted Aptian ammonites of the Russian Platform. In: Klug C, Fuchs D (eds) Intern Symp Cephalopods present and past in combination with the 5th Intern Symp Coleoid cephalopods through time Abstracts and program. Univ Zurich
Doguzhaeva LA, Kabanov GK (1988) Muscle scars in ammonoids. Dokl Akad Nauk USSR 301:210–212
Doguzhaeva LA, Mikhailova IA (1991) New data on muscle system of heteromorph ammonites. Dokl Akad Nauk USSR 318:981–984
Doguzhaeva LA, Mikhailova IA (2002) The jaw apparatus of the heteromorphic ammonite *Australiceras whitehouse*, 1926 (Mollusca: Cephalopoda) from the Aptian of the Volga Region. Dokl Akad Nauk USSR 382:38–40
Doguzhaeva LA, Mutvei H (1989) *Ptychoceras* - a heteromorphic lytoceratid with truncated shell and modified ultrastructure (Mollusca: Ammonoidea). Palaeontogr A 208:91–121
Doguzhaeva LA, Mutvei H (1990) Radula, aptychi and counter-aptychi in Cretaceous ammonite *Aconeceras*. Dokl Akad Nauk USSR 313:192–195
Doguzhaeva LA, Mutvei H (1991) Organization of the soft body in *Aconeceras* (Ammonitina), interpreted on the basis of shell morphology and muscle scars. Palaeontogr A 218:17–33
Doguzhaeva LA, Mutvei H (1992) Radula of Early Cretaceous ammonite *Aconeceras* (Mollusca: Cephalopoda). Palaeontogr A 223:167–177
Doguzhaeva LA, Mutvei H (1993a) Shell ultrastructure, muscle scars, and buccalapparatus in ammonoids. Geobios 15:111–119
Doguzhaeva LA, Mutvei H (1993b) Structural features in Cretaceous ammonoids indicative of semi-internal or internal shells. In: House MH (ed) The Ammonoidea Environment, Ecology, and Evolutionary Change. Syst Ass Special Vol 47: 99–114. Clarendon Press Oxford
Doguzhaeva LA, Mutvei H (1996) Attachment of the body to the shell in Ammonoids. In: Landman NH, Tanabe K, Davis RA (eds) Ammonoid Paleobiology, Topics in Geobiology 13:43–63. Plenum Press New York and London
Doguzhaeva LA, Mutvei H, Summesberger H & Dunca E (2004) Bituminous soft body tissues in Late Triassic ceratitid Austrotrachyceras Mitt Geol-Paläont Inst Univ Hamburg 88:37–50
Doguzhaeva LA, Mutvei H (2015) The additional external shell layers indicative of "endocochleate experiments" in some ammonoids. This volume
Doguzhaeva LA, Summesberger H (2012) Pro-ostraca of Triassic belemnoids (Cephalopoda) from Northern Calcareous Alps, with observations on their mode of preservation in an environment of northern Tethys which allowed for carbonization of non-biomineralized structures. N Jahrb Geol Paläont Abh 266:31–38
Doguzhaeva LA, Mapes RH, Mutvei H (2002) Early Carboniferous coleoid *Hematites* Flower & Gordon, 1959 (Hematitida ord. nov.) from Midcontinent (USA). Abh Geol B-A 57:299–320
Doguzhaeva LA, Mapes RH, Mutvei H (2003) The shell and ink sac morphology and ultrastructure of the Late Pennsylvanian cephalopod *Donovaniconus* and its phylogenetic significance. Berl Paläobiol Abh 3:61–78

Doguzhaeva LA, Mutvei H, Summesberger H, Dunca E (2004) Bituminous soft tissues in the body chamber of the Late Triassic ceratitid *Austrotrachyceras* from the Austrian Alps. Mitt Geol-Paläont Inst Univ Hamburg 88:37–50

Doguzhaeva LA, Summesberger H, Mutvei H (2006) An unique Upper Triassic coleoid from the Austrian Alps reveals pro-ostracum and mandible ultrastructure. Acta Univ Carolinae Geol 49:69–82

Doguzhaeva LA, Summesberger H, Mutvei H, Brandstaetter F (2007a) The mantle, ink sac, ink, arm hooks and soft body debris associated with the shells in Late Triassic coleoid cephalopod *Phragmoteuthis* from the Austrian Alps. Palaeoworld 16:272–284

Doguzhaeva LA, Mapes RH, Summesberger H, Mutvei H (2007b) The preservation of body tissues, shell, and mandibles in the ceratitid ammonoid *Austrotrachyceras* (Late Triassic), Austria. In: Landman NH, Davis RA, Mapes RH (eds) Cephalopods present and past new insights and fresh perspectives. Springer, Dordrecht

Doguzhaeva LA, Bengtson SM, Mutvei H (2010) Structural and morphological indicators of mode of life in the Aptian lytoceratid ammonoid *Eogaudryceras*. In: Tanabe K, Shigeta Y, Sasaki T, Hirano H (eds) Cephalopods—present and past. Tokai University Press, Tokyo

Drushchits VV, Doguzhaeva LA (1981) Ammonoids in electron microscope (internal shell structure and systematics of Mesozoic phylloceratids, lytoceratids and six families of the Early Cretaceous ammonitids). Moscow State University Press, Moscow

Drushchits VV, Doguzhaeva LA, Mikhailova IA (1978) Unusual coating layers in ammonoids. Paleont Zh 2:36–44 [In Russian]

Griffith J (1977) The Upper Triassic fishes from Polzberg bei Lunz. Zool J Linnean Soc 60:1–93

Griffin LE (1900) The anatomy of *Nautilus pompilius*. Mem Nat Acad Sci 8:101–230

Haug E (1898) Études sur les goniatites. Mém Soc Geol France, Paléont 7(4). Mém 18:1–112

Henderson RA (1984) A muscular attachment proposal for septal function in Mesozoic ammonites. Palaeontology 27:461–486

Hewitt RA, Checa A, Westermann GEG, Zaborski PMP (1991) Chamber growth in ammonites inferred from colour markings and naturally etched surfaces of Cretaceous vascoceratids from Nigeria. Lethaia 24:271–284

Hoffmann R (2010) New insights on the phylogeny of the Lytoceratoidea (Ammonitina) from the septal lobe and its functional interpretation. Rev Paléobiol 29(1):1–156

Ivanov AN (1975) Late ontogeny in ammonoids, and its specific features in macro-, micro- and megaconchs. Scientific notes Yaroslavl Pedagogical Inst, Geol and Paleont 87:5–57 [In Russian]

Jacobs DK, Chamberlain JA Jr (1996) Buoyancy and hydrodynamics in ammonoids. In: Landman NH, Tanabe K, Davis RA (eds) Ammonoid Paleobiology, Topics in Geobiology 13:169–224

Jones DL (1961) Muscle attachment impressions in a Cretaceous ammonite. J Paleont 35:502–504

Jordan R (1968) Zur Anatomie mesozoischer Ammoniten nach den Strukturelementen der Gehäuse-Innenwand. Beih Geol Jahrb 77:1–64

Kakabadze MV, Sharikadze MZ (1993) On the mode of life of hetemmorph ammonites (heterocone, ancylocone, ptychocone). Geobios 15:209–215

Kar AJ, Briggs DEG, Donovan DT (1995) Decay and fossilization of non-mineralized tissue in coleoid cephalopods. Palaeontology 38:105–131

Kennedy WJ, Cobban WA (1976) Apects of ammonite biology, biogeography, and biostratigraphy. Spec Pap Palaeon 17:1–64

Kennedy WJ, Cobban WA, Klinger HC (2002) Muscle attachment and mantle-related features in Upper Cretaceous *Baculites* from the United States Western Interior. Abh Geol B-A 57:89–112

Keupp H (2000) Ammoniten. Paläobiologische Erfolgsspiralen. Thorbecke, Stuttgart

Kier WM (1988) The arrangement and function of molluscan muscle. In: Trueman ER, Clarke MR (eds) The Mollusca. vol 11. Form and Function. Academies Press, San Diego

Kier WM, Thompson JT (2003) Muscle arrangement, function and specialization in Recent coleoids. In: Warnke K, Keupp H, Boletzky S v (eds) Coleoid cephalopods through time. Berl Paläobiol Abh 3:141–162

Klug C (2004) Mature modifications, the black band, the black aperture, the black stripe, and the periostracum in cephalopods from the Upper Muschelkalk (Middle Triassic, Germany). Mitt Geol-Paläont Inst Univ Hamburg 88:63–78

Klug C, Korn D, Richter U, Urlichs M (2004) The black layer in cephalopods from the German Muschelkalk (Middle Triassic). Palaeontolology 47:1407–1425

Klug C, Montenari M, Schultz H, Ulrichs M (2007) Soft-tissue attachment of Middle Triassic Ceratitida from Germany. In: Landman NH, Davis RA, Mapes RH (eds) Cephalopods present and past. New Insights and fresh perspectives. 6th intern symp cephalopods—present and past. Springer, Dordrecht

Klug C, Meyer E, Richter U, Korn D (2008) Soft-tissue imprints in fossil and Recent cephalopod septa and septum formation. Lethaia 41:477–492

Korn D, Klug C (2007) Conch form analysis, variability, morphological disparity, and mode of life of the Frasnian (Late Devonian) ammonoid *Manticoceras* from Coumiac (Montagne Noire, France). In: Landman NH, Davis RA, Mapes RH (eds) Cephalopods—present and past. New insights and fresh perspectives. 6th International Symposium. Springer, Dordrecht

Kröger B (2002) On the ability of withdrawing of some Jurassic ammonoids. Abh Geol B-A 57:199–204

Kyuma Y, Nishida T (1987) *Akiyoshiceras*, a new neoicoceratid ammonoid genus from the Upper Carboniferous of Akiyoshi. Bull Akiyoshi-dai Mus Nat Hist 22:23–41

Kulicki C, Doguzhaeva LA (1994) Development and calcification of the ammonitella shell. Acta Palaeontol Pol 39:17–44

Kulicki C, Landman NH, HeaneyMJ, Mapes RH, Tanabe K (2002) Morphology of the early whorls of *Goniatites* from the Carboniferous Buckhorn Asphalt (Oklahoma) with aragonitic preservation. Abh Geol B-A 57:205–224

Landman NH, Bandel K (1985) Internal structures in the early whorls of Mesozoic ammonites. Am Mus Novit 2823:1–21

Landman NH, Waage KM (1993) Scaphitid ammonites of the Upper Cretaceous (Maastrichtian) Fox Hills formaion in South Dakota and Wyoming. Bull Amer Mus Nat Hist 215:1–257

Landman NH, Tanabe K, Shigeta Y (1996) Ammonoid embryonic development. In: Landman NH, Tanabe K, Davis RA (eds) Ammonoid Paleobiology, Topics in Geobiology 13:343–405

Landman NH, Mapes RH, Tanabe K (1999a) Internal features of the embryonic shells of Late Carboniferous Goniatitina. In: Oloriz F, Rodriguez-Tovar FJ (eds) Advancing research on living and fossil cephalopods. Kluwer Academic Press, New York

Landman NH, Lane J, Cobban WA, Jorgensen SD, Kennedy WJ, Larson NL (1999b) Impressions of the attachment of the soft body to the shell in Late Cretaceous pachydiscid ammonites from the Western Interior of the United States. Amer Mus Nov 3273:1–31

Landman NH, Polizzotto K, Mapes RH, Tanabe K (2006) Cameral membranes in prolecanitid and goniatitid ammonoids from the Permian Arcturus Formation, Nevada, USA. Lethaia 39:365–379

Landman NH, Johnson RO, Garb MP, Edwards LE, Kyte FT (2007) Cephalopods from the Cretaceous/Tertiary boundary interval on the Atlantic coastal plain, with a description of the highest ammonite zones in North America. Part 3. Manasquan River Basin, Monmouth country, New Jersey. Bull Amer Mus Nat Hist 303:1–122

Landman NH, Mapes RH, Cruz C (2010) Jaws and soft tissues in ammonoids from the Lower Carboniferous (Upper Mississippian) Bear Gulch Beds, Montana, USA. In: Tanabe K, Shigeta Y, Sasaki T, Hirano H (eds) Cephalopods—present and past. Tokai University Press, Tokyo

Matsuoka A, Anso J, Nakada K, Terabe K, Sato T (2010) Biometrical analysis on primary rib number of the middle Jurassic ammonoids *Pseudoneuqueniceras yokoyamai* (Kobayashi & Fukuda) and its allied forms. In: Tanabe K, Shigeta Y, Sasaki T, Hirano H (eds) Cephalopods—present and past. Tokai University Press, Tokyo

Mitta VV, Starodubceva IA (2000) Stchirowsky and study of the Mesozoic in Alatyr-Kurmush area (Basin of the Middle Volga). VernadskyMus-Nov 5:1–20

Monks N, Young JR (1998) Body position and the functional morphology of Cretaceous heteromorph ammonites. Palaeontol Electronica 1.1.1A. http://palaeo-electronica.org/1998_1/monks/text.pdf

Münster GGz (1839) Beiträge zur Petrefaktenkunde mit XVIII nach der Natur gezeichneten Tafeln der Herren Hermann v. Meyer und Professor Rudolph Wagner. Buchner'sche Buchhandlung, Bayreuth

Mutvei H (1957) On the relations of the principal muscles to the shell in *Nautilus* and some fossil nautiloids. Arkiv Mineral Geol 2:219–254

Mutvei H (1964) Remarks on the anatomy of Recent and fossil Cephalopoda. Stockholm Contr Geol 11:79–102

Mutvei H (2013) Characterization of nautiloid orders Ellesmerocerida, Oncocerida, Tarphycerida, Discosorida and Ascocerida: new superorder Multiceratoidea. GFF 135:171–183. doi: 10.1080/11035897.2013.801034

Mutvei H, Doguzhaeva LA (1997) Shell ultrastructure and ontogenetic growth in *Nautilus pompilius* L. (Mollusca: Cephalopoda). Palaeontogr A 246:33–52

Mutvei H, Dunca E (2007) Connecting ring ultrastructure in the Jurassic ammonoid *Quenstedtoceras* with discussion on mode of life of ammonoids. In: Landman NH, Davis RA, Mapes RH (eds) Cephalopods present and past. New insights and fresh perspectives. Springer, Dordrecht

Mutvei H, Reyment RA (1973) Buoyancy control and siphuncle function in ammonoids. Palaeontology 16:623–636

Mutvei H, Arnold JM, Landman NH (1993) Muscles and attachment of the body to the shell in embryos and adults of *Nautilus belauensis* (Cephalopoda). Amer Mus Nov 3059:1–15

Obata I, Futakami M, Kawashita Y, Takahashi T (1978) Apertural features in some Cretaceous ammonites from Hokkaido. Bull Nat Sci Mus C 4:139–155

O'Dor RK, Forsythe J, Webber DM, Wells J, Wells MJ (1990) Activity levels of *Nautilus* in the wild. Nature 362:626–627

Rakús M (1978) Sur l'existence de deux types distincts d'empreintes de muscles rétracteurs chez les ammonites. Bull Soc Vaudoise Sci Nat 354(74):139–145

Richter U (2002) Gewebeansatz-Strukturen auf Steinkernen von Ammonoideen. Geol Beitr Hannover 4:1–113

Richter U, Fischer R (2002) Soft tissue attachment structures on pyritized internal moulds of ammonoids. Abh Geol B-A 57:139–149

Ruzhencev VE (1956) Early Permian ammonoids from south Urals. II. The ammonoids of the Artinskya Stage. Trudy Paleont Inst USSR 60:1–275 [in Russian]

Ruzhencev VE (1962) General Part. In: Ruzencev VE (ed) Superorder Ammonoidea. Mollusca, Cephalopoda I. Osnovy Paleontologii. USSR Academy of Sciences, Moscow [In Russian]

Sasaki T, Shigeno S, Tanabe K (2010) Anatomy of living *Nautilus*: Reevaluation of primitiveness and comparison with Coleoidea. In: Tanabe K, Shigeta Y, Sasaki T, Hirano H (eds) Cephalopods—present and past. Tokai University Press, Tokyo

Saunders WB, Shapiro EA (1986) Calculation and simulation of ammonoid hydrostatics. Paleobiol 12:64–79

Saunders WB, Work DM (1997) Evolution of shell morphology and sutures complexity in Paleozoic prolecanitids, the rootstock of Mesozoic ammonoids. Paleobiol 23:301–325

Sharikadze MZ, Lominadze TA, Kvantaliani IV (1990) Systematische Bedeutung von Muskelabdrucken spätjurassischer Ammonoideae. Z geol Wiss 18:1031–1039

Shigeta Y, Weitschat W (2004) Origin of the Ammonitina (Ammonoidea) inferred from the internal shell structures. Mitt Geol-Paläontol Inst Univ Hamburg 88:179–194

Shigeta Y, Izukura M (2013) The earliest Cenomanian ammonoid *Tanabeceras yezoense* (Shigeta) from the Hobetsu area, Hokkaido. Bull Hobetsu Mus 28:1–6

Shulga-Nesterenko MI (1925) Internal shell structure of the Artinskian ammonites. Bull Moskovskogo obsh'estva ispytatelej prirody. Otd Geol 4(1–2):81–100

Stenzel HB (1964) Living *Nautilus*. In: Moore R (ed) Treatise on invertebrate paleontology, Mollusca 3: K59-K93. Geological Society and University of Kansas Press, Lawrence, KS

Suess E (1865) Über Ammoniten. Sitzungsber kaiserl Akad Wiss, Wien 52:3–19

Tanabe K, Landman NH, Mapes RH (1998) Muscle attachment scars in a Carboniferous goniatite. Paleont Res 2(2):130–136

Trueman AE (1941) The ammonite body-chamber, with special reference to the buoyancy and mode of life of the living ammonite. Q J Geol Soc 384:339–383

Vogel KP (1959) Zwergwuchs bei Polyptychiten (Ammonoidea). Geol Jb 76:469–540

Weitschat W (1986) Phosphatisierte Ammonoideen aus der Mittleren Trias von Central-Spitzbergen. Mitt Geol-Paläont Inst Univ Hamburg 61:249–279

Weitschat W, Bamdel K (1991) Organic components in phragmocones of Boreal Triassic ammonoids: implications for ammonoid biology. Paläontol Z 65:269–303

Wells MJ (1988) The mantle muscle and mantle cavity in cephalopods. In: Truman ER, Clarke MR (eds) The Mollusca, form and function 11. Academies Press, London

Wells MJ, O'Dor RK (1991) Jet propulsion and the evolution of cephalopods. Bull Mat Sci 49:419–432

Westermann GEG (1996) Ammonoid life and habitat. In: Landman NH, Tanabe K, Davis RA (eds) Ammonoid Paleobiology. Topics in Geobiology 13: 607–707. Springer US

Westermann GEG, Tsujita J (1999) Life habits of ammonoids. In: Savazzi E (ed) Morphology of invertebrate skeleton. Wiley, Chichester

Zaborski PMP (1986) Internal mould markings in a Cretaceous ammonite from Nigeria. Palaeontology 29:725–738

Zakharov JD (1974) New data on internal shell structures in Carboniferous, Triassic and Cretaceous ammonoids. Paleontol Zh 8(1):25–36

Zakharov JD (1978) Lower Triassic ammonoids of East USSR. Nauka, Moscow

Chapter 15
The Additional External Shell Layers Indicative of "Endocochleate Experiments" in Some Ammonoids

Larisa A. Doguzhaeva and Harry Mutvei

15.1 Introduction

In ectocochleate cephalopods, the shell wall typically consists of the outer prismatic, nacreous and inner prismatic layers; this standard shell wall pattern is well illustrated by some orthoconic shells from the Pennsylvanian (Late Carboniferous) Buckhorn Asphalt of the USA (Erben et al. 1969; Ristedt 1971; Mutvei 1972; Blind 1988) and the Early Cretaceous orthocerid shell from the north-western Caucasus (Doguzhaeva 1994). The body chamber may be lined with the wrinkle layer, which, coated the surface of a segment of the adjacent whorl beyond the shell aperture (Tozer 1972; Doguzhaeva 1980, 1981; Korn 2000; Kulicki et al. 2001; Klug et al. 2007). In ammonoids, the standard shell wall pattern was usually reinstalled during the repair of after shell damage (Doguzhaeva et al. 2010; Doguzhaeva 2012). In the shells secreted in an unfavourable environment, the shell wall shows pathological disturbances, particularly within the nacreous layer, but no aberrant shell wall structures (Doguzhaeva 2002). The aberrant shell wall pattern formed by adding extra prismatic and nacreous layers to the inside of the main shell wall; this was revealed in Jurassic ammonites of the family Dactylioceratidae (Guex 1970; Howarth 1975). This additional portion of shell wall "*lines the whole of the lateral and ventral parts of the main shell, while its front edge is 1/8–1/2 whorl behind the mouth border and is similar in shape*" (Howarth 1975, p. 47). The aberrant shell wall pattern, characterized by extra layers on the surface of the standard shell wall was recorded in the Late Cretaceous lytoceratid *Gaudryceras tenuiliratum* Yabe 1903 (Figs 15.1, 15.2; Drushchits et al. 1978; Birkelund 1981; Kulicki et al. 2001),

L. A. Doguzhaeva (✉) · H. Mutvei
Department of Palaeobiology, Swedish Museum of Natural History, Stockholm, Sweden
e-mail: larisa.doguzhaeva@nrm.se

H. Mutvei
e-mail: harry.mutvei@nrm.se

C. Klug et al. (eds.), *Ammonoid Paleobiology: From Anatomy to Ecology*,
Topics in Geobiology 43, DOI 10.1007/978-94-017-9630-9_15

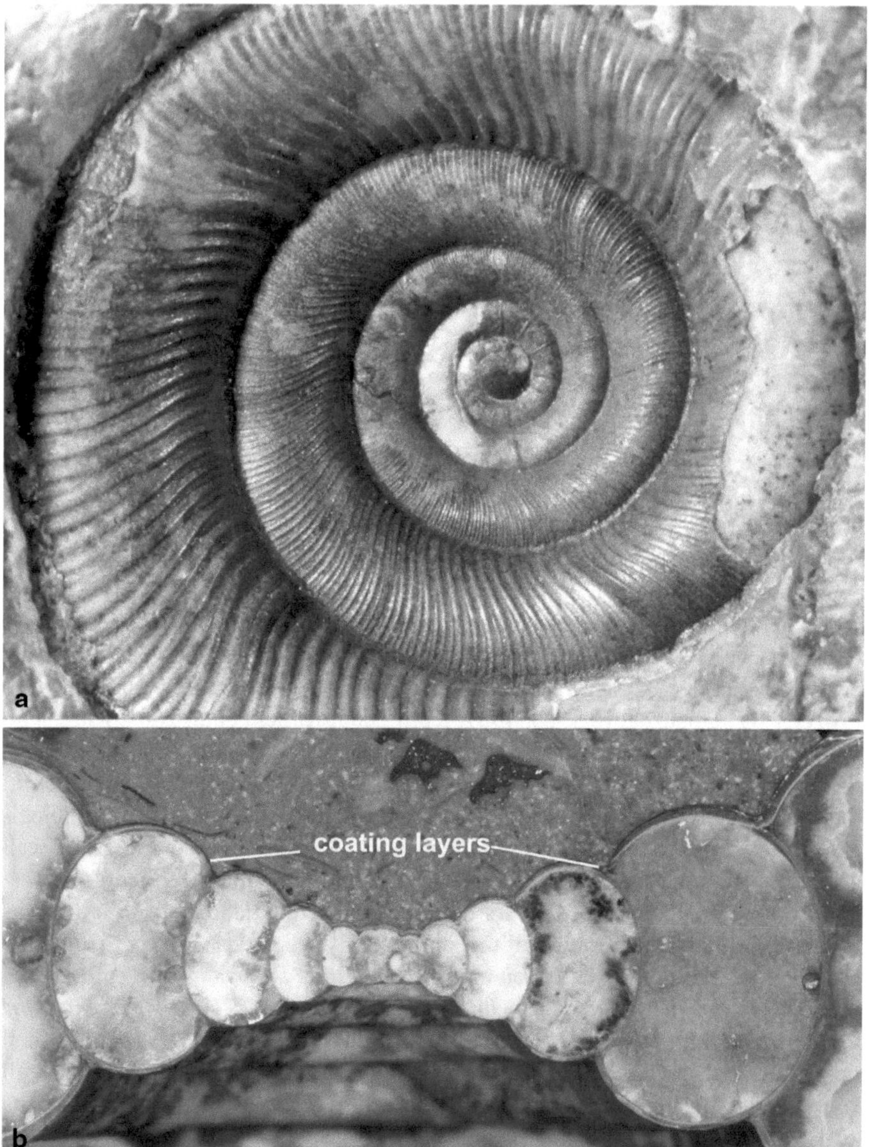

Fig. 15.1 *Gaudryceras tenuiliratum*. Turonian-Campanian, Upper Cretaceous; island of Sakhalin, Russia. **a** NRM–PZ Mo 167766, ribbed shell with a broad shallow umbilicus, x 9. **b** NRM–PZ Mo 167767, section of the shell through the protoconch; x 7

the Late Jurassic perisphinctid *Indosphinctes (Elatmites) submutatus* Nikitin 1881 (Figs 15.3, 15.4; Doguzhaeva 2012), the Early Cretaceous ammonitid *Aconeceras*

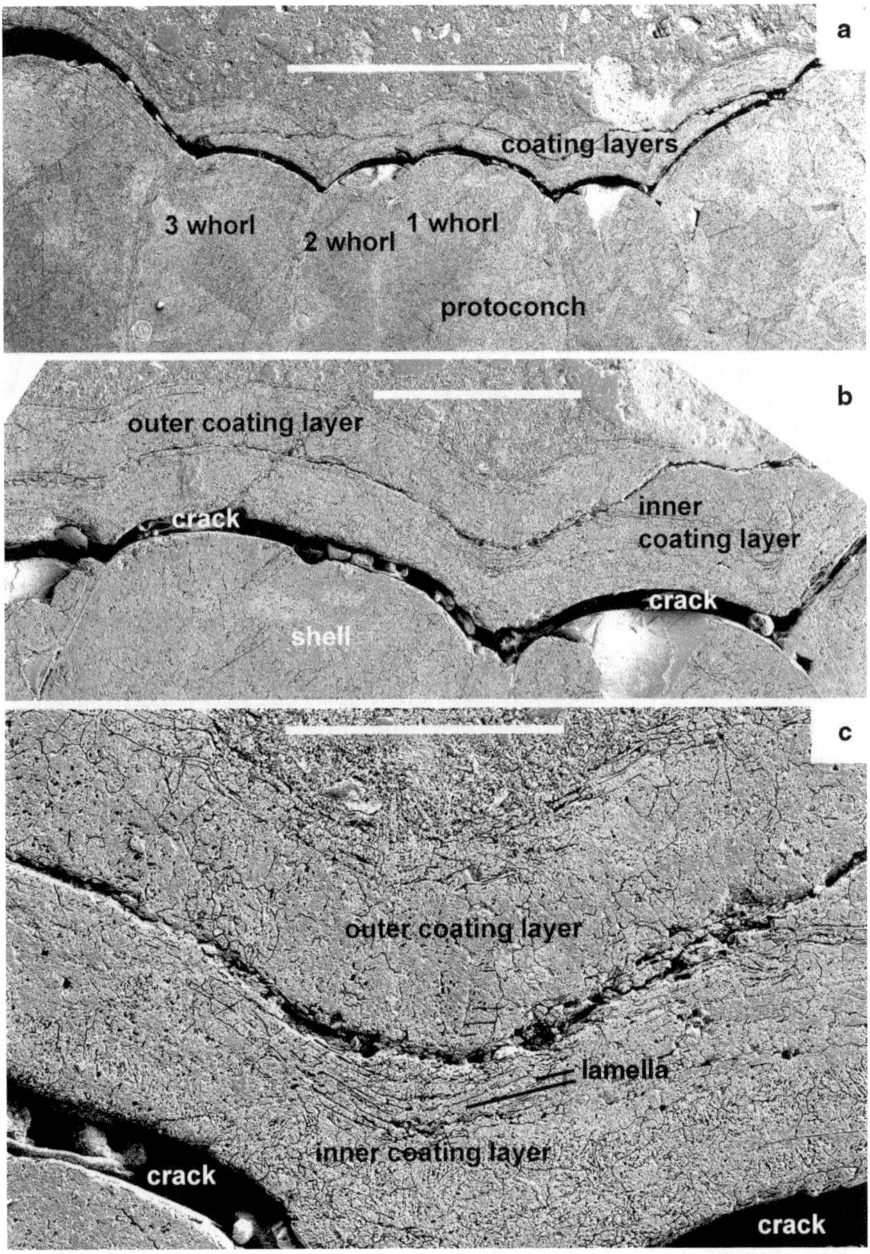

Fig. 15.2 *Gaudryceras tenuiliratum*. NRM–PZ Mo. 167767. Turonian-Campanian, Upper Cretaceous; island of Sakhalin, Russia. **a** cross shell section through the protoconch to show two thick coating layers on the surface of the protoconch and first three whorls; scale bar = 1.0 mm. **b**, **c** close-up of (**a**); scale bars = 100 μm and 200 μm, respectively

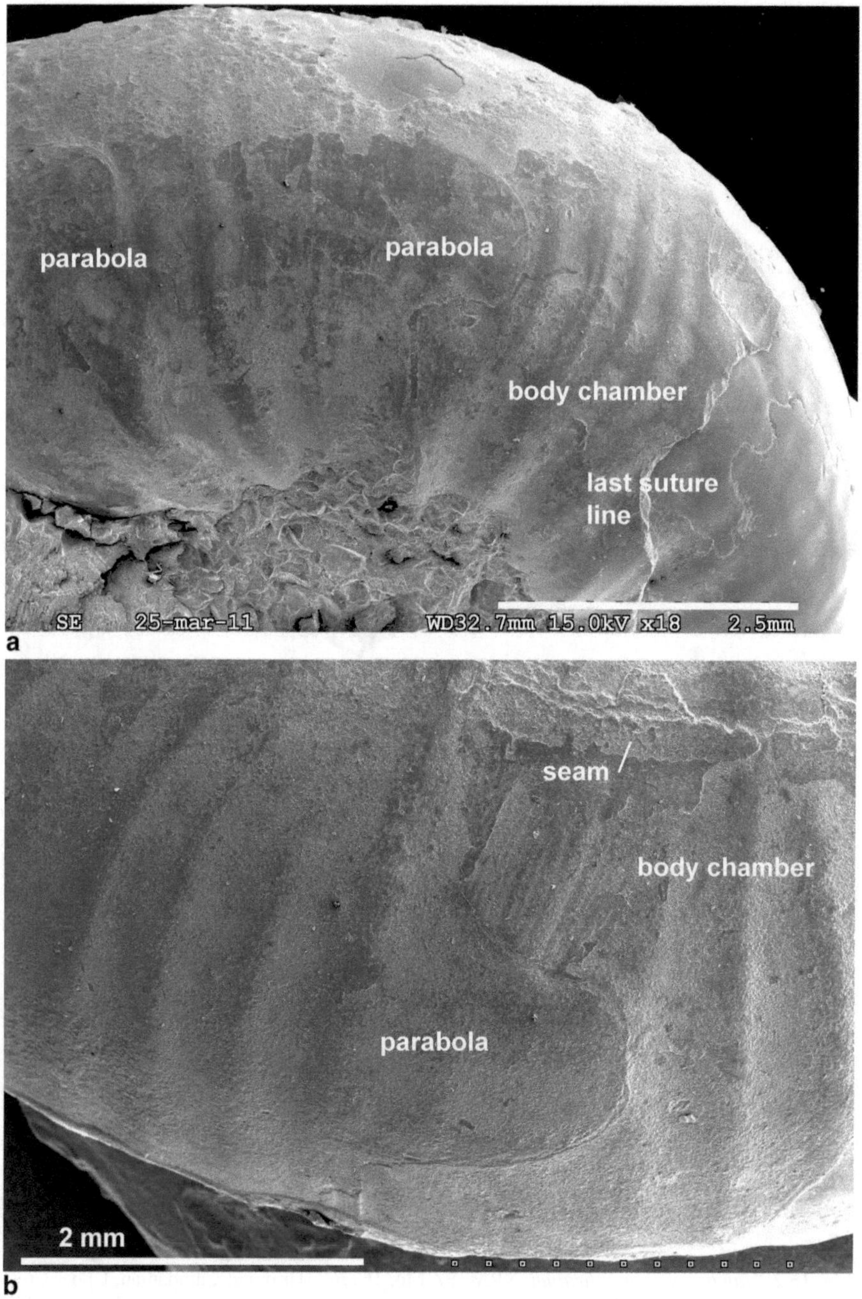

Fig. 15.3 *Indosphinctes (E.) mutatus*. Callovian, Middle Jurassic; Ryazan Region, Central Russia. Two shells showing parabolae. **a** NRM–PZ Mo. 167641; a parabola on the external surface of the body chamber; the last suture line is exposed where the shell wall is broken. **b** NRM–PZ Mo. 167640; the seam on the top of the image shows that the last whorl was broken; the parabola is therefore exposed on the dorsal wall here

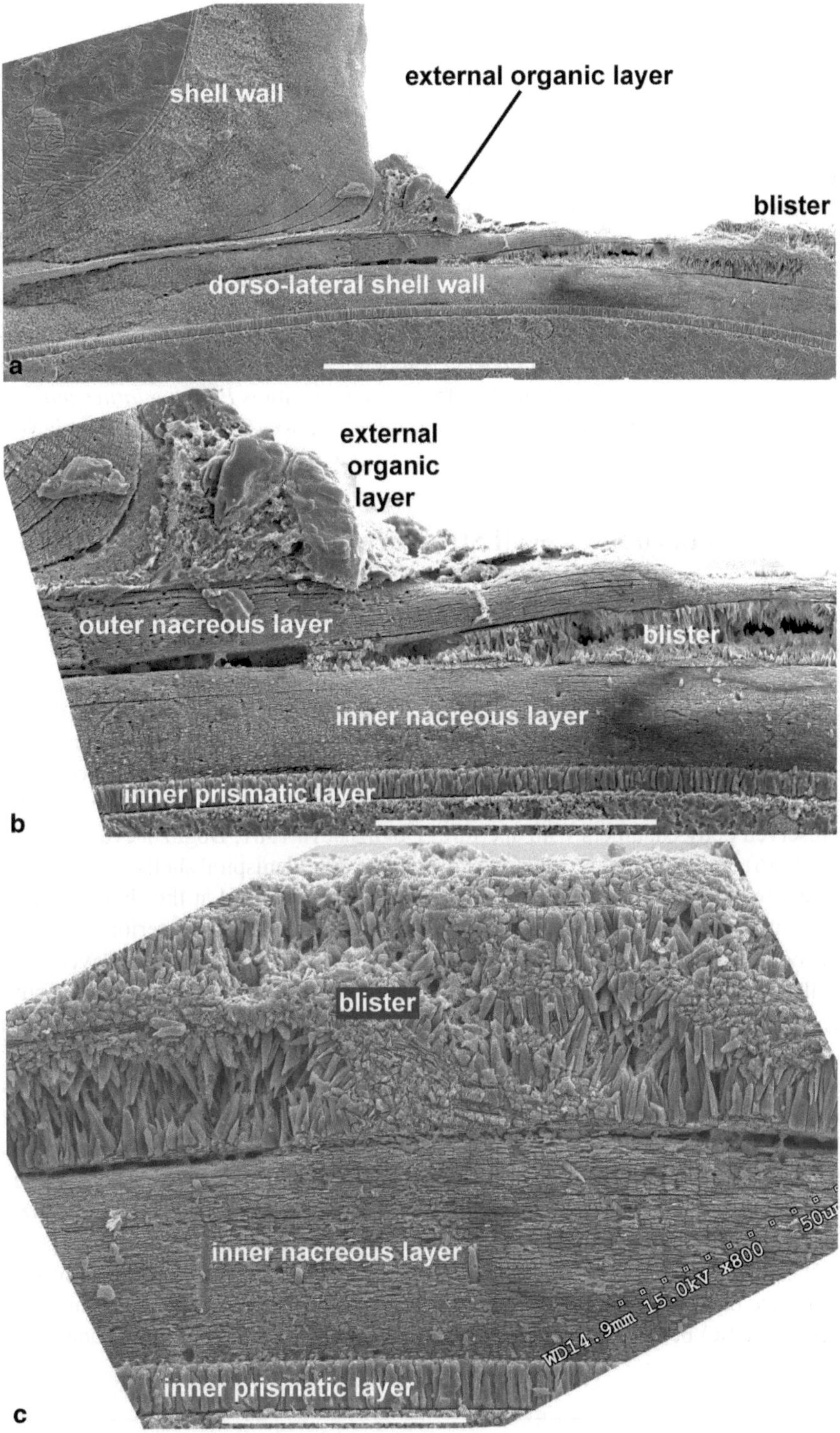

Fig. 15.4 *Indosphinctes (E.) mutatus*. NRM–PZ Mo. 167637. Callovian, Middle Jurassic; Ryazan Region, Central Russia. **a** cross section of the shell to show the dorso–lateral shell wall with remnants of the external organic layer, and two nacreous layers with a lens like blister; scale bar

trautscholdi (Sinzow 1870) (Figs 15.5-15.7; Doguzhaeva and Mutvei 1991), and the heteromorphs *Ptychoceras renngarteni* Egoian 1969, *P. levigatum* Egoian 1969 and *P. parvum* Egoian 1969 (Figs 15.8-15.10, 15.12; Doguzhaeva and Mutvei 1989). The aberrant shell wall structures, characterized by extra layers on the surface of the standard shell wall, supposedly indicate the "*endocochleate experiments*". This expression was introduced by Turek and Manda (2012) and applied by these authors for the Silurian straight-shelled cephalopod *Sphooceras truncatum* (Barrande 1860). The relevant data on the umbilical membrane of the Devonian ammonoid *Prolobites* Karpinsky 1885 and the helicolateral deposits of the Carboniferous goniatite *Clystoceras globosum* Nassichuk 1967 are discussed as well as the wrinkle layer on the external shell surface of the goniatites *Platygoniatites molaris* Ruzhencev 1956 (Carboniferous) and *Erfoudites rugosus* Korn 2000 (Devonian).

15.2 Standard Shell Wall Structure of Ammonoids

The embryonic shell or the ammonitella (term introduced by Drushchits and Khiami 1970) has a prismatic shell wall; it fades out at the nacreous primary varix underlying the aperture of the ammonitella (Drushchits and Doguzhaeva 1974, 1981; Howarth 1975; Kulicki and Doguzhaeva 1994; Kulicki 1996). The shell wall layers of the hatchlings lie on the inner surface of the primary varix (Drushchits and Doguzhaeva 1974; Howarth 1975; Birkelund 1981). The secreting sequence of the shell wall layers in adult shells has been established by tracing the layers in fully preserved body chambers with aperture (Doguzhaeva 1981; Doguzhaeva and Mutvei 1986). In the ventral and lateral walls of the adult planispiral shells, three of four layers (outer prismatic, nacreous and wrinkle) were secreted at the shell aperture; the fourth layer, the inner prismatic layer, was secreted in the posterior portion of the body chamber where it covers the wrinkle layer. The inner prismatic layer gradually wedged out in the anterior direction. In phylloceratids and lytoceratids it covers a short posterior portion of the body chamber approximately equal to the length of the last two or three chambers. In ammonitids this layer covers a longer posterior portion of the body chamber equal to half of the body chamber length (Drushchits and Doguzhaeva 1974; Drushchits et al. 1978; Doguzhaeva 1981; Doguzhaeva and Mutvei 1986). The wrinkle layer was secreted on the inner surface of the nacreous layer in the body chamber and also on the surface of the adjacent whorl outside the body chamber (Ruzhencev 1962; Walliser 1970; House 1971; Senior 1971; Tozer 1972; Kulicki 1979; Doguzhaeva 1981; Doguzhaeva and Mutvei 1986; Korn 2000; Kulicki et al. 2001). The adjacent shell wall of the consecutive whorls shows a sequence of six layers: four of these layers (outer prismatic, nacreous, wrinkle, inner prismatic) belong to the ventral wall of previous whorl, and the additional two

=200 μm. **b** close-up of (**a**) to show remnants of the external organic layer and lens-like blister that corresponds to the parabolic notch; scale bar =100 μm. **c** close-up of (**a**) to show the structure of the blister composed of loosely mineralized prisms and spherulites; scale bar =50 μm

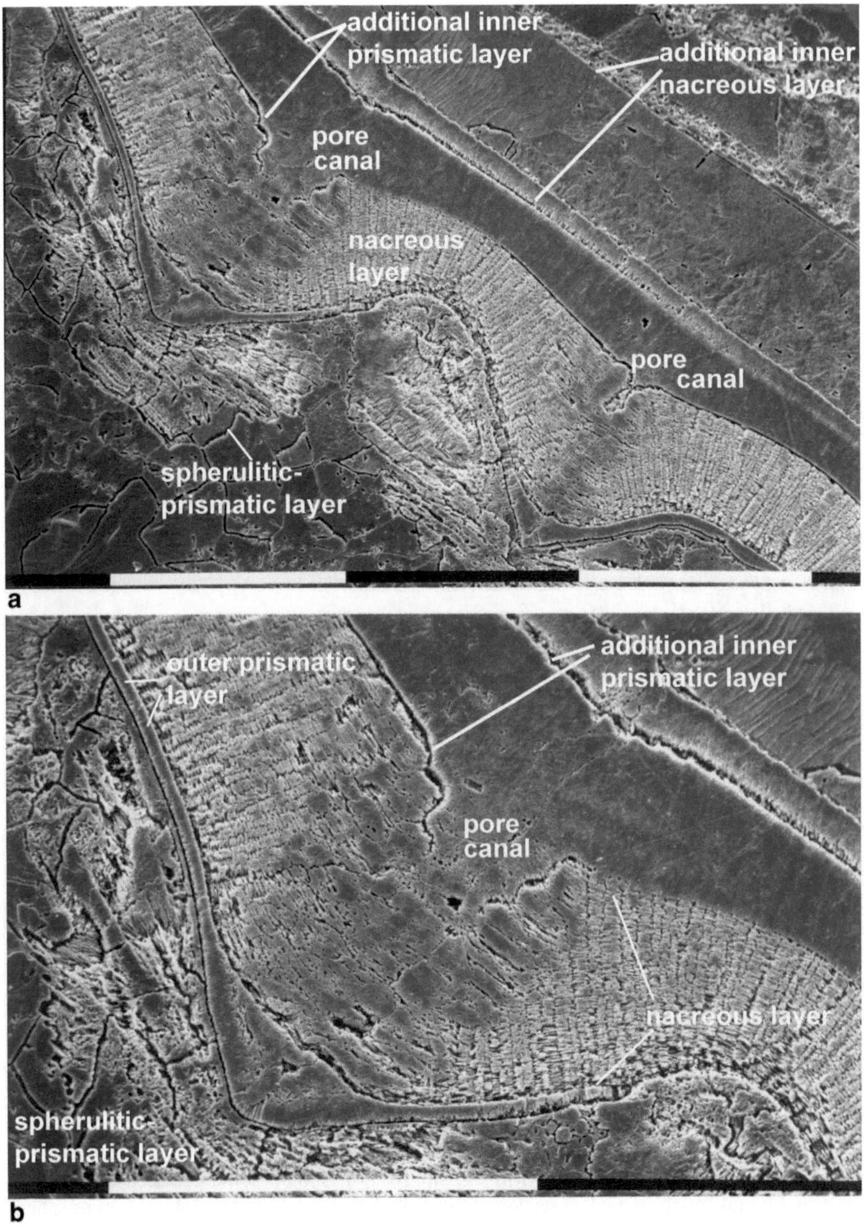

Fig. 15.5 *Aconeceras trautscholdi*. NRM–PZ Mo. 167769. Aptian, Lower Cretaceous; Ulyanovsk Region, Central Russia. **a** cross section of the shell to show the additional outer and inner prismatic layers in the shell wall, and the spherulitic-prismatic layer around the keel. **b** close-up of (**a**) to show irregularly mineralized, porous spherulitic-prismatic layer around the keel

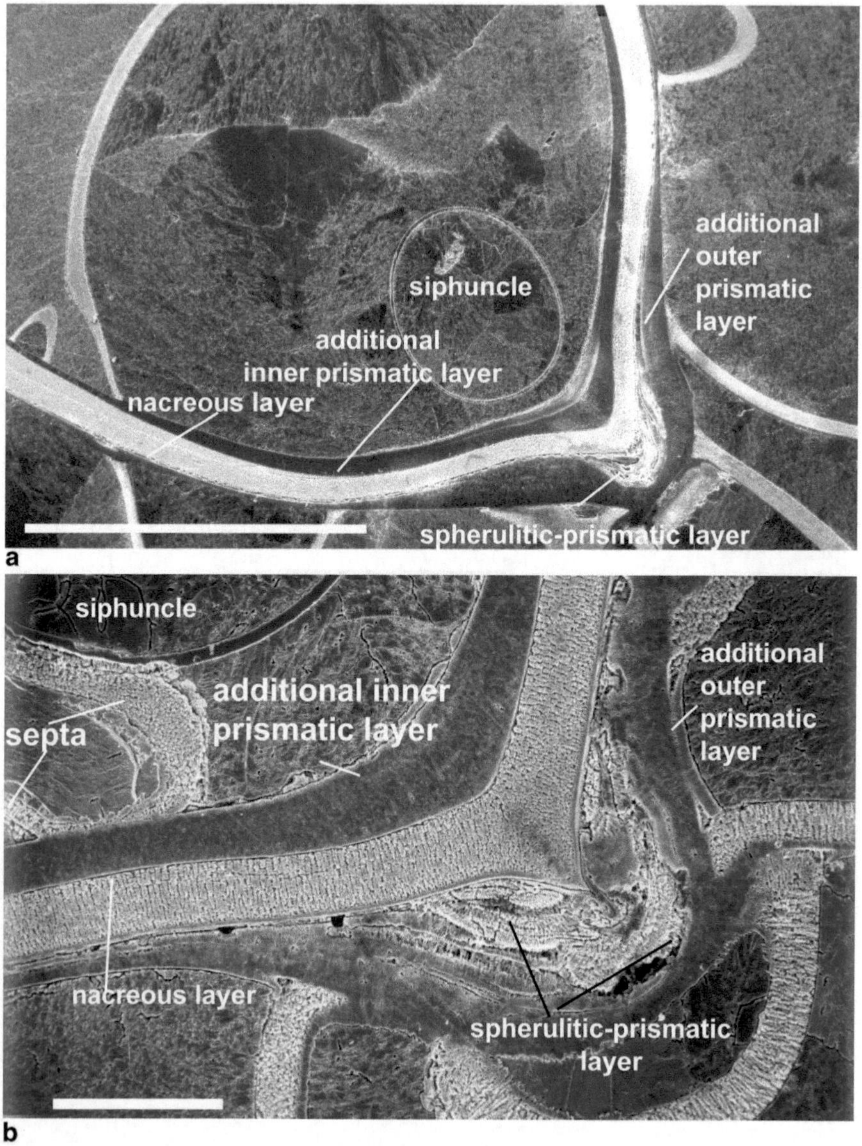

Fig. 15.6 *Aconeceras trautscholdi*. NRM–PZ Mo. 167770. Aptian, Lower Cretaceous; Ulyanovsk Region, Central Russia. **a** median section of the ventral shell wall to show a thick, porous spherulitic-prismatic layer which covers the serrate keel from outside, the additional inner prismatic and nacreous layers, and the pore canals of the keel. **b** close-up of (**a**) to show the exposed depression at the periphery of the pore canal and dark, supposedly organic material within the nacreous layer around the pore canal; scale bars = 150 μm and 450 μm, respectively

Fig. 15.7 *Aconeceras trautscholdi*. NRM–PZ Mo. 167768. Aptian, Lower Cretaceous; Ulyanovsk Region, Central Russia. Lateral view on a fully-grown shell exhibiting a short body chamber (less than a half of the whorl long) and a large muscle scar; x3

layers (wrinkle, inner prismatic) to the dorsal wall of the next whorl (Doguzhaeva 1980, 1981, Fig. 1). In the heteromorphic ammonoids with open whorls, the shell wall seems to be composed of equal number of layers around the whorls (Doguzhaeva and Mikhailova 1982). The periostracal layers consisted of perishable organic matter and have not yet been demonstrated to occur in ammonoids.

15.3 Aberrant Shell Wall Structures with Extra Layers

15.3.1 Extra Layers of the Cretaceous Gaudryceras tenuiliratum

Drushchits et al. (1978) observed the coating layers in about fifteen shells of *G. tenuiliratum* from the Turonian-Campanian of the Island Sakhalin, Russia. These black layers coat the umbilical parts of the shell and hide the umbilical seals (Fig. 15.1, 15.2). They are formed by two layers, which are similar in structure and have about equal thickness (Fig. 15.2). The inner layer encapsulates the first three whorls, and the outer layer covers the inner coating layer and extends to the surface of the fourth and fifth whorls. In the umbilical part of shells the total thickness of coating layers exceeds three to seven times the rest of the shell wall. The coating layers are well mineralized and show a fine lamination (Fig. 15.2c) similar to that of an organic material like chitin. The lamellae are deposited parallel to the shell surface. The black colour indicates that the coating layers are rich in organic matter. The coating layers of *G. tenuiliratum* were also observed by Birkelund who found that in the specimens of *G. tenuiliratum* from Turonian-Campanian of Sakhalin,

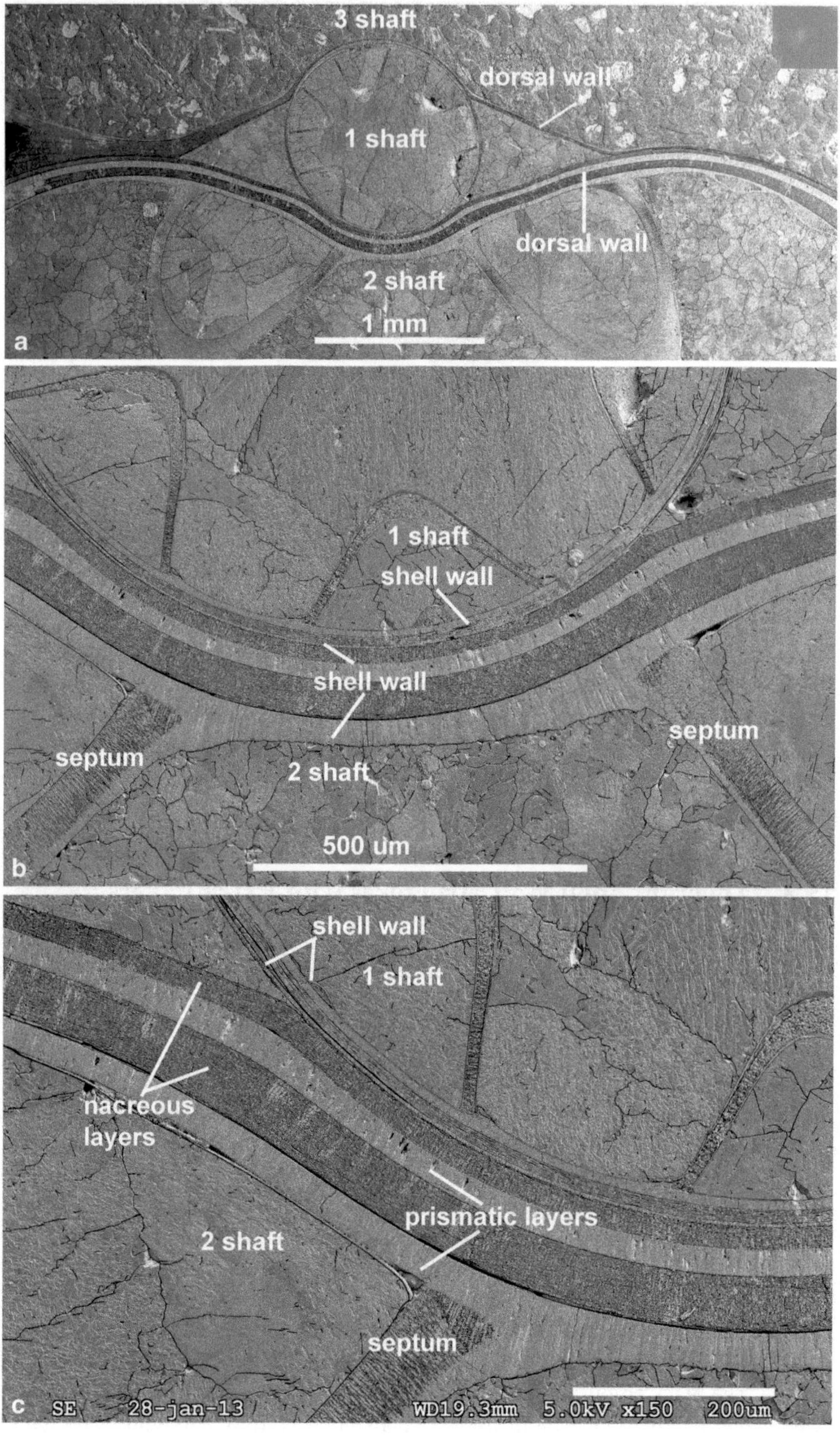

Fig. 15.8 *Ptychoceras renngarteni*; NRM–PZ Mo. 167773. Aptian, Lower Cretaceous; NW Caucasus, Adygeya Republik, Russia. **a** cross section of the first (in the centre), second (bottom) and third (top) shafts. **b**, **c**, enlargements of the dorsal walls of the first and second shafts

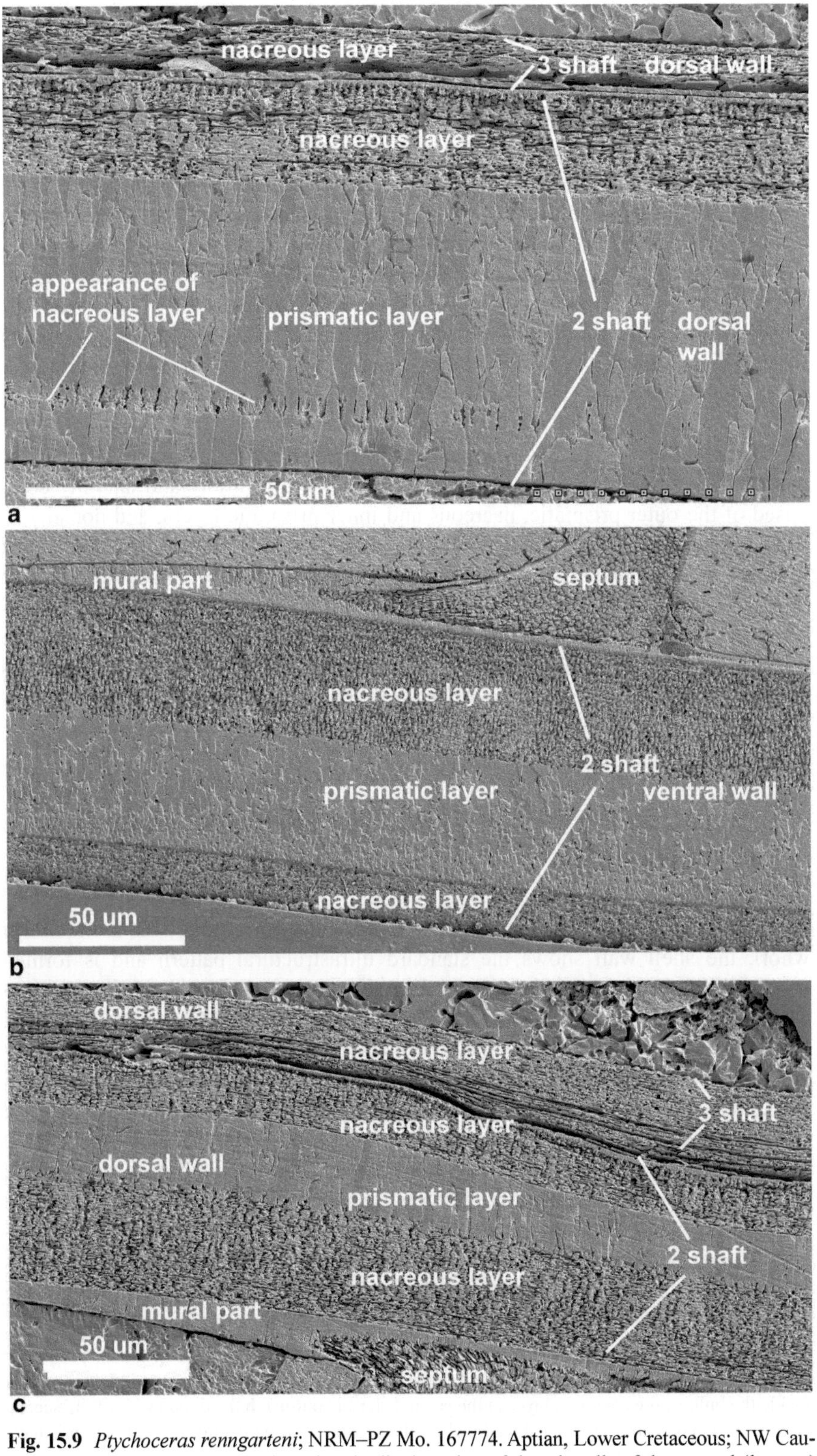

Fig. 15.9 *Ptychoceras renngarteni*; NRM–PZ Mo. 167774. Aptian, Lower Cretaceous; NW Caucasus, Adygeya Republik, Russia. **a** longitudinal section of dorsal walls of the second (bottom) and third (top) shafts after truncation of the first shaft; note that there are two nacreous layers along

Russia, these layers completely encrust the third whorl (Birkelund 1981, p. 202). The coating layers are also visible in the umbilical parts of the shells of *G. tenuiliratum* from Sakhalin studied by Hirano (1975, pl. 26, Fig. 1, 6, 10). Kulicki et al. (2001) described the coating layer in *G. tenuiliratum* from Hokkaido, Japan.

The post-mortem or diagenetic origin of the coating layers of the *G. tenuiliratum* has been refuted for the following reasons: (1) these layers are only present in *G. tenuiliratum* but missing in the co-occurring ammonoid shells from the same locality; (2) they are present not only on the external shell surfaces but also between the whorls; (3) they have a micro-laminated ultrastructure. If they were diagenetically formed layers they would have a predominantly spherulitic-prismatic or irregular prismatic ultrastructure (Drushchits et al. 1978). The formation of the coating layers in *G. tenuiliratum* probably took place as follows.

The ammonitella and first three whorls were secreted as in typical ammonoids, and their shell wall has a standard structure in being ventrally and laterally composed of the outer prismatic, nacreous and inner prismatic layers, and dorsally of the outer prismatic layer. After formation of the third whorl, the shell was coated by the mantle extensions that provided secretion of the coating layers around the third whorl including its ventral and lateral sides. The fourth and the fifth whorls were secreted in the ordinary manner and have a standard shell wall structure. The repeated mantle expansion occurred when the formation of the fourth whorl was complete. The second coating layer covered the outer surface of the first coating layer (Drushchits et al. 1978).

15.3.2 Extra Layers of the Jurassic Indosphinctes (Elatmites) submutatus

At juvenile stages of *I. (E.) submutatus,* approximately from the first to the fourth whorl, the shell wall shows the standard ultrastructural pattern and is formed ventrally and laterally by the inner prismatic, nacreous and outer prismatic layers (Doguzhaeva 2012). The parabolae were possibly not formed at this stage, but appear at later stages (Fig. 15.3a, b). Through ontogeny the nacreous layer gradually increases in thickness, and in adult shells, it forms the bulk of the shell wall thickness. The prismatic layers gradually decrease in thickness and fade out. The outermost, perishable, thick, non-biomineralized layer is partially preserved in the umbilical corners and between the whorls where it is relatively thin (Fig. 15.4a, b). The shell cross-sections through the parabolae show blisters at the sites of the parabolic notches (Fig. 15.4b). They are situated on the inner portion of the nacreous layer and coated with the additional nacreous layer; therefore, the blisters form lens-like intrusions within the nacreous layer (Fig. 15.4a, b). The blisters are

---

the contact between the second and third shafts, and that the additional prismatic layer appears inside the bulk of the nacreous layer in the second shaft (bottom). **b** the dorsal wall of the second shaft with two nacreous layers. **c** dorsal wall of the second shaft with a thick, additional prismatic layer; scale bars = 50 μm

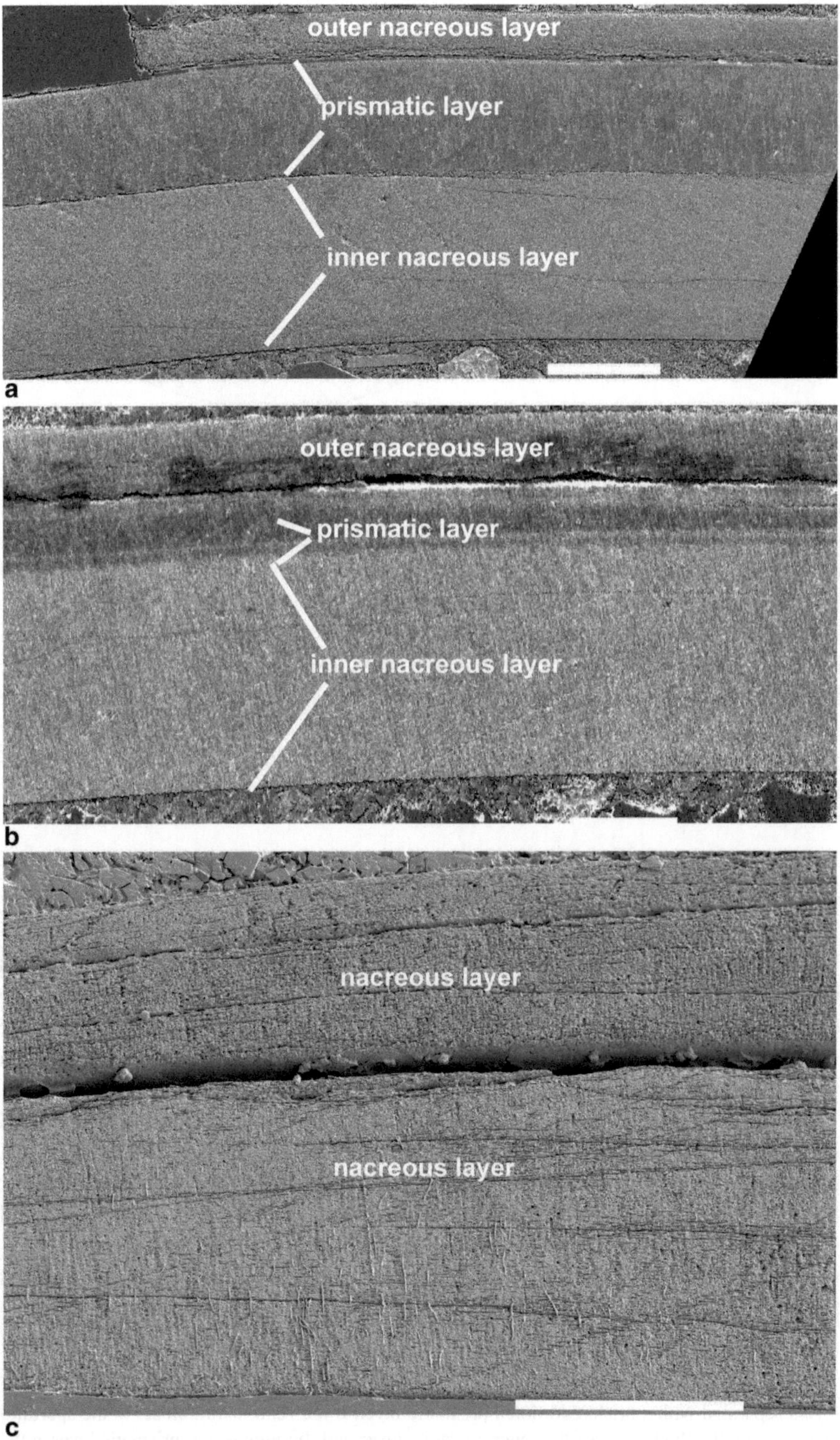

Fig. 15.10 *Ptychoceras renngarteni*; NRM–PZ Mo. 167775. Aptian, Lower Cretaceous; NW Caucasus, Adygeya Republik, Russia. **a** ventral wall structure of the third shaft with two nacreous

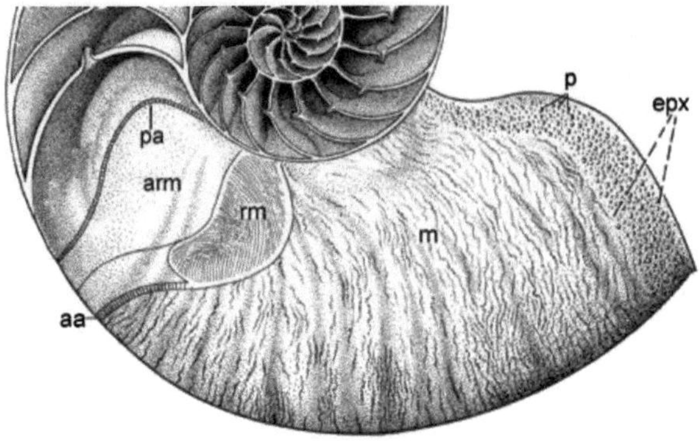

Fig. 15.11 *Nautilus* (modified from Mutvei and Doguzhaeva 1997). Schematic presentation of the body to show mantle surface (*m*), sites of the origin of the retractor muscles (*rm*) and the mantle and septal attachment zones (*aa, pa*), mantle attachment zone to the shell aperture (*exp*) with finger like epithelial extensions (*p*)

formed by oblique irregularly mineralized prismatic layers, with the spherulites and nacreous chips in the spaces between the layers (Fig. 15.4c). The aberrant shell wall structure in *I. (E.) submutatus* is characterized by two outermost extra-layers, one of which is the outermost nacreous layer and the other is the non-biomineralized layer on the surface of the former. The secretion of the aberrant shell wall structure in *I. (E.) submutatus* required that the mantle be expanded on the outer surface of the body chamber. The mantle expansion was probably long and attached to the inner and outer surface of the body chamber along the parabolae (for details see Doguzhaeva 2012). Thus, contrary to *G. tenuiliratum* which has two coating layers of similar ultrastructure, one of the two coating layers in *I. (E.) submutatus* is non-biomineralized and the second, internal layer, is nacreous.

15.3.3 Extra Layers of the Cretaceous **Aconeceras trautscholdi**

The standard ammonoid shell wall formed by the outer and inner prismatic and nacreous layers was developed at the early ontogenetic stages characterized by the evolute shell. Then the shell became involute and the shell wall was modified in the way that the outer prismatic layer decreased considerably in thickness and is even wedged out in places (Doguzhaeva and Mutvei 1993b).

At the ontogenetic stages when the outer prismatic layer essentially decreased in thickness or is wedged out, the outer portion of the nacreous layer was in several places transformed into a platy sub-layer that differs from the rest of the nacreous layer by consisting of varying sizes and orientation of calcareous plates separated by distinct interspaces (Doguzhaeva and Mutvei 1993b). The thickness of the platy sub-layer is variable. The interspaces between the plates may have been formed from post-mortem destruction of the perishable organic matter between the plates.

layers separated with a thick, additional prismatic layer. **b** the ventral wall closer to the apertural margin; note the thickening of the inner nacreous layer and thinning of the prismatic layer. **c** the ventral wall still closer to the shell aperture; note that the prismatic layer is missing and the shell wall consists of the nacreous layer; scale bars: **a**, **b** 150 μm, **c** 100 μm

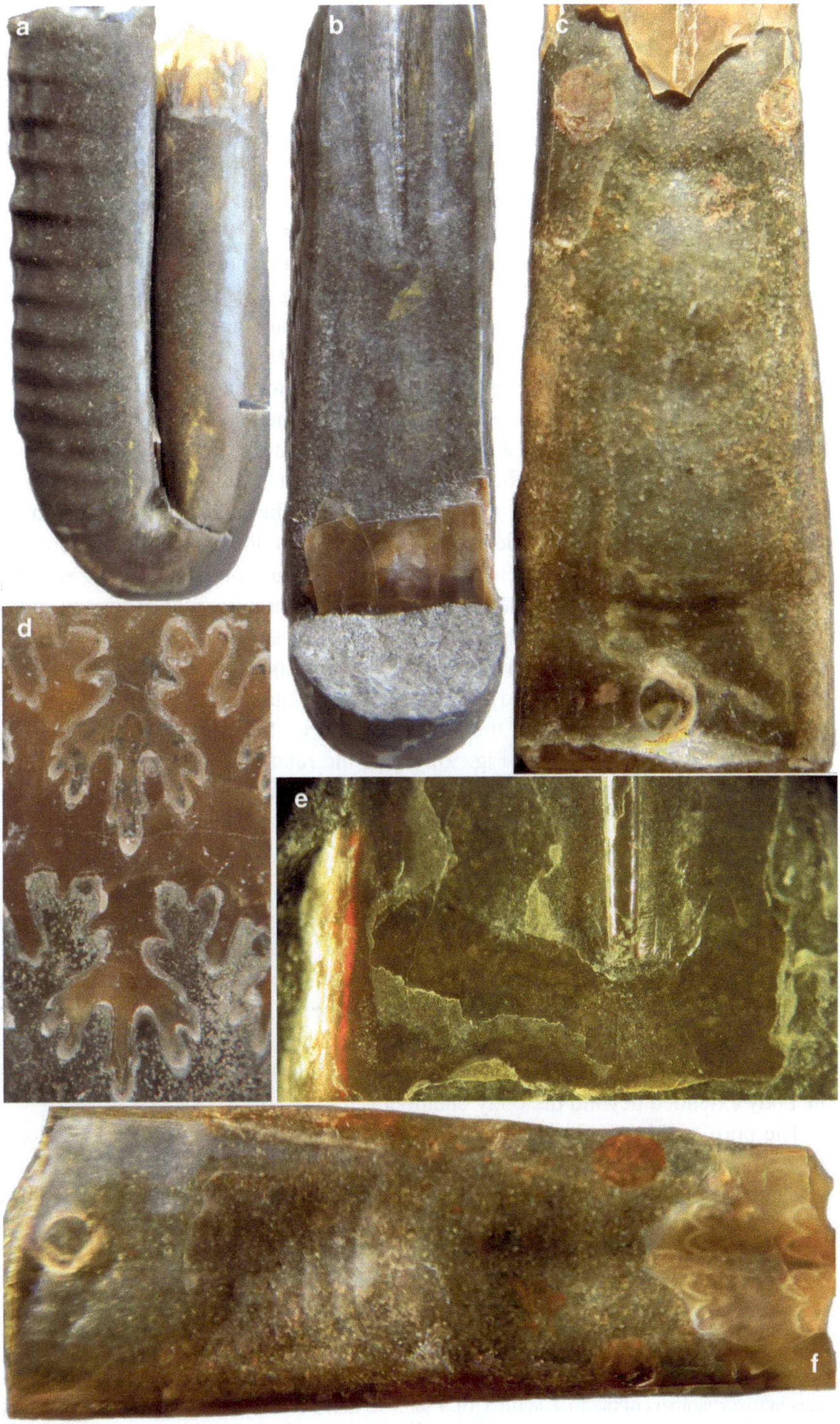

Fig. 15.12 *Ptychoceras sp.* **a–d**, **f**: NRM–PZ Mo. 167771; **e**: NRM–PZ Mo 167772. Aptian, Lower Cretaceous; NW Caucasus, Adygeya Republik, Russia. **a** lateral view on the body chamber

This would indicate increasing organic content in the shell wall caused by covering of the shell by the mantle extension (= 'shifting of the shell inside the mantle fold').

Two thick additional layers, nacreous and inner prismatic, or only the inner prismatic, invest the ventral and ventro-lateral portions of the body chamber (Fig. 15.5, 15.6; Doguzhaeva and Mutvei 1991, Fig. 2). The distribution of these layers corresponds to the location of the ventro-lateral muscle scars.

This layer surrounds the keel (Fig. 15.5, 15.6; Doguzhaeva and Mutvei 1991, Fig. 2; pl. 1) and is composed of spherulites of varying sizes (range of the diameter is 5–100 µm), and more or less continuous prismatic lamellae separated by numerous larger or smaller interspaces. The interspaces apparently contained a perishable organic matter during life. The porous spherulitic-prismatic layer around the keel may have been secreted by the mantle extension over the external shell surface. It is covered from the outside by an additional outer prismatic layer.

The length of the body chamber at early growth stages, at least at the shell diameter of 10–12 mm, is about two-thirds of the whorl. At the succeeding growth stages the length of the body chamber decreases (Fig. 15.7), attaining a length of about one half of the whorl at a shell diameter of 27–50 mm. At early ontogenetic stages (shell diameter is about 6 mm), the apertural margin forms thin, nearly transparent, mid-lateral lappets and a prominent, acute, serrate ventral rostrum. It is logical to assume that the shortened body chamber housed the posterior portion of the body, whereas the cephalic portion was located beyond the body chamber and supported by the rostrum on the ventral side and the lappets on the ventro–lateral sides (Fig. 15.7; Doguzhaeva and Mutvei 1991, Fig. 9). While the relative shortening of the body chamber in ontogeny of ammonoids is a common phenomenon, the extreme shortening of the body chamber, like that of *A. trautscholdi*, is to our knowledge extraordinary (see Doguzhaeva and Mapes 2015).

A. trautscholdi has large, ventro-lateral muscle scars about as long as 0.3–0.5 of body chamber length and as broad as 0.3–0.5 of body chamber height (Fig. 15.6; Doguzhaeva and Mutvei 1991, pl. 2, figs 2–4; pl. 3, Fig. 1, 2; pl. 4, Fig. 1, 2; Doguzhaeva and Mutvei 1993a, Fig. 1a, b, c, d; Doguzhaeva 2014; Doguzhaeva and Mapes 2015). Muscle scars of similar size and shape are also known in several other ammonoid genera (see Appendix 2 in Doguzhaeva and Mapes 2015). The association of the large ventro-lateral muscle scars and a short body chamber suggests that the body extended beyond the body chamber.

The pore canals regularly traverse the total thickness of the shell wall along the keel (Fig. 15.7). The canals are perpendicular to the shell surface and extend to the outer shell surface at pointed elevations or serrae (Doguzhaeva and Mutvei

occupying half of the second (to the right) and the third (to the left) shafts; x 2.3. **b** the specimen on (**a**) after splitting; dorsal view on the third shaft to show the groove at place of the removed first shaft; x 2.7. **c** the posterior portion of the body chamber showing umbilical muscle scars and W-shaped pallial line in front of the last suture line, and unpaired round dorsal scar at the end of the second shaft (bottom); x 4. **d** enlarged detail of (**c**) (top) to show small flask-shaped dorsal scars in two chambers of the first shaft; x 10. **f** same as in (**c**), rotated at 90°, to show repetition of the umbilical muscle scars; x 4.5. **e** the truncated first shaft on the dorsal side of the second shaft; x 3.5. Abbreviations: *bc* body chamber, *d* dorsal scar, *ls* last suture line, *p* pallial line, *sh1, sh2, sh3* the first, second and third shafts, *u* umbilical muscle scar

1993b, Fig. 6.6, 6.7). The pore canals possibly housed finger-like microvilli of the mantle that helped to temporarily attach the external portion of mantle to the outer shell surface (Doguzhaeva and Mutvei 1993b). For comparison, in extant *Nautilus* the pore canals in the near-apertural margin serve for shell/mantle attachment to the inner surface of the body chamber (Mutvei and Doguzhaeva 1997, pl. 7, Fig. 1–3).

15.3.4 Aberrant Shell Wall Structure of the Cretaceous Heteromorph **Ptychoceras**

15.3.4.1 Standard Shell Wall Structure of *P. minimum* Rouchadze 1933

P. minimum is characterized by smaller size of its shell in comparison with that in three other, co-occurring species *P. levigatum*, *P. renngarteni and P. parvum*. Only in *P. minimum* does the apical shell portion show the non-destructed protoconch surrounded by planispiral whorl with a primary constriction and primary varix (=ammonitella), and the first chambers of the straight shaft. The ammonitella and the first shaft have a standard shell wall structure: the shell wall of the protoconch and the first whorl (up-to the primary varix) is prismatic; the nacreous layer first appears in the primary varix; the shell wall between the primary varix and approximately the half of the first shaft is composed of the outer prismatic, nacreous and inner prismatic layers (for details see Doguzhaeva and Mutvei 1989).

15.3.4.2 Aberrant Shell Wall Structure of *Ptychoceras*

All examined specimens of *P. levigatum, P. renngarteni* and *P. parvum* show regular truncation of the initial portion of shell and have an aberrant shell wall structure (Fig. 15.8, 15.9, 15.10). The first aberrant change of the shell wall structure was a pronounced reduction of the outer prismatic layer that is about two to three nacreous laminae thick, or 35 times thinner than the nacreous layer in fully grown shells. Approximately in the middle of the first shaft the nacreous layer became the bulk of the shell wall thickness. The second aberrant feature is the appearance of the additional nacreous layer in the second shaft. The shell wall is here and up–to the beginning of the third shaft composed of two nacreous layers separated by a prismatic layer (Fig. 15.10a). The prismatic layer wedged out along the third shaft towards the aperture (Fig. 15.10b). The two nacreous layers fused with each other towards the aperture (Fig. 15.10c). The third aberrant feature is the formation of an additional, thin, outermost nacreous layer at the aperture in fully-grown shells. This layer forms prominent growth lines that parallel the shell surface but have a distinct angle to the growth laminae of the ordinary nacreous layers. The different orientation of the growth laminae indicates that the thin outermost nacreous layer was secreted on the shell surface by the mantle epithelium that covered the apertural region of the shell from the outside (Fig. 15.11).

15.3.4.3 Shell Truncation in *Ptychoceras*

The truncation of the initial shell portion took place in all examined specimens of *P. renngarteni*, *P. levigatum* and *P. parvum*. The truncated portion regularly comprises the ammonitella and the next 15–17 chambers of the phragmocone (Fig. 15.12b, e). Numerous fragments of the destroyed initial portion of shell are often preserved around the shell wall at the place of truncation. Moreover, comparatively large pieces of the truncated portion of the shell, sometimes four-five chambers, seem to have been transported for a comparatively long distance from their original location, and then embedded between the second and third shaft (Doguzhaeva and Mutvei 1989, pl. 2, Figs 2, 3; pl. 5, Fig. 1). The broken end of the retained portion of the first shaft lacked a plug or any other structure that would seal the opening of the phragmocone. The latter remained open till the dorsal wall of the second shaft was secreted at this place and partly sealed the opening due to its thickening. At the place of the truncation the dorsal wall of the second shaft is also thickened.

15.3.4.4 Muscle Scars in *Ptychoceras*

At early ontogenetic stages small, unpaired, dorsal scars are present within the dorsal lobes (Fig. 15.12d; Šarikadze et al. 1990, Fig. 14). A pair of large rounded umbilical muscle scars is situated in front of the last suture line as well (Fig. 15.12c, f; Šarikadze et al. 1990, Fig. 6). There is also a dorsal rounded muscle scar placed at the end of the second shaft in the middle length of the body chamber (Fig. 15.12c, f). In front of the umbilical scars there is an incised line on the internal mold of the body chamber, that probably was a w-shaped pallial line (Fig. 15.12c; compare with the mantle attachment zone in *Nautilus* on Fig. 15.11). In front of the pallial line, the shell was elevated as expressed by the relief of the internal mold (Fig. 15.12c). In ammonoids, the dorsal scar is usually located near the last septum. In *Ptychoceras,* the dorsal muscle scar at the terminal body chamber is far away from the last suture line and in this respect *Ptychoceras* differs from other ammonoids (compare with Fig. 15.7) but similar to heteromorph *Audouliceras* (Doguzhaeva 2014; Doguzhaeva and Mapes, 2015).

15.4 Helicolateral Deposits of the Carboniferous *Clystoceras globosum*

The globose shell of *Clystoceras globosum*, with deep narrow umbilicus and narrow whorls, reveals an unusual shell element called helicolateral deposits that are unknown in other ammonoids (Nassichuk 1967). The helicolateral deposits look like head-phones as they represent a pair of relatively large rounded structures on the umbilicus of both sides of the nearly spherical shells (Nassichuk 1967, pl. 28, Fig. 1–11). The helicolateral deposits fill the umbilicus and coat the adjacent portions of the shell. According to Nassichuk (1967, p. 240), the shell wall does not continue into the helicolateral deposits. The wrinkle layer observed in the body chamber (Nassichuk 1967, pl. 28, Fig. 1–2) makes the boundary between the shell and helicolateral deposits distinct. The helicolateral deposits exhibit "*Evenly spaced lateral wave-like ridges and troughs* […] *suggesting an even periodicity or pulsation of growth*" (Nassichuk 1967, p. 240). The umbilical portion of the suture was influenced by a dorsolateral cameral flexure, which resulted from "*overriding*" of the dorsolateral helicolateral deposits by the dorsum of the body chamber (Nassichuk 1967, p. 240). The original composition and ultrastructure of the helicolateral deposits are so far unknown.

15.5 Umbilical Membrane of the Devonian *Prolobites*

In *Prolobites,* the subglobular to subdiscoidal shell has a broad umbilicus at early ontogenetic stages that is closed by the umbilical membrane at later ontogenetic stages (Bogoslovsky 1969, p. 180, Fig. 49, 52). The protection of the umbilical portion of the shell supposedly enforced the mechanical strength of the entire shell. The original composition and ultrastructure of the umbilical membrane are so far unknown.

15.6 Wrinkle Layer on the Outer Shell Surface

The expansion of the wrinkle layer upon the previous whorl of the shell in front of the shell aperture has been observed in different Paleozoic and Mesozoic ammonoids (Fig. 15.13; Ruzhencev 1962; House 1971; Doguzhaeva 1981; Korn 1985, 2000; Kulicki et al. 2001; Klug 2007). The shell segment in front of the aperture showing the wrinkle layer is observed to be about 60 ° long (Tozer 1972, Fig. 3a), but in the Late Devonian goniatite *Erfoudites rugosus* Korn 2000, this segment is about 180 ° of the body chamber length (Korn 2000, Fig. 2, 3a, 4, 5). In *E. rugosus,* the wrinkles are oriented mostly parallel to the apertural margin. The wrinkles on the surface of the body chamber have been interpreted as secreted by the mantle "expanded on the conch" (Korn 2000, p. 26).

15.7 Endocochleate Experiments as a Pathway to Protect the Phragmocone

The additional external layers in the shell wall of the Late Cretaceous *Gaudryceras tenuiliratum*, Late Jurassic *Indosphinctes (Elatmites) submutatus,* Early Cretaceous *Aconeceras trautscholdi* and *Ptychoceras renngarteni, P. levigatum* and *P. parvum*

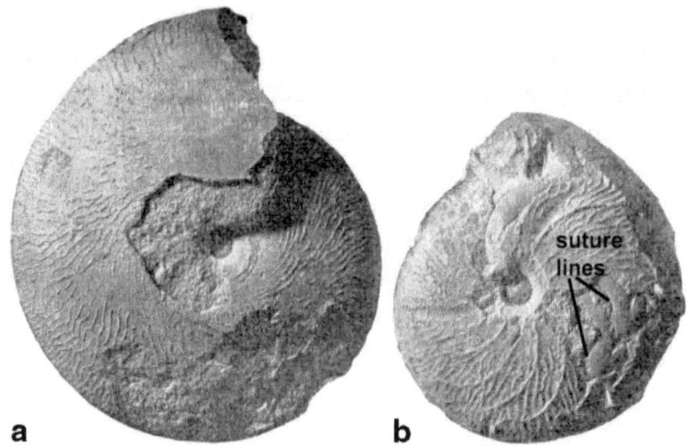

Fig. 15.13 *Platigoniatites molaris* Ruzhencev 1956 (modified from Ruzhencev 1962). **a** wrinkle layer covering the outer shell surface. **b** wrinkle layer on the inner surface of the phragmocone; x 2

provide evidences that the mantle of these ammonoids coated the external shell surface and secreted additional external layers. The number of ammonoids that underwent the "endocochleate experiment" seems not to have been restricted to the four genera referred to above in which the shell preservation allowed us to carry out ultrastructural investigations of the shell. For instance, Arkell et al (1957, p. L244, Fig. 270, 5) published a photo of the *Tmaegophioceras* Spath 1925 *(Arietites laevis* Geyer 1886) showing the supposed coating layers hiding the umbilical seams. The figured specimen (Geyer 1886) illustrated in the Treatise on Invertebrate Paleontology is not any longer available (see Gerry 1973). The supposed coating layers of *Tmaegophioceras* are similar to the coating layers on the photo of *G. tenuiliratum* (Hirano 1975) published before these layers were recognized for the first time by Drushchits et al. (1978). *Tmaegophioceras* falls in the superfamily Psilocerataceae comprising the earliest Jurassic ammonites in many of which the shell has a broad umbilicus. Therefore, this superfamily might be a potential site to search for the ammonites with coating layers. In the ammonoids with a broad umbilicus, like in *Gaudryceras,* the coating layers might be secreted and be notably thicker than the rest of the shell wall. Because these layers coat a broad shallow umbilicus, they would reinforce the shell strength; this might be important for preventing shell crushing in deep–water environment. The shell of *G. tenuiliratum* apparently allowed diurnal migration to deep–water environment. It has a very narrow lace-like siphuncle; it is strongly attached to the shell wall with the aid of long septal necks; the long cuffs tightly attach the chitinous connecting rings to the inner surface of septal neck. The diameter of the siphuncle is from 0.04 to 0.27 of the whorl height on the first–fifth whorls. The septal necks achieve a maximum length at the third–fourth whorls where they are about 0.3–0.5 chamber length. The cuffs are about 0.3–0.5 septal neck length in the third–fourth whorls. Moreover, the shell wall shows regular thickenings of the nacreous layer (up to five in a whorl), associated with the collars (Drushchits and Doguzhaeva 1981). The ammonoids with the small siphuncular diameter had supposedly a higher resistance against hydrostatic pressure and could live in deeper habitats than the majority of ammonitids (see Westermann 1982, 1996). The evolute shell with a comparatively long body chamber of *Gaudryceras* would have had an unstable orientation because the centres of buoyancy and gravity were situated close together (see Raup and Chamberlain 1967). However, the collars seem to have significantly increased the stability and also shell buoyancy. The bases of the collars form a thickening of the shell wall which increased the mechanical strength of the shell against hydrostatic pressure. The collars apparently contributed to the diurnal migrations to deep-water environment as well (Doguzhaeva et al. 2010). In *Gaudryceras*, like in all other lytoceratids (Hoffmann 2010), septal lobes of each septum are attached to the preceding septum; therefore, the shell/body attachment was strengthened as well. All these morphological elements, together with the coating layers, seem to be needed for the deep-water environment and daily vertical migrations from surface to larger depths. Our conclusion, based on the functional morphology of the shell, is supported by the sedimentological data on Santonian–Campanian of Sakhalin—that yielded the studied shells of *G. tenuiliratum*—suggesting a deep water slope to basinal environment for formation of these sediments (see Yazykova 2002). The shells of G. tenuiliratum collected from the outer shelf deposits of northwestern Hokkaido, Japan, gave an opportunity to examine the shell parameters of the lytoceratid ammonoid dwelled in deep-water environment (Ikeda and Wani 2012). The authors demonstrated that ratio of whorl breadth and shell diameter shows that *G. tenuiliratum* migrated at different depths on the outer shelf environment (Ikeda and Wani 2012). Thus, the Late Cretaceous lytoceratid *Gaudryceras tenuiliratum* possesses: (1) thick coating layers; (2) a narrow evolute shell with long body chamber and broad shallow umbilicus; (3) a narrow siphuncle with relatively long septal necks and cuffs in the third-fifth whorls; (4) septa attached to each other by means of septal lobes, and (5) collars associated with thickenings of the nacreous layer in the shell wall. These features suggest that the shell of *Gaudryceras* was designed for deep-water habitats, diurnal vertical migrations, hovering and drifting in mid–water.

In *Aconeceras trautscholdi,* the aberrant shell wall structure is associated with a short (about 180 ° in length) body chamber, deep apertural lateral sinuses and pronounced ventral keel; thin semi–transparent apertural wall, large (up to 1/3 of body chamber length) paired muscle scars, and a perforated edge of the keel. This morphological association leads to the conclusion that in *A. trautscholdi,* the body was protected by the body chamber only posteriorly while anteriorly it was free from the shell. The body that supposedly exceeded the body chamber size would need strong muscle–retractors; in *A. trautsholdi,* the enlarged lateral attachment scars were apparently indicative of such muscles. The pore canals of the keel possibly served for housing of the mantle papillae, which may have enforced the conjunction of the body and shell along the ventral side.

In *Ptychoceras,* the aberrant shell wall structure is associated with the (a) very thin apertural shell wall characterized either by the extremely thin outer prismatic layer that is about 0.03 of the thickness of the nacreous layer (= thickness of 2–3 nacreous laminas), or its complete absence on the external surfaces of the third shaft, (b) long curved body chamber occupying the second half of the second shaft and the third one; (c) specific system of enlarged muscle scars which form a long acute–

angled triangle on the dorsal side of the body chamber, and (d) mode of truncation resulted in temporary open first shaft and reservation of shell fragments at place of truncation. In *Ptychoceras,* the repaired injuries of the shell wall were observed by Kakabadze and Sharikadze (1993, Fig. 5). These authors schematically illustrated the restoration of the lost broken part and shown that the patch of the shell wall was added from the innner side of the broken shell wall in a similar way as it happened in monomorph ammonoids. This observation was interpreterd as evidence that in *Ptychoceras*, the mantle could not coat the shell (Kakabadze and Sharikadze 1993, p. 214). For the interpretation of the illustrated repairation, it is important to know what side of the shell is shown on Fig. 15.7a, b (this information is missing in the paper). The outline of the shell on Fig. 15.7a allows suggesting that this is the dorsal side; if it is so, than the restoration of the dorsal wall is shown on Fig. 15.7b. In *Ptychoceras*, the dorsal wall tightly touches the dorsal wall of the next shaft, and the mantle could not coat the dorsal side of the shell; thereby, the shown shell wall restoration resembles this in monomorph ammonoids. Besides, in *Ptychoceras*, like in *Spirula* (Chun 1898, 1899), the outermost part of the shell wall was secreted from the outside but the inner part from the inside. This points out that the interpretation of the shell wall restoration requests the sections of the total shell wall which could reveal the relationship between the different layers (for comparison, see Doguzhaeva 2012). Moreover, the fact that the fractured fragments of the truncated shell wall were regularly, rather than occasionally, retained in the shells should be explained from the author's viewpoint.

Judging on the umbilical membrane of the Devonian anarcestid ammonoid *Prolobites* (Bogoslovsky 1969) and the helicolateral deposits of the Late Carboniferous goniatitid *Clystoceras globosum* (Nassichuk 1967), the attempts to "coat" the umbilical parts of the shell and to protect them by means of secretion of additional layers above the standard shell wall occurred already in the early stages of the ammonoid evolution. This would be hardly possible without the capability to stretch the mantle upon the extrenal shell surface. The expansions of the wrinkle layer from the body chamber on the external shell surface (Korn 2000; Klug 2007; Klug et al 2007) make this assumption highly probable.

Thus, the implication of the additional external shell layers and the "endocochleate experiments" can be clarified due to the analysis of the associated morphological traits such as body chamber length, shell aperture, shell geometry, siphuncle characteristics, wrinkle layer distribution and muscle scars. Further study on the aberrant shell wall structures seems to be essential to get better understanding of locomotion, habits, and other aspects of the ammonoid paleobiology.

Acknowledgements This study was carried out thanks to support by the Royal Swedish Academy of Sciences and personally by Prof. Stefan Bengtson, Department of Palaebiology of the Swedish Museum of Natural History. Prof. Royal H. Mapes (Ohio University, Athens, USA) and Rene Hoffman (Freie University, Berlin, Germany) are the reviewers of the paper. We are indebted for the help to all these people.

References

Arkell WJ, Kummel B, Wright CW (1957) Systematic descriptions. In: Moore RC (ed) Treatise on invertebrate paleontology, Part L:129–465. Geol Soc Amer and Univ Kansas Press

Barrande J (1860) Troncature normale ou périodique de la coquille dans certains céphalopodes paléozoïques. Bull Soc Géol France, sér 2 17:573–601

Birkelund T (1981) Ammonoid shell structure. In: MR House (ed) Syst Ass Spec Vol 12:177–214. Academic, London

Blind W (1988) Comparative investigations on the shell morphology and structure of *Nautilus pompilius, Orthoceras* sp., *Pseudorthoceras* sp., and *Kionoceras* sp. In: Wiedmann J, Kullmann J (eds) Cephalopods—present and past. Schweizerbart, Stuttgart

Bogoslovsky BI (1969) Devonian ammonoids. I. Agoniatitida Tr Paleont Inst Akad Nauk USSR 124:1–341

Chun C (1898–1899) The Cephalopoda. German Deep Sea Expedition 18:1–435

Doguzhaeva LA (1980) New data on shell wall structure in ammonoids. Dokl Akad Nauk USSR 254:745–748

Doguzhaeva LA (1981) Wrinkle layer in ammonoid shell. Paleont Zh 1:38–48

Doguzhaeva LA (1994) An Early Cretaceous orthocerid cephalopod from North-Western Caucasus. Palaeontology 37:889–899

Doguzhaeva LA (2002) Pre-mortem septal crowding and pathological shell wall ultrastructure of ammonite younglings from the lower Aptian of Central Volga (Russia). In: M Wagreich (ed) Aspects of Cretaceous Stratigraphy and Palaeobiogeography. Österreich Akad Wiss Schrift Erdwiss Komm 15:171–185. Graz Austria

Doguzhaeva LA (2012) Functional significance of parabolae, interpreted on the basis of shell morphology, ultrastructure and chamical analyses of the Callovian ammonite *Indosphinctes* (Ammonoidea: Perisphinctidae), Central Russia. Rev Paléobiol vol Spec 11:89–101

Doguzhaeva LA, Mikhailova IA (1982) The genus *Luppovia* and the phylogeny of Cretaceous heteromorphic ammonoids. Lethaia 15:55–65

Doguzhaeva LA (2014) Muscle and mantle attachment marks as well as body chamber lengths indicative of diverse life styles of coexisted Aptian ammonites of the Russian Platform. In: C Klug et al (eds) Intern Symp Cephalopods present and past in combination with the 5th Intern Symp Coleoid cephalopods through time Abstracts and program. Univ Zurich

Doguzhaeva LA, Mapes RH (2015) Body Chamber Length and Muscle and Mantle Attachments in Ammonoids. This volume

Doguzhaeva LA, Mutvei H (1986) Functional interpretation of inner shell layers in Triassic ceratid ammonites. Lethaia 19:195–209

Doguzhaeva LA, Mutvei H (1989) *Ptychoceras*—a heteromorphic lytoceratid with truncated shell and modified ultrastructure (Mollusca: Ammonoidea). Palaeontogr A 208:91–121

Doguzhaeva LA, Mutvei H (1991) Organization of the soft body in *Aconeceras* (Ammonitina), interpreted on the basis of shell morphology and muscle scars. Palaentogr A 218:17–33

Doguzhaeva LA, Mutvei H (1993a) Structural features in Cretaceous ammonoids indicative of semi-internal or internal shells. In: House MR (ed) The Ammonoidea: Environment, Ecology, and Evolutionary Change. Syst Ass Spec Special Vol. 47:99–114. Clarendon Press Oxford

Doguzhaeva LA, Mutvei H (1993b) Shell ultrastructure, muscle scars, and buccal apparatus in ammonoids. Geobios 15:111–119

Doguzhaeva LA, Mapes RH (2015) Evidences and implications of mantle and muscle attachments in ammonoid body chamber. This volume

Doguzhaeva LA, Bengtson S, Mutvei H (2010). Structural and morphological indicators of mode of life in the Aptian lytoceratid ammonoid *Eogaudryceras*. In: Tanabe K, Shigeta Y, Sasaki T, Hirano H (eds) Cephalopods—present and past. Tokai University Press, Tokyo

Drushchits VV, Khiami N (1970) Structure of the septa, protoconch walls and initial whorls in Early Cretaceous ammonites. Paleont Zh 1:26–38

Drushchits VV, Doguzhaeva LA (1974) About some features of morphogenesis of Phylloceratida and Lytoceratida (Ammonoidea). Paleont Zh 1:42–53
Drushchits VV, Doguzhaeva LA (1981) Ammonites in the scanning electron microscope (Internal shell structures and systematics of Mesozoic phylloceratids, lytoceratids and six families of the Early Cretaceous ammonitids). Moscow State University, Moscow
Drushchits VV, Doguzhaeva LA, Mikhailova IA (1978) Unusual coating layers in ammonoids. Paleont Zh 2:36–44
Egoian VL (1969) The ammonites from the Clansayesian deposits in West Caucasus. Tr Krasnodar Fil Research Oil Inst 19(126–188):264–315
Erben HK, Flajs G, Siehl A (1969) Die frühontogenetische Entwicklung der Schalenstruktur ectocochleater cephalopoden. Palaeontogr A 132:1–54
Gerry TA (1973) A revision of the generic classification of the family Echioceratidae (Cephalopoda, Ammonoidea) (Lower Jurassic). Paleontol Contr 63:1–32
Geyer G (1886) Lieber die liassischen Cephalopoden der Hierlatz bei Hallstatt. KK Geol Reichsanst Wien Abh 12:213–287
Guex J (1970) Sur les moules interne des Dactyliocératides. Bull Lab Geol Miner Geophys Mus Geol Univ Lausanne 70(182):1–7
Hirano H (1975) Ontogenetic study of Late Cretaceous *Gaudryceras tenuiliratum*. Mem Fac Sci Kyushu Univ Ser D Geol 22:165–192
Hoffmann R (2010) New insights on the phylogeny of the Lytoceratoidea (Ammonitina) from the septal lobe and its functional interpretation. Rev Paléobiol Genève 29(1):1–156
House M (1971) The goniatite wrinkle layer. Smiths Contr Paleobiol 3:23–32
Howarth M (1975) The shell structure of the Liassic ammonite family Dactylioceratidae. Bull Brit Mus (NH) Geol 26:45–67
Ikeda Y, Wani R (2012) Different modes of migration among Late Cretaceous ammonoids in Northwestern Hokkaido, Japan: evidence from the analyses of shell whorls. J Paleont 86:605–615
Kakabadze MV, Sharikadze MZ (1993) On the mode of life of heteromorph ammonites (heterocone, ancylocone, ptychocone). Geobios 15:209–215
Karpinsky AP (1885) Geological study in Southern Urals during summer 1884. Izv Geol Comm 4:1–323
Klug C (2007) Sublethal injuries in Early Devonian cephalopod shells from Morocco. Acta Palaeont Pol 52:749–759
Klug C, Brühwiler T, Korn D, Schweigert G, Brayard A, Tilsley J (2007) Ammonoid shell structures of primary organic composition. Palaeontology 50:1463–1478
Korn D (1985) Runzelschicht und Ritzstreifung bei Clymenien. N Jahrb Geol Paläont Mh 1985:533–541
Korn D (2000) Mantle expansion upon the conch in the Late Devonian ammonoid *Erfoudites*. Acta Geol Pol 50:21–27
Kulicki C (1979) The ammonite shell: its structure, development and biological significance. Palaeont Pol 39:97–142
Kulicki C (1996) Ammonoid shell microstructure. In: Landman NH, Tanabe K, Davis RA (eds) Ammonoid paleobiology (pp. 65–101). Plenum, New York
Kulicki C, Doguzhaeva LA (1994) Development and calcification of the ammonitella shell. Acta Palaeont Pol 39:17–44
Kulicki C, Tanabe K, Landman NH, Mapes RH (2001) Dorsal shell wall in ammonoids. Acta Palaeont Pol 46:23–42
Mutvei H (1972) Ultrastructural studies on cephalopod shells. Part II. Orthoconic cephalopods from the Pennsylvanian Buckhorn Asphalt. Bull Geol Inst Univ Uppsala NS 3:263–272
Mutvei H, Doguzhaeva LA (1997) Early ontogeny of shell ultrastructure in *Nautilus pompilius*. Palaeontogr A 246:33–52
Nassichuk WW (1967) A morphological character new to ammonoids portrayed by *Clistoceras* gen. nov. from Pennsylvanian of Arctic Canada. J Paleont 41:237–242
Nikitin S (1881) Jurassic deposits in-between Rybinsk, Mologa, and Myshkin. Materials for Geology of Russia 10(3):200–332

Raup DM, Chamberlain JA (1967) Equations for volume and center of gravity in ammonoid shells. J Paleont 41:566–574
Ristedt H (1971) Zum Bau der orthoceriden Cephalopoden. Palaeontogr A 137:155–195
Rouchadze J (1933) Les Ammonites aptiennes de la Géorgie occidentale. Bull Inst Géol Géorgie Tiflis 1(3):1–273
Ruzhencev VE (1956) About some new genera of ammonoids. Dokl Akad Nauk USSR 107(1):158–161
Ruzhencev VE (1962) General part. In: Ruzencev VE (ed) Superorder Ammonoidea. Osnovy Paleontologii. Mollusca, Cephalopoda I (pp. 243–302). USSR Akad Sci, Moscow
Senior JR (1971) Wrinkle-layer structures in Jurassic ammonites. Palaeontology 14:107–113
Šarikadze MZ, Lominadze NF, Kvantaliani IV (1990) Systematische Bedeutung von Muskelabdrücken spätjurassischer Ammonoidea. Z Geol Wiss Berlin 18:1031–1039
Sinzow IF (1870) Studies on geology of Saratov Province. Zap Imp St-P Miner Soc Ser 2(5):1–55
Spath LF (1925) The collection of fossils and rocks from Somaliland. Ammonites and aptychi. Monogr Geol Dep Hunter Mus 1:111–164
Tozer ET (1972) Observations on the shell structures of Triassic ammonoids. Palaeontology 15:637–654
Turek V, Manda S (2012) "An endocochleate experiment" in the Silurian straight-shelled cephalopod *Sphooceras*. Bull Geosci 87:767–813
Yazykova E (2002) Ammonite and inoceramid radiations after the Santonian–Campanian bioevent in Sakhalin, Far East Russia. Lethaia 35:51–60
Yabe H (1903) Cretaceous Cephalopoda from the Hakkaido (Yesso). J Coll Sci Imp Univ Japan 18(2):1–55
Walliser OH (1970) Über die Runzelschicht bei Ammonoidea. Göttinger Arb Geol Paläont 5:115–126
Westermann GEG (1982) The connecting rings in *Nautilus* and Mesozoic ammonoids: implications for ammonoid bathymetry. Lethaia 5:373–384
Westermann GEG (1996) Ammonoid life and habitat. In: Landman NH, Tanabe K, Davis RA (eds) Ammonoid paleobiology. Plenum, New York

Part IV
Habit and Habitats

Chapter 16
Ammonoid Buoyancy

René Hoffmann, Robert Lemanis, Carole Naglik and Christian Klug

16.1 Introduction

In this chapter we summarize the current knowledge on the buoyancy apparatus of ammonoids. We also discuss the ability of ammonoids to use their phragmocone to obtain neutral buoyancy. A major line of reasoning is the actualistic comparison with living nautilids (e.g., Ward 1979, 1982, 1986, 1987) and other phragmocone-bearing cephalopods (e.g., Ward and Boletzky 1984; Warnke et al. 2010). The upward acting buoyant force is a product of the volume and density of the displaced liquid (in this case seawater, since all cephalopods are marine) which is equivalent to the volume of the submerged object, its magnitude is equal to the weight of the displaced fluid (Archimedes principle, see, e.g., Heath 1897).

The buoyant force acts through a definable point called the centre of buoyancy which is the centre of mass of the displaced fluid (e.g., Trueman 1941). The magnitude of the buoyant force depends only on the volume of the submerged object. Two objects of equivalent volume will experience the same buoyant force regardless of their respective masses. The net buoyancy which predicts whether an object will rise, sink, or stabilize in a column of water, is the difference between the weight of the submerged object and the buoyant force. Some animals evolved mechanisms to

R. Hoffmann (✉) · R. Lemanis
Department of Earth Sciences, Institute of Geology, Mineralogy, and Geophysics,
Ruhr-Universität Bochum, Universitätsstraße 150, 44801 Bochum, Germany
e-mail: rene.hoffmann@rub.de

R. Lemanis
e-mail: robert.lemanis@rub.de

C. Naglik · C. Klug
Paläontologisches Institut und Museum, University of Zurich,
Karl Schmid-Strasse 6, CH-8006 Zurich, Switzerland
e-mail: carole.naglik@pim.uzh.ch

C. Klug
e-mail: chklug@pim.uzh.ch

C. Klug et al. (eds.), *Ammonoid Paleobiology: From anatomy to ecology*,
Topics in Geobiology 43, DOI 10.1007/978-94-017-9630-9_16

change their effective weight while retaining their volume (e.g., *Argonauta* see Finn and Norman 2010).

Changing the spatial distribution of mass causes a simultaneous shift of the respective centre of mass (coincident with the centre of gravity). Mineralized hard parts and muscle tissue of active marine animals are denser than seawater and have negative buoyancy when separated from the rest of the organism. Macroscopic animals spending large parts of their lifetime in the free water column evolved two general solutions to this problem that can be broadly divided into active and passive methods. Active methods are characterized by propelling a volume of water in the direction opposite to the direction of motion (i.e. swimming). Certain animals, such as the flatfish, tuna and some sharks, depend on the force generated through swimming to counteract their sinking due to negative buoyancy. Passive methods are seen in organisms that have evolved gas floats or incorporate low density tissue or fluids in their bodies in order to negate the need to expend energy to counteract sinking. These gas floats, tissues and fluids have a density below the density of water and provided the animal with a total density equal or very close to that of water, a condition referred to as neutral buoyancy (this can also be expressed as the point in which the total weight of the animal is equal in magnitude to the buoyant force).

The ammonia-containing tissues of some squids (Denton et al. 1969) or the swim bladder of some fish (Denton 1962) are examples of tissues and organs which reduce weight and thereby enable the entire organism to reach neutral buoyancy. Similar to the swim bladder, cephalopods evolved a chambered shell (the phragmocone) which acts as a gas float for buoyancy control. It is well known that cephalopods with chambered shells use their phragmocone to regulate their actual weight which also causes a shift of the centre of gravity, to become neutrally buoyant (Ward 1987).

Buoyancy control is applied very regularly by cephalopods with chambered shells such as *Nautilus, Sepia* and *Spirula*, for example in order to facilitate their vertical migrations in *Nautilus* (Crick 1988; Clarke 1969; Denton and Gilpin-Brown 1961b; Dunstan et al. 2011). Changes between negatively, neutrally and positively buoyant animals were observed by Bidder (1962) for aquarium reared *Nautilus* suggesting that liquid can move into and out of the phragmocone chambers.

In addition to the extinct ammonoids and some coleoids, a number of extinct Paleozoic taxa and the Mesozoic and Cenozoic nautilid clade have such chambered shells (Fig. 16.1). The actual weight of an ammonoid is the sum of the shell (conch, septa), soft parts (soft body in the body chamber, siphuncle and pellicle in the phragmocone), liquids and gas in the chambers—the latter being negligible. It was demonstrated that in modern cephalopods with chambered shells the newly built chamber was initially filled with liquid (Ward 1987). The liquid was subsequently pumped out of the chambers by the siphuncle rendering the shell less dense than sea water. Thereby the shell compensates the actual weight of the animal resulting in neutral buoyancy. This mechanism allows for flotation without expending muscular energy just to stay in the water column.

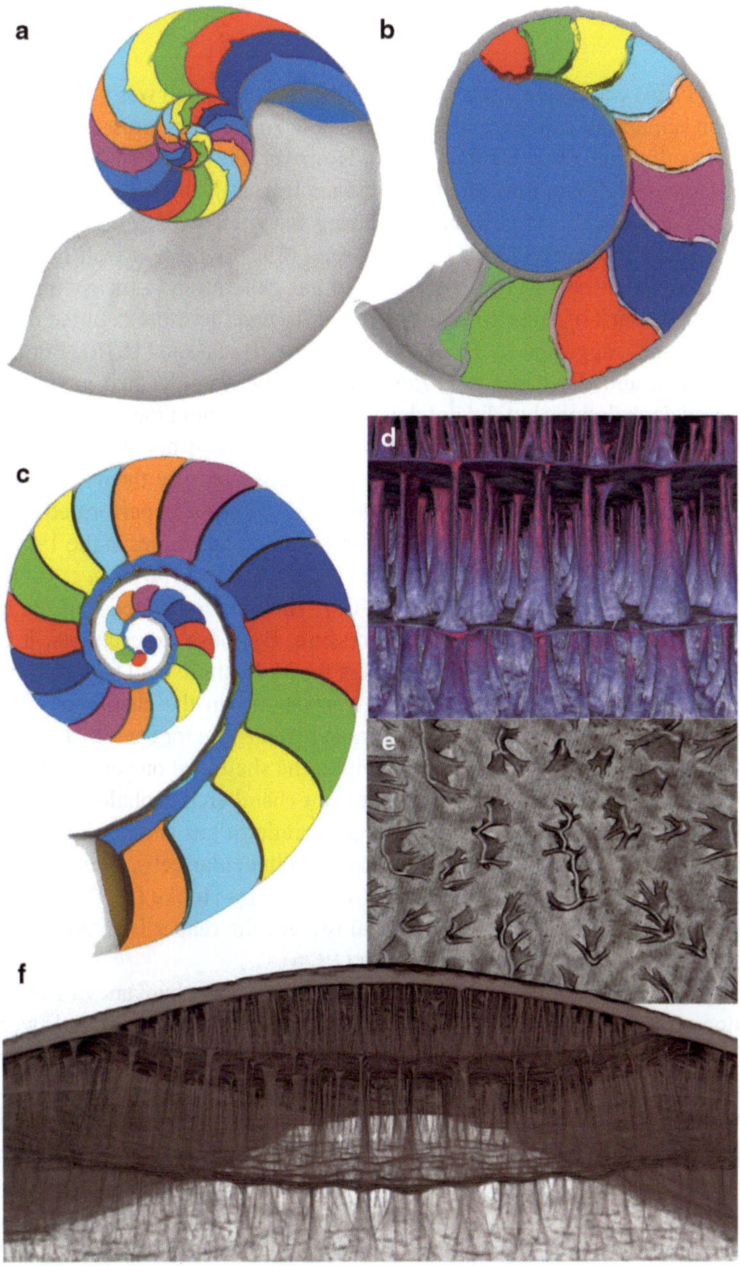

Fig. 16.1 **a–f** section through the shells of chambered cephalopods. **a** *Nautilus pompilius* with the body chamber in grey and the phragmocone chambers coloured, note the slight septal crowding between the last and the penultimate septum. **b** juvenile *Cadoceras* sp. with its protoconch (in *blue*) and the earliest chambers of the phragmocone in different colours. **c** *Spirula spirula* with the siphuncle (*blue*) running through the complete shell. **d–f** *Sepia officinalis*. **d** a close up of f showing the perpendicular pillars between the chamber walls of the phragmocone. **e** a top view of the same pillars showing the complex structure that is reminiscent of ammonoid septa but do not subdivide the chamber. **f** cross section through the cuttlebone of a juvenile *Sepia officinalis* with the thick dorsal shield on top and three subsequent chambers and pillars

16.2 History

Derham (1726) was the first to suggest that the *Nautilus* shell was a buoyancy compensation apparatus and suggested the possible role of gas pressure in expelling water from the shell (*fide* Derham 1726). However, the understanding of the function of the nautilid phragmocone remained unclear until the early 1960s. A simple test to prove Hooke's suggestion, puncturing the shell while immersed in water, was suggested by Owen (1832, 1878). Owen (1832), who was well familiar with the vascularisation necessary for pressurised gas generation in fish swim bladders, recognized the lack of those structures in *Nautilus* and therefore doubted the idea of pressurized gas suggested by Hooke (1695). Only Buckland (1836), Pfaff (1911), Spath (1919), and Westermann (1956) accepted the non-elevated gas-pressure in chambered cephalopods but did not develop new ideas about the chamber emptying process. Arkell (1957), the preeminent ammonitologist of his time, attacked Westermann's (1956) objections and supported the gas pressure theory. Surprisingly, the simple experiment suggested by Owen (1832, 1878) was performed about 130 years later in the framework of functional and anatomical studies in the 1960–1970s by Denton (1962, 1971, 1974), Denton and Gilpin-Brown (1961a, b, c, 1966, 1971, 1973), as well as Denton et al. (1961, 1967). They demonstrated that the nautilid shell does not contain an elevated gas pressure. By these experiments, it became clear that the cephalopod shell does support the hydrostatic load resulting from the overlying column of water at atmospheric pressures in the chambers. They also demonstrated that the siphuncle was the organ responsible for osmotically emptying the cameral liquid and that the pressure within the shell is at one atmosphere or below. Approximately neutral buoyancy in extant chambered cephalopods (*Nautilus*, *Sepia* and *Spirula*) is achieved by changing the ratio of gas and liquid in the phragmocone chambers by osmotic pressure in the siphuncular cells. The hypothesis of pre-septal gas as a buoyancy mechanism invented by Hooke (1695) was rejected by Bidder (1962). For a detailed historical review, the reader is referred to Jacobs (1992) as well as Jacobs and Chamberlain (1996).

First extensive trials for quantitative calculations of the buoyancy of chambered cephalopods were made by means of mathematical models. Such models were published by Moseley (1838), Trueman (1941), and later by Currie (1957); the latter two models were corrected by Heptonstall (1970), Reyment (1958, 1973), Mutvei and Reyment (1973), Westermann (1977), Saunders and Shapiro (1986), Tsujita and Westermann (1998), Westermann (1998b), Kröger (2000) and the latest much more complex calculations were done by Kröger (2002), Hammer and Bucher (2006) and Longridge et al. (2009).

Recently, empirical data have been employed by some authors in order to reconstruct the buoyancy of ammonoids (Tajika et al. 2014; Naglik et al. 2014). These empirical data were produced by both technically intricate, energetically expensive and time-consuming methods such as grinding tomography, computer tomography and synchrotron tomography (for overviews see Sutton et al. 2001; Hoffmann and Zachow 2011; Hoffmann et al. 2014; Sutton et al. 2014).

16.3 Ammonoid Life Habits

One of the oldest and still lasting debates among ammonoid researchers is related with the life habits of ammonoids including whether they could actively swim or not (Naglik et al. 2015). Several attempts were made (e.g., Trueman 1941; Heptonstall 1970) to calculate the buoyancy of fossil cephalopods in order to reconstruct their mode of life. Already Trueman (1941) faced the difficulties of mathematical calculation of masses and volumes of extinct animals by applying geometrical approximations (see below). Since no perfect mathematical method is available and the results of buoyancy calculations seem to support either positive, occasionally neutral or sometimes negative buoyancy for ammonoids, this debate seemed to never end. By contrast, new empirical studies will put an end to these debates, hopefully.

In general, the discussion is twofold. On the one hand, ammonoid life habit is compared to benthic gastropods and thus, this idea is dubbed the "benthic crawler" hypothesis. That all ammonoids were gastropod-like benthic crawlers was repeatedly stressed by Ebel (1983, 1985, 1990, 1992, 1993, 1999) and Rein (1999) for ceratitid ammonoids. These authors even included the shape of the soft body in this hypothesis and reconstructed the head-foot-complex of ammonoids with a gastropod-like foot. Similarly, *Nipponites* was, for a long time, the single known example of an ectocochleate cephalopod referred to a sessile, benthonic habit (Diener 1912; Schmidt 1930; Moore et al. 1952; Tasch 1973). This hypothesis was later rejected by Ward and Westermann (1977) based on new buoyancy calculations in favour for a planktic mode of life. Ebel's (1983) theory was criticized by Westermann (1993, 1996, 1998a), Engeser (1996), Jacobs and Chamberlain (1996), and Kröger (2001) and most other ammonoid researchers. Shigeta (1993) reported on positively buoyant hatchlings becoming markedly negatively buoyant at 2.0–2.5 mm diameter (see Jacobs and Chamberlain 1996, p. 185 for further explanations).

These other researchers including Denton and Gilpin-Brown (1961a, 1973), Ward and Martin (1978) and many more now favour a swimming mode of life of extinct ammonoids similar to their extant relatives and assume that ammonoids were well capable of attaining neutral buoyancy. Neutral buoyancy was suggested by Westermann (1990) and Kröger (2000, 2001) for Jurassic and Cretaceous ammonoids and for Late Paleozoic ammonoids by Swan and Saunders (1987). Heptonstall (1970) calculated a density for ammonites well below that of seawater. A planktic mode of life for heteromorphic ammonoids with a U-shaped body chamber has been considered by Schmidt (1925), Berry (1928), Donovan (1964), Packard (1972), Klinger (1981), and Westermann (1996). Furthermore, Crick (1988), exemplified for nautiloids, outlined the problem of cephalopods not as a problem of increasing buoyancy but as a problem of reducing it. This point of view was supported by the calculations of Westermann (1998a) for earliest plectronocerid cephalopods. Endosiphonal or endocameral deposits in orthoconic cephalopods point in a similar direction (e.g. Westermann 2013). These precipitations act as ballast in the apical part of the phragmocone to offset the strong buoyancy of the phragmocone and to control the attitude of the shell (Kröger 2001). Denton and Gilpin-Brown (1971) reported for *Spirula* that the earliest 1–5 chambers of the phragmocone were filled with liquid. Liquid ballast also occurs in the cuttlebone of extant *Sepia* (Denton and Gilpin-Brown 1961a) as a solution (without carbonate precipitation) to maintain neutral buoyancy and/or a horizontal swimming position. Recently, Westermann (2013) could successfully demonstrate for the orthoconic *Baculites* that the apical part of the phragmocone must have been filled with liquid if a horizontal swimming position is accepted. Seilacher (1960) described an ammonoid (*Buchiceras*) that was not only able to control its buoyancy to some extend but could also compensate a significant change of density that had occurred due to encrustations by oysters during lifetime. The same phenomenon was demonstrated by Klug et al. (2004) for Triassic *Ceratites* which reacted on *syn vivo Placunopsis*-overgrowth by shortening the body chamber. Seilacher's (1960) example was used by Heptonstall (1970) to demonstrate that this *Buchiceras* must have had about 25 % of its phragmocone volume filled with liquid as removable ballast (which is more or less the case in Recent nautilids and was demonstrated by Tajika et al. 2014 for a Jurassic ammonite). Similarly heavy *syn vivo* encrustations by oysters and Cirripedia were shown by Keupp et al. (1999). Other indicators not directly related with buoyancy calculations point to a swimming habit of ammonoids as well. These are the presence of ammonites in black shale deposits representing oceanic anoxic events with oxygen depleted bottom water and the lack of trace fossils. A bottom crawling habit of ammonites with a strong muscular foot comparable to gastropods would favour the preservation of these soft parts in Konservat-Lagerstätten like La Voulte (France; Charbonnier 2009) or Herefordshire (UK; Sutton et al. 2006). However, not a single report about ammonoid trace fossils or massively phosphatized soft body is known to the authors (only poorly carbonized parts; Klug et al. 2012; Klug and Lehmann 2014).

During the last two decades, a more differentiated approach for ammonoid life habits based on shell shape (Westermann 1996, 1998b; Ritterbush and Bottjer 2012), analyses of stable isotopes (Moriya et al. 2003; Lukeneder et al. 2010; Kruta et al. 2014) and sedimentological data (Reboulet et al. 2005) became available.

Universal ammonoid behaviour was challenged by recent observations of depth distribution of *Nautilus pompilius*. Dunstan et al. (2011) document a continuous nightly movement between 130–700 m water depth and a daytime behaviour of either stasis in relatively shallow 160–225 m depth or active foraging in depths between 489–700 m. Additionally, Dunstan et al. (2011) demonstrated the same depth distribution for juveniles and daytime feeding behaviour. These data document a more complex vertical migration pattern dictated by optimal feeding, avoidance of daytime predators, requirements for resting periods at around 200 m to regain nearly neutral buoyancy, upper temperature limits of 25 °C and implosion depths of 800 m. These are important facts that should be considered in future discussions about ammonoid distribution and behaviour.

16.4 Pumping Mechanism

In order to understand buoyancy control and the mode of life of fossil cephalopods, it is most important to understand the ecology and functional morphology of their close living relatives with functional phragmocones, i.e. *Nautilus*, *Spirula* and *Sepia*.

Bruun (1943), who worked on *Spirula*, was the first to mention an osmotic mechanism in cephalopods acting to remove the cameral liquid. However, the osmotic difference between fresh and salt water produces an osmotic pressure of about 25 atmospheres. This value is achieved in 240 m water depth. At greater depths, the hydrostatic pressure will exceed osmotic pressure and chambers would be re-filled with liquid causing negative buoyancy. Due to the occurrence of *Spirula* far below 240 m, Bruun (1943, 1950) rejected the osmotic pump hypothesis. Further observations by Clarke (1970) on the depth distribution confirmed Bruun's (1943) earlier report. Clarke (1970) found high concentrations of juvenile *Spirula* in depths between 1000–1750 m and larger specimens between 600–700 m. Depth distribution during ontogeny is fortunately also recorded in the stable oxygen isotopes of these cephalopod shells. Based on the shells of two live caught specimens of *Spirula*, Warnke et al. (2010) documented ontogenetic changes in terms of depth distribution. Juveniles hatch at around 800 m water depth and subsequently migrate upwards to around 350–400 m; as adults, they sink again down to about 550–600 m (Warnke et al. 2010).

Similar discrepancies between the actual depth distribution (Dunstan et al. 2011) and limitation by the osmotic mechanisms were observed for *Nautilus*. Ward and Martin (1978) demonstrated that below 250 m water depths the osmotic pressure gradient between the siphuncle tissue and phragmocone chambers of *Nautilus* causes an influx of liquid into the chambers, resulting in negative buoyancy. Furthermore, it is possible for *Nautilus* to empty its chambers and regain neutral buoyancy only at depths less than 250 m. Again, the limitation of a simple osmotic mechanism is contradicted by the observation of juvenile *Nautilus pompilius* at 703 m depth reported by Dunstan et al. (2011). After migrating below a water depth of 250 m, a resting and re-equilibration period at around 200 m becomes necessary which has been documented by Dunstan et al. (2011). Note that diving depth of *Nautilus* is limited by temperature, chamber re-filling and shell implosion with the greatest depth being around 800 m (Kanie et al. 1980; Ward et al. 1980a).

After Bruun's (1943, 1950) scepticism, the mechanisms of buoyancy control of extant chambered cephalopods (*Nautilus*, *Sepia* and *Spirula*) have been clarified by the work of Denton and Gilpin-Brown (1961a, b, c, 1966, 1971) and Denton et al. (1961, 1967). Denton and Gilpin-Brown (1961a, b, c) as well as Denton et al. (1961) were not aware of the experiment suggested by Owen (1832), so they started their extensive survey to explore the buoyancy mechanism in cephalopods with the cuttlebone of the cuttlefish *Sepia*. According to the old theory of Hooke (1695), they expected to find pressurized gas within the chambers of freshly caught specimens.

Surprisingly, gas did not emanate from the live caught cuttlefish as a consequence of the sudden ascent and decompression. Gas was also not detected in the tissue surrounding the cuttlebone. When punctured under water, fluid flows into the chambers because phragmocone chamber gas pressure is usually around 0.8 atmospheres. It was observed that more fluid enters the recently formed chambers of the cuttlebone than into older ones, thus indicating a continuously increasing gas pressure. Gas within the chambers was composed of primarily nitrogen, much less oxygen (Bert 1867) and more carbon dioxide compared to atmospheric conditions and the fluid was principally a solution of sodium chloride. Gas pressure in older chambers was in equilibrium with the gases in the blood. *Sepia* varies in density as a result from a change in the density of the cuttlebone being negatively buoyant during day time and neutrally or positively buoyant at night. Quick changes of density and thus buoyancy were arranged by changing the amount of liquid and gas volume in the cuttlebone but not under pressurized conditions. Accordingly, Denton and Gilpin-Brown (1961a, b, c) as well as Denton et al. (1961) assumed an osmotic mechanism to pump cameral liquid into or out of the chambers. While the chamber liquid can be hyposmotic to sea water (when caught in greater depths), the cuttlefish blood is almost isosmotic with sea water. Changes of cameral liquid must be carried out by exchanges in salt or water across the siphuncular membrane in order to balance the hydrostatic pressure by an osmotic force. The siphuncular membrane covering the siphuncular wall acts as a barrier; through active transport of ions across that membrane, water follows due to the higher salt concentration compared to the cameral liquid. Contrary, concentration changes of the liquid within each chamber depend on diffusion of salts either towards or away from the region of the siphuncular membrane. When the cameral liquid of a newly built chamber was completely emptied, gas migrated under very low pressure by passive diffusion from the blood of the cuttlefish into that chamber.

Nautilus and *Spirula*, were also studied by Denton and Gilpin-Brown (1966, 1971) as well as Denton et al. (1967). Both taxa share a siphuncular tube which runs through all chambers of the phragmocone, while in *Sepia*, the siphuncle is almost flat (siphuncular field). Only through the permeable siphuncle, exchanges of liquids and substances between chambers and tissues can take place (Denton and Gilpin-Brown 1966; Collins and Minton 1967; Chamberlain and Moore 1982). However, the results they obtained show very great similarities with their findings in *Sepia*. Gas pressure in the chambers of *Spirula* was lowest in newly built chambers but remains below normal atmospheric pressure at sea-level (around 0.8 atmospheres). Concerning the gas pressure, the same holds true for *Nautilus* and consists mainly of nitrogen (Vrolik 1843). Contrary to *Sepia*, the liquid of a newly formed chamber in *Nautilus* was isosmotic with sea water at first. Concentration of salts within the liquid of a new chamber decreases to about one fifth of that of sea water and the animal's blood before water begins to leave and before the first gas bubble enters the chamber. During later ontogenetic stages, the earliest chambers will become refilled with liquid that gradually changes from hyposmotic to isosmotic to sea water (Denton 1971). For *Nautilus* with its external shell, the general physiology was found to be very similar to those of *Spirula* and *Sepia*. One could argue that these three living genera represent an extant phylogenetic bracket *sensu* Witmer (1995), thus corroborating similar buoyancy regulation in ammonoids. Nevertheless, there are many neutrally buoyant living coleoids that employ different mechanisms. Unlike *Sepia*, both *Nautilus* and *Spirula* are as close to neutral buoyancy as possible.

For *Nautilus*, it has been demonstrated that cameral liquid first closely resembles sea water in its ion content, but with sufficient differences to assume that cameral liquid is a secreted body fluid and not unaltered sea water (Denton and Gilpin-Brown 1966; Greenwald and Ward 1982; Mangum and Towle 1982). Prior to removing the cameral liquid, some NaCl was removed by the siphuncle; subsequently, true emptying including salt and water transport out of the chamber starts (Denton and Gilpin-Brown 1966; Ward 1979). Salt concentration progressively decreases at the beginning after 50% of the cameral liquid was removed; the remaining liquid lost direct contact with the siphuncle and salt concentration increases (Ward 1979; Ward et al. 1981). At the same time when cameral liquid is removed from the chamber, a vacuum is created into which various blood gases (nitrogen, argon, carbon dioxide, oxygen) diffuse. After about 50% of the cameral liquid in the final chamber was pumped out, a new chamber formation cycle starts (Ward et al. 1981). The cameral liquid of the newest chamber initially acts as a brace against external hydrostatic pressure (Collins et al. 1980) and as a damper (buffer) to avoid explosion of the siphuncle (not yet calcified) due to blood pressure. Denton and Gilpin-Brown (1966) as well as Ward and Martin (1978) stressed that emptying is a slow and irreversible process in *Nautilus*. Ward and Greenwald (1982) and Greenwald and Ward (1987) managed to demonstrated that emptying, i.e. chamber refilling, can also be reversed.

Finally, Denton (1971, 1974) summarized that the mechanism of pumping salts and water is basically the same in *Sepia*, *Nautilus* and *Spirula*. Later, in the context of fossil cephalopods (nautiloids, ammonoids, and belemnoids), Denton and Gilpin-Brown (1973) assumed a similar mechanism for chambered fossil cephalopods with a siphuncle.

However, even after the osmotic mechanism had been adequately clarified, the same conflict between observed depth distribution of chambered cephalopods and the assumed depth limitations for the osmotic pump remain unclear; the largest difference in osmolarity between chamber liquid and sea water could produce a pressure equalled at 240 m water depth (Denton and Gilpin-Brown 1966; Greenwald et al. 1980, 1984). By contrast, Clarke (1970) has shown that *Spirula* spends only a small fraction of its life at depths with less pressure. Denton (1971) described the observation that dead *Spirula* does not fill with liquid when exposed to high hydrostatic pressures.

16.4.1 Decoupling

One possible explanation for the conflict between osmotic pumping and observed habitat was the decoupling argument (Denton and Gilpin-Brown 1966; Denton et al. 1967). It was suggested that *Nautilus* and *Spirula* emptied cameral liquid above 250 m water depths and when migration into greater depths takes place, the chamber liquid was decoupled. That means that chamber liquid was not in direct contact with the siphuncle. The decoupling argument has been applied by ammonite workers to explain chamber, septal and suture complexity of ammonoids (e.g., Gottobrio and Saunders 2005).

This assumption is contradicted by the report of Ward et al. (1984) and Dunstan et al. (2011) for *Nautilus* and by Clarke (1970) for *Spirula*. Ward et al. (1984) documented the vertical distribution of one *Nautilus* specimen being tracked for seven days which was commonly below 240 m and Dunstan et al. (2011) recorded the deepest encounters of living *Nautilus pompilius* at 703 m. However, as stressed by Jacobs (1992, 1996) as well as Jacobs and Chamberlain (1996), decoupling would not prevent liquid to flow back into the chambers when hydrostatic pressure exceeds the osmotic pressure. The presence of the pellicle, a wetted organic sheet that covers the septal surface of *Nautilus* chambers with the potential for fluid transport, makes it highly unlikely that fluid can be successfully isolated. The pellicle acts as a wick or blotting paper. As a consequence, decoupling of liquid from the siphuncle was rejected in favour of local osmosis, where organisms can concentrate salts in intracellular structures (Diamond and Bossert 1967). Maybe this rejection was related to a misinterpretation of the idea of Denton and Gilpin-Brown (1966) as well as Denton et al. (1967) and the decoupling model. In his final summary, Denton (1974) already mentioned for *Nautilus* that the chalky tube and the pellicle still allow liquid to be drawn towards the permeable part of the siphuncle after losing direct connection. Pumping is still possible but at a lower rate. Similar cases are reported for *Sepia* and *Spirula*, both with a very small region through which liquids can be exchanged between living tissue and the chambers.

The theory of decoupled liquid was revived by Kröger (2003) and is agreed here in its narrow sense. It was pointed out that the decoupled theory is based on decoupled spaces used by cephalopods like buffer zones while changing their position in the water column. The decoupled spaces thereby protect the animal against liquid flowing back into the phragmocone chambers. The theory was based on the empirically observed phenomenon of decoupled spaces in *Sepia*. It was stressed that any decoupling space that creates a system with different salinity acts like the standing gradient model proposed by Diamond and Bossert (1967). It is more energetically efficient to transport liquid along an osmotic gradient than with no or against an osmotic gradient. Decoupled spaces therefore should be much smaller than the entire chamber, because ion concentration is more rapid and easier to maintain. These kinds of spaces are especially useful in species with an active use of their buoyancy apparatus (Kröger 2003).

16.4.2 Local Osmosis

The second explanation discussed by Denton and Gilpin-Brown (1966) as well as Denton et al. (1967) was that localized high concentrations of solutes could be built up in a "*pumping*" epithelium by a mechanism similar to that proposed by Diamond and Bossert (1967, 1968) as "*standing gradient model*" and used to extract liquids from the chambers of the shell by simple osmosis.

Later, Denton and Gilpin-Brown (1973) cited a study of hyperosmotic pumping observed for sea birds suggesting a similar system for cephalopods in depths below 250 m. Such a pumping mechanism was described by Greenwald et al. (1982, 1984) for the *Nautilus* siphuncle. The siphuncular epithelium (Fig. 16.2) has microvilli at the site of putative NaCl absorption into the cell as well as elaborate folds of the basement membrane. These folds are located at the site of NaCl extrusion from the cell extending throughout the cytoplasm as a network of blind ending sacs (canaliculi). The canaliculi were lined with numerous mitochondria (Greenwald et al. 1980, 1982, 1984; Ward et al. 1980b; Mangum and Towle 1982; Tanabe et al. 2000). That arrangement is typical for tissues that carry out NaCl solute and water transport against osmotic or hydrostatic pressure (Berridge and Oschman 1972) and was found in a single ammonoid siphuncle of *Akmilleria* so far (Tanabe et al. 2000). This transport system is known as local osmosis system (Diamond and Bossert 1967), in which mitochondria supply ATP (energy) to salt ion pumps (= hyperosmotic pump of Jacobs 1996). The hyperosmotic pump actively transported Na^+ ions across cell membranes into microscopic intracellular channels, while Cl^- ions follow passively to maintain the charge balance. The resulting locally elevated solute concentration causes water to diffuse from the cytoplasm of the cell (local osmosis) into the microscopic channels (Fig. 16.2). In turn, these high pressures in the intracellular channels exceed the hydrostatic pressure and water thus flows into the channels and therefore into the siphuncle. This water leaves the cell through the openings of the folds. In the case of the channels, their geometry plays an important role. Water diffuses readily across the relatively large surface area of the channel, but flow down the channel and into the siphuncular lumen is much more rapid than diffusive processes.

The ultrastructure of the *Nautilus* siphuncle epithelium (Mangum and Towle 1982; Tanabe et al. 2000) and the presence of the sodium transport enzyme Na^+/K^+-ATPase (Bonting 1970) localized at the folds supports the assumption of the presence of a local osmosis system in the siphuncle of *Nautilus*. The salt of the basolateral labyrinth was transported by Na-K-ATPase into the intracellular system of canaliculi. A sufficiently high salt concentration generates a high osmotic concentration that draws liquid out of the chambers against either osmotic or hydrostatic gradients (Greenwald et al. 1984). These canaliculi systems were observed for the twelve most recent chambers but not from the third or fourth oldest chambers. It therefore appears likely that the earliest chambers did not contain living siphuncular tissue. Thus, intracellular channels can transport fluid against very high pressures as long as ions are present to be pumped.

The above described local osmosis model allows for cameral liquid emptying below 240 m water depths and could pump against almost any pressure. Accordingly, the limitation of osmotic pumping is now the concentration of ions that could be maintained in the tube instead of the ion concentration of the ambient sea water (Greenwald et al. 1982). Decoupling in its wider sense becomes needless for the

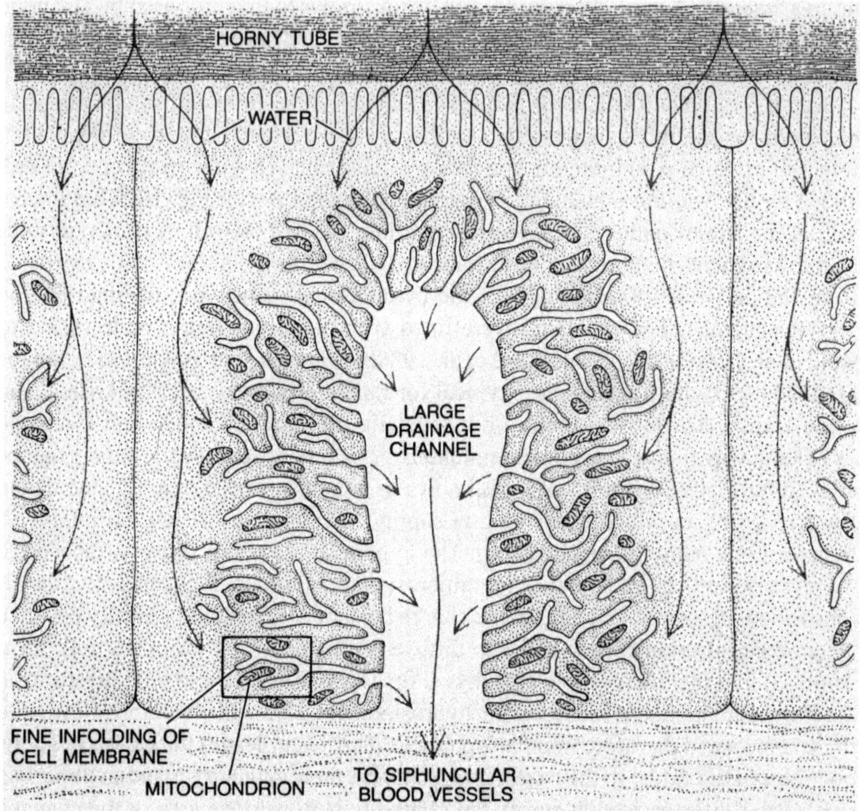

AQ1

Fig. 16.2 Diagrammatic representation of a single transporting cell of siphuncular epithelium. Water molecules (*arrows*) move from horny tube, across the brush border, and into the cell. Fine infoldings within the cell communicate with a central drainage canal. The water ultimately reaches the siphuncular blood vessels, and is then carried to the body (reproduction from Ward et al. 1980b, by courtesy Peter D. Ward, Adelaide)

explanation of depth distribution in the presence of local osmosis in extant cephalopods. Due to the finding of soft parts in the *Akmilleria* siphuncle with preserved ultrastructure including canaliculi, it is now likely that local osmosis was the active pumping mechanism in ammonoids (Tanabe et al. 2000). However, presence of canaliculi for *Spirula* has not been objectively demonstrated (Chun 1915).

16.4.3 Preseptal Gas

Since Hooke (1695) compared the buoyancy mechanism in *Nautilus* with the fish bladder (Derham 1726), the gas pressure hypothesis became widely accepted without any justification (see Chap. 16.2). Meigen (1870) and Schmidt (1925) supposed that *Nautilus* used preseptal gas to control its density. The gas pressure hypothesis then was also applied to fossil cephalopods but was rejected by Bidder (1962) for *Nautilus*.

The explanation of an internal pressure-driven buoyancy regulation with the last septum remaining uncalcified and functioning analogue to a fish swim bladder, i.e. the Cartesian diver model (Seilacher and LaBarbera 1995, Seilacher and Gishlick 2015), was rejected by Jacobs (1996), Kröger (2002) and Klug et al. (2008) for various reasons. No fossil indications such as muscle scars were found supporting the proposed muscular diaphragm enclosing a gas bubble and other mechanical implausibilities. Guex (2005) stressed the presence of a preseptal cavity in cephalopods recognized by the Haftband-Struktur, without providing new evidence. That cavity was suggested to allow adult ammonites to continue with growth, even if the phragmocone did not contain any cameral liquid by passive diffusion of gas into the preseptal cavity. In our opinion, this idea is currently lacking any support.

16.4.4 Role of the Siphuncle

The structure and detailed explanation of the function of the ammonoid siphuncle will be given elsewhere (Kulicki 1996; Tanabe and Landman 1996; Tanabe et al. 2000, 2015). Ward (1982) demonstrated the proportional dependence between the siphuncular surface area and the chamber emptying rate in *Nautilus* (see also Kröger 2003). Because siphuncular epithelia of *Nautilus*, *Sepia*, *Spirula* and ammonoids are histologically similar (Denton and Gilpin-Brown 1973; Drushchits and Doguzhaeva 1981; Barskov 1990, 1996, 1999; Tanabe et al. 2000), it was assumed that pumping power and morphology of siphuncular epithelia, the place of osmotic pumping, remained constant during cephalopod evolution (Ward 1982; Kröger 2003). A detailed structural description for the *Nautilus* siphuncle was presented by Ward (1987) as well as Tanabe et al. (2000), for *Virgatites* (Perisphinctidae) by Barskov (1996) and for *Akmilleria* (Prolecanitida) by Tanabe et al. (2000). Besides osmotic or hydrostatic pressures, the osmotic pump mechanism responsible for emptying and refilling of cameral liquid depends on the diameter and ultrastructure of siphuncle and surrounding tissues. Recently, Westermann et al. (2002) localized putative neurotransmitters in the mantle and siphuncle of *Nautilus pompilius* with the assumed function of blood circulation as well as complex secretory mechanisms. The siphuncle diameter itself is limited by strength requirements (Westermann 1971; Ward 1982; Jacobs 1992; Hewitt 1996; Hewitt and Westermann 1997). Depth distribution of chambered cephalopods is further limited by shell- and siphuncular strength. Westermann (1971, 1982) introduced the siphuncular strength index. Concerning the siphuncle, tubes of smaller diameter are stronger but have a smaller active surface for pumping which will lower the emptying rate. According to Westermann (1971, 1982), lytoceratids and phylloceratids have a greater siphuncular strength index, thus indicating that these forms lived at greater depths compared to other ammonites. Ward and Martin (1978) reported on chamber emptying rates of a maximum of 1.0 ml/day/chamber in living *Nautilus*.

By contrast, Ward and Greenwald (1982) reported a maximal refilling rate of 100 µl/h in response to a sudden buoyancy increase in living *Nautilus*. Based on the observed maximum emptying rate of 1 ml/day and maximum refilling rate of 2 ml/day, Ward and Greenwald (1982) concluded that *Nautilus* can only fine tune its buoyancy in order to ensure that the animal is always slightly negatively buoyant (which human scuba-divers also do). But these rates are not rapid enough to serve for the observed diurnal migrations, i.e. daily buoyancy-aided ascents and descents as suggested by Hooke (1695), Willey (1902), and Heptonstall (1970). Consequently, the siphuncle of *Nautilus* is only capable of long term but not of short term or sudden buoyancy regulations.

Due to their large siphuncular surface area relative to chamber volume, only sepiids and potentially some Paleozoic cephalopods such as actinocerids and endocerids are able to adjust their buoyancy on a daily basis (Ward 1982; Kröger 2003). For ammonoids, Ward (1982) excluded the possibility of diurnal migration (see below).

Based on the calibrated pumping rate of siphuncular tissue, it turned out that the actively pumping siphuncular area is too small relative to the chamber volume. Instead, these cephalopods, including nautilids, may have accommodated vertical movements through active locomotion as observed for extant *Nautilus* (O'Dor et al. 1993).

Jacobs and Chamberlain (1996) argued that structural and energetic reasons prevent the development of large siphuncular tubes in cephalopods allowing for rapid density changes. The structural argument is based on the fact that the siphuncular strength is inversely related to diameter. This limitation was also observed in cuttlefish. Ward and Boletzky (1984) reported that shallow water forms have a flat cuttlebone with a large siphuncular area, while forms from deeper water have a cylindrical cuttlebone with a smaller surface area. Shallow water sepiids apparently can change their density faster compared to deeper forms (which might be linked with physical necessities). Energetic limitations are due to energy costs of pumping and maintaining empty chambers. These costs are proportional to the siphuncular surface area and will increase with water depths (hydrostatic pressure).

By contrast, Bandel and Stinnesbeck (2006) stressed that the permeable zone within the siphuncular tube of *Spirula* is as long as one chamber (Fig. 16.3). As a

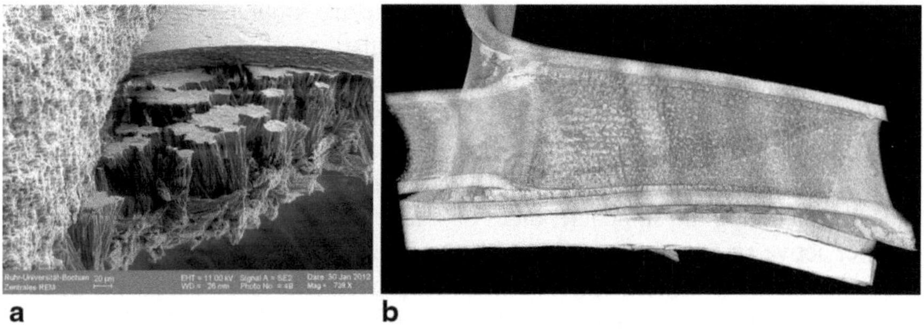

Fig. 16.3 **a** Siphuncle of *Nautilus pompilius*. **b** Siphuncle of *Spirula spirula* with the porous zone all along the inside of the septal neck

consequence, most of the actively pumping siphuncle epithelium is in contact with the cameral liquid and the ratio of pumping area to chamber volume is increased. Due to its small size and the precise buoyancy adjustment with its largest chamber volume of about 2 ml (Denton 1971), buoyancy may change fairly rapidly (Bandel and Boletzky 1979) and is mainly supported by pumping cameral liquid in or out of the chambers.

Following Mutvei and Reyment (1973), Kröger (2003) rejected Ward's (1982) idea that the relation of siphuncular surface area to chamber volume (si = siphuncle surface area index) was too small to support daily buoyancy adjustments in ammonoids. Kröger (2003) compared 250 cephalopods including nautiloids, ammonoids and coleoids. He demonstrated that cephalopods with high but also small si values were capable of rapid buoyancy adjustments. Already Heptonstall (1970) stressed that in ammonoids, the siphuncle with its extreme ventral position has the greatest possible length and hence is in the most efficient position for cameral liquid exchange.

Exceptional with respect to the position of the siphuncle is the group of Late Devonian clymeniids with the siphuncle in a dorsal position. Superficially almost indistinguishable from other ammonoids in terms of shell geometry and sutural complexity, clymeniids differ by their two to three times larger siphuncle diameter and up to 33% thicker shells. With its dorsal position, the siphuncle surface and volume was decreased by about 50% compared to narrower ventral siphuncles (Gottobrio and Saunders 2005). The reduction of siphuncular active pumping area was partially or fully compensated by its increased thickness. According to Westermann (1971, 1982), this restricted clymeniids to a shallow marine habitat.

Several authors focused on the refill ability of the siphuncle (Keupp 1997, 2000, 2012; Daniel et al. 1997; Kröger 2002). Based on observations of repaired injuries accompanied by shell loss, these authors speculated on a high rate of liquid refill. That refill had to compensate for the decreased weight by shell loss before the animals reached the surface.

16.4.5 Role of Cameral Liquid

In all three extant cephalopods that built a chambered shell (*Nautilus*, *Spirula*, and *Sepia*), the liquid that fills a newly formed chamber differs in composition from sea water and is recognized as a body fluid (Denton 1974). During ontogeny, the role of liquid for supporting buoyancy adjustment decreases with the growing total volume of the animal. Juvenile *Nautilus* retains about 32% cameral liquid while adults keep 12% of liquid compared to the phragmocone volume (Ward 1980, 1987). By perforating a single chamber directly above the hood (i.e. about 270° behind the aperture) of a *Nautilus* of 200 mm diameter, Tsujino and Shigeta (2012) demonstrated that *Nautilus* loses its neutral buoyancy due to water logging. The exact volume of water causing the permanent benthic position of *Nautilus* was not measured (pers. com. Tsujino). Nevertheless, the specimen managed to swim actively by producing

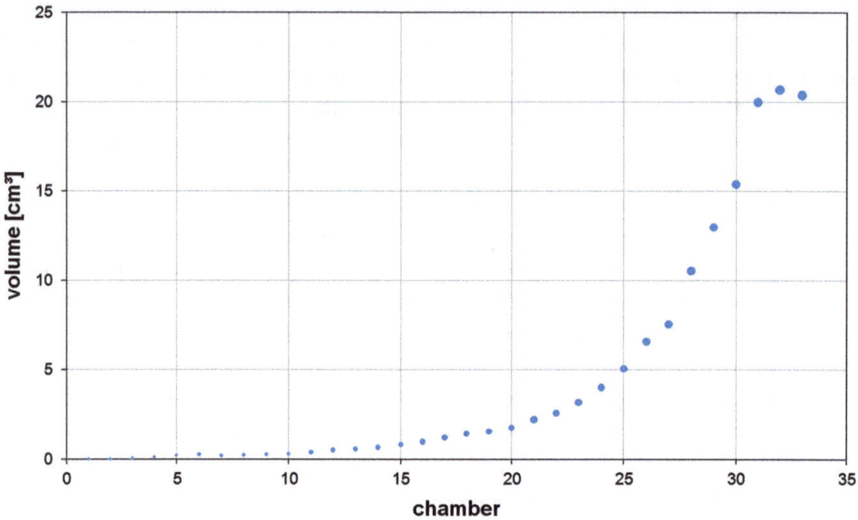

Fig. 16.4 Chamber volume for the 33 chambers of the *Nautilus pompilius* shell shown in Fig. 16.1A. Septal crowding can be recognised by decreased chamber volume

a jet over a short time close to the sediment surface. Furthermore, the siphuncle was cut and afterwards, the adapertural part of the siphuncle decomposed. This demonstrated that perforation of the phragmocone and siphuncle is not necessarily a fatal injury.

Denton et al. (1967) found small amounts of liquid in the last three chambers of a *Spirula* phragmocone made of 30 chambers, followed by 18 completely emptied chambers. In the subsequent chambers, the amount of liquid rises with the five smallest chambers being completely filled with liquid. When the animal grows, the earliest, smallest chambers of an adult animal have only a small fraction of volume of the latest chambers (Fig. 16.4). Kröger (2003) speculated that the high energy cost of emptying these chambers and maintenance is inversely related to their importance of buoyancy regulation. This may be the reason for chamber refill of the earliest, smallest chambers observed in *Spirula* (Denton et al. 1967).

Retained liquid was interpreted as ballast water for buoyancy regulation, e.g., to balance density increase due to shell and tissue growth, epizoa growing on the outer shell, injuries that cause partial water logging and to support vertical migration (Heptonstall 1970; Denton 1974; Keupp et al. 1999). At the time of sexual maturity, when growth ended, most or all cameral liquid was removed in living *Nautilus* (Collins et al. 1980).

The pellicle, the organic sheet covering the inner surface of phragmocone chambers of *Nautilus* (Denton and Gilpin-Brown 1966), *Spirula* (Appellöf 1893; Denton and Gilpin-Brown 1973) and ammonoids (Weitschat and Bandel 1991), transfers the cameral liquid from the more remote parts of the chambers to the siphuncle. As exemplified by Kröger (2003), the relative inner surface of the chambers represents the potential of liquid refill rate under the assumption that the ammonoid chambers were also covered with the pellicle. For comparison of the ability of cephalopods to compensate slow or rapid buoyancy increase (e.g. due to shell loss by predation), Kröger (2003) introduced the inner chamber area index (*cai*) which is the inner chamber surface against its volume.

16.5 Buoyancy Calculations for Ammonoids

Due to comparable shell architecture and the presence of a siphuncle, the observed pumping mechanisms in extant relatives (*Nautilus*, *Spirula*, and *Sepia*) like osmotic pumping and local osmosis are assumed to act in ammonoids as well. Besides, it remains unclear if and how much liquid was kept in the ammonoid phragmocone (fill fraction in the sense of Kröger 2002), original shell density, ratio of soft body and body chamber volume. Further issues related with the calculation of the buoyancy for extinct animals like the relationship of density and volume and simplifying assumptions are discussed below.

16.5.1 Issues of Buoyancy Calculation

Trueman (1941) conducted the first extensive research to calculate masses and volumes of ammonoid taxa. For extant *Nautilus*, Trueman (1941) calculated the volume of the shell by the displaced amount of water and determined the volume of each chamber by cutting the shell along its median plan and fill the chambers with liquid. For extinct ammonoids, Trueman (1941) measured shell and septal thickness and applied the density of crystal aragonite for the shell. He assumed that the siphuncle contributed about 1% to the total weight of the shell while the septa contributed 10% (see his tab. 1, p. 360 ff). Trueman's (1941) and Currie's (1957) data were re-evaluated by Heptonstall (1970) using modern estimates for cephalopod tissue and shell densities. Mainly the equations of Raup and Chamberlain (1967) were applied to buoyancy calculations with additional corrections and extrapolations for different shell parameters such as ribs introduced by Kröger (2002). However, coiling of the shell is not dictated by mathematical parameters and does not follow a perfect logarithmic spiral. Shigeta (1993) could demonstrate that the planispiral whorls show a polyphasic allometric growth pattern (see also Klug 2001 and references therein). Coiling parameters and shapes may also change during ontogeny as is possible for all other morphological features and subsequently derived ratios (see Chap. 9 for intraspecific variability). Therefore, geometrical approximation of actual shapes of structures such as the whorl section as circles or polygons or the complex shape of septa bias the results. Calculations based on the statement that volumes in general are approximated by the visual matching of geometrical triangles, rectangles or trapezes (Ebel 1983) have therefore been rejected (e.g., Jacobs and Chamberlain 1996).

16.5.2 Volumes

Another point is the over- or underestimation of body chamber lengths and the volume of the soft body. Complete body chambers are rarely reported for ammonoids and nautiloids (Klug et al. 2014b). In earlier approaches, body chamber lengths were calculated for a range of ammonoids assuming neutral buoyancy. Therefore, information about volumes and densities of all negatively buoyant structures (shell and soft body) are required. A major error in buoyancy calculations is related with the unknown, and therefore approximated, body chamber length and the resulting soft body volume (Kröger 2002). Additionally, the pellicle and cameral sheets have to be taken into account with approximately 14 % of the chamber volume and a density of 1.055 g/cm³ as assumed by Hewitt and Westermann (1996).

Additional problems arise, because the actual distribution of shell material in the ammonoid phragmocone is not trivial to quantify in detail. Hammer and Bucher (2006) calculated for the ammonoid *Intornites* that the shell material makes up about 18 % (about 11.8 g) of the total mass (65.56 g). In their empirical study, Tajika et al. (2014) found that in *Normannites*, the shell mass is about 15 % (5.9 g) of the entire mass (24.36 g with aptychi). These discrepancies in shell-weight proportion of total weight are related with the fact that shell wall thickness is not constant but increases during ontogeny and wedges out or may strengthen near the adult aperture. Shell wall thickness also varies in different positions such as on the umbilicus, flanks or venter (Kröger 2002). Also, ornamentation like ribs, nodes or clavi contribute to the total shell weight but may increase or decrease in expression and density during ontogeny. In order to take ribs into account, Magnin in Delanoy et al. (1991) introduced a factor of sculpture (F_s) under the assumption of a sinusoidal curvature and was subsequently used by Kröger (2002). However, ribs neither are exactly radial nor represent an ideal sinusoidal curvature and their expression may also vary along the shell surface, becoming more prominent with a stronger curvature of the conch wall (Guex et al. 2003; Hammer and Bucher 2005).

Averaged values for septal thickness do not account for the fact that septal thickness decreases to the shell margin (Westermann 1975; Hewitt 1985; Hewitt and Westermann 1987). For example, estimations for septal thickness were expressed as ratio of the whorl height. Shigeta (1993) applied 2 % of the whorl height as being the septal- and conch wall thickness, while Westermann (1993) found 1.3–1.8 % in several Late Cretaceous ammonite genera and in epicontinental ammonoids, he measured a ratio of 1 % of whorl height. Westermann (1993) states that fluted septa thin out from the centre to the margin at about the same rate as fluting increases area; consequently, fluting does not add extra shell and weight in such way that similar total weights were obtained independent of the degree of fluting, assuming a simple plate of constant thickness equalling that of the centre of the septum. This has never been proven.

Finally, Saunders and Shapiro (1986) employed thickness terms for the shell and septa derived from digitized sections of the specimens that should lead to more reliable results. Nevertheless, the distribution of shell material did not resemble the three dimensional distribution of shell material. By contrast, a simple circular whorl section was assumed by Saunders and Shapiro (1986). For compressed shells, this assumption is likely to produce an overestimate of shell volume relative to shell mass and an overestimate of body chamber length.

16.5.3 Shell Density

Another problem related with the shell is not only the approximation of its volume but the assumed density of the ammonoid shell material. Trueman (1941) applied the highest value with 2.94 g/cm³ which is the density of the aragonite crystal and was also used by Raup and Chamberlain (1967) as well as Crick (1988). These authors overlooked the values presented as relative density by Schwartz (1894) with 2.68 g/cm³ and Kelly (1901) with 2.688 g/cm³. After Trueman (1941), Reyment (1958) introduced a second lighter value of 2.62 g/cm³ due to the recognized incorporation of high amounts of organic components into the mollusc shell. Kelly (1901) implied 15% organic components within the nacre of *Nautilus* and *Spirula* nacre due to the different densities between crystal and nacre. For the organic components, Kelly (1901) assumed a density of 1.1–1.2 g/cm³. That value (2.62 g/cm³) was used by most of the subsequent worker such as Heptonstall (1970), Reyment (1973), Tanabe (1975), Ward and Westermann (1977), Saunders and Shapiro (1986), Okamoto (1988), Shigeta (1993), Kröger (2002), Hammer and Bucher (2006). Other density values used so far given in chronological order are: 2.6 g/cm³ by Westermann (1977), 2.53 g/cm³ by Collins et al. (1980) and Ebel (1983), 2.67 g/cm³ by Chamberlain et al. (1981) and Longridge et al. (2009), 2.7 g/cm³ for the shell *Nautilus* by Greenwald and Ward (1987), 2.5958 g/cm³ by Ebel (1993), 2,69 g/cm³ by Hewitt and Westermann (1996) and Hewitt et al. (1999) and finally 2.65 g/cm³ by Westermann (2013) for nacre density.

Applied shell densities vary from 2.53–2.94 g/cm³. Variation of shell density within the shell was mentioned by Mutvei (1983) due to incorporation of different amounts of organic matter. For a *Nautilus* shell, collected some thirty years ago, Hoffmann and Zachow (2011) documented different shell densities for the conch wall (2.57 g/cm³) and the septa (2.51 g/cm³). Hewitt and Westermann (1996) reported that the density of the *Nautilus* phragmocone with a value of 2.668 g/cm³ differs from that of the adult body chamber with a reduced density of 2.582 g/cm³. On the contrary, the same density was assumed mostly for the conch wall and septa of ammonoids (e.g. Longridge et al. 2009), thus neglecting the small contribution of 2–6% (Trueman 1941) of the septa documented for Mesozoic ammonoids or 6% of septa and siphuncle (Raup and Chamberlain 1967) to the total shell volume. While Saunders and Shapiro (1986) report values of 5–14.6% for septa and siphuncle of the volume of the entire shell, Kröger (2002) decided to choose a mean value of 5.7%.

16.5.4 Soft Body Density

Imprecise calculations for the shell may cause significant errors in the estimations of the body chamber length and thus the soft body volume and could also lead to wrong assumptions about the paleobiology of ammonoids. Trueman (1941) calculated a soft body density of 1.13 g/cm³ under the assumption that the body chamber was completely filled with the soft body. Later, it was generally accepted that the ammonoid soft body had the same density as that of *Nautilus* (1.06 g/cm³) and that the soft body completely filled the body chamber (e. g. Reyment 1973; Tanabe 1975; Ward and Westermann 1977; Greenwald and Ward 1987; Monks and Young 1998). The latter was criticized by Kröger (2002) as too high. Up to 15 % has to be subtracted from the body chamber volume for the mantle cavity which is used by the animal for respiration and jet propulsion (Chamberlain 1987; Wells 1990). This mantle cavity ratio remains constant throughout ontogeny (Chamberlain 1987). Recently, Westermann (2013) suggested that only two thirds of the body chamber were occupied by the soft body with the remainder being filled with sea water. It can be speculated that ammonoids with long body chambers had similar or higher values.

Other density values for the soft body are given in chronological order with 1.068 g/cm³ applied by Denton and Gilpin-Brown (1961a, b, c, 1966) as well as Heptonstall (1970), 1.067 g/cm³ applied by Raup and Chamberlain (1967), Okamoto (1988) and Shigeta (1993), instead, Ward and Westermann (1977) applied 1.05–1.07 g/cm³, Ward (1986) differentiated the *Nautilus* soft body in coelomic fluid with a density of 1.026 g/cm³ and the tissues with 1.068 g/cm³, 1.0568 g/cm³ was applied by Ebel (1993), while Saunders and Shapiro (1986), Hewitt and Westermann (1996), Hewitt et al. (1999), Hammer and Bucher (2006) as well as Longridge et al. (2009) applied 1.055 g/cm³, Hoffmann and Zachow (2011) applied a mean value of 1.047 g/cm³ based on the two values presented by Ward (1986) and most recently Westermann (2013) applied 1.065 g/cm³ taking the buccal mass including the aptychus into account. Tajika et al. (2014) separately considered soft tissue density and jaw density.

For the buccal mass of the cephalopod head region, including the radula and the chitinous jaw apparatus, an average density of *Nautilus* jaw elements of 1.655 g/ml was reported by Hewitt and Westermann (1993). Both the volume of siphuncle has to be subtracted from chamber volume and the density of soft tissue has to be applied to the siphuncle volume. To be precise, one should include the pellicle, the wettable tissue covering the inner surface of each chamber, in buoyancy calculations, although its contribution in terms of total weight might be low.

For *Nautilus*, Ward et al. (1977) reported a maximum crop content of 76 g (21.7 % of the soft body mass). This crop content was responsible for a mass increase (negative buoyancy) of the animal in sea water of 0.9 g, while the same animal weighed under water with its shell was only 0.3 g. Accordingly, the crop content needs to be considered due to its nontrivial mass.

16.5.5 Sea Water Density

There was a more or less uniform use for the values of sea water density with 1.025 g/cm³ for sea water at the sea surface at room temperature (Greenwald and Ward 1987) or 1.026 g/cm³ used by Raup and Chamberlain (1967), Heptonstall (1970), Ward and Westermann (1977), Saunders and Shapiro (1986), Ebel (1993), Kröger (2002), as well as Hammer and Bucher (2006). Only Tanabe (1975) applied a density of 1.00 g/cm³ for Late Cretaceous sea water. This is of importance for buoyancy calculations to estimate the weight of the sea water volume displaced by the animal's total volume.

16.5.6 Model Approach

In order to avoid mathematical assumptions, approximations or speculations, Mutvei and Reyment (1973), Monks and Young (1998) as well as Westermann (2013) successfully carried out flotation experiments on more or less exact plastic models of key ammonoid morphotypes including three kinds of heteromorphs. The correct specific weight was obtained by electroplating the models internally and externally with metal. The major outcome of the experiments was that ammonoids must have retained much more cameral liquid in the chambers compared to the extant *Nautilus*. Additionally, ammonoids were possibly better adapted for vertical movements with a continuous adjustment to pressure gradients. The latter statement is supported by the wide occurrence of ammonoids like vascoceratids in shallow marine deposits. They possibly did not need larger adjustments of buoyancy which might be reflected in their simplified suture line although this is speculative (compare Klug et al. 2008). By contrast, ammonoids with highly complex suture lines can be speculatively interpreted as actively using their buoyancy apparatus for vertical movement (see also decoupling).

Due to its high density compared to the soft body, it turned out that the exact determination of shell volume was one of the main issues in earlier attempts for buoyancy calculation. Shell thickness must be measured for different ontogenetic stages. Thickness of the conch wall must be distinguished from thickness of the septa due to different densities (Hoffmann and Zachow 2011). Loss of shell volume during diagenesis was stressed by Reyment (1958) and Heptonstall (1970). Longridge et al. (2009) for the reason of diagenesis reduced the volume of their model by 7.6% before they included it in the calculations. It has to be taken into account that this assumption of material loss is based on the suggestion that during diagenesis, such volume changes occurred on a regular basis. We think that this is incorrect, because in most cases, shell replacement occurs after sediment consolidation, thus inhibiting volume alterations. In most cases, the measurable volumes even in calcitic replacement shells will probably be more or less accurate (e.g., Tajika et al. 2014; Naglik et al. 2014). It remains arguable whether limits of error from neutral buoyancy in seawater of +/−10% (Westermann 1993; Kröger 2002) are acceptable or not, especially when volumes are measured empirically using any kind of

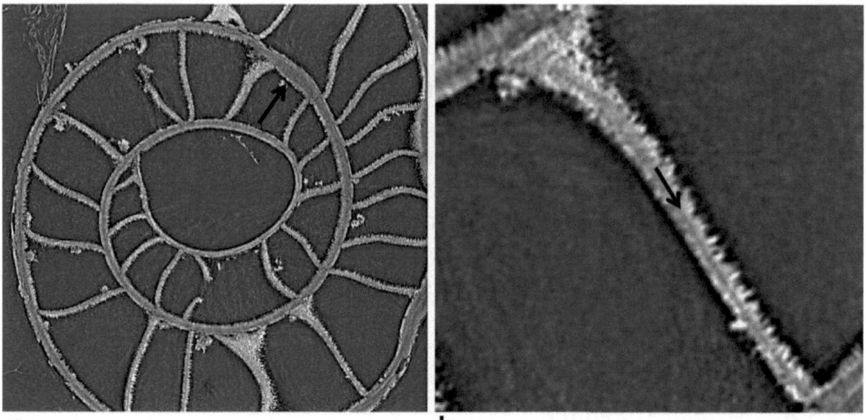

Fig. 16.5 juvenile *Cadoceras* sp. (Callovian of Russia) with secondary calcite crystals covering the inner surface of its chambers. **a** black arrow points to the nepionic constriction. **b** close up of a single septum with crystals growing on both sides and are in total about four to eight times thicker compared to the septa nevertheless this specimen was positively buoyant in fresh water, the specimen was scanned with a spatial resolution of 0.74 μm isotropic voxel size, length of the lower image edge is 1.4 mm

tomography (Fig. 16.5). According to Kröger (2002), an exact estimation of the soft body volume represents a major challenge because a few degrees of change in the length of the body chamber can alter the entire shell weight significantly.

16.6 New Approaches for Improved Buoyancy Calculations

16.6.1 Computer Tomography

Attempts to reconstruct ammonoid buoyancy had largely depended on mathematical modelling as discussed above. These models have not only simplified geometry and aspects of shell ornamentation but have often ignored changes in shell and septa thickness (for example: Trueman 1941, Westermann 1990, Ebel 1992, Kröger 2002) as well as allometric growth etc. Modern technology has provided novel approaches to address the buoyancy of ammonites as well as broader questions of functional morphology in the form of tomography including X-ray (CT), synchrotron and grinding tomography.

CT generates a series of X-ray images of the rotated object that are used to reconstruct a stack of 2D images. These images reveal the internal and external morphology of the object of interest in a non-invasive manner, whereas grinding tomography is destructive (Sutton et al. 2001; Tajika et al. 2014; Naglik et al. 2014). The details visible in CT images largely depend on (1) how the X-rays interact with the materials in the field of view (i.e. density differences and the absorption properties of the material) and (2) the contrast of material properties between two adjacent materials (Stock 2009). If two adjacent materials possess the same material properties distinction between the different materials may be impossible. CT applications in biological and paleontological sciences often utilize micro-CT and synchrotron micro-CT (e.g., Dumont et al. 2005; Jones et al. 2012; Schmidt et al. 2013) which, compared to common medical CT scanners, produce scans with a much higher-resolution of up to fractions of a micron (for a more detailed review of these methods and their underlying science see Stock 2009 or Sutton et al. 2014).

These methods can be used to reconstruct a 3D model of a scanned specimen from the image stack using a variety of commercial software (e.g., Avizo, Amira, VGstudiomax, Geomagic) and freeware (e.g., S.P.I.E.R.S., YaDIV). Hoffmann (2010) illustrated this method by creating 3D models of two ammonite specimens, namely *Argonauticeras besairiei* and *Gaudryceras* sp. These 3D models, provided the datasets possess sufficient resolution, can be used to perform 3D measurements such as volume and surface area calculations which can then be applied to functional analysis of the buoyancy apparatus of cephalopods as shown in Longridge et al. (2009), Hoffmann and Zachow (2011) and Tajika et al. (2014). The benefit of these empirical 3D models over mathematical models is their ability to accurately maintain the original geometry of the shell and thereby eliminating errors due to oversimplification and the use of average shell properties used by previous methods of buoyancy reconstruction. However, depending on the resolution of the scans and the visibility of fine structures, there still are limitations in accuracy.

The creation of these 3D models from CT data can be a relatively straightforward process in extant animals as the specimens scanned are usually complete (i.e. undamaged), show no post-depositional deformation exhibited by many fossils, and possess no sediment infill that can make the separation of the shell material from sedimentary or diagenetic infill much more difficult and very time-consuming. However, the modelling process for both extant and extinct specimens is principally the same; if preservation and image-quality permits, the shell material can be isolated automatically by the thresholding algorithms of the program from the surrounding medium (i.e. rock or air) in the CT scans and any shelly material not selected can be added or subtracted manually (in the case of fossil specimens, some parts of the surrounding rock may have been included in the threshold; Fig. 16.6). Once the shell material is selected, measurements relevant to reconstruction of the function of the buoyancy apparatus, such as the volume of the chambers and surface area of the siphuncle, become possible; see Hoffmann and Zachow (2011) for an example of volumetric analysis using CT data for *Nautilus pompilius* (Fig. 16.4) and for examples of ammonoids, for which grinding tomography was used, see Tajika et al. (2014) and Naglik et al. (2014) (Fig. 16.7).

It is important to note that tomographic methods have their own sets of complications that can arise. One of the main benefits of this method is the preservation of the original geometry of the shell, but this is only true if the CT scanner used has a sufficient resolution relative to the size of the specimen. If the resolution of the

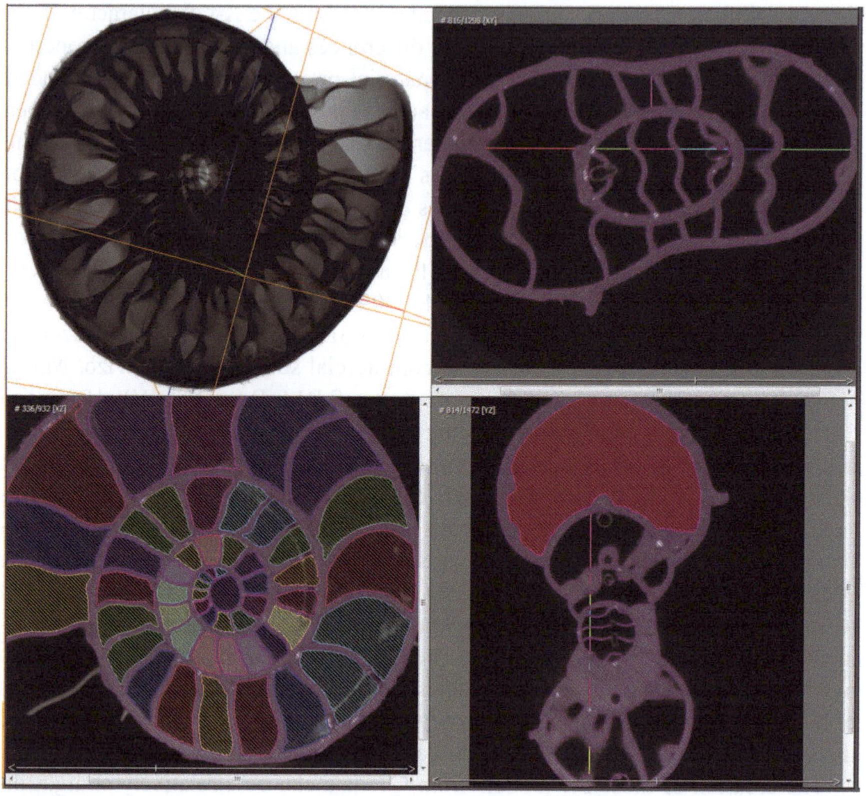

Fig. 16.6 Computer tomographic dataset of a scanned ammonoid visualized with the 3D software Avizo Fire 7.0, the specimen is being segmented, i.e., the selection of different materials of interest such as air, shell, sediment, scan artefacts

scanner is much larger than the structure of interest then some relevant geometric detail will be lost. For instance, if the resolution of the scanner is 1 mm but the width of the septa is only 1 μm, then a reliable reconstruction of the septa will be impossible. If the specimen is filled with material that has the same absorption properties as the specimen, such as a carbonate ammonite shell full of carbonate sediments, then the differentiation of the shell material and the surrounding medium will be possible only with grinding tomography and impossible with conventional CT. Advanced CT techniques such as phase contrast CT which is sensitive to changes in refractive index, can increase the differentiation between materials with similar absorption properties. In depth discussion of such techniques as well as additional experimental CT methods currently being explored that can mitigate this limitation can be found in Sutton et al. (2014).

The extraction of a complete surface from the data can be relatively quick for extant specimens, on the order of one to several days. By contrast, fossil specimens may require a much greater amount of work due to the complications explained above. A single ammonite specimen can take several months to complete depending

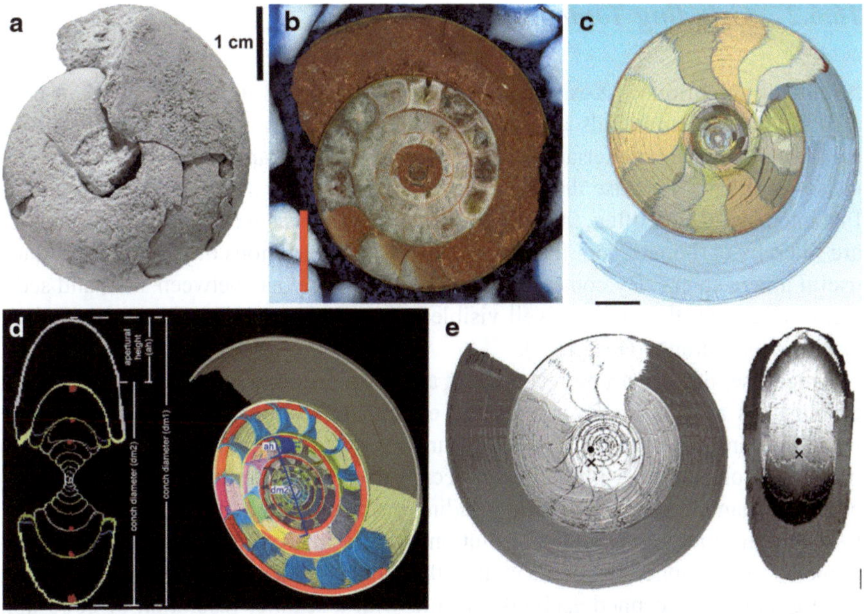

Fig. 16.7 The process from the fossil to the volume model (modified from Naglik et al. 2014, in press.): **a** *Diallagites lenticulifer*, Eifelian, Devonian, Hamar Laghdad (Morocco); the now pulverized specimen. **b** One of the 334 scans of the specimen in A; note that missing parts were manually reconstructed; the colour information is preserved, but the various shell structures are not uniformly well visible. **c** virtual reconstruction of the shell using only every fourth slice; the outer shell is here transparent, to show the color-labelled phragmocone chambers. **d** two different sections showing the labelled shell parts. E, reconstruction of shell orientation using the volume model (x—center of mass; o—center of buoyancy)

on the detail, quality of the images and contrast between shell and internal sediment or mineral filling.

Errors can be introduced by what we refer to as the partial volume effect (PVE). CT data is generated in voxels which can be thought of as 3D pixels. A single value is assigned to each voxel. This value controls the shade (from black to white) of each voxel shown on the CT slice. Because each voxel is represented only by a single value, the shade of a voxel represents an average of the material properties, if two or more materials are neighbouring the same voxel. This means that, when segmenting different materials, you may accidentally include a certain percentage of the surrounding materials due to the difficulty of detecting the exact boundary between adjacent materials. The PVE can be mitigated by scanning an object with similar absorption properties as the material of interest but with a known volume or density that can be used as a standard to accurately segment the specimen. This procedure is called quantitative computed tomography (qCT) in the medical literature, where the standard is referred to as a "phantom" (Kalender et al. 1995). It should be noted that this error is not unique to CT data and is also present in all scanned images represented by pixels.

16.6.2 Grinding Tomography

As already mentioned, X-ray tomography and synchrotron tomography can only display limits of materials in a scanned object, if there is a difference in physical properties of the materials under consideration. In ammonoids, this commonly causes problems, because the shell is carbonatic, even if it is diagenetically modified into calcite, and often, the phragmocone chambers as well as the body chamber are filled by carbonate, too. In such cases, CTs and variations of it will not produce useful image stacks. By contrast, if a colour contrast exists between shell and sediment or cement, this will be well visible in scans of ground surfaces obtained by grinding tomography (Fig. 16.7).

Of course, there is no perfect method and thus, grinding tomography (serial sectioning) has also its disadvantages and errors. For example, depending on image resolution and the spacing of ground surfaces, the partial volume effect will also produce errors. Another problem can occur, when the material is clayey. This may cause mechanical problems in the grinding process and thus bad scans. In contrast to synchrotron tomography, the resolution is lower in all three dimensions. The ammonoids tomographed by Tajika et al. (2014) and Naglik et al. (2014), for example, were ground and scanned each 60 - 48 μm with a scanner resolution of 2000 dpi. Therefore, a much lower resolution is obtained, but colour information is gained. The specimen gets destroyed by this method, so it is reasonable to apply it only to specimens of common taxa. In consequence, the application of this method makes sense for specimens smaller than 30 cm (depending also on the dimensions of the structure of interest that should be at least twice the resolution), where only a moderate resolution is needed to produce moderately well resolved volume models. Depending on the scientific question, this might well make sense and produce stimulating results based on empirical data (Tajika et al. 2014; Naglik et al. 2014; Naglik et al. in press).

16.6.3 Synchrotron and Neutron Tomography

In recent years, the use of synchrotron tomography has become fashionable in paleontology (e.g., Sutton et al. 2014). This method has the great advantage of producing image stacks of incredibly high resolution, thus revealing finest details (e.g., Kruta et al. 2011, Hoffmann et al. 2014 and references therein) and a small partial volume effect. At the same time, the application of this method is energy consuming, it requires enormous apparatuses of very limited access to researchers, it produces data sets of gigantic volume using a lot of computer disc space, data processing requires fast hardware with large RAM and often expensive software, sample size (usually <10 cm) is much more limited than in CT, and no colour information is preserved. In contrast to grinding tomography, however, the specimens are not lost.

Using neutron sources has been tested by us. This has a number of disadvantages such as a low resolution (>0.1 mm), the risk of producing radioactive nuclides in the sample and the low permeability of materials containing a lot of hydrogen and oxygen. Perhaps, this method might be of use in very special kinds of preservation, where the sample contains low amounts of hydrogen and a lot of iron or else.

16.7 Efficiency of the Ammonoids Hydrostatic Apparatus

Ward (1982) doubted that ammonoids used the buoyancy apparatus to support short term vertical migration comparable to the diurnal migration of the extant *Nautilus*. His interpretation was mainly based on the active siphuncular area available for liquid pumping. If we accept that there are no substantial differences in all known cephalopod siphuncular epithelia (Tanabe et al. 2000), the ratio between chamber volume and siphuncular volume should be a measure of short time buoyancy change ability (see above). On the contrary, Kröger (2002) convincingly demonstrated the efficiency of the ammonoid hydrostatic apparatus. He analyzed the five greatest sublethal injuries found in Mesozoic ammonoids and calculated the loss of shell mass that has occurred. It turned out that the maximum tolerated shell loss was four times larger than in extant *Nautilus*. Furthermore, his findings implied a buoyancy compensation mechanism that includes a rapid liquid refill of the phragmocone chambers. *Nautilus* could compensate for a maximum shell loss requiring a liquid refill of 3 % of the chamber volume (Ward 1987) equaling a shell loss of 5 % of the total shell mass (Ward 1986).

Daily density changes of *Sepia* are maintained by a compensatory liquid exchange of about 6 % of the chamber volume (Denton and Gilpin-Brown 1961b, 1973). Kröger (2002) expected a possibly higher compensatory liquid refill in ammonoids than the 6 % of *Sepia*. He could also demonstrate that the capability of compensation of drastic shell mass loss in ammonoids was significantly higher than in *Nautilus* and exceeds the observed values for normal daily density changes of *Sepia*. Mesozoic ammonites compensated a maximum shell loss requiring a liquid refill of more than 10 % of the chamber volume with an observed maximum of 18 %. If we assume that the shell loss has to be compensated before the animal rose to the sea surface, the compensation mechanism must have acted very rapidly. The above mentioned necessary liquid refill to compensate for shell loss mainly depends on the active siphuncular surface in relationship with the chamber volume. This relationship, the connecting ring surface area in mm^2/ chamber volume in ml, is the *si* of Kröger (2003). He calculated the following values: *Lithacoceras*—0.043 (Kröger 2002), adult *Spirula*—0.01, adult *Nautilus*—0.01, *Sepia*—0.1 (Ward 1982). Based on five specimens a mean *si* of 0.028 was presented for Paleozoic ammonoids and Kröger (2003) concluded that the ammonoid buoyancy apparatus was much more involved in active buoyancy regulation than in *Nautilus* and *Spirula*.

Liquid refill into the chambers is the only mechanism to compensate for a sudden increase of buoyancy. While Ward and Greenwald (1982) and Ward (1986) found an upper limit of liquid refill at approximately 3 % of the chamber volume, it was much higher (9–18 %) and therefore efficient in ammonoids (Kröger 2002).

This high ability of refill in ammonoids was explained by the more complex architecture of the phragmocone of the ammonoids with folded septa compared to the simple septa in the *Nautilus* phragmocone. In both groups, the inner surfaces of the chambers are lined with the pellicle (Weitschat and Bandel 1991). The pellicle is hydrophilic and plays an important role in the transportation of liquid out of and into the chambers by the use of capillary forces. The effect or power of the capillary forces for rapid refill of liquid into the chambers were demonstrated by the refloatation experiments of dry *Nautilus pompilius* shells by Hewitt and Westermann (1996). As for the refilling of liquid into the phragmocone, Wani et al. (2005) found through field experiments that waterlogging of the phragmocone in modern nautili does not occur until the mantle tissue detaches from the shell due to decomposition. Kröger (2002) assumed a direct relationship between the volume of capillary tissue (pellicle), ability of capillary fluid transport and of fluid storage (decoupling in its narrow sense). Due to higher capillary forces in the small pockets of complexly folded septa, chamber liquid will be stored in these spaces (Kulicki 1979; Weitschat and Bandel 1991; Checa 1996). A large pellicle volume is related to a large surface lined by the pellicle that is directly linked with the complexity of septa. Therefore, the highest ability of liquid transport can be assumed for ammonoids with the most complex septa such as, e.g., lytoceratids and phylloceratids. Daniel et al. (1997) were the first to consider the refill ability in conjunction with septal complexity. Mesozoic ammonoids survived a significantly higher shell loss than *Nautilus* or Paleozoic ammonoids (Keupp 2012) with much simpler suture lines. It can be assumed that survived shell loss is positively correlated with suture line complexity, although this requires further testing.

16.8 Evolution of the Hydrostatic Apparatus

Possibly, the phragmocone is the most important autapomorphy of the Cephalopoda (for cephalopod ancestry and origin see Klug et al. 2015a; Kröger et al. 2011; Ax 2001). With respect to cephalopod buoyancy, a hypothesis suggested by Dzik (1981) has to be mentioned: According to Pojeta (1980), the siphuncle of the earliest cephalopods, i.e. the Plectronoceridae, might be homologous to the snorkel-like process of monoplacophoran Yochelcionellidae. Dzik (1981) then suggested that stemgroup cephalopods or derived monoplacophorans have produced a hyposaline liquid, stored it in the apex of the shell, thereby reducing their buoyancy and becoming able to swim. This is in accordance with the osmotic process involved in the removal of cameral liquid in ammonoids and other chambered cephalopods. In the opinion of Dzik (1981), the septa then evolved by a repeated change of apical liquid- and carbonate secretion, followed by gas diffusion which was enabled by the growing partial pressures produced by the ionic pump in the chambers.

In the course of cephalopod evolution, the buoyancy apparatus underwent many modifications including thick (endocerid) and thin (orthocerid) siphuncles, simple (actinocerid) and complex (ammonoid) septa, straight (orthocerid) and coiled (ammonoid) phragmocones. Undoubtedly, the major innovations characterizing most Ammonoidea are the ventral siphuncle, the tightly coiled shell and more or less strongly folded septa. In how far morphological evolutionary innovations represent adaptations to improve its function to regulate buoyancy is nearly impossible to prove. In any case, we do not doubt that buoyancy control was the phragmocone's principal function because it persisted throughout the entire phylogeny and underwent an increase in complexity which was partially reversed only in very few taxa (e.g., Cenomanian *Neolobites*).

Above, we have mentioned the line of reasoning that the increasing sutural frilling might serve to increase the surface of the phragmocone (see Chap. 3). This complexity is largely limited to the part of the septa, where they are in contact with the outer shell (the suture), while much of the septum remains gently folded. This arrangement is not the optimal construction to maximize the pellicle-covered surface of the inside of the chamber. This might indicate a different meaning of the suture frilling which has been suggested by Klug et al. (2008). Accordingly, the complexity increased the length of attachment of the organic preseptum prior to mineralization and might have helped to allow an earlier start of the emptying process in newly formed chambers. In both cases, the increase in sutural frilling is linked with a rise in efficiency of the buoyancy apparatus, thus supporting the likely great importance of buoyancy regulation in ammonoids. In turn, this points at the possibly immense meaning of the control of vertical movements in the life of ammonoids. Taking the implosion depths of ammonoids of perhaps above 300 m into account (e.g., Westermann 1973, 1996), the importance of buoyancy control becomes essential for survival. Phragmocone differentiation could be speculatively explained as a reflection of specializations to certain depths and/ or distinct modes of vertical migration. With the increase in accuracy and number of buoyancy models of ammonoid species, we will hopefully learn more about their habitats and perhaps also behavior in terms of vertical migration.

Acknowledgements CK and CN thank the Swiss National Science Foundation (SNF project numbers 200021-113956/ 1, 200020-25029, and 200020-132870) and RH and RL thank the Deutsche Forschungsgemeinschaft (DFG project numbers HO 4674/2-1) for financial support of their research, especially for the grinding tomography. We greatly appreciate the work of the members of the Heidelberg grinding tomography lab, namely Stefan Götz, who died much too young, Enrique Pascual-Cebrian, and Dominik Hennhöfer (all Heidelberg).

References

Appellöf A (1893) Die Schalen von *Sepia, Spirula* und *Nautilus*—Studien über den Bau und das Wachstum. Kongl Svenska Vetensk Akad Handl 25:1–106

Arkell WJ (1957) Sutures and septa in Jurassic ammonite systematic. Geol Mag 94:235–248

Ax P (2001) Das System der Metazoa. Fischer, Stuttgart

Bandel K, von Boletzky S (1979) A comparative study of the structure, development and morphological relationships of chambered cephalopod shells. Veliger 21:313–354

Bandel K, Stinnesbeck W (2006) *Naefia* Wetzel 1930 from the Quiriquina Formation (Later Maastrichtian, Chile): relationship to modern *Spirula* and ancient Coleoidea (Cephalopoda). Acta Univ Carol Geol 49:21–32

Barskov IS (1990) Internal structure of siphuncle of the Late Jurassic ammonite *Virgatites virgatus*. Trans Paleontol Inst 243:127–132

Barskov IS (1996) Phosphatized blood vessels in the siphuncle of Jurassic ammonites. Bull Inst Océanogr, (Monaco, special) 14:335–341

Barskov IS (1999) Why ammonoids have complex septa and sutures? In: Rozanov AY, Shevyrev AA (eds) Fossil cephalopods: recent advances in their study. Russian Academy of Science, Moscow

Berridge MJ, Oschman JL (1972) Transporting epithelia. Academic Press, New York

Berry E (1928) Cephalopod adaptations—the record and its interpretations. Q Rev Biol (Baltimore) 3:92–108

Bert P (1867) Mémoire sur la physiologie de la Seiche. Mem Soc Sci Phys Nat Bordeaux 5:114–138

Bidder AM (1962) Use of the tentacles, swimming and buoyancy control in the Pearly *Nautilus*. Nature 196:451–454

Bonting SL (1970) Sodium-potassium activated adenosine triphosphatase and cation transport. In: Bittar I (ed) Membranes and Ion Transport. Wiley, New York

Bruun AF (1943) The biology of *Spirula spirula* (L.). Dana Rep 4:1–46.

Bruun AF (1950) New light on the biology of *Spirula*, a mesopelagic cephalopod (Essays on the natural sciences in honour of Captain Allan Hancock). University of Southern California Press, Los Angeles, pp 61–72

Buckland W (1836) Geology and mineralogy considered with reference to natural theology, vol 1. William Pickering, London

Chamberlain Jr JA (1987) Locomotion of *Nautilus*. In: Saunders WB, Landman NH (eds) *Nautilus*—the biology and paleobiology of a living fossil. Springer, Dordrecht

Chamberlain JA Jr, Moore WA Jr (1982) Rupture strength and flow rate of *Nautilus* siphuncular tube. Paleobiology 8:408–425

Chamberlain JA Jr, Ward PD, Weaver JS (1981) Post-mortem ascent of *Nautilus* shells: implications for cephalopod paleobiogeography. Paleobiology 7:494–509

Charbonnier S (2009) Le Lagerstätte de la Voulte un environment bathyal au Jurassique. Mém Mus Natl Hist Nat 199:1–272

Checa A (1996) Origin of intracameral sheets in ammonoids. Lethaia 29:61–75

Chun C (1915) The Cephalopoda part 1: Oegopsida, part 2: Myopsida, Octopoda—text and atlas. Scientific results of the German deepsea expedition on board the steamship “Valdivia” 1898–1899

Clarke MR (1969) Cephalopoda collected on the Sond Cruise. J Mar Biol Assoc UK 49:961–976

Clarke MR (1970) Growth and development of *Spirula spirula*. J Mar Biol Assoc UK 50:53–64

Collins DH, Minton P (1967) Siphuncular tube of *Nautilus*. Nature 216:916–917

Collins DH, Ward, PD, Westermann GEG (1980) Function of cameral water in *Nautilus*. Paleobiology 6:168–172

Crick RE (1988) Buoyancy regulation and macroevolution in nautiloid cephalopods. Senck Leth 69:13–42

Currie ED (1957) The mode of life of certain goniatites. Trans Geol Soc Glasg 22:169–186

Daniel TL, Helmuth BS, Saunders WB, Ward PD (1997) Septal complexity in ammonoid cephalopods increased mechanical risk and limited depth. Paleobiology 23:470–481

Delanoy G, Magnin A, Sélébran M, Sélébran J (1991) *Moutoniceras nodosum* d’Orbigny, 1850 (Ammonoidea, Ancyloceratina), une très grande ammonite hétéromorphe du Barrémien inférieur. Rev Paléobiol 10:229–245.

Denton EJ (1962) Some recently discovered buoyancy mechanisms in marine animals. Proc R Soc Lond B 265:366–370

Denton EJ (1971) Examples of the use of active transport of salts and water to give buoyancy in the sea. Phil Trans R Soc Lond B 262:277–287

Denton EJ (1974) On buoyancy and the lives of modern and fossil cephalopods. Proc R Soc Lond B 185:273–299

Denton EJ, Gilpin-Brown JB (1961a) The buoyancy of the cuttlefish *Sepia officinalis* (L.). J Mar Biol Assoc UK 41:319–342

Denton EJ, Gilpin-Brown JB (1961b) The effect of light on the buoyancy of the cuttlefish. J Mar Biol Assoc UK 41:343–350

Denton EJ, Gilpin-Brown JB (1961c) The distribution of gas and liquid within the cuttlebone. J Mar Biol Assoc UK 41:365–381

Denton EJ, Gilpin-Brown JB (1966) On the buoyancy of the pearly *Nautilus*. J Mar Biol Assoc UK 46:723–759

Denton EJ, Gilpin-Brown JB (1971) Further observations on the buoyancy of *Spirula*. J Mar Biol Assoc UK 51:363–373

Denton EJ, Gilpin-Brown JB (1973) Floatation mechanisms in modern and fossil cephalopods. Adv Mar Biol 11:197–268

Denton EJ, Gilpin-Brown JB, Howarth JV (1961) The osmotic mechanism of the cuttlebone. J Mar Biol Assoc UK 41:351–364

Denton EJ, Gilpin-Brown JB, Howarth JV (1967) On the buoyancy of *Spirula spirula*. J Mar Biol Assoc UK 47:181–191

Denton EJ, Gilpin-Brown JB, Shaw TI (1969) A buoyancy mechanism found in cranchid squid. Proc R Soc Lond B 174:271–279

Derham W (1726) Philosophical experiments and observations of the late eminent Dr. Robert Hooke, Derham, London

Diamond JM, Bossert WH (1967) Standing gradient osmotic flow—a mechanism for coupling water and solute transport in epithelia. J Gen Physiol 50:2061–2083

Diamond JM, Bossert WH (1968) Functional consequences of ultra-structural geometry in "backwards" fluid-transporting epithelia. J Cell Biol 37:694–702

Diener C (1912) Lebensweise und Verbreitung der Ammoniten. Neues Jahrb Miner Geol Palaontol 2:67–89

Donovan D (1964) Cephalopod phylogeny and classification. Biol Rev 39:259–287

Drushchits VV, Doguzhaeva LA (1981) Ammonites under the electron microscope (internal shell structure and systematics of Mesozoic Phylloceratidae, Lytoceratidae and 6 families of Early Cretaceous Ammonitidae). Moscow University, Moscow

Dumont ER, Piccirillo J, Grosse IR (2005) Finite-element analysis of biting behavior and bone stress in the facial skeletons of bats. Anat Rec 283A:319–330

Dunstan AJ, Ward PD, Marshall NJ (2011) Vertical Distribution and Migration Patterns of *Nautilus pompilius*. PLoS One 6:e16312

Dzik J (1981) Origin of the Cephalopoda. Acta Palaeont Pol 26:161–91

Ebel K (1983) Berechnungen zur Schwebfähigkeit von Ammoniten. Neues Jahrb Geol Paläontol (MMonatshefte) 1983:614–640

Ebel K (1985) Gehäusespirale und Septenform bei Ammoniten unter der Annahme vagil benthischer Lebensweise. Paläontol Z 59:109–123

Ebel K (1990) Swimming abilities of ammonites and limitations. Paläontol Z 64:25–37

Ebel K (1992) Mode of life and soft body shape of heteromorph ammonites. Lethaia 25:179–193

Ebel K (1993) Negative buoyancy of ammonoids—reply. Lethaia 26:260

Ebel K (1999) Hydrostatics of fossil ectocochleate cephalopods and its significance for the reconstruction of their lifestyle. Paläontol Z 73:277–288

Engeser, T (1996) The position of the Ammonoidea within the Cephalopoda. In: Landman NH, Tanabe K, Davis RA (eds) Ammonoid paleobiology. Topics in Geobiology 13. Plenum, New York

Finn, JK, Norman, MD (2010) The argonaut shell: gas-mediated buoyancy controlin a pelagic octopus. Proc R Soc B 277:2967–2971

Gottobrio WE, Saunders WB (2005) The clymeniid dilemma: functional implications of the dorsal siphuncle in clymeniid ammonoids. Paleobiology 31:233–252

Greenwald L, Ward PD (1982) On the source of cameral liquid in the chambered *Nautilus*. Veliger 25:169–170.

Greenwald L, Ward PD (1987) Buoyancy in *Nautilus*. In: Saunders BW, Landman NH (eds) *Nautilus*—the biology and paleobiology of a living fossil. Springer, Dordrecht
Greenwald L, Ward PD, Greenwald OE (1980) Cameral liquid transport and buoyancy control in the chambered nautilus (*Nautilus macromphalus*). Nature 286:55–56
Greenwald L, Cook CB, Ward PD (1982) The structure of the chambered *Nautilus* siphuncle: the siphuncular epithelium. J Morphol 172:5–22
Greenwald L, Verderber G, Singley C (1984) Localization of Na-K ATPase activity in the *Nautilus* siphuncle. J Exp Zool 229:481–484
Guex J (2005) Buoyancy control and growth rates in ammonoids: new preliminary remarks about an old Red Herring. Bull Géol Lausanne 365:1–4
Guex J, Koch A, O'Dogherty L, Bucher H (2003) A morphogenetic explanation of Buckman's law of covariation. Bull Soc Géol Fr 174:603–606
Hammer Ø, Bucher H (2005) Buckman's law of covariation—a case of proportionality. Lethaia 38:67–72
Hammer Ø, Bucher H (2006) Generalized ammonoid hydrostatics modelling, with application to *Intornites* and intraspecific variation in *Amaltheus*. Paleontol Res 10:91–96
Heath TL (1897) The works of Archimedes. Clay and Sons, Cambridge University Press, Warehouse, London
Heptonstall WB (1970) Buoyancy control in ammonoids. Lethaia 3:317–328.
Hewitt RA (1985) Numerical aspects of sutural ontogeny in the Ammonitina and Lytoceratina. Neues Jahrb Geol Palaontol Abh 170:273–290
Hewitt RA (1996) Architecture and strength of the ammonoid shell. In: Landman NH, Tanabe K, Davis RA (eds) Ammonoid paleobiology. Plenum, New York.
Hewitt RA, Westermann GEG (1987) Function of complexly fluted septa in ammonoid shells 2. Septal evolution and conclusions. Neues Jahrb Geol Palaontol Abh 174:135–169
Hewitt RA, Westermann GEG (1993) Growth rates of ammonites estimated from aptychi. Geobios Mem Spec 15:203–208
Hewitt RA, Westermann GEG (1996) Post-mortem behaviour of Early Paleozoic nautiloids and paleobathymetry. Paläontol Z 70:405–424
Hewitt RA, Westermann GEG (1997) Mechanical significance of ammonoid septa with complex sutures. Lethaia 30:205–212
Hewitt RA, Westermann GEG, Judd RL (1999) Buoyancy calculations and ecology of Callovian (Jurassic) cylindroteuthid belemnites. Neues Jahrb Geol Paläont Abh 211:89–112
Hoffmann R (2010) New insights on the phylogeny of the Lytoceratoidea (Ammonitina) from the septal lobe and its functional interpretation. Rev Paléobiol 29:1–156
Hoffmann R, Zachow S (2011) Non-invasive approach to shed new light on the buoyancy business of chambered cephalopods (Mollusca). IAMG 2011 publication, Salzburg. doi:10.5242/iamg.2011.0163:506-516
Hoffmann R, Schultz JA, Schellhorn R, Rybacki E, Keupp H, Gerden SR, Lemanis R, Zachow S (2014) Non-invasive imaging methods applied to neo- and paleontological cephalopod research. Biogeosciences 11: 2721–2739. doi:10.5194/bg-11-2721-2014
Hooke R (1726) Philosophical experiments and observations. In: Derham W (ed) Printers to the Royal Society 8:807–810
Jacobs DK (1992) The support of hydrostatic load in cephalopod shells—adaptive and ontogenetic explanations of shell form and evolution from Hooke 1695 to the present. In: Hecht MK, Wallace B, MacIntyre RJ (eds) Evolutionary biology, 26, Plenum, New York
Jacobs DK (1996) Chambered cephalopod shells, buoyancy, structure and decoupling: history and red herrings. Palaios 11:610–614
Jacobs DK, Chamberlain JA Jr (1996) Buoyancy and hydrodynamics in ammonoids. In: Landman NH, Tanabe K, Davis RA (eds) Ammonoid paleobiology. Topics in geobiology 13. Plenum, New York
Jones D, Evans AR, Siu KWK (2012) The sharpest tool in the box? Quantitative analysis of conodont element functional morphology. Proc R Soc Biol 279:2849–2854

Kalender W, Felsenberg D, Genant HK (1995) The European Spine Phantom—a tool for standardization and quality control in spinal bone mineral measurements by DXA and QCT. Eur J Radiol 20:83–92

Kanie Y, Fukuda Y, Nakahara K, Seki K, Hattori H (1980) Implosion of living *Nautilus* under increased pressure. Paleobiology 6:44–47

Kelly A (1901) Beiträge zur mineralogischen Kenntnis der Kalkausscheidungen im Tierreich. Jenä Z 35:429–494

Keupp H (1997) Paläopathologische Analyse einer "Population" von *Dactylioceras athleticum* (Simpson) aus dem Unter-Toarcium von Schlaifhausen/Oberfranken. Berl Geowiss Abh 25:243–267

Keupp H (2000) Ammoniten—Paläobiologische Erfolgsspiralen. Thorbecke, Stuttgart

Keupp H (2012) Atlas zur Paläopathologie der Cephalopoden. Berl Paläobiol Abh 10:1–390

Keupp H, Röper M, Seilacher A (1999) Paläobiologische Aspekte von syn vivo-besiedelten Ammonoideen im Plattenkalk des Ober-Kimmeridgiums von Brunn in Ostbayern. Berl Geowiss Abh 30:121–145

Klinger HC (1981) Speculations on buoyancy control and ecology in some heteromorph ammonites. Syst Assoc Spec Vol 18:337–355

Klug C (2001) Life-cycles of Emsian and Eifelian ammonoids (Devonian). Lethaia 34:215–233

Klug C, Lehmann J (2015) Soft part anatomy of ammonoids: reconstructing the animal based on exceptionally preserved specimens and actualistic comparisons (this volume)

Klug C, Korn D, Richter U, Urlichs M (2004) The black layer in cephalopods from the German Muschelkalk (Middle Triassic). Palaeontology 47:1407–1425

Klug C, Meyer E, Richter U, Korn D (2008) Soft-tissue imprints in fossil and Recent cephalopod septa and septum formation. Lethaia 41:477–492

Klug C, Riegraf W, Lehmann J (2012) Soft-part preservation in heteromorph ammonites from the Cenomanian-Turonian Boundary Event (OAE 2) in the Teutoburger Wald (Germany). Palaeontology 55:1307–1331

Klug C, Kröger B, Vinther J, Fuchs D, De Baets K (2015a) Ancestry, origin and early evolution of ammonoids. (this volume)

Klug C, Zatoń M, Parent H, Hostettler B, Tajika A (2015b) Mature modifications and sexual dimorphism. (this volume)

Kröger B (2000) Schalenverletzungen an jurassischen Ammoniten—ihre paläobiologische und palökologische Aussagefähigkeit. Berl Geowiss Abh 33:1–97

Kröger B (2001) Discussion—comments on Ebel's benthic-crawler hypothesis for ammonoids and extinct nautiloids. Paläontol Z 75:123–125

Kröger B (2002) On the efficiency of the buoyancy apparatus in ammonoids: evidences from sublethal shell injuries. Lethaia 35:61–70

Kröger B (2003) The size of the siphuncle in cephalopod evolution. Senckenberg Lethaea 83:39–52

Kröger B, Vinther J, Fuchs D (2011) Cephalopod origin and evolution: a congruent picture emerging from fossils, development and molecules. Bioessays 12 pp. doi:10.1002/bies.201100001

Kruta I, Landman NH, Rouget I, Cecca F, Tafforeau P (2011) The role of ammonites in the Mesozoic marine food web revealed by exceptional jaws preservation. Science 331(70):70–72

Kruta I, Landman NH, Cochran JK (2014) A new approach for the determination of ammonite and nautilid habitats. PLoS One 9:e87479 doi:10.1371/journal.pone.0087479

Kulicki C (1979) The ammonite shell, its structure, development and biological significance. Palaeontol Pol 39:97–142

Kulicki C (1996) Ammonoid shell microstructure. In: Tanabe K, Davis RA (eds) Ammonoid Paleobiology. Plenum, New York

Longridge LM, Smith PL, Rawlings G, Klaptocz V (2009) The impact of asymmetries in the elements of the phragmocone of early Jurassic ammonites. Palaeontol Electron 12:1–15

Lukeneder A, Harzhauser M, Müllegger S, Piller WE (2010) Ontogeny and habitat change in Mesozoic cephalopods revealed by stable isotopes (delta^{18}O, delta^{13}C). Earth Planet Sci Lett 296:103–114 doi:10.1016/ j.epsl.2010.04.053

Mangum CP, Towle DW (1982) The *Nautilus* siphuncle as an ion pump. Pac Sci 36:273–282

Meigen W (1870) Über den hydrostatischen Apparat bei *Nautilus pompilius*. Arch Naturgesch 36:1–36

Monks N, Young JR (1998) Body position and the functional morphology of Cretaceous heteromorph ammonites. Palaeontogr Electron, http:/www-odp.tamuedu/paleo/1998_1/toc.htm. Accessed 17 Jan 2015

Moore R, Lalicker C, Fischer A (1952) Invertebrate fossils. McGraw-Hill Co., New York

Moriya K, Nishi H, Kawahata H, Tanabe K, Takayanagi Y (2003) Demersal habitat of Late Cretaceous ammonoids: evidence from oxygen isotopes for the campanian (Late Cretaceous) northwestern Pacific thermal structure. Geology 31:167–170

Moseley H (1838) On the geometrical form of turbinated and discoid shells. Phil Trans R Soc Lond 1838:351–370

Mutvei H (1983) Flexible nacre in the nautiloid *Isorthoceras*, with remarks on the evolution of cephalopod nacre. Lethaia 16:233–240

Mutvei H, Reyment RA (1973) Buoyancy control and siphuncle function in ammonoids. Palaeontology 16:623–636

Naglik C, Monnet C, Götz S, Kolb C, De Baets K, Klug C (2015) Growth trajectories in chamber and septum volumes in major subclades of Paleozoic ammonoids. Lethaia: DOI:10.1111/let.12085. Accessed 17 Jan 2015

Naglik C, Rikhtegar F, Klug C (2014) Buoyancy of some Palaeozoic ammonoids and their hydrostatic properties based on empirical 3D-models. Lethaia: ca. 14 pp.

O'Dor RK, Forsythe J, Webber DM, Wells J, Wells MJ (1993) Activity levels of *Nautilus* in the wild. Nature 362:626–628

Okamoto T (1988) Changes in life orientation during the ontogeny of some heteromorph ammonites. Paleontology 31:281–294

Owen R (1832) Memoir of the Pearly Nautilus (*Nautilus Pompilius*, Linn.). London. pp 1–68

Owen R (1878) On the relative positions to their constructions of the chambered shells of cephalopods. Proc Zool Soc Lond 1878:955–975

Packard A (1972) Cephalopods and fish: the limits of convergence. Biol Rev 47:241–307

Pfaff E (1911) Über Form und Bau der Ammonitensepten und ihre Beziehungen zur Suturlinie. Jahresber Niedersächs Geol Ver (Geol Abt Naturhist Ges Hannover) 4:207–223

Pojeta J Jr (1980) Molluscan phylogeny. Tulane Stud Geol Paleontol 16:55–80

Raup DM, Chamberlain JA Jr (1967) Equations for volume and center of gravity in ammonoids shells. J Paleontol 41:566–574

Reboulet S, Giraud F, Proux O (2005) Ammonoid abundance variations related to changes in trophic conditions across the Oceanic Anoxic Event 1d (Latest Albian, SE France). Palaios 20:121–141

Rein S (1999) On the swimming abilities of *Ceratites* De Haan and *Germanonautilus* Mojsisovics from the Upper Muschelkalk (Middle Triassic). Freiber Forschungsheft C481:39–47

Reyment RA (1958) Some factors in the distribution of fossil Cephalopods. Acta Univ Stockh—Stockh Contrib in Geol 1:97–184

Reyment RA (1973) Factors in the distribution of fossil cephalopods. Part 3: experiments with exact models of certain shell types. Bull Geol Inst Univ Uppsala N. S. 4:7–41

Ritterbush KA, Bottjer DJ (2012) Westermann Morphospace displays ammonoid shell shape and hypothetical paleoecology. Paleobiology 38:424–446

Saunders WB, Shapiro EA (1986) Calculation and simulation of ammonoid hydrostatics. Paleobiology 12:64–79

Schmidt M (1925) Ammonitenstudien. Fortschr Geol Palaeontol 10:75–363

Schmidt H (1930) Über die Bewegungsweise der Schalencephalopoden. Paläontol Z 12:194–208

Schmidt DN, Rayfield ER, Cocking A (2013) Linking evolution and development: synchrotron radiation X-ray tomographic microscopy of planktic foraminifers. Palaeontology 56:741–749

Schwarz EHL (1894) The Aptychus. Geol Mag, N S(Decade IV) 1:454–459

Seilacher A (1960) Epizoans as a key to ammonoid ecology. J Paleontol 34:183–193

Seilacher A, Gishlick AD (2015) Morphodynamics. CRC Press Taylor & Francis Group

Seilacher A, Labarbera M (1995) Ammonites as Cartesian Divers. Palaios 10:493–506

Shigeta Y (1993) Post-hatching early life history of Cretaceous Ammonoidea. Lethaia 26:133–146

Spath LF (1919) Notes on ammonites. Geol Mag 56:26–58, 65–74, 115–122, 170–177, 220–225

Stock SR (2009) MicroComputed tomography: methodology and applications. CRC Press, London

Sutton MD, Briggs DEG, Siveter DJ et al (2001) Methodologies for the Visualization and Reconstruction of Three-dimensional Fossils from the Silurian Herefordshire Lagerstätte. Palaeontol Electron 4:1–17

Sutton MD, Briggs DEG, Siveter DJ, Siveter DJ (2006) Fossilized soft tissues in a Silurian platyceratid gastropod. Proc Royal Soc B 273(1590):1039–1044

Sutton MD, Rahman IA, Garwood RJ (2014) Techniques for virtual palaeontology. Wiley, New York. doi:10.1002/9781118591192

Swan ARH, Saunders WB (1987) Function and shape in late Paleozoic (mid-Carboniferous) ammonoids. Paleobiology 13:297–311

Tajika A, Naglik C, Morimoto N, Pascual-Cebrian E, Hennhöfer DK, Klug C (2014) Empirical 3D-model of the conch of the Middle Jurassic ammonite microconch *Normannites*, its buoyancy, the physical effects of its mature modifications and speculations on their function. Hist Biol, 27:181–191. Accessed 17 Jan 2015

Tanabe K (1975) Functional morphology of *Otoscaphites puerculus* (Jimbo), an Upper Cretaceous ammonite. Trans Proc Palaeont Soc Jpn, N S 99:109–132

Tanabe K, Landman NH (1996) Septal neck—siphuncular complex of ammonoids. In: Landman NH, Tanabe K, Davis RA (eds) Ammonoid paleobiology, Plenum, New York

Tanabe K, Mapes RH, Sasaki T, Landman NH (2000) Soft part anatomy of the siphuncle in Permian prolecanitid ammonoids. Lethaia 3:83–91

Tanabe K, Sasaki T, Mapes RH (2014) Soft-part anatomy of the siphuncle in ammonoids (this volume)

Tasch P (1973) Paleobiology of invertebrates. Wiley, New York.

Trueman AE (1941) The ammonite body chamber with special reference to the buoyancy and mode of life of the living ammonite. Quart J Geol Soc Lond 96:339–383

Tsujino Y, Shigeta Y (2012) Biological response to experimental damage of the phragmocone and siphuncle in *Nautilus pompilius* Linnaeus. Lethaia 45:443–449

Tsujita CJ, Westermann GEG (1998) Ammonoid habitats and habits in the Western Interior Seaway: a case study from the Upper Cretaceous Bearpaw Formation of southern Alberta, Canada. Palaeogeogr Palaeoclim Palaeoecol 144:135–160

Vrolik W (1843) On the Anatomy of the Pearly *Nautilus*. Ann Mag Nat Hist 12:173–175

Wani R, Kase T, Shigeta Y, De Ocampo R (2005) New look at ammonoid taphonomy, based on field experiments with modern chambered nautilus. Geology 33:849–852

Ward PD (1979) Cameral liquid in *Nautilus* and ammonites. Paleobiology 5:40–49

Ward PD (1980) Restructuring the chambered *Nautilus*. Paleobiology 6: 247–249

Ward PD (1982) The relationship of siphuncle size to emptying rates in chambered cephalopods: implications for cephalopod paleobiology. Paleobiology 8:426–433

Ward PD (1986) Rates and processes of compensatory buoyancy change in *Nautilus macromphalus*. Veliger 28:356–368

Ward PD (1987) The Natural History of *Nautilus*. Allen & Unwin, Boston

Ward PD, von Boletzky S (1984) Shell implosion depth and implosion morphologies in three species of *Sepia* (Cephalopoda) from the Mediterranean Sea. J Mar Biol Assoc UK 64:955–966

Ward PD, Greenwald L (1982) Chamber refilling in *Nautilus*. J Mar Biol Assoc UK 62:469–475

Ward PD, Martin AW (1978) On the buoyancy of the Pearly *Nautilus*. J Exp Zool 205:5–12

Ward PD, Westermann GEG (1977) First occurrence, systematics, and functional morphology of *Nipponites* (Cretaceous Lytoceratina) from the Americas. J Paleontol 51:367–372

Ward PD, Stone R, Westermann GEG, Martin A (1977) Notes on animal weight, cameral fluids, swimming speed, and color polymorphism of the Cephalopod *Nautilus pompilius* in the Fiji Islands. Paleobiology 3:377–388

Ward PD, Greenwald L, Rougerie F (1980a) Shell implosion depth for living *Nautilus macromphalus* and shell strength of extinct cephalopods. Lethaia 13:182

Ward PD, Greenwald L, Greenwald OE (1980b) The buoyancy of the chambered *Nautilus*. Sci Am 243:190–203

Ward PD, Greenwald L, Magnier Y (1981) The chamber formation cycle in *Nautilus macromphalus*. Paleobiology 7:481–493

Ward PD, Carlson B, Weekley M, Brumbaugh B (1984) Remote telemetry of daily vertical and horizontal movement by *Nautilus* in Palau. Nature 309:248–250

Warnke KM, Oppelt A, Hoffmann R (2010) Stable isotopes during ontogeny of *Spirula* and derived hatching temperatures. Ferrantia 59:191–201

Weitschat W, Bandel K (1991) Organic components in phragmocones of boreal Triassic ammonoids: implications for ammonoid biology. Paläontol Z 65: 269–303

Wells M (1990) The dilemma of the jet set. New Sci 1704:44–47

Westermann GEG (1956) Phylogenie der Stephanocerataceae und Perisphinctaceae des Dogger. Neues Jahrb Geol Paläont Abh 103:233–279

Westermann GEG (1971) Form, structure and function of shell and siphuncle in coiled mesozoic ammonoids. Life Sci Contrib, R Ont Mus 78:1–39

Westermann GEG (1973) Strength of concave septa and depth limits of fossil cephalopods. Lethaia 6:383–403

Westermann GEG (1975) Model for origin, function and fabrication of fluted cephalopod septa. Paläontol Z 49:235–253

Westermann GEG (1977) Form and function of orthoconic cephalopod shells with concave septa. Paleobiology 3:300–321

Westermann GEG (1982) The connecting rings of *Nautilus* and Mesozoic ammonids: implications for ammonite bathymetry. Lethaia 15:373–384

Westermann GEG (1990) New developments in ecology of Jurassic-Cretaceous ammonoids. In: Pallini G, Cresta S, Santantonio M (eds) Fossili, Evolutione, Ambiente. Atti II Convenio Internationale Pergola 1987

Westermann GEG (1993) On alleged negative buoyancy of ammonoids. Lethaia 26:246

Westermann GEG (1996) Ammonoid life and habitat. In: Landman NH, Tanabe K, Davis RA (eds) Ammonoid paleobiology, Plenum, New York

Westermann GEG (1998a) Life habits of nautiloids. In: Savazzi E (ed) Functional morphology of the invertebrate skeleton. Wiley, Chichester

Westermann GEG (1998b) Life habits of ammonoids. In: Savazzi E (ed) Functional morphology of the invertebrate skeleton. Wiley, Chichester

Westermann GEG (2013) Hydrostatics, propulsion and life-habits of the Cretaceous ammonoid *Baculites*. Rev Paléobiol 32:249–265

Westermann B, Beuerlein K, Hempelmann G, Schipp R (2002) Localization of putative neurotransmitters in the mantle and siphuncle of the mollusc *Nautilus* L. (Cephalopoda). Histochem J 34:435–440

Willey A (1902) Contributions to the natural history of the Pearly Nautilus. In: Wiley A (ed.) Zoological results part 6, Cambridge University Press, Cambridge

Witmer LM (1995) The extant phylogenetic bracket and the importance of reconstructing soft tissues in fossils. In: Thomason JJ (ed) Functional morphology in vertebrate paleontology. Cambridge University Press, Cambridge

Chapter 17
Ammonoid Locomotion

Carole Naglik, Amane Tajika, John Chamberlain and Christian Klug

17.1 Introduction

The locomotor capacity of ammonoids is still a matter of much debate. This question is intimately linked with questions concerning ammonoid habitat and buoyancy (Ritterbush et al. 2014). Aspects of buoyancy were reviewed by Hoffmann et al. (2015). Based on theoretical models of ammonoid buoyancy (e.g., Trueman 1941; Saunders and Shapiro 1986) in combination with the latest empirical studies on volume models of ammonoids (Tajika et al. 2015; Naglik et al. 2015), we can now confidently reject the hypothesis of an obligatorily benthic mode of life for most ammonoids advocated by Ebel 1983 (see also Westermann 1993, 1996; Kröger 2001 or Jacobs and Chamberlain 1996 for views contrasting Ebel's ideas). The function of the phragmocone as a buoyancy device has been corroborated by a great number of studies (see Hoffmann et al. 2015 and references therein) including the latest volume models of ammonoid shells and the linked buoyancy calculations (Tajika et al. 2015; Naglik et al. 2015), most mathematical models of buoyancy (Hoffmann et al. 2015), the convergent evolution of an upward orientation of the aperture in many

C. Naglik (✉) · A. Tajika · C. Klug
Paläontologisches Institut und Museum, University of Zurich, Karl Schmid-Strasse 6, 8006 Zurich, Switzerland
e-mail: carole.naglik@pim.uzh.ch

A. Tajika
e-mail: amane.tajika@pim.uzh.ch

C. Klug
e-mail: chklug@pim.uzh.ch

J. Chamberlain
Department of Earth and Environmental Sciences, Brooklyn College of CUNY, Brooklyn, NY 11210, USA

Doctoral Programs in Biology and Earth and Environmental Sciences, CUNY Graduate Center, New York, NY 10016, USA
e-mail: JohnC@brooklyn.cuny.edu

C. Klug et al. (eds.), *Ammonoid Paleobiology: From Anatomy to Ecology*, Topics in Geobiology 43, DOI 10.1007/978-94-017-9630-9_17

major ammonoid lineages (e.g., Raup and Chamberlain 1967; Bayer and McGhee 1984; Saunders and Shapiro 1986; Klug 2001; Korn and Klug 2003; Monnet et al. 2011) and the increase of phragmocone complexity through ammonoid evolution (Saunders 1995; Saunders and Work 1996, 1997; Daniel et al. 1997; Saunders et al. 1999). This is significant because *syn vivo* shell orientation can be used as an indicator of locomotion and habitat preference, although there are great limits for the accuracy of such conclusions.

Several authors have sought information on habitat depth and swimming speed in the physical properties of ammonoid shells. For example, shell implosion depths suggest limits for maximum diving depths (e.g., Westermann 1973, 1996; Saunders and Wehman 1977; Jacobs 1992a; Hewitt 1996; Batt 2007). Inferences on diving depth and ammonoid behavior have also been drawn based on siphuncle properties (e.g., Westermann 1971, 1996; Mutvei and Reyment 1973; Mutvei 1975; Chamberlain and Moore 1982; Ward 1982; Hewitt 1996). Oxygen isotopes have also been used to approximate diving/living depths of ammonoids (Moriya et al. 2003; Lukeneder et al. 2010, Lukeneder 2015; Moriya 2015).

Similarly, streamlining and drag have been quantified for a wide range of shell shapes (Kummel and Lloyd 1955; Westermann 1971, 1996; Reyment 1973; Chamberlain 1976, 1981; Chamberlain and Westermann 1976; Jacobs 1992b, Jacobs et al. 1994; Monnet et al. 2011; Ritterbush and Bottjer 2012; Ritterbush et al. 2014), and the results related to mode of life. Mutvei and Reyment (1973) as well as Mutvei (1975) argued that muscle attachment was too small and weak to allow ammonoids to swim well. However, there is some indication suggesting that ammonoids may have powered their locomotion with a muscular mantle not firmly attached to the shell (Jacobs and Landman 1993; Jacobs and Chamberlain 1996). Muscle attachment is discussed in detail in Doguzhaeva and Mapes (2015).

Sedimentary facies in which ammonoids are preserved may provide some information about lifestyle and habitat (Wang and Westermann 1993; Westermann 1996; Tsujita and Westermann 1998; Westermann and Tsujita 1999), although post mortem transport can complicate the picture (e.g., Kennedy and Cobban 1976; Tanabe 1979: Westermann 1996 and references therein). Nevertheless, the broad range of facies types in which ammonoid remains occur in combination with the great disparity in shell morphology supports a wide variety of life habitats and habits for these animals that in principle relate to differing locomotor capabilities as exemplified by Jacobs et al (1994).

Another line of evidence comes from sublethal injuries. It was especially Keupp (review in Keupp and Hoffmann 2015), who, in a series of articles (Keupp 1984, 1985, 1992, 1996, 1997, 2000, 2006, 2008, 2012), proposed that several types of injuries commonly recorded in ammonoid shells were inflicted by benthic crustaceans. If that is correct, this would support at least a temporarily demersal habitat for ammonoids showing such injuries; other injuries related to nektic predators, however, corroborate a nektic mode of life for at least some ammonoid groups (compare Ritterbush et al. 2014; Keupp and Hoffmann 2015).

Finally, *syn vivo* epizoans also provide some information on swimming direction and orientation of the shell (Keupp et al. 1999). However, such cases of epizoans that can be interpreted in that respect are rare (Seilacher 1960, 1982a, b; Keupp et al. 1999; Seilacher and Keupp 2000; Hauschke et al. 2011). De Baets et al. (2015b) review information obtained from epizoans attached to ammonoids as a function of orientation (e.g., Seilacher 1960). Their results support the swimming orientations discussed herein.

Because ammonoids apparently did not produce unequivocal trace fossils of their movements *syn vivo*, no evidence from this source is available to help interpret ammonoid locomotion.

17.2 Limits of Research on Ammonoid Locomotion

Because ammonoids are extinct, we cannot provide direct, observational evidence on their swimming ability from study of the living creatures. There is no direct way to measure such parameters as maximal swimming speed, maneuverability, or the efficiency of the musculature in extinct animals like ammonoids. Thus, in this paper we attempt to reconstruct ammonoid swimming ability and maneuverability using indirect evidence that can be gleaned from the fossil record; from analogy to the performance of modern relatives; and from awareness of the uncertainties inherent in such an effort.

There are only a few aspects of ammonoid locomotion, which at the outset appear highly plausible to us:

1. Ammonoids generally were able to produce neutral buoyancy by means of their buoyancy apparatus.
2. Most ammonoids were not fully benthic since they did not leave any traces in the sediment, had often upward pointing apertures and were preyed upon by nektic organisms or only from below by benthic organisms.
3. Most ammonoids were probably capable of swimming movement powered by jet propulsion, arm beating, or other mode of propulsion.
4. Locomotor capabilities were not uniform across all ammonoid taxa since they had sometimes quite large differences in shell orientation, hyponomic sinus, body size or shell shape.

17.3 Shell Orientation

17.3.1 Mathematical Models

With his pioneering work on ammonoid shell geometry and buoyancy, Trueman (1941) initiated a line of investigation that continues down to the present day, in which palaeontologists utilize mathematical modeling techniques to gain insight on ammonoid buoyancy and shell orientation. These models usually employ the parameters used by Raup and Chamberlain (1967), namely W (expansion rate), K (area of last generating curve) and R (distance from the coiling axis). However, such models are predicated on a number of simplifications (e.g., Trueman 1941; Raup (1967); Raup and Chamberlain (1967); Ebel 1983; Saunders and Shapiro 1986; Shapiro and Saunders 1987; Okamoto 1988, 1996; Klug 2001; Korn and Klug 2003). Commonly, these models include the assumption of self-similar (gnomonic), logarithmic shell growth, uniform shell thickness independent of position on the whorl section and the presence of a stable coiling axis. None of these simplifying assumptions necessarily coincide with actual ammonoid shells, i.e., shell growth in ammonoids was not perfectly logarithmic (e.g., Okamoto 1996; Klug 2001; Korn 2012; Tajika et al. 2015; Naglik et al. 2015), shell thickness varies and the coiling axis can permanently change its position throughout ontogeny (e.g., Urdy et al. 2010a, b).

Most authors, who produced mathematical models of shell geometry (Trueman 1941; Raup 1967; Raup and Chamberlain 1967; Saunders and Shapiro 1986; Okamoto 1988, 1996), tested their models, usually with data from Recent nautilids (Packard et al. 1980; Chamberlain 1987; Ward 1987; Jacobs and Landman 1993), and found reasonably good agreement between their results and the modeled attributes of the living animal. According to these models, the orientation of the aperture largely depended on the whorl expansion rate and ranged between about 30° and 110° from the vertical direction in normally coiled ammonoids with planispiral shells (Saunders and Shapiro 1986). In straight bactritoids (Fig. 17.1) and other heteromorph ammonoids, the aperture may have faced more or less downward, for example in more or less orthoconic forms (without counterbalancing options) such as baculitids or in some early ammonoids with very loosely coiled shells such as *Metabactrites* (e.g. Klug and Korn 2004), or in subadult *Anisoceras*, turrilitids and other heteromorphs. As shown in Fig. 17.1, shell orientation may have varied quite strongly throughout ontogeny.

As shown by Westermann (1996), the majority of Mesozoic ammonoids had body chamber lengths between 200° and 300° (Fig. 17.2). According to him and the model by Saunders and Shapiro (1986), this would coincide with an apertural orientation of about 80° to 100°, i.e. with the aperture oriented more or less horizontally. Only forms with extremely high or extremely low whorl expansion rates and body chambers shorter than half a whorl or exceeding one whorl in length would have had an aperture oriented below 50° from vertical.

17.3.2 Mechanical Models

In addition to mathematical modeling, some authors have employed mechanical, i.e. physical, models to help reconstruct shell orientation in ammonoids. Among the first to use such models were Mutvei and Reyment (1973), who built metal-coated, plastic shell models, vacuum molded from real ammonoid and *Nautilus* shells, to investigate the buoyancy and floating position of the animals thus modeled. These authors were later followed in using physical models by, e.g., Elmi (1991, 1993), Seki et al. (2000), Klug and Korn (2004), Westermann (2013) as well as Parent et al. (2014).

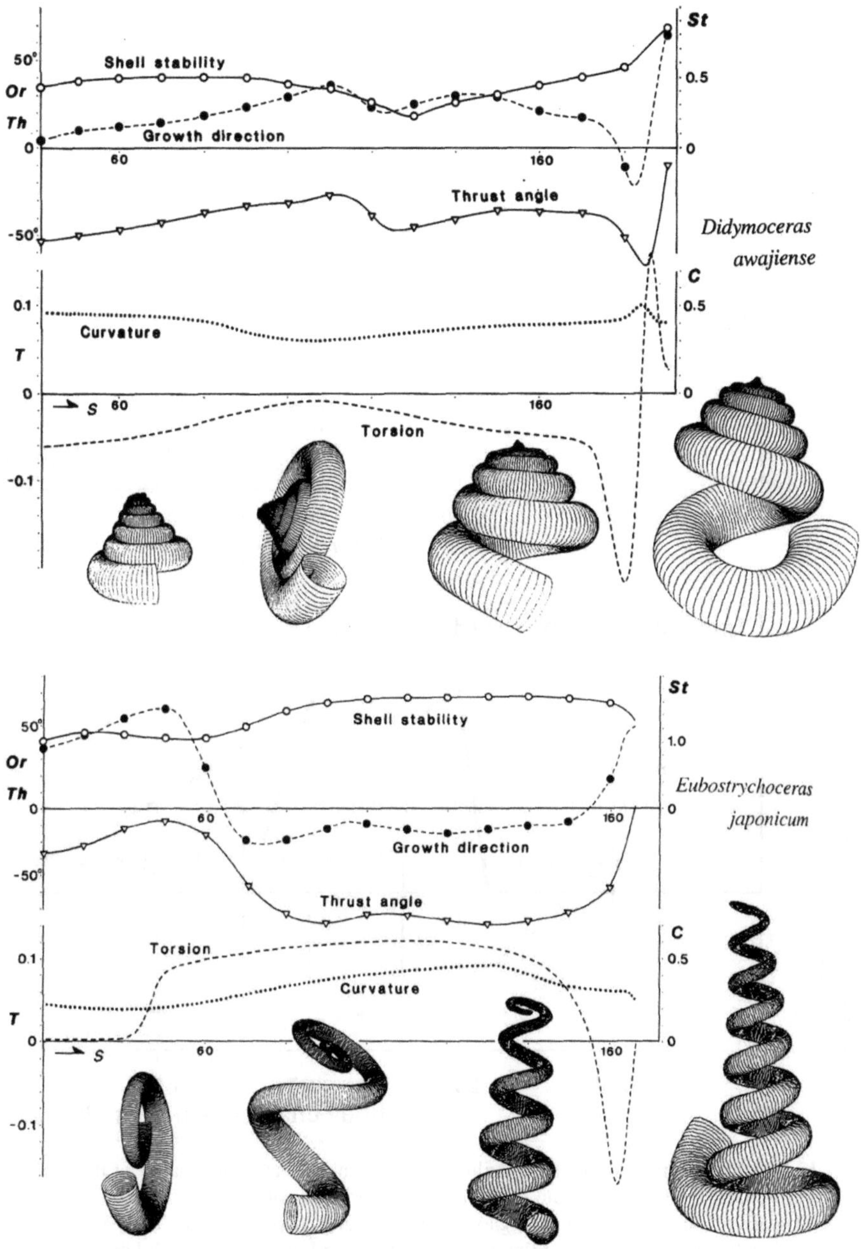

Fig. 17.1 Shell curvature, torsion, growth direction as well as hydrodynamic characters such as hydrodynamic stability, orientation of the aperture and hyponome jet thrust angle. (Source: Okamoto 1996)

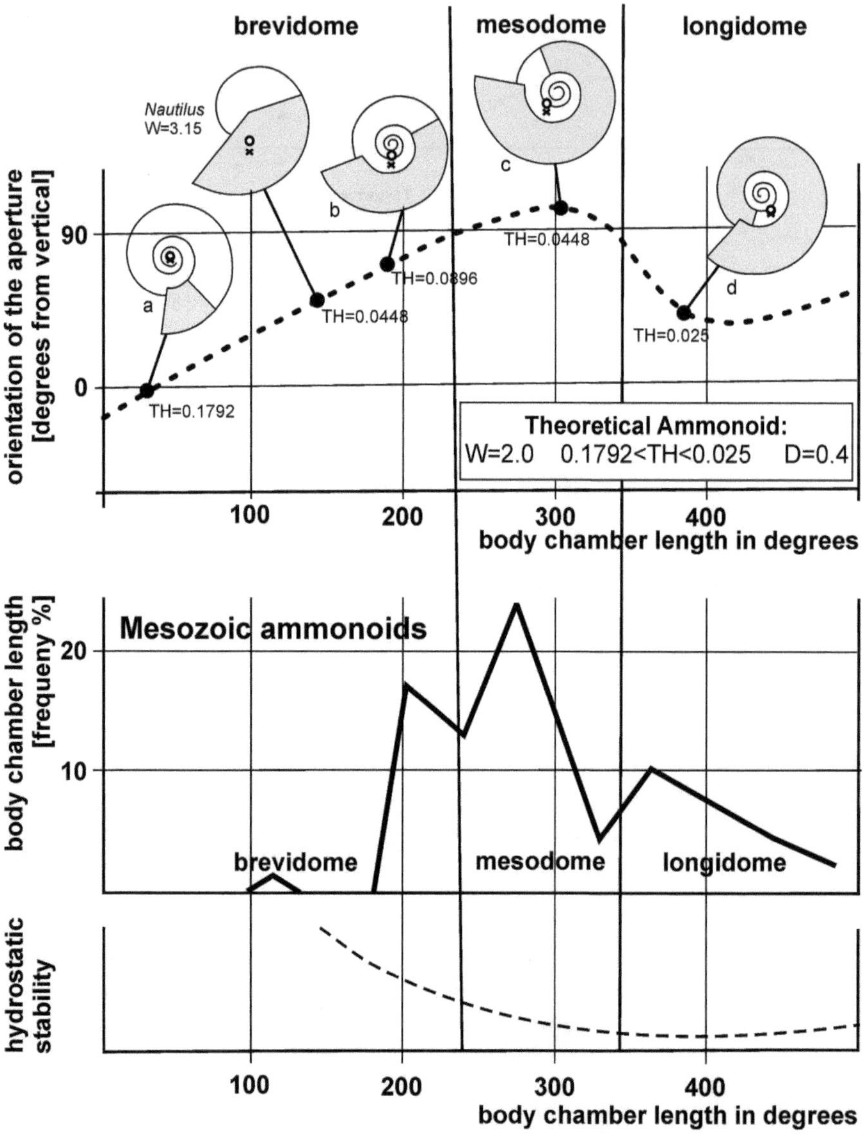

Fig. 17.2 Relationships between shell coiling, body chamber length, orientation of the aperture and hydrostatic stability of Mesozoic ammonoids. *W* whorl expansion rate, *TH* relative shell thickness, *D* relative distance between coiling axis and generating curve. Note the three peaks in body chamber length abundance and that these peaks coincide with a commonly sub-horizontal apertural margin and thus upward facing aperture. (Modified after Westermann (1996) incorporating a graph of Saunders and Shapiro (1986))

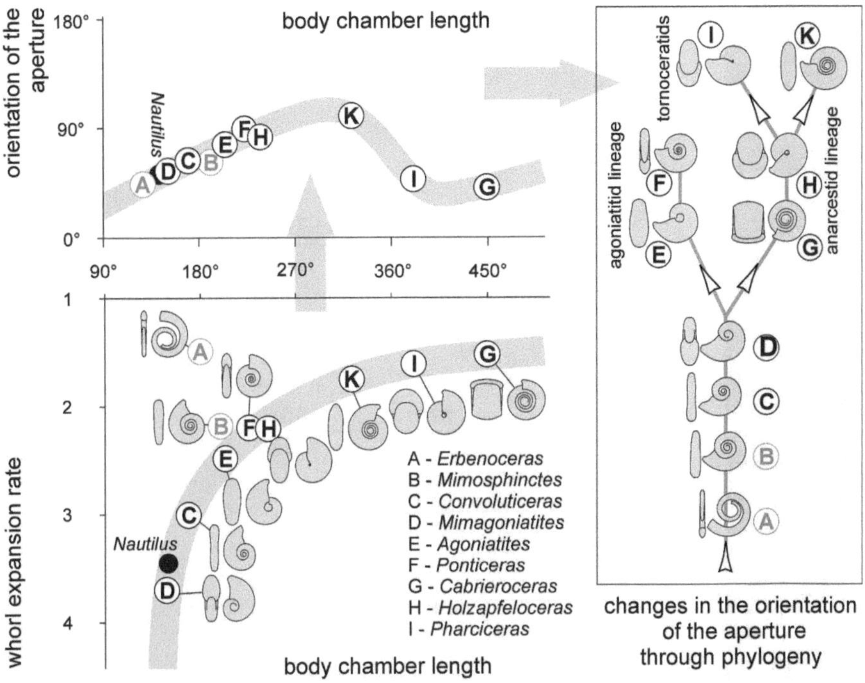

Fig. 17.3 Relationships between shell coiling, body chamber length, orientation of the aperture throughout the evolution of Devonian ammonoids. Note that three of the most important clades more or less independently evolved horizontal apertures early in the evolution of ammonoids. (Source: Klug and Korn 2004)

Klug and Korn (2004) showed, how shell orientation changed from facing downward in Orthocerida and Bactritida with orthoconic shells to oblique downward in Ammonoidea with loosely coiled shells, to oblique upward in less loosely coiled forms, to more or less horizontally upward in fully coiled shells (Fig. 17.3, 17.4). A progression of this type in aperture orientation is associated with iterative evolutionary trends (Fig. 17.3) in the major Devonian ammonoid clades (Mimosphinctoidea, Mimagoniatitoidea, Agoniatitoidea; Korn and Klug 2003) and even in parallel in two Devonian clades (Auguritidae and Pinacitidae; Monnet et al. 2011).

The question of whether ectocochleate cephalopods with orthoconic shells were capable of bringing their shell and body into a horizontal position is of long interest (e.g., Schmidt 1930; Ward 1976). Using physical models, Westermann (2013) demonstrated that a horizontal position in baculitid ammonoids could have been achieved by accumulating liquid in the most apical chambers. Such a horizontal position of cephalopods with orthoconic shells may also have been achieved with apical intracameral or intrasiphuncular deposits (e.g., Actinocerida, Endocerida) or chamber liquid (Westermann 1977, 2013; House 1981). For baculitids, Westermann (2013) suggested a vertical orientation of the shells of juveniles and a nearly horizontal orientation of subadults and adults because he assumed that juveniles had

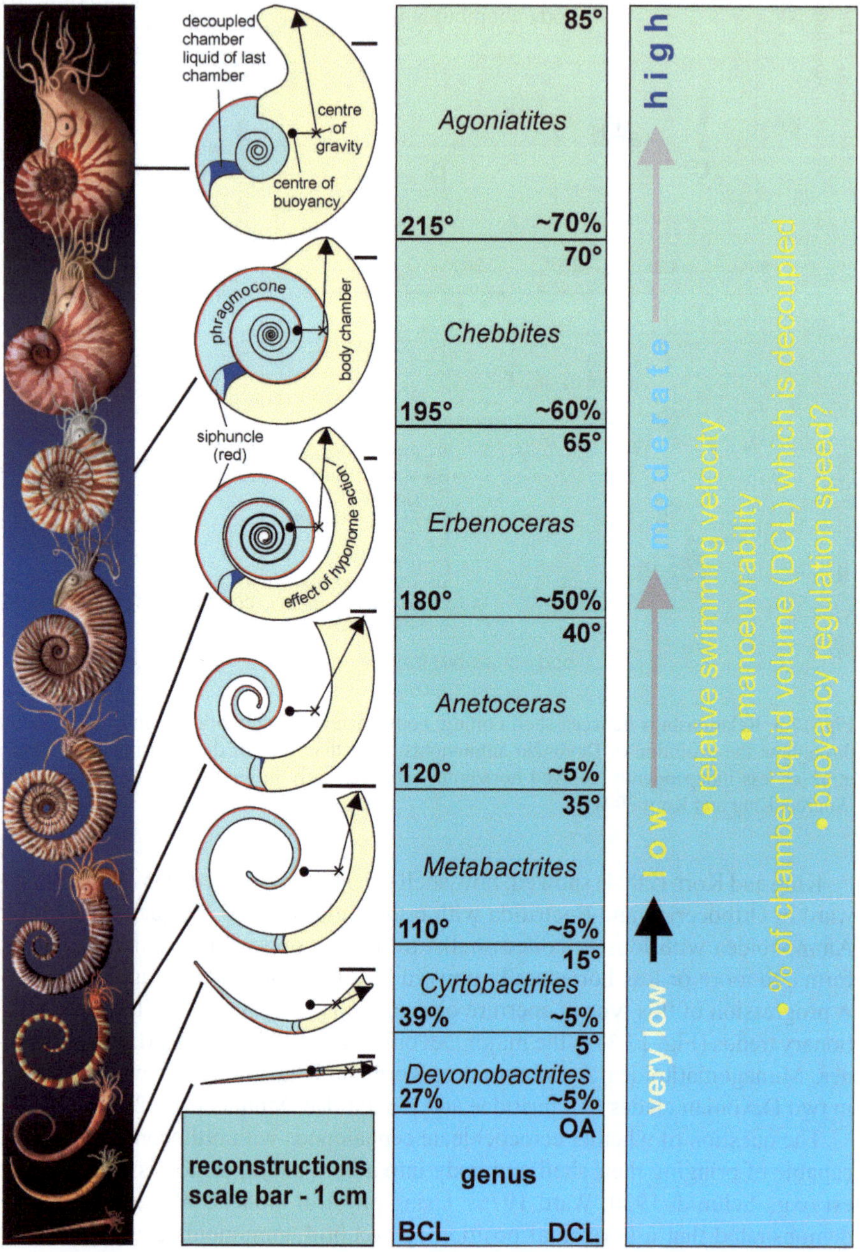

Fig. 17.4 Evolution of coiled ammonoid shells from straight bactritid shells and the consequences for body chamber length, aperture orientation, thrust angle of the hyponome jet, hydrodynamic stability and interpretations for swimming capabilities throughout evolution. (Modified after Klug and Korn 2004 as well as Klug et al. 2008). *BCL* body chamber length, *OA* orientation of the aperture, *DCL* decoupled chamber liquid.

phragmocones with more or less uniformly distributed liquid while in adults, the chamber liquid accumulated apically as a counterweight. The spatial distribution of chamber liquid might have also played a role in other ammonoids (Ward 1979, 1982; Kaplan 2002; Klug et al. 2008), although quantitative evidence on liquid distribution in ammonoid phragmocones has not yet been obtained.

Parent et al. (2014) experimented with a physical model comprised of weights and levers to assess the possible effect of the position of aptychi in aptychophoran ammonites on shell orientation. They concluded that in cases where the aptychus contains a sufficient mass and density relative to the animal's soft tissue and shell, the forward and backward movement of the buccal mass would have affected the orientation of the shell. Ammonites, such as some aspidoceratids, could have altered shell orientation in such a way that the aperture was lowered to $<25°$ from the vertical position.

Earlier physical models suggested capability of certain heteromorph ammonites (particularly the so-called "*shaft and hook shaped body chamber*" ammonoids; Kaplan 2002) to change their shell orientation (Kakabadzé and Sharikadzé 1993; Monks and Young 1998) by displacement of fluid and gas in the phragmocone (Kakabadzé and Sharikadzé 1993), or by moving the soft body of the animal within the living chamber, assuming that the animal was much smaller than its body chamber (Monks and Young 1998).

17.3.3 Empirical Models

We use the term "*empirical models*" to mean three-dimensional physical models of ammonoid shells constructed from stacks of cross-sections cut through a real shell. A similar approach was first employed by Chamberlain (1969), who built Plexiglas shell models from computer-produced topographic cross-sections of hypothetical ammonoid shells, which he then used for hydrodynamic experimentation (Chamberlain 1976, 1980, 1981). More recently, tomographic techniques have been developed, which greatly advance our skill to more confidently reconstruct *syn vivo* shell orientation(e.g., Longridge et al. 2009; Hoffmann and Zachow 2011; Hoffmann et al. 2013; Tajika et al. 2015; Naglik et al. 2015).These models are based on image stacks produced by different tomographic methods. Attempts to obtain image stacks by computer tomography often failed due to the lack of density contrast. This is probably the reason for the relatively late appearance of tomographic images of the interior of ammonoids in the scientific literature. Accordingly, tomographic data were sometimes obtained by serial sectioning (Tajika et al. 2015; Naglik et al. 2015). The latter method has the advantage that the images provide colour information and lack certain artifacts occurring in CT-data such as ring artifacts (see Hoffmann et al. 2013). In any case, these empirical models (Fig. 17.5) largely corroborate the results of mathematical modeling: forms with body chambers $<100°$ or $>360°$ have low apertures while the majority of shell forms with body chambers of 200° to 300° have more or less horizontally arranged apertures facing upward (Tajika et al. 2015; Naglik et al. in press).

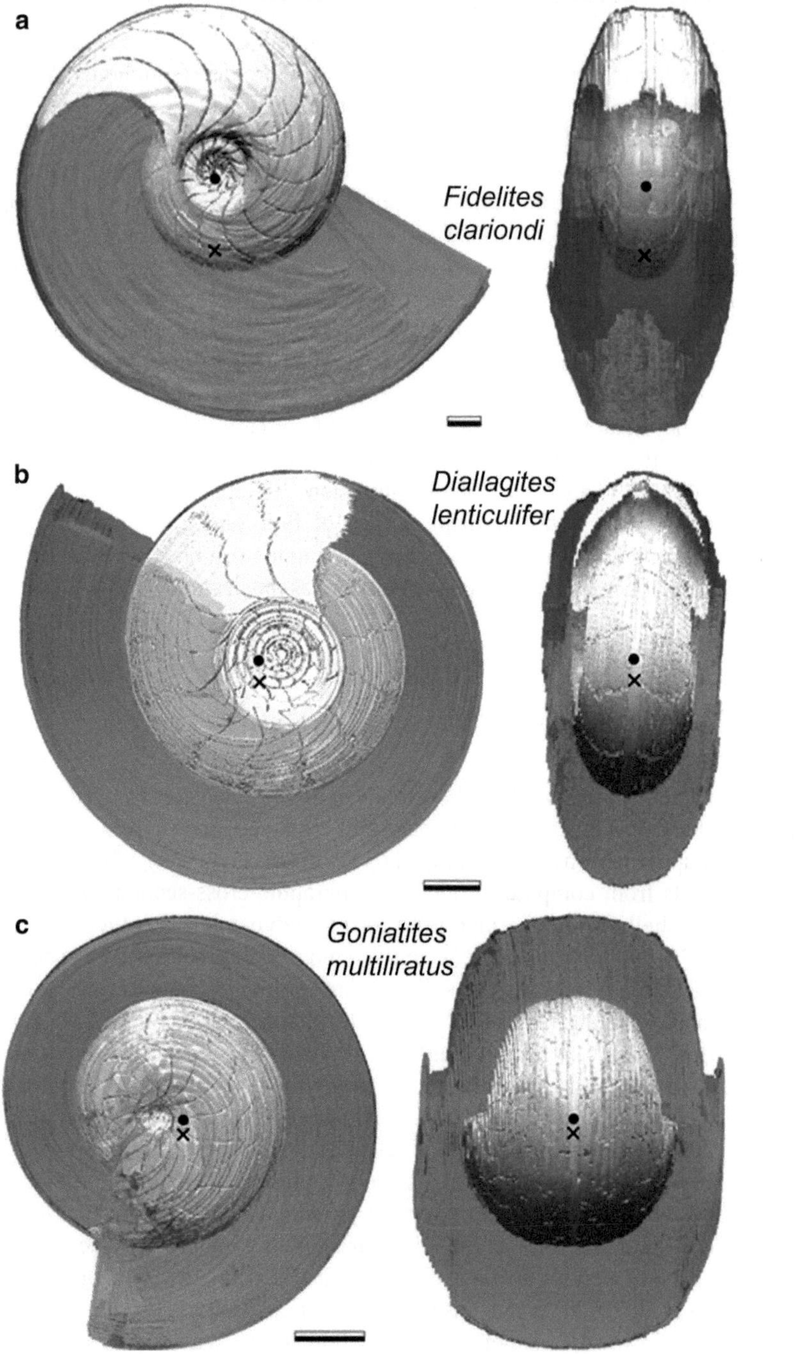

Fig. 17.5 Three Paleozoic ammonoids that have been subjected to grinding tomography in order to produce virtual 3D-reconstructions. Based on the image stacks, centers of mass (x) and buoyancy (o) were established based on these empirical models. (From Naglik et al. 2015). Scale bar: 0.5 cm

17.4 Muscles, Drag and Power

Because we cannot directly observe live ammonoids swimming, testing hypotheses on ammonoid swimming speed and swimming behavior presents obvious challenges to researchers interested in such matters. A wide variety of approaches to these issues are possible, but so far interest has centered primarily on muscles used to generate propulsion, drag, and power.

17.4.1 Muscles

Muscle attachment structures are reviewed by Doguzhaeva and Mapes (2015). According to them, jet-powered swimming could have been possible for forms with a body chamber length of one whorl or less and muscle attachments that would permit some muscles to extend straight across the body chamber and attach to the head and to the funnel.

A basic question here is the following: which of the extant cephalopods, if any, have a propulsive muscular system and mode of locomotion similar to that of ammonoids? The following kinds of muscular systems characterize modern cephalopods:

1. *Nautilus*-like: Because Recent nautilids are the only extant ectocochleate cephalopods, they have commonly been used as paradigms to understand the paleobiology of ammonoids. In modern nautilids, the large cephalic retractor muscles, which are attached to the inner shell wall of the body chamber, pull the head complex back into the body chamber, thus compressing the mantle cavity and expelling a propulsive jet of water out of the hyponome (Packard et al. 1980; Chamberlain 1981, 1987). The occurrence of apparent retractor muscle attachment scars in some ammonoids, as pointed out in Doguzhaeva and Mapes (2015), suggests that such ammonoids may have powered themselves by a "piston-pump" system not unlike what is seen in *Nautilus*.
2. Squid-like: Taking cephalopod phylogeny into account, ammonoids are more closely related to coleoids than to nautilids (Jacobs and Landman 1993; Kröger et al. 2011), and perhaps one may thus expect some similarities in coleoid and ammonoid propulsion systems. Modern squids use their muscular mantle (Bone et al. 1981) to pressurize mantle cavity water, which is then ejected through the funnel. However, the mantle in squids is not surrounded by shell as in *Nautilus*, or ammonoids, and does not function in shell secretion. It is generally considered that ammonoids also used their mantle to secrete the shell, as do ectocochleate cephalopods in general. Whether this necessarily implies that the ammonoid mantle was attached to the inner shell surface, and was secretory rather than muscular and incapable of compressing the mantle cavity is open to debate. In this regard, Jacobs and Landman (1993) as well as Jacobs and Chamberlain (1996) have suggested that the absence of large lateral muscle scars in some ammonoids

may mean that such animals used a squid-like system of propulsion involving the mantle. It has even been suggested that some ammonoids may actually have internalized shells (Doguzhaeva and Mutvei 1991, 1993), but in such cases, the mantle cavity was still located inside the body chamber of the shell. The *Nautilus* hyponome is also muscular and can direct and further compress the propulsive flow, but unlike the tubular funnel in coleoids, the *Nautilus* hyponome is a flap of tissue with folded, overlapping edges. Westermann (2013) proposed that some ammonoids may have had a powerful, coleoid-like tubular hyponome that was the main source of propulsive power. Nevertheless, the fins (probably not present in ammonoids) sometimes also play a role in squid locomotion (Packard 1972; Well 1995; Boyle and Rodhouse 2005)

3. *Argonauta*-like: Females of the octobrachian *Argonauta* produce an egg-case that is used both to shelter the eggs and to pick up air at the water surface in order to regulate buoyancy (Finn and Norman 2010). The shell differs from ammonoid shells in the absence of chambers and the fact that the *Argonauta* shell is secreted by two modified arms; other characters also differ (compare Hewitt and Westermann 2003). The mantle is not firmly attached to the shell and propulsion is carried out by means of the mantle as in other octobrachians (Young 1960; Finn and Norman 2010; Rosa and Seibel (2010).It is highly unlikely that ammonoids propelled themselves in a way analogous to that of a female *Argonauta*.
4. *Vampyroteuthis/Octopus*-like: *Vampyroteuthis* and several octobrachians can swim by contracting their arms with the velar skins, thus expelling water (Boyle and Rodhouse 2005). Since hardly anything is known about ammonoid arms (Klug and Lehmann 2015), it is currently impossible to conclude if such a mode of locomotion occurred in ammonoids.

In our view, it is likely, but not proven, that many ammonoids used longitudinal muscles to power jet propulsion. Evidence for the use of arms, velar webs, fins and mantle muscles in ammonoid locomotion is still poor or lacking. The possible efficiency, energy requirements and energy consumption associated with ammonoid propulsion are discussed in Chap. 17.4.3 below.

17.4.2 Drag

Drag is a physical term, which describes the forces that counteract the motion of an object moving through a fluid, namely seawater in the case of ammonoids. Drag is the product of the inertial and viscous forces acting on such an object, and thus depends on size, shape, and speed of the object, and on the density and viscosity of the fluid. Drag is one of the physical aspects of ammonoids that can be measured directly, even on fossil specimens (Schmidt 1930; Kummel and Lloyd 1955; Chamberlain 1976, 1980, 1981; Chamberlain and Westermann 1976; Jacobs 1992b; Jacobs et al. 1994; Jacobs and Chamberlain 1996). Drag force is usually measured directly, as was done in the studies noted immediately above. In situations where separated flow occurs, as would normally be the case for medium-sized and large ammonoids moving relatively fast, drag can also be calculated from the following equation.

$$F_D = \tfrac{1}{2}\rho\, v^2\, C_d A$$

F_D—drag force; ρ—density of the medium (seawater); v—velocity of the object (ammonoid); C_d—drag coefficient of the object (a dimensionless number which can be thought as representing the shape of the moving object); A—an area representative of the size of the moving object.

For objects, such as ammonoids, which have complex shapes, and thus complex flow interactions, shell volume raised to the two-thirds power ($V^{2/3}$) is generally the areal parameter of choice (Chamberlain 1976; Vogel 1981). It is important to understand that C_d is not a constant; it is a coefficient that varies widely for a given object depending on flow conditions. Flow state around an object, like an ammonoid, is described in terms of the Reynolds number.

$$\mathrm{Re} = \mathrm{dm}\, v / \upsilon \quad \text{with} \quad \upsilon = \gamma / \rho$$

Re—Reynolds number; dm—specimen shell diameter in the direction of motion; v—velocity of the object; υ—kinematic viscosity of seawater (viscosity [γ] divided by seawater density [ρ]). When Re is low ($\mathrm{Re} < 1000$ approximately), flow is attached to the object (this is often referred to as Stokes Flow); drag is due entirely to surface friction; C_d is very high, often more than 100, and varies directly with velocity and Re. Spherical objects generate the least drag because they have the smallest surface area, and hence least frictional drag per unit volume. For ammonoids, these conditions would hold for small ammonoids swimming slowly. When Re exceeds approximately 10,000, flow is separated to some degree from the object (this is often referred to as separated or non-Stokes flow); drag is due to a combination of friction and an adverse pressure gradient created by the separation; and C_d is low and often constant, or nearly so, as Re and velocity change, fusiform objects generate the least drag because they have the smallest possible pressure drag component (fusiform shapes minimize the extent and magnitude of separation and the posterior low pressures that derive from separation). For many objects operating in separated flow, a large reduction in C_d occurs when the character of the fluid boundary layer lying on the surface of the object converts from laminar to turbulent conditions. For ammonoids, separated flow would hold for large ammonoids swimming quickly. At intermediate values of Re ($1000 > \mathrm{Re} < 10{,}000$ approximately), flow is unstable and can vary from a separated to attached state depending on such factors as object shape and surface features. This would apply to ammonoids of intermediate size moving at intermediate speeds.

Several authors examined drag using ammonoid models (Schmidt 1930; Kummel and Lloyd 1955; Chamberlain 1976; Jacobs 1992b; Jacobs and Chamberlain 1996). Modeling focused on the shell only (Kummel and Lloyd 1955; Chamberlain 1976); the shell and attached prostheses imitating extruding soft parts (Chamberlain 1980); or shell and artificial surface sculpture (Chamberlain 1981). For models

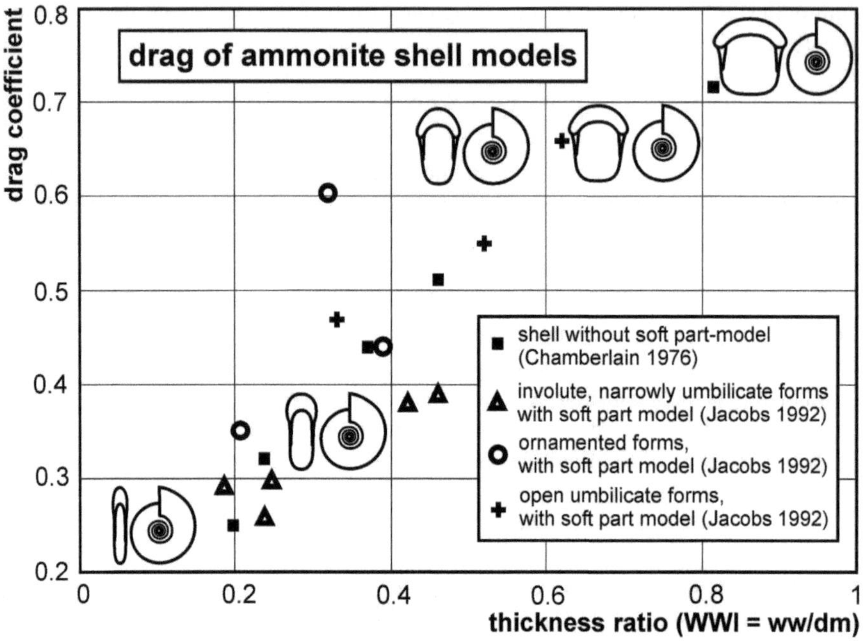

Fig. 17.6 Relationship between the thickness ratio and the drag coefficient, depending to a lesser degree on other factors such as umbilical width and ornament strength. These data (Chamberlain 1976; Jacobs 1992b) were obtained from models in a water tank (modified after Jacobs and Chamberlain 1996)

with representative values of Raup's W and D values (Raup 1966, 1967; Raup and Chamberlain 1967), Chamberlain (1976) determined drag coefficients in separated flow (i.e. for higher velocities and larger shells) where pressure drag is the key hydrodynamic factor. These experiments on models revealed that narrower shells had lower drag values. It appears to be mainly shell thickness and umbilical width, which play important role in generating drag in such flow conditions (Fig. 17.6).

In a later study, Jacobs (1992b) focused on drag for ammonoids of small size and low velocity (Re below about 25000), where frictional drag is the key hydrodynamic factor (see also Jacobs and Chamberlain 1996). Some results of Jacobs (1992b) are reproduced in Fig. 17.7. Note that in each graph in Fig. 17.7, the curves for wide and narrow forms cross at a point between Reynolds numbers of 5000 and 10,000. At Re less than the crossing value, the wider shells have lower drag coefficients (less frictional drag in Stokes flow) than the narrow shells, but at Re greater than the crossing value, the narrow shells have lower drag coefficients (less pressure drag in separated flow). This implies that different shell morphologies are more efficient at different sizes and swimming speeds (Table 17.1). Narrow forms produce less drag than wide forms at higher Reynolds numbers (faster speeds, larger size), while wide shells generate less drag at low Reynolds numbers (slower speeds, smaller size) than do narrow shells. This situation implies that the com-

Table 17.1 Possible swimming behavior of ammonoids in dependence of their shell shape. (Modified after Jacobs and Chamberlain 1996). For *Baculites*, we used the interpretation of Westermann (2013). Additional information comes from Klinger (1981) and Seki et al. (2000)

Shell shape	Slow, continuous swimming	Fast, continuous swimming	Acceleration	Vertical
Compressed involute				
Oxyconic (e.g., *Sphenodiscus*)	Poor	Good	Excellent	Moderate
Platyconic with rounded venter (e.g., *Oppelia*)	Moderate	Excellent	Good	Moderate
Platyconic with tabulate venter (e.g., *Anahoplites*)	Good?	Good	Moderate	Moderate
Moderately compressed				
Platyconic, moderately evolute (e.g., *Mesobeloceras*)	Moderate	Good	Moderate	Moderate
Involute juvenile (e.g., *Scaphites*)		Moderate	Moderate	Moderate
Evolute, rounded whorls (e.g., *Lytoceras*)	Moderate	Moderate	Poor	Moderate
Compressed evolute	Good	Moderate	Moderate	Good
Depressed				
Sphaeroconic involute (e.g., *Goniatites*)	Moderate	Poor	Poor	Moderate
Cadiconic, evolute (e.g., *Cabrieroceras, Gabbioceras*)	Moderate	Poor	Poor	Moderate
Heteromorphic				
Orthoconic (e.g., *Baculites*)	Moderate	Moderate	Excellent?	Moderate
Torticonic (e.g., *Turrilites*)	Poor	Poor	Poor	Good
Loosely coiled in three dimensions (e.g., *Nipponites, Didymoceras*)	Poor	Poor	Poor	Good

mon ammonoid ontogenetic change in shell morphology from depressed juvenile whorls to more compressed whorl shape near maturity could be linked with this flow state dependent change in drag coefficient (Jacobs and Chamberlain 1996). The latter authors also suggested that morphologic change related to hydrodynamic factors operating in the evolution of ammonoid clades should be linked with different host facies. This link was examined by various authors (e.g., Ziegler 1967; Batt 1989; Bayer and McGhee 1984; Marchand 1992; Courville and Thierry 1993;

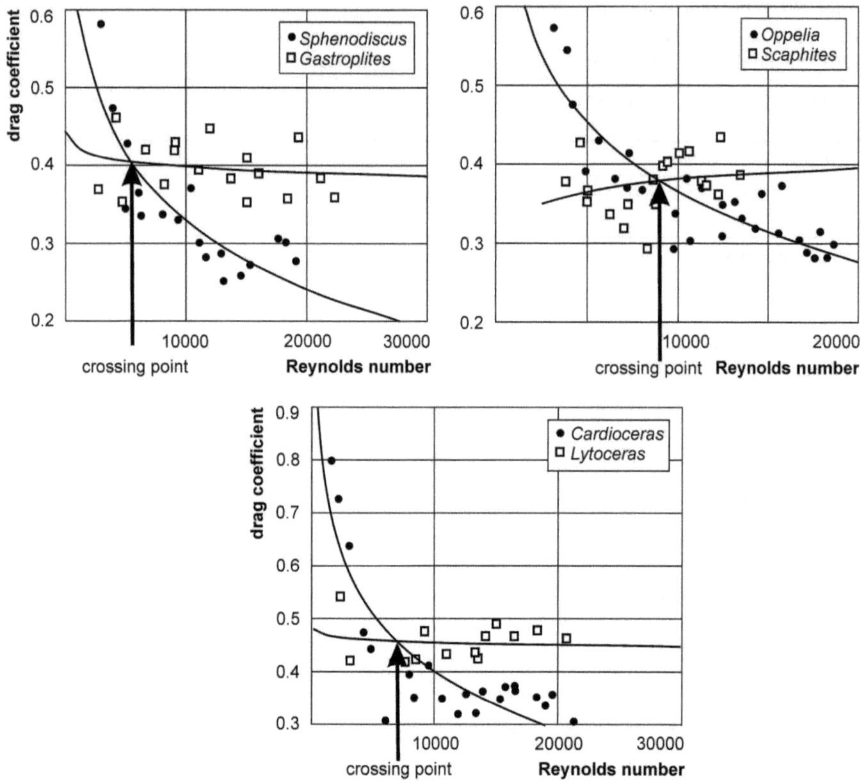

Fig. 17.7 Relationships between drag coefficients and Reynolds number (Re) of three different pairs of Jurassic and Cretaceous ammonoids. In each pair, one form has a narrow shell (*dots*), and one has a wide shell (*open squares*). Note that in each graph the curves for the two forms cross at a point between Reynolds numbers of 5000 and 10000. At Re less than the crossing value, the wider shells have lower drag coefficients (less frictional drag in Stokes flow) than the narrow shells, but at Re greater than the crossing value, the narrow shells have lower drag coefficients (less pressure drag in separated flow)

Jacobs et al. 1994; Klug 2002; Kawabe 2003). Such studies are hampered by the possibility that ammonoid shells were transported post mortem and the imperfect knowledge of habitat depth, because the sedimentary context in which ammonoids are preserved mainly informs about the energy in the water column and the volume of sediment that is delivered in combination with accommodation space. It is possible that ammonoids could have lived in more quiet waters near the sea-floor or in more agitated waters near the surface uncharacteristic of the sedimentary context of the rock itself. Additional factors, such as time-averaging might also complicate straight forward interpretations (compare De Baets et al. 2015a).

In any case, the measurable disparity of ammonoids throughout ontogeny and evolution as well as the recurrent ontogenetic change in shell shape indicate that minimizing drag played an important role in ammonoid evolution. It also indicates that different forms were possibly specialized for different modes of life with correspondingly different swimming abilities.

17.4.3 Power

The use and availability of power for swimming in ammonoids cannot be measured directly and thus has to be addressed based on actualistic comparisons with living organisms (e.g., Trueman and Packard 1968).

The physical term, power, simply describes the ratio between the work, W, expended in a time interval, t:

$$P = \Delta W / \Delta t$$

Assuming constant velocity during the time interval in question, this can be modified to the following equation using drag force F_D and velocity v:

$$P = F_D\ v$$

Power consumption during swimming thus depends directly on drag coefficient and can be estimated from the relationship between drag coefficient and Reynolds number, and thus with respect to size and velocity (Jacobs 1992b; Jacobs and Chamberlain 1996). In order to assess the differences in power consumption as a function of shell form, size and velocity, Jacobs (1992b) produced drag data for the thick genus *Gastroplites* (ww/dm = 0.42) and the thin genus *Sphenodiscus* (ww/dm = 0.19). His results are reproduced here in Fig. 17.8. According to Fig. 17.8, *Gastroplites* would require less power at sizes below 10 cm and velocities below 50 cm/s. At a shell size of 10–100 cm and speeds below 15 cm/s, the two shell shapes would require about the same power. At higher speeds and sizes exceeding 10 cm, *Sphenodiscus* would need less power and swim more economically. Whether these ammonoids could actually produce the power necessary to swim at these speeds cannot be inferred from such data, however.

Knowledge of swimming speed in fossil ammonoids requires knowledge of the power output generated by live ammonoids. The power produced by live ammonoids is unknown. However, one can gain useful insight into this matter by applying to this question data on power output of modern swimmers, particularly modern cephalopods. Of primary interest is the power output of modern analogues in sustained swimming (powered by aerobic muscle contraction), and in burst swimming (powered by anaerobic muscle contraction). Also of interest is metabolic scope, i.e., the difference between the power requirements during inactivity and periods of maximum activity. Unsurprisingly, power output and metabolic scope differ strongly between living cephalopods such as *Nautilus* with very low metabolic rates and the active squid *Illex* with a high metabolic scope (O'Dor 1982, 1988a, b; Chamberlain 1987; O'Dor and Wells 1990; O'Dor et al. 1990, 1993; O'Dor and Webber 1991). Even among squids, metabolic rates can vary strongly depending on

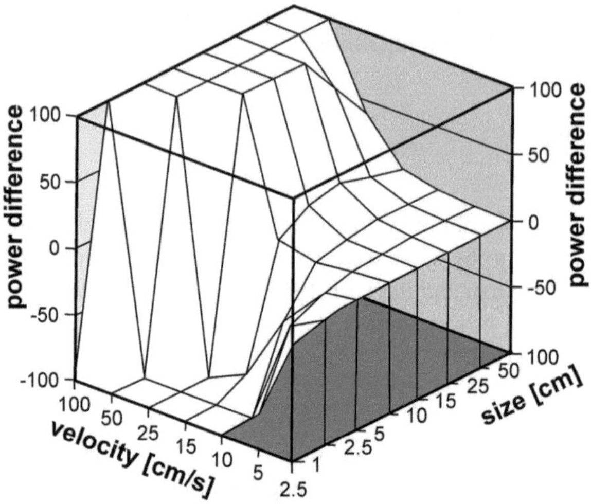

Fig. 17.8 Differences in power consumption (in ergs/s/cm^3) in a broad, depressed form (*Gastroplites*) and in a narrow, laterally compressed form (*Sphenodiscus*). Power difference was calculated by subtracting the power required per unit volume in *Sphenodiscus* from that of *Gastroplites*. Depending on this ratio, one obtains positive or negative values: when the values of power difference are negative, *Gastroplites* requires less power. The greatest power difference is seen at low sizes and high velocities. Jacobs and Chamberlain (1996) considered these differences as so profound that they appear to be biologically significant. Power differences > 100 ergs/s/cm^3 are not shown (modified after Jacobs and Chamberlain 1996)

their habitats (Seibel et al. 1997). For instance, the deep-sea squid *Vampyroteuthis infernalis* has a metabolic rate a hundred times lower than the shallow water *Gonatus onyx* (Seibel et al. 1997).

Estimates of power production in ammonoids depend on whether Recent nautilids are considered the better model organisms with their similarly constructed external shell or whether coleoids should rather be used as paradigms because they are more closely related to ammonoids. Several authors (e.g., Trueman 1941; Swan and Saunders 1987; Jacobs and Landman 1993; Kröger et al. 2011) have argued in favor of coleoids rather than nautilids on the basis of shell form and phylogeny. In order to estimate sustainable swimming speeds in ammonoids, Jacobs (1992b) argued that a metabolic rate of 200 ml of oxygen per kilogram per hour, which is close to that of *Sepia* (O'Dor and Webber 1991), probably represents a reasonable figure for most ammonoids. He also advocated that for ammonoids, sepiids represent the most meaningful model organisms among coleoids because like ammonoids, they have a large chambered phragmocone, which greatly limits the relative proportion of propulsive muscle (and soft tissue generally) to total volume of the animal (see also Chamberlain 1981, 1990, 1992, 1993). By comparison, squids like *Illex*, pack their bodies much more fully with propulsive muscle. O'Dor and Webber (1991) found that the metabolic scope of the highly active *Illex* was four times larger than in *Sepia* and additionally, the efficiency of their muscles exceeds that of sepiids. In

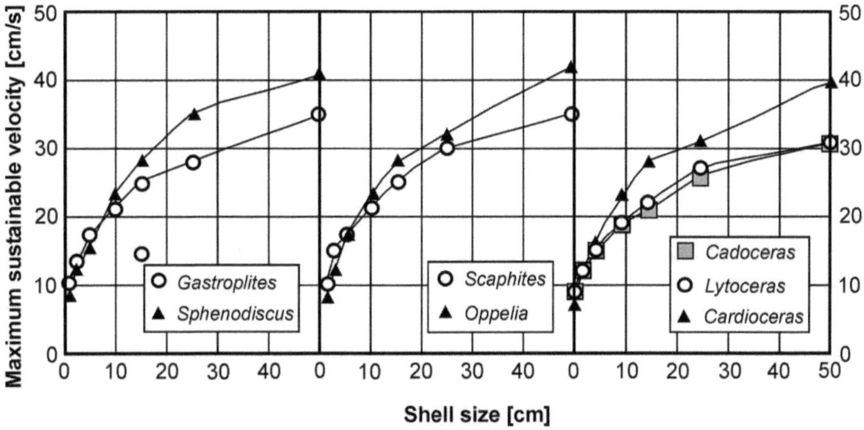

Fig. 17.9 Maximum sustainable swimming velocities in seven ammonoid genera. These are arranged in groups of two or three, always comprising a genus with a more compressed and one with a more depressed shell form. The velocity values are based on the assumption that the maximum power availability was 400 ergs/s/cm³. Overall, the curves resemble each other and in the curve pairs, they cross each other at a size of 5 to 10 cm (modified after Jacobs (1992b) as well as Jacobs and Chamberlain (1996))

consequence, power output is ten times higher in *Illex*, thus making *Sepia* the better actualistic model organism for ammonoids (Jacobs and Chamberlain 1996).

O'Dor and Webber (1991) observed swimming speeds of maximally 65 cm/s (2.3 km/h), which required a power output of 1000 µJ/s/cm³. Jacobs (1992b) as well as Jacobs and Chamberlain (1996) concluded that in ammonoids, this figure would probably not have exceeded 600 µJ/s/cm³ because only about 40% of the organism's volume is occupied by soft parts. The maximum swimming speeds of some ammonoid species, which are based on these assumptions, are depicted in Fig. 17.9. Maximum swimming speeds of large ammonoids like *Sphenodiscus* with a shell diameter of 25 cm would not have exceeded 100 cm/s (3.6 km/h). *Gastroplites* of the same size would have a speed of about 70 cm/s (2.5 km/h). The latter velocity corresponds to the maximum in *Sepia* (Jacobs and Chamberlain 1996). As a lower limit of energy availability, *Nautilus* can be used as model. Nautilus can activate up to 100 µJ/s/cm³, i.e. a tenth of that of *Sepia.* Using this figure, a 25 cm *Gastroplites* could reach 40 cm/s (0.54 km/h) and *Sphenodiscus* would have been able to swim 55 cm/s (1.98 km/h). These results are similar to swimming velocity estimates based primarily on drag considerations made by Chamberlain (1981, Fig. 17.8).

In summary, Jacobs (1992b) as well as Jacobs and Chamberlain (1996) found that swimming speed of ammonoids likely depended on various factors including shell shape (e.g. Table 17.1), body chamber angle, size, energy availability and power consumption. For large size, ammonoids with compressed shell form (low ww/dm ratio) could swim faster than those with depressed shells (high ww/dm ratio); while at small size this relationship is reversed.

17.4.4 Acceleration

Accelerating an object in a fluid involves accelerating fluid entrained in the object's wake and also fluid in direct contact with the surface of the object, i.e. in the boundary layer. In the case of swimming organisms, this also applies and in order to estimate swimming speeds and energy requirements, this added mass has to be taken into account (Chamberlain 1987; Jacobs 1992b; Jacobs and Chamberlain 1996). The force required to accelerate this added mass can be quantified by the following equation, which was introduced by Daniel (1984):

$$G = -\text{ar V} \left(\text{du}/\text{dt}\right)$$

G—acceleration reaction force; a—added mass coefficient (a function of thickness ratio ww/dm); r—density of the fluid; V—volume of the object/ammonoid; du/dt—acceleration.

The acceleration reaction force occurs both in acceleration and deceleration (Daniel 1984; Chamberlain 1987; Jacobs 1992b; Jacobs and Chamberlain 1996). For ammonoids, the symmetry of the shell in swimming direction, shell shape, differences in acceleration and deceleration processes as well as the formation of vortices in the wake play a role.

In cephalopods, acceleration is produced by a series of water expulsions from the hyponome with interim phases of water intake into the mantle cavity. The animal accelerates when the propulsive muscles contract forcing water from the mantle cavity and decelerates during the recovery phase of the propulsive cycle when water is taken into the mantle cavity in preparation for the next mantle cavity contraction. When an organism starts swimming, energy is mainly invested in acceleration while at higher speeds when velocity is more constant, the energetic cost of drag rises. Acceleration force also depends on the width of the ammonoid shell (ww/dm ratio; Fig. 17.10) and it roughly doubles from ww/dm = 0.2 to a value of 0.4 (Jacobs and Chamberlain 1996). According to Daniel (1984, 1985), the ratio of energetic costs of drag to that of acceleration varies from 48 % in a small squid accelerating from 0 to 2000 cm/s^2 to 62 % in a medusa accelerating to 700 cm/s^2 to 92 % in a salp accelerating to 23 cm/s^2. These values show that the faster an organism accelerates to a higher velocity, the lower the relative energy investment into added mass and the higher the investment into overcoming drag.

This relationship points to a potentially multiple functions of shell shape in ammonoids. While some shell morphologies reduced drag, other shell morphologies, such as oxycones with a small umbilicus, would have reduced the energetic cost invested in added mass (Jacobs 1992b; Jacobs and Chamberlain 1996). In that respect, ammonoids with narrow oxyconic shells would resemble ambush predators among fish (e.g. pike, barracuda) whose body geometry is too elongate to be purely adapted to reduce drag. Instead, their long and narrow shape strongly reduces acceleration reaction force and enables them to accelerate strongly from a standing start. While there is no corroboratory evidence for an ambush predator strategy

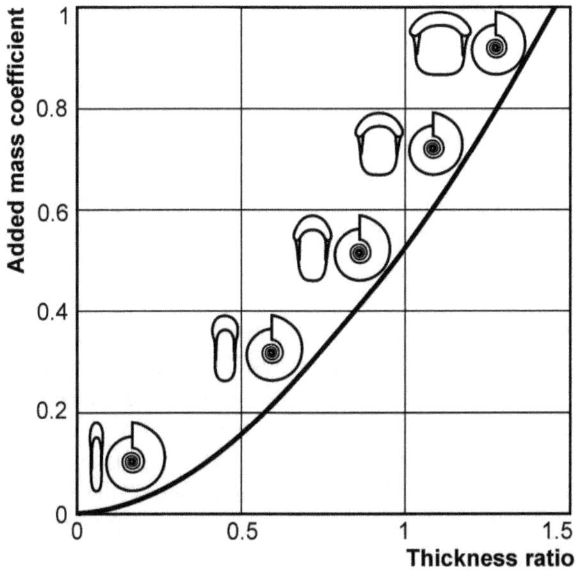

Fig. 17.10 Depending on shell shape and ornament, differing amounts of added mass of water accelerated with the ammonoid in the boundary layer and the wake can be expected. The acceleration reaction is a linear function velocity change (acceleration) and a function of the added mass coefficient, which -in turn- depends on shell shape and orientation relative to the direction of acceleration. According to these relationships, ammonoids with laterally compressed shells had substantially less added mass than ammonoids with depressed shells (modified after Jacobs (1992b) as well as Jacobs and Chamberlain (1996))

in oxyconic ammonoids, the fact that oxyconic shell form evolved many times iteratively and sometimes even in parallel (e.g., Bayer and McGhee 1984; Klug and Korn 2002; Monnet et al. 2011) shows that this shell shape may indeed have had a positive adaptive benefit for ammonoids.

17.4.5 Cost of Transportation

The cost of transportation (COT) is a metric that describes the energetic cost of locomotion. COT has been defined in a variety of ways. For example, in his comparison of the energy cost of different styles of animal locomotion Schmidt-Nielsen (1972) defined COT as (metabolic rate/(body weight and speed). O'Dor (1988a) and O'Dor and Webber (1991) in their study of squid locomotion, and Chamberlain (1990) in his study of *Nautilus* locomotion, determined COT by calculating metabolic output from oxygen consumption data for swimming animals. In all such approaches the aim has been to express COT in terms of the propulsive power produced by a swimming animal relative to some measure of its size, speed, and distance travelled. COT is thus simply stated in terms of propulsive power per unit of animal size per unit of speed or distance traveled where animal size is represented by weight or volume.

The power produced by a swimming ammonoid can be expressed as follows:

$$P = W/t = (F\ d)/t = F\ v$$

where P is the metabolic output (power) used to produce locomotion; W is the work needed for locomotion; t is the time interval over which the locomotion occurs; F is the force or thrust developed by the swimming ammonoid and is assumed to be constant over the interval t; d is the distance traveled; and v is the animal's velocity, also assumed to be constant.

Jacobs (1992b) and Jacobs and Chamberlain (1996) used the power-required data and the efficiency assumptions of Jacobs (1992b) to evaluate COT for a few representative ammonoids. Following Jacobs (1962b), they calculated COT as propulsive power per unit of total shell volume per unit of distance traveled. Their results are diagrammed in Fig. 17.11. The upper panel in this figure indicates that, assuming *Sepia* metabolic output, *Gastroplites* COT depends on size. Larger animals have lower COTs for a given velocity than smaller ones. This is largely the result of larger animals operating in separated flow where drag coefficients are smaller while small animals operate in Stokes flow where drag coefficient is much higher for objects of the same shape. The upper panel also indicates that if we assume *Gastroplites* had a lower metabolic output equivalent to that of *Nautilus*, its COT would also be lower. Perhaps the most interesting observation to be made from Fig. 17.11 is that for each curve there is a specific velocity for which COT is minimal. If energy conservation in swimming ammonoids mirrors that of flying animals, where flight speed usually reflects minimal COT, and there is no reason why it should not, this may mean that this minimal COT speed represents the usual swimming speed for the ammonoid to which the curve applies. The steepness of the curve on either side of the minimum COT speed implies that there would be considerable gain in cost to the animal in moving away from this optimum speed. The lower panel in Fig. 17.11 shows that the modern swimmers plotted here, both coleoids and fish, have COT-velocity curves much less steeply inclined as velocity increases above the minimum COT speed. This means that these modern animals are not nearly so constrained in terms of COT in varying their swimming speed than is the case for the ammonoids plotted here as well. Swimming over a range of velocities does not greatly influence their COT. It would appear that these modern swimmers have a much more flexible swimming repertoire than did fossil ammonoids.

Jacobs (1992b) as well as Jacobs and Chamberlain (1996) suggested that due to their neutral buoyancy, ammonoids, like *Nautilus*, may have had a low use of energy at rest and that the cost of transportation in ammonoids was accordingly low at low velocities. Alternatively, if ammonoids were closer to sepiids in their metabolic rates, the cost of transportation would have been lower at higher swimming speeds depending on their size (Fig. 17.11). Jacobs and Chamberlain (1996, p. 210) summarized this idea as follows: "*ammonoids may not have been pursuit predators, comparable to tuna or some squids, that spend long periods of time chasing down prey at high speed. This would deny the utility of the neutrally buoyant shell in limiting energetic expenditure. However, life styles that require only intermittent bursts*

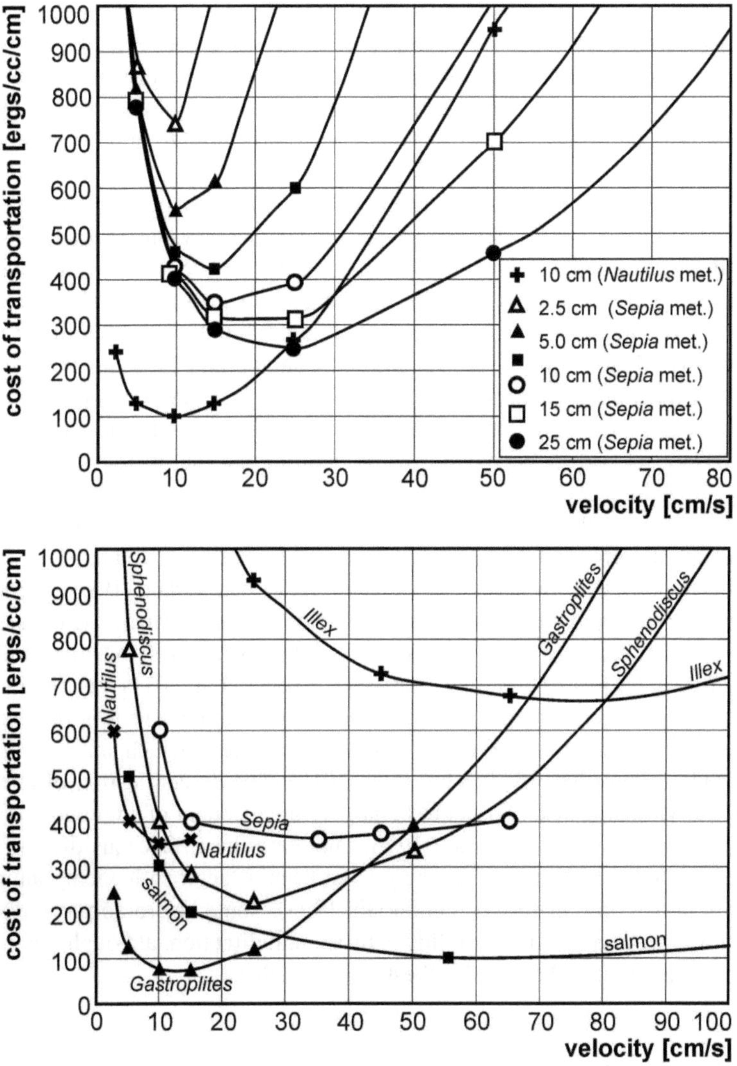

Fig. 17.11 Cost of transportation (*COT*) in relation to velocity depends on shell size and metabolic rate (upper diagram) and differs between modern animal groups (lower diagram). Modified after Jacobs (1992b) as well as Jacobs and Chamberlain (1996). The efficiency of energy conversion into propulsion force is estimated to be 10 %. The upper diagram shows the COT for various sizes of *Gastroplites*, assuming metabolic rates (O'Dor and Webber 1991) of *Sepia* (3900 ergs/s/cm^3) and in one case of *Nautilus* (1/7th of *Sepia*). With increasing size, less energy is required for locomotion. The lower diagram shows the COT of *Sphenodiscus* and *Gastroplites* in comparison to various recent cephalopods and a fish (O'Dor and Webber 1991). Resting metabolic rates were estimated for *Sphenodiscus* to resemble that of *Sepia* and for *Gastroplites* to resemble that of *Nautilus*. At higher velocities, the costs rise much faster in the shelled swimmers than in fishes and squids. However, the ammonoid curves are based on a series of estimates for the metabolic rates, added mass and other modes of locomotion (fins in *Sepia*)

of energy, such as ambush predation, seem possible, and oxyconic shell shape [...] *may have been conducive to such a mode of life*." It should be remembered that high speed is not required for successful predation. A predator must only move faster than its prey. If its prey is slow, a predator can be slow also. Oxycones would not need the fast burst speed of *Illex* or *Sphyraena* (barracuda) to prey on slower moving ammonoids.

Jacobs (1992b) and Jacobs and Chamberlain (1996) also pointed out that energy used for transport is energy that cannot be used in other ways; there is a trade-off between these costs and the energetic cost of other life functions. Nautilids have a low metabolism and can fast over lengthy time spans. In such a case, slow swimming speeds (O'Dor et al. 1990) are advantageous in promoting prolonged food searches (Wells 1987; Chamberlain 1990; Jacobs and Chamberlain 1996), as is the case for *Nautilus* (Ward and Wicksten 1980). Wells and O'Dor (1991) thought that other ectocochleates such as ammonoids may have pursued a similar low energy mode of life. They supported this hypothesis by pointing out that increasing numbers of fish occupied high energy nektonic habitats (for these macroecological changes, see Signor and Brett 1984; Bambach 1999; Kröger 2005; Klug et al. 2010) and would have competitively excluded most ammonoids from these habitats. The problem with this hypothesis is twofold: (1) As Jacobs and Chamberlain (1996) pointed out, ammonoids are more closely related to coleoids (some of which use considerable energy in relation to body size and also swim at high velocities) than they are to low energy nautilids (Jacobs and Landman 1993; Kröger et al. 2011). (2) The radiation of gnathostome fish in the Silurian and Devonian, a major event in the evolution and history of diversification of fishes, was also a time in which ammonoids originated and rapidly diversified (Klug et al. 2010). The diversification of teleostean fish in the Mesozoic also appears to be largely independent of ammonoid diversity changes (Jacobs and Chamberlain 1996), although the Cretaceous diversification of deep-bodied acanthopterygians may have been a factor influencing ammonoid diversity late in their history (Chamberlain 1993). Some heteromorphs might have been slower swimmers than nautilids in horizontal direction, although this requires further research (e.g., Ward 1979; Westermann 1996).

17.4.6 The Role of Ornament

As in sharks (Reif 1982; Oefner and Lauder 2012) and golf balls, a fine regular surface ornament can reduce drag by forcing conversion of the boundary layer around an ammonoid shell from laminar to turbulent flow at lower Reynolds numbers than would normally be the case. Boundary layer conversion reduces the scale of the turbulent wake and the pressure drag that results from it. Chamberlain and Westermann (1976) and Chamberlain (1981) examined this phenomenon and concluded that it could have a positive effect for some ammonoids by bringing lower drag and more efficient swimming into the velocity range of some ammonoids. Nevertheless, the lowering of the coefficient of drag would have been significant at Reynolds numbers exceeding 40,000, a figure that could potentially only be achieved in large ammonoids moving at relatively high velocities (Chamberlain 1981).

Jacobs and Chamberlain (1996) speculated that in cadicones, the coarse ribs or nodes as in *Cabrieroceras*, *Gastrioceras* or *Teloceras* might have caused the formation of vortices covering the entire umbilicus. Similarly, they suggested that, in forms with tabulate venter (or with ventral band as in Devonian forms such as *Gyroceratites*, *Armatites* or *Kosmoclymenia*), the water might have been divided into two fields, thus maintaining flow attachment and reducing turbulence in their wake, at least at certain velocities and sizes. They also reasoned that ribs tend to be the largest near the aperture and to be oriented in swimming direction, thus stabilizing the shell orientation during backward swimming in forms, which are more or less involute and carry moderately strong ribs such as *Cardioceras*. Westermann (1966) even speculated that this might be a driving force behind Buckman's law of covariation, although this law can be conveniently explained by morphogenetic processes (Monnet et al. 2015) without an adaptive interpretation (compare Hewitt 1996 for an alternative functional explanation). In contrast, strong ornament significantly increased drag (Chamberlain 1976; Jacobs 1992b; Hewitt 1996; Jacobs and Chamberlain 1996), thus supporting indirectly its possibly defensive function shell sculpture (e.g., Ward 1981).

17.4.7 Hydrodynamics Through Ammonoid Development

As discussed in Hoffmann et al. (2015), the flow regime in which ammonoid swimming took place changed through ontogeny as ammonoids grew in body size and shell diameter. Ontogenetic size increase covered two orders of magnitude or more in most ammonoid taxa. While embryonic shells (Landman et al. 1983; De Baets et al. 2012) vary about one order in magnitude in size between the earliest forms (>5 mm) and several derived Mesozoic forms (ca. 0.5 mm), the adult shells vary from less than 1 cm to over 2 m. Because small individuals have less power in relation to drag, adult ammonoids could probably swim one to two orders of magnitude faster than hatchlings (Jacobs and Chamberlain 1996).

In hatchlings, much of the energy invested in locomotion will be absorbed by skin friction drag. Jacobs and Chamberlain (1996) guessed that a hatchling of 1 mm diameter might have attained a swimming speed of 1 cm/s, which corresponds to a Reynolds number near 10. Accordingly, shells with a high whorl width index would have been favorable. In that light, the common decrease in whorl width index, which occurs at that size, appears less surprising (Fig. 17.12). Jacobs and Chamberlain (1996) assumed that added mass and acceleration was more important for small than for large individuals. Consequently, early ontogenetic stages would have profited more from compressed shell shapes, which would have reduced the energetic cost of the acceleration reaction. Taking limited energy resources into account, it becomes clear that hatchlings and early juveniles were limited in most cases to a rather passive, probably planktonic mode of life. Residence of early ontogenetic stages in the water column is evidenced by ammonitellae and early ontogenetic stages of ammonoids in black shale deposits (e.g., Landman 1988; Mapes and Nützel 2008) and other lines of evidence (Landman et al. 1996; Ritterbush et al. 2014; De Baets et al. 2015c).

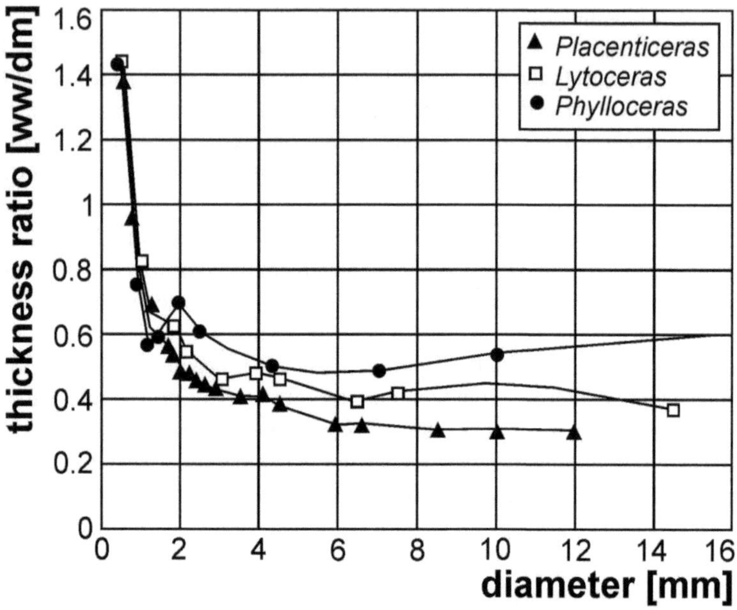

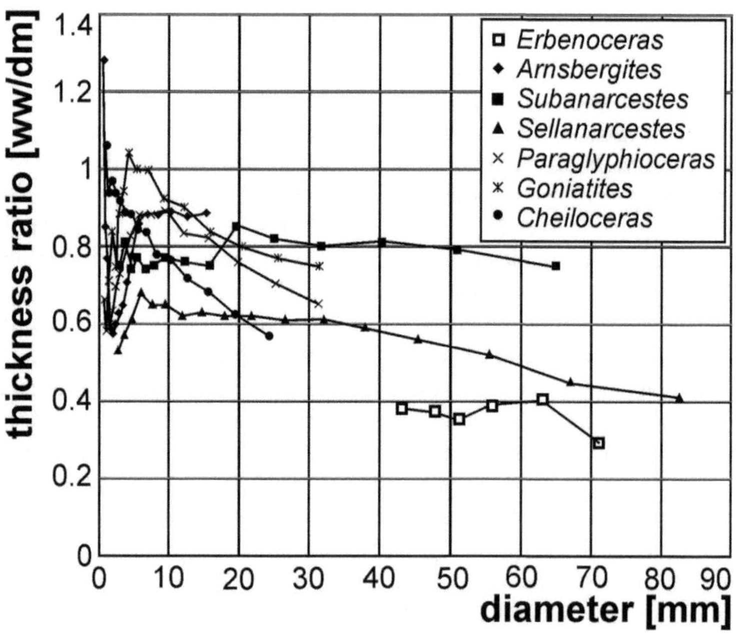

Fig. 17.12 Thickness ratio and shell size in Mesozoic (*top*) and Paleozoic ammonoids (*bottom*) through ontogeny. Modified after Jacobs (1992b) as well as Jacobs and Chamberlain (1996) with new data first reported here. Between hatching (dm < 5mm) and the end of the neanic stage (ca. 10 mm), ammonoids moved only slowly and had wide shells and thus swam at low Reynolds numbers. In all ammonoids, whorl width is reduced after the neanic stage, in Mesozoic forms to values between 0.3 and 0.6 and in Paleozoic forms to values between 0.3 and 0.8. These observations suggest that the ontogenetic late neanic change in shell shape may be an adaptation reflecting the change in hydrodynamic flow conditions

Such accumulations of early ontogenetic stages have often, although not exclusively, been found from the Devonian to the Cretaceous in strata, where benthic life was strongly limited (compare De Baets et al. 2015c). Jacobs (1992b) suggested that the serpenticonic shell shape commonly found in ammonoids (Raup 1967) permitted ammonoids to optimize shell shape for swimming as Reynolds number increased during growth. In contrast, most nautilids (except the aturiids) avoided the smallest size-range for their juveniles, which would have forced the juveniles into a passive planktonic mode of life and similarly, serpenticonic shell shapes are absent in post-Paleozoic Nautilida. Because of these poor locomotory capabilities of ammonoid hatchlings, Jacobs and Chamberlain (1996) considered the possibility of brood care in ammonoids, which finds some support in the occasionally extreme size-dimorphism among ammonoids (e.g., in scaphitids; compare De Baets et al. 2012; 2015c). Walton et al. (2010) speculated on brood care in the Late Devonian genus *Prolobites* based on the extremely low body chamber and terminal aperture, but in this case perhaps outside of the shell of the brooding adult.

Independent of the presence or absence of brood care in ammonoids, the profound morphologic changes that occur around hatching, at the end of the neanic stage, and at maturity (e.g., Westermann 1996; Klug 2001; Korn and Klug 2003) likely had effects on the physical framework for locomotion. It is also striking that commonly, morphologic changes occur at shell diameters between 1 and 2 cm, i.e., when active swimming became feasible for the young ammonoids.

17.5 Information from Epizoans

Some sessile organisms are known to attach themselves in an oriented way depending on the prevailing current direction. Ammonoid shells are well-known to have been inhabited by numerous different invertebrates *syn vivo* (Seilacher 1960; Davis et al. 1999). Some of these epizoans have accordingly been used to interpret the predominant swimming direction of ammonoids. For example, Seilacher (1960) showed bivalve overgrowth on *Buchiceras*, which supported an oblique upward orientation of the aperture of this Cretaceous ammonite.

Keupp et al. (1999) Seilacher and Keupp (2000) as well as Keupp (2012) described a Tithonian aspidocerid inhabited by numerous cirripeds. These epizoans likely attached themselves to the shell of the living ammonite because its aptychi are still in the body chamber and the cirripeds are well articulated. The feeding appendages point in the direction of the aperture, thus suggesting forward swimming, i.e. not backward, as it is usually done by modern cephalopods. This is consistent with the interpretations of Parent et al. (2014) regarding the effect of the aptychi in this genus on swimming speed and swimming direction. Forward swimming would have the advantage that the low hydrodynamic stability of many ammonoids would not have played a big role, because the ammonite shell would have followed the propellent.

Hauschke et al. (2011) described the oriented attachment of a cirripede (goose neck barnacle) on a baculitid. Their findings support forward swimming, but there is also some indication for an approximately horizontal shell orientation during swimming of this orthoconic ammonite. Westermann (2013) contradicts this interpretation, arguing that these cirripedes might actually have colonized shells without a clear preference of orientation and because he thinks that the apical parts of the phragmocones were largely free of chamber water at such early ontogenetic stages. In addition with the rather long body chambers, it would have made young baculitids swim with their shells in a more or less vertical position.

17.6 Facies of the Host Rock and Habitats

It is one of the classical arguments in cephalopod paleobiology as to whether the host rock facies of a cephalopod fossil can be considered as an indicator of habitat in the live animal. The main reason for doubting the usefulness of studies on the rocks that contain ammonoids is the likelihood of post mortem transport (e.g., Kennedy and Cobban 1976; Tanabe 1979; Marchand 1984). Post mortem transport of nautilids over thousands of kilometers has been shown by various authors (Iredale 1944; Hamada 1964; Stenzel 1964; Toriyama et al. 1965; House 1973, 1987; Chirat 2000). In contrast, Chamberlain et al. (1981) argued that the strong pressure gradient between phragmocone chambers and ambient pressure in modern *Nautilus* leads to rapid post mortem waterlogging of the shell in animals dying within the normal depth range of the live animals (100–300 m). This would rapidly produce negative buoyancy and cause the empty shell to sink, thus precluding significant postmortem drift (Maeda and Seilacher 1996). Animals dying at shallow depths would have a much greater chance of reaching the ocean surface and drifting significantly from their original habitat. Independent of the correctness of the preceding opinion, some recurring patterns have been found where the same taxa have been discovered in different regions in similar facies (Fig. 17.13 and 17.14; e.g., Westermann 1996). In such cases, one could argue that the same ammonoid taxa may have lived in the same part of a transgression or regression, which thus produced fossils in similar rock types. Especially when ammonoids are found in small basins with restricted connections to the oceans, the probability of extended distances of drift is lower. Naturally, even within small basins, a great range of habitats existed.

Ammonoids were probably not capable of long distance high speed swimming like some modern decabrachian squids or certain pelagic fishes such as tuna. For that reason, Jacobs and Chamberlain (1996) suggested that ammonoids either lived in conditions with slow currents or currents like ocean gyres or in a demersal habitat in regions with slow or absent bottom currents. In one way or the other, ammonoids had to be able to remain in a habitat with favorable conditions, i.e., sufficient food, oxygen, and also mating partners. In turn, it can be expected to find ammonoid remains more commonly in sediments typical for moderate to low water currents (Jacobs 1992b; Jacobs et al. 1994), although not in the deep sea as their shells would have imploded there, or dissolved if below the carbonate compensation depth.

There are several studies, which examined relationships between ammonoid shell shapes and sedimentary facies. For example, Batt (1989, 1993) used shell morpholo-

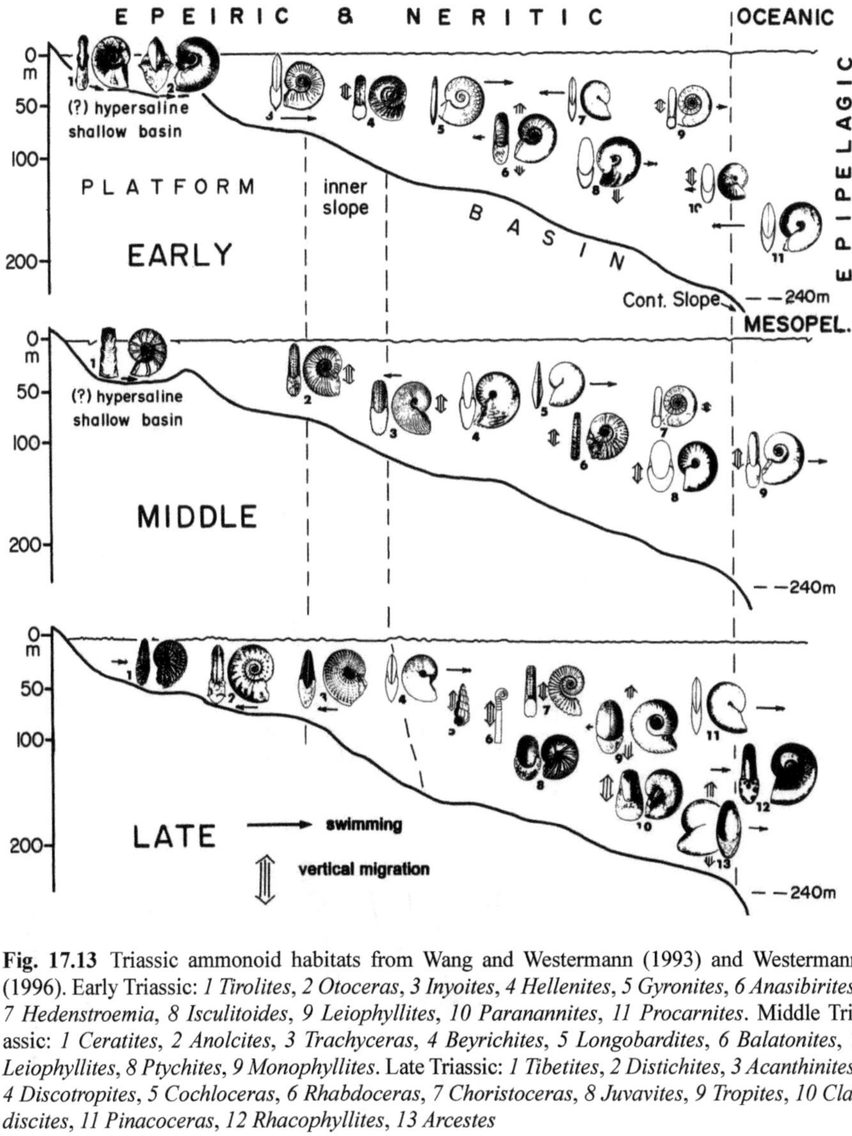

Fig. 17.13 Triassic ammonoid habitats from Wang and Westermann (1993) and Westermann (1996). Early Triassic: *1 Tirolites*, *2 Otoceras*, *3 Inyoites*, *4 Hellenites*, *5 Gyronites*, *6 Anasibirites*, *7 Hedenstroemia*, *8 Isculitoides*, *9 Leiophyllites*, *10 Paranannites*, *11 Procarnites*. Middle Triassic: *1 Ceratites*, *2 Anolcites*, *3 Trachyceras*, *4 Beyrichites*, *5 Longobardites*, *6 Balatonites*, *7 Leiophyllites*, *8 Ptychites*, *9 Monophyllites*. Late Triassic: *1 Tibetites*, *2 Distichites*, *3 Acanthinites*, *4 Discotropites*, *5 Cochloceras*, *6 Rhabdoceras*, *7 Choristoceras*, *8 Juvavites*, *9 Tropites*, *10 Cladiscites*, *11 Pinacoceras*, *12 Rhacophyllites*, *13 Arcestes*

gies to interpret oxygen availability near the sea-floor. In his opinion, heteromorphs like baculitids and loosely coiled forms lived in the water column, while the more tightly coiled heteromorphs and the normally coiled ammonoids occupied a more demersal habitat. Therefore, if the latter group is missing, this might be an indicator of hypoxic to anoxic conditions near the sea-floor (e.g., Monnet and Bucher 2007). Bayer and McGhee (1984) as well as McGhee et al. (1991) employed a more evolutionary approach. They documented how, in the parts of Middle Jurassic transgressive-regres-

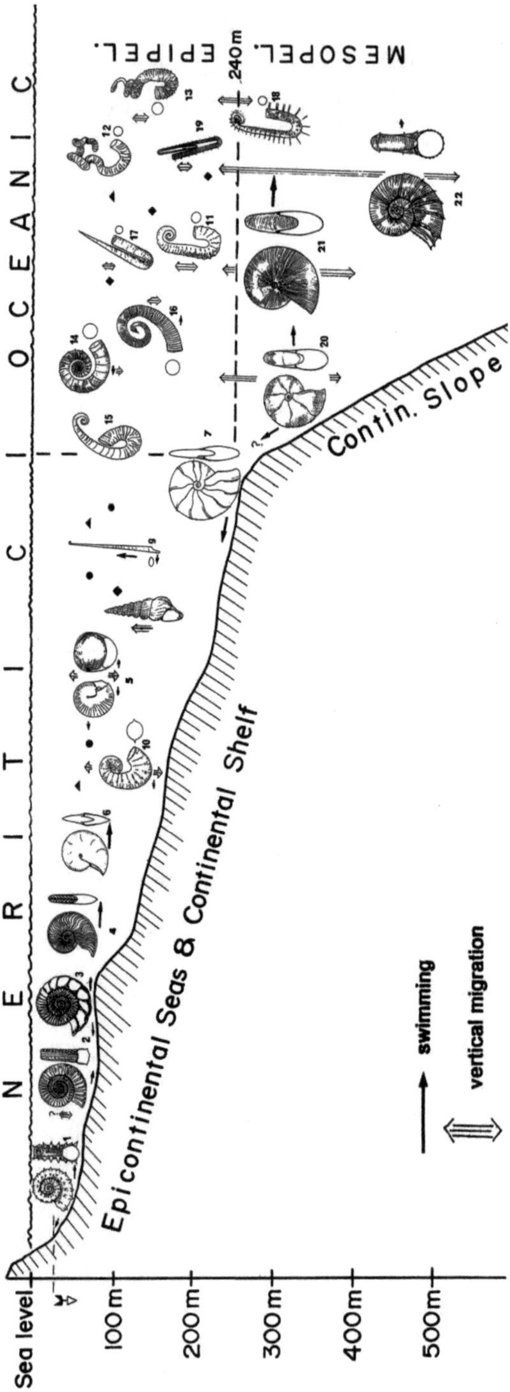

Fig. 17.14 Jurassic and Cretaceous ammonoid habitats from Westermann (1990) and Westermann (1996): *1 Peltoceras*, *2 Arietites*, *3 Perisphinctes*, *4 Harpoceras*, *5 Sphaeroceras*, *6 Oxycerites*, *7 Barremites*, *8 Turrilites*, *9 Baculites*, *10 Scaphites*, *11 Ancyloceras*, *12 Nipponites*, *13 Didymoceras*, *14 Crioceratites*, *15 Labeceras*, *16 Glyptoxoceras*, *17 Hamulina*, *18 Anisoceras 19 Pseudoxybeloceras*, *20 Holcophylloceras*, *21 Phylloceras*, *22 Lytoceras*

sive "Klüpfel cycles" with higher water energy, more involute and compressed shell forms evolved in the Leioceratinae and Graphoceratinae iteratively. Landman and Waage (1993) found that the lineages of the genera *Hoploscaphites* and *Jeletzkytes* both evolved more compressed representatives while the facies changed from the deeper water Pierre Shale to the shallower water sandy Fox Hills Formation. Jacobs et al. (1994) found that more compressed, lower drag shell morphs of *Scaphites whitfieldi* are associated with sandy facies in the Cretaceous Carlisle Shale of the American Western Interior while thicker higher drag shell morphs of the same species occur in finer grain facies. A similar pattern was reported by Courville and Thierry (1993) from *Thomasites* but inverse patterns are sometimes also found in strongly ornamented taxa such as *Schloenbachia* (Wilmsen and Mosavinia 2011), which can complicate interpretations (Ritterbush et al. 2014; De Baets et al. 2015a). Westermann (1996) listed a great number of examples for several marine basins, where he assigned certain ammonoid groups to distinct habitats (Fig. 17.13, 17.14). Klug (2002) suggested that Early and Middle Devonian anarcestids and agoniatitids, which mainly differ in whorl expansion rate and umbilical width, had different ecological preferences since he found them in more clayey or more limey facies, respectively. However, this study used low specimen numbers, thus leading to a low statistical power.

In the Early and Middle Devonian, two such lineages evolved in parallel as shown by Monnet et al. (2011). In the Auguritidae and Pinacitidae, oxyconic shell forms with closed umbilicus evolved independently, and in both lineages, the most derived forms occurred in carbonates that were probably deposited under shallower water conditions than those associated with the ancestral forms. Most of the studies listed in the preceding paragraphs appear to coincide with the interpretations of Jacobs (1992b) as well as Jacobs and Chamberlain (1996), but there are not many such studies, their statistical power tends to be low, and the causality between habit, habitat and shell morphology is difficult to establish with certainty; this can only be achieved by combining multiple lines of evidence including analysis of shell shape, facies and geographic distribution, isotope analysis, etc. (e.g., Tsujita and Westermann 1998; Ritterbush et al. 2014).

A different approach to identify habitat depth is discussed in detail in Chaps. 17.1 and 2. In these studies, stable isotopes of oxygen have been used to assess the habitat depth of various Cretaceous ammonoids (Moriya et al. 2003; Lukeneder et al. 2010). Unfortunately, the error sources of such studies are often large and the number of these studies is still too low. Examination of oxygen isotopes in ammonoid shells is still one of the most promising methods to reveal new information on ammonoid habitats.

17.7 Swimming Modes

Taking the uncertain knowledge of ammonoid soft parts into account, most interpretations of the 'ammonoid power plant' are based on actualistic comparisons. Packard et al. (1980) examined the swimming modes in Recent nautilids (see also

Crick 1898; Chamberlain 1987, 1990, 1992). In *Nautilus*, very slow movement can be produced by the water expelled through the hyponome during aeration of the gills. Normal swimming speeds are produced by mantle cavity water expelled by contraction of the cephalic retractor muscles and funnel muscles. The animal moves forward, backward, up, or down, depending on the orientation of the highly flexible hyponome (Johanson et al. 1972; Ward et al. 1977; Packard et al. 1980; Chamberlain 1981; 1987; Wells and Wells 1985; Webber and O'Dor 1986; Wells and O'Dor 1991).

Many squids including *Sepia* have lateral fins, which function in thrust production and in turning in some squid locomotor behaviors. Octobrachians and some decabrachians use their arms, sometimes connected with velar skins, to swim by expelling water entrained within their arm crowns with rhythmic beating of their arms, in the style of medusoid cnidarian bells. These two modes of locomotion appear unlikely to have been present in ammonoids, or at least, there is no evidence at all yet to support their occurrence among ammonoids.

The most likely mode of swimming is by contracting the mantle cavity, although it is not clear, which muscles were responsible for this task in ammonoids. There are several alternatives, namely the mantle musculature (as in coleoids), the cephalic retractors (as in nautilids) or potentially also other longitudinal muscles (not realized in Recent forms) in combination with the hyponome musculature. It appears also likely that the water was expelled through a hyponome, since hyponomic sinuses are present in many ammonoids. Hyponomes have not yet been found fossilized in ammonoids. Thus, another open question is the flexibility of the ammonoid hyponome. Was it long and flexible enough to point backwards and allow forward swimming?

Recently, Westermann (2013) revived a hypothesis earlier introduced by Schmidt (1930). This "*Twin nozzle-Hypothesis*" roots in the fact that many Mesozoic ammonoids have a more or less long ventral projection (e.g., in *Amaltheus*) combined with a probably more or less horizontal aperture. This would be an adverse combination of character states for straight backwards swimming, because the mentioned ventral apertural projection would have interfered with movement of the hyponome. Therefore, these authors suggested that the hyponome had evolved two openings, one on each side of the ventral projection. Both hyponome parts could be moved independently according to them. This is an interesting idea but so far, it is not supported by fossil evidence.

Monnet et al. (2011) discussed the peculiar way, in which the umbilicus was closed in some Devonian Auguritidae and Pinacitidae. The most derived representatives of both families have largely covered the umbilicus with a projection of the lateral shell over the umbilicus. This projection formed umbilical sinuses, which might have been horizontally aligned with the hyponome sinus. It would have allowed these species to take in water from the swimming direction into the mantle cavity (in *Nautilus*, water is taken in at the same place according to Packard et al. 1980), accelerating the water by compressing the mantle cavity, and expelling it out of the hyponome. This means that these forms potentially sucked water into the mantle cavity after completion of a hyponome jet.

Acknowledgements We greatly appreciate the financial support by the Swiss National Science foundation (project numbers 200021-113956/1, 200020-25029, -132870, and -149120). We greatly appreciate the effort the reviewers Kenneth De Baets (Erlangen) and Benjamin J. Linzmeier (University of Wisconsin-Madison) have put into their reviews, thereby helping us to improve our manuscript.

References

Bambach RK (1999) Energetics in the global marine fauna: a connection between terrestrial diversification and change in the marine biosphere. Geobios 32:131–144

Batt RJ (1989) Ammonite shell morphospace distribution in the Western Interior Greenhorn Sea and some paleoecological implications. Palaios 4:32–43

Batt RJ (1993) Ammonite shell morphotypes as indicators of oxygenation in a Cretaceous epicontinental sea. Lethaia 26:49–63

Batt RJ (2007) Sutural amplitude of ammonite shells as a paleoenvironmental indicator. Lethaia 24:219–225

Bayer U, McGhee GR Jr (1984) Iterative evolution of Middle Jurassic ammonite faunas. Lethaia 17:1–16

Bone Q, Pulsford A, Chubb AD (1981) Squid mantle muscle. J Mar Biol Assoc UK 61:327–342

Boyle P, Rodhouse P (2005) Cephalopods: ecology and fisheries. Wiley, Oxford

Chamberlain JA Jr (1969) Technique for scale modeling of cephalopod shells. Palaeontology 12:48–55

Chamberlain JA Jr (1976) Flow patterns and drag coefficients of cephalopod shells. Palaeontology 19:539–563

Chamberlain JA Jr (1980) The role of body extension in cephalopod locomotion. Palaeontology 23:445–461

Chamberlain JA Jr (1981) Hydromechanical design of fossil cephalopods. In: House MR, Senior JR (eds) The Ammonoidea. Syst Assoc Spec, vol 18. Academic, London

Chamberlain JA Jr (1987) Locomotion of *Nautilus*. In: Saunders WB, Landman NH (eds) *Nautilus*-The biology and paleobiology of a living fossil. Plenum, New York

Chamberlain JA Jr (1990) Jet propulsion of *Nautilus*: a surviving example of early Paleozoic locomotor design. Can J Zool 68:806–814

Chamberlain JA Jr (1992) Cephalopod locomotor design and evolution: the constraints of jet propulsion. In: Rayner MV, Wootton RJ (eds) Biomechanics and evolution. Cambridge University Press, Cambridge

Chamberlain JA Jr (1993) Locomotion in ancient seas: constraint and opportunity in cephalopod adaptive design. Geobios Spec Mem 15:49–61

Chamberlain JA Jr, Moore WA (1982) Rupture strength and flow rate of *Nautilus* siphuncular tube. Paleobiology 8:408–425

Chamberlain JA Jr, Westermann GEG (1976) Hydrodynamic properties of cephalopod shell ornament. Paleobiology 2:316–331

Chamberlain JA Jr, Ward PD, Weaver JS (1981) Post-mortem ascent of Nautilus shells: implications for cephalopod paleobiogeography. Paleobiology 7:494–509

Chirat R (2000) The so-called 'cosmopolitan palaeobiogeographic distribution' of tertiary Nautilida of the genus *Aturia* Bronn 1838: the result of post-mortem transport by oceanic palaeocurrents. Palaeogeogr Palaeoclim Palaeoecol 157:59–77

Courville P, Thierry J (1993) Nouvelles données biostratigraphiques sur les dépôts cénomanoturoniens du Nord-Est du fossé de ia Bénoué (Nigeria). Cretaceous Research 14(4–5):385–396

Crick GS (1898) On the muscular attachment of the animal to the shell in some fossil Cephalopoda (Ammonoidea). Trans Linn Soc NY 7:71–113

Daniel TL (1984) The unsteady aspects of locomotion. Am Zool 24:121–134

Daniel TL (1985) Cost of locomotion: unsteady medusan swimming. J Exp Biol 119:149–164
Daniel TL, Helmuth BS, Saunders WB, Ward PD (1997) Septal complexity in ammonoid cephalopods increased mechanical risk and limited depth. Paleobiology 23:470–481
Davis RA, Mapes RH, Klofak SM (1999) Epizoa on externally shelled cephalopods. In: Rozanov AY, Shevyrev AA (eds) Fossil cephalopods: recent advances in their study. Russian Academy of Sciences, Palaeontological Institute. Moskva
De Baets K, Klug C, Korn D, Landman NH (2012) Evolutionary trends in ammonoid embryonal development. Evolution 66:1788–1806
De Baets K, Bert D, Hofmann R, Monnet C, Yacobucci MM, Klug C (2015a) Ammonoid intraspecific variation. This volume
De Baets K, Keupp H, Klug C (2015b) Parasitism in ammonoids. This volume
De Baets K, Landman NH, Tanabe K (2015c) Ammonoid embryonic development. This volume
Doguzhaeva LA, Mapes RH (2015) Muscle scars in ammonoid shells. This volume
Doguzhaeva LA, Mutvei H (1991) Organization of the soft body in *Aconeceras*(Ammonitina), interpreted on the basis of shell morphology and muscle scars. Palaeontogr A 218:17–33
Doguzhaeva LA, Mutvei H (1993) Structural features in Cretaceous ammonoids indicative of semi-internal or internal shells. In: House MR (ed) The Ammonoidea: environment, ecology, and evolutionary change. Syst Assoc Spec, vol 47. Clarendon Press, Oxford
Ebel K (1983) Berechnungen zur Schwebefähigkeit von Ammoniten. N Jb Geol Paläont Mh 1983:614–640
Elmi S (1991) Données expérimentales sur l'architecture fonctionelle de la coquille des ammonodes Jurassiques. Géobios, Mémoire Spécial 13:155–160
Elmi S (1993) Loi des aires, couche-limite et morphologie fonctionnelle de la coquille des Céphalopodes (Ammonoides). Geobios 26(Suppl 1):121–138
Finn JK, Norman MD (2010) The argonaut shell: gas-mediated buoyancy control in a pelagic octopus. Proc Ro Soc B 277(1696):2967–2971. doi:10.1098/rspb.2010.0155
Gaillard C (1977) Cannelures d'érosion et figures d'impact dues à des coquilles d'ammonites à épines (Oxfordien supérieur du Jura français). Eclogae Geol. Helvetiae 70:701–715
Hamada T (1964) Notes on drifted *Nautilus* in Thailand. Sci Pap Coll Gen Educ Univ Tokyo 14:255–277
Hauschke N, Schöllmann L, Keupp H (2011) Oriented attachment of a stalked cirripede on an orthoconic heteromorph ammonite—implications for the swimming position of the latter. N Jahrb Geol Paläont Abh 202:199–212
Hewitt RA (1996) Architecture and strength of the ammonite shell. In: Landman NH, Tanabe K, Davis RA (eds) Ammonoid paleobiology. Plenum, New York
Hewitt RA, Westermann GEG (2003) Recurrences of hypotheses about ammonites and *Argonauta*. J Paleontol 77:792–795
Hoffmann R, Zachow S 2011 Non-invasive approach to shed new light on the buoyancy business of chambered cephalopods (Mollusca). Extended Abstract IAMG Salzburg 2011:1–9
Hoffmann R, Schultz JA, Schellhorn R, Rybacki E, Keupp H, Gerden SR, Lemanis R, Zachow S (2013) Non-invasive imaging methods applied to neo- and paleontological cephalopod research. Biogeosciences Discuss 10:18803–18851:2013. doi:10.5194/bgd-10-18803-2013
Hoffmann R, Lemanis R, Naglik C, Klug C (2015) Ammonoid buoyancy. This volume
House MR (1973) An analysis of Devonian goniatite distributions. In: Hughes NF (ed) Organisms and continents through time. Spec Pap Palaeont 12:305–317
House MR (1981) On the origin, classification and evolution of the early Ammonoidea. In: House MR, Senior JR (eds) The Ammonoidea: the evolution, classification, mode of life and geological usefulness of a major fossil group. Academic, London
House MR (1987) Geographic distribution of *Nautilus* shells. In: Saunders WB, Landman NH (eds) *Nautilus*. The biology and paleobiology of a living fossil. Plenum, New York
Iredale T (1944) Australian pearly *Nautilus*. Austr. Zool 10:294–298
Jacobs DK (1992a) The support of hydrostatic load in cephalopod shells-adaptive and ontogenetic explanations of shell form and evolution from Hooke 1695 to the present. In: Hecht MK, Wallace B, Macintyre RJ (eds) Evolutionary biology, vol 26. Plenum, New York

Jacobs DK (1992b) Shape, drag, and power in ammonoid swimming. Paleobiology 18:203–220
Jacobs DK, Chamberlain JA (1996) Buoyancy and hydrodynamics in ammonoids. In: Landman NH, Tanabe K, Davis RA (eds) Ammonoid paleobiology. Topics in geobiology13. Plenum, New York
Jacobs DK, Landman NH (1993) Is *Nautilus* a good model for the function and behavior of ammonoids? Lethaia 26:101–110
Jacobs DK, Landman NH, Chamberlain JA Jr (1994) Ammonite shell shape covaries with facies and hydrodynamics: iterative evolution as a response to changes in basinal environment. Geology 22:905–908
Johansen W, Soden PD, Trueman ER (1972) A study in jet propulsion: an analysis of the motion of the squid, *Loligo vulgaris*. J Exp Biol 56:155–156
Kakabadzé MV, Sharikadzé MZ (1993) On the mode of life of heteromorph ammonites (heterocone, ancylocone, ptychocone). Geobios 26(Suppl 1):209–215
Kaplan P (2002) Biomechanics as a test of functional plausibility: testing the adaptive value of terminal-countdown heteromorphy in Cretaceous ammonoids. Abh Geol B-A 57:181–197
Kawabe F (2003) Relationship between mid-Cretaceous (upper Albian–Cenomanian) ammonoid facies and lithofacies in the Yezo forearc basin, Hokkaido, Japan. Cret Res 24:751–763
Kennedy WJ, Cobban WA (1976) Aspects of ammonite biology, biogeography, and biostratigraphy. Spec Pap Palaeontol 17:1–94
Keupp H (1984) Pathologische Ammoniten—Kuriositäten oder paläobiologische Dokumente? (Teil 1). Fossilien 1(6):258–262, 267–275
Keupp H (1985) Pathologische Ammoniten—Kuriositäten oder paläobiologische Dokumente? (Teil 2). Fossilien 2(1):23–35
Keupp H (1992) Rippenscheitel bei Ammoniten-Gehäusen. Fossilien 5:283–290
Keupp H (1996) Paläopathologische Analyse einer Ammoniten-Vergesellschaftung aus der Mittleren Volga-Stufe des subpolaren Urals. Fossilien 1:45–54
Keupp H (1997) Paläopathologische Analyse einer "Population" von *Dactylioceras athleticum* (Simpson) aus dem Unter-Toarcium von Schlaifhausen/Oberfranken. Berliner geowiss Abh E 25:243–267
Keupp H (2000) Ammoniten—paläobiologische Erfolgsspiralen. Thorbecke, Stuttgart
Keupp H (2006) Sublethal punctures in body chambers of Mesozoic ammonites (forma aegra fenestra n.f.), a tool to interpret synecological relationships, particularly predator-prey interactions. Paläontol Z 80:112–123
Keupp H (2008) Wer hat hier zugebissen? Ammoniten-Prädation. Fossilien 2008(2):109–112
Keupp H (2012) Atlas zur Paläopathologie der Cephalopoden. Berliner geowiss Abh E 12:1–390
Keupp H, Hoffmann R (2015) Ammonoid paleopathology. This volume
Keupp H, Röper M, Seilacher A (1999) Paläobiologische Aspekte von syn vivo- besiedelten Ammonoideen im Plattenkalk des Ober-Kimmeridgiums von Brunn in Ostbayern. Berliner geowiss Abh E 30:121–145
Klinger HC (1981) Speculation on buoyancy control and ecology in some heteromorph ammonites. In: House MR, Senior JR (eds) The Ammonoidea. Syst Assoc, Spec, vol 18. Academic, London
Klug C (2001) Life-cycles of Emsian and Eifelian ammonoids (Devonian). Lethaia 34:215–233
Klug C (2002) Quantitative stratigraphy and taxonomy of late Emsian and Eifelian ammonoids of the eastern Anti-Atlas (Morocco). Cour Forschungsinst Senck 238:1–109
Klug C, Korn D (2002) Occluded umbilicus in the Pinacitinae (Devonian) and its palaeoecological implications. Palaeontology 45:917–931
Klug C, Korn D (2004) The origin of ammonoid locomotion. Acta Palaeont Pol 49:235–242
Klug C, Lehmann J (2015) Soft-part anatomy of ammonoids: reconstructing the animal based on exceptionally preserved specimens and actualistic comparisons. This volume
Klug C, Meyer E, Richter U, Korn D (2008) Soft-tissue imprints in fossil and Recent cephalopod septa and septum formation. Lethaia 41:477–492
Klug C, Kröger B, Kiessling W, Mullins GL, Servais T, Frýda J, Korn D, Turner S (2010) The Devonian nekton revolution. Lethaia 43:465–477

Korn D (2012) Quantification of ontogenetic allometry in ammonoids. Evol Dev 14:501–514. doi:10.1111/ede.12003

Korn D, Klug C (2003) Morphological pathways in the evolution of Early and Middle Devonian ammonoids. Paleobiology 29:329–348

Kröger B (2001) Comments on Ebel's benthic-crawler hypothesis for ammonoids and extinct nautiloids. Paläontol Z 75:123–125

Kröger B (2005) Adaptive evolution in Paleozoic coiled cephalopods. Paleobiology 31:253–268

Kröger B, Vinther J, Fuchs D (2011) Cephalopod origin and evolution: a congruent picture emerging from fossils, development and molecules. Bioessays 12. doi:10.1002/bies.201100001

Kummel B, Lloyd RM (1955) Experiments on the relative streamlining of coiled cephalopod shells. J Paleontol 29:159–170

Landman NH (1988) Early ontogeny of Mesozoic ammonites and nautilids. In: Wiedmann J, Kullmann J (eds) Cephalopods-present and past. Schweizerbart, Stuttgart

Landman NH, Cobban WA (2007) Ammonite touch marks in Upper Cretaceous (Cenomanian-Santonian) deposits of the Western Interior Sea. In: Landman NH, Davis RA, Mapes RH (eds) Cephalopods present and past: new insights and fresh perspectives. Springer, Dordrecht

Landman NH, Waage KM (1993) Scaphitid ammonites of the Upper Cretaceous (Maastrichtian) fox hills formation in South Dakota and Wyoming. Bull Am Mus Nat Hist 215:1–257

Landman NH, Rye DM, Shelton KL (1983) Early ontogeny of *Eutrephoceras* compared to recent *Nautilus*and Mesozoic ammonites: evidence from shell morphology and light stable isotopes. Paleobiology 9:269–279

Landman NH, Tanabe K, Shigeta Y (1996) Ammonoid Embryonic Development. In: (Eds) Landman, N.H., Tanabe, K., Davis, R.A. *Ammonoid Paleobiology. Vol. 13, Topics in Geobiology.* 343–405. Plenum Press, New York

Longridge LM, Smith PL, Rawlings G, Klaptocz V (2009) The impact of asymmetries in the elements of the phragmocone of early Jurassic ammonites. Palaeontol Electron 12(1A):1–15

Lukeneder A (2015) Ammonoid habitats and life history. This volume

Lukeneder A, Harzhauser M, Müllegger S, Piller WE (2010) Ontogeny and habitat change in Mesozoic cephalopods revealed by stable isotopes (delta^{18}O, delta^{13}C). Earth and Planetary Science Letters 296:103–111. doi:10.1016/j.epsl.2010.04.053

Maeda H, Seilacher A (1996) Ammonoid taphonomy. In: Landman NH, Tanabe K, Davis RA (eds) Ammonoid paleobiology. Plenum, New York

Mapes RH, Nützel A (2008) Late Palaeozoic mollusc reproduction: cephalopod egg-laying behavior and gastropod larval palaeobiology. Lethaia 42:341–356

Marchand D (1984) Ammonites et paléoenvironnements; une nouvelle approche. Geobios Mém. spécial 8:101–107

Marchand D (1992) Ammonites et paléoprofondeur: les faits, les interprétations. Paleovox 1:49–68

McGhee GC, Bayer U, Seilacher A (1991) Biological and evolutionary responses to transgressive-regressive cycles. In: Einsele G, Ricken W, Seilacher A (eds) Cycles and events in stratigraphy. Springer, Berlin

Monks N, Young JR (1998) Body position and the functional morphology of Cretaceous heteromorph ammonites. Palaeontol Electron 1:15

Monnet C, Bucher H (2007) European ammonoid diversity questions the spreading of anoxia as primary cause for the Cenomanian/Turonian (Late Cretaceous) mass extinction. Swiss J Geosci 100:137–144

Monnet C, Klug C, De Baets K (2011) Parallel evolution controlled by adaptation and covariation in ammonoid cephalopods. BMC Evol Bio 11(115):1–21

Monnet C, De Baets K, Yacobucci MM (2015) Buckman's rules of covariation. In Klug C, Korn D, De Baets K, Kruta I, Mapes RH (eds): Ammonoid Paleobiology, Vol.2: From macroevolution to biogeography. Springer, Dordrecht

Moriya K (2015) Isotope signature of ammonoid shells. This volume

Moriya K, Nishi H, Kawahata H, Tanabe K, Takayanagi Y (2003) Demersal habitat of Late Cretaceous ammonoids: evidence from oxygen isotopes for the Campanian (Late Cretaceous) northwestern Pacific thermal structure. Geology 31:167–170

Mutvei H (1975) The mode of life in ammonoids. Paläontol Z 49:196–206
Mutvei H, Reyment RA (1973) Buoyancy control and siphuncle function in ammonoids. Palaeontology 16:623–636
Naglik C, Monnet C, Götz S, Kolb C, De Baets K, Klug C (2015) Growth trajectories in chamber and septum volumes in major subclades of Paleozoic ammonoids. Lethaia 48(1):29–46
Naglik C, Rikhtegar F, Klug C (in press) Buoyancy of some Palaeozoic ammonoids and their hydrostatic properties based on empirical 3D-models. Lethaia 10pp. DOI 10.1111/let.12125
O'Dor RK (1982) Respiratory metabolism and swimming performance of the squid, *Loligo opalescens*. Can J Fish Aquat Sci 39:580–587
O'Dor RK (1988a) The energetic limits on squid distributions. Malacologia 29:113–119
O'Dor RK (1988b) The forces acting on swimming squid. J Exp Biol 137:421–442
O'Dor RK, Webber DM (1991) Invertebrate athletes: trade-offs between transport efficiency and power density in cephalopod evolution. J Exp Biol 160:93–112
O'Dor RK, Wells MJ (1990) Performance limits of "antique" and "state-of-the-art" cephalopods, *Nautilus* and squid. Am Malacol Union Prog Abstr. 56th Ann Meeting, 52
O'Dor RK, Wells MJ, Wells J (1990) Speed jet pressure and oxygen consumption relationships in free-swimming *Nautilus*. J Exp Biol 154:383–396
O'Dor RK, Forsythe J, Webber DM, Wells J, Wells MJ (1993) Activity levels of *Nautilus* in the wild. Nature 362:626–627
Oeffner J, Lauder GV (2012) The hydrodynamic function of shark skin and two biomimetic applications. J Exp Biol 215:785–795
Okamoto T (1988) Analysis of heteromorph ammonoids by differential geometry. Palaeontology 31:35–52
Okamoto T (1996) Theoretical modeling of ammonoid morphology. In: Landman NH, Tanabe K, Davis RA (eds) Ammonoid paleobiology. Topics in geobiology 13. Plenum, New York
Packard A (1972) Cephalopods and fish: the limits of convergence. Biol Rev 47:241–307
Packard A, Bone Q, Hignette M (1980) Breathing and swimming movements in a captive *Nautilus*. J Mar Biol Assoc UK 60:313–327
Parent H, Westermann GEG, Chamberlain JA Jr (2014) Ammonite aptychi: functions and role in propulsion. Geobios 47:45–55
Raup DM (1966) Geometric analysis of shell coiling: general problems. J Paleont 40:1178–1190
Raup DM (1967) Geometric analysis of shell coiling: coiling in ammonoids. J Paleontol 41:43–65
Raup DM, Chamberlain JA Jr (1967) Equations for volume and center of gravity in ammonoid shells. J Paleontol 41:566–574
Reif WE (1982) Morphogenesis and function of the squamation in sharks. 1. Comparative functional morphology of shark scales, and ecology of scales. N Jahrb Geol Paläont Abh 164:172–183
Reyment RA (1973) Factors in the distribution of fossil cephalopods. Part 3: experiments with exact models of certain shell types. Bull Geol lnst Univ Uppsala N S 4:7–41
Ritterbush K, Bottjer DJ (2012) Westermann Morphospace displays ammonoid shell shape and hypothetical paleoecology. Paleobiology 38:424–446. doi:10.1666/10027.1
Ritterbush K, De Baets K, Hoffmann R, Lukeneder A (2014) Pelagic Palaeoecology: the importance of recent constraints on ammonoid palaeobiology and life history. J Zool. doi:10.1111/jzo.12118
Rosa R, Seibel BA (2010) Voyage of the argonauts in the pelagic realm: physiological and behavioural ecology of the rare paper nautilus, *Argonauta nouryi*. ICES J Mar Sci J du Conseil 67:1494–1500
Rothpletz A (1909) Über die Einbettung der Ammoniten in die Solnhofener Schichten. Abh math-phys Kl der königl Bayr Akad der Wiss München 24(2):313–337
Saunders WB (1995) The ammonoid suture problem: relationship between shell and septal thickness and sutural complexity in Paleozoic ammonoids. Paleobiology 21:343–355
Saunders WB, Shapiro EA (1986) Calculation and simulation of ammonoid hydrostatics. Paleobiology 12:64–79

Saunders WB, Wehman DA (1977) Shell strength of *Nautilus* as a depth limiting factor. Paleobiology 3:83–89
Saunders WB, Work DM (1996) Shell morphology and suture complexity in Upper Carboniferous ammonoids. Paleobiology 22:189–218
Saunders WB, Work DM (1997) Evolution of shell morphology and suture complexity in Paleozoic prolecanitids, the rootstock of Mesozoic ammonoids. Paleobiology 23:301–325
Saunders WB, Work DM, Nikolaeva SV (1999) Evolution of complexity in Paleozoic ammonoids. Science 286:760–763
Schmidt H (1930) Ueber die Bewegungsweise der Schalencephalopoden. Paläontol Z 12:194–208
Schmidt-Nielsen K (1972) Locomotion: energy cost of swimming, flying and running. Science 177:222–228
Seibel BA (2007) On the depth and scale of metabolic rate variation: scaling of oxygen consumption rates and enzymatic activity in the class Cephalopoda (Mollusca). J Exp Biol 210:1–11
Seibel BA, Thuesen EV, Childress JJ, Gorodezky LA (1997) Decline in pelagic Cephalopod metabolism with habitat depth reflects differences in locomotory efficiency. Biol Bull 192:262–278
Seilacher A (1960) Epizoans as a key to ammonoid ecology. J Paleont 34:189–193
Seilacher A (1963) Umlagerung und Rolltransport von Cephalopodengehäusen. N Jahrb Geol Paläont Mh 11:593–615
Seilacher A (1982a) Ammonite shells as habitats in the Posidonia shales of Holzmaden—floats or benthic islands? N Jahrb Geol Paläont Mh 1982:98–114
Seilacher A (1982b) Ammonite shells as habitats—floats or benthic islands? In Einsele G, Seilacher A (eds) Cyclic and event in stratification. Springer, Berlin. doi:10.1007/978-3-642-75829-4_38
Seilacher A, Keupp H (2000) Wie sind Ammoniten geschwommen? Fossilien 5:310–313
Seki K, Tanabe K, Landman NH, Jacobs DK (2000) Hydrodynamic analysis of Late Cretaceous desmoceratine ammonites. Rev Paléobiol Vol spéc 8:141–155
Shapiro EA, Saunders WB (1987) *Nautilus* shell hydrostatics. In: Saunders WB, Landman NH (eds) *Nautilus*—The biology and paleobiology of a living fossil. Plenum, New York
Signor PW III, Brett CE (1984) The mid-Paleozoic precursor to the Mesozoic marine revolution. Paleobiology 10:229–245
Stenzel HB (1964) Living *Nautilus*. In: Moore RC (ed) Treatise on invertebrate paleontology part K (Mollusca 3). Geological Society of America and University of Kansas Press, Lawrence, pp. K59–K93
Summesberger H, Jurkivsek B, Kolar-Jurkovsek T (1999) Rollmarks of soft parts and a possible crop content of Late Cretaceous ammonites from the Slovenian karst. In: Olóriz F, Rodríguez-Tovar FJ (eds) Advancing research on living and fossil Cephalopods. Kluwer Academic/Plenum, New York
Swan RTH, Saunders, WB (1987) Function and shape in late Paleozoic (mid-carboniferous) ammonoids. Paleobiology 13:297–311
Tajika A, Naglik C, Morimoto N, Pascual-Cebrian E, Hennhöfer DK, Klug C (2015) Empirical 3D-model of the conch of the Middle Jurassic ammonite microconch *Normannites*, its buoyancy, the physical effects of its mature modifications and speculations on their function. Historical Biology: An International Journal of Paleobiology, 27(2):181–191. DOI: 10.1080/08912963.2013.872097
Tanabe K (1979) Palaeoecological analysis of ammonoid assemblages in the Turonian *Scaphites* facies of Hokkaido, Japan. Palaeontology 22:609–630
Toriyama R, Sato T, Hamada T, Komalarjun P (1965) *Nautilus pompilius* drifts on the west coast of Thailand. Jpn J Geol Geog. 36:149–161
Trammer J, Niechwedowicz M (2007) Hydrodynamically controlled anagenetic evolution of Famennian goniatites from Poland. Acta Palaeont Pol 52:63–75
Trueman AE (1941) The ammonite body chamber, with special reference to the buoyancy and mode of life of the living ammonite. Q J Geol So. 96:339–383

Trueman ER, Packard A (1968) Motor performances of some cephalopods. J Exp Biol 49:495–507
Tsujita CJ, Westermann GEG (1998) Ammonoid habitats and habits in the Western Interior Seaway: a case study from the Upper Cretaceous Bearpaw Formation of southern Alberta, Canada. Palaeogeogr Palaeoclim Palaeoecol 144:135–160
Urdy S, Goudemand N, Bucher H, Chirat R (2010a) Allometries and the morphogenesis of the molluscan shell: a quantitative and theoretical model. J Exp Biol B 314:280–302
Urdy S, Goudemand N, Bucher H, Chirat R (2010b) Growth-dependent phenotypic variation of molluscan shells: implications for allometric data interpretation. J Exp Biol B 314:303–26
Vogel S (1981) Life in moving fluids: the physical biology of flow. Princeton University Press, Princeton
Walton S, Korn D, Klug C (2010) Size distribution of the Late Devonian ammonoid *Prolobites*: indication for possible mass spawning events. Swiss J of Geosci 103:475–494
Wang Y, Westermann GEG (1993) Paleoecology of triassic ammonoids. Geobios Mem Spec 15:373–392
Ward PD (1976) Stratigraphy, paleoecology and functional morphology of heteromorph ammonites of the Upper Cretaceous Nanaimo Group, British Columbia and Washington. PhD thesis McMaster University Library, Thesis QE788134 (39005047235555), Hamilton, Canada
Ward P (1979) Functional morphology of Cretaceous helically-coiled ammonite shells. Paleobiology 5:415–422
Ward PD (1981) Shell sculpture as a defensive adaptation in ammonoids. Paleobiology 7:96–100
Ward PD (1982) The relationship of siphuncle size to emptying rates in chambered cephalopods: implications for cephalopod paleobiology. Paleobiology 8:426–433
Ward PD (1987) The natural history of *Nautilus*. Allen and Unwin, Winchester
Ward PD, Wicksten MK (1980) Food sources and feeding behavior of *Nautilus macromphalus*. Veliger 23:119–124
Ward PD, Stone R, Westermann GEG, Martin A (1977) Notes on animal weight, cameral fluids, swimming speed, and colour polymorphism of the cephalopod, *Nautilus pompilius*, in the Fiji Islands. Paleobiology 3:377–388
Webber DM, O'Dor RK (1986) Monitoring the metabolic rate and activity of free-swimming squid with telemetered jet pressure. J Exp Biol 126:205–224
Wells MJ (1987) Ventilation and oxygen extraction by *Nautilus*. In: Saunders WB, Landman NH (eds) *Nautilus*-The biology and paleobiology of a living fossil. Plenum, New York
Wells MJ (1995) The evolution of a racing snail. Mar Freshw Behav Physiol 25:1–12
Wells MJ, O'Dor RK (1991) Jet propulsion and the evolution of Cephalopods. Bull Mar Sci 49:419–432
Wells MJ, Wells J (1985) Ventilation and oxygen uptake by *Nautilus*. J Exp Biol 118:297–312
Westermann GEG (1966) Covariation and taxonomy of the Jurassic ammonite *Sonninia adicra* Waagen. N Jb Geol Paläont Abh 124:289–312
Westermann GEG (1971) Form, structure and function of shell and siphuncle in coiled Mesozoic ammonoids. Life Sci Contrib R Ont Mus 78:1–39
Westermann GEG (1973) Strength of concave septa and depth limits of fossil cephalopods. Lethaia 6:383–403
Westermann GEG (1977) Form and Function of orthocone cephalopod shells with concave septa. Paleobiology 3:300–321
Westermann GEG (1990) New developments in ecology of Jurassic-Cretaceous ammonoids. In: Pallini G, Cecca F, Cresta S, Santantonio M (eds) Fossili, evoluzione, ambiente. Atti II Conv Int Pergola 1987. Tecnostampa, Ostra Vetere
Westermann GEG (1993) On alleged negative buoyancy of ammonoids. Lethaia 26:246. doi:10.1111/j.1502–3931.1993.tb01526.x
Westermann GEG (1996) Ammonoid life and habitat. In: Landman NH, Tanabe K, Davis RA (eds) Ammonoid paleobiology. Plenum, New York
Westermann GEG (2013) Hydrostatics, propulsion and life-habits of the Cretaceous ammonoid *Baculites*. Rev Paléobiol 32:249–265

Westermann GEG, Tsujita CJ (1999) Life habits of ammonoids. In: Savazzi E (ed) Functional morphology of the invertebrate skeleton. Wiley, Hoboken

Wilmsen M, Mosavinia A (2011) Phenotypic plasticity and taxonomy of *Schloenbachia varians* (J. Sowerby, 1817) (Cretaceous Ammonoidea). Paläontol Z 85:169–184

Young JZ (1960) Observations on *Argonauta* and especially its method of feeding. Proc Zool Soc London 133:471–479

Ziegler B (1967) Ammonitenökologie am Beispiel des Oberjura. Geol Rundsch 56:439–446

Chapter 18
Ammonoid Habitats and Life History

Alexander Lukeneder

18.1 Introduction

Ammonoid paleobiology has been a topic of interest for many years and an enormous amount of knowledge and data on extant (*Nautilus*, *Octopus*, *Sepia*, *Spirula*) and fossil cephalopods (ammonoids, nautiloids, belemnoids, teuthoids) has been compiled in recent decades (Arkell 1957; House and Senior 1981; Wiedmann and Kullmann 1988; House 1993a; Landman et al. 1996a; Payne et al. 1998; Olóriz and Rodríguez–Tovar 1999; Summesberger et al. 2002; Landman et al. 2007a; Tanabe et al. 2010a; Ritterbush et al. 2014). Reconstructing the life history and habitats of extinct groups is one of the classic problems. Ammonoids are an extinct group within the Cephalopoda with an external aragonitic shell (ectocochleate). Various shell morphologies exist, ranging from planispiral to heteromorphic shells with straight, openly coiled, helicospiral shells or even various combinations of these. The shell sculpture varies from smooth ('Leiostraca'; e.g., *Phylloceras*) to strongly ribbed ('Trachyostraca'; e.g., *Trachyceras*), with tubercles or even long spines (Fig. 18.1).

Ammonoid researchers assumed that the great majority of ammonoids lived close to the epicontinental (epeiric) sea floor, termed either 'nektobenthic' or 'benthopelagic/demersal' (Westermann 1990, 1996). Westermann (1996) concluded in his work on ammonoid habitats and life that probably more ammonoids were nektic (=nektonic=active swimming) or were members of planktic communities (=planktonic=passively drifting) rather than being only nektobenthic (=demersal=near-bottom swimming). Such general assumptions are nonetheless highly speculative: they are not based on evidence or data sets, yet may represent good ideas and hypotheses.

A. Lukeneder (✉)
Department of Geology and Paleontology, Natural History Museum, 1010 Vienna, Austria
e-mail: alexander.lukeneder@nhm-wien.ac.at

C. Klug et al. (eds.), *Ammonoid Paleobiology: From Anatomy to Ecology*,
Topics in Geobiology 43, DOI 10.1007/978-94-017-9630-9_18

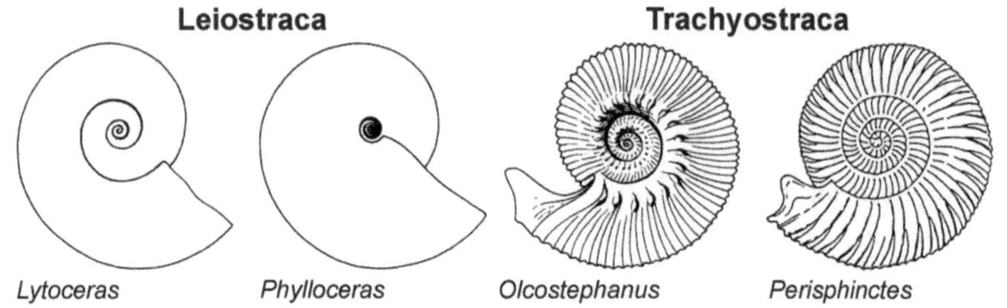

Fig. 18.1 Main shell sculpture morphogroups in planispirals, from smooth to fine-ribbed 'Leiostraca' with *Phylloceras* and *Lytoceras* to strongly ribbed 'Trachyostraca' with *Olcostephanus* and *Perisphinctes*. Schematic sketches not to scale

Knowledge about life cycles, ecology and ontogeny of fossil cephalopods and especially of ammonoids is still poor and often speculative. While ammonoids are frequently found in Paleozoic and Mesozoic marine sediments worldwide from the tropic-subtropics via Boreal/Austral to Arctic/Antarctic zones, information on their habitat and ecology is scarce and imprecise. Their habitat is suggested to be epipelagic, mesopelagic or epibenthic. Ammonoids probably spawned in benthic, demersal or even midwater habitats (Mapes and Nützel 2009) in the neritic to oceanic zone above the shelf areas and upper slopes (Westermann 1996). Females are thought to have laid 100–500000 eggs on the sea floor (*r-strategy*) or spawned egg masses in the water column. At the latest after hatching, hatchlings become a part of the plankton (e.g., Landman et al. 1983; Tanabe et al. 1993b; Mapes and Nützel 2009; De Baets et al. 2012; epiplankton after Westermann 1996). This strategy is comparable to most oceanic coleoids, except for the *K-strategist* nautilids and sepiids. Although it is still under debate, female ammonoids returned to primary habitats (home-grounds), spawned and died afterwards, assuming a semelparous strategy (single reproductive event) at least for some taxa (Callomon 1980, Landman et al. 2003; Stephen et al. 2012), as often observed in neritic, extant coleoids (sepiids or loliginids). Ammonoids perhaps undertook vertical diurnal migrations, as is characteristic for many planktic ocean dwellers such as the deep-water squid *Spirula* (Clarke 1969; Lukeneder et al. 2008, 2010; Doghuzhaeva et al. 2010). Note, however, that a contrasting model arguing for a permanent demersal habitat in adult stages of late Cretaceous ammonoids from Japan has been presented by Moriya et al. (2003) and Moriya (2015).

Since the last detailed reviews on ammonoid habitats and life histories published by Kennedy and Cobban (1976) and Westermann (1996), extensive work has been done on ammonoid autecology as well as on the synecology of ammonoids and Recent relatives. Ammonoids probably started developing from a planktic hatchling stage (Kennedy and Cobban 1976; Shigeta 1993; Mapes and Nützel 2009), subsequently followed by a differentiation in the mode of life as continued planktic or nektonic or nektobenthic (demersal). The earliest juvenile stages were interpreted by Kennedy and Cobban (1976) as mostly being benthic due to their notable substrate affinities while Mapes and Nützel (2009) suggested a planktic life for some

Paleozoic forms. Sexually dimorphic pairs (i.e., macroconch females, microconch males) might have been separated into different habitats (e.g., *Acanthodiscus*: Reboulet 1996; *Olcostephanus*: Lukeneder and Harzhauser 2003), as is the case in maturity differentiation (i.e., juvenile to adults; *Scaphites*: Kennedy and Cobban 1976).

While the functional morphology of ammonoids has been discussed starting more than half a century ago (Trueman 1941; Arkell 1957), advanced analyses on buoyancy and shell orientation, for example, started in the 1980s (Saunders and Shapiro 1986; Hewitt and Westermann 1987; Hewitt and Westermann 1988; Jacobs 1992a). Functional morphological research on ammonoids has focused on three major paleobiological questions: 1) buoyancy, 2) mobility, and 3) habitat depth. Although some ambiguities remain, most previous studies (excluding Ebel 1983, 1992) estimated that living ammonoids had a density almost equivalent to that of seawater, making them neutrally buoyant (Hoffmann et al. 2015; Tajika et al. 2015). The neutral buoyancy hypothesis was supported by theoretical morphological considerations on the mode of coiling and life orientation of some Cretaceous heteromorph ammonites (Okamoto 1988a, b, c). Recently, empirical models corroborated Okamoto's findings (Hoffmann et al. 2015; Tajika et al. 2015; see Westermann 1993a).

A second aspect is swimming and mobility in coiled and chambered ammonoids. It has been shown that shell stability (determined by the distance between the centers of gravity and buoyancy) is influenced by pressure and hence an expression of the water depth in which the animal lived. Shell size and drag coefficient are key factors in estimating swimming ability (Chamberlain 1976; Jacobs 1992a; Jacobs and Chamberlain 1996; Naglik et al. 2015). One of the outcomes of these studies is that a thinner shell shape is more advantageous in a flow at higher Reynolds numbers, while a thicker shape is more advantageous at low Reynolds numbers (Jacobs 1992a). Another aspect of mobility is (diurnal) vertical movement. Living chambered cephalopods (e.g., *Nautilus*, *Sepia*, *Spirula*) have long been considered to move vertically by controlling the amount of cameral liquid using osmotic pressure of the blood vessels (Denton and Gilpin-Brown 1973). The water depth in which efficient pumping using a simple osmotic mechanism can be accomplished is shallower than 240 m (Greenwald et al. 1980). Accordingly, ammonoids from a deep-water habitat might have often ascended to shallow waters to pump out the cameral liquid (Westermann 1989). Direct observations of vertical movements of *Nautilus* with remote telemetry (Carlson et al. 1984; Ward et al. 1984) support this idea. Greenwald et al. (1982), however, studied the *Nautilus* siphuncle at the ultrastructural level and documented the presence of structures associated with a hyperosmotic pump. This would enable the animals to discharge cameral liquid in deep water. Consequently, the argument for ammonoid vertical movement by means of osmotic pressure lost some of its power. Emptying and refilling the chambers to control buoyancy has been considered unlikely in *Nautilus* for some time (Ward 1986a). It has also been considered unlikely for ammonoids, although that is being debated (e.g., Mutvei and Dunca 2007).

Thirdly, the depths that ammonoids might have inhabited were also estimated based on mechanical properties of the shell against ambient hydrostatic pressure.

Based on the wide variation in the mechanical strength of shell materials among various Mesozoic ammonoid morphotypes, Westermann (1996) suggested that Mesozoic ammonoids had various habitat depths. Nonetheless, the implosion depth of a *Nautilus* shell does not represent an actual depth limit of living animals. Direct observations using a remote camera and capture records using baited traps have demonstrated that the optimal habitat depth of *Nautilus* ranges from 150 to 300 m in Palau and from 300 to 500 m in Fiji (Saunders 1984; Hayasaka et al. 1987), whereas the shell implosion depth of *Nautilus* is about 800 m (Kanie et al. 1980). Although these analyses in functional morphology provided valuable suggestions and limitations for considering ammonoid ecology, reasonable and unanswered questions remain for future discussion.

Functional morphology describes the relationship of a single individual, group of individuals, or distinct ammonoid species with its environment (e.g., all parameters of seawater). Accordingly, shell features such as composition, thickness, siphuncle strength and shape are extremely important for understanding the ammonoid/habitat relationship. Additional investigations on the life and habitats of Recent genera such as *Nautilus*, *Spirula* and *Sepia* shed light on the paleoecology and lifestyle of fossil cephalopods. These approaches conclude that most morphogroups were poor swimmers. New findings of soft parts (e.g., stomach content, buccal masses) have enhanced the picture of ammonoids, their hydrodynamics and diets (Kruta et al. 2011).

Ammonoids are marine animals; their lives are conditioned by abiotic parameters such as salinity, with most groups inhabiting euryhaline conditions. A few probably lived in superhaline waters, while others could tolerate subhaline or even brackish waters (Zaborski 1982; Kennedy et al. 1998). Temperature is also a crucial parameter (calculated by $\delta^{18}O$ values) together with oxygenation (oxygenated to dysoxic). As noted by Westermann (1996), oxygenation and salinity covary in the water column. Water layers (in epeiric seaways) above an oxygen-minimum zone may have had reduced salinity, enabling only euryhaline organisms (certain fishes) to survive and preventing abundant ammonoid populations. Hydrostatic pressure at corresponding water depths are reflected by adapted conchs, septa and siphuncles of ectocochleates, helping to estimate rupture (Chamberlain and Moore 1982) or implosion depths. Biotic changes in trophic conditions (predator-prey relationships), food supply (diet differs in different water layers), and natural enemies (cephalopods, fishes, marine reptiles) triggered changes in populations and morphotypes. The sum of the abiotic parameters describes and defines the habitat or environment in which ammonoids lived. Biotic factors relate to the environments, also termed as ecological parameters influencing the mode of life.

Numerous papers of the last decades (Scott 1940; Ziegler 1967; Donovan 1985; Batt 1987, 1989; Westermann 1990, 1996 and references therein) have related shell morphotypes and structural features to depth as determined by sedimentary facies and/or submarine topography. The detection of changes in water depth is important to understand habitat changes in ammonoids.

Did ammonoids actively change their depth-dependent habitats during ontogeny, or did similar ammonoid groups (i.e., morphogroups) migrate into different habitats during long-term evolution? Are ammonoids able to adapt their habits to long-term environmental changes or do certain conditions or events result in a dead end in the evolution of certain lineages?

Geological analyses from a sedimentological and lithological view also brought clues to understand ammonoid ecology. Westermann's (1990, 1996) investigations on ammonoid autecology emphasized the interrelations between sediment and biofacies. Facies changes were triggered by sea-level fluctuations caused by eustasy, tectonics or subsidence. Kennedy and Cobban (1976) assumed that there is no consistent pattern between ammonoid occurrence and facies. Nonetheless, there are morphotypes or taxa that exhibit significant correlations to sedimentary facies, excluding transport by postmortem drift or redeposition (e.g., Scott 1940; Ziegler 1967; Bayer and McGhee 1984; Donovan 1985; Batt 1987, 1989). Maeda and Seilacher (1996) and Westermann (1996) assumed that postmortem sinking occurred mostly in specimens living in greater depths, especially in mesopelagic habitats, as well as in juveniles and small species. Postmortem ascent, followed by surface drift, was thought to occur mostly in shallow habitats and in adults of larger ammonoids, as is known from extant *Nautilus*, *Spirula*, and some sepiids. This hypothesis implies that if ammonoids inhabited various depths within the water column (including shallower water), carcasses would have been widely dispersed by postmortem drift. This would be reflected by each morphology showing no coherency to sedimentary facies. Contrary to this expectation, many previous studies on faunal distributions of ammonoid assemblages have shown that spectra of faunal compositions are closely related to lithofacies, as mentioned above. These facies-dependences suggest that ammonoid carcasses were not transported over long distances by postmortem drift, assuming that ammonoids did not inhabit shallower parts of the water column or that they were at greater depths when they died.

Although many studies have assessed ammonoid ecology, the main problem in understanding ammonoid life and habitat is that they are extinct. Investigations on Recent relatives comprising the ectocochleate *Nautilus* as well as the endocochleate *Spirula* and *Sepia* shed some light on the paleoecology and lifestyle of fossil cephalopods. Nonetheless, the systematic distance between extant and extinct cephalopods prevents a direct comparison in anatomical and ethological aspects. Ammonoid soft parts (e.g., arms, eyes) are in most cases unknown except stomach contents and buccal masses. Taxonomic concepts in paleontology can be based only on morphology and not on genetic principles; this hampers determining exact ammonoid lineages and separating points for species or morphotypes.

Besides these morphological and sedimentological considerations, stable isotope data ($\delta^{13}C$, $\delta^{18}O$, $^{87}Sr/^{86}Sr$; clumped isotopes) extracted from original shell material (i.e., aragonite) of ammonoids could yield new insights in ammonoid lifestyle, ontogeny and habitats (Jordan and Stahl 1970; Brand 1986; Anderson et al. 1994; Fatherree et al. 1998; Tsujita and Westermann 1998; Cochran et al. 2003, 2010a, b; Moriya et al. 2003; Lukeneder et al. 2010; Dennis et al. 2013; Stevens et al. 2015). If we can neglect vital effects in calcification of ammonoid (Urey et al. 1951; Moriya et al. 2003) and Recent *Nautilus* (Landman et al. 1994) shells, then the physicochemical properties, especially $\delta^{18}O$, would provide independent and solid evidence for calcification temperature (and salinity). When the ammonoid calcification temperatures are compared with the thermal structure of the water column, we can identify the habitat depth of ammonoids (Anderson et al. 1994; Moriya et al. 2003). Furthermore, a combination of $\delta^{13}C$ and $\delta^{18}O$ profiles provides information on the life history of ammonoids, such as sexual maturity (Zakharov et al. 2005, 2006, 2011; Lécuyer and Bucher 2006; Lukeneder et al. 2010). The observed seasonal variations in oxygen isotope data of ammonoid shells potentially enables an age estimation for ammonoid maturity. What has been difficult to measure and evaluate in stable isotope data from ammonoid shells is the possible annual migration over hundreds or thousands of kilometers, a phenomenon observed in numerous Recent cephalopods. It is even more complicated when cephalopods die after spawning in habitats that are quite different and distant from those in which they spent most of their lives (Westermann 1996).

Consequently, a key question is how can we compare the circumstances for fossil, extinct ammonoids and Recent, extant cephalopods when ammonoids in most cases show no resemblance to Recent cephalopods. Also, the phylogenetic relationship is likely closer to coleoids than to *Nautilus* (Jacobs and Landman 1993; Hewitt and Westermann 2003; Warnke et al. 2003; Warnke and Keupp 2005). This fact is a powerful argument for investigating ammonoid paleoecology through the record itself, because modern analogues are not the best. Catastrophic events led to extinctions at the Cretaceous/Paleogene boundary about 66 million years ago, but orders such as Nautilida, Spirulida or members of the Coleoida survived this big mass extinction. Factors such as the differentiation of ecological niches and different spawning strategies may have triggered different histories of extinct and extant cephalopod groups. A difference can be determined between quantity (*r*) with small but numerous offspring in 'opportunistic habitats' (ammonoids) versus quality (*K*) with bigger but fewer egg capsules in more stable 'consistent habitats' (nautilids).

The state of the art is reviewed and summarized, new methods and insights are given resulting in perspectives within this promising field of cephalopod paleontology. The current compilation is based on the inspiring work of Westermann (1996), a pioneer in the field of habitat and life of ammonoids.

18.2 Morphology, Architecture, Mobility and Stability of Ammonoid Conchs

The most important terms for shell morphology and ontogenetic stages are summarized below. These are needed to understand interpretations on habitat and lifestyle given in the literature and herein. Morphological terms follow Batt (1989, 1991), Westermann (1996), Wright et al. (1996), and Klug et al. (2015a). Westermann (1996) identified two major morphogroups in ammonoids, the planispirals ('normal coiled') and the heteromorphs (varying in coiling-axis).

Planispirals include 12 morphogroups (Westermann 1996, Fig. 18.2) with spherocone (planktic), discocone (planktic-nektic), cadicone (planktic), oxycone

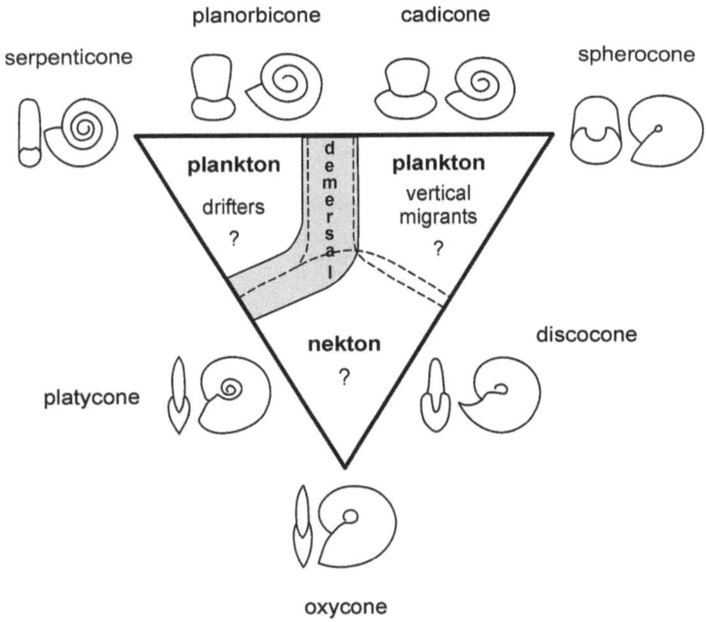

Fig. 18.2 Basic planispiral ammonoid shell shapes and their principal assumed habitats, adapted from the original figure in Westermann (1996, Fig. 1, p. 611) and modified after Ritterbush et al. (2014, Fig. 3). Distinct morphogroups appear in the hypothetical diagram with overlapping areas (dashed lines). Grey-shaded region was interpreted by Westermann (1996) as being characteristic for a demersal (nektobenthic) life mode

(nektic), platycone (demersal), planorbicone (planktic-demersal), serpenticone (planktic), 'Leiostraca', 'Trachyostraca', brevidome, mesodome, and longidome.

Heteromorphs include 7 morphogroups with orthocone (planktic-demersal), cyrtocone (quasiplanktic), gyrocone (quasiplanktic), torticone ('trochospiral', planktic), ancylocone (planktic vertical migrants), hamiticone (planktic vertical migrants), and vermicone (planktic drifters; Fig. 18.3).

The basic planispiral and heteromorph shell shapes were shown in ternary diagrams by Westermann (1996). The majority of heteromorphs was suggested to be planktic, mostly vertical migrants (Westermann 1996). Small gyrocone forms were interpreted as pseudoplanktic in floating algal mats, but detailed evidence is still missing. The main morphotypes and hypothesized modes of life were measured and computed on ternary diagrams to estimate the ammonoid ecospace in the water column (Westermann Morphospace Method; Ritterbush and Bottjer 2012; Ritterbush et al. 2014; Fig. 18.2, 18.3).

Klinger (1980) speculated that torticones and ancylocones were demersal or planktic floaters. Additionally, orthocones were suggested to have buoyancy strategies more similar to *Sepia* or *Spirula* rather than to *Nautilus* (Klinger 1980). Numerous authors (e.g., Batt 1989, 1991, 1993; Westermann 1996; Ritterbush and Bottjer 2012) argued that each shell shape or at least morphogroup had its own

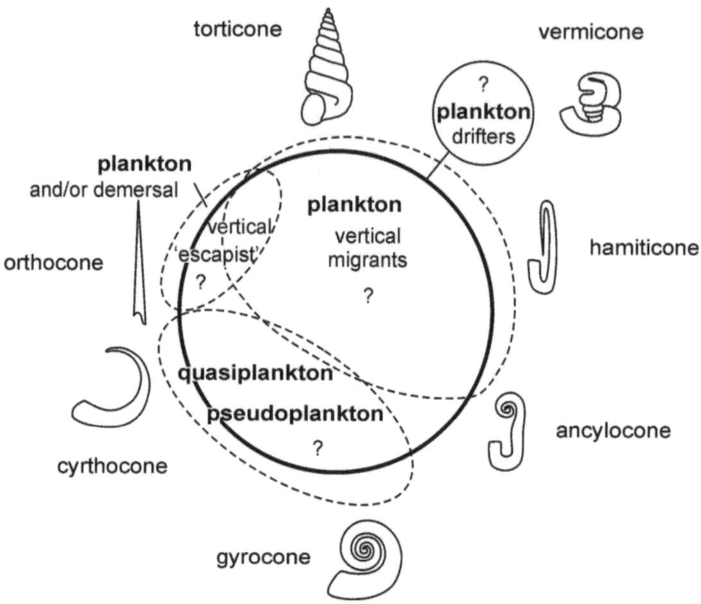

Fig. 18.3 Basic heteromorph ammonoid shell shapes and their principal assumed habitats, adapted from the original figure in Westermann (1996, Fig. 2, p. 612). Distinct morphogroups appear in the hypothetical diagram with overlapping areas (dashed lines)

special habitats (e.g., demersal, planktonic) and motility (e.g., swimming, passive drifting). Hence, shell morphologies have been interpreted as being directly dependent on habitat. Furthermore, habitat and water depth was argued as being reflected in the lithology, geochemistry, and biofacies (assemblages). Such assumptions, however, are often speculative and based on spatial distribution (e.g., Cretaceous Western Interior Seaway; Batt 1989, 1991, 1993). Many workers calculated the *syn vivo* density of ammonoids based on shell thickness, the volume ratio between phragmocone and body chamber (Fig. 18.4), and the density of supposed soft parts, which commonly is assumed to be equal to that of *Nautilus* (i.e., neutrally buoyant), for understanding the presumed habitat.

Besides density and buoyancy, the hydrodynamics of ammonoid shells were also examined by many authors. For example, Chamberlain (1980) suggested that a nektobenthic mode of life was predominant among ammonoids. Jacobs et al. (1994) suggested that compressed and depressed morphotypes in Cretaceous scaphitids from the USA had advantages for swimming at higher and lower velocity, respectively. These results are consistent with the fact that compressed morphotypes are found in shallower (more energetic) and depressed morphotypes in deeper (less energetic) sedimentary facies. For more compressed morphotypes, a higher swimming velocity was assumed to be essential for life in shallow waters, whereas the depressed morphotypes are thought to inhabit deeper (offshore) areas (Jacobs et al. 1994).

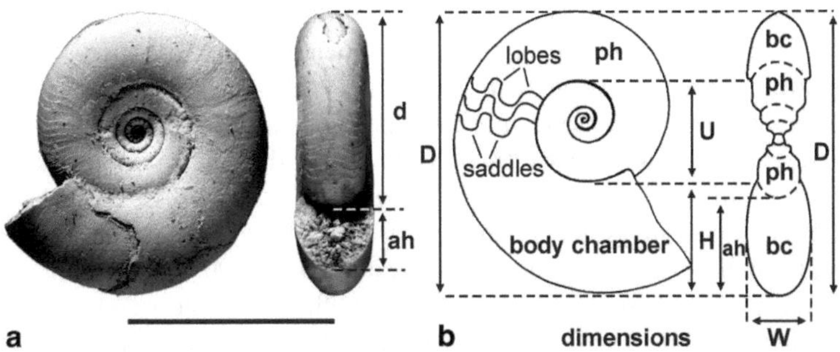

Fig. 18.4 Explanation of dimensions and conch parameters given in the text. **a** lateral (left) and apertural (right) views of an ammonoid (*Kasimlarceltites krystyni*, NHMW 2012z0133/0262; adapted from Lukeneder and Lukeneder 2014). **b** schematic ammonoid-sketch with indicated conch parameters. ah aperture height, bc body chamber, D diameter, d diameter exclusive of last whorl, H whorl height, ph phragmocone, U umbilical width, W whorl breadth. Scale bar: 1 cm

Landman et al. (2012) examined septal and siphuncle strength (Hewitt 1993; 1996; Tanabe 1979) combined with stable isotope data (Chap. 18.5; Cochran et al. 2010a, b) of *Hoploscaphites* from the Cenomanian–Maastrichtian of the Western Interior Seaway (USA). They concluded that habitat depth was < 100 m and near the sea floor. The apertural angle excludes a nektobenthic mode of life for these heteromorph morphotypes, at least in adult specimens (Landman et al. 2012, Chap. 18.11). Stable isotope data for the relative habitat depths of *Hoploscaphites* and *Baculites* strengthened the supposed shallower environment for *Scaphites* (Cochran et al. 2010a, b; Landman and Klofak 2012; Henderson and Price 2012). Touch marks on the sea floor attributed to these scaphitids were formed by postmortem bouncing or rolling shells; hence, they do not provide evidence for a demersal lifestyle (Seilacher 1963; Landman and Cobban 2007). Yamada and Wani (2013) measured the thickness ratios of shells (whorl width/diameter) of *Scaphites* from the lower middle Turonian of Hokkaido (Japan) in order to determine the migration mode: the ratios of *Scaphites planus* differed significantly between distinct localities. They argued that different populations did not frequently migrate between such areas. Based on hatchling sizes, they concluded that thickness ratios became manifested after hatching due to limited migration within a nektobenthic habitat. That study suggests that scaphitid ammonoids became nektobenthic with limited migration at an earlier stage than previously thought (i.e., not during the transition from normal to heteromorph coiling; Yamada and Wani 2013).

Oxycone ammonoids such as *Sphenodiscus* from the Western Interior Seaway were used to define migration paths during the Maastrichtian (Ifrim and Stinnesbeck 2010). Sphenodiscids were reported to be characteristic morphotypes for shallow-water environments near the wave base (Batt 1989; Jacobs 1992a; Jacobs and Chamberlain 1996; Ifrim and Stinnesbeck 2010). Ifrim and Stinnesbeck (2010) assumed a connection between near-shore environments and the occurrence of *Sphenodiscus*, at least for North America.

Many hypotheses about the mode of life of distinct morphogroups are largely speculative. A demersal mode of life (nektobenthic) was proposed for several heteromorph groups by Wiedmann (1973), Vašíček and Wiedmann (1994), and Klinger (1980). The same mode of life was postulated for small heteromorph forms (*Karsteniceras*) by Lukeneder (2003, 2005) during dysoxic environmental conditions in the Early Cretaceous from the Northern Calcareous Alps (Austria). Ebel (1985, 1990, 1992) suggested a benthic, gastropod-like (see rebuttal comments in Kröger 2001) lifestyle for heteromorphs. Westermann (1990, 1996) summarized the imaginable life habits from active swimming (with varying steerage) to a passively drifting planktic mode, and from horizontal to vertical migratory types. Westermann (1996) separated different strategies within the heteromorphs, a distinct demersal mode of life for cyrtocone forms (e.g., *Protancyloceras*), a pseudoplanktic one for gyrocone forms (e.g., *Crioceratites*), and a planktic mode for vermicone ones (e.g., *Nipponites*; see also Klinger 1980). According to Westermann (1990, 1996), various life modes existed in photic and aphotic zones in the ocean. Stable isotope data confirm this scenario (Lukeneder et al. 2010). Hence, the $\delta^{18}O$ values of the forms with strong ribbing such as *Hypacanthoplites*, *Nowakites* and *Perisphinctes* were confirmed to have been inhabitants of the epipelagic zone in the neritic province. *Hypacanthoplites* (discocone deshayesitid with moderately coarse ribbing) possibly inhabited the photic zone of the uppermost 50–100 m with a planktic to nektobenthic, mobile lifestyle. *Nowakites* (platycone-discocone pachydiscid with strong ribs) appears to have preferred the transition from the epipelagic to the mesopelagic zone, with a mobile to sluggish mode of life and vertical migrations. Heteromorphs like *Baculites* potentially preferred a similar water depth (Henderson and Price 2012). In contrast, the much more positive $\delta^{18}O$ values of the sub-sphaerical and depressed *Cadoceras* indicate a habitat in the cooler and deeper mesopelagic zone. Based on this low number of species, the general rule of thumb that strongly ribbed forms dominated above depths of 100 m in the neritic epipelagic zone, whereas smooth forms dominated in the deeper oceanic mesopelagic zone appears to be valid.

A major difference in the modes of life of many ammonoids compared to the modern cephalopods *Nautilus*, *Sepia* and *Spirula* is evident in their adult stage. All extant examples tend to retreat into the deepest environments as adult animals (e.g., *Spirula*, *Sepia*) or at least to remain there throughout their post-juvenile phase (e.g., *Nautilus*). In contrast, all measured ammonoids, except for strongly ornamented *Perisphinctes*, display a clear tendency to migrate into shallower environments in their latest ontogenetic stage (Chap. 18.5).

Constraints from phylogenetic analyses are needed in the case of arm number, which is mostly given as ten (Mehl, 1984; Jacobs and Landman, 1993; Westermann 1996; Lukeneder 2012; Klug and Lehmann 2015). Moreover, the extent to which the shell was covered by the mantle in different morphogoups remains unclear (Doguzhaeva and Mutvei 1989, 1991, 1992, 1993). Those authors suggested that the shells of the Cretaceous genera *Sinzovia*, *Gaudryceras*, and *Ptychoceras* were semi-internal or even fully covered by mantle. Full coverage in *Ptychoceras* was rejected by Westermann (1996) due to buoyancy and soft tissue estimations. Shell-wall

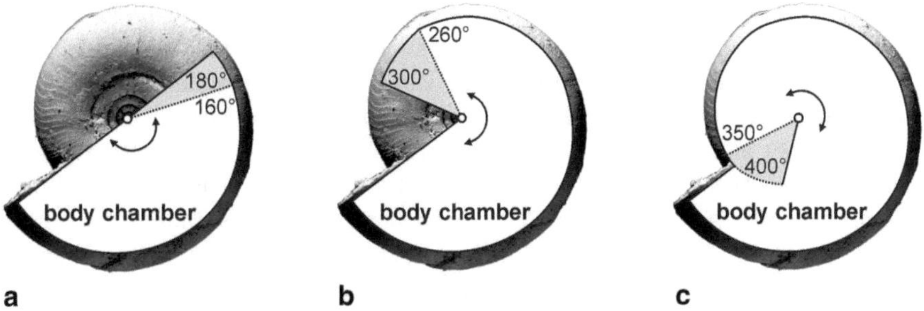

Fig. 18.5 Main classes of body chamber length (angles) in planispiral ammonoids, measured from the aperture. **a** Brevidome (160–180°), **b** mesodome (260–300°), and **c** longidome. (350–400°; adapted from Westermann 1996)

duplication was interpreted by Birkelund (1981) in *Gaudryceras* as indication that its juvenile shell was exceptionally covered by the mantle. Additionally, *in situ* fossil buccal masses are commonly located near the centre of the body chamber, supporting a jaw-function rather than an operculum-function (Schindewolf 1958; Lehmann 1976, 1980, 1988; Kennedy and Cobban 1976; Tanabe 1983; Morton and Nixon 1987; Tanabe and Fukuda 1987; Mapes 1987; Bandel 1988; Nixon 1988, 1996; Seilacher 1993; Westermann 1996; Kennedy et al. 2002; Wippich and Lehmann 2004; Doguzhaeva et al. 2007; Keupp 2007; Landman et al. 2007b, 2010, 2012). As shown by Wani (2007), *in situ* preservation of buccal masses in the body chamber of ammonoids coincides with the *in situ* deposition of the ammonoid shell itself (Chamberlain et al. 1981; Wani et al. 2005).

In planispiral ammonoid shells, the static orientation of the aperture (apertural angle) is interrelated with hydrodynamic stability (distance between the centers of buoyancy and mass; Westermann 1996). These centers depend mostly on body chamber length (Fig. 18.4, 18.5), causing differences in soft body volume (i.e., soft body mass). The consistency and density of ammonoid soft bodies are imprecisely known (Hoffmann et al. 2015; Klug and Lehmann 2015). This hampers estimations (Saunders and Swan 1984; Saunders and Shapiro 1986: Shapiro and Saunders 1987; Swan and Saunders 1987) of buoyancy, bathymetry (depth limits), swimming style etc. Assuming neutral buoyancy, Saunders and Swan (1984), Saunders and Shapiro (1986), Shapiro and Saunders (1987), as well as Swan and Saunders (1987) calculated the orientation and stability of Carboniferous goniatitids and prolecanitids. Three body chamber groupings were established in conjunction with their supposed swimming-potential and maneuverability in ammonoids. The criteria were shell shape, aperture orientation, and stability (Westermann 1996; adapted after Saunders and Shapiro 1986), and the groupings were brevidome (<220° body chamber length), mesodome (220–320°), and longidome (<320°; Fig. 18.5). Subsequently, the relation between shell thickness ratio (whorl width/diameter) and body chamber length is combined with the apertural orientation. Brevidome, mesodome and longidome ammonoid shell hydrodynamic stability and steerage mechanisms (Fig. 18.5) were discussed by several authors (Davis et al. 1969;

Davis 1972; Hengsbach 1978; Chamberlain 1980, 1991; Ward 1981, 1986b; Bayer 1982; Saunders and Swan 1984; Callomon 1985; Saunders and Shapiro 1986; Westermann and Callomon 1988; Checa and Westermann 1989; O'Dor and Wells 1990; Westermann 1990, 1996; Doguzhaeva and Mutvei 1992; Jacobs 1992a; Jacobs and Landman 1993; Dagys and Weitschat 1993; Elmi 1993; Jacobs and Landman 1993, 1996; Jacobs et al. 1994). Heteromorph shell types and their swimming performance were also discussed in numerous papers (Mapes 1979; Ward 1976a; Ward and Westermann 1977; Westermann 1977; Klinger 1980; Okamoto 1984, 1988a, b, c; Batt 1993; Fig. 18.3). A pelagic, quasi-planktic life mode was suggested for Cenomanian openly coiled *Allocrioceras* by Wippich and Lehmann (2004) conbased on *in situ* aptychi and stomach contents.

18.3 Conch Parameters as Proxies for Ecology

Numerous studies drew attention to the relation between the ecology of individual ammonoid taxa and morphological shell parameters (Fig. 18.4) such as rib numbers, diameter (D), siphuncle diameter (sd), siphuncle wall thickness (swt) whorl width-diameter ratios (W/D) or whorl expansion rates (WER; e.g., Kant 1975; Westermann 1996, Nikolaeva 1999; Sarti 1999; Klug 2002, Korn and Klug 2001, 2002, 2012; Korn 2010; Korn et al. 2004; Matsuoka et al. 2010; Ritterbush and Bottjer 2012; Yahama and Wani 2013). Kant (1975) found allometric growth changes in Carboniferous ammonoids. Septal complexity was also investigated biometrically (Kahn and Kant 1975) to gain ideas on ammonoid ontogeny. The same authors showed that measurements on ammonoid sutures can potentially help to estimate growth rate and mode. Ballentine (2007) mathematically analyzed the index of sutural complexity (ISC) used to classify ammonoid shells. He concluded that this index has the same value for suture patterns of very different shapes. Accordingly, it should be combined with other shell characteristics (siphuncle strength). Fernandez-López et al. (1999) demonstrated a relation between ornamentation or conch shape (coiling) and habitat for *Tmetoceras* (Middle Jurassic, Spain). Evolute morphotypes inhabited basinal areas, whereas involute poplulations were dominant in epicontinental, relatively shallow platforms.

Korn and Klug (2001, 2002, 2012) and Klug (2001, 2002) published conch parameters of Devonian ammonoids (Agoniatitina and Anarcestina) from Morocco, which yielded results on ontogeny, taxonomy, covariation of D, WER, and relations to sea-level fluctuations. Analyses of lithology, grain size and microfacies of the host sediment allowed a correlation with global sea-level curves, although on a weak statistical basis (Klug 2002). Klug (2002) found smaller conchs with lower WER (<2.0; Anarcestina) in deeper areas and/or trangressive phases and larger morphotypes (Agoniatina; WER >2.2) with increasing W/D ratios in more shallow areas and/or regressive phases. W/D ratios were also important in detecting sexual dimorphism (Sarti 1999). Klug (2002) suggested that well-preserved assemblages comprising all growth stages of ammonoids indicate reproduction within the region.

Biometric estimations (morphospace) by Neige et al. (1997) on shell parameters including whorl expansion rate, diameter, whorl height and whorl width in Callovian ammonoids from France were used to relate morphology to sea-level changes. A clear trend was detected from subserpenticone (i.e., evolute, depressed, nektopelagic; Fig. 18.2) to suboxycone (i.e., involute, compressed, nektobenthic) morphotypes during a transgression. This indicates a potential usefulness of ammonoids as paleodepth indicators because morphology might change with depth (Navarro et al. 2005). Those authors performed multivariate analyses on Middle Jurassic Cardioceratidae, which made morphospace calculations less intuitive, increasing the reliability of their results and conclusions drawn based on ammonoid habitats. Morphological diversity during immigration phases increased because the animals inhabited distinct niches, a phenomenon probably valid for all ammonoid taxa (see Westermann 1993b).

Biometric analyses are useful to detect differences in shell parameters and their ratios. Celtitids from the Carnian of Turkey have quite similar shell morphologies. Taxonomic separation is mainly based on sculptural differences. Several ontogenetic stages could be distinguished by biometric analyses (WER, W/D) in *Kasimlarceltites* (Lukeneder and Lukeneder 2014). Embryonic to early juvenile stages start with high WER values (2.16–2.17, spherical), which drastically decreases in older juveniles (1.75, strongly depressed early whorls), then increases markedly in mid-age, followed by unstable mid-aged phases of (WER 1.89–1.79), and finally peaks in high values of the preadult to adult stages (2.05–2.17, compressed). This morphological change mirrors a trend from from planktic to nektic or active swimming from the second life-phase on Investigations on the ammonoid shell shape in morphospace of the serpenticone *K. krystini* pointed to a hypothetical planktic lifestyle of ceratitid ammonoids (calculated by K. Ritterbush 2013, Fig. 18.2 18.6). A mainly planktic lifestyle is assumed based on the calculations for the small ceratitid ammonoid (max. size 33 mm; Lukeneder and Lukeneder (2014).

Despite of the huge amount of data and articles on ammonoid biometry, the question remains whether there is a relation between conch shape and the physical and chemical parameters of the environment (depth, pressure, salinity, temperature, light etc.). Most studies seem to support such a relationship.

18.4 Ontogenetic Stages in Ammonoids

The terminology of growth stages used herein is based on ontogenetic stages defined by numerous authors (House 1985; Kant 1975; Kullmann and Scheuch 1970, 1972, Westermann 1954, 1958; Landman 1987; Kullmann 1981; Hewitt 1988; Kant and Kullmann 1988; Landman 1988; Bucher et al. 1996; Klug 2001; Korn and Klug 2002, 2007, 2012; Etches et al. 2009; Lukeneder et al. 2010; De Baets et al. 2012, 2013, 2015; Laptikhovsky et al. 2013; Lukeneder and Lukeneder 2014). The exact definition of the ontogenetic stages is crucial for reconstructing life cycles with coeval changes of habitats.

The ammonitella (=embryonic stage; embryo in Lukeneder et al. 2010) consists of the subelliptical protoconch or initial chamber and three-fourths to one-and-a-quarter whorls with body chamber phragmocone chambers. The ammonitella is usually delimited by the primary (nepionic) constriction (Birkelund 1981; Landman et al. 1996b, Klofak et al. 1999, 2007; Etches et al. 2009). After that stage, a change in ornamentation and shape occured (Currie 1942, 1944; Burnaby 1966, Lehmann 1966; Kulicki 1974, 1979, 1996; Bandel et al. 1982; Ward and Bandel 1987; Landman 1987, 1988, Landman et al. 1996b, 2007a; Maeda 1993; Kulicki and Doguzhaeva 1994; Bucher et al. 1996; Doguzhaeva 2002; Kulicki et al. 2002; Sprey 2002; Korn and Klug 2007; Tanabe et al. 2010b). The corresponding term in nautiloids is nauta (Laptikhovsky et al. 2013). According to Laptikhovsky et al. (2013), the average ammonitella diameter was 0.54–2.6 mm, decreasing in mean size from the Devonian (up to >5 mm; De Baets et al. 2015) via the Carboniferous with 0.6–1.4 mm to the latest Cretaceous with 0.7–1.0 mm. Laptikhovsky et al. (2013) concluded that seawater temperatures were the key factor provoking historical changes in ammonoid and nautiloid evolution. The authors argued that eggs were larger in species from temperate than from equatorial areas and also during global warming. The negative relationship between egg size and environmental temperatures is a also known as Thorson-Rass rule (Laptikhovsky et al. 2013).

The neanoconch (=neanic stage of Hyatt 1894, hatchling to early juvenile; juvenile in Lukeneder et al. 2010) comprises 2.5–3.3 additional whorls. It grew mainly in height, is often planorbiconic and weakly sculpted with 3–5 mm diameter (Westermann 1958).

The juvenile phase (=juvenile stage; juvenile to mid-age in Lukeneder et al. 2010) comprises the late juvenile plus adolescent/immature or preadult phase with several additional whorls. These differ from the preceding ones in various growth parameters, e.g., abrupt growth of width and sculpture, roughly to half adult size.

The adult phase (=adult stage or adulthood in Bucher et al. 1996; adult in Lukeneder et al. 2010; maturity) is fully grown and thus mature. The post-juvenile shell has 3 or more whorls, comprising the final body chamber with different coiling, cross section, sculpture, and/or peristome (Davis et al. 1996; Klug et al. 2015b).

As noted by several authors, Recent cephalopods exhibit two principal spawning strategies (Nesis 1987; Boletzky 1987; Hewitt 1988; Westermann 1996; De Baets et al. 2012, 2015). A benthic *K*-type strategy with few big eggs and a planktic *r*-type strategy with numerous small eggs. Most ammonoid eggs (Etches et al. 2009; Tajika and Wani 2011; Stephen et al. 2012; Laptikhovsky et al. 2013; Yamada and Wani 2013; De Baets et al. 2012, 2015), as inferred from the ammonitella, resembled those of the small to smallest eggs of present-day cephalopods. Environmental influences such as warm water and good oxygenation enhanced egg growth (Westermann 1996; Laptikhovsky et al. 2013). Accordingly, aerobic conditions were necessary for full development. Most taxa possibly encased their eggs in a light gel, singly or in masses (Westermann 1996; Mapes and Nützel 2009). Alternatively, the female may have carried the eggs, as in Recent pelagic octopods (e.g., *Argonauta*). Recent *Nautilus* produces less than 10 eggs of a diameter of c. 20 mm

(Tanabe et al. 1993b; Kröger 2005), laying the egg capsule directly on the sea floor (Saunders and Landman 1987, 2010).

Differing reproductive strategies in nautiloids and ammonoids in early ontogenies are reported by Stephen and Stanton (2002), Kröger (2005) and Klug (2007). Fossil nautiloids produced fewer, larger eggs than ammonoids (Landman et al. 1996b; Chirat 2001; Klug 2001, 2007). A typical ammonoid egg measured 1–2 mm (c. 100 to 500 000 per mother animal; Klug 2001, 2007; Korn and Klug 2007, 2012; De Baets et al. 2012, 2015; Laptikhovsky et al. 2013). In contrast, the hatchling size for Cretaceous nautiloids varies from 9 to 35 mm (Wani et al. 2011).

It may be speculated that the often small ammonoid hatchling size (compared to nautiloids) enabled them to produce more ammonoid offspring (*r*-strategists, type III survivorship curve). They drifted as plankton in the water column (Klug 2001, 2007; Landman 1985; Landman et al. 1996b) and were part of planktic food webs (Laptikhovsky et al. 2013). Further information is obscured by the lack of knowledge about where ammonoids laid their eggs (Westermann 1996). As noted by Klug (2001) as well as Korn and Klug (2007), floating ammonoid egg masses in the water column (see also Tanabe et al. 1993b, Mapes and Nützel 2009) would change our understanding of ammonoid ecology and habitat preference.

The ecology of hatchlings and early juveniles (neanic stage) of living cephalopods is well known (e.g., Boletzky 1974, 1992; Nesis 1987; Westermann 1996; Stephen and Stanton 2002; Korn and Klug 2007; Laptikhovsky et al. 2013). The hatchling stage in ammonoids began at 3–5 mm diameter (Landman 1987; Landman and Waage 1993; Okamoto 1988a, b; Shigeta 1993; Tanabe and Ohtsuka 1985; Tanabe et al. 1981, 1995, 2003; Bucher et al. 1996; De Baets et al. 2015), thus categorizing the hatchlings and neanic stages as microplankton. The early post-hatching interval in at least Cretaceous ammonoids was assumed to be planktic, changing into nektobenthic or nektoplanktic after reaching a diameter of 2–2.5 mm (Shigeta 1993). Westermann (1996) dismissed this based on difficulties in measuring ammonitella volumes. The most common habitat of the neanic stage was probably the deep midwater of hypoxic, epeiric basins, perhaps somewhat above or below the dysaerobic/aerobic interface (e.g., Schindewolf 1959; Landman 1982; Tanabe et al. 1993b, 1995; Westermann 1996).

Juvenile specimens are usually rare in ammonoid assemblages of the aerobic facies (Westermann 1954, 1996; Kennedy and Cobban 1976; Callomon 1985; Landman 1987). They appear more frequent in anaerobic and dysaerobic black shales (Westermann 1996), being either pelagic or pseudoplanktic or, as postulated by Westermann (1996), developing from planktic, vertical migrants into demersal swimmers.

In adult stages (mature, fully grown), ammonoids were most probably generalists populating almost every single niche in the marine environment environment (Fig. 18.7),, with the ability to migrate between and within several depth zones. Adult stages are mainly detected based on apertural collars (macroconchs, females) or lappets (microconchs, males) reflecting sexual dimorphism, and based on changes in ribbing or on final septal crowding (Callomon 1980; Davis 1972; Elmi 1993; Davis et al. 1996; Lukeneder 2004; Klug et al. 2015b). Ontogenetically controlled migrations mirror long-term habits, whereas foraging and preying would reflect daily behaviors in ammonoids. According to Becker (1993) and Westermann (1996), even brackish waters could be inhabited by at least some ammonoid groups (e.g., some goniatids, *Placenticeras*, perhaps *Dactylioceras*). Westermann (1996) considered the lifestyle flexibility of ammonoids versus extant cephalopods as follows: "*except for their reduced depth, speed, and, hence, range of migration, ammonoids presumably had similar ecologic diversity*".

Adult modifications such as coarsening sculpture, shell thickening or uncoiling were discussed by numerous authors (Tanabe and Shigeta 1987; Jacobs and Landman, 1993; Jacobs et al. 1994; Jacobs and Chamberlain 1996; Westermann 1996; Davis et al. 1996; Klug et al. 2015b). Based on estimated shape and hydrodynamic properties, a mainly nektobenthic life was suggested for adult ammonoids by Chamberlain (1980). As summarized by Westermann (1996), the main late ontogenetic trends in ammonoid life were (a) from planktic (or demersal vertical migrant) to (b) sluggish nektic or (c) demersal, passive drifters or active swimmers. Late ontogeny was perhaps accompanied by an increased diurnal vertical migration of planktic forms (see Doghuzhaeva et al. 2010), increased speed and acceleration of nektic or demersal forms, from sluggish to mobile, as in many of the smaller microconchs from demersal to nektic.

Post-neanic ammonoids resemble *Nautilus* in the roughly linear growth rate (Saunders 1984; Hewitt et al. 1994; Bucher et al. 1996). As noted by Westermann (1971, 1977, 1996), maximum growth rates of ammonoid shells probably depended more on chamber growth than on apertural growth. The neanic growth rate may have differed greatly from that of the juvenile, especially if there was a major habitat change, e.g., from planktic to demersal (Westermann 1996). Growth rate might have decreased with increasing habitat depth (e.g., increasing pressure reduced the osmotic emptying rate, decreasing temperature and food supply, thus slowing the metabolism). As speculated by Westermann (1996), the lifespan may have varied from 1 to 2 years for small shallow-water ammonoids (<50 mm diameter) to 5 to 10 years for most epeiric and inner-neritic ammonoids (see Bucher et al. 1996), to as much as 50 or 100 years (overestimate?) in mesopelagic giants such as large *Lytoceras* (Stevens 1988). In general, growth rates were higher and size larger in shallow platform dwellers than in basin dwellers (Elmi and Benshili 1987). Ontogenetic stable isotope analyses on baculitid shells from the Campanian from South Dakota (USA) were conducted to assess $\delta^{18}O$ data concerning marine paleotemperature. Our understanding of ammonoid growth rate will be enhanced in the future by the analysis of isotope composition, which changes during migration and habitat change. No exact estimates (c. 33 cm/a) were presented by Fatherree et al. (1998) for *Baculites*, suggesting a quite rapid growth. More detailed ontogenetic analyses of stable isotopes could probably help to solve that problem. Detailed stable isotope studies with the goal of a better reconstruction of seasonal variations will help to estimate at least the duration from embryonic to full growth (Chap. 18.5).

No evidence has been presented for the exact habitat where ammonoids laid their eggs and where spawning took place. Stable isotope data (Fig. 18.8, 18.9, 18.10, 18.11) are still too imprecise to reconstruct these aspects using protoconchs and ammontellae shell. Specifically, the volumes required for carbonate analyses

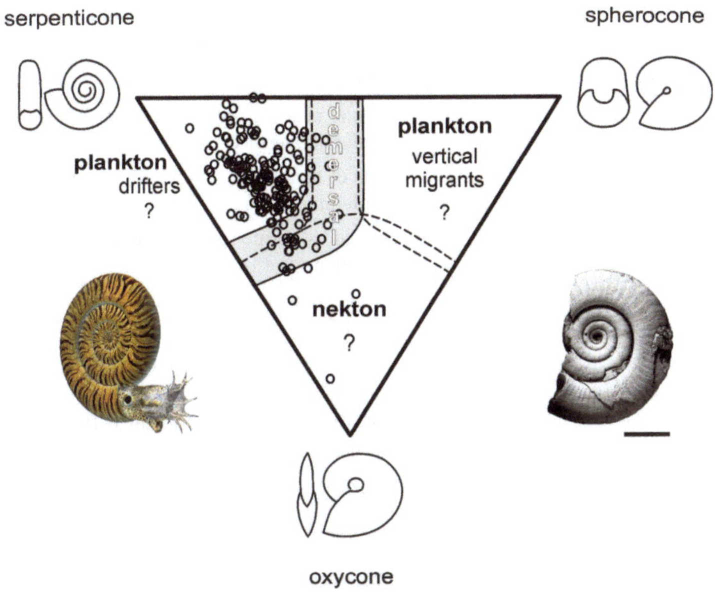

Fig. 18.6 The serpenticone Upper Triassic (Carnian) ammonoid shell shape and the principal assumed habitats in *Kasimlarceltites krystini* (calculated by K. Ritterbush 2013). Distinct morphogroups appear in the hypothetical diagram with overlapping areas. For details see Fig. 18.2 herein. A mainly planktic life-style is assumed from the calculations for the small ceratitid ammonoid (max. diameter 33 mm; Lukeneder and Lukeneder 2014). Reconstruction of *K. krystini* (left, by 7reasons Media Company) and the holotype of *K. krystini* (right, NHMW 2012z0133/0014; Lukeneder and Lukeneder 2014). Scale bar: 1 cm

(e.g., Keil Carbonate Device) are large (3050 µg) and drill bits (30–50 µm) are thicker than ammonitella shell walls (10–20 µm; Kulicki and Doguzhaeva 1994). Large differences in embryonic stages are visible only when comparing different ammonoid groups. This provides no information on where the eggs were stored (i.e., water column or benthic). All ammonoid hatchlings are suggested to have started passively drifting as plankton (Kennedy and Cobban 1976; Tajika and Wani 2011), the duration of that phase being unknown. The mechanisms of dispersal of the hatchlings and neanic stages were likely entirely passive and depended on the presence and strength of marine currents (Westermann 1996). Westermann (1990, 1996) concluded that dispersal distance ranged from tens of kilometers (enclosed epeiric seas) over a few hundred kilometers (shelf seas) up to 1000–2000 km in the open ocean. These early ontogenetic stages likely formed an important constituent of the Mesozoic plankton in shelf seas and oceans (Fig. 18.7, Table. 18.1).

Recently the probable ecology of eggs was discussed by Etches et al. (2009), Tajika and Wani (2011), Stephen et al. (2012), and Laptikhovsky et al. (2013). Etches et al. (2009) reported eggs enclosed in egg sacks (see also Lehmann 1966, 1976; Müller 1969) from the Upper Jurassic (Kimmeridgian) of the UK. Some were still attached to ammonoids (possible parentage), e.g., *Aulacostephanus* and *Pectinatites*. Etches et al. (2009) suggested a firm substrate where eggs were anchored below the storm wave base. It is controversial whether such egg sacks still contained eggs

Table 18.1 Terms used in the text and figures. Zonation of oceans, habitats of ammonoids, morphological terms, ontogenetic stages and hypothetical mode of life in ammonoids

Habitat and marine zonation	Ammonoids
Benthic	On or near the sea floor, bottom and water column interface
Epibenthic	On or near the sea floor, bottom and water column interface
Pelagic	Includes neritic and oceanic zones, 0 m down to 6000 m
Neritic	Near shore to marine areas above the continental shelves
Epipelagic	Uppermost part of the marine oceanic province, down to approx. 200 m
Mesopelagic	Marine water column 200 m to 1000 m in the oceanic province
Morphology	–
Outer shell	–
Planispiral	Coiling in a single horizontal plane, whorl height increasing from the coiling-axis
Heteromorph	Open coiled, coiling in various planes or with irregular coiling
Leiostraca	Smooth to fine ribbed shells
Trachyostraca	Strong ribbing of shells
Body chamber	–
Brevidome	Body chamber length from 160° to 180°
Mesodome	Body chamber length from 260° to 300°
Longidome	Body chamber length from 350° to 400°
Planispirals	–
Cadicone	Evolute, subglobular shell with angular umbilicus and depressed shell whorls
Discocone	Involute, with ovate and slightly compressed whorls
Oxycone	Involute, with subtriangular whorls
Planorbicone	Evolute, with subcircular to depressed shell whorls
Platycone	Involute to moderate evolute, with subrectangular compressed whorls, keeled
Serpenticone	Evolute to advolute, with circular to depressed whorls
Spherocone	Involute to convolute, subglobular shell with ovate shell whorls
Heteromorphs	–
Ancylocone	Closed or open, planar or helical spire followed by a hook
Cyrthocone	Open, curved with less than one or one circular to ovate whorl

Table 18.1 (continued)

Morphology	
Gyrocone	Open spire, with more than one circular to ovate whorl
Hamiticone	Two or more straight shafts, with final hook
Orthocone	Straight, with circular to ovate whorl sections
Torticone	Trochospiral or helical, with loose or close coiling
Vermicone	Worm-like, with irregular coiling
Ontogenetical stages	–
Protoconch	subspherical, egg shaped initial chamber
Ammonitella	Embryonic stage, starting with protoconch, ending at primary constriction
Neanoconch	Neanic stage, hatchling to early juvenile, 2.5–3.3 additional whorls
Juvenile	Juvenile stage, late juvenile plus adolescent , plus several additional whorls
Adult	Adult stage, mature with modified ribbing, final lappets or collars at peristome
Mode of Life	–
Demersal	Live and feed on or near the sea floor
Nektic	Active and free swimming, independet of ocean currents
Planktic	Passively floating and drifted by ocean currents
Nektobenthic	Live and feed on or near the sea floor, synonymous to demersal

in the fossil record, and thus whether they were deposited *in situ* (Lehman 1976). As speculated by Callomon (1980) and Etches et al. (2009), some groups were gregarious, migrating over long distances and depths. The latter interpretation finds some support in the isotope analyses of Lukeneder et al. (2010, Fig. 18.9). In some Mesozic ammonoids, the sexes were perhaps spatially separated as preadults (Davis et al. 1996; Lukeneder and Harzhauser 2003; Lukeneder 2004), met for mating and separated during spawning. As suggested by Callomon (1980), female ammonoids might have returned to their primary habitats ("home-grounds" in Callomon 1980) for oviposition (Lukeneder et al. 2011), spawned and died afterwards (i.e., semelparous strategy; see also Stephen et al. 2012). The overall strategy might have resembled that of Recent neritic coleoids such as sepiids or loliginids (2009).

As argued by Tajika and Wani (2011), who studied the Cretaceous ammonoids of Japan, egg masses or ammonitellae (with given protoconch and ammonitellae size ranges) were drifted over long distances at speeds of approx. 110 km within a minimum of 5 days, from spawning places by ocean currents. This is similar to extant coleoids (2009). Tajika and Wani (2011) based their calculations on modern oceanogeography, under the assumption that paleocurrents were similar to modern ocean

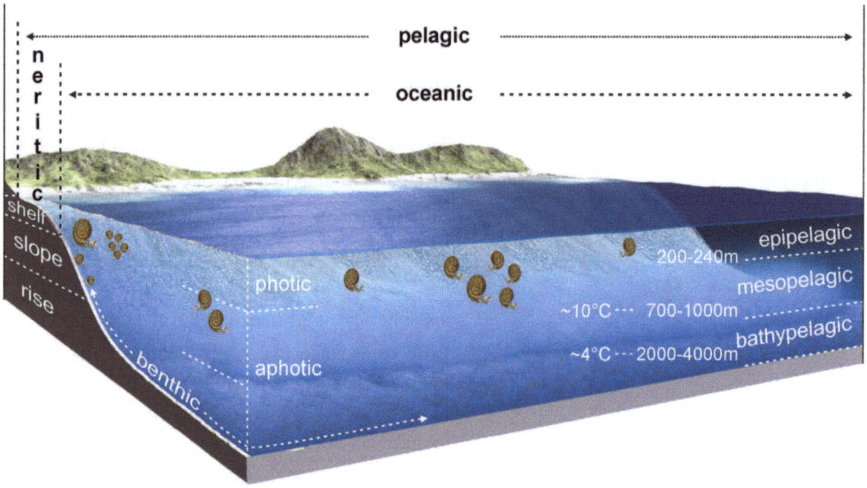

Fig. 18.7 Schematic diagram showing the pelagic zone and its horizontal layers, marine environments inhabited by ammonoids. For details and explanations see the text. Ammonoid size and paleogeographic distances not to scale

currents around Hokkaido. Eggs and hatchlings of extant coleoids (e.g., *Illex*) can be drifted by water currents for 2–3 months, during which they still show the same size range (Haimovici et al. 1998; Tajika and Wani 2011). Laptikhovsky et al. (2013) studied reproduction strategies in the Cretaceous. Based on egg size measurements, they concluded that the evolution of smaller eggs in ammonoids permitted them to occupy new habitats, whereas nautiloids survived despite their inability to migrate and settle new ecological niches. These arguments may also be valid for the ammonoids from the latest Cretaceous (latest Maastrichtian to K/P boundary), which were oceanic to suboceanic (distal neritic), deep epipelagic to mesopelagic nekton and plankton (Ward 1987, 1990a, b; Wiedmann 1988b; Westermann 1996; Fig. 18.6, Table. 18.1). The deep-water character of these last faunas shows low evolutionary rates and a high cosmopolitanism, both characteristic of oceanic taxa (Ward and Signor 1983). The fact that the nautilids survived this event indicates that the main difference might lie in the habitat differences (Westermann 1996) from mostly pelagic Maastrichtian ammonoids to demersal nautilids like Recent *Nautilus*. An additional cause for the survival of nautiloids was noted by Wani et al (2011): the much larger hatchling size of Cretaceous nautiloids contrasts with those in ammonoids (Ritterbush et al. 2014).

As noted by Klug et al. (2010), ammonoids presumably began to occupy the water column during the Devonian nekton revolution as nektic organisms, already exhibiting at this early phase various life strategies such as demersal, planktic and nektic (Klug 2001).

Based on Carboniferous anoxic–dysoxic occurrences, Boston and Mapes (1991) speculated that the very young ammonoids might have been benthic (sessile or

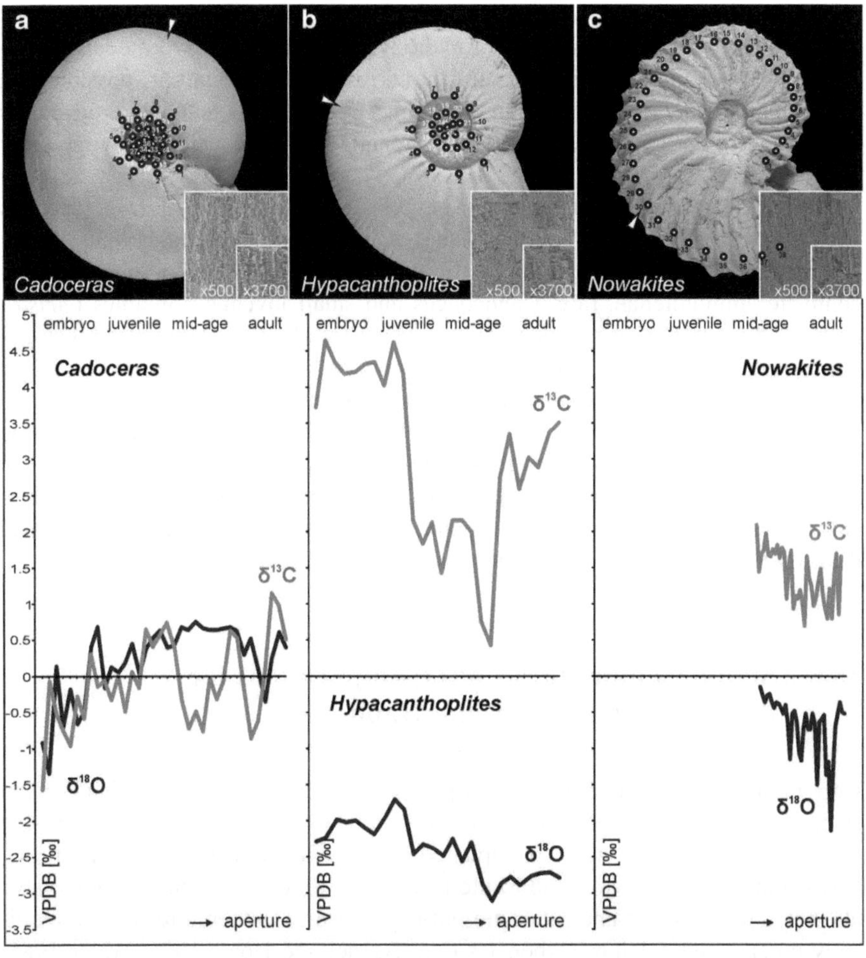

Fig. 18.8 Lateral view with numbered samples in ontogenetic direction; growth direction with indicated SEM images of aragonitic ultrastructure, ×500 and ×3700. **a** Jurassic (Callovian) *Cadoceras*, **b** Cretaceous (Aptian) Hypacanthoplites and **c** Cretaceous (Santonian) *Nowakites* shells. Arrows mark the position of last suture, the beginning of the body chamber. Stable isotope curves ($\delta^{18}O$ in black and $\delta^{13}C$ in grey) of the corresponding ammonoids *Cadoceras*, *Hypacanthoplites*, and *Nowakites* (adapted from Lukeneder et al. 2010)

vagile) and were unable to escape periodic benthic anoxia. In the past decades, a sessile mode of life was rejected for all ontogenetic stages in ammonoids by most ammonoid specialists. Ammonoids of larger growth stages in higher, photic zone depths (Boston and Mapes 1991) were less vulnerable to predators than the very small juveniles or were able to survive brief hypoxic events like living *Nautilus* (Wells et al. 1992). As noted by Stephens et al. (2012), mass mortalities of ammonoids could occur if they were semelparous (i.e, reproduction mass mortality), or mass mortalities of their ammonitellae or juveniles were mostly driven by environmental catastrophes (e.g., anoxic or toxic events) or reflect taphonomic processes such as stratigraphic condensation or hydrodynamic trapping. Two models for the formation of ammonoid egg masses and the corresponding habitats are possible: egg masses floating in the water column or benthic egg masses. Potential hydrodynamic processes (e.g., accumulation by ocean currents) were not considered by the latter authors.

Abundant embryonic ammonoid assemblages were described from deposits from the USA and Canada (Kansas; Mapes et al. 1992; Tanabe et al. 1993b, 1994a, b; Alberta; Pamenter 1956; Schindewolf 1959; Becker 1993). Complete, and incomplete ammonitellae, rare neanoconchs and small juvenile goniatites (*Aristoceras*, *Vidrioceras*) occur in thick layers. Mapes et al. (1992) concluded that egg masses floated above the bottom and were not planktic, or that they were slightly transported from nearby spawning sites. Westermann (1996) suggested a second hypothesis for such egg masses: neanoconchs were floating planktic-quasiplanktic in moderately deep basins, killed from time to time by storms mixing anoxic waters into those areas of the sea; larger juveniles and adults lived higher up in the water column and were unaffected (Westermann 1996). Mapes and Nützel (2009) revised these opinions and suggested floating egg masses and planktic juveniles.

18.5 Stable Isotopes ($\delta^{13}C$, $\delta^{18}O$)-Implications for Ammonoid Ontogeny and Habitat

Urey (1947), Urey et al. (1951) and Epstein et al. (1953) discovered that oxygen stable isotope ratios of calcium carbonates correlate with the temperature of the water from which they were precipitated and developed a method to determine paleotemperatures (Rexfort and Mutterlose 2006). Since then, many authors have attempted to assess the ancient or present-day seawater temperatures from biologically precipitated calcium carbonates (e.g. Bandel and Hoefs 1975; Wefer 1985; Wefer and Berger 1991). The aragonitic composition of the external shells and septa in ammonoids makes them suitable for isotopic measurements (Jordan and Stahl 1970; Brand 1986; Anderson et al.1994; Moriya et al. 2003; Zakharov et al. 2005, 2006; Lécuyer and Bucher 2006; Lukeneder et al. 2010; Henderson and Price 2012; Stevens et al. 2015), when the aragonite is pristine. Purity of shell material should be tested using modern scanning electron microscopy (SEM), computed tomography (CT) and cathodoluminescence (CL; Lukeneder et al. 2010). The fractionation and isotopic composition in oxygen and carbon isotopes ($^{16}O/^{18}O$, $^{12}C/^{13}C$) refers to a change in stable isotope ratios, reflecting chemical and physical processes (Hoefs 2004) in a cephalopod's life.

Oxygen isotope composition ($\delta^{18}O$) in carbonate minerals precipitated in water is a function of $\delta^{18}O$ of water and precipitated temperature. $\delta^{18}O$ of the carbonate reflects the $\delta^{18}O$ of the water precipitated under isotopic equilibrium. Therefore, if the $\delta^{18}O$ of water can be estimated, the temperature during precipitation can be calculated from $\delta^{18}O$ of the carbonate according to the formula

$$T\left({}^{\circ}C\right) = 21.8 - 4.69\left(\delta^{18}O_{aragonite} - \left[\delta^{18}O_{water} - 0.2\right]\right)$$

(Grossman and Ku 1986; Goodwin et al. 2003; Kobashi et al. 2003; Lécuyer et al. 2004; Kim et al. 2007, 2010), where $\delta^{18}O_{aragonite}$ is $\delta^{18}O$ of carbonate analyzed in VPDB scale, $\delta^{18}O_{water}$ is $\delta^{18}O$ of ambient water in VSMOW scale. A shift of one per mill in the oxygen isotope ratio corresponds to a temperature change of approximately 4.3 °C. Lécuyer et al. (2004) assumed that the isotope ratio is related to

$$\delta^{18}O_{water} = -9.986 + 0.3 * S$$

(GEOSECS Executive Committee 1987; see also Railsbeck et al. 1989; Geary et al. 1992) where *S* is the salinity of about 35‰ at a depth of around 300–500 m according to data by Auclair et al. (2004) and Watanabe et al. (2003). Considering that the analytical error in isotope measurements is usually $< \pm 0.1$‰ (practically, error would be slightly $> \pm 0.1$‰), at least 1 °C changes in calcification temperature should be detectable and discussable. Biogenic carbonates, however, are sometimes secreted under disequilibrium, which is termed vital effect (Urey et al. 1951). The vital effect is the most confusing bias in the stable isotope records of biologically precipitated carbonates. This effect may be greater in C isotopes than in O isotopes (Wefer and Berger 1991). Only minor or no vital effects are usually observed in the O isotopic compositions of molluscs (Wefer and Berger 1991). In contrast, it is practically impossible to assess the degree of vital effect in ammonoids. Cephalopods (*Nautilus*, *Sepia*) precipitate their shells in an O isotopic equilibrium (Landman et al. 1994; Bettencourt and Guerra 1999). One possibility to resolve such problems would be to compare the stable isotope and clumped isotope data to other organisms in the same assemblage.

Carbon isotope composition ($\delta^{13}C$) of synthetic carbonate is predominantly a function of $\delta^{13}C$ of dissolved inorganic C and carbonate ion concentration within the solution. However, $\delta^{13}C$ in biological carbonate may show a greater vital effect than $\delta^{18}O$, as mentioned above, and it is generally more difficult to assess the specific controls (McConnaughey and Gillikin 2008). Among a variety of biological activities, changes in metabolic rate and incorporation of carbon from food would be major sources of vital effects in molluscs, except for photosymbiotic species (e.g., Jones et al. 1986; Tanaka et al. 1986; Romanek et al. 1987; Henderson and Price 2012). In some cases, abrupt changes of $\delta^{13}C$ profiles may be used to identify changes in metabolic rate, such as sexual maturity. As noted by Henderson and Price (2012), studies of *Nautilus* (Auclair et al. 2004) and the deep-water squid *Spirula spirula* (Lukeneder et al. 2008; Price et al. 2009) show increasingly depleted carbon values in shell carbonate due to metabolic effects related to changes in rates of growth or food sources.

18.5.1 Recent Cephalopods-Learning for the Past

$\delta^{18}O$ is a proxy for water temperature in aragonitic and calcitic shells such as those of extant (*Nautilus*, *Spirula*, *Sepia*) and ancient cephalopods (ammonoids, nautiloids, belemnoids, teuthoids; Tarutani et al. 1969; Grossmann and Ku 1986; Lukeneder et al. 2008, 2010; Zakharov et al. 2006; Stevens et al. 2015; Fig. 18.8, 18.9, 18.10, 18.11). Food webs and food selection can be examined based on carbon ($\delta^{13}C$) and nitrogen ($\delta^{15}N$) isotopes (Minagawa and Wada 1984; Cherel and Hobson 2005; Hobson and Cherel 2006). In addition to these techniques, nitrogen isotopes in amino acids are a powerful tool for analysing trophic level in modern species (Chikaraishi et al. 2009; Kashiyama et al. 2010). Consensus exists amongst scientists working with fossil and extant cephalopods on the interpretation of $\delta^{18}O$ data. By contrast, $\delta^{13}C$ trends are more complicated to interpret in fossil cephalopods and molluscs in general.

Sclerochronologic (i.e., hard-part measurements in chronological order) $\delta^{13}C$ and $\delta^{18}O$ records from numerous *Nautilus* species were provided by several authors (Eichler and Ristedt 1966a, b; Cochran et al. 1981; Taylor and Ward 1983; Wefer 1985; Landman et al. 1983, 1994; Auclair et al. 2004; Zakharov et al. 2006). $\delta^{13}C$ and $\delta^{18}O$ in *Sepia* cuttlebones were also measured (Longinelli 1966; Longinelli and Nuti 1973; Hewitt and Stait 1988; Bettencourt and Guerra 1999; Hobson and Cherel 2006; Rexfort and Mutterlose 2006; Cherel et al. 2009). Stable isotope composition of *Spirula* was presented by Lukeneder et al. (2008), Price et al. (2009) and Warnke et al. (2010). Chitinous *Octopus* beaks were analysed by Cherel and Hobson (2005) for their $\delta^{13}C$ and $\delta^{15}N$ composition.

Ontogeny-related stable isotope patterns in the shells of Recent cephalopods reveal two main lifecycle types (Fig. 18.9). One is represented by *Nautilus* and *Sepia*, which start in warm shallower waters as juveniles and migrate into cooler and deeper waters later in ontogeny. This strategy is contrasted with *Spirula*, which starts in cold deep waters, subsequently migrates into warmer habitats, and finally inhabits deeper waters in adult stages (Lukeneder et al. 2008). Time averaging of short-term (diurnal) migration does not significantly influence stable isotope trends within the shell. In all taxa, a roughly parallel trend of C isotopes suggests a concomitant change in diet and/or water chemistry due to habitat change.

The comparison of Recent *Nautilus*, *Sepia* and *Spirula* allows quite different modes of life to be deciphered based on stable isotope signatures. Applying these methods to Mesozoic ammonoids sheds light on strategies and environmental requirements of fossil cephalopods. Due to its unusual morphology, *Spirula* is used as a key genus in paleontological papers that attempt to interpret the mode of life of Mesozoic ammonoids (Lukeneder et al. 2010).

18.5.2 *Spirula*

The $\delta^{13}C$, $\delta^{18}O$ data reported by Lukeneder et al. (2008) for the deep-water squid *Spirula spirula* revealed three ontogenetic stages, including a major shift from positive to negative values corresponding to sexual maturation, the initiation of reproduction, and concomitant changes in diet (Fig. 18.9). An embryonic stage with $\delta^{18}O$ values at +3.6 to +3.1‰ is followed by a steady decrease from +3.5 to +3.0‰ to +1.7 to +2.0‰ in juvenile stages and an increase to +2.2 to +2.6‰ in adults (Lukeneder et al. 2008, see also Price et al. 2009 and Warnke et al. 2010). After hatching at depths >1000 m (4–6 °C), the squid migrates into shallower, warmer waters at depths of 400–600 m (12–14 °C; Levitus 1994). Subsequently, the animals migrate back into somewhat cooler, deeper habitats at 500–600 m (mean 9.1 °C; Lukeneder et al. 2008), identified for all major oceans. The importance of knowing water temperatures in certain depths is shown by the correlation of known temperatures (Levitus 1994) and stable isotope data of extant molluscs. Little is known about the composition of paleooceanic water stratification (Hay 2008), hence the arrangement of temperature zonation is ambiguous for ancient seas. This makes it much more complicated to obtain reliable conclusions on habitats of fossil organisms than for Recent relatives. As noted by Hay (2008), the circulation of the Cretaceous ocean may have been very different from that of extant conditions, for example by lacking a well-developed pycnocline (e.g., stratification, convection, thermodynamic heat transport) and dominant tropical/subtropical gyres.

18.5.3 Sepia

Sepia officinalis hatches in warm waters (>20°C) at depths of <20 m and descends to greater depths (below thermocline) in accordance with a change in lifestyle from nektobenthic to nektic/nektobenthic (Rexfort and Mutterlose 2006; Lukeneder et al. 2008). Wild-caught specimens displayed $\delta^{18}O$ values from −1.3 to +3.0‰ during ontogeny, comparable to calcification temperatures of 21 °C in adolescent and 5 °C in adult specimens (i.e., cuttlebones; Fig. 18.9). This ontogenetic trend of migration from warm water into deeper cooler waters was also documented by isotope studies of Longinelli and Nuti (1973) and Bettencourt and Guerra (1999). Compared to *Nautilus*, sepiids grow much faster, maturing in 1–2 years and dying immediately after reproduction (Packard 1972; Wells 1983; Landman and Cochran 1987, 2010). Contrastingly, *Nautilus* shows slower growth but lives for several years after maturing (Saunders 1983, 1984). *Sepia* attains sexual maturity in 200 days (Rexfort and Mutterlose 2006) to 2 years (Bettencourt and Guerra 2001; Challier et al. 2005; Ceriola and Milone 2007), whereas ammonoids might have taken from 5–15 years to mature, as determined by growth line measurements, encrustation patterns and stable isotope data (Kennedy and Cobban 1976; Bucher et al. 1996; Westermann 1996; Fatherree et al. 1998; Lukeneder et al. 2010). Interestingly, sexually dimorphic ammonoids seem to reach maturity at different ages (Kennedy and Cobban 1976), earlier in males and later in females. Estimates were based on whorl number (max. size) with fewer in males and more in females. Stable isotope analyses performed on micro- and macroconchs could help to strengthen or refute these assumptions because $\delta^{13}C$ and $\delta^{18}O$ data generate details on seasonality and ontogeny.

18.5.4 Nautilus

$\delta^{18}O$ analyses of the shell of various species of *Nautilus* show more negative values in the embryonic stage and more positive values in the postembryonic stage, indicating migration from warmer shallower into cooler deeper environments during growth (Fig. 18.9; Eichler and Ristedt 1966a, b; Cochran et al. 1981; Taylor and Ward 1983; Landman et al. 1983, 1994; Landman 1988; Oba et al. 1992; Auclair et al. 2004; Zakharov et al. 2006; Kruta et al. 2014). Due to the isochronous timing of the isotope shift from negative to positive values and the decrease in septal spacing and position of the nepionic constriction, this isotope shift is considered to represent hatching of the animal (Cochran et al. 1981; Oba et al. 1992).

Initially, there was debate as to whether the more negative $\delta^{18}O$ values in *Nautilus'* embryonic stage resulted from the precipitation of embryonic shells in isotopic disequilibrium with the ambient water in eggshells (Taylor and Ward 1983). Crocker et al. (1985) showed that the $\delta^{18}O$ of the egg water of two *Nautilus* species is depleted by approximately 1‰ relative to ambient water. However, the eggs analysed by Crocker et al. (1985) did not contain developing embryos, so the fractionation between the egg water and embryonic shell was not evaluated. Landman et al. (1994) finally succeeded to analyze the $\delta^{18}O$ of shells of *N. belauensis* raised under controlled temperatures in the Waikiki Aquarium from embryonic and early postembryonic stages. The shells of both stages were secreted under isotopic equilibrium with the *in situ* water.

Based on O isotope data, nautilid embryonic development takes place at 22–30 °C and depths of 100–200 m, depending on the species. As inferred from aquarium observations, hatching occurs after about 1 year of development (Landman et al. 1994; Uchiyama and Tanabe 1999); in nature, juveniles then migrate into cooler deeper waters (150–400 m; 14–16 °C). Consequently, Taylor and Ward (1983) defined two ontogenetic stages via stable isotope data, the embryonic stage and the free-swimming stage. The main $\delta^{18}O$ shift corresponds with the formation of the 7th to 8th septum, reflecting hatching after the embryonic stage. The embryonic stage has $\delta^{18}O$ values below c. –1.0‰ (>20 °C), the free-swimming stage values above –1.0‰ (<20 °C). Variations from a $\delta^{13}C$ value of –1.3 to a value of +1.5‰ are correlated with changes from the embryonic to juvenile-adult stages (Taylor and Ward 1983) and the change in habitat (Fig. 18.8).

Nautilus macromphalus was found to reach sexual maturity in c. 2.5–6 years (Martin et al. 1978; Landman and Cochran 2010; Collins and Ward 2010) and 3 years (Zakharov et al. 2006). For *Nautilus belauensis*, maturity was attained with 15 years (Saunders 1983) and 10 years (Landman and Cochran 1984, 2010). Typical embryonic $\delta^{18}O$ values range around –1.07 to –3.0‰. After c. 269 to 362 days (Uchiyama and Tanabe 1999), hatching takes place and a migration into cooler, deeper waters (150–400 m; 14–16 °C) starts. This is documented by increasing $\delta^{18}O$ values of +0.4 to +1.21‰.

This migrational behaviour is characteristic for Nautilidae since at least the Cretaceous because similar $\delta^{18}O$ patterns, from more negative to more positive values after hatching, occurred in fossil *Eutrephoceras* (Landman et al. 1983; Landman 1988).

In *Eutrephoceras*, $\delta^{18}O$ shifts occur between septa 2 and 4, with a magnitude of the shift varying from 1.6‰ to 2.9‰. Calcification temperatures at embryonic and postembryonic stages correspond to 22–23 °C and 14–20 °C, respectively. Thereafter, *Eutrephoceras* descended into deeper waters with c. 14 °C (Landman et al. 1983).

Isotopic analyses on the nautiloid *Aturia* from the Cenozoic (Eocene, Miocene) of Slovakia indicate a nektobenthic lifestyle with temperatures of 13–16 °C and 14–18 °C for juvenile and adult stages, respectively (Schlögl et al. 2011; Fig. 18.8). Schlögl et al. (2011) suggested that newly hatched juveniles and adults inhabited the same water at c. 240–330 m. The constant habitat depth through ontogeny in *Aturia* contrasts to the drastic change in *Nautilus* (Zakharov et al. 2006).

18.5.5 $\delta^{15}N$-Amino Acids in Nautilus

The $\delta^{15}N$ of amino acids incorporated in shell carbonates of wild *Nautilus pompilius* specimens captured in the Philippines provide new insight into its trophic level (Kashiyama et al. 2010). Values in bulk $\delta^{15}N$ decrease from embryonic to postembryonic stages. Three trophic stages in ontogeny were detected. Trophic levels calculated from compound-specific amino acid $\delta^{15}N$ (Chikaraishi et al. 2009) decrease from 12–14‰ (embryonic) to 9–11‰ (post hatching), and finally to 10–12‰ (juvenile to mature). The higher trophic level values in the embryonic versus postembryonic stage is explained by consumption of egg yolk produced by the adult parent; i.e., the offspring is eating parts of their parents. Kashiyama et al. (2010) also applied this method to the Cretaceous (Albian) nautiloid *Cymatoceras* from Madagascar. In contrast to the stepwise decreasing $\delta^{15}N$ values in the extant *Nautilus*, $\delta^{15}N$ values in the fossil nautiloid material gradually increased in ontogeny from 1.2‰ to 3.8‰, probably reflecting a change in diet. The authors, however, did not trust their results because "*a potential pitfall of the current method is that we cannot exclude the possibility that organic matter was added to or formed within the samples during postmortem degradation*" (Kashiyama et al. 2010). While the applicability of this method to fossil materials might be limited, future new techniques would help to understand the life history of extinct animals.

18.5.6 Ammonoid Ontogeny-Implications from Stable Isotopes

In early O isotope paleothermometry work on ammonids, the focus was often on understanding marine paleotemperatures (Triassic and Jurassic ammonoid shells, Kaltenegger 1967; Fabricius et al. 1970; Kaltenegger et al. 1971; Jeletzky and Zapfe 1976). For this purpose, belemnoids were also frequently utilized since Lowenstam and Epstein (1954). This method generated insights into ancient ocean water temperatures at different Mesozoic ages (Kaltenegger 1967; Stahl and Jordan 1969; Tourtelot and Rye 1969; Jordan and Stahl 1970; Fabricius et al. 1970; Kaltenegger et al. 1971; Jeletzky and Zapfe 1976; Forester et. al. 1977; Marshall 1981; Landman et al. 1983; Whittaker et al. 1987; Fatherree et al. 1998; Cochran et al. 2003, 2010a, b; Moriya et al. 2003; He et al. 2005; Ifrim and Stinnesbeck 2010; Henderson and Price 2012; Zakharov et al. 2005, 2006, 2001, 2012). Similar investigations were made on the endocochleate belemnites (Lowenstein and Epstein 1954; Spaeth et al. 1971; Ditchfield 1997; Podlaha et al. 1998; Price et al. 2000; Niebuhr and Jochiamski 2002; McArthur et al. 2004; Price and Mutterlose 2004; Dutton et al. 2007; Zakharov et al. 1999, 2005, 2011; Stevens et al. 2015). Additional analyses were performed on nacreous shells of Carboniferous nautiloids from Oklahoma (Seuss et al. 2012).

Most work on isotope paleothermometry involving ammonoids utilized only the shells (outer shell, septa), with more recent efforts using calcitic aptychi and rhyncholites (Kruta et al. 2014) to discuss paleooceanography or ammonoid ecology. If we assume that ammonoids were mobile organisms living in the water column, one must be cautious when discussing isotope results based on ammonoid shells. When $\delta^{18}O$ data of a single ammonoid individual show a sinusoidal ontogenetic pattern, we have two potential explanations for it: (1) the temperature of the inhabited water fluctuated seasonally or (2) the analyzed individual successively migrated into warmer/cooler water masses during ontogeny.

A pioneering work on ammonoid isotope paleothermometry (Stahl and Jordan 1969) used this approach to discuss ammonoid ecology, especially growth rate. Since then, the number of ontogenetic stable isotope analyses on ammonoid shells has increased. Ontogenetic sampling is especially useful for paleotemperature estimations because $\delta^{18}O$ data of single cephalopod shells can range over 2‰, spanning a temperature range of almost 8–10 °C (Fig. 18.9, 18.10, 18.11). Single point measurements (Kaltenegger 1967; Fabricius et al. 1970; Kaltenegger et al. 1971; Jeletzky and Zapfe 1976) will only snapshot ocean water temperatures at a specific point of development (Lukeneder et al. 2010).

Life habitat throughout ontogeny was reconstructed for several ammonoids by Lukeneder et al. (2010). Entire ammonoid specimens were measured in the spiral direction (embryonic to adult aperture) to gain information on ecology and habitat preferences. $\delta^{18}O$ data from a Jurassic *Cadoceras* and Cretaceous *Hypacanthoplites* and *Nowakites* were chosen due to the primary aragonitic shell preservation (Fig. 18.8), and compared with Recent *Nautilus*, *Spirula*, and *Sepia* as well as Cenozoic *Aturia*, which possess equivalent or comparable hard parts (Fig. 18.9). The most suitable approach to assess diagenetic processes and the value of shell alteration (in all molluscs) is to check the purity of the shell material (i.e., aragonite; see methods listed above).

The $\delta^{13}C$ values revealed three ontogenetic stages in *Cadoceras* and *Haypacanthoplites*, including two major shifts from positive to negative and from negative to positive values (Fig. 18.8). These probably correspond to sexual maturation, the start of reproduction, and concomitant changes in diet (Lukeneder et al. 2010). $\delta^{13}C$ and $\delta^{18}O$ records both suggest separated phases corresponding to ontogenetically controlled vertical migrations within the water column. The values mark three to four phases in ontogeny: embryonic, juvenile, mid-aged (preadult) and adult.

The $\delta^{18}O$ values of juvenile *Hypacanthoplites* shells increase from juvenile –2.29‰ to an adult value of –2.80‰, with a mid-aged minimum of –3.11‰ (Fig. 18.8). This reflects an ontogenetic migration from warm shallow marine environments to even warmer environments (27–28 °C; Fig. 18.9). The $\delta^{18}O$ values of juvenile *Cadoceras* shells decrease from juvenile (–0.93‰) to adult (+0.39‰), with a juvenile minimum of –1.35‰ and a mid-aged minimum of +0.75‰ (Fig. 18.8). This points to a shallow, c. 21 °C warm marine habitat of juvenile *Cadoceras*. Later, the mid-aged animal preferred slightly cooler and deeper environments (12–16 °C). Finally, the adult *Cadoceras* migrated back to slightly warmer and shallower environments (c. 17 °C; Fig. 18.9).

The resulting maximum temperature range to which the ammonoids were exposed during ontogeny is about 10 °C in *Cadoceras* and 6 °C in *Hypacanthoplites*. Compared to recent seawater temperatures (Levitus 1994; NOAA 2014), this suggests a bathymetric range of 50–700 m in *Cadoceras* and of only 60 m in *Hypacanthoplites*. With respect to the probably unstratified ocean system in the Jurassic and Cretaceous (Hay 2008), the observed data might point to even larger ranges.

From the isotope results, ontogenetic strategies for the morphologic groups 'Leiostraca' and 'Trachyostraca' were postulated by Lukeneder et al. (2010). The *wcw*-type (warm–cool–warm type) of *Cadoceras* resembles the patterns in *Nautilus* and *Sepia*, starting (a) as early juvenile in shallow waters migrating (b) as juvenile to mid-aged into deeper environments and (c) as adult back to shallow and thus warm waters. The *cww*-type (cool–warm–warmer type) of *Hypacanthoplites*, represented in modern cephalopods by *Spirula* (*cwc*-type), migrates (a) as a juvenile from deeper environments into (b) shallower ones and then (c) as an adult back into deeper habitats. The latter trend is slightly modified in *Hypacanthoplites*: it experiences a higher temperature plateau as an adult. During this adult phase from 28 to 29 °C (e.g., late mid-aged to adult), a slightly decreasing temperature trend occurs with age (28.9 °C to aperture with 27.5 °C; Fig. 18.9).

Data on *Perisphinctes* were published by Lécuyer and Bucher (2006), although they did not cover the earliest ontogenetic stages (Fig. 18.9). A threefold post-embryonic development is signaled by the O isotope values: a shallow and warm late juvenile to mid-aged phase at c. 23–24 °C is contrasted by slightly cooler and deeper juvenile and adult stages at 20–21 °C (Lukeneder et al. 2010). This pattern is similar to that of *Spirula*, although its deep mesopelagic habitat differs from the epipelagic environment of *Perisphinctes*. The C pattern of *Perisphinctes*, however, has little in common with *Spirula*, but is strongly reminiscent of that of *Nautilus*: very low embryonic and early juvenile stages and a considerable shift towards positive values thereafter. As this shift is not reflected in the $\delta^{18}O$ values, the $\delta^{13}C$ shift was probably induced by a major change in diet or the onset of sexual maturity rather than a marked habitat change. The animals probably changed their prey preference after attaining a certain size (Lukeneder et al. 2010). The impulse for the ontogenetic migrations in both groups (i.e., 'Leiostraca' and 'Trachyostraca') may parallel that in modern cephalopods. Firstly, the change in diets and, secondly, the mating–spawning phase in mid-aged to adult ammonoids.

As summarized by Lukeneder et al. (2010) and reviewed herein, stable isotope data on ammonoids (mostly single point measurements) are available for the Triassic *Arcestes*, *Carnites*, *Sagenites* and *Rhacophyllites*, the Jurassic *Amaltheus*, *Cadoceras*, *Leioceras*, *Parkinsonia*, *Perisphinctes*, *Phlyseogrammoceras*, *Quenstedtoceras* and *Staufenia* and the Cretaceous *Acanthoceras*, *Acanthoscaphites*, *Anagaudryceras*, *Baculites*, *Canadoceras*, *Chimbuites*, *Cleoniceras*, *Demesites*, *Desmoceras*, *Didymoceras*, *Discoscaphites*, *Douvilleiceras*, *Eotetragonites*, *Euomphaloceras*, *Eupachydiscus*, *Exiteloceras*, *Gaudryceras*, *Hauericeras*, *Hoploscaphites*, *Hypacanthoplites*, *Hypophylloceras*, *Hypoturrilites*, *Jeletzkytes*, *Kossmaticeras*, *Menuites*, *Mesopuzosia*, *Nowakites*, *Otohoplites*, *Oxytropidoceras*, *Pachydiscus*, *Phyllopachyceras*, *Polyptychoceras*, *Scalarites*, *Scaphites*, *Sciponoceras*, *Simbirskites*, *Sphenodiscus*, *Tetragonites*, *Tragodesmoceras*, *Turrilites* and *Yokoyamaoceras*.

Cochran et al. (2003) measured Maastrichtian cephalopods (e.g., *Eutrephoceras*, *Jeletzkytes*) from the Western Interior Sea in North America to define various depositional settings. $\delta^{18}O$ values of the fossils show decreases from the marine (ca. 35‰) to brackish (ca. 20‰) biofacies consistent with increasing temperatures (ca. 13 to 23 °C) or, if temperatures were relatively constant, to a decrease in the $\delta^{18}O$ of the water in which the shell formed (Cochran et al. 2003). The latter interpretation is consistent with less-than-fully marine salinities in the nearshore biofacies, but changes in both the temperature and the isotopic composition of the water may have occurred in this environment. Stahl and Jordan (1969) and Jordan and Stahl (1970) found a seasonal variation of about 8–9 °C in the Callovian based on *Quenstedtoceras* and *Staufenia*. Similarly, Lécuyer and Bucher (2006) analyzed Jurassic *Perisphinctes* from Madagascar to obtain seasonal surface water temperatures of the southern hemisphere. Zakharov et al. (2005, 2011) suggested seasonal temperature fluctuations of 1.7 °C in *Anagaudryceras* during the Coniacian from Russia and in Albian ammonoid shells (e.g., *Cleoniceras*, *Douvilleiceras*) from Madagascar. Numerous values in Zakharov et al. (2011) may be uncertain because the percent aragonite (e.g., 88%) is lower, presumably due to diagenesis. Stable isotope data from the Albian of Madagascan belemnoids (e.g., *Parahibolites*, *Tetrabelus*), ammonoids (e.g., *Eotetragonites*, *Cleoniceras*, *Desmoceras*, *Douvilleiceras*) and nautiloid shells (e.g., *Cymatoceras*) given by Zakharov et al. (2011) were interpreted as seasonal signals. Seasonality is probably over-interpreted in that study because the measurements were not performed ontogenetically (insufficient number of samples), only encompassed small parts of the shell (Zakharov et al. 2005) and the spacing between sampling points was probably too wide. These data more likely reflect habitat changes during ontogeny as suggested by Lukeneder et al. (2010), and not seasonal variations as assumed by Zakharov et al. (2011). Nevertheless, the interpretation of paleotemperature of the Albian Madagascan sea (tropical–subtropical zone) appears to be correct, as are habitat inferences from upper mesopelagic to lower epipelagic for the ammonoids (Fig. 18.7, 18.10, Table 18.1). *Eotetragonites* might have lived within a paleotemperature range of 13.3–16.4 °C, *Cleoniceras* of 16.4–19.4 °C, *Desmoceras* of 15.5–21.1 °C, and *Douvilleiceras* of 20.2–21.6 °C.

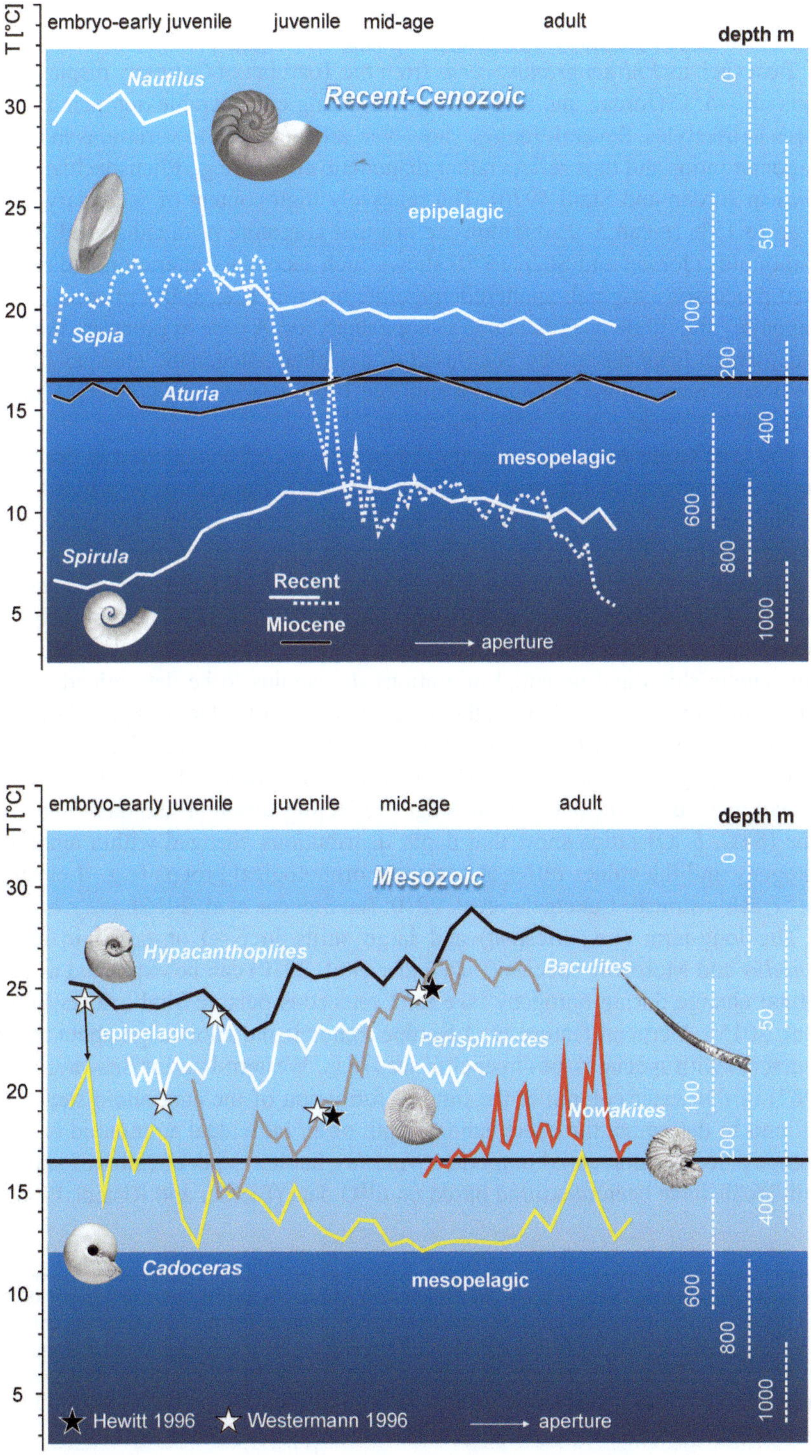

Fig. 18.9 Ontogeny in Recent *Nautilus*, *Sepia*, *Spirula*, and Cenozoic *Aturia* (above). Ontogenetic migrations in the ammonoids *Baculites*, *Cadoceras*, *Hypacanthoplites*, *Nowakites*, and

The only nautilod, *Cymatoceras*, hints at epi– to mesopelgic water layers of 16.4–17.2 °C (only 2 samples in Zakharov et al. 2011).

Taxa such as *Phlyseogrammoceras* from the Toarcian of Germany display variations of 3–4 °C (Jordan and Stahl 1970), indicating considerable ontogentic differences in lifestyles. Several 'ranges', however, are attributed to variations in calcite-aragonite ratios and thus reflect rather diagenetic effects (e.g., Pliensbachian *Amaltheus* in Jordan and Stahl 1970). The relatively high content of secondary calcite of up to 15% (mean 5–9%) within the original aragonite of the analyzed Jurassic ammonoids (Jordan and Stahl 1970) shows such secondary alteration and replacement processes. Original, unaltered aragonite material (i.e., 100% pristine primary aragonite) should be used in stable isotope analyses. A case in point are the baculitids, which have repeatedly been used to decipher ontogenetic changes of these Cretaceous heteromorphs (Tourtelot and Rye 1969; Forester et al. 1977; Whittaker et al. 1987; Fatherree et al. 1998; Fig. 18.9).

The Late Cretaceous heteromorph *Polyptychoceras pseudogaltinum* from Hokkaido was analyzed by Okamoto and Shibata (1997). The authors concluded, based on buoyancy and mode of growth investigations, either a demersal mode of life almost touching the sea bottom or a nekto–planktic mode in the water column. A demersal lifestyle was proposed by Moriya et al. (2003) for several taxa from the Upper Cretaceous of the northwestern Pacific (Fig. 18.11). The relatively short shell-sections utilized for that study, however, did not include a full record of the ontogenetic shifts and potential migrations. It remains to be determined whether data from Moriya et al. (2003) on the wide range of $\delta^{18}O$ values in *Hypophylloceras* (i.e., Hs1 to Hs2) or in *Polyptychoceras* (i.e., Pp1 to Pp2) reflect slight differences in stratigraphy and variations in lithology (i.e., concretions, Moriya et al. 2003) or instead mark different ontogenetic stages reflecting different habitats. Stable isotope ($\delta^{13}C$, $\delta^{18}O$) ratios show that depth distributions changed within ammonoid ontogeny and the values differ in various morphological groups (e.g. 'Leiostraca' vs. 'Trachyostraca'; Lukeneder et al. 2010; see Stevens et al. 2015).

The long-term (i.e., lifespan) and large shifts in $\delta^{18}O$ observed in modern *Nautilus* and Mesozoic species (Lukeneder et al. 2010) can be explained as major habitat change during ontogeny (see for Cretaceous pelagic Simbirskites Stevens et al. 2015). A crucial feature in O isotope trends is the knowledge about thermal structures with special issues in epeiric seas (e.g., Western Interior Seaway; Wright 1987) of the water column at the time of formation of the ammonoid shells. It is difficult to determine the exact habitat depth of an individual ammonoid based on isotope paleothermometry. For modern *Nautilus*, habitat depth of embryos and adult individuals have been measured based on $\delta^{18}O$ data (Eichler and Ristedt 1966a, b;

Perisphinctes (below). Calculated water temperature curves in growth direction, based on $\delta^{18}O$ and depth distribution of the cephalopods investigated compared with additional Recent *Sepia*, *Spirula*, *Nautilus*, Cenozoic *Aturia*, and Mesozoic *Perisphinctes* and *Baculites* (all literature data; see text for explanation). Adapted and extended after Lukeneder et al. (2010). Maximal depth range estimates based on siphuncle strength index (SiSI) from Hewitt (1996, black stars) and Westermann (1996, white stars) in corresponding colours to temperature curves

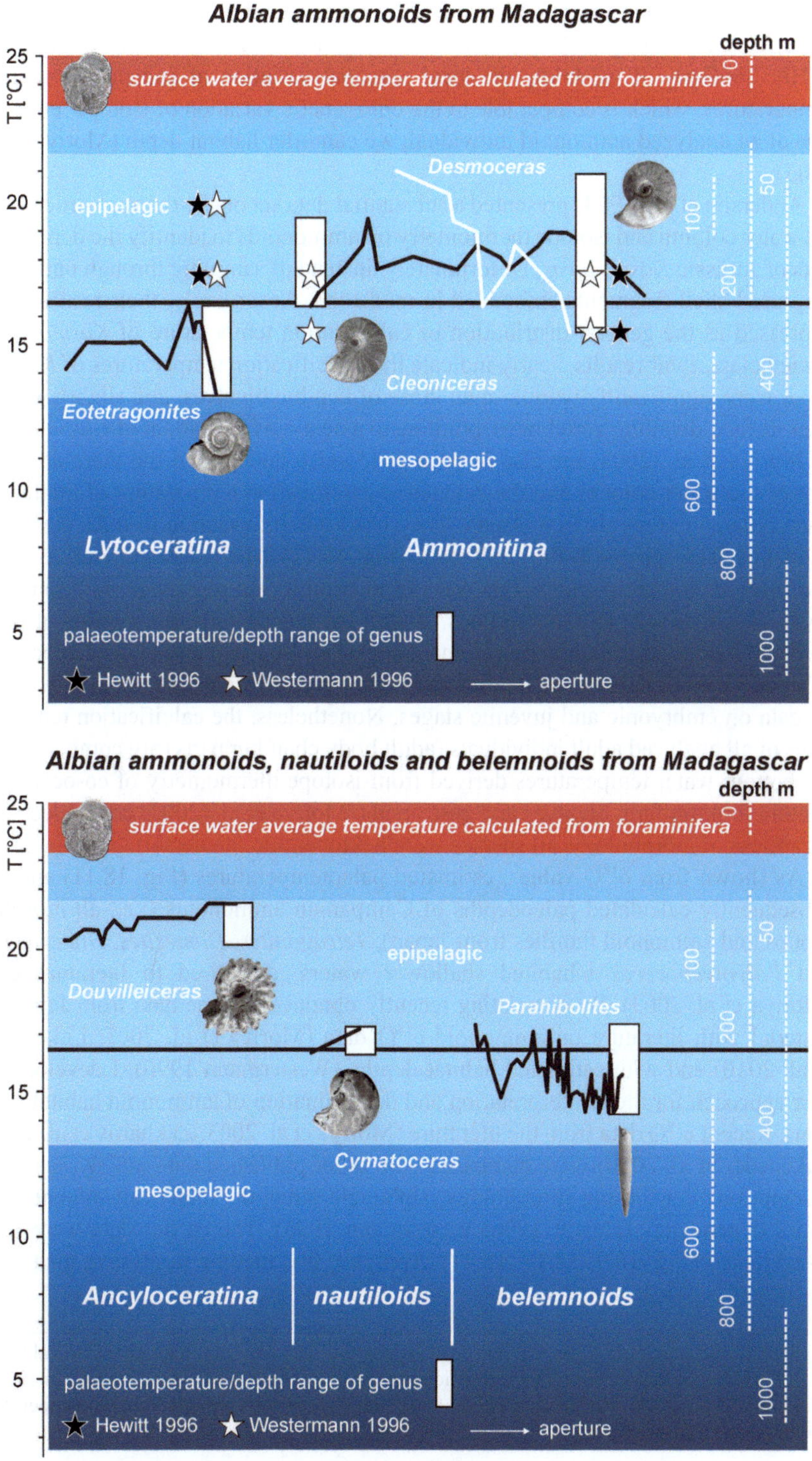

Fig. 18.10 Paleotemperatures estimated based on stable isotope data ($\delta^{18}O$) from the Lower Cretaceous (Albian) of Madagascar. Ammonoid (*Eotetragonites, Cleoniceras, Desmoceras,*

Cochran et al. 1981; Taylor and Ward 1983; Landman et al. 1983, 1994; Landman 1988; Oba et al. 1992; Moriya et al. 2003; Auclair et al. 2004; Lukeneder et al. 2010). Similarly, if we know the local seasonal variation of ancient sea surface temperatures, which is comparable to the ontogenetic variation of isotope temperature of an analyzed ammonoid individual, we can infer habitat depth (Moriya et al. 2003).

Anderson et al. (1994) presented a substantial data set on the thermal structure of the water column and isotope thermometry on ammonoids to identify the depth habitats of Jurassic *Kosmoceras*. Unfortunately, instead of sampling through ontogeny, they used shell fragments embedded in mudstone. Accordingly, their results were expressed as the general distribution of calcification temperature of *Kosmoceras*. Nonetheless, their results clearly indicate that calcification temperatures of *Kosmoceras* were significantly warmer than those of benthic bivalves, and slightly cooler than surface-dwelling vertebrates, pointing to a near-surface habitat of *Kosmoceras*.

Moriya et al. (2003; see also Moriya 2015) also determined the thermal structure of the water column and the calculated calcification temperature of numerous Campanian ammonoids from Japan. They used this information to infer the depth habitat of each species in the adult stage (Fig. 18.11). The paleotemperature range and calcification temperature data were from benthic and planktonic foraminifera, bivalves, gastropods and ammonoids. Within the ammonoids, the authors utilized 15 discrete individuals of 9 species and 4 superfamilies, which include discocones, oxycones, planorbicones and hamitocones. However, Moriya et al. (2003) provided no data on embryonic and juvenile stages. Nonetheless, the calcification temperatures of all analyzed adult individuals (adult body chamber parts) are comparable to the bottom water temperatures derived from isotope thermometry of co-occurring benthic foraminifera, bivalves and gastropods. Moriya et al. (2003) concluded that, regardless of morphology, all species were nektobenthic (demersal).

As shown from $\delta^{18}O$ values, estimated paleotemperatures (Fig. 18.11) and the subsequently calculated paleodepths of Campanian ammonoids (i.e., all morphogroups and ammonoid families from Japan), *Tetragonites*, *Damesites*, *Hauericeras* and *Polyptychoceras* inhabited shallower waters, compared to literature data (Moriya et al. 2003). By correlating recently obtained isotope data from Japanese material with literature on ammonoid $\delta^{18}O$ data (Moriya et al. 2003; Lukeneder et al. 2010) and with estimated habitat depths (Westermann 1996), I developed a new approach for the reinterpretation and determination of ammonoid/habitat relations. Recent $\delta^{18}O$ data from the literature (Moriya et al. 2003; Zakharov et al. 2005; Lukeneder et al. 2010) were compared with other published data mainly based on assumptions concerning morphology, siphuncle strength, shell wall thickness or facies dependency (Hewitt 1996; Westermann 1996). The earlier data were then re-evaluated (Fig. 18.9, 18.10, 18.11). Applying this method to all taxa measured

Douvilleiceras), nautiloid (*Cymatoceras*) and belemnoid data based on data given by Zakharov et al. (2011). Maximal depth range estimations based on Siphuncle Strength Index from Hewitt (1996, black stars) and Westermann (1996, white stars) in corresponding colours to temperature curves (see text for explanation) in Moriya et al. (2003) yields a more detailed range of 70–350 m (Fig. 18.11), exclusively from the lower to middle epipelagic, with a clear dominance in the lower epipelagic areas. The exception are the Phylloceratina (*Hypophylloceras*): they also inhabited the upper mesopelagic, ranging up to the middle epipelagic from 100–300 m. This contrasts with much greater depths given in the literature (down to 600 m; Westermann 1996). Accordingly, the ranges of estimated paleodepths and habitat boundaries of Cretaceous ammonoids from Japan (Moriya et al. 2003) are narrower than assumed based on earlier data (Hewitt 1996; Westermann 1996). Nonetheless, a shallowing gradient is evident from smooth to fine ribbed types (i.e., 'Leiostraca') of Phylloceratina (*Hypophylloceras*, *Phyllopachyceras*) and Lytoceratina (*Tetragonites*, *Gaudryceras*), over stronger ribbed types (i.e., 'Trachoystraca') of the Ammonitina (*Eupachydiscus*, *Yokoyamaoceras*) to heteromorph Ancyloceratina (*Polyptychoceras*). At least *Polyptychoceras* (i.e., Pp1 to Pp2 in Moriya et al. 2003) appears to have migrated in ontogeny from lower epipelagic at 200 m to higher epipelagic zones at 60 m (Fig. 18.11). Japanese *Gaudryceras* is suggested to have inhabited depths of 80–210 m, similar to estimated 100–180 m for *Gaudryceras* from eastern Russia given by Zakharov et al. (2005; only last whorl; Fig. 18.11; see also Zakharov et al. 2006). Note that this would follow the hypothesis given in Lukeneder et al. (2010) suggesting that numerous groups migrate within the water column during ontogeny. This is in contrast with Moriya et al. (2003; also pers. comm. K. Moriya 2013), who concluded that their Late Cretaceous ammonoids from Japan did not migrate during ontogeny, rather being nektobenthic (demersal) for their entire lives. One important distinction between the arguments in Moriya et al. (2003) and proposals of Lukeneder et al. (2010) and additional data (herein) is the duration of suggested migrations. Diurnal migrations are not ruled out by Lukeneder et al. (2010) and Moriya et al. (2003). In fact, they are likely. Moriya et al. (2003) argued for a constant long-term mean depth (e.g., average depth over months to years) that does not change in ontogeny. This argument is not necessarily unequivocal because only few individuals have been sampled to date. Including the data from the scheme outlined by Lukeneder et al. (2010) for ammonoid and extant cephalopod habitats, combined with data from Hewitt (1996) and Westermann (1996), indicates that a migration of at least a few ammonoid species from Japan is possible or even likely in the Late Cretaceous sea (Fig. 18.11). New geochemical results indicate that classic assumptions about the habitat depth of each ammonoid morphotype based on functional morphology should be reassessed or partly revised based on physicochemical evidence.

Possible habitat depths of ammonoids, according to literature data and more recent stable oxygen isotope data ($\delta^{18}O$), range from approx. 20–500 m (Fig. 18.9, 18.10, 18.11). Marcinowski and Wiedmann (1988) and Wiedmann (1988a) already stated that bathymetry and climate (i.e., temperature) are the main motors and controlling factors for ammonoid distribution. Depth data (implosion depth) estimated from siphuncle and shell wall strengths (Tanabe 1979; Hewitt 1993, 1996; Westermann 1971, 1975a, b, 1996) are reliable for ammonoid paleobathymetry. Earlier data based on such estimations were demonstrated as being correct or at

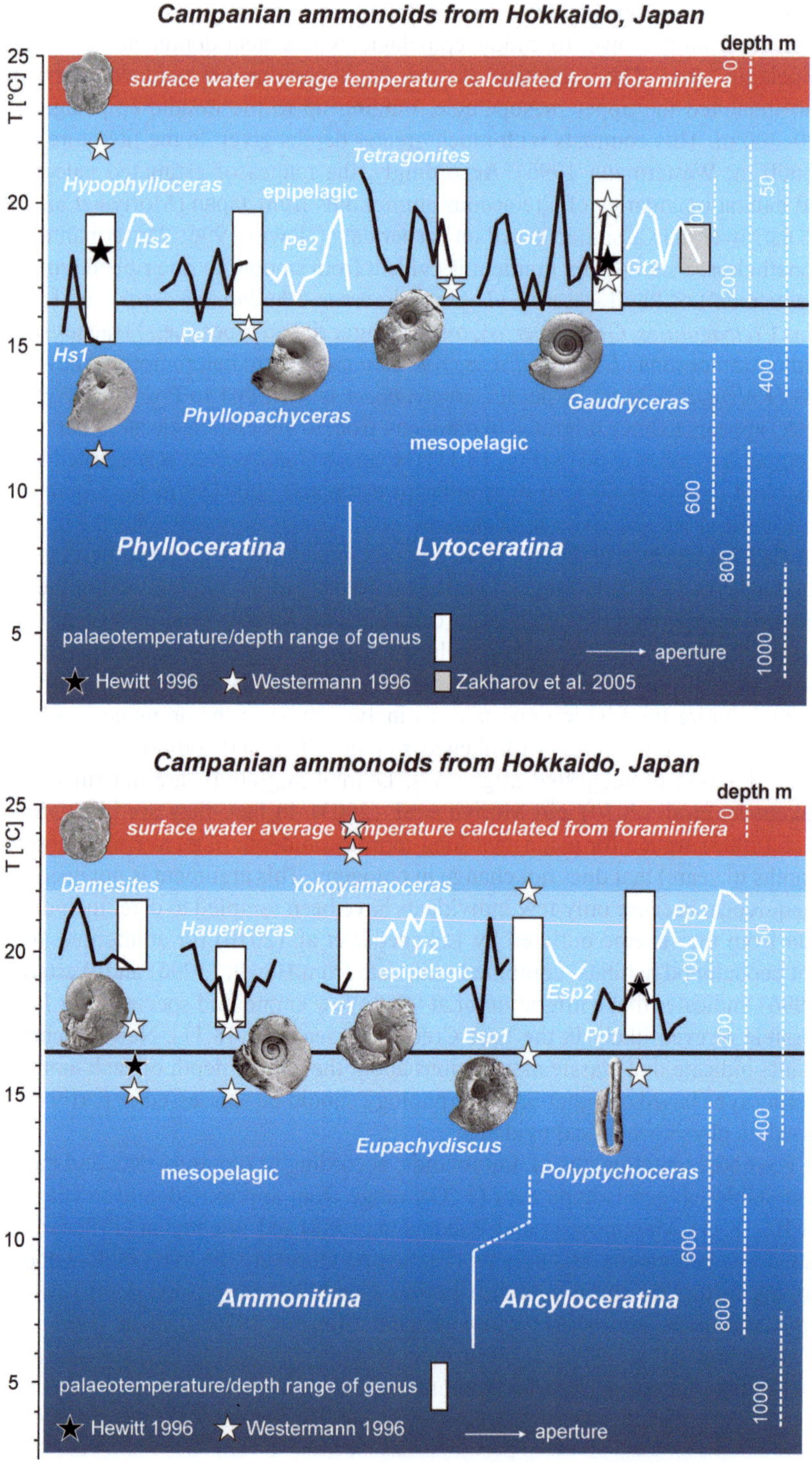

Fig. 18.11 Paleotemperature data estimated using stable isotope data ($\delta^{18}O$) from Late Cretaceous (Campanian) ammonoids from Hokkaido, Japan (Moriya et al. 2003). Light blue shading

least valuable in most cases (i.e., by comparing with $\delta^{18}O$-paleotemperature data). Nonetheless, several values given in the literature seem to be imprecise.

Lukeneder et al. (2010) described a bathymetric range of 50–700 m for *Cadoceras* starting in the lower epipelagic and migrating deeper during ontogeny to the upper mesopelagic realm. With a water-depth range from 10–60 m in *Hypacanthoplites*, ontogeny starts in deeper epipelgic layers, followed by shallower epipelagic stages. *Nowakites* is interpreted to have lived from 50–300 m, implying migrations from deeper mesopelagic layers to shallower epipelagic habitats (Fig. 18.9). The maximal depth range given for *Baculites* (Cochran et al. 2010a, b; Landman and Klofak 2012, Henderson and Price 2012) could be specified to be relatively shallow from 50–100 m in the epipelagic zone. *Persiphinctes* apparently inhabited almost the same middle epipelagic zone (50–100 m), but migrated to deeper areas as adults (comparable to early juveniles).

Stable isotope data from Albian ammonoids of Madagascar (e.g., *Eotetragonites*, *Cleoniceras*, *Desmoceras*, *Douvilleiceras*) were given by Zakharov et al. (2011). The ontogenetic rows (as indicated by shell diameters) enable an interpretation on paleotemperature and habitat: tropical-subtropical, upper mesopelagic to lower epipelagic (Fig. 18.7, 18.10) with a depth range from 70–400 m. Zakharov et al. (2006) concluded that belemnites underwent ontogenetic changes, such as in *Nautilus*, but argued against an ontogenetic migration in ammonoids. Price et al. (2011) described a habitat differentiation within different belemnite groups from the Early Cretaceous based on $\delta^{18}O$ data combined with Mg/Ca ratios (Dutton et al. 2007; see Stevens et al. 2015). Alberti et al. (2011) measured 61 belemnite rostra and reconstructed a paleotemperature model based on the $\delta^{18}O$ data from the Middle to Upper Jurassic of western India. The authors proposed a migration at least during spawning times from cooler areas into warmer strata. The conclusions drawn by Zakharov et al. (2006) are contrasted by their own figured distribution patterns of Cretaceous ammonoids and by their assumptions that large pachydiscids (*Canadoceras*, *Menuites*, *Eupachydiscus*) were able to migrate to colder waters during ontogeny (shown herein and in Lukeneder et al. 2010). The numerous ammonoid data given by Zakharov et al. (2006) point to a range of at least 7 °C, which, in their opinion, is explainable rather by seasonal fluctuations than ontogenetic trends. That range, however, is too wide for seasonal temperature variations at 100–200 m depth (for temp. see Levitus 94); hence, migration during ontogeny is indicated for most Albian ammonoids. Consideration must also be given to the paleogeographic position of Madagascar during the Albian at middle latitudes of 30–40°S (Stampfli and Borel 2002; Zakharov et al. 2006; Lukeneder et al. 2013).

represents the seasonal range of bottom water temperature estimated from oxygen isotope thermometry of co-occurring benthic organisms. Red shading represents annual mean sea surface water temperature estimated from oxygen isotope thermometry of surface-dwelling planktic foraminifera. Maximal depth range estimations based on the Siphuncle Strength Index (SiSI) from Hewitt (1996, black stars) and Westermann (1996, white stars) in corresponding colours to temperature curves (see text for explanation). Black and white lines for distinct specimens. White bars indicate the depth/paletemperature range of distinct genera given in Moriya et al. (2003). Grey bar indicates a *Gaudryceras* measured by Zakharov et al. (2005) for comparison

The Lytoceratina (*Eotetragonites*) appear to be the only inhabitants of the upper mesopelagic layers (200–400 m; Fig. 18.10). Contrastingly, Westermann (1996) suggested a shallower epipelagic range (100–180 m) based on siphuncle index estimates. Albian members of the Ammonitina lived within the lower epipelagic zone. The finely ribbed forms *Cleoniceras* and *Desmoceras* are a case in point: the former ranges from 100–200 m (max. 250 m in Westermann 1996) with an ontogenetic trend from deeper waters, into shallower epipelagic layers and back (*cwc*–type), the latter from 70–250 m (min. 180 m in Westermann 1996). The ontogenetic trend in *Desmoceras* shows a major habitat change from higher epipelagic waters into the upper mesopelagic waters (*wc*–type), i.e. much deeper in adult stages. Strongly ribbed Ancyloceratina like *Douvilleiceras* are relatively shallow-water inhabitants of the middle epipelagic at depths from 70–90 m, migrating into slightly shallower waters during ontogeny (Fig. 18.10). The Albian nautiloid *Camytoceras* inhabited depths around 180–200 m (not ontogenetically measured), equivalent to modern *Nautilus*. Scott (1940) was among the first to present estimations on comparable Albian faunas (e.g., *Desmoceras*, *Douvilleiceras*, *Hypacanthoplites*) from Texas. He suggested that ammonoids dominated in the infraneritic zone (20–100 fathoms in Scott 1940, = 37–183 m). Depth was estimated mostly based on morphological shell features and facies: 36.6–183 m in *Hypacanthoplites*, 146–183 m in *Douvilleiceras*, 183–200 m in *Desmoceras*. Recent estimates from isotope data show much shallower depth values than already given in the historic literature for all members: 10–60 m for *Hypacanthoplites*, 70–90 m for *Douvilleiceras* and 70–250 m for *Desmoceras* (Fig. 18.9, 18.10).

$\delta^{18}O$ data are available from Cenomanian ammonoids of Australia (e.g., *Euompaloceras*, *Acanthoceras*; Henderson and Price 2012). The reconstructed paleotemperatures ranged up to 34 °C in accordance with the subtropical paleolatitude and the thermal maximum of the northern Australian area during the early Late Cretaceous (Henderson and Price 2012). Similar isotope data from benthic gastropods and bivalves as well as the straight-shelled baculitid *Sciponoceras* show calculated paleotemperatures of 21 °C, assuming a nektobenthic mode of life. Heteromorph (helically coiled) ammonoids such as *Hypoturrilites* and *Turrilites* with $\delta^{18}O$ from −2.9‰ to −0.8‰ were interpreted to inhabit bottom- and midwater depths, excluding surface waters (Henderson and Price 2012; Batt 1989). Henderson and Price (2012) argued that the relatively positive $\delta^{13}C$ values (0.0‰ to 2.8‰) of the Australian helical ammonoids are typical of normal marine waters, but may also point to a sluggish mode of life and slow metabolic rate. More negative $\delta^{18}O$ data were derived (Henderson and Price 2012) from the planispiral, strongly ornamented *Acanthoceras* and *Euomphaloceras* (−4.9‰ to −1.6‰). These shell morphologies suggested pelagic and nektobenthic life habits (Stilwell and Henderson 2002). Henderson and Price (2012) interpreted these ammonoids as growing in near-surface waters based on the most negative $\delta^{18}O$ values. This conflicts with the interpretation of a deep marine nektobenthic habit and with data from Campanian planispiral taxa recorded by Moriya et al. (2003), as is also challenged herein (see above discussion of the interpretation of data in Moriya et al. 2003). Based on $\delta^{18}O$ paleotemperature records relative to associated benthos, those authors considered a taxonomically diverse assemblage of planispiral forms to be nektobenthic. Note that the Campanian forms in Moriya et al. (2003) differ from those in Henderson and Price (2012) in morphology as they are, except *Eupachydiscus*, smooth to lightly ornamented and lack ventral tubercules. Based on $\delta^{18}O$ values for ammonoids, the general rule of thumb that 'Trachyostraca' dominated above 200 m in the neritic epipelagic zone whereas 'Leiostraca' dominated in the lowermost epipelgic and deeper mesopelagic zones appears valid.

18.6 Morphological Assessment of Bathymetry in Ammonoids

Ecological conclusions drawn on habitat, bathymetry and life history of ammonoids are mostly speculative. Hewitt (1993) stated that ammonoids and nautiloids are directly adapted to a maximum water depth, in either pelagic or benthic habitats. A maximal depth of approximately 240 m was given for ammonoids (Denton 1974; Hewitt 1993) based on the maximal depth for cameral fluid osmosis.

Information about the paleobiolgy of ammonoid assemblages can be gained in several ways. The first is a synoptic examination of the literature on ammonoid ecology dealing with morphology–habitat relations (Ziegler 1963, 1967; Kauffmann 1977; Donovan 1985; Batt 1989, 1993; Jacobs et al. 1994; Westermann 1996; Neige et al. 1997; Navarro et al. 2005). A second approach is to analyze the relation between the ammonoid fauna and the embedding facies (Chap. 18.7; literature reviewed by Westermann 1996). An additional possibility is to regard conch parameters (shell morphology, siphuncle strength, etc.) as being directly linked to and hence reflecting primary habitat conditions (depth, temperature etc.) as displayed by morphotypes and hypothesized life modes (Hewitt 1993; Westermann 1996; Klug 2002; Mutvei and Dunca 2007; Ritterbush and Bottjer 2012).

The main structural elements increasing the strength of an ammonoid shell against hydrostatic pressure and predators are the shell wall and the septa (form and spacing). Their stremgth is thus a main proxy for the mode of life (Westermann 1971; Denton 1974; Ebel 1983; Saunders and Swan 1984; Saunders and Shapiro 1986; Shapiro and Saunders 1987; Swan and Saunders 1987; Hewitt 1993; Shigeta 1993). According to Westermann (1996), decreasing shell thickness correlates with an increase in body chamber length, a change in aperture orientation (more horizontally) and a reduced hydrostatic stability. Nonetheless, Westermann (1996) stated that the siphuncle strength (SiSI) is a more reliable proxy than shell thickness for estimating depth and bathymetry. The septa are truncated by the siphuncle from the first, embryonic chamber to the last septum at the border to the body chamber (Westermann 1958, 1996; Bayer 1977; Jacobs 1992b; Hewitt and Westermann 1987; Korn 1992).

External shell features such as shell strength (thickness) and whorl cross section are crucial for the hydrostatic stability of ammonoid shells in the water column. The most pressure-resistant shells have circular to semicircular sections, followed by elliptical and ovate ones (Westermann 1958, 1975b, 1996; Ward and Westermann 1985; Jacobs 1992b; Hewitt and Westermann 1987). In contrast, shallow-water cephalopods required thick shells mainly for protection from predators (Westermann 1977; Bond and Saunders 1989). Mathematical and statistical estimations of shell thickness and internal features (see below) that draw conclusions about implosion depth based on the maximal pressure resistance point of an ammonoid shell remain theoretical and provide no direct proof. This calls for additional methods (e.g., stable isotope data, assemblage analyses).

Internal shell features such as septa (thickness, spacing), septal necks (length, diameter, orientation), siphuncles (thickness, diameter) and sutures (complexity) were often used to infer habitat depth (Westermann 1996; Oloriz and Palmqvist 1995; Mutvei and Dunca 2007; Klug and Hoffmann 2015). Strength calculations were based on septal flutes (Hewitt 1996). The septal flute strength index (SFSI, function of maximum septal thickness, minimum whorl radius, and tensile strength; Hewitt and Westermann 1987, 1997; Wang and Westermann 1993; Lewy 2002), combined with statistical methods, help to estimate habitat depth limits (Hewitt and Westermann 1988; 1990a). The relation between septal shape, suture complexity (shape and amplitude; Checa and Garcia–Ruiz 1996), septal spacing and the resulting assumptions about habitat depth (grade of resistance againts seawater pressure) were discussed in several articles (Kahn and Kant 1975; Westermann 1958, 1971, 1975b, 1996; Westermann and Ward 1980; Ward and Westermann 1985; Doghuzaeva 1988; Batt 1989; Hirano et al. 1990; Jacobs 1992b; Boyajian and Lutz 1992; Olóriz and Palmqvist 1995; Saunders 1995; Hewitt and Westermann 1997, 2003; Checa 2003; Hassan et al. 2002; Lewy 2002, 2003).

Olóriz and Palmqvist (1995) questioned whether sutural complexity is a real indicator for habitat depth or whether it is only an artifact in ammonoid ecology. The underlying question is whether sutures are adaptive or non-adaptive elements. They concluded, based on Jurassic ammonoids, that suture complexity is related more to shell coiling and whorl sections than to bathymetry (Olóriz et al. 1999). Ward and Westermann (1985) favoured siphuncle strength as reliable function for paleodepth estimations.

Concomitant changes in habitat and shell structure reflect life cycles and changes in ecology (Westermann 1996). Changes in siphuncle strength (SiSI) and shell/septa thickness were reported to be reliable proxies for habitat depth estmates (Hewitt and Westermann 1988; Hewitt 1996; Westermann 1996). High pumping rates would have been limited to ammonoids with large siphuncular radii that are found only among shallow-water forms with small SiSI, such as the Upper Cretaceous large ammonoid *Placenticeras* (Westermann 1996). Shell growth was accelerated in shallow water, resulting in thinner, more economic, and lighter septa (Westermann 1996). The relative septal thickness, the major variable relating to septal strength (Hewitt and Westermann 1988; Hewitt 1993), was therefore either ecologically controlled or evolved rapidly in the new habitat. Both are good evidence for autochthonous populations. Thus, the exceptionally large difference between the septal and siphuncular strengths is indicative of the rapid habitat change from the usual oceano–mesopelagic to the new epeiro–epipelagic habitat (Westermann 1996; see

Fig. 18.7). Hewitt (1993) suggested hatching in epipelagic zones and dispersal in deeper mesopelagic areas after hatching; this was based on septal strength estimation, assuming that the last septum is the weakest part against hydrostatic pressure in ammonoids. Fossil nautiloids were believed to have hatched at shallower benthic sites (Hewitt 1993). The author noted that assumptions on epibenthonic or epipelagic habitats could not be made by the estimated water pressure based on septal strength. Sutural complication has been used in bathymetry, assuming that it improves shell strength against ambient pressure and, hence, varies with habitat depth (Westermann 1996).

Contrastingly, Daniel et al. (1997) demonstrated that there was no positive effect of septal complexity on habitat depth, and even suggested a shallower position for ammonoids with complex sutures. They concluded that selective pressures from predators and buoyancy control were the determining factors: depth did not cause increasing septa complexity. Assumptions based on computed models, simulated and estimated from hypothetical chambers and septal shapes, and values filled in formulae, do not sufficiently mirror the real life of ammonoids and should only be used in combination with other bathymetric indicators (e.g., foraminifera, facies, ammonoid assemblage, stable isotopes) to avoid overestimations.

Westermann (1996) referred to an additional internal shell component believed to influence the possible maximal habitat depth: the septal neck (Doguzhaeva 1988; Tanabe et al. 1993a; Hewitt 1996; Tanabe and Landman 1996). Estimations on bathymetry were calculated by correlating the length of septal necks and the siphuncular strength index (SiSI; Westermann 1971, 1973, 1975b, 1982; Hewitt and Westermann 1988). The congruent ontogenetic reduction of SiSI and septal flute strength indicates that connecting rings weaken during ammonoid ontogeny (Westermann 1971, 1996). Mutvei and Dunca (2007) rejected the siphuncle strength index as an appropriate tool for paleobathymtric estimations; they noted that the index should only be used as an indicator for relative differences in distinct ammonoids, in combination with connecting ring permeability data.

As noted by Mutvei (1975), Tanabe (1977), Westermann (1971, 1996), Hewitt and Westermann (1988), Geraghty and Westermann (1994), Hewitt (1996) and Ballentine (2007), the siphuncle strength and connecting ring ultrastructure (Mutvei and Dunca 2007) can directly be connected to implosion depth for ammonoids. Doghuzhaeva et al. (2010) noted that conchs with a small siphuncle diameter and long septal necks, as in most phylloceratids and lytoceratids, could resist hydrostatic pressure and thus inhabit deeper waters. The main parameters are the siphuncle diameter (sd) and the siphuncle wall thickness (swt), yielding an estimated siphuncle strength index (SiSI). Westermann (1971) incorporated these parameters in the formula $SiSI = h/r \times 100$ (h siphuncle wall thickness, r siphuncle radius), adapted by Hewitt (1996) to

$$\mathrm{SiSI} = \mathrm{d}' \times 100/\mathrm{r} - (\mathrm{d}'/2)$$

(d′ wall thickness, r inner radius). An adapted formula,

$$\mathrm{SiSI} = \mathrm{swt} \times 100/\mathrm{r} - (\mathrm{swt}/2)$$

is applied to Late Triassic ceratitids from Turkey (swt = siphuncle wall thickness; Lukeneder and Lukeneder 2014). The resulting SiSI values for *Kasimlarceltites* are 10–12.6, which, following Geraghty and Westermann (1994), are equivalent to a maximal depth of 200–252 m. Westermann (1971) also noted that ammonoids with ceratitic or pseudoceratitic sutures like *Kasimlarceltites* probably inhabited even shallower waters. This would tentatively strengthen the habitat suggestions as epipelagic at shallow, upper mid ramp for *Kasimlarceltites* as discussed above (Lukeneder et al. 2012; Lukeneder and Lukeneder 2014; Lukeneder and Mayrhofer 2014; see Fig. 18.7).

18.7 Allochthonous Versus Autochthonous Ammonoid Assemblages

Based on the considerations in Westermann (1996), this chapter discusses the relation of ammonoid associations and biofacies using Paleozoic to Recent case studies. All scientists working on ammonoid assemblages, regardless of the time slice, have to deal with the same problem: recognizing whether ammonoid shells are autochthonous or allochthonous. Postmortem drift could have been controlled by sea surface currents (wind) or by bottom currents (density differences). Paleobiostratinomy deals with the period from the ammonoid's death to its final deposition on the sea floor before burial. The main feature causing postmortem drift in cephalopod shells is the presence of a gas-filled phragmocone.

Postmortem drift of empty cephalopod shells can be observed in Recent *Nautilus* (Kennedy and Cobban 1976; Saunders and Landman 1987; Ward 1987; Reyment 2008). Hamada (1964, 1984) and Toriyama et al. (1964) report similar shell concentrations for modern *Nautilus*. Despite Reyment's (1958) doubts, that empty shells of fossil cephalopods floated like those of *Spirula* (mostly washed on shore) and modern *Nautilus*, a 'surface-drift-theory' for the postmortem drifts of the Cenozoic nautilid *Aturia* shells is favoured (Lukeneder and Harzhauser 2002). The studies of Hamada (1964, 1984) and Toriyama et al. (1964) on the necroplanktic features of recent *Nautilus* shells provide evidence for postmortem driftings of 3000 km and floats of more than 1 year, which are reflected in bioerosion and epifaunal settlement. Similarly, Saunders and Spinosa (1979) recorded a postmortem drift of a *Nautilus* shell over a distance of 1000 km (between Palau and Mindanao) in a maximum of 138 days, an average of approximately 7 km per day. The maximum drifting distance (so far) involved a *Nautilus* specimen that drifted ashore after 11 years (Ishii 1981). Postmortem drift behaviour and mechanisms for Recent cephalopods such as *Nautilus* and *Spirula* were summarized by Reboulet et al. (2005) and Reyment (2008).

Sedimentological and paleoecological studies (Lukeneder and Harzhauser 2002) on a Miocene section from Austria revealed significant shell accumulations of the nautilid *Aturia aturi*. Each of these allochthonous occurrences containing abundant nautiloid shells within a littoral to shallow sublittoral mollusc fauna reflected current-induced, postmortem drifts of the otherwise rare cephalopod. Taphonomic features (including broken shells with broken phragmocones, broken body chambers and fragmented specimens without body chamber) strengthen the idea of prolonged transport. This is comparable to abundant *Nautilus* washed ashore after stormy weather (own observations Susanne Mayrhofer, Andaman and Nocobar Islands, Indian Ocean, August 2006). As noted by Hamada (1964), ca. 80% of the *Nautilus* shells washed ashore had lost their body chamber.

According to Ward (1987), nautiloids are never abundant in Cenozoic strata but are not rare either. However, the Austrian occurrences provide evidence that specimens of *Aturia* can be abundant in special cases and may even form shell concentrations in environments distinctly different from their habitat. The described shell accumulations are restricted to littoral to shallow sublittoral environments along the coast of the Paratethys. Since the rocky shores and shallow embayments cannot be expected to have been a suitable habitat for a nectonic, deep-water dwelling cephalopod, the shell accumulations are linked to current-induced, postmortem drifts of *Aturia* shells (Lukeneder and Harzhauser 2002).

Such open seasurface drift-routes are associated with currents, whereas nearshore drifts are mainly induced by wind. Although the actualistic ecological comparison of *Nautilus* with the fossil *Aturia* is complicated by differences in shell shape and suture lines, which may reflect adaptation to different habits (see Ward 1980, 1987), the general drift-behaviour is considered to be identical. This hypothesis is also strongly supported by the observations of Kobayashi (1954), who related the various Cenozoic, nearshore occurrences of *Aturia* in Japan with the proto-Kuroshiwo current. As shown by Kobayashi (1954) by bottle-float experiments, it is not possible to reconstruct the necroplanktic history of shells. During one of his experiments, one bottle arrived after 10 months, whereas a second bottle reached the same locality after 2 months. This difference is explained by the complicated interplay of oceanic currents and countercurrents, tidal currents and winds. Correspondingly, the unknown interplay of these factors, resulting in the formation of shell accumulations, prevent a detailed reconstruction of paleocurrent-induced drift routes.

Similar to the Austrian shell accumulations (Lukeneder and Harzhauser 2002), most of the Japanese findings were classified by Kobayashi (1954) as embayment type occurrences, representing necroplanktic floats. He also stated that the animals did not inhabit these embayments, but does conclude that the natural habitat was probably not very far from these localities. This might also be valid for the Austrian aturiids. The usually excellent preservation of the shell surfaces (as seen on plaster moulds) indicates a short floating period. Epifauna as described by Hamada (1964, 1984) on *Nautilus* shells and by Seilacher (1960) on ammonoid shells, being characteristic for long floatings, is completely missing on the material from Obermarkersdorf and Unternalb. Similarly, heavy exfoliation or abrasion can be excluded: even delicate growth lines are well preserved. Thus, the shells are interpreted to have been transported over a rather short distance from the adjacent Molasse Basin. According to Teichert and Matsumoto (1987), the endogastric position of the siphuncle in *Aturia* indicates adaptations to deep rather than shallow-water environments, pointing at the deep Miocene Molasse Basin as its habitat.

The shallow embayments along the coast acted as traps for postmortem drifts (Lukeneder and Harzhauser 2002).

The transgression, which affected the coasts in that region during the Miocene (Roetzel et al. 1999) resulted in a rapid sediment covering of the *Aturia*-bearing strata and probably explains the preservation of the littoral to sublittoral taphocoenoses. Moreover, the slightly heterochronous (i.e., up to 1000 years) deposition of the two shell accumulations seems to be linked to this transgression. Hence, the reported mass-occurrences of *Aturia* are always bound to littoral environments, whereas the sublittoral fauna of the distal section Unternalb bears only a single specimen. This specimen consists solely of the early parts of the phragmocone and displays some breakage of the shell surface. As demonstrated by Toriyama et al. (1964), Recent *Nautilus* floats easily if the camerate portion of the shell is in a good state of preservation, even when the body chamber is broken. By contrast, during floating experiments, the *Nautilus* shells sank rather abruptly if the phragmocone was broken. The single fragmented *Aturia* from Unternalb therefore had little drifting capacity. This *Aturia* occurrence coincides with a thin coquina of the archaeogastropod *Diloma* (*Paroxystele*) *amedei*. Kroh and Harzhauser (1999) interpreted this layer to represent an allochthonous occurrence of shells, which were transported into the shallow basin from an adjoining littoral environment during a storm. This nautilid shell was probably damaged during this high-energy event and then transported from the coast to the outer bay. The energetic conditions along the coast of the investigated area of the Bohemian Massif resulted in a fair percentage of breakage of the body chambers, but the good preservation of shell surfaces shows that the fragmentation usually did not affect the phragmocones.

Nautiloid accumulations are well known in the literature, but are interpreted in different ways. *Cymatoceras* accumulations from Cretaceous deposits of Russia (Shimansky 1975) were related to turbidites. *Aturia* accumulations from the Miocene of western USA (Moore 1984) and from the Miocene of Slovakia were considered to be deposited autochthonously (Schlögl et al. 2011), *Aturia* from the Eocene of Antarctica (Zinsmeister 1987) to be deposited in high-energy environments (e.g., shoreline). *Hercoglossa* from the Paleogene of Argentina are interpreted by Casadío and Concheyro (1992) to be shore-drifted specimens. Grunert et al. (2010) reported an *Aturia* accumulation from the Miocene as being triggered by several consecutive biostratinomic mechanisms. The authors suggested a deep-water environment as its habitat, followed by transport to the shoreline and redeposition in the dysoxic-anoxic basin by storm events (> 100 m depth). Schlögl et al. (2011) interpreted the accumulation of *Aturia* shells (juvenile–adult) as being living and hence deposited autochthonously in dysoxic environments in 240–330 m depth (Chap. 18.5). These authors excluded transport and drift based on the presence of related jaws and the shell diameter range (7–38 mm). An exceptional storm accumulation of Early Cretaceous nautiloids was reported by Cichowolski et al. (2011) from Argentina. This occurrence was deposited at deeper environments near mid-ramp, accumulated by storm-induced drifts, trapping on sea floor and probably reflecting gregareous lifestyle. Nonetheless, recent *Nautilus* can be accumulated in deep water areas, as reported by Roux (1990 in Maeda et al. 2003), Roux et al. (1991) and Mapes et al. (2010a, b).

Drifted nautiloids and ammonoids can change the relative proportions of ammonid assemblages (Stevens 1997). Nonetheless, in practice the effect of post-mortem drift of ammonoid shells, becoming nekroplanktic, is thought to be minimal (Chamberlain et al. 1981; Cecca 1992; Reboulet 2001; Westermann 1990, 1996; Chirat 2000). For detailed factors in the distribution of fossil cephalopods and exact floating orientations, see Reyment (1958, 1973, 1980; 2008).

The main and probably the starting point of ammonoid biostratinomy is why, how and when they sank. For details on waterlogging, implosion, buoyancy of empty ammonoid shells, pressure of the cameral gas, surfacing to never surfacing and drifting behaviour of ammonoids, see Maeda and Seilacher (1996) and Wani and Grupta (2015). As noted by Chamberlain et al. (1981) for *Nautilus*, postmortem drift is enhanced when predators or scavengers remove the soft body. This considerably increases buoyancy and the shell drifts easily. In contrast, when shells become negatively buoyant at certain depths (critical at 50 m in *Nautilus*), they sink. The critical depth of surfacing and dispersal patterns were discussed by Maeda and Seilacher (1996). The depth limit, at which ammonoid shells float up to the surface, marks a boundary in the water column at a specific hydrostatic pressure. Below, ammonoid shells would not have surfaced, above, they would have drifted.

Sinking history can also be detected by analyzing the encrusting community and the distribution of the latter on the shell (Reyment 1973, 2008; Seilacher 1982). The path of an ammonoid shell after the animal died is controlled by several environmental factors such as depth at death, the initial buoyancy of the empty conch, the morphological and physiological parameters (e.g., morphology, size, weight) of the conch, and the rate of seawater influx into the phragmocone chambers (Kennedy and Cobban 1976; Chamberlain and Weaver 1978; Chamberlain et al. 1981; Olóriz et al. 1996; Chirat 2000; Reboulet et al. 2003, 2005).

The cephalopod/aptychi ratio provides a useful criterion to evaluate the postmortem transport of ammonoids (Olóriz et al. 1996; Lukeneder and Tanabe 2002; Reboulet et al. 2003; Doguzhaeva et al. 2007; Schlögl et al. 2011). Accumulated ammonoid shells can be affected by resedimentation processes generating fragmentation (*sensu* Fernández–López 1991, 1997; Fernández–López and Meléndez 1994, 1995; 2004; Olóriz et al. 1996; Fernández-López et al. 1999; Landman and Klofak 2012). Resedimentation does not necessarily imply significant lateral transport. The described mechanisms for postmortem drift and the biostratinomic processes in fossil cephalopods show the complexity and the difficulties for a detailed assignment of a distinct ammonoid assemblage to a specific habitat or water depth.

18.8 Ammonoid Associations and Facies

Ammonoids have been used extensively by cephalopod researchers to correlate shell morphologies with environmental conditions (habitats), and distinct morphological groups were assigned to certain sea level changes, reflected by the lithology and facies (Ziegler 1963, 1967; Tanabe 1979; Enay 1980; Donovan 1985; Marchand

and Thierry 1986, Marchand 1992, Hantzpergue 1995, Westermann 1996, Neige et al. 1997; Klug 2002). Proxies for estimations related to shell morphology and the development of morphospace are ornament strength and whorl cross-section (e.g., circular, subtriangular, elliptic; depressed vs compressed etc.).

Abel (1916) was the first to show a correspondence and interaction between the environment and the newly evolving morphologies of cephalopods. That seminal paper on the paleobiology of cephalopods underlined that cephalopod evolution is closely related with changing environmental conditions. Evolution moulds the genetic programming of cephalopods and the potential for adaptation (Jacobs et al. 1994; Young et al. 1998; Yacobucci 1999). Adaptation is one major motor for evolution, a situation recognized by Abel (1916) when he erected the new field paleobiology. Paleobiology shows how important the animal–environment interaction is for promoting evolution. Do ammonoids speciate profusely because internal factors enhance variability and reproductive success? Or do ammonoids respond passively to environmental changes and therefore react after changes of the environment? Problems associated with these major issues in cephalopod research have been highlighted by diverse papers expressing different points of view. The adoption of new habits interacts with long-lasting morphological change and therefore appears as a new evolutionary trend. Evolutionary trends show the main directions and pathways, but are only descriptive mirrors for more important processes that more cephalopod workers should recognize (House and Senior 1981; House 1993b). Spectacular evolutionary radiations mostly took place when the environment changed drastically. New forms evolved due to environmental changes or due to adaptions to the preferred habitat. The adaptive strategy is reflected in the change of morphology in the fossil record and the embedding fauna.

There are still numerous open questions on the ammonoid–facies relation. The confusion in the literature calls for clarifying that ammonoids cannot be dependent on a facies: they can merely show dependency on primary environments. Environments are characterized by several parameters such as depth, light, pressure, oxygen, salinity and others. All these factors together led to the formation of a certain sediment, yielding a special lithology or facies. The appearance–disappearance or presence–absence of ammonoid taxa within different facies was interpreted to be related to specific life habits and habitats (Kennedy and Cobban 1976; Westermann 1996). Kennedy and Cobban (1976) noted the wide range of ammonoid-facies relations from little relation indicating nektic to planktic habits over strong relation reflecting a benthic mode to almost no relation in many taxa. No relation, or independence, is indicated when a certain ammonoid taxon appears in every single facies, contrasted by high relation when a taxon appears in a single facies only. As noted by Kennedy and Cobban (1976), relative abundances can hint at different habitat–ammonoid relations, if postmortem drift can be excluded.

An extensive review of the relation of ammonid assemblages to the corresponding facies was given in Westermann (1996). As stated by Kennedy and Cobban (1976) in their paper on Late Cretaceous ammonoids (N America), no clear patterns of an ammonoid-facies relationship were found (Reyment 1958; Ziegler 1967). Kennedy and Cobban (1976) referred to another problem. It involves the assumption by several authors (e.g., Böse 1928; Sornay 1955; Collignon 1963, 1964) that micromorph ammonoids (tiny or dwarfed forms) are exclusively linked to shaly, offshore deep-water facies in the Cretaceous. As noted by the latter authors and observations below, such pyritized small ammonoids are often the nuclei (inner, juvenile whorls), hence synonymous forms of large specimens from shallower areas. As argued by Keupp (1997), highly morphologically variable species refute the paradigm of Westermann (1996) that explicitly requires a dependency of shell morphology on the mode of life. The Bajocian *Sonninia* apparently shows 64 different morphologies (from oxycone to serpenticone) in a single layer, all living in the same habitat (Keupp 1997). Keupp (1997) also quoted the apertural injuries in *Dactylioceras* inflicted by benthic crustaceans on the sea floor (Liassic, Germany). *Dactylioceras* was assumed by Westermann (1996) to be a planktic form, hence should not have been attacked by benthic organisms. Polymorphism from a single horizon was shown by Wilmsen and Mosavinia (2011) for *Schloenbachia* from the Cenomanian. The morphotypes were distributed within a gradient from shallow-water areas (proximal) to deeper environments (distal). Strongly sculptured morphotypes dominate the shallow environments; deep-water areas were dominated by compressed, weakly ornamented morphotypes (Wilmsen and Mosavinia 2011) causally induced by water velocity and predation pressure.

Less attention was given in the literature to the ontogenetic variation within one species or even within a single specimen. Such information, however, is crucial for understanding the life and habitat changes in ammonoids. New methods and data on stable isotope analyses (Chap. 18.5.) have shifted our knowledge on paleodepths, habitats, and lifestyles summarized in Westermann (1996).

As evident in the literature (Hewitt and Westermann 1987; Batt 1991) there are also different points of view about the complex relation between suture complexity and habitat and its proxies such as depth (i.e., pressure, water density). Commonly used terms such as "*probably planktic*", "*presumably pelagic*" and "*may have lived*" show the insecurity in asigning habitats or lifestyles.

The same applies to the connection at various systematic levels to particular facies. Westermann (1996) still separated ammonoid assemblages and biofacies types from the typical environments such as the ocean floor, the ocean margin, the hypoxic cratonic seas, the epeiric basins and the epeiric reef slopes and carbonate platforms. He summarized Paleozoic and Mesozoic fossil assemblages with the intention of showing a relation between ammonoid association and facies (e.g., litho- and biofacies). He was also the first to figure dioramas and possible scenarios of ammonoid habitats, summarized (for adult stages) related to the corresponding facies (tab. 1, p 684–686 in Westermann 1996). However, most of these scenarios are speculative and need testing with geochemical and stable isotope data.

A crucial fact is that ammonoids lived in the water column above the sediment in which they finally became embedded, although postmortem transport is possible. In most cases, it remains unknown, hence highly speculative, in which depth the ammonoid lived. Siphuncle strength index (SiSI) and suture amplitude index (SAI) could be used in combination with stable isotope shell analyses (Fig. 18.8, 18.9, 18.10, 18.11). Nonetheless, Westermann (1996) correctly assumed that ammonoids were probably capable of inhabiting most parts of the sea.

In ocean floor ammonoid assemblages (inclusive aptychi), Westermann (1996) included associations from the Jurassic to Cretaceous of the North American Basin and the Central Pacific (Jansa et al. 1979; Renz 1972, 1973, 1978, 1979). In fact, these occurrences merely show that ammonoids lived in the water column from the bottom to the surface.

18.8.1 Sculpture and Bathymetry

Morphological parameters (e.g., ornamentation, conch geometry) are meant to reflect the lifestyle of separated morphogroups such as the 'Leiostraca' (i.e., smooth shells in deeper water; Fig. 18.1, Table 18.1) and the 'Trachyostraca' (i.e., strong ribbing in shallow water; Westermann 1996; Lukeneder et al. 2010). Characteristic 'Leiostracans' are *Lytoceras*, *Phylloceras*, *Desmoceras* or *Haploceras*, comparable to recent *Nautilus*. Typical 'Trachyostracans' are *Trachyceras*, *Stephanoceras* or *Deshayesites*. The ratio between the 'Leiostraca' and the 'Trachyostraca' within an assemblage (deep water with up to 90% 'Leiostraca', platform environments with 0–10% 'Leiostraca') is often used to conclude on the inhabited paleoenvironments and water depths. The ratio was given for Asian examples from the Cretaceous (Hokkaido, Japan; Obata and Futakami 1977; Tanabe et al. 1978; Tanabe 1979; Matsumoto et al. 1981; Westermann 1996; Kawabe 2003) and the Triassic (Lukeneder and Lukeneder 2014). North American examples were given for the Cretaceous in the Western Interior (USA) by Scott (1940) and Batt (1989). Examples from Europe characterized by that ratio were mentioned by Kennedy (1971) for the Late Cretaceous of England, and by Lukeneder and Reháková (2004) for the Early Cretaceous of Austria. Assumptions on ammonoid assemblages were mostly drawn from the percentages of their constituents (Marchand 1984, 1992; Company 1987; Reboulet et al. 1992; Cecca et al. 1993; Reboulet 1996; Reboulet and Atrops 1997; Cecca 1998) and from absolute abundance variations (Reboulet et al. 2000, 2003).

Especially deep water areas (e.g., deep structural rises, basins and the distal shelves to upper slopes) from the Triassic to the Cretaceous were characterized by the presence/absence rate of smooth 'Leiostraca' (Oloriz 1976; Geczy 1982, 1984; Sandoval 1983; Clari et al. 1984; Bruna and Martire 1985; Galacz and Horvath 1985; Fourcade et al. 1991; Oloriz et al. 1991; Cecca 1992; Westermann 1996). Accordingly, the proportion of 'Leiostraca' among the ammonoids increases with depth (Westermann 1996).

Cecca (1992) assumed, based on Tithonian ammonoid associations, the 'Leiostraca'/'Trachyostraca' ratio and the suture complexity of the Mediterranean, that most ammonoids were demersal swimmers (neritic nektobenthos, epipelagic; e.g., *Aspidoceras*, *Simoceras*, *Pseudolissoceras*) at depths from about 100–200 m. Additionally, pelagic benthic habitats (i.e., mesopelagic; e.g., *Haploceras*, *Ptychophylloceras*, *Protetragonites*) or intermediate habitats (Cecca 1992) were considered for other ammonoids, showing the wide range of presumed habitats. Late Cretaceous ammonoid assemblages from Hokkaido (Japan) show a gradient

('Leiostraca' vs. 'Trachyostraca') correlating to a neritic (above 200 m) to oceanic (below 200 m) gradient (Tanabe 1977, 1979; Westermann 1996). The baculitids were suggested to be planktic, hamiticones as epipelagic, scapihitids as epipelagic (at least juveniles), desmoceratids mostly as mesopelagic, phylloceratids as nektic and lytoceratids as planktic (Tanabe 1977, 1979; Fig. 18.7). 'Leiostraca' (desmoceratids only 1% in shallow areas) vs. 'Trachyostraca' (hoplitids up to 99% in shallow areas) show the same ratio in the Albian from Poland (Marcinowski and Wiedmann 1985). Shallow-water hoplitids were suggested to be demersal and desmoceratids to be pelagic with planktic heteromorphs. Westermann (1996) doubted the benthic lifestyle suggested for these heteromorphs by Marcinowski and Wiedmann (1985). According to Westermann (1996), the planktic lifestyle was also supported by the synonymy with planktic morphogroups (e.g., morphogroup 16; see Fig. 18.2) given by Batt (1989). The literature shows that the morphotype groups given in Batt (1989) and Westermann (1996) were often cited without being questioned. This led to numerous interpretations without testing. Analyzing the data and figures given in Batt (1989) reveals that a real morphotype–facies relation is rare (e.g., *Calycoceras*, *Sciponoceras*) or absent in most genera (e.g., *Watinoceras*, *Collignoniceras*). This is why Batt (1989) designated the habitats as "*possible*". Nonetheless, stronger ribbing appears to dominate in shallower areas. A change in environmental conditions (e.g., oxygenation), as given by Batt (1993) and thus an ecologically dependent change in ammonoid assemblage composition, is more likely. This hampers different groups from inhabiting distinct zones (e.g., near sea floor; Monnet 2009, Chap. 18.8.3). Other forms might be able to live above such hostile conditions due to a different mode of life.

Numerous authors (Ziegler 1967; Kennedy and Cobban 1976; Westermann 1990, 1996; Cecca 1992; Fernandez-Lopez and Melendez 1996) have drawn attention to the ecological differences between these two major groups of ammonoids: the streamlined, smooth-shelled 'Leiostraca' and the strongly ornamented 'Trachyostraca' (Fig. 18.1). The quantitative composition of the faunal assemblages is important both for paleoecologic and paleobiogeographic studies. The fluctuation in relative abundance and diversity of the various morphotypes reflects environmental changes related to sea-level changes (Reboulet and Atrops 1995; Reboulet 1996, Reboulet et al. 2005).

The 'Leiostraca' have been declared to indicate pelagic to deep-shelf habitats (ca. 250–300 m depth; Westermann 1990, 1996; Cecca et al. 1990; Mouterde and Elmi 1991). In contrast, the 'Trachyostraca' are generally considered to have inhabited neritic shallow-water environments (ca. 30–100 m depth; Westermann 1990; 1996). Kennedy and Cobban (1976) described local abundances of 'Leiostraca' in shallow-water deposits, uncover the weakly ornamented 'Leiostraca' as not exclusively deep-water inhabitants. Proportions of 'Leiostraca' vs. 'Trachyostraca' were often used as indicators for habitat depth (Ziegler 1967). Early Cretaceous ammonoid assemblages of Austria (Lukeneder and Reháková 2004) are represented by means of spectra that illustrate the proportions of 'Leiostraca' and 'Trachyostraca'. Ziegler (1967) proposed that the dominance of 'Leiostraca' at a specific stratigraphic level was indicative of deep-water conditions (ca. 300–400 m depth). However, as cautioned by Kennedy and Cobban (1976), 'off–shore' rather than 'deep-water' *per se* may be the more appropriate appellation. Westermann (1990, 1996) and Cecca (1992) regarded 'Leiostraca' as pelagic, capable of occupying deep-water habitats, in contrast to the neritic shallow-water 'Trachyostraca'. Fernandez-Lopez and Melendez (1996) have correlated the relative abundance of 'Leiostraca' with changes accompanying eustatic rises in sea-level (see also Gygi 1986).

These percentages may be interpreted as indicating either the presence of offshore (pelagic) influences or local postmortem drift of the otherwise pelagic ammonoids (Tanabe 1979; Cecca 1992, Batt 1989, 1993; Lukeneder and Reháková 2004). Stratigraphic units with percentages of 'Leiostraca' ranging from 5 to 10% reflect environments from the shelf or upper slope, as indicated by the ammonoid genera analysis from the Vocontian Trough (SE-France; Bulot 1993; Reboulet 1996). Similar fluctuations in the proportions of 'Leiostraca' and 'Trachyostraca', presumably reflecting a comparable interplay of environmental factors, have been documented by Sarti (1986a, b) from Northern Italy, by Christ (1960), Wendt (1963) and D'Arpa and Meléndez (2004) from Western Sicily and Hungary, by Cecca et al. (1990), Cecca (1988, 1992), Fernandez-Lopez and Melendez (1996) from the Western Tethys, and by Stevens (1997) from New Zealand.

As stated in Buckman's Law of Covariation, the ammonoid shell shape and ornamentation are typically correlated. Compressed, involute forms display a fine ornamentation, while more depressed, evolute morphotypes show stronger ornamentation (Yacobucci 2004). Such covariation implies a link between the morphogenesis of shell shape and ornamentation rather than an ecological control. In contrast, there is evidence that ornament growth is controlled by genetics and that shell shape is merely influenced by environmental factors. This is known as Buckman's Paradox. Yacobucci (2004) suggested that shell shape and rib growth are controlled by different processes. Ribbing appears to be more constrained than shell shape, consistent with the view that ornamentation is more tightly controlled by the genetic growth programme in ammonoids (Yacobucci 2004).

More recently, Doghuzhaeva et al. (2010) suggested a middle pelagic habitat for lytoceratids (i.e., *Eogaudryceras*) from the Aptian of Russia, also suggesting that they were capable of diurnal vertical migration and hovering (see also Mutvei and Dunca 2007). These suggestions were based on narrow siphuncles with long septal necks and shell analyses (e.g., wrinkle layer, myostracum) combined with external features such as numerous thick collars stabilising the conch. Prominent collars feign a strong ribbing of an otherwise almost smooth to fine-ribbed form. Similarities in lifestyle were proposed in *Eogaudryceras* with the deep-water *Spirula* (Doghuzhaeva et al. 2010, Chap. 18.5.2).

18.8.2 Sexual Dimorphism and Sea-Level

Sexual dimorphism occurs with sex ratios up to 100:1 for either macroconchs or microconchs (Callomon 1980; Howarth 1992; Davis et al. 1996; Westermann 1996; Davis et al. 1996; Klug et al. 2015b). The sex ratios may diverge considerably even between closely related species in the same assemblage (e.g., *Stephanoceras*; Westermann 1996). In such cases, the dimorphs might have lived in segregated swarms. In normal occurrences of adult ammonoids (both dimorphs), dispersed or abundant, the animals died of old age (Westermann 1996). If sexual dimorphism can be observed in a single species from adjoining areas, one can speculate that adult females left hypoxic, cold basinal waters to spawn on the warm, solid, and oxygenated bottom of the adjoining platform, and died afterwards. A considerable amount of ammonoid taxa was, at least sporadically, separated according to sex before mating and/or during spawning. Callomon (1980) also suggested that female ammonoids return to their primary habitats. A sexual and age (juvenile and adult) separation seems to have existed in numerous ammonoid taxa (Kennedy and Cobban 1976). These authors suggested that earlier ontogenetic stages inhabited offshore waters, contrasted by shallower areas in adults, hence exhibiting an ontogenetic change of habitats. This is somewhat supported by stable isotope data (Lukeneder et al. 2010; Fig. 18.9).

Two Early Cretaceous ammonoid taxa from Austria (Northern Calcareous Alps) were examined with respect to the evolution of shape as well as morphology and environment (Lukeneder and Harzhauser 2003; Lukeneder 2004). Both ammonoids outnumber other ammonoid taxa and were deposited in mass-occurrences: the planispiral *Olcostephanus* (98%; Lukeneder 2004) and the criocone *Karsteniceras* (91% Lukeneder 2003; see Chap. 18.8.4).

The antidimorphs of the Valanginian *Olcostephanus* lived in two different environments and adapted morphologiclly to these somewhat different environments (shallow to deep gradient; Lukeneder 2004). Lithological differences around the *Olcostephanus* Level are due to an altered paleooceanography and reflect sea-level fluctuations (transgressive facies) during the Early Cretaceous (Hoedemaeker 1990; Lukeneder 2004); Lukeneder and Reháková 2004. Faunal turnovers, mass-occurrences, and migrations have often been considered to be controlled by transgressive/regressive cycles in various ammonoid groups (Rawson 1981, Hoedemaeker 1990; Lukeneder 2004). Studies conducted in SE France provide evidence of ammonoid distribution linked to facies and point to faunal assemblage variations between basin and outer shelf (Bulot 1993; Reboulet 1996; Reboulet and Atrops 1997). In *Olcostephanus*, several species are restricted to the outer shelf facies while others occur in basinal facies (Bulot 1993). Data from the western Tethys confirm this facies-link of *Olcostephanus* (N Caucasus in Kvantaliani and Sakharov 1986; Spain in Company 1987; Switzerland in Bulot 1989, 1992) and underline a general trend for the entire Tethyan Realm (Bulot and Company 1990).

Variations in the ratios between 'Leiostraca' and 'Trachyostraca' (Fig. 18.1) in Austria were investigated by Lukeneder and Reháková (2004). The *Olcostephanus* Level (upper bathyal or deep sublittoral) shows a value from 5–10% of 'Leiostraca' (lytoceratids, phylloceratids). The changes in ammonoid faunal spectra from the Steinmühl Formation (Lukeneder and Reháková 2004) to the Schrambach Formation reflect a complex of changes. This includes an altered pelagic influence, the sedimentological change related to the prograding development of a fan system, eustatic changes in sea level, and changes in bioproductivity (Lukeneder and Harzhauser 2003; Lukeneder and Reháková 2004).

18.8.3 Dysaerobic to Anaerobic Environments

As noted by Westermann (1996), a crucial component in correlating ammonoid occurrences is the oxygen content of the water column. Ammonoids are widely distributed in well-oxygenated water, rare in dysoxic layers and absent in anoxic environments. Such facies are characterized by the trace fossil content, sulphur values (e.g., pyrite formation) and total organic carbon contents. Hypoxic (oxygen-poor or dysaerobic, $O_2 = 0.3–1.0$ ml/l; anaerobic $O_2 < 0.3$ ml/l) sea water conditions are important in benthic habitats (mostly bivalves) and for pelagic organisms (e.g., ammonoids) throughout the Phanerozoic (Westermann 1996). The geochemical and biotic conditions are important in the seawater, at the seawater–sediment interface, and in the sediment of such areas (Boucot 1981; Morris 1979, 1980; Wilde and Berry 1984; Wignall 1987, 1990; Jenkyns 1988; Brett et al. 1991; Oschmann 1991; Savrda and Bottjer 1991; Tyson and Pearson 1991; Wignall and Hallam 1991, 1993; Kauffman et al. 1992; Monnet 2009).

Ammonoids, as indicators for environmental conditions, reacted on changes in water geochemistry. Anoxic, and in parts dysoxic, conditions, induced the deposition of black shales. These are commonly characterized by the absence of ammonoids, depending on where the oxygen-depleted layers were located in the water column (e.g., near-bottom or in the water column). If bottom waters were anoxic, demersal and benthic forms were absent; only nektopelagic or planktic ammonoids appear in the sediment or laminated facies. The vanishing of demersal forms during anoxic to dysoxic conditions was shown by Batt (1993; see also Monnet 2009). The high metabolism of ammonoids, especially of larger forms, excluded most of them from dysoxic–anoxic conditions. Exceptions were reported repeatedly (Rieber 1977; Donovan 1993; Vašíček and Wiedmann 1994; Cecca 1997, 1998; Lukeneder 2003, 2005, 2007; Lukeneder and Smrečková 2006) for small heteromorphs. They were probably adapted to such conditions, at least for short times for preying or during migration phases. Such small demersal ammonoids were perhaps able to survive briefly in dysoxic waters as reported for the Cenozoic nautilid *Aturia* (Schlögl et al. 2011) and the Recent *Nautilus* (Wells et al. 1992). A similar strategy was suggested for several species of *Baculites* (Westermann 1977, 1996): speculatively, they normally inhabited aerobic waters but could dive shortly into dysoxic zones to feed near the sea floor. Well-oxygenated waters probably enhanced the growth of eggs. Accordingly, full oxygenation of the primary habitat is required for early ontogenetic development (Westermann 1996; Laptikhovsky et al. 2013). Adult female ammonoids are thought to have spawned in well-oxygenated habitats.

During the Devonian nekton revolution (Klug et al. 2010), ammonoids started to explore various ecological strategies (demersal, planktic, nektic) and to occupy various habitats in the water column. This also depended on the oxygen content and is reflected in black shale deposition, affecting the presence of distinct taxa. Boston and Mapes (1991) postulated for Carboniferous anoxic–dysoxic occurrences that very young ammonoids were benthic inhabitants, unable to escape the periodic benthic anoxia, but Mapes and Nützel (2009) revised this view profoundly. Adult ammonoids lived in the water column or were capable of surviving brief hypoxic events (Boston and Mapes 1991), like extant *Nautilus* (Wells et al. 1992).

Late Cretaceous black shales from Hokkaido (offshore shelves and deep basins) were deposited in the deep waters of the continental margin (Hirano 1986, 1993; Westermann 1996). They yield abundant desmoceratids, with either planktic *Desmoceras* (inflated) or nektic *Pachydesmoceras* (compressed). Abundant *Desmoceras* were overcome by deepwater benthic anoxia, indicating a demersal habitat (Westermann, 1990, 1996). Heteromorphs like *Turrilites* were suggested as being planktic vertical migrants (Westermann 1996). Hirano (1993) reported that ammonoid populations were affected by oceanic anoxic conditions prevailing during the Cenomanian-Turonian boundary in Japan. *Desmoceras japonicum* died out, enabling the spread of *Tragodesmoceroides subcostatus*. The latter species arose from the former by migrating into shallow, oxygenated waters. Thus, the possible escape strategies of Cretaceous ammonoids (Hirano 1988, 1993) were a) to survive in other niches of the outer shelf to upper slope environments, b) to undergo phyletic evolution and c) to adapt to changing oxygen conditions. More recently, Kawabe (2003) showed, for Albian-Cenomanian assemblages from Japan, the correlation from depressed (low-energy, offshore) and compressed shell morphologies (high-energy regimes) in desmoceratids (*Desmoceras*) and gaudryceratids (*Zelandites*, nektic, nearshore) from the same localities (inner shelf, compressed vs. slope, depressed forms). Kawabe (2003) also noted that the shell ornamentation does not depend on lithofacies within these faunas, but that whorl section varies with lithofacies, reflecting environmental conditions (e.g., high or low energy, waves, currents; see Jacobs et al. 1994).

Unstable bottom-water conditions prevailed in some more or less hypoxic epeiric basins (ca. 50–100 m depth), marked by the deposition of black-shale facies (Little et al. 1991; Oschmann, 1991; Wignall and Hallam 1991; Westermann 1996). Westermann (1996) noted a major second component in correlating ammonoid occurrences and oxygenation of the water column, namely the salinity. Nektobenthic ammonoids could not survive when oxygen levels became low or salinity changed drastically. The range of tolerable salinity in modern cephalopods is 16–41‰ (Forsythe and Van Heukelem 1987; Cochran et al. 2003).

The Givetian in the USA (New York, Pennsylvania) shows that the total number of species increased when the oxygenation of bottom waters improved (Kammer et al. 1986; Brett et al. 1990, 1991). Within such facies, the presence of *in situ* aptychi can be important evidence for the autochthonous deposition of ammonoids (Frye and Feldman 1991). Cephalopod abundance peaks within a dysaerobic biofacies (House and Price 1985). The presence of all size classes and exceptional preservation suggested to House and Price (1985) a demersal swimming habit, with the animals preying on epibenthos (Westermann 1996). Additionally, bactritids were planktic (vertical migrants), living in midwaters (Westermann 1996).

Triassic dysoxic conditions prevailed in Carnian times (e.g., Reingrabner facies). A remarkable site for that is the Polzberg locality described by Glaessner (1931) and Krystyn (1991) in Austria. Krystyn (1991) suggested a relatively shallow intraplatform basin with dysoxic or even occasional anoxic conditions, and termed it a Konservat Lagerstätte based on the enormous amounts of well-preserved fishes, conodont clusters, fecal pellets, coleoids and mass occurrences of ammonoids (*Austrotrachyceras*) with *in situ* buccal masses (Doghuzaeva et al. 2007). Benthic life is represented by mass occurrences of the bivalve *Halobia* (comparable to *Posidonia* in the Toarcian *Posidonia* Shale) and rare crustaceans. Nektic life is dominated by the almost monospecific ammonoid fauna with *Austrotrachyceras* ($n > 1000$, all size classes), accompanied by bony fishes and sharks. The ammonoids and members of other nektic groups were able to live above the oxygen-deficient layers present throughout the sedimentary succession of this locality. *Austrotrachyceras* was preyed upon by fishes and ichthyosaurs, as is evident from coprolites with numerous incorporated specimens (own observations).

The Toarcian *Posidonia* Shales are characteristic for NW Europe and other parts of the world (Jenkyns 1988). These shales consist of laminated black shales, marly shales and bioturbated (hypoxic) carbonates (Urlichs et al. 1979). Abundant oxyconic harpoceratids (max. habitat depth c. 50 m, Hewitt 1996; Westermann 1996), platyconic hildoceratids (>60 m; Hewitt 1996; Westermann 1996), typical serpenticonic dactylioceratids (c. 70 m; Hewitt 1996; Westermann 1996), dominate over *Phylloceras* and *Lytoceras* (i.e., 'Leiostraca'). Squid predation (in the upper water column) is indicated by clusters of harpoceratinid fragments (Mehl 1978b). Abundance peaks of *Dactylioceras* (c. 70 m, Hewitt 1996; Westermann 1996) might be traces of mass mortalities by anoxic admixing during storms (Westermann 1996). Harpoceratids and hildoceratids frequently containing *in situ* aptychi hint at hypoxic habitats or rapid burial under such conditions (Westerman 1996). Normally coiled dactylioceratids such as *Peronioceras* were rare in this highly bituminous facies, suggesting that they might have been more bottom-dependent than *Dactylioceras* (Schmidt–Effing 1972; Loh et al. 1986; Westermann 1996). *Posidonia* Shales of Germany were deposited in shallow seas characterized by varying salinities (low salinity due to freshwater input) and periodic anoxic events (Brumsack 1991; Röhl et al. 2001; Schmid–Röhl and Röhl 2003). Few groups tolerated short anoxic or dysoxic pulses. Among the adapted organisms were fishes and rare ammonoid taxa: they survived such environmental changes (i.e., salinity and oxygen; Westermann 1996). Facies distributions and faunal ammonoid assemblages confirm a planktic lifestyle of *Dactylioceras* (Tintant et al. 1982). A euryhaline habit was suggested (Westermann 1996) for *Dactylioceras* (e.g., surface-water habitat and dispersal across shallow shelves; Elmi and Almeras 1984). More probably, these dactylioceratids were midwater drifters as indicated by their shell and siphuncle strength, which is average for Ammonitina (Westermann 1996). The global distribution of *Dactylioceras* species in all facies (Schmidt–Effing 1972) supports a planktic habit in epeiric and neritic seas as well as in the oceans. This is in contrast to Keupp's (1997, 2000) interpretation based on the huge number of apertural injuries in Dactylioceras (e.g., caused by crustaceans on the sea floor), assuming a demersal life. Harpoceratinae and Hildoceratinae were suggested as being nektic (Westermann 1996).

A black shale sequence was reported from the Middle Jurassic (Bathonian–Callovian) Los Molles Formation in Argentina (e.g., Neuquén Basin; Riccardi and

Westermann 1991; Riccardi et al. 1992). The invertebrate fauna consists mainly of abundant adult ammonoids with very low diversity but high intraspecific variation (Westermann 1996). Adult sphaeroceratids (*Eurycephalites*, *Stenocephalites*) and reineckeiids (*Neuqueniceras*) were interpreted as adults that died normally, whereas the juveniles were killed by single events (e.g., anoxia) in masses. Young stages probably lived in deeper waters that were exposed to periodic upwelling and, after death, were drifted and accumulated by bottom-water currents (Westermann 1996). In the basin center (max. depth), abundant small *Ptychophylloceras* (oceanic to mesopelagic; Westermann 1996) occur. O isotope data in similar forms of phylloceratins (*Hypophylloceras*, *Phyllopachyceras*) from the Campanian of Japan yielded depth ranges for that ammonoid group of 100–300 m (Chap. 18.5.6, Fig. 18.11). This is the upper mesopelagic to middle epipelagic zone. From estimates of habitat paleodepth in the Neuquén Basin, *Ptychophylloceras* inhabited shallower areas (Westermann 1996; Hewitt 1993, 1996).

The Callovian to Tithonian black shales (Lower Oxford Clay and the Kimmeridge Clay of England; Cope 1967, 1974; Callomon 1985; Duff 1975; Hudson and Martill 1991; Morris 1979; Oschmann 1991; Wignall 1990; Wignall and Hallam 1991) yield monospecific ammonoid assemblages with *Kosmoceras*. These are usually adult and strongly dimorphic with a high intraspecific variation. Depth estimates for the Kimmeridgean area are 30–50 m (Westermann 1996). Assuming normal salinity, the sluggish ammonoids lived at 17–20 °C (max. 16–23 °C), inhabiting deeper, slightly oxygen-deficient (upper dysaerobic) waters, away from predators (O'Dor and Wells 1990). The more serpenticonic and less sculptured kosmoceratids are found in the deepest facies (Marchand et al. 1985), which does not imply exlusively demersal habitats (Westermann 1996).

The Kimmeridgian to Tithonian of the Antarctic Peninsula, the anaerobic Nordenskjold Formation, was deposited in a deep basin (Doyle and Whitham 1991). The small (10–110 mm) ammonoids there belong to coarsely sculptured, serpenticonic perisphinctids. 70% of all ammonoids are infested by bivalves *syn vivo*. Mass mortality events occured, caused by storm-induced admixing of anoxia (Westermann 1996). Higher in the section, small, smooth, discoconic haploceratids lack epizoans, suggesting that the perisphinctids were planktic drifters in the upper dysaerobic zone. The 'leiostracan' *Haploceras* questionably swam in the aerobic zone, as assumed previously based on its cosmopolitan distribution (Westermann 1996). It was immune to drag-increasing epizoans (smooth surface; perhaps a special periostracum). In deep turbiditic sedimentation areas, the ammonoid fauna consists of dominantly finely costate *Virgatosphinctes* and *Lithacoceras*. Accessories were costate, platyconic berriasellids, planorbiconic olcostephanids, as well as haploceratids. Again, only perisphinctids carry epizoans. These perisphinctids are mesodomic, compressed planorbicones to discocones that were presumably moderately mobile and perhaps bottom-feeders.

The Albian Mowry Shale of USA (Western Interior) includes hoplitid assemblages (*Neogastroplites*; Reeside and Cobban 1960; Batt 1989). It appears that the immature, sluggish hoplitids lived in midwater of the shallow, epeiric sea above dysaerobic bottom waters (Landman and Waage 1993). More recently, Monnet (2009;

inferred by Batt 1989) showed that, in the Late Cenomanian of the Western Interior Seaway (USA), nektobenthic forms such as *Acanthoceras* and *Mammites* (strongly ornamented) were the first to vanish during anoxic/hypoxic bottom conditions (Oceanic Anoxic Event 2). Subsequently, ancyloconic heteromorphs (*Hamites*, *Allocrioceras*) of the middle pelagic zone were eliminated by an upward migration of the oxygen minimum zone (OMZ). Nonetheless, Monnet (2009) showed that this is not a worldwide phenomenon and thus not the rule or main mechanism for ammonoid extinctions. This was illustrated in examples from Tunisia contradicting the 'rising-anoxia' model. In fact, Monnet (2009) reported an increase in ammonoid taxa during the formation of black shales in Tunisia during the late Cenomanian.

In contrast, an oxygen-dependent lifestyle was indicated by Landman et al. (2012) for scaphitids, e.g., from the Campanian Western Interior (Pierre Shale, USA). *Hoploscaphites* and benthic fauna did not occur during phases of dysoxic bottom waters: only *Placenticeras* and *Baculites* were present, both inhabiting higher zones in the water (see also Bearpaw Shale in Canada, Tsujita and Westermann 1998). During phases of good oxygenation, *Hoploscaphites* (juveniles and adults) recolonized the basinal areas, co-occurring with other ammonoids, except during spawning events, subsequently followed by mass mortality of females (Landman et al. 2003). A deeper habitat for *Hoploscaphites* was also supported by stable isotope data, in contrast to the shallower habitat of *Baculites* (Cochran et al. 2010a, b; Kruta et al. 2014; Fig. 18.9). Henderson and Price (2012) challenged that shallow habitat interpretation after investigating baculitids (*Sciponoceras*) from the Late Cretaceous of Australia. The authors argued for a nektobenthic mode of life in this baculitid, based on relatively positive $\delta^{18}O$ data (mean calculated paleotemperature 21 °C; Henderson and Price 2012). These data contrasted with those for bivalves and gastropods. The authors proposed an adaptation among baculitids as specialists in demersal predation (see also Kruta et al. 2011).

Santonian to Campanian bituminous, laminated shales of the Bearpaw Formation and coeval Pierre Formation (both Western Interior, USA) contain abundant ammonoids. The main faunal components are *Baculites* and large *Placenticeras*. Lithofacies characterized by diverse benthos indicates better-oxygenated bottom water, yielding scaphitids and accumulations of neanic to juvenile baculitids (Landman 1987; Westermann 1996). *Placenticeras meeki* is strongly attacked in shallow, surface waters by predators (e.g., mosasaurs; 10–20% of all specimens; Kauffman 1990, Hewitt and Westermann 1990b; Westermann 1996). Oxygen and carbon isotope analyses (Chap. 18.5) revealed markedly reduced salinities of the surface waters inhabited by *P. meeki* (Westermann 1996). This clearly indicates that the large predator *Placenticeras* occasionally lived in brackish surface waters (e.g., <30 m below surface; Westermann 1996). This contrasts with the deeper habitats assumed by Ziegler (1967, 1981) and Westermann (1990) above wave base. Shell characteristics of *P. meeki* are a complicated suture, low sutural amplitude index (Batt 1991), low resistance to hydrostatic pressure, and a high predation rate (Hewitt and Westermann 1990b). *Baculites* was also commonly attacked by mosasaurs, evident from several punctures (Kauffman 1990; Westermann 1996; Tsujita and Westermann 2001). This points to a pelagic habitat (Westermann 1977, 1996) in the aerobic waters. *Baculites* is therefore suggested as being capable of diving rapidly (5–100 m deep into dysoxic zones), feeding on the bottom. The maximal depth range for *Baculites* was corroborated by isotope data (Fatherree et al. 1998; Lukeneder et al. 2010; Fig. 18.9): relatively shallow from 50–100 m in epipelagic seas (Fig. 18.7, Table 18.1). *Baculites* apparently shifted into shallower waters after reaching maturity (Fatherree et al. 1998; Lukeneder et al. 2010). In contrast to the pure surface-water dweller *Hypancanthoplites* (Lukeneder et al. 2010), it thus changed its depth preference and seems to have started as a juvenile in the lower epipelagic zone. Similar depths for *Baculites* were given (40–120 m) based on SiSI values (Hewitt 1996; Westermann 1996). Abundant but smaller species and juveniles of the giant species were suggested to be vertical migrants (Klinger 1980; Batt 1989; Westermann 1990), probably mostly in midwater. The giant adults (>0.8–1 m) of *B. cuneatus*, in turn, became mainly demersal swimmers.

18.8.4 R-Strategy in Oxygen-Depleted Habitats

The Barremian heteromorph ammonoid *Karsteniceras* (N Calcareous Alps, Austria) evolved under intermittent oxygen-depleted conditions associated with stable, salinity-stratified water masses; it exhibited mass occurrences (Lukeneder 2003, 2005, 2007; Lukeneder and Smrečková 2006). Lithological and geochemical analyses (e.g., dark laminated shales; $CaCO_3$, TOC, S) were combined with investigations of trace fossils, microfossils and macrofossils. This led to the assumption of an invasion of an opportunistic *Karsteniceras* community (with sexual dimorphism, most probably applicable to the whole leptoceratoid group) during unfavorable conditions (Lukeneder 2003, 2007; Lukeneder and Tanabe 2002). The significant effect of autecological stress, caused either by changes in abiotic environmental factors (oxygen content, salinity, depth) or by synecological stress associated with biotic competitors occupying the same environment was shown by Lukeneder (2003, 2007). Jaw apparatuses are found *in situ* or isolated but are associated with a mass-occurrence of the genus *Karsteniceras*, reflecting a fast and autochthonous deposition. Low energy on the sea floor (absence of bottom currents) and dysaerobic conditions prevented predators from isolating the shells from the jaw apparatuses and led to the unusual preservation of the ammonoid conch-jaw association. The positions of the lower jaw in the body chamber of *Karsteniceras* indicate that the dead ammonoid bodies were not subjected to long postmortem drift; they rapidly became waterlogged and sank to the sea floor. These exceptional preservational features are typical for 'Konservat–Lagerstätten' (Seilacher et al. 1976), which always show exceptional preservation of articulated hard parts and/or soft body parts.

Terms often used in ammonoid ecology such as allochthonous and autochthonous, must be used with caution when linked with fossil cephalopods because shells could have been subject to major postmortem drift. In most cases, the water depth or primary habitat is unknown. A certain likelihood concerning the autochthonous deposition of ammonoid specimens, hence the primary depositional area, is warranted by the preservation of *in situ* buccal masses located in the body chamber

(Schindewolf 1958; Bandel 1988; Frye and Feldman 1991; Summesberger et al. 1996; Lukeneder and Tanabe 2002).

Karsteniceras inhabited areas of stagnant water with low dissolved oxygen, showing abundance peaks during times of oxygen depletion, which partially hindered other invertebrates from settling. An invasion of the *r*-strategist *Karsteniceras* assemblage during unfavorable conditions over the sea bed was proposed. The host layers do not reflect catastrophic mass mortality. The specimens are not concentrated in single horizons, but are abundant in layers of 3–10 cm thickness, reflecting 100,000 of years (calculations based on ammonoid zonation).

Taking the speculations on ammonoid life-habitats into account, demersal forms feeding from the sea floor should be rare or absent in the anoxic to dysoxic levels of the water column (Batt 1993). Vašíček and Wiedmann (1994) noted the possibility that the biotope of the Leptoceratoidinae was close to stagnant, poorly oxygenated environments. This has been interpreted to reflect the opportunistic bahavior of some heteromorphs (bochianitids, leptoceratoids) in unfavorable environments (Cecca 1997, 1998). Ecology, paleobiology and habitats of small heteromorphs such as leptoceratoids and spiroceratids were interpreted to have occupied various habitats (Thieuloy 1966) such as nektic above an anoxic bottom (Rieber 1977), on the distal shelf (Westermann 1990, 1996), in turbiditic environments (Vašíček and Wiedmann 1994), and demersal, capable of life in dysoxic waters (Lukeneder 2003, 2005).

18.8.5 Oxygenated Shallow-Water Habitats

Carbonate platforms are a special environment for ammonoid occurrences. It is relatively shallow and well oxygenated, mostly well agitated by bottom or surface currents. Some early ammonoid taxa like Devonian clymeniids were suggested to appear within such environments. They exhibited a morphological depth gradient with coarse ribbing in shallow and fine ribbing in deeper habitats (Korn 1986, 1988, 1992; Westermann 1996).

Middle Triassic ceratitids were the main inhabitants of shallow, epeiric platforms. They probably tolerated a wide range of salinities (euryhaline). Sutural simplification occurred in several immigrant clades from the deeper Tethys in the German Muschelkalk (Urlichs and Mundlos 1985; Klug et al. 2005). Coarse sculpture is also commonly associated with a basin-slope habitat, whereas a basinal pelagic habitat is usually indicated by smooth or weakly sculptured discocones. A few taxa (*Monophyllites*, *Psilosturia*) probably lived on the continental slope of the oceanic region (Westermann 1996), but based on their shell strength were still epipelagic. In the Late Triassic, radiation affected almost all morphs. The proportion of oceanic ‘Leiostraca’ greatly increased, and the first heteromorphic ammonoids appeared with torticone, orthocone and gyrocone morphotypes (Westermann 1996).

A more Recent study on Triassic ceratitid ammonoids was performed on a mass-occurrence (*Kasimlarceltites*; $n >> 300$ million; M and m, embryonic to adults)

Fig. 18.12 The gregarious ceratitid ammonoid *Kasimlarceltites krystyni* (max. diameter 35 mm) in assumed spawning grounds and the nektic actinopterygiid fish *Saurichthys* (max. length 1 m) in shallow waters of a Triassic carbonate platform. Reconstruction (by 7reasons Media Company) based on studies of Upper Triassic mass occurrences of the ammonoids in the Taurus Mountains (SW Turkey; Lukeneder and Lukeneder 2014)

from Turkey (Lukeneder and Lukeneder 2014; Lukeneder et al. 2014; Lukeneder and Mayrhofer 2014). *Kasimlarceltites* is morphologically similar (i.e., serpenticone, highly evolute; Fig. 18.6, 18.12) and taxonomically closely related to other celtitid ammonoids including *Aplococeras*, *Lecantites* and *Celtites*. These genera were interpreted by Assereto (1969) as shallow 'platform faunal associations' from Latemar or Marmolada platforms (Dolomites, Italy). The celtitids are rare to absent in deeper basinal environments (Assereto 1969). The ammonoid-facies relation was interpreted to be controlled by ecological factors (Brack and Rieber 1993, 1996). *Kasimlarceltites* probably inhabited environments on the inner- to mid-ramp, adjacent to the Carnian platform (Lukeneder and Lukeneder 2014; Lukeneder and Mayrhofer 2014; Fig. 18.12). They were influenced by this environment and may have migrated near the sea floor, comparable to extant *Nautilus*. Assumptions on a facies or depth gradient dependency in celtitids (*Kasimlarceltites*) fit well into the interpretation of the Turkish material. For ammonoid assemblage-facies relations in the Mesozoic Era see Westermann (1996). Numerous layers are dominated by small ammonoids ($<$30 mm) accompanied by shallow-water faunal components (transported ammonoids, bivalves, corals, gastropods, sponges) derived from the platform to inner ramp. They were therefore interpreted as being redeposited from shallower areas down to a shelf margin or mid-ramp position (Lukeneder et al. 2012). In contrast to Assereto (1969), additional findings of celtitids were presented by Manfrin et al. (2005) from adjacent basinal series surrounding the Latermar platform. The faunas from ammonoid layers were described as storm deposits. These were transported and comprise mixed faunal elements (pelagic ammonoids and benthic gastropods; Manfrin et al. 2005). This is almost identical to the assemblages found at Aşağiyaylabel. The use of such assemblages as characteristic elements for platform deposition or even primary habitat of celtitids is still arguable and needs more taxonomic and correlation work. Hence, a facies dependency of *Kasimlarceltites* in the carbonate member at Aşağiyaylabel must be considered with reservation. Taphonomic analyses rather point at allochthonous deposition after considerable transport (current, turbidites, mass flow) within ammonoid-mass occurrences (Lukeneder et al. 2014). The occurrence of single layers comprising mostly transported conchs of juvenile or exclusively adult shells of *Kasimlarceltites*, and beds bearing mostly adults, support the hypothesis that different ontogenetic stages inhabited different habitats (Lukeneder and Mayrhofer 2014; Lukeneder et al. 2014).

Similar hypotheses on the relation of closely resembling morphotypes versus habitat and lifestyle of serpenticone morphotypes were given for *Paraceltites* (Permian; Spinosa et al. 1975), *Psiloceras* (Jurassic; Westermann 1996), and *Celtites* (Triassic; Rieber 1973, 1975). These were presented and commented upon in Westermann (1996). Spinosa et al. (1975) proposed a "*tropical platform limestone*" depositional area for such serpenticone ammonoid types and a presumable planktic lifestyle was assumed (Fig. 18.6, 18.12) for such costate, celtitid morphotypes (Rieber 1973, 1975; Westermann 1996). As postulated by Klug (2002) and Klug and Korn (2004), ontogenetic alterations in morphology reflect, at least in early ammonoid taxa, a change in mode of life. This is because these changes appear several times independently.

18.9 Autecology in Ammonoids

Neumayr (1883), Uhlig (1911), Scott (1940), Ziegler (1967, 1981), Kennedy and Cobban (1976), and, more recently, Westermann (1990, 1996) have studied the autecology of ammonoids and related ecological factors. This relation is denoted by the variation and ratio of shell morphologies and assemblages in general, related to tectonically caused or eustatic sea level fluctuations (Enay 1980; Enay and Mangold 1982; Marchand and Thierry 1986, 1997; Lominadze and Sakharov 1988; Wiedmann 1988a; Cecca 1992; Bulot 1993; Rawson 1993; Hantzpergue 1995; Reboulet and Atrops 1995; Reboulet 1996; Enay and Cariou 1997; Pucéat et al. 2003). Such environmental changes are accompanied by changes in many other ecological parameters. This has consequences for the organisms living in the affected environment. Neige et al. (1997) stressed the close relation between disparity and (eustatic) sea-level changes. These authors presented a new approach based on Callovian ammonoid faunas of France (Lominadze and Sakharov 1988, Callovian of Caucasus). Here, the morphospace calculations reflected the most important conch parameters (diameter, whorl height and width, etc). Mathematical estimations by Neige et al. (1997) resulted in a more precise use of ammonoid morphotypes as paleodepth indicators.

Oxygen is a biolimiting element for metazoans and one of the most important factors influencing species diversity and abundance in the marine realm. Batt (1993) suggested that the relationship between ammonoid shell morphotypes and position within the water column provides a useful tool in interpreting fluctuations in O levels in marine depositional systems of the Cretaceous Western Interior Seaway (N America). Suggested habitats in Batt (1993) were based on siphuncle strength index (SiSI), septal amplitude index (SAI) and septal complexity (Batt 1991, 1993; Ballentine 2007). As noted by Batt (1991) such indices are only valuable in combination with other shell morphological features (e.g., ribbing). The author concluded that the presence of nektobenthic morphotypes (e.g., *Scaphites*, *Baculites*) indicates oxygenation of the sea floor and lower parts of the water column, while that of pelagic morphotypes (e.g., *Watinoceras*, *Borrisiakoceras*) or shallow-water ammonoids (e.g., *Placenticeras*) may be used to interpret oxygenation higher in the water column. A nektobenthic mode in *Scaphites* is probably indicated by the highly injured specimens noted from Landman and Waage (1986; see also Landman 1986). *Baculites* was given with a depth range from 40–120 m based on SiSI values (Hewitt 1996; Westermann 1996; Fig. 18.9). The occurrence of shallow middle- and upper water column pelagic ammonoids in deeper-water stratigraphic intervals yielding no nektobenthic ones appears to be a useful indicator of oxygenated conditions within the upper part of the water column at a time when benthic environments were anoxic. The additional absence of ancyloconic heteromorphs in sediments deposited in water deeper than 100 m that contain upper-water pelagics indicates that anoxia extended to within 100 m of the surface (Batt 1993). Changes in ammonoid morphotype representation during upward migration of a poorly oxygenated water mass, hence fully oxygenated water column appears with all morphotypes (pelagic and nektobenthic). Bottom near anoxic conditions first eliminated nektobenthic ammonoids, and continued with anoxic conditions up to 100 m of the surface hampered life of pelagic forms (megaplanktic ancyloconic heteromorphs). Finally, only pelagic ammonoids from the upper water column are represented in sediments (Batt 1993; Reboulet et al. 2000, 2005; Lukeneder and Grunert 2013).

18.10 Epizoans Versus Postmortem Epicoles in Ammonoids

Studying the spatial relationships among organisms that lived with each other provides insight into the synecology, autecology, and biostratinomic processes of these organisms. It also sheds light on their paleoenvironment and paleocommunity structure. Epibiosis is one of the few well-preserved biotic interactions in the fossil record dating back to the Early Paleozoic (Palmer 1982). A crucial aspect is to detect if infestation occurred before or after death. This determines whether an autecological or synecological relation is indicative for paleodepth, behavioral or habitat assumptions. Organisms that infested ammonoid shells within the water column are also referred to as pseudoplankton (Wignall and Simms 1990). If infestation occured postmortem, it can be speculated whether shell was infested while drifting or after it sunk to the sediment surface.

Davis et al. (1999) introduced terms to distinguish between the different aspects of epizoans. According to them, an organism that settles an lives on another one is an epizoan. By contrast, if the substrate object or organism is dead, Davis et al. (1999) suggested to call the animal living on it an epicole. These terms will be used subsequently. Countless examples of epizoa and epicoles on ammonoids have been described and some examples will be listed.

Ammonoid assemblages from the Early Cretaceous of Austria (Chap. 18.8.2; Lukeneder and Harzhauser 2003; Lukeneder 2004) yield abundant specimens of *Olcostephanus guebhardi* showing unique encrustations by placopsilinid foraminifera, bryozoans, serpulids, and crinoids (*Phyllocrinus*). The pattern of infestation clearly documents a 100% preference of these taxa for the inner shell surface, characterizing them as an epicole cavity-dwelling paleocommunity (Lukeneder and Harzhauser 2003). The encrustation in the body chamber furnishes evidence for postmortem settlement. *O. guebhardi* is suggested to have preferred shallow water and was most abundant in depths of 40–100 m (Bulot pers. comm. 2002). Major redeposition of the Austrian ammonoid shells is evident based on a mixture of olcostephanids with encrustations on their upper side and specimens with epibionts on their lower side within single layers. The shells sunk to the sea floor after death and became partly filled with sediment on the side of the shell facing the sediment. Thus, epicoles could not settle these areas of the body chamber. Only the 'sediment free' upper shell parts could be inhabited (*Placopsilina*, bryozoans, serpulids). This mode of infestation indicates that encrustation during postmortem drift can be excluded (Lukeneder and Harzhauser 2003). *Phyllocrinus* is one of the most frequent species associated with *Olcostephanus*. The calices of this sessile crinoid are often found in the sediment-infill of the ammonoids and the adherent holdfasts are exclusively found on the upper, outer side of the shell. In this soft bottom environment, the empty shells of *Olcostephanus* served as hardgrounds.

A more complex history is reported from the Valanginian to Aptian deposits of the Dolomites (N Italy). These deposits yield a rich ammonoid fauna showing unique epifaunal encrustations by an ahermatypic, solitary scleractinian coral (Lukeneder 2008). The coral encrusted only the outer shell surfaces; the inner surface remained unaffected. Shells of dead ammonoids sank to the bottom and became colonized by the coral (and serpulid) larvae, as documented by the location of the epibionts only on the upper side. Encrustation by the bivalve *Placunopsis* represents a different situation because both sides of the ammonoid shells were affected. This points to encrustation of floating or repeatedly turned (dead) or swimming (alive) ammonoids (Lukeneder 2008; Misaki et al. 2013). Such long-term infestation in the water column contrasts with coral settlement on the sea floor. Ammonoid specimens encrusted by *Placunopsis* never exhibit encrustation by corals (Merkt 1966; Meischner 1968).

The so-called coelobites are a somewhat neglected part of the vast spectrum of epibioses. They are defined by Ginsburg and Schroeder (1973) as cavity-dwelling organisms, often found in cryptic habitats. Studies on such habitats of modern environments have focused on coral reefs and reef-associated submarine caves (Jackson and Winston 1982; Rasmussen and Brett 1985; Meesters et al. 1991). Despite the smaller scale and limited spatial resource of the ammonoid shells, several parallels with modern cryptic habitats are obvious. As emphasized by Wunsch and Richter (1998) and Richter and Wunsch (1999), cavities and other cryptic habitats are "*spatially confined habitats*", which provide living space for low-light adapted organisms. Among these, especially cryptic suspension feeders predominate. Cryptic habitats provide shelter for sessile species from predation and physical disturbance (Gischler and Ginsburg 1996). In modern cryptic habitats, the coelobite's mode of life is mainly adopted by sponges, algae, serpulids, tunicates, bryozoans, and foraminifera. Several taxa such as placopsilinid foraminifera and the bryozoan *Stomatopora* have been shown to be characteristic constituents of all these paleocommunities. These taxa were also recorded adhering to ammonoid shells (Lukeneder and Harzhauser 2003).

Can such encrustation yield information on paleodepth and ammonoid habitats? The ecological depth ranges of the encrusting organism groups are enormous. The main encrusters on ammonoids are bivalves (e.g., oysters, *Exogyra*, *Placunopsis*, *Gervillia*, *Anomia*; Larson 2007; Misaki et al. 2013), gastropods (limpets), brachiopods (discinids; Seilacher 1982), serpulids, tunicates, solitary corals, bryozoans, and foraminifera. Larval stages often settle on secondary hard grounds (ammonoid shells). Unfortunately, synecological relationships (i.e., infestation during life) of ammonoid and encrusting taxa also can rarely be used to conclude on habitat or depth. Oysters (Merkt 1966), for example, inhabit depths from almost 0–850 m (Rooij et al. 2010), the bivalve *Placunopsis* hundreds of meters, serpulids can live in thousands of meters. Serpulids were able to grow with ammonoids in a synecological manner during life, on the ventral side of the shell. This makes them useful to estimate ammonoid growth rates and ages (Schindewolf 1934; Andrew et al. 2011). Andrew et al. (2011) suggest a distinct infestation by serpulids at different habitat depths for the Jurassic, mostly of living, juvenile ammonoids (*Promicroceras*) from the UK (Dorset). The dependency or at least preference of several serpulid species for shallower waters (6–12 m) led to the assumption that juvenile *Promicroceras* inhabited shallower areas: they were more highly infested than adults. Andrew et al. (2011) concluded that *Promicroceras* was negatively influenced by the parasitic serpulids, by the higher body weight and the increasing drag, dying at smaller diameters at an age of 2–3 years.

Synecological infestation by foraminifera potentially points to paleodepths of 15–70 m for *Placopsilina* attached to the ammonoid *Olcostephanus* (Lukeneder and Harzhauser 2003). Correspondingly, *O. guebhardi* was suggested to have preferred shallower shelf areas and was most abundant at depths of 40–100 m (pers. comm. Bulot 2002). Home scars of patellogastropods (limpets) on ammonoid shells were shown by Kase et al. (1994, 1998). Limpets, when settled postmortem on ammonoids, point to a drifting phase because limpets inhabit shallow areas and therefore attached near the sea surface (see Westermann and Hewitt 1995).

Maeda and Seilacher (1996) reported *syn vivo* and postmortem epibiont assemblages. The encrustation by inoceramids, oysters or *Gervillia* of both the flanks and venter in living ammonoids is remarkable (e.g., Jurassic *Lytoceras* and *Harpoceras*; Cretaceous *Buchiceras*; Seilacher 1960; Heptonstall 1970; Seilacher 1982; Maeda and Seilacher 1996; Larson 2007). One-side-only infestation by oysters was also reported as being caused by horizontal drifting of ammonoid shells on the water surface (e.g., *Pectinatites*; Donovan 1989). Sinking history can also be detected by analyzing the encrusting community and its position (Reyment 1973, Kauffmann 1978; Seilacher 1982). Kauffmann (1978) postulated the 'benthic island' infestation on sea floor for the Jurassic *Posidonia* Shale ammonoids, but this was refuted by Seilacher (1982) and Maeda and Seilacher (1996) based on the encrustation of both flanks (e.g., *Lytoceras*) in the water column. Further examples of *syn vivo* epizoans are discussed in Naglik et al. (2015).

18.11 Trophic Level

The trophic level of ammonoids was perhaps some kind of a mid-order omnivore level. They exhibit a range of feeding strategies from planktotrophic to microphagous (i.e. lower trophic levels). They were preyed upon by vertebrates (ichthyosaurs, mosasaurs, sharks, fishes; Fig. 18.13) and invertebrates (other cephalopods, crustaceans). Ammonoids ate possibly everything they got hold of from the lower trophic levels, leading to the interpretation as demersal herbivores, scavengers via microphagous carnivores to active, nektonic carnivores (Kennedy and Cobban 1976).

As noted by Westermann (1996), numerous feeding strategies can be envisioned for neanic and juvenile ammonoids. These speculatively fed on mobile prey with ejectile tentacles, visually on essentially planktic prey (ammonoid hatchlings, pseudoplanktic and planktic microorganisms such as ostracods or microgastropods), with non-ejectile tentacles or velar webs, and tactile or chemosensory feeding by 'pseudoscavenging' of organic particles floating and sinking at all depths (Reyment 1988; Westermann 1996). Reyment (1988) imagined a more scavenging and browsing mode, concluding that especially orthocones and heteromorph forms were incapable of active predation. The preying behavior is crucial for reconstructing the ammonoid habitat because ammonoids might have actively followed prey. This can cause ontogenetic or diurnal migrations, preying on different resources during their lifespan, or preying at night when some other predators sleep, comparable to Recent *Nautilus*.

Most recently, spectacularly preserved specimens of the Late Cretaceous heteromorph *Baculites* from South Dakota (USA) were reported by Kruta et al. (2011). They discovered buccal masses with radulae containing plankton found within. This represents an important indication for the trophic habits of these heteromorph ammonoids. One of the specimens revealed a planktic snail and three tiny planktic crustaceans (Isopoda) in its buccal mass (Kruta et al. 2011). As already stated by Westermann (1996), knowledge on stomach or crop contents is scarce (Lehmann and Weitschat 1973; Lehmann 1975, 1976, 1985, 1988; Summesberger 2000; Keupp 2007). See Klug and Lehmann (2015) for a review.

Numerous problems related to the dietary and feeding habits for the diversely shaped buccal organs are still unsolved, even for some extant cephalopods (Nesis 1986, 1987; Lehmann 1988). Westermann (1996) noted that besides the micro and mesophagy documented by crop/stomach contents, macrophagy and even duraphagous strategies might have also been present in ammonoids (Schindewolf 1958; Lehmann et al. 1980; Tanabe 1983; Nesis 1986; Tanabe and Fukuda 1987; Lehmann and Kulicki 1990; Hewitt 1993; Seilacher 1993; Keupp 2007). Westermann (1996) suggested that most ammonoids belonged to the pelagic food-web. Ammonoids fed directly on microplankton, perhaps at the base of the photic zone, themselves falling prey to larger ammonoids and vertebrates. An important feature for the ammonoid/prey interaction was therefore the beginning formation of a distinct stratification including a well-defined pycnocline in Cretaceous oceans (Hay 2008), as in Recent oceans. Other groups were vertical migrants in deep water, the mesopelagic zone. There, some fed on mesopelagic organisms including sluggish juvenile ammonoids, whereas others might have caught zooplankton with arms connected by velar skins (Westermann 1996). Only relatively few consumed soft-bodied epibenthos or preyed like *Nautilus* on shelly benthos.

The relation between changes in ammonoid assemblages (radiation of new taxa, appearance of additional morphogroups, size variations) and evolutionary events was interpreted by several authors (Elmi and Alméras 1984; Reboulet 1996, 1998, 2001; Reboulet and Atrops 1997; Cecca 1997, 1998; Lehmann 2000; Guex 2001; Reboulet et al. 2005). They considered the relation to be caused or at least influenced by changes of trophic levels (food, nutrients, energy, etc.). Reboulet et al (2005), however, noted that the assumptions on ammonoid trophic levels in the former papers were speculative. Reboulet et al. (2000, 2003) interpreted Valanginian assemblages by performing a quantitative and integrated high-resolution analysis of macrofauna, nannofossils, microfacies, trace fossils, and total organic carbon. The resulting absolute abundance variations reflected the trophic resources in the oceanic waters.

A trophic level specifies the position of an organism in the food web, often generated by the nutrient supply into that system. Accordingly, the water mass can be characterized as oligotrophic (i.e., less nutrients, less organic production, high oxygen content), mesotrophic (i.e., increasing nutrients, increasing productivity and enough oxygen), and eutrophic (i.e., high nutrient values, high organic production, decreasing oxygen). Nutrient supply creates energy, whose availability alters biodiversity (Tittensor et al. 2011). Less nutrients are available at great distances from land and in deeper waters in contrast to nearshore and shallow areas.

Abundance data on ammonoids have been collected (Reboulet et al. 2005) from the Albian of the Vocontian Basin (France) across an anoxic interval (OAE 1d, Breistroffer interval). Reboulet et al. (2005) suggested that these were influenced by trophic changes. The Breistroffer deposits do not record eutrophication of surface waters associated with expansion of the oxygen minimum zone. Rather, the deposits record changes from mesotrophic to more oligotrophic conditions in surface waters (Reboulet et al. 2000). Ammonoids are abundant and morphologically diverse (i.e., seven morpho-groups) in a section from proximal areas (platform environments) and the pelagic realm (open-marine water column). Ammonoid morphology was interpreted (Reboulet et al. 2005) to indicate aspects of their modes of life (Westermann 1996; Cecca 1997; Klug 2001), although all were attributed to the epipelagic zone. For example, *Hypophylloceras* (Phylloceratina) and *Tetragonites* (Lytoceratina) were placed by Reboulet et al. (2005) in the deepest layers of the epipelagic zone within the oceanic domain (150–200 m depth). Recent data on oxygen isotopes (Moriya et al. 2003; reinterpreted herein, Fig. 18.11) yielded wider depth ranges for both genera: 100–300 m (upper mesopelagic to middle epipelagic) in *Hypophylloceras* and 80–180 m in *Tetragonites* (lower to middle epipelagic). *Tetragonites* probably inhabited slightly shallower layers, as also figured by Reboulet et al. (2005).

Ammonoid taxa can be linked to changes in trophic conditions as inferred from a study on calcareous nannofossils (Reboulet et al. 2005). Heteromorphs, which were dominant during mesotrophic conditions, could have been more competitive than involute/evolute planispirals (normally coiled) when environmental conditions became more unstable. *Lechites* (orthocone) was assigned to an epipelagic habitat (i.e., distal paleoenvironments); it was interpreted to be a vertical migrant, able to move with increasing nutrients to meso- and eutrophic layer conditions in surface waters, and also down in order to avoid oligotrophic surface waters, preying on food-rich layers in deeper waters (Reboulet et al. 2005). More proximal paleoenvironments were suggested for *Scaphites*, which maybe was a vertical migrant. Quasiplanktic *Turrilitoides* and *Mariella* (torticones) mainly inhabited neritic paleoenvironments, occupying more distal paleoenvironments with mesotrophic conditions (increasing nannofossil content) in the surface water column. *Anisoceras* and *Hamites* (quasiplanktic U-shaped heteromorphs) lived in distal, epipelagic habitats. They were more competitive when oligotrophic conditions prevailed in surface waters. *Mortoniceras* (planispiral) was interpreted to inhabit the lower part of the epipelagic zone with a deep nektonic mode of life (Reboulet et al. 2005).

Early Cretaceous size variations (on species level) along a depth gradient are known from S France (Bulot 1993, 1995; Reboulet 1996, 2001; Reboulet and Atrops 1997). These ammonoids exhibited larger conchs in Valanginian shallow-water facies (Provence platform) and smaller ones in deepwater facies (Vocontian basin). Reboulet (2001) noted a size dependency on higher hydrostatic pressure (i.e., emptying rate of cameral fluid; Ward 1987) and cooler waters. This led to dwarfed forms in the basinal areas. Reboulet (2001) assigned the Valanginian ammoinoid taxa (macro- and microconchs; *Acanthodiscus*, *Busnardoites*, *Karakaschiceras*, *Leopoldia*, *Lytoceras*, *Neocomites*, *Neolissoceras*) to a deep nektonic to nektobenthic mode of life. *Acanthodiscus radiatus* is much more common on the shallow platform than in the basin (Reboulet 1996). It was segregated by its dimorphs: macroconchs dominate on the platform whilst microconchs are abundant in both habitats. *A. radiatus* shows the same size in shallow and deep-water facies, probably reflecting a migrant lifestyle between different environments. *Neocomites peregrinus* appears in equal numbers in both environments and is thus thought to be more tolerant of environmental changes. Reboulet (2001) proposed that the main factor in ammonoid segregation was water temperature, as in Recent *Nautilus* (Ward 1987) and coleoids (Mangold 1989). Thus, temperature may have triggered sexual maturity and the corresponding growth; food resources were also suggested to have influenced size differentiation. The important factor of food availability was also noted by Bucher et al. (1996). Speculating on the assumptions above, nutrient supply is higher on shallow platforms (see Cecca 1998). Accordingly, planktic activity increased and more food was available, so that ammonoid size increased in such environments (Reboulet 2001).

Trends in ammonoid diversity, habitat, and trophic levels were compared by Lukeneder and Grunert (2013) from the Trento Plateau in Italy and from the Betic Cordillera in Spain (Company et al. 2005). This yielded new insights into the paleoenvironments, which were determined by trophic fluctuations. Both data sets agree in the strongly increasing diversity of epipelagic vertical migrants and planktic drifters during the Late Hauterivian (*Pseudothurmannia mortilleti* and *P. picteti* subzones; Faraoni Level). Company et al. (2005) and Lukeneder and Grunert (2013) related this distinct turnover to phytoplankton (radiolarian) bloom during the Faraoni event and during the coeval 2nd-order peak transgression in the lower *P. mortilleti* Zone. The increase in trophic resources favored the diversification of planktic ammonoids. In the Puez area, similar blooms in productivity are expressed as radiolarian wackestones. This fact, however, does not explain why this habitat group remains diverse in the aftermath of the Faraoni event in the upper *P. mortilleti* and *P. picteti* subzones when oligotrophic conditions were established (Lukeneder and Grunert in 2013). These contrasting trends show the difficulties in interpreting ammonoid habitat groups and feeding strategies. Company et al. (2005) explained the extinction of deep nektic groups around the Faraoni Level by an oxygen-minimum zone that developed due to phytoplankton blooms. Trends in radiolarian and nannoconid abundances suggest a turnover in the trophic structure of the surface water from eutrophication around the Faraoni event to oligotrophic conditions. These changes in primary productivity, however, are also reflected in the organic matter with values up to 7.0% TOC. Company et al. (2005) developed a classification of ammonoids into six life-habitat groups: planktic drifters, epipelagic vertical migrants, epipelagic nektic ammonoids, mesopelagic vertical migrants, mesopelagic nektic ammonoids, and nektobenthic ammonoids. Members of all these groups occur in the studied interval (Lukeneder and Grunert 2013). Planktic drifters are represented by *Crioceratites*, *Pseudothurmannia*, *Paracostidiscus*, *Paraspiticeras*, *Karsteniceras*, *Hamulinites*, *Sabaudiella* and *Megacrioceras*; *Honnoratia*, *Hamulina*, and *Anahamulina* were epipelagic vertical migrants. *Lytoceras* and *Protetragonites* were possibly mesopelagic vertical migrants, and the mesopelagic nektic ammonites include *Phylloceras*, *Phyllopachyceras*, *Kotetishvilia*, and *Neolissoceras*. *Barremites*, *Plesiospitidiscus*, and *Discoidellia* were epipelagic nektic ammonoids. Finally, *Abrytusites* and *Astieridiscus* are nektobenthic forms.

The analysis of ammonoid diversity and life-habitat groups (Westermann 1996; Company et al. 2005) suggests a strong influence of sea-level on the plateau. While epi- and mesopelagic ammonoids occur commonly in the *Crioceratites krenkeli*, *P.*

Fig. 18.13 Reconstruction of a Lower Jurassic ichthyosaur hunting on ammonoids, biting into the shell of *Dactylioceras* in the epipelagic water zone of the Liassic sea (based on studies of Posidonia shales in SW Germany). Upper Cretaceous mosasaur bite marks (max. puncture diameter 3 cm; lower right corner) on a Campanian *Placenticeras* shell (cast replica, max. diameter 30 cm) from the Western Interior Seaway in South Dakota

mortilleti, and *P. picteti* subzones, a severe reduction in all groups occurred during the sea-level lowstand of the *Spathicrioceras seitzi* and *Pseudothurmannia ohmi* subzones. The Faraoni event is heralded by an ammonoid faunal turnover characterized by an increasing diversity of epipelagic ammonoids. Although a similar trend was observed in the Betic Cordillera in Spain, the trend differs in the common occurrence of deep nektic ammonoids, related to a better oxygenated environment on the Trento Plateau.

18.12 Implications for Ecology from Traces of Predation

Shell damage is an indicator for predator-prey interactions (Kowalewski 2002) involving predators within distinct and characteristic ammonoid habitats. Ammonoids are both predators and prey, foraging and being attacked in the same habitat. Mapes and Chaffin (2003) summarized the sublethal and lethal shell damage from recent and fossil cephalopods (Keupp and Hoffmann 2015). Predation in the water column can involve ichthyosaurs (Fig. 18.13), mosasaurs, nothosaurs, sharks or bony fishes, and invertebrates including other ammonoids, nautiloids, coleoids (Landman and Waage 1986; Kröger 2000, 2002a, b, c; Keupp 2000, 2006; Larson 2003, 2007; Tichy and Urbanek 2004; Andrew et al. 2010; Landman et al. 2012; Kauffman and Sawdo 2013) and arthropods (Keupp and Hoffmann 2015). As noted by Klompmaker et al. (2009) for Mesozoic and Andrew et al. (2010) for Jurassic ventral bite marks, they are most probably caused by predatory attacks on living animals by teuthoids (Yomogida and Wani 2013). These authors argued that the combination of an intact aperture and the absence of shell chips (i.e., broken parts of corresponding injury) excludes scavengers as producers of such injuries.

Recently, remarkable bite marks on both flanks (i.e., by both jaws of a predatory fish) were reported on Jurassic *Oxycerites* (Richter 2009). Some ammonoids are thought to be demersal feeders, scavenging on the sea floor (Reyment 1988): there, they were attacked by crustaceans (Radwansky 1996; Kröger 2002b, c; Keupp 2006), causing sublethal to lethal injuries (Keupp and Hoffmann 2015). Keupp (2000) also attributed the apertural injuries in *Dacytlioceras* to crustaceans on the sea floor (Keupp 1997), assuming a demersal habitat. This significantly contrasts with Westermann (1996), who proposed that *Dactylioceras* was a planktic form. The same former result was presented for the entire Callovian assemblage (perisphinctids, kosmoceratids) from northern France (Keupp 1992): all ammonoids were assumed to be nektobenthic to benthic (habitat dependency in accordance to frequency of injuries). That author used sublethal shell injuries of ammonoid shells, caused by crustaceans to draw conclusions on paleopathology. Keupp and Schobert (2011) even suggested a special crab (*Palaeopagurus*) as the predator on Liassic ammonoids (*Amaltheus, Amauroceras*) from Germany, which were assumed to have been demersal forms inhabiting near-bottom areas. *Palaeopagurus* is as the hermit crab also known to inhabit the body chambers of Jurassic and Cretaceous ammonoid shells on the sea floor (Germany and UK; Fraaije 2003; Jagt et al 2006).

As noted by Keupp (2000), shell damage on the rocky sea floor caused by the animal itself can be ruled out in most injuries (Bayer 1970; Checa and Westermann 1989; Zatoń 2010). A predator-prey dependency (Vermeij 1977; Keupp 2000a, 2006) puts pressure on the prey and probably causes it to evolve new strategies such as coarser ribbing, a more complex suture or increasing coiling (i.e., from straight to involute). Nonetheless, ecological pressure can also cause strategic changes, as seen in Recent *Nautilus*: it migrates at night to prey on various resources, and may then prey when its main enemies such as many fishes sleep (Carlson et al. 1984; Saunders 1984; Ward et al. 1984; Dunstan et al. 2011).

Shell damage by marine reptiles was documented from Middle Jurassic *Kosmoceras* (Ward and Hollingworth 1990), Late Cretaceous *Placenticeras* and *Sphenodiscus* (Kauffman and Kiesling 1960; Kauffman 1990; Hewitt and Westermann 1990b; Stewart and Carpenter 1990, 1999), and the nautiloid *Eutrephoceras* (Kauffman and Sawdo 2013). Kauffman and Sawdo (2013) argued that reports of mosasaur predation on Early Maastrichtian nautiloids from the Western Interior are rare because of the prey's predominantly deep, epibenthic habitat. They deduced this from modern *Nautilus* and applied it to the Cretaceous *Eutrephoceras*. Home scars of patellogastropods (i.e., round marks in a line) located on ammonoid shells have been used to falsify the mosasaur bite theory by Kase et al. (1994; 1998). In several cases, however, there appears to be no doubt about the mosasaur predation (Keupp 1991; Hewitt and Westermann 1990b; Fig. 18.13). Interestingly, patellogastropods, when attached to ammonoids (after their death) show a drifting history: limpets live in shallow areas and must have attached at the sea surface. Plesiosaurs were reported to prey on ammonoids by Sato and Tanabe (1998; see also Keupp 2000) from the Cretaceous of Japan (stomach content). Fish predators have been inferred (Ward 1981) and documented in Late Jurassic Haploceratidae by aptychi in holostean feces (Mehl 1978a). Jurassic ammonoids from the shallow sea in Solnhofen were preyed upon by teleosts such as *Gyrodus* or *Dapedium* (Keupp 2000). They broke parts off the aperture to expose the soft body, as observed in Recent *Nautilus* preyed upon by parrotfishes (Saunders et al. 1987), triggerfishes, and groupers (Mapes and Chaffin 2003). Injuries inflicted by fishes can be sublethal or lethal. Sublethal injuries can also reflect octopod attacks (e.g., borings in the shell; Mapes and Chaffin 2003). A more peculiar 'attack' is observed in extant *Nautilus* during mating (Arnold 1985). In this case, the partner bites and damages the apertural edge, which is also visible in fossil conchs. The mechanisms that damage extant *Nautilus* (sediment loading, within-sediment transport, collision during floating, predation) were analyzed by Wani (2004). He concluded that these mechanisms produce distinct and characteristic shell damage, hence being comparable to biostratinomic and taphonomic features in fossil materials.

Stomach contents of plesiosaurs from the Late Cretaceous of Japan represent important evidence for lethal injuries (Massare 1987; Sato and Tanabe 1998). They contain pieces of broken shell material from *Scaphites* and other ammonoids.

Keupp (2000) noted healed injuries caused by fish attacks on *Kranaosphinctes* and by crustaceans on *Prososphinctes* (Oxfordian, Madagascar). Teuthoid predators were suspected to have produced clusters of uniformly sized shell fragments of Early Jurassic harpoceratids and Late Jurassic *Gravesia* (Mehl 1978b). They probably also caused most of the peristomal mantle injuries in Jurassic to Cretaceous ammonoids. Unless these were lethal, they caused the well-known shell abnormalities. Evidence for predation (predator not specified) of Early Cretaceous ammonoids is also given by Reboulet and Rard (2008). Klug (2007) reported sublethal injuries on Devonian ammonoids that lived in the water column; that damage was likely caused by other cephalopods or phyllocarids. Shark attacks were reported as a possible cause for goniatid injuries in Carboniferous ammonoids (USA; Mapes and Hansen 1984; Mapes et al. 1995).

A remarkable ammonoid occurrence was documented by Mapes and Dalton (2002) from the Mississippian of Arkansas (USA). Ammonoid accumulations (e.g., *Emstites*, *Cluthoceras*) in carbonate concretions form halos around body chambers of *Rayonnoceras* (Actinocerida) and were interpreted as prey of nautiloid predators (based on stomach contents); scavenging was excluded by Mapes and Dalton (2002). Those assumptions were mainly based on missing body chambers, the random orientation of ammonoids, and the rarity of ammonitellae in such concretions. My own interpretation of that material is that ammonoids were redeposited into the broken phragmocone and body chambers of *Rayonnoceras* (fossil trap).

As noted by several authors, increased coiling and sculpture might have enhanced shell resistance towards breakage during predation attempts (Ward 1981; Klug 2007). Tight coiling implies an increased resistance against breakage of entire whorls when compared to loosely coiled shells (Nützel and Frýda 2003 and references therein). Consequently, a simple change in morphology had a major impact on ecological fitness, partially explaining the evolutionary success of ammonoids (Kröger 2005). Coiling probably enabled even early forms to increase their maximum swimming velocity compared to cephalopods with orthoconic shells (e.g., Jacobs 1992a; Jacobs and Chamberlain 1996; Westermann 1996; Korn and Klug 2002; Klug and Korn 2004). Predation and damage in relation to coiling in Carboniferous nautiloids and ammonoids was treated by Mapes and Chaffin (2003): predation levels were much higher in the latter. Swimming velocity was assumed to be higher in nearshore morphotypes (compressed) of *Scaphites* (Cretaceous, USA). Higher velocities are assumed to be essential for life in shallower areas (Jacobs et a. 1994), contrasting to the decreased velocity values of the depressed morphs of deeper areas (offshore). The more compressed morphotypes can better escape potential predators.

18.13 Conclusions

Ammonoid migrations within and between different habitats are ontogenetically induced and detectable by stable isotope data ($\delta^{13}C$, $\delta^{18}O$) and morphology. Ammonoids generally start with a perhaps planktic embryonic stage (floating egg masses), followed by the hatchlings, which differentiated in mode of life; they lived either planktic, nektic, or nektobenthic (demersal). Antidimorphic pairs (female/male, macroconch/microconch) were apparently separated into different habitats, at least during spawning periods. Females are thought to have laid 100–500000 eggs (*r*-strategy) on the sea floor or in the water column; ammonoid eggs and hatchlings were thus an important constituent of the plankton. Eggs were perhaps enclosed in egg sacks, anchored (in pouches) below the storm wave base or floated as egg masses in the upper water column. Hatchlings had a nektoplanktic lifestyle in the water column.

Several taxonomic groups were probably gregarious, some migrated over long distances and across depth levels. This conclusion has been strengthened by stable isotope analyses. Female ammonoids appear to have returned to their home-grounds for spawning and died afterwards (i.e., semelparous). This resembles the situation in extant neritic coleoids. Planktic egg masses and hatchlings/young juveniles floated for months within the water column, driven by ocean currents (surface or bottom-water currents). During the nektoplanktic juvenile phase and subsequent stages, they fed on various nutrients including nanno- and micro-plankton members as trophic resources. Taxonomic groups can therefore appear or vanish due to changes in trophic levels. Ammonoids inhabited most known parts of ancient seas (lack of knowledge from paleo-oceans), from euryhaline to brackish waters, and probably tolerated brief hypoxic events like extant cephalopods such as *Nautilus* and *Spirula*.

The best-estimate for habitat depths of ammonoids is between c. 20–500 m according to the literature and recent stable isotope data ($\delta^{18}O$). Maximal depth (implosion depth) estimated from siphuncle and shell wall strengths are now recognized to be reliable for application to ammonoid paleobathymetry. Nonetheless, some literature values are apparently too deep and the ranges inaccurate.

Bathymetric conclusions were drawn from $\delta^{18}O$ values for several ammonoid morphogroups. The range in *Cadoceras* (Jurassic, Callovian) is 50–700 m. It therefore started in the lower epipelagic and migrated deeper during ontogeny into the upper mesopelagic. With a depth range of 10–60 m, the ontogeny of *Hypacanthoplites* (Cretaceous, Aptian) starts in deeper epipelagic layers, followed by shallower epipelagic stages. *Nowakites* (Cretaceous, Santonian) is interpreted to have inhabited depths from 50–300 m. This implies migrations from deeper mesopelagic layers to shallower epipelagic habitats. The depth range given in the literature for *Baculites* (Cretaceous, Campanian) was specified as being relatively shallow (50–100 m) in the epipelagic zone. *Perisphinctes* (Jurassic, Oxfordian) apparently inhabited almost the same habitat in the middle epipelagic zone (50–100 m), although they migrated to deeper areas as young juveniles and inhabit again cooler waters as adults.

Two main groups with different ontogenetic strategies based on the $\delta^{18}O$ data were established by Lukeneder et al. (2010). The first is the *wcw*–type (warm–cool–warm type) of *Cadoceras*. Its lif strategy resembles those of *Nautilus* and *Sepia*, which migrate from shallow into deeper environments and back in ontogeny. The second is the *cw*–type (cool–warm type) of *Hypacanthoplites*. It resembling the first two migration phases of the *cwc*–type of *Spirula* (cool–warm-cool type), which migrates from deeper into shallower waters and back again.

$\delta^{18}O$ values as proxies for paleotemperature and paleohabitat-depth estimates show that Campanian taxa from Japan of all morphogroups and ammonoid families (*Tetragonites*, *Damesites*, *Hauericeras*, *Polyptychoceras*) inhabited shallower waters than indicated earlier. This contrasts with the opinion of Moriya et al. (2003), who assumes a bottom-dwelling demersal lifestyle for all Late Cretaceous ammonoids they measured. They therefore suggested that the ammonoids from Japan undertook no long- or short-term vertical migrations. They measured only the body chambers of the specimens. This hinders any conclusions on life-history trends. Applying the scheme of Lukeneder et al. (2010) on ammonoid and extant cephalopod habitats, water depth and temperature yielded a somewhat divergent picture: dispersal during ontogeny is evident. The range for all included taxa from the Cretaceous ammonoids of Hokkaido would then be 70–350 m, i.e. exclusively lower to middle epipelagic (the middle epipelagic zone dominates). The exception is the Phylloceratina (*Hypophylloceras*). They were restricted from the upper mesopelagic to middle epipelagic, a range from 100–300 m; this contrasts with the much deeper values given in the literature (up to 600 m). This means that the difference in depth and habitat between Phylloceratina, Lytoceratina, Ammonitina and Ancyloceratina are not as large as previously thought. Nonetheless, a shallowing gradient is evident from smooth to fine-ribbed types of Phylloceratina (*Hypophylloceras*, *Phyllopachyceras*) and Lytoceratina (*Tetragonites*, *Gaudryceras*), over stronger ribbed types of the Ammonitina (*Eupachydiscus*, *Yokoyamaoceras*) to heteromorph

Ancylocetratina (*Polyptychoceras*). At least *Polyptychoceras* (i.e., Pp1 to Pp2) appears to have migrated during ontogeny from the lower epipelagic at 200 m to shallower epipelagic zones at 60 m.

The conclusion in Moriya et al. (2003) that all species of the taxonomically diverse assemblage investigated were entirely demersal throughout their lives was challenged by $\delta^{18}O$ data presented by Henderson and Price (2012) for Late Cretaceous ammonoids (*Acanthoceras*, *Euomphaloceras*) from Australia. The highly negative $\delta^{18}O$ values were interpreted as indicating the growth of strongly ornamented morphogroups in near-surface waters. This conflicts with the interpretation of a deep marine nektobenthic life habit and with data from Late Cretaceous (Campanian) planispiral taxa recorded by Moriya et al. (2003).

Stable isotope data from Cretaceous (Albian) ammonoids of Madagascar (*Eotetragonites*, *Cleoniceras*, *Desmoceras*, *Douvilleiceras*) were given by Zakharov et al. (2011). Concerning the data as ontogenetic rows, the interpretation regarding the paleotemperature and habitat of the Albian Madagascan sea (tropical–subtropical zone) reflects habitats from upper mesopelagic to lower epipelagic for these ammonoids, with a general depth range from 70–400 m. The Lytoceratina with *Eotetragonites* appear to be the only inhabitants of the upper mesopelagic layers at 200–400 m. Contrastingly, Westermann (1996) suggested a shallower epipelagic range from 100–180 m based on siphuncle index estimations. The Albian members of the Ammonitina lived within the lower epipelagic zone. There are two finely ribbed forms: *Cleoniceras* shows a depth range from 100–200 m (max. depth 250 m in Westermann 1996) with a constant ontogenetic trend, whereas *Desmoceras* ranges from 70–250 m (min. depth 180 m in Westermann 1996). The ontogenetic trend in *Desmoceras* shows a major habitat change, i.e., a deepening in adult stages. Strongly ribbed Ancyloceratina like *Douvilleiceras* are relatively shallow-water inhabitants of the middle epipelagic at depths of 70–90 m, slightly increasing in ontogeny. The Albian nautiloid *Camytoceras* inhabited a depth of c. 180–200 m (not ontogenetically measured) in the range of extant *Nautilus*.

Based on $\delta^{18}O$ values for ammonoids, the rule of thumb that 'Trachyostraca' dominated above 200 m depth in the neritic epipelagic zone, whereas 'Leiostraca' dominated in the lowermost epipelagic and deeper mesopelagic zones, appears to be valid. There were, however, exceptions. In general, ammonoids were mostly adapted to neritic areas associated with the continental shelf. Crucial for future work on stable isotope trends in mobile and migrating ammonoids is an understanding of ocean stratification during the Jurassic and Cretaceous (Hay 2008). Knowledge about the presence or absence of a pycnocline and hence the thermal structure of the ocean bodies will improve our understanding of $\delta^{18}O$ data from cephalopod shells in the Mesozoic.

A major difference in the modes of life of many ammonoids compared to the extant cephalopods *Nautilus*, *Sepia*, and *Spirula* occured in the latest adult stage. All extant examples tend to retreat into the deepest environments as mature adult animals (*Spirula*, *Sepia*) or remain there throughout their post-juvenile phase (*Nautilus*). In contrast, all measured ammonoids except for *Perisphinctes* display a clear tendency to migrate into shallower environments in their latest ontogenetic stage (mature, fully grown; Lukeneder et al. 2010). Sexually dimorphic ammonoids might have reached maturity at different ages, earlier in males and later in females. Additional stable isotope analyses of antidimorphic pairs will help to test this hypothesis.

Caution should be exercised when estimating or assuming an ammonoid–facies dependency. Ammonoids were probably dependent on a specific environment, reflected today in the lithology/facies of the host rock. Ammonoids were probably capable of migrating between different habitats, or at least of escaping from hostile environmental changes (e.g., oxygen depletion, anoxia), giving rise to new populations. Such new faunal constituents are marked by distinct morphological adaptations to shallow or deeper environments, predominantly triggered by water energy and hydrostatic changes, and less bound to facies. Highly morphologically variable species challenge the paradigms that explicitly require a dependency of shell morphology and a special mode of life. Single species exhibiting numerous morphotypes, for example from oxycone to serpenticone detected in a single rock-layer were also able to inhabit the same habitat (pelagic areas) without any separation into morphogroups reflecting special facies dependence. An expansion of shallow-water areas can, however, influence morphological evolution of ammonoid shells, mainly during sea level transgressions onto shelves.

The main prey for ammonoids was most likely zooplankton from different water layers, from mid-water to benthic areas. Ostracods, foraminifera, tiny ammonoids (e.g., mandibles of juveniles hint at cannibalism ; Summesberger 2000) and planktic crinoids have been detected in stomach/crop content (Jäger and Fraaye 1997; Keupp 2007; see also Milson 1994), indicating that they were important food sources. The feeding strategies of ammonoids suggest that they probably ate everything they could reach, foraging as demersal herbivores, scavengers or even as active nektic carnivores.

The implications from ammonoid shell damage caused by predators are twofold. Predation can take place directly in the water column by mosasaurs, sharks, bony fishes and invertebrates (other cephalopods) or near the sea floor by, e.g., crustaceans, causing sublethal or lethal injuries. Abundant injuries caused by crustaceans hint at a (at least temporal) demersal life or benthic resting times. This recognition significantly changed the opinions presented in the past. A more demersal, nektobenthic to benthic life (indicated by the frequency of injuries) was suggested by some authors. Some ammonoids speculatively undertook vertical diurnal migrations, crossing different oceanic strata of the epipelagic to upper mesopelagic zones. This is characteristic for many planktic ocean dwellers and for nektobenthic genera such as *Nautilus* and *Spirula*.

Different ammonoid groups show major differences (size, morphology) in embryonic stages, not assessing where the eggs were stored (i.e., water column or benthic areas). The spawning strategy in ammonoids (*r*-strategy, 100–500 000 eggs per mother animal) contrasts with that of extant nautiloids (*K*-strategy, < 10 eggs or capsulae). The dispersal mechanisms of the planktic hatchlings and subsequent neanic ammonoids were entirely passive and depended on currents. Dispersal distance ranged from tens of kilometers in enclosed epeiric seas, over a few hundred kilometers on shelf seas, to up to 1000–2000 km in the open ocean. Planktic stages drifted passively, making up a large part of the Mesozoic plankton in shelf seas and adjoining oceans.

The postmortem history of ammonoids can change our picture of an assemblage in diverse ways. Shells that reach the sea surface after death can drift over long distances or long times. Postmortem drifting is enhanced when predators or scavengers remove the soft body from the shell, increasing buoyancy. This obscures the primary habitat and depth of death. In contrast, those shells that became negatively buoyant at a certain depth and sank to sea floor reveal a more reliable nearly autochthonous picture of the primarily deeper habitat. The path of a shell after the animal's death is controlled by several environmental factors such as depth at death, the initial buoyancy of the empty shell, the morphological and physiological parameters of the shell, and the rate of sea water influx into the phragmocone. Sinking history can also be detected by analyzing the encrusting community and its position. Buccal masses found within body chambers provide a useful criterion for evaluating postmortem transport. *In situ* buccal masses (upper and lower jaws) are suggested to reflect an *in situ* deposition of the ammonoid.

New stable isotope data ($\delta^{13}C$, $\delta^{18}O$) extracted from original shell materials (aragonite) revealed different ammonoid lifestyles (e.g., *wcw*–type, *cw*–type, *wc*-type), ontogenies, and varying habits in different habitats. Additional work on these isotopes and on the composition of ammonoid shells will enhance our knowledge on lifestyles and habitats. Exact ontogenetic measurements of $\delta^{13}C$ and $\delta^{18}O$ from neanoconchs to adult stages, from ammonitellae to terminal apertures, are crucial in future investigations on ammonoids.

Acknowledgements Studies were enhanced by the data generated from projects (P22109–B17, P20018–N10) financially supported by the Austrian Science Fund (FWF). The author wishes to thank Kazuyoshi Moriya (University of Kanazawa) for his extensive review of the manuscript and Christian Klug (University of Zurich) for his constructive comments and review. The author is grateful to Susanne Mayrhofer and Mathias Harzhauser (both Natural History Museum, Vienna) for discussions on specific problems on cephalopods and the life and habitat of ammonoids. The author thanks Kathleen A. Ritterbush (University of Southern California) for the calculation of a morphospace diagram. For important information and comments, I am grateful to Neil H. Landman (American Museum of Natural History, New York), Benjamin J. Linzmeier (University of Wisconsin-Madison) and two anonymous reviewers. Special thanks for editorial work and fruitful discussions on ammonoid life and habitat go to Christian Klug (University of Zurich) and Dieter Korn (Natural History Museum, Berlin).

References

Abel O (1916) Paläobiologie der Cephalopoden. Aus der Gruppe der Dibranchiaten. Verlag Gustav Fischer, Jena

Alberti M, Fürsich FT, Pandey DK, Ramkumar M (2011) Stable isotope analyses of belemnites from the Kachchh Basin, western India: paleoclimatic implications for the Middle to Late Jurassic transition. Facies 58:261–278

Anderson TF, Popp BN, Williams AC, Ho LZ, Hudson JD (1994) The stable isotopic record of fossils from the Peterborough Member, Oxford Clay Formation (Jurassic), UK: Palaeoenvironmental implications. J Geol Soc Lond 151:125–138

Andrew C, Hoew P, Paul CRC, Donovam SK (2010) Fatally bitten ammonoids from the lower Lias Group (Lower Jurassic) of Lyme Regis, Dorset. Proc York Geol Soc 58:81–94

Andrew C, Hoew P, Paul CRC, Donovam SK (2011) Epifaunal worm tubes on Lower Jurassic (Lower Lias) ammonites from Dorset. Proc Geol Assoc 34–46

Arkell W (1957) Introduction to Mesozoic Ammonoidea. In: Moore RC (ed) Treatise on Invertebrate Paleontology Part L Mollusca 4 Cephalopoda Ammonoidea. The University of Kansas Press, New York, pp. L81–L129

Arnold JM (1985) Shell growth, trauma and repair as an indicator of life history for *Nautilus*. Veliger 27:386–396

Assereto R (1969) Sul significato stratigrafico della 'Zona ad *Avisianus*' del Trias medio delle Alpi. Boll Soc Geol Italia 88:123–145

Auclair A-C, Lecuyer C, Bucher H, Sheppard SMF (2004) Carbon and oxygen isotope composition of *Nautilus macromphalus:* a record of thermocline waters off New Caledonia. Chem Geol 207:91–100

Ballentine CM (2007) A mathematical analyses of some indices used to classify ammonite shells. Lethaia 40:197–198

Bandel K (1988) Operculum and buccal mass of ammonites. In: Wiedmann J, Kullmann J (eds) Cephalopods–present and past. Schweizerbart, Stuttgart

Bandel K, Hoefs J (1975) Die Isotopenzusammensetzung von Molluskenschalen am Beispiel der Gastropoden. Neues Jahrb Geol Paläontol Monatshefte 1975:1–11

Bandel K, Landman NH, Waage KM (1982) Micro-ornament on early whorls of Mesozoic ammonites: implications for early ontogeny. J Paleontol 56:386–391

Batt RJ (1987) Pelagic biofacies of the Western Interior Greenhorn Sea (Cretaceous). Evidence from ammonites and planktonic foraminiferas [Ph.D. Thesis]: University of Colorado, p 778

Batt RJ (1989) Ammonite shell morphotype distribution in the Western Interior Greenhorn sea and some paleoecological implication. Palaios 4:32–42

Batt RJ (1991) Sutural amplitude of ammonite shells as a paleoenvironmental indicator. Lethaia 24:219–225

Batt RJ (1993) Ammonite shell morphotypes as indicators of oxygenation in ancient epicontinental seas: example from Late Cretaceous Greenhorn Cyclothem (U.S.A.). Lethaia 26:49–64

Bayer U (1970) Anomalies in Aalenian and Bajocian ammonites as clues to their mode of life. Neues Jahrb Geol Paläontol Abh 135:19–41

Bayer U (1977) Cephalopod septa I. Constructional morphology of the ammonite septum. Neues Jahrb Geol Paläontol Abh 154:290–366

Bayer U (1982) Ammonite maneuverability-a new look at the functions of shell geometry. Neues Jahrb Geol Paläontol Abh 164:154–156

Bayer U, McGhee GR Jr (1984) Iterative evolution of Middle Jurassic ammonite faunas. Lethaia 17:1–16

Becker RT (1993) Anoxia, eustatic changes, and Upper Devonian to lowermost Carboniferous global ammonoid diversity. In: House MR (ed) The Ammonoidea: Environment, Ecology, and Evolutionary Change. The Systematics Association, special vol 47. Clarendon Press, Oxford

Bettencourt V, Guerra A (1999) Carbon- and oxygen-isotope composition of the cuttlebone of *Sepia officinalis:* a tool for predicting ecological information? Marine Biol 133:651–657

Bettencourt V, Guerra A (2001) Age studies based on daily growth increments in statoliths and growth lamellae in cuttlebone of cultured *Sepia officinalis*. Marine Biol 139:327–334

Birkelund T (1981) Ammonoid shell structure. In: House MR, Senior J (eds) The Ammonoidea. Systematics Association, special vol 18. Academic Press, London

Boletzky S v (1974) The "larvae" of Cephalopoda: a review. Thalassia Jugosl 10:45–76

Boletzky SV (1987) Juvenile behaviour. In: Boyle PR (ed) Cephalopod life cycles, vol II. Academic Press, London

Boletzky SV (1992) Evolutionary aspects of development, life style, and reproductive mode in incirrate octopods (Molluscs, Cephalopoda). Rev Suisse Zool 99:755–770

Bond PN, Saunders WB (1989) Sublethal shell repair in Upper Mississippian ammonoids. Paleobiol 15:414–428

Böse E (1928) Cretaceous ammonites from Texas and northern Mexico. Bull Univ Tex 2748:143–312

Boston WB, Mapes RH (1991) Ectocochleate cephalopod taphonomy. In: Donovan SK (ed) Processes of fossilization. Bellhaven Press, London

Boucot AJ (1981) Principles of benthic marine paleoecology. Academic Press, New York

Boyajian G, Lutz T (1992) Evolution of biological complexity and its relation to taxonomic longevity in the ammonoidea. Geology 20:983–986

Brack P, Rieber H (1993) Towards a better definition of the Anisian/Ladinian boundary: new biostratigraphic data and correlations of boundary sections from the Southern Alps. Eclogae Geol Helvetiae 86:415–527

Brack P, Rieber H (1996) The 'High resolution Middle Triassic ammonoid standard scale' proposed by Triassic researchers from Padova–a discussion of the Anisian/Ladinian boundary interval. Albertiana 17:42–50

Brand U (1986) Paleoenvironmental analysis of Middle Jurassic (Callovian) ammonoids from Poland: trace elements and stable isotopes. J Paleontol 60:293–301

Brett CE, Miller KB, Baird GC (1990) A temporal hierarchy of paleoecologic processes within a Middle Devonian epeiric sea. In: Miller W (ed) Paleocommunity Temporal Dynamics: The Longterm Development of Multispecles Assemblages. Paleontol Soc Spec Publ 5:178–203

Brett CE, Dick VB, Baird GC (1991) Comparative taphonomy and paleoecology of Middle Devonian dark gray and black shale facies from western New York. State Mus Bull 469:5–36

Brumsack HJ (1991) Inorganic geochemistry of the German 'Posidonia Shale': Palaeoenvironmental consequences. In: Tyson RV, Pearson TH (eds) Modern and ancient continental shelf anoxia. Geol Soc Spec Publ 58:353–362

Bruna GD, Martire L (1985) La successione giurassica (Pliensbachiano-Kimmeridgiano) delle Alpi Feltrine (Belluno). Riv Ital Paleontol Stratigr 91:15–62

Bucher H, Landman NH, Klofak SM, Guex J (1996) Mode and rate of growth in ammonoids. In: Landman NH, Tanabe K, Davis RA (eds) Ammonoid Paleobiology. Topics in Geobiology 13, Plenum, New York

Bulot L (1989) Les Olcostephaninae (Ammonitina, Cephalopoda) dans le Crétacé inférieur du Jura Suisse et Francais. In: Museé d'Histoire Naturelle de Genève Résumés (ed) Réunion Commune APF-SPS, 21–22 Octobre 1989, Geneva, p 5

Bulot L (1992) Les Olcostephaninae Valanginiens et Hauteriviens (Ammonitina, Cephalopoda) du Jura Franco-Suisse: systematique et interet biostratigraphique. Rev Paléobiol 11:149–166

Bulot L (1993) Stratigraphic implications of the relationship between ammonites and facies: examples taken from the Lower Cretaceous (Valanginian-Hauterivian) of the western Tethys. In: House MR (ed) The Ammonoidea: Environment, Ecology, and Evolutionary Change, Systematics Association, special vol 47. Clarendon Press, Oxford

Bulot L (1995) Les formations à ammonites du Crétacé inférieur dans le SE de la France (Berriasien-Hauterivien): Biostratigraphie, paléontologie et cycles sédimentaires [PhD thesis]: Paris, University Paris, p. 374

Bulot L, Company M (1990) Les *Olcostephanus* du groupe *atherstoni* (Ammonitina, Cephalopoda): potential d'utilisation pour les correlations biostratigraphiques à longue distance. In: VI J Soc Esp Paleontol rés, p. 34

Burnaby TP (1966) Allometric growth of ammonoid shells: a generalization of the logarithmic spiral. Lett Nat 209:904–906

Callomon JH (1980) Dimorphism in ammonoids. In: House MR, Senior JR (eds) The Ammonoidea, Systematics Association, special vol 18. Academic Press, London

Callomon JH (1985) The evolution of the Jurassic ammonite family Cardioceratidae. Spec Pap Palaeontol 33:49–90

Carlson BA, McKibben, JN, deGruy, MV (1984) Telemetric investigation of vertical migration of *Nautilus belauensis*. Pac Sci 38:183–188

Casadío S, Concheyro A (1992) Facies y ambientes de sedimentación en el límite Cretácico-Terciario de La Pampa, Argentina. Act VIII Cong Latin Geol Salamanca 4:30–34

Cecca F (1988) Ammonites méditerranéennes du Tithonique inférieur de l'Ardèche (Sud-Est de la France): analyse des afflux téthysiens. Geobios 21:169–186

Cecca F (1992) Ammonite habitats in the Early Tithonian of Western Tethys. Lethaia 25:257–267

Cecca F (1997) Late Jurassic and Early Cretaceous uncoiled ammonites: trophism-related evolutionary processes. Comptes Rendus Acad Sci Paris, Ser II 325:629–634

Cecca F (1998) Hypothesis about the role of the trophism in the evolution of the uncoiled ammonites: the adaptive radiations of the Ancyloceratina (Ammonoidea) at the end of the Jurassic and in the Lower Cretaceous. Rend Fis Acc Lincei 9:213–226

Cecca F, Cresta S, Pallini G, Santantonio M (1990) Il Giurassico di Monte Nerone (Appennino marchigiano, Italia Centrale): biostratigrafia, litostratigrafia ed evoluzione paleogeografica. In: Pallini G, Cecca F, Cresta S, Santantonio M (eds) Fossili, Evoluzione e Ambiente. Atti de Secondo Convegno Internazionale, Pergola ottobre 1987, Ed Comitato Centenario R Piccinini, Ancona

Cecca F, Fozy I, Wierzbowski A (1993) Ammonites et paléoécologie: étude quantitative d'associations du Tithonien inférieur de la Tethys occidentale: In: Elmi S, Mangold C, Alméras Y (eds) 3ème Symposium International, Céphalopodes Actuels et Fossiles (Lyon, 1990), Geobios Mém Spéc 15:39–48

Ceriola L, Milone N (2007) Cephalopods age determination by statolith reading: a technical manual. Adria Med Technical Documents 22, Rome, 78 p

Challier L, Dunn M R, Robin J-P (2005) Trends in age-at-recruitment and juvenile growth of cuttlefish, *Sepia officinalis*, from the English Channel. ICES J Mar Sci 62:1671–1682

Chamberlain JA (1976) Flow patterns and drag coefficients of cephalopod shells. Paleobiology 19:539–563

Chamberlain JA Jr (1980) Hydrodynamical design of fossil cephalopods. In: House MR, Senior JR (eds) The Ammonoidea: the evolution, classification, mode of life and geological usefulness of a major fossil group. The Systematics Association, special, vol 18. Academic Press, London

Chamberlain JA Jr (1991) Cephalopod locomotor design and evolution: the constraints of jet propulsion. In: Rayner JMV, Wooton RJ (eds) Biomechanics and evolution. Cambridge University Press, Cambridge

Chamberlain JA Jr, Moore WA Jr (1982) Rupture strength and flow rate of *Nautilus* siphuncular tube. Paleobiology 8:408–5452

Chamberlain JA Jr, Weaver JS (1978) Equations of motion for post-mortem sinking of cephalopod shells. Math Geol 10:673–689

Chamberlain JA Jr, Ward PD, Weaver JS (1981) Postmortem ascent of *Nautilus* shells: implications for cephalopod paleogeography. Paleobiology 7:494–509

Checa AG (2003) Fabrication and function of ammonite septa—comment on Lewy. J Paleontol 77:790–791

Checa AG, Garcia-Ruiz JM (1996) Morphogenesis of the septum in ammonoids. In: Landman NH, Tanabe K, Davis RA (eds) Ammonoid Paleobiology. vol 13 of Topics in Geobiology, Plenum, New York

Checa A, Westermann GEG (1989) Segmental growth in planulate ammonites: inferences on costae function. Lethaia 22:95–100

Cherel Y, Hobson KA (2005) Stable isotopes, beaks and predators: a new tool to study the trophic ecology of cephalopods, including giant and colossal squids. Proc Biol Soc 272:1601–1607

Cherel Y, Ridoux V, Spitz J, Richard P (2009) Stable isotopes document the trophic structure of a deep-sea cephalopod assemblage including giant octopod and giant squid. Biol Lett 5:364–367

Chikaraishi Y, Ogawa NO, Kashiyama Y, Takano Y, Suga H, Tomitani A, Miyashita H, Kitazato H, Ohkouchi N (2009) Determination of aquatic food-web structure based on compound-specific nitrogen isotopic composition of amino acids: Liminology and Oceanography. Methods 7:740–750

Chirat R (2000) The so-called "cosmopolitan palaeobiogeographic distribution" of Tertiary Nautilida of the genus *Aturia* Bronn 1838: the result of post-mortem transport by oceanic palaeocurrents. Palaeogeogr Palaeoclimatol Palaeoecol 157:59–77

Chirat R (2001) Anomalies of embryonic shell growth in post-Triassic Nautilida. Paleobiology 27:485–499

Christ HA (1960) Beiträge zur Stratigraphie und Paläontologie des Malm von Westsizilien. Mém Soc Paléont Suisse 77:1–138

Cichowolski M, Pazos PJ, Tunik MA, Aguirre-Urreta MB (2011) An exceptional storm accumulation of nautilids, Lower Cretaceous, Neuquén Basin, Argentina. Lethaia 45:121–138

Clari PA, Marini P, Pastorini M, Pavia G (1984) Il Rosso Ammonitico Inferiore (Baiociano-Calloviano) nei Monti Lessini Settentrionali (Verona). Riv Ital Paleontol Stratigr 90:15–86

Clarke MR (1969) Cephalopoda collected on the SOND cruise. J Mar Biol Assoc UK 49:961–976

Cochran JK, Rye DM, Landman NH (1981) Growth rate and habitat of *Nautilus pompilius* inferred from radioactive and stable isotope studies. Paleobiology 7:469–480

Cochran JK, Landman NH, Turekian KK, Michard A, Schrag DP (2003) Paleoceanography of the Late Cretaceous (Maastrichtian) Western Interior Seaway of North America: evidence from Sr and O isotopes. Palaeogeogr Palaeoclimatol Palaeoecol 191:45–64

Cochran JK, Kallenberg K, Landman NH, Harries PJ, Weinreb D, Turekan KK, Beck AJ, Cobban WA (2010a) Effect of diagensis on the Sr, O and C isotopic composition of Late Cretaceous mollusks from the Western Interior of North America. Am J Sci 310:69–88

Cochran JK, Landman NH, Harries PJ, Larson NL, Garb MP, Klofak SM, Myers C, Brezina J (2010b) Stable isotopes in well-preserved shells of a methane seep fauna from the Upper Cretaceous (Campanian) Pierre Shale, U.S. Western Interior. 8th International Symposium, Cephalopods—Present and Past, 30 August–3 September, 2010 Dijon, Abstract Volume, 119

Collignon M (1963) Atlas des fossils caractéristiques de Madagascar (Ammonites), Albien. Serv Géol Tananarive 10:184

Collignon M (1964) Atlas des fossils caractéristiques de Madagascar (Ammonites), Cenomanien. Serv Géol Tananarive 11:152

Collins D, Ward PD (2010) Adolescent growth and maturity in *Nautilus*. In: Saunders WB, Landman NH (eds) *Nautilus*. the biology and paleobiology of a living fossil. Topics in Geobiology, 6, Springer Press, New York (reprinted from Saunders WB, Landman NH (eds) 1987 with additions) pp. 421–432

Company M (1987) Los Ammonites del Valanginiense del sector oriental de las Cordilleras Béticas (SE de Espana) [PhD thesis]: Universidad de Granada, Granada, p 294

Company M, Aguado R, Sandoval J, Tavera JM, Jimenez de Cisneros C, Vera JA (2005) Biotic changes linked to a minor anoxic event (Faraoni Level, Latest Hauterivian, Early Cretaceous). Palaeogeogr Palaeoclimat Palaeoecol 224:186–199

Cope JCW (1967) The palaeontology and stratigraphy of the lower part of the Upper Kimmeridge Clay of Dorset. Bull Br Mus Nat Hist Geol 15:1–79

Cope JCW (1974) Upper Kimmeridge ammonite faunas of wash area and a subzonal scheme from the lower part of the Upper Kimmeridgian. Bull Geol Surv GB 1974:29–37

Crocker KC, DeNio, MJ, Ward, PD (1985) Stable isotopic investigations of early development in extant and fossil chambered cephalopods I. Oxygen isotopic composition of eggwater and carbon isotopic composition of siphuncle organic matter in *Nautilus*. Geochim Cosmochim Acta 49:2527–2532.

Currie ED (1942) Growth stages in the ammonite *Promicroceras marstonense* Spath. Proc R Soc Edinb B 61:344–367

Currie, ED. (1944) Growth stages in some Jurassic ammonites. Trans R Soc Edinb 61:171–198

Dagys AS, Weitschat W (1993) Extensive intraspecific variation in a Triassic ammonoid from Siberia. Lethaia 26:113–122

Daniel, TL, Helmuth BS, Sunders WB, Ward PD (1997) Septal complexity in ammonoid cephalopods increased mechanical risk and limited depth. Paleobiology 23: 470–481

D'Arpa C, Meléndez G. (2004) Oxfordian biostratigraphy and ammonite associations from west Sicily: biostratigraphic succession of genus *Gregoryceras* ad correlation with Tethyan Perisphinctid scale. Riv Ital Paleontol Strati 110:225–267

Davis RA (1972) Mature modification and dimorphism in selected Late Paleozoic ammonoids. Bull Am Paleontol 62:26–130

Davis RA, Furnish WM, Glenister BF (1969) Mature modification and dimorphism in Late Paleozoic ammonoids. In: Westermann GEG (ed) Sexual dimorphism in fossil metazoa and taxonomic implications. Schweizerbart, Stuttgart

Davis RA, Landman NH, Dommergues J-L, Marchand D, Bucher H (1996) Mature modifications and dimorphism in ammonoid cephalopods. In: Landman NH, Tanabe K, Davis RA (eds) Ammonoid paleobiology. Plenum, New York

Davis RA, Mapes RH, Klofak SM (1999) Epizoa on externally shelled cephalopods. In: Rozanov AY, Shevyrev AA (eds) Fossil Cephalopods: recent advances in their study. Russian Academy of Sciences, Palaeontological Institute, Moskva

De Baets K Klug C Korn D Landman NH. (2012) Early evolutionary trends in ammonoid embryonic development. Evolution 66:1788–1806

De Baets K Klug C Monnet C (2013) Intraspecific variability through ontogeny in early ammonoids. Paleobiology 39: 75–94

De Baets K, Landman NH, Tanabe K (2015) Ammonoid embryonic development. This volume

Dennis KJ, Cochran JK, Landman NH, Schrag DP (2013) The climate of the Late Cretaceous: new insights from the application of the carbonate clumped isotope thermometer to Western Interior Seaway macrofossil. Earth Planet Sci Lett 362:51–65

Denton EJ (1974) On the buoyancy and the lives of modern and fossil cephalopods. Proc R Soc Lond B 185:273–299

Denton EJ, Gilpin-Brown JB (1973) Floatation mechanisms in modern and fossil cephalopods. Adv Mar Biol 11:197–268

Ditchfield PW (1997) High Northern palaeolatitude Jurassic-Cretaceous palaeotemperature variation: new data from Kong Karls Land, Svalbard. Palaeogeogr Palaeoclimatol Palaeoecol 61:237–254

Doguzhaeva LA (1988) Siphuncular tubes and septal necks in ammonite evolution. In: Wiedmann J, Kullmann J (eds) Cephalopods–present and past. Schweizerbart, Stuttgart

Doguzhaeva LA (2002) Adolescent bactritoid, orthoceratoid, ammonoid and coleoid shells from the Upper Carboniferous and Lower Permian of the South Urals. In: Summesberger H, Histon K, Daurer A (eds) Cephalopods–present and past. Abh Geol B-A 57: 9–55

Doguzhaeva LA, Mutvei H (1989) *Ptychoceras*—a heteromorphic lytoceratid with truncated shell and modified ultrastructure (Mollusca: Ammonoidea). Palaeontogr Abt A 208:91–121

Doguzhaeva LA, Mutvei H (1991) Organization of the soft body in *Aconeceras* (Ammonitina), interpreted on the basis of shell morphology and muscle scars. Palaeontogr Abt A Palaeoz Stratigr 218:17–33

Doguzhaeva LA, Mutvei H. (1992) Radula of the Early Cretaceous ammonite *Aconeceras* (Mollusca: Cephalopoda). Palaeontogr A 223:167–177

Doguzhaeva LA, Mutvei H (1993) Structural features in Cretaceous ammonoids indicative of semiinternal or internal shells. In: House MR (ed) The Ammonoidea: Environment, Ecology, and Evolutionary Change. Systematics Association, special vol 47. Clarendon Press, Oxford

Doguzhaeva LA, Mapes RH, Summesberger H, Mutvei H (2007) The preservation of body tissues, shell, and mandibles in the ceratitid Ammonoid *Austrotrachyceras* (Late Triassic), Austria. In: Landman NH, Davis RA, Mapes RH (eds) Cephalopods Present and Past: new insights and fresh perspectives. Springer, Dordrecht

Doguzhaeva LA, Bengtson S, Mutvei H (2010) Structural and morphological indicators of mode of life in the Aptian lytoceratid ammonoid *Eogaudryceras*. In: Tanabe K, Shigeta Y, Sasaki T, Hirano H (eds) Cephalopods Present and Past. Tokai University Press, Tokyo

Donovan DT (1985) Ammonite shell form and transgression in the British Lower Jurassic. In: Bayer U, Seilacher A (eds) Sedimentary and evolutionary cycles, vol. 1. Springer–Verlag, Berlin

Donovan SK (1989) Taphonomic significance of the encrustation of the dead shell of recent *Spirula spirula* (Linneé) (Cephalopoda: Coleoidea) by *Lepas anatifera* Linné (Cirripedia: Thoracia). J Paleontol 63:698–702

Donovan DT (1993) Ammonites in 1991. In: House MR (ed) The Ammonoidea. environment, ecology and evolutionary change. Systematics Association, special vol. 47. Clarendon Press, Oxford

Doyle P, Whitham AG (1991) Palaeoenvironments of the Nordenskjold Formation: an Antarctic Late Jurassic–Early Cretaceous black shale-tuff sequence. In: Tyson RV, Pearson TH (eds) Modern and ancient continental shelf anoxia. Geol Soc Spec Publ 58:397–414

Duff KL (1975) Palaeoecology of a bituminous shale—the Lower Oxford Clay of central England. Palaeontology 18:443–482

Dunstan A, Bradshaw CJA, Marshall J (2011) *Nautilus* at Risk—Estimating population size and demography of *Nautilus pompilius*. PLoS ONE 6(2):e16716. doi:10.1371/journal.pone.0016716

Dutton A, Huber BT, Lohmann KC, Zinsmeister WJ (2007) High-resolution stable isotope profiles of a dimitobelid belemnite: implications for a paleodepth habitat and late Maastrichtian climate seasonality. Palaios 22:642–650

Ebel K (1983) Calculations on the buoyancy of ammonites. Neues Jahrb Geol Paläontol Monatshefte 1983:614–640

Ebel K (1985) Gehäusespirale und Septenformen bei Ammoniten unter Annahme vagil benthischer Lebensweise. Paläontol Z 59:109–123

Ebel K (1990) Swimming abilities of ammonites and limitations. Paläontol Z 64:25–37

Ebel K (1999) Hydrostatics of fossil ectochochleate cephalopods and its significance for the reconstruction of their lifestyle. Paläontol Z 73:277–288

Ebel K (1992) Mode of life and soft body shape of heteromorph ammonites. Lethaia 25:179–193

Eichler R, Ristedt H (1966a) Isotopic evidence on the Earth Life History of *Nautilus pompilius* (Linné). Science 153:734–736

Eichler R, Ristedt H (1966b) Untersuchungen zur Frühontogenie von *Nautilus pompilius* (Linné). Paläontol Z 40:173–191

Elmi S (1993) Area-rule, boundary layer and functional morphology of cephalopod shells (Ammonoids). Geobios Mem Spéc 15:121–138

Elmi S, Alméras Y (1984) Physiography, palaeotectonics and palaeoenvironment as control of changes in ammonite and brachiopod communities (an example from the Early and Middle Jurassic of western Algeria). Pelaeogeogr Palaeoclimatol Palaeoecol 47:347–360

Elmi S, Benshili K (1987) Relations entre la structuration tectonique, la composition des peuplements et l'évolution; exemple du Toarcian du Moyen–Atlas meridional (Maroc). Boll Soc Paleontol Ital 26:47–62

Enay R (1980) Paléobiogéographie et ammonites jurassiques: "rythmes fauniques" et variations du niveau marin; voies d'échanges, migrations et domaines biogéographiques. Mém Hors sér Soc Géol France 10:261–81

Enay R, Cariou E (1997) Ammonite faunas andpalaeobiogeography of the Himalayan belt during the Jurassic: initiation of a Late Jurassic austral ammonite fauna. Palaeogeogr Palaeoclimat Palaeoecol 134:1–38

Enay R, Mangold C (1982) Dynamique biogéographique et évolution des faunes d'ammonites au Jurassique. Bull Soc Géol France 24:1025–1046

Epstein S, Buchsbaum R, Lowenstam HA, Urey HC (1953) Revised carbonate-water isotopic temperature scale. Bull GSA 64:1315–1326

Etches S, Clarke J, Callomon J (2009) Ammonite eggs and ammonitellae from the Kimmeridge Clay Formation (Upper Jurassic) of Dorset, England. Lethaia 42:204–217

Fabricius F, Friedrichsen H, Jacobshagen V (1970) Paläotemperaturen und Paläoklima in Obetrias und Lias der Alpen. Int J Earth Sci 59:805–826

Fatherree JW, Harries PJ, Quinn TM (1998) Oxygen and carbon isotopic "dissection" of *Baculites compressus* (Mollusca: Cephalopoda) from Pierre Shale (Upper Campanian) of South Dakota: implications for paleoenvironmental reconstructions. Palaios 13:376–385

Fernandez-López S (1991) Taphonomic concepts for a theoretical biochronology. Rev Esp Paleontol 6:37–49

Fernández-López S (1997) Ammonites, taphonomic cycles and stratigraphic cycles in carbonate epicontinental platforms. Cuad Geol Iber 23:95–136

Fernández-López S, Meléndez G (1994) Abrasion surfaces on internal moulds of ammonites as palaeobathymetric indicators. Palaeogeogr Palaeoclimat Palaeoecol 110:29–42

Fernández-López S, Meléndez G (1995) Taphonomic gradients in Middle Jurassic ammonites of the Iberian Range (Spain). Geobios Mém Spéc 18:155–165

Fernandez-López S, Melendez G (1996) Phylloceratina ammonoids in the Iberian Basin during the Middle Jurassic: a model of biogeographical and taphonomic dispersal related to relative sea-level changes. Palaeogeogr Palaeoclimat Palaeoecol 120:291–302

Fernandez-López S, Melendez G (2004) Fossilization of ammonites and sedimentary events in deep environments of carbonate platform (highest Middle to lowest Upper Oxfordian, Iberian Range, Spain). Riv It Paleontol Strati 110:219–230

Fernandez-López SR, Henriques MH, Linares A, Sandoval J, Ureta MS (1999) Aalenian *Tmetoceras* (Ammonoidea) from Iberia. taxonomy, habitats and evolution In: Olóriz F, Rodríguez-Tovar FJ (eds) Advancing Research on Living and Fossil Cephalopods. Plenum, New York

Forester RW, Caldwell WGE, Oro FH (1977) Oxygen and carbon isotopic study of ammonites from Late Cretaceous Bearpaw Formation in southwestern Saskatchewan. Can J Earth Sci 14:2086–2110

Forsythe JW, Van Heukelem WF (1987) Growth. In: Boyle PR (ed) Cephalopod life cycles, Vol. 2. Academic Press, New York

Fourcade E, Azema J, Cecca F, Bonneau M, Peybernes B, Dercourt J (1991) Essai de reconstitution cartographique de la Paléogéographie et des Paléoenvironnements de la Tethys au Tithonique supérieur (138 Ã 135 Ma). Bull Soc Géol Fr 162:1197–1208

Fraaije RHB (2003) The oldest *in situ* hermit crab from the Lower Cretaceous of Speeton, UK. Palaeontology 46:53–58

Frye CJ, Feldman RM (1991) North American Late Devonian cephalopod aptychi. Kirtlandia 49:49–71

Galacz A, Horwath E (1985) Sedimentary and structural evolution of the Bakony Mountains (Transdanubian Central Range, Hungary): Paleogeographic implications. Acta Geol Hung 28:85–100

Geary D, Brieske TA, Bemis B (1992) The influence and interaction of temperature, salinity, and upwelling on the stable isotopic profiles of strombid gastropod shells. Palaios 7(1):77–85

Geczy B (1982) The Davoi Zone in the Bakony Mountains, Hungary. Ann Univ Sci Budap Sect Geol 21:1–11

Geczy B (1984) Provincialism of Jurassic ammonites; examples from Hungarian faunas. Acta Geol Hung 27:379–389

Geosecs Atlantic, Pacific and Indian Ocean Expeditions (1987) Shore-based data and graphics GEOSECS Executive Committee IDOE National Science Foundation 7

Geraghty M, Westermann GEG (1994) Composition and origin of Jurassic ammonite concretions from Alfeld, Germany: a biogenic alternative. Paläontol Z 68:473–490

Ginsburg RN, Schroeder JH (1973) Growth and submarine fossilization of algal cup reefs, Bermuda. Sedimentology 20:575–614

Gischler E, Ginsburg RN (1996) Cavity dwellers (coelobites) under coral rubble in southern Belize barrier and atoll reefs. Bull Mar Sci 58:570–589

Glaessner M (1931) Eine Crustaceenfauna aus den Lunzer Schichten Niederösterreichs. Jb Geol BA 81:467–486

Goodwin DH, Schöne BR, Dettman DL (2003) Resolution and fidelity of oxygen isotopes as palaeotemperature proxies in bivalve mollusk shells: models and observations. Palaios 18:110–125

Greenwald L, Ward PD, Greenwald OE (1980) Cameral liquid transport and buoyancy control in chambered *Nautilus* (*Nautilus macromphalus*). Nature 286:55–56.

Greenwald KP, Cook CB, Ward P (1982) The structure of the chambered *Nautilus* siphuncle: the siphuncular epithelium. J Morphol 172:5–22

Grossman EL, Ku T (1986) Oxygen and carbon isotope fractionation in biogenic aragonite: temperature effects. Chem Geol 59:59–74

Grunert P, Harzhauser M, Rögl F, Sachsenhofer R, Gratzer R, Soliman A, Piller WE (2010) Oceanographic conditions triggering the formation of an Early Miocene (Aquitanian) Konservat-Lagerstätte in the Central Paratethys Sea. Palaeogeogr Palaeoclimatol Palaeoecol 292:425–442

Guex J (2001) Environmental stress and atavism in ammonoid evolution. Eclogae Geol Helv 94:321–328

Gygi RA (1986) Eustatic sea level changes of the Oxfordian (Late Jurassic) and their effect documented in sediments and fossil assemblages of an epicontinental sea. Eclogae Geol Helv 79:455–491

Haimovici M, Brunetti NE, Rodhouse PG, Csirke J, Leta RH (1998) *Illex argentinus*. In: Rodhouse PG, Dawe EG, O'Dor RK (eds) Squid recruitment dynamics. The genus *Illex* as a model. The commercial *Illex* species. Infiuences on variability. FAO Fish Tech Pap 376:27–58
Hamada T (1964) Notes on the drifted *Nautilus* in Thailand. Contr Geol Palaeont Southeast Asia, 21, Sci Pap Coll Geb Educ 14:255
Hamada T (1984) Further notes on *Nautilus* drifts. Contr Geol Palaeont SE Asia 25:263
Hantzpergue P (1995) Faunal trends and sea-level changes: biogeographic patterns of Kimmeridgian ammonites on the Western European Shelf. Geol Rundsch 84:245–54
Hassan MA, Westermann GEG, Hewitt RA, Dokainish MA (2002) Finite-element analysis of simulated ammonoid septa (extinct Cephalopoda): septal and sutural complexities do not reduce strength. Paleobiology 28:113–126
Hay WW (2008) Evolving ideas about the cretaceous climate and ocean circulation. Cretac Res 29:725–753
Hayasaka S, Oki K, Tanabe K, Saisho T, Shinomiya A (1987) On the habitat of *Nautilus pompilius* in taon strait (Philippines) and the Fiji Islands. In: Saunders W B, Landman N H (eds) *Nautilus*: The Biology and Paleobiology of a Living Fossil. Plenum Press, New York
He S, Kyser TK, Caldwell GE (2005) Paleoenvironment of the Western Interior Seaway inferred from $\delta^{18}O$ and $\delta^{13}C$ values of molluscs from Cretaceous Bearpaw marine cyclothem. Palaeogeogr Palaeoclimatol Palaeoecol 217:67–85
Henderson RA, Price GD (2012) Paleoenvironment and paleoecology inferred from oxygen and carbon isotopes of subtropical mollusks from the late Cretaceous (Cenomanian) of Bathurst Island, Australia. Palaios 27(9):617–626
Hengsbach R (1978) Bemerkungen über das Schwimmvermögen der Ammoniten und die Funktion der Septen. Sitzungsberichte Ges Naturforschungs Freunde Berl Neue Folge 18:105–117
Heptonstall WB (1970) Buoyancy control in ammonoids. Lethaia 3:317–328
Hewitt RA (1988) Significance of early septal ontogeny in ammonoids and other ectocochliates. In: Wiedmann J, Kullmann J (eds) Cephalopods Present and Past. Schweizerbart, Stuttgart
Hewitt RA (1993) Relation of shell strength to evolution. In: House MR (ed) The Ammonoidea: Environment, Ecology, and Evolutionary Change. Systematics Association, special vol. 47. Clarendon Press, Oxford
Hewitt RA (1996) Architecture and strength of the ammonoid shell. In: Landman NH, Tanabe K, Davis RA (eds) Ammonoid paleobiology. Plenum, New York
Hewitt RA, Stait B (1988) Seasonal variations in septal spacing of *Sepia officinalis* and some Ordovician actinocerid nautiloids. Lethaia 21:283–394
Hewitt RA, Westermann GEG (1987) Functions of complexly fluted septa in ammonoid shells. II. Septal evolution and conclusions. Neues Jahrb Geol Paläontol Abh 174:135–169
Hewitt RA, Westermann GEG (1988) Stress and strain in *Nautilus* shells: some limitations on the buoyancy control and vertical migration of ectocochliates. In: Wiedmann J, Kullmann J (eds) Cephalopods Present and Past. Schweizerbart, Stuttgart
Hewitt RA, Westermann GEG (1990a) *Nautilus* shell strength variance as an indicator of habitat depth limits. Neues Jahrb Geol Paläontol Abh 179:73–97
Hewitt RA, Westermann GEG (1990b) Mosasaur tooth marks on the ammonite *Placenticeras* from the Upper Cretaceous Bearpaw Formation of Alberta. Can J Earth Sci 27:469–472
Hewitt RA, Westermann GEG (1997) Mechanical signifcance of ammonoid septa with complex sutures. Lethaia 30:205–212
Hewitt RA, Westermann GEG (2003) Recurrences of hypotheses about ammonites and *Argonauta*. J Paleontol 77:792–795
Hewitt RA, Westermann GEG, Checa A, Zaborski PM (1994) Growth rates of ammonoids estimated from aptychi. Geobios Mém Spéc 15:203–208
Hirano H (1986) Cenomanian and Turonian biostratigraphy of the off-shore facies of the Northern Pacific-an example of the Oyubari area, central Hokkaido, Japan. Bull Sci Eng Res Lab Waseda Univ 113:6–20
Hirano H (1988) Evolutionary mode of some late cretaceous ammonites in offshore waters. In: Wiedmann J, Kullmann J (eds) Cephalopods Present and Past, Schweizerbart, Stuttgart

Hirano H (1993) Phyletic evolution of desmoceratine ammonoids through the Cenomanian–Turonian oceanic anoxic event. In: House MR (ed) The Ammonoidea: Environment, Ecology, and Evolutionary Change, Systematics association, special vol 47. Clarendon Press, Oxford
Hirano H, Okamoto T, Hattori K (1990) Evolution of some Late Cretaceous desmoceratine ammonoids. Trans Proc Palaeontol Soc Jpn NS 157:382–411
Hobson KA, Cherel Y (2006) Isotopic reconstruction of marine food webs using cephalopod beaks: new insight from captively raised Sepia officinalis. Can J Zool 84:766–770
Hoedemaeker PJ (1990) The Neocomian boundaries of the Tethyan Realm based on the distribution of ammonites. Cretac Res 11:331–342
Hoefs J (2004) Stable isotope geochemistry. (Fifth revised and updated edition). Springer Verlag, New York, p. 244
Hoffmann R, Lemanis R, Naglik C, Klug C (2015) Ammonoid buoyancy. This volume
House MR (1985) A new approach to an absolute time scale from measurements of orbital cycles and sedimentary microrhythms. Nature 316:721–725
House MR (ed) (1993a) The Ammonoidea: Environment, Ecology, and Evolutionary Change. Systematics Association, special vol 47. Clarendon Press, Oxford, p 354
House MR (1993b) Fluctuations in ammonoid evolution and possible environmental causes. In: House MR (ed) The Ammonoidea: Environment, Ecology, and Evolutionary Change. Systematics Association, special vol 47. Clarendon Press, Oxford
House MR, Price JD (1985) New Late Devonian genera and species of tornoceratid goniatites. Palaeontology 28:159–188
House MR, Senior JR (eds) (1981) The Ammonoidea: the evolution, classification, mode of life and geological usefulness of a major fossil group. Systematics Association, special vol 18. Academic Press, London
Howarth MK (1992) The ammonite family Hildoceratidae in the Lower Jurassic of Britain. Palaeontogr Soc Monogr 590(146):1–200
Hudson JD, Martill DM (1991) The lower Oxford Clay: Production and preservation of organic matter in the Callovian (Jurassic) of central England. In: Tyson RV, Pearson TH (eds) Modern and ancient continental shelf anoxia. Geol Soc Spec Pap 58:363–379
Hyatt A (1894) Phylogeny of an acquired characteristic. Am Philos Soc 32 143:349–647
Ifrim C, Stinnesbeck W (2010) Migration pathways of the late Campanian and Maastrichtian shallow facies ammonite *Sphenodiscus* in North America. Palaeogeogr Palaeoclimatol Palaeoecol 292(1):96–102
Ishii T (1981) Shells of *Nautilus* drifted ashore after an interval of 11 years. J Malac Soc Jpn 12:37–39
Jackson JBC, Winston JE (1982) Ecology of cryptic coral reef communities. I. Distribution and abundance of major groups of encrusting organisms. J Exper Mar Biol Ecol 57:135–147
Jacobs DK (1992a) Shape, drag, and power in ammonoid swimming. Paleobiology 18:203–220
Jacobs DK (1992b) The support of hydrostatic load in cephalopod shells. Adaptive and ontogenetic explanations of shell form and evolution from Hooke 1695 to the present. In: Hecht MK, Wallace B, MacIntyre RJ (eds) Evolutionary biology, vol 26. Plenum, New York
Jacobs DK, Chamberlain JA (1996) Buoyancy and hydrodynamics in ammonoids. In: Landman NH, Tanabe K, Davis RA (eds) Ammonoid Paleobiology. Plenum, New York
Jacobs DK, Landman NH (1993) *Nautilus*—a poor model for the function and behavior of ammonoids? Lethaia 26:101–112
Jacobs DK, Landman NH, Chamberlain JA (1994) Ammonite shell shape covaries with facies and hydrodynamics: iterative evolution as a response to changes in basinal environment. Geology 22:905–908
Jäger M, Fraaye R (1997) The diet of the Early Toarcian ammonite *Harpoceras falciferum*. Palaeontology 40:557–574
Jagt JWM, Bakel BWM van, Fraaije RHB, Neumann C (2006) *In situ* fossil hermit crabs (Paguroidea) from northwest Europe and Russia. Preliminary data on new records. Rev Mex Cienc Geol 23:364–369

Jansa LF, Emos P, Tcholke BE, Gradstein E, Sheridan RE (1979) Mesozoic-Cenozoic sedimentary formations of the North American Basin, western North Atlantic. In: Talman M, Hay W, Ryan WBE (eds) Deep drilling results in the Atlantic Ocean Continental Margins and Paleoenvironment, Maurice Ewing Series 3, American Geophysical Union, Washington, DC
Jeletzky JA, Zapfe H (1976) Coleoid and Orthocerid Cephalopods of the Rhaetian Zlambach Marl from the Fischerwiese near Aussee, Styria (Austria). Ann Naturhist Mus Wien 71:69–106
Jenkyns H (1988) The early Toarcian (Jurassic) anoxic event. Stratigraphic, sedimentary, and geochemical evidence. Am J Sci 288:101–151
Jones DS, Williams DF, Romanek CS (1986) Life history of symbiont-bearing giant clams from stable isotope profiles. Science 231:46–48
Jordan R, Stahl W (1970) Isotopische Paläotemperatur-Bestimmungen an jurassischen Ammoniten und grundsätzliche Voraussetzungen für diese Methode. Geol Jahrb 89:33–62
Kahn P, Kant R (1975) Biometrische Untersuchungen zur Lobenentwicklung. Paläontol Z 49:287–297
Kaltenegger W (1967) Paläotemperaturbestimmungen an aragonitischen Dibranchiatenrostren der Trias. Naturwissenschaften 54:515
Kaltenegger W, Preisinger A, Rögl F (1971) Palaeotemperature determinations of aragonitic mollusks from the Alpine Mesozoic. Palaeogeogr Palaeoclimatol Palaeoecol 10:273–285
Kammer TW, Brett, CE, Boardman DR IS, Mapes RH (1986) Ecologic stability of the dysaerobic biofacies during the Late Paleozoic. Lethaia 19:109–121
Kanie Y, Fukuda Y, Nakayama H, Seki K, Hattori M (1980) Implosion of living *Nautilus* under increased pressure: Paleobiology 6(1):44–47
Kant R (1975) Biometric analysis of ammonoid shells. Paläontol Z 49:203–220
Kant R, Kullmann J (1988) Changes in conch form in the Paleozoic ammonoids. In: Wiedmann J, Kullmann J (eds) Cephalopods Present and Past. Schweizerbart, Stuttgart
Kase T, Shigeta F, Futakami M (1994) Limpet home depressions in Cretaceous ammonites. Lethaia 25:49–58
Kase T, Johnson PA, Seilacher A, Boyce JP (1998) Alleged mosasaur bite marks on late Cretaceous ammonites are limpet (patellogastropod) home scars. Geology 26:947–950
Kashiyama Y, Ogawa NO, Chikaraishi Y, Kashiyama N, Sakai S, Tanabe K, Ohkouchi N (2010) Reconstructing the life history of modern and fossil nautiloids based on the nitrogen isotopic composition of shell organic matter and amino acids. In Tanabe K, Shigeta Y, Sasaki T, Hirano H, (eds.) Cephalopods Present and Past. Tokai Unive Press, Tokyo
Kauffman EG (1977) Geological and Biological Overview: Western Interior Cretaceous Basin, U.S.A. In: Kauffman EG (ed) Field Guide: North American Paleontological Convention II; Cretaceous facies, faunas, and paleoenvironments across the Western Interior Basin: The Mountain geologist 14:75–79
Kauffman EG (1978) Benthic environments and paleoecology of the Posidonienschiefer (Toarcian). Neues Jahrb Geol Paläontol Abh 157:18–36
Kauffman EG (1990) Mosasaur predation on ammonites during the Cretaceous–an evolutionary history. In: Boucot AJ (ed) Evolutionary paleobiology of behaviour and coevolution. Elsevier, New York
Kauffman EG, Kiesling RV (1960) An Upper Cretaceous ammonite bitten by a mosasur. Contr Michigan Univ Mus Paleontol 15:193–248
Kauffman EG, Sawdo JK (2013) Mosasaur predation on a nautiloid from the Maastrichtian Pierre Shale, Central Colorado, Western Interior Basin, United States. Lethaia 46:180–187
Kauffman EG, Villamil T, Harries PJ, Meyer C (1992) The flat clam controversy: where did they come from? Where did they go? Paleontol Soc Spec Publ 6:159
Kawabe F (2003) Relationship between mid-Cretaceous (upper Albian-Cenomanian) ammonoid facies and lithofacies in the Yezo forearc basin, Hokkaido, Japan. Cretac Res 24:751–763
Kennedy WJ (1971) Cenomanian ammonites from southern England. Spec Pap Palaeont 8:1–133
Kennedy WJ, Cobban WA (1976) Aspects of ammonite biology, biogeography, and biostratigraphy. Spec Pap Palaeont 17:1–94

Kennedy WJ, Landman NH, Christensen WK, Cobban WA, Hancock JM (1998) Marine connections in North America during the Late Maastrichtian: palaeogeographic and palaeobiogeographic significance of *Jeletzkytes nebrascensis* Zone cephalopod fauna from the Elk Butte Member of the Pierre Shale, SE South Dakota and NE Nebraska. Cretac Res 19:745–775
Kennedy WJ, Landman NH, Cobban WA, Larson NL (2002) Jaws and radulae in *Rhaeboceras*, a late Cretaceous ammonite. In: Summesberger H, Histon K, Daurer A (eds) Cephalopods Present and Past B-A 57:113–132
Keupp H (1991) Bissmarken oder postmortale Implosionsstrukturen? Fossilien 1992:141–146
Keupp H (1992) Paläopathologie der Ammonitenfauna aus dem Obercallovium der Normandie und ihre palökologische Interpretation. Berl Geowiss Abh 3:171–189
Keupp H (1997) Paläopahologische Analyse einer „Population" von *Dactylioceras athleticum* (Simpson) aus dem Unter-Toarcium von Schlaifhausen/Oberfranken. Berl Geowiss Abh E 25:243–267
Keupp H (2000) Ammoniten. Paläobiologische Erfolgsspiralen. Thorbecke, Stuttgart
Keupp H (2006) Sublethal punctures in body chambers of Mesozoic ammonites (forma aegrafenestra n. f.), a tool to interpret synecological relationships, particularly predator–prey interactions. Paläontol Z 80:112–123
Keupp H (2007) Complete ammonoid jaw apparatuses from the Solnhofen plattenkalks: implications for aptychi function and microphagous feeding of ammonoids. N Jahrb Geol Paläontol Abh 245:93–101
Keupp H, Hoffmann R (2015) Ammonoid paleopathology. This volume
Keupp H, Schobert J (2011) Gehäuseanomalien bei Klein-Ammoniten von Buttheim/Oberfranken. Fossilien 3:164–174
Kim S-T, O'Neil JR, Hillaire-Marcel C, Mucci A (2007) Oxygen isotope fractionation between synthetic aragonite and water: Influence of temperature and Mg^{2+} concentration. Geochim Cosmochim Acta 71:4704–4715
Klinger HC (1980) Speculations on buoyancy control and ecology in some heteromorph ammonites. In: House MR, Senior JR (eds) The Ammonoidea: The evolution, classification, mode of life and geological usefulness of a major fossil group. Systematics Association, special vol 18. Academic Press, London
Klofak SM, Landman NH, Mapes RH (1999) Embryonic development of primitive ammonoids and the morphology of the Ammonoidea. In: Olóriz F, Rodríguez-Tovar FJ (eds) Advancing Research on Living and Fossil Cephalopods. Plenum, New York
Klofak SM, Landman NH, Mapes RH (2007) Patterns of embryonic development in early to middle Devonian ammonoids. In: Landman NH, Davis RA, Mapes RH (eds) Cephalopods Present and Past: new insights and fresh perspectives. Springer, Dortrecht
Klompmaker AA, Waljaard, Fraaije RHB (2009) Ventral bite marks in Mesozoic ammonoids. Palaeogeogr Palaeoclimatol Palaeoecol 280:245–257
Klug C (2001) Life-cycle of some Devonian ammonoids. Lethaia 34:215–233
Klug C (2002) Conch parameters and habitats of Emsian and Eifelian ammonoids from the Tafilalt (Morocco) and their relation to global events. Abh Geol B-A 57:523–538
Klug C (2007) Sublethal injuries in Early Devonian cephalopod shells from Morocco. Acta Palaeont Polonica 52:749–759
Klug C, Hoffmann R (2015) Ammonoid septa and sutures. This volume
Klug C, Korn D (2004) The origin of ammonoid locomotion. Acta Palaeont Pol 49:235–242
Klug C, Lehmann J (2015) Soft-part anatomy of ammonoids: reconstructing the animal based on exceptionally preserved specimens and actualistic comparisons. This volume
Klug C, Schatz W, Korn D, Reisdorf A (2005) Morphological fluctuations of ammonoid assemblages from the Muschelkalk (Middle Triassic) of the Germanic Basin—indicators of their ecology, extinctions, and immigrations. Palaeogeogr, Palaeoclim Palaeoeco 221:7–34
Klug C, Kröger B, Kiessling W, Mullins GL, Servais T, Frýda J, Korn D, Turner S (2010) The Devonian nekton revolution. Lethaia 43:465–477
Klug C, Korn D, Landman NH, Tanabe K, De Baets K, Naglik C (2015a) Describing ammonoid conchs. This volume

Klug C, Zatoń M, Parent H, Hostettler, Tajika A (2015b) Mature modifications and sexual dimorphism. This volume

Kobashi T, Grossman EL (2003) The oxygen isotopic record of seasonality in *Conus* shells and its application to understanding late middle Eocene (38 Ma) climate. Paleontol Res 7:343–355

Kobayashi T (1954) A contribution toward Palaeo-Flumenology, Science of the Oceanic Current in the Past, with a description of a new Miocene *Aturia* from Central Japan. Jap J Geol Geogr Transact 25:35–59

Korn D (1986) Ammonoid evolution in late Famennian and early Tournaisian. Ann Soc Géol Belg 109:49–54

Korn D (1988) Oberdevonishe Goniatiten mit dreieckigen Innenwindungen. N Jahrb Geol Palaont Mh 1988:605–610

Korn D (1992) Relationship between shell form, septal construction and suture line in clymeniid cephalopods (Ammonoidea: Upper Devonian). N Jahrb Geol Paläontol Abh 185:115–130

Korn D (2010) A key for the description of Palaeozoic ammonoids. Fossil Rec 13:5–12

Korn D, Klug C (2001) Biometric analyses of some Palaeozoic ammonoid conchs. Berl Geowiss Abh E 36:173–187

Korn D, Klug C (2002) Ammoneae Devonicae. In Riegraf (ed). Fossilium Catalogus I: Animalia. Backhuys Publishers, Leiden, 138:375

Korn D, Klug C (2007) Conch form analysis, variability, morphological disparity, and mode of life of the Frasnian (Late Devonian) ammonite *Manticoceras* from Coumiac (Montagne nOire, France). In: Landman NH, Davis RA, Mapes RH (eds) Cephalopods Present and Past: new insights and fresh perspectives. Springer, Dortrecht

Korn D, Klug C (2012) Ammoneae Devonicae. In: Riegraf W (ed) Fossilium Catalugus I: Animalia. Blackhuys Publishers, Leiden

Korn, D, Klug, C, Mapes RH (2004) Cuboid Carboniferous ammonoids. Mitt Geol Palaöont Inst Univ Hambg 88:79–98

Kowalewski M (2002) The fossil record of predation: an overview of analytical methods. In: Kowalewski M, Kelley PH (eds) The fossil record of predation. Paleontol Soc Pap 8:1–42

Kröger B (2000) Schalenverletzungen an jurassischen Ammoniten—ihre paläobiologische und paläoökologische Aussagefähigkeit. Berl Geowiss Abh E 33:1–97

Kröger B (2001) Comments on Ebel's benthic–crawler hypothesis for ammonoids and exticnt nautiloids. Paläontol Z 75:123–125

Kröger B (2002a) On the effciency of the buoyancy apparatus in ammonoids: evidences from sublethal shell injuries. Lethaia 35:61–70

Kröger B (2002b) On the ability of withdrawing of some Jurassic ammonoids. In: Summesberger H, Histon K, Daurer K (eds) Cephalopods Present and Past. Abh Geol B-A 67:199–244

Kröger B (2002c) Antipredatory traits of the ammonoid shell—Indications from Jurassic ammonoids with sublethal injuries. Paläontol Z 76:223–234

Kröger B (2005) Adaptive evolution in Paleozoic coiled cephalopods. Paleobiology 31:253–268

Kroh A, Harzhauser M (1999) An echinoderm fauna from the Lower Miocene of Austria: paleoecology and implications for Central Paratethys Paleobiogeography. Ann Nat Mus Wien 101A:145–191

Kruta I, Landman N, Rouget I, Cecca F, Tafforeau P (2011) The role of ammonites in the Mesozoic marine food web revealed by jaw preservation. Science 2011:70–72

Kruta I, Landman N, Cochran JK (2014) A new approach for the determination of ammonite and nautilid habitats. PLos ONE 9(1):e87479. doi:10.1371/journal.pone.0087479

Krystyn L (1991) Die Fossillagerstätten der alpinen Trias. In: Nagel D, Rabeder G (eds) Exkursionen im Jungpaläozoikum und Mesozoikum Österreichs. Österr Paläontol Ges, Wien

Kulicki C (1974) Remarks on the embryogeny and postembryonal development of ammonites. Acta Palaeontol Pol 19:201–224

Kulicki C (1979) The ammonite shell: Its structure, development and biological significance. Palaeontol Pol 39:97–142

Kulicki C (1996) Ammonoid shell microstructures. In: Landman NH, Tanabe K, Davis RA (eds) Ammonoid Paleobiology. Plenum, New York

Kulicki C, Doguzhaeva L (1994) Development and calcification of the ammonitella shell. Acta Palaeont Pol 39:17–44

Kulicki C, Landman NH, Heaney MJ, Mapes RH, Tanabe K (2002) Morphology of the early whorls of goniatites from the Carboniferous Buckhorn Asphalt (Oklahoma) with aragonitic preservation. In: Summesberger H, Histon K, Daurer A (eds) Cephalopods—present and past. Abh Geol B-A 57:225–255

Kullmann J (1981) Carboniferous goniatites. In: House MR, Senior JR (eds) The Ammonoidea. Systematics Association, special vol 18. Academic Press, London

Kullmann J, Scheuch J (1970) Wachstums-Änderungen in der Ontogenese paläozoischer Ammonoideen. Lethaia 3:397–412

Kullmann J, Scheuch J (1972) Absolutes und relatives Wachstum bei Ammonoideen. Lethaia 5:129–146

Kvantaliani IV, Sakharov AS (1986) Valanginian ammonites of the Northern Caucasus (Russ.). Geol Balc 163:55–68

Landman NH (1982) Embryonic shells of *Baculites*. J Paleontol 56:1235–1241

Landman NH (1985) Preserved ammonitellas of *Scaphites* (Ammonoidea, Ancyloceratina). Am Mus Novit 2815:1–10

Landman NH (1986) Shell abnormalities in scaphitid ammonites. Lethaia 19:211–224

Landman NH (1987) Ontogeny of Upper Cretaceous (Turonian-Santonian) scaphitid ammonites from the Western Interior of North America: systematics, developmental patterns, and life history. Bull Am Mus Nat Hist 185:117–241

Landman NH (1988) Early ontogeny of Mesozoic ammonites and nautilids. In: Wiedmann, J, Kullmann J (eds) Cephalopods Present and Past. Schweizerbart, Stuttgart

Landman NH, Cobban WA (2007) Ammonite touch marks in Upper Cretaceous (Cenomanian-Santonian) deposits of the Western Interior Seaway. In: Landman NH, Davis RA, Mapes RH (eds) Cephalopods Present and Past: new insights and fresh perspectives. Springer, Dortrecht

Landman NH, Cochran JK (1987) Growth and longevity of *Nautilus*. In: Saunders WB, Landman NH (eds) *Nautilus*: The Biology and Paleobiology of a Living Fossil. Topics in Geobiology, vol 6. Springer, New York

Landman NH, Cochran, JK (2010) Growth and longevity of *Nautilus*. In: Saunders WB, Landman NH (eds) *Nautilus*: The Biology and Paleobiology of a Living Fossil. Topics in Geobiology, vol 6. Springer, New York

Landman NH., Klofak, SM (2012). Anatomy of a concretion: life, death, and burial in the Western Interior Seaway. Palaios 27:672–693

Landman NH, Waage KM (1986) Shell abnormalities in scaphitid ammonites. Lethaia 19:211–224

Landman NH, Waage KM (1993) Scaphitid ammonites of the Upper Cretaceous (Maastrichtian) Fox Hills Formation in South Dakota and Wyoming. Bull Am Mus Nat Hist 215:l–257

Landman NH, Rye DM, Shelton KL (1983) Early ontogeny of *Eutrephoceras* compared to Recent *Nautilus* and Mesozoic ammonites: evidence from shell morphology and light stable isotopes. Paleobiol 9:269–279

Landman NH, Saunders WB, Winston JE, Harries PJ (1987) Incidence and kinds of epizoans on the shells of living *Nautilus*. In Saunders WB, Landman NH (eds) *Nautilus*: The Biology and Paleobiology of a Living Fossil. Plenum, New York

Landman NH, Cochran, JK, Rye DM, Tanabe K, Arnold JM (1994) Early life history of *Nautilus:* evidence from isotopic analysis of aquarium reared specimens. Paleobiology 20:40–51

Landman NH, Tanabe K, Davis RA (eds) (1996a) Ammonoid Paleobiology. Plenum Press, New York, pp 857

Landman NH, Tanabe K, Shigeta Y (1996b) Ammonoid embryonic development. In: Landman NH, Tanabe K, Davis RA (eds) Ammonoid Paleobiology. Plenum, New York

Landman NH, Klofak SM, Sarg KB (2003) Variation in adult size of scaphitid ammonites from the Upper Cretaceous Pierre Shale and Fox Hill Formation. In: Harries PJ (ed) Approaches in high-resolution stratigraphic paleontology. Kluwer Academic, Dordrecht

Landman NH, Davis RA, Mapes RH (eds) (2007a) Cephalopods Present and Past: new insights and fresh perspectives. Springer, Dortrecht

Landman NH, Larson NL, Cobban WA (2007b) Jaws and radula of *Baculites* from the Upper Cretaceous (Campanian) of North America. In: Landman NH, Davis RA, Mapes RH (eds) Cephalopods Present and Past: new insights and fresh perspectives. Springer, Dortrecht

Landman NH, Mapes RH, Cruz C (2010) Jaws and soft tissues in ammonoids from Lower Carboniferous (Upper Mississippian) Cear Gulch Beds Montan, USA. In: Tanabe K, Shigeta Y, Sasaki T, Hirano H (eds) Cephalopods Present and Past. Tokai Univ Press, Tokyo

Landman NH, Cobban WA, Larson NL (2012) Mode of life and habitat of scaphitid ammonites. Geobios 45:87–98

Laptikhovsky VL, Rogov, MA, Nikolaeva SE, Arkhipkin AI (2013) Environmental impact on ectocochleate cephalopod reproductive strategies and the evolutionary significance of cephalopod egg size. Bull Geosci 88:83–93

Larson NL (2003) Predation and pathologies in the Late Cretaceous ammonite family Scaphitidae. Mid-Am Paleontol Soc 26:1–30

Larson NL (2007) Deformities in the Late Callovian (Late Middle Jurassic) ammonite fauna from Saratov, Russia. In: Landman NH, Davis RA, Mapes RH (eds) Cephalopods Present and Past: new insights and fresh perspectives. Springer, Dortrecht

Lécuyer C, Bucher H (2006) Stable isotope composition of late Jurassic ammonite shell: a record of seasonal surface water temperatures in the southern hemisphere? eEarth Discuss 1:1–7

Lécuyer C, Reynard B, Martineau F (2004) Stable isotope fractionation between mollusc shells and marine waters from Martinique Island. Chem Geol 213:293–305

Lehmann U (1966) Dimorphismus bei Ammoniten der Ahrensburger Lias-Geschiebe. Paläontol Z 40:26–55

Lehmann U (1975) Über Nahrung und Ernährungsweise der Ammoniten. Paläontol Z 49:187–195

Lehmann U (1976) Ammoniten-Ihr Leben und Ihre Umwelt. Enke, Stuttgart

Lehmann U (1980) Ammonite jaws and soft parts. In: House MR, Senior JR (eds) The Ammonoidea: the evolution, classification, mode of life and geological usefulness of a major fossil group. Systematics Association, special vol 18, Academic Press, London

Lehmann U (1985) Zur Anatomie der Ammoniten: Tintenbeutel, Kiemen, Augen. Paläontol Z 59:99–108

Lehmann U (1988) On the dietary habits and locomotion of fossil cephalopods. In: Wiedmann J, Kullmann J (eds) Cephalopods Present and Past. Schweizerbart, Stuttgart

Lehmann J (2000) Upper Albian ammonites from ODP leg 171B off northern Florida. Palaeontology 43:41–61

Lehmann U, Kulicki C (1990) Double function of aptychi (Ammonoidea) as jaw elements and opercula. Lethaia 23:325–331

Lehmann U, Weitschat W (1973) Zur Anatomie und Okologie der Ammoniten. Funde von Kropf und Kiemen. Paläontol Z 47:69–76

Lehmann U, Tanabe K, Kanie Y, Fukuda Y (1980) Über den Kieferapparat der Lytoceraten (Ammonoidea). Paläontol Z 54:319–329

Levitus (1994) World Ocean Atlas, an atlas of objectively analyzed fields of major ocean parameters at the annual, seasonal, and monthly time scales. http://iridl.ldeo.columbia.edu/SOURCES/.LEVITUS94/. Accessed 18 June 2012

Lewy Z (2002) The function of ammonite fluted septal margins. J Paleontol 76:63–69

Lewy Z (2003) Reply to Checa and to Hewitt and Westermann. J Paleontol 77:796–798

Little R, Baker DR, Leythaeuser D, Rullkottner J (1991) Keys to the depositional history of the Posidonia Shale (Toarcian) in the Hills Syncline, northern Germany. In: Tyson RV, Pearson TH (eds) Modern and ancient continental margin anoxia. Geol Soc Spec Publ 58:311–333

Loh H, Maul B, Prauss M, Riegel W (1986) Primary production, marl formation and carbonate species in the Posidonia Shale of NW Germany. Mitt Geol Palaont Inst Univ Hambg 60:397–421

Lominadze T, Sakharov AS (1988) Ecology of caucasian callovian ammonitida. In: Wiedmann J, Kullmann J (eds) Cephalopods—present and past. Schweizerbart, Stuttgart

Longinelli A (1966) Ratios of oxygen-18: oxygen-16 in phosphate and carbonate from living and fossil marine organisms. Nature 211:923–927

Longinelli A., Nuti S (1973) Revised phosphate-water isotopic temperature scale. Earth Planet Sci Lett 19:373–376

Lowenstein HA, Epstein S (1954) Paleotemperatures of the post-Albian Cretaceous as determined by the oxygen isotope method. J Geol 62:207–248

Lukeneder A (2003) The *Karsteniceras* level: dysoxic ammonoid beds within the Early Cretaceous (Barremian, Northern Calcareous Alps, Austria). Facies 49:87–100

Lukeneder A (2004) The *Olcostephanus* level: an Upper Valanginian ammonoid mass-occurrence (Lower Cretaceous, Northern Calcareous Alps, Austria). Acta Geol Pol 54:23–33

Lukeneder A (2005) The Early Cretaceous *Karsteniceras* level in the Vienna Woods (Northern Calcareous Alps, Lower Austria). Geol Carpathica 56:307–315

Lukeneder A (2007) Cephalopod evolution: a new perspective—implications from two Early Cretaceous ammonoid suborders (Northern Calcareous Alps, Upper Austria). Denisia 20:395–404

Lukeneder A (2008) The ecological significance of solitary coral and bivalve epibionts on lower Cretaceous (Valangianian–Aptian) ammonoids from the Italian dolomites. Acta Geol Pol 58:425–436

Lukeneder A (2012) Computed 3D visualisation of an extinct cephalopod using computer tomographs. Comput Geosci 45:68–74

Lukeneder A, Grunert P (2013) Palaeoenvironmental evolution of the Southern Alps across the Faraoni Level equivalent: new data from the Trento Plateau (Upper Hauterivian, Dolomites, N. Italy). Acta Geol Pol 63:89–104

Lukeneder A, Harzhauser M (2002) Shell accumulations of the Nautilidae *Aturia (Aturia) aturi* (Bast.) in the Lower Micocene Paratethys (Lower Austria). In: Summesberger H, Histon K, Daurer A (eds) Cephalopods Present and Past. Abh Geol B-A 57:459–466

Lukeneder A, Harzhauser M (2003) *Olcostephanus guebhardi* as cryptic habitat for an Early Cretaceous coelobite community (Valanginian, Northern Calcareous Alps, Austria). Cretac Res 24:477–485

Lukeneder S, Lukeneder A (2014) A new ammonoid fauna from the Carnian (Upper Triassic) kasimlar formation of the Taurus Mountains (Anatolia, Turkey). Palaeontology 57:357–396

Lukeneder A, Mayrhofer S (2014) Taphonomic implications from Upper Triassic mass flow deposits: 2-dimensional reconstructions of an ammonoid mass occurrence (Carnian, Taurus Mountains, Turkey). Geol Carpath 65/5:339–364

Lukeneder A, Reháková D (2004) Lower Cretaceous section of the Ternberg Nappe (Northern Calcareous Alps, Upper Austria): facies-changes, biostratigraphy and paleoecology. Geol Carpathica 55:227–237

Lukeneder A, Smrečková M (2006) An Early Cretaceous radiolarian assemblage: palaeoenvironmental and palaeoecological implications for the Northern Calcareous Alps (Barremian, Lunz Nappe, Lower Austria). Ann Naturhist Mus Wien 107A:23–57

Lukeneder A, Tanabe K (2002) In situ finds of aptychi in the Barremian of the Alpine Lower Cretaceous (Barremian, Northern Calcareous Alps, Upper Austria). Cretac Res 23:15–24

Lukeneder A, Harzhauser M, Müllegger S, Piller W (2008) Stable isotopes ($\delta^{18}O$ and $\delta^{13}C$) in *Spirula spirula* shells from three major oceans indicate developmental changes paralleling depth distributions. Mar Biol 154:175–182

Lukeneder A, Harzhauser M, Müllegger S, Piller WE (2010) Ontogeny and habitat change in Mesozoic cephalopods revealed by stable isotopes ($\delta^{18}O$, $\delta^{13}C$). Earth Planet Sci Lett 296:103–114

Lukeneder S, Lukeneder A, Harzhauser M., Islamoglu Y, Krystyn L, Lein R (2012) A delayed carbonate factory breakdown during the Tethyan-wide Carnian Pluvial Episode along the Cimmerian terranes (Taurus, Turkey). Facies 58:279–296

Lukeneder A, Suttner TJ, Bertle RJ (2013) New ammonoid taxa from the Lower Cretaceous Giumal Formation of the Tethyan Himalaya (Northern India). Palaeontology 56:991–1028

Lukeneder S, Lukeneder A, Weber GW (2014) Computed reconstruction of spatial ammonoid-shell orientation captured from digitized grinding and landmark data. Comput Geosci 64:104–114

Maeda H (1993) Dimorphism of Late Cretaceous false-Puzosiinae ammonites, *Yokoyamaoceras* Wright and Matsumoto, 1954 and *Neopuzosia* Matsumoto, 1954. Trans Proc Palaeont Soc Japan N S 169:97–128
Maeda H, Seilacher A (1996) Ammonoid taphonomy. In: Landman NH, Tanabe K, Davis RA (eds) Ammonoid paleobiology. Plenum, New York
Maeda H, Mapes RH, Mapes G (2003) Taphonomic features of a Lower Permian beached cephalopod assemblage from Central Texas. Palaios 18:421–434
Manfrin S, Mietto P, Preto N (2005) Ammonoid biostratigraphy of the Middle Triassic Latemar platform (Dolomites, Italy) and its correlation with Nevada and Canada. Geobios 38:477–504
Mangold K (1989) Reproduction, croissance et durée de vie. In: Grassé PP (ed) Traité de Zoolgie, 5 (4). Anatomie, Systématique, Biologie. Céphalopodes. Masson, Paris
Mapes RH (1979) Carboniferous and Permian Bactritoidea (Cephalopoda) in North America. Univ Kans Paleontol Contrib Artic 64:1–75
Mapes RH (1987) Upper Paleozoic cephalopod mandibles: frequency of occurrence, modes of preservation, and paleoecological implications. J Paleontol 61:521–538
Mapes RH, Chaffin DT (2003) Predation on cephalopods. a genaral overview with case study from the Upper Carboniferous of Texas. In: Kelley PH, Kowalewski M, Hansen TA (eds). Predator-Prey Interactions in the Fossil record. Kluwer Academic, New York
Mapes RH, Dalton RB (2002) Scavening or Predation?—Mississippian ammonoid accumulations in carbonate concretion halos around *Rayonnoceras* (Actinoceratoidea—Nautiloidea) body chambers. In: Summesberger H, Histon K, Daurer A (eds) Cephalopods Present and Past. Abh der Geol B-A 57:407–422
Mapes RH, Hansen MC (1984) Pennsylvanian shark–cephalopod predation: a case study. Lethaia 17:175–183
Mapes RH, Nützel A (2009) Late Palaeozoic mollusc reproduction: cephalopod egg-laying behavior and gastropod larval palaeobiology. Lethaia 42:341–356
Mapes RH, Tanabe K, Landman NH, Faulkner CJ (1992) Upper Carboniferous ammonoid shell clusters: transported accumulations or *in situ* nests? Paleontol Soc Spec Pub 6:196
Mapes RH, Sims MS, Boardman DR II (1995) Predation on the Pennsylvanian ammonoid *Gonioloboceras* and ist implications for allochthonous vs. autochthonous accumulations of goniatites and other ammonoids. J Palaeontol 69:441–446
Mapes RH, Hembree DI, Rasor BA, Stigall A, Goirand C, de Forges BR (2010a) Modern *Nautilus* (Cephalopoda) taphonomy in a subtidal to backshore environment, Lifou (Loyalty Islands). Palaios 25:656–670
Mapes RH, Landman NH, Cochran K, Goiran C, Richer De Forges B, Renfro A (2010b) Early taphonomy and significance of naturally submerged *Nautilus* shells from the New Caledonia Region. Palaios 25:597–610
Marchand T (1984) Ammonites et paleoenvironnements: une nouvelle approche. Geobios Mém Spéc 8:101–107
Marchand D (1992) Ammonites et paléoprofondeur: les faits, les interprétations. Paleovox 1:49–68
Marchand D, Thierry J (1986) Relations entre les évènements calloviens et l'évolution des peuplements d'ammonites en Europe occidentale. Bull Centr Rech Elf Aquitaine 10:383–92
Marchand T, Thierry J (1997) Enregistrement des variations morphologiques et de la composition des peuplements d'ammonites durant le cycle régressif/transgressif de 2^{éme} ordre Bathonien inférieur–Oxfordien inférieur en Europe occidentale. Bull Soc Géol Fr 168:121–132
Marcinowski R, Wiedmann J (1985) The Albian ammonite fauna of Poland and its paleogeographical significance. Acta Geol Pol 35:199–218
Marcinowski R, Wiedmann J (1988) Paleogeographic implications of the Albian ammonite faunas of Poland. In: Wiedmann J, Kullmann J (eds) Cephalopods Present and Past. Schweizerbart, Stuttgart
Marchand D, Thierry J, Tintant H (1985) Influence des seuls et des hauts–fonds sur la morphology et l'évolution des ammonites. Inst Sci Terre Univ Dijon Bull Sect Sci 9:191–202
Marshall JD (1981) Zoned calcites in Jurassic ammonite chambers: trace elements, isotopes and neomorphic origin. Sedimentology 28:867–887

Martin AW, Catala-Stucki I, Ward DP (1978) The growth rate and reproductive behaviour of *Nautilus* macromphalus. N Jahrb Geol Paläontol Abh 156:207–225

Massare JA (1987) Tooth morphology and prey preference of Mesozoic marine reptiles. J Vert Paleontol 7:121–137

Matsumoto T, Futakami M, Tanabe K, Obata I (1981) Upper Turonian ammonite assemblages in the Pombets area, central Hokkaido. Bull Kitakyushu Mus Nat Hist 3:1–10

Matsuoka A, Anso J, Nakada K, Terabe K, Sato T (2010) Biometrical analysis on primary rib number of the Middle Jurassic ammonoid *Pseudoneuqueniceras yokoyamai* (Kobayashi & Fukada) and its allied forms. In: Tanabe K, Shigeta Y, Sasaki T, Hirano H (eds) Cephalopods Present and Past. Tokai University Press, Tokyo

McArthur JM, Mutterlose J, Price GD, Rawson PF, Ruffell A., Thirlwall MF (2004) Belemnites of Valanginian, Hauterivian and Barremian age: Sr-isotope stratigraphy, composition ($^{87}Sr/^{86}Sr$, $\delta^{13}C$, $\delta^{18}O$, Na, Sr, Mg), and palaeo-oceanography. Palaeogeogr Palaeoclimatol Palaeoecol 202:253–272

McConnaughey TA, Gillikin DP (2008) Carbon isotopes in mollusk shell carbonates. Geo-Mar Lett 28:287–299

Meesters E, Knijn R, Pennarts R, Roebers G, van Soest RWM (1991) Sub-rubble communities of Curacao and Bonaire coral reefs. Coral Reefs 10:189–197

Mehl J (1978a) Ein Koprolith mit Ammoniten-Aptychen aus dem Solnhofer Plattenkalk. Jahresb Wetterau Ges Naturkunde 129–130:85–89

Mehl J (1978b) Anhäufungen scherbenartiger Fragmente von Arnmonitenschalen im süddeutschen Lias und Malm und ihre Deutung als Frassreste. Ber Naturforsch Ges Freib Breisgau 68:75–93

Mehl J (1984) Radula and arms of *Michelinoceras* sp. from the Silurian of Bohemia. Paläontol Z 58:211–229

Meischner D (1968) Perniciöse, Epökie von *Placunopsis* auf *Ceratites*. Lethaia 1:156–174

Merkt J (1966) Über Austern und Serpeln als Epöken auf Ammonitengehäusen. N Jahrb Geol Palöont Abh 125:467–479

Milson CV (1994) *Saccocoma*, a benthic crinoid from the Jurassic Solnhofen Limestone, Germany. Palaeontology 37:121–130

Minagawa M, Wada E (1984) Stepwise enrichment of ^{15}N along food chains: further evidence and the relation between delta ^{15}N and animal age. Geochim Cosmochim Acta 48:1135–1140

Misaki A, Maeda H, Kumagae T, Ichida M (2013) Commensal anomiid bivalves on Late Cretaceous heteromorph ammonites from south-west Japan. Palaeontology 57:77–95

Monnet C (2009) The Cenomanian–Turonian boundary mass extinction (Late Cretaceous): new insights from ammonoid biodiversity patterns of Europe, Tunesia and the Western Interior (North America). Palaeogeogr Palaeoclimatol Palaeoecol 282:88–104

Moore EJ (1984) Molluscan paleontology and biostratigraphy of the lower Miocene upper part of the lincoln creek formation in southwestern Washington. Contr Sci Nat Hist Mus Los Angeles City 351:1–39

Moriya K (2015) Isotope signature of ammonoid shells. This volume

Moriya K, Nishi H, Kawahata H, Tanabe K, Takayanagi Y (2003) Demersal habitat of Late Cretaceous ammonoids: evidence from oxygen isotopes for the Campanian (Late Cretaceous) northwestern Pacific thermal structure. Geology 31:167–170

Morris KA (1979) A classification of Jurassic marine shale sequences: an example from the Toarcian (Lower Jurassic) of Great Britain. Palaeogeogr Palaeoclimatol Palaeoecol 26:117–126

Morris KA (1980) Comparison of major regions of organic–rich mud deposition in the British Jurassic. J Geol Soc Lond 137:157–170

Morton N, Nixon M (1987) Size and function of ammonoid aptychi in comparison with buccal masses in modern cephalopods. Lethaia 20:231–238

Mouterde R, Elmi S (1991) Caractères différentiels des faunes d'ammonites du Toarcien des bordures de la Téthys. Signification paléogéographique. Bull Soc Géol Fr 162:1185–1195

Müller AH (1969) Ammoniten mit 'Eierbeutel' und die Frage nach dem Sexual-Dimorphismus der Ceratiten (Cephalopoda). Mber Dt Akad Wiss Berlin 11:411–420

Mutvei H (1975) The mode of life in ammonoids. Paläontol Z 49:196–202
Mutvei H, Dunca E (2007) Connecting ring ultrastructure in the Jurassic ammonoid *Quenstedtoceras* with discussion on mode of life of ammonoids. In: Landman NH, Davis RA, Mapes RH (eds) Cephalopods Present and Past: new insights and fresh perspectives. Springer, Dortrecht
Naglik C, Tajika A, Chamberlain JA, Klug C (2015) Ammonoid locomotion. This volume
Navarro N, Neige P, Marchand D (2005) Faunal invasion as a source of morphological constraints and innovations? The diversification of the early Cardioceratidae (Ammonoidea; Middle Jurassic). Palaeobiology 31:98–116
Neige P, Marchand D, Bonnot A (1997) Ammonoid morphological signal versus sea-level changes. Geol Mag 134:261–264
Nesis KN (1986) On the feeding and causes of extinction of certain heteromorph ammonites. Paleontol Zh 1986:8–15 (Engl transl, Paleontol J 20:5–11)
Nesis KN (1987) Cephalopods of the world, squids, cuttlefishes, octopuses and allies. (transl from Russian). TFH Publications, Neptune City, p. 351
Neumayr M (1883) Über Klimatische Zonen während der Jura und Kreidezeit: Denk K Akad Wiss Math-Nat Kl 48:57–142
Niebuhr S, Jochiamski MM (2002) Stable isotope and trace element geochemistry of Upper Cretaceous carbonates and belemnite rostra (Middle Campanian, North Germany). Géobios 35:51–64
Nikolaeva SV (1999) Morphological diversity of ammonoids from the Lower Namutian of Central Asia. In: Olóriz F, Rodríguez-Tovar FJ (eds) Advancing research on living and fossil cephalopods. Plenum, New York
Nixon M (1988) The feeding mechanism and diets of cephalopods-living and fossil. In: Wiedmann J, Kullmann J (eds) Cephalopods Present and Past. Schweizerbart, Stuttgart
Nixon M (1996) Morphology of the jaws and radula in ammonoids. In: Landman NH, Tanabe K, Davis RA (eds) Ammonoid Paleobiology. Plenum, New York
NOAA National Oceanic and Atmospheric Administration (2014) National oceanographic data center (NODC). United States department of commerce. http://www.nodc.noaa.gov. Accessed 12 April 2014
Nützel A, Frýda, J (2003) Paleozoic plankton revolution: Evidence from early gastropod ontogeny. Geology 31:829–831
Oba T, Kai M, Tanabe K (1992) Early life history and habitat of *Nautilus pompilius* inferred from oxygen isotope examinations. Mar Bio 113:211–217
Obata I, Futakami M (1977) The Cretaceous sequence of the Manji Dome. Palaeontol Soc Jpn Spec Pap 21:23–30
O'Dor RK, Wells MJ (1990) Performance limits of 'antique' and 'state-of-the-art' cephalopods, *Nautilus* and squid. Am Mahcologicnl Union Program Abstr Annu Meet 56:52
Okamoto T (1984) Theoretical morphology of *Nipponites* (a heteromorph ammonite). Fossils (Kaseki) Palaeont Soc Jpn 36:37–51
Okamoto T (1988a) Analysis of heteromorph ammonoids by differential geometry. Palaeontology 31:35–52
Okamoto T (1988b) Changes in life orientation during the ontogeny of some heteromorph ammonoids. Palaeontology 31:281–294
Okamoto T (1988c) Developmental regulation and morphological saltation in the heteromorph ammonite *Nipponites*. Paleobiology 14:273–286
Okamoto T, Shibata M (1997) A cyclic mode of shell growth and its implications in a Late Cretaceous heteromorph ammonite *Polyptychoceras pseudogaultinum* (Yokoyama). Palaeontol Res 1:29–46
Olivero EB, Zinsmeister WJ (1989) Large heteromorph ammonites from the Upper Cretaceous of Seymour Island, Antarctica. J Paleontol 63:626–635
Olóriz E (1976) Kimmeridgiano-Tlithonico inferior en el sector central de las Cordilleras Béticas (Zona Subbética). Paleontologia, Bioestratigrafía [PhD thesis]. Universidad de Granada, Granada

Olóriz F, Palmqvist P (1995) Sutural complexity and bathymetry in ammonites: fact or artefact? Lethaia 28:167–170

Olóriz F, Rodríguez-Tovar FJ (eds) (1999) Advancing Research on Living and Fossil Cephalopods. Kluwer Academic, New York

Olóriz E, Marques B, Rodriguez-Tovar FJ (1991) Eustatism and faunal associations. examples from the south Iberian margin during the Late Jurassic (Oxfordian-Kimmeridgian). Eclogae Geol Helv 84:83–106

Olóriz, F, Caracuel, JE, Ruiz-Heras JJ, Rodríguez-Tovar FJ, Marques B (1996) Ecostratigraphic approaches, sequence stratigraphy proposals and block tectonics: examples from epioceanic swell areas in south and east Iberia. Palaeogeogr Palaeoclimat Palaeoecol 121:273–295

Olóriz F, Palmqvist P, Pérez-Claros A (1999) Recent advances in morphometric approaches to co-variation of shell features and the complexity of suture lines in Late Jurassic ammonites, with reference to the major environments colonized. In: Olóriz F, Rodríguez-Tovar FJ (eds) Advancing Research on Living and Fossil Cephalopods. Kluwer Academic, New York,

Oschmann W (1991) Distribution, dynamics and palaeontology of Kimmeridgian (Upper Jurassic) shelf anoxia in western Europe. In: Tyson RV, Pearson TH (eds) Modern and ancient continental shelf anoxia. Geol Soc Spec Pap 58:381–395

Packard, A (1972) Cephalopods and fish: the limits of convergence. Biol Rev 47:241–307

Palmer TJ (1982) Cambrian to Cretaceous changes in hardground communities. Lethaia 15:309–323

Pamenter CB (1956) *Imitoceras* from the Exshaw Formation of Alberta. J Paleont 30:965–966

Payne AIL, Lipiński MR, Clarke MR, Roeleveld MAC (1998) Cephalopod biodiversity, ecology and evolution. South American Journal of Marine Science 20. Sea Fisheries Department of Environmental Affairs and Tourism, Cape Town, p 469

Podlaha OG, Mutterlose J, Veizer J (1998) Preservation of $\delta^{18}O$ and $\delta^{13}C$ in belemnite rostra from the Jurassic/Early Cretaceous successions. Am J Sci 298:324–347

Price GD, Mutterlose J (2004) Isotopic signals from late Jurassic-early Cretaceous (Volgian-Valanginian) sub-Arctic belemnites, Yatra River, Western Siberia. J Geol Soc Lond 161:959–968.

Price GD, Ruffel AH, Jones CE, Kalin RM, Mutterlose J (2000) Isotopic evidence for temperature variation during the Early Cretaceous (late Ryazanian-mid Hauterivian). J Geol Soc Lond 157:335–343

Price GD, Twitchett RJ, Smale Ch, Marks V (2009) Isotopic analysis of the life history of the enigmatic squid *Spirula spirula*, with implications for studies of fossil cephalopods. Palaios 24:273–279

Price GD, Főzy I, Janssen NMM, Pálfy (2011) Late Valanginian-Barremian (Early Cretaceous) palaeotemperatures inferred from belemnite stable isotope and Mg/Ca ratios from Bersek Quarry (Gerecse Mountains, Transdanubian Range, Hungary). Palaeogeogr Palaeoclimatol Palaeoecol 305:1–9

Pucéat E, Lecuyer C, Sheppard SMF, Dromart G, Reboulet S, Grandjean P (2003) Thermal evolution of Cretaceous Tethyan marine waters inferred from oxygen isotope composition of fish tooth enamels. Paleoceanography 18:7–12

Radwansky A (1996) The predation upon, and the extinction of, the latest Maastrichtian populations of the ammonite species *Hoploscaophites constrictus* (J. Sowerby, 1817) from the Middle Vistula Valley, Central Poland. Acta Geol Pol 46:117–135

Railsback BL, Anderson TF, Ackerly SC, Cisne JL (1989) Paleoceanographic modeling of temperature-salinity profiles from stable isotopic data. Paleoceanography 4(5):585–591

Rasmussen K, Brett C (1985) Taphonomy of Holocene cryptic biotas from St. Croix, Virgin Islands: information loss and preservational biases. Geology 13:551–553

Rawson PF (1981) Early Cretaceous ammonite biostratigraphy and biogeography. In: House MR, Senior JR (eds) The Ammonoidea. Systematics Association, special vol 18, Academic Press, London

Rawson PF (1993) The influence of sea level changes on the migration and evolution of early Cretaceous (pre-Aptian) ammonites. In: House MR (ed) The Ammonoidea: Environment,

Ecology, and Evolutionary Change. Systematics Association, special vol 47. Oxford Science Publications, Oxford
Reboulet S (1996) L'évolution des ammonites du Valanginien–Hauterivien inférieur du bassin vocontien et de la plate-forme provençale (Sud-Est de la France): relations avec la stratigraphie séquentielle et implications biostratigraphique. Doc Labor Géol Lyon 137:371
Reboulet S (1998) Diversification des ammonites hétéromorphes. In: Gayet M, Otero O (eds) Paleodiversifications, Land and Sea Compared: International Symposium, Lyon, France (July 6–8, 1998), Abstr Vol, p. 54
Reboulet S (2001) Limiting factors on shell growth, mode of life and segregation of Valanginian ammonoid populations: evidence from adult-size variations. Geobios 34:423–435
Reboulet S, Atrops F (1995) Rôle du climat sur les migrations et la composition des peuplements d'ammonites du Valanginien supérieur du bassin vocontien (S-E de la France). Geobios Mém Spéc 18:357–365
Reboulet S, Atrops F (1997) Quantitative variations of the Valanginian ammonite fauna of the Vocontian Basin (southeastern France) between limestone-marls and within parasequence sets. Palaeogeogr Palaeoclimat Palaeoecol 135:145–155
Reboulet S, Rard A (2008) Double alignments of ammonoid aptychi from the Lower Cretaceous of Southeast France: Result of a post–mortem transport or bromalites? Acta Palaeontol Pol 53:261–274
Reboulet, S, Atrops F, Ferry S, Schaaf A (1992) Renouvellement des ammonites en fosse vocontienne à la limite Valanginien-Hauterivien. Geobios 25:469–476
Reboulet S, Proux O, Giraud F, Baudin F, Olivero D, Aucour AM (2000) Characterization and significance of ammonoid and nannoplankton assemblages during an "anoxic" event: the Breistroffer level (Upper Albian, SE France). In: Summesberger H, Kollmann H (eds) 6th International Cretaceous Symposium, Vienna, Austria (August 27–September 4), Abstr vol, p 111
Reboulet S, Mattioli, E, Pittet B, Baudin F, Olivero D, Proux O (2003) Ammonoid and nannoplankton abundance in Valanginian (early Cretaceous) limestone-marl successions from the southeast France Basin: carbonate dilution or productivity? Palaeogeogr Palaeoclimat Palaeoecol 201:113–139
Reboulet S, Giraud F, Proux O (2005) Ammonoid abundance variations related to changes in trophic conditions across the Oceanic Anoxic Event 1d (Latest Albian, SE France). Palaios 20:121–141
Reeside JB, Cobban WA (1960) Studies of the Mowry Shale (Cretaceous) and contemporary formations in the United States and Canada. US Geol Surv Prof Pap 355:1–126
Renz O (1972) Aptychi (Ammonoidea) from the Upper Jurassic and Lower Cretaceous of the western North Atlantic (site 105, leg 11, DSDP). In: Holister CD, Ewing JI et al (eds) Initial reports DSDP, No. 11, US Government Printing Office, Washington DC
Renz O (1973) Two lamellaptychi (Ammonoidea) from the Magellan Rise in the central Pacific. In: Winterer EL, Hewing JL (eds) Initial reports DSDP, No 17, US Government Printing Office, Washington DC
Renz O (1978) Aptychi (Ammonoidea) from the early cretaceous of the Blake–Bahama Basin, leg 44, hole 391c, DSDP. In: Benson WE, Sheridan RE (eds) Initial reports DSDP, No. 44, US Government Printing Office, Washington DC
Renz O (1979) Aptychi (Ammonoidea) and ammonites from the Lower Cretaceous of the western Bermuda Rise, leg 43, site 387, DSDP. In: Tucholke BE, Vogt PR (eds) Initial reports DSDP, No. 43, US Government Printing Office, Washington DC
Rexfort A, Mutterlose J (2006) Stable isotope records from *Sepia officinalis*—a key to understand the ecology of belemnites? Earth Planet Sci Lett 247:212–221
Reyment RA (1958) Some factors in the distribution of fossil cephalopods. Stock Contr Geol 1:97–184
Reyment RA (1973) Factors in the distribution of fossil cephalopods. Part 3: experiments wth exact models of certain shell types. Bull Geol Inst Univ Uppsala 4(2):7–41
Reyment RA (1980) Floating orientations of cephalopod shell models. Palaeontology 23:931–936.

Reyment RA (1988) A foraging model for shelled cephalopods. In: Wiedmann J, Kullmann J (eds) Cephalopods Present and Past. Schweizerbart, Stuttgart

Reyment RA (2008) Reyment, Richard A. 2008. A review of the post-mortem dispersal of cephalopod shells. Palaeontol Electr 11:12A:13p. http://palaeo-electronica.org/2008_3/148/index.html

Riccardi AC, Westermann GEG (1991) Middle Jurassic ammonite fauna of the Argentine-Chilean Andes, III: Bajocian-Callovian Eurycephalitinae, Stephanocerataceae. Palaeontogr A 216:1–110

Riccardi AC, Gulisano CA, Mojica J, Palacios O, Schubert C, Thomson MRA (1992) Western South America and Antarctica. In: Westermann GEG (ed) The Jurassic of the Circum-Pacific. Cambridge University Press, New York

Richter AE (2009) Ammoniten-Gehäuse mit Bissspuren. Berl Paläontol Abh 10:297–305

Richter C, Wunsch M (1999) Cavity-dwelling suspension feeders in coral reefs-a new link in reef trophodynamics. Mar Ecol Prog Ser 188:105–116

Rieber H (1973) Ergebnisse paläontologisch-stratigraphischer Untersuchungen in der Grenzbitumenzone (Mittlere Trias) des Monte San Giorgio. Eclogae Geol Helv 66:667–685

Rieber H (1975) Der Posidonienschiefer (Oberer Lias) von Holzmaden und die Grenzbitumenzone (Mittlere Trias) des Monte San Giorgio (Kanton Tessin, Schweiz). Jahrb Ges Naturkd Württemb 130:163–190

Rieber H (1977) Eine Ammonitenfauna aus der oberen Maiolica der Breggia-Schlucht (Tessin/Schweiz). Eclogae Geol Helv 70:777–787

Ritterbush KA, Bottjer DJ (2012) Westermann Morphospace displays ammonoid shell shape and hypothetical paleoecology. Paleobiology 38:424–446

Ritterbush KA, Lukeneder A, Hoffmann R, De Baets K (2014) Pelagic palaeoecology: the importance of recent constraints on ammonoid palaeobiology and life history. J Zool 292:229–241

Roetzel R, Mandic O, Steininger FF (1999) Lithostratigraphie und Chronostratigraphie der tertiären Sedimente im westlichen Weinviertel und angrenzenden Waldviertel. Arb Geol BA 1999:38–54

Röhl H, Schmid-Röhl A, Oschmann W, Frimmel A, Schwark L (2001) The Posidonia Shale (Lower Toarcian) of SW-Germany: an oxygen depleted ecosystem controlled by sea level and palaeoclimate. Palaeogeogr Palaeoclimat Palaeoeco 165:27–52

Romanek CS, Jones DS, Williams DF, Krantz DE, Radtke R (1987) Stable isotopic investigation of physiological and enviromnental changes recorded in shell carbonate from the giant clam *Tridacna maxima*. Mar Biol 94:385–393

Rooij D van, Mol L de, Guilloux E le, Wisshak M, Huvenne VAI, Moeremans R, Henriet JP (2010) Environmental setting of deep-water oysters in the Bay of Biscay. Deep Sea Res Part I: Oceanogr Res Pap 57:1561–1572

Roux M (1990) Underwater observations of *Nautilus macromphalus* off New Caledonia. Chamb Nautil Newsl 60:1

Roux M, Bouchet P, Bourseau J-P, Gaillard C, Grandperrin R, Guille A, Laurin B, Monniot C, Richer de Forges B, Rio M, Segonzac M, Vacelet J, Zibrowius H (1991) L'environnement bathyal au large de la Nouvelle-Calédonie: résultats préliminaires de la campagne CALSUB et conséquences paléoécologiques. Bull Soc Géol Fr 162:675–685

Sandoval J (1983) Bioestratigrafia y paleontologia (Stephanocerataceae y Perisphinctaceae) del Bajocense y Bathoniense en las Cordilleras Beticas [PhD thesis]. Universidad de Granada, Granada, p. 613

Sarti C (1986a) Faune e biostratigrafia del Rosso Ammonitico del Trentino centrale (Kimmeridgiano-Titoniano). Boll Soc Paleontol Ital 23:473–514

Sarti C (1986b) Considerazioni sul Rosso Ammonitico Veronese del Col Santino (M. Pasubio) e raffronti con altre successioni del Trentino. In: Pallini G (ed) Atti I Conv Int Fossili Evoluzione Ambiente. Pergola, ottobre 1984

Sarti C (1999) Whorl width in the body chamber of ammonites as a sign of dimorphism. In: Olóriz F, Rodríguez–Tovar FJ (eds) Advancing Research on Living and Fossil Cephalopods. Kluwer Academic, New York

Sato T, Tanabe K (1998) Cretaceous plesiosaurs ate ammonites. Nature 394:629–630

Saunders WB (1983) Natural rates of growth and longevity of *Nautilus belauensis*. Paleobiology 9:280–288
Saunders WB (1984) *Nautilus belauensis* growth and longevity: evidence from marked and recaptured animals. Science 224:990–992
Saunders WB (1995) The ammonoid suture problem: relationships between shell septum thickness and suture complexity in Paleozoic ammonoids. Paleobiology 21:343–355
Saunders WB, Landman NH (eds) (1987) *Nautilus*. The Biology and Paleobiology of a Living Fossil. Topics in Geobiology 6, Springer Press, New York
Saunders WB, Landman NH (eds) (2010) ditto (reprinted from 1987 with additions)
Saunders WB, Shapiro EA (1986) Calculation and simulation of ammonoid hydrostatics. Paleobiology 12:64–79
Saunders WB, Spinosa, C (1979) *Nautilus* movement and distribution in Palau, Western Caroline Islands. Science 204:1199–1201
Saunders WB, Swan RH (1984) Morphology and morphologic diversity of mid-Carboniferous (Namurian) ammonoids in time and space. Paleobiol 10:195–228
Saunders WB, Spinosa C, Davies LE (1987) Predation on *Nautilus*. In: Saunders WB, Landman NH (eds) (2010) *Nautilus*. The Biology and Paleobiology of a Living Fossil. Topics in Geobiology 6, Springer Press, New York (reprinted from Saunders WB, Landman NH (eds) 1987 with additions), pp. 201–212
Savrda CE, Bottjer DJ (1991) Oxygen-related biofacies in marine strata: an overview and update. In: Tyson RV, Pearson TH (eds) Modern and ancient continental shelf anoxia. Geol Soc Spec Pap 58:201–219
Schindewolf OH (1934) Über Epöken auf Cephalopoden-Gehäusen. Paläontol Z 16:15–31
Schindewolf OH (1958) Über Aptychen (Ammonoidea). Palaeontogr A 111:1–46
Schindewolf OH (1959) Adolescent cephalopods from the Exshaw Formation of Alberta. J Paleontol 33:971–976
Schlögl J, Chirat R, Balter V, Joachimski M, Hudáčková N, Quillévéré F (2011) *Aturia* from the Miocene Paratethys: an exceptional window on nautilid habitat and lifestyle. Palaegeogr Palaeoclimat Palaeoecol 308:330–338
Schmid-Röhl A, Röhl H-J (2003) Overgrowth on ammonite conchs: environmental implications for the Lower Toarcian Posidonia Shale. Palaeontology 46:339–352
Schmidt-Effing R (1972) Die Dactylioceratidae, eine Ammoniten-Familie des unteren Jura. Münst Forsch Geol Paläontol 25/26:1–254
Scott G (1940) Paleoecological factors controlling the distribution and mode of life of Cretaceous ammonoids in the Texas area. J Paleontol 14:299–323
Seilacher A (1960) Epizoans as a key to ammonoid ecology. J Paleontol 34:189–193
Seilacher A (1963) Umlagerung und Rolltransport von Cephalopoden-Gehäusen. Neues Jahrb Geol Paläontol Monatshefte 1963:593–615
Seilacher A (1982) Ammonites as habitats in the Posidonia Shale—floats or benthic islands? Neues Jahrb Geol Paläontol Monatshefte 1982:98–114
Seilacher A (1993) Ammonite aptychi: how to transform a jaw into an operculum. Am J Sci 293A:20–32
Seilacher A, Analib F, Dietl G, Gocht H (1976) Preservational history of compressed ammonites from southern Germany. Neues Jahrb Geol Paläontol Abh 152:307–356
Seuss B, Titschack J, Seifert S, Neubauer J, Nützel A (2012) Oxygen and stable carbon isotopes from a nautiloid from the middle Pennsylvanian (Late Carboniferous) impregnation Lagerstätte 'Buckhorn Asphalt Quarry'—Primary paleo-environmental signals versus diagenesis. Palaeogeogr Palaeoclimat Palaeoecol 319/320:1–15
Shapiro EA, Saunders WB (1987) *Nautilus* shell hydrostatics. In: Saunders WB, Landman NH (eds) *Nautilus*. Plenum, New York
Shigeta Y (1993) Post–hatching early life history of Cretaceous Ammonoidea. Lethaia 26:23–46
Shimansky VN (1975) Cretaceous nautiloids. Acad Sci USSR Trans Palaeontol Inst 150:1–288 [In Russian]

Sornay J (1955) Ammonites nouvelles du Crétacé de la région des Monts du Mellègue (Constantine). Bull Serv Carte Geol Algér 1 ser Paléontol 18:1–40

Spaeth C, Hoefs J, Vetter U (1971) Some aspects of isotope composition of belemnites and related paleotemperature. Geol Soc Amer Bull 82:3139–3150

Spinosa PL, Furnish WM, Glenister GE (1975) The Xenodiscidae, Permian ceratitoid ammonoids. J Paleontol 49:239–283

Sprey AM (2002) Early ontogeny of three Callovian ammonite genera (*Binatisphinctes, Kosmoceras* (*Spinikosmoceras*) and *Hecticoceras*) from Ryazan (Russis). In: Summesberger H, Histon K, Daurer A (2002) Cephalopods Present and Past. Abh Geol B-A 57:225–255

Stahl W, Jordan R (1969) General considerations on isotopic paleotemperature determinations and analysis on Jurassic ammonites. Earth Planet Sci Lett 6:173–178

Stampfli GM, Borel GD (2002) A plate tectonic model for the Paleozoic and Mesozoic constrained by dynamic plate boundaries and restored synthetic oceanic isochrons. Earth Planet Sci Lett 196:17–33

Stephen DA, Stanton RJ (2002) Impact of reproductive strategy on cephalopod evolution. In: Summesberger H, Histon K, Daurer A (eds) Cephalopods Present and Past. Abh Geol B-A 57:151–155

Stephen DA, Bylund KG, Garcia P, McShinsky RD, Carter HJ (2012) Taphonomy of dense concentrations of juvenile ammonoids in the Upper Cretaceous Mancos Shale, east-central Utah, USA. Geobios 45:121–128

Stevens GR (1988) Giant ammonites: a review. In: Wiedmann J, Kullmann J (eds) Cephalopods Present and Past. Schweizerbart, Stuttgart

Stevens GR (1997) The Late Jurassic ammonite fauna of New Zealand. Institute of Geological and Nuclear Sciences monograph 18. N Z Geol Surv Paleontol Bull 74:1–217

Stevens K, Mutterlose J, Wiedenroth (2015) Stable isotope data ($\delta^{18}O$, $\delta^{13}C$) of the ammonite genus Simbirskites—implications for habitat reconstructions of extinct cephalopods. Palaeogeogr Palaeoclimatol Palaeoecol 417:164–175

Stevens K, Mutterlose J, Wiedenroth (2015) Stable isotope data ($\delta^{18}O$, $\delta^{13}C$) of the ammonite genus *Simbirskites*—implications for habitat reconstructions of extinct cephalopods. Palaeogeogr Palaeoclimatol Palaeoecol 417:164–175

Stewart JD, Carpenter K (1990) Examples of vertebrate predation on cephalopods in the Late Cretaceous of the Western Interior. In: Boucot AJ (ed) Evolutionary paleobiology of behavior and coevolution. Elsevier, Amsterdam

Stewart JD, Carpenter K (1999) Examples of vertebrate predation on cephalopods in the Late Cretaceous of the Western Interior. Bull S Cal Paleontol Soc 31:66–73

Stilwell JD, Henderson RA (2002) Description and paleobiogeographic significance of a rare Cenomanian molluscan faunule from Bathurst Island, northern Australia. J Paleontol 76:447–471

Summesberger H (2000) Ammoniten als Kannibalen. Das Naturhistorische 1, p. 10

Summesberger H, Jurkovsek B, Kolar-Jurkovsek T (1996) Aptychi associated with ammonites from the lipica-formation (Upper Cretaceous, Slovenia). Ann Naturhist Mus Wien 97A:1–19

Summesberger H, Histon K, Daurer A (2002) Cephalopods Cephalopods Present and Past. Abh Geol B-A 57:1–569

Swan RH, Saunders WB (1987) Function and shape in Late Paleozoic (mid-Carboniferous) ammonoids. Paleobiology 13:297–311

Tajika A, Wani R (2011) Intraspecific variation of hatchling size in Late Cretaceous ammonoids from Hokkaido, Japan: implication for planktic duration at early ontogenetic stage. Lethaia 44: 287–298

Tajika A, Naglik C, Morimoto N, Pascual-Cebrian E, Hennhöfer DK, Klug C (2015): Empirical 3D-model of the conch of the Middle Jurassic ammonite microconch *Normannites,* its buoyancy, the physical effects of its mature modifications and speculations on their function. Hist Biol, 11 pp. doi:10.1080/08912963.2013.872097

Tanabe K (1977) Functional evolution of *Otoscaphites puerculus* (Jimbo) and *Scaphites planus* (Yabe), Upper Cretaceous ammonites. Mem Fac Sci Kyushu Univ Ser D Geol 23:367–407

Tanabe K (1979) Palaeoecological analysis of ammonoid assemblages in the Turonian *Scaphites* facies of Hokkaido, Japan. Palaeontology 22:609–630

Tanabe K (1983) The jaw apparatus of Cretaceous desmoceratid ammonites. Palaeontology 26:677–689

Tanabe K, Fukuda Y (1987) The jaw apparatus of the Cretaceous ammonite *Reesidites*. Lethaia 20:41–48

Tanabe K, Landman NH (1996) Septal neck-siphuncular complex of ammonoids. In: Landman NH, Tanabe K, Davis RA (eds) Ammonoid Paleobiology. Plenum, New York

Tanabe K, Ohtsuka Y (1985) Ammonoid early internal shell structure: Its bearing on early life history. Paleobiology 11:310–322

Tanabe K, Shigeta Y (1987) Ontogenetic shell variation and streamlining of some Cretaceous ammonites, Trans Proc Palaeonr Soc Jpn N S 147:165–179

Tanabe K, Obata I, Futakami M (1978) Analysis of ammonoid assemblages in the Upper Turonian of the Manji area, central Hokkaido. Bull Nat Sci Mus Tokyo C 4:37–62

Tanabe K, Obata I, Futakami M (1981) Early shell morphology in some Upper Cretaceous heteromorph ammonites. Trans Proc Palaeontol Soc Jpn NS 124:215–234

Tanabe K, Landman NH, Weitschat W (1993a) Septal necks in mesozoic ammonoidea: structure, ontogenetic development and evolution. In: House MR (ed) The Ammonoidea. Environment, Ecology, and Evolutionary Change. Systematic Association special vol 47, Clarendon Press, Oxford

Tanabe K, Landman NH, Mapes, RH, Faulkner CJ (1993b) Analysis of a Carboniferous embryonic ammonoid assemblage-implications for ammonoid embryology. Lethaia 20:215–224

Tanabe K, Landman NH, Mapes RH (1994a) Early shell features of some Late Paleozoic ammonoids and their systematic implications. Trans Proc Palaeontol Soc Jpn NS 173:383–400

Tanabe K, Landman NH, Mapes RH, Faulkner CJ (1994b) Analysis of a Carboniferous embryonic ammonoid assemblage from Kansas. U.S.A.—implications for ammonoid embryology. Lethaia 26:215–224

Tanabe K, Shigeta Y, Mapes, RH (1995) Early life history of Carboniferous ammonoids inferred from analysis of fossil assemblages and shell hydrostatics. Palaios 10:80–86

Tanabe K, Landman NH, Yoshioka Y (2003) Intra- and interspecific variations in the early internal shell features of some Cretaceous ammonoids. J Paleontol 77:876–887

Tanabe K, Shigeta Y, Sasaki T, Hirano H (eds) (2010a) Cephalopods Present and Past. Tokai University Press, Tokyo

Tanabe K, Kulicki C, Landman NH, Kaim A (2010b) Tuberculate micro-ornamentation on embryonic shells of Mesozoic ammonoids: microstructure, taxonomic variation, and morphogenesis In: Tanabe K, Shigeta Y, Sasaki T, Hirano H (eds) Cephalopods Present and Past. Tokai University Press, Tokyo

Tanaka N, Monaghan MC, Rye DM (1986) Contribution of metabolic carbon to mollusc and barnacle shell carbonate. Nature 320:520–523

Tarutani T, Clayton RN, Mayeda TK (1969) The effects of polymorphism and magnesium substitution on oxygen isotope fractionation between calcium carbonate and water. Geochim Cosmochim Acta 33:987–996

Taylor BE, Ward PD (1983) Isotopic studies of *Nautilus macromphalus* Sowerby (New Caledonia) and *Nautilus pompilius* L. (Fiji). Palaeogeogr Palaeoclimatol Palaeoecol 41:1–16

Teichert C, Matsumoto T (1987) The ancestry of the genus *Nautilus*. In: Saunders WB, Landmann NH (eds) Nautilus: The Biology and Paleobiology of a Living Fossil. Topics in Geobiology 6, Springer, New York

Thieuloy JP (1966) Leptocères berriasiens du massif de la Grande-Chartreuse. Trav Lab Géol Fac Sci Univ Grenoble 42:281–295

Tichy G, Urbanek E (2004) Biss-Spuren eines Sauriers an *Pinacoceras parma* Mojsisovics, ein Ammonit der Halleiner Obetrias. GeoAlp 1:87–90

Tintant H, Marchand D, Mouterde R (1982) Relations entre les milieux marins et l'evolution des Ammonoides: Les radiations adaptives du Lias. Bull Soc Géol Fr 24:951–961

Tittensor DP, Rex MA, Stuart CT, McClain CR, Smith CR (2011) Species-energy relationships in deep-sea molluscs. Biol Letters 23:718–722

Toriyama R, Sato T, Hamada T, Komalarjun P (1964) *Nautilus pompilius* drifts on the west coast of Thailand. Jap J Geol Geogr 36: 149–161

Tourtelot HA, Rye RO (1969) Distribution of oxygen and carbone isotopes in the fossil of Late Cretecaous age, Western Interior region of North America. Geol Soc Am Bull 80:1903–1922

Trueman AE (1941) The ammonoid body chamber with special reference to the buoyancy and mode of life of the living ammonite. Quart J Geol Soc Lond 96: 339–383

Tsujita CJ, Westermann GEG (1998) Ammonoid habitats and habits in the Western Interior Seaway: a case study from the Upper Cretaceous Bearpaw Formation of southern Alberta, Canada. Palaeogeogr Palaeoclimat Palaeoecol 144:135–160

Tsujita CJ, Westermann GEG (2001) Were limpets or mosasaurs responsible for the perforations in the ammonite *Placenticeras*? Palaeogeogr Palaeoclimatol Palaeoecol 169:245–270

Tyson RV, Pearson TH (1991) Modern and ancient continental shelf anoxia: an overview. In: Tyson RV, Pearson TH (eds) Modern and ancient continental shelf anoxia. Geol Soc Spec Pub 58:l–26

Uchiyama K, Tanabe K (1999) Hatching of *Nautilus macromphalus* in the Toba aquarium, Japan. In: Olóriz F, Rodríguez-Tovar FJ (eds) Advancing Research on Living and Fossil Cephalopods. Kluwer Academic, New York

Uhlig V (1911) Über die sogen. borealen Typen des südandinen Reiches. Cent Mineral Geol Paläontol 15, 16, 17:483–490, 517–522, 536–548

Urey HC (1947) The thermodynamic properties of isotopic substances. J Chem Soc 1947:562–581

Urey HC, Lowenstam HA, Epstein S, McKinney CR (1951) Measurements of paleotemperatures and temperatures of the Upper Cretaceous, Denmark, and southeastern United States. GSA Bull 62:399–416

Urlichs M, Mundlos R (1985) Immigration of cephalopods into the German Muschelkalk basin and its influence on the suture lines. In: Bayer U, Seilacher A (eds) Sedimentary and Evolutionary Cycles. Lecture Notes in Earth Sciences. Springer, Berlin

Urlichs M, Wild R, Ziegler B (1979) Fossilien aus Holzmaden. Stuttg Beitr Naturkunde Ser C 11:1–34

Vašíček Z, Wiedmann J (1994) The Leptoceratoidinae: small heteromorph ammonites from the Barremian. Palaeontology 37:203–239

Vermeij GJ (1977) The Mesozoic marine revolution: evidence from snails, predators and grazers. Paleobiology 3:245–258

Wang Y, Westermann GEG (1993) Paleoecology of Triassic ammonoids. Geobios Mem Spec 15:373–392

Wani R (2004) Experimental fragmentation pattersn of modern *Nautilus* shells and the implications for fossil cephalopod taphonomy. Lethaia 37:113–123

Wani R (2007) How to recognize *in situ* fossil cephalopods: evidence from experiments with modern *Nautilus*. Lethaia 40:305–311

Wani R, Gupta NS (2015). Ammonoid taphonomy. In: Klug C, Korn D, De Baets K, Kruta I, Mapes RH (eds) Topics in Geobiology, vol 43. Ammonoid Paleobiology: From macroevolution to paleogeography. Vol 2, Part III - Ammonoids through time. Springer, Dordrecht

Wani R, Kase T, Shigeta Y, De Ocampo R (2005) New look at ammonoid taphonomy, based on field experiments with modern chambered *Nautilus*. Geology 33:849–852

Wani R, Kurihara K, Ayyasami K (2011) Large hatchling size in Cretaceous nautiloids persists across the end-Cretaceous mass extinction: new data of Hercoglossidae hatchlings. Cretac Res 32:618–622

Ward PD (1976a) Stratigraphy, paleoecology and functional morphology of heteromorph ammonites in the Upper Cretaceous Nanaimo Group, British Columbia and Washington. Dissertation, McMaster University of Hamilton, Ontario.

Ward PD (1980) Comparative shell shape distributions in Jurassic-Cretaceous ammonites and nautilids. Paleobiology 6:32–43

Ward PD (1981) Shell sculpture as a defensive adaptation in ammonoids. Paleobiology 7:96–100.

Ward PD (1986a) Rates and processes of compensatory buoyancy change in *Nautilus macromphalus*. Veliger 1986:356–368
Ward PD (1986b) Cretaceous ammonite shell shapes. Malacologia 27:3–28
Ward PD (1987) The natural history of *Nautilus*. Allen and Unwin, Boston
Ward PD (1990a) A review of Maastrichtian ammonite ranges. GSA Spec Pap 247:519–530
Ward PD (1990b) The Cretaceous/Tertiary extinctions in the marine realm: A 1990 perspective. GSA Spec Pap 247:425–432
Ward PD, Bandel K. (1987) Life history strategies in fossil cephalopods. In: life cycles. Academic Press, London
Ward DJ, Hollingworth NTJ (1990) The first record of a bitten ammonite from the Middle Oxford Clay (Callovian, Middle Jurassic) of Bletchley, Buckingshamshire. Mesozoic Res 2:153–161
Ward PD, Signor PW III (1983) Evolutionary tempo in Jurassic and Cretaceous ammonites. Paleobiology 9:183–198
Ward PD, Westermann GEG (1977) First occurrence, systematics and functional morphology of *Nipponites* from the Americas. J Paleontol 51:367–372
Ward PD, Westermann GEG (1985) Cephalopod paleoecology. In: Bottjer DJ, Hickman CS, Ward PD (org), Broadhead TW (ed) Mollusks, Notes for a short course. Univ Tenn Geol Sci Stud 13:1–18
Ward PD, Carlson B, Weekly M, Brumbaugh B (1984) Remote telemetry of daily vertical and horizontal movement of *Nautilus* in Palau. Nature 309:248–250
Warnke K, Keupp H (2005) *Spirula*—A window to the embryonic developments of ammonoids? Morphological and molecular indications for a palaeontological hypothesis. Facies 51:60–65
Warnke K, Plötner J, Santana JI, Rueda MJ, Llinas O (2003) Reflections on the phylogenetic position of *Spirula* (Cephalopoda): preliminary evidence from the 18S ribosomal RNA gene. Berl Paläobiol Abh 3:253–260
Warnke K, Oppelt A, Hoffmann R (2010) Stable isotopes during ontogeny of *Spirula* and derived hatching temperatures. Ferrantia 59:191–201
Watanabe T, Gagan MK, Correge T, Scott-Gagan H, Cowley J, Hantoro WS (2003) Oxygen isotope systematics in *Diploastrea heliopora*: new coral archive of tropical paleoclimate. Geochim Cosmochim Acta 67:1349–1358
Wefer G (1985) Die Verteilung stabiler Isotope in Kalkschalen mariner Organismen. Geol Jahrb 82:3–111.
Wefer G, Berger, WH (1991) Isotope paleontology: growth and composition of extant calcareous species. Mar Geol 100:207–248
Wells MJ (1983) Cephalopods do it different. New Sci 100:332–338
Wells MJ, Wells J, O'Dor RK (1992) Life at low oxygen tensions: the behaviour and physiology of *Nautilus pompilius* and the biology of extinct forms. J Mar Biol Assoc UK 72:313–328
Wendt J (1963) Stratigraphisch-paläontologische Untersuchungen im Dogger Westsiziliens. Boll Soc Paleontol Ital 2:57–145
Westermann GEG (1954) Monographie der Otoitidae (Ammonoidea). Geol Jahrb Beih 15:1–364.
Westermann GEG (1958) The significance of septa and sutures in Jurassic ammonite systematics. Geol Mag 45:441–455
Westermann GEG (1971) Form, structure and function of shell and siphuncle in coiled Mesozoic ammonoids. Life Sci Contrib R Ont Mus 78:1–39
Westermann GEG (1973) Strength of concave septa and depth limits of fossil cephalopods. Lethaia 6:373–403
Westermann GEG (1975a) Architecture and buoyancy of simple cephalopod phragmocones and remarks on ammonites. Paläontol Z 49:221–234
Westermann GEG (1975b) A model for origin, function and fabrication of fluted cephalopod septa. Paläontol Z 49:235–253
Westermann GEG (1977) Form and function of orthoconic cephalopod shells with concave septa. Paleobiology 3:300–321
Westermann GEG (1982) The connecting rings of *Nautilus* and Mesozoic ammonoids: Implications for ammonoid bathymetry. Lethaia 15:374–384

Westermann GEG (1989) New developments in ecology of Jurassic-Cretaceous ammonoids, in Pallini, G., Cecca, F., and Cresta, S., (eds), Fossili, evoluzione, ambiente, Att II Conventione Internationale Pergola 1987: Otra Vetere, Italy, Tecnostampa
Westermann GEG (1990) New developments in ecology of Jurassic-Cretaceous ammonoids. In: Pallini G, Cecca E, Cresta S, Santantonio M (eds) Fossili, Evolutione, Ambiente. Atti II Conv Int F E A Pergola 1987. Com Cent Raff Piccinini, Ostra Vetere
Westermann GEG (1993a) On alleged negative buoyancy in ammonoids. Lethaia 26:246
Westermann GEG (1993b) Global bio-events in mid-Jurassic ammonites controlled by seaways. In: House MR (ed) The Ammonoidea. Environment, Ecology and Evolutionary Change. Systematics Association, special volume 47, Clarendon Press, Oxford
Westermann GEG (1996) Ammonoid life and habitat. In: Landman NH, Tanabe K, Davis RA (eds) Ammonoid Paleobiology. Plenum, New York
Westermann GEG, Callomon JH (1988) The Macrocephalitinae and associated Bathonian and early Callovian (Jurassic) ammonoids of the Sula lslands and New Guinea. Palaeontogr A 203:1–90
Westermann GEG, Hewitt RA (1995) Do limpet pits indicate that desmoceratacean ammonites lived mainly in surface waters? Lethaia 28:24
Westermann GEG, Ward P (1980) Septum morphology and bathymetry in cephalopods. Paleobiol 6:48–50
Whittaker SG, Kyser TK, Caldwell GE (1987) Paleoenvironmental geochemistry of the Claggett marine Cyclothem in south-central Saskatchewan. Can J Earth Sci 24:967–984
Wiedmann J (1973) Evolution or revolution of ammonoids at Mesozoic system boundaries. Biol Rev 48:159–194
Wiedmann J (1988a) Plate tectonics, sea level changes, climate and the relationship to ammonite evolution, provincialism, and mode of life. In: Wiedmann J, Kullmann J (eds) Cephalopods Present and Past. Schweizerbart, Stuttgart
Wiedmann J (1988b) Ammonite extinction and the Cretaceous Tertiary boundary event. In: Wiedmann J, Kullmann J (eds) Cephalopods Present and Past. Schweizerbart, Stuttgart
Wiedmann J, Kullmann J (eds) (1988) Cephalopods Present and Past. Schweizerbart, Stuttgart
Wignall PB (1987) A biofacies analysis of the *Gastrioceras cumbriense* Marine Band (Namurian) of the Central Pennines. Proc York Geol Soc 46:111–121
Wignall PB (1990) Observations on the evolution and classification of dysaerobic communities. In: Miller W (ed) Paleocommunity Temporal Dynamics: The Long-term Development of Multispecies Assemblies. Paleontol Soc Spec Publ 5:99–111
Wignall PB, Hallam A (1991) Biofacies, stratigraphic distribution and depositional models of British onshore Jurassic black shales. In: Tyson RV, Pearson TH (eds) Modern and ancient continental shelf anoxia. Geol Soc Spec Pap 58:291–309
Wignall PB, Hallam A (1993) Griesbachian (earliest Triassic) palaeoenvironmental changes in the Salt Range, Pakistan and southeast China and their bearing on the Permo-Triassic mass extinction. Palaeogeogr Palaeoclimatol Palaeoecol 102:215–237
Wignall PB, Simms MJ (1990) Pseudoplankton. Palaeontology 33:359–378
Wilde P, Berry WBN (1984) Destabilization of the oceanic density structure and its significance to marine "extinction" events. Palaeogeogr Palaeoclimatol Palaeoecol 48:143–162
Wilmsen M, Mosavinia A (2011) Phenotypic plasticity and taxonomy of *Schloenbachia varians* (J. Sowerby, 1817). Paläontol Z 85:169–184
Wippich MGE, Lehmann J (2004) *Allocrioceras* from the Cenomanian (Mid-Cretaceous) of the Lebanon and its bearing on the palaeobiological interpretation of heteromorphic ammonites. Palaeontology 47:1093–1107
Wright CW, Calloman JH, Howarth MK (1996) Treatise on Invertebrate Paleontology, Part L, Mollusca 4 revised (Cretaceous Ammonoidea), GSA, Boulder. University of Kansas Press, Lawrence
Wright EK (1987) Stratification and paleocirculation of the Late Cretaceous Western Interior Seaway of North America. GSA Bull 99:480–490

Wunsch M, Richter C (1998) The CaveCam–an endoscopic underwater videosystem for the exploration of cryptic habitats. Mar Ecol Prog Ser 169:277–282
Yacobucci MM (1999) Plasticity of developmental timing as underlying cause of high speciation rates in ammonoids. In: Olóriz F, Rodríguez-Tovar FJ (eds) Advancing Research on Living and Fossil Cephalopods. Kluwer Academic, Plenum, New York
Yacobucci M (2004) Buckman's Paradox: variability and constraints on ammonoid ornament and shell shape. Lethaia 37:57–69
Yahada H., Wani R (2013) Limited migration of scaphitid ammonoids: evidence from the analyses of shell whorls. J Paleontol 87(3):406–412
Yomogida S, Wani R (2013) Higher risk of fatality by predatory attacks in earlier ontogenetic stages of modern *Nautilus pompilius* in the Philippines: evidence from the ontogenetic analyses of shell repairs. Lethaia 46:317–330
Young RE, Vecchione M, Donovan DT (1998) The evolution of coloid cephalopods and their present biodiversity and ecology. In: Payne AIL, Lipinski MR, Clarke MR, Roeleveld MAC (eds) Cephalopod diversity, Ecology and Evolution. Afr J Mar Sci 20:393–420
Zaborski PMP (1982) Campanian and Maastrichtian sphenodiscid ammonites from southern Nigeria. Bull Br Mus Nat Hist (Geol) 36:303–332
Zakharov YD, Boriskina, N.G., Ignatyev AV, Tanabe K, Shigeta Y, Popov AM, Afanasyeva TB, Maeda H (1999) Palaeotemperature curve for the Late Cretaceous of the northwestern circum-Pacific. Cretac Res 20:685–997
Zakharov YD, Smyshlyaeva OP., Tanabe K, Shigeta Y, Maeda H, Ignatiev AV, Velivetskaya TA, Afanasyeva TB, Popov AM, Golozubov VV, Kolyadae AA, Cherbadzhi AK, Moriya K (2005) Seasonal temperature fluctuations in the high northern latitudes during the Cretaceous Period: isotopic evidence from Albian and Coniacian shallow-water invertebrates of the Talovka River Basin, Koryak Upland, Russian Far East. Cretac Res 26:113–132
Zakharov YD, Shigeta Y, Smyshlyaeva OP, Popov AM, Ignatiev AV (2006) Relationship between $\delta^{13}C$ and $\delta^{18}O$ values of the Recent *Nautilus* and brachiopod shells in the wild and the problem of reconstruction of fossil cephalopod habitat. Geosci J 10:331–345
Zakharov YD, Shigeta Y, Nagendra R, Safronov PP, Popov AM, Velivetskaya TA, Afanasyeva TB (2011) Cretaceous climate oscillations in the southern palaeolatitudes: new stable isotope evidence from India and Madagascar. Cretac Res 32:623–645
Zakharov YD, Melnikov ME, Popov AM, Pletnev SP, Khudik VD, Punina TA (2012) Cephalopod and brachiopod fossils from the Pacific: Evidence from the Upper Cretaceous of the Magellan Seamounts. Geobios 45:145–156
Zatoń M (2010) Sublethal injuries in Middle Jurassic ammonite shells from Poland. Geobios 43:365–375
Ziegler B (1963) Ammoniten als Faziesfossilien. Paläontol Z 37:96–102
Ziegler B (1967) Ammoniten-Ökologie am Beispiel des Oberjura. Geol Rundsch 56:439–464
Ziegler B (1981) Ammonoid biostratigraphy and provincialism: Jurassic-Old World. In: House MR, Senior JR (eds) The Ammonoidea: Systematics Assocoation, speial volume 18. Academic Press, London
Zinsmeister WJ (1987) Unusual nautilid occurrence in the upper Eocene La Meseta Formation, Seymour Island, Antarctica. J Paleontol 61:724–726

Chapter 19
Isotope Signature of Ammonoid Shells

Kazuyoshi Moriya

19.1 Introduction

From the middle Paleozoic to the end of Mesozoic, ammonoids represent one of the major constituents of macro-organisms that are preserved as fossils in the epicontinental seas. Their rich fossil records and remarkable diversification of shell forms in space and time indicate that ammonoids are one of the best examples for understanding relationships between evolution of marine biota and changes in paleoenvironment and/or paleoceanography. The biotic and abiotic components in the ancient ocean are, obviously, closely connected through the chemical cycles and energy flow in the ancient eco-system. Therefore, for understanding the cause and/or effect of diversity change or extinction events in the ancient earth- and eco-system, identifying ecological niche, such as habitat, food habit, life history strategy, etc., of ammonoid species is extremely important.

The mode of life of ammonoids has hitherto been investigated based mainly on functional morphology of their shells, as well as synpaleoecological analyses of fossil assemblages. The previous approaches on the basis of functional morphology are roughly divided into three aspects; (1) buoyancy of the animal including both a shell and a supposed soft part, (2) swimming and mobility and (3) depth habitat.

Many workers (Trueman 1941; Heptonstall 1970; Ebel 1983; Saunders and Swan 1984; Saunders and Shapiro 1986; Ebel 1992; Shigeta 1993; and others) calculated density of living ammonoids based on shell thickness, volume ratio between a phragmocone and a body chamber, and the estimated density of the hypothetized soft part. Although there are some ambiguities, such as accuracy of shell thickness calculations based on theoretical models, quantity of cameral liquid in the phragmocone and the supposed soft part density, most previous studies, excluding Ebel (1983, 1992) and Shigeta (1993), estimated that living ammonoids had a density

K. Moriya (✉)
Faculty of Natural System, Institute of Science and Engineering, Kanazawa University, Kakuma-machi, Kanazawa 920-1192, Japan
e-mail: kmoriya@staff.kanazawa-u.ac.jp

C. Klug et al. (eds.), *Ammonoid Paleobiology: From Anatomy to Ecology*,
Topics in Geobiology 43, DOI 10.1007/978-94-017-9630-9_19

almost equivalent to that of seawater (= neutral buoyancy). The neutral buoyancy hypothesis was subsequently supported by theoretical morphological considerations on the mode of coiling and life orientation of some Cretaceous heteromorph ammonites (Okamoto 1988a, b).

It has been shown that shell stability (determined by the distance between the centers of gravity and buoyancy), shell size and drag coefficient are the important factors in estimating swimming ability of ammonoids (Chamberlain 1976; Jacobs 1992; Jacobs and Chamberlain 1996). One of the outcomes of these studies is that, a laterally compressed morphology is more advantageous in a flow at higher Reynolds numbers, while a thicker morphology is more advantageous at low Reynolds numbers (Jacobs 1992). This correlation between relative shell thickness and Reynolds number, which is proportional to current velocity, is partly supported by a field observation (Bayer and McGhee 1984; Jacobs et al. 1994; Kawabe 2003). Another aspect of mobility is (diurnal) vertical migrations. Living chambered cephalopods (*Nautilus*, *Sepia* and *Spirula*) had long been considered to move vertically by controlling the amount of cameral liquid using osmotic pressure of the blood vessels (Denton and Gilpin-Brown 1973). In fact, direct observations of vertical movements of *Nautilus* with remote telemetry (Carlson et al. 1984; Ward et al. 1984), and detailed isotope record (Oba et al. 1992) seem to support this idea. This observation in modern *Nautilus* has been proposed for ammonites (e.g. Westermann 1996). Because the water depth in which efficient pumping using the simple osmotic mechanism can be accomplished is estimated to be shallower than 240 m (Greenwald et al. 1980), ammonoids which inhabited deeper part of the water column might have often needed to ascend to shallow waters to pump out the cameral liquid (Westermann 1989). However, Greenwald et al. (1982) studied the *Nautilus* siphuncle at the ultrastructural level and documented the presence of structures associated with a hyperosmotic pump, which enables them to discharge cameral liquid in deep water. Consequently, the argument for the ammonoid vertical movement by means of pumping out the cameral liquid with simple osmotic pressure lost the positive evidence.

A third aspect is habitat depth of ammonoid animals. The habitat depth of ammonoids has been estimated based on mechanical strength of the shell against ambient hydrostatic pressure. Such indices as Siphunclar Strength Index (Westermann 1971), Septal Flute Strength (Hewitt and Westermann 1987), Septal Amplitude Index (Batt 1991), etc. have been proposed to estimate the mechanical strength of the shell, and those indices are calibrated as absolute habitat depth by the implosion depth of *Nautilus* shell. Based on the fact that these indices show wide variation among various Mesozoic ammonoid morphotypes, Westermann (1996) suggested that Mesozoic ammonoids had various habitat depths. However, the implosion depth of the shell of living *Nautilus* does not represent an actual preferred depth of the animal. Direct observations using a remote camera and capture records using baited traps have demonstrated that the optimal habitat depth of *Nautilus* ranges from 150 and 300 m deep in Palau and between 300 and 500 m deep in Fiji (Saunders 1984; Hayasaka et al. 1987; Dunstan et al. 2011), whereas the shell implosion depth of *Nautilus* is comparable to 800 m deep (Kanie et al. 1980). Although these analyses in functional morphology gave us many great suggestions for considering ammonoid ecology, reasonable questions unanswered are still left on a table for future discussion.

Neutral buoyancy and swimming by jet propulsion of water in ammonoids have been suggested by a number of indirect evidence. On the other hand, inferences of vertical movements and the habitat depths appear to be highly suspicious. If ammonoids were distributed at various depths in the water column (including shallow water), carcasses of ammonoids would have been widely dispersed by post-mortem drift (Maeda and Seilacher 1996). Contrary to this expectation, some previous studies on faunal distributions of ammonoid assemblages have shown that spectra of faunal compositions are closely related to lithofacies (Scott 1940; Tanabe 1979; Cecca 1992; Olóriz et al. 1993). These facies-dependent occurrences of ammonoids suggest that ammonoid carcasses were not transported for a long distance by the post-mortem drift, except water column was shallow, which implies those ammonoids did not inhabit shallower part of the water column.

Recently, new data on ammonoid anatomy and/or ecology have been revealed from investigations of samples showing exceptional preservations (Lagerstätten) (Tanabe et al. 2000; Klug et al. 2012 and references therein), and a new extremely high resolution imaging technique (Kruta et al. 2011). While most of the soft part preservations are reported on buccal masses, sometimes including radulae, Tanabe et al. (2000) found and investigated a phosphatized siphuncle of Permian *Akmilleria electraensis*. They showed that microanatomy of the siphuncle of *Akmilleria* is almost identical to that of living cephalopods, indicating that "ammonoids could control their buoyancy by transferring liquid osmotically between chambers and blood vessels of the siphuncle (Tanabe et al. 2000)". Kruta et al. (2011) reported potential food remains in a buccal mass of Cretacesous *Baculites*. They found both benthic (demersal) and palnktic micro-organisms in the buccal mass, and concluded "these ammonites fed on small organisms in the water column, rather than capturing and eating large prey on the ocean bottom, as exemplified by living nautilus (Kruta et al. 2011)". This evidence implies that aptychophoran ammonoids were detritus and/or suspension feeders, rather than scavenger like modern nautilus. These new findings from fossils showing exceptional preservations present direct evidence on ecology and/or physiology of ammonoids. While these examples are especially important for discussing ammonoid ecology, very unfortunately, occurrences of these examples are rare.

In addition to analyses of the physical evidence of fossil materials, geochemical signatures of ammonoid shells potentially provide us an additional piece of information on ammonoid ecology. Although behavior and/or physiology of ammonoids have became better understood with examinations of extraordinarily preserved fossils and/or cladistics (e.g. Jacobs and Landman 1993; Tanabe et al. 2000), morphological analyses still need some assumptions (e.g. morphology of soft part, physiological processes, etc.) On the other hand, geochemical signatures preserved in ammonoid shells are, ideally, controlled by chemical and/or physicochemical processes. Therefore, geochemical signatures of both modern and fossil biominerals are encoded by the same processes, which reduces number of assumptions necessary for understanding ecology of ancient animals secreting calcareous hard parts. Chemical compositions of aragonite, comprising a shell of ammonoids, precipitated in the ocean record a variety of physical and chemical qualities of the ocean, such as ocean circulation, pH, productivity, salinity, temperature, etc. (cf. Austin and James 2008; Hillaire-Marcel and de Vernal 2008). Among these properties, temperature is especially informative for understanding ammonoid ecology. Sinusoidal temperature fluctuations with ontogeny, which might be attributable to seasonal seawater temperature variations, would provide the growth rate of the individuals (Stahl and Jordan 1969; Jordan and Stahl 1970; Fatherree et al. 1998). When the ammonoid calcification temperatures are compared with the thermal structure of the water column that those ammonoids inhabited, the habitat depth of each ammonoid species and ontogenetic migration could be determined (Anderson et al. 1994; Moriya et al. 2003; Lukeneder et al. 2010). Examples of these geochemical studies on ammonoid shells are summarized in Chap. 3.5 of this volume. In this chapter, the nature of geochemical signatures of ammonoid shells is summarized from the practical point of view with a little information of background knowledge and potential pitfalls.

19.2 Geochemical Signatures Recorded on Ammonoid Shells

Ammonoids, known as a group of extinct ectocochleate cephalopods, secrete calcareous hard parts composed of aragonite. Aragonite, which is often found as a biomineral in many groups of animals in many geological periods, is orthorhombic calcium carbonate. Biologically precipitated marine aragonite includes not only calcium, but also many other metal elements as a solid solution and/or interstitial contaminations, for example, boron, magnesium, strontium, cadmium, barium, uranium, and many others. In addition to carbon and oxygen isotopic compositions of carbonate, abundance of these metals relative to calcium and isotopic ratio of these metals are empirically used for paleoenvironmental proxies. For example, the relative abundance of Sr (Sr/Ca), which can replace Ca in the lattice as a substitutional solid solution, is commonly used as a paleotemperature proxy (Gagan et al. 2000; Henderson 2002 and many others). However, since those trace metals could be preferentially dissolved and altered through diagenesis (e.g. Cochran et al. 2010), using trace metal proxies on fossil material would be difficult in some cases. Diagenesis of geochemical signals in fossil materials will be discussed below.

On the other hand, carbon and oxygen isotopic ratios of marine biogenic aragonite are much more promising geochemical signatures. Compared with the oxygen isotopic composition, which is a function of calcification temperature and oxygen isotopic ratio of the solution (seawater), the carbon isotope signal is a little more complicated. In this section, theoretical and empirical backgrounds of these isotopic signatures are briefly summarized.

19.2.1 Oxygen Isotope Thermometry

Since Urey (1947), MacCrea (1950), and Epstein et al. (1953) discovered that the stable oxygen isotopic ratios of calcium carbonates correlate with the temperature of the water in which they are precipitated, and developed a method to evaluate paleotemperatures, many authors have attempted to assess ancient or present day seawater temperatures from biologically precipitated calcium carbonates (e.g. Wefer 1985; Wefer and Berger 1991 and references therein). The oxygen isotope paleothermomety is especially well-developed in the field of paleoceanography for estimation of sea surface temperature and deep-sea temperature in both geological and seasonal time scales in the Earth history (e.g. Emiliani 1954; Shackleton and Opdyke 1973; Shackleton and Kennett 1975; McCulloch et al. 1994; Leder et al. 1996; Gagan et al. 1998; Suzuki et al. 1999; Suzuki et al. 2001; Zachos et al. 2001; Zachos et al. 2008; Watanabe et al. 2011; Friedrich et al. 2012).

19.2.1.1 Concepts and Equations for Oxygen Isotope Thermometry

On the basis of theoretical prediction of Urey (1947) that the fractionation of oxygen isotopes between calcium carbonate and water should vary with the temperature of the water, many authors determined equilibrium fractionation factors $\alpha_{carbonate\text{–}water}$ and/or temperature equations in both synthetic and biogenic carbonates (cf. Grossman 2012). $\alpha_{carbonate\text{–}water}$ is the fractionation factor between carbonates and water, and described as follows:

$$\alpha_{\text{carbonate-water}} = \frac{\left(\frac{^{18}\text{O}}{^{16}\text{O}}\right)_{\text{carbonate}}}{\left(\frac{^{18}\text{O}}{^{16}\text{O}}\right)_{\text{water}}} \tag{19.1}$$

While most of the studies of the temperature dependency of the fractionation factor α have been made on calcite, Kim et al. (2007) finally made a systematic work on synthetic aragonite (Table 19.1). The temperature dependency of $\alpha_{carbonate\text{–}water}$ they proposed over the temperature range of 0–40°C is described as follows:

$$1000 ln\alpha_{\text{aragonite-water}} = 17.88 \times \left(\frac{10^2}{\text{T}}\right) - 31.14 \tag{19.2}$$

where T is temperature in Kelvin. Instead of reporting the temperature dependency of the fractionation factor, Grossman and Ku (1986) reported the empirical and conventional temperature scale over the temperature range of 2.6–22°C on the basis of oxygen isotope analyses of modern aragonitic benthic foraminifers and molluscs.

Table 19.1 Oxygen isotopic fractionation equations and conventional isotopic temperature scales for molluscan aragonite

$\delta^{18}O_c$ (VPDB)	$\delta^{18}O_w$ (VSMOW)	Equation (T in °K)	Conventional temperature scale (T in °C)	Temperature calcurated (°C)	Temperature range (°C)	Reference
1.0	−1.0	$1000\ln\alpha=18.04(10^3/T)-31.12$	$T=19.74-4.37(\delta^{18}O_c(VPDB)-\delta^{18}O_w(VSMOW))$	10.7	2.6–22	Grossman and Ku (1986); Hudson and Anderson (1989); Kim et al. (2007)
1.0	−1.0	$1000\ln\alpha=16.74(10^3/T)-26.39$	$T=21.36-4.83(\delta^{18}O_c(VPDB)-\delta^{18}O_w(VSMOW))$	11.4	8–24	White et al. (1999)
1.0	−1.0	$1000\ln\alpha=17.88(10^3/T)-31.14$	–	9.7	0–40	Kim et al. (2007)
−1.0	−1.0	$1000\ln\alpha=18.04(10^3/T)-31.12$	$T=19.74-4.37(\delta^{18}O_c(VPDB)-\delta^{18}O_w(VSMOW))$	19.9	2.6–22	Grossman and Ku (1986); Hudson and Anderson (1989); Kim et al. (2007)
−1.0	−1.0	$1000\ln\alpha=16.74(10^3/T)-26.39$	$T=21.36-4.83(\delta^{18}O_c(VPDB)-\delta^{18}O_w(VSMOW))$	21.4	8–24	White et al. (1999)
−1.0	−1.0	$1000\ln\alpha=17.88(10^3/T)-31.14$	–	18.9	0–40	Kim et al. (2007)
−3.0	−1.0	$1000\ln\alpha=17.88(10^3/T)-31.14$	–	28.8	0–40	Kim et al. (2007)

For temperature calculation, $\alpha(CO_2\text{-aragonite})$ and $\alpha(CO_2\text{-}H_2O)$ are assumed as 1.01025 and 1.0412, respectively for Grossman and Ku (1986) and White et al. (1999), 1.01063 and 1.0412, respectively, for Kim et al (2007)

They reported temperature dependency of ($\delta^{18}O_c$–$\delta^{18}O_w$), where $\delta^{18}O_c$ is oxygen isotopic composition of a carbonate analyzed; $\delta^{18}O_w$ is oxygen isotopic composition of seawater (Table 19.1).

Oxygen isotopic ratio is reported in delta notation relative to international reference standard, which are VPDB (Vienna Peedee Belemnite) and VSMOW (Vienna Standard Mean Ocean Water) (IAEA 1995). The delta value is described as follows:

$$\delta^{18}O\ (‰) = \left(\frac{\left(\frac{^{18}O}{^{16}O}\right)_{\text{carbonate}}}{\left(\frac{^{18}O}{^{16}O}\right)_{\text{standard}}} - 1 \right) \times 1000 \qquad (19.3)$$

Therefore, $\delta^{18}O$ should now always be reported relative to VPDB ($\delta^{18}O_{(VPDB)}$) or VSMOW ($\delta^{18}O_{(VSMOW)}$). However, since Grossman and Ku (1986) reported oxygen isotopic compostion of seawater relative to"average marine water", and they used 0.2‰ for conversion of $\delta^{18}O$ of CO_2 equilibrated to seawater into $\delta^{18}O$ of CO_2 evolved from the reactin between carbonates and phosphoric acid, the equation is revised by Hudson and Anderson (1989) for $\delta^{18}O_{c(VPDB)}$ and $\delta^{18}O_{w(VSMOW)}$ (see Epstein et al. 1953; Grossman and Ku 1986; Grossman 2012 for details). However, a precise method for obtaining the revised equation was not described in Hudson and Anderson (1989), and they only made a revision to equation (1) of Grossman and Ku (1986) (the equation for all aragonitic tests including benthic foraminifers and molluscs), so all three conventional temperature scales reported in Grossman and Ku (1986) are reassessed herein (Fig. 19.1, Table 19.1). The regressions shown in red solid lines represent reduced major axis regressions obtained herein (Fig. 19.1b, c, d). All data used in this reassessment are from Grossman and Ku (1986). As mentioned in Grossman and Ku (1986), since both *in situ* temperatures (y axis) and oxygen isotopic compositions (x axis) include analytical error, a type II regression is employed. The regressions yield the following equations:

$$T = 19.74 - 4.37 \times \left(\delta^{18}O_{c(VPDB)} - \delta^{18}O_{w(VSMOW)}\right) \qquad (19.4)$$

$$T = 19.17 - 4.21 \times \left(\delta^{18}O_{c(VPDB)} - \delta^{18}O_{w(VSMOW)}\right) \qquad (19.5)$$

$$T = 20.70 - 4.57 \times \left(\delta^{18}O_{c(VPDB)} - \delta^{18}O_{w(VSMOW)}\right) \qquad (19.6)$$

where T is temperature in Celsius; $\delta^{18}O_{c(VPDB)}$ is oxygen iosotpic composition of the aragonaite analyzed; $\delta^{18}O_{w(VSMOW)}$ is oxygen isotopic composition of seawater. Equations (19.4), (19.5), and (19.6) are for all data including benthic foraminifers and molluscs, live benthic foraminifers, and molluscs (gastropods and scaphopods), respectively. 95% confidential and prediction intervals of least square regres-

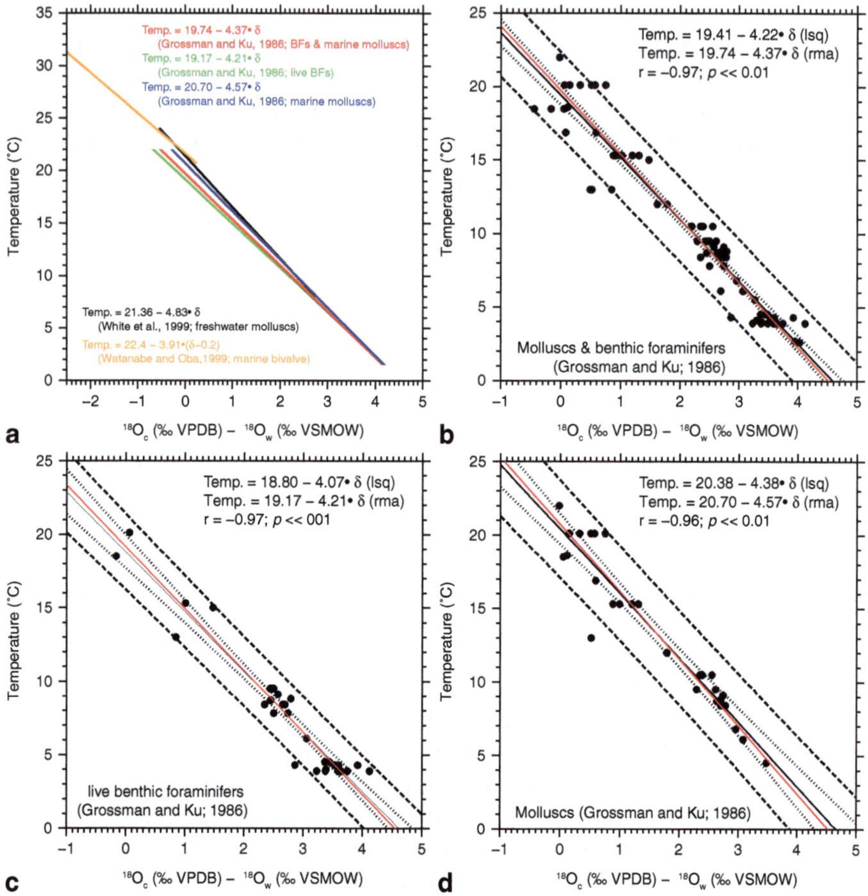

Fig. 19.1 Oxygen isotopic paleotemperature equations for mollucan biogenic aragonite. (**a**) Intercomparison of temperature scales published by Grossman and Ku (1986) (reassessed in this study), Watanabe and Oba (1999), and White et al. (1999). Since Watanabe and Oba (1999) determined the temperature scale against [($\delta^{18}O_c$ against VPDB)–($\delta^{18}O_w$ against Average Marine Water)], 0.2 is subtracted from δ (see Grossman (2012) for detail). (**b**) Revised temperature scale for aragonitic molluscs and benthic foraminifers. Black solid, dotted, and dashed lines represent the least square regression, 95 % confidential interval, and 95 % prediction interval for the regression, respectively. Red solid line shows the reduced major axis regression determined in this study. All data shown in black solid circles are cited from Grossman and Ku (1986). (**c**) Revised temperature scale for aragonitic live benthic foraminifers. Black solid, dotted, and dashed lines represent the least square regression, 95 % confidential interval, and 95 % prediction interval for the regression, respectively. Red solid line shows the reduced major axis regression determined in this study. All data shown in black solid circles are cited from Grossman and Ku (1986). (**d**) Revised temperature scale for aragonitic molluscs. Black solid, dotted, and dashed lines represent the least square regression, 95 % confidential interval, and 95 % prediction interval for the regression, respectively. Red solid line shows the reduced major axis regression determined in this study. All data shown in black solid circles are cited from Grossman and Ku (1986). lsq; least square regression. rma; reduced major axis regression. $\delta^{18}O_c$; oxygen isotopic composition of carbonate analyzed. $\delta^{18}O_w$; oxygen isotopic composition of water. $\delta=(\delta^{18}O_c-\delta^{18}O_w)$

sions are also shown in Fig. 19.1b, c, and d. In additon to the pionieering work of Grossman and Ku (1986), White et al. (1999) determined the temperature dependency of fractionation factor $\alpha_{\text{aragonite–water}}$ and a conventional temperature scale on the basis of oxygen isotopic analyses of non-marine gastropods. However, since they determined those conventional temperature scales in relatively lower temperature regime (2.6–22 °C), those scales are not applicable for materials obtained in the warm climate interval and tropical regions. On the other hand, Watanabe and Oba (1999) reported a conventional temperature scale in a higher temperature regime with oxygen isotopic analyses of a shallow marine tridacnid bivalve (Fig. 19.1a). The conventional temperature scales reported in these studies are slightly different from each other. However, considering the prediction intervals of these equations shown in Fig. 19.1b, c, and d, equations reported in Grossman and Ku (1986) and White et al. (1999) are indistinguishable each other (Fig. 19.1). Additionally, these biogenic aragonite–water fractionation curves are also statistically indistinguishable from the abiotic aragonite–water fractionation curve reported by Kim et al. (2007), indicating that these biogenic aragonite shells are precipitated at or very close to isotopic equilibrium.

Although the equation reported by Watanabe and Oba (1999) at the higher temperature regime is also similar to the extrapolation of the equation of Grossman and Ku (1986), slopes of the equation are different from each other. These results imply that extrapolating the equation of Grossman and Ku (1986) to the higher temperature regime or that of Watanabe and Oba (1999) to the lower temperature regime might produce artificial errors on paleotemperature calculations. In fact, even in an inter-comparison in the lower temperature regime, each equation gives slightly different paleotemperature estimation (Table 19.1). Considering the temperature range in which each equation is applicable, it should be emphasized that using the equation reported by Kim et al. (2007) would be appropriate for paleotemperature estimation on ammonoids, especially in the greenhouse interval.

19.2.1.2 Vital Effect

For oxygen isotope thermometry, one of the most important things one should be aware of is that the biogenic aragonite as an analyte be precipitated under isotopic equilibrium. However, biogenic aragonite can be precipitated under isotopic disequilibrium with ambient water because of many biological processes involved, such as growth rate, precipitation rate of aragonite, feeding, symbiosis, etc. The biogenic process making isotope disequilibrium is called a vital effect (Urey et al. 1951). The vital effect on biogenic aragonite precipitation can be found in many groups of animals (Veizer 1983; Wefer and Berger 1991; Veizer 1992). In comparison with carbon isotopes, which will be mentioned below, while the vital effect might not be prominent on oxygen isotopes (Wefer and Berger 1991), some examples have been known hitherto in corals, echinoderms, foraminifers, etc. (e.g. Weber and Raup 1966; Williams et al. 1981; McConnaughey 1989; Bemis et al. 1998). However, in some animals, a significant correlation between $\delta^{18}O$ and ambient temperature has been observed, even when they show a vital effect for $\delta^{18}O$

(Bemis et al. 1998; Gagan et al. 2000). In modern ocean, ($\delta^{18}O_c$–$\delta^{18}O_w$) and *in situ* temperature can be directly measured and an original temperature scale for each species of animals can be established as mentioned above. However, it is impossible to assess the presence or degree of vital effect in extinct fossil animals including ammonoids.

If we could assume that biomineralization processes of cephalopods, including proteins and enzymes involved, are conservative through their evolutional history, we might be able to find a clue from a phylogenetic position of ammonoids. On the basis of cladistics analyses, Ammonoidea is a sister group of Coleoidea (e.g. Ruppert et al. 2004). Since both modern nautiloids and coleoides secrete their shells at or very close to isotopic equilibrium (Landman et al. 1994; Bettencourt and Guerra 1999), we assume that ammonoids also secrete their shells at or very close to isotopic equilibrium. As mentioned in the previous section, both marine and non-marine molluscs often secrete their shells under isotopic equilibrium. Additionally, neither *Tridacna*, bearing photosymbionts, nor *Calyptogena*, bearing chemosymbionts, show a sign of vital effects on oxygen isotopic compositions (cf. Kulm et al. 1986; Romanek and Grossman 1989; Grossman 1993). Therefore, although there is no direct evidence, we hypothesize that ammonoids secrete their shells at or very close to isotopic equilibrium for oxygen isotopes.

19.2.1.3 The Effect of Diagenesis on Isotopic Signals

The other process that can alter the original isotopic signals on ammonoid shells is diagenesis of fossil materials. During the diagenesis, metastable aragonite, comprising ammonoid shells, can be partly or completely dissolved and replaced by neomorphic aragonite or calcite, depending on Mg ion concentration of the solution. Since original geochemical signals are altered by diagenesis, pristine aragonitic shells are required for geochemical analyses. Diagenesis, including dissolution of the original materials, can happen in both marine and meteoric water.

The fully marine epicontinental ocean, where many ammonoid fossils have been buried, is saturated in aragonite and calcite in the period of relatively low atmospheric carbon dioxide concentration, so abiotic secondary aragonite or calcite would be precipitated within and/or on the surface of shell materials. The precipitation of abiotic secondary aragonite is well known in modern corals (Enmar et al. 2000; Sayani et al. 2011). When Mg/Ca ratio in the ocean water is relatively low (calcite sea), calcite would be precipitated. On the other hand, the dissolution of aragonite shells might happen when the atmospheric carbon dioxide concentration is relatively high, as is predicted in future ocean acidification.

When sedimentary rocks containing ammonoid fossils are delivered to a subaerial condition, fossils may be exposed to meteoric water. Because aragonite is easily dissolved in undersaturated meteoric water, original shell materials are dissolved and neomorphic calcite can be precipitated (Veizer 1992). This meteoric digenesis is possibly more important, because ammonoid fossils for geochemical analyses are usually obtained from outcrops.

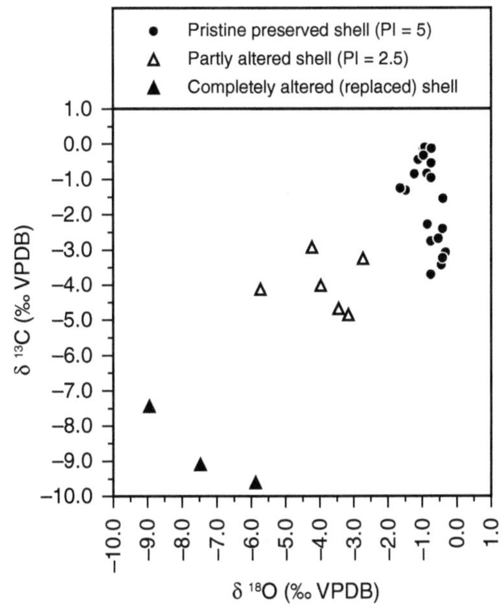

Fig. 19.2 Carbon and oxygen isotopic compositions of ammonoid fossils obtained from the late Cretaceous Yezo Group distributed in Hokkaido, Japan. All data are cited from Moriya (2008). Solid circles indicate results from a pristine preserved shell showing Preservation Index (PI) = 5. Open triangles represent results from a partly altered shell showing PI = 2.5. Solid triangles are results from a completely altered shell (PI is not defined for the completely altered shell). PI is defined by Cochran et al. (2010) (see text and Fig. 19.4 for details). Both $\delta^{13}C$ and $\delta^{18}O$ from the altered shell represent significantly negative values, comprising an end member of a diagenetic isotopic mixing line

Because of the oxygen isotopic fractionation between water vapor and liquid water, precipitation and ground water are usually depleted in ^{18}O (Gat et al. 2001; Geyh 2001). Dissolved inorganic carbon within meteoric water is also depleted in heavier species because of enhanced ^{12}C input from the decomposition of organic matters. Therefore, $\delta^{13}C$ and $\delta^{18}O$ of diagenetic carbonates are lower (more negative) than those in original biogenic carbonate (Brand and Veizer 1981; Mitchell et al. 1997; Moriya 2008; Cochran et al. 2010) (Fig. 19.2). As a result, $\delta^{13}C$ and $\delta^{18}O$ of altered materials lie on a covariant mixing line (Mitchell et al. 1997; Cochran et al. 2010) (Fig. 19.2), and are described with conventional mass balance equation:

$$\delta_{ori} \times A + \delta_{dia} \times (100 - A) = \delta_{total} \tag{19.7}$$

where δ_{ori} is δ value of original pristine aragonite; δ_{dia} is δvalue of diagenetic carbonate; δ_{total} is δ value of analytes; A is relative amount of original aragonite in wt%. The result of this conventional mass balance calculation is shown in Fig. 19.3, assumming δ value of original aragonite equals to zero. When δ_{dia} is −10‰, contamination of only 5 wt% diagenetic carbonate makes 0.5‰ shift in δ_{total}. According to temperature scales discribed above (Table 19.1), 0.5‰ shift in $\delta^{18}O$ is comparative

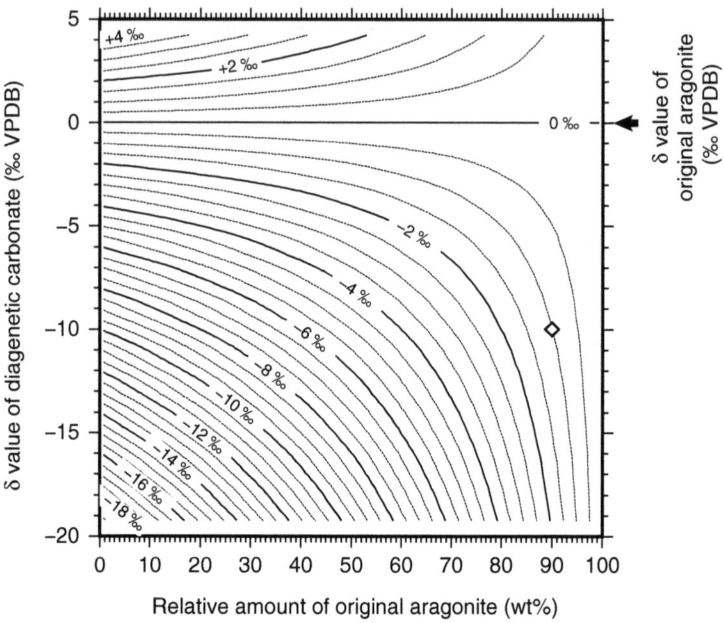

Fig. 19.3 Contour diagram showing the mass balance modeling of diagenetic alteration of isotopic value against relative amount of diagenetic carbonate in wt %. δ value of original unaltered aragonite is estimated as 0 ‰ (VPDB). The vertical and horizontal axes indicate δ value of diagenetic carbonate and relative amount of original aragonite in wt %, respectively. Contours represent δ value of carbonate minerals to be analyzed. For example, when carbonate as an analyte has contamination of 10 wt% diagenetic calcite of −10‰, δ value of carbonate analyzed will be −1‰, which is shown by an open diamond

to approximately 2 °C shift in temperature estimation. Therefore, choosing pristine preserved materias is especially important for isotope analyses.

Many criteria for selecting well preserved materials have been proposed to identify degree of diagenesis on fossil materials. Among them, mineralogical and elemental analyses are especially informative for assessing the effects of diagenetic alteration on original aragonitic shell materials (Brand and Veizer 1980; Brand and Veizer 1981; Popp et al. 1986; Anderson et al. 1994; Cochran et al. 2010). Because aragonitic fossils are easily dissolved in meteoric water and neomorphic calcite will be precipitated, mineralogical analyses on X-ray diffractometer is commonly employed (e.g. Davies and Hooper 1963; Anderson et al. 1994; Moriya et al. 2003). In addition to mineralogical analyses, trace element compositions are also revealing. Comparing to marine water, meteoric water is depleted in Mg and Sr concentrations, so Mg/Ca and Sr/Ca tend to decrease through diagenesis. On the other hand, since meteoric water is enriched in Fe and Mn concentrations, Fe/Ca and Mn/Ca increase with diagenesis and showing positive correlation (Brand and Veizer 1980; Popp et al. 1986; Tucker and Wright 1990; Anderson et al. 1994; Moriya et al. 2003). However, in a specific geological background, Sr/Ca would increase with diagenesis (Cochran et al. 2010). Because trace element compositions are altered from their original value in any case, it is one of promising indices for evaluating diagenesis on ammonoid fossils. Although it is relatively qualitative, instead of quantitative, observation of shell microstructure under scanning electron microscope is also useful (Moriya 2008; Cochran et al. 2010). Because diagenesis on geochemical signatures of aragonitic fossils is always accompanied by dissolution and/or precipitation of neomorphic carbonates, visual observations showing pristine shell microstructure assure the preservation of geochemical signature (Cochran et al. 2010; Sayani et al. 2011).

Compared to isotopic compositions, trace element compositions are more sensitive to diagenesis (Brand and Veizer 1980; Popp et al. 1986; Anderson et al. 1994; Cochran et al. 2010). In particular, although both Sr/Ca and $\delta^{18}O$ are proxies for paleotemperature, Cochran et al. (2010) showed that $\delta^{18}O$ is more durable against diagenesis in terms of SEM preservation index (PI). According to Cochran et al. (2010) PIs are defined as follows;

- PI=5 (Excellent; Fig. 19.4a, b): Surface clean and unetched; internal nacreous tablets distinct from adjacent layers and well-defined. Samples at this level of preservation are indistinguishable from nacreous shell structure in modern mollusks (for example, Nautilus),
- PI=4 (Very Good; Fig. 19.4c, d): Surface clean and unetched; nacreous tablets well defined, but their surfaces are slightly irregular and boundaries between adjacent tablets are slightly less distinct than in excellent preservation,
- PI=3 (Good; Fig. 19.4e, f): Surface good, but shows some signs of etching; nacreous tablets are visible, but show the onset of fusion with adjacent tablets. Specimens in this group may show a range of preservation, such that SEM images from different parts of the shell show variation in preservation,
- PI=2 (Fair; Fig. 19.4g, h): Surface shows etching; nacreous tablets are discernible, but show fusion with adjacent layers,
- PI=1 (Poor; Fig. 19.4i, j): Surface is significantly etched; nacreous tablets are indistinct and fused with adjacent layers.

In addition to the Sr concentrations in orthorhombic aragonite, selective alteration of Mg in trigonal rhombohedral calcite has also been reported (Barker et al. 2005). Even in the modern ocean, Mg/Ca ratio of foraminiferal tests is altered because of the preferential dissolution of high Mg/Ca regions within the test. Comparing to the pristine calcite composed of $CaCO_3$, the regions containing Mg in the lattice are assumed to be less crystallized. Therefore, in the process of the initial dissolution, those less crystallized regions including more Mg would be preferentially dissolved (Bassinot et al. 2004; Barker et al. 2005). The same scenario would be true for Sr/Ca in aragonite fossils. In fact, while $\delta^{18}O$ stays constant from 5 through 2.5 in PI, Sr/Ca becomes altered from 4 in PI (Cochran et al. 2010). To conclude, metal/Ca proxies would be less promising in comparison to $\delta^{13}C$ and $\delta^{18}O$, except for the materials showing exquisite preservation.

Using the weakness of trace element compositions for diagenesis, trace metal compositions can be used as an index of diagenetic alteration (Brand and Veizer 1980; Popp et al. 1986; Anderson et al. 1994; Cochran et al. 2010). Although trace element compositions are usually determined by inductively coupled plasma–atomic emission spectrometry (ICP-AES), inductively coupled plasma–mass

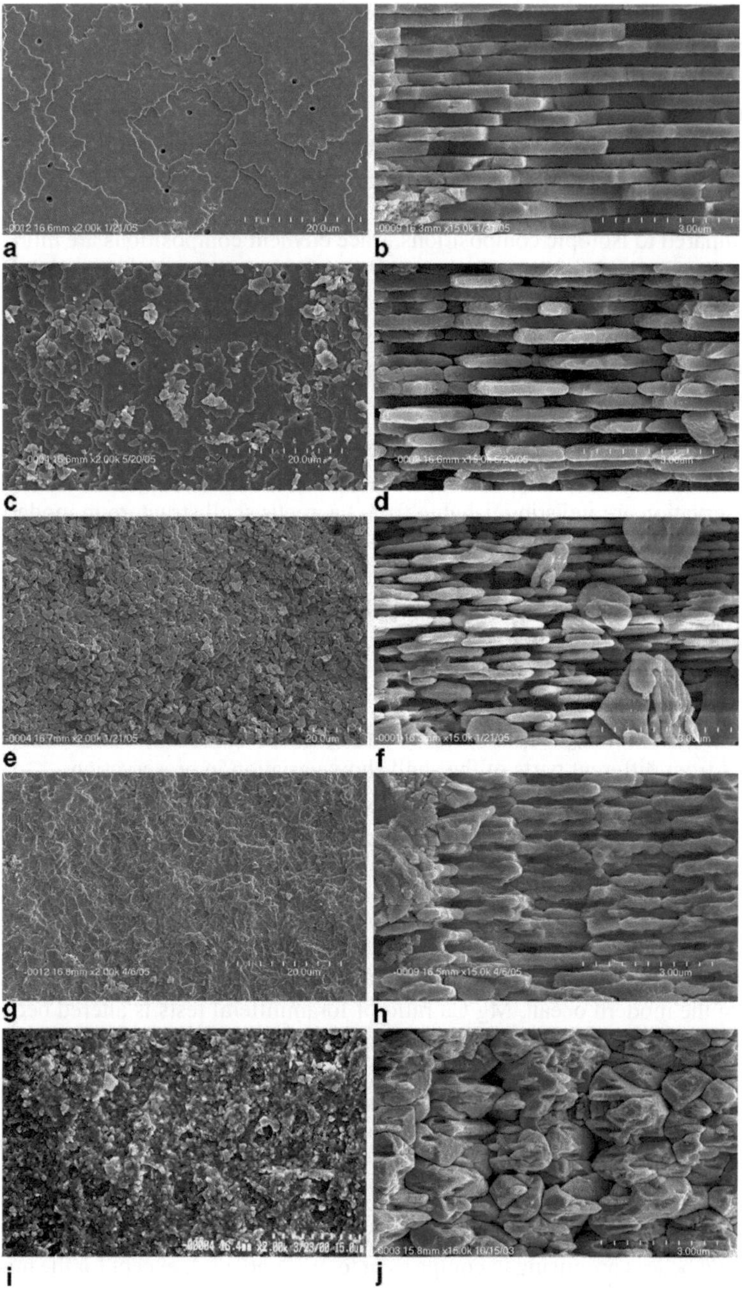

Fig. 19.4 Scanning electron micrographs showing different degrees of preservation (expressed as a Preservation Index, PI) of aragonitic nacreous layer of fossils from the late Cretaceous Western Interior Seaway. (**a, b**) "Excellent" preservation (PI = 5) in *Baculites compressus*: **a**, surface; **b**, cross-section. (**c, d**) "Very Good" preservation (PI = 4) in *Baculites compressus*: **c**, surface; **d**, cross-section. (**e, f**) "Good" preservation (PI = 3) in *Placenticeras meeki*: **e**, surface; **f**, cross-section. (**g, h**) "Fair" preservation (PI = 2) in *Hoploscaphites brevis*: **g**, surface; **h**, cross-section. (**i, j**) "Poor" preservation (PI = 1) in *Placenticeras meeki*: **i**, surface; **j**, cross-section. Scale bars are shown in each panels. Redrawn from Cochran et al. (2010) with permission of American Journal of Science

spectrometry (ICP-MS) and/or electron probe microanalyses (EPMA), cathodeluminescence (CL) are also knows as useful techniques for identifying diagenesis (Tucker and Wright 1990). In contrast to dissolution of Mg and/or Sr, Mn and/or Fe will be incorporated into neomorphic carbonates during diagenesis (Brand and Veizer 1980; Tucker and Wright 1990). Since Mn and Fe commonly intensify luminescence and produce yellow to orange emission, CL provides nondestructive qualitative results (Machel 2000). Table 19.2 shows variety of techniques used for identifying diagenesis of ammonoid fossils. According to Cochran et al. (2010), when PI is more than 4, δ^{18}O should be unaltered. Although a set of analyses of SEM, ^{87}Sr/^{86}Sr ratios, Sr/Ca ratios and X-ray diffraction (XRD) would ideally be required, observation of PI under SEM is useful to eliminate potentially altered specimens (Cochran et al. 2010). Although, Stahl and Jordan (1969) and Jordan and Stahl (1970) did not provide any scanning electron micrographs, they presented XRD results. However, some individuals contain a little amount of diagenetic calcite (up to 9 %), indicating δ^{18}O may partly be altered. While Anderson et al. (1994) did not show scanning electron micrographs as well, they provided precise results of trace element compositions which are used for selecting unaltered samples. Since they used samples composed of approximately 100 wt% aragonite showing very low Mn concentration for isotopic analyses, there results should be secured. Moriya et al. (2003) used CL, SEM and EPMA for calcite fossils, and SEM and XRD for aragonite fossils. Those calcite and aragonite fossils show extremely low Mn and Fe and well preserved shell microstructures, and well preserved shell microstructures and approximately 100 wt% aragonite, respectively. Additionally, the scanning electron micrographs indicate 5 in PI, which assure the preservation of those fossils. Lécuyer and Bucher (2006) made ICP, SEM and XRD analyses. Although it is not clearly shown, scanning electron micrographs seem to show 5 in PI, indicating the specimen used in their analyses may be well preserved. Lukeneder et al. (2010) provided CL, SEM, EPMA and XRD results. While most of their samples seem to show well preservation, shell microstructure of *Nowakites* may show 2 in PI (shown in a scanning electron micrograph in Appendix A-B-C in Lukeneder et al. 2010).

19.2.1.4 Other Factors Affecting δ^{18}O of Ammonoid Shells

Besides the vital effect and diagenetic alteration mentioned above, the original oxygen isotopic composition of biogenic aragonite can be affected by the oxygen isotopic composition of seawater and concentration of bicarbonate ions within seawater in which the shell materials are precipitated.

Because both salinity and δ^{18}O of sea surface water are modified by evaporation, they usually show a strong correlation. A salinity increase of 2 resulted from enhanced local evaporation is equivalent to an approximately 0.5 ‰ increase of the oxygen isotopic composition of sea surface water in the present day equatorial Pacific (Fairbanks et al. 1997). As mentioned above, 0.5 ‰ shifts in δ^{18}O corresponds to approximately 2 °C shifts in paleotemperature estimation. Considering the meridional gradient of net evaporation rate from the equator to the pole, Zachos et al. (1994) discussed a method for adjusting δ^{18}O of sea surface water as a function of

Table 19.2 Criteria used for identifying diagenetic alteration of fossil materias in literatures cited in this chapter

Literature	Materials for oxygen isotoic thermometry	Cathodo-luminescence	Scanning electron microscopy	$^{87}Sr/^{86}Sr$	Trace element composition	X-ray diffraction	Preservation Index (PI)[a]
Stahl and Jordan (1969); Jordan and Stahl (1970)	Ammonites	NA.	NA	N.A.	N.A.	◎	N.A.
Anderson et al. (1994)	Ammonites Belemnites Bivalves Vertebrate remains	NA.	NA	N.A.	ICP-AES	◎	N.A.
Moriya et al. (2003)	Ammonites Bivalves Foraminifers Gastropod	◎	◎	N.A.	WDS	◎	5
Lécuyer and Bucher (2006)	Ammonites Bivalves	NA.	◎	N.A.	ICP-AES, ICP-MS	◎	5?
Cochran etal. (2010)	Ammonites Nautiloides	NA.	◎	◎	ICP-MS	N.A.	–†
Lukeneder et al. (2010)	Ammonites	◎	◎	N.A.	EDS	◎	2?–5

[a] =Preservaton Index Is defined by Cochran et al. (2010). †=Not applicable because PI Is defined In this litreture. ◎=done. N.A.=not applicable ICP-AES = Inductively Coupled Plasma – Atomic Emission Spectrometry, ICP-MS = Inductively Coupled Plasma – Mass Spectrometry WDS=Wavelength Dispersive X-ray Spectrometry, EDS = Energy Dispersive X-ray Spectrometry,?=not clearly identified

latitude in the modern ocean. They proposed an equation between the oxygen isotopic composition of present day sea surface water and latitude as follows:

$$y = 0.576 + 0.041 \times L - 0.0017 \times L^2 + 1.35 \times 10^{-5} \times L^3 \qquad (19.8)$$

where y is the oxygen isotopic composition of sea surface water, and L is the absolute latitude in the range of 0° to 70°. However, since the relationship between salinity and $\delta^{18}O$ of sea surface water varies regionally, seasonally and/or in longer timescales (Rohling and Bigg 1998; Bigg and Rohling 2000; Signorini and McClain 2012), prediction of local $\delta^{18}O$ of seawater in the geological past is rather difficult, especially in the epicontinental ocean. In fact, because of reduced equilibrium fractionation between water vapor and liquid water, $\delta^{18}O$ of precipitation and sea surface water in the greenhouse Earth would be significantly different from modern values (Ufnar et al. 2004; Zhou et al. 2008). Additionally, $\delta^{18}O$ of local water would also be different from the zonal mean sea surface $\delta^{18}O$ (Zhou et al. 2008). Therefore, while we have known that there is a meridional gradient in $\delta^{18}O$ of sea surface water, the precise method for applying this to the ancient ocean is still unclear. When comparing isotopic temperature from many different localities in different latitudes, we should be kept in mind this meridional $\delta^{18}O$ gradient in sea surface water.

The other aspect of the salinity effect on $\delta^{18}O$ of seawater is associated with river input. Since precipitation and river runoff are significantly depleted in ^{18}O, $\delta^{18}O$ of brackish water in a restricted basin, such as modern Chesapeake Bay, show negative $\delta^{18}O$ value comparing to the open ocean. As a result, $\delta^{18}O$ of aragonite shells also become significantly negative value. One potential example is reported by Tsujita and Westermann (1998), showing significantly negative $\delta^{18}O$ in the late Cretaceous ammonoids from northern Western Interior. Since Western Interior became more enclosed and restricted in the late Cretaceous (Dean and Arthur 1998), salinity and $\delta^{18}O$ of seawater/brackish water in northern Western Interior might be decreased. If we assume that $\delta^{18}O$ of background marine water and river runoff are 0‰ and −10‰, respectively, contamination of 20% of runoff lowers $\delta^{18}O$ of carbonate precipitated by 2‰ (cf. Fig. 19.3). Salinity of the water will be 27.2, assuming the salinity of background marine water is 34. A modern example of the molluscan biocalcification in this kind of blackish water has been described by Elliot et al. (2003). They showed that *Mercenaria* bivalves secrete their shells under isotopic equilibrium even in blackish water.

The effect of bicarbonate ion concentration in seawater on the isotopic ratio of carbonates has been described by Spero et al. (1997) and Zeebe (1999). According to their results, bicarbonate ion concentration, pH, and isotopic composition of seawater were discussed in the Cretaceous greenhouse by Zeebe (2001). The $\delta^{18}O$ of marine carbonate increase with the decrease of pH. For example, on the assumption that pH of the Cretaceous seawater is about 7.7 (Zeebe 2001), the oxygen isotopic composition of the Cretaceous marine carbonate should be more positive than that of the modern marine carbonate by 0.7‰. Although it is uncertain if the increased atmospheric carbon dioxide concentration would affect the oxygen isotopic compositions of surface and intermediate waters more prominently, or the whole water column (Zeebe 2001), this effect may decrease the paleotemperature estimated by about 3 °C regardless of latitude. Therefore, differences between surface water temperatures and bottom water temperatures might be underestimated in the greenhouse interval.

19.2.2 Carbonate Clumped Isotope Thermometry

In addition to the conventional oxygen isotope thermometry mentioned in the previous section, a new method of isotope thermometry using carbonate has been developed in this decade (e.g. Eiler and Schauble 2004; Wang et al. 2004; Schauble et al. 2006; Eiler 2007; Hill et al. 2014). In the conventional stable isotope analyses of carbonate (analyses of $\delta^{13}C$ and $\delta^{18}O$), compositions of isotopic species containing one rare isotope have been measured. For example, in the analyses of CO_2, $^{13}C^{16}O_2$ and $^{12}C^{18}O^{16}O$ are measured except for very rare isotope species containing ^{17}O. Then, isotopic species containing multiple rare isotopes (e.g. $^{13}C^{18}O^{16}O$, $^{12}C^{18}O_2$, etc.) are actually neglected in the conventional isotope analyses. However, since these multiply-substituted isotopologues, which are called clumped isotopic species, also have distinct physical and chemical properties, analyses of these clumped isotope species may provide new and useful information (Eiler 2007).

One of the most important physico-chemical dynamics of stable isotope geochemistry is that substituting a heavy isotope for a light isotope in a chemical bond reduces the vibration frequencies of that bond and its zero-point energy (e.g. Urey 1947). For example, the vibration energy of C–O bonds in $^{18}O–^{12}C–^{16}O$ (CO_2) is slightly lower than that of $^{16}O–^{12}C–^{16}O$. Because of this difference, ^{13}C is preferentially substitute ^{12}C in $^{18}O–^{12}C–^{16}O$, making $^{18}O–^{13}C–^{16}O$ (Eiler 2007). The important point is that there is a thermodynamic driving force to promote clumping of heavy isotopes into multiply-substituted isotopologues. The details of mechanisms and thermodynamics of isotope clumping are described in Wang et al (2004), Schauble et al. (2006), Eiler (2007) and some other.

If each isotopologue is stochastically distributed, abundance of each clumped isotope species is determined by a simple function of the symmetry numbers and stoichiometry coefficients of reactant and product molecules. However, because clumped isotope species are more stable than the theoretical prediction in lower temperature region (for example, lower than ca. 500 °C) (e.g. Urey 1947; Eiler 2007), clumped isotope species become more enriched at lower temperature region. On the other hand, enrichment of clumped isotope species comparing to stochastic distribution approaches to zero with temperature increase. In clumped isotope analyses, this difference between actual abundance of clumped isotope species and stochastic distribution is measured and used as a paleothermometer.

In calcium carbonate, the homologous equilibrium below is expected at thermodynamic equilibrium (Schauble et al. 2006; Eiler 2007);

$$Ca^{12}C^{18}O^{16}O_2 + Ca^{13}C^{16}O_3 = Ca^{13}C^{18}O^{16}O_2 + Ca^{12}C^{16}O_3 \tag{19.9}$$

As mentioned above, clumped isotopologue of $Ca^{13}C^{18}O^{16}O_2$ becomes more abundant comparing to the stochastic distribution with decreasing temperature. Thus, by analyzing relative abundance of all isotopologues in calcium carbonate concerning with reaction (19.9), one can estimate calcification temperature. This approach is referred as carbonate clumped isotope thermometry.

Carbonate clumped isotope thermometry is significantly advantageous comparing to conventional oxygen isotope thermometry which is based on a fractionation of ^{18}O between carbonate and water. As mentioned in the previous section, accurate temperature estimation is achieved when $\delta^{18}O$ of ambient water is known, which is practically difficult in some cases at a given geological age. On the other hand, carbonate clumped isotope thermometry is based on stochastic distribution of each isotopologues, and every information needed to estimate calcification temperature is preserved in carbonate minerals.

Practically, carbonate is reacted with anhydrous phosphatic acid, and CO_2 evolved is measured on isotopic ratio mass spectrometer (Ghosh et al. 2006; Came et al. 2007; Wacker et al. 2013). Carbonate clumped isotope thermometry uses value of Δ_{47} to denote the excess of isotopologue of 47, which is mainly composed of $^{13}C^{18}O^{16}O$ (Eiler 2007), ralative to the stochastic distribution. The Δ_{47} value of CO_2 is determined as follows:

$$\Delta_{47} = \left[\left(\frac{R^{47}}{R^{47*}} - 1\right) - \left(\frac{R^{46}}{R^{46*}} - 1\right) - \left(\frac{R^{45}}{R^{45*}} - 1\right)\right] \times 1000 \tag{19.10}$$

where R^{47}, R^{46}, R^{45} are abundance ratio of massed 47, 46, 45 relative to mass 44. These R^{i} values are determined by comparison with a reference standard of which isotopic value is previously known. R^{47*}, R^{46*}, R^{45*} are the expected stochastic distribution of a sample calculated from $\delta^{13}C_{(VPDB)}$ and $\delta^{18}O_{(SMOW)}$ of the sample analyzed. The temperature dependency of Δ_{47} has been calibrated for synthetic calcite (Ghosh et al. 2006; Zaarur et al. 2013), aragonitic corals (Ghosh et al. 2006), natural inorganic calcite from soil and aragonitic otoliths (Ghosh et al. 2007), aragonitic molluscs and calcitic brachiopods (Came et al. 2007), calcitic foraminifers and coccoliths (Tripati et al. 2010) and others. Regardless of their mineralogy and origin (synthetic or biotic), all of these results fall on a uniform calibration, implying that mineral-specific fractionations and vital effects are relatively unimportant for this system (Eiler 2007). The temperature calibrations of Δ_{47} are determined as follows;

$$\Delta 47 = \frac{0.0592 \times 10^6}{T^2} - 0.02 \text{ (Ghosh et al. 2006)} \tag{19.11}$$

$$\Delta \cdots = \frac{0.0526 \times 10^6}{T^2} - \tag{19.12}$$

Considering the analytical precision of Δ_{47} (ca. ±0.005 %) (Came et al. 2007; Eiler 2007) and temperature sensitivity of calibrations, expected error in temperature estimation is approximately ± 1 °C at earth-surface temperatures.

With respect to the carbonate clumped isotope thermometry of cephalopods, Came et al. (2007), Dennis et al. (2013) and Price and Passey (2013) provided some results. Among them, Dennis et al. (2013) showed a substantial data set of modern and fossil cephalopods including modern *Nautilus* and *Sepia* and Cretaceous belemnites and ammonoids. They used wild and aquarium cultured *Nautilus* and *Sepia* of which habitat temperatures and $\delta^{18}O$ of ambient seawater are known to test applicability of carbonate clumped isotope thermometry on cephalopods. However, while carbonate clumped isotope thermometry is expected to be robust against vital effects (Eiler 2007), modern *Nautilus* and *Sepia* show significant vital effects on Δ_{47} (Dennis et al. 2013). This is actually very confusing because conventional bulk oxygen isotope ($\delta^{18}O$) is under isotopic equilibrium, showing very accurate temperatures. Then, Dennis et al. (2013) discussed that "the apparent discrepancy between Δ_{47} and $\delta^{18}O$ seen in modern cephalopods could be explained by: (i) inadequate knowledge of growth temperatures and/or $\delta^{18}O$ of ambient seawater, (ii) problems with the calibration of the carbonate clumped isotope thermometer, and/or (iii) methodological artifacts (Dennis et al. 2013)". While the exact reason for this discrepancy is still unknown, one should be very careful when applying carbonate clumped isotope thermometry on fossil cephalopods. Even worse, degree of vital effects on *Nautilus* and *Sepia* seems to be different from each other. *Nautilus* and *Sepia* show vital effects of 0.063 ± 0.019 and 0.047 ± 0.020 in Δ_{47}, corresponding 14.5 °C and 11.0 °C, respectively, if calibration of Zaarur et al. (2013) is applied. As mentioned in the previous section, conventional oxygen isotopic thermometry of fossil cephalopods is based on the assumption that fossil species precipitated their shell materials under isotopic equilibrium, which is inferred from the principle of parsimony (see 2.1.2 Vital effect). However, if modern *Nautilus* and *Sepia* do have independent vital effects on Δ_{47}, it is easy to imagine that other fossil genera or species show different degree of vital effects, which is commonly found in $\delta^{18}O$ of corals and foraminifers (e.g. Watanabe et al. 2003; Ishimura et al. 2012). While Dennis et al. (2013) applied 0.059 ± 0.019 (average of all modern *Nautilus* and *Sepia* analyzed, corresponding to 13.6 °C) for correcting vital effects on Δ_{47} measured from Cretaceous (Campanian and Maastrichtian) ammonoids in Western Interior Seaway, applying that average value for correcting Δ_{47} in fossil ammonoids might not be valid.

An example of potential vital effect on the Early Cretaceous (Berriasian and Valanginian) belemnites, which is collected at 60–65 °N in paleolatitude, is shown in Price and Passey (2013). Their results indicate that paleotemperature at that time in that high latitude was 10–20 °C, which is significantly warmer than $\delta^{18}O$ thermometry of benthic foraminifers showing ca. 12 °C in the Aptian (Friedrich et al. 2012). Considering that climate in Berriasian was expected to be cooler than Aptian (Pucéat et al. 2003), and benthic foraminiferal temperature might be comparative to sea surface water temperature at high latitudes, 10–20 °C at 60–65 °N in the Berriasian and Valanginian might be too warm. In fact, Price and Passey (2013) mentioned "We cannot exclude the possibility that the belemnite growth temperatures are seasonally biased due to high shell accretion rates during the summer or migration from warmer-water regions (Price and Passey 2013)". While this seasonal bias would be a candidate, another possibility is isotopic disequilibrium in Δ_{47}. If the average vital effect of 13.6 °C on cephalopod proposed by Dennis et al. (2013) is applied on results of Price and Passey (2013), some of their results become freezing temperatures in normal marine salinity, implying that vital effects on cephalopods might not be uniform.

Although there are some ambiguities in vital effects, both Dennis et al. (2013) and Price and Passey (2013) discussed $\delta^{18}O$ of seawater in the Cretaceous. Because carbonate clumped isotope thermometry does not require $\delta^{18}O$ of seawater, by combining carbonate clumped isotope thermometry and conventional oxygen isotope thermometry, $\delta^{18}O$ of seawater can be estimated. Especially, Dennis et al. (2013) discussed salinity of fresh, brackish and marine waters in and around the Cretaceous Western Interior Seaway. Since even apparently well preserved, fossils sometimes show unrealistically high temperatures in the epicontinental sea implying contamination of less saline water (Tsujita and Westermann 1998; Cochran et al. 2003, 2010; Dennis et al. 2013), this approach is significantly advantageous in those settings.

Finally, very surprisingly, Δ_{47} is astonishingly robust against diagenesis (Dennis et al. 2013). While it is very easy to imagine that precipitation of neomorphic carbonate provides significant alteration on all of $\delta^{13}C$, $\delta^{18}O$, and Δ_{47}, initial dissolution might not change Δ_{47} (Dennis et al. 2013). Although $\delta^{13}C$ and $\delta^{18}O$ are significanly altered when Preservation Index (PI) becomes less than 3 (Cochran et al. 2010), there is no trend between Δ_{47} and PI, suggesting alteration has not changed the clumped isotope signature (Dennis et al. 2013). However, since $\delta^{13}C$ and $\delta^{18}O$ are altered when the preservation of samples becomes less than 3 in PI, those samples can not be use for calculating $\delta^{18}O$ of seawater. While there may be many unknown and/or unrecognized issues on carbonate clumped isotope thermometry of cephalopods, this technique would explore new frontiers of ammonoid paleoecology in future.

19.2.3 Carbon and Nitrogen Isotope Signatures

The carbon isotope composition of synthetic carbonate is predominantly a function of $\delta^{13}C$ of dissolved inorganic carbon and carbonate ion concentration within the solution. However, $\delta^{13}C$ in biogenic carbonate may show a greater vital effect than does $\delta^{18}O$ as mentioned above (Veizer 1983; Wefer and Berger 1991). Among the variety of biological activities, changes in metabolic rate of the organisms and incorporation of carbons in food would be major candidates for vital effect in molluscs (e.g. Jones et al. 1986; Tanaka et al. 1986; Romanek et al. 1987; McConnaughey and Gillikin 2008).

In some cases, abrupt changes of $\delta^{13}C$ profiles may be used for identification of changes in metabolic rate accompanied by sexual maturity (Jones et al. 1986; Romanek et al. 1987). The cause of abrupt changes in $\delta^{13}C$ profile is explained by the trade-off of carbon distribution between growth of the individual and gametogenesis. Carbon from foods could also be incorporated in a biogenic carbonate. Tanaka et al. (1986) discussed that a large percentage of carbon in calcareous hard parts of bivalves and gastropods analyzed is metabolic carbon, indicating that $\delta^{13}C$ of these shells can't be used to predict $\delta^{13}C$ of dissolved inorganic carbon in the ancient ocean. The incorporation of metabolic carbon into biogenic carbonates is also discussed in calcitic planktic foraminiferal tests (Spero and Lea 1996).

In combination with $\delta^{13}C$, nitrogen isotopic composition ($\delta^{15}N$) indicates a trophic level of the individual analyzed (Minagawa and Wada 1984). This technique has been widely applied in many groups of marine and non-marine aquatic organisms, including modern cephalopods (Cherel and Hobson 2005; Hobson and Cherel 2006; Kashiyama et al. 2010). $\delta^{15}N$ of soft tissue of living animals refracts $\delta^{15}N$ of their diet. $\delta^{15}N$ increases by ~3 ‰ in every trophic level, thus the $\delta^{15}N$ of soft tissue of predators is ~3 % higher than that of prey in carnivores (Minagawa and Wada 1984; Zanden and Rasmussen 2001).

This method can be utilized not only for soft tissue, but also for inter- and intra-crystalline organic compounds within hard parts. Kashiyama et al. (2010) reported ontogenetic variation of $\delta^{15}N$ of organic compounds within shells of modern *Nautilus*. Their results indicate that $\delta^{15}N$ of embryonic shells is higher than that of post embryonic shells by up to approximately 3 ‰, indicating that pre-hatching animals have been consuming (or eating) the yolk precipitated by adult animals.

In addition to this traditional bulk $\delta^{15}N$ technique, nitrogen isotopes in amino acids are know to be a powerful tool for analyzing the trophic level in modern species (Chikaraishi et al. 2007, 2009; Kashiyama et al. 2010; Ohkouchi et al. 2012). Ohkouchi et al. (2012) showed $\delta^{15}N$ of 10 species of amino acids in both soft tissue and hard parts of three species of modern *Sepia*, and in soft tissue of modern *Spirula*. The $\delta^{15}N$ of amino acids from hard parts of *Sepia* shows significant correlation with that from soft tissue, verifying the applicability of the method to calcareous hard parts. Kashiyama et al. (2010) also presented $\delta^{15}N$ of amino acids incorporated in shell carbonates of wild *Nautilus* specimens captured in Philippine. In addition to the apparent decrease in bulk $\delta^{15}N$ from embryonic to post-embryonic stages, the Amino acid Trophic Level (ATL), calculated from compound specific amino acid $\delta^{15}N$ (cf. Chikaraishi et al. 2009), decreases from 4.5 in the embryonic stage to 3.8 in the post-embryonic stage. The fact that the embryonic stage shows a higher trophic level than the post-embryonic stage is explained by consumption of egg yolk produced by an adult animal as mentioned above. They also applied this method to fossil *Cymatoceras*. However, they were less confident on their results because "a potential pitfall of the current method is that we cannot exclude the possibility that organic matter was added to or formed within the samples during post-mortem degradation (Kashiyama et al. 2010)". While applicability of this method to fossil materials might be limited, development of new techniques, including physical evidence such as the data presented by Kruta et al. (2011), would help to understand life history of extinct animals.

19.2.4 Keys to Understand Isotopic Signatures on Ammonoid Shells

19.2.4.1 Precipitation of Shell Materials: Continuous or Intermittent?

While oxygen isotopic thermometry and other geochemical analyses of ammonoid shells are powerful tools to investigate ammonoid paleoecology, those results should be treated with caution for further discussion. Because we analyze shell materials secreted by ammonoids, obviously, only the signals recorded in shell materials will be obtained. More precisely, the signals in the interval when the ammonoid analyzed ceased secreting shell materials will never be recorded.

An example of this intermittent shell precipitation can be found in the oxygen isotopic records of modern *Natutilus* (Moriya et al. 2003) (Fig. 19.5). Figure 19.5 shows oxygen isotopic records of an outer whorl at the apertural margin and adult septa of wild *Nautilus pompilius* collected at Tañon Strait, Philippine. Since modern *Nautilus* is known to conduct diurnal vertical migration within a water column, both warmer (shallower) and cooler (deeper) oxygen isotopic temperatures should be recorded within shell materials. In fact, oxygen isotopic temperatures of the outer whorl were scattered in a wider temperature range from 21.5–26.5 °C (Moriya et al. 2003) (Fig. 19.5). These temperatures are comparable to depth of 150–50 m at Tañon Strait (Hayasaka et al. 1982), which roughly agrees with the range of habitat depth of *N. pompilius* on the basis of trapping experiments (Hayasaka et al. 1982).

On the other hand, isotopic temperatures of adult septa of the same individual show little variation around 20 °C throughout its ontogeny (Moriya et al. 2003) (Fig. 19.5). Oba et al. (1992) also presented isotopic records of *N. pompilius* collected at Tañon Strait, confirming that isotopic temperature of all septal materials show approximately 20 °C. If each septum had been secreted continuously with growth of the animal, isotopic tempters should corresponds to average of the habitat temperature, namely approximately 23 °C. Oba et al. (1992) showed quasi-sinusoidal patterns in the $\delta^{18}O$ profile within a single septum as well, which is explained as the evidence of diurnal vertical migration of the individual. However, isotopic temperatures of the serial samples within a single septum are comparative to 18.5–21.5°C, which is significantly cooler than the highest temperatures obtained from the outer whorl (Fig. 19.5). These lines of evidence imply that septa of the individuals analyzed were secreted only in the deeper part of their habitat on the assumption that both outer whorls and septa were precipitated under isotopic equilibrium, which was confirmed by Landman et al. (1994).

Unlike some bivalves, shells of ammonoids and *Nautilus* lack of apparent chronologically regulated growth lines on the section surface except for lirae on the shell surface. In some studies, those ornamentations on the shell surface were used for estimating the growth rate of the animal, on the assumption that those ornamentations were formed at a constant rate (Bucher et al. 1996 and references therein). However, it is somewhat doubtful that the ornamentations including "growth lines" on the surface of ammonoid shells are periodic (cf. Landman 1983). Therefore,

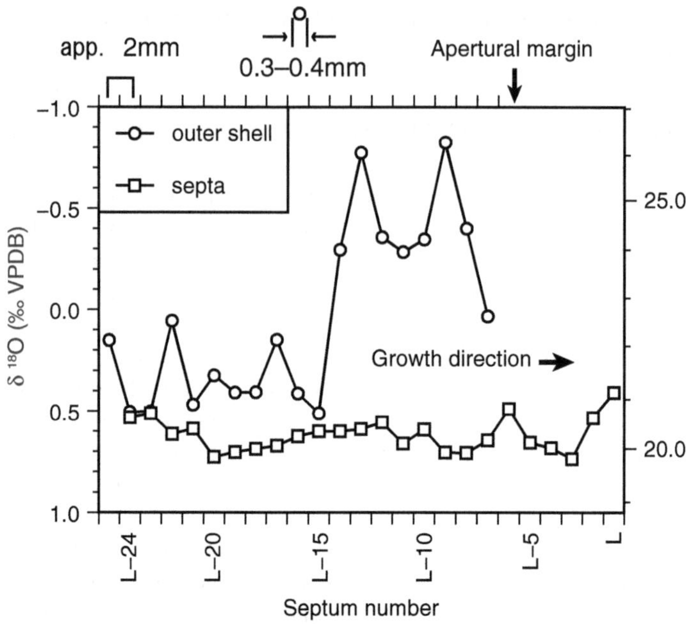

Fig. 19.5 Oxygen isotopic records of outer whorl and septum materials of wild *Nautilus pompilius* caught in Tañon Stait, Fiji. All data are cited from Moriya et al. (2003). Isotopic temperatures are calculated using a equation and $\delta^{18}O$ of seawater for *N. pumpilius* in Tañon Strait published by Oba et al. (1992). Open circles show $\delta^{18}O$ of outer whorl materials. These samples are scraped from ventral margin of shell surface with micro-drill of 0.3 mm in diameter. Considering that the rate of apertural growth is approximately 0.1–0.3 mm/day in adult specimens (Landman and Cochran 1987), each analysis shows the average for one to three days records. Samples are scraped approximately every 2 mm from apertural margin toward adapical direction. Mineralogy of the sample scraped with micro-drill is analyzed with X-ray diffractmeter, and confirmed to be composed of 100% aragonite. Open squares indicate $\delta^{18}O$ of septal materials in the post hatching stage. Each symbol represents results from each septum. A broken piece of septa is milled with an agate mortar and pestle for isotopic analyses. L; last septum of the individual. L−n (n=5, 10, 15, 20, and 24); n-th septum counting from the last septum

sclerochronologic works cannot be achieved on ammonoid shells, making identification of a growth increment per day/year difficult. On the other hand, Cochran and Landman (1984) showed excellent results on growth rate of modern *Nautilus* on the basis of radiogenic isotopic analyses of shell materials (Cochran et al. 1981; Cochran and Landman 1984; Landman et al. 1988; Cochran and Landman 1993). Although these techniques, unfortunately, can be applied only on modern species, they indicate that rate of apertural growth is 0.12–0.21 mm/day, and time of septal formation ranges from approximately 70–180 days in wild specimens (Landman and Cochran 1987; Landman et al. 1989). On the other hand, Oba et al. (1992) described that they found 40–100 cycles in the $\delta^{18}O$ profile within a single septum of 1.53 mm thick. Although the septa are assumed to be secreted only in the deeper part of their habitat as mentioned above, the septal formation of the specimen analyzed by Oba et al. (1992) possible took for 40–100 days, which is comparable to the results shown in Landman and Cochran (1987) and Landman et al. (1989).

19.2.4.2 Nature as a Data Logger: Stationary or Mobile?

Calcareous hard harts of aquatic organisms can be compared with an environmental data logger such as a conductivity and temperature probe in the water column. As there are two types of methods for a water quality monitoring, namely stationary or mobile, organisms with calcareous hard parts are classified into two types in terms of their mobility. The first group is composed of animals without significant mobility, for example, benthic foraminifers, bivalves, and corals. Geochemical records in calcareous hard parts of these organisms provide excellent sets of Eulerian data. Paleoceanography in the geological timescale, and seasonal changes in sea surface water quality have been elucidated using benthic foraminifers, and coral skeletons, respectively (e.g. Gagan et al. 2000; Zachos et al. 2001; Watanabe et al. 2011; Friedrich et al. 2012). According to the growth rate and longevity of the organisms analyzed, many different types of data have been extracted from geochemical analyses. While planktic foraminifers are mobile organisms within the water column in the strict sense, some species inhabited the mixed layer and can be used as stationary examples, because they always represent conditions of sea surface water (e.g. Fairbanks et al. 1980; Moriya et al. 2007).

The other group consists of mobile organisms, such as fish (otolith), cephalopods, etc. These organisms are actively moving within a water column, so their geochemical records can be compared with, rather than proper Lagrangian data, the data obtained by Argo floats which is drifting robotic probes ascending and descending within a water column on its own. Since geochemical temperature proxy records of hard parts of those animals show a history of temperature that the animals have experienced, ontogenetic analyses of these proxy records thus reveal a history of ontogenetic migration of the individuals analyzed (e.g. Oba et al. 1992; Tsukamoto et al. 1998; Tsukamoto and Arai 2001; Auclair et al. 2004; Zakharov et al. 2006). This idea has been applied to living *Nautilus*, and provided convincing results, which will be discussed below.

Since ammonoids are expected to be mobile organisms within a water column, their shells can be likened to Argo floats, instead of stationary monitoring posts. As shown in previous studies on fish otoliths and *Nautilus* shells, oxygen isotopic temperatures on ammonoid shells could provide temperature records of the habitat that they experienced. In a simple gedankenexperiment, potential ammonoid habitat can be divided into five groups as follows (Fig. 19.6):

- Type I: surface dweller within a mixed layer (planktic/nektic)
- Type II: subsurface dweller within a thermocline (planktic/nektic)
- Type III: bottom dweller (demersal)
- Type IV: vertical migrant (planktic/nektic)
- Type V: bottom migrant (demersal)

On the assumption that ammonoids continuously secrete their shells under isotopic equilibrium, Fig. 19.7 shows isotopic temperature expected for each type of ammonoids. The isotopic temperatures of ammonoids inhabited the mixed layer, i.e. Type I, show sinusoidal patterns which must represent seasonal sea surface tem-

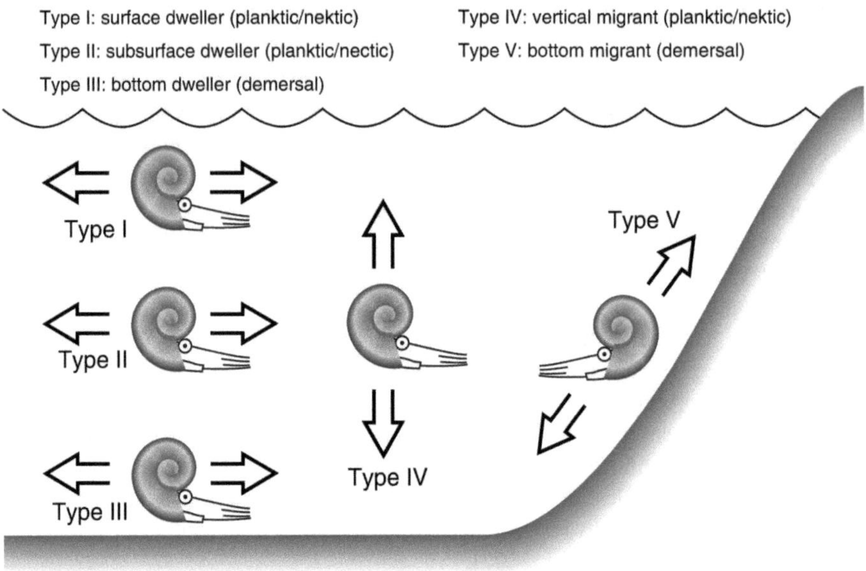

Fig. 19.6 Schematic view of possible habitat and mode of life of ammonoids. Type I; permanent surface dweller within a mixed layer. Type II; permanent subsurface dweller within a thermocline layer. Type III; permanent bottom dweller (demersal) on and very close to seafloor. Type IV; planktic/nektic vertical migrant. Type V; demersal bottom migrant. The migration period for Types IV and V can be diurnal, seasonal, annual, biennial, etc. Even unidirectional migration from shallower/deeper to deeper/shallower, and ontogenetic migration from shallower to deeper, then back into shallower in the end of ontogeny are considered as examples of Types IV and V

perature fluctuations (Type i in Fig. 19.7). Similar to planktic foraminifers, Type I ammonoids can be considered as an example of stationary monitoring, because they always represent sea surface temperature unless they migrate substantial distance in their life. Because sinusoidal patterns represent annual cycles, growth rate and longevity could be discussed if the patterns are confirmed as true seasonal signals (cf. Urey et al. 1951; Stahl and Jordan 1969; Jordan and Stahl 1970). In modern coral skeletons, since daily and annual growth lines are visible under a soft X-ray radiograph, sinusoidal patterns in geochemical signals are compared to the annual growth increments, allowing identification of annual cycles in δ^{13}C and δ^{18}O. However, it is practically very difficult to discern annual cycles in δ^{18}O of ammonoid shells because of fundamental lack of growth lines as mentioned above.

Types II and III would show almost constant temperature throughout their life (Types ii and iii in Fig. 19.7), because seasonal temperature fluctuations below the mixed layer are much less prominent than those in the mixed layer, for example less than 2–4 °C in the modern Pacific midlatitudes (http://dx.doi.org/10.12770/1282383d-9b35-4eaa-a9d6-4b0c24c0cfc9). In other words, when the isotopic results of the individual analyzed indicate Types ii or iii (Fig. 19.7), the habitat depth of the individual is expected to be thermocline or deeper (e.g. Moriya et al. 2003).

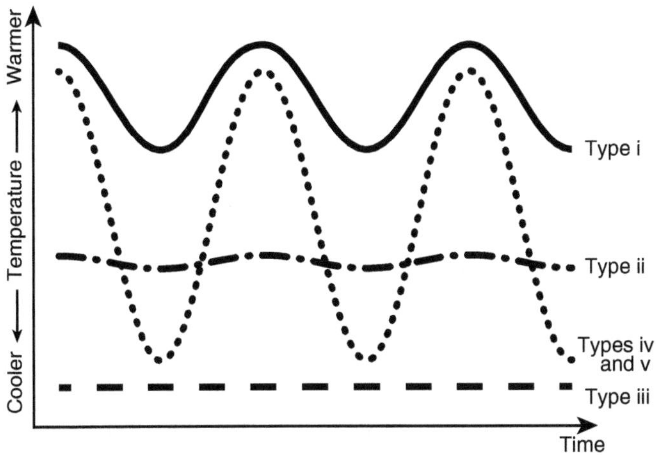

Fig. 19.7 Schematic view of possible patterns of isotopic temperature recorded on ammonoid shells on the assumption that calcareous shells have been secreted continuously. Type i; isotopic records expected for Type I ammonoids. Sinusoidal fluctuations represent seasonal sea surface water temperature. Type ii; isotopic records expected for Type II ammonoids. Much less prominent sinusoidal fluctuations than those in Type i might be identified. Type iii; isotopic records expected for Type III ammonoids. Significant sinusoidal pattern might not be observed. Types iv and v; isotopic records expected for Types IV and V ammonoids. Sinusoidal patterns would represent diurnal, seasonal, annual, biennial, etc. migration of the individual. The patterns can be composed of only a half-period or one period, representing an unidirectional shift from warmer/cooler to cooler/warmer, or a warmer-cooler-warmer cycle

Although Types iv and v would also indicate sinusoidal patterns, a period and amplitude would be different from Type i, i.e. the scales of abscissas between Type i and Types iv and v are different from each other. While the cycle should be one per year, for Type i it can be one per day (diurnal migration), one per season (seasonal migration), one per year (annual migration), one per two years (biennal migration) and so on for Types iv and v. When Types IV and V ammomoids conduct diurnal migration and each sample for isotopic analyses averages more than one day, oxygen isotopic signal would be widely scattered from warmer to cooler as shown in the outer whorl of modern *Nautilus* (Moriya et al. 2003; Auclair et al. 2004) (Fig. 19.5). On the other hand, when they have biannual, annual, or biennial migration, the results would be comparable to Type i, except for the amplitude and cycle/year. Types iv and v also include the unidirectional ontogenetic migration, which is well known in septum records of modern *Nautilus* and *Sepia* (e.g. Eichler and Ristedt 1966; Taylor and Ward 1983; Oba et al. 1992; Rexfort and Mutterlose 2006). While some species of these modern cephalopods return to the depth in which they hatched for reproduction, those signals might not be recorded on shell materials because of lack of calcification (cf. Rexfort and Mutterlose 2006).

Unlike Argo floats, geochemical records on ammonoid shells include fundamental demerits for estimating life and habitat of ammonoids; lack of depth and time in a data set. When the ammonoid analyzed shows sinusoidal cycles in $\delta^{18}O$, Types

I, IV and V would be candidates. However, since there is no independent proxy indicating time on ammonoid shells, it is virtually impossible to estimate the period of each cycle. While radiogenic isotopes have been utilized for estimating time in shells of modern *Nautilus* (Cochran et al. 1981; Cochran and Landman 1984), this technique could hardly be applied to fossil martial millions of years old as mentioned above.

Besides time, the other major difficulty in discussion of geochemical signatures on ammonoid shells is estimation of calcification depth. Since the thermal structure of the water column, especially in epicontinental seas where most of ammonoid fossils occur, shows significant regional and temporal variations, there is no apparent correlation between calcification depth and calcification temperature in a global scale. In fact, sea surface temperature, deep water temperature and meridional gradient in sea surface temperature in the greenhouse interval are considerably different from those in modern ocean (e.g. Zachos et al. 2001; Bice and Norris 2002; Huber et al. 2002; Moriya et al. 2007; Moriya 2011; Friedrich et al. 2012). Although warmer calcification temperature would represent shallower calcification depth within a limited area, it is unfortunately impossible to estimate absolute depth from calcification temperature. Therefore, when ammonoid analyzed shows almost constant temperature through its ontogeny, Type ii (planktic/nektic) is indistinguishable from Type iii (demersal) without an external thermal scale. The same confusion would happen between Type i and Types iv and v, when Types IV and V ammonoids represent annual migration.

In modern examples on *Nautilus* and *Sepia*, since the thermal structure of the water column and $\delta^{18}O$ of ocean water are directly measured, oxygen isotopic temperature on shell materials can be compared with the thermal structure of the water column to identify habitat depth of the individual analyzed (Oba et al. 1992; Landman et al. 1994; Bettencourt and Guerra 1999). Similar to these examples, calcification temperature of ammonoid shells could be converted to calcification depth, if the thermal structure of the water column is identified by the other independent criteria. Anderson et al. (1994) and Moriya et al. (2003) used this idea to determine the thermal structure of the local water column from oxygen isotopic analyses of apparent surface dwelling and benthic organisms. The isotopic temperatures of ammonoid analyzed were superimposed onto the thermal structure of the water column to identify the depth habitat of ammonoids. Anderson et al. (1994) presented oxygen isotopic temperatures of surface dwelling vertebrates (e.g. fishes) and benthic bivalves. The isotopic temperatures of ammonoids show significantly warmer than those of benthic bivalves, but slightly cooler than those of surface dwelling vertebrates, indicating sub-surface dwelling nature of those ammonoids. Unfortunately, since Anderson et al. (1994) analyzed bulk shell fragments, rather than serial samples, cyclicity in $\delta^{18}O$ was not discussed. Moriya et al. (2003) showed oxygen isotopic temperatures of planktic and benthic foraminifers, and benthic molluscs. All isotopic temperatures obtained from ammonoids with serial sampling did not show any significant cyclicity (Moriya et al. 2003), indicating those ammonoids could be Types II or III. In addition to this result, isotopic temperatures of ammonoids are comparable to those of benthic foraminifers and benthic molluscs, suggesting those ammonoids were Type III demersal organisms. However, in the strict sense, complete thermal structure can not be determined from isotopic analyses of apparent planktic and benthic organisms. When the depth of the basin is deeper than the depth of the thermocline, isotopic temperatures of all subthermocline dwelling animals are comparatable to the local bottom water temperature. In this case, if Type II ammonoids were subthermocline species, their isotopic results can not be distinguished from that of Type III. The details of results shown in Anderson et al. (1994) and Moriya et al. (2003) will be discussed below.

Fundamental lack of proxies for time (growth rate) and pressure (depth) on ammonoid shells makes interpretation of isotopic signals more confusing. Nonetheless, when the data set of oxygen isotopic thermometry is supplemented by another independent data set, for example, the thermal structure of the water column and/or food habit, life and habitat of ammonoid fossils would potentially be better understood. For instance, Types i and ii are distinguished from Types iv and iii, respectively, when $\delta^{18}O$ of ammonoids is compared with surface, intermediate and bottom water temperatures. Additionally, when it is confirmed that the individual analyzed feeding on benthic/demersal preys, Type v would be distinguished from Type iv. Individual examples for these analyses will be summarized in the next chapter.

19.3 Growth and Habitat of Ammonoids Inferred from Geochemical Signatures

19.3.1 Growth Rate

As mentioned in the previous sections, it is well known that some skeletal organisms show daily and/or annual growth bands on their calcareous hard parts. With counting the number of annual growth lines and daily growth lines within an annual increment, growth rate and longevity of the individual and even the number of days within a year have been discussed (e.g. Wells 1963; Buddemeier et al. 1974; Jones and Quitmyer 1996; Schöne et al. 2004). However, very unfortunately, no apparent annual growth line has been observed on ammonoid shells. While chronological analyses of ammonoid shells are especially challenging, some estimations have been made on the assumption that particular morphological features were secreted at a constant rate (Bucher et al. 1996 and references therein).

Instead of morphological patterns, geochemical signatures have also been utilized for chronological analyses of cephalopod shells. As mentioned above, Oba et al. (1992) reported quasi-sinusoidal patterns in $\delta^{18}O$ within a single septum. Although the septa might be precipitated only in the deeper part of their habitat, if the sinusoidal patterns in septal $\delta^{18}O$ represent the part of diurnal migration, time for precipitating a single septum is estimates for 40–100 days. The number of days for precipitating a single septum has also been documented with analyses of radionuclides within septal materials (Cochran et al. 1981; Cochran and Landman 1984;

Landman et al. 1988). Those results indicate time of septal formation is 67–92 and 184 days in immature and submature wild *Nautilus*, which is comparable to the estimation from $\delta^{18}O$ analyses (Oba et al. 1992). Although, as mentioned above, the radioisotopic technique can't be utilized on fossil materials, oxygen isotopic temperatures showing quasi-sinusoidal patterns have also been reported from the Jurassic ammonoids, *Staufenia staufensis* and *Quenstedtoceras* sp. (Stahl and Jordan 1969; Jordan and Stahl 1970; Bucher et al. 1996). The sinusoidal patterns in isotopic records have also been found in the Jurassic belemnite rostrum in the pioneering work by Urey et al. (1951). Urey et al. (1951) argued that the sinusoidal patterns in $\delta^{18}O$ represent seasonal seawater temperature (Type i in Fig. 19.7), and the individual analyzed lived at least for three years. Stahl and Jordan (1969) employed the same logic to estimate the growth rate of the Jurassic ammonites. They analyzed $\delta^{18}O$ of septal materials, and concluded that *Staufenia* and *Quenstedtoceras* secrete 12 and 5 septa/year on the assumption that sinusoidal patterns in $\delta^{18}O$ represent seasonal seawater temperatures. However, they also mentioned that the individuals analyzes include 6–9 or 2–4 % diagenetic calcite for *Staufenia* and *Quenstedtoceras*, respectively. Some of their data points show abnormally high temperature, implying the presence of meteoric water diagenesis of the materials they analyzed. Therefore, their conclusions would require further verification.

The other example is presented by Lécuyer and Bucher (2006). They collected *Perisphinctes* and a benthic bivalve (*Asarte*) from shallow marine deposits of which bathymetry was shallower than 50 m deep. In addition to the fact that the amplitude of sinusoidal patterns in $\delta^{18}O$ of *Perisphinctes* is comparable to that of the bivalve, the depth of the basin assure that the sinusoidal patterns in $\delta^{18}O$ of *Perisphinctes* should be the Type i signal, if the individual inhabited the basin throughout its life. They estimated growth rate of the outer whorl of the individual analyzed as 0.7 mm/day. This study is outstanding from two points of view; $\delta^{18}O$ of *Perisphinctes* was compared to that of the benthic bivalve co-occurring, and the depth of the basin was estimated from sedimentology. When the $\delta^{18}O$ records on ammonoid shells are confirmed to show Type i, isotopic records on ammonoid shells would be very informative.

19.3.2 Habitat Depth

Besides estimating the growth rate from Type i records, habitat depth of planktic/nektic organisms has been the other target for geochemical analyses. In fact, segregation of habitat depth of planktic foraminifers within a water column has widely been discussed with isotopic analyses (e.g. Saito and Donk 1974; Fairbanks et al. 1980; Mortyn and Charles 2003). Although morphological analyses of mechanical strength of shell materials may provide a limitation on the maximum depth of habitat to some extent (Westermann 1996, Chap. 3.5 in this volume), the actual depth of habitat of the animal might be much shallower than the maximum explosion depth estimated from the mechanical regulations. In fact, depth habitat of modern

Nautilus is much shallower than the shell explosion depth (cf. Kanie et al. 1980; Saunders and Ward 2010). On the other hand, as shown in exquisite examples on isotopic analyses of modern *Nautilus*, oxygen isotopic thermometry would enhance our knowledge on ammonoid habitat.

Isotopic works on cephalopod shells, in terms of discussion on the mode of life of the animal, have began from the pioneering work by Eichler and Ristedt (1966). They found that both carbon and oxygen isotopic compositions of modern *Nautilus* show abrupt change through their ontogeny. Because the timing of oxygen isotope shift from negative to positive value is comparable to the decrease in septal spacing and position of the nepionic constriction, this isotope shift is considered to represent hatching of the animal (Eichler and Ristedt 1966; Cochran et al. 1981; Oba et al. 1992).

However, in the early phase of the history of isotope analyses of *Nautilus*, there was a debate whether more negative $\delta^{18}O$ value in the embryonic stage could result from precipitation under isotopic disequilibrium to ambient water in eggshells (cf. Taylor and Ward 1983). Crocker et al. (1985) also showed that $\delta^{18}O$ of eggwaters of two species of *Nautilus* is depleted by approximately 1 ‰ to ambient water. However, the eggs analyzed by Crocker et al. (1985) didn't contain any developing embryo, so the fractionation between the eggwater and embryonic shell was not discussed. After these discussions, Landman et al. (1994) finally succeeded in analyzing $\delta^{18}O$ of shells of *N. belauensis* raised under controlled temperature in Waikiki Aquarium. Their results indicate that both embryonic and post-embryonic shells are secreted under isotope equilibrium, showing *in situ* temperature.

According to these lines of discussion, it is now usually accepted that $\delta^{18}O$ shift from negative in the embryonic stage to positive in the post-embryonic stage shows that *Nautilus* laid their eggs at shallower depth than the habitat depth of adult individuals. On the basis of direct measurements of isotopic composition of seawater and thermal structure of the water column, it is revealed that *N. pompilius* in Tañon Strait, Philliphine, and Kandavu Passage, Fiji, laid their eggs at 50 and 300 m deep and the hatchlings migrating into 150 and 500 m deep, respectively (Oba et al. 1992). These results indicate that each population of *N. pompilius* has slightly different habitat depth within water columns in each locality.

In this context, the depth habitat of ammonoids can be investigated from isotopic records of their shell materials. Lukeneder et al. (2010) presented that similar set of data from three ammonoids, Jurassic *Cadoceras*, and Cretaceous *Hypacanthoplites* and *Nowakites*. They showed that oxygen isotopic temperatures in juvenile and mature stages within an individual of *Cadoceras* are warmer than that of adolescent stage, indicating that individual analyzed migrated from warmer water to cooler water, and then back into warmer water at the mature stage. On the other hand, *Hypacanthoplites* show a unidirectional change from cooler to warmer. Since both genera show $\delta^{18}O$ shifts through their ontogeny, their isotopic records are identified as Types iv or v (Fig. 19.8). However, since Lukeneder et al. (2010) did not provide sea surface and/or bottom water temperature of the water column which those specimens inhabited, habitat depth cannot be determined.

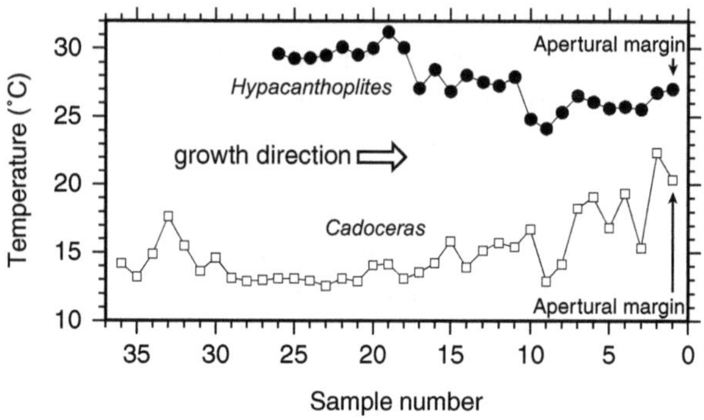

Fig. 19.8 Ontogenetic trends of oxygen isotopic temperatures of the Jurassic *Cadoceras* and Cretaceous *Hypacanthoplites*. All data are cited from Appendix A-C of Lukeneder et al. (2010). Oxygen isotopic temperatures are recalculated using the equation published by Kim et al. (2007). Acid fractionation factor for calcite and aragonite are estimated as 1.00878 and 1.00909 at 70 °C, respectively. Note that these two individuals had never inhabited the same water column in the same age. Although the growth direction is the wrong way around against Fig. 5 of Lukeneder et al. (2010), all data are plotted as shown in Fig. 1 and Appendix A-C of Lukeneder et al. (2010). *Cadoceras* and *Hypoacanthoplites* show a cyclic shift of warmer–cooler–warmer and a unidirectional shift from cooler to warmer, respectively (Lukeneder et al. 2010). These two isotopic trends can be classified into Types iv and v. However habitat depth cannot be identified because of lack of thermal structure of the water column

Although some data on ammonoid habitat, such as the potential shift of the habitat from warmer to cooler, can be deduced from isotopic records on ammonoid shells, there are fundamental problems for discussion on the ammonoid habitat, namely lack of information on $\delta^{18}O$ of seawater and thermal structure of a water column. Most of works on isotope thermometry on ammonoids predominantly utilized only ammonoid shells for discussing paleoceanography or ammonoid ecology. However, it is, unfortunately, unattainable to determine the depth of habitat of the individual without data of the thermal structure of a local water column.

From this point of view, Anderson et al. (1994) presented a substantial data set for sea surface and bottom water temperatures of the water column and isotope thermometry on ammonoids to identify habitat depth of Jurassic *Kosmoceras* (Fig. 19.9). Average precipitation temperatures of two benthic bivalves and surface dwelling vertebrates depict the temperature difference between sea surface and bottom water of the local water column. Unfortunately, instead of using a serial sampling method from a discrete individual, they used bulk shell fragments of *Kosmoceras* for isotope analyses, so their results were expressed as the general distribution of calcification temperature of *Kosmoceras*. Figure 19.9 shows isotopic temperatures of each group of animal analyzed by Anderson et al. (1994). Their results clearly indicate that distribution of calcification temperature of *Kosmoceras* is significantly warmer than those of benthic bivalves, and slightly cooler than surface dwelling vertebrates, indicating subsurface or intermediate dwelling habitat of *Kosmoceras*. Although ontogenetic variation in $\delta^{18}O$ cannot be discussed from their bulk isotope results, these results imply that *Kosmoceras* would be identified

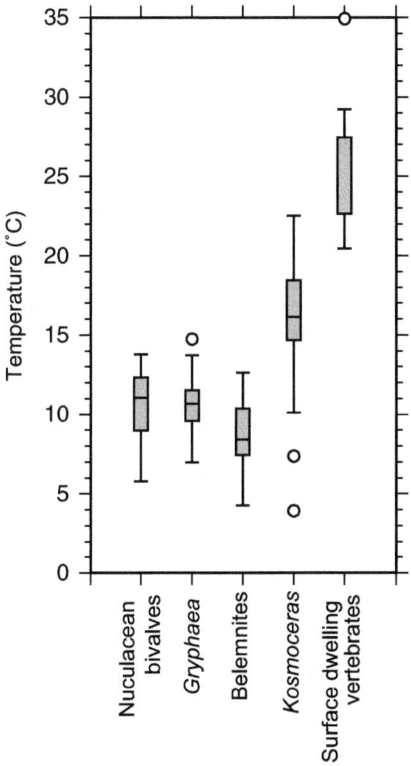

Fig. 19.9 Distribution of isotopic temperatures of belemnoids (N=67), benthic molluscs (N=44 for *Grypaera*, N=31 for nuculacea), surface dwelling vertebrates (N=10), and Jurassic *Kosmoceras* (N=78) recovered from the Jurassic Oxford Clay Formation, UK. Horizontal bars within boxes show median temperatures. Top and bottom of boxes represent 25 % and 75 % quantiles, respectively. Top and bottom whiskers indicate maximum and minimum values, respectively. Open circles show outliers. All data are cited from Anderson et al. (1994). Isotopic temperatures of two benthic bivalves are significantly cooler than that of surface dwelling vertebrates, indicating deference between sea surface and bottom water temperatures of the water column. Isotopic temperature of *Kosmoceras* is significantly warmer than benthic bivalves and slightly cooler than surface dwelling vertebrates, showing subsurface dwelling nature of *Kosmoceras*

as Type II. These results emphasize the importance of information of the thermal structure of the local water column.

Moriya et al. (2003) also provided the water column temperature and isotopic temperatures of the late Cretaceous ammonoids for discussing depth habitat of each species in adult stage (Fig. 19.10a). They utilized 15 discrete individuals of 9 species and 4 superfamilies, which include discocones, oxycones, planorbicones and hamitocones. Although they did not provide any data on embryonic and juvenile stage, isotopic results from serial sampling of adult outer whorls indicate that calcification temperatures of all adult individuals analyzed are comparable to the bottom water temperature derived from isotopic thermometry of co-occurred benthic foraminifers, bivalves and gastropods (Moriya et al. 2003) (Fig. 19.10a). They also

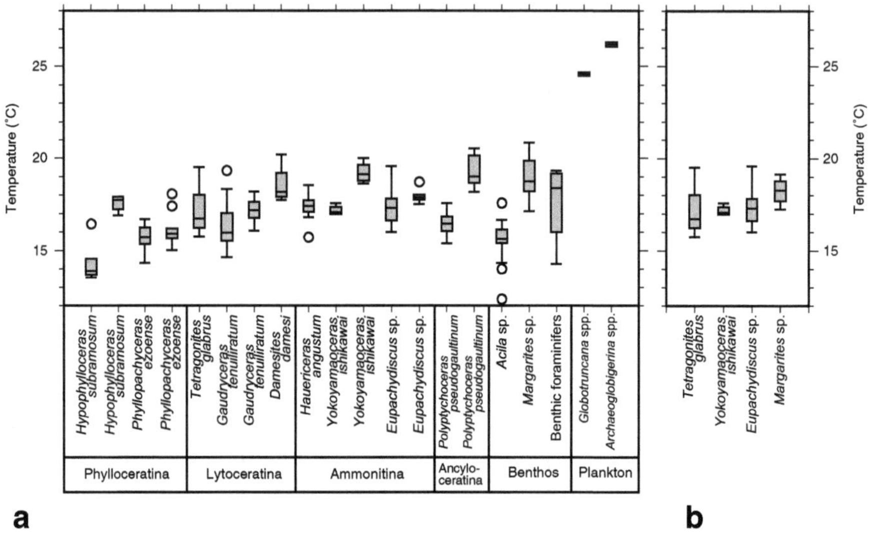

Fig. 19.10 Distribution of isotopic temperatures of ammonoids, benthic organisms (molluscs and foraminifers), and planktic foraminifers recovered from the late Cretaceous Yezo Group, Japan. Horizontal bars within boxes show median temperatures. Top and bottom of boxes represent 25 % and 75 % quantiles, respectively. Top and bottom whiskers indicate maximum and minimum values, respectively. Open circles show outliers. All data are cited from Moriya et al. (2003). Isotopic temperatures for molluscs and foraminifers are recalculated with an equation published by Kim et al. (2007) and Bemis et al. (1998), respectively. $\delta^{18}O$ of seawater is estimated to be −1 ‰ (VSMOW). (**a**) Isotopic temperatures of all materials analyzed. Isotopic temperatures of all ammonoids are comparable to those of benthic organisms which is clearly distinguished from sea surface temperature, showing permanent bottom dwelling nature of all ammonoid analyzed. (**b**) Distribution of isotopic temperatures of materials yielded from a single concretion. Although isotopic temperatures of ammonoids in panel (**a**) scatter from 13–20 °C, individuals obtained from a single concretion always show almost the same temperature as show in panel (**b**)

provided sea surface, and subsurface water temperatures by analyses of surface (*Archaeoglobigerina*) and subsurface (*Globotruncana*) dwelling planktic foraminifers (Fig. 19.10a). Considering that (i) deep dwelling (deep thermocline and/or subthermocline) planktic foraminifers have not been found in this basin, (ii) the depth of the basin is estimated to be shallower than 400m deep (Moriya et al. 2003) and (iii) predictions of the Cretaceous climate models indicating thermocline depth is deeper than 400 m deep in mid latitudes (e.g. Bice and Norris 2002), temperature of the water column might show gradual decrease from sea surface to bottom water temperature. Under this water column condition estimated, none of ammonoid data from serial sampling exceeds the range of bottom water temperature, indicating that there is no evidence for vertical migration for all individuals analyzed. On the basis of isotopic results from serial sampling of outer whorls and superimposing ammonoid calcification temperatures onto the local thermal structure, regardless of their morphology, demersal habitat (Type III) of all ammonoid analyzed were clearly documented. Although it looks that some individual show rather cooler or warmer temperatures (Fig. 19.10a), this is an artifact of sampling (Moriya et al. 2003). They used specimens from a few discrete concretions which were embedded in the host rock a few meters apart from each other. Therefore, while all individuals, including benthic gastropods, recovered from a single concretion show exactly the same temperature (Fig. 19.10b), individuals from different concretions experienced slightly different bottom water temperature because of secular changes in bottom water temperature (Moriya et al. 2003). Judging from the average sedimentation rate estimated in this basin (ca. 30 cm/kyr), these concretions could be separated by more than 10 kyr from earth other. Nonetheless, significant difference between sea surface temperature from $\delta^{18}O$ of *Archaeoglobigerina* spp. and *Globotruncana* spp. and calcification temperatures of ammonoids clearly indicate their bottom-dwelling nature. While all individuals analyzed were recovered from epicontinental silty facies, indicating that those individuals inhabited an epicontinental sea, absolute depth of habitat cannot be discussed from isotopic records as mentioned above.

19.4 Conclusions

Geochemical analyses of ammonoid fossils are potentially very powerful tools for discussing paleoecology of ammonoids. Many proxies for seawater qualities can be measured within well preserved aragonite shells of ammonoids. However, geochemical signals on ammonoid shells might be degraded by diagenetic alteration in some cases. One should be very careful to assess diagenetic alteration because degree of alteration on geochemical signals may be independent from the relative abundance of diagenetic minerals in wt%. When diagenetic minerals have exactly the same chemical composition to that of the original shell materials, original geochemical signals will never be altered after diagenesis (Fig. 19.2). On the other hand, a more plausible scenario in the natural environment is that the chemical composition of diagenetic minerals are significantly different from that of original shell materials. In some cases, only a tiny amount of contamination of diagenetic minerals would substantially modify the original geochemical signals (Fig. 19.2).

When we find fossil materials showing exquisite preservation, those ammonoid shells would be utilized for geochemical analyses. Among many proxies for water quality, such as B/Ca, Sr/Ca, etc., oxygen isotope thermometry is one of the most informative and robust tools for analyzing ammonoid life and ecology. In addition to these conventional oxygen isotope thermometry, carbonate clumped isotope thermometry has a grate potential to be a robust and clearer paleotemperature proxy. Since carbonate clumped isotope thermometry does not require $\delta^{18}O$ of ambient seawater, the combination of $\delta^{18}O$ analyses and clumped isotope analyses would provide both temperature and $\delta^{18}O$ of seawater at a same time. However, there might be some issues, for example vital effects, need to be solved before applying this method to fossil materials at a given age (Dennis et al. 2013).

When annual cycles in seawater temperature have been found in serial $\delta^{18}O$ analyses, growth rate and longevity of the individual would be discussed (Lécuyer and Bucher 2006). However, even if sinusoidal cycles have been found in $\delta^{18}O$, the cycles might not indicate seasonal seawater temperature changes but represent biannual, biennial, etc. migration of the animal (Fig. 19.6, 19.7). In this case, growth rate and/or longevity of the individual would be misidentified.

Besides growth rate of ammonoids, habitat depth and ontogenetic migration may also be determined from oxygen isotopic thermometry. Very unfortunately, since a proxy for paleo-pressure within a water column has not been found for biological carbonates so far, calcification depth cannot be discussed from geochemical records on ammonoid shells. Although calcification temperature of ammonoid shells indicates relative calcification depth, namely shallower (warmer) or deeper (cooler), it is hard to discuss exact depth of habitat of individuals analyzed (Fig. 19.6, 19.7). However, when ammonoid calcification temperature is compared with the thermal structure of the water column, which can be decoded from oxygen isotopic thermometry of apparent planktic and benthic organisms, the habitat of the individual may be identified. The results shown by Anderson et al. (1994), Moriya et al. (2003), and Lécuyer and Bucher (2006) emphasized the importance of the external criteria for evaluating isotopic data obtained from ammonoid shells (Fig. 19.9, 19.10). When identifying the thermal structure of the water column, if possible, a set of isotopic analyses of apparent surface, subsurface, thermocline, subthermocline and benthic organisms provide more accurate results. However, when the depth of the basin, which ammonoid specimens are obtained from, is deeper than the depth of the thermocline, bottom dwelling habitat and subthermocline dwelling habitat cannot be distinguished.

In addition to oxygen isotopic thermometry, carbon and nitrogen isotope compositions of inter- and intra-crystalline organic compounds would be a potential tool for elucidating the food habit of ammonoids. Since vertical migrants (Types IV and iv in Fig. 19.6, 19.7) and demersal migrants (Types V and v in Fig. 19.6, 19.7) cannot be identified from the thermal records, independent evidence for the food habit, for instance, predatory or detritivore, will enhance our knowledge on ammonoid life and habitat. However, because of diagenetic alteration, carbon and nitrogen isotopic analyses of fossil organic compounds would be more difficult than carbonate isotope analyses (Kashiyama et al. 2010).

For getting better insights on ammonoid life and habitat, physical evidences on fossils from Lagerstätten (Tanabe et al. 2000; Klug et al. 2012 and references therein) and/or a new imaging techniques (Kruta et al. 2011) would be required as well. In fact, if detritivore habitat is indicated from physical evidences, Type v records can be distinguished from Type iv records. Needless to say, morphological analyses on ammonoid shells are also highly valuable. When the fundamental knowledge on habitat of ammonoids is established from geochemical and physical evidences, functional morphology on a specific aspect within the habitat estimated would greatly improve our knowledge. In conclusion, while geochemical analyses of ammonoid fossils provide independent and objective data set, the data set will become more informative when those analyses are compared with morphological and physical evidence.

Acknowledgement I would like to express the deepest appreciation to Prof. Hiromichi Hirano of Waseda University and Prof. Kazushige Tanabe of the University of Tokyo for their continuous encouragement and guidance. Without their help, this paper would not have been possible. I acknowledge Prof. J. Kirk Cochran and Benjamin Linzmeier for critical reading of the manuscript and for useful discussion. I also thank Profs. Hodaka Kawahata, and Kazuyoshi Endo of the University of Tokyo, Prof. Tatsuo Oji of Nagoya University, Prof. Hiroshi Nishi of Tohoku University, and Dr. Neil Landman of American Museum of Natural History for their fruitful discussion. Dr. Toyoho Ishimura of Ibaraki National College of Technology kindly provided some comments on oxygen isotopic thermometry. All figures except for Fig. 19.4, 19.6 and 19.7 were generated with Generic Mapping Tools (Wessel et al. 2013). Statistic analyses were made with R (Ihanka and Gentleman 1996).

References

Anderson TF, Popp BN, Williams AC, Ho L-Z, Hudson JD (1994) The stable isotopic record of fossils from the Peterborough Member, Oxford Clay Formation (Jurassic), UK: palaeoenvironmental implications. J Geol Soc Lond 151:125–138

Auclair A-C, Lecuyer C, Bucher H, Sheppard SMF (2004) Carbon and oxygen isotope composition of *Nautilus macromphalus*: a record of thermocline waters off New Caledonia. Chem Geol 207:91–100

Austin WEN, James RH (2008) Biogeochemical controls on palaeoceanographic environmental proxies. Geological Society, London

Barker S, Cacho I, Benway H, Tachikawa K (2005) Planktonic foraminiferal Mg/Ca as a proxy for past oceanic temperatures: a methodological overview and data compilation for the Last Glacial Maximum. Quat Sci Rev 24:821–834

Bassinot FC, Mélières F, Gehlen M, Levi C, Labeyrie L (2004) Crystallinity of foraminifera shells: A proxy to reconstruct past bottom water $CO_3^{=}$ changes? Geochem Geophys Geosy 10.1029/2003gc000668

Batt RJ (1991) Sutural amplitude of ammonite shells as a paleoenvironmental indicator. Lethaia 24:219–225

Bayer U, McGhee GR (1984) Iterative evolution of Middle Jurassic ammonites faunas. Lethaia 17:1–16

Bemis BE, Spero HJ, Bijima J, Lea DW (1998) Reevaluation of the oxygen isotopic composition of planktonic foraminifera: Experimental results and revised paleotemperature equations. Paleoceanogr 13:150–160

Bettencourt V, Guerra A (1999) Carbon- and oxygen-isotope composition of the cuttlebone of *Sepia officinalis*: a tool for predicting ecological information? Mar Biol 133:651–657

Bice KL, Norris RD (2002) Possible atmospheric CO_2 extremes of the Middle Cretaceous (late Albian-Turonian). Paleoceanogr 17:22-1–22-17

Bigg GR, Rohling EJ (2000) An oxygen isotope data set for marine waters. J Geophys Res 105(C4):8527–8535

Brand U, Veizer J (1980) Chemical Diagenesis of a multicomponent carbonate system –1: trace elements. J Sediment Petrol 50:1219–1236

Brand U, Veizer J (1981) Chemical diagensis of a multicomponent carbonate system –2: stable isotopes. J Sediment Petrol 51:987–997

Bucher H, Landman NH, Klofak SM, Guex J (1996) Mode and rate of growth in ammonoids. In: Landman NH, Tanabe K, Davis RA (eds) Ammonoid Paleobiology. Plenum, New York

Buddemeier RW, Maragos JE, Knutson DW (1974) Radiographic studies of reef coral exoskeletons: Rates and patterns of coral growth. J Exp Mar Biol Ecol 14:179–199

Came RE, Eiler JM, Veizer J, Azmy K, Brand U, Weidman CR (2007) Coupling of surface temperatures and atmospheric CO_2 concentrations during the Palaeozoic era. Nature 449:198–U193

Carlson BA, McKibben JN, deGruy MV (1984) Telemetric investigation of vertical migration of *Nautilus belauensis*. Pac Sci 38:183–188

Cecca F (1992) Ammonite habitats in the Early Tithonian of Western Tethys. Lethaia 25:257–267

Chamberlain JA (1976) Flow patterns and drag coefficients of cephalopod shells. Paleobiology 19:539–563

Cherel Y, Hobson KA (2005) Stable isotopes, beaks and predators: a new tool to study the trophic ecology of cephalopods, including giant and colossal squids. Proceedings. Biol sci/ Royal Soc 272:1601–1607

Chikaraishi Y, Kashiyama Y, Ogawa NO, Kitazato H, Ohkouchi N (2007) Metabolic control of nitrogen isotope composition of amino acids in macroalgae and gastropods: implications for aquatic food web studies. Mar Ecol Prog Ser 342:85–90

Chikaraishi Y, Ogawa NO, Kashiyama Y, Takano Y, Suga H, Tomitani A, Miyashita H, Kitazato H, Ohkouchi N (2009) Determination of aquatic food-web structure based on compound-specific nitrogen isotopic composition of amino acids. limin Oceanogr Methods 7:740–750

Cochran JK, Landman NH (1984) Radiometric determination of the growth-rate of *Nautilus* in nature. Nature 308:725–727

Cochran JK, Landman NH (1993) Using radioisotopes: to determine growth rates of marine organisms. J Chem Educ 70:749–754

Cochran JK, Rye DM, Landman NH (1981) Growth rate and habitat of *Nautilus pompilius* inferred from radioactive and stable isotope studies. Paleobiology 7:469–480

Cochran JK, Landman NH, Turekian KK, Michard A, Schrag DP (2003) Paleoceanography of the Late Cretaceous (Maastrichtian) Western Interior Seaway of North America: evidence from Sr and O isotopes. Palaeogeogr Palaeocl 191:45–64

Cochran JK, Kallenberg K, Landman NH, Harries PJ, Weinreb D, Turekian KK, Beck AJ, Cobban WA (2010) Effect of diagenesis on the Sr, O and C isotope composition of Late Cretaceous mollusks from the Western Interior Seaway of North America. Am J Sci 310:69–88

Crocker KC, DeNio MJ, Ward PD (1985) Stable isotopic investigation of early development in extant and fossil chambered cephalopods I. Oxygen isotopic composition of eggwater and carbon isotopic composition of siphuncle organic matter in *Nautilus*. Geochim Cosmochim Acta 49:2527–2532

Davies TT, Hooper PR (1963) The determination of the calcite: aragonite ratio in mollusc shells by X- ray diffraction. Mineral Mag 33:608–612

Dean WED, Arthur MA (1998) Stratigraphy and paleoenvironments of the Cretaceous Western Interior Seaway, USA. SEPM, Tusa

Dennis KJ, Cochran JK, Landman NH, Schrag DP (2013) The climate of the Late Cretaceous: new insights from the application of the carbonate clumped isotope thermometer to Western Interior Seaway macrofossil. Earth Planet Sc Lett 362:51–65

Denton EJ, Gilpin-Brown JB (1973) Floatation mechanisms in modern and fossil cephalopods. Adv Mar Biol 11:197–268

Dunstan AJ, Ward PD, Marshall NJ (2011) Vertical Distribution and Migration Patterns of *Nautilus pompilius*. Plos One 6:e16311

Ebel K (1983) Berechnungen zur Schwebefähigkeit von Ammoniten. Neues Jahrb Geol Paläontl, Mh 1983:614–640

Ebel K (1992) Mode of life and soft body shape of heteromorph ammonites. Lethaia 25:179–193

Eichler R, Ristedt H (1966) Isotopic evidence on the early life history of *Nautilus pompilius* (Linné). Science 153:734–736

Eiler JM, Schauble E (2004) $^{18}O^{13}C^{16}O$ in Earth's atmosphere. Geochim Cosmochim Acta 68:4767–4777

Eiler JM (2007) "Clumped-isotope" geochemistry—the study of naturally-occurring, multiply-substituted isotopologues. Earth Planet Sc Lett 262:309–327

Elliot M, deMenocal PB, Linsley BK, Howe SS (2003) Environmental controls on the stable isotopic composition of *Mercenaria mercenaria*: Potential application to paleoenvironmental studies. Geochem Geophy Geosy doi:10.1029/2002GC000425

Emiliani C (1954) Temperatures of Pacific bottom water and polar superficial waters during the Tertiary. Science 119:853–855

Enmar R, Stein M, Bar-Matthews M, Sass E, Katz A, Lazar B (2000) Diagenesis in live corals from the Gulf of Aqaba. I. The effect on paleo-oceanography tracers. Geochim Cosmochim Acta 64:3123–3132

Epstein S, Buchsbaum R, Lowenstam HA, Urey HC (1953) Revised carbonate-water isotopic temperature scale. Geol Soc Am Bull 64:1315–1325

Fairbanks RG, Wiebe PH, Bé AWH (1980) Vertical distribution and isotopic compositoiin of living planktonic foraminifera in the wetern North Atlantic. Science 207:61–63

Fairbanks RG, Evans MN, Rubenstone JL, Mortlock RA, Broad K, Moore MD, Charles CD (1997) Evaluating climate indices and their geochemical proxies measured in corals. Coral Reefs 16:16S93–S100

Fatherree JW, Harries PJ, Quinn TM (1998) Oxygen and carbon isotopic "dissection" of *Baculites compressus* (Mollusca: Cephalopoda) from the Pierre Shale (Upper Campanian) of South Dakota: implications for paleoenvironmental reconstructions. Palaios 13:376–385

Friedrich O, Norris RD, Erbacher J (2012) Evolution of middle to Late Cretaceous oceans—A 55 m.y. record of Earth's temperature and carbon cycle. Geology 40:107–110

Gagan MK, Ayliffe LK, Hopley D, Cali JA, Mortimer GE, Chappell J, McCulloch MT, Head MJ (1998) Temperature and surface-ocean water balance of the mid-Holocene tropical Western Pacific. Science 279:1014–1018

Gagan MK, Ayliffe LK, Beck JW, Cole JE, Druffel ERM, Dunbar RB, Schrag DP (2000) New views of tropical paleoclimates from corals. Quat Sci Rev 19:45–64

Gat JR, Mook WG, J. MHA (2001) Environmental isotopes in the hydrological cycle: principles and applications, Volume II atmospheric water. International Atomic Energy Agency and United Nations Educational, Scientific and Cultural Organization, Paris

Geyh M (2001) Environmental isotopes in the hydrological cycle: principles and applications, Volume IV groundwater saturated and undersaturated zone. International Atomic Energy Agency and United Nations Educational, Scientific and Cultural Organization, Paris

Ghosh P, Adkins J, Affek H, Balta B, Guo WF, Schauble EA, Schrag D, Eller JM (2006) ^{13}C–^{18}O bonds in carbonate minerals: a new kind of paleothermometer. Geochim Cosmochim Acta 70:1439–1456

Ghosh P, Eiler J, Campana SE, Feeney RF (2007) Calibration of the carbonate 'clumped isotope' paleothermometer for otoliths. Geochim Cosmochim Acta 71:2736–2744

Greenwald L, Ward PD, Greenwald OE (1980) Cameral liquid transport and buoyancy control in chambered *Nautilus* (*Nautilus macromphalus*). Nature 286:55–56

Greenwald KP, Cook CB, Ward P (1982) The structure of the chambered *Nautilus* siphuncle: The siphuncular epithelium. J Morphol 172:5–22

Grossman EL (1993) Evidence that inoceramid bivalves were benthic and harbored chemosynthetic symbionts: comment and reply. Geology 21:94–95

Grossman EL (2012) Oxygen isotope stratigraphy. In: Gradstein FM, Ogg JG, Schmitz MD, Ogg GM (eds) The geologic time scale 2012. Elsevier, Oxford

Grossman EL, Ku TL (1986) Oxygen and carbon isotope fractionation in biogenic aragonite: temperature effects. Chem Geol 59:59–74

Hayasaka S, Saisho T, Kakinuma Y, Shinomiya A, Oki K, Hamada T, Tanabe K, Kanie Y, Hattori M, Vande Vusse F, Alcala L, Cordero PA, Cabrera J, J., Garcia R, G. (1982) Field study on the habitat of *Nautilus* in the environments of Cebu and Negros Islands, the Philippines. Mem Kagoshima Univ Res Cent S Pac 3:67–137

Hayasaka S, Oki K, Tanabe K, Saisho T, Shinomiya A (1987) On the habitat of *Nautilus pompilius* in Tañon Strait (Philippines) and the Fiji Islands. In: Saunders WB, Landman NH (eds) *Natilus*: the biology and paleobiology of a living fossil. Plenum, New York

Henderson GM (2002) New oceanic proxies for paleoclimate. Earth Planet Sc Lett 203:1–13

Heptonstall WB (1970) Buoyancy control in ammonoids. Lethaia 3:317–328

Hewitt RA, Westermann GEG (1987) Function of complexly fluted septa in ammonoid shells II. Septal evolution and conclusions. Neues Jahrb Geol und Paontol, Ab 174:135–169

Hill PS, Tripati AK, Schauble EA (2014) Theoretical constraints on the effects of pH, salinity, and temperature on clumped isotope signatures of dissolved inorganic carbon species and precipitating carbonate minerals. Geochim Cosmochim Acta 125:610–652
Hillaire-Marcel C, de Vernal A (2008) Proxies in Late Cenozoic paleoceanography. Elsevier, Oxford
Hobson KA, Cherel Y (2006) Isotopic reconstruction of marine food webs using cephalopod beaks: new insight from captively raised Sepia officinalis. Can J Zool 84:766–770
Huber BT, Norris RD, MacLeod KG (2002) Deep-sea paleotemperature record of extreme warmth during the Cretaceous. Geology 30:123–126
Hudson JD, Anderson TF (1989) Ocean temperatures and isotopic compositions through time. Trans Royal Soc Edinb Earth Sci 80:183–192
Ihaka R, Gentleman R (1996) R: A language for data analysis and graphics. J Comput Graph Stat 5:299–314
International Atomic Energy Agency (IAEA) (1995) Reference and intercomparison materials for stable isotopes of light elements. IAEA, Vienna
Ishimura T, Tsunogai U, Hasegawa S, Nakagawa F, Oi T, Kitazato H, Suga H, Toyofuku T (2012) Variation in stable carbon and oxygen isotopes of individual benthic foraminifera: tracers for quantifying the magnitude of isotopic disequilibrium. Biogeosciences 9:4353–4367
Jacobs D, Landman NH (1993) Nautilus—a poor model for the function and behavior of ammonids? Lethaia 26:101–111
Jacobs DK (1992) Shape, drag, and power in ammonoid swimming. Paleobiology 18:203–220
Jacobs DK, Chamberlain JA (1996) Buoyancy and hydrodynamics in ammonoids. In: Landman NH, Tanabe K, Davis RA (eds) Ammonoid paleobiology. Plenum, New York
Jacobs DK, Landman NH, Chamberlain JA (1994) Ammonite shell shape covaries with facies and hydrodynamics: iterative evolution as a response to changes in basinal environment. Geology 22:905–908
Jones DS, Quitmyer IR (1996) Marking time with bivalve shells: oxygen isotopes and season of annual increment formation. Palaios 11:340–346
Jones DS, Williams DF, Romanek CS (1986) Life history of symbiont-bearing giant clams from stable isotope profiles. Science 231:46–48
Jordan R, Stahl W (1970) Isotopische Paläotemperatur Bestimmungen an jurassischen Ammoniten and grundsätzliche Voraussetzungen für diese Methode. Geol Jahrb 89:33–62
Kanie Y, Fukuda Y, Nakayama H, Seki K, Hattori M (1980) Implosion of living *Nautilus* under increased pressure. Paleobiology 6:44–47
Kashiyama Y, Ogawa NO, Chikaraishi Y, Kashiyama N, Sakai S, Tanabe K, Ohkouchi N (2010) Reconstructing the life history of modern and fossil nautiloids based on the nitrogen isotopic composition of shell organic matter and amino acids. In: Tanabe K, Shigeta Y, Sasaki T, Hirano H (eds) Cephalopods—present and past. Tokai University Press, Tokyo
Kawabe F (2003) Relationship between mid-Cretaceous (upper Albian–Cenomanian) ammonoid facies and lithofacies in the Yezo forearc basin, Hokkaido, Japan. Cretaceous Res 24:751–763
Kim ST, O'Neil JR, Hillaire-Marcel C, Mucci A (2007) Oxygen isotope fractionation between synthetic aragonite and water: Influence of temperature and Mg^{2+} concentration. Geochim Cosmochim Acta 71:4704–4715
Klug C, Riegraf W, Lehmann J (2012) Soft-part preservation in heteromorph ammonites from the Cenomanian-Turonian Boundary Event (OAE 2) in north-west Germany. Palaeontology 55(6):1307–1331
Kruta I, Landman N, Rouget I, Cecca F, Tafforeau P (2011) The role of ammonites in the Mesozoic marine food web revealed by jaw preservation. Science 331:70–72
Kulm LD, Suess E, Moore JC, Carson B, Lewis T, Ritger SD, Kadko DC, Thornburg TM, Embley RW, Rugh WD, Massoth GJ, Langseth MG, Cochrane GR, Scamman RL (1986) Oregon subduction zone: venting, fauna, and carbonates. Science 231:561–231
Landman NH (1983) Ammonoid growth rhythms. Lethaia 16:248–248
Landman NH, Cochran JK (1987) Growth and longevity of Nautilus. In: Saunders WB, Landman NH (eds) Nautilus: the biology and paleobiology of a living fossil. Plenum, New York

Landman NH, Druffel ERM, Cochran JK, Donahue DJ, Jull AJT (1988) Bomb-produced radiocarbon in the shell of the chambered Nautilus—rate of growth and age at maturity. Earth Planet Sc Lett 89:28–34

Landman NH, Cochran JK, Chamberlain JA, Hirschberg DJ (1989) Timing of septal formation in 2 species of Nautilus based on radiometric and aquarium data. Mar Biol 102:65–72

Landman NH, Cochran JK, Rye DM, Tanabe K, Arnold JM (1994) Early life history of *Nautilus*: evidence from isotopic analyses of aquarium-reared specimens. Paleobiology 20:40–51

Lécuyer C, Bucher H (2006) Stable isotope compositions of a late Jurassic ammonite shell: a record of seasonal surface water temperatures in the southern hemisphere? eEarth 1:1–7

Leder JJ, Swart PK, Szmant AM, Dodge RE (1996) The origin of variations in the isotopic record of scleractinian corals: I. Oxygen. Geochim Cosmochim Acta 60:2857–2870

Lukeneder A, Harzhauser M, Müllegger S, Piller WE (2010) Ontogeny and habitat change in Mesozoic cephalopods revealed by stable isotopes (δ^{18}O, δ^{13}C). Earth Planet Sci Lett 296:103–114

Maeda H, Seilacher A (1996) Ammonoid taphonomy. In: Landman NH, Tanabe K, Davis RA (eds) Ammonoid Paleobiology. Plenum, New York

Machel H (2000) Application of cathodoluminescence to carbonate diagenesis. In: Pagel M, Barbin V, Blanc P, Ohnenstetter D (eds) Cathodoluminescence in geosciences. Springer, Berlin

McConnaughey T (1989) ^{13}C and ^{18}O isotopic disequilibrium in biological carbonates: I. Patterns. Geochim Cosmochim Acta 53:151–162

McConnaughey TA, Gillikin DP (2008) Carbon isotopes in mollusk shell carbonates. Geo-Mar Lett 28:287–299

McCrea JM (1950) On the isotopic chemistry of carbonates and a paleotemperature scale. J Chem Phys 18:849–856

McCulloch MT, Gagan MK, Mortimer GE, Chivas AR, Isdale PJ (1994) A high-resolution Sr/Ca and δ^{18}O coral record from the Great Barrier Reef, Australia, and the 1982–1983 El Niño. Geochim Cosmochim Acta 58:2747–2754

Minagawa M, Wada E (1984) Stepwise enrichment of ^{15}N along food chains: further evidence and the relation between δ^{15}N and animal age. Geochim Cosmochim Acta 48:1135–1140

Mitchell SF, Ball JD, Crowley SF, Marshall JD, Paul CRC, Veltkamp CJ, Samir A (1997) Isotope data from Cretaceous chalks and foraminifera: environmental or diagenetic signals? Geology 25:691–694

Moriya K (2008) Diagenetic processes for biogenic carbonates; implications for paleotemperature analyses in the geological age. Mon Chikyu 30:329–337

Moriya K (2011) Development of the Cretaceous greenhouse climate and the oceanic thermal structure. Paleontol Res 15:77–88

Moriya K, Nishi H, Kawahata H, Tanabe K, Takayanagi Y (2003) Demersal habitat of Late Cretaceous ammonoids: evidence from oxygen isotopes for the Campanian (Late Cretaceous) northwestern Pacific thermal structure. Geology 32:167–170

Moriya K, Wilson PA, Friedrich O, Erbacher J, Kawahata H (2007) Testing for ice sheets during the mid-Cretaceous greenhouse using glassy foraminiferal calcite from the mid-Cenomanian tropics on Demerara Rise. Geology 35:615–618

Mortyn PG, Charles CD (2003) Planktonic foraminiferal depth habitat and δ^{18}O calibrations: plankton tow results from the Atlantic sector of the Southern Ocean. Paleoceanography. doi:10.1029/2001PA000637

Oba T, Kai M, Tanabe K (1992) Early life history and habitat of *Nautilus pompilius* inferred from oxygen isotope examinations. Mar Biol 113:211–217

Ohkouchi N, Tsuda R, Chikaraishi Y, Tanabe K (2012) A preliminary estimate of the trophic position of the deep-water ram's horn squid Spirula spirula based on the nitrogen isotopic composition of amino acids. Mar Biol 160:773–779

Okamoto T (1988a) Changes in life orientation during the ontogeny of some heteromorph ammonoids. Palaeontol 31:281–294

Okamoto T (1988b) Develipmental regulation and morphological saltation in the heteromorph ammonite *Nipponites*. Paleobiology 14:272–286

Olóriz F, Rodriguez-Tovar FJ, Marques B, Caracuel JE (1993) Ecostratigraphy and sequence stratigraphy in high frequency sea level fluctuations: Examples from Jurassic macroinvertebrate assemblages. Palaeogeogr Palaeoclimatol Palaeoecol 101:131–145

Popp BN, Anderson TF, Sandberg PA (1986) Brachiopods as indicators of original isotopic compositions in some Paleozoic limestones. Geol Soc Am Bull 97:1262–1269

Price GD, Passey BH (2013) Dynamic polar climates in a greenhouse world: evidence from clumped isotope thermometry of Early Cretaceous belemnites. Geology 41:923–926

Pucéat E, Lécuyer C, Sheppard SMF, Dromart G, Reboulet S, Grandjean P (2003) Thermal evolution of Cretaceous Tethyan marine waters inferred from oxygen isotope composition of fish tooth enamels. Paleoceanography 18. doi:10.1029/2002PA000823

Rexfort A, Mutterlose J (2006) Stable isotope records from *Sepia officinalis*—a key to understanding the ecology of belemnites? Earth Planet Sci Lett 247:212–221

Rohling EJ, Bigg GR (1998) Paleosalinity and $\delta^{18}O$: a critical assessment. J Geophys Res: Oceans 103:1307–1318

Romanek CS, Grossman EL (1989) Stable isotope profiles of *Tridacna maxima* as environmental indicators. Palaios 4:402–413

Romanek CS, Jones DS, Williams DF, Krantz DE, Radtke R (1987) Stable isotopic investigation of physiological and environmental changes recorded in shell carbonate from the giant clam *Tridacna maxima*. Mar Biol 94:385–393

Ruppert EE, Fox RS, Barnes RD (2004) Invertebrate zoology: a functional evolutionary approach. Brooks Cole, California

Saito T, Donk JV (1974) Oxygen and carbo isotope measurements of Late Cretaceous and Early Tertiary foraminifera. Micropaleontol 20:152–177

Saunders WB (1984) The role and status of *Nautilus* in its natural habitat: evidence from deep-water remote camera photosequences. Paleobiology 10:469–486

Saunders WB, Swan RH (1984) Morphology and morphologic diversity of mid-Carboniferous (Namurian) ammonoids in time and space. Paleobiology 10:195–228

Saunders WB, Shapiro EA (1986) Calculation and simulation of ammonoid hydrostatics. Paleobiology 12:64–79

Saunders WB, Ward P (2010) Ecology, distribution, and population characteristics of Nautilus. In: Saunders WB, Landman N (eds) Nautilus: biology and paleobiology of a living fossil. Plenum, New York

Sayani HR, Cobb KM, Cohen AL, Elliott WC, Nurhati IS, Dunbar RB, Rose KA, Zaunbrecher LK (2011) Effects of diagenesis on paleoclimate reconstructions from modern and young fossil corals. Geochim Cosmochim Acta 75:6361–6373

Schauble EA, Ghosh P, Eiler JM (2006) Preferential formation of ^{13}C–^{18}O bonds in carbonate minerals, estimated using first-principles lattice dynamics. Geochim Cosmochim Acta 70:2510–2529

Schöne BR, Oschmann W, Tanabe K, Dettman D, Fiebig J, Houk SD, Kanie Y (2004) Holocene seasonal environmental trends at Tokyo Bay, Japan, reconstructed from bivalve mollusk shells—implications for changes in the East Asian monsoon and latitudinal shifts of the Polar Front. Quat Sci Rev 23:1137–1150

Scott G (1940) Paleoecological factors controlling distribution and mode of life of Cretaceous ammonoids in Texas. Journal of Paleontology 14:1164–1203

Shackleton NJ, Opdyke ND (1973) Oxygen isotope and palaeomagnetic stratigraphy of Equatorial Pacific core V28–238: Oxygen isotope temperatures and ice volumes on a 105 year and 106 year scale. Quaternary Res 3:39–55

Shackleton NJ, Kennett JP (1975) Paleotemperature history of the Cenozoic and the initiation of Antarctic glaciation: oxygen and carbon isotopic analysis in DSDP sites 277, 279 and 281. In: Kennett JP, Houtz RE (eds) Initial reports of the Deep Sea Drilling Project. U.S. Government Printing Office, Washington D.C.

Shigeta Y (1993) Post-hatching early life history of Cretaceous Ammonoidea. Lethaia 26:133–146

Signorini SR, McClain CR (2012) Subtropical gyre variability as seen from satellites. Remote Sens Lett 3:471–479

Spero HJ, Lea DW (1996) Experimental determination of stable isotope variability in *Globigerina bullides*: Implications fro paleoceanographic reconstructions. Mar Micropaleontol 28:231–246
Spero HJ, Bijma J, Lea DW, Bemis BE (1997) Effect of seawater carbonate concentration on foraminiferal carbon and oxygen isotopes. Nature 390:497–500
Stahl W, Jordan R (1969) General considerations on isotopic paleotenperature deternimations and analyses of Jurassic ammonites. Earth Planet Sci Lett 6:173–178
Suzuki A, Yukino I, Kawahata H (1999) Temperature-skeletal δ^{18}O relationship of *Porites australiensis* from Ishigaki Island, the Ryukyus, Japan. Geochem J 33:419–428
Suzuki A, Gagan MK, De Deckker P, Omura A, Yukino I, Kawahata H (2001) Last Interglacial coral record of enhanced insolation seasonality and seawater ^{18}O enrichment in the Ryukyu Islands, northwest Pacific. Geophys Res Lett 28:3685–3688
Tanabe K, Mapes RH, Sasaki T, Landman NH (2000) Soft-part anatomy of the siphuncle in Permian prolecanitid ammonoids. Lethaia 33:83–91
Tanaka N, Monaghan MC, Rye DM (1986) Contribution of metabolic carbon to mollusc and barnacle shell carbonate. Nature 320:520–523
Taylor BE, Ward PD (1983) Stable isotope studies of *Nautilus macromphalus* Sowerby (New Caledonia) and *Nautilus pompilius* L. (Fiji). Palaeogeogr Palaeoclimatol Palaeoecol 41:1–16
Tripati AK, Eagle RA, Thiagarajan N, Gagnon AC, Bauch H, Halloran PR, Eiler JM (2010) ^{13}C–^{18}O isotope signatures and 'clumped isotope' thermometry in foraminifera and coccoliths. Geochim Cosmochim Acta 74:5697–5717
Trueman AE (1941) The ammonite body-chamber, with special reference to the buoyancy and mode of life of the living ammonite. Quart J Geol Soc Lond 96:339–383
Tsujita CJ, Westermann GEG (1998) Ammonoid habitats and habits in the Western Interior Seaway: A case study from the Upper Cretaceous Bearpaw Formation of southern Alberta, Canada. Palaeogeogr Palaeoclimatol Palaeoecol 144:135–160
Tsukamoto K, Arai T (2001) Facultative catadromy of the eel *Anguilla japonica* between freshwater and seawater habitats. Mar Ecol Prog Ser 220:265–276
Tsukamoto K, Nakai I, Tesch WV (1998) Do all freshwater eels migrate? Nature 396:635–636
Tucker ME, Wright VP (1990) Carbonate sedimentology. Blackwell Science, Oxford
Ufnar DF, Ludvigson GA, González LA, Brenner RL, Witzke BJ (2004) High latitude meteoric δ^{18}O compositions: paleosol siderite in the Middle Cretaceous Nanushuk Formation, North Slope, Alaska. Geol Soc Am Bull 116:463
Urey HC (1947) The thermodynamic properties of isotopic substances. J Chem Soc 1947:562–581
Urey HC, Lowenstam HA, Epstein S, Mckinney CR (1951) Measurements of paleotemperatures and temperatures of the Upper Cretaceous of England, Demmark, and southeastern Unites States. Geol Soc Am Bull 62:399–416
Veizer J (1983) Trace elements and isotopes in sedimentary carbonates. In: Reeder RJ (ed) Carbonates: mineralogy and chemistry. Mineralogical Society of America, Michigan
Veizer J (1992) Depositional and diagenetic history of limestones: stable and radiogenic isotopes. In: Clauer N, Chaudhuri S (eds) Isotopic signatures and sedimentary records. Springer-Verlag, Berlin
Wacker U, Fiebig J, Schoene BR (2013) Clumped isotope analysis of carbonates: comparison of two different acid digestion techniques. Rapid Commun Mass Spectrom 27:1631–1642
Wang Z, Schauble EA, Eiler JM (2004) Equilibrium thermodynamics of multiply substituted isotopologues of molecular gases. Geochim Cosmochim Acta 68:4779–4797
Ward PD, Carlson B, Weekly M, Brumbaugh B (1984) Remote telemetry of daily vertical and horizontal movement of *Nautilus* in Palau. Nature 309:248–250
Watanabe T, Oba T (1999) Daily reconstruction of water temperature from oxygen isotopic ratios of a modern *Tridacna* shell using a freezing microtome sampling technique. J Geophys Res 104:20667–20674
Watanabe T, Gagan MK, Corrége T, Scott-Gagan H, Cowley J, Hantoro WS (2003) Oxygen isotope systematics in *Diploastrea heliopora*: new coral archive of tropical paleoclimate. Geochim Cosmochim Acta 67:1349–1358

Watanabe T, Suzuki A, Minobe S, Kawahima T, Kameo K, Minoshima K, Aguilar YM, Wani R, Kawahata H, Sowa K, Nagai T, Kase T (2011) Permanent El Niño during the Pliocene warm period not supported by coral evidence. Nature 471:209–211

Weber JN, Raup DM (1966) Fractionation of the stable isotopes of carbon and oxygen in marine calcareous organisms-the Echinoidea. Part II. Environmental and genetic factors. Geochim Cosmochim Acta 30:705–736

Wefer G (1985) Die Verteilung stabiler Isotope in Kalkschalen mariner Organismen. Geologisches Jahrb, Reihe A 82:3–111

Wefer G, Berger WH (1991) Isotope paleontology: growth and composition of extant calcareous species. Mar Geol 100:207–248

Wells JW (1963) Coral growth and geochronometry. Nature 197:948–950

Wessel P, Smith WHF, Scharroo R, Luis J, Wobbe F (2013) Generic mapping tools: improved version released. EOS Transac Am Geophys Union 94:409–410

Westermann GEG (1971) Form, structure and function of shell and siphuncle in coiled Mesozoic ammonoids. Life Sci Contrib R Ont Mus 78:1–39

Westermann GEG (1989) New developments in ecology of Jurassic-Cretaceous ammonoids. In: Pallini G, Cecca F, Cresta S (eds) Fossili, evoluzione, ambiente, Att II Conventione Internationale Pergola 1987. Tecnostampa, Otra Vetere, Italy

Westermann GEG (1996) Ammonoid life and habitat. In: Landman N, Tanabe K, Davis RA (eds) Ammonoid paleobiology. Plenum, New York

White RMP, Dennis PF, Atkinson TC (1999) Experimental calibration and field investigation of the oxygen isotopic fractionation between biogenic aragonite and water. Rapid Commun Mass Spectrom 13:1242–1247

Williams DF, Rötter R, Schmaljohann R, Keigwin L (1981) Oxygen and carbon isotopic fractionation and algal symbiosis in the benthic foraminifera *Heterostegina depressa*. Palaeogeogr Palaeoclimatol Palaeoecol 33:231–251

Zachos JC, Stott LD, Lohmann KC (1994) Evolution of early Cenozoic marine temperatures. Paleoceanogr 9:353–387

Zachos JC, Pagani M, Sloan L, Thomas E, Billups K (2001) Trends, rhythms, and abrrations in global climate 65 Ma to present. Science 292:686–693

Zachos JC, Dickens GR, Zeebe RE (2008) An early Cenozoic perspective on greenhouse warming and carbon-cycle dynamics. Nature 451:279–283

Zanden MJV, Rasmussen JB (2001) Variation in δ^{15}N and δ^{13}C trophic fractionation: Implications for aquatic food web studies. Limnol Oceanogr 46:2061–2062-2066

Zakharov YD, Shigeta Y, Smyshlyaeva OP, Popov AM, Ignatiev AV (2006) Relationship between delta C-13 and delta O-19 values of the recent Nautilus and brachiopod shells in the wild and the problem of reconstruction of fossil cephalopod habitat. Geosci J 10:331–345

Zaarur S, Affek HP, Brandon MT (2013) A revised calibration of the clumped isotope thermometer. Earth Planet Sc Lett 382:47–57

Zeebe RE (1999) An explanation of the effect of seawater carbonate concentration on foraminiferal oxygen isotopes. Geochim Cosmochim Acta 63:2001–2007

Zeebe RE (2001) Seawater pH and isotopic paleotemperatures of Cretaceous oceans. Palaeogeogr Palaeoclimatol Palaeoecol 170:49–57

Zhou J, Poulsen CJ, Pollard D, White TS (2008) Simulation of modern and middle Cretaceous marine δ^{18}O with an ocean-atmosphere general circulation model. Paleoceanography. doi:10.1029/2008PA001596

Chapter 20
Parasites of Ammonoids

Kenneth De Baets, Helmut Keupp and Christian Klug

20.1 Introduction

Parasites in many groups of extant cephalopods are still poorly investigated (Pascual et al. 2007) and even more so in fossil species, in spite of evidence that parasitism can provide important information on the ecology (i.e., diet, mode of life), phylogeny, and the evolutionary history of their hosts. There is currently no documented record of parasite body fossils from externally shelled cephalopods, which is probably related to the paucity of soft-tissue preservation in externally shelled forms like ammonoids, the small size of parasites (at least at some life stages), the residence of endoparasites within the host for the large part of their life cycle, and the rarity of preservable (hard) tissues in many parasites (Conway-Morris 1981, 1990; Littlewood and Donovan 2003; De Baets et al. 2011). The main evidence for parasitism in fossil cephalopods like ammonoids is indirect and in the form of shell pathologies (Hengsbach 1991a, 1996; Keupp 2000, 2012).

Shell "*malformations*" in ammonoids, coined "*paleopathies*" by Hengsbach (1996), have been known since the nineteeth century. The recognition that parasites might be responsible for at least some paleopathies came comparatively late

K. De Baets (✉)
GeoZentrum Nordbayern, Fachgruppe PaläoUmwelt, Friedrich-Alexander University Erlangen-Nuremberg, Loewenichstraße 28, 91054 Erlangen, Germany
e-mail: kenneth.debaets@fau.de

H. Keupp
Institute of Geological Sciences, Branch Palaeontology, Freie Universität Berlin, 12249 Berlin, Germany
e-mail: keupp@zedat.fu-berlin.de

C. Klug
Paläontologisches Institut und Museum, University of Zurich, Karl Schmid-Strasse 6, 8006 Zürich, Switzerland
e-mail: chklug@pim.uzh.ch

C. Klug et al. (eds.), *Ammonoid Paleobiology: From Anatomy to Ecology*, Topics in Geobiology 43, DOI 10.1007/978-94-017-9630-9_20

in paleopathological research on ammonoids (compare Hengsbach 1991a). Engel (1894) was the first to suspect parasites and other pathogens ("Schmarotzer") to be responsible for some of these pathologies. Hölder (1956) introduced a classification of pathologies based on their morphological expressions into '*forma*' types, each of which encompasses a disease and/ or healing pattern. He was also the first to attribute ammonoid gigantism to bacterial infection following reports from pathological gigantism in gastropods. Rieber (1963) was the first to explain asymmetrical growth of suture lines and shells by parasitism. Wetzel (1964) dedicated an entire article to parasitism in the ammonoid shell, which was later reinterpreted as bioerosion by fungi that might have occurred post-mortem (Keupp 2012). Significant contributions to the study of ammonoid parasites were published by Hengsbach (1979a, 1986a, 1986b, 1991a, 1996) and Keupp (1976, 1977, 1979, 1986, 1995, 2000, 2012). Additional contributions were made by various authors (e.g., House 1960; Schindewolf 1962, 1963; Rieber 1963; Bayer 1970; Hölder 1970; Morton 1983; Landman and Waage 1986; Rein 1989; Manger et al. 1999; Kröger 2000; Seltzer 2001, 2009; Larson 2007; De Baets et al. 2011, 2013b; Mironenko 2012; Rakociński 2012).

Proving a parasitic origin of these shell structures is not straightforward (Hengsbach 1991a, 1996). Among extant *Nautilus*, only parasitic copepods have been reported in natural environments (e.g., Ho 1980), which are of little help to the interpretation of these structures since they are not known to influence shell growth. A variety of parasites have been described from extant coleoids (Hochberg 1983, 1989, 1990; Castellanos-Martínez and Gestal 2013; Keupp and Hoffmann 2014), the closest relatives of ammonoids (Kröger et al. 2011), but comparatively little research has been carried out on pathological reactions on their shells (Keupp 2012). Initially, we will briefly discuss the definition of parasitism and their recognition in both extant and extinct cephalopods. Then, we demonstrate the importance of comparative work on pathologies in other shelled mollusks, particularly bivalves and gastropods, to identify causes and host reactions to parasites. We also discuss how pathologies caused by parasitic infestations can be recognized.

20.2 Definitions

Parasitism is usually defined as a symbiotic relationship whereby one individual, the parasite, derives benefits (e.g., energy, matter, nutrition, protection) at the expense of another, the host, by means of close or long-term association (Conway-Morris 1990; Rohde 2005); parasitism often tends to result in demonstrable negative effects on the host (Kinne 1980). An organism can be parasitic for its whole life (holoparasitism) or only part of its life cycle (meroparasitism). The (metabolic) dependence of the parasite may be facultative or obligate. Ectoparasites live externally on the host body, while endoparasites live inside cells (intracellular), between cells (intercellular) or within alimentary tracts, cavities, kidneys, and other open spaces inside the host (extracellular; Kinne 1980). Most species of parasites are obligate parasites which means that they need a host for survival at least during certain stages of their life cycle. Parasites can have a simple or direct life cycle with a single host, or a complex or indirect one with various additional intermediate hosts (Rohde 2005). Intermediate hosts harbor immature, developing stages, while definitive or final hosts harbor the sexually mature stage of a parasite. Paratenic hosts harbor larval forms that do not develop within the host (Rohde 2005) and are therefore not crucial to the development of a particular species of parasite; but these hosts can serve an important role in maintaining the life cycle of that parasite (Combes 2001). The ecological definition of parasitism sometimes grades into other symbioses (Zapalski 2011) and different long-term interactions between organisms including epizoism (Hengsbach 1991a) or (*in-vivo*) bioerosion (Wisshak and Tapanila 2008) which can also harm or influence the growth of their hosts (Keupp 2012). In some cases, *in vivo* epizoism can be detrimental to both the epizoa and their hosts (e.g., Meischner 1968, Hengsbach 1991a, Larson 2007; Keupp 2012; Keupp and Hoffmann 2014). In the fossil record, parasitism can mostly only be recognized by negative effects on its host and by comparisons with extant parasite-host associations.

20.3 Parasites of Extant Cephalopods

Measured by the number of times it evolved independently in several lineages and how many parasitic species are presently in existence, parasitism is one of the most successful modes of life displayed by living organisms (Poulin and Morand 2000). Although research has advanced significantly in the last decades, parasites of many groups of extant cephalopods are still poorly known (Hochberg 1983, 1989, 1990; Pascual et al. 1996; 2007; Gonzalez et al. 2003; Castellanos-Martinez and Gestal 2013). Adult specimens of commercially exploited species which have been the most investigated, are all known to harbor multiple groups of parasites (Pascual et al. 2007). Morphological determination of the systematic affinity of these parasites is not always straight forward, because of their high degree of specialization compared with their free-living relatives, high degree of reduced/ simplified morphology related with their parasitic lifestyle, and sometimes complex life cycles consisting of morphologically distinct ontogenetic stages (Brooks and McLennan 1993). An anecdotal example for cephalopods is the original description of the hectocotylus or modified arm of some male cephalopods as a parasitic helminth (Della Chiaje 1825; Cuvier 1829, 1830). The heterogeneous distribution and differing prevalence of parasites (and host reactions) within the infected tissues, within certain individuals and/ or regions, might also hamper their discovery when species have been poorly investigated or sampled. Molecular techniques have become useful complementary taxonomic tools to parasite diagnosis and specific identification in many groups including cephalopods (Pascual et al. 2007), but are of course unavailable in long extinct fossil forms.

The only parasites known from the extant externally shelled cephalopods (*Nautilus* and *Allonautilus*) in the wild (natural environments) are copepods living on the gills (Willey 1897; Haven 1972; Ho 1980; Hochberg 1983, 1990; Carlson 1987; Ward 1987). A possible exception might be a bacterial infection and a nematode infestation reported by Barord et al. (2012) in a captive *Nautilus* soon after it was captured from the wild. Many different kinds of parasites have been reported from extant coleoids, which are now generally accepted to be more closely related to ammonoids than to nautilids (e.g., Jacobs and Landman 1993; Kröger et al. 2011). Parasites have been recovered from almost all tissues and organs of these cephalopods (Hochberg 1983, 1989, 1990). They are most commonly located on gills, in digestive tracts, the excretory organs, mantle cavities, gonads, and in the musculature (Hochberg 1989, 1990; Pascual et al. 2007; Castellanos-Martinez and Gestal 2013). Pathogens of extant coleoids (Hanlon and Forsythe 1990; Hochberg 1983, 1989, 1990; Castellanos-Martinez and Gestal 2013) include bacteria, fungi, Sarcomastigophora, Apixcomplexa, Ciliophora, Dicyemida, parasitic flatworms (Monogenea, Trematoda, Cestoda), Acanthocephala, Nematoda, Annelida (Hirudinae, Polychaeta), and various crustacean Arthropoda (Copepoda, Isopoda, Branchiura, Malacostraca). Parasites can actively penetrate the host's body or be passively obtained through feeding or respiration. Many cephalopod species serve as primary hosts for protozoans, dicyemids, helminths, and crustaceans, but more commonly serve as secondary, third or reservoir (paratenic) hosts for intermediate (larval) stages of helminths such as digeneans, cestodes, and nematodes (Hochberg 1983, 1989, 1990). Cephalopods thus play a vital role in the transfer of parasites through the food web of marine ecosystems to final hosts such as elasmobranchs, fishes, and marine mammals (Hochberg 1983, 1989, 1990; Pascual et al. 2007). Crustaceans are macroparasites that mainly inhabit the mantle cavity and gills of cephalopods (Castellanos-Martinez and Gestal 2013), although they can also be found on external surfaces of the body such as on the arms or the head (Hochberg 1990).

Many parasites are acquired through contact with hosts in their habitat or through feeding, so that they might even give information about the feeding grounds or ecology of extinct cephalopods. Parasites cannot only provide information on cephalopod predator-prey interactions, but also on their distribution in the water column. Benthic and coastal cephalopods usually have different parasites then pelagic and oceanic species (compare Hochberg 1983; Pascual et al. 1996; Gonzalez et al. 2003).

20.4 Parasites of Fossil Cephalopods

Despite the ubiquity of the parasitic life style, its independent appearance in several unrelated lineages, and hypothesized ancient origins, relatively few clues can be derived from the fossil record. This is related with the fact that parasites (at least in some stages) are commonly small (+/− microscopic), frequently lack hard parts, often live within the (soft parts of the) host for the longest portion of their lifecycle (endoparasites), and/or might be isolated from their host post-mortem (Conway-Morris 1981; De Baets et al. 2011). These phenomena weigh against the preservation and recognition of fossil parasitoses, and as a consequence, the detection of parasitic biota is rare and mostly restricted to localities with exceptional preservation such as Konservat-Lagerstätten (Littlewood and Donovan 2003; Boucot and Poinar 2010). Additionally, even when preserved and detected, the quality of preservation will hamper precise identification of such fossil parasites. Soft part preservation is also rare in ammonoids (e.g., Klug et al. 2012; Klug and Lehmann 2015) and therefore, no direct evidence for parasitism in the form of body fossils has yet been found.

Ammonoid workers (like many other palaeontologists), therefore, have to rely on host shell pathologies which are characteristic of parasitism. Similar pathologies can however be caused by different agents that are difficult to assign to a particular extant parasite lineage and in some cases might belong to extinct lineages of parasites. Just as with trace fossils, different parasites can create similar pathologies when infesting similar tissues or behaving in the same way. The same parasite can induce different pathologies depending on their abundance, behavior, as well as the state of and position within the host.

20.5 Identifying Parasitism as A Cause for Ammonoid Pathologies

Identifying a parasitic cause for pathologies is not straight forward, but some general guide lines can be given (compare also Hengsbach 1991a, 1996; Keupp 2000, 2012; De Baets et al. 2011):

1. Preservation of parasite remains: The perfect evidence for parasitism in ammonoids would be the remains of the parasites preserved with the ammonoids showing these pathologies. Unfortunately, such cases have not been documented. The closest thing so far are blister pearls, which are interpreted to have enclosed the dead parasite remains and partially reproduced its shape as well as size. Examples include organic tubes preserved in some Housean pits in the Devonian (De Baets et al. 2011) and elliptical (now recrystallized) egg-shape inclusions in Jurassic ammonoid blister pearls (Keupp 1986, 2000, 2012). It should also be noted that traces of small soft-bodied parasites are known to vanish during pearl formation (Lauckner 1983).
2. Comparison with pathologies in extant externally shelled mollusks (Fig. 20.1): Interpretation of parasitism should be biologically plausible (Hengsbach 1991a), so that similar, characteristic pathologies in extant shelled mollusks could indicate that a related parasite or a parasite behaving in the same way might have caused these injuries in ammonoids. One must remain cautious if alternative interpretations for these phenomena also exist. Gigantism could be related to parasitic castration (which is often caused by parasitic flatworms in extant gastropods and bivalves), but can also be explained by different pathologies, a natural rarity of larger individuals, and/ or considerable intraspecific variation in adult size, which is not unusual in extant cephalopods. Therefore, gigantic size of individuals on its own is insufficient to prove a parasitic infestation (Klug et al. 2014; De Baets et al. 2015).

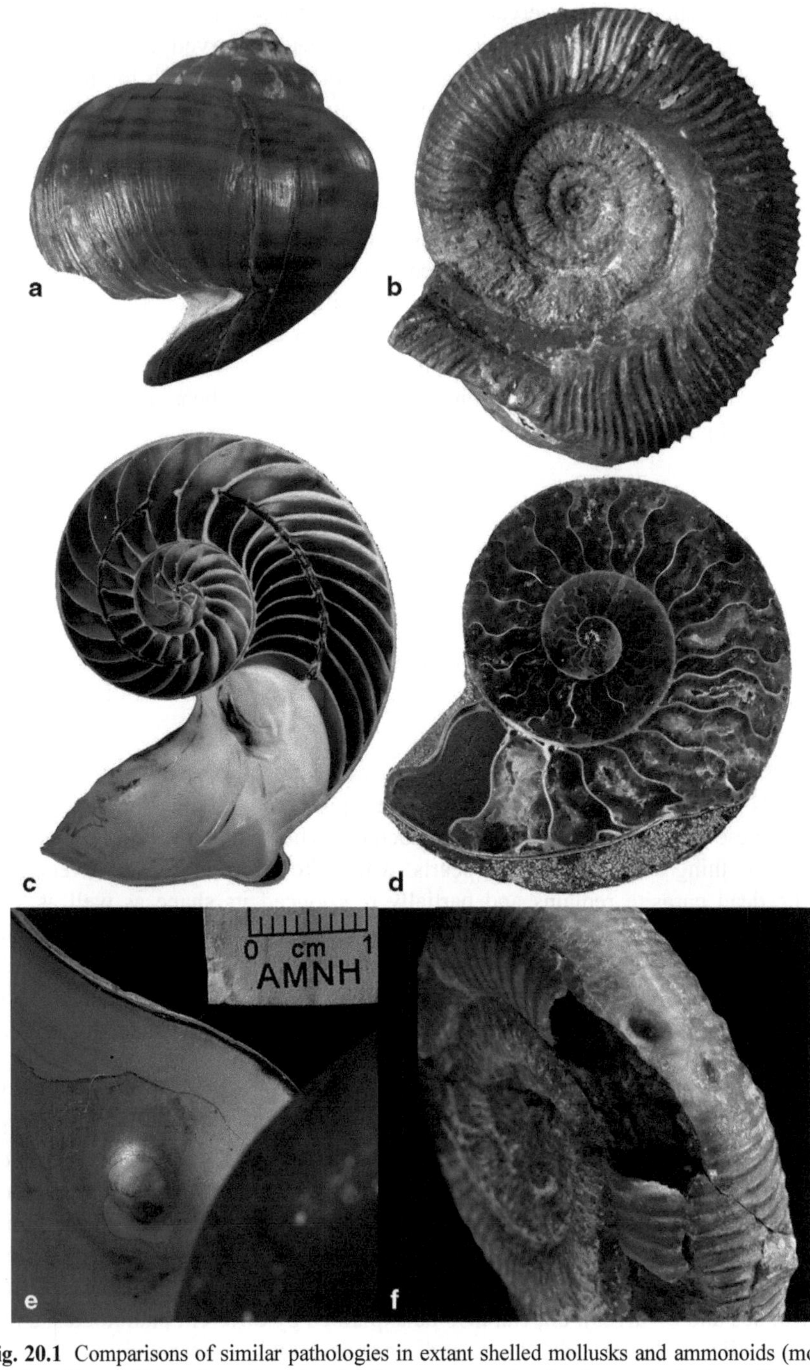

Fig. 20.1 Comparisons of similar pathologies in extant shelled mollusks and ammonoids (modified from Keupp 2012 unless otherwise stated). **a**, **b** progressive development of a deep slit-shaped recess in the apertural margin attributed to parasitic infestation in *Pila* sp. (**a**) Recent, Egypt and *forma umbilicata* in *Dactylioceras anguinum* (**b**) Early Toarcian (Jurassic), Altdorf near Nürnberg (Germany), SHK PA-1694a, dm 39 mm. **c**, **d** development of shell lamellae related with a local

3. Misinterpretation of normal shell features as pathologies: Parasitism should be excluded if features misidentified as pathologies are part of the normal growth or anatomy (e.g., soft-tissue). In some cases, muscle attachment scars might be confused with certain pathologies (*forma umbilicata* of Kröger 2000; Fig. 20.1b), but are restricted to internal moulds and always orientated towards the back of the body chamber (Keupp 2012).
4. Lack of external injury: Parasitic or other endogenous causes are often put forward in absence of external injuries or epizoa. In rare cases, parasitic infestations or infections might have happened preferentially after soft-tissue was exposed due to injury causing short-term swellings such as *forma inflata* (compare Keupp 1976, 2000, 2012), although the injury and the pathology are not directly related. Lack of proof as an argument for endogenous cause on its own cannot be accepted (Zapalski 2011), as one never knows if such an absence is caused by a true absence or only that it is impossible to identify it (soft parts not preserved).
5. Characteristics of the structures: The position, distribution, size, and morphology of these structures might also reveal a parasitic origin, particularly, when similar structures are known from extant shelled molluscs. Even when similar structures are absent in extant shelled molluscs, the position of these pathologies deep in the soft-tissue and/or far away from the aperture which could only be reached and/or inhabited by parasitic organisms feeding on tissues or benefitting from the host in different ways, can still corroborate their parasitic nature.
   a. Asymmetric development: Deviations from bilateral symmetry or symmetropathies (Hengsbach 1991b) have been attributed to parasitism including asymmetry of the suture line or shell sculpture and whorl section. In several taxa, the position of the lateral lobe as well as siphon are tightly constrained and asymmetry of suture line can even be species-specific and not of pathological origin (Keupp 2012). A progressive development of asymmetry of whorl section or suture line could potentially relate to the asymmetrical swellings which could be caused by local infections and parasitic infestations (Hengsbach 1991a, 1996). However, a gradual development of asymmetry of the suture line might also be related with non-pathological asymmetrical growth of soft-tissues during ontogeny, particularly towards maturity (e.g., Yacobucci and Manship 2011). The low prevalence and irregular appearance, variation in its development between individuals, sudden to temporary development in post-embryonic stages, and lack of traces of external injuries or epizoa are therefore of key importance to corroborate a parasitic origin of these phenomena (Hengsbach 1991a, 1996; Keupp 2000, 2012).

---

detachment of the mantle in a captive *Nautilus pompilius*, Recent (1990–1993), aquarium of the Jura-Museum Eichstätt, SHK PN-12, dm 110 mm (**c**) and a specimen of *Cleoniceras besairiei*, Albian (Cretaceous), Ambatolafia (Madagascar), PA-33582–1, dm 85 mm (**d**). **e, f** blister pearl in *Nautilus macromphalus*, Recent, New Caledonia, AMNH, max dm of pearl: 6 mm (**e**) and casts of two blister pearls on an internal mold of *Dactylioceras anguinum* **f** Toarcian (Jurassic), Altdorf near Nürnberg (Germany), SHK PA-643, dm 59 mm. The formation of blister pearls is typically induced by irritants (organic material, parasites, sediment grains, epizoa, etc.), which get lodged between the mantle and the shell

b. Prevalence of pathologies within population: The prevalence of parasites (and associated pathologies) in their host populations can be quite variable (from about 1 % to over 70 % or higher, but usually significantly below 100 %) and hard to predict. Typically, not all specimens of a species within a population are infested and even less might develop deformities indicative of parasitism. According to some authors, the number of pathological specimens attributable to parasitism should be consistently low (1–10 %: compare Keupp 2012). Higher prevalence (70 % or more) might be related with a specialist among parasites (e.g., Hengsbach 1991a, 1996; but compare Keupp 2000, 2012 for a different viewpoint). In certain cases, one kind of parasitosis might be particularly common in some regions, while in other areas, they might be rare or absent (even in large samples of the same taxon with similar preservation). Some authors have interpreted a high prevalence of a pathology as a sign that a certain area supported a high population of parasites (Morton 1983; Keupp 2000, 2012), but this could also be related with higher infection rates (in response to different feeding or living habits in separate regions), different immune responses or to ammonoids being accidental hosts (which do not normally harbor the parasite) in these environments. If the phenomenon occurs in 100 % of the representatives of a taxon and if its appearance is consistent from one individual to another (same side affected, same degree of development), parasitism is unlikely and genetic causes can be suspected. It can be difficult to classify a feature as pathology if all specimens have it (Keupp 2012) and most postnatal causes with the exception of obligate symbiosis can be ruled out (Hengsbach 1996). However, symbiosis is hard to verify in the fossil record (Zapalski 2011).

6. Appearance after hatching and later in ontogeny: In most cases, species are infested directly or by taking in infested food after hatching, which means that the pathologies should at the earliest develop after hatching or probably much later, when the ammonoid came in contact with the parasite or their larvae (intermediate stages). Structures being present at or before birth might be related with teratological causes (e.g., congenital disorders). The blister pearls in certain species of Paleozoic ammonoids occur early in ontogeny (but always clearly after the embryonic shell) and more importantly, appear at different points in the ontogeny of different individuals within the same species. In the case of parasitism, such variability and no correlation with growth of the organism (e.g., ontogenetic stages, septal spacing) is expected (Hengsbach 1996; De Baets et al. 2013b).
7. Host specificity: Not all ammonoid taxa at a locality show identical pathologies; parasites are often quite host-specific and might cause different reactions in separate taxa, so that the pathologies might often be restricted to a single species, clade or lineage. Sometimes, ecology might play a bigger role than phylogeny, so that certain parasitic infestations can be present in one or multiple (not necessarily closely related) taxa with a particular ecology (food source, mode of life, habitat).
8. Gradual development of irregular, ontogenetically long-lasting pathologies: Rieber (1963) interpreted a pathological ammonoid individual with a progressive displacement of the keel and the siphuncle after at least three normal whorls

as the results of parasite infestation, which has been generally accepted as a plausible explanation. This probably reflects a struggle between the host and the infesting parasite(s). After the death of the parasites, the malformation might remain this way or gradually disappear depending on the damage as well as the need and available mechanisms to return to normal morphology (Hengsbach 1991a, 1996). In cases of swellings, the pathologies might disappear as suddenly as they have developed.

These guidelines can be used to evaluate the plausibility and certainty of several pathologies attributed to parasites in ammonoids. Due to the difficulties in identifying the potential parasitic culprits; we will discuss possible cases of parasitism in ammonoids grouped by the morphological expression of pathologies or paleopathies.

Several structures and pathologies have been related with parasitism in shelled cephalopods with various degrees of certainty (Hengsbach 1991a, 1996; Keupp 2000, 2012; De Baets et al. 2011; Fig. 20.2). These pathologies range from (blister) pearls, symmetropathies (asymmetry of shell tube and suture line), temporal increase in shell volume and ornamentation, anomalies in shell secretion to various other pathologies. Parasitic flatworms have most commonly been implicated in several pathologies, including pathological gigantism (Manger et al. 1999), blister pearls (Keupp 1986; De Baets et al. 2011, Fig. 20.1f), and certain perturbations in shell growth (e.g., *forma umbilicata*: Keupp 2000, 2012; Fig. 20.1b) in analogy with prevalence of these parasites in extant cephalopods and cause of similar pathologies in extant shelled molluscs (Fig. 20.1). These hypotheses are not unlikely as parasitic flatworms have probably been around since the Ordovician based on the extrapolation of extant parasite-host relationships (Littlewood 2006), but so far, direct fossil evidence is lacking prior to the Permian (Dentzien-Dias et al. 2013). Earlier records of parasitic flatworms are still debated (Upeniece 2001); no remains have been found directly associated with ammonoids. It seems likely that other parasites were also present including forms that were unable to leave any trace (e.g., parasitic crustaceans and bacterial infections: compare Hölder 1956) and some traces might even relate to now extinct groups of parasites (Boucot and Poinar 2010).

20.5.1 Disturbances in Apertural Shell Growth

Keupp (1979) described a pathology, where the shell formation is delayed near the umbilicus leading to the progressive development of a deep slit-shaped recess (and ribbing vertices similar to *forma verticata*) in the apertural margin of some specimens of Jurassic *Dactylioceras* (Fig. 20.1b). In the absence of external injuries, he attributed this to parasitism based on the protected position of the paleopathies and comparisons to other specimens in the local population. Superficially similar structures ("Rippenscheitelungen") can also be formed in association with external injuries, traditionally described as *forma verticata* and *forma semiverticata* (compare Hengsbach 1991a, 1996; Keupp 2000, 2012; Zatoń 2010). Kröger (2000) introduced the term *forma umbilicata* for these pathologies in absence of external injuries. Their parasitic origin is corroborated by a similar phe-

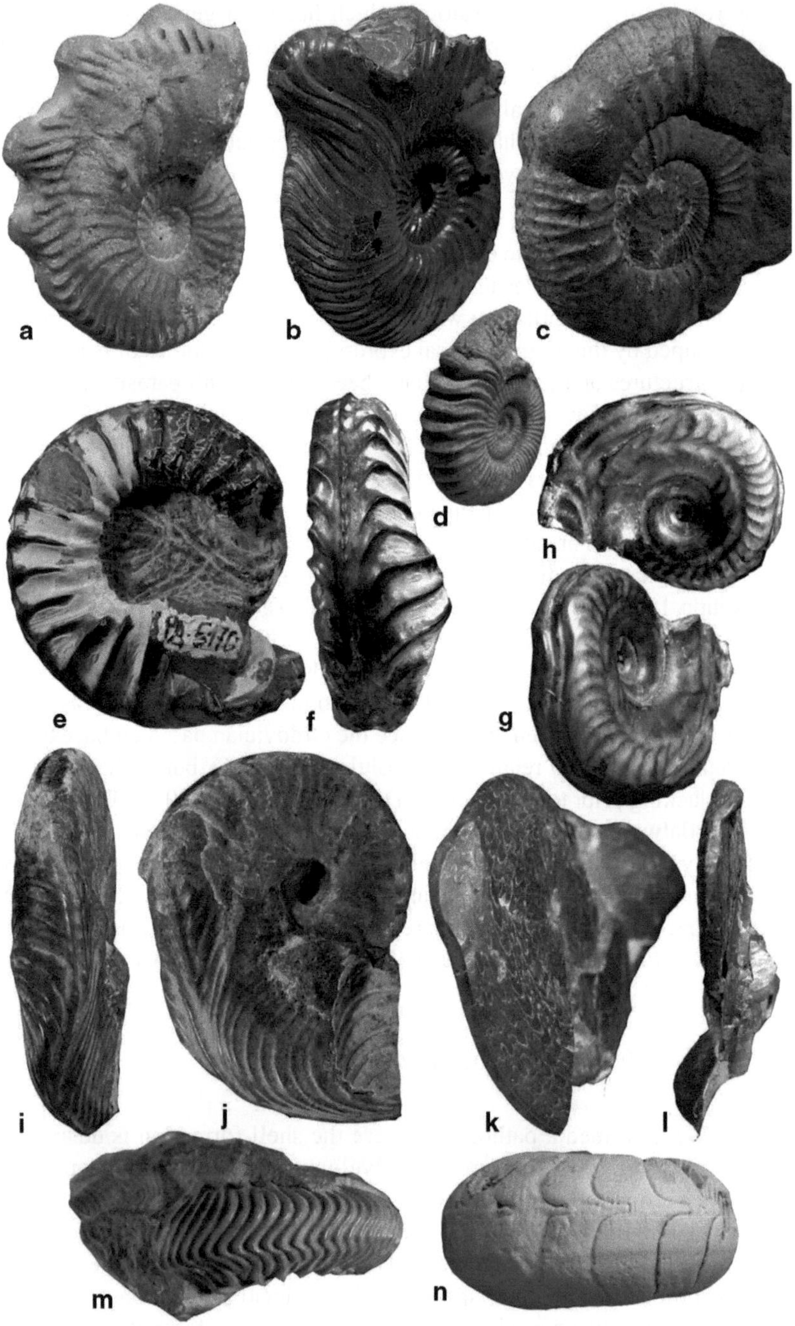

Fig. 20.2 Examples of different pathologies attributed to parasitic infestations with various degrees of certainty from the Devonian to Cretaceous (modified from Keupp 2012; unless stated otherwise): **a** *Amoeboceras alternans* with progressive enlargement of the ventral ornamentation (*forma augata* Kröger 2000), Oxfordian (Jurassic), Kucha near Hersbruck (Germany), SHK PA-785, dm 12 mm. **b** *Quenstedtoceras leachi* with a conspicuous, temporary bulbous swelling of the ventral shell and

nomenon in the freshwater gastropod *Pila* (Keupp 2000, 2012; Fig. 20.1a), where the infestation with intermediate stages of trematode flatworms in the mantle tissue of the apertural margin leads to the formation of similar abnormal slit-shaped openings in the aperture, which becomes progressively broader with the increasingly infested tissue. These structures should not be confused with muscle scars ("*Muskelleisten*": compare Keupp 2012) which are restricted to internal moulds from the back of the body chamber to the phragmocone (as opposed to the front of the body chamber on both the shell and internal mould).

These pathologies have also been reported from rare specimens of *Epivirgatites* from the Tithonian and *Quenstedtoceras* from the Callovian of Russia (Keupp 2012, pp. 223–224). These specimens demonstrate that the infestations are not always restricted to the umbilical margin and can vary even within the same taxon which probably depends on the position of the parasite; some specimens were able to subsequently close the recess with smooth or irregular shell material (Keupp 2012). So far, the structures are only confidently known from these three genera of Jurassic Ammonitida (Keupp 2012). These pathologies are best known from the Early Toarcian of Altdorf near Nürnberg (Germany), where they are restricted to dactylioceratids (Fig. 20.1b) and absent from all other ammonoid taxa (Harpoceratidae, Hildoceratidae, Phylloceratidae, Lytoceratidae). This might indicate that these parasites are host specific. These structures are rare in dactylioceratids in other regions, which might suggest that the ecology (mode of life, predator-prey relationships) of these taxa in this region also increased their infection risk. The prevalence is quite high in the Toarcian of Altdorf where up to 5 % of *Dactylioceras* display these structures which Keupp (2012) attributed to a high regional population of these parasites.

20.5.2 Pathological Gigantism

Parasites are well known to alter the behavior, growth, and morphology of their hosts (Miura et al. 2006). Gigantism is one of the most striking modifications in

ornamentation (*forma augata* Kröger 2000), Callovian (Jurassic), Dubki near Saratov (Russia), SHK PA-20114, dm 58 mm. **c** *Orthosphinctes* with conspicuous, temporary bulbous swelling of the shell without ornamentation behind the aperture (*forma inflata* Keupp 1976), Kimmeridgian (Jurassic), Hartmannshof (Germany), SHK PA-1871, dm 60 mm. **d** *Dactylioceras athleticum* with a progressive development of shovel-like ribs (*forma augata* Kröger 2000), Kimmeridgian (Jurassic), Hartmannshof (Germany), SHK PA-1871, dm 60 mm. **e, f** *Pleuroceras spinatum* with a unilateral thickening of the body chamber resulting in a significant *left-right* asymmetry of the whorl section, SHK PA-5170, Pliensbachian (Jurassic), Unterstürmig (Germany), dm 41 mm. **g, h** *Hildoceras bifrons* showing multiple oscillations of the keel around its normal ventral position (*forma undaticarinata*), SHK PA-6245, Toarcian (Jurassic), Grand Causses (France), dm 23 mm. **i, j** *Cleviceras elegans* with Morton's syndrome (sensu Landman and Waage 1986), SHK PA-543, Toarcian (Jurassic), Altdorf near Nürnberg (Germany), dm 55 mm. **k, l** *Pseudosageceras multilobatum* with Morton's syndrome (sensu Landman and Waage 1986), SHK PA-9204, Early Triassic, Vikinghøgda south of Sassendalen (Spitzbergen), dm 34 mm. **m** *Amoeboceras sp.* with a progressive asymmetric shift of the keel (*forma juxtacarinata* Hölder 1956; similar to the case described by Rieber 1973), SHK PA-786, Oxfordian (Jurassic), Scarborough (United Kingdom), dm 20 mm (previously unpublished). **n** *Latanarcestes noeggerathi* with an exentrical position of the external lobe (*forma juxtalobata* Hölder 1956), Emsian (Devonian), Tafilalt (Morocco), dm 27 mm morphology and growth, which can be induced by parasitic organisms and can significantly alter the size-structure, use of resources, and intraspecific competitive interactions of the host population. It is one of the common paleopathies to be linked with parasitic infestations in ammonoids (Hölder 1956; Hengsbach 1991a, 1996; Manger et al. 1999; Keupp 2000, 2012). Hölder (1956) was the first to suspect that parasite-caused gigantism might have occurred in ammonoids based on studies of the present-day terrestrial snail *Zebrina* by Boettger (1953a, 1953b), who documented parasite-caused delay in sexual maturation, corresponding castration, and retardation in growth.

Pathological gigantism should only be used to refer to rare specimens of a certain taxon which have abnormally large sizes (Tasnadi-Kubacska 1962) that were triggered by pathogens, hormonal disorders or other endogenic causes. Hengsbach (1996) introduced the term *forma gigantea* to refer to pathological gigantism in ammonoids (which effects only a small fraction of ammonoid populations). Other causes for size differences such as continuous and discontinuous (e.g., sexual dimorphism) intraspecific variability in adult size or ecological causes should be ruled out (De Baets et al. 2015; Keupp and Hoffmann 2015). So far, pathological gigantism caused by parasitic castration has only been suggested by Manger et al. (1999) to explain rare exceptionally larger individuals of some Carboniferous ammonoid and nautiloid taxa. Manger et al. (1999) suggest that infestation of the gonads by trematode larvae and castration might have been responsible. Several gastropod species exhibit growth to abnormally large sizes following infection by trematodes or other parasites (Sousa 1983; Sorensen and Minchella 2001), which can be caused by enhanced growth, sexual retardation, or even castration. Castration in extant bivalves and gastropods is most commonly caused by parasitic flatworms (Lafferty et al. 2009; but see Boettger 1953a, b for an example with a bacterial cause). However, not all infestations or even castrations result in abnormally larger sizes or noticeable effects on (shell) growth in molluscs. Some specimens can become mature before castration or continue to grow after castration. Additionally, gigantism has only been documented in short-lived and primarily fresh water and terrestrial gastropods, while studies on long-lived marine gastropod species have found that trematodes have either no effect on growth or even stunt growth (Sousa 1983; Sorensen and Minchella 2001; Miura et al. 2006). Castration and pathological gigantism have so far not been reported from extant cephalopods, which are commonly infested by parasitic flatworms and other parasites, highlighting the need for independent evidence for parasitic infestation of these specimens other than size.

However, there is no direct proof in the form of shell reactions or preserved parasite remains (compare Hengsbach 1996; De Baets et al. 2011; Klug et al. 2014 for a review) to corroborate the views of pathological gigantism caused by parasitism in ammonoids. Indications used to support this interpretation comprise the large size of these cephalopod fossils (two to four times as big as “normal” specimens from the same layers), the scarcity of specimens of this size belonging to members of a species that are common, and the absence of indications for adulthood. Ivanov (1971, 1975) introduced the term *“megaconch”* to refer to rare abnormally large specimens of Jurassic ammonoid species which show no signs of maturation, although he did not specifically discuss parasitism as a cause for this phenomenon and these criteria might be hard to interpret in incomplete specimens. Rare larger specimens in semelparous accumulations might reflect the general paucity of larger individuals as well as multiple spawning seasons and areas in cephalopods (Rocha et al. 2001). Castrated ammonoids are usually interpreted to not have reached maturity and/or grew beyond their normal size (Hölder 1956; Manger et al. 1999). The absence of signs of adulthood could also be explained by various other factors, including incomplete preservation of the body chamber of the specimens, a common taphonomic and collection bias in larger specimens, and the lack of no clear signs of maturity in the shell or continuous growth until death which is not uncommon in some taxa of ammonoids (compare Davis et al. 1996; Bucher et al. 1996). Furthermore, some commonly caused adult modifications such as septal crowding might be useless in some cases as they can also be induced by environmental stress (Kraft et al. 2008). Although it is difficult to impossible to prove the absence of pathogens or the normal levels of hormones in these giants, we suggest that the most likely explanation for many cases of supposedly pathologic gigantism might just be highly variable adult sizes (not uncommon in extant cephalopods and can sometimes be discontinuous: compare De Baets et al. 2015) and/or reproductive strategies, where only a few individuals achieved maturity out of a thousand eggs. More evidence is necessary to ascribe giant size of rare specimens of a species to pathologies and even more to attribute this to parasitism, although it is biologically plausible (Klug et al. 2015).

20.5.3 Pearl Formation

The formation of blister and free pearls in bivalves has often been linked with parasitic flatworms (Conway-Morris 1981; Combes 2001; Littlewood and Donovan 2003; Boucot and Poinar 2010). Nevertheless, various other parasitic organisms such as fungi, unicellular organisms, nematodes or arthropods (and their eggs) as well as various other irritants including inorganic material which get stuck between mantle and shell can induce pearl formation (Götting 1974, 1979; Lauckner 1983). Nevertheless, the morphology of some parasite-induced structures in extant bivalves are believed to be very specific to parasitic flatworms such as pits in shells and igloo-shaped shell secretions or to polychaete worms such as borings and can be traced back in the fossil record (Liljedahl 1985, Ruiz and Lindberg 1989; Ituarte et al. 2001, 2005; Huntley 2007; De Baets et al. 2011). Blister pearls (Fig. 20.1f) in ammonoids have therefore often been linked with parasitic infestations (House 1960; Keupp 1986; Hengsbach 1996; Davis and Mapes 1999; De Baets et al. 2011, 2013b). Furthermore, the fact that blister pearls (cephalopods, bivalves, and gastropods) and free pearls (bivalves, gastropods) are known from a wide variety of fossil and extant taxa (Binder 2002; Boucot and Poinar 2010 and references therein) suggest that every shelled mollusc is capable of forming such structures (Landman et al. 2001). In the fossil record, these pearls can be preserved as pits on internal moulds (steinkerns).

House (1960) was the first to describe more or less regularly distributed pits on internal moulds of Early to Middle Devonian ammonoids (Fig. 20.3, 20.4, 20.5). Many authors (Chlupáč and Turek 1983; Becker and House 1994; Klug 2002a,

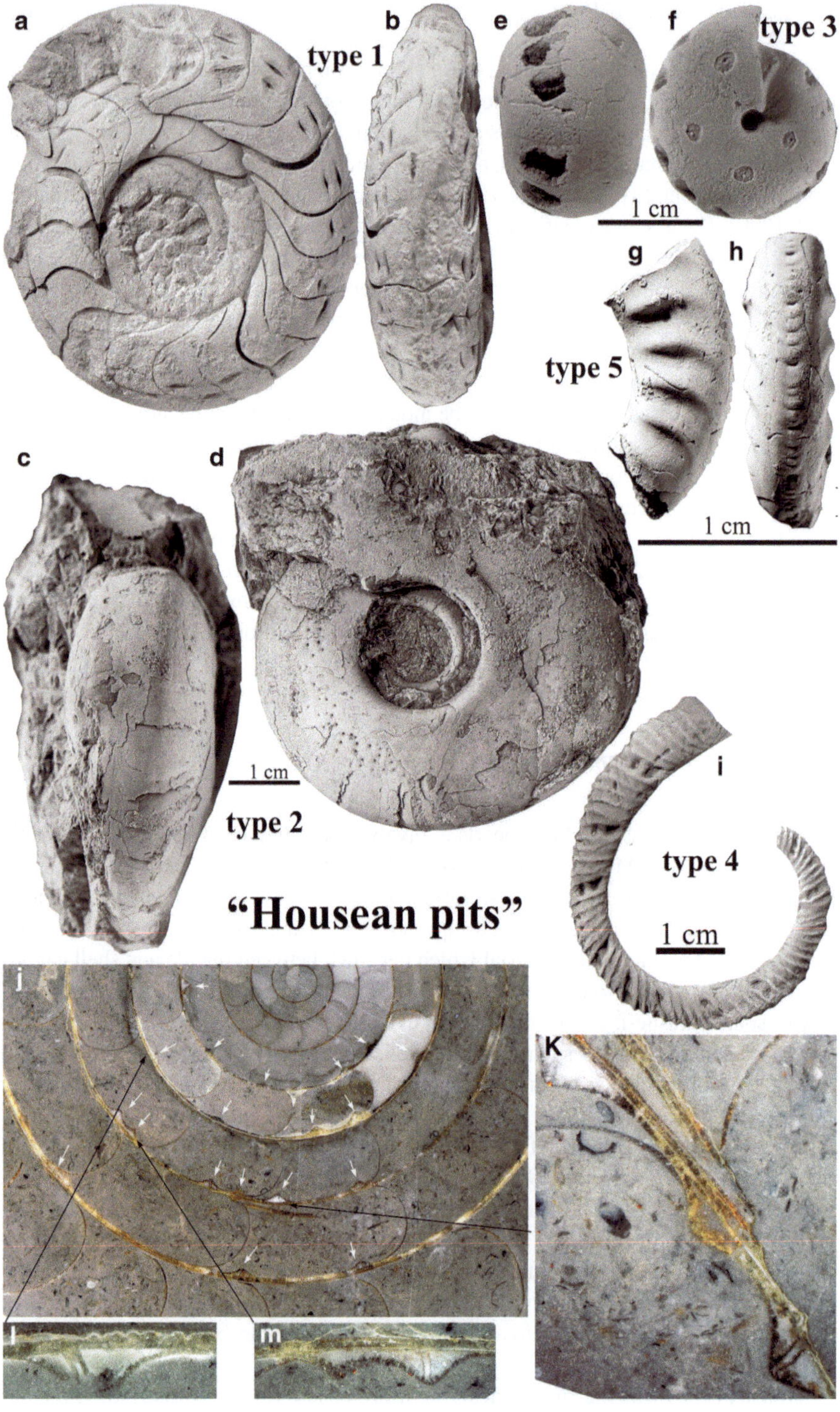

Fig. 20.3 Types of Housean pits and morphological details. Images from De Baets et al. (2011). **a**, **b** lateral and ventral views of *Sellanarcestes* cf. *ebbighauseni*, PIMUZ 28582, Sellanarcestes wenkenbachi Zone, Emsian, Jebel Ouaoufilal, Tafilalt, Morocco. **c**, **d** ventral and lateral view of *Crispoceras tureki* Klug, 2002, PIMUZ 28591, Pinacites jugleri Zone, Eifelian, Jebel Ouaou-

2002b, 2007; Bockwinkel et al. 2009; Ebbighausen et al. 2011) published specimens showing these pits, but without discussing them in detail. House (1960) and others concluded that these pits represent the moulds of pearls that must have formed during the ammonoid's life on the inside of the shell within the body chamber, i.e., the part of the shell that contained the soft parts and lacked chambers. These pits/pearls are commonly arranged spirally, radially or chaotically on the flanks and venter of these ammonoids and are oval to circular in outline, occurring in varying numbers. House (1960) lacked sufficient information to decide whether these pits represent traces of a parasite or particles that entered between mantle and shell, causing an irritation and the formation of a pearl-like structure. Davis and Mapes (1999) named these features "*Housean pits*" in honor of House and favoured a parasitic origin. Hengsbach (1991a, 1996) included these pathological shell concretions together with more isolated blister pearls (e.g., Keupp 1986) in a paleopathology he dubbed *forma concreta*.

De Baets et al. (2011) reviewed the morphology and distribution of these structures and distinguished five types of pits on Devonian ammonoids (Fig. 20.4):

- Type 1: fairly large oval pits, which are lengthened in a longitudinal direction, predominately ordered in large spiral rows and often paired. They are common in *Anarcestes* and *Sellanarcestes*.
- Type 2: multiple, small pits, arranged in more or less radial rows or more chaotically. They were initially thought to be restricted to the late Early Devonian to Middle Devonian (Late Emsian to Givetian: *Anarcestes*, *Afromaenioceras*, *Crispoceras*, *Sobolewia* and *Subanarcestes*), but were recently also reported from the Late Devonian (*Felisporadoceras*: Rakociński 2012).
- Type 3: rare circular pits with a central deepening, which have so far only been reported from five specimens of *Sobolewia* from the Givetian of Algeria. Although similar pits might also be present in Aulatornoceratinae (Jürgen Bockwinkel, personal communication 2013)
- Type 4 (Opitzian pits): flat, large, radially arranged, paired pits which have so far only been reported from *Ivoites* from the early Emsian Hunsrück Slate (De Baets et al. 2013b) and absent from the same taxon at other localities (De Baets et al. 2013a)
- Type 5: kidney-shaped pits in the middle of the venter, which are found in certain taxa of Early Emsian ammonoids (*Chebbites*, *Gracilites*, *Lenzites*). These pits usually coincide with the most posterior points of the hyponomic sinus of their growth lines (compare Becker and House 1994; Klug 2001; Korn and Klug 2002; Klug et al. 2008)

filal, Tafilalt, Morocco. **e**, **f** *Sobolewia nuciformis* (Whidborne 1889), three specimens kept under the same number (MNHN–R.08459), Givetian, Redjel Iamrad, Algeria, Jacques Follot coll., **g**, **h** *Chebbites reisdorfi* Klug, 2001, PIMUZ 7484. **i** *Ivoites opitzi*, Lehmann col., early Emsian, Bundenbach, Germany. **j-k** Longitudinal section through PIMUZ 28583 of *Sellanarcestes* spp., *wenkenbachi* Zone, Emsian, Oufrane (S of Tata), Morocco. **j** median section displaying many "Housean pits" (*arrows*), most with internal tube; overview. **k** three closely spaced pits, two with internal tubes, note the continuous ammonoid shell layer covering pits and septum. **l** corroded pit with tube, note the continuation of the innermost ammonoid shell layer. **m** two fused pits, the right pit shows the delicate internal tube, mural part of septum on the left

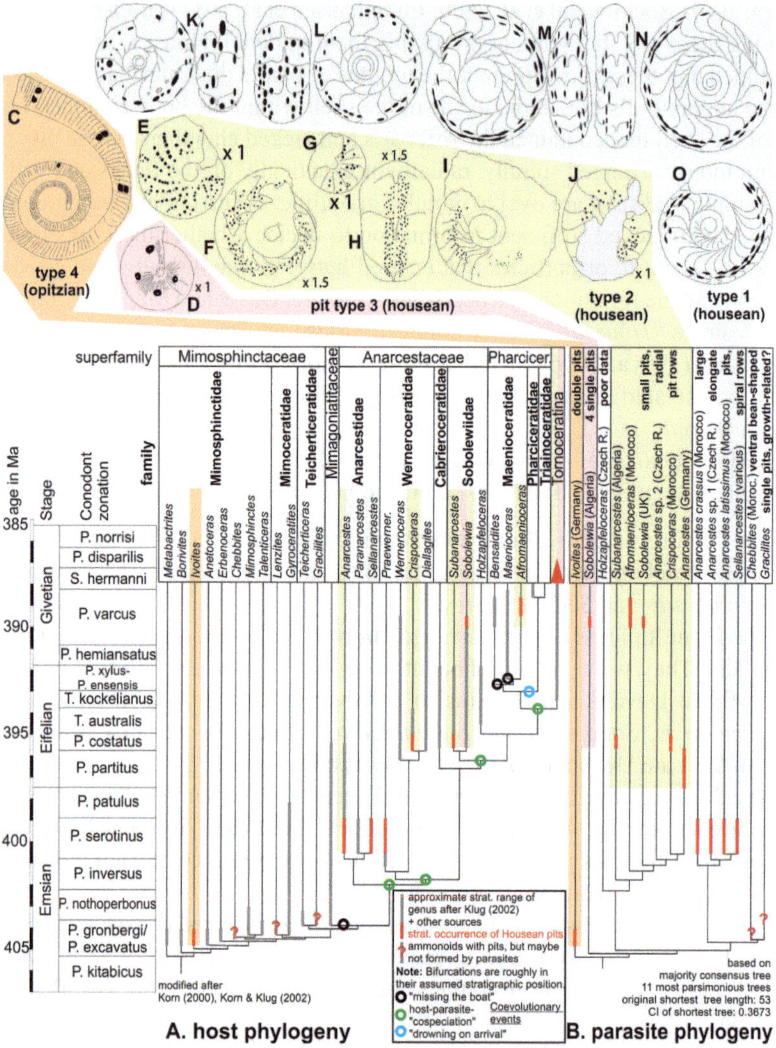

Fig. 20.4 Morphology and Distribution of Opitzian (type 4), Housean (type 1–3) and other pits (type 5) known from Devonian ammonoids updated from De Baets et al. (2011) to include Late Devonian records discussed in the text: *red arrow* list the extended range of type 2 (*green*: Rakocinski 2012) and type 3 (*pink*) pits in Tornoceratina. Note the coevolution of Devonian ammonoids and their parasites as reflected in the arrangement and shape of the blister pearls. **a** Host phylogeny. **b** Parasite phylogeny. The ammonoid phylogeny is based on a majority consensus tree consisting of the11 most parsimonious trees original shortest tree length (compare Korn 2001; Korn and Klug 2002; De Baets et al. 2011). **c** *Ivoites opitzi* (early Emsian, middle Kaub Formation, Hunsrück, Germany: from De Baets et al. 2013b). **d** *Sobolewia nuciformis* (Givetian, Redjel Iamrad, Algeria). **e** *Subanarcestes* sp. (Eifelian, Erg El Djemel, Algeria; after House 1960). F. *Afromaenioceras sulcatostriatum* (Givetian, Jebel Ouaoufilal, Morocco). G. *Sobolewia* aff. *nuciformis* (Givetian, Cornwall). H. *Anarcestes* sp. (late Emsian, Koněprusy, Czech Republic). I. *Crispoceras tureki* (Eifelian, Jebel Ouaoufilal, Morocco). J. *Anarcestes* sp. (Eifelian, Wissenbacher Schiefer, Germany). K. *Anarcestes* sp. (late Emsian, Jebel Mech Agrou, Morocco). L. *Anarcestes latissimus* (late Emsian, Hassi Moudaras, Morocco). M. *Sellanarcestes* cf. *ebbighauseni* (late Emsian, Jebel Ouaoufilal, Morocco). N. *Sellanarcestes ebbighauseni* (late Emsian, northern Jebel Amessoui, Morocco). O. *Sellanarcestes neglectus* (late Emsian, southern Jebel Mech Agrou, Morocco)

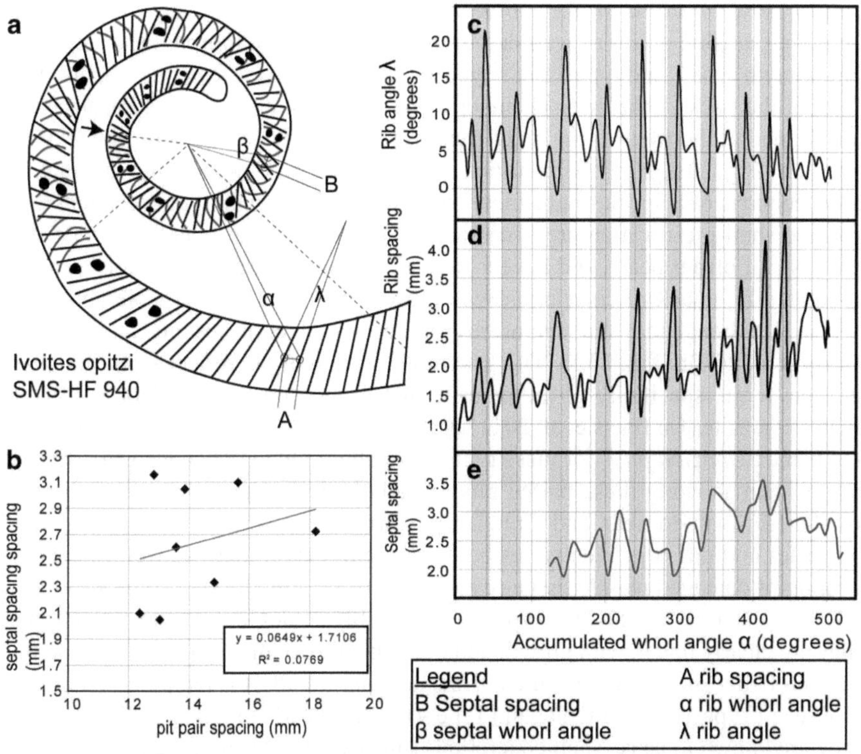

Fig. 20.5 Specimen of *Ivoites opitzi* from the Hunsrück Slate showing the morphology of Opitzian pits and their local influence on rib angle and spacing. Note, the lack of a general correlation of pit pair spacing with growth rhythms or septal spacing (modified from De Baets et al. 2013b). **a** Schematic drawing of the retrodeformed specimen with position of pits, ribs, and septa and analysed parameters. The *arrow* marks the starting point of the analyses (0 on the graphs), which proceed counter-clockwise. **b** Correlation between pit pair and septal spacing. **c** Plot of rib angle λ versus accumulated whorl angle Σ α. Note the drop in the angle between the rib before and following the pit pair and the subsequent rise between the rib and the second rib after the pit pair. **d** Plot of rib spacing A versus accumulated whorl angle Σ α. Note the rise in rib spacing A coinciding with the position of the pit pair. **e** Plot of septal spacing A versus accumulated whorl angle Σ α. Note a drop in septal spacing which more or less follows on each pit pair. The septal spacing has been translated for a rib angle corresponding to the body chamber length of about 115° (cf. Bucher et al. 1996)

De Baets et al. (2011) demonstrated that the Housean pits were the casts of blister pearls at least for type 1 pits which were overgrowing organic tubes, because they found these structures in specimens where the associated shell and the blister pearls were preserved (Fig. 20.3). The tubes are interpreted to be the remains of parasitic organisms which lodged themselves between mantle and shell due to their position far away from the aperture (other symbiotic relationships or long-term guests such as epizoa are therefore unlikely) and which were overgrown by shell material (probably after the death of the parasite). No borings in the shell or inorganic particles were found inside of the pearls. Tubes are also present in type 3 pits suggesting a similar parasitic cause for these blister pearls. Type 2 is tentatively interpreted as being of parasitic origin as well due to their similarities in morphology and distribution with type 1 and type 3 pits. Type 1–3 are typical Housean pits because they form at the back of the body chamber. Type 4 pits are an exception because they are formed at the front of the body chamber and locally affect shell growth at the aperture (Fig. 20.4, 20.5). Similarities with the other pit types and the fact that not all specimens of a species have them suggest a similar origin, but specimens with pristine shell preservation are necessary to directly test the hypotheses that they are casts of blister pearls and have a parasitic origin. A parasitic origin is also corroborated by a lack of correlation with growth rhythms (long-term rib and septal spacing), although they did temporarily and locally affect rib spacing and angle (De Baets et al. 2013b; Fig. 20.5). This apertural location might reflect an evolutionary change in site specificity of the parasite that formed pit types 1–3 or a different type of parasite (De Baets et al. 2011). De Baets et al. (2013b) therefore introduced the name "*Opitzian pits*" for these type 4 pits after Opitz (1932), who first figured a specimen showing these paired pits. A pathological origin is also corroborated by the fact that Housean and Opitzian pits are not known from all specimens of a species and the ratio even varies between regions (Fig. 20.6) as well as signs of healing of damaged mantle tissue (e.g., spiral traces: Fig. 20.3) in some specimens (De Baets et al. 2011, 2013b). Type 5 pits probably do not have a parasitic origin as they are clearly linked with growth and occur in all specimens of these taxa, so that they probably should not be called "*Housean pits*" at all. It is tempting to attribute Housean (Type 1–3) and Opitzian (Type 4) blister pearls to parasitic flatworms as they are known to cause shell concretions (mostly after death of the parasite: compare Lauckner 1983) in living bivalves, but so far no identical structures are known from extant cephalopods and the identity of these Devonian parasites remains a mystery. Other structures like igloo-shaped concretions can be formed when both the bivalve and parasite are alive (Ituarte et al. 2001, 2005).

The types of pearls correspond well with large groups of ammonoids suggesting a certain degree of parasite-host coevolution (De Baets et al. 2011; Fig. 20.4). The pits become rarer during the Givetian (Fig. 20.6) which might also explain why they have initially been overlooked in the Late Devonian (House 1960; De Baets et al. 2011). Rakociński (2012) subsequently described type 2 pearls in *Felisporadoceras* from the Famennian of Poland extending their range into the latest Devonian. They might also be present in other Late Devonian taxa (e.g., Frasnian aulatornoceratids: Jürgen Bockwinkel, personal communication 2013). Superficially similar pits in *Cymaclymenia* figured by Schindewolf (1934) are probably borings (House 1960). Housean pits seem to have disappeared at the end of the Devonian which might indicate the extinction of this particular lineage of parasites or that ammonoids found countermeasures against them or that the parasites stopped inducing the formation of pearls. It is unclear if the absence in derived ammonoids reflect changes in the complexity of food webs, because pearls might suggest the presence of complex parasite life cycles and food webs involving bivalves as well as jawed vertebrates (compare De Baets et al. 2011). Clusters of blister pearls (or their possible casts) have also been reported from Silurian nautiloids (Stridsberg and Turek 1997; Manda

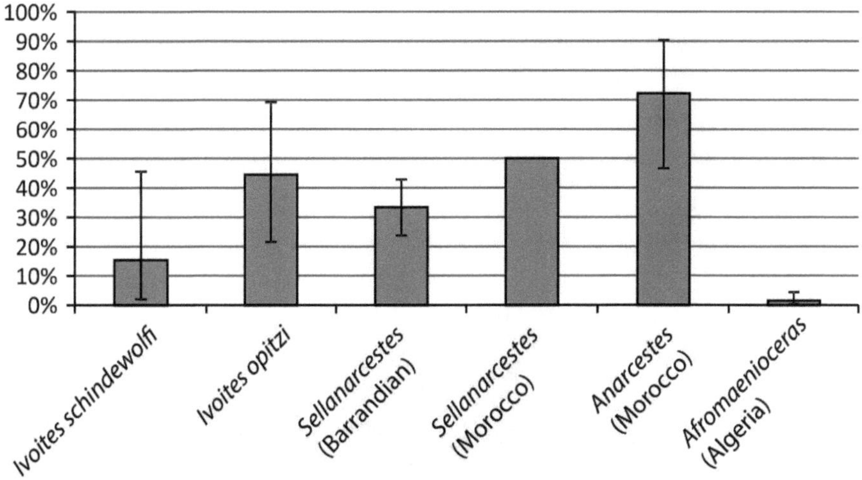

Fig. 20.6 Prevalence of Opitzian and Housean Pits in Early to Middle Devonian ammonoids. *Ivoites opitzi* and *I. schindewolfi* derive from the early Emsian, *Sellanarcestes* from the late Emsian; *Anarcestes* from the late Emsian to early Eifelian; *Afromaenioceras* from the Givetian. Data compiled from Chlupáč and Turek (1983) and De Baets et al. (2011). Note the large variability between taxa and within *Sellanarcestes* depending on the region (left: Barrandian, Czech Republic, right: Anti-Atlas, Morocco) as well as the low prevalence of Housean pits in Givetian *Afromaenioceras*. The 95% binomial confidence intervals (following Raup 1991 and De Baets et al. 2012) were calculated using the binom.confint function of the Binom Package in R (using the exact approach)

and Turek 2009; Turek and Manda 2010) and Late Jurassic ammonoids (Mironenko 2012), although they are never long-lasting and not regularly arranged in rows as with Devonian Housean or Opitzian pits (De Baets et al. 2011). It is unclear what causes their repeated and rather regular appearance in the Devonian, however, it might be linked to an episodic release of new parasitic stages as these rhythms are not related with the growth rhythm of the ammonoid itself (e.g., apertural growth or septal spacing: De Baets et al. 2011, 2013b). Furthermore, the parasite might target certain tissues or organs which possibly contribute to their regular position or paired arrangement (compare De Baets et al. 2011).

Isolated blister pearls and their casts have been reported from the Devonian (*Cheiloceras*: Keupp 2012), Triassic (*Ceratites*: Kirchner 1927), and Jurassic (Keupp 1986, 2000, 2012; Mironenko 2012), but not all might be of parasitic origin. Blister pearls are also reported from extant *Nautilus* (Landman et al. 2001; Fig. 20.1e) and fossil nautiloids (Kieslinger 1926; Manda and Turek 2009), although their causes are not well investigated. They are mostly located in a lateral (Fig. 20.1e) or dorsal position on the shell (compare Kieslinger 1926). On internal moulds, casts of blister pearls could be confused with borings or dissolved epicoles which grew postmortem on the inner part of the shell (e.g., Miller 1938; House 1960; Keupp 2012). A boring in the shell would, however, result in a positive structure in the internal mould rather than in a negative structure (De Baets et al. 2013b). The best evidence

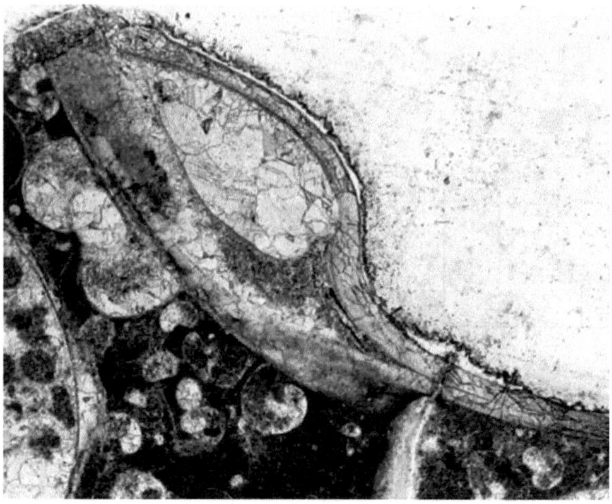

Fig. 20.7 Cross section through a blister pearl from a *Dactylioceras anguinum*, Toarcian (Jurassic), Altdorf near Nürnberg (Germany), PA-1696 (modified from Keupp 2012). Note, the egg shaped cavity (dm 3 mm) within the blister pearl, now filled with cement, interpreted to be the original shape of overgrown parasite (compare Keupp 1986, 2012)

for pearls derives from specimens where the shell, the blister and potentially even the irritant are preserved together. Inorganic material and fossilized parasite hard parts can be preserved in the nucleus of these pearls, while soft-bodied remains of parasites can be erased during the pearl formation process (Lauckner 1983).

Keupp (1986) was the first to describe blister pearls in ammonoids and to discuss their parasitic nature for at least some blister pearls in the Jurassic. He described similar indentations along with the adhering shell and egg-like concretions in two specimens of *Dactylioceras anguinum* from the Toarcian of Germany (Fig. 20.1, 20.7). The strange elliptical concretions are attributed to an overgrown parasite (Keupp 2012), potentially parasitic flatworms (Keupp 1986). These particular types of blister pearls were recently also reported from the Tithonian of Russia (*Kachpurites*: Mironenko 2012) which extended their range from the Early into the Late Jurassic.

20.5.4 Volume and Ornamentation-Enlarging Pathologies

Temporary increases in the volume of the shell and ornamentation have been linked with parasitism by Keupp (1976) who described all these phenomena as *forma inflata*. Subsequent authors have consistently attributed these temporary swellings to mantle tissue infections and infestations caused by parasites (Keupp 1976, 2000, 2012; Hengsbach 1979b, 1991a, 1996; Kröger 2000). Nevertheless, this needs to be further corroborated by studies on extant bivalves or other shelled molluscs. Since the work of Kröger (2000) two main types have been defined: shell volume-enlarging pathologies (*forma inflata* Keupp 1976; Fig. 20.2c) and ornamentation-enlarging pathologies (*forma augata* Kröger 2000; Fig. 20.2a, b, d).

Keupp (1976, 2000) described a pathological specimen of *Amoeboceras* (Fig. 20.1a) in which the crenulated keel temporally and progressively developed several large protuberances (superficially similar to the keel in *Creniceras*), which he attributed to a short-term infestation of the ventral apertural mantle tissue by parasites. Keupp (1984, 1997, 2000) described a similar phenomenon in *Dactylioceras* (Fig. 20.1d); in the material described therein, the ribs suddenly and progressively developed into large shovel-like bands on the venter which subsequently return back to normal ribs.

Such malformations which progressively develop and return to normal, have only been reported from single specimens of taxa from the Early and Late Jurassic which have been related to the parasitic infestation of the ventral mantle epithelium (Keupp 1976; Hengsbach 1996). The appearance of this anomaly seems to be dependent on the type of ornamentation and its development program (e.g., transformation of the crenulated keel to large protuberances in *Amoeboceras* or the enlargement of ventral ribs to successive shovel-like bands in *Dactylioceras*; compare Keupp 2000). The increase in development of ornamentation might give rare specimens a polygonal outline (Keupp 2012), although this should not be confused with other types of pathologies resulting in a polygonal shape possibly linked to endogenic causes (*forma polygonia* Hüne and Hüne 2006) or taxa where a polygonal whorl is normal and all specimens have it (e.g., triangular *Soliclymenia* from the Late Devonian: Korn et al. 2005b or the tetrangular *Entogonites* from the Carboniferous: Korn et al. 2005a). Fernandez-Lopez (1987, pl. 1, Fig. 1) figured a specimen of *Bajocisphinctes* with a similar pathology which can be attributed to the struggle between ammonoid and parasite (Hengsbach 1996). Keupp (2012) described phenomena resulting from the single swelling of ornamentation from the Late Triassic (*Arcestes*), Early Jurassic (*Pleuroceras*), Middle Jurassic (*Quenstedtoceras*; Fig. 20.2b), and Late Jurassic (*Orthospinctes*). The normal ornamentation surrounding these structures suggests that the mantle tissue at the apertural margin was probably not infected. These phenomena were originally described as *forma inflata*, but Kröger (2000) introduced the *forma augata* to refer to this ornamentation-enlarging phenomena, particularly in case of multiple swellings. Further examples of *forma augata* in Quenstedtoceratinae were also figured by Keupp (1985), Seltzen (2001, 2009) and Larson (2007).

The *forma inflata* should be restricted to conspicuous, temporary bulbous swellings of the shell behind the aperture which are typically associated with the regeneration of external injuries (Kröger 2000; Keupp 1995, (2006, 2012; Fig. 20.2c). The swellings are mostly smooth indicating that the mantle tissue at the aperture was not involved in their formation, although in rare cases, damages to the apertural mantle tissue might have resulted in phenomena similar to *forma augata* during further growth. Such temporary swellings were first figured by Lehmann (1975) which he related to the mantle protruding considerably outside of the shell after some injuries behind the aperture. Keupp (2012) pointed out that Lehmann's explanation would be rather unlikely because of the consistency of the mantle musculature as well as the offset between the injuries and the bulbous swellings. According to Keupp (1976, 1995, 2000, 2006, 2012), the temporarily exposed mantle tissue (as consequence of an injury) was more prone to infection and infestation by parasites, thus resulting in the temporary swelling. The *forma inflata* is known from a wide variety of taxa (Lehmann 1975; Keupp 1976, 1995, 2000, 2006, 2012; Hengsbach 1996; Kröger 2000) from the Early to Late Jurassic (e.g., *Dactylioceras*, *Pleuroceras*, *Rehmannia*, *Divisosphinctes*, *Orthosphinctes*: Fig. 20.2c) and the Late Cretaceous (e.g., "*Jeletzkytes*", now synonymized with *Hoploscaphites* by Landman et al. 2010).

Keupp (1995, 2012) also described additional volume increasing phenomena which might be linked to parasitism. They correspond with a gradual thickening of one side of the whorl resulting in significant left-right asymmetry of the whorl section which could be interpreted as volume enlargement of the soft-tissue along one side of the body in response to a parasitic infestation and/or tumor-like swelling (extant cephalopods are not believed to develop real tumors: Sparks 1972). So far, these unilaterial swellings have been reported from a *Pleuroceras* (Early Jurassic: Keupp 1995; Fig. 20.2e, f) and a "*Jeletzkytes*" (Late Cretaceous: Keupp 2012).

20.5.5 Symmetropathies

Hengsbach (1991b) introduced the term "symmetropathy" to refer to pathological asymmetry or deviations from the plane of bilateral symmetry. Several symmetropathies in both the phragmocone and conotheca in absence of external injuries or epizoa have been attributed to parasitism with various degrees of certainty (Hengsbach 1991a, 1996; Keupp 2000), including asymmetry in the position of the keel or the entire shell as well as asymmetry of the siphuncle and ventral lobe.

Rieber (1963) was the first to interpret a parasitic cause for a paleopathy he observed in *Cardioceras* with an asymmetrically situated keel (*forma juxtacarinata* Hölder 1956; compare Fig. 20.2m) and siphuncle (*forma juxtalobata*). The formation of the septa by the mantle is spatially and temporarily separated from the formation of shell material at the aperture, so that pathologies that affected both the phragmocone and the aperture are usually rare (Keupp 2012). Rieber (1963) interpreted this lateral displacement of the keel as being the result of an infestation of the ventral mantle by parasites. A parasitic infestation seems plausible as the *Cardioceras* specimen showed a lateral displacement of both medial elements after at least three normal whorls without indications for external injuries and with a progressively increasing degree of deviation of the keel (and the siphuncle) adaperturally. Several authors have discussed similar cases and followed Rieber (1963) in attributing similar deviations in symmetry of the ventral elements of the conotheca to parasitic infestations (Bayer 1970; Hölder 1970; Hengsbach 1991a, 1996; Keupp 2000, 2012).

Heller (1958) described a *Pleuroceras* with an asymmetrical keel that shows multiple oscillations and only normalizes towards the end of the body chamber. Heller (1958) dubbed this phenomenon *forma undaticarinata* (Fig. 20.2g, h). It can probably be best explained by a temporary parasite infestation of the ventral mantle epithelium. This is supported by the fact that it can be associated with a flattening and asymmetrical appearance of the keel crenulation, which is reminiscent

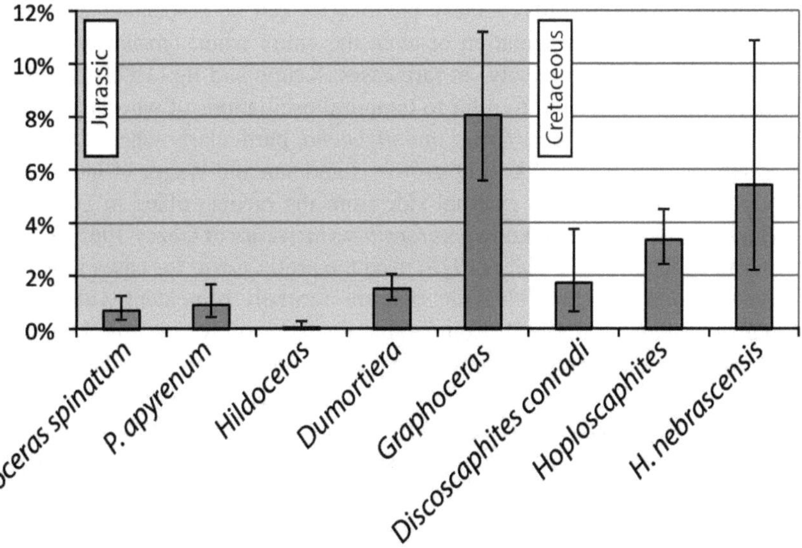

Fig. 20.8 Prevalence of Morton's syndrome and similar phenomena (*forma undaticarinata*) in Early Jurassic to Late Cretaceous ammonoids (in chronological order). Data compiled from Morton (1983), Landman and Waage (1986) and Keupp (2012). The 95% binomial confidence intervals (following Raup 1991 and De Baets et al. (2012) were calculated using the binom.confint function of the Binom Package in R (using the exact approach).

of pathologies dubbed *forma cicatricocarinata* (Heller 1964) caused by injuries to the ventral mantle epithelium as well as unilateral temporary disappearance of the groove surrounding the keel. This phenomenon is not only known from the Amaltheidae, but also from the Harpoceratidae and Hildoceratidae (Fig. 20.2g, h) as well as potentially other taxa with keeled or sharp venters (Keupp 2012). Rare quantitative analyses (Fig. 20.8) show a 10 times higher prevalence of this pathology in *Pleuroceras* than in *Hildoceras* (compare Keupp 2012). The keel can maximally deviate about 90° from the planispiral position and if it exceeds this value, it can result in one or multiple reestablishments of the keel and an associated chaotic, zigzag pattern of the ornamentation (*forma choatica* of Keupp 1977).

Landman and Waage (1986) introduced the term Morton's Syndrome to refer to forms of shell asymmetry in which the midline of the shell venter is deflected to the right or left of the plane of symmetry (Fig. 20.2i, j, k, l, 20.8). The name was based on the work of Morton (1983), who described a high proportion (8.1%) of both macro- and microconchs of *Graphoceras* from the Aalenian, whose whorls grew over to one side resulting in unilateral deformation of the whorl cross section and an excentrical position of the planispiral plane after initial normal growth. Morton (1983) attributed this displacement to parasitic infestation or disease, an opinion shared by Hengsbach (1991b, 1996) and Keupp (2000, 2012). It can result in a bowl-shaped morphology which could be described with the term *forma excentrica* (Hölder 1956) and can also be caused by *in vivo* encrustations of epizoa.

According to Keupp (2012), these pathologies can be associated with counteroscillations of the ornamentation or even the entire whorl (*forma undatacarinata* or *undaticoncha*, respectively) in rare cases. Keupp and Ilg (1992) introduced the term *forma undatispirata* to refer to temporal oscillations of whorls, while Hengsbach (1996) used the term *forma undaticoncha*, particularly when associated with oscillations of the ventral keel or groove. Landman and Waage (1986) described a deviation of the flattened external side from the median plane in *Discoscaphites* and *Hoploscaphites* (also known as *forma juxtasulcata* of Gezcy 1965) as Morton's syndrome (Fig. 20.8). Hölder (1970) used the prefix *juxta-* for cases in which there is a separation of medial elements that are normally coincident with one another (compare Hengsbach 1996). Comparable deviations have become known from a wide variety of taxa (reviewed by Keupp 2012) with mostly planulate to discocone conchs ranging from the Early to Middle Triassic (*Pseudosageceras*, *Columbites*, *Tropigastrites*), over to the Early to Middle Jurassic (Amaltheidae, Graphoceratidae such as *Graphoceras*, Hildoceratidae such as *Cleviceras, Dumortiera* or *Pleydellia*, Cardioceratidae such as *Quenstedtoceras*) to Cretaceous (*Deshayesites*: Doguzhaeva et al. 1990; *Saynoceras:* Ploch 2007). Similar pathologies might be present in the Devonian as well (compare Bockwinkel et al. 2013, Fig. 6D for a report of a specimen of *Pseudoprobeloceras pernai* with possible Morton's Syndrome), but the deviations from normal coiling of the inner whorls in this case might also be related with external injuries or epizoa (compare Klug and Korn 2001) which are now overgrown and not detectable anymore.

The proportion of these pathologies is variable between populations and localities (0.05– 8.1%; Fig. 20.8). The high proportion of these structures in some populations (e.g., *Graphoceras* of the Isle of Skye: Morton 1983; *Hoploscaphites nebrascensis*: Landman and Waage 1986; Fig. 20.8) might point to a large parasite population at certain localities (e.g., Keupp 2012), although they might also be related with other factors such as a particular ecology (food, mode of life) or an oversensitivity of a population or that these ammonoids were false or facultative hosts at some sites.

In most ammonoid taxa, the midventral position of the siphuncle and of the ventral lobe of the suture line is constant, however, in some taxa the ventral placement can be quite variable or its lateral displacement is even species-specific (Keupp 2012; Fig. 20.9) and it can also change towards the end of ontogeny through the asymmetrical growth of organs (Yacobucci and Manship 2011). This phenomenon has been documented from the Devonian to the Cretaceous (Ziegler 1958; Kemper 1961; Hengsbach 1977a, 1977b, 1977c, 1978, 1979a, 1986a, 1986b; Landman and Waage 1986) with differing prevalences in separate taxa and populations. It also shows no clear boundary (smooth transition) between pathological and normal development of this asymmetry (Keupp 2012; Fig. 20.9). Hölder (1956) introduced the term *forma juxtalobata* (Fig. 20.2n) to refer to the pathological ventral displacement of the siphuncle and the ventral lobe of the suture to one side which should only affect a certain portion of the ammonoid population. According to Keupp (2012), only a small fraction of ammonoid specimens should be affected by parasites (~1%) and show pathologies, while others like Hengsbach (1996) still attribute a significantly higher prevalence of these pathologies (up to 70%) to specialized parasitic infection. In the absence of external injuries

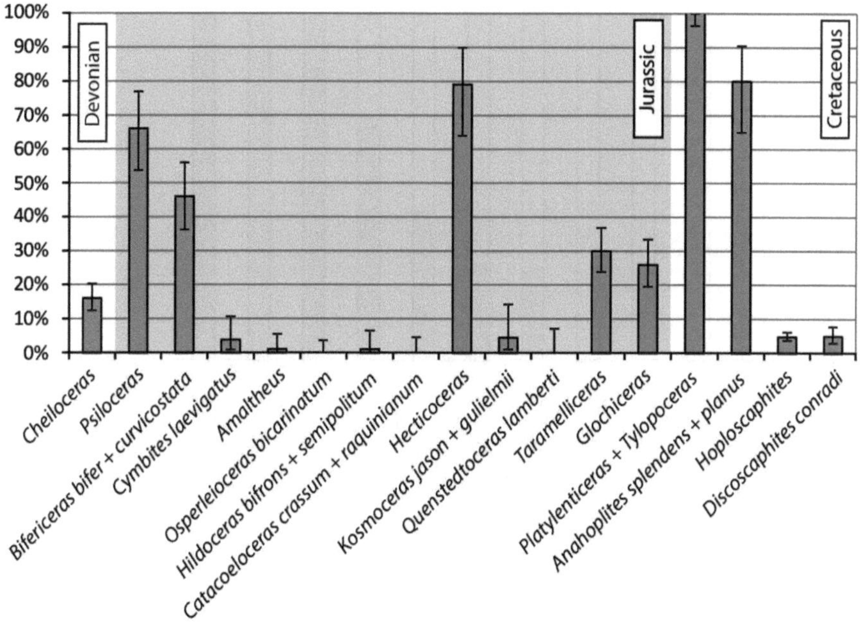

Fig. 20.9 Prevalence of asymmetry of the ventral lobe and siphuncle (not restricted to pathological cases: *forma juxtalobata*: see discussion in text) in Devonian to Cretaceous ammonoids (in chronological order). Data derive from Ziegler (1958), Hengsbach (1976, 1977a, 1977b, 1977c, 1978, 1979a, 1980, 1986a, 1986b) and Landman and Waage (1986); fide Keupp (2012). The 95% binomial confidence intervals (following Raup 1991 and De Baets et al. 2012) were calculated using the binom.confint function of the Binom Package in R (using the exact approach).

or developmental disorders, the *forma juxtalobata* could potentially be related with various endogenic diseases such as those caused by parasitic infestations (Hengsbach 1991b, 1996; Keupp 2012). Hengsbach (1986a, 1986b, 1991a, 1996) argued that at least some sutural asymmetry-paleopathies in Jurassic taxa were caused by parasitism. He attributed them to an infestation of the septal mantle at or near the siphuncle of young ammonoids (potentially resulting in an asymmetrical swelling) which caused a displacement of the septal root and hence the ventral lobe. According to Keupp (2000, 2012), a pathological cause for the asymmetry of septa in several groups of Jurassic ammonoids with a high proportion of asymmetry (including the cases discussed by Hengsbach) are still speculative.

This is related to the fact that the direction and degree of asymmetry of the suture line seems to be constant (genetically fixed?) in some taxa to highly variable in other taxa (Keupp 2012). Furthermore, a parasitic (or a different pathological) cause appears rather unlikely in many cases as the asymmetry of the suture often lacks a gradual development and can show a continuum between a very low to very high prevalence within some genera (Keupp 2000, 2012; Fig. 20.9, 20.10). Hengsbach (1986a) reported a 26% prevalence of asymmetry in *Glochiceras*, while Keupp (2012, p. 253), based on the material of Ziegler (1958), stated that the prevalence within this taxon differs between species from 0–100% (compare Keupp 2012, p. 253; Fig. 20.10).

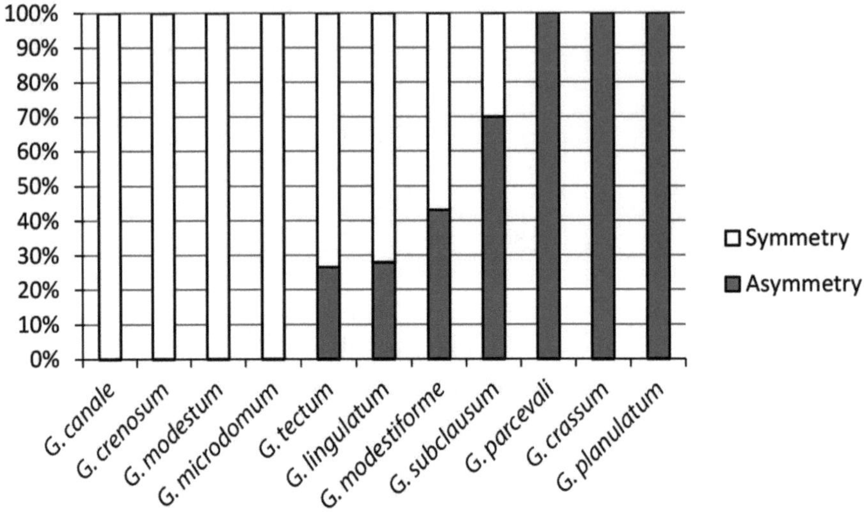

Fig. 20.10 Prevalence of asymmetry of the ventral lobe and siphuncle in *Glochiceras* as listed in Keupp (2012; compare Ziegler 1958)

Exceptions include the specimen described by Rieber (1963) and potential similar cases, where both the siphuncle and the keel progressively develop a lateral displacement after a certain amount of normal coiling as discussed above. Extant *Nautilus* is of little help to interpret the asymmetry of the ventral lobe because its siphuncle is located centrally, so that the parasitic cause remains unproven.

Parasitism remains attractive to explain asymmetry, even in the buccal mass, where Schweigert (2009) suggested parasites to be responsible for assymmetrically arranged deformations in anaptychi without an apparent orientation preference (e.g., Schweigert and Dietl 2001). Kruta and Landman (2008) investigated injuries and anomalies in *Nautilus* jaws which are probably mostly related to diet or mating behavior. They hypothesized that one type of anomaly might potentially be linked to parasitism, however, only parasitic copepods have been reported from *Nautilus* in the wild (see 3.6.3) which are not known to cause these pathologies.

20.5.6 Other Pathologies attributed to Parasitism

Some other pathologies without clear signs for external injuries or epizoa have also been attributed to parasitism or more generally to endogenic causes (see Hengsbach 1996; Keupp 2012 for a more general reviews). Shell lamellae (*forma aptycha* of Keupp 1977; *forma conclusa* of Rein 1989; Fig. 20.1d) have also been discussed to encapsulate foreign bodies (Lehmann 1990, p. 194) such as parasites (Rein 1989) or an injured or diseased area. Such secondary shell lamellae (Keupp 1977; Rein 1989; Lehmann 1990; Rein 1994; Keupp 1994, 1996, 1998, 2000, 2012) have been reported from several ammonoid taxa ranging from the Triassic (*Ceratites, Nevadites*) to the Jurassic (*Chondroceras, Dactylioceras, Elatmites, Kosmoceras, Pleuroceras, Sigaloceras, Virgatites*) and in to the Cretaceous (*Audouliceras, Pavlovia/Strajevskya*).

Similar lamellae have been reported from extant captive *Nautilus* (Keupp and Riedel 1995; Keupp 2012, p. 231–234; Fig. 20.1c), where the shell material is secreted by the mantle and a local detachment of the mantle and shell occurs (Keupp 2012). No parasitic cause is necessary to explain these structures and according to Keupp (2012), they are mostly related with external injuries. However, parasitic or other endogenic causes cannot be completely ruled out if no obvious injury can be found (compare Keupp 2012, p. 161).

The temporary loss or diminution of ornamentation (described by Lange 1941 as *forma cacoptycha*) might also be related to endogenic causes (Keupp 2012), including parasitism or metabolic-physiological disorders (Hengsbach 1996); however, similar phenomena are also known during regeneration of injuries. Hengsbach(1996) suggested that parasitism could have been involved in cicatrisations of the crenulated keel (*forma cicatricocarinata* of Heller 1964) and Keupp (2012) suggested that endogenic causes might be involved in keel-like raised cicatrices along the venter and flanks of some ammonoids (*forma pseudocarinata* of Fernandez-Lopez 1987), although more comparative studies and data are necessary to corroborate these hypotheses.

20.5.7 Negative Effects of Bioerosion and Epizoa

Parasitism should not be confused with other symbiotic or long-term associations such as epizoa and bioerosion which can also affect the growth of ammonoids (but do necessarily bring advantage to the bioeroding or encrusting agents). Epizoa (Fig. 20.11, 20.12) can form long-term associations and may cause damage or influence growth of ammonoids (e.g., Davis et al. 1999; Klug and Korn 2001; Checa et al. 2002; Larson 2007; Keupp 2012). They are, however, not necessarily parasites since many grew on shells of both living and dead ammonoids as sclerobionts or epicoles (Fig. 20.11, 20.12; Keupp and Hoffman 2015), colonizing their floating shell or their shell when it is already deposited on the seafloor forming a benthic island (Seilacher 1982; Paul and Simms 2012; De Baets et al. 2013b). We follow the terminology suggested by Davis et al. (1999) to refer to epizoa as organisms living on other organisms while both are alive, while we will use epicoles to refer to organisms that live on a hard substrate or shells also when the host is already dead and gone. *In vivo* encrusters of ammonoid shells are known from the Devonian (Klug and Korn 2001) to the Cretaceous (reviewed by Keupp 2012) and include algae, foraminifers, tabulate corals, bryozoans, brachiopods, annelids, lepadomorphs, gastropods, bivalves, and crinoids (see also Rakociński 2011). Various other organisms ranging from fungi to cystoids have also been documented to encrust ammonoid shells, but so far there is no evidence that this happened during their life. Some re-

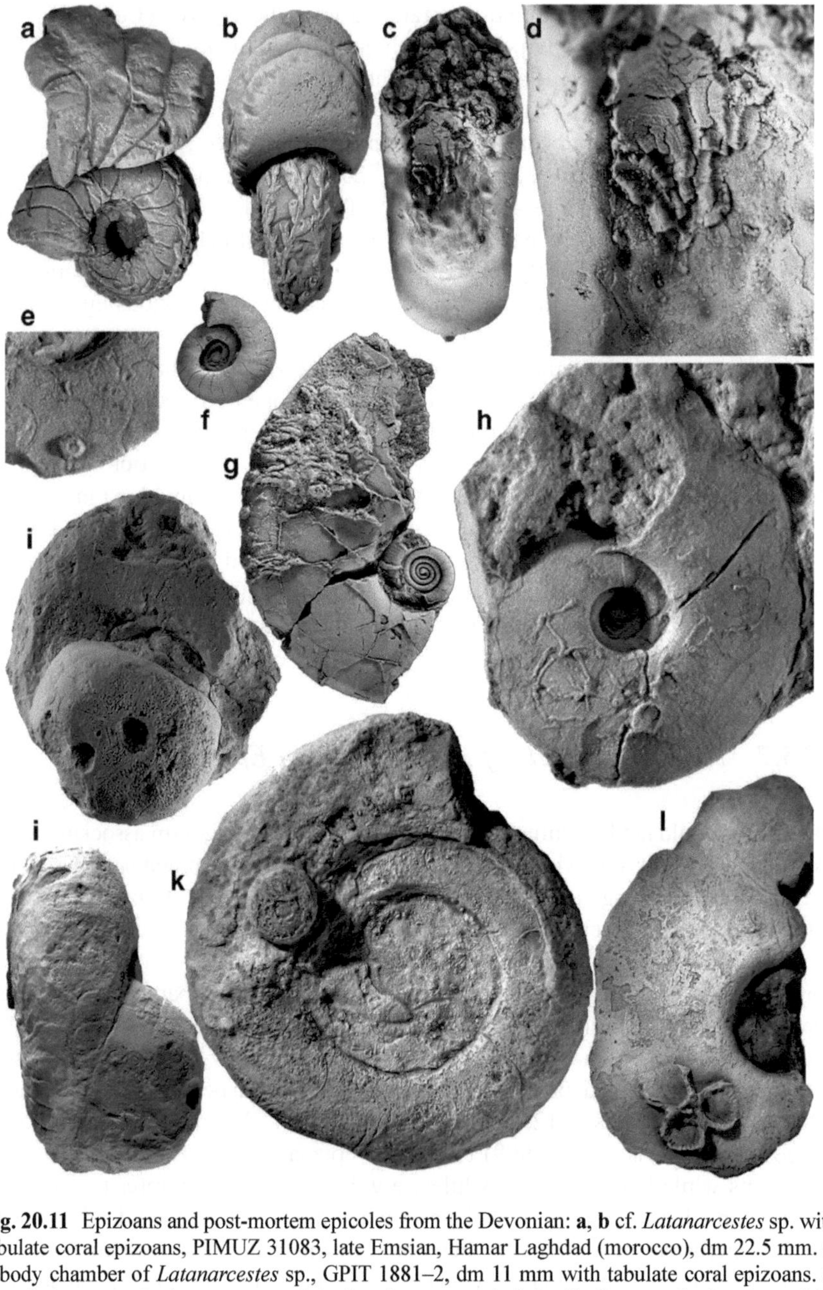

Fig. 20.11 Epizoans and post-mortem epicoles from the Devonian: **a**, **b** cf. *Latanarcestes* sp. with tabulate coral epizoans, PIMUZ 31083, late Emsian, Hamar Laghdad (morocco), dm 22.5 mm. **c**, **d** body chamber of *Latanarcestes* sp., GPIT 1881–2, dm 11 mm with tabulate coral epizoans. **c**, dorsal view, note the last septum and the imprint zone, x 4. **d** detail of **c**, note the imprints of the overgrown corals. **e**, detail of a *Endosiphonites muensteri* with a crinoid epicole, GPIT 1850–10, late Famennian, Ouidane Chebbi (Morocco), dm 72 mm, x 1. **f** *Paranarcestes chalix*, elliptical coiling due to epizoans, GPIT 1871–206, late Emsian, Ouidane Chebbi (Morocco), dm 7.9 mm. **g** *Rherisites tuba* with tabulate coral epicoles, GPIT 1869–7, late Emsian, Jebel Mech Agrou, Tafilalt (Morocco), dm 52.2 mm. **h** *Cymaclymenia involvens* with auloporid coral epicole (? *Cladochonus*

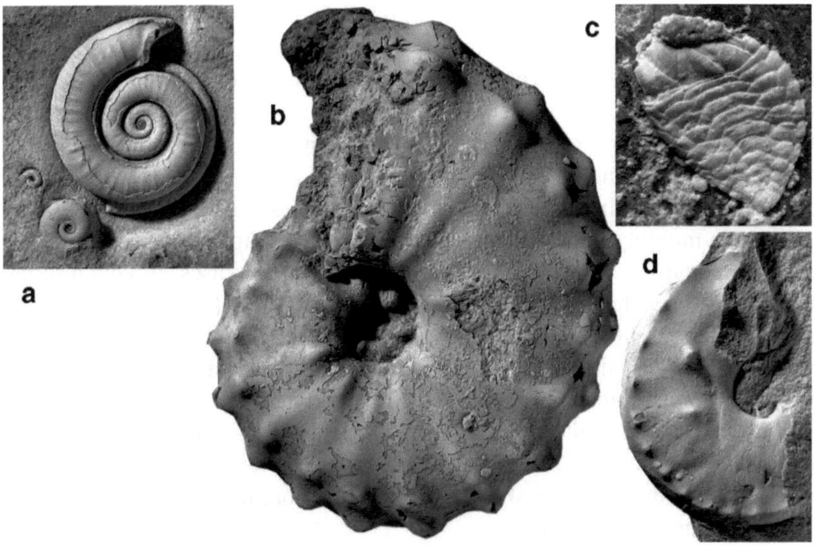

Fig. 20.12 Epizoans and post-mortem epicoles from the Mesozoic: **a** *Arnioceras miserabile*, with a serpulid epizoan, which forced the ammonite to alter its shell morphology, Early Sinemurian, Semicostatum Zone, Charmouth, Dorset (UK), dm 30 mm. **b**, **c** *Mammites nodosoides* with four specimens of the crustacean epicole *Stramentum* sp., Turonian, Goulmima (Morocco), dm 75 mm. **d** *Ceratites pulcher* with 15 juvenile brachiopod epicoles of *Discinisca discoides*, PIMUZ 31080, Anisian, Triassic, dm 37 mm

ports of epicoles might be actually conellae (inorganic growths) which have sometimes been confused and wrongly described (Keupp 2012) as barnacle-like epicoles (Maubeuge 1949, Gerasimov 1955) or limpet gastropods (Quenstedt 1884; Busse 1976). When correctly recognized, epizoa can provide important information on ammonoid growth (Bucher et al. 1996) and ecology (Seilacher 1960; Keupp et al. 1999) including shell orientation in the water column (Hauschke et al. 2011; Ritterbush et al. 2014). *In vivo* epizoism can be easily recognized when epizoa are attached on both sides or overgrown by the ammonoid shell (Paul and Simms 2012) which often influence the growth of the ammonoid resulting in changes in shell morphology such as asymmetry of the whorl section and deviations from normal coiling (Checa et al. 2002). When the final size is reached, *in vivo* epizoism can only be recognized by a preferential orientation of epizoa to currents (Seilacher 1960). Multiple generations of epizoa showing a clear size gradation and exceptional preservation of ammonoid remains (presence of buccal mass) might also point to the

sp.), GPIT 1850–22, late Famennian, Madene El Mrakib (Morocco), dm 60 mm. **i**, **j** lateral and dorsal view of *Chlupacites praeceps* with the cystoid epicole *Eucystis* sp., GPIT 1881–5, late Emsian, Tazoulait, Morocco, dm 46 mm. **k** *Sellanarcestes* cf. *tenuior* with the cystoid epicole *Eucystis* sp., late Emsian, Filon 12 (Morocco), dm 64 mm. **l** *Cymaclymenia* sp. with a coral epicole (Cleistoporidae gen. et sp. indet.), late Famennian, Lambidia (Morocco), dm 60 mm. Images **c** to **j** from Klug and Korn (2001) fact that organisms encrusted the ammonoid shell during its lifetime (Keupp et al. 1999). Multiple generations of epizoa would take a while to grow superseding the phase of post-mortem drift. After a long postmortem drift or slow burial, one would not expect the exceptional preservation with *in situ* buccal masses (compare Wani 2007; Keupp 2012) and renal concrements or uroliths.

The type of relationship between the epizoa and the ammonoids probably varied from taxon to taxon. In some cases, only the ammonoid might have had disadvantages as his mobility was constrained by increased drag, additional weight, and the influence on growth because of the epizoa. In some cases, *in vivo* infestation brought negative effects for both the epizoa and the host (compare Larson 2007; Keupp 2012), for which Meischner (1968) introduced the term "*perniciöse Epökie*". This was particularly the case for epizoa which could not change their relative position after fixation on the substrate or which grew longitudinally, because they soon ended up away from their preferred current orientation and were eventually overgrown by the ammonoid shell (Keupp 2012, p. 183). The epizoa in turn influenced ammonoid growth as well as shell shape and therefore the orientation and drag in the water column of the shells.

Whether or not carbonate boring fungi, found as trace fossils (Wetzel 1954; Schindewolf 1962, 1963; Wetzel 1964; Keupp 2012) and more rarely as body fossils (Weitschat 1986; Lehmann 1990) in ammonoid shells, already infested ammonoids during their life and can be treated as parasites is still debated (Keupp 2012). Algae and some parasites are also known to be actively involved in bioerosion of shelled mollusks (e.g., the foraminifer *Hyrrokin*: Beuck et al. 2008). These and other microborers could already have infested the ammonoid during their lifetime (Schindewolf 1962, 1963; Wetzel 1964), but often do so after death (Dullo 1981; Keupp 2012). The traces described by Schindewolf (1962) as *Mycelites* from ammonoid shells are definitively microborings, although this ichnogenus and its ichnospecies are no longer applied following the invention of the cast-embedding technique (Wisshak and Tapanila 2008). The study of microborings has greatly advanced and a lot of ichnotaxa have been erected that were formerly subsumed under terms like *Mycelites*. Most of the traces are reminiscent of *Orthogonum lineare* (Glaub 1994) which are produced by a heterotrophic organism based on its distribution down to aphotic depths and fungi have been tentatively assigned to be the most likely producer. Not all traces reported by Schindewolf are produced by heterotrophic organisms (Max Wisshak, personal communication 2012). However, the hypothesis of Schindewolf (1962, 1963) and Wetzel (1964) of a *syn vivo* infestation of the microborers is not inconceivable and has been reported from oysters, balanids, and serpulids (Max Wisshak, personal communication 2012). The periostracum might, however, serve as a barrier for many microendoliths, but not for all. Some microborings have been described, which even specifically penetrate or entirely dwell within the periostracum (e.g, in the deep-sea bivalve *Bathymodiolus*: Hook and Golubic 1988, 1992). Bioerosion already starts during the animal's lifetime in extant *Nautilus* (Seuss et al. 2015). So far, no direct evidence for a *syn vivo*-infestation of ammonoids (such as active countermeasures) have been documented, so it appears more reasonable to assume that in many cases, the boring traces were mostly formed postmortem unless demonstrated otherwise (Keupp 2012).

20.6 Conclusions and Future Perspectives

Pathologies in ammonoid shells show that they were frequently infested- and affected by parasites of various kinds. Among the most convincing manifestations of such parasitoses are blister pearls (Devonian-Jurassic: Keupp 1986; De Baets et al. 2011), asymmetry of the shell in absence of external injuries or epizoa (Triassic-Cretaceous: Rieber 1963; Morton 1983; Keupp 2012), and disturbances in shell growth (Jurassic: Keupp 1979; Kröger 2000) which is corroborated by comparative studies on extant and fossil shelled mollusks. Pathological gigantism (Carboniferous: Manger et al. 1999) and pathological enlargement of shell volume or ornamentation (Triassic-Cretaceous: Keupp 1976; Kröger 2000) could also be related with parasitism, although more evidence is necessary to further confirm these hypotheses. Only in some cases (e.g., Rieber 1963), asymmetry of the suture line (Devonian-Cretaceous: Keupp 2012), could potentially also be related with parasitoses. Parasitic infestations were more widespread than suspected from counting such paleopathies in the ammonoid shell because most soft-tissue parasites did not leave direct traces in the fossil record and only a fraction caused shell growth pathologies in their hosts. For the same reasons, the identity of the parasites causing these pathologies is so far mostly unknown. Parasitic flatworms were among the likely suspects which is based on extant parasite-host relationships, their high prevalence in extant coleoids, and studies of similar pathologies they cause in other externally shelled mollusks today. Various other extant and even extinct lineages of parasites with similar behavior might also have been responsible. Furthermore, many common cephalopod parasites like bacteria, helminths, as well as crustaceans living in gills of both Recent coleoids and *Nautilus* have a low fossilization potential. The high concentration or restriction of these anomalies to certain taxa or timeframes (Housean pits in Devonian Anarcestina, Pharciceratina and Tornoceratina: De Baets et al. 2011, "*forma umbilicata*" in Dactylioceratidae of the Early Toarcian: Keupp 1979, 2000, 2012; particular cases of "*forma augata*" in Quenstedtoceratidae of the Early Callovian: Seltzer 2001, 2009; Larson 2007; Keupp 2012) suggests parasitic culprits with a high degree of host specificity and makes them potentially an important tool for investigating coevolution in deep time (e.g., De Baets et al. 2011). It has been suggested that the presence of blister pearls in the earliest ammonoids might also indicate the presence of complex parasite life cycles and food webs already in the Devonian because extant parasitic flatworms inducing pearls are transmitted from intermediate host to final hosts by feeding (compare Fig. 20.13). Keupp (2000) suggested that the prevalence of possible trematode-induced pearls in Jurassic *Dactylioceras* might corroborate their planktic lifestyle as intermediate hosts today are mostly mollusks feeding on plankton. However, more studies on the distribution of pathologies are necessary to confirm such hypotheses, as they might not only give important information on the evolutionary history of their hosts but also on their mode of life and predator-prey relationships. Such hypotheses can only be adequately corroborated by finding exceptionally preserved parasites in ammonoid soft parts which is highly implausible, but not impossible (cf. Klug et al.

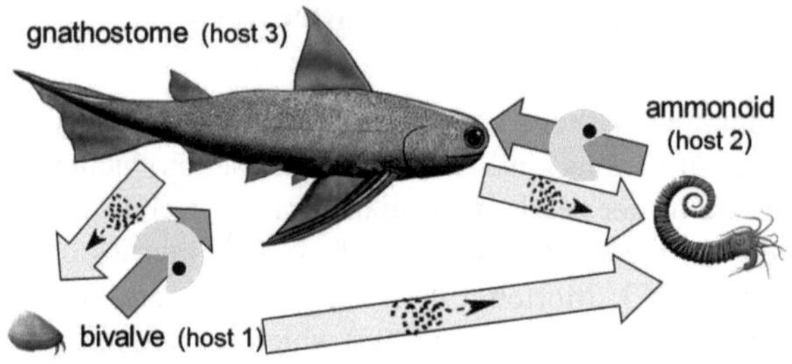

Fig. 20.13 Possible complexity of parasite life cycles and food webs in the Devonian. Infestation of bivalves and ammonoids is based on the presence of blister pearls, possibly induced by intermediate life stages parasitic flatworms or other parasites with similar behaviours (De Baets et al. 2011). Infestation of gnathostomes is indicated by the presence of parasitic attachment organs in gill region and abdomen of Late Devonian acanthodians and placoderms attributed to parasitic flatworms (Upeniece 2001, 2011)

2012). Further comparative work on pathologies in extant and fossil cephalopods and other molluscs with an accretionary shell, particularly bivalves and gastropods, can also be useful to identify their parasitic causes and the behavior of the culprit. The influence of *in vivo* epizoa and bioerosion during the lifetime of ammonoids also needs to be further investigated.

Acknowledgements Some of the materials and studies used for this contribution resulted from research which was carried out in the research projects with the numbers 200021–113956⁄ 1, 200020–25029, and 200020–132870 funded by the Swiss National Science Foundation SNF. We would like to thank: Matthias López Correa (Geozentrum Nordbayern, Erlangen), Barbara Seuß (Geozentrum Nordbayern, Erlangen) and Max Wisshak (Senckenberg am Meer, Wilhelmshaven) for discussions on bioerosion in extant molluscs; Carlo Romano (Palaontologisches Institut und Museum, Zürich) for helping with obtaining literature; Aleksandr Mironenko (Moscow) and Mikhail Rogov (Geological Institute of Russian Academy of Sciences, Moscow) for help with Russian literature. We thank the reviewers Arnaud Brayard (Université de Bourgogne, Dijon), Isabella Kruta (AMNH, New York) and Joshua Slattery (University of South Florida) for their constructive comments and suggestions. John Huntley (University of Missouri) proofread an earlier draft of the manuscript.

References

Barord GJ, Ju C, Basil JA (2012) First report of a successful treatment of a mucodegenerative disease in the chambered nautilus (*Nautilus pompilius*). J Zoo Wildl Med 43:636–639

Bayer U (1970) Anomalien bei Ammoniten des Aaleniums und Bajociums und ihre Beziehung zur Lebensweise. Neues Jahrb Geol Paläontol Abh 135:19–41

Becker RT, House MR (1994) International devonian goniatite zonation, Emsian to Givetian, with new records from Morocco. Cour Forschungsinst Senckenb 169:79–135

Beuck L, Correa M, Freiwald A (2008) Biogeographical distribution of *Hyrrokkin* (Rosalinidae, Foraminifera) and its host-specific morphological and textural trace variability. In: Wisshak M, Tapanila L (eds) Current developments in bioerosion. Springer, Berlin

Binder H (2002) Fossile Perlen aus dem Karpatium des Korneuburger Beckens (Österreich, Untermiozän). Beiträge zur Paläontologie 27:259–271

Bockwinkel J, Becker RT, Ebbighausen V (2009) Upper Givetian ammonoids from Dar Kaoua (Tafilalt, SE Anti-Atlas, Morocco). Berl Paläobiol Abh 10:61–128

Bockwinkel J, Becker RT, Ebbighausen† V (2013) Late Givetian ammonoids from Hassi Nebech (Tafilalt Basin, Anti-Atlas, southern Morocco). Foss Rec 16:5–65

Boettger CR (1953a) Größenwachstum und Geschlechtsreife bei Schnecken und pathologischer Riesenwuchs als Folge einer gestörten Wechselwirkung beider Faktoren. Zoologischer Anzeiger (Suppl 17):468–487

Boettger CR (1953b) Riesenwuchs der Landschnecke *Zebrina* (*Zebrina*) *detrita* (Müller) als Folge parasitärer Kastration. Archiv für Molluskenkunde 82:151–152

Boucot AJ, Poinar GO Jr (2010) Fossil behavior compendium. CRC Press, Boca Raton

Brooks DR, McLennan DA (1993) Parascript: parasites and the language of evolution. Smithsonian Institution Press, Washington

Bucher H, Landman NH, Klofak SM, Guex J (1996) Mode and rate of growth in ammonoids. In: Landman NH, Tanabe K, Davis RA (eds) Ammonoid paleobiology. Plenum, New York

Busse E (1976). Eine Napfschnecke (Gastropoda, Cyclobranchia, Patellaceae) im Oberen Muschelkalk (Mittlere Ceratitenschichten/Ladin) Niederhessens. Geol Jahrb Hessen 104:5–7

Carlson BA (1987) Collection and aquarium maintenance of nautilus. In: Saunders WB, Landman NH (eds) Nautilus. Springer, Dordrecht

Castellanos-Martínez S, Gestal C (2013) Pathogens and immune response of cephalopods. J Exp Mar Biol Ecol 447:14–22

Checa AG, Okamoto T, Keupp H (2002) Abnormalities as natural experiments: a morphogenetic model for coiling regulation in planispiral ammonites. Paleobiology 28:127–138

Chlupáč I, Turek V (1983) Devonian goniatites from the Barrandian area. Rozpr Ustred Ust Geol 46:1–159

Combes C (2001) Parasitism: the ecology and evolution of intimate interactions. University of Chicago Press, Chicago

Conway Morris S (1981) Parasites and the fossil record. Parasitology 83:489–509

Conway Morris S (1990) Parasitism. In: Briggs DEG, Crowther PR (eds) Palaeobiology: a synthesis. Blackwell Science, Oxford

Cuvier G (1829) Le règne animal distribué d'après son organisation. Tome I. Imprimerie D'Hippolyte Tilliard, Paris

Cuvier G (1830) Considérations sur les mollusques, et en particulier sur les céphalopodes. Ann Sci Nat 19:241–259.

Davis RA, Mapes RH (1999) Pits in internal molds of cephalopods. Ber Geol BA 46:31

Davis RA, Landman NH, Dommergues J-L, Marchand D, Bucher H (1996) Mature Modifications and dimorphism in ammonoid cephalopods. In: Landman NH, Tanabe K, Davis RA (eds) Ammonoid paleobiology. Plenum, New York

Davis RA, Klofak SM, Landman NH (1999) Epizoa on externally shelled cephalopods. In: Rozanov AY, Shevyrev AA (eds) Fossil cephalopods: recent advances in their study. Russian Academy of Sciences Paleontological Institute, Moscow

De Baets K, Klug C, Korn D (2011) Devonian pearls and ammonoid-endoparasite co-evolution. Acta Palaeontol Pol 56:159–180

De Baets K, Klug C, Korn D, Landman NH (2012) Early evolutionary trends in ammonoid embryonic development. Evolution 66:1788–1806

De Baets K, Goolaerts S, Jansen U, Rietbergen T, Klug C (2013a) The first record of Early Devonian ammonoids from Belgium and their stratigraphic significance. Geol Belg 16:148–156

De Baets K, Klug C, Korn D, Bartels C, Poschmann M (2013b) Emsian Ammonoidea and the age of the Hunsrück Slate (Rhenish Mountains, Western Germany). Palaeontogr A 299:1–113

De Baets K, Bert D, Hoffmann R, Monnet C ,Yacobucci MM, Klug C (2015) Ammonoid intraspecific variability. In: Klug C, Korn D, De Baets K, Kruta I, Mapes RH (eds) Ammonoid Paleobiology: From anatomy to ecology. Springer, Dordrecht

Della Chiaje S (1825) Memorie sulla storia e notomia degli animali senza vertebre del regno di Napoli, 2. Napoli. Stamperia Della Societá Tipografica 1–4:185–444

Dentzien-Dias PC, Poinar G Jr, de Figueiredo AEQ, Pacheco ACL, Horn BLD, Schultz CL (2013) Tapeworm Eggs in a 270 Million-Year-Old Shark Coprolite. PlosOne 8:1–4, DOI: 10.1371/journal.pone.0055007

Doguzhaeva LA, Michailova IA, Kabanov GK (1990) Irregular forms of *Deshayesites* (Ancyloceratina) from the Volga Region near Uljanowsk. Trans Paleontol Inst 243:120–127 [in Russian]

Dullo C (1981) Zur Erhaltung und Mikrostruktur der Ammonitenschalen aus Unterstürmig (Ofr.). Geol Nordost-Bayern 31:75–83

Ebbighausen V, Becker RT, Bockwinkel J (2011) Emsian and Eifelian ammonoids from Oufrane, eastern Dra Valley (Anti-Atlas, Morocco)—taxonomy, stratigraphy and correlation. Neues Jahrb Geol Paläontol Abh 259:313–379

Engel MS (1894) Über kranke Ammonitenformen im Schwäbischen Jura. Nova Acta Leopold 61:327–384

Fernández-López SR (1987) Necrocinesis y colonización posmortal en *Bajocisphinctes* (Ammonoidea) de la Cuenca Ibérica: implicaciones paleoecológicas y paleobatimétricas. Bol Real Soc Esp His Nat Secc Geol 82:151–184

Geczy B (1965) Pathologische jurassische Ammoniten aus dem Bakony-Gebirge. Ann Univ Sci Budap Rolando Eötvös Nominatae Sect Geol 9:31–37

Gerasimov PA (1955) Guide-fossils of Mesozoic of the central regions of the European part of USSR. Part II. Echinodermata, Crustaceans, Vermes, Bryozoans and corals from the Jurassic deposits. Gosgeolyekhizdat, Moscow [in Russian]

Glaub I (1994) Mikrobohrspuren in ausgewählten Ablagerungsräumen des europäischen Jura und der Unterkreide (Klassifikation und Palökologie). Cour Forsch Senckenberg 174:1–324

González AF, Pascual S, Gestal C, Abollo E, Guerra A (2003) What makes a cephalopod a suitable host for parasite? The case of Galician waters. Fish Res 60:177–183. doi:10.1016/s0165-7836(02)00059–0

Götting K-J (1974) Malakozoologie. Grundriss der Weichtierkunde. Fischer, Stuttgart

Götting K-J (1979) Durch Parasiten induzierte Perlbildung bei *Mytilus edulis* L. (Bivalvia). Malacologia 18:563–567

Hanlon RT, Forsythe JW (1990) Diseases caused by microorganisms In: Kinne O (ed) Diseases of Marine Animals. Vol. III. Biologische Anstalt Helgoland, Hamburg

Hauschke N, Schöllmann L, Keupp H (2011) Oriented attachment of a stalked cirripede on an orthoconic heteromorph ammonite—implications for the swimming position of the latter. Neues Jahrb Geol Paläontol Abh 262:199–212

Haven N (1972) The ecology and behavior of *Nautilus pompilius* in the Philippines. Veliger 15:75–81

Heller F (1958) Gehäusemißbildungen bei Amaltheiden. Ein neuer Fund aus dem fränkischen Jura. Geol Blätter für Nordost-Bayern und angrenzende Gebiete 8:66–71

Heller F (1964) Neue Fälle von Gehäuse-Mißbildungen bei Amaltheiden. Paläontol Z 38:136–141

Hengsbach R (1976) Über Sutur-Asymmetrie bei *Cymbites laevigatus* (Ammonoidea; Jura). Senckenb Lethaea 56:463–468

Hengsbach R (1977a) Cheiloceraten (Ammon., Devon) mit asymmetrischen Phragmocon. Sitzungsber Ges Natur Freunde Berlin (N F) 17:69–72

Hengsbach R (1977b) Über die Sutur-Asymmetrie einiger Psiloceraten. Sitzungsber Ges Natur Freunde Berlin (N F) 17:59–67

Hengsbach R (1977c) Zur Sutur-Asymmetrie bei *Platylenticeras.* Zool Beitr 23:459–468

Hengsbach R (1978) Zur Sutur-Asymmetrie bei *Anahoplites.* Senck leth 59:377–385

Hengsbach R (1979a) Zur kenntnis der Asymmetrie der Ammoniten-Lobenlinie. Zool Beitr 25:107–162

Hengsbach R (1979b) Weitere Anomalien an Amaltheen-Gehäusen (Ammonoidea; Lias). Senck leth 60:243–251

Hengsbach R (1980) Über die Sutur-Asymmetrie bei *Hecticoceras* (Ammonoidea; Jura). Senck leth 60:463–473

Hengsbach R (1986a) Zur Kenntnis der Asymmetrie der Sutur-Asymmetrie bei Ammoniten. Senck leth 67:119–149
Hengsbach R (1986b) Ontogenetisches Auftreten und Entwicklung der Sutur-Assymmetrie bei einigen Psilocerataceae (Ammonoidea; Jura). Senck leth 67:323–330
Hengsbach R (1990) Studien zur Paläopathologie der Invertebraten. 1: Die Paläoparasitologie, eine Arbeitsrichtung der Paläobiologie. Senck leth 70:439–461
Hengsbach R (1991a) Studien zur Paläopathologie der Invertebraten III: Parasitismus bei Ammoniten. Paläontol Z 65:127–139
Hengsbach R (1991b) Studien zur Paläopathologie der Invertebraten II: Die Symmetropathie—ein Beitrag zur Erforschung sogenannter Anomalien. Senck leth 71:339–366
Hengsbach R (1996) Ammonoid pathology. In: Landman NH, Tanabe K, Davis RA (eds) Ammonoid paleobiology. Plenum, New York
Ho JS (1980) Anchicaligus *nautili* (Willey), a caligid copepod parasitic on *Nautilus* in Palau, with discussion of *Caligulina* Heegaard, 1972. Publ Seto Mar Biol Lab 25:157–165
Hochberg FG (1983) The parasites of cephalopods: a review. Mem Natl Mus Vic 44:109–145
Hochberg FG (1989) Les parasites. In: Mangold K (ed) Traité de Zoologie, Tome V (Cephalopodes), Fascicule 4. Masson, Paris
Hochberg FG (1990) Diseases of mollusca: cephalopoda. In: Kinne O (ed) Diseases of marine animals, Vol. III. Biologische Anstalt Helgoland, Hamburg
Hölder H (1956) Über Anomalien an jurassischen Ammoniten. Paläontol Z 30:95–107
Hölder H (1970) Anomalien an Molluskenschalen, insbesondere Ammoniten, und deren Ursachen. Paläontol Z 44:182–195
Hook JE, Golubic S (1988) Mussel periostracum from deep-sea redox communities as a microbial habitat: the scalloping periostracum borer. Mar Ecol 9:347–364
Hook JE, Golubic S (1992) Mussel periostracum from deep-sea redox communities as a microbial habitat: 3. Secondary inhabitants. Mar Ecol 13:119–131
House MR (1960) Abnormal growths in some Devonian goniatites. Palaeontology 3:129–136
Hüne L, Hüne P (2006). Des phénomènes paléopathologiques chez une faune d'ammonites du Callovien superieur de Bénerville-sur-Mer (Calvados, France). L'ècho des falaises 10: 33–37
Huntley JW (2007) Towards establishing a modern baseline for paleopathology: trace-producing parasites in a bivalve host. J Shellfish Res 26:253–259
Ituarte CF, Cremonte F, Deferrari G (2001) Mantle-shell complex reactions elicited by digenean metacercariae in *Gaimardia trapesina* (Bivalvia: Gaimardiidae) from the Southwestern Atlantic Ocean and Magellan Strait. Dis Aquat Org 48:47–56
Ituarte C, Cremonte F, Zelaya DG (2005) Parasite-mediated shell alterations in Recent and Holocene sub-antarctic bivalves: the parasite as modeler of host reaction. Invertebr Biol 124:220–229. doi: 10.1111/j.1744–7410.2005.00021.x
Ivanov AN (1971) Problems of the periodization of ontogeny in ammonites. Yarosl Ped Inst Uch Zap Geol i Paleont 87:76–119 [in Russian]
Ivanov AN (1975) Late ontogeny in ammonites and its characteristics in micro-, macro- and megaconches. Yarosl Ped Inst. Sb Nauchn Trudy 142:5–57 [in Russian]
Jacobs DK, Landman NH (1993) *Nautilus*-a poor model for the function and behavior of ammonoids? Lethaia 26:101–111
Kemper E (1961) Die Ammonitengattung *Platylenticeras* (= *Garnieria*). Beihefte zum Geologischen Jahrbuch 47:1–195
Keupp H (1976) Neue Beispiele für den Regenerationsmechanismus bei verletzten und kranken Ammoniten. Paläontol Z 50:70–77
Keupp H (1977) Paläopathologische Normen bei Amaltheiden (Ammonoidea) des Fränkischen Lias. Jahrb Coburg Landes-Stiftung 1977:263–280
Keupp H (1979) Nabelkanten-präferenz der *forma verticata* Hölder 1956 bei Dactylioceraten (ammonoidea, Toarcien). Paläontol Z 53:214–219
Keupp H (1984) Pathologische Ammoniten, Kuriositäten oder paläobiologische Dokumente? (Teil 1). Fossilien 1984: 258–262

Keupp H (1986) Perlen (Schalenkonkretionen) bei Dactylioceraten aus dem fränkischen Lias. Natur und Mensch 1986:97–102

Keupp H (1994) Volumenvermindernde Gehäuse-Anomalien bei Jura-Ammoniten. Fossilien 1994:38–44

Keupp H (1995) Volumenvergrößernde Anomalien bei Jura-Ammoniten. Fossilien 1995:54–59

Keupp H (1996) Paläopathologische Analyse einer Ammoniten-Vergesellschaftung aus dem Ober-Jura Westsibiriens. Fossilien 1996:45–54

Keupp H (1997) Paläopathologische Analyse einer „Population“ von *Dactylioceras athleticum* (Simpson) aus dem Unter-Toarcium von Schlaifhausen/Oberfranken. Berl Geowiss Abh E 25:243–267

Keupp H (1998) Mundsaumverletzungen bei *Pleuroceras* (Ammonoidea). Fossilien 1998:37–42

Keupp H (2000) Ammoniten—Paläobiologische Erfolgsspiralen. Thorbecke, Stuttgart

Keupp H (2006) Sublethal punctures in body chambers of Mesozoic ammonites (*forma aegra fenestra* n. f.), a tool to interpret synecological relationships, particularly predator-prey interactions. Paläontol Z 80:112–123

Keupp H (2012) Atlas zur Paläopathologie der Cephalopoden. Berl Paläobiol Abh 12:1–390

Keupp H, Hoffmann R (2015) Ammonoid paleopathology. In: Klug C, Korn D, De Baets K, Kruta I, Mapes RH (eds) Ammonoid Paleobiology: From anatomy to ecology. Springer, Dordrecht

Keupp H, Ilg A (1992) Paläopathologie der Ammoniten fauna aus dem Obercallovium der Normandie und ihre palökologische Interpretation. Berl Geowiss Abh E 3:171–189

Keupp H, Riedel F (1995) *Nautilus pompilius* in captivity: a case study of abnormal shell growth. Berl Geowiss Abh E 16 2:663–681

Keupp H, Röper M, Seilacher A (1999) Paläobiologische Aspekte von syn vivo-besiedelten Ammonoideen im Plattenkalk des Ober-Kimmeridgiums von Brunn in Ostbayern. Berl Geowiss Abh E 30:121–145

Kieslinger A (1926) Untersuchungen an triadischen Nautiloideen. Paläontol Z 7:101–122

Kinne O (1980) Diseases of marine animals: general aspects. In Kinne O (ed) Diseases of marine animals, volume I: general aspects, protozoa to gastropoda. Biologische Anstalt Helgoland, Hamburg

Kirchner H (1927) Perlbildung bei einem Ceratiten. Zentra Miner Geol Paläont Abt B Geol Paläont 4:148–150

Klug C (2001) Early Emsian ammonoids from the eastern Anti-Atlas (Morocco) and their succession. Paläontol Z 74:479–515

Klug C (2002a) Quantitative stratigraphy and taxonomy of late Emsian and Eifelian ammonoids of the eastern Anti-Atlas (Morocco). Cour Forschungsinst Senckenb 238:1–109

Klug C (2002b) Conch parameters and ecology of Emsian and Eifelian ammonoids from the Tafilalt (Morocco) and their relation to global events. Ber Geol BA 57:523–538

Klug C (2007) Sublethal injuries in Early Devonian cephalopod shells from Morocco. Acta Palaeontol Pol 52:749–759

Klug C, Korn D (2001) Epizoa and post-mortem epicoles on cephalopod shells—examples from the Devonian and Carboniferous of Morocco. Berl Geowiss Abh E 36:145–155

Klug C, Lehmann J (2015) Soft part anatomy of ammonoids: reconstructing the animal based on exceptionally preserved specimens and actualistic comparisons. In: Klug C, Korn D, De Baets K, Kruta I, Mapes RH (eds) Ammonoid Paleobiology: From anatomy to ecology. Springer, Dordrecht

Klug C, De Baets K, Kröger B, Bell MA, Korn D, Payne JL (2015) Normal giants? Temporal and latitudinal shifts of Palaeozoic marine invertebrate gigantism and global change. Lethaia 48:267–288. DOI: 10.1111/let.12104

Klug C, Kröger B, Korn D, Rucklin M, Schemm-Gregory M, De Baets K, Mapes RH (2008) Ecological change during the early Emsian (Devonian) in the Tafilalt (Morocco), the origin of the Ammonoidea, and the first African pyrgocystid edrioasteroids, machaerids and phyllocarids. Palaeontogr A 283:83–176

Klug C, Riegraf W, Lehmann J (2012) Soft–part preservation in heteromorph ammonites from the Cenomanian–Turonian Boundary Event (OAE 2) in north–west Germany. Palaeontology 55:1307–1331
Korn D, Klug C (2002) Ammoneae Devonicae. In: Riegraf W (ed) Fossilium Catalogus 1: Animalia, 138. Backhuys, Leiden
Korn D, Klug C, Mapes RH (2005a) The lazarus ammonoid family Goniatitidae, the tetrangularly coiled Entogonitidae, and Mississippian biogeography. J Paleontol 79:356–365
Korn D, Niedzwiedzki R, Posieczek JB (2005b) Age, distribution, and phylogeny of the peculiar late Devonian ammonoid *Soliclymenia.* Act Geol Pol 55:99–108
Kraft S, Korn D, Klug C (2008) Patterns of ontogenetic septal spacing in Carboniferous ammonoids. N Jahrb Geol Paläont Abh 250:31–44
Kröger B (2000) Schalenverletzungen an jurassischen Ammoniten—ihre paläobiologische und paläoökologische Aussagefähigkeit. Berl Geowiss Abh E 33:1–97
Kröger B, Vinther J, Fuchs D (2011) Cephalopod origin and evolution: a congruent picture emerging from fossils, development and molecules. Bioessays 33:602–613
Kruta I, Landman NH (2008) Injuries on *Nautilus* jaws: implications for the function of ammonite aptychi. Veliger 50:241–247
Lafferty KD, Kuris AM (2009) Parasitic castration: the evolution and ecology of body snatchers. Trends Parasitol 25:564–572
Lange W (1941) Die Ammonitenfauna der Psiloceras-Stufe Norddeutschlands. Palaeontogr A 93:1–186
Lauckner G (1983) Diseases of mollusca: Bivalvia. In: Kinne O (ed) Diseases of marine animals, vol II. Biologische Anstalt Helgoland, Hamburg
Landman NH, Waage KM (1986) Shell abnormalities in scaphitid ammonites. Lethaia 19:211–224
Landman NH, Mikkelsen PM, Bieler R Bronson B (2001) Pearls: a natural history. Abrams, New York.
Landman NH, Kennedy WJ, Cobban WA, Larson NL (2010) *Scaphites* of the "Nodosus Group" from the Upper Cretaceous (Campanian) of the Western Interior of North America. Bull Am Mus Nat Hist:1–242
Larson N (2007) Deformities in the Late Callovian (late Middle Jurassic) ammonite Fauna from Saratov, Russia. In: Landman NH, Davis RA, Mapes RH (eds) Cephalopods—Present and Past: new insights and fresh perspectives. Springer, Netherlands
Lehmann U (1975) Über Biologie und Gehäusebau bei *Dactylioceras* (Ammonoidea) aufgrund einer Fraktur-Analyse. Mitt Geol-Paläontol Inst Univ Hamburg 44:195–206
Lehmann U (1990) Ammonoideen: leben zwischen Skylla and Charybdis. Enke, Stuttgart
Liljedahl L (1985) Ecological aspects of a silicified bivalve Fauna from the Silurian of Gotland. Lethaia 18:53–66
Littlewood DTJ (2006) The evolution of parasitism in flatworms. In: Maule AG, Marks NJ (eds) Parasitic flatworms: molecular biology, biochemistry, immunology and physiology. CABI, Wallingford
Littlewood DTJ, Donovan SK (2003) Fossil parasites: a case of identity. Geol Today 19:136–142
Manda S, Turek V (2009) Minute Silurian oncocerid nautiloids with unusual colour patterns. Acta Palaeontol Pol 54:503–512
Manger WL, Meeks LK, Stephen DA (1999) Pathologic gigantism in middle Carboniferous cephalopods, Southern Midcontinent, United States. In: Olóriz F, Rodríguez-Tovar FJ (eds) Advancing research on living and fossil cephalopods. Springer, Dordrecht
Maubeuge PL (1949) Sur la nature des "conelles" (Quenstedt). Bull Soc Sci Nancy 1:1–3.
Meischner D (1968) Perniciöse Epökie von *Placunopsis* auf *Ceratites*. Lethaia 1:156–174
Miller AI (1938) Devonian ammonoids of America. GSA Spec Pap 14:1–294
Mironenko AA 2012. Traces of lifetime damage on the shells of Upper Jurassic (upper Volgian) *Kachpurites* (Craspeditidae, Ammonoidea). In: Leonova TB, Barskov IS, Mitta VV (eds) Modern problems of studying of cephalopod molluscs: morphology, taxonomy, evolution, ecology, biostratigraphy. Paleont Inst Russian Acad Sci, Moscow [in Russian]

Miura O, Kuris AM, Torchin ME, Hechinger RF, Chiba S (2006) Parasites alter host phenotype and may create a new ecological niche for snail hosts. Proc R Soc Lond B 273:1323–1328
Morton N (1983) Pathologically deformed *Graphoceras* (Ammonitina) from the Jurassic of Skye, Scotland. Palaeontology 26:443–453
Opitz R (1932) Bilder aus der Erdgeschichte des Nahe-Hunsrück-Landes Birkenfeld. Enke, Birkenfeld
Pascual S, Gestal C, Estévez JM, Rodríguez H, Soto M, Abollo E, Arias C (1996) Parasites in commercially-exploited cephalopods (Mollusca, Cephalopoda) in Spain: an updated perspective. Aquaculture 142:1–10
Pascual S, Gonzáez A, Guerra A (2007) Parasites and cephalopod fisheries uncertainty: towards a waterfall understanding. Rev Fish Biol Fish 17:139–144
Paul CRC, Simms MJ (2012) Epifauna on ammonites from the Lower Jurassic of the Severn basin, southern England, and their palaeoenvironmental and taphonomic significance. Proc Geol Assoc 123:508–519
Ploch I (2007) Intraspecific variability and problematic dimorphism in the Early Cretaceous (Valanginian) ammonite *Saynoceras verrucosu*m (d'Orbigny, 1841). Act Geologica Sin 81:877–882
Poulin R, Morand S (2000) The diversity of parasites. The Quart Rev Biol 75:277–293. doi:10.2307/2665190
Quenstedt FA (1884) Petrefaktenkunde Deutschlands, Erste Abtheilung, B, 1. Band Fues´s, Leipzig
Rakociński M. (2011) Sclerobionts on Upper Famennian cephalopods from the Holy cross mountains, Poland. Palaeobiol Palaeoenviron 91:63–73, doi 10.1007/s12549–010-0045-x
Rakociński M (2012) The youngest Devonian record of "Housean pits" in ammonoids. Geol Q 56:387–390
Raup DM (1991) The future of analytical paleobiology. Short Course Paleontol 4:207–216
Rein S (1989) Über das Regenerationsvermögen der germanischen Ceratiten (Ammonoidea) des Oberen Muschelkalks (Mitteltrias). Veröff Naturhistorischen Mus Schleus 4:47–54
Rein S (1994) Sekundärschalenbildungen (*forma conclusa*) bei germanischen Ceratiten. Fossilien 1994:372–376
Rieber H (1963) Ein *Cardioceras* (Ammonoidea) mit asymmetrischer Lage von Phragmokon und Kiel. Neues Jahrb Geol Paläontol Mh 1963:289–294
Ritterbush KA, Hoffmann R, Lukeneder A, De Baets K (2014) Pelagic palaeoecology: the importance of recent constraints on ammonoid palaeobiology and life history. J Zool 292:229–241
Rocha F, Guerra A, Gonzalez AF (2001) A review of reproductive strategies in cephalopods. Biol Rev 76:291–304
Rohde K (2005) Definitions, and adaptations to a parasitic way of life. In: Rohde K (ed) Marine parasitology. CSIRO, Oxon
Ruiz GM, Lindberg DR (1989) A fossil record for trematodes: extent and potential uses. Lethaia 22:431–438
Schindewolf OH (1934) Über Epöken auf Cephalopoden-Gehäusen. Paläontol Z 16:15–31
Schindewolf OH (1962) Parasitäre Thallophyten in Ammoniten-Schalen. Paläontol Z 36:206–215
Schindewolf OH (1963) Pilze in oberjurassischen Ammoniten-Schalen. N Jahrb Geol Paläont Abh 118:177–181
Schweigert G (2009) First three-dimensionally preserved in situ record of an aptychophoran ammonite jaw apparatus in the Jurassic and discussion of the function of aptychi. Berl Paläobiol Abh 10:321–330
Schweigert G, Dietl G (2001) Die Kieferelemente von *Physodoceras* (Ammonitina, Aspidoceratidae) im Nusplinger Plattenkalk (Oberjura, Schwäbische Alb). Berl Geowiss Abh E 36:131–143
Seilacher A (1960) Epizoans as a key to ammonoid ecology. J Paleont 34:189–193
Seilacher A (1982) Ammonite shells as habitats—floats or benthic islands? (Abstract). In: Einsele G, Seilacher A (eds) Cyclic and event stratification. Springer, Berlin
Seltzer VB (2001) About anomal shells from the callovian ammonites. Transactions of the scientific research geological institute of the N. G. Chernyshevskii Saratov State Univ 8:29–45

Seltzer VB (2009) Anomalous phragmocones of the Late Callovian Cardioceratidae (Ammonoidea) In: Leonova TB, Barskov IS, Mitta VV (eds) Contributions to current cephalopod research: morphology, systematics, evolution, ecology and biostratigraphy. PIN RAS, Moscow

Seuss B, Wisshak M, Mapes RH, Landman NH (2015) Syn-Vivo Bioerosion of *Nautilus* by Endo- and Epilithic Foraminiferans (New Caledonia and Vanuatu). *PLoS ONE* 10:e0125558

Stridsberg S, Turek V (1997) A revision of the Silurian nautiloid genus *Ophioceras* Barrande. GFF 119:21–36

Sousa WP (1983) Host life history and the effect of parasitic castration on growth: a field study of *Cerithidea californica* Haldeman (Gastropoda: Prosobranchia) and its trematode parasites. J Exp Mar Biol Ecol 73:273–296

Sorensen RE, Minchella DJ (2001) Snail-trematode life history interactions: past trends and future directions. Parasitology 123(7):S3–S18

Sparks AK (1972) Invertebrate Pathology: Non-communicable Diseases. Academic Press, New York

Tasnádi-Kubacska A (1962) Paläopathologie: Pathologie der Vorzeitlichen Tiere: Vol. 1. Fischer, Jena

Turek V, Manda S (2010) Variability of colour pattern and shell abnormalities in Silurian nautiloid *Peismoceras* Hyatt, 1884. J Natl Mus (Prague), Nat Hist Ser 179:171–178

Upeniece I (2001) The unique fossil assemblage from the Lode quarry (Upper Devonian, Latvia). Foss Rec 4:101–119

Upeniece I (2011) Palaeoecology and juvenile Individuals of the Devonian placoderm and acanthodian fishes from Lode Site, Latvia. University of Latvia, Riga

Wani R (2007) How to recognize in situ fossil cephalopods: evidence from experiments with modern *Nautilus*. Lethaia 40:305–311

Ward PD (1987) The natural history of *Nautilus*. Allen and Unwin, Winchester

Weitschat W (1986) Phosphatisierte Ammonoideen aus der Mittleren Trias von Central-Spitzbergen. Mitt Geol-Paläont Inst Univ Hamburg 61:249–279

Wetzel W (1954) Untersuchung eines großen Jurafindlings von besonderem sedimentologischen und paläobiologischen Interesse. Palaeontogr A 105:133–165

Wetzel W (1964) Schalen-Parasitismus bei Ammoniten (aufgrund schleswig-holsteinischer Funde). Meyniana 14:66–69

Willey A (1897) Zoological observations in the South Pacific. Q J Microscopical Sci (N. S.) 39:219–231

Wisshak M, Tapanila L (2008) Current developments in Bioerosion. In: Freiwald A (ed) Erlangen earth conference series. Springer, Heidelberg

Yacobucci MM, Manship LL (2011) Ammonoid septal formation and suture asymmetry explored with a geographic information systems approach. Palaeontol Electron 14(1):3A:17p

Zapalski MK (2011) Is absence of proof a proof of absence? Comments on commensalism. Palaeogeogr Palaeoclimatol Palaeoecol 302:484–488

Zatoń M (2010) Sublethal injuries in middle Jurassic ammonite shells from Poland. Geobios 43:365–375.

Ziegler B (1958) Monographie der Ammonitengattung *Glochiceras* im epikontintalen Weißjura Mitteleuropas. Palaeontogr A 110:93–164

Chapter 21
Ammonoid Paleopathology

René Hoffmann and Helmut Keupp

21.1 Introduction

21.1.1 Definition of (Paleo-)Pathological Phenomena

"*Normality is fiction*" (Hölder 1956) and is best understood as arithmetic average. Deviations from the "*normal*" phenotype of a given population that are caused by endogene or exogene growth anomalies are called pathologies. Paleopathologies, as a rule, are constrained to growth anomalies of mineralized hardparts that passed down in the fossil record. In case of ammonoids these hardparts are the conch and subordinate jaw elements, especially the calcareous aptychi, often found isolated from their host.

21.1.2 Historical Aspects

During early scientific time, pathological ammonoids were regarded as monstrous curiosities till the end of the nineteenth century, e.g., d'Orbigny 1842–1851; Fraas 1863; Quenstedt 1885–1888. Modern scientific treatment appeared at the beginning of the twentieth century with the publication of Engel (1894) who conducted research regarding the causes of such phenomena. Finally, the summarizing paleopathological descriptions of Moodie (1926) and Tasnadi-Kubacska (1962) focused

R. Hoffmann (✉)
Department of Earth Sciences, Institute of Geology, Mineralogy, and Geophysics,
Ruhr-Universität Bochum, 44780 Bochum, Germany
e-mail: rene.hoffmann@rub.de

H. Keupp
Institute of Geological Sciences, Branch Palaeontology, Freie Universität Berlin,
12249 Berlin, Germany
e-mail: keupp@zedat.fu-berlin.de

C. Klug et al. (eds.), *Ammonoid Paleobiology: From Anatomy to Ecology*,
Topics in Geobiology 43, DOI 10.1007/978-94-017-9630-9_21

on vertebrate paleontology and introduced the field of paleopathology as an independent paleobiological discipline (Hengsbach 1991).

Engel (1894) mentioned four factors causing paleopathological phenomena: (1) "*Krüppel ab ovo*" (cripple within the egg), (2) "*Bastarde bzw. senile Formen*" (bastards respectively senile forms), (3) "*kranke Formen*" (ill forms), (4) "*verletzte Formen*" (injured forms). Today, according to Lehmann (1976), anomalies are understood as conch modifications that only affect single or few specimens of a species and can be attributed to diseases, injuries or other interactions between organisms, e.g., parasites, epizoa (Keupp 2012). Disease that affect the whole population (="*Überindividuelle krankhafte Erscheinungen*" of Schwegler 1939), and higher taxonomic units e.g., heteromorph ammonoids, specific modification of the aperture, or ontogenetic changes in ornamentation patterns, that were described by former authors as "*ill*", "*abnormal*" or "*degenerated*" (Quenstedt 1858; Engel 1894; Pompeckj 1894; Tornquist 1896; Schindewolf 1929), do not represent pathological phenomena.

For paleontologists, confirmation or identification of conch anomalies that are caused by genetic dysfunctions (Engel 1894: "*ab ovo*" or crossing) is difficult because the mutagenic character hinders a differentiation between intraspecific variability or speciation (development of new species). In many cases traumatic events during an early/juvenile ontogenetic stage that caused phenotypical modifications also in later ontogenetic stages were misinterpreted as being caused by genetic dysfunctions. Such cases are the source of taxonomic confusion and phylogenetic misconceptions in paleontology (e.g. Maubeuge 1957). Based on singular pathological cases, several new species have been raised; Keupp (2012) lists 40 such cases. Two cases will be explored in more detail in the following: (1) fastigate ceratitids from the German Muschelkalk, and (2) ring-ribbed pleuroceratitids from the middle Lower Jurassic.

The species "*Ammonites fastigatus*" was established by Credner (1875) based on a single ceratitid ammonite from the German Muschelkalk, whose strong ventral ribs extended over the otherwise smooth ventral side (=ring ribs; Fig. 21.1). Afterwards, additional specimens of fastigate ceratitids from different species were found (e.g., Eck 1879; Zimmermann 1883; Blankenhorn 1887; Philippi 1901; Riedel 1916; Bülow 1918; Böttcher 1938; Müller 1954, 1970a, b, c; Busse 1954; Wenger 1957; Weyer 1964; Mayer 1966, 1974, 1978, 1981; Rummel 1973; Rein 1989, 1991; Claus 1992). In most cases, the preservation as internal moulds prevents the recognition of the causing injury of the juvenile ceratitid aperture. Only the more or less pronounced asymmetry of the ornamentation perturbance itself indicates a traumatic event (Keupp 1985). The misinterpretation of these anomalies as genetic dysfunction led to new "*fastigo-species*" (Rothe 1949, 1955; Wunsch 1957; Mundlos 1963). Müller (1976) interprets fastigate ceratitids as "*prologism*", i.e. single specimen genetically implement ornamentation patterns that become regular later during the Jurassic. Considering the fact that ceratitids represent a blind ending group of ammonoids that is not phylogenetically connected to the younger groups of Jurassic and Cretaceous ammonoids, demonstrates the unreliability of that hy-

Fig. 21.1 **a** Reproduction of "*Ceratites fastigatus*" from Credner (1875). **b** fastigate *Ceratites nodosus* (Schlotheim) from Oberstetten/Mainfranken, already described by Eck (1897), Philippi (1901) and Wenger (1957) (original housed at Staatliches Museum für Naturkunde Stuttgart, Nr. 21076)

pothesis. Hence, fastigate ceratitids are the result of traumatic events that caused a so-called "*compensations ornamentale*" (Guex 1967).

Two Lower Jurassic pleuroceratitids with ring-ribs were described in a similar way. The lack of the typical "*Zopfkiel*" was caused by an injury of the shell-secreting, ventral mantle epithelium during an earlier ontogenetic stage, but was misinterpreted as genetic dysfunction by Kolb (1955). The resulting phenotype resembles some Aegoceratidae, the phylogenetically ascending keel-less group from which the Amaltheidae arise. Kolb (1955) therefore argued that this phenomenon might represent atavism. Maubeuge (1957) re-examined the same two specimens. He draws the same conclusion as he did before for similar cases (Maubeuge 1949a), that these ring-ribbed pleuroceratitids represent independent valid taxa, which he transferred to the genus *Androgynoceras* and *Oistoceras*, respectively. Both genera are restricted to the Lower Pliensbachian.

The morphological variability of anomalies causing ring-ribs is reflected in the assignment to two different aegoceratid genera. Strength of expression of that anomaly depends on the intensity of the injury and degree of regeneration of the injured epithelia; both allow for differing growth rates of the ventral part of the conch (Fig. 21.2). Hölder (1956) introduced for these anomalous ring-ribbed ammonites the forma aegra circumdata (forma aegra will be abbreviated as f. a. onwards) as one of his "standardized anomalies" (see below). That was based on his obersavtions of ring-ribbed *Schlotheimia* that was earlier described by Martin (1858) as "*Ammonites circumdatus*"

Fig. 21.2 Three specimens of *Pleuroceras spinatum* (Bruguière) from the Upper Pliensbachian of Buttenheim/Germany (leg. J. Schobert): **a** normal conch morphology, **b** pathological specimen (f. a. circumdata), that after a traumatic loss of its keel resembles *Androgynoceras*, **c** after loss of its keel during early ontogeny this ring-ribbed *Pleuroceras* resembles *Oistoceras*

21.1.3 Taxonomic Handling: Symptom versus Syndrome

Hölder (1956) introduced an open nomenclature below the species level for characteristic, so-called "*standardized*" anomalies that occur in different ammonoid taxa and cause similar phenomena (= forma aegra). In many cases different causes, partly endogenic or exogenic underlie the descriptive-phenomenologically defined types of forma aegra. Therefore in most cases an etiology of the phenomenon cannot be clarified. A summarizing table of all hitherto known 42 types of forma aegra defined for ammonoids was presented by Keupp (2012) and is reproduced here (Tab. 21.1).

Two different ways of treatment has been established in order to handle cases where several anomalous phenomena (= symptoms) due to larger perturbances of the shell geometry occur. One option is to describe the sequence of anomalous morphologies by application of the correct f. a. types (e.g., Keupp 1976, 1984). On the other hand, several anomalies can be combined and described as a complex medical syndrome under its own name (e.g., "*Morton's syndrome*" by Landman and Waage 1986). As a rule, only the reaction of the organism, here the ammonoids, to the disturbing factor should be considered to characterize a pathological phenomenon. However, in some cases also the potential trigger of growth disturbances like the configuration of shell injuries have been determined in a nomenclatorical manner (e.g., characteristic bite marks; f. a. seccata Hölder 1956, f. a. mordata Hengsbach 1996). Accordingly, Kröger (2000) suggested in dependence to Ward (1987) three different types of characteristic injuries of the peristome. These are "*typus stupidus*", "*typus acutus*" and "*typus parvus*".

Table 21.1 Alphabetic list of "standardized anomalies" (=forma aegra) modified after Keupp (2012)

Forma aegra	Author	Primarily thought causality	Description of the phenomenon	Remarks
abrupta	(Stahl 1824) Hölder 1956	Injury	Abrupt sculptural perturbance along a distinct line	Collective term describing f. verticata Hölder 1956 and f. substructa Hölder 1973
augata	Kröger 2000	Parasitism	Temporarely enlargement of single sculptural elements	Ex forma inflata Keupp 1976
alternospinata	Keupp 1977	Asymmetry of the peristome	Alternating nodes (or ribs) on the flanks	The phenomenon affected also parabolic nodes.
aptycha	Keupp 1977	Mostly injuries	Internal, horizontal sheets of the body chamber reducing its inner space	=forma conclusa Rein 1989
bovicornuta	Keupp 1977	Injury of the peristome	Anomalous orientation (mostly fore-wards) of marginal spines	Special phenomenon connected with the "com-pensations ornamentale" Guex, 1967
cacoptycha	Lange 1941	Endogenic cause	Temporarely loss or reduction of sculptural elements	Similar phenomena can be observed also during regeneration of sublethal injuries
calcar	(Zieten 1830) Hölder 1956	Predominantly injuries of the peristome	Coalescence of two primarily marginal nodes to a single median element (incl. Parable nodes)	cf. Keupp 1993: special case of "compensations ornamentale" Guex 1967
chaotica	Keupp 1977	Endo- and exogenic causes	Iterative zigzag-like dislocation of the keel from extern median to the flanks	Compare with "chaotische Skulptur" (chaotic sculpture) Hölder 1970
cicatricocari-nata	Heller 1964	Injury of the peri-stome, partly with endogenic causes	Irregular scars and grooves instead of a median keel	
circumdata	(Martin 1858) Hölder 1956	Predominantly injuries of the peristome	Anomalous continuation of lateral ribs over the otherwise smooth or covered with other elements ventral side.	Special case of "compensations ornamentale" (Guex 1967)=forma fastigata (Credner 1875) Rein 1989

Table 21.1 (continued)

Forma aegra	Author	Primarily thought causality	Description of the phenomenon	Remarks
complicata	Keupp and Riedel 1995	Endogenic perturbance	Folding of the shell at the peristome	
conclusa	Rein 1989; junior syn. of f. aptycha Keupp 1977	Predominantly injuries	Calcareous sheets within the body chamber	
concreta	Hengsbach 1996	Parasitism	Internal shell concretion (blister? pearls)	see Keupp 1987
disseptata	Hölder 1956	Endogenic perturbance	Single incomplete or lack of septa	
duplicarinata	Keupp 1976	Injury of the peristome	Temporary development of a double keel due to laterally dislocated new insertions of a regenerative keel	Special case of "compensations ornamentale" (Guex 1967)
excentrica	Hölder 1956	Endogen and exogen	Anomalous trochospiral conch	Phenomen of the "Morton-syndrom", epizoa: Merkt 1966, Keupp 1984
fastigata	(Credner 1875) Rein 1989	Injury of the peristome often misinterpreted as genetic dysfunction (e.g., Müller 1976)	Lateral ribs cross the otherwise smooth venter (term predominantly used for ceratitids)	Special case of "compensatio ornamentale" (Guex 1967) = forma circumdata (Martin 1858) Hölder 1956
fenestra	Keupp 2006	Injury	Window-like shell breakage caused a perturbance of shell sculpture in all directions	
gigantea	Hengsbach 1996	Parasitism	Pathological gigantism	="megaconchs": Ivanov 1975
inflata	Keupp 1976	Parasitism	Anomalous enlarged sculptural elements and blister-like shell outgrowth (protuberance)	According to Kröger 2000 only blister-like structures. anomalous enlarged sculptural elements represent forma aegra augata Kröger 2000

Table 21.1 (continued)

Forma aegra	Author	Primarily thought causality	Description of the phenomenon	Remarks
intracarinata	Keupp 2000b	Endogenic	Spiral-furrows visible on the internal mould caused by anomalous strengthened muscle attachments	
heterumbilicata	Keupp 2007c	Injury of the peristome	Enlargement of the umbilical width caused by the regenerative backwards dislocation of the umbilical edge on one side of the conch	
juxtacarinata	Hölder 1956	Injury or endogenic	Dislocation of the median keel to the lateral side of the shell	Analogue for ammonoids with a ventral furrow: forma juxtasulcata Geczy 1965
juxtalobata	Hölder 1956	Endogenic	Asymmetric phragmocone with dislocation of the external lobe and siphuncle to the side	For a few taxa this phenomenon has been recognized as being species specific and not pathologic
juxtasulcata	Geczy 1965	Injury or endogenic	Dislocation of the median furrow to the lateral side of the shell	Analogue to the forma juxtacarinata Hölder 1956 reported for ammonoids with a median keel
mediosulcata	Hengsbach 1986a	No pathology, within the range of intraspecific variability	Formation of ventral ridges for muscle attachment	Bayer 1977a, b; Dagys and Keupp 1998
mordata	Hengsbach 1996	Lethal injury	Bite marks of vertebrates	Without any traces of regeneration
pexa	Hölder 1973	Injury of the peristome	Thickening of growth lines as a late effect of a forma aegra verticata	Special phenomenon that belongs to the forma verticata Hölder 1956
polygonia	Hüne and Hüne 2006	Endogenic causes	Polygonal cross section of the whorl	Compare with Keupp and Richter 2010
pseudocarinata	Fernandez-López 1987	Injury of the peristome	Forma aegra verticata (ribbing vertex) developed as keel-like ridge	Special phenomenon that belongs to the forma verticata Hölder 1956

Table 21.1 (continued)

Forma aegra	Author	Primarily thought causality	Description of the phenomenon	Remarks
refecta	Rein 1994a	Injury of the peristome	Regenerated shell which underpine the breaking edge is bulged outwards above the normale shell niveau	Junior synonym of forma substructa Hölder 1973
seccata	Hölder 1956	Lethal attack of crustaceans	Fragmentation of the shell by the activity of crustaceans	See: Roll 1935; Hölder 1955
semiverticata	Hölder 1977	Injury of the peristome	Forma aegra verticata (Rippenschei-tel) on only one side due to the posi-tioning parallel to the umbilical edge	Special phenomenon that belongs to the forma aegra verticata Hölder 1956, similar phenomena can be triggered by parasitism (Keupp 1979 2000b): Kröger 2000: f. umbilicata
septaedefor-mata	Rein 1990	Endogenic: disloca-tion of the rear soft body within the body chamber	Septal inclination followed a disseptie	Rein 1993; 2000
substructa	Hölder 1973	Injury of the peristome	Regeneration of a shell injury by underpinning the breakage edge with new shell material	Junior synonym: forma refecta Rein 1994a
superstructa	Keupp 1998	Injury of the peristome	Regeneration due to overlapping of regenerating shell material	
syncosta	Hengsbach 1976	"Genetic"	Two former independent ribs growth together in amaltheids to a ventro-lateral spine; however this represents part of the normal Bauplan of spine bearing taxa	See Hengsbach 1996; according to Keupp and Ilg 1992 a phenomenon of forma aegra substructa Hölder 1973 after an injury of the peristome (see also Kröger 2000)
umbilicata	Kröger 2000	Parasitism	Flattening and delay of shell secretion along the umbilical edge, similar to forma aegra semiverticata Hölder 1977	See: "Nabelkanten-Anomalie" (umbilical edge anomaly by Keupp 1979

Table 21.1 (continued)

Forma aegra	Author	Primarily thought causality	Description of the phenomenon	Remarks
undaticari-nata	Heller 1958	Endogenic,? Parasitism	Iterative swinging and partly buckling of the median keel	Similar phenomena can be triggered during the aftermath of the regeneration of injuries see: Keupp 1985
undaticari-nata-undati-concha	Hengsbach 1996	Parasitism	Temporarely swinging of the whorl in combination with the swinging of ventral keel or—furrow	See: “Mortons’s syndrome” Landman and Waage 1986; Morton 1983; junior synonym of forma aegra undatispirata Keupp and Ilg 1992: cf. Kröger 2000
undatispirata	Keupp and Ilg 1992	Epizoans and endogenic	Temporary swinging of the whorls	Junior synonym: forma aegra undaticoncha Hengsbach 1996
verticata	Hölder 1956	Injury of the peristome	Crest-like splitting of growth lines and ribs (ribbing vertex) caused by the regeneration of injured punctua-tion of shell secreting epithelia	See: Keupp 1992a: different morphologies: f. pexa Hölder 1973, semiverticata Hölder 1977; abrupta Hölder 1956 (pars), pseudocarinata Fernandez-López 1987

21.2 Exogenic Reasons

For practical reasons, pathological phenomena caused by exogenic factors and recognizable on the shell (injuries, epibiosis) are differentiated from endogenic disturbances (illness, parasitoses). For the first case, intensity of anomalies is constantly decreasing during regeneration. Contrary, intensity of anomalies is constantly increasing for the latter case of intrinsic growth irregularities.

21.2.1 Injuries

The majority of conch injuries were caused by predator-prey interactions. While injuries of lethal attacks are hard to distinguish from post-mortem shell damages (Roll 1935; Mayr 1967; Keller 1977; Lehmann 1990; Radwanski 1996; Keupp 2008, 2012; Wani et al. 2012), the post-traumatic reaction of shell growth document the syn-vivo character of sublethal injuries. Mainly the interplay of four factors controls the phenotype of anomalies caused by injuries:

- Position at the conch (peristome, rear part of the body chamber, phragmocone)
- Configuration of the injury which depends on the predator
- Intensity of injury (shell only or additionally the shell-secreting mantle epithelium)
- Genetically-fixed Bauplan (whorl parameter, sculpture, etc.) of a certain ammonoid.

Injuries which did not affect the shell secreting epithelium, caused only short-term anomalies during the post-traumatic phase (Fig. 21.3). Oblique anomalies of the

Fig. 21.3 Simple, regenerated injuries of the former peristome only affecting the shell not the shell secreting epithelium (f. a. substructa Hölder 1973): **a** clymeniid from the Upper Devonian of the south Urals (from Keupp 2012), **b, c** *Labeceras bryani* (Whitehouse), Upper Aptian of Walsh River, Australia. Injuries often depend on the construction and configuration of their producer. Vertebrates, due to their isolated tooths, can cause partly characteristic bite marks (Fig. 21.4; Kauffman and Kesling 1960; Mapes and Hansen 1984; Sims et al. 1987; Hansen and Mapes 1990; Keupp 1985; 1991, 2000b, Kauffman 1990; Martill 1990; Hewitt and Westermann 1990; Mapes et al. 1995; Tsujita and Westermann 1998; King 2009; Richter 2009a, b), and partly non-specific u-shaped exposures along the peristome (Keupp 2012)

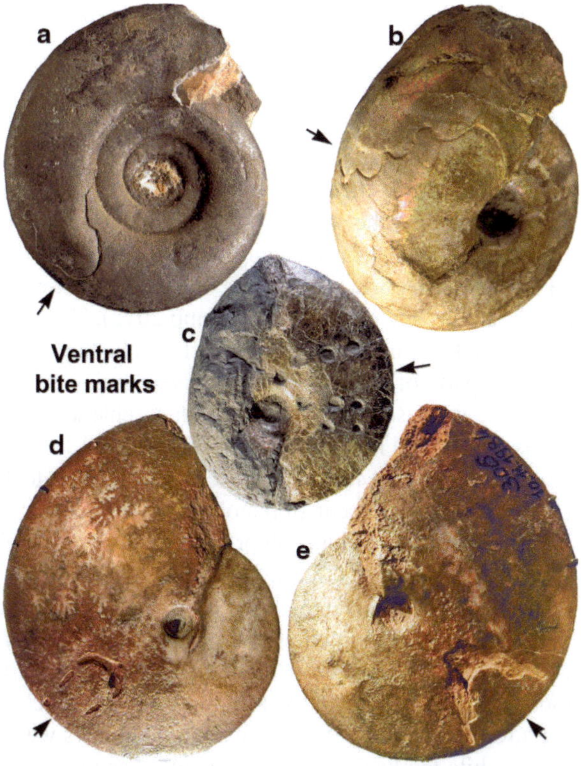

Fig. 21.4 Vertebrate bite marks, regenerated (**a**, **b**) and lethal (**c**, **d**, **e**): **a** u-shaped exposure caused by a fish (*Gymnites* sp., Anisian of Epidauros, Greece), **d, e** pycnodontid bite mark on both sides (*Oxycerites*, Upper Bajocian, Sengenthal, southern Germany), **b** *Desmoceras latidorsatum* Michelin from the Lower Albian of Ambatolafia, Madagascar with typical bite mark of a semionotid fish, **c** *Placenticeras* sp. from the Upper Cretaceous of South Dakota with presumed bite marks of a mosasaur (original: Ruhrland-Museum, Essen, Germany, coll. no. RE 551 763 333 A 3073)

sculpture of the regenerated shell are caused due to the underpinning of the break edge by the peristomal epithelium and depend on the size of the injury. In adoral direction obliqueness of the sculpture decreases and merges into the normal sculptural pattern. Regenerated shell portions of window-like injuries of the body chamber behind the peristome show a distorted sculpture if the position of the puncture allows for a generation of the injury by the retracted peristomal epithelium. Shell areas of regenerated punctures are smooth without growth lines when these were repared by the mantle epithelium of the rear part of the body chamber.

Injuries along the phragmocone were predominantly lethal, especially for brevidomic ammonoids (Kröger 2000). For nautilids, some cases of small injuries of the phragmocone were reported which did not cause a lethal fluting of the shell but could be successfully sealed from the outside. One prerequisite is the position of the injury within the zone influenced by the soft body (Stumbur 1960; Kröger and Keupp 2004; Tsujino and Shigeta 2012).

21.2.2 Characteristic Phenomena and their Producers

Arthropoda: Benthic crustacea (malacostraca) which radiated during the Mesozoic played an important role as predators. With their specialized limbs they could produce characteristic injuries. On the one hand, they could cause point-like injuries with their claws at the peristome which usually also perforated the peristomal epithelium. On the other hand, they could cut the body chamber starting at the peristome and produce far-reaching back "*band cuts*" (Kröger 2000). Usually broad band cuts are reported for Lower Paleozoic ammonoids; their producers are to be sought among the chelicerates (Keupp 2012). The stomatopods with their specialized angled pair of maxillipeds, which can be thrown with high active force against the hard parts of potential prey, evolved during the Upper Paleozoic. They crush preferred molluscan shells and produce characteristic window-like holes in the shells (Baluk and Radwanski 1996; Radwanski 1996; Keupp 2006).Thylacocephalan arthropods that ranged from Early Cambrian to Late Cretaceous had similar raptorial limbs as recent stomatopods. (Schram et al. 2003). They are too are hypothesised to prey as ambush predators, throwing the limbs with high active force stabbing the prey (Fig. 21.5).

Nautilids/ coleoids: One basic character of cephalopods is the chitinous, parrot-like jaw apparatus, which allows nautilids and coleoids a predatory lifestyle. Only certain ammonoids, due to their microphageous and partly planktotrophic diet, modified the jaw apparatus (Keupp 2000b; Kruta et al. 2011; Tanabe 2011). With the advent of the aptychus some ammonoid jaws were no longer able to cause significant bite marks. The most frequently occurring nautilid and coleoid bite marks during the early Paleozoic are typical trigonal in their configuration (Fig. 21.6).

If in addition to the conch, the shell secreting mantle is injured, long-term post-traumatic conch anomalies with decreasing intensity during the process of regeneration of the epithelium are the results. However, in many cases a complete regeneration of the shell-forming function was not possible, so that anomalous phenomena were built permanently. The mechanism of the "*compensations ornamentale*" (Guex 1967) causes significant deformation of conch features after a partial failure of the peristomal epithelium. The intact, not included in the injury parts of the epithelium were dragged under strain at the site of injury, in order to take over the shell-forming function. Single parts of the ammonoid peristomal epithelium seem to be strictly genetically programmed in favour of a realisation of a narrow part of the sculpture pattern. Therefore, this regeneration mechanism results in a shift and distortion of certain sculptural elements during the post-traumatic shell formation. In the case the ventral epithelium of the keel-bearing ammonite migrates towards the flank, the keel was not built at the median position, but dislocated by the amount of dislocation of the epithelium. Usually the keel was not only dislocated but deformed as well (= f. a. juxtacarinata Hölder 1956; Fig. 21.7). Another characteristic phenomenon of "*compensations ornamentale*" is the f. a. calcar (Zieten; see Hölder 1956). The latter two rows of marginal nodes merge to a single row of median cusps by shortening of the ventral epithelium. Accordingly, Zieten (1830) has described a

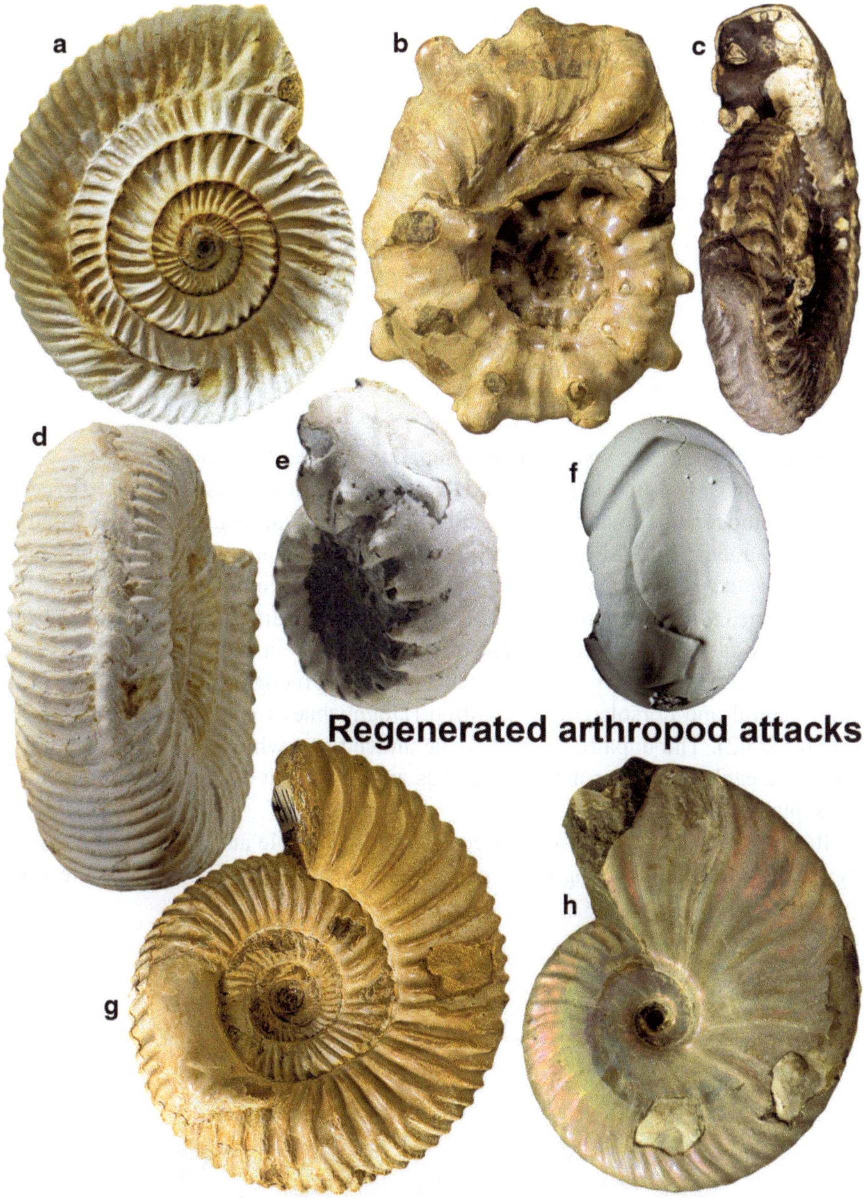

Fig. 21.5 Regenerated arthropod attacks: **a**, **b** ribbing vertex (f. a. verticata Hölder 1956) *Divisosphinctes* sp., Upper-Oxfordian of Sakaraha, SW-Madagascar (**a**), and *Douvilleiceras inaequinodosum* Parona and Bonarelli from the Lower Albian of the Mahajanga Basin, Madagascar. **c**, **d** "Band-cuts" *Clionites acutocostatum* (Klipstein), Ladinian of Nifukokko, Timor (**c**) and "*Kranaosphinctes*" sp. from the Oxfordian of Sakaraha, Madagascar (**d**). **e** tongue-shaped anomaly of dwarf and juvenile amaltheids caused by paginurid crustaceans (Keupp and Schobert 2011). **f** broad "*Band-cuts*" are characteristic injuries of Devonian ammonoids: cheiloceratid ammonoid, Upper Devonian, Tafilalt, Morocco. **g**, **h** stomatopod attacks left window-like punctures on the shell of Mesozoic ammonoids (Keupp 2006): "*Divisosphinctes*" sp. from the Upper Oxfordian of Sakaraha, Madagascar (**g**), *Cleoniceras besairiei* Collignon from the Lower Albian of Ambatolafia, Madagascar

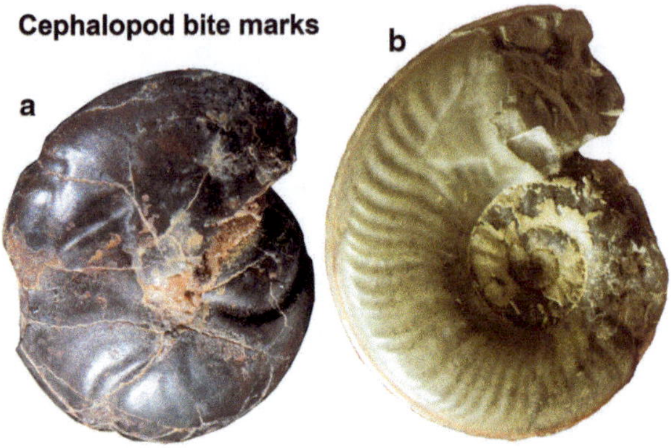

Fig. 21.6 Regenerated cephalopod bite marks with trigonal outline. **a** cheiloceratid from the Upper Devonian of the Tafilalt, Morocco (Ø 8 mm). **b** *Pelekodites* sp. from the Bajocian of Gerzen, Germany (Ø 2.5 cm)

Distichoceras from the Callovian as a separate species "*Ammonites calcar*". Such cases are not only found in ammonoid taxa with marginal rows of nodes (e.g. ceratitids, kosmoceratids, mammitids, scaphitids), but is analogous and transferable to parabolic nodes (Fig. 21.8; Keupp 1993, 2012). Prerequisite for the functioning of the "*compensatio ornamentale*" is a targeted and differentiated mobility of the mantle edge. A differentiated dislocation of epithelia in ammonoids may have been favored by the complex attachment of the soft body in the shell. Thus, at least three independent paired retractor systems have been recognized, which are ventral, ventrolateral and dorsolateral in position (Doghuzhaeva and Mutvei 1991, 1996; Richter 2002). The unpaired dorsal tissue attachment, which also occurs in endocochleate cephalopods, probably served as an attachment of the septal shell sac to the septa.

It seems that nautiloids, by their comparatively simple attachment to the conch by means of the annular ring and a pair of large ventro-lateral muscles (Keferstein

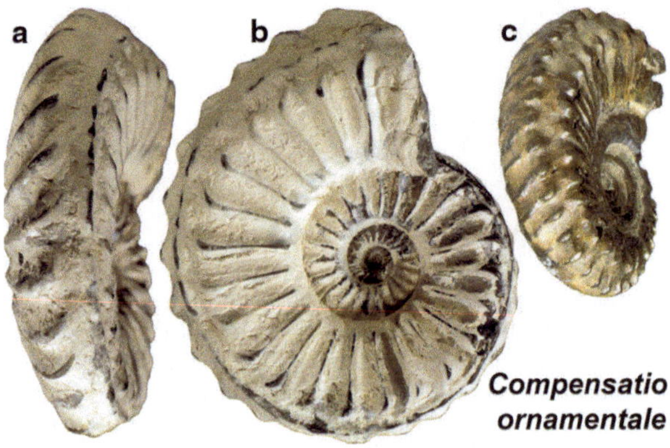

Fig. 21.7 Examples for "*compensations ornamentale*": **a**, **b** *Pleuroceras spinatum* f. **a**. juxtacarinata Hölder 1956 from the Upper Pliensbachian of Unterstürmig, southern Germany, **c** *Kosmoceras* sp. f.a. juxtasulcata Geczy 1965 from the Callovian of Rijasan, Russia

Fig. 21.8 Examples for "*compensatio ornamentale*": **a**, **b**, **d** f.a. calcar (Zieten 1830), *Distichoceras bicostatum* (Stahl) from the Callovian of Thanheim, Württemberg (**a**), *Mammites* sp. from the Lower Turonian of Asfla, Morocco (**d**), Discoscaphites sp. from the Pierre Shale (Upper Cretaceous), Fox Hills, South Dakota (**c**) analogue to the shortening of the ventral epithelium in a perisphinctid causes clasp-like ventrally furcating ribs due to the lack of marginal nodes (Upper Oxfordian, Sakaraha, southwest Madagascar), **e** analogue phenomena of parabolic nodes of a perisphinctid (Lower Kimmeridgian, Hartmannsdorf, southern Germany) do also belong to the f. a. calcar

1861; Mutvei 1957), were not capable of the necessary mantle mobility. Herewith, the mechanism of the "*compensations ornamentale*" is regarded as species-specific in ammonoids.

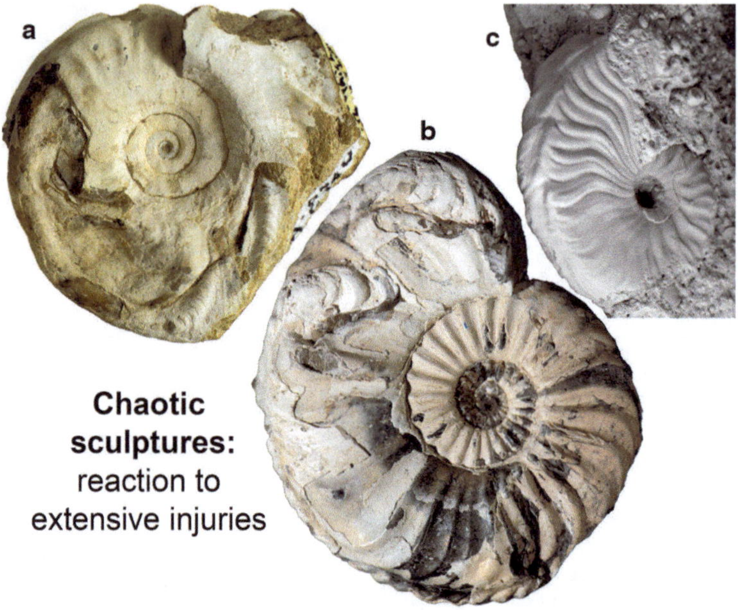

Fig. 21.9 Chaotic sculptures triggered by large-scaled injuries **a** *Eleganticeras elegantulum* (Young and Bird), *Lower* Toarcian, "*Ahrensburger Geschiebegruppe*" i.e. wellestablished glacial erratic boulder association) versus endogenic dysfunction. **b** *Pleuroceras solare* Phillips, Upper Pliensbachian of Buttenheim, southern Germany. **c** *Cleviceras exaratum* (Young and Bird), Lower Toarcian of Altdorf, southern Germany)

21.2.3 Chaotic Sculptures

A continuous shell growth obviously was not possible after a disruption of the lateral shell-secreting epithelium, due to a large scale injury or endogenic reasons that exceeds the maximal tolerable dislocation. The result is a "*chaotic*" sculptural phenomenon characterized by a repeated redeployment of shell formation. The specific mode of regeneration is particularly evident in keel-bearing ammonites. A sawtooth-like course of the keel was achieved by the iterative attachment of parts of the shell partly in order to bring the keel back to an approximately normal position but sliding back to the maximum tolerable juxtaposition. A case study of 286 pleuroceratids from the middle Lower Jurassic of southern Germany bearing an anomalous keel position demonstrated two different scenarios. First, a lateral dislocation of the keel by 20 % of the whorl height usually results in a permanent keel-dislocation; contrary dislocations of about 25–50 % of the whorl height cause usually a chaotic course of the keel (Fig. 21.9, 21.10 Keupp 2012).

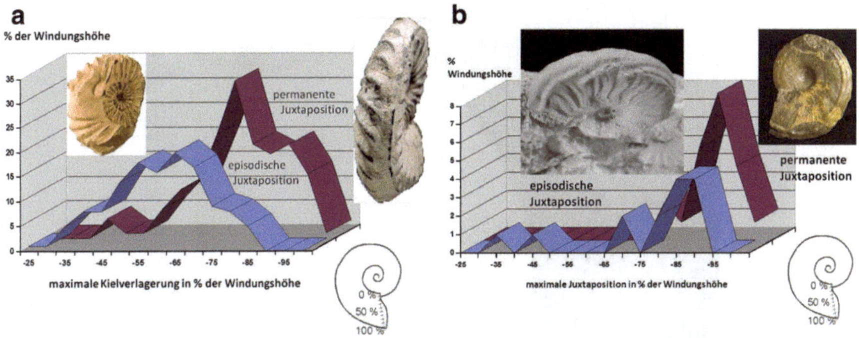

Fig. 21.10 Tolerance of permanent dislocation of the keel varies between taxonomic groups. **a** *Pleuroceras spinatum* (Bruguière), Lower Jurassic of Unterstürmig, Germany ($n = 286$) tolerates permanent dislocation of the keel (= f. a. juxtacarinata Hölder 1956) up to 20 % of the corresponding whorl height, above that dislocation will cause chaotic sculptures. **b** Contrary, the limit of tolerance in hildoceratids from the Lower Toarcian of Altdorf, Germany ($n = 35$) is only 90–95 % of the whorl height (after Keupp 2012)

21.2.4 Internal Growth-Disturbations

Damage of the aperture/peristome sometimes lead to secondary phenomena. These secondary phenomena could only be recognized in the internal mould within the body chamber. Thus, the retraction of the peristomal epithelium could occasionally cause a detachment of mantle segments from the inner surface of the conch. That detachment triggered the formation of one or more mineralized internal sheets/lamellae showing nacre ultrastructure (Fig. 21.11). Keupp (1977) introduced for

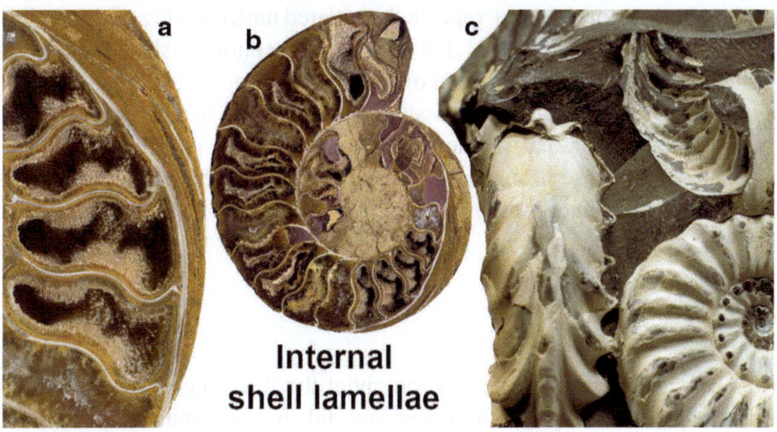

Fig. 21.11 Internal shell lamellae (f. a. aptycha Keupp 1977) significantly restrict the body chamber: **c** *Pleuroceras spinatum* Brug. from the Upper Pliensbachian of Unterstürmig, southern Germany, **a**, **b** median section showing the phragmocone of *Cleoniceras besairiei* Coll. (Lower Albian of Ambatolafia, Madagascar) with a multiple insertion of internal lamellae and fitting of septa afterwards

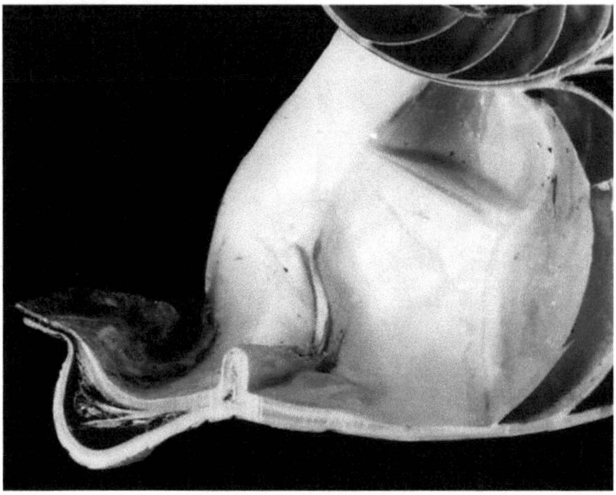

Fig. 21.12 Recent *Nautilus pompilius* Linné from the aquarium of the Jura-Museums Eichstätt, with anomalous shell bulge bridged by multiple internal shell lamellae. The lacking continuation of body chamber growth causes an extreme shortening of the body chamber due to continuous septal formation

such phenomena the f. a. type aptycha. For the same phenomenon observed for Middle Triassic (Muschelkalk) *Ceratites*, Rein (1989) suggested the f. a. conclusa (Lehmann 1990; Keupp 1994). The principle mechanism of formation of internal shell lamellae was observed in modern *Nautilus*, which were aquarium-reared at unnatural water chemistry by Keupp and Riedel (1995). Under these artificial conditions the peristome grew outwards in a bead-like form. The resulting gap between soft body and shell is like an open shell injury bridged by underpinning with a regenerative shell. A multiphased, repeatedly initiated underpinning of the shell-bulge prevents growth progress (Fig. 21.12) and causes a continuously shortening of the body chamber due to the progression of septal formation.

Injuries of shell area related to tissue attachment site usually cause, similar to intrinsic disruption, internal problems, especially for the attachment of the large retractor-muscles. These problems generated ridge-like structure at the inner shell surface (f. a. intracarinata Keupp 2000a; Fig. 21.13).

21.2.5 Aptychi

There is a long controversial discussion about the function of the double-valved, calcareous aptychi interpreted as lower jaw, lid, or a combination of both (e.g., Schindewolf 1958; Lehmann 1972, 1990; Morton 1981; Lehmann and Kulicki 1990; Seilacher 1993; Keupp and Veit 1996; Keupp 2003; Schweigert 2009; Kruta et al. 2011; Tanabe 2011; Keupp and Mitta 2013). Regenerated injuries causing

Fig. 21.13 Anomalous internal ridges serving as attachment sites of the large retractor muscle; **a** dorsolateral muscle of *Divisosphinctes* (Oxfordian, Sakaraha, SW-Madagascar), **b** ventrolateral muscle of *Amaltheus gibbosus* (Pliensbachian, Buttenheim, Germany)

partly long-term growth anomalies like growth line vertices or longitudinal bulges are also recorded for aptychi (Schindewolf 1958; Keupp et al. 1999; Engeser and Keupp 2002, see Fig. 21.14). Usually these anomalies start along the aboral margin of the aptychus pointing towards the inner body. Due to the exposed position for potential predators (usually crustaceans) of the rear edge of the aptychus when in the locked-position, injuries of the aptychus were used by Keupp et al. (1999) and Keupp (2012) as an argument for a potential lid-function of that structure. Contrary, Schweigert and Dietl (2001) and Schweigert (2009) assume a specialized jaw function for the aptychi. These authors interpret growth anomalies as a possible consequence of parasites. Damage of the lower jaw by the hard parts of their potential

Fig. 21.14 Growth anomalies of *Laevaptychus latus* from the Kimmeridgian presumably caused by injuries of the aboral edge: **a**, **b** growth line vertices, aptychus from Nusplingen. **c** longitudinal bulges, aptychus in living position, from Kirchheim, Ries, southern Germany

prey, as described by Kruta and Landman (2007) for modern nautilids, is unlikely for aptychi-bearing ammonoids (Aptychophora—see Engeser and Keupp 2002) due to the mostly microphageous diet (Keupp 2000b; Kruta et al. 2011; Tanabe 2011).

21.2.6 Paleobiological Aspects of Sublethal Injuries

The analysis of shell injuries allows for new insights in aut- and synecological aspects. One autecologic aspect, i.e., group-specific variable tolerance to compensate for growth asymmetries, was exemplified for the permanent dislocation of the median keel in succession of the "*compensations ornamentale*" (see above). In general, it seems likely that ammonoids that lived as demersal in low-energy water had a higher tolerance of shell asymmetries compared to taxa that lived close to the surface water in a high energetic milieu. Injuries are also useful to identify constructural limits of survival. Two examples below illustrate that issue

21.2.6.1 Efficiency of the Hydrostatic Apparatus

The amount of liquid, that was retained within the nautilid phragmocone in order to balance weight changes, is negligibly small (Ward 1979) with a maximum of 7 g (=about 2% of the total chamber volume). Therefore, shell loss of more than 5 g (=4% of the total shell mass) exceeds the limits of compensation (Ward 1986) and is lethal for *Nautilus* due to buoying upwards to the surface waters.

Kröger (2000, 2002a) calculated for Jurassic ammonoids the sudden shell loss of large-scaled (up to 20% of the total shell mass) but re-generated, i.e. survived, shell breakages (Fig. 21.15). Thereby, he recognized that ammonites could survive shell loss (decreasing weight) being five times higher compared to modern *Nautilus*. Accordingly, Kröger (2000, 2002a) concluded that ammonoids had flooded a larger part of their phragmocone.

21.2.6.2 Mobility of the Soft Body in the Body Chamber

Kröger (2002b) demonstrated for meso- and longidomic ammonites (cf. Westermann 1996), that shell injuries extending from the aperture to approximately half the length of the body chamber could be survived and repaired. Similar cases were reported by Keupp (2012, p. 109) for brevidomic taxa. The regenerated shell, in particular of those species with long body chambers (dactylioceratids, perisphinctids), shows growth lines and sculptural elements. They prove that the peristomal epithelium could withdraw up to the aboral edge of injury. The required flexibility of the ammonoid soft body, was facilitated due to the differentiated retractor muscle system consisting of the paired dorsolateral, ventrolateral, and attachment sites, as already pointed out for "*compensations ornamentale*". The ability of the potential

Fig. 21.15 Regenerated shell area of *"Lithacoceras" torquatiforme* Collignon from the Upper Oxfordian of Sakaraha, SW-Madagascar (67 mm in diameter, coll. Keupp, PA-10675), equates a relative weight proportion of 20 % of the complete shell (Kröger 2000), demonstrating that ammonoids could at least compensate fivetimes the amount of shell loss compared to modern *Nautilus*

retraction of the soft body reveals the ammonoid body chamber fundamentally had a volume that significantly exceeds that of the soft body.

Comparative quantitative analysis of the abundance of injuries and types of injuries observed for ammonoid taphocoenosis also provides interesting insight into synecological aspects, in particular predator-prey relationships. Using this detour may help to elucidate different life habits of ammonoids with different shell morphologies. Accordingly, Keupp and Ilg (1992) interpret decreasing abundances of regenerated injuries caused by benthic crustaceans, observed for an ammonite fauna from the Upper Callovian of Normandie, as indicator for different frequencies of demersal contacts. Therefore, Keupp and Ilg (1992) assumed a demersal lifestyle for the coarse ribbed peltoceratids (rate of injuries ca. 14 %), a habitat in the middle of the water column for cosmoceratids and *Quenstedtoceras* (rate of injuries 3–4 %), and for hecticoceratids a habitat close to the sea surface (rate of injuries 1 %). A similar interpretation can be drawn from the analysis of Upper Oxfordian ammonoids from Madagascar (Fig. 21.13). The strongly ornamented perisphinctids show the highest rate of injuries (13 %). About 1/3 of these injuries were caused by benthic crustaceans. Contrary, less sculptured lytoceratids (Protetragonites) and phylloceratids show significant lower rates of injuries (1.25–2.21 %) are presumed fish attacks.

The analysis of a total number of 1190 weakly ornamented ("*leiostracans*" and 328 strongly ornamented ("*trachyostracans*") ammonoids from the Upper Triassic of SW-Timor (Keupp 2012, Fig. 7) shows significant differences between the rates of injuries of smooth shelled species (2–6 %) and sculptured species (8–17 %). These differences confirmed that the phylogenetic megatrend within the ammonoids towards sculptured species, were a defensive strategy (Ward 1981), particularly

Taxon	Individuals	Nr. of sublethal injuries	Portion of crab attacks (f. a. verticata)	Rate of injuries (%)
1 *Taramelliceras* cf. *externnodosum*	440	5	3	1.14
2 *Protetragonites fraasi*	400	5	0	1.25
3 Phylloceratidae div. sp.	1356	30	0	2.21
4 *Aspidoceras* sp.	864	31	11 (=35%)	3.6
5 *Epaspidoceras jeanneti*	215	5	1	2.3
6 Perisphinctidae div. sp.	2593	337	113 (=33.5%)	13

Fig. 21.16 Different rates of injuries within taphocoenoses ($n = 5868$) from the *Upper*-Oxfordian of Sakaraha, Madagascar

since the earliest Mesozoic, in order to counteract the increasing predation pressure in the benthic regime (Vermeij 1977, 1982, 1983, 1987). The increasing predatory pressure was predominantly due to the radiation of malacostracan crustaceans since the beginning of the Triassic (Walker and Brett 2002).

Kröger (2002c), on the basis of another quantitative analysis of 1500 specimens of Jurassic ammonites with sublethal injuries, postulated a connection between characteristic shell features (shell form, body chamber length and sculpture) and susceptibility for injuries. His results show that the antipredatory traits were in adaptational conflict with traits demanded for high manoeuverability and streamlining.

21.2.7 Epizoans and Reactions of Shell Growth

The external shells of ammonoids were a welcome hard substrate for a plethora of different colonizers. Two categories can be distinguished within the group of epizoans: On the one hand, sessile suspension feeder, attached with their body to the shell and being partly pseudoplanctonic as free-rider on a mobile raft that ensures a varied diet. On the other hand, mobile benthic-organisms, occupying the shell and use it as a temporarily feeding ground, consuming organic particles on the shell surface or incorporated in the shell. While the first group usually colonize the shell surface during the lifetime of the ammonoid animal (="*true epoecy*" *sensu* Linck 1956), but also colonize empty shells floating in the water column or lying on the ground as substrate islands (="*false epoecy*" *sensu* Linck 1956). Contrary, the second group of grazing animals attack the shells usually post-mortem (Schindewolf 1962; Wetzel 1964; Kase et al. 1998; Seilacher 1998; Keupp and Richter 2010). Following Davis et al. (1999) we summarize, as an umbrella term, the colonization of the shell surface by filter feeding organisms, independently whether the emplacement took place during the lifetime of the host or after its death, as Epicoles. The subset of epicoles, that colonize the shell surface during the lifetime of the host are called Epizoa (=Epoecy).

In particular syn vivo epizoa are of interest in connection with paleopathological studies, since the affected ammonoids usually show characteristic reactions documented in their shell growth.

Evidence of *syn-vivo* colonization can be drawn in two ways:

1. By means of morphological reactions of the host during shell growth (cf. Lange 1932; Merkt 1966; Hölder 1970; Keupp 1984, 1992b, 1996; Davis et al. 1999; Klug and Korn 2001).
2. With the help of shell-morphological reactions of the epizoans.

A prerequisite to document noticeable reactions in shell growth of ammonoids is the colonization of a juvenile, actively growing shell. Significantly reactions are the overgrowth of free riders, which were settled in the overlapping area of the whorls. The process of overgrowth causes a whorl-bend (Philippi 1897; Ilovaisky 1917; Sornay 1955; Hölder 1970; Keupp 1984/1985, 1992b, 1996, 1997, 2000b, 2012; Klug and Korn 2002; Checa et al. 2002). Such morphologies can mimic the shape of heteromorphy ammonoids due to the loss of the epizoa by taphonomy (Quenstedt 1858, 1886/1887; Hyatt 1889; Vadász 1908; Lange 1932; Schindewolf 1934; Tasnadi-Kubacska 1962; Mitta et al. 1999; Keupp 1992b, 2000b, 2012; Keupp and Schweigert 2009; Wannemacher 2010; Frerichs 2011). On the other side, laterally attached epizoa caused counter-steering in growth direction in order to maintain balance. A trend towards a trochospiral shell was activated in some cases (Merkt 1966; Keupp 1984), but predominantly caused an oscillation of the whorls around the median plane ((Fig. 21.17) Keupp and Ilg 1992; Checa et al. 2002).

The morphological reactions of the epizoans are mainly related with their positioning at the ammonoid shell and their response to the spiral growth of the

Fig. 21.17 Reactions in shell growth due to epizoans: **a** *Owenites koeneni* (Hyatt and Smith), Upper Skythian, Crittenden Springs, Nevada with a bending whorl. **b** *Quenstedtoceras* sp. colonized with oysters, Upper Callovian, Dubki near Saratov, Russia. **d** Apparently gyrocone shell caused by diagenetically triggered Serpula-epoecy on *Gagaticeras neglectum* (Simpson), Sinemurian of Whitby, United Kingdom. **c** Eccentric growth in order to balance for laterally attached epizoa: perisphinctid from the Upper Bajocian of Sengenthal, southern Germany. **e** Apparently ancylocone shell of a *Pleuroceras spinatum* (Bruguière), Upper Pliensbachian of Aubächle near Blumberg, southern Germany. **f** *Douvilleiceras albense* Spath (Lower Albian of Mahajanga, Madagascar) showing oscillating whorls due to infestination with a soft bodied epizoan (Keupp 2012)

ammonoid shell (Seilacher 1954, 1960; Meischner 1968, 2002; Baird et al. 1989; Keupp et al. 1999; Hauschke et al. 2011; Keupp et al. 2012). For example epizoa orientate during their settlement as larvae streamlined in terms of their own suspension feeding activities. During growth of the ammonoid shell, bivalves, brachiopods and other organisms attached to the shell, rotate away from their preferred position (="*perniciöse Epökie*" *sensu* Meischner 1968).

Especially serpulids (Annelida) could offset the rotations from their preferred position by synchronizing the growth of their living tubes with the growth of the ammonoid shell (Lange 1932; Schindewolf 1934; Keupp et al. 2012; Fig. 21.14).

21.2.8 Paleobiological Aspects of Anomalies Caused by Epizoans

The interpretation of close organismic interactions between the now extinct ammonoids and their epizoa, whose modern representatives allow us to draw some conclusions about their needs, open up a broad range of interesting paleobiological aspects, for example:

- habit and life style of ammonoids (position of the peristome, predominant swimming direction)
- age
- efficiency of the hydrostatic apparatus

The nature of organismic interactions between host and epizoans whether symbiosis, commensalism, or parasitism, usually remains speculative (Zapalski 2011). In most cases the alliance between host and epizoan was to the detriment of both (="*perniciöse Epökie*": Meischner 1968). The interaction between *Serpula raricosta* Quenstedt and various evolute ammonoids is largely restricted to the Lower Jurassic (Lange 1932; Schindewolf 1934; Merkt 1966; Müller 1982; Jäger 1991; Hungerbühler 1992; Keupp 1992b, 2000b, 2012), and indicates a possible adaptative relationship due to a single strategy. This strategy implies that the larvae of Serpula infested the umbilical area of juvenile ammonoids. During growth the shortest route was chosen to bring their living tube to the flow-exposed, ventral site of the shell. In order to keep its position the tubeworm could compensate further growth of the ammonite shell by lengthwise growth of its tube in the same direction. Older portions of the Serpula tube become overgrown by the spiral shell of the ammonite and incorporated between two whorls forming a double helix of same growth direction (Fig. 21.18 right).

Accumulations of colonized ammonoid shells were mainly reported from the margins of marine basins, due to the distribution of suspensions feeders that attached their shells to the substrate (here the ammonoid shell), in coastal, high-energetic shallow marine settings (e.g., ceratitids of the German Muschelkalks: Philippi 1897; Linck 1956; Wenger 1957; Mayer 1975; Hagdorn and Simon 1985; Rein 1996; Suchopar 1997; Knoch 1989; Meischner 1968, 2002; *Quenstedtoceras* from the Callovian of Saratov, Russia: Seltzer 2001; Larson 2007; Keupp 2012; or aspidoceratids from the Kimmeridgian of Brunn, southern Germany, Bavaria: Keupp et al. 1999, 2012; Seilacher and Keupp 2000 and others). The colonization of the ventral center of mass by various epizoa (Seilacher 1960; Keupp et al. 1999, 2011; Seilacher and Keupp 2000) shows two things: the position of the peristome during the life time of the ammonoids and that the affected individuals had an "*epidemersal*" life style (Westermann 2013), otherwise the downward hanging epizoa

Fig. 21.18 *Left:* Concentration of epizoa attached to the centre of mass of a *Physodoceras* sp. (cirripeds: *Pollicipes*) and serpulid tube growing synchronous with the spiral shell of an ammonoid, both indicating syn-vivo infestination and an upward orientation of the aperture; Upper Kimmeridgian, platy limestones of Brunn, southern Germany, Bavaria. *Right:* Concordant growth direction of the ammonoid shell and serpulid tubes attached to the shell of *Schlotheimia angulosa* Lange (Hettangian of Holzen, Germany) suggests a possible adaptation of the worm to its favoured substrate (Keupp 2000b)

(cirripeds, serpulids, bivalves) would be stifled. Studies of the Upper Kimmeridgian *Physodoceras* from the platy limestones of Brunn (southern Germany) show that epizoans (such as cirripeds, serpulids, and bivalves) tend to be concentrated on the ventral center of mass indicating the peristome had an orientation of about 70–80° upwards (Fig. 21.15; Keupp et al. 1999). This is consistent with calculated values by Trueman (1941) derived from the ratio of phragmocone to body chamber length and demonstrates their basic orientation in the water column. Also the settling of *Stramentum* (cirripeds) at the centre of mass of the orthocone *Sciponoceras* (Upper Cretaceous) points toward a horizontal orientation of this heteromorph ammonoid with an upwardly directed peristome (Fig. 21.19; Hauschke et al. 2011). Another argument comes from mathematical modelling of some *Pavlovia* shells from the Tithonian of Russia. The whorls of these specimens bend and oscillate due to attached oysters and serpulids. The models show that the deformation of the shell served to keep a constant position of the peristome during shell growth/ ontogeny (Fig. 21.21; Checa et al. 2002).

Cirripeds always orient their feeding apparatus against the water current (Seilacher and Keupp 2000). The ammonoid shells infested with cirripeds clearly show a preferred orientation towards the peristome (Keupp et al. 1999; Hauschke et al. 2011). This proves that ammonoids preferred a frontal water current hitting the peristome first. Accordingly, Seilacher and Keupp (2000) assumed a predominantly forward migration pattern for all Mesozoic ammonoids. It reveals that, in accordance with the loss of the ventral sinus, since the beginning of the Triassic the jet propelling mechanisms plays no important role for movement.

There are several attempts to reconstruct the potential period of growth and age of ammonoids based on the growth rate of epizoans. Although such calculations are

Fig. 21.19 Position of *Stramentum pulchellum* (Sowerby), a cirriped, at the ventral centre of mass of a Sciponoceras sp. shell from the Upper Cenomanian of Lengerich, NW-Germany indicates a horizontal swimming position with upwardly directed peristome (from: Hauschke et al. 2011)

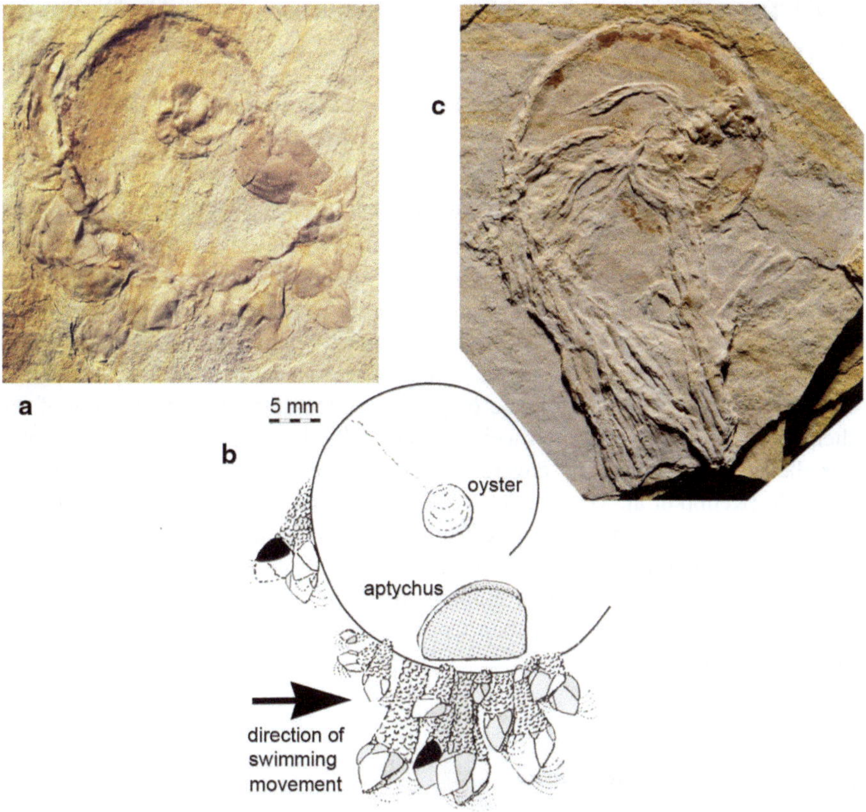

Fig. 21.20 Concentration of Epizoa on a *Physodoceras* sp. shell from the Upper Kimmeridgian of Brunn, southern Germany (Bavaria), platy limestone, marking the centre of mass and points to a 70–80° upward orientation of the peristome. The forward orientation of the filter apparatus of these suspension feeders like *Pollicipes* sp. (cirripeds in **a** and **b**) and serpulids (**c**) indicate a predominantly forward migration pattern of these ammonoids

Fig. 21.21 Mathematical simulation made by T. Okamoto of bending whorls caused by epizoans for *Pavlovia* cf. *iatrensis* Ilovaisky from the Tithonian of the polar Ural; it shows that during the phase of anomalous shell growth the peristome was orientated constantly in its normal position (from Keupp 2000b)

subject to many uncertainties, Schindewolf (1934); Seilacher (1960); Merkt (1966), and Meischner (1968, 2002), independently come to the same conclusion, that an average-sized ammonite had several years of growth followed by at least 3 more years as adults. The life expectancy of ammonoids thus corresponds approximately with that what is known for the modern *Nautilus* (Landman and Cochran 1987) reaching sometimes an age of about 20 years (Dunstan et al. 2011).

While modern nautilids tolerate only small amounts of epizoa (3–16 % of the shell surface or 1 % of the shell weight; Zann 1985; Landman et al. 1987), ammonite shells nearly completely covered with epizoa seem to keep their epi-demersal life-style (Keupp et al. 1999). It proves that ammonoids, as compared to modern *Nautilus*, possessed a much more effective hydrostatic apparatus. Kröger (2000), based on a number of large sublethal injuries, calculated that Jurassic ammonoids could handle significantly larger amounts of liquid within the phragmocone (see above) and could tolerate more than five times the shell losses compared with the extant *Nautilus*.

21.3 Endogenic Reasons

Anomalies that are not caused by injuries or epizoans are best summarized as endogenic or intrinsic. External causes for this type of pathologies cannot be recognized. Intrinsic pathologies are characterized by an usually increasing intensity of the anomaly.

Most endogenous ammonoid shell anomalies, attributed to dysfunctions of the soft body due to parasite infestation or at least the possibility of a parasitic infestation, is discussed (e.g., Rieber 1963; Keupp 1976, 1979, 1984/1985, 1995, 1997; Hengsbach 1990, 1991, 1996; Kröger 2000; Larson 2007; De Baets et al. 2011, 2015).

21.3.1 Gigantism and Dwarfism

Extraordinarily large ammonoid species repeatedly occur during transgressive phases of Earth history (Wiedmann 1973; Johnson 1984; Stephens 1988; Klug 1999). For example, some Upper Devonian manticoceratids and Permian metalegoceratids achieve shell diameters up to 60 cm, while some Upper Cretaceous parapuzosiids reached up to 3 m in shell diameter (Krüger 1984; Stephens 1988). Other cases of genetically controlled size differentiation within a certain ammonoid species can be seen in pronounced sexual dimorphism (Callomon 1963; Makowski 1963; Lehmann 1990; Keupp 2000b).

For pathological cases of gigantism, that affected only single specimens of otherwise growth-restricted species, Ivanov (1975) introduced the name megaconch. They can be seen as a pathological deregulation of the species specific growth limitation. Therefore, the shells of these megaconchs generally show no signs of growth limitation like septal crowding or deterministic differentiation of the aperture.

Manger et al. (1999) list a larger number of Upper Paleozoic species, the phenomenon of pathological gigantism. They also reported that affected individuals exceed the shell diameter by two to four times as compared to their normal counterparts with whom these are associated in the same beds (Stephen 1997). As a potential reason for pathological gigantism Manger et al. (1999) discussed a possible infestation of the gonads with trematod larvae. Analogous to modern cases of gigantism in terrestrial, tropical gastropods, trematod larvae prevent the advent of sexual maturity, and with it the hormone-controlled growth stop (Hölder 1956, 1960; Bucher et al. 1996; Davis et al. 1996).

The term dwarfism for dwarfish ammonites was coined by Tasch (1953). Ager (1963) distinguishes between genetically-fixed dwarfism and ecologically-related growth limitations ("*stunting*"). The former, partly representing phylogenetic end members of a neotenic development, e.g., *Cymbites laevigatus* (Sowerby) from the Pliensbachian (Donavan 1957), *Trochleiceras magneti* Collignon from the Albian, or partly representing a pronounced sexual dimorphism (e.g., *Oecoptychius* from the Callovian; Schweigert and Dietze 1999) are not related with pathological phenomena.

The so-called "*Miniconchs*", which Matyja (1986) described for Callovian *Quenstedtoceras*, possibly documents a hormonally controlled early onset of sexual maturity in the sense of "*Microgerontie*" (Schmidt 1926). Miniconchs usually grow no larger than 1/3 of the normal sized microconchs and 1/8 of macroconchs. Keupp (2012) discussed a possible dwarfism for *Pleuroceras apyrenum* (Buckman), since no sexual dimorphism was detected for pleuroceratids (Fig. 21.22).

Fig. 21.22 Possible "Miniconch" of *Pleuroceras apyrenum* compared with normal-sized specimen (middle Lower Jurassic, Buttenheim, Germany). The small specimen shows characteristic features of an adult shell like final flattening of the ribs, and the depression of the ventral crenulated keel

21.3.2 Disturbance of the Septal Apparatus

The chamber formation in ammonoids was performed by the rear of the mantle sack. Septal formation was, most likely, temporally, spatially, and functionally independent of the growth processes at the peristome (Keupp and Riedel 1995). A regular and episodic phenomenon was observed on pre-adult shells, which ensured the regulation of weight equilibration of the growing and weight gaining animal. Ammonoid septal formation, however, was subject of a high flexibility in their species-specific but also individual configuration (Keupp 1995). Thus, orientation and symmetry of the septa within the conotheca (conch), as well as the relative distance from each other, vary considerably from specimen to specimen (Bayer 1977b).

The reactive flexibility to spontaneous changes in the geometry of the tube-shaped outer shell, which represents the frame in which the septum was formed, was considerable but independent from phylogenetic and ontogenetic controlled complexity. Perhaps, the punctual attachment of the organically preformed septa within the tube promoted that flexibility (Seilacher 1975). Sculptured ammonoids with a corrugated-metal-like shell for example were able to adapt their septa to constricted intercostals and widened costal sections of the conotheca. In addition, ammonoids were also able to cope with extreme changes of the whorl cross-section, due to pathological processes and proceed with the septal formation. This applies to both reduced and expanded cross-sections through anomalous internal shell lamellae (f. a. aptycha Keupp 1977) or bubble-like outgrowths (f. a. inflata Keupp 1976) (Fig. 21.23; Keupp 1994, 1995, 2000b, 2012). The extreme adaptability during the formation of the septal apparatus ensures its hydrostatic function and can be assumed as an essential survival strategy.

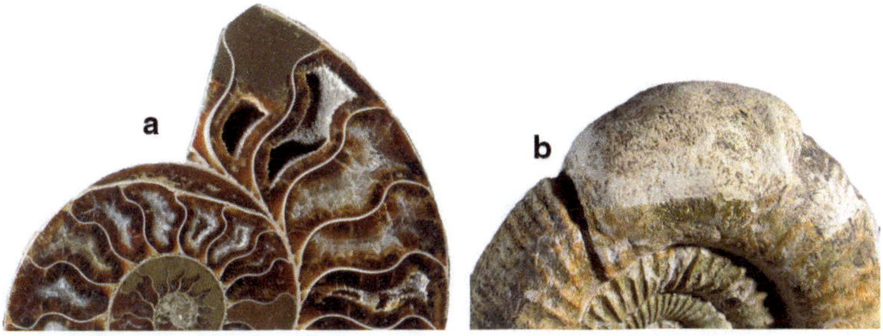

Fig. 21.23 Two examples showing the high flexibility of septal formation after anomalous changes of the whorl cross section. **a** *Cleoniceras besairiei* Collignon (Lower Albian, Mahajanga Basin, Madagascar), despite the reduction of the whorl height by 35 % caused by the f. a aptycha it could form new septa. **b** *Orthosphinctes* sp. (Lower Kimmerdigian of Neumarkt, southern Germany) shows complexly folded suture lines also within its bubble-like extension of the whorl cross-section (f. a. inflata Keupp 1976)

First evidences of dysfunctions of the septal apparatus of fossil cephalopods with an external shell caused by pathologies go back to Kieslinger (1926). He described incompletely preserved septa for Mesozoic nautilids. However, the nature of the phenomenon described by Kieslinger remains controversial. It either could represent primary incompletely formed septa or was due to partial diagenetic dissolution (Hölder 1956).

For ammonoids a number of different phenomena of anomalous septal formation can be distinguished based on numerous findings:

21.3.2.1 Anomalous Septal Spacing and Orientation

In most cephalopods a growth stop is correlated with the onset of sexual maturity. Septal crowding indicates the final growth phase for pre-adult to adult ammonoids. Pre-adult septal crowding were partly triggered by rhythmic deceleration of growth due to specific and regular modifications of the sculpture (e.g., *Horrioceras*, Callovian; Bayer 1977b; Keupp 2012), and partly due to pathological growth anomalies during the regeneration of injuries or due to intrinsic or ecological reasons (Eichler and Ristedt 1966). Septal spacing can be reduced to almost zero in extreme cases, which results in the impression that the septa are in direct contact or bifurcate (Fig. 21.24 Right; Blind and Jordan 1979).

The opposite of septal crowding is the anomalous enlargement of septal spacing, which gives the impression of a septal loss (= f. a. dissepta Hölder 1956). Although most of these cases are better attributed to selective diagenetic secondary dissolution of the septa (Hölder 1956; Keupp 2012), individual cases in goniatites and ceratites seem to be primary in nature (Becker et al. 2000; Müller 1978; Rein 1990, 1997). One-sided enlargements of septal spacing can also by caused by anomalous inclination of individual septa (Fig. 21.25).

- Perturbance of the septal morphology

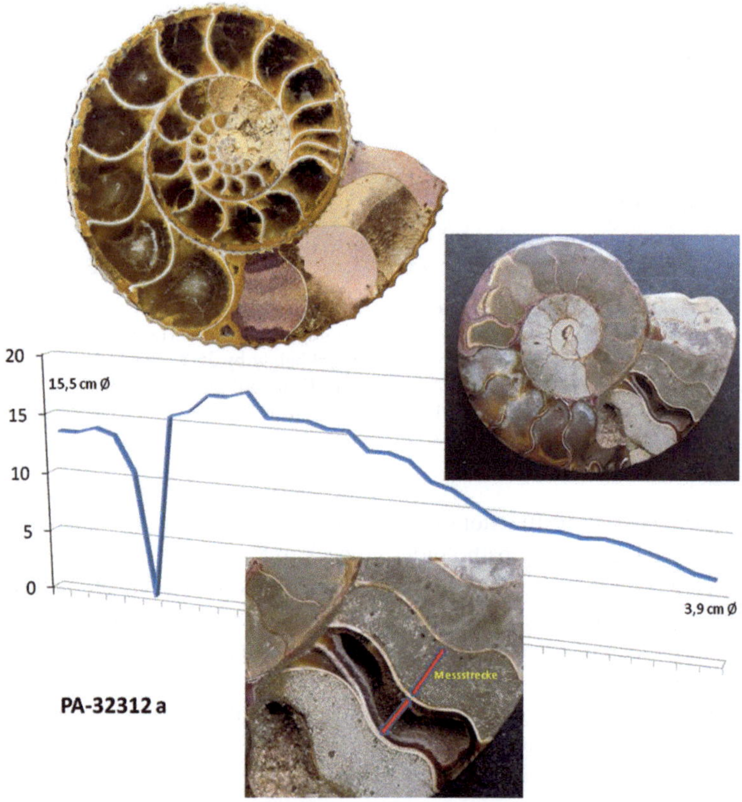

Fig. 21.24 *Left:* Median section through the phragmocone of a perisphinctid (Upper Oxfordian, SW-Madagascar, coll. Keupp, coll. no. PA-32896) with anomalous, pre-adult (endogenic) septal crowding, *Right: Cleoniceras besairiei Collignon* (Lower Albian, Mahajanga Basin, Madagascar) with apparently bifurcated septum caused by extreme pre-adult septal crowding

The general flexibility of the septal formation enabled many individuals, even if the whorl cross section was modified, e.g., deformation of the outline, increasing or decreasing volumes (Keupp 1994, 1995), to form their septa morphologically modified. This could cause asymmetric septal surfaces (="*adaptive septal-malformation*" Keupp 2012; Fig. 21.26).

21.3.2.2 Disturbance of Suture Line Symmetry

A symmetropathy of the body chamber usually does not cause a corresponding asymmetry of the phragmocone due to the independent formation mechanisms and functions of the conotheca and phragmotheca (Nicolesco 1921, p. 49). Thus, only a few individual observations of a coupled asymmetry of conotheca and phragmotheca exist (e.g., Fraas 1863, pl. 1: 1.3; Rieber 1963).

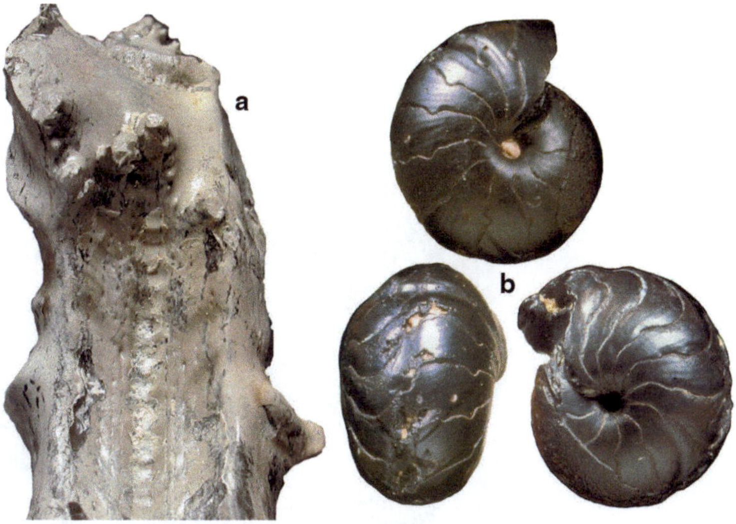

Fig. 21.25 Anomalous inclination of septa shown for *Pleuroceras salebrosum* (Hyatt) from the Upper Pliensbachian of Buttenheim, southern Germany (**a**) and for *Cheiloceras undulosum* (Münster) from the Upper Devonian of Fezzou/Morocco (**b**)

If one side of the rear soft body is deformed, and causes an asymmetric position within the shell, the described flexibility during the septal attachment would allow a modification of the septal formation. Thereby the original inclination is reflected by the asymmetric suture line. That process often results in different degrees of complexity of the suture line on both sides (Schindewolf 1961; Hengsbach 1976; Keupp 2012).

A frequently encountered phenomenon since the Paleozoic is the offset of the siphuncle from its ventral, median position (Fig. 21.27). Accordingly, the external lobe was displaced towards the right or left flank. At the same time the bilateral symmetry of the conotheca was retained. Other parts of the suture line, including the internal lobe are usually not affected by the asymmetry (Hölder 1956; Wiedmann 1972; Hengsbach 1986a). However, an analogue asymmetry was observed for the attachment site of the ventro-lateral retractor musculature (Landman and Waage 1986). New calculations of the effect of an offset siphuncle of *Badouxia columbiae* (Frebold) from the Sinemurian (Lower Jurassic) by Longridge et al. (2009) have shown that the effect of the eccentric position of the siphuncle is negligible for the position and the balance of the ammonoid shell within the water column due to its small mass as compared to the soft body of the animal. The asymmetric position of the siphuncle simply forced expansion of the septal suture line and septal folds on the non-siphuncle side of the shell but no significant counterbalance was in effect.

Within the ammonoids, we can distinguish three different groups in terms of the preferred constancy of the position of the external lobe:

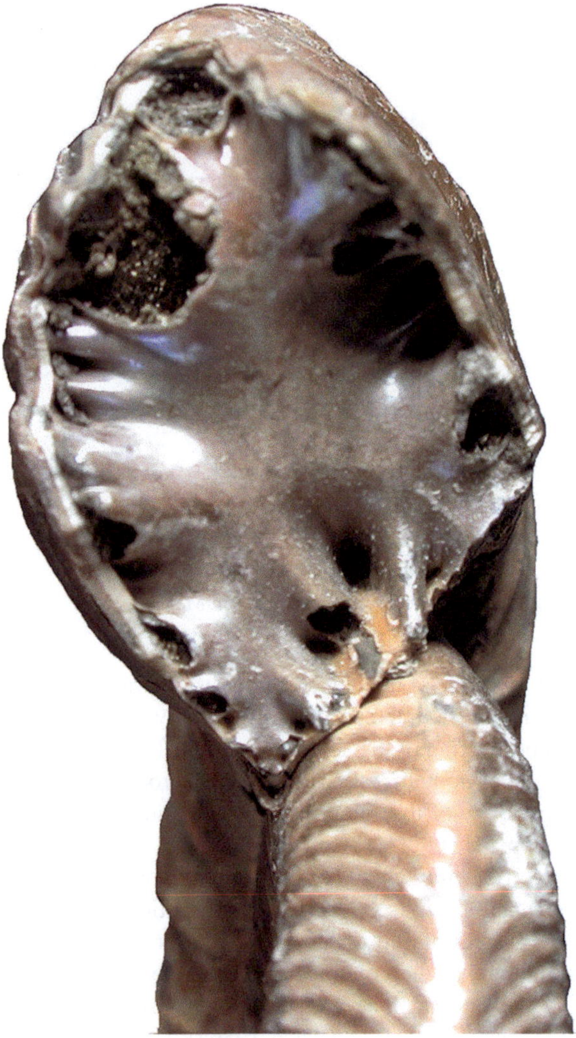

Fig. 21.26 Adaptive septal malformation caused by perniciöse epoecy deforming the conch tube (conotheca). *Quenstedtoceras* sp., Upper Callovian of Dubki near Saratov, Russia

- A species-specific constant eccentric position of siphuncle and external lobe for example characterizes the Lower Cretaceous platylenticeratids (particularly in the Boreal Realm, Kemper 1961) and *Anahoplites* (Hölder 1956; Hengsbach 1978).
- In some taxa [e.g., cheiloceratids and armatids (Upper Devonian), psiloceratids (Blind 1963) and schlotheimiids (Hettangian), arnioceratids (Sinemurian), Cymbites (Pliensbachian), *Hecticoceras* (Callovian), glochiceratids (Ziegler 1958), *Physodoceras* (Upper Jurassic), different hoplitid taxa, as well as numerous scaphitids (Upper Cretaceous)], the onset of an unstable positioning of the external lobe after the ammonitella stage results in frequent deviation from symmetry of larger portions of the population (Lange 1929, 1941; Hölder 1956;

Fig. 21.27 *Latanarcestes noeggerathi* (Buch), Lower Devonian, Tafilalt, southern Morocco with eccentric position of the siphuncle and the external lobe (= forma aegra juxtalobata Hölder 1956; from Keupp 2012)

Schindewolf 1961; Hengsbach 1976, 1977a, 1977b, 1980, 1986a, b). The degree of asymmetry varies between negligible and significant without a preferred direction (left or right).

- For *Hoploscaphites nicoletti* from the Upper Cretaceous, Landman and Waage (1986) have shown that asymmetries predominantly occur during the final growth phase of macroconchs. About 86% of the affected specimens indicate a deviation to the right.
- Most Mesozoic ammonoid taxa are characterized by a constant ventral, median position of the siphuncle and the external lobe. Deviations are pathological and affect single specimens of a population.
- Hölder (1956) introduced the f. a. juxtalobata based on his observations of *Harpoceras* (Toarcian) and *Taramelliceras* (Upper Jurassic) for such cases.

21.3.2.3 Suture-Inversion

To date the question to what extent the different orientation of the septal concavity between nautiloids and ammonoids refers to different pressure regimes during the septal formation, is controversial. Ward (1987) demonstrates for the modern *Nautilus* that low/negative pressure exists after the preforming organic phragma of the septum was mineralized. That negative pressure creates an aboral oriented tension to the corresponding concave septum. The last chamber is still filled with liquid during the septal formation process and initial mineralization processes. Therefore, a uniformly directed pressure to all sides exists and the septum itself is formed without facing directed stress. Several authors postulated for ammonoids with generally convex, arched (towards the peristome) septa (Westermann 1975; Bayer 1977a, 1977b; Checa and Garcia-Ruiz 1996), a septal formation under high pressure or stressed conditions. In this case, stress was caused by muscular tension (Seilacher 1975; Seilacher and LaBarbera 1995). Bayer (1978) assumes that high pressure conditions within the ammonoid chambers are necessary for the morphogenesis of the septa that are punctually attached to the conotheca ("*Pneu-Modell*"). According to this concept the configuration of the suture line, with its rounded saddle endings and pointed lobe endings, represent the expression of that tension. Any reversal of the suture line morphology, i.e., the formation of pointed saddles and rounded lobes, was therefore excluded categorically by Bayer (1978). However, Ward and Westermann (1976) and Henderson et al. (2002) report such a suture line inversion for the two heteromorph ammonoid genera *Glyptoxoceras* and *Baculites*. For the former the suture line inversion, including pointed saddles and rounded lobe endings, only occur temporarily during the juvenile stage (septa 2–6). During subsequent growth stages, the suture line returned to normal. For *Baculites* from the Santonian of the Western Interior, the overall morphology of the suture line and septal spacing was similar to its conspecifics. However, a continuous modification with pointed saddles and rounded lobe endings was reported. The authors conclude, from these extremely rare cases of suture line inversion, that the configuration and formation of the ammonoid septa was carried out under uniform hydrostatic pressure and therefore comparable to the modern *Nautilus*. The reason for the suture line inversion reported for *Baculites* is seen by these authors as a genetic defect. Thus, the rear visceral mass including the surrounding mantle has been shaped independently. On the other hand, particularly for *Baculites*, the difference between pointed lobe- and rounded saddle endings are less significant. Following the "pneu-model", the tension and the postulated high pressure in the last chamber would already have been small. The assumption that straight shells had flooded a large part of their phragmocone with liquid in order to ensure a more or less horizontal swimming position (Hauschke et al. 2011) supports the idea of a more or less uniform pressure load on the forming septum. Therefore, the restricted observation of suture line inversions just for heteromorph ammonoids supports the "*pneu model*" of septal formation of normal planspirally coiled ammonoid shells.

21.3.2.4 Anomalous Simplification of Sutures

In this context the description of a *Brasilia* from the Middle Jurassic of Dorset with a pathological simplification of the suture by Rieber (1979) is interesting to be mentioned (Fig. 21.28). The specimen shows an inclined septum with one side unusually distant from the preceding septum. Its configuration shows simplified saddles and lobes. The non-subdivided lateral lobe appears broadly rounded and the rudimentary umbilical saddles have rather pointed protuberances. Rieber (1979) interprets the anomaly due to the asymmetric detachment of the rear subepithelial musculature and thus a shift of the "*Muralleiste*" (=mural ridge, *sensu* Blind 1975).

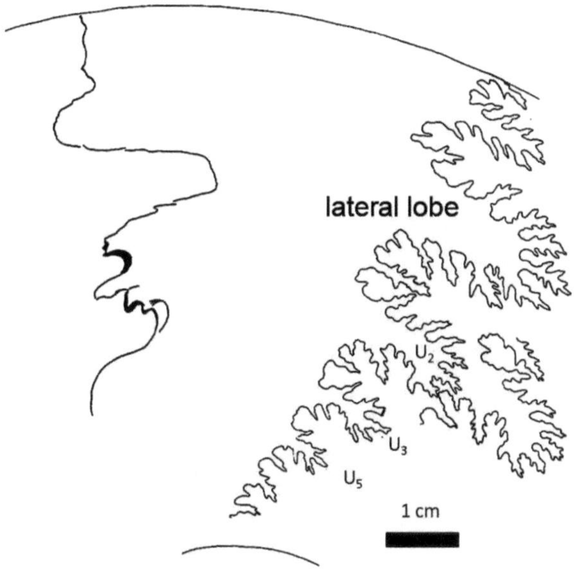

Fig. 21.28 Anomalous simplified and inverted suture line of *Brasilia decipiens* (Buckman) from the Middle Jurassic of Horn Park near Beaminster, diameter 17.7 cm. Drawing after Rieber (1979, Fig. 2; from Keupp 2012)

The mural ridge as forming organ for the septal wall no longer takes the usual course. An alternative interpretation can be drawn from the fact that the suture line is not only simplified but also inverted. This can be explained by a detachment of the aboral body portion that causes an anomalous low pressure and subsequently a collapse of the final, exceedingly large, chamber.

21.3.2.5 Pneumosepta

The term "*pneumosepta-syndrom*" was coined by Keupp and Mitta (2004) based on a *Quenstedtoceras* (Upper Callovian). The anomalous forward projected septum shows that during its formation there was a high pressure, which had already deformed the organic pre-septum. The reason for that septal projection is a breakaway of the siphuncle from its ventral position and could be visualized using computed tomography images. Thus, the foramen (point of passage of the siphuncle through the septum) was displaced dorsally due to the anomalous bulged septum. Another case of a "*pneumoseptum*" was reported by Keupp (2012) for *Douvilleiceras* (Lower Albian).

Occurrence of pneumosepta within planispiral ammonoids again support the "*pneu-model*" for chamber formation (Bayer 1978). The examples (Fig. 21.29) have shown that at least for individual cases the chamber formation and suture line configuration can also take place under adorally directed tension. Thus, it becomes conceivable that ammonoids, contrary to the modern *Nautilus*, form their chambers regularly under slightly higher pressure conditions.

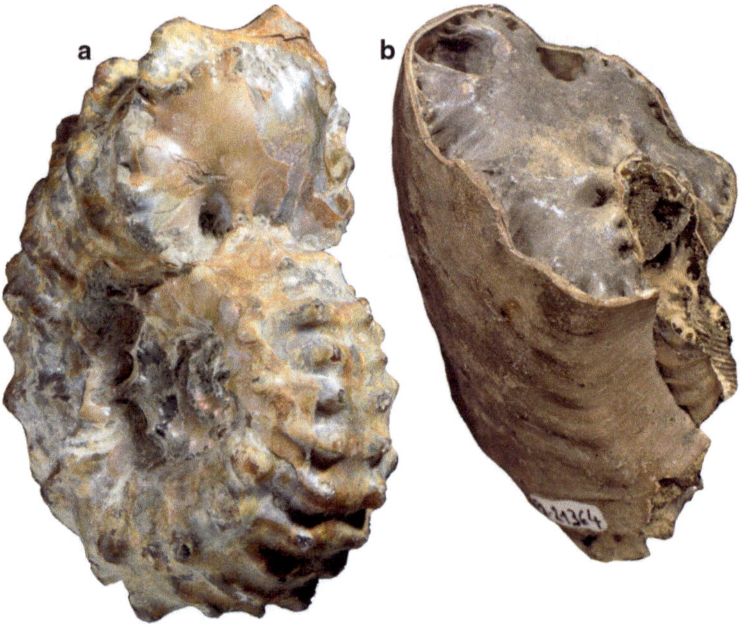

Fig. 21.29 Anomalous bulged septa indicate high pressure conditions within the chambers during the septal formation process ("pneumosepta-syndrom"). **a** *Quenstedtoceras* sp. from the Upper Callovian of Dubki near Saratov, Russia (from Keupp and Mitta 2004), **b** *Douvilleiceras mammillatum* (Schlotheim) from the Lower Albian, Mahajanga Basin, Madagascar (from Keupp 2012)

21.4 Anomalies Caused by Physiological Disturbations

Only a few descriptions about physiological malfunctions in succession of diseases (e.g., virulent tumors) are available for extant cephalopods (e.g., *Octopus* see Hanlon and Forsyte 1990; Rungger et al. 1971). In fact, these are fairly common in other molluscs (e.g., bivalves see Sparks 1972). In contrast, multiple disruptions of shell formation, especially the increased incorporation of black lines due to ecological stress especially of aquarium reared nautilids, were described (Carlson 1987). Adult nautilids develop a marginal thickening of their peristome and a black band of conchiolin which is of individual thickness in their natural habitat (Ward 1987). Almost all captive nautiloids show a typical response to stressed conditions in their shell, namely the reduced growth of the conotheca. The slowdown which simulates a complete growth stop for short episodes causes the occurrence of a sequence of pre-adult black bands. These pre-adult black bands will be incorporated into the shell during the next growth phase. The result is a shell with a high proportion of black polymeric proteins incorporated within the strong growth lines (Saunders and Landman 1987). Potentially, the stretch zones that were reported by Landman and Waage (1986) for ammonoids represent similar phenomena. Growth stress during the morphological transformation of the final, hook-shaped living chamber of

Fig. 21.30 *Kranaosphinctes rabei* Collignon with a stretch zone *sensu* Landman and Waage (1986) in connection with the development of parabolic ribs (Upper Oxfordian, Sakaraha, SW-Madagascar)

Upper Cretaceous scaphitids causes an anomalous sculpture at the rear living chamber according to Landman and Waage (1986). That sculpture consists of an anomalous ventral depression and is reduced to densely packed growth lines.

Comparable temporary growth retardations can also be found in perisphinctids in connection with the formation of parabolic ribs (Fig. 21.30). Stretch zones are not to be confused with the final weakening of the sculpture, e.g., on the adult body chamber, which were described as so-called "senile-sculptures" in *Pleuroceras* from the Lower Jurassic (Frentzen 1937). Pre-adult stretch zones, without changes of the shell, rather indicate a physiological or ecological stress.

"*Kümmerwuchs*" (stunting *sensu* Ager 1963) has been postulated for various assemblages. Stunted populations of a certain species or species associations are due to certain environmental parameters. Therefore, these stunted forms represent ecophenotypes and thus, they are not pathological. However, it is difficult to distinguish between ecologically-induced stunting and true pathologies. Stunted ammonoid fauna, in which all representatives are smaller compared to other localities have usually been described as "*dwarf fauna*" (e.g., Sturani 1971). Stunted fauna can be caused by specific environmental situations. These parameters, including food availability, oxygen depletion, hazardous substances, or geographical restriction of the habitat, could have negatively influenced the growth of ammonoids (Tasch 1953; Vogel 1959; Hallam 1965; Clausen 1968; Wendt 1971; Krystyn et al. 1971; Wagenplast 1972; Mignot 1993; Mignot et al. 1993; Besnosov and Mitta 1996; Marchand et al. 2002). Another example comes from the Upper Triassic of Southern Tyrol with its peculiar miniaturized mollusc and brachiopod fauna. Richthofen (1860) and Laube (1869) discussed unfavourable life conditions within the Cassianian Basin, while Fürsich and Wendt (1977) interpreted them as remains of a sea-weed meadow. However, due to the presence of primary small ammonoids (e.g., *Lobites*), preservation of numerous juvenile forms (Fuchs 1871) and due to taphonomic phenomena, a stunting fauna was in some cases partly faked. Only the fully chambered inner whorls and juvenile specimens were recorded within the marly facies (Urlichs 2001, 2004). Comparative studies of septal spacing between the Cassian fauna and the contemporaneous Hallstatt limestones showed that no septal crowding occurred in trachyceratids, proarcestids, lecantitids, and megaphyllitids from the Cassianian Basin, i.e. a true stunting was not verifiable for these taxa. For *Lobites* it seems apparent that the amount of very small "stunted" form is rather artificial and related to the intense collecting of the larger specimens (Urlichs 1994).

21.5 Conclusions

The presented pathological phenomena in ammonoids, which are related with shell anomalies and disturbances in the growth of aptychi, are caused partly exogenously (injuries, epibiosis), and partly endogenously ("*diseases*", parasitism). It turns out that repeatedly occurring morphologically similar "*standardized*" anomalies are often overinterpreted and led to incorrect taxonomic and phylogenetic evaluations. For these "*standardized*" anomalies Hölder (1956) introduced in open nomenclature the term forma aegra which currently is in common usage. The etiology of anomalies and the reaction of affected ammonoids open a broad range of paleobiological interpretations. This applies equally to aut- and synecological aspects. As examples of autecological aspects we discussed: the hydrostatic efficiency of the phragmocone, the mobility of the soft body in the living chamber, life- and feeding habits, and the lifespan of the ammonites. Synecological aspects were developed on the basis of possible coevolutionary processes of parasites and epizoa, and especially of predator-prey relationships.

References

Ager DV (1963) Principles of paleontology. McGraw Hill, New York

Baird GC, Brett CE, Frey R (1989) "Hitchhiking" epizoans on orthoconic cephalopods: preliminary review of the evidence and its implications. Senck leth 69:39–465

Baluk W, Radwanski A (1996) Stomatopod predation upon gastropods from the Korytnica Basin, and from other classical Miocene localities in Europe. Acta Geol Pol 46:279–304

Bayer U (1977a) Cephalopoden-Septen, Teil I: Konstruktionsmorphologie des Ammoniten-Septums. N Jb Geol Paläont Abh 154:290–366

Bayer U (1977b) Cephalopoden-Septen, Teil II: Regelmechanismen im Gehäuse- und Septenbau der Ammoniten. N Jb Geol Paläont Abh 155:162–215

Bayer U (1978) The impossibility of inverted suture lines in ammonites. Lethaia 11:307–313

Becker RT, House MR (1994) International Devonian goniatitic zonation, Emsian to Givetian, with new records from Morocco. Courier Forschungsinst Senck (Willi Ziegler Festschrift II) 169:79–135

Besnosov NV, Mitta VV (1996) "Dwarf" ammonites from the Calloviense Zone of the Great Balkhan (Callovian, Western Turkmenistan), their ecological and taphonomic environments. Paleontol J 30(4):389–395

Blankenhorn M (1887) Über Ceratiten des oberen deutschen Muschelkalks. Verh Naturhist Ver preuß Rheinl Sitz Ber 44:28

Blind W (1963) Die Ammoniten des Lias alpha aus Schwaben, vom Fonsjoch und Breitenberg (Alpen) und ihre Entwicklung. Palaeontogr A 121:38–131

Blind W (1975) Über die Entstehung und Funktion der Lobenlinie bei Ammonoideen. Paläontol Z 49:254–267

Blind W, Jordan R (1979) "Septen-Gabelung" an einer *Dorsetensia romani* (Oppel) aus dem nordwestdeutschen Dogger. Paläontol Z 53 (3/4):137–141

Böttcher J (1938) Versteinerungen des Oberen Muschelkalkes bei Ohrdorf als aufschlussreiche Dokumente für die Geschichte des deutschen Muschelkalkmeeres. Beitr Geol Thüringen 5: 99–105

Bucher H, Landman NH, Klofak SM, Guex J (1996) Mode and rate of growth in ammonoids. In: Landman NH, Tanabe K, Davis RA (eds) Ammonoid paleobiology. Plenum, New York

Bülow Ev (1918) Über einige abnorme Formen bei den Ammoniten. Z dtsch geol Ges, Monatsber 69:132–139

Busse E (1954) Profil der Unteren und Mittleren Ceratitenschichten vom Eisenberg bei Hessisch-Lichtenau und Walburg. Notizblatt hess LA Bodenforsch 80(3):118–137

Callomon JH (1963) Sexual dimorphism in Jurassic ammonites. Trans Leicester Liter Philos Soc 57:21–56

Carlson BA (1987) Collection and aquarium maintenance of *Nautilus*. In: Saunders, WB, Landman NH (eds) *Nautilus*, the biology and paleobiology of a living fossil. Plenum, New York

Checa A, García-Ruiz JM (1996) Morphogenesis of the septum in ammonites. In: Landman NH, Tanabe K, Davis RA (eds) Ammonoid paleobiology. Plenum, New York

Checa AG, Okamoto T, Keupp H (2002) Abnormalities as natural experiments: a morphogenetic model for coiling regulation in planspiral ammonites. Paleobiology 28:127–138

Claus W (1992) Der Obere Muschelkalk des Coburger Landes. Fossilien 1992(2):122–127

Clausen CD (1968) Oberdevonische Cephalopoden aus dem Rheinischen Schiefergebirge, I Orthocerida, Bactritida. Palaeontogr A 128: 1–86

Credner GR (1875) *Ceratites fastigatus* und *Salenia texana*. Z Ges Naturwiss 46:105–116

Dagys AS, Keupp H (1998) Internal ventral keels in Triassic ceratid ammonoids: description and functional interpretation as muscle scars. Z dtsch geol Ges 149(1):81–89

Davis RA, Landman NH, Dommergues JL, Marchand D, Bucher H (1996) Mature modifications and dimorphism in ammonoid cephalopods. In: Landman NH, Tanabe K, Davis RA (eds) Ammonoid paleobiology. Plenum, New York

Davis RA, Mapes RH, Klofak SM (1999) Epizoa on externally shelled cephalopods. In: Rozanov AY, Shevyrev AA (eds) Fossil cephalopods: recent advances in their study. Palaeontological Institute, Moscow

De Baets K Klug C Korn D (2011) Devonian pearls and ammonoid-endoparasitic co-evolution. Acta Paleontol Pol 56:159–180

De Baets K, Keupp H, Klug C (2015) Parasites of Ammonoids. In: Klug C, Korn D, De Baets K, Kruta I, Mapes RH (eds) Ammonoid paleobiology: from anatomy to ecology. Springer, Dordrecht

Doguzhaeva LA, Mutvei R (1991) Organization of the soft body in *Aconeceras* (Ammonitina), interpreted on the basis of shell morphology and muscle-scars. Paleontogr A 218:17–33

Doguzhaeva LA, Mutvei H (1996) Attachment of the body to the shell in ammonoids. In: Landman NH, Tanabe K, Davis RA (eds) Ammonoid paleobiology. Plenum, New York

Donavan DT (1957) Note on the species *Cymbites laevigatus* (J. de C. Sowerby) and on the genus Cymbites Neumayr. Geol Mag 94(5):413–420

Dunstan A, Bradshaw CJA, Marshall J (2011) *Nautilus* at risk-estimating population size and demography of *Nautilus pompilius*. Plos ONE 6(2):1–9 (e16716)

Eck H (1879) Ueber einige Trias-Versteinerungen. Z dtsch Geol Ges 31:254–281

Eichler R, Ristedt H (1966) Isotopic evidence on the early life history of *Nautilus pompilius* (Linné). Science 153:734–736

Engel T (1894) Ueber kranke Ammonitenformen im schwäbischen Jura. Nova Acta Kgl Leop Carol dtsch Akad Naturf 61(5):326–384

Engeser T, Keupp H (2002) Phylogeny of the aptychi-posessing Neoammonoidea (Aptychophora nov., Cephalopoda). Lethaia 34:79–96

Fernández-López S (1987) Necrocinesis y colonización postmortal en Bajosphinctes (Ammonoidea) de la Cuenca Ibérica. Implicationes paleoecológicas y paleobatimétricas. Bol R Soc Española Hist Nat (Geol) 82(1–4):151–184

Fraas O (1863) Abnormitäten bei Ammoniten. Jahresverh Ver vaterländ Naturkunde Wrttbg 19:111–113

Frentzen K (1937) Ontogenie, Phylogenie und Systematik der Amaltheen des Lias Delta Südwestdeutschlands. Abh Heidelberger Akad Wiss mathem naturwiss Kl 23:1–136

Frerichs U (2011) Epöken: Über Siedler und "Piraten". Fossilien 2011(5):295–298

Fuchs T (1871) Über lokale Anreicherung kleiner Organismen und insbesondere über die Fauna von St. Cassian. Verh k k geol Reichsanstalt 1871:204–206

Fürsich F, Wendt J (1977): Biostratinomy and palaeoecology of the Cassian Formation (Triassic) of the Southern Alps. Palaeogeogr Palaeoclim Palaeoecol 22:257–323

Geczy B (1965) Pathologische jurassische Ammoniten aus dem Bakony-Gebirge. Ann Univ Sci Budapest Sec Geol 9:31–37

Guex J (1967) Contribution ã l'etude des blessures chez les ammonites. Bull Lab Géol Univ Lausanne 165:1–16

Hagdorn H, Simon T (1985) Geologie und Landschaft des Hohenloher Landes. Thorbecke, Sigmaringen

Hallam A (1965) Environmental causes of stunting in living and fossil marine benthonic invertebrates. Paleontology 8:132–155

Hanlon RT, Forsyte JW (1990) Diseases of Mollusca: Cephalopoda. 1.1 Diseases caused by microorganisms. In: Kinne O (ed) Diseases of marine animals, 3. Biologische Anstalt Helgoland, Hamburg

Hansen MC, Mapes RH (1990) A predator-prey relationship between sharks and cephalopods in the Late Paleozoic. In: Boucot A (ed) Evolutionary paleobiology of behaviour and coevolution. Elsevier, Amsterdam

Hauschke N, Schöllmann N, Keupp H (2011) Oriented attachment of a stalked cirripede on an orthoconic heteromorph ammonite—implications for the swimming position of the latter. N Jahrb Geol Paläont Abh 262(2):199–212

Heller F (1958) Gehäusemißbildungen bei Amaltheiden. Geol Bl NO-Bayern 8:66–71

Heller F (1964) Neue Fälle von Gehäusemißbildungen bei Amaltheiden. Paläontol Z 38(3/4): 136–141

Henderson RA, Kennedy WJ, Cobban WA (2002) Perspectives of ammonite paleobiology from shell abnormalities in the genus Baculites. Lethaia 35:215–230

Hengsbach R (1976) Über Sutur-Asymmetrie bei *Cymbites laevigatus* (Ammonoidea; Jura). Senck leth 56(6):463–468

Hengsbach R (1977a) Cheiloceraten (Ammon., Devon) mit asymmetrischem Phragmocon. Sitzungsberichte Ges Naturf Freunde Berlin (NF) 17:69–72

Hengsbach R (1977b) Über die Sutur-Asymmetrie einiger Psiloceraten. Sitzungsber Ges Naturf Freunde Berlin (NF) 17:59–67

Hengsbach R (1978) Zur Sutur-Asymmetrie bei *Anahoplites* (Ammonoidea, Kreide). Senck leth 59:377–385

Hengsbach R (1980) Über die Sutur-Asymmetrie bei *Hecticoceras* (Ammonoidea; Jura). Senck leth 60(4/6):463–473

Hengsbach R (1986a) Zur Kenntnis der Sutur-Asymmetrie bei Ammoniten. Senck leth 67(1/4):119–149

Hengsbach R (1986b) Ontogenetisches Auftreten und Entwicklung der Sutur-Asymmetrie bei einigen Psiloceratacea (Ammonoidea; Jura). Senck leth 67(1/4):323–330

Hengsbach R (1990) Studien zur Paläopathologie I: Die Paläoparasitologie, eine Arbeitsrichtung der Paläobiologie. Senck leth 70(4/6):439–461

Hengsbach R (1991) Die Symmetropathie, ein Beitrag zur Erforschung sogenannter Anomalien. Senck leth 71(3/4):339–366

Hengsbach R (1996) Ammonoid pathology. In: Landman NH, Tanabe K, Davis RA (eds) Ammonoid paleobiology. Plenum, New York

Hewitt RA, Westermann GEG (1990) Mosasaur tooth marks on the ammonite *Placenticeras* from the Upper Cretaceous of Alberta, Canada. Can J Earth Sci 27:469–472

Hölder H (1955) Die Ammonitengattung *Taramelliceras* im südwestdeutschen Unter- und Mittelmalm. Paleontogr A 106:37–153

Hölder H (1956) Über Anomalien an jurassischen Ammoniten. Paläontol Z 30:95–107

Hölder H (1960) Zur Frage des Wachstumsendes bei Ammoniten. Paläontol Z 34:61–68

Hölder H (1970) Anomalien an Molluskenschalen, insbesondere Ammoniten, und deren Ursachen. Paläontol Z 44:182–195

Hölder H (1973) Miscellanea cephalopodica. Münster Forsch Geol Paläont 29:39–76

Hölder H (1977) Zwei ungewöhnliche Erscheinungsformen anomaler Jura-Ammoniten der forma aegra verticata. Paläontol Z 51(3/4):254–257

Hüne L, Hüne P (2006) Des phénomènes paléopathologiques chez une faune d'Ammonites du Callovien supérieur de Bénerville-sur-Mer (Calvados, France). L'Echo des falaises 2006(7):33–37

Hungerbühler A (1992) Fossilien. Kosmos, Stuttgart

Hyatt A (1889) Genesis of the Arietidae. Smithson Contr Knowledge 26, Art. II + XI: 673 + 238 pp.

Ilovaisky D (1917) Les ammonites du Jurassique supérieur du pays de Liapine. Ouvrage Sect Géol Soc Imper Am Sci Nat 1(1–2):1–180

Ivanov AN (1975) Late ontogeny in ammonites and its characteristics in micro-, macro- and megaconchs. Yarosl Pedagog Inst Sb Nauchn Tr 142:5–57

Jäger M (1991) Nicht alltäglich: Ein Ammonit mit zwei Spiralen. Fossilien 1991(6):351–353

Johnson JG (1984): Temperature and biotic crisis in the marine realm. Geology 12(12):741

Kase T, Johnston PA, Seilacher A, Boyce J (1998) Alleged mosasaur bite marks on Late Cretaceous ammonites are limpet (patellogastropod) home scars. Geology 26(10):947–950

Kauffman EG (1990) Mosasaur predation on ammonites during the Cretaceous—a evolutionary history. In: Boucot A (ed) Evolutionary paleobiology of behavior and coevolution. Elsevier, Amsterdam

Kauffman EG, Kesling RV (1960) An Upper Cretaceous ammonite bitten by a mosasaur. Contr Michigan Univ Mus Paleont 15:193–248

Keferstein W (1861) Kopftragende Weichthiere (Malacozoa cephalophora). In: Bronn's Klassen und Ordnungen der Weichthiere, Bd. 3. C.F. Winter'sche Verlagshandlung, Leipzig

Keller T (1977) Fraßreste im süddeutschen Posidonienschiefer. Jahresh Ges Naturkde Württ 132:117–134

Kemper E (1961) Die Ammonitengattung *Platylenticeras* (= *Garnieria*). Beih Geol Jb 47:1–195

Keupp H (1976) Neue Beispiele für den Regenerationsmechanismus bei verletzten und kranken Ammoniten. Paläontol Z 50(1/2):70–77

Keupp H (1977) Paläopathologische Normen bei Amaltheiden (Ammonoidea) des Fränkischen Lias. Jb Coburger Landesstiftung 1977:263–280

Keupp H (1979) Nabelkanten-Präferenz der forma verticata Hölder 1956 bei Dactylioceraten (Ammonoidea, Toarcien). Paläontol Z 53:214–219

Keupp H (1984/85) Pathologische Ammoniten, Kuriositäten oder paläobiologische Dokumente, Teil 1 und 2. Fossilien 1984/6:258–262, 267–275; 1985/1: 23–35

Keupp H (1985) Das "Fastigatus"-Problem bei Ceratiten des Germanischen Muschelkalks. In: Hagdorn H (ed) Geologie und Paläontologie im Hohenloher Land. Symposium 100. Geburtstag von Georg Wagner. Künzelsau

Keupp H (1987) Perlen (Schalenkonkretionen) bei Dactylioceraten aus dem fränkischen Lias. Jahresmitt Naturhist Ges Nürnberg 1986:97–102

Keupp H (1991) Bißmarken oder postmortale Implosionsstrukturen? Fossilien 1991(5):275–280

Keupp H (1992a) Rippenscheitel bei Ammoniten-Gehäusen. Fossilien 1992(5):283–290

Keupp H (1992b) Wachstumsstörungen bei *Pleuroceras* und anderen Ammonoidea durch Epökie. Berl geowiss Abh E 3:113–119

Keupp H (1993) Fälle der forma aegra calcar (Zieten) bei mesozoischen Ammoniten. Fossilien 1993(3):140–144

Keupp H (1994) Volumenvermindernde Gehäuse-Anomalien bei Jura-Ammoniten. Fossilien 1994(1):38–44

Keupp H (1995) Volumenvergrößernde Anomalien bei Jura-Ammoniten. Fossilien 1995/1:54–59

Keupp H (1996) Paläopathologische Analyse einer Ammoniten-Vergesellschaftung aus dem Oberjura Westsibiriens. Fossilien 1996/1:45–54

Keupp H (1997) Paläopathologische Analyse einer "Population" von *Dactylioceras athleticum* (Simpson) aus dem Unter-Toarcium von Schlaifhausen/Oberfranken. Berl geowiss Abh E 25:243–267

Keupp H (1998) Mundsaumverletzungen bei *Pleuroceras* (Ammonoidea). Fossilien 1998(1):37–42

Keupp H (2000a) Anomale Muskelleisten bei Ammoniten. Berl Geowiss Abh E 34:279–289

Keupp H (2000b) Ammoniten- Paläobiologische Erfolgsspiralen. Thorbecke, Stuttgart

Keupp H (2003) Aptychen: Kiefer, Deckel oder beides? Fossilien 2003(2):104–110

Keupp H (2006) Sublethal punctures in body chambers of Mesozoic ammonites (forma aegra fenestra n.f.), a tool to interpret synecological relationships, particularly predator-prey interactions. Paläontol Z 80(2):112–123

Keupp H (2007) Anomale Nabelweiten bei Ammoniten und Nautiliden. http://www.leitfossil.de/Abonnenten/Lehrreiches

Keupp H (2008) Wer hat hier zugebissen? Ammoniten-Prädation. Fossilien 2008(2):109–112

Keupp H (2012) Atlas zur Paläopathologie der Cephalopoden. Berl Paläobiol Abh 12:1–390

Keupp H, Ilg A (1992) Paläopathologie der Ammonitenfauna aus dem Obercallovium der Normandie und ihre palökologische Interpretation. Berl geowiss Abh E 3:171–189

Keupp H, Riedel F (1995) *Nautilus pompilius* in captivity: a case study of abnormal shell growth. Berl geowiss Abh E 16/2:663–681

Keupp H, Veit R (1996) Ein *Phylloceras* mit Anaptychus. Fossilien 1996(6):343–351

Keupp H, Mitta V (2004) Septenbildung bei *Quenstedtoceras* (Ammonoidea) von Saratov (Russland) unter anomalen Kammerdruckbedingungen. Mitt Geol Paläont Inst Univ Hamburg 88:51–62

Keupp H, Schweigert G (2009) Scheinbare heteromorphe Ammoniten. Fossilien 2009(6):358–362

Keupp H, Richter AE (2010) Napfschnecken-Fraßkuhlen als Ursache von blasenartigen Steinkernstrukturen bei meozoischen Ammoniten (*Kranaosphinctes*, *Eupachydiscus*) von Madagaskar. http://www.leitfossil.de/Abonnenten/Lehrreiches

Keupp H, Schobert J (2011) Gehäuseanomalien bei Klein-Ammoniten von Buttenheim/Oberfranken. Fossilien 2011(3):164–174

Keupp H, Mitta V (2013) Cephalopod jaws from the Middle Jurassic of the Unzha-Basin, Kostroma Region, Central Russia. N Jahrb Geol Paläont Abh 270(1):23–54

Keupp H, Röper M, Seilacher A (1999) Paläobiologische Aspekte von syn vivo-besiedelten Ammonoideen im Plattenkalk des Ober-Kimmeridgiums von Brunn in Ostbayern. Berl geowiss Abh E 30:121–145

Keupp H, Röper M, Rothgaenger M (2012) Serpuliden-Epökie auf Ammoniten aus dem Brunner Plattenkalk (Ober-Kimmeridgium) in Ostbayern. Archaepteryx 29:1–12.

Kieslinger A (1926) Untersuchungen an triadischen Nautiloideen. Paläontol Z 7:101–122

King SD (2009) The ability of Mosasaurs to produce unique puncture marks on ammonite shells. Master Thesis, Graduate College at Bowling Green State University, 144 pp.

Klug C (1999) Devonian ammonoid biometry and global events. preliminary results. In: Histon K (ed) V Intern. Symp. Cephalopods—Present and Past. Ber Geol BA 46:60

Klug C, Korn D (2001) Epizoa and post-mortem epicoles on cephalopod shells—Devonian and Carboniferous examples from Morocco. Berl geowiss Abh E 36:145–155

Klug C, Korn D (2002) Occluded umbilicus in the Pinacitinae (Devonian) and its paleoecological implications. Paleontology 45:917–931

Knoch U (1989) *Placunopsis* auf einem Ceratiten. Fossilien 1989(6):280

Kolb A (1955) Über zwei abnorme Ammonitengehäuse der Gattung *Paltopleuroceras*. Geol Bl NO-Bayern 5:148–150

Kröger B (2000) Schalenverletzungen an jurassischen Ammoniten – ihre palöäobiologische und paläoökologische Aussagefähigkeit. Berl Geowiss Abh E 33:1–97

Kröger B (2002a) On the efficiency to the buoancy apparatus in ammonoids: evidences from sublethal shell injuries. Lethaia 35:61–70
Kröger B (2002b) On the ability of withdrawing of some Jurassic ammonoids. Berl Geol BA 57:199–204
Kröger B (2002c) Antipretatory traits of the ammonoid shell—interactions from Jurassic ammonoids with sublethal injuries. Paläontol Z 76:223–234
Kröger B, Keupp H (2004) A paradox survival—report of a repaired syn vivo perforation in a nautiloid phragmocone. Lethaia 37:439–444
Krüger FJ (1984) Professor Landois und die Riesenammoniten. Fossilien 1984(6):283–285
Kruta I, Landman NH (2007) Injuries on *Nautilus* jaws: implications for the function of ammonite aptychi. Veliger 50(3):241–247
Kruta I, Landman N, Rouget I, Cecca F, Tafforeau P (2011) The role of ammonites in the Mesozoic marine food web revealed by jaw preservation. Science 331:70–72
Krystyn L, Schäfer G, Schlager W (1971) Über die Fossil-Lagerstätten in den triadischen Hallstätter Kalken der Ostalpen. N Jb Geol Paläont Abh 137:284–304
Landman NH, Waage KM (1986) Shell abnormalities in scaphitid ammonites. Lethaia 19:211–224
Landman NH, Cochran JK (1987) Growth and longevity of *Nautilus*. In: Saunders WB, Landman NH (eds) *Nautilus*—the biology and paleobiology of a living fossil. Plenum, New York
Landman NH, Saunders WB, Winston JE, Harries PJ (1987) Incidence and kinds of epizoans on the shells of live *Nautilus*. In: Saunders WB, Landman NH (eds) *Nautilus*—the biology and paleobiology of a living fossil. Plenum, New York
Lange W (1929) Zur Kenntnis des Oberdevons am Enkeberg und bei Balve (Sauerland). Abh Preuß Geol LA, NF 119:1–132
Lange W (1932) Über Symbiosen von *Serpula* mit Ammoniten im unteren Lias Nordwestdeutschlands. Z dtsch geol Ges 84:229–234
Lange W (1941) Die Ammonitenfauna der *Psiloceras*-Stufe Norddeutschlands. Paleontogr A 93:1–186
Larson N (2007) Deformities in the Late Callovian (Late Middle Jurassic) ammonite fauna from Saratov, Russia. In: Landman NH, Tanabe K, Mapes RH (eds) Cephalopods—present and past: new insights and fresh perspectives. Springer, New York
Laube GC (1869) Die Fauna der Schichten von St. Cassian. Denkschr kaiserl Akad Wiss mathem naturwiss Kl 30:1–106
Lehmann U (1972) Aptychen als Kieferelemente der Ammoniten. Paläontol Z 46:34–48
Lehmann U (1976) Ammoniten, ihr Leben und ihre Umwelt. Enke, Stuttgart
Lehmann U (1990) Ammonoideen, Leben zwischen Skylla und Charybdis. Enke, Stuttgart
Lehmann U, Kulicki C (1990) Double function of aptychi (Ammonoidea) as jaw elements and opercula. Lethaia 23:325–331
Linck O (1956) Echte und unechte Besiedler (Epöken) des deutschen Muschelkalk-Meeres. Aus der Heimat 64:161–173
Longridge LM, Smith PL, Rawlings G, Klaptocz V (2009) The impact of asymmetries in the elements of the phragmocone of Early Jurassic ammonites. Paleontol Electron 12(1):1A,15 pp. http://paleo-electronica.org/2009
Makowski H (1963) Problem of sexual dimorphism in ammonites. Paleontol Pol 12:1–92
Manger WL, Meeks LK, Stephen DA (1999) Pathological gigantism in Middle Carboniferous cephalopods, southern midcontinent, United States. In: Oloriz F, Rodriguez-Tovar (eds) Advancing research on living and fossil cephalopods. Kluwer, Plenum, New York
Mapes RH, Hansen M (1984) Pennsylvanian shark-cephalopod predation: a case study. Lethaia 17:175–182
Mapes RH, Sims MS, Boardman DRII (1995) Predation on the Pennsylvanian ammonoid *Gonioloboceras* and its implications for allochthonous vs. autochthonous accumulations of goniatites and other ammonoids. J Paleontol 69:441–446
Marchand D, Courville P, Bonnot A, Rossi J, Scoufflaire Q (2002) Very small ammonites (Micromorphs) from Lower Oxfordian Marls (Mariae Zone). Abh Geol BA 57:467–478

Martill DM (1990) Predation on *Kosmoceras* by semionotid fish in the Middle Jurassic Lower Oxford Clay of England. Palaeontology 33:739–742

Martin J (1858) Notice paléontologique et stratigraphique du Lias inférieur de la Côte d'Or, de l'Yonne et de Luxembourg. Bull Congr Sci France Auxerre 1858:377

Matyja BA (1986) Developmental polymorphism in Oxfordian ammonites. Acta Geol Pol 36:37–68

Maubeuge PL (1949a) Sur quelques échantillons anormoux d'Ammonites jurassiques. Arch Inst Grand-Ducal Luxembourg NS 18:127–147

Maubeuge PL (1949b) Sur la nature des "conelles" (Quenstedt). Bull Soc Sci Nancy 1: Not Paléont AS:1–3

Maubeuge PL (1957) Deux ammonites nouvelles du Lias moyen de l'Allemagne septentrionale. Bull Soc Sci Nancy 1957:1–6

Mayer G (1966) Ein *Ceratites* (*Ceratites?*) *sublaevigatus* Wenger fastigatus und zwei weitere Ceratiten mit fastigatus-Merkmalen. Der Aufschluss 17:295–298

Mayer G (1974) Ein anomaler *Ceratites* (*Acanthoceratites*) *spinosus spinosus* Philippi aus dem mittleren Hauptmuschelkalk von Zuzenhausen (Kraichgau). Der Aufschluss 25:191–192

Mayer G (1975) Ein bemerkenswerter Ceratit aus dem Oberen Muschelkalk von Schöningen/ Elm. Der Aufschluss 26:304–305

Mayer G (1978) Ceratiten mit Skulpturanomalien aus dem süddeutschen und französischen Muschelkalk. Der Aufschluss 29:71–75

Mayer G (1981) Zwei fastigate Ceratiten aus dem mainfränkischen Muschelkalk. Der Aufschluss 32:345–347

Mayr FX (1967) Paläobiologie und Stratinomie der Plattenkalke der Altmühlalb. Erlanger Geol Abh 67:1–40

Meischner D (1968) Perniciöse Epökie von *Placunopsis* auf *Ceratites*. Lethaia 1(2):156–174

Meischner D (2002) Aufwuchs von Muscheln auf Ceratiten – Ein Ärgernis? Nat Museum 132(1):23–29

Merkt J (1966) Über Austern und Serpeln als Epöken auf Ammonitengehäusen. N Jb Geol Paläont Abh 125:467–479

Mignot Y (1993) Un problem de paléobiologie chez les ammonoids (Cephalopoda): Croissance et miniaturisation en liaison avev les environnements. Documents des Laboratoires de Géologie Lyon 124:1–113

Mignot Y, Elmi S, Domergues JL (1993) Croissance et miniaturisation de quelques *Hildoceras* (Cephalopoda) en liaison avec l'environnement contraignants de la Téthys Toarcienne. Geobios Mém Spec 15:305–312

Mitta VV, Michailova IA, Sumin DL (1999) Unusual Volgian scaphitoid ammonites from Central Russia. Paleontol J 33(6):614–619

Moodie RL (1926) La paléopathologie des mammiferès du pléistocène. Biol Médicale 16(9):431–440

Morton N (1981) Aptychi: the myth of the ammonite operculum. Lethaia 14:57–61

Morton N (1983) Pathologically deformed *Graphoceras* (Ammonitina) from the Jurassic of Skye, Scotland. Paleontology 26(2):443–453

Müller AH (1954) Zur Entwicklungsgeschichte der Ceratiten des Germanischen Muschelkalks, mit einigen Bemerkungen über Abnormitäten. Geologie 3:28–41

Müller AH (1970a) Über die Fastigatus-Anomalie der Ceratiten (Ammonoidea, Cephalopoda) des Germanischen oberen Muschelkalkes. Monatsber dtsch Akad Wiss 12:303–321

Müller AH (1970b) Weitere Aberrationen bei Ceratiten (Ammonoidea, Cephalopoda) aus dem germanischen oberen Muschelkalk und Bemerkungen zur Originalschale der Ceratiten. Monatsber dtsch Akad Wiss 12:219–231

Müller AH (1970c) Neue Funde seltener Ceratiten aus dem germanischen Muschelkalk und Keuper. Monatsber dtsch Akad Wiss 12(8):632–642

Müller AH (1976) Über einen besonderen Typ phylogenetisch deutbarer Aberrationen fossiler Tiere. Biologische Rundschau 14:190–204

Müller AH (1978) Über Ceratiten mit fehlenden oder unvollständigen Kammerscheidewänden (Septen) und die Frage nach der Lebensweise der Ammonoidea (Cephalopoda). Freiberger Forschungsheft C 334:69–83

Müller AH (1982) Zur Morphologie, Taxonomie und Ökologie fossiler und rezenter Serpulimorpha (Polychaeta). Biologische Rundschau 20:330–351

Mundlos R (1963) Fundgrube Schöningen (Braunschweig). Der Aufschluß 14:76–80

Mutvei H (1957) On the relation of the principal muscles to the shell in *Nautilus* and some fossil nautiloids. Arkiv Mineralogy an Geology. Stockholm 2:219–254

Nicolesco CP (1921) Étude sur la Dissymmetrie de certaines Ammonites. Paris

Orbigny A de (1842–1851) Paléontologie Francaise, Terrain Jurassique, Tome 1: Cephalopodes. Masson, Paris

Philippi E (1897) Ein interessantes Vorkommen von *Placunopsis ostracinna* v.Schloth. sp. Z dt geol Ges 51:67–69

Philippi E (1901) Die Ceratiten des oberen deutschen Muschelkalks. Paleontol Abh NF 4:347–458

Pompeckj JF (1894) Über Ammonoideen mit anormaler Wohnkammer. Jh Ver vaterl Naturkunde Württ 49:220–290

Quenstedt FA (1858) Der Jura. Schweizerbart, Tübingen

Quenstedt FA (1885–1888) Die Ammoniten des Schwäbischen Jura, I-III. Schweizerbart, Stuttgart

Radwanski A (1996) The predation upon, and the extinction of, the latest Maastrichtian populations of the ammonite species *Hoploscaphites constrictus* (J. Sowerby 1817) from the Middle Vistula Valley, Central Poland. Acta Geol Pol 46:117–135

Rein S (1989) Über das Regenerationsvermögen der germanischen Ceratiten (Ammonoidea) des Oberen Muschelkalks (Mitteltrias). Veröff Naturhist Mus Schleus 4:47–54

Rein S (1990) Über Ceratiten (Cephalopoda, Ammonoidea) mit “fehlenden” Septen. Veröff Naturhist Mus Schleus 5:22–25

Rein S (1991) Die fastigaten Ceratiten in den Sammlungen des Erfurter Naturkundemuseums. Veröff Naturkundemuseum Erfurt 10:66–79

Rein S (1993) Zur Biologie und Lebensweise der germanischen Ceratiten. In: Hagdorn H, Seilacher A (eds) Muschelkalk, Schöntaler Symposium 1991. Goldschneck, Korb

Rein S (1994) Über eine interessante Gehäuseregenerierung der Ceratiten. Veröff Naturkundemuseums Erfurt 13:91–100

Rein S (1996) Über Epöken und das Schwimmvermögen der Ceratiten. Veröff Naturhist Mus Schleus 11:65–75

Rein S (1997) Biologie und Lebensweise von *Germanonautilus* Mojsisovics 1902: Teil I: Das Schwimmvermögen von *Germanonautilus*. Veröff Naturhist Mus Schleus 12:43–51

Rein S (2000) Zur Lebensweise von *Ceratites* und *Germanonautilus* im Muschelkalkmeer. Veröff Naturhist Mus Schleus 15:25–40

Richter AE (2009a) Ammoniten-Gehäuse mit Bissspuren. Berl paläobiol Abh 10:297–305

Richter AE (2009b) Geologische Notizen: Ammonitengehäuse mit Bissspuren. Natur und Mensch, Jahresmitt der Naturhist Ges Nürnberg 2008:169–175

Richter U (2002) Gewebeansatz-Strukturen auf pyritisierten Steinkernen von Ammonoideen. Geol Beitr Hannover 4:1–113

Richthofen Fv (1860) Geognostische Beschreibung der Umgebung von Predazzo, Sanct Cassian und der Seiser Alpe in Süd-Tyrol. Gotha

Rieber H (1963) Ein *Cardioceras* (Ammonoidea) mit asymmetrischer Lage von Phragmokon und Kiel. N Jahrb Geol Paläont Mh 1963:289–294

Rieber H (1979) Eine abnorme, stark vereinfachte Lobenlinie bei *Brasilia decipiens* (Buckman). Paläontol Z 53:230–236

Riedel A (1916) Beiträge zur Paläontologie und Stratigraphie der Ceratiten des deutschen oberen Muschelkalks. Jb Kgl preuß geol LA 37:1–116

Roll A (1935) Über Fraßspuren an Ammoniten. Zbl Min Geol Paläont Abt B 1935:120–124

Rothe HW (1949) Zum Problem des *Ceratites fastigatus* Credn. mit Beispielen von thüringischen Fundorten. Hallesches Jb mitteldt Erdgesch 1:27–32

Rothe P (1955) Die Ceratiten und die Ceratitenschichten des Oberen Muschelkalks (Trias) im Thüringer Becken. Beitr Geol Thüringen 8:255–323

Rummel O (1973) Fund eines dritten ringrippigen Ceratiten im Steinbruch Luxwinkel bei Schöningen. Der Aufschluss 24:367–368

Rungger D, Rastelli M, Braendle E, Malsberger RG (1971) A virus like particle associated with lesions in the muscles of *Octopus vulgaris*. J Invertebr Pathol 17:72–80

Saunders WB, Landman NH (eds) (1987) *Nautilus*—The biology and paleobiology of a living fossil. Plenum, New York

Schindewolf OH (1929) Vergleichende Studien zur Phylogenie, Morphogenie und Terminologie der Ammoneenlobenlinie. Abh preuß geol LA, NF 115:1–102

Schindewolf OH (1934) Über Epöken auf Cephalopoden-Gehäusen. Paläontol Z 16(1/2):15–31

Schindewolf OH (1958) Über Aptychen (Ammonoidea). Paleontogr A 111:1–46

Schindewolf OH (1961) Die Ammoniten-Gattung *Cymbites* im deutschen Lias. Paleontogr A 117:193–232

Schindewolf OH (1962) Parasitäre Thallophyten in Ammoniten-Schalen. Paläontol Z H. Schmidt-Festband: 206–215

Schmidt H (1926) Neotenie und beschleunigte Entwicklung bei Ammoneen. Paläontol Z 7:197–205

Schram FR, Hof CJ, Steeman FA (2003) Thylacocephala (Arthropoda: Crustacea ?) from the Cretaceous of Lebanon and implications fpr Thylacocephalan systemetics. Palaeontology 42:769–797

Schwegler E (1939) Eine merkwürdige Krankheitserscheinung bei einem Belemniten aus dem Braunen Jura epsilon Schwabens und ihre Deutung. Zentralbl Min Geol Paläont 1939 Abt.B:74–80

Schweigert G (2009) Doppelt gescheitelt: Aus der Schatzkiste eines Sammlers. Fossilien 2009(4):250–252

Schweigert G, Dietze V (1999) *Oecoptychius* – des Rätsels Lösung. Fossilien 1999(1):51–58

Schweigert G, Dietl G (2001) Die Kieferelemente von *Physodoceras* (Ammonitina, Aspidoceratidae) im Nusplinger Plattenkalk (Oberjura, Schwäbische Alb). Berl geowiss Abh E 36:131–143

Seilacher A (1954) Ökologie der triassischen Muschel *Lima lineata* (Schloth.) und ihre Epöken. N Jahrb Geol Paläont Mh 1954:163–183

Seilacher A (1960) Epizoans as a key to ammonoid ecology. J Paleontol 34(1):189–193

Seilacher A (1975) Mechanische Simulation und funktionelle Evolution des Ammoniten-Septums. Paläontol Z 49:268–286

Seilacher A (1993) Ammonite aptychi: how to transform a jaw into an operculum? Am J Sci 293(AS):20–32

Seilacher A (1998) Mosasaurs, limpets or diagenesis: How *Placenticeras* shells got punctured. Mitt Mus Naturkde Berlin. Geowiss Reihe 1:93–102

Seilacher A, La Barbera M (1995) Ammonites as Cartesian divers. Palaios 10:493–506

Seilacher A, Keupp H (2000) Wie sind Ammoniten geschwommen? Fossilien 2000(5):310–313

Seltzer VB (2001) About anomal shells from the Callovian ammonites. Transact Sci Res Geol Inst Saratov State Univ, NS 8:29–45

Sims MS, Boston WB, Mapes RH (1987) Predation on an Upper Carboniferous ammonoid *Gonioloboceras*. Geol Surv Am 19(Abstracts with Program):57–58

Sornay J (1955) Nautile fossile ã coquille anormale. Muséum Nat Hist Bull, ser 2 27:260–261

Sparks AK (1972) Invertebrate pathology. Academic, New York

Stahl KF (1824) Verzeichnis der Versteinerungen Württembergs. Correspondenzbl kgl württemb landwirtsch Ver 6:1–91

Stephen DA (1997) Possible reproductive mass mortality and pathologic gigantism in Middle Carboniferous (Chesterian-Morrawan) cephalopod assemblages, southern Midcontinent. Oklahama Geol 57(3):113

Stephens GR (1988) Giant ammonites: a review. In: Wiedmann J, Kullmann J (eds) Cephalopods—present and past. Schweizerbart, Stuttgart

Stumbur KA (1960) Life-time injuries in some nautiloid shells. Paleontol J 4:133–135

Sturani C. (1971) Ammonites and stratigraphy of the "*Posidonia alpina*" beds of the Venetian Alps. Memorie degli Institutid di Geologia e Mineralogia dell'Universitá di Padova 28:1–190
Suchopar J (1997) Die Ceratiten der semipartitus-Zone von Gänheim. Fossilien 1997(4):222–224
Tanabe K (2011) The feeding habits of ammonites. Science 331:37–38
Tasch P (1953) Causes and paleontological significance of dwarfed fossil marine invertebrates. J Paleontol 27:356–444
Tasnadi-Kubacska A (1962) Paläopathologie, Pathologie der vorzeitlichen Tiere. Fischer, Jena
Tornquist A (1896) Die degenerierten Perisphinctiden des Kimmeridge von le Havre. Abh der Schweiz Paläont Ges 23:1–43
Trueman AE (1941) The ammonite body-chamber, with special reference to the buoyancy and mode of life of the living ammonite. Quart J Geol Soc London 94(4):339–383
Tsujino Y, Shigeta Y (2012) Biological response to experimental damage of the phragmocone and siphuncle in *Nautilus pompilius* Linnaeus. Lethaia 45:443–449
Tsujita CJ, Westermann GEG (1998) Ammonoid habitats and habits in the Western Interior Seaway: a case study from the Upper Cretaceous Bearpaw Formation of southern Alberta, Canada. Paleogeogr Paleoclim Paleoecol 144:135–160
Urlichs M (1994) *Trachyceras* Laube 1869 (Ammonoidea) aus dem Unter-Karn (Obertrias) der Dolomiten (Italien). Stuttg Beitr Natkde B 217:1–55
Urlichs M (2001) Die Zwergfauna aus der Obertrias von St. Kassian (Dolomiten). In: Weidert K (ed) Klassische Fundstellen der Paläontologie, Bd. IV. Goldschneck, Korb
Urlichs M (2004) Kümmerwuchs bei *Lobites* Mojsisovics 1902 (Ammonoidea) aus dem Unter-Karnium der Dolomiten (Ober-Trias, Italien) mit Revision der unterkarnischen Arten. Stuttg Beitr Natkde B 344:1–37
Vadász E (1908) Über eine oberliassische Lytocerasart mit aufgelöster Wohnkammer. Z Ung geol Ges 38:131–136
Vermeij GJ (1977) Mesozoic marine revolutions. Paleobiology 3:245–258
Vermeij GJ (1982) Gastropod shell form, breakage, and repair in relation to predation by the crab Calappa. Malacologia 23:1–12
Vermeij GJ (1983) Traces and trends of predation with special reference to bivalve animals. J Paleontol 26:455–465
Vermeij GJ (1987) Evolution and escalation: an ecological history of life. Princeton University Press, Princeton
Vogel KP (1959) Zwergwuchs bei Polyptychiten (Ammonoidea). Geol Jahrb 76:469–540
Wagenplast P (1972) Ökologische Untersuchungen der Fauna aus Bank- und Schwammfazies des Weißen Jura der Schwäbischen Alb. Arbeiten aus dem Geologisch-Paläontologischen Institut an der Universität Stuttgart. NF 67:1–99
Walker, S.E. and Brett, C. (2002) Post-Paleozoic patterns in marine predation: was there a Mesozoic and Cenozoic marine predatory revolution? Paleontol Soc Pap 8:119–193
Wani R, Jenkins RG, Mapes RH (2012) Preferential predatory peeling: ammonoid vs. nautiloid shells from the Upper Carboniferous of Texas, USA. Geobios 45(1):129–137
Wannenmacher N (2010) Aus der Form geraten: Pathologische Lias-Ammoniten aus Schwaben. Fossilien 2010(4):225–228
Ward PD (1979) Cameral liquid in *Nautilus* and ammonites. Paleobiology 5(1):40–49
Ward PD (1981) Shell sculpture as an adaptation in ammonoids. Paleobiology 7:96–100
Ward PD (1986) Rates and processes of compensatory buoyancy change in *Nautilus macromphalus*. Veliger 28 (4):356–368.
Ward PD (1987) The natural history of *Nautilus*. Allen and Unwin, Boston
Ward PD, Westermann GEG (1976) Sutural inversion in a heteromorphy ammonite and its implications for septal formation. Lethaia 9:357–361
Wenger R (1957) Die germanischen Ceratiten. Paleontogr A 108:57–129
Wendt J (1971) Genese und Fauna submariner sedimentärer Spaltenfüllungen im mediterranen Jura. Paleontogr A 136:121–192
Westermann GEG (1975) Model for origin, function and fabrication of fluted cephalopod septa. Paläontol Z 49:235–253

Westermann GEG (1996) Ammonoid life and habit. In: Landman NH, Tanabe K, Davis RS (eds) Ammonoid paleobiology. Plenum, New York

Westermann GEG (2013) Hydrostatics, propulsion and life-habits of the Cretaceous ammonoid *Baculites*. Revue de Paléobiologie 32(1):249–265

Wetzel W (1964) Schalen-Parasitismus bei Ammoniten (aufgrund schleswig-holsteinischer Funde). Meyniana 14:66–69

Weyer D (1964) Ein ungewöhnlicher fastigater Ceratit aus dem germanischen Oberen Muschelkalk. Geologie 13(4):478–481

Wiedmann J (1972) Ammoniten-Nuclei aus Schlämmproben der nordalpinen Obertrias, ihre stammesgeschichtliche und stratigraphische Bedeutung. Mitteilungen Ges Geol Bergbaustudent 21:561–622

Wiedmann J (1973) Evolution or revolution of ammonoids at Mesozoic system boundaries. Biol Rev 48(2):159–194

Wunsch P (1957) Ein typischer *Ceratites fastigiotenuis* Rothe aus der Umgebung von Göttingen. Geol Jb 73:557–560

Zapalski MK (2011) Is absence of proof a proof of absence? Comments on commensalism. Palaeogeogr Palaeoclim Palaeoecol 302:484–488

Zann L (1985) The rhythmic activity of *Nautilus pompilius*, with notes on its ecology and behavior in Fiji. Veliger 27(1):19–28

Ziegler B (1958) Monographie der Ammonitengattung *Glochiceras* im epikontinentalen Weißjura Mitteleuropas. Paleontogr A 110:93–164

Zieten CHv (1830–1833) Die Versteinerungen Württembergs. Schweizerbart, Stuttgart

Zimmermann E (1883) Über einen neuen Ceratiten aus dem Grenzdolomit Thüringens und über Glacialerscheinungen bei Klein-Pörthen zwischen Gera und Zeitz. Z dtsch geol Ges 35: 382–384

Index

C. Klug et al. (eds.), *Ammonoid Paleobiology: From Anatomy to Ecology*, Topics in Geobiology 43, DOI 10.1007/978-94-017-9630-9

T

U

MIX
Papier aus verantwortungsvollen Quellen
Paper from responsible sources
FSC® C105338

If you have any concerns about our products,
you can contact us on
ProductSafety@springernature.com

In case Publisher is established outside the EU,
the EU authorized representative is:
Springer Nature Customer Service Center GmbH
Europaplatz 3, 69115 Heidelberg, Germany

Printed by Libri Plureos GmbH
in Hamburg, Germany